$$[\nabla^2 \mathbf{v}]_x \equiv [\nabla \cdot \nabla \mathbf{v}]_x = \frac{\partial^2 v_x}{\partial x^2} + \frac{\partial^2 v_x}{\partial y^2} + \frac{\partial^2 v_x}{\partial z^2}$$

$$[\nabla^2 \mathbf{v}]_y \equiv [\nabla \cdot \nabla \mathbf{v}]_y = \frac{\partial^2 v_y}{\partial x^2} + \frac{\partial^2 v_y}{\partial y^2} + \frac{\partial^2 v_y}{\partial z^2}$$

$$[\nabla^2 \mathbf{v}]_z \equiv [\nabla \cdot \nabla \mathbf{v}]_z = \frac{\partial^2 v_z}{\partial x^2} + \frac{\partial^2 v_z}{\partial y^2} + \frac{\partial^2 v_z}{\partial z^2}$$

$$[\mathbf{v} \cdot \nabla \mathbf{v}]_x = v_x \frac{\partial v_x}{\partial x} + v_y \frac{\partial v_x}{\partial y} + v_z \frac{\partial v_x}{\partial z}$$

$$[\mathbf{v} \cdot \nabla \mathbf{v}]_y = v_x \frac{\partial v_y}{\partial x} + v_y \frac{\partial v_y}{\partial y} + v_z \frac{\partial v_y}{\partial z}$$

$$[\mathbf{v} \cdot \nabla \mathbf{v}]_z = v_x \frac{\partial v_z}{\partial x} + v_y \frac{\partial v_z}{\partial y} + v_z \frac{\partial v_z}{\partial z}$$

$$[\nabla \cdot \mathbf{v}\mathbf{v}]_x = \frac{\partial (v_x v_x)}{\partial x} + \frac{\partial (v_y v_x)}{\partial y} + \frac{\partial (v_z v_x)}{\partial z}$$

$$[\nabla \cdot \mathbf{v}\mathbf{v}]_y = \frac{\partial (v_x v_y)}{\partial x} + \frac{\partial (v_y v_y)}{\partial y} + \frac{\partial (v_z v_y)}{\partial z}$$

$$[\nabla \cdot \mathbf{v}\mathbf{v}]_z = \frac{\partial (v_x v_z)}{\partial x} + \frac{\partial (v_y v_z)}{\partial y} + \frac{\partial (v_z v_z)}{\partial z}$$

$$[\nabla \cdot \boldsymbol{\tau}]_x = \frac{\partial \tau_{xx}}{\partial x} + \frac{\partial \tau_{yx}}{\partial y} + \frac{\partial \tau_{zx}}{\partial z}$$

$$[\nabla \cdot \boldsymbol{\tau}]_y = \frac{\partial \tau_{xy}}{\partial x} + \frac{\partial \tau_{yy}}{\partial y} + \frac{\partial \tau_{zy}}{\partial z}$$

$$[\nabla \cdot \boldsymbol{\tau}]_z = \frac{\partial \tau_{xz}}{\partial x} + \frac{\partial \tau_{yz}}{\partial y} + \frac{\partial \tau_{zz}}{\partial z}$$

$$(\boldsymbol{\tau} : \nabla \mathbf{v}) = \tau_{xx} \frac{\partial v_x}{\partial x} + \tau_{xy} \frac{\partial v_x}{\partial y} + \tau_{xz} \frac{\partial v_x}{\partial z}$$

$$+ \tau_{yx} \frac{\partial v_y}{\partial x} + \tau_{yy} \frac{\partial v_y}{\partial y} + \tau_{yz} \frac{\partial v_y}{\partial z}$$

$$+ \tau_{zx} \frac{\partial v_z}{\partial x} + \tau_{zy} \frac{\partial v_z}{\partial y} + \tau_{zz} \frac{\partial v_z}{\partial z}$$

Note: the differential operations may *not* be simply generalized to curvilinear coordinates; see Tables A.7-2 and A.7-3.

Transport Phenomena

Second Edition

R. Byron Bird
Warren E. Stewart
Edwin N. Lightfoot

Chemical Engineering Department
University of Wisconsin-Madison

John Wiley & Sons, Inc.
New York / Chichester / Weinheim / Brisbane / Singapore / Toronto

Acquisitions Editor	*Wayne Anderson*
Marketing Manager	*Katherine Hepburn*
Senior Production Editor	*Petrina Kulek*
Director Design	*Madelyn Lesure*
Illustration Coodinator	*Gene Aiello*

This book was set in Palatino by UG / GGS Information Services, Inc. and printed and bound by Hamilton Printing. The cover was printed by Phoenix.

This book is printed on acid free paper. ∞

Library of Congress Cataloging-in-Publication Data
Bird, R. Byron (Robert Byron), 1924–
    Transport phenomena / R. Byron Bird, Warren E. Stewart, Edwin N. Lightfoot.—2nd ed.
      p. cm.
    Includes indexes.
    ISBN 0-471-41077-2 (cloth : alk. paper)
     1. Fluid dynamics. 2. Transport theory.   I. Stewart, Warren E., 1924–   II. Lightfoot, Edwin N., 1925–   III. Title.
    QA929.B5 2001
    530.13′8—dc21   2001023739

ISBN 0-471-41077-2

Printed in the United States of America

10 9 8 7 6 5 4 3

Preface

While momentum, heat, and mass transfer developed independently as branches of classical physics long ago, their unified study has found its place as one of the fundamental engineering sciences. This development, in turn, less than half a century old, continues to grow and to find applications in new fields such as biotechnology, microelectronics, nanotechnology, and polymer science.

Evolution of transport phenomena has been so rapid and extensive that complete coverage is not possible. While we have included many representative examples, our main emphasis has, of necessity, been on the fundamental aspects of this field. Moreover, we have found in discussions with colleagues that transport phenomena is taught in a variety of ways and at several different levels. Enough material has been included for two courses, one introductory and one advanced. The elementary course, in turn, can be divided into one course on momentum transfer, and another on heat and mass transfer, thus providing more opportunity to demonstrate the utility of this material in practical applications. Designation of some sections as optional (○) and other as advanced (●) may be helpful to students and instructors.

Long regarded as a rather mathematical subject, transport phenomena is most important for its physical significance. The essence of this subject is the careful and compact statement of the conservation principles, along with the flux expressions, with emphasis on the similarities and differences among the three transport processes considered. Often, specialization to the boundary conditions and the physical properties in a specific problem can provide useful insight with minimal effort. Nevertheless, the language of transport phenomena is mathematics, and in this textbook we have assumed familiarity with ordinary differential equations and elementary vector analysis. We introduce the use of partial differential equations with sufficient explanation that the interested student can master the material presented. Numerical techniques are deferred, in spite of their obvious importance, in order to concentrate on fundamental understanding.

Citations to the published literature are emphasized throughout, both to place transport phenomena in its proper historical context and to lead the reader into further extensions of fundamentals and to applications. We have been particularly anxious to introduce the pioneers to whom we owe so much, and from whom we can still draw useful inspiration. These were human beings not so different from ourselves, and perhaps some of our readers will be inspired to make similar contributions.

Obviously both the needs of our readers and the tools available to them have changed greatly since the first edition was written over forty years ago. We have made a serious effort to bring our text up to date, within the limits of space and our abilities, and we have tried to anticipate further developments. Major changes from the first edition include:

- transport properties of two-phase systems
- use of "combined fluxes" to set up shell balances and equations of change
- angular momentum conservation and its consequences
- complete derivation of the mechanical energy balance
- expanded treatment of boundary-layer theory
- Taylor dispersion
- improved discussions of turbulent transport

- Fourier analysis of turbulent transport at high Pr or Sc
- more on heat and mass transfer coefficients
- enlarged discussions of dimensional analysis and scaling
- matrix methods for multicomponent mass transfer
- ionic systems, membrane separations, and porous media
- the relation between the Boltzmann equation and the continuum equations
- use of the "Q+W" convention in energy discussions, in conformity with the leading textbooks in physics and physical chemistry

However, it is always the youngest generation of professionals who see the future most clearly, and who must build on their imperfect inheritance.

Much remains to be done, but the utility of transport phenomena can be expected to increase rather than diminish. Each of the exciting new technologies blossoming around us is governed, at the detailed level of interest, by the conservation laws and flux expressions, together with information on the transport coefficients. Adapting the problem formulations and solution techniques for these new areas will undoubtedly keep engineers busy for a long time, and we can only hope that we have provided a useful base from which to start.

Each new book depends for its success on many more individuals than those whose names appear on the title page. The most obvious debt is certainly to the hard-working and gifted students who have collectively taught us much more than we have taught them. In addition, the professors who reviewed the manuscript deserve special thanks for their numerous corrections and insightful comments: Yu-Ling Cheng (University of Toronto), Michael D. Graham (University of Wisconsin), Susan J. Muller (University of California-Berkeley), William B. Russel (Princeton University), Jay D. Schieber (Illinois Institute of Technology), and John F. Wendt (Von Kármán Institute for Fluid Dynamics). However, at a deeper level, we have benefited from the departmental structure and traditions provided by our elders here in Madison. Foremost among these was Olaf Andreas Hougen, and it is to his memory that this edition is dedicated.

Madison, Wisconsin

R. B. B.
W. E. S.
E. N. L.

Contents

vi   Contents

Chapter 16 Energy Transport by Radiation   487

Part III Mass Transport

Chapter 17 Diffusivity and the Mechanisms of Mass Transport   513

Chapter 18 Concentration Distributions in Solids and Laminar Flow   543

Chapter 19 Equations of Change for Multicomponent Systems   582

Chapter 0

The Subject of Transport Phenomena

§0.1 What are the transport phenomena?

§0.2 Three levels at which transport phenomena can be studied

§0.3 The conservation laws: an example

§0.4 Concluding comments

The purpose of this introductory chapter is to describe the scope, aims, and methods of the subject of transport phenomena. It is important to have some idea about the structure of the field before plunging into the details; without this perspective it is not possible to appreciate the unifying principles of the subject and the interrelation of the various individual topics. A good grasp of transport phenomena is essential for understanding many processes in engineering, agriculture, meteorology, physiology, biology, analytical chemistry, materials science, pharmacy, and other areas. Transport phenomena is a well-developed and eminently useful branch of physics that pervades many areas of applied science.

§0.1 WHAT ARE THE TRANSPORT PHENOMENA?

The subject of transport phenomena includes three closely related topics: fluid dynamics, heat transfer, and mass transfer. Fluid dynamics involves the transport of *momentum*, heat transfer deals with the transport of *energy*, and mass transfer is concerned with the transport of *mass* of various chemical species. These three transport phenomena should, at the introductory level, be studied together for the following reasons:

- They frequently occur simultaneously in industrial, biological, agricultural, and meteorological problems; in fact, the occurrence of any one transport process by itself is the exception rather than the rule.

- The basic equations that describe the three transport phenomena are closely related. The similarity of the equations under simple conditions is the basis for solving problems "by analogy."

- The mathematical tools needed for describing these phenomena are very similar. Although it is not the aim of this book to teach mathematics, the student will be required to review various mathematical topics as the development unfolds. Learning how to use mathematics may be a very valuable by-product of studying transport phenomena.

- The molecular mechanisms underlying the various transport phenomena are very closely related. All materials are made up of molecules, and the same molecular motions and interactions are responsible for viscosity, thermal conductivity, and diffusion.

The main aim of this book is to give a balanced overview of the field of transport phenomena, present the fundamental equations of the subject, and illustrate how to use them to solve problems.

There are many excellent treatises on fluid dynamics, heat transfer, and mass transfer. In addition, there are many research and review journals devoted to these individual subjects and even to specialized subfields. The reader who has mastered the contents of this book should find it possible to consult the treatises and journals and go more deeply into other aspects of the theory, experimental techniques, empirical correlations, design methods, and applications. That is, this book should not be regarded as the complete presentation of the subject, but rather as a stepping stone to a wealth of knowledge that lies beyond.

§0.2 THREE LEVELS AT WHICH TRANSPORT PHENOMENA CAN BE STUDIED

In Fig. 0.2-1 we show a schematic diagram of a large system—for example, a large piece of equipment through which a fluid mixture is flowing. We can describe the transport of mass, momentum, energy, and angular momentum at three different levels.

At the *macroscopic level* (Fig. 0.2-1a) we write down a set of equations called the "macroscopic balances," which describe how the mass, momentum, energy, and angular momentum in the system change because of the introduction and removal of these entities via the entering and leaving streams, and because of various other inputs to the system from the surroundings. No attempt is made to understand all the details of the system. In studying an engineering or biological system it is a good idea to start with this macroscopic description in order to make a global assessment of the problem; in some instances it is only this overall view that is needed.

At the *microscopic level* (Fig. 0.2-1b) we examine what is happening to the fluid mixture in a small region within the equipment. We write down a set of equations called the "equations of change," which describe how the mass, momentum, energy, and angular momentum change within this small region. The aim here is to get information about velocity, temperature, pressure, and concentration profiles within the system. This more detailed information may be required for the understanding of some processes.

At the *molecular level* (Fig. 0.2-1c) we seek a fundamental understanding of the mechanisms of mass, momentum, energy, and angular momentum transport in terms of mol-

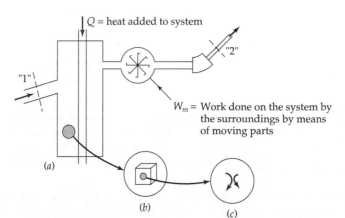

Fig. 0.2-1 (a) A macroscopic flow system containing N_2 and O_2; (b) a microscopic region within the macroscopic system containing N_2 and O_2, which are in a state of flow; (c) a collision between a molecule of N_2 and a molecule of O_2.

ecular structure and intermolecular forces. Generally this is the realm of the theoretical physicist or physical chemist, but occasionally engineers and applied scientists have to get involved at this level. This is particularly true if the processes being studied involve complex molecules, extreme ranges of temperature and pressure, or chemically reacting systems.

It should be evident that these three levels of description involve different "length scales": for example, in a typical industrial problem, at the macroscopic level the dimensions of the flow systems may be of the order of centimeters or meters; the microscopic level involves what is happening in the micron to the centimeter range; and molecular-level problems involve ranges of about 1 to 1000 nanometers.

This book is divided into three parts dealing with

- Flow of pure fluids at constant temperature (with emphasis on viscous and convective momentum transport)—Chapters 1–8

- Flow of pure fluids with varying temperature (with emphasis on conductive, convective, and radiative energy transport)—Chapters 9–16

- Flow of fluid mixtures with varying composition (with emphasis on diffusive and convective mass transport)—Chapters 17–24

That is, we build from the simpler to the more difficult problems. Within each of these parts, we start with an initial chapter dealing with some results of the molecular theory of the transport properties (viscosity, thermal conductivity, and diffusivity). Then we proceed to the microscopic level and learn how to determine the velocity, temperature, and concentration profiles in various kinds of systems. The discussion concludes with the macroscopic level and the description of large systems.

As the discussion unfolds, the reader will appreciate that there are many connections between the levels of description. The transport properties that are described by molecular theory are used at the microscopic level. Furthermore, the equations developed at the microscopic level are needed in order to provide some input into problem solving at the macroscopic level.

There are also many connections between the three areas of momentum, energy, and mass transport. By learning how to solve problems in one area, one also learns the techniques for solving problems in another area. The similarities of the equations in the three areas mean that in many instances one can solve a problem "by analogy"—that is, by taking over a solution directly from one area and, then changing the symbols in the equations, write down the solution to a problem in another area.

The student will find that these connections—among levels, and among the various transport phenomena—reinforce the learning process. As one goes from the first part of the book (momentum transport) to the second part (energy transport) and then on to the third part (mass transport) the story will be very similar but the "names of the players" will change.

Table 0.2-1 shows the arrangement of the chapters in the form of a 3×8 "matrix." Just a brief glance at the matrix will make it abundantly clear what kinds of interconnections can be expected in the course of the study of the book. We recommend that the book be studied by columns, particularly in undergraduate courses. For graduate students, on the other hand, studying the topics by rows may provide a chance to reinforce the connections between the three areas of transport phenomena.

At all three levels of description—molecular, microscopic, and macroscopic—the *conservation laws* play a key role. The derivation of the conservation laws for molecular systems is straightforward and instructive. With elementary physics and a minimum of mathematics we can illustrate the main concepts and review key physical quantities that will be encountered throughout this book. That is the topic of the next section.

Table 0.2-1   Organization of the Topics in This Book

Type of transport	Momentum	Energy	Mass
Transport by molecular motion	1 Viscosity and the stress (momentum flux) tensor	9 Thermal conductivity and the heat-flux vector	17 Diffusivity and the mass-flux vectors
Transport in one dimension (shell-balance methods)	2 Shell momentum balances and velocity distributions	10 Shell energy balances and temperature distributions	18 Shell mass balances and concentration distributions
Transport in arbitrary continua (use of general transport equations)	3 Equations of change and their use [isothermal]	11 Equations of change and their use [nonisothermal]	19 Equations of change and their use [mixtures]
Transport with two independent variables (special methods)	4 Momentum transport with two independent variables	12 Energy transport with two independent variables	20 Mass transport with two independent variables
Transport in turbulent flow, and eddy transport properties	5 Turbulent momentum transport; eddy viscosity	13 Turbulent energy transport; eddy thermal conductivity	21 Turbulent mass transport; eddy diffusivity
Transport across phase boundaries	6 Friction factors; use of empirical correlations	14 Heat-transfer coefficients; use of empirical correlations	22 Mass-transfer coefficients; use of empirical correlations
Transport in large systems, such as pieces of equipment or parts thereof	7 Macroscopic balances [isothermal]	15 Macroscopic balances [nonisothermal]	23 Macroscopic balances [mixtures]
Transport by other mechanisms	8 Momentum transport in polymeric liquids	16 Energy transport by radiation	24 Mass transport in multi-component systems; cross effects

§0.3   THE CONSERVATION LAWS: AN EXAMPLE

The system we consider is that of two colliding diatomic molecules. For simplicity we assume that the molecules do not interact chemically and that each molecule is homonuclear—that is, that its atomic nuclei are identical. The molecules are in a low-density gas, so that we need not consider interactions with other molecules in the neighborhood. In Fig. 0.3-1 we show the collision between the two homonuclear diatomic molecules, A and B, and in Fig. 0.3-2 we show the notation for specifying the locations of the two atoms of one molecule by means of position vectors drawn from an arbitrary origin.

Actually the description of events at the atomic and molecular level should be made by using quantum mechanics. However, except for the lightest molecules (H_2 and He) at

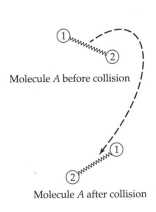

Molecule *A* before collision

Molecule *A* after collision

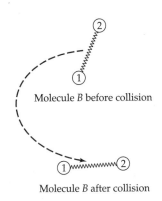

Molecule *B* before collision

Molecule *B* after collision

Fig. 0.3-1 A collision between homonuclear diatomic molecules, such as N_2 and O_2. Molecule *A* is made up of two atoms *A*1 and *A*2. Molecule *B* is made up of two atoms *B*1 and *B*2.

temperatures lower than 50 K, the kinetic theory of gases can be developed quite satisfactorily by use of classical mechanics.

Several relations must hold between quantities before and after a collision. Both before and after the collision the molecules are presumed to be sufficiently far apart that the two molecules cannot "feel" the intermolecular force between them; beyond a distance of about 5 molecular diameters the intermolecular force is known to be negligible. Quantities after the collision are indicated with primes.

(a) According to the *law of conservation of mass*, the total mass of the molecules entering and leaving the collision must be equal:

$$m_A + m_B = m'_A + m'_B \qquad (0.3\text{-}1)$$

Here m_A and m_B are the masses of molecules *A* and *B*. Since there are no chemical reactions, the masses of the individual species will also be conserved, so that

$$m_A = m'_A \quad \text{and} \quad m_B = m'_B \qquad (0.3\text{-}2)$$

(b) According to the *law of conservation of momentum* the sum of the momenta of all the atoms before the collision must equal that after the collision, so that

$$m_{A1}\dot{\mathbf{r}}_{A1} + m_{A2}\dot{\mathbf{r}}_{A2} + m_{B1}\dot{\mathbf{r}}_{B1} + m_{B2}\dot{\mathbf{r}}_{B2} = m'_{A1}\dot{\mathbf{r}}'_{A1} + m'_{A2}\dot{\mathbf{r}}'_{A2} + m'_{B1}\dot{\mathbf{r}}'_{B1} + m'_{B2}\dot{\mathbf{r}}'_{B2} \qquad (0.3\text{-}3)$$

in which $\mathbf{r}_{A1}$ is the position vector for atom 1 of molecule *A*, and $\dot{\mathbf{r}}_{A1}$ is its velocity. We now write $\mathbf{r}_{A1} = \mathbf{r}_A + \mathbf{R}_{A1}$ so that $\mathbf{r}_{A1}$ is written as the sum of the position vector for the

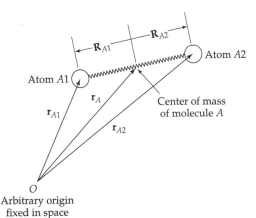

Fig. 0.3-2 Position vectors for the atoms *A*1 and *A*2 in molecule *A*.

center of mass and the position vector of the atom with respect to the center of mass, and we recognize that $\mathbf{R}_{A2} = -\mathbf{R}_{A1}$; we also write the same relations for the velocity vectors. Then we can rewrite Eq. 0.3-3 as

$$m_A \dot{\mathbf{r}}_A + m_B \dot{\mathbf{r}}_B = m_A \dot{\mathbf{r}}'_A + m_B \dot{\mathbf{r}}'_B \tag{0.3-4}$$

That is, the conservation statement can be written in terms of the molecular masses and velocities, and the corresponding atomic quantities have been eliminated. In getting Eq. 0.3-4 we have used Eq. 0.3-2 and the fact that for homonuclear diatomic molecules $m_{A1} = m_{A2} = \frac{1}{2}m_A$.

(c) According to the *law of conservation of energy*, the energy of the colliding pair of molecules must be the same before and after the collision. The energy of an isolated molecule is the sum of the kinetic energies of the two atoms and the interatomic potential energy, ϕ_A, which describes the force of the chemical bond joining the two atoms 1 and 2 of molecule A, and is a function of the interatomic distance $|\mathbf{r}_{A2} - \mathbf{r}_{A1}|$. Therefore, energy conservation leads to

$$\left(\tfrac{1}{2}m_{A1}\dot{r}^2_{A1} + \tfrac{1}{2}m_{A2}\dot{r}^2_{A2} + \phi_A\right) + \left(\tfrac{1}{2}m_{B1}\dot{r}^2_{B1} + \tfrac{1}{2}m_{B2}\dot{r}^2_{B2} + \phi_B\right) =$$
$$\left(\tfrac{1}{2}m'_{A1}\dot{r}'^2_{A1} + \tfrac{1}{2}m'_{A2}\dot{r}'^2_{A2} + \phi'_A\right) + \left(\tfrac{1}{2}m'_{B1}\dot{r}'^2_{B1} + \tfrac{1}{2}m'_{B2}\dot{r}'^2_{B2} + \phi'_B\right) \tag{0.3-5}$$

Note that we use the standard abbreviated notation that $\dot{r}^2_{A1} = (\dot{\mathbf{r}}_{A1} \cdot \dot{\mathbf{r}}_{A1})$. We now write the velocity of atom 1 of molecule A as the sum of the velocity of the center of mass of A and the velocity of 1 with respect to the center of mass; that is, $\dot{\mathbf{r}}_{A1} = \dot{\mathbf{r}}_A + \dot{\mathbf{R}}_{A1}$. Then Eq. 0.3-5 becomes

$$\left(\tfrac{1}{2}m_A \dot{r}^2_A + u_A\right) + \left(\tfrac{1}{2}m_B \dot{r}^2_B + u_B\right) = \left(\tfrac{1}{2}m'_A \dot{r}'^2_A + u'_A\right) + \left(\tfrac{1}{2}m'_B \dot{r}'^2_B + u'_B\right) \tag{0.3-6}$$

in which $u_A = \tfrac{1}{2}m_{A1}\dot{R}^2_{A1} + \tfrac{1}{2}m_{A2}\dot{R}^2_{A2} + \phi_A$ is the sum of the kinetic energies of the atoms, referred to the center of mass of molecule A, and the interatomic potential of molecule A. That is, we split up the energy of each molecule into its kinetic energy with respect to fixed coordinates, and the internal energy of the molecule (which includes its vibrational, rotational, and potential energies). Equation 0.3-6 makes it clear that the kinetic energies of the colliding molecules can be converted into internal energy or vice versa. This idea of an interchange between kinetic and internal energy will arise again when we discuss the energy relations at the microscopic and macroscopic levels.

(d) Finally, the *law of conservation of angular momentum* can be applied to a collision to give

$$([\mathbf{r}_{A1} \times m_{A1}\dot{\mathbf{r}}_{A1}] + [\mathbf{r}_{A2} \times m_{A2}\dot{\mathbf{r}}_{A2}]) + ([\mathbf{r}_{B1} \times m_{B1}\dot{\mathbf{r}}_{B1}] + [\mathbf{r}_{B2} \times m_{B2}\dot{\mathbf{r}}_{B2}]) =$$
$$([\mathbf{r}'_{A1} \times m'_{A1}\dot{\mathbf{r}}'_{A1}] + [\mathbf{r}'_{A2} \times m'_{A2}\dot{\mathbf{r}}'_{A2}]) + ([\mathbf{r}'_{B1} \times m'_{B1}\dot{\mathbf{r}}'_{B1}] + [\mathbf{r}'_{B2} \times m'_{B2}\dot{\mathbf{r}}'_{B2}]) \tag{0.3-7}$$

in which $\times$ is used to indicate the cross product of two vectors. Next we introduce the center-of-mass and relative position vectors and velocity vectors as before and obtain

$$([\mathbf{r}_A \times m_A\dot{\mathbf{r}}_A] + \mathbf{l}_A) + ([\mathbf{r}_B \times m_B\dot{\mathbf{r}}_B] + \mathbf{l}_B) =$$
$$([\mathbf{r}'_A \times m_A\dot{\mathbf{r}}'_A] + \mathbf{l}'_A) + ([\mathbf{r}'_B \times m_B\dot{\mathbf{r}}'_B] + \mathbf{l}'_B) \tag{0.3-8}$$

in which $\mathbf{l}_A = [\mathbf{R}_{A1} \times m_{A1}\dot{\mathbf{R}}_{A1}] + [\mathbf{R}_{A2} \times m_{A2}\dot{\mathbf{R}}_{A2}]$ is the sum of the angular momenta of the atoms referred to an origin of coordinates at the center of mass of the molecule—that is, the "internal angular momentum." The important point is that there is the possibility for interchange between the angular momentum of the molecules (with respect to the origin of coordinates) and their internal angular momentum (with respect to the center of mass of the molecule). This will be referred to later in connection with the equation of change for angular momentum.

The conservation laws as applied to collisions of monatomic molecules can be obtained from the results above as follows: Eqs. 0.3-1, 0.3-2, and 0.3-4 are directly applicable; Eq. 0.3-6 is applicable if the internal energy contributions are omitted; and Eq. 0.3-8 may be used if the internal angular momentum terms are discarded.

Much of this book will be concerned with setting up the conservation laws at the microscopic and macroscopic levels and applying them to problems of interest in engineering and science. The above discussion should provide a good background for this adventure. For a glimpse of the conservation laws for species mass, momentum, and energy at the microscopic and macroscopic levels, see Tables 19.2-1 and 23.5-1.

§0.4   CONCLUDING COMMENTS

To use the macroscopic balances intelligently, it is necessary to use information about interphase transport that comes from the equations of change. To use the equations of change, we need the transport properties, which are described by various molecular theories. Therefore, from a teaching point of view, it seems best to start at the molecular level and work upward toward the larger systems.

All the discussions of theory are accompanied by examples to illustrate how the theory is applied to problem solving. Then at the end of each chapter there are problems to provide extra experience in using the ideas given in the chapter. The problems are grouped into four classes:

Class A: Numerical problems, which are designed to highlight important equations in the text and to give a feeling for the orders of magnitude.

Class B: Analytical problems that require doing elementary derivations using ideas mainly from the chapter.

Class C: More advanced analytical problems that may bring ideas from other chapters or from other books.

Class D: Problems in which intermediate mathematical skills are required.

Many of the problems and illustrative examples are rather elementary in that they involve oversimplified systems or very idealized models. It is, however, necessary to start with these elementary problems in order to understand how the theory works and to develop confidence in using it. In addition, some of these elementary examples can be very useful in making order-of-magnitude estimates in complex problems.

Here are a few suggestions for studying the subject of transport phenomena:

- Always read the text with pencil and paper in hand; work through the details of the mathematical developments and supply any missing steps.

- Whenever necessary, go back to the mathematics textbooks to brush up on calculus, differential equations, vectors, etc. This is an excellent time to review the mathematics that was learned earlier (but possibly not as carefully as it should have been).

- Make it a point to give a physical interpretation of key results; that is, get in the habit of relating the physical ideas to the equations.

- Always ask whether the results seem reasonable. If the results do not agree with intuition, it is important to find out which is incorrect.

- Make it a habit to check the dimensions of all results. This is one very good way of locating errors in derivations.

We hope that the reader will share our enthusiasm for the subject of transport phenomena. It will take some effort to learn the material, but the rewards will be worth the time and energy required.

QUESTIONS FOR DISCUSSION

1. What are the definitions of momentum, angular momentum, and kinetic energy for a single particle? What are the dimensions of these quantities?
2. What are the dimensions of velocity, angular velocity, pressure, density, force, work, and torque? What are some common units used for these quantities?
3. Verify that it is possible to go from Eq. 0.3-3 to Eq. 0.3-4.
4. Go through all the details needed to get Eq. 0.3-6 from Eq. 0.3-5.
5. Suppose that the origin of coordinates is shifted to a new position. What effect would that have on Eq. 0.3-7? Is the equation changed?
6. Compare and contrast angular velocity and angular momentum.
7. What is meant by internal energy? Potential energy?
8. Is the law of conservation of mass always valid? What are the limitations?

Part One

Momentum Transport

Chapter 1

Viscosity and the Mechanisms of Momentum Transport

§1.1 Newton's law of viscosity (molecular momentum transport)

§1.2 Generalization of Newton's law of viscosity

§1.3 Pressure and temperature dependence of viscosity

§1.4° Molecular theory of the viscosity of gases at low density

§1.5° Molecular theory of the viscosity of liquids

§1.6° Viscosity of suspensions and emulsions

§1.7 Convective momentum transport

The first part of this book deals with the flow of viscous fluids. For fluids of low molecular weight, the physical property that characterizes the resistance to flow is the *viscosity*. Anyone who has bought motor oil is aware of the fact that some oils are more "viscous" than others and that viscosity is a function of the temperature.

We begin in §1.1 with the simple shear flow between parallel plates and discuss how momentum is transferred through the fluid by viscous action. This is an elementary example of *molecular momentum transport* and it serves to introduce "Newton's law of viscosity" along with the definition of viscosity μ. Next in §1.2 we show how Newton's law can be generalized for arbitrary flow patterns. The effects of temperature and pressure on the viscosities of gases and liquids are summarized in §1.3 by means of a dimensionless plot. Then §1.4 tells how the viscosities of gases can be calculated from the kinetic theory of gases, and in §1.5 a similar discussion is given for liquids. In §1.6 we make a few comments about the viscosity of suspensions and emulsions.

Finally, we show in §1.7 that momentum can also be transferred by the bulk fluid motion and that such *convective momentum transport* is proportional to the fluid density ρ.

§1.1 NEWTON'S LAW OF VISCOSITY (MOLECULAR TRANSPORT OF MOMENTUM)

In Fig. 1.1-1 we show a pair of large parallel plates, each one with area A, separated by a distance Y. In the space between them is a fluid—either a gas or a liquid. This system is initially at rest, but at time $t = 0$ the lower plate is set in motion in the positive x direction at a constant velocity V. As time proceeds, the fluid gains momentum, and ultimately the linear steady-state velocity profile shown in the figure is established. We require that the flow be laminar ("laminar" flow is the orderly type of flow that one usually observes when syrup is poured, in contrast to "turbulent" flow, which is the irregular, chaotic flow one sees in a high-speed mixer). When the final state of steady motion

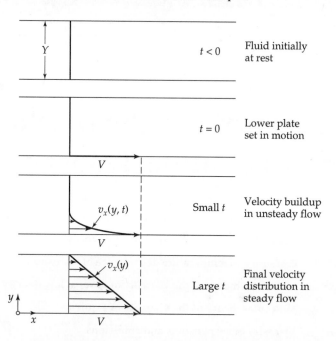

Fig. 1.1-1 The buildup to the steady, laminar velocity profile for a fluid contained between two plates. The flow is called "laminar" because the adjacent layers of fluid ("laminae") slide past one another in an orderly fashion.

has been attained, a constant force F is required to maintain the motion of the lower plate. Common sense suggests that this force may be expressed as follows:

$$\frac{F}{A} = \mu \frac{V}{Y} \tag{1.1-1}$$

That is, the force should be proportional to the area and to the velocity, and inversely proportional to the distance between the plates. The constant of proportionality μ is a property of the fluid, defined to be the *viscosity*.

We now switch to the notation that will be used throughout the book. First we replace F/A by the symbol τ_{yx}, which is the force in the x direction on a unit area perpendicular to the y direction. It is understood that this is the force exerted by the fluid of lesser y on the fluid of greater y. Furthermore, we replace V/Y by $-dv_x/dy$. Then, in terms of these symbols, Eq. 1.1-1 becomes

$$\tau_{yx} = -\mu \frac{dv_x}{dy} \tag{1.1-2}[1]$$

This equation, which states that the shearing force per unit area is proportional to the negative of the velocity gradient, is often called *Newton's law of viscosity.*[2] Actually we

---

[1] Some authors write Eq. 1.1-2 in the form

$$g_c \tau_{yx} = -\mu \frac{dv_x}{dy} \tag{1.1-2a}$$

in which τ_{yx} [=] lb_f/ft^2, v_x [=] ft/s, y [=] ft, and μ [=] $lb_m/ft \cdot s$; the quantity g_c is the "gravitational conversion factor" with the value of 32.174 poundals/lb_f. In this book we will always use Eq. 1.1-2 rather than Eq. 1.1-2a.

[2] Sir **Isaac Newton** (1643–1727), a professor at Cambridge University and later Master of the Mint, was the founder of classical mechanics and contributed to other fields of physics as well. Actually Eq. 1.1-2 does not appear in Sir Isaac Newton's *Philosophiae Naturalis Principia Mathematica*, but the germ of the idea is there. For illuminating comments, see D. J. Acheson, *Elementary Fluid Dynamics*, Oxford University Press, 1990, §6.1.

should not refer to Eq. 1.1-2 as a "law," since Newton suggested it as an empiricism[3]—the simplest proposal that could be made for relating the stress and the velocity gradient. However, it has been found that the resistance to flow of all gases and all liquids with molecular weight of less than about 5000 is described by Eq. 1.1-2, and such fluids are referred to as *Newtonian fluids*. Polymeric liquids, suspensions, pastes, slurries, and other complex fluids are not described by Eq. 1.1-2 and are referred to as *non-Newtonian fluids*. Polymeric liquids are discussed in Chapter 8.

Equation 1.1-2 may be interpreted in another fashion. In the neighborhood of the moving solid surface at $y = 0$ the fluid acquires a certain amount of x-momentum. This fluid, in turn, imparts momentum to the adjacent layer of liquid, causing it to remain in motion in the x direction. Hence x-momentum is being transmitted through the fluid in the positive y direction. Therefore τ_{yx} may also be interpreted as the *flux of x-momentum in the positive y direction*, where the term "flux" means "flow per unit area." This interpretation is consistent with the molecular picture of momentum transport and the kinetic theories of gases and liquids. It also is in harmony with the analogous treatment given later for heat and mass transport.

The idea in the preceding paragraph may be paraphrased by saying that momentum goes "downhill" from a region of high velocity to a region of low velocity—just as a sled goes downhill from a region of high elevation to a region of low elevation, or the way heat flows from a region of high temperature to a region of low temperature. The velocity gradient can therefore be thought of as a "driving force" for momentum transport.

In what follows we shall sometimes refer to Newton's law in Eq. 1.1-2 in terms of forces (which emphasizes the mechanical nature of the subject) and sometimes in terms of momentum transport (which emphasizes the analogies with heat and mass transport). This dual viewpoint should prove helpful in physical interpretations.

Often fluid dynamicists use the symbol ν to represent the viscosity divided by the density (mass per unit volume) of the fluid, thus:

$$\nu = \mu/\rho \tag{1.1-3}$$

This quantity is called the *kinematic viscosity*.

Next we make a few comments about the units of the quantities we have defined. If we use the symbol $[=]$ to mean "has units of," then in the SI system $\tau_{yx} [=] \text{N/m}^2 = \text{Pa}$, $v_x [=] \text{m/s}$, and $y [=] \text{m}$, so that

$$\mu = -\tau_{yx}\left(\frac{dv_x}{dy}\right)^{-1} [=] (\text{Pa})[(\text{m/s})(\text{m}^{-1})]^{-1} = \text{Pa} \cdot \text{s} \tag{1.1-4}$$

since the units on both sides of Eq. 1.1-2 must agree. We summarize the above and also give the units for the c.g.s. system and the British system in Table 1.1-1. The conversion tables in Appendix F will prove to be very useful for solving numerical problems involving diverse systems of units.

The viscosities of fluids vary over many orders of magnitude, with the viscosity of air at 20°C being 1.8×10^{-5} Pa · s and that of glycerol being about 1 Pa · s, with some silicone oils being even more viscous. In Tables 1.1-2, 1.1-3, and 1.1-4 experimental data[4] are

---

[3] A relation of the form of Eq. 1.1-2 does come out of the simple kinetic theory of gases (Eq. 1.4-7). However, a rigorous theory for gases sketched in Appendix D makes it clear that Eq. 1.1-2 arises as the first term in an expansion, and that additional (higher-order) terms are to be expected. Also, even an elementary kinetic theory of liquids predicts non-Newtonian behavior (Eq. 1.5-6).

[4] A comprehensive presentation of experimental techniques for measuring transport properties can be found in W. A. Wakeham, A. Nagashima, and J. V. Sengers, *Measurement of the Transport Properties of Fluids*, CRC Press, Boca Raton, Fla. (1991). Sources for experimental data are: Landolt-Börnstein, *Zahlenwerte und Funktionen*, Vol. II, 5, Springer (1968–1969); *International Critical Tables*, McGraw-Hill, New York (1926); Y. S. Touloukian, P. E. Liley, and S. C. Saxena, *Thermophysical Properties of Matter*, Plenum Press, New York (1970); and also numerous handbooks of chemistry, physics, fluid dynamics, and heat transfer.

Table 1.1-1 Summary of Units for Quantities Related to Eq. 1.1-2

	SI	c.g.s.	British
τ_{yx}	Pa	dyn/cm^2	poundals/ft^2
v_x	m/s	cm/s	ft/s
y	m	cm	ft
μ	Pa·s	gm/cm·s = poise	lb$_m$/ft·s
ν	m^2/s	cm^2/s	ft^2/s

Note: The pascal, Pa, is the same as N/m^2, and the newton, N, is the same as kg·m/s^2. The abbreviation for "centipoise" is "cp."

Table 1.1-2 Viscosity of Water and Air at 1 atm Pressure

Temperature T (°C)	Water (liq.)[a]		Air[b]	
	Viscosity μ (mPa·s)	Kinematic viscosity ν (cm^2/s)	Viscosity μ (mPa·s)	Kinematic viscosity ν (cm^2/s)
0	1.787	0.01787	0.01716	0.1327
20	1.0019	0.010037	0.01813	0.1505
40	0.6530	0.006581	0.01908	0.1692
60	0.4665	0.004744	0.01999	0.1886
80	0.3548	0.003651	0.02087	0.2088
100	0.2821	0.002944	0.02173	0.2298

[a] Calculated from the results of R. C. Hardy and R. L. Cottington, *J. Research Nat. Bur. Standards*, **42**, 573–578 (1949); and J. F. Swidells, J. R. Coe, Jr., and T. B. Godfrey, *J. Research Nat. Bur. Standards*, **48**, 1–31 (1952).

[b] Calculated from "Tables of Thermal Properties of Gases," *National Bureau of Standards Circular* **464** (1955), Chapter 2.

Table 1.1-3 Viscosities of Some Gases and Liquids at Atmospheric Pressure[a]

Gases	Temperature T (°C)	Viscosity μ (mPa·s)	Liquids	Temperature T (°C)	Viscosity μ (mPa·s)
i-C$_4$H$_{10}$	23	0.0076[c]	(C$_2$H$_5$)$_2$O	0	0.283
SF$_6$	23	0.0153		25	0.224
CH$_4$	20	0.0109[b]	C$_6$H$_6$	20	0.649
H$_2$O	100	0.01211[d]	Br$_2$	25	0.744
CO$_2$	20	0.0146[b]	Hg	20	1.552
N$_2$	20	0.0175[b]	C$_2$H$_5$OH	0	1.786
O$_2$	20	0.0204		25	1.074
Hg	380	0.0654[d]		50	0.694
			H$_2$SO$_4$	25	25.54
			Glycerol	25	934.

[a] Values taken from N. A. Lange, *Handbook of Chemistry*, McGraw-Hill, New York, 15th edition (1999), Tables 5.16 and 5.18.

[b] H. L. Johnston and K. E. McCloskey, *J. Phys. Chem.*, **44**, 1038–1058 (1940).

[c] *CRC Handbook of Chemistry and Physics*, CRC Press, Boca Raton, Fla. (1999).

[d] *Landolt-Börnstein Zahlenwerte und Funktionen*, Springer (1969).

Table 1.1-4 Viscosities of Some Liquid Metals

Metal	Temperature T (°C)	Viscosity μ (mPa · s)
Li	183.4	0.5918
	216.0	0.5406
	285.5	0.4548
Na	103.7	0.686
	250	0.381
	700	0.182
K	69.6	0.515
	250	0.258
	700	0.136
Hg	−20	1.85
	20	1.55
	100	1.21
	200	1.01
Pb	441	2.116
	551	1.700
	844	1.185

Data taken from *The Reactor Handbook*, Vol. 2, Atomic Energy Commission AECD-3646, U.S. Government Printing Office, Washington, D.C. (May 1955), pp. 258 *et seq.*

given for pure fluids at 1 atm pressure. Note that for gases at low density, the viscosity *increases* with increasing temperature, whereas for liquids the viscosity usually *decreases* with increasing temperature. In gases the momentum is transported by the molecules in free flight between collisions, but in liquids the transport takes place predominantly by virtue of the intermolecular forces that pairs of molecules experience as they wind their way around among their neighbors. In §§1.4 and 1.5 we give some elementary kinetic theory arguments to explain the temperature dependence of viscosity.

EXAMPLE 1.1-1

Calculation of Momentum Flux

Compute the steady-state momentum flux τ_{yx} in lb_f/ft^2 when the lower plate velocity V in Fig. 1.1-1 is 1 ft/s in the positive x direction, the plate separation Y is 0.001 ft, and the fluid viscosity μ is 0.7 cp.

SOLUTION

Since τ_{yx} is desired in British units, we should convert the viscosity into that system of units. Thus, making use of Appendix F, we find $\mu = (0.7 \text{ cp})(2.0886 \times 10^{-5}) = 1.46 \times 10^{-5} \, lb_f \, s/ft^2$. The velocity profile is linear so that

$$\frac{dv_x}{dy} = \frac{\Delta v_x}{\Delta y} = \frac{-1.0 \text{ ft/s}}{0.001 \text{ ft}} = -1000 s^{-1} \tag{1.1-5}$$

Substitution into Eq. 1.1-2 gives

$$\tau_{yx} = -\mu \frac{dv_x}{dy} = -(1.46 \times 10^{-5})(-1000) = 1.46 \times 10^{-2} \, lb_f/ft^2 \tag{1.1-6}$$

§1.2   GENERALIZATION OF NEWTON'S LAW OF VISCOSITY

In the previous section the viscosity was defined by Eq. 1.1-2, in terms of a simple steady-state shearing flow in which v_x is a function of y alone, and v_y and v_z are zero. Usually we are interested in more complicated flows in which the three velocity components may depend on all three coordinates and possibly on time. Therefore we must have an expression more general than Eq. 1.1-2, but it must simplify to Eq. 1.1-2 for steady-state shearing flow.

This generalization is not simple; in fact, it took mathematicians about a century and a half to do this. It is not appropriate for us to give all the details of this development here, since they can be found in many fluid dynamics books.[1] Instead we explain briefly the main ideas that led to the discovery of the required generalization of Newton's law of viscosity.

To do this we consider a very general flow pattern, in which the fluid velocity may be in various directions at various places and may depend on the time t. The velocity components are then given by

$$v_x = v_x(x, y, z, t); \qquad v_y = v_y(x, y, z, t); \qquad v_z = v_z(x, y, z, t) \tag{1.2-1}$$

In such a situation, there will be nine stress components τ_{ij} (where i and j may take on the designations x, y, and z), instead of the component τ_{yx} that appears in Eq. 1.1-2. We therefore must begin by defining these stress components.

In Fig. 1.2-1 is shown a small cube-shaped volume element within the flow field, each face having unit area. The center of the volume element is at the position x, y, z. At

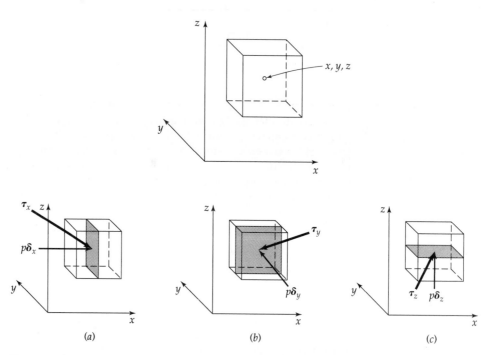

Fig. 1.2-1 Pressure and viscous forces acting on planes in the fluid perpendicular to the three coordinate systems. The shaded planes have unit area.

[1] W. Prager, *Introduction to Mechanics of Continua*, Ginn, Boston (1961), pp. 89–91; R. Aris, *Vectors, Tensors, and the Basic Equations of Fluid Mechanics*, Prentice-Hall, Englewood Cliffs, N.J. (1962), pp. 30–34, 99–112; L. Landau and E. M. Lifshitz, *Fluid Mechanics*, Pergamon, London, 2nd edition (1987), pp. 44–45. **Lev Davydovich Landau** (1908–1968) received the Nobel prize in 1962 for his work on liquid helium and superfluid dynamics.

any instant of time we can slice the volume element in such a way as to remove half the fluid within it. As shown in the figure, we can cut the volume perpendicular to each of the three coordinate directions in turn. We can then ask what force has to be applied on the free (shaded) surface in order to replace the force that had been exerted on that surface by the fluid that was removed. There will be two contributions to the force: that associated with the pressure, and that associated with the viscous forces.

The pressure force will always be perpendicular to the exposed surface. Hence in (a) the force per unit area on the shaded surface will be a vector $p\boldsymbol{\delta}_x$—that is, the pressure (a scalar) multiplied by the unit vector $\boldsymbol{\delta}_x$ in the x direction. Similarly, the force on the shaded surface in (b) will be $p\boldsymbol{\delta}_y$, and in (c) the force will be $p\boldsymbol{\delta}_z$. The pressure forces will be exerted when the fluid is stationary as well as when it is in motion.

The viscous forces come into play only when there are velocity gradients within the fluid. In general they are neither perpendicular to the surface element nor parallel to it, but rather at some angle to the surface (see Fig. 1.2-1). In (a) we see a force per unit area $\boldsymbol{\tau}_x$ exerted on the shaded area, and in (b) and (c) we see forces per unit area $\boldsymbol{\tau}_y$ and $\boldsymbol{\tau}_z$. Each of these forces (which are vectors) has components (scalars); for example, $\boldsymbol{\tau}_x$ has components τ_{xx}, τ_{xy}, and τ_{xz}. Hence we can now summarize the forces acting on the three shaded areas in Fig. 1.2-1 in Table 1.2-1. This tabulation is a summary of the forces per unit area (*stresses*) exerted within a fluid, both by the thermodynamic pressure and the *viscous stresses*. Sometimes we will find it convenient to have a symbol that includes both types of stresses, and so we define the *molecular stresses* as follows:

$$\pi_{ij} = p\delta_{ij} + \tau_{ij} \qquad \text{where } i \text{ and } j \text{ may be } x, y, \text{ or } z \qquad (1.2\text{-}2)$$

Here δ_{ij} is the *Kronecker delta*, which is 1 if $i = j$ and zero if $i \neq j$.

Just as in the previous section, the τ_{ij} (and also the π_{ij}) may be interpreted in two ways:

$\pi_{ij} = p\delta_{ij} + \tau_{ij} =$ force in the j direction on a unit area perpendicular to the i direction, where it is understood that the fluid in the region of lesser x_i is exerting the force on the fluid of greater x_i

$\pi_{ij} = p\delta_{ij} + \tau_{ij} =$ flux of j-momentum in the positive i direction—that is, from the region of lesser x_i to that of greater x_i

Both interpretations are used in this book; the first one is particularly useful in describing the forces exerted by the fluid on solid surfaces. The stresses $\pi_{xx} = p + \tau_{xx}$, $\pi_{yy} = p + \tau_{yy}$, $\pi_{zz} = p + \tau_{zz}$ are called *normal stresses*, whereas the remaining quantities, $\pi_{xy} = \tau_{xy}$, $\pi_{yz} = \tau_{yz}$, ... are called *shear stresses*. These quantities, which have two subscripts associated with the coordinate directions, are referred to as "tensors," just as quantities (such as velocity) that have one subscript associated with the coordinate directions are called

Table 1.2-1   Summary of the Components of the Molecular Stress Tensor (or Molecular Momentum-Flux Tensor)[a]

Direction normal to the shaded face	Vector force per unit area on the shaded face (momentum flux through shaded face)	Components of the forces (per unit area) acting on the shaded face (components of the momentum flux through the shaded face)		
		x-component	y-component	z-component
x	$\boldsymbol{\pi}_x = p\boldsymbol{\delta}_x + \boldsymbol{\tau}_x$	$\pi_{xx} = p + \tau_{xx}$	$\pi_{xy} = \tau_{xy}$	$\pi_{xz} = \tau_{xz}$
y	$\boldsymbol{\pi}_y = p\boldsymbol{\delta}_y + \boldsymbol{\tau}_y$	$\pi_{yx} = \tau_{yx}$	$\pi_{yy} = p + \tau_{yy}$	$\pi_{yz} = \tau_{yz}$
z	$\boldsymbol{\pi}_z = p\boldsymbol{\delta}_z + \boldsymbol{\tau}_z$	$\pi_{zx} = \tau_{zx}$	$\pi_{zy} = \tau_{zy}$	$\pi_{zz} = p + \tau_{zz}$

[a] These are referred to as components of the "molecular momentum flux tensor" because they are associated with the molecular motions, as discussed in §1.4 and Appendix D. The additional "convective momentum flux tensor" components, associated with bulk movement of the fluid, are discussed in §1.7.

"vectors." Therefore we will refer to $\boldsymbol{\tau}$ as the *viscous stress tensor* (with components τ_{ij}) and $\boldsymbol{\pi}$ as the *molecular stress tensor* (with components π_{ij}). When there is no chance for confusion, the modifiers "viscous" and "molecular" may be omitted. A discussion of vectors and tensors can be found in Appendix A.

The question now is: How are these stresses τ_{ij} related to the velocity gradients in the fluid? In generalizing Eq. 1.1-2, we put several restrictions on the stresses, as follows:

- The viscous stresses may be linear combinations of all the velocity gradients:

$$\tau_{ij} = -\textstyle\sum_k\sum_l \mu_{ijkl}\frac{\partial v_k}{\partial x_l} \qquad \text{where } i, j, k, \text{ and } l \text{ may be 1, 2, 3} \qquad (1.2\text{-}3)$$

  Here the 81 quantities μ_{ijkl} are "viscosity coefficients." The quantities x_1, x_2, x_3 in the derivatives denote the Cartesian coordinates x, y, z, and v_1, v_2, v_3 are the same as v_x, v_y, v_z.

- We assert that time derivatives or time integrals should not appear in the expression. (For viscoelastic fluids, as discussed in Chapter 8, time derivatives or time integrals are needed to describe the elastic responses.)

- We do not expect any viscous forces to be present, if the fluid is in a state of pure rotation. This requirement leads to the necessity that τ_{ij} be a symmetric combination of the velocity gradients. By this we mean that if i and j are interchanged, the combination of velocity gradients remains unchanged. It can be shown that the only symmetric linear combinations of velocity gradients are

$$\left(\frac{\partial v_j}{\partial x_i} + \frac{\partial v_i}{\partial x_j}\right) \quad \text{and} \quad \left(\frac{\partial v_x}{\partial x} + \frac{\partial v_y}{\partial y} + \frac{\partial v_z}{\partial z}\right)\delta_{ij} \qquad (1.2\text{-}4)$$

- If the fluid is isotropic—that is, it has no preferred direction—then the coefficients in front of the two expressions in Eq. 1.2-4 must be scalars so that

$$\tau_{ij} = A\left(\frac{\partial v_j}{\partial x_i} + \frac{\partial v_i}{\partial x_j}\right) + B\left(\frac{\partial v_x}{\partial x} + \frac{\partial v_y}{\partial y} + \frac{\partial v_z}{\partial z}\right)\delta_{ij} \qquad (1.2\text{-}5)$$

  We have thus reduced the number of "viscosity coefficients" from 81 to 2!

- Of course, we want Eq. 1.2-5 to simplify to Eq. 1.1-2 for the flow situation in Fig. 1.1-1. For that elementary flow Eq. 1.2-5 simplifies to $\tau_{yx} = A\, dv_x/dy$, and hence the scalar constant A must be the same as the negative of the *viscosity* μ.

- Finally, by common agreement among most fluid dynamicists the scalar constant B is set equal to $\frac{2}{3}\mu - \kappa$, where κ is called the *dilatational viscosity*. The reason for writing B in this way is that it is known from kinetic theory that κ is identically zero for monatomic gases at low density.

Thus the required generalization for Newton's law of viscosity in Eq. 1.1-2 is then the set of nine relations (six being independent):

$$\tau_{ij} = -\mu\left(\frac{\partial v_j}{\partial x_i} + \frac{\partial v_i}{\partial x_j}\right) + \left(\tfrac{2}{3}\mu - \kappa\right)\left(\frac{\partial v_x}{\partial x} + \frac{\partial v_y}{\partial y} + \frac{\partial v_z}{\partial z}\right)\delta_{ij} \qquad (1.2\text{-}6)$$

Here $\tau_{ij} = \tau_{ji}$, and i and j can take on the values 1, 2, 3. These relations for the stresses in a Newtonian fluid are associated with the names of Navier, Poisson, and Stokes.[2] If de-

---

[2] C.-L.-M.-H. Navier, *Ann. Chimie*, **19**, 244–260 (1821); S.-D. Poisson, *J. École Polytech.*, **13**, Cahier 20, 1–174 (1831); G. G. Stokes, *Trans. Camb. Phil. Soc.*, **8**, 287–305 (1845). **Claude-Louis-Marie-Henri Navier** (1785–1836) (pronounced "Nah-vyay," with the second syllable accented) was a civil engineer whose specialty was road and bridge building; **George Gabriel Stokes** (1819–1903) taught at Cambridge University and was president of the Royal Society. Navier and Stokes are well known because of the Navier–Stokes equations (see Chapter 3). See also D. J. Acheson, *Elementary Fluid Mechanics*, Oxford University Press (1990), pp. 209–212, 218.

sired, this set of relations can be written more concisely in the vector-tensor notation of Appendix A as

$$\boxed{\boldsymbol{\tau} = -\mu(\nabla\mathbf{v} + (\nabla\mathbf{v})^\dagger) + (\tfrac{2}{3}\mu - \kappa)(\nabla\cdot\mathbf{v})\boldsymbol{\delta}}$$

(1.2-7)

in which $\boldsymbol{\delta}$ is the *unit tensor* with components δ_{ij}, $\nabla\mathbf{v}$ is the *velocity gradient tensor* with components $(\partial/\partial x_i)v_j$, $(\nabla\mathbf{v})^\dagger$ is the "transpose" of the velocity gradient tensor with components $(\partial/\partial x_j)v_i$, and $(\nabla\cdot\mathbf{v})$ is the *divergence* of the velocity vector.

The important conclusion is that we have a generalization of Eq. 1.1-2, and this generalization involves not one but two coefficients[3] characterizing the fluid: the viscosity μ and the dilatational viscosity κ. Usually, in solving fluid dynamics problems, it is not necessary to know κ. If the fluid is a gas, we often assume it to act as an ideal monoatomic gas, for which κ is identically zero. If the fluid is a liquid, we often assume that it is incompressible, and in Chapter 3 we show that for incompressible liquids $(\nabla\cdot\mathbf{v}) = 0$, and therefore the term containing κ is discarded anyway. The dilational viscosity is important in describing sound absorption in polyatomic gases[4] and in describing the fluid dynamics of liquids containing gas bubbles.[5]

Equation 1.2-7 (or 1.2-6) is an important equation and one that we shall use often. Therefore it is written out in full in Cartesian (x, y, z), cylindrical (r, θ, z), and spherical (r, θ, ϕ) coordinates in Table B.1. The entries in this table for curvilinear coordinates are obtained by the methods outlined in §§A.6 and A.7. It is suggested that beginning students not concern themselves with the details of such derivations, but rather concentrate on using the tabulated results. Chapters 2 and 3 will give ample practice in doing this.

In curvilinear coordinates the stress components have the same meaning as in Cartesian coordinates. For example, τ_{rz} in cylindrical coordinates, which will be encountered in Chapter 2, can be interpreted as: (i) the viscous force in the z direction on a unit area perpendicular to the r direction, or (ii) the viscous flux of z-momentum in the positive r direction. Figure 1.2-2 illustrates some typical surface elements and stress-tensor components that arise in fluid dynamics.

The shear stresses are usually easy to visualize, but the normal stresses may cause conceptual problems. For example, τ_{zz} is a force per unit area in the z direction on a plane perpendicular to the z direction. For the flow of an incompressible fluid in the convergent channel of Fig. 1.2-3, we know intuitively that v_z increases with decreasing z; hence, according to Eq. 1.2-6, there is a nonzero stress $\tau_{zz} = -2\mu(\partial v_z/\partial z)$ acting in the fluid.

Note on the Sign Convention for the Stress Tensor   We have emphasized in connection with Eq. 1.1-2 (and in the generalization in this section) that τ_{yx} is the force in the positive x direction on a plane perpendicular to the y direction, and that this is the force exerted by the fluid in the region of the *lesser y* on the fluid of *greater y*. In most fluid dynamics and elasticity books, the words "lesser" and "greater" are interchanged and Eq. 1.1-2 is written as $\tau_{yx} = +\mu(dv_x/dy)$. The advantages of the sign convention used in this book are: (a) the sign convention used in Newton's law of viscosity is consistent with that used in Fourier's law of heat conduction and Fick's law of diffusion; (b) the sign convention for τ_{ij} is the same as that for the convective momentum flux $\rho\mathbf{vv}$ (see

---

[3] Some writers refer to μ as the "shear viscosity," but this is inappropriate nomenclature inasmuch as μ can arise in nonshearing flows as well as shearing flows. The term "dynamic viscosity" is also occasionally seen, but this term has a very specific meaning in the field of viscoelasticity and is an inappropriate term for μ.

[4] L. Landau and E. M. Lifshitz, *op. cit.*, Ch. VIII.

[5] G. K. Batchelor, *An Introduction to Fluid Dynamics*, Cambridge University Press (1967), pp. 253–255.

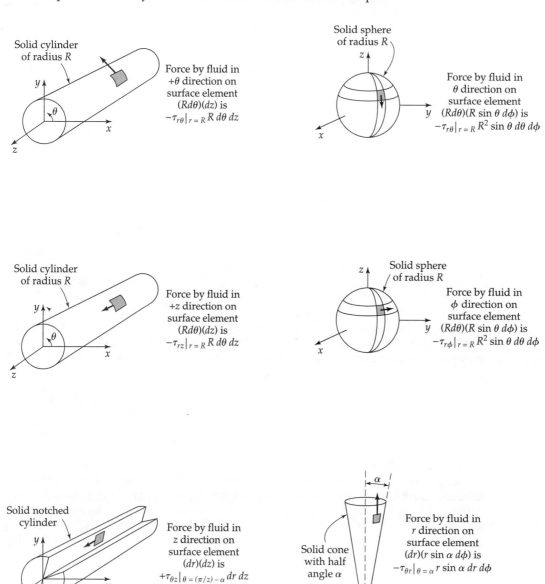

Fig. 1.2-2 (*a*) Some typical surface elements and shear stresses in the cylindrical coordinate system. (*b*) Some typical surface elements and shear stresses in the spherical coordinate system.

§1.7 and Table 19.2-2); (c) in Eq. 1.2-2, the terms $p\delta_{ij}$ and τ_{ij} have the same sign affixed, and the terms p and τ_{ii} are both positive in compression (in accordance with common usage in thermodynamics); (d) all terms in the entropy production in Eq. 24.1-5 have the same sign. Clearly the sign convention in Eqs. 1.1-2 and 1.2-6 is arbitrary, and either sign convention can be used, provided that the physical meaning of the sign convention is clearly understood.

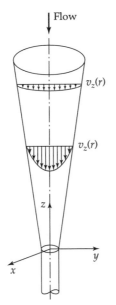

Flow

$v_z(r)$

$v_z(r)$

Fig. 1.2-3 The flow in a converging duct is an example of a situation in which the normal stresses are not zero. Since v_z is a function of r and z, the normal-stress component $\tau_{zz} = -2\mu(\partial v_z/\partial z)$ is nonzero. Also, since v_r depends on r and z, the normal-stress component $\tau_{rr} = -2\mu(\partial v_r/\partial r)$ is not equal to zero. At the wall, however, the normal stresses all vanish for fluids described by Eq. 1.2-7 provided that the density is constant (see Example 3.1-1 and Problem 3C.2).

§1.3   PRESSURE AND TEMPERATURE DEPENDENCE OF VISCOSITY

Extensive data on viscosities of pure gases and liquids are available in various science and engineering handbooks.[1] When experimental data are lacking and there is not time to obtain them, the viscosity can be estimated by empirical methods, making use of other data on the given substance. We present here a *corresponding-states correlation*, which facilitates such estimates and illustrates general trends of viscosity with temperature and pressure for ordinary fluids. The principle of corresponding states, which has a sound scientific basis,[2] is widely used for correlating equation-of-state and thermodynamic data. Discussions of this principle can be found in textbooks on physical chemistry and thermodynamics.

The plot in Fig. 1.3-1 gives a global view of the pressure and temperature dependence of viscosity. The reduced viscosity $\mu_r = \mu/\mu_c$ is plotted versus the reduced temperature $T_r = T/T_c$ for various values of the reduced pressure $p_r = p/p_c$. A "reduced" quantity is one that has been made dimensionless by dividing by the corresponding quantity at the critical point. The chart shows that the viscosity of a gas approaches a limit (the low-density limit) as the pressure becomes smaller; for most gases, this limit is nearly attained at 1 atm pressure. The viscosity of a gas at low density *increases* with increasing temperature, whereas the viscosity of a liquid *decreases* with increasing temperature.

Experimental values of the critical viscosity μ_c are seldom available. However, μ_c may be estimated in one of the following ways: (i) if a value of viscosity is known at a given reduced pressure and temperature, preferably at conditions near to those of

[1] J. A. Schetz and A. E. Fuhs (eds.), *Handbook of Fluid Dynamics and Fluid Machinery*, Wiley-Interscience, New York (1996), Vol. 1, Chapter 2; W. M. Rohsenow, J. P. Hartnett, and Y. I. Cho, *Handbook of Heat Transfer*, McGraw-Hill, New York, 3rd edition (1998), Chapter 2. Other sources are mentioned in fn. 4 of §1.1.

[2] J. Millat, J. H. Dymond, and C. A. Nieto de Castro (eds.), *Transport Properties of Fluids*, Cambridge University Press (1996), Chapter 11, by E. A. Mason and F. J. Uribe, and Chapter 12, by M. L. Huber and H. M. M. Hanley.

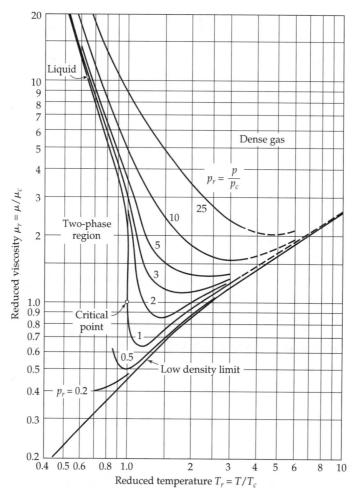

Fig. 1.3-1 Reduced viscosity $\mu_r = \mu/\mu_c$ as a function of reduced temperature for several values of the reduced pressure. [O. A. Uyehara and K. M. Watson, *Nat. Petroleum News, Tech. Section,* **36**, 764 (Oct. 4, 1944); revised by K. M. Watson (1960). A large-scale version of this graph is available in O. A. Hougen, K. M. Watson, and R. A. Ragatz, *C. P. P. Charts,* Wiley, New York, 2nd edition (1960)].

interest, then μ_c can be calculated from $\mu_c = \mu/\mu_r$; or (ii) if critical p-V-T data are available, then μ_c may be estimated from these empirical relations:

$$\mu_c = 61.6(MT_c)^{1/2}(\tilde{V}_c)^{-2/3} \quad \text{and} \quad \mu_c = 7.70M^{1/2}p_c^{2/3}T_c^{-1/6} \qquad (1.3\text{-}1a, b)$$

Here μ_c is in micropoises, p_c in atm, T_c in K, and $\tilde{V}_c$ in cm^3/g-mole. A tabulation of critical viscosities[3] computed by method (i) is given in Appendix E.

Figure 1.3-1 can also be used for rough estimation of viscosities of mixtures. For N-component fluids with mole fractions x_α, the "pseudocritical" properties[4] are:

$$p_c' = \sum_{\alpha=1}^{N} x_\alpha p_{c\alpha} \qquad T_c' = \sum_{\alpha=1}^{N} x_\alpha T_{c\alpha} \qquad \mu_c' = \sum_{\alpha=1}^{N} x_\alpha \mu_{c\alpha} \qquad (1.3\text{-}2a, b, c)$$

That is, one uses the chart exactly as for pure fluids, but with the pseudocritical properties instead of the critical properties. This empirical procedure works reasonably well

---

[3] O. A. Hougen and K. M. Watson, *Chemical Process Principles,* Part III, Wiley, New York (1947), p. 873. **Olaf Andreas Hougen** (pronounced "How-gen") (1893–1986) was a leader in the development of chemical engineering for four decades; together with K. M. Watson and R. A. Ragatz, he wrote influential books on thermodynamics and kinetics.

[4] O. A. Hougen and K. M. Watson, *Chemical Process Principles,* Part II, Wiley, New York (1947), p. 604.

unless there are chemically dissimilar substances in the mixture or the critical properties of the components differ greatly.

There are many variants on the above method, as well as a number of other empiricisms. These can be found in the extensive compilation of Reid, Prausnitz, and Poling.[5]

EXAMPLE 1.3-1	Estimate the viscosity of N_2 at 50°C and 854 atm, given $M = 28.0$ g/g-mole, $p_c = 33.5$ atm, and $T_c = 126.2$ K.
Estimation of Viscosity from Critical Properties	**SOLUTION**

Using Eq. 1.3-1b, we get

$$\mu_c = 7.70(28.0)^{1/2}(33.5)^{2/3}(126.2)^{-1/6}$$
$$= 189 \text{ micropoises} = 189 \times 10^{-6} \text{ poise} \tag{1.3-3}$$

The reduced temperature and pressure are

$$T_r = \frac{273.2 + 50}{126.2} = 2.56; \qquad p_r = \frac{854}{33.5} = 25.5 \tag{1.3-4a, b}$$

From Fig. 1.3-1, we obtain $\mu_r = \mu/\mu_c = 2.39$. Hence, the predicted value of the viscosity is

$$\mu = \mu_c(\mu/\mu_c) = (189 \times 10^{-6})(2.39) = 452 \times 10^{-6} \text{ poise} \tag{1.3-5}$$

The measured value[6] is 455×10^{-6} poise. This is unusually good agreement.

§1.4   MOLECULAR THEORY OF THE VISCOSITY OF GASES AT LOW DENSITY

To get a better appreciation of the concept of molecular momentum transport, we examine this transport mechanism from the point of view of an elementary kinetic theory of gases.

We consider a pure gas composed of rigid, nonattracting spherical molecules of diameter d and mass m, and the number density (number of molecules per unit volume) is taken to be n. The concentration of gas molecules is presumed to be sufficiently small that the average distance between molecules is many times their diameter d. In such a gas it is known[1] that, at equilibrium, the molecular velocities are randomly directed and have an average magnitude given by (see Problem 1C.1)

$$\bar{u} = \sqrt{\frac{8\kappa T}{\pi m}} \tag{1.4-1}$$

in which κ is the Boltzmann constant (see Appendix F). The frequency of molecular bombardment per unit area on one side of any stationary surface exposed to the gas is

$$Z = \tfrac{1}{4}n\bar{u} \tag{1.4-2}$$

---

[5] R. C. Reid, J. M. Prausnitz, and B. E. Poling, *The Properties of Gases and Liquids*, McGraw-Hill, New York, 4th edition (1987), Chapter 9.

[6] A. M. J. F. Michels and R. E. Gibson, *Proc. Roy. Soc.* (London), **A134**, 288–307 (1931).

[1] The first four equations in this section are given without proof. Detailed justifications are given in books on kinetic theory—for example, E. H. Kennard, *Kinetic Theory of Gases*, McGraw-Hill, New York (1938), Chapters II and III. Also E. A. Guggenheim, *Elements of the Kinetic Theory of Gases*, Pergamon Press, New York (1960), Chapter 7, has given a short account of the elementary theory of viscosity. For readable summaries of the kinetic theory of gases, see R. J. Silbey and R. A. Alberty, *Physical Chemistry*, Wiley, New York, 3rd edition (2001), Chapter 17, or R. S. Berry, S. A. Rice, and J. Ross, *Physical Chemistry*, Oxford University Press, 2nd edition (2000), Chapter 28.

The average distance traveled by a molecule between successive collisions is the *mean free path* λ, given by

$$\lambda = \frac{1}{\sqrt{2}\pi d^2 n} \tag{1.4-3}$$

On the average, the molecules reaching a plane will have experienced their last collision at a distance a from the plane, where a is given very roughly by

$$a = \tfrac{2}{3}\lambda \tag{1.4-4}$$

The concept of the mean free path is intuitively appealing, but it is meaningful only when λ is large compared to the range of intermolecular forces. The concept is appropriate for the rigid-sphere molecular model considered here.

To determine the viscosity of a gas in terms of the molecular model parameters, we consider the behavior of the gas when it flows parallel to the xz-plane with a velocity gradient dv_x/dy (see Fig. 1.4-1). We assume that Eqs. 1.4-1 to 4 remain valid in this non-equilibrium situation, provided that all molecular velocities are calculated relative to the average velocity $\mathbf{v}$ in the region in which the given molecule had its last collision. The flux of x-momentum across any plane of constant y is found by summing the x-momenta of the molecules that cross in the positive y direction and subtracting the x-momenta of those that cross in the opposite direction, as follows:

$$\tau_{yx} = Zmv_x\big|_{y-a} - Zmv_x\big|_{y+a} \tag{1.4-5}$$

In writing this equation, we have assumed that all molecules have velocities representative of the region in which they last collided and that the velocity profile $v_x(y)$ is essentially linear for a distance of several mean free paths. In view of the latter assumption, we may further write

$$v_x\big|_{y\pm a} = v_x\big|_y \pm \tfrac{2}{3}\lambda \frac{dv_x}{dy} \tag{1.4-6}$$

By combining Eqs. 1.4-2, 5, and 6 we get for the net flux of x-momentum in the positive y direction

$$\tau_{yx} = -\tfrac{1}{3}nm\bar{u}\lambda \frac{dv_x}{dy} \tag{1.4-7}$$

This has the same form as Newton's law of viscosity given in Eq. 1.1-2. Comparing the two equations gives an equation for the viscosity

$$\mu = \tfrac{1}{3}nm\bar{u}\lambda = \tfrac{1}{3}\rho\bar{u}\lambda \tag{1.4-8}$$

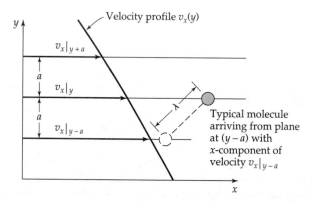

Fig. 1.4-1 Molecular transport of x-momentum from the plane at $(y-a)$ to the plane at y.

or, by combining Eqs. 1.4-1, 3, and 8

$$\mu = \frac{2}{3} \frac{\sqrt{m\kappa T/\pi}}{\pi d^2} = \frac{2}{3\pi} \frac{\sqrt{\pi m \kappa T}}{\pi d^2} \qquad (1.4\text{-}9)$$

This expression for the viscosity was obtained by Maxwell[2] in 1860. The quantity πd^2 is called the *collision cross section* (see Fig. 1.4-2).

The above derivation, which gives a qualitatively correct picture of momentum transfer in a gas at low density, makes it clear why we wished to introduce the term "momentum flux" for τ_{yx} in §1.1.

The prediction of Eq. 1.4-9 that μ is independent of pressure agrees with experimental data up to about 10 atm at temperatures above the critical temperature (see Fig. 1.3-1). The predicted temperature dependence is less satisfactory; data for various gases indicate that μ increases more rapidly than $\sqrt{T}$. To better describe the temperature dependence of μ, it is necessary to replace the rigid-sphere model by one that portrays the attractive and repulsive forces more accurately. It is also necessary to abandon the mean free path theories and use the Boltzmann equation to obtain the molecular velocity distribution in nonequilibrium systems more accurately. Relegating the details to Appendix D, we present here the main results.[3,4,5]

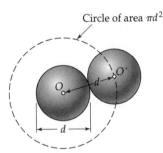

Circle of area πd^2

Fig. 1.4-2 When two rigid spheres of diameter d approach each other, the center of one sphere (at O') "sees" a circle of area πd^2 about the center of the other sphere (at O), on which a collision can occur. The area πd^2 is referred to as the "collision cross section."

---

[2] **James Clerk Maxwell** (1831–1879) was one of the greatest physicists of all time; he is particularly famous for his development of the field of electromagnetism and his contributions to the kinetic theory of gases. In connection with the latter, see J. C. Maxwell, *Phil. Mag.*, **19**, 19, Prop. XIII (1860); S. G. Brush, *Am. J. Phys*, **30**, 269–281 (1962). There is some controversy concerning Eqs. 1.4-4 and 1.4-9 (see S. Chapman and T. G. Cowling, *The Mathematical Theory of Non-Uniform Gases*, Cambridge University Press, 3rd edition 1970), p. 98; R. E. Cunningham and R. J. J. Williams, *Diffusion in Gases and Porous Media*, Plenum Press, New York (1980), §6.4.

[3] **Sydney Chapman** (1888–1970) taught at Imperial College in London, and thereafter was at the High Altitude Observatory in Boulder, Colorado; in addition to his seminal work on gas kinetic theory, he contributed to kinetic theory of plasmas and the theory of flames and detonations. **David Enskog** (1884–1947) (pronounced, roughly, "Ayn-skohg") is famous for his work on kinetic theories of low- and high-density gases. The standard reference on the Chapman–Enskog kinetic theory of dilute gases is S. Chapman and T. G. Cowling, *The Mathematical Theory of Non-Uniform Gases*, Cambridge University Press, 3rd edition (1970); pp. 407–409 give a historical summary of the kinetic theory. See also D. Enskog, *Inaugural Dissertation*, Uppsala (1917). In addition J. H. Ferziger and H. G. Kaper, *Mathematical Theory of Transport Processes in Gases*, North-Holland, Amsterdam (1972), is a very readable account of molecular theory.

[4] The Curtiss–Hirschfelder[5] extension of the Chapman–Enskog theory to multicomponent gas mixtures, as well as the development of useful tables for computation, can be found in J. O. Hirschfelder, C. F. Curtiss, and R. B. Bird, *Molecular Theory of Gases and Liquids*, Wiley, New York, 2nd corrected printing (1964). See also C. F. Curtiss, *J. Chem. Phys.*, **49**, 2917–2919 (1968), as well as references given in Appendix E. **Joseph Oakland Hirschfelder** (1911–1990), founding director of the Theoretical Chemistry Institute at the University of Wisconsin, specialized in intermolecular forces and applications of kinetic theory.

[5] C. F. Curtiss and J. O. Hirschfelder, *J. Chem. Phys.*, **17**, 550–555 (1949).

A rigorous kinetic theory of monatomic gases at low density was developed early in the twentieth century by Chapman in England and independently by Enskog in Sweden. The Chapman–Enskog theory gives expressions for the transport properties in terms of the *intermolecular potential energy* $\varphi(r)$, where r is the distance between a pair of molecules undergoing a collision. The intermolecular force is then given by $F(r) = -d\varphi/dr$. The exact functional form of $\varphi(r)$ is not known; however, for nonpolar molecules a satisfactory empirical expression is the *Lennard-Jones (6–12) potential*[6] given by

$$\varphi(r) = 4\varepsilon \left[\left(\frac{\sigma}{r} \right)^{12} - \left(\frac{\sigma}{r} \right)^{6} \right] \tag{1.4-10}$$

in which σ is a characteristic diameter of the molecules, often called the *collision diameter* and ε is a characteristic energy, actually the maximum energy of attraction between a pair of molecules. This function, shown in Fig. 1.4-3, exhibits the characteristic features of intermolecular forces: weak attractions at large separations and strong repulsions at small separations. Values of the parameters σ and ϵ are known for many substances; a partial list is given in Table E.1, and a more extensive list is available elsewhere.[4] When σ and ε are not known, they may be estimated from properties of the fluid at the critical point (c), the liquid at the normal boiling point (b), or the solid at the melting point (m), by means of the following empirical relations:[4]

$$\varepsilon/\kappa = 0.77T_c \qquad \sigma = 0.841\tilde{V}_c^{1/3} \quad \text{or} \quad \sigma = 2.44(T_c/p_c)^{1/3} \tag{1.4-11a, b, c}$$
$$\varepsilon/\kappa = 1.15T_b \qquad \sigma = 1.166\tilde{V}_{b,\text{liq}}^{1/3} \tag{1.4-12a, b}$$
$$\varepsilon/\kappa = 1.92T_m \qquad \sigma = 1.222\tilde{V}_{m,\text{sol}}^{1/3} \tag{1.4-13a, b}$$

Here ε/κ and T are in K, σ is in Ångström units (1 Å = 10^{-10} m), $\tilde{V}$ is in cm³/g-mole, and p_c is in atmospheres.

The viscosity of a pure monatomic gas of molecular weight M may be written in terms of the Lennard-Jones parameters as

$$\mu = \frac{5}{16} \frac{\sqrt{\pi m \kappa T}}{\pi \sigma^2 \Omega_\mu} \quad \text{or} \quad \mu = 2.6693 \times 10^{-5} \frac{\sqrt{MT}}{\sigma^2 \Omega_\mu} \tag{1.4-14}$$

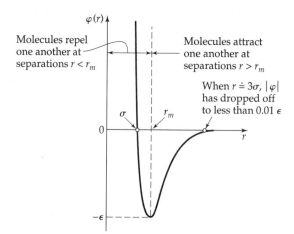

Molecules repel one another at separations $r < r_m$

Molecules attract one another at separations $r > r_m$

When $r \doteq 3\sigma$, $|\varphi|$ has dropped off to less than $0.01\ \epsilon$

Fig. 1.4-3 Potential energy function $\varphi(r)$ describing the interaction of two spherical, nonpolar molecules. The Lennard-Jones (6–12) potential, given in Eq. 1.4-10, is one of the many empirical equations proposed for fitting this curve. For $r < r_m$ the molecules repel one another, whereas for $r > r_m$ the molecules attract one another.

[6] J. E. (Lennard-)Jones, *Proc. Roy. Soc.*, **A106**, 441–462, 463–477 (1924). See also R. J. Silbey and R. A. Alberty, *Physical Chemistry*, Wiley, 2nd edition (2001), §§11.10, 16.14, and 17.9; and R. S. Berry, S. A. Rice, and J. Ross, *Physical Chemistry*, Oxford University Press, 2nd edition (2000), §10.2.

In the second form of this equation, if $T \, [=] \, K$ and $\sigma \, [=] \, Å$, then $\mu \, [=] \, g/cm \cdot s$. The dimensionless quantity Ω_μ is a slowly varying function of the dimensionless temperature $\kappa T / \varepsilon$, of the order of magnitude of unity, given in Table E.2. It is called the "collision integral for viscosity," because it accounts for the details of the paths that the molecules take during a binary collision. If the gas were made up of rigid spheres of diameter σ (instead of real molecules with attractive and repulsive forces), then Ω_μ would be exactly unity. Hence the function Ω_μ may be interpreted as describing the deviation from rigid-sphere behavior.

Although Eq. 1.4-14 is a result of the kinetic theory of monatomic gases, it has been found to be remarkably good for polyatomic gases as well. The reason for this is that, in the equation of conservation of momentum for a collision between polyatomic molecules, the center of mass coordinates are more important than the internal coordinates [see §0.3(b)]. The temperature dependence predicted by Eq. 1.4-14 is in good agreement with that found from the low-density line in the empirical correlation of Fig. 1.3-1. The viscosity of gases at low density increases with temperature, roughly as the 0.6 to 1.0 power of the absolute temperature, and is independent of the pressure.

To calculate the viscosity of a gas mixture, the multicomponent extension of the Chapman–Enskog theory can be used.[4,5] Alternatively, one can use the following very satisfactory semiempirical formula:[7]

$$\mu_{\text{mix}} = \sum_{\alpha=1}^{N} \frac{x_\alpha \mu_\alpha}{\sum_\beta x_\beta \Phi_{\alpha\beta}} \tag{1.4-15}$$

in which the dimensionless quantities $\Phi_{\alpha\beta}$ are

$$\Phi_{\alpha\beta} = \frac{1}{\sqrt{8}} \left(1 + \frac{M_\alpha}{M_\beta} \right)^{-1/2} \left[1 + \left(\frac{\mu_\alpha}{\mu_\beta} \right)^{1/2} \left(\frac{M_\beta}{M_\alpha} \right)^{1/4} \right]^2 \tag{1.4-16}$$

Here N is the number of chemical species in the mixture, x_α is the mole fraction of species α, μ_α is the viscosity of pure species α at the system temperature and pressure, and M_α is the molecular weight of species α. Equation 1.4-16 has been shown to reproduce measured values of the viscosities of mixtures within an average deviation of about 2%. The dependence of mixture viscosity on composition is extremely nonlinear for some mixtures, particularly mixtures of light and heavy gases (see Problem 1A.2).

To summarize, Eqs. 1.4-14, 15, and 16 are useful formulas for computing viscosities of nonpolar gases and gas mixtures at low density from tabulated values of the intermolecular force parameters σ and ε/κ. They will not give reliable results for gases consisting of polar or highly elongated molecules because of the angle-dependent force fields that exist between such molecules. For polar vapors, such as H_2O, NH_3, $CHOH$, and $NOCl$, an angle-dependent modification of Eq. 1.4-10 has given good results.[8] For the light gases H_2 and He below about 100K, quantum effects have to be taken into account.[9]

Many additional empiricisms are available for estimating viscosities of gases and gas mixtures. A standard reference is that of Reid, Prausnitz, and Poling.[10]

---

[7] C. R. Wilke, *J. Chem. Phys.*, **18**, 517–519 (1950); see also J. W. Buddenberg and C. R. Wilke, *Ind. Eng. Chem.*, **41**, 1345–1347 (1949).

[8] E. A. Mason and L. Monchick, *J. Chem. Phys.*, **35**, 1676–1697 (1961) and **36**, 1622–1639, 2746–2757 (1962).

[9] J. O. Hirschfelder, C. F. Curtiss, and R. B. Bird, *op. cit.*, Chapter 10; H. T. Wood and C. F. Curtiss, *J. Chem. Phys.*, **41**, 1167–1173 (1964); R. J. Munn, F. J. Smith, and E. A. Mason, *J. Chem. Phys.*, **42**, 537–539 (1965); S. Imam-Rahajoe, C. F. Curtiss, and R. B. Bernstein, *J. Chem. Phys.*, **42**, 530–536 (1965).

[10] R. C. Reid, J. M. Prausnitz, and B. E. Poling, *The Propeties of Gases and Liquids*, McGraw-Hill, New York, 4th edition (1987).

EXAMPLE 1.4-1

Computation of the Viscosity of a Pure Gas at Low Density

Compute the viscosity of CO_2 at 200, 300, and 800 K and 1 atm.

SOLUTION

Use Eq. 1.4-14. From Table E.1, we find the Lennard-Jones parameters for CO_2 to be $\varepsilon/\kappa = 190$ K and $\sigma = 3.996$ Å. The molecular weight of CO_2 is 44.01. Substitution of M and σ into Eq. 1.4-14 gives

$$\mu = 2.6693 \times 10^{-5} \frac{\sqrt{44.01T}}{(3.996)^2 \Omega_\mu} = 1.109 \times 10^{-5} \frac{\sqrt{T}}{\Omega_\mu} \tag{1.4-17}$$

in which μ [=] g/cm · s and T [=] K. The remaining calculations may be displayed in a table.

				Viscosity (g/cm · s)	
T (K)	$\kappa T/\varepsilon$	Ω_μ	$\sqrt{T}$	Predicted	Observed[11]
200	1.053	1.548	14.14	1.013×10^{-4}	1.015×10^{-4}
300	1.58	1.286	17.32	1.494×10^{-4}	1.495×10^{-4}
800	4.21	0.9595	28.28	3.269×10^{-4}	$\cdots$

Experimental data are shown in the last column for comparison. The good agreement is to be expected, since the Lennard-Jones parameters of Table E.1 were derived from viscosity data.

EXAMPLE 1.4-2

Prediction of the Viscosity of a Gas Mixture at Low Density

Estimate the viscosity of the following gas mixture at 1 atm and 293 K from the given data on the pure components at the same pressure and temperature:

Species α	Mole fraction, x_α	Molecular weight, M_α	Viscosity, μ_α (g/cm · s)
1. CO_2	0.133	44.01	1462×10^{-7}
2. O_2	0.039	32.00	2031×10^{-7}
3. N_2	0.828	28.02	1754×10^{-7}

SOLUTION

Use Eqs. 1.4-16 and 15 (in that order). The calculations can be systematized in tabular form, thus:

α	β	M_α/M_β	μ_α/μ_β	$\Phi_{\alpha\beta}$	$\sum_{\beta=1}^{3} x_\beta \Phi_{\alpha\beta}$
1.	1	1.000	1.000	1.000	
	2	1.375	0.720	0.730	0.763
	3	1.571	0.834	0.727	
2.	1	0.727	1.389	1.394	
	2	1.000	1.000	1.000	1.057
	3	1.142	1.158	1.006	
3.	1	0.637	1.200	1.370	
	2	0.876	0.864	0.993	1.049
	3	1.000	1.000	1.000	

[11] H. L. Johnston and K. E. McCloskey, *J. Phys. Chem.*, **44**, 1038–1058 (1940).

Eq. 1.4-15 then gives

$$\mu = \frac{(0.1333)(1462)(10^{-7})}{0.763} + \frac{(0.039)(2031)(10^{-7})}{1.057} + \frac{(0.828)(1754)(10^{-7})}{1.049}$$

$$= 1714 \times 10^{-7} \, \text{g/cm} \cdot \text{s}$$

The observed value[12] is $1793 \times 10^{-7} \, \text{g/cm} \cdot \text{s}$.

§1.5   MOLECULAR THEORY OF THE VISCOSITY OF LIQUIDS

A rigorous kinetic theory of the transport properties of monatomic liquids was developed by Kirkwood and coworkers.[1] However this theory does not lead to easy-to-use results. An older theory, developed by Eyring[2] and coworkers, although less well grounded theoretically, does give a qualitative picture of the mechanism of momentum transport in liquids and permits rough estimation of the viscosity from other physical properties. We discuss this theory briefly.

In a pure liquid at rest the individual molecules are constantly in motion. However, because of the close packing, the motion is largely confined to a vibration of each molecule within a "cage" formed by its nearest neighbors. This cage is represented by an energy barrier of height $\Delta \tilde{G}_0^{\ddagger}/\tilde{N}$, in which $\Delta \tilde{G}_0^{\ddagger}$ is the molar free energy of activation for escape from the cage in the stationary fluid (see Fig. 1.5-1). According to Eyring, a liquid at rest continually undergoes rearrangements, in which one molecule at a time escapes from its "cage" into an adjoining "hole," and that the molecules thus move in each of the

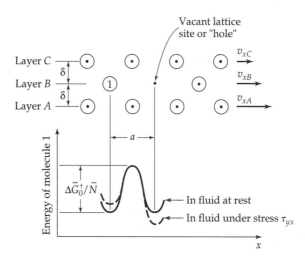

Fig. 1.5-1 Illustration of an escape process in the flow of a liquid. Molecule 1 must pass through a "bottleneck" to reach the vacant site.

[12] F. Herning and L. Zipperer, *Gas- und Wasserfach*, **79**, 49–54, 69–73 (1936).

[1] J. H. Irving and J. G. Kirkwood, *J. Chem. Phys.*, **18**, 817–823 (1950); R. J. Bearman and J. G. Kirkwood, *J. Chem. Phys*, **28**, 136–146 (1958). For additional publications, see John Gamble Kirkwood, *Collected Works*, Gordon and Breach, New York (1967). **John Gamble Kirkwood** (1907–1959) contributed much to the kinetic theory of liquids, properties of polymer solutions, theory of electrolytes, and thermodynamics of irreversible processes.

[2] S. Glasstone, K. J. Laidler, and H. Eyring, *Theory of Rate Processes*, McGraw-Hill, New York (1941), Chapter 9; H. Eyring, D. Henderson, B. J. Stover, and E. M. Eyring, *Statistical Mechanics*, Wiley, New York (1964), Chapter 16. See also R. J. Silbey and R. A. Alberty, *Physical Chemistry*, Wiley, 3rd edition (2001), §20.1; and R. S. Berry, S. A. Rice, and J. Ross, *Physical Chemistry*, Oxford University Press, 2nd edition (2000), Ch. 29. **Henry Eyring** (1901–1981) developed theories for the transport properties based on simple physical models; he also developed the theory of absolute reaction rates.

coordinate directions in jumps of length a at a frequency ν per molecule. The frequency is given by the rate equation

$$\nu = \frac{\kappa T}{h} \exp(-\Delta \tilde{G}_0^+ / RT) \tag{1.5-1}$$

In which κ and h are the Boltzmann and Planck constants, $\tilde{N}$ is the Avogadro number, and $R = \tilde{N}\kappa$ is the gas constant (see Appendix F).

In a fluid that is flowing in the x direction with a velocity gradient dv_x/dy, the frequency of molecular rearrangements is increased. The effect can be explained by considering the potential energy barrier as distorted under the applied stress τ_{yx} (see Fig. 1.5-1), so that

$$-\Delta \tilde{G}^+ = -\Delta \tilde{G}_0^+ \pm \left(\frac{a}{\delta}\right)\left(\frac{\tau_{yx}\tilde{V}}{2}\right) \tag{1.5-2}$$

where $\tilde{V}$ is the volume of a mole of liquid, and $\pm(a/\delta)(\tau_{yx}\tilde{V}/2)$ is an approximation to the work done on the molecules as they move to the top of the energy barrier, moving *with* the applied shear stress (plus sign) or *against* the applied shear stress (minus sign). We now define ν_+ as the frequency of forward jumps and ν_- as the frequency of backward jumps. Then from Eqs. 1.5-1 and 1.5-2 we find that

$$\nu_\pm = \frac{\kappa T}{h} \exp(-\Delta \tilde{G}_0^+ / RT) \exp(\pm a\tau_{yx}\tilde{V}/2\delta RT) \tag{1.5-3}$$

The net velocity with which molecules in layer A slip ahead of those in layer B (Fig. 1.5-1) is just the distance traveled per jump (a) times the *net* frequency of forward jumps ($\nu_+ - \nu_-$); this gives

$$v_{xA} - v_{xB} = a(\nu_+ - \nu_-) \tag{1.5-4}$$

The velocity profile can be considered to be linear over the very small distance δ between the layers A and B, so that

$$-\frac{dv_x}{dy} = \left(\frac{a}{\delta}\right)(\nu_+ - \nu_-) \tag{1.5-5}$$

By combining Eqs. 1.5-3 and 5, we obtain finally

$$-\frac{dv_x}{dy} = \left(\frac{a}{\delta}\right)\left(\frac{\kappa T}{h} \exp(-\Delta \tilde{G}_0^+ / RT)\right)(\exp(+a\tau_{yx}\tilde{V}/2\delta RT) - \exp(-a\tau_{yx}\tilde{V}/2\delta RT))$$

$$= \left(\frac{a}{\delta}\right)\left(\frac{\kappa T}{h} \exp(-\Delta \tilde{G}_0^+ / RT)\right)\left(2 \sinh \frac{a\tau_{yx}\tilde{V}}{2\delta RT}\right) \tag{1.5-6}$$

This predicts a nonlinear relation between the shear stress (momentum flux) and the velocity gradient—that is, *non-Newtonian flow*. Such nonlinear behavior is discussed further in Chapter 8.

The usual situation, however, is that $a\tau_{yx}\tilde{V}/2\delta RT \ll 1$. Then we can use the Taylor series (see §C.2) $\sinh x = x + (1/3!)x^3 + (1/5!)x^5 + \cdots$ and retain only one term. Equation 1.5-6 is then of the form of Eq. 1.1-2, with the viscosity being given by

$$\mu = \left(\frac{\delta}{a}\right)^2 \frac{\tilde{N}h}{\tilde{V}} \exp(\Delta \tilde{G}_0^+ / RT) \tag{1.5-7}$$

The factor δ/a can be taken to be unity; this simplification involves no loss of accuracy, since $\Delta \tilde{G}_0^+$ is usually determined empirically to make the equation agree with experimental viscosity data.

It has been found that free energies of activation, $\Delta \tilde{G}_0^+$, determined by fitting Eq. 1.5-7 to experimental data on viscosity versus temperature, are almost constant for a given fluid and are simply related to the internal energy of vaporization at the normal boiling point, as follows:[3]

$$\Delta \tilde{G}_0^{\ddagger} \approx 0.408 \, \Delta \tilde{U}_{vap} \tag{1.5-8}$$

By using this empiricism and setting $\delta/a = 1$, Eq. 1.5-7 becomes

$$\mu = \frac{\tilde{N}h}{\tilde{V}} \exp (0.408 \, \Delta \tilde{U}_{vap}/RT) \tag{1.5-9}$$

The energy of vaporization at the normal boiling point can be estimated roughly from Trouton's rule

$$\Delta \tilde{U}_{vap} \approx \Delta \tilde{H}_{vap} - RT_b \cong 9.4RT_b \tag{1.5-10}$$

With this further approximation, Eq. 1.5-9 becomes

$$\mu = \frac{\tilde{N}h}{\tilde{V}} \exp(3.8T_b/T) \tag{1.5-11}$$

Equations 1.5-9 and 11 are in agreement with the long-used and apparently successful empiricism $\mu = A \exp(B/T)$. The theory, although only approximate in nature, does give the observed decrease of viscosity with temperature, but errors of as much as 30% are common when Eqs. 1.5-9 and 11 are used. They should not be used for very long slender molecules, such as n-$C_{20}H_{42}$.

There are, in addition, many empirical formulas available for predicting the viscosity of liquids and liquid mixtures. For these, physical chemistry and chemical engineering textbooks should be consulted.[4]

EXAMPLE 1.5-1

Estimation of the Viscosity of a Pure Liquid

Estimate the viscosity of liquid benzene, C_6H_6, at 20°C (293.2 K).

SOLUTION

Use Eq. 1.5-11 with the following information:

$$\tilde{V} = 89.0 \text{ cm}^3/\text{g-mole}$$
$$T_b = 80.1°C$$

Since this information is given in c.g.s. units, we use the values of Avogadro's number and Planck's constant in the same set of units. Substituting into Eq. 1.5-11 gives:

$$\mu = \frac{(6.023 \times 10^{23})(6.624 \times 10^{-27})}{(89.0)} \exp\left(\frac{3.8 \times (273.2 + 80.1)}{293.2}\right)$$
$$= 4.5 \times 10^{-3} \text{ g/cm} \cdot \text{s} \quad or \quad 4.5 \times 10^{-4} \text{ Pa} \cdot \text{s} \quad or \quad 0.45 \text{ mPa} \cdot \text{s}$$

§1.6   VISCOSITY OF SUSPENSIONS AND EMULSIONS

Up to this point we have been discussing fluids that consist of a single homogeneous phase. We now turn our attention briefly to two-phase systems. The complete description of such systems is, of course, quite complex, but it is often useful to replace the suspension or emulsion by a hypothetical one-phase system, which we then describe by

---

[3] J. F. Kincaid, H. Eyring, and A. E. Stearn, *Chem. Revs.*, **28**, 301–365 (1941).

[4] See, for example, J. R. Partington, *Treatise on Physical Chemistry*, Longmans, Green (1949); or R. C. Reid, J. M. Prausnitz, and B. E. Poling, *The Properties of Gases and Liquids*, McGraw-Hill, New York, 4th edition (1987). See also P. A. Egelstaff, *An Introduction to the Liquid State*, Oxford University Press, 2nd edition (1994), Chapter 13; and J. P. Hansen and I. R. McDonald, *Theory of Simple Liquids*, Academic Press, London (1986), Chapter 8.

Newton's law of viscosity (Eq. 1.1-2 or 1.2-7) with two modifications: (i) the viscosity μ is replaced by an *effective viscosity* μ_{eff}, and (ii) the velocity and stress components are then redefined (with no change of symbol) as the analogous quantities averaged over a volume large with respect to the interparticle distances and small with respect to the dimensions of the flow system. This kind of theory is satisfactory as long as the flow involved is steady; in time-dependent flows, it has been shown that Newton's law of viscosity is inappropriate, and the two-phase systems have to be regarded as viscoelastic materials.[1]

The first major contribution to the theory of the *viscosity of suspensions of spheres* was that of Einstein.[2] He considered a suspension of rigid spheres, so dilute that the movement of one sphere does not influence the fluid flow in the neighborhood of any other sphere. Then it suffices to analyze only the motion of the fluid around a single sphere, and the effects of the individual spheres are additive. The *Einstein equation* is

$$\frac{\mu_{eff}}{\mu_0} = 1 + \frac{5}{2}\phi \tag{1.6-1}$$

in which μ_0 is the viscosity of the suspending medium, and ϕ is the volume fraction of the spheres. Einstein's pioneering result has been modified in many ways, a few of which we now describe.

For *dilute suspensions of particles of various shapes* the constant $\frac{5}{2}$ has to be replaced by a different coefficient depending on the particular shape. Suspensions of elongated or flexible particles exhibit non-Newtonian viscosity.[3,4,5,6]

For *concentrated suspensions of spheres* (that is, ϕ greater than about 0.05) particle interactions become appreciable. Numerous semiempirical expressions have been developed, one of the simplest of which is the *Mooney equation*[7]

$$\frac{\mu_{eff}}{\mu_0} = \exp\left(\frac{\frac{5}{2}\phi}{1 - (\phi/\phi_0)}\right) \tag{1.6-2}$$

in which ϕ_0 is an empirical constant between about 0.74 and 0.52, these values corresponding to the values of ϕ for closest packing and cubic packing, respectively.

---

[1] For dilute suspensions of rigid spheres, the linear viscoelastic behavior has been studied by H. Fröhlich and R. Sack, *Proc. Roy. Soc.*, **A185**, 415–430 (1946), and for dilute emulsions, the analogous derivation has been given by J. G. Oldroyd, *Proc. Roy. Soc.*, **A218**, 122–132 (1953). In both of these publications the fluid is described by the Jeffreys model (see Eq. 8.4-4), and the authors found the relations between the three parameters in the Jeffreys model and the constants describing the structure of the two-phase system (the volume fraction of suspended material and the viscosities of the two phases). For further comments concerning suspensions and rheology, see R. B. Bird and J. M. Wiest, Chapter 3 in *Handbook of Fluid Dynamics and Fluid Machinery*, J. A. Schetz and A. E. Fuhs (eds.), Wiley, New York (1996).

[2] **Albert Einstein** (1879–1955) received the Nobel prize for his explanation of the photoelectric effect, not for his development of the theory of special relativity. His seminal work on suspensions appeared in A. Einstein, *Ann. Phys. (Leipzig)*, **19**, 289–306 (1906); erratum, *ibid.*, **24**, 591–592 (1911). In the original publication, Einstein made an error in the derivation and got ϕ instead of $\frac{5}{2}\phi$. After experiments showed that his equation did not agree with the experimental data, he recalculated the coefficient. Einstein's original derivation is quite lengthy; for a more compact development, see L. D. Landau and E. M. Lifshitz, *Fluid Mechanics*, Pergamon Press, Oxford, 2nd edition (1987), pp. 73–75. The mathematical formulation of multiphase fluid behavior can be found in D. A. Drew and S. L. Passman, *Theory of Multicomponent Fluids*, Springer, Berlin (1999).

[3] H. L. Frisch and R. Simha, Chapter 14 in *Rheology*, Vol. 1, (F. R. Eirich, ed.), Academic Press, New York (1956), Sections II and III.

[4] E. W. Merrill, Chapter 4 in *Modern Chemical Engineering*, Vol. 1, (A. Acrivos, ed.), Reinhold, New York (1963), p. 165.

[5] E. J. Hinch and L. G. Leal, *J. Fluid Mech.*, **52**, 683–712 (1972); **76**, 187–208 (1976).

[6] W. R. Schowalter, *Mechanics of Non-Newtonian Fluids*, Pergamon, Oxford (1978), Chapter 13.

[7] M. Mooney, *J. Coll. Sci.*, **6**, 162–170 (1951).

Another approach for concentrated suspensions of spheres is the "cell theory," in which one examines the dissipation energy in the "squeezing flow" between the spheres. As an example of this kind of theory we cite the *Graham equation*[8]

$$\frac{\mu_{\text{eff}}}{\mu_0} = 1 + \frac{5}{2}\phi + \frac{9}{4}\left(\frac{1}{\psi(1 + \frac{1}{2}\psi)(1 + \psi)^2}\right) \tag{1.6-3}$$

in which $\psi = 2[(1 - \sqrt[3]{\phi/\phi_{\text{max}}})/\sqrt[3]{\phi/\phi_{\text{max}}}]$, where ϕ_{max} is the volume fraction corresponding to the experimentally determined closest packing of the spheres. This expression simplifies to Einstein's equation for $\phi \to 0$ and the Frankel–Acrivos equation[9] when $\phi \to \phi_{\text{max}}$.

For *concentrated suspensions of nonspherical particles*, the *Krieger–Dougherty* equation[10] can be used:

$$\frac{\mu_{\text{eff}}}{\mu_0} = \left(1 - \frac{\phi}{\phi_{\text{max}}}\right)^{-A\phi_{\text{max}}} \tag{1.6-4}$$

The parameters A and ϕ_{max} to be used in this equation are tabulated[11] in Table 1.6-1 for suspensions of several materials.

Non-Newtonian behavior is observed for concentrated suspensions, even when the suspended particles are spherical.[11] This means that the viscosity depends on the velocity gradient and may be different in a shear than it is in an elongational flow. Therefore, equations such as Eq. 1.6-2 must be used with some caution.

Table 1.6-1   Dimensionless Constants for Use in Eq. 1.6-4

System	A	ϕ_{max}	Reference
Spheres (submicron)	2.7	0.71	*a*
Spheres (40 μm)	3.28	0.61	*b*
Ground gypsum	3.25	0.69	*c*
Titanium dioxide	5.0	0.55	*c*
Laterite	9.0	0.35	*c*
Glass rods (30 × 700 μm)	9.25	0.268	*d*
Glass plates (100 × 400 μm)	9.87	0.382	*d*
Quartz grains (53–76 μm)	5.8	0.371	*d*
Glass fibers (axial ratio 7)	3.8	0.374	*b*
Glass fibers (axial ratio 14)	5.03	0.26	*b*
Glass fibers (axial ratio 21)	6.0	0.233	*b*

[a] C. G. de Kruif, E. M. F. van Ievsel, A. Vrij, and W. B. Russel, in *Viscoelasticity and Rheology* (A. S. Lodge, M. Renardy, J. A. Nohel, eds.), Academic Press, New York (1985).

[b] H. Giesekus, in *Physical Properties of Foods* (J. Jowitt et al., eds.), Applied Science Publishers (1983), Chapter 13.

[c] R. M. Turian and T.-F. Yuan, *AIChE Journal*, **23**, 232–243 (1977).

[d] B. Clarke, *Trans. Inst. Chem. Eng.*, **45**, 251–256 (1966).

[8] A. L. Graham, *Appl. Sci. Res.*, **37**, 275–286 (1981).

[9] N. A. Frankel and A. Acrivos, *Chem. Engr. Sci.*, **22**, 847–853 (1967).

[10] I. M. Krieger and T. J. Dougherty, *Trans. Soc. Rheol.*, **3**, 137–152 (1959).

[11] H. A. Barnes, J. F. Hutton, and K. Walters, *An Introduction to Rheology*, Elsevier, Amsterdam (1989), p. 125.

For *emulsions* or *suspensions of tiny droplets*, in which the suspended material may undergo internal circulation but still retain a spherical shape, the effective viscosity can be considerably less than that for suspensions of solid spheres. The viscosity of dilute emulsions is then described by the *Taylor equation*:[12]

$$\frac{\mu_{\text{eff}}}{\mu_0} = 1 + \left(\frac{\mu_0 + \frac{5}{2}\mu_1}{\mu_0 + \mu_1}\right)\phi \tag{1.6-5}$$

in which μ_1 is the viscosity of the disperse phase. It should, however, be noted that surface-active contaminants, frequently present even in carefully purified liquids, can effectively stop the internal circulation;[13] the droplets then behave as rigid spheres.

For *dilute suspensions of charged spheres*, Eq. 1.6-1 may be replaced by the *Smoluchowski equation*[14]

$$\frac{\mu_{\text{eff}}}{\mu_0} = 1 + \frac{5}{2}\,\phi\left(1 + \frac{(D\zeta/2\pi R)^2}{\mu_0 k_e}\right) \tag{1.6-6}$$

in which D is the dielectric constant of the suspending fluid, k_e the specific electrical conductivity of the suspension, ζ the electrokinetic potential of the particles, and R the particle radius. Surface charges are not uncommon in stable suspensions. Other, less well understood, surface forces are also important and frequently cause the particles to form loose aggregates.[4] Here again, non-Newtonian behavior is encountered.[15]

§1.7 CONVECTIVE MOMENTUM TRANSPORT

Thus far we have discussed the *molecular transport* of momentum, and this led to a set of quantities π_{ij}, which give the flux of j-momentum across a surface perpendicular to the i direction. We then related the π_{ij} to the velocity gradients and the pressure, and we found that this relation involved two material parameters μ and κ. We have seen in §§1.4 and 1.5 how the viscosity arises from a consideration of the random motion of the molecules in the fluid—that is, the random molecular motion with respect to the bulk motion of the fluid. Furthermore, in Problem 1C.3 we show how the pressure contribution to π_{ij} arises from the random molecular motions.

Momentum can, in addition, be transported by the bulk flow of the fluid, and this process is called *convective transport*. To discuss this we use Fig. 1.7-1 and focus our attention on a cube-shaped region in space through which the fluid is flowing. At the center of the cube (located at x, y, z) the fluid velocity vector is $\mathbf{v}$. Just as in §1.2 we consider three mutually perpendicular planes (the shaded planes) through the point x, y, z, and we ask how much momentum is flowing through each of them. Each of the planes is taken to have unit area.

The volume rate of flow across the shaded unit area in (*a*) is v_x. This fluid carries with it momentum $\rho\mathbf{v}$ per unit volume. Hence the momentum flux across the shaded area is $v_x\rho\mathbf{v}$; note that this is the momentum flux from the region of lesser x to the region

[12] G. I. Taylor, *Proc. Roy. Soc.*, **A138**, 41–48 (1932). **Geoffrey Ingram Taylor** (1886–1975) is famous for Taylor dispersion, Taylor vortices, and his work on the statistical theory of turbulence; he attacked many complex problems in ingenious ways that made maximum use of the physical processes involved.

[13] V. G. Levich, *Physicochemical Hydrodynamics*, Prentice-Hall, Englewood Cliffs, N.J. (1962), Chapter 8. **Veniamin Grigorevich Levich** (1917–1987), physicist and electrochemist, made many contributions to the solution of important problems in diffusion and mass transfer.

[14] M. von Smoluchowski, *Kolloid Zeits.*, **18**, 190–195 (1916).

[15] W. B. Russel, *The Dynamics of Colloidal Systems*, U. of Wisconsin Press, Madison (1987), Chapter 4; W. B. Russel, D. A. Saville, and W. R. Schowalter, *Colloidal Dispersions*, Cambridge University Press (1989); R. G. Larson, *The Structure and Rheology of Complex Fluids*, Oxford University Press (1998).

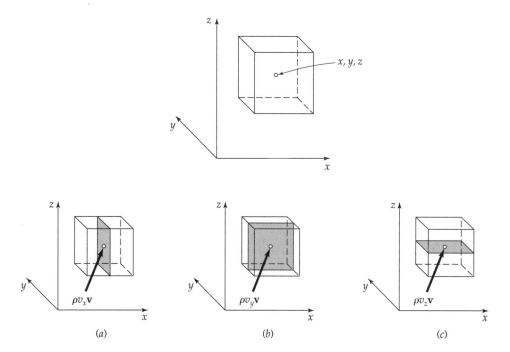

Fig. 1.7-1 The convective momentum fluxes through planes of unit area perpendicular to the coordinate directions.

of greater x. Similarly the momentum flux across the shaded area in (b) is $v_y\rho\mathbf{v}$, and the momentum flux across the shaded area in (c) is $v_z\rho\mathbf{v}$.

These three vectors—$\rho v_x\mathbf{v}$, $\rho v_y\mathbf{v}$, and $\rho v_z\mathbf{v}$—describe the momentum flux across the three areas perpendicular to the respective axes. Each of these vectors has an x-, y-, and z-component. These components can be arranged as shown in Table 1.7-1. The quantity $\rho v_x v_y$ is the convective flux of y-momentum across a surface perpendicular to the x direction. This should be compared with the quantity τ_{xy}, which is the molecular flux of y-momentum across a surface perpendicular to the x direction. The sign convention for both modes of transport is the same.

The collection of nine scalar components given in Table 1.7-1 can be represented as

$$\rho\mathbf{vv} = (\Sigma_i\boldsymbol{\delta}_i\rho v_i)\mathbf{v} = (\Sigma_i\boldsymbol{\delta}_i\rho v_i)(\Sigma_j\boldsymbol{\delta}_j v_j)$$
$$= \Sigma_i\Sigma_j\boldsymbol{\delta}_i\boldsymbol{\delta}_j\rho v_i v_j \tag{1.7-1}$$

Since each component of $\rho\mathbf{vv}$ has two subscripts, each associated with a coordinate direction, $\rho\mathbf{vv}$ is a (second-order) tensor; it is called the *convective momentum-flux tensor*. Table 1.7-1 for the convective momentum flux tensor components should be compared with Table 1.2-1 for the molecular momentum flux tensor components.

Table 1.7-1   Summary of the Convective Momentum Flux Components

Direction normal to the shaded surface	Flux of momentum through the shaded surface	Convective momentum flux components		
		x-component	y-component	z-component
x	$\rho v_x\mathbf{v}$	$\rho v_x v_x$	$\rho v_x v_y$	$\rho v_x v_z$
y	$\rho v_y\mathbf{v}$	$\rho v_y v_x$	$\rho v_y v_y$	$\rho v_y v_z$
z	$\rho v_z\mathbf{v}$	$\rho v_z v_x$	$\rho v_z v_y$	$\rho v_z v_z$

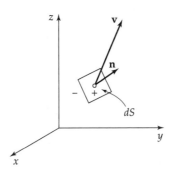

Fig. 1.7-2 The convective momentum flux through a plane of arbitrary orientation **n** is $(\mathbf{n} \cdot \mathbf{v})\rho\mathbf{v} = [\mathbf{n} \cdot \rho\mathbf{vv}]$.

Next we ask what the convective momentum flux would be through a surface element whose orientation is given by a unit normal vector **n** (see Fig. 1.7-2). If a fluid is flowing through the surface dS with a velocity **v**, then the volume rate of flow through the surface, from the minus side to the plus side, is $(\mathbf{n} \cdot \mathbf{v})dS$. Hence the rate of flow of momentum across the surface is $(\mathbf{n} \cdot \mathbf{v})\rho\mathbf{v}dS$, and the convective momentum flux is $(\mathbf{n} \cdot \mathbf{v})\rho\mathbf{v}$. According to the rules for vector-tensor notation given in Appendix A, this can also be written as $[\mathbf{n} \cdot \rho\mathbf{vv}]$—that is, the dot product of the unit normal vector **n** with the convective momentum flux tensor $\rho\mathbf{vv}$. If we let **n** be successively the unit vectors pointing in the x, y, and z directions (i.e., $\boldsymbol{\delta}_x$, $\boldsymbol{\delta}_y$, and $\boldsymbol{\delta}_z$), we obtain the entries in the second column of Table 1.7-1.

Similarly, the *total molecular momentum flux* through a surface of orientation **n** is given by $[\mathbf{n} \cdot \boldsymbol{\pi}] = p\mathbf{n} + [\mathbf{n} \cdot \boldsymbol{\tau}]$. It is understood that this is the flux from the minus side to the plus side of the surface. This quantity can also be interpreted as the force per unit area exerted by the minus material on the plus material across the surface. A geometric interpretation of $[\mathbf{n} \cdot \boldsymbol{\pi}]$ is given in Problem 1D.2.

In this chapter we defined the *molecular transport* of momentum in §1.2, and in this section we have described the *convective transport* of momentum. In setting up shell momentum balances in Chapter 2 and in setting up the general momentum balance in Chapter 3, we shall find it useful to define the *combined momentum flux*, which is the sum of the molecular momentum flux and the convective momentum flux:

$$\boldsymbol{\phi} = \boldsymbol{\pi} + \rho\mathbf{vv} = p\boldsymbol{\delta} + \boldsymbol{\tau} + \rho\mathbf{vv} \tag{1.7-2}$$

Keep in mind that the contribution $p\boldsymbol{\delta}$ contains no velocity, only the pressure; the combination $\rho\mathbf{vv}$ contains the density and products of the velocity components; and the contribution $\boldsymbol{\tau}$ contains the viscosity and, for a Newtonian fluid, is linear in the velocity gradients. All these quantities are second-order tensors.

Most of the time we will be dealing with components of these quantities. For example the components of $\boldsymbol{\phi}$ are

$$\phi_{xx} = \pi_{xx} + \rho v_x v_x = p + \tau_{xx} + \rho v_x v_x \tag{1.7-3a}$$
$$\phi_{xy} = \pi_{xy} + \rho v_x v_y = \tau_{xy} + \rho v_x v_y \tag{1.7-3b}$$

and so on, paralleling the entries in Tables 1.2-1 and 1.7-1. The important thing to remember is that

ϕ_{xy} = the combined flux of y-momentum across a surface perpendicular to the x
   direction by molecular and convective mechanisms.

The second index gives the component of momentum being transported and the first index gives the direction of transport.

The various symbols and nomenclature that are used for momentum fluxes are given in Table 1.7-2. The same sign convention is used for all fluxes.

Table 1.7-2   Summary of Notation for Momentum Fluxes

Symbol	Meaning	Reference
$\rho\mathbf{vv}$	Convective momentum-flux tensor	Table 1.7-1
$\boldsymbol{\tau}$	Viscous momentum-flux tensor[a]	Table 1.2-1
$\boldsymbol{\pi} = p\boldsymbol{\delta} + \boldsymbol{\tau}$	Molecular momentum-flux tensor[b]	Table 1.2-1
$\boldsymbol{\phi} = \boldsymbol{\pi} + \rho\mathbf{vv}$	Combined momentum-flux tensor	Eq. 1.7-2

[a] For viscoelastic fluids (see Chapter 8), this should be called the viscoelastic momentum-flux tensor or the viscoelastic stress tensor.

[b] This may be referred to as the molecular stress tensor.

QUESTIONS FOR DISCUSSION

1. Compare Newton's law of viscosity and Hooke's law of elasticity. What is the origin of these "laws"?
2. Verify that "momentum per unit area per unit time" has the same dimensions as "force per unit area."
3. Compare and contrast the molecular and convective mechanisms for momentum transport.
4. What are the physical meanings of the Lennard-Jones parameters and how can they be determined from viscosity data? Is the determination unique?
5. How do the viscosities of liquids and low-density gases depend on the temperature and pressure?
6. The Lennard-Jones potential depends only on the intermolecular separation. For what kinds of molecules would you expect that this kind of potential would be inappropriate?
7. Sketch the potential energy function $\varphi(r)$ for rigid, nonattracting spheres.
8. Molecules differing only in their atomic isotopes have the same values of the Lennard-Jones potential parameters. Would you expect the viscosity of CD_4 to be larger or smaller than that of CH_4 at the same temperature and pressure?
9. Fluid A has a viscosity twice that of fluid B; which fluid would you expect to flow more rapidly through a horizontal tube of length L and radius R when the same pressure difference is imposed?
10. Draw a sketch of the intermolecular force $F(r)$ obtained from the Lennard-Jones function for $\varphi(r)$. Also, determine the value of r_m in Fig. 1.4-2 in terms of the Lennard-Jones parameters.
11. What main ideas are used when one goes from Newton's law of viscosity in Eq. 1.1-2 to the generalization in Eq. 1.2-6?
12. What reference works can be consulted to find out more about kinetic theory of gases and liquids, and also for obtaining useful empiricisms for calculating viscosity?

PROBLEMS

1A.1 Estimation of dense-gas viscosity. Estimate the viscosity of nitrogen at 68°F and 1000 psig by means of Fig. 1.3-1, using the critical viscosity from Table E.1. Give the result in units of $lb_m/ft \cdot s$. For the meaning of "psig," see Table F.3-2.

Answer: $1.4 \times 10^{-5}\ lb_m/ft \cdot s$

1A.2 Estimation of the viscosity of methyl fluoride. Use Fig. 1.3-1 to find the viscosity in $Pa \cdot s$ of CH_3F at 370°C and 120 atm. Use the following values[1] for the critical constants: $T_c = 4.55°C$, $p_c = 58.0$ atm, $\rho_c = 0.300$ g/cm^3.

[1] K. A. Kobe and R. E. Lynn, Jr., *Chem. Revs.* **52**, 117–236 (1953), see p. 202.

1A.3 Computation of the viscosities of gases at low density. Predict the viscosities of molecular oxygen, nitrogen, and methane at 20°C and atmospheric pressure, and express the results in mPa · s. Compare the results with experimental data given in this chapter.

Answers: 0.0202, 0.0172, 0.0107 mPa · s

1A.4 Gas-mixture viscosities at low density. The following data[2] are available for the viscosities of mixtures of hydrogen and Freon-12 (dichlorodifluoromethane) at 25°C and 1 atm:

Mole fraction of H_2:	0.00	0.25	0.50	0.75	1.00
$\mu \times 10^6$ (poise):	124.0	128.1	131.9	135.1	88.4

Use the viscosities of the pure components to calculate the viscosities at the three intermediate compositions by means of Eqs. 1.4-15 and 16.

Sample answer: At 0.5, $\mu = 0.01317$ cp

1A.5 Viscosities of chlorine–air mixtures at low density. Predict the viscosities (in cp) of chlorine–air mixtures at 75°F and 1 atm, for the following mole fractions of chlorine: 0.00, 0.25, 0.50, 0.75, 1.00. Consider air as a single component and use Eqs. 1.4-14 to 16.

Answers: 0.0183, 0.0164, 0.0150, 0.0139, 0.0130 cp

1A.6 Estimation of liquid viscosity. Estimate the viscosity of saturated liquid water at 0°C and at 100°C by means of **(a)** Eq. 1.5-9, with $\Delta \hat{U}_{vap} = 897.5$ Btu/lb$_m$ at 100°C, and **(b)** Eq. 1.5-11. Compare the results with the values in Table 1.1-2.

Answer: **(b)** 4.0 cp, 0.95 cp

1A.7 Molecular velocity and mean free path. Compute the mean molecular velocity $\bar{u}$ (in cm/s) and the mean free path λ (in cm) for oxygen at 1 atm and 273.2 K. A reasonable value for d is 3 Å. What is the ratio of the mean free path to the molecular diameter under these conditions? What would be the order of magnitude of the corresponding ratio in the liquid state?

Answers: $\bar{u} = 4.25 \times 10^4$ cm/s, $\lambda = 9.3 \times 10^{-6}$ cm

1B.1 Velocity profiles and the stress components τ_{ij}. For each of the following velocity distributions, draw a meaningful sketch showing the flow pattern. Then find all the components of τ and ρvv for the Newtonian fluid. The parameter b is a constant.

(a) $v_x = by$, $v_y = 0$, $v_z = 0$
(b) $v_x = by$, $v_y = bx$, $v_z = 0$
(c) $v_x = -by$, $v_y = bx$, $v_z = 0$
(d) $v_x = -\frac{1}{2}bx$, $v_y = -\frac{1}{2}by$, $v_z = bz$

1B.2 A fluid in a state of rigid rotation.

(a) Verify that the velocity distribution (c) in Problem 1B.1 describes a fluid in a state of pure rotation; that is, the fluid is rotating like a rigid body. What is the angular velocity of rotation?

(b) For that flow pattern evaluate the symmetric and antisymmetric combinations of velocity derivatives:

(i) $(\partial v_y / \partial x) + (\partial v_x / \partial y)$
(ii) $(\partial v_y / \partial x) - (\partial v_x / \partial y)$

(c) Discuss the results of (b) in connection with the development in §1.2.

1B.3 Viscosity of suspensions. Data of Vand[3] for suspensions of small glass spheres in aqueous glycerol solutions of ZnI_2 can be represented up to about $\phi = 0.5$ by the semiempirical expression

$$\frac{\mu_{eff}}{\mu_0} = 1 + 2.5\phi + 7.17\phi^2 + 16.2\phi^3 + \cdots \quad (1B.3\text{-}1)$$

Compare this result with Eq. 1.6-2.

Answer: The Mooney equation gives a good fit of Vand's data if ϕ_0 is assigned the very reasonable value of 0.70.

1C.1 Some consequences of the Maxwell–Boltzmann distribution. In the simplified kinetic theory in §1.4, several statements concerning the equilibrium behavior of a gas were made without proof. In this problem and the next, some of these statements are shown to be exact consequences of the Maxwell–Boltzmann velocity distribution.

The Maxwell–Boltzmann distribution of molecular velocities in an ideal gas at rest is

$$f(u_x, u_y, u_z) = n(m/2\pi\kappa T)^{3/2} \exp(-mu^2/2\kappa T) \quad (1C.1\text{-}1)$$

in which **u** is the molecular velocity, n is the number density, and $f(u_x, u_y, u_z)du_x du_y du_z$ is the number of molecules per unit volume that is expected to have velocities between u_x and $u_x + du_x$, u_y and $u_y + du_y$, u_z and $u_z + du_z$. It follows from this equation that the distribution of the molecular speed u is

$$f(u) = 4\pi n u^2 (m/2\pi\kappa T)^{3/2} \exp(-mu^2/2\kappa T) \quad (1C.1\text{-}2)$$

(a) Verify Eq. 1.4-1 by obtaining the expression for the mean speed $\bar{u}$ from

$$\bar{u} = \frac{\displaystyle\int_0^\infty u f(u) du}{\displaystyle\int_0^\infty f(u) du} \quad (1C.1\text{-}3)$$

(b) Obtain the mean values of the velocity components $\bar{u}_x$, $\bar{u}_y$, and $\bar{u}_z$. The first of these is obtained from

$$\bar{u}_x = \frac{\displaystyle\int_{-\infty}^{+\infty}\int_{-\infty}^{+\infty}\int_{-\infty}^{+\infty} u_x f(u_x, u_y, u_z) du_x du_y du_z}{\displaystyle\int_{-\infty}^{+\infty}\int_{-\infty}^{+\infty}\int_{-\infty}^{+\infty} f(u_x, u_y, u_z) du_x du_y du_z} \quad (1C.1\text{-}4)$$

What can one conclude from the results?

---

[2] J. W. Buddenberg and C. R. Wilke, *Ind. Eng. Chem.* **41**, 1345–1347 (1949).

[3] V. Vand, *J. Phys. Colloid Chem.*, **52**, 277–299, 300–314, 314–321 (1948).

(c) Obtain the mean kinetic energy per molecule by

$$\tfrac{1}{2}m\overline{u^2} = \frac{\displaystyle\int_0^\infty \tfrac{1}{2}mu^2 f(u)\,du}{\displaystyle\int_0^\infty f(u)\,du} \tag{1C.1-5}$$

The correct result is $\tfrac{1}{2}m\overline{u^2} = \tfrac{3}{2}\kappa T$.

1C.2 The wall collision frequency. It is desired to find the frequency Z with which the molecules in an ideal gas strike a unit area of a wall from one side only. The gas is at rest and at equilibrium with a temperature T and the number density of the molecules is n. All molecules have a mass m. All molecules in the region $x < 0$ with $u_x > 0$ will hit an area S in the yz-plane in a short time Δt if they are in the volume $Su_x\Delta t$. The number of wall collisions per unit area per unit time will be

$$Z = \frac{\displaystyle\int_{-\infty}^{+\infty}\int_{-\infty}^{+\infty}\int_0^{+\infty} (Su_x\Delta t)f(u_x, u_y, u_z)\,du_x\,du_y\,du_z}{S\Delta t}$$

$$= n\left(\frac{m}{2\pi\kappa T}\right)^{3/2}\left(\int_0^{+\infty} u_x \exp(-mu^2/2\kappa T)\,du_x\right)$$

$$\left(\int_{-\infty}^{+\infty}\exp(-mu^2/2\kappa T)\,du_y\right)\left(\int_{-\infty}^{+\infty}\exp(-mu^2/2\kappa T)\,du_z\right)$$

$$= n\sqrt{\frac{\kappa T}{2\pi m}} = \tfrac{1}{4}n\overline{u} \tag{1C.2-1}$$

Verify the above development.

1C.3 Pressure of an ideal gas.[4] It is desired to get the pressure exerted by an ideal gas on a wall by accounting for the rate of momentum transfer from the molecules to the wall.

(a) When a molecule traveling with a velocity $\mathbf{v}$ collides with a wall, its incoming velocity components are u_x, u_y, u_z, and after a specular reflection at the wall, its components are $-u_x, u_y, u_z$. Thus the net momentum transmitted to the wall by a molecule is $2mu_x$. The molecules that have an x-component of the velocity equal to u_x, and that will collide with the wall during a small time interval Δt, must be within the volume $Su_x\Delta t$. How many molecules with velocity components in the range from u_x, u_y, u_z to $u_x + \Delta u_x$, $u_y + \Delta u_y$, $u_z + \Delta u_z$ will hit an area S of the wall with a velocity u_x within a time interval Δt? It will be $f(u_x, u_y, u_z)du_x\,du_y\,du_z$ times $Su_x\Delta t$. Then the pressure exerted on the wall by the gas will be

$$p = \frac{\displaystyle\int_{-\infty}^{+\infty}\int_{-\infty}^{+\infty}\int_0^{+\infty} (Su_x\Delta t)(2mu_x)f(u_x, u_y, u_z)\,du_x\,du_y\,du_z}{S\Delta t} \tag{1C.3-1}$$

Explain carefully how this expression is constructed. Verify that this relation is dimensionally correct.

(b) Insert Eq. 1C.1-1 for the Maxwell–Boltzmann equilibrium distribution into Eq. 1C.3-1 and perform the integration. Verify that this procedure leads to $p = n\kappa T$, the ideal gas law.

1D.1 Uniform rotation of a fluid.

(a) Verify that the velocity distribution in a fluid in a state of pure rotation (i.e., rotating as a rigid body) is $\mathbf{v} = [\mathbf{w} \times \mathbf{r}]$, where $\mathbf{w}$ is the angular velocity (a constant) and $\mathbf{r}$ is the position vector, with components x, y, z.

(b) What are $\nabla\mathbf{v} + (\nabla\mathbf{v})^\dagger$ and $(\nabla \cdot \mathbf{v})$ for the flow field in (a)?

(c) Interpret Eq. 1.2-7 in terms of the results in (b).

1D.2 Force on a surface of arbitrary orientation.[5] (Fig. 1D.2) Consider the material within an element of volume $OABC$ that is in a state of equilibrium, so that the sum of the forces acting on the triangular faces $\triangle OBC$, $\triangle OCA$, $\triangle OAB$, and $\triangle ABC$ must be zero. Let the area of $\triangle ABC$ be dS, and the force per unit area acting from the minus to the plus side of dS be the vector $\boldsymbol{\pi}_n$. Show that $\boldsymbol{\pi}_n = [\mathbf{n} \cdot \boldsymbol{\pi}]$.

(a) Show that the area of $\triangle OBC$ is the same as the area of the projection $\triangle ABC$ on the yz-plane; this is $(\mathbf{n} \cdot \boldsymbol{\delta}_x)dS$. Write similar expressions for the areas of $\triangle OCA$ and $\triangle OAB$.

(b) Show that according to Table 1.2-1 the force per unit area on $\triangle OBC$ is $\boldsymbol{\delta}_x\pi_{xx} + \boldsymbol{\delta}_y\pi_{xy} + \boldsymbol{\delta}_z\pi_{xz}$. Write similar force expressions for $\triangle OCA$ and $\triangle OAB$.

(c) Show that the force balance for the volume element $OABC$ gives

$$\boldsymbol{\pi}_n = \sum_i \sum_j (\mathbf{n} \cdot \boldsymbol{\delta}_i)(\boldsymbol{\delta}_j\pi_{ij}) = \left[\mathbf{n} \cdot \sum_i \sum_j \boldsymbol{\delta}_i\boldsymbol{\delta}_j\pi_{ij}\right] \tag{1D.2-1}$$

in which the indices i, j take on the values x, y, z. The double sum in the last expression is the stress tensor $\boldsymbol{\pi}$ written as a sum of products of unit dyads and components.

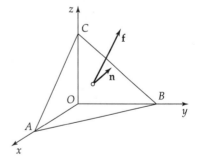

Fig. 1D.2 Element of volume $OABC$ over which a force balance is made. The vector $\boldsymbol{\pi}_n = [\mathbf{n} \cdot \boldsymbol{\pi}]$ is the force per unit area exerted by the minus material (material inside $OABC$) on the plus material (material outside $OABC$). The vector $\mathbf{n}$ is the outwardly directed unit normal vector on face ABC.

---

[4] R. J. Silbey and R. A. Alberty, *Physical Chemistry*, Wiley, New York, 3rd edition (2001), pp. 639–640.

[5] M. Abraham and R. Becker, *The Classical Theory of Electricity and Magnetism*, Blackie and Sons, London (1952), pp. 44–45.

Chapter 2

Shell Momentum Balances and Velocity Distributions in Laminar Flow

In this chapter we show how to obtain the velocity profiles for laminar flows of fluids in simple flow systems. These derivations make use of the definition of viscosity, the expressions for the molecular and convective momentum fluxes, and the concept of a momentum balance. Once the velocity profiles have been obtained, we can then get other quantities such as the maximum velocity, the average velocity, or the shear stress at a surface. Often it is these latter quantities that are of interest in engineering problems.

In the first section we make a few general remarks about how to set up differential momentum balances. In the sections that follow we work out in detail several classical examples of viscous flow patterns. These examples should be thoroughly understood, since we shall have frequent occasions to refer to them in subsequent chapters. Although these problems are rather simple and involve idealized systems, they are nonetheless often used in solving practical problems.

The systems studied in this chapter are so arranged that the reader is gradually introduced to a variety of factors that arise in the solution of viscous flow problems. In §2.2 the falling film problem illustrates the role of gravity forces and the use of Cartesian coordinates; it also shows how to solve the problem when viscosity may be a function of position. In §2.3 the flow in a circular tube illustrates the role of pressure and gravity forces and the use of cylindrical coordinates; an approximate extension to compressible flow is given. In §2.4 the flow in a cylindrical annulus emphasizes the role played by the boundary conditions. Then in §2.5 the question of boundary conditions is pursued further in the discussion of the flow of two adjacent immiscible liquids. Finally, in §2.6 the flow around a sphere is discussed briefly to illustrate a problem in spherical coordinates and also to point out how both tangential and normal forces are handled.

The methods and problems in this chapter apply only to *steady flow*. By "steady" we mean that the pressure, density, and velocity components at each point in the stream do not change with time. The general equations for unsteady flow are given in Chapter 3.

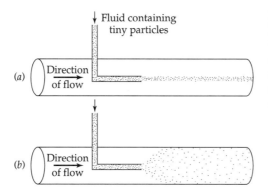

Fig. 2.0-1 (*a*) Laminar flow, in which fluid layers move smoothly over one another in the direction of flow, and (*b*) turbulent flow, in which the flow pattern is complex and time-dependent, with considerable motion perpendicular to the principal flow direction.

This chapter is concerned only with *laminar flow*. "Laminar flow" is the orderly flow that is observed, for example, in tube flow at velocities sufficiently low that tiny particles injected into the tube move along in a thin line. This is in sharp contrast with the wildly chaotic "turbulent flow" at sufficiently high velocities that the particles are flung apart and dispersed throughout the entire cross section of the tube. Turbulent flow is the subject of Chapter 5. The sketches in Fig. 2.0-1 illustrate the difference between the two flow regimes.

§2.1   SHELL MOMENTUM BALANCES AND BOUNDARY CONDITIONS

The problems discussed in §2.2 through §2.5 are approached by setting up momentum balances over a thin "shell" of the fluid. For *steady flow*, the momentum balance is

$$
\begin{Bmatrix} \text{rate of} \\ \text{momentum in} \\ \text{by convective} \\ \text{transport} \end{Bmatrix} - \begin{Bmatrix} \text{rate of} \\ \text{momentum out} \\ \text{by convective} \\ \text{transport} \end{Bmatrix} + \begin{Bmatrix} \text{rate of} \\ \text{momentum in} \\ \text{by molecular} \\ \text{transport} \end{Bmatrix} - \begin{Bmatrix} \text{rate of} \\ \text{momentum out} \\ \text{by molecular} \\ \text{transport} \end{Bmatrix} + \begin{Bmatrix} \text{force of gravity} \\ \text{acting on system} \end{Bmatrix} = 0 \quad (2.1\text{-}1)
$$

This is a restricted statement of the law of conservation of momentum. In this chapter we apply this statement only to one component of the momentum—namely, the component in the direction of flow. To write the momentum balance we need the expressions for the convective momentum fluxes given in Table 1.7-1 and the molecular momentum fluxes given in Table 1.2-1; keep in mind that the molecular momentum flux includes both the pressure and the viscous contributions.

In this chapter the momentum balance is applied only to systems in which there is just one velocity component, which depends on only one spatial variable; in addition, the flow must be rectilinear. In the next chapter the momentum balance concept is extended to unsteady-state systems with curvilinear motion and more than one velocity component.

The procedure in this chapter for setting up and solving viscous flow problems is as follows:

- Identify the nonvanishing velocity component and the spatial variable on which it depends.

- Write a momentum balance of the form of Eq. 2.1-1 over a thin shell perpendicular to the relevant spatial variable.

- Let the thickness of the shell approach zero and make use of the definition of the first derivative to obtain the corresponding differential equation for the momentum flux.

- Integrate this equation to get the momentum-flux distribution.
- Insert Newton's law of viscosity and obtain a differential equation for the velocity.
- Integrate this equation to get the velocity distribution.
- Use the velocity distribution to get other quantities, such as the maximum velocity, average velocity, or force on solid surfaces.

In the integrations mentioned above, several constants of integration appear, and these are evaluated by using "boundary conditions"—that is, statements about the velocity or stress at the boundaries of the system. The most commonly used boundary conditions are as follows:

a. At *solid–fluid* interfaces the fluid velocity equals the velocity with which the solid surface is moving; this statement is applied to both the tangential and the normal component of the velocity vector. The equality of the tangential components is referred to as the "no-slip condition."

b. At a *liquid–liquid* interfacial plane of constant x, the tangential velocity components v_y and v_z are continuous through the interface (the "no-slip condition") as are also the molecular stress-tensor components $p + \tau_{xx}$, τ_{xy}, and τ_{xz}.

c. At a *liquid–gas* interfacial plane of constant x, the stress-tensor components τ_{xy} and τ_{xz} are taken to be zero, provided that the gas-side velocity gradient is not too large. This is reasonable, since the viscosities of gases are much less than those of liquids.

In all of these boundary conditions it is presumed that there is no material passing through the interface; that is, there is no adsorption, absorption, dissolution, evaporation, melting, or chemical reaction at the surface between the two phases. Boundary conditions incorporating such phenomena appear in Problems 3C.5 and 11C.6, and §18.1.

In this section we have presented some guidelines for solving simple viscous flow problems. For some problems slight variations on these guidelines may prove to be appropriate.

§2.2   FLOW OF A FALLING FILM

The first example we discuss is that of the flow of a liquid down an inclined flat plate of length L and width W, as shown in Fig. 2.2-1. Such films have been studied in connection with wetted-wall towers, evaporation and gas-absorption experiments, and applications of coatings. We consider the viscosity and density of the fluid to be constant.

A complete description of the liquid flow is difficult because of the disturbances at the edges of the system ($z = 0, z = L, y = 0, y = W$). An adequate description can often be

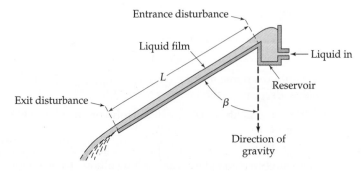

Fig. 2.2-1 Schematic diagram of the falling film experiment, showing end effects.

obtained by neglecting such disturbances, particularly if W and L are large compared to the film thickness δ. For small flow rates we expect that the viscous forces will prevent continued acceleration of the liquid down the wall, so that v_z will become independent of z in a short distance down the plate. Therefore it seems reasonable to *postulate* that $v_z = v_z(x)$, $v_x = 0$, and $v_y = 0$, and further that $p = p(x)$. From Table B.1 it is seen that the only nonvanishing components of $\boldsymbol{\tau}$ are then $\tau_{xz} = \tau_{zx} = -\mu(dv_z/dx)$.

We now select as the "system" a thin shell perpendicular to the x direction (see Fig. 2.2-2). Then we set up a z-momentum balance over this shell, which is a region of thickness Δx, bounded by the planes $z = 0$ and $z = L$, and extending a distance W in the y direction. The various contributions to the momentum balance are then obtained with the help of the quantities in the "z-component" columns of Tables 1.2-1 and 1.7-1. By using the components of the "combined momentum-flux tensor" $\boldsymbol{\phi}$ defined in 1.7-1 to 3, we can include all the possible mechanisms for momentum transport at once:

rate of z-momentum in across surface at $z = 0$	$(W\Delta x)\phi_{zz}\|_{z=0}$	(2.2-1)
rate of z-momentum out across surface at $z = L$	$(W\Delta x)\phi_{zz}\|_{z=L}$	(2.2-2)
rate of z-momentum in across surface at x	$(LW)(\phi_{xz})\|_{x}$	(2.2-3)
rate of z-momentum out across surface at $x + \Delta x$	$(LW)(\phi_{xz})\|_{x+\Delta x}$	(2.2-4)
gravity force acting on fluid in the z direction	$(LW\,\Delta x)(\rho g \cos \beta)$	(2.2-5)

By using the quantities ϕ_{xz} and ϕ_{zz} we account for the z-momentum transport by all mechanisms, convective and molecular. Note that we take the "in" and "out" directions in the direction of the positive x- and z-axes (in this problem these happen to coincide with the directions of z-momentum transport). The notation $\|_{x+\Delta x}$ means "evaluated at $x + \Delta x$," and g is the gravitational acceleration.

When these terms are substituted into the z-momentum balance of Eq. 2.1-1, we get

$$LW(\phi_{xz}\|_x - \phi_{xz}\|_{x+\Delta x}) + W\Delta x(\phi_{zz}\|_{z=0} - \phi_{zz}\|_{z=L}) + (LW\,\Delta x)(\rho g \cos \beta) = 0 \qquad (2.2\text{-}6)$$

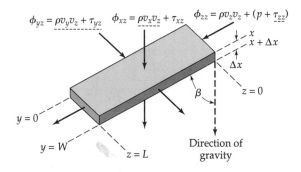

$\phi_{yz} = \rho v_y v_z + \tau_{yz}$ $\phi_{xz} = \rho v_x v_z + \tau_{xz}$ $\phi_{zz} = \rho v_z v_z + (p + \tau_{zz})$

Fig. 2.2-2 Shell of thickness Δx over which a z-momentum balance is made. Arrows show the momentum fluxes associated with the surfaces of the shell. Since v_x and v_y are both zero, $\rho v_x v_z$ and $\rho v_y v_z$ are zero. Since v_z does not depend on y and z, it follows from Table B.1 that $\tau_{yz} = 0$ and $\tau_{zz} = 0$. Therefore, the dashed-underlined fluxes do not need to be considered. Both p and $\rho v_z v_z$ are the same at $z = 0$ and $z = L$, and therefore do not appear in the final equation for the balance of z-momentum, Eq. 2.2-10.

When this equation is divided by $LW\,\Delta x$, and the limit taken as Δx approaches zero, we get

$$\lim_{\Delta x \to 0}\left(\frac{\phi_{xz}|_{x+\Delta x} - \phi_{xz}|_x}{\Delta x}\right) - \frac{\phi_{zz}|_{z=0} - \phi_{zz}|_{z=L}}{L} = \rho g \cos \beta \tag{2.2-7}$$

The first term on the left side is exactly the definition of the derivative of ϕ_{xz} with respect to x. Therefore Eq. 2.2-7 becomes

$$\frac{\partial \phi_{xz}}{\partial x} - \frac{\phi_{zz}|_{z=0} - \phi_{zz}|_{z=L}}{L} = \rho g \cos \beta \tag{2.2-8}$$

At this point we have to write out explicitly what the components ϕ_{xz} and ϕ_{zz} are, making use of the definition of $\boldsymbol{\phi}$ in Eqs. 1.7-1 to 3 and the expressions for τ_{xz} and τ_{zz} in Appendix B.1. This ensures that we do not miss out on any of the forms of momentum transport. Hence we get

$$\phi_{xz} = \tau_{xz} + \rho v_x v_z = -\mu \frac{\partial v_z}{\partial x} + \rho v_x v_z \tag{2.2-9a}$$

$$\phi_{zz} = p + \tau_{zz} + \rho v_z v_z = p - 2\mu \frac{\partial v_z}{\partial z} + \rho v_z v_z \tag{2.2-9b}$$

In accordance with the postulates that $v_z = v_z(x)$, $v_x = 0$, $v_y = 0$, and $p = p(x)$, we see that (i) since $v_x = 0$, the $\rho v_x v_z$ term in Eq. 2.2-9a is zero; (ii) since $v_z = v_z(x)$, the term $-2\mu(\partial v_z/\partial z)$ in Eq. 2.2-9b is zero; (iii) since $v_z = v_z(x)$, the term $\rho v_z v_z$ is the same at $z = 0$ and $z = L$; and (iv) since $p = p(x)$, the contribution p is the same at $z = 0$ and $z = L$. Hence τ_{xz} depends only on x, and Eq. 2.2-8 simplifies to

$$\boxed{\frac{d\tau_{xz}}{dx} = \rho g \cos \beta} \tag{2.2-10}$$

This is the differential equation for the momentum flux τ_{xz}. It may be integrated to give

$$\tau_{xz} = (\rho g \cos \beta)x + C_1 \tag{2.2-11}$$

The constant of integration may be evaluated by using the boundary condition at the gas–liquid interface (see §2.1):

B.C. 1:   at $x = 0$, $\tau_{xz} = 0$ \hfill (2.2-12)

Substitution of this boundary condition into Eq. 2.2-11 shows that $C_1 = 0$. Therefore the momentum-flux distribution is

$$\tau_{xz} = (\rho g \cos \beta)x \tag{2.2-13}$$

as shown in Fig. 2.2-3.

Next we substitute Newton's law of viscosity

$$\tau_{xz} = -\mu \frac{dv_z}{dx} \tag{2.2-14}$$

into the left side of Eq. 2.2-13 to obtain

$$\frac{dv_z}{dx} = -\left(\frac{\rho g \cos \beta}{\mu}\right)x \tag{2.2-15}$$

which is the differential equation for the velocity distribution. It can be integrated to give

$$v_z = -\left(\frac{\rho g \cos \beta}{2\mu}\right)x^2 + C_2 \tag{2.2-16}$$

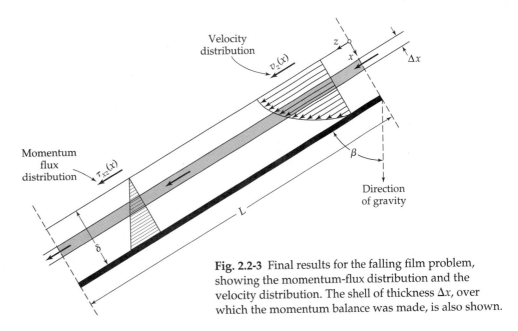

Fig. 2.2-3 Final results for the falling film problem, showing the momentum-flux distribution and the velocity distribution. The shell of thickness Δx, over which the momentum balance was made, is also shown.

The constant of integration is evaluated by using the no-slip boundary condition at the solid surface:

B.C. 2   at $x = \delta$, $v_z = 0$ (2.2-17)

Substitution of this boundary condition into Eq. 2.2-16 shows that $C_2 = (\rho g \cos \beta / 2\mu)\delta^2$. Consequently, the velocity distribution is

$$v_z = \frac{\rho g \delta^2 \cos \beta}{2\mu}\left[1 - \left(\frac{x}{\delta}\right)^2\right]$$ (2.2-18)

This parabolic velocity distribution is shown in Fig. 2.2-3. It is consistent with the postulates made initially and must therefore be a *possible* solution. Other solutions might be possible, and experiments are normally required to tell whether other flow patterns can actually arise. We return to this point after Eq. 2.2-23.

Once the velocity distribution is known, a number of quantities can be calculated:

(i) The *maximum velocity* $v_{z,max}$ is clearly the velocity at $x = 0$; that is,

$$v_{z,max} = \frac{\rho g \delta^2 \cos \beta}{2\mu}$$ (2.2-19)

(ii) The *average velocity* $\langle v_z \rangle$ over a cross section of the film is obtained as follows:

$$\langle v_z \rangle = \frac{\int_0^W \int_0^\delta v_z\,dx\,dy}{\int_0^W \int_0^\delta dx\,dy} = \frac{1}{\delta}\int_0^\delta v_z\,dx$$

$$= \frac{\rho g \delta^2 \cos \beta}{2\mu}\int_0^1\left[1 - \left(\frac{x}{\delta}\right)^2\right]d\left(\frac{x}{\delta}\right)$$

$$= \frac{\rho g \delta^2 \cos \beta}{3\mu} = \tfrac{2}{3}v_{z,max}$$ (2.2-20)

The double integral in the denominator of the first line is the cross-sectional area of the film. The double integral in the numerator is the volume flow rate through a differential element of the cross section, $v_z\,dx\,dy$, integrated over the entire cross section.

(iii) The *mass rate of flow* w is obtained from the average velocity or by integration of the velocity distribution

$$w = \int_0^W \int_0^\delta \rho v_z\,dxdy = \rho W \delta \langle v_z \rangle = \frac{\rho^2 g W \delta^3 \cos \beta}{3\mu} \tag{2.2-21}$$

(iv) The *film thickness* δ may be given in terms of the average velocity or the mass rate of flow as follows:

$$\delta = \sqrt{\frac{3\mu \langle v_z \rangle}{\rho g \cos \beta}} = \sqrt[3]{\frac{3\mu w}{\rho^2 g W \cos \beta}} \tag{2.2-22}$$

(v) The force per unit area in the z direction on a surface element perpendicular to the x direction is $+\tau_{xz}$ evaluated at $x = \delta$. This is the force exerted by the fluid (region of lesser x) on the wall (region of greater x). The z-component of the *force* $\mathbf{F}$ *of the fluid on the solid surface* is obtained by integrating the shear stress over the fluid–solid interface:

$$F_z = \int_0^L \int_0^W (\tau_{xz}|_{x=\delta})dy\,dz = \int_0^L \int_0^W \left(-\mu \frac{dv_z}{dx}\bigg|_{x=\delta}\right)dy\,dz$$

$$= (LW)(-\mu)\left(-\frac{\rho g \delta \cos \beta}{\mu}\right) = \rho g \delta L W \cos \beta \tag{2.2-23}$$

This is the z-component of the weight of the fluid in the entire film—as we would have expected.

Experimental observations of falling films show that there are actually three "flow regimes," and that these may be classified according to the *Reynolds number*,[1] Re, for the flow. For falling films the Reynolds number is defined by Re $= 4\delta \langle v_z \rangle \rho / \mu$. The three flow regime are then:

laminar flow with negligible rippling   Re < 20
laminar flow with pronounced rippling   20 < Re < 1500
turbulent flow   Re > 1500

The analysis we have given above is valid only for the first regime, since the analysis was restricted by the postulates made at the outset. Ripples appear on the surface of the fluid at all Reynolds numbers. For Reynolds numbers less than about 20, the ripples are very long and grow rather slowly as they travel down the surface of the liquid; as a result the formulas derived above are useful up to about Re = 20 for plates of moderate length. Above that value of Re, the ripple growth increases very rapidly, although the flow remains laminar. At about Re = 1500 the flow becomes irregular and chaotic, and the flow is said to be turbulent.[2,3] At this point it is not clear why the value of the

---

[1] This dimensionless group is named for **Osborne Reynolds** (1842–1912), professor of engineering at the University of Manchester. He studied the laminar-turbulent transition, turbulent heat transfer, and theory of lubrication. We shall see in the next chapter that the Reynolds number is the ratio of the inertial forces to the viscous forces.

[2] G. D. Fulford, *Adv. Chem. Engr.*, **5**, 151–236 (1964); S. Whitaker, *Ind. Eng. Chem. Fund.*, **3**, 132–142 (1964); V. G. Levich, *Physicochemical Hydrodynamics*, Prentice-Hall, Englewood Cliffs, N.J. (1962), §135.

[3] H.-C. Chang, *Ann. Rev. Fluid Mech.*, **26**, 103–136 (1994); S.-H. Hwang and H.-C. Chang, *Phys. Fluids*, **30**, 1259–1268 (1987).

Reynolds number should be used to delineate the flow regimes. We shall have more to say about this in §3.7.

This discussion illustrates a very important point: theoretical analysis of flow systems is limited by the postulates that are made in setting up the problem. It is absolutely necessary to do experiments in order to establish the flow regimes so as to know when instabilities (spontaneous oscillations) occur and when the flow becomes turbulent. Some information about the onset of instability and the demarcation of the flow regimes can be obtained by theoretical analysis, but this is an extraordinarily difficult subject. This is a result of the inherent nonlinear nature of the governing equations of fluid dynamics, as will be explained in Chapter 3. Suffice it to say at this point that experiments play a *very* important role in the field of fluid dynamics.

EXAMPLE 2.2-1 *Calculation of Film Velocity*	An oil has a kinematic viscosity of 2×10^{-4} m^2/s and a density of 0.8×10^3 kg/m^3. If we want to have a falling film of thickness of 2.5 mm on a vertical wall, what should the mass rate of flow of the liquid be?

SOLUTION

According to Eq. 2.2-21, the mass rate of flow in kg/s is

$$w = \frac{\rho g \delta^3 W}{3\nu} = \frac{(0.8 \times 10^3)(9.80)(2.5 \times 10^{-3})^3 W}{3(2 \times 10^{-4})} = 0.204 W \tag{2.2-24}$$

To get the mass rate of flow one then needs to insert a value for the width of the wall in meters. This is the desired result provided that the flow is laminar and nonrippling. To determine the flow regime we calculate the Reynolds number, making use of Eqs. 2.2-21 and 24

$$\text{Re} = \frac{4\delta \langle v_z \rangle \rho}{\mu} = \frac{4w/W}{\nu\rho} = \frac{4(0.204)}{(2 \times 10^{-4})(0.8 \times 10^3)} = 5.1 \tag{2.2-25}$$

This Reynolds number is sufficiently low that rippling will not be pronounced, and therefore the expression for the mass rate of flow in Eq. 2.2-24 is reasonable.

EXAMPLE 2.2-2 *Falling Film with Variable Viscosity*	Rework the falling film problem for a position-dependent viscosity $\mu = \mu_0 e^{-\alpha x/\delta}$, which arises when the film is nonisothermal, as in the condensation of a vapor on a wall. Here μ_0 is the viscosity at the surface of the film and α is a constant that describes how rapidly μ decreases as x increases. Such a variation could arise in the flow of a condensate down a wall with a linear temperature gradient through the film.

SOLUTION

The development proceeds as before up to Eq. 2.2-13. Then substituting Newton's law with variable viscosity into Eq. 2.2-13 gives

$$-\mu_0 e^{-\alpha x/\delta} \frac{dv_z}{dx} = \rho g x \cos \beta \tag{2.2-26}$$

This equation can be integrated, and using the boundary conditions in Eq. 2.2-17 enables us to evaluate the integration constant. The velocity profile is then

$$v_z = \frac{\rho g \delta^2 \cos \beta}{\mu_0} \left[e^\alpha \left(\frac{1}{\alpha} - \frac{1}{\alpha^2} \right) - e^{\alpha x/\delta} \left(\frac{x}{\alpha\delta} - \frac{1}{\alpha^2} \right) \right] \tag{2.2-27}$$

As a check we evaluate the velocity distribution for the constant-viscosity problem (that is, when α is zero). However, setting $\alpha = 0$ gives $\infty - \infty$ in the two expressions within parentheses.

This difficulty can be overcome if we expand the two exponentials in Taylor series (see §C.2), as follows:

$$(v_z)_{\alpha=0} = \frac{\rho g \delta^2 \cos \beta}{\mu_0} \lim_{\alpha \to 0} \left[\left(1 + \alpha + \frac{\alpha^2}{2!} + \frac{\alpha^3}{3!} + \cdots \right) \left(\frac{1}{\alpha} - \frac{1}{\alpha^2} \right) \right.$$

$$\left. - \left(1 + \frac{\alpha x}{\delta} + \frac{\alpha^2 x^2}{2! \delta^2} + \frac{\alpha^3 x^3}{3! \delta^3} + \cdots \right) \left(\frac{x}{\alpha \delta} - \frac{1}{\alpha^2} \right) \right]$$

$$= \frac{\rho g \delta^2 \cos \beta}{\mu_0} \lim_{\alpha \to 0} \left[\left(\frac{1}{2} + \frac{1}{3} \alpha + \cdots \right) - \left(\frac{1}{2} \frac{x^2}{\delta^2} - \frac{1}{3} \frac{x^3}{\delta^3} \alpha + \cdots \right) \right]$$

$$= \frac{\rho g \delta^2 \cos \beta}{2 \mu_0} \left[1 - \left(\frac{x}{\delta} \right)^2 \right] \tag{2.2-28}$$

which is in agreement with Eq. 2.2-18.

From Eq. 2.2-27 it may be shown that the average velocity is

$$\langle v_z \rangle = \frac{\rho g \delta^2 \cos \beta}{\mu_0} \left[e^\alpha \left(\frac{1}{\alpha} - \frac{2}{\alpha^2} + \frac{2}{\alpha^3} \right) - \frac{2}{\alpha^3} \right] \tag{2.2-29}$$

The reader may verify that this result simplifies to Eq. 2.2-20 when α goes to zero.

§2.3   FLOW THROUGH A CIRCULAR TUBE

The flow of fluids in circular tubes is encountered frequently in physics, chemistry, biology, and engineering. The laminar flow of fluids in circular tubes may be analyzed by means of the momentum balance described in §2.1. The only new feature introduced here is the use of cylindrical coordinates, which are the natural coordinates for describing positions in a pipe of circular cross section.

We consider then the steady-state, laminar flow of a fluid of constant density ρ and viscosity μ in a vertical tube of length L and radius R. The liquid flows downward under the influence of a pressure difference and gravity; the coordinate system is that shown in Fig. 2.3-1. We specify that the tube length be very large with respect to the tube radius, so that "end effects" will be unimportant throughout most of the tube; that is, we can ignore the fact that at the tube entrance and exit the flow will not necessarily be parallel to the tube wall.

We postulate that $v_z = v_z(r)$, $v_r = 0$, $v_\theta = 0$, and $p = p(z)$. With these postulates it may be seen from Table B.1 that the only nonvanishing components of τ are $\tau_{rz} = \tau_{zr} = -\mu(dv_z/dr)$.

We select as our system a cylindrical shell of thickness Δr and length L and we begin by listing the various contributions to the z-momentum balance:

rate of z-momentum in across annular surface at $z = 0$	$(2\pi r \Delta r)(\phi_{zz})\|_{z=0}$	(2.3-1)
rate of z-momentum out across annular surface at $z = L$	$(2\pi r \Delta r)(\phi_{zz})\|_{z=L}$	(2.3-2)
rate of z-momentum in across cylindrical surface at r	$(2\pi r L)(\phi_{rz})\|_r = (2\pi r L \phi_{rz})\|_r$	(2.3-3)
rate of z-momentum out across cylindrical surface at $r + \Delta r$	$(2\pi (r + \Delta r) L)(\phi_{rz})\|_{r+\Delta r} = (2\pi r L \phi_{rz})\|_{r+\Delta r}$	(2.3-4)
gravity force acting in z direction on cylindrical shell	$(2\pi r \Delta r L)\rho g$	(2.3-5)

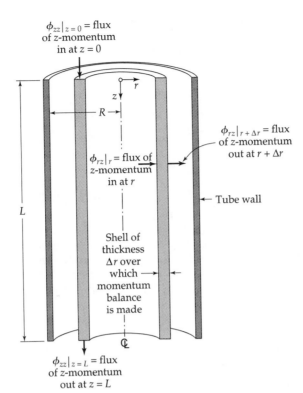

Fig. 2.3-1 Cylindrical shell of fluid over which the z-momentum balance is made for axial flow in a circular tube (see Eqs. 2.3-1 to 5). The z-momentum fluxes ϕ_{rz} and ϕ_{zz} are given in full in Eqs. 2.3-9a and 9b.

$\phi_{zz}|_{z=0}$ = flux of z-momentum in at $z = 0$

$\phi_{rz}|_{r+\Delta r}$ = flux of z-momentum out at $r + \Delta r$

$\phi_{rz}|_r$ = flux of z-momentum in at r

← Tube wall

Shell of thickness Δr over which → ← momentum balance is made

$\phi_{zz}|_{z=L}$ = flux of z-momentum out at $z = L$

The quantities ϕ_{zz} and ϕ_{rz} account for the momentum transport by all possible mechanisms, convective and molecular. In Eq. 2.3-4, $(r + \Delta r)$ and $(r)|_{r+\Delta r}$ are two ways of writing the same thing. Note that we take "in" and "out" to be in the positive directions of the r- and z-axes.

We now add up the contributions to the momentum balance:

$$(2\pi r L\phi_{rz})|_r - (2\pi r L\phi_{rz})|_{r+\Delta r} + (2\pi r \Delta r)(\phi_{zz})|_{z=0} - (2\pi r\Delta r)(\phi_{zz})|_{z=L} + (2\pi r \Delta r L)\rho g = 0 \quad (2.3\text{-}6)$$

When we divide Eq. (2.3-8) by $2\pi L\Delta r$ and take the limit as $\Delta r \to 0$, we get

$$\lim_{\Delta r \to 0}\left(\frac{(r\phi_{rz})|_{r+\Delta r} - (r\phi_{rz})|_r}{\Delta r}\right) = \left(\frac{\phi_{zz}|_{z=0} - \phi_{zz}|_{z=L}}{L} + \rho g\right)r \quad (2.3\text{-}7)$$

The expression on the left side is the definition of the first derivative of $r\phi_{rz}$ with respect to r. Hence Eq. 2.3-7 may be written as

$$\frac{\partial}{\partial r}(r\phi_{rz}) = \left(\frac{\phi_{zz}|_{z=0} - \phi_{zz}|_{z=L}}{L} + \rho g\right)r \quad (2.3\text{-}8)$$

Now we have to evaluate the components ϕ_{rz} and ϕ_{zz} from Eq. 1.7-1 and Appendix B.1:

$$\phi_{rz} = \tau_{rz} + \rho v_r v_z = -\mu\frac{\partial v_z}{\partial r} + \rho v_r v_z \quad (2.3\text{-}9a)$$

$$\phi_{zz} = p + \tau_{zz} + \rho v_z v_z = p - 2\mu\frac{\partial v_z}{\partial z} + \rho v_z v_z \quad (2.3\text{-}9b)$$

Next we take into account the postulates made at the beginning of the problem—namely, that $v_z = v_z(r)$, $v_r = 0$, $v_\theta = 0$, and $p = p(z)$. Then we make the following simplifications:

(i) because $v_r = 0$, we can drop the term $\rho v_r v_z$ in Eq. 2.3-9a; (ii) because $v_z = v_z(r)$, the term $\rho v_z v_z$ will be the same at both ends of the tube; and (iii) because $v_z = v_z(r)$, the term $-2\mu \partial v_z / \partial z$ will be the same at both ends of the tube. Hence Eq. 2.3-8 simplifies to

$$\frac{d}{dr}(r\tau_{rz}) = \left(\frac{(p_0 - \rho g 0) - (p_L - \rho g L)}{L}\right)r \equiv \left(\frac{\mathcal{P}_0 - \mathcal{P}_L}{L}\right)r \tag{2.3-10}$$

in which $\mathcal{P} = p - \rho g z$ is a convenient abbreviation for the sum of the pressure and gravitational terms.[1] Equation 2.3-10 may be integrated to give

$$\tau_{rz} = \left(\frac{\mathcal{P}_0 - \mathcal{P}_L}{2L}\right)r + \frac{C_1}{r} \tag{2.3-11}$$

The constant C_1 is evaluated by using the boundary condition

B.C. 1:   at $r = 0$, $\tau_{rz} =$ finite $\tag{2.3-12}$

Consequently C_1 must be zero, for otherwise the momentum flux would be infinite at the axis of the tube. Therefore the momentum flux distribution is

$$\boxed{\tau_{rz} = \left(\frac{\mathcal{P}_0 - \mathcal{P}_L}{2L}\right)r} \tag{2.3-13}$$

This distribution is shown in Fig. 2.3-2.

Newton's law of viscosity for this situation is obtained from Appendix B.2 as follows:

$$\tau_{rz} = -\mu \frac{dv_z}{dr} \tag{2.3-14}$$

Substitution of this expression into Eq. 2.3-13 then gives the following differential equation for the velocity:

$$\frac{dv_z}{dr} = -\left(\frac{\mathcal{P}_0 - \mathcal{P}_L}{2\mu L}\right)r \tag{2.3-15}$$

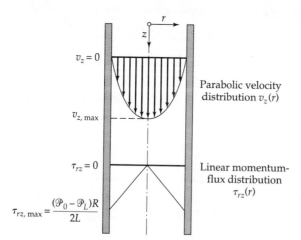

$v_z = 0$

Parabolic velocity distribution $v_z(r)$

$v_{z,\,max}$

$\tau_{rz} = 0$

Linear momentum-flux distribution $\tau_{rz}(r)$

$\tau_{rz,\,max} = \dfrac{(\mathcal{P}_0 - \mathcal{P}_L)R}{2L}$

Fig. 2.3-2 The momentum-flux distribution and velocity distribution for the downward flow in a circular tube.

---

[1] The quantity designated by $\mathcal{P}$ is called the *modified pressure*. In general it is defined by $\mathcal{P} = p + \rho g h$, where h is the distance "upward"—that is, in the direction opposed to gravity from some preselected reference plane. Hence in this problem $h = -z$.

This first-order separable differential equation may be integrated to give

$$v_z = -\left(\frac{\mathcal{P}_0 - \mathcal{P}_L}{4\mu L}\right)r^2 + C_2 \qquad (2.3\text{-}16)$$

The constant C_2 is evaluated from the boundary condition

B.C. 2:   at $r = R$, $v_z = 0$ (2.3-17)

From this C_2 is found to be $(\mathcal{P}_0 - \mathcal{P}_L)R^2/4\mu L$. Hence the velocity distribution is

$$\boxed{v_z = \frac{(\mathcal{P}_0 - \mathcal{P}_L)R^2}{4\mu L}\left[1 - \left(\frac{r}{R}\right)^2\right]} \qquad (2.3\text{-}18)$$

We see that the velocity distribution for laminar, incompressible flow of a Newtonian fluid in a long tube is parabolic (see Fig. 2.3-2).

Once the velocity profile has been established, various derived quantities can be obtained:

(i)   The *maximum velocity* $v_{z,\text{max}}$ occurs at $r = 0$ and is

$$v_{z,\text{max}} = \frac{(\mathcal{P}_0 - \mathcal{P}_L)R^2}{4\mu L} \qquad (2.3\text{-}19)$$

(ii) The *average velocity* $\langle v_z \rangle$ is obtained by dividing the total volumetric flow rate by the cross-sectional area

$$\langle v_z \rangle = \frac{\displaystyle\int_0^{2\pi}\int_0^R v_z\, r\, dr\, d\theta}{\displaystyle\int_0^{2\pi}\int_0^R r\, dr\, d\theta} = \frac{(\mathcal{P}_0 - \mathcal{P}_L)R^2}{8\mu L} = \tfrac{1}{2}v_{z,\text{max}} \qquad (2.3\text{-}20)$$

(iii) The *mass rate of flow* w is the product of the cross-sectional area πR^2, the density ρ, and the average velocity $\langle v_z \rangle$

$$w = \frac{\pi(\mathcal{P}_0 - \mathcal{P}_L)R^4\rho}{8\mu L} \qquad (2.3\text{-}21)$$

This rather famous result is called the *Hagen–Poiseuille*[2] *equation*. It is used, along with experimental data for the rate of flow and the modified pressure difference, to determine the viscosity of fluids (see Example 2.3-1) in a "capillary viscometer."

(iv) The z-component of the *force, F_z, of the fluid on the wetted surface* of the pipe is just the shear stress τ_{rz} integrated over the wetted area

$$F_z = (2\pi RL)\left(-\mu \frac{dv_z}{dr}\right)\Bigg|_{r=R} = \pi R^2(\mathcal{P}_0 - \mathcal{P}_L)$$

$$= \pi R^2(p_0 - p_L) + \pi R^2 L\rho g \qquad (2.3\text{-}22)$$

This result states that the viscous force F_z is counterbalanced by the net pressure force and the gravitational force. This is exactly what one would obtain from making a force balance over the fluid in the tube.

---

[2] G. Hagen, *Ann. Phys. Chem.*, **46**, 423–442 (1839); J. L. Poiseuille, *Comptes Rendus*, **11**, 961 and 1041 (1841). **Jean Louis Poiseuille** (1799–1869) (pronounced "Pwa-zø´-yuh," with ø is roughly the "oo" in book) was a physician interested in the flow of blood. Although Hagen and Poiseuille established the dependence of the flow rate on the fourth power of the tube radius, Eq. 2.3-21 was first derived by E. Hagenbach, *Pogg. Annalen der Physik u. Chemie*, **108**, 385–426 (1860).

The results of this section are only as good as the postulates introduced at the beginning of the section—namely, that $v_z = v_z(r)$ and $p = p(z)$. Experiments have shown that these postulates are in fact realized for Reynolds numbers up to about 2100; above that value, the flow will be turbulent if there are any appreciable disturbances in the system—that is, wall roughness or vibrations.[3] For circular tubes the Reynolds number is defined by $Re = D\langle v_z\rangle\rho/\mu$, where $D = 2R$ is the tube diameter.

We now summarize all the assumptions that were made in obtaining the Hagen–Poiseuille equation.

(a) The flow is laminar; that is, Re must be less than about 2100.

(b) The density is constant ("incompressible flow").

(c) The flow is "steady" (i.e., it does not change with time).

(d) The fluid is Newtonian (Eq. 2.3-14 is valid).

(e) End effects are neglected. Actually an "entrance length," after the tube entrance, of the order of $L_e = 0.035D\,Re$, is needed for the buildup to the parabolic profile. If the section of pipe of interest includes the entrance region, a correction must be applied.[4] The fractional correction in the pressure difference or mass rate of flow never exceeds L_e/L if $L > L_e$.

(f) The fluid behaves as a continuum—this assumption is valid, except for very dilute gases or very narrow capillary tubes, in which the molecular mean free path is comparable to the tube diameter (the "slip flow region") or much greater than the tube diameter (the "Knudsen flow" or "free molecule flow" regime).[5]

(g) There is no slip at the wall, so that B.C. 2 is valid; this is an excellent assumption for pure fluids under the conditions assumed in (f). See Problem 2B.9 for a discussion of wall slip.

EXAMPLE 2.3-1

Determination of Viscosity from Capillary Flow Data

Glycerine ($CH_2OH \cdot CHOH \cdot CH_2OH$) at 26.5°C is flowing through a horizontal tube 1 ft long and with 0.1 in. inside diameter. For a pressure drop of 40 psi, the volume flow rate w/ρ is 0.00398 ft³/min. The density of glycerine at 26.5°C is 1.261 g/cm³. From the flow data, find the viscosity of glycerine in centipoises and in Pa · s.

SOLUTION

From the Hagen–Poiseuille equation (Eq. 2.3-21), we find

$$\mu = \frac{\pi(p_0 - p_L)R^4}{8(w/\rho)L}$$

$$= \frac{\pi\left(40\,\dfrac{lb_f}{in.^2}\right)\left(6.8947 \times 10^4\,\dfrac{dyn/cm^2}{lb_f/in.^2}\right)\left(0.05\,in. \times \dfrac{1}{12}\,\dfrac{ft}{in.}\right)^4}{8\left(0.00398\,\dfrac{ft^3}{min} \times \dfrac{1}{60}\,\dfrac{min}{s}\right)(1\,ft)}$$

$$= 4.92\,\text{g/cm} \cdot \text{s} = 492\,\text{cp} = 0.492\,\text{Pa} \cdot \text{s} \qquad (2.3\text{-}23)$$

---

[3] A. A. Draad [Doctoral Dissertation, Technical University of Delft (1996)] in a carefully controlled experiment, attained laminar flow up to Re = 60,000. He also studied the nonparabolic velocity profile induced by the earth's rotation (through the Coriolis effect). See also A. A. Draad and F. T. M. Nieuwstadt, *J. Fluid. Mech.*, **361**, 207–308 (1998).

[4] J. H. Perry, *Chemical Engineers Handbook*, McGraw-Hill, New York, 3rd edition (1950), pp. 388–389; W. M. Kays and A. L. London, *Compact Heat Exchangers*, McGraw-Hill, New York (1958), p. 49.

[5] **Martin Hans Christian Knudsen** (1871–1949), professor of physics at the University of Copenhagen, did key experiments on the behavior of very dilute gases. The lectures he gave at the University of Glasgow were published as M. Knudsen, *The Kinetic Theory of Gases*, Methuen, London (1934); G. N. Patterson, *Molecular Flow of Gases*, Wiley, New York (1956). See also J. H. Ferziger and H. G. Kaper, *Mathematical Theory of Transport Processes in Gases*, North-Holland, Amsterdam (1972), Chapter 15.

To check whether the flow is laminar, we calculate the Reynolds number

$$\text{Re} = \frac{D\langle v_z\rangle\rho}{\mu} = \frac{4(w/\rho)\rho}{\pi D\mu}$$

$$= \frac{4\left(0.00398\ \frac{\text{ft}^3}{\text{min}}\right)\left(2.54\ \frac{\text{cm}}{\text{in.}} \times 12\ \frac{\text{in.}}{\text{ft}}\right)^3\left(\frac{1}{60}\ \frac{\text{min}}{\text{s}}\right)\left(1.261\ \frac{\text{g}}{\text{cm}^3}\right)}{\pi\left(0.1\ \text{in.} \times 2.54\ \frac{\text{cm}}{\text{in.}}\right)\left(4.92\ \frac{\text{g}}{\text{cm} \cdot \text{s}}\right)}$$

$$= 2.41\ \text{(dimensionless)} \tag{2.3-24}$$

Hence the flow is indeed laminar. Furthermore, the entrance length is

$$L_e = 0.035D\ \text{Re} = (0.035)(0.1/12)(2.41) = 0.0007\ \text{ft} \tag{2.3-25}$$

Hence, entrance effects are not important, and the viscosity value given above has been calculated properly.

EXAMPLE 2.3-2

Compressible Flow in a Horizontal Circular Tube[6]

Obtain an expression for the mass rate of flow w for an ideal gas in laminar flow in a long circular tube. The flow is presumed to be isothermal. Assume that the pressure change through the tube is not very large, so that the viscosity can be regarded a constant throughout.

SOLUTION

This problem can be solved *approximately* by assuming that the Hagen–Poiseuille equation (Eq. 2.3-21) can be applied over a small length dz of the tube as follows:

$$w = \frac{\pi\rho R^4}{8\mu}\left(-\frac{dp}{dz}\right) \tag{2.3-26}$$

To eliminate ρ in favor of p, we use the ideal gas law in the form $p/\rho = p_0/\rho_0$, where p_0 and ρ_0 are the pressure and density at $z = 0$. This gives

$$w = \frac{\pi R^4}{8\mu}\frac{\rho_0}{p_0}\left(-p\frac{dp}{dz}\right) \tag{2.3-27}$$

The mass rate of flow w is the same for all z. Hence Eq. 2.3-27 can be integrated from $z = 0$ to $z = L$ to give

$$w = \frac{\pi R^4}{16\mu L}\frac{\rho_0}{p_0}(p_0^2 - p_L^2) \tag{2.3-28}$$

Since $p_0^2 - p_L^2 = (p_0 + p_L)(p_0 - p_L)$, we get finally

$$w = \frac{\pi(p_0 - p_L)R^4\rho_{\text{avg}}}{8\mu L} \tag{2.3-29}$$

where $\rho_{\text{avg}} = \frac{1}{2}(\rho_0 + \rho_L)$ is the average density calculated at the average pressure $p_{\text{avg}} = \frac{1}{2}(p_0 + p_L)$.

§2.4   FLOW THROUGH AN ANNULUS

We now solve another viscous flow problem in cylindrical coordinates, namely the steady-state axial flow of an incompressible liquid in an annular region between two coaxial cylinders of radii κR and R as shown in Fig. 2.4-1. The fluid is flowing upward in

---

[6] L. Landau and E. M. Lifshitz, *Fluid Mechanics*, Pergamon, 2nd edition (1987), §17, Problem 6. A perturbation solution of this problem was obtained by R. K. Prud'homme, T. W. Chapman, and J. R. Bowen, *Appl. Sci. Res*, **43**, 67–74 (1986).

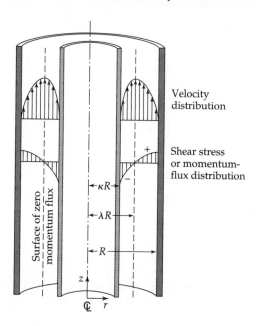

Fig. 2.4-1 The momentum-flux distribution and velocity distribution for the upward flow in a cylindrical annulus. Note that the momentum flux changes sign at the same value of r for which the velocity has a maximum.

Velocity distribution

Shear stress or momentum-flux distribution

the tube—that is, in the direction opposed to gravity. We make the same postulates as in §2.3: $v_z = v_z(r)$, $v_\theta = 0$, $v_r = 0$, and $p = p(z)$. Then when we make a momentum balance over a thin cylindrical shell of liquid, we arrive at the following differential equation:

$$\frac{d}{dr}(r\tau_{rz}) = \left(\frac{(p_0 + \rho g 0) - (p_L + \rho g L)}{L}\right)r \equiv \left(\frac{\mathcal{P}_0 - \mathcal{P}_L}{L}\right)r \tag{2.4-1}$$

This differs from Eq. 2.3-10 only in that $\mathcal{P} = p + \rho g z$ here, since the coordinate z is in the direction opposed to gravity (i.e., z is the same as the h of footnote 1 in §2.3). Integration of Eq. 2.4-1 gives

$$\tau_{rz} = \left(\frac{\mathcal{P}_0 - \mathcal{P}_L}{2L}\right)r + \frac{C_1}{r} \tag{2.4-2}$$

just as in Eq. 2.3-11.

The constant C_1 cannot be determined immediately, since we have no information about the momentum flux at the fixed surfaces $r = \kappa R$ and $r = R$. All we know is that there will be a maximum in the velocity curve at some (as yet unknown) plane $r = \lambda R$ at which the momentum flux will be zero. That is,

$$0 = \left(\frac{\mathcal{P}_0 - \mathcal{P}_L}{2L}\right)\lambda R + \frac{C_1}{\lambda R} \tag{2.4-3}$$

When we solve this equation for C_1 and substitute it into Eq. 2.4-2, we get

$$\tau_{rz} = \frac{(\mathcal{P}_0 - \mathcal{P}_L)R}{2L}\left[\left(\frac{r}{R}\right) - \lambda^2\left(\frac{R}{r}\right)\right] \tag{2.4-4}$$

The only difference between this equation and Eq. 2.4-2 is that the constant of integration C_1 has been eliminated in favor of a different constant λ. The advantage of this is that we know the geometrical significance of λ.

We now substitute Newton's law of viscosity, $\tau_{rz} = -\mu(dv_z/dr)$, into Eq. 2.4-4 to obtain a differential equation for v_z

$$\frac{dv_z}{dr} = -\frac{(\mathcal{P}_0 - \mathcal{P}_L)R}{2\mu L}\left[\left(\frac{r}{R}\right) - \lambda^2\left(\frac{R}{r}\right)\right] \tag{2.4-5}$$

Integration of this first-order separable differential equation then gives

$$v_z = -\frac{(\mathscr{P}_0 - \mathscr{P}_L)R^2}{4\mu L}\left[\left(\frac{r}{R}\right)^2 - 2\lambda^2 \ln\left(\frac{r}{R}\right) + C_2\right]$$ (2.4-6)

We now evaluate the two constants of integration, λ and C_2, by using the no-slip condition on each solid boundary:

B.C. 1:   at $r = \kappa R$, $v_z = 0$ (2.4-7)

B.C. 2:   at $r = R$, $v_z = 0$ (2.4-8)

Substitution of these boundary conditions into Eq. 2.4-6 then gives two simultaneous equations:

$$0 = \kappa^2 - 2\lambda^2 \ln \kappa + C_2; \quad 0 = 1 + C_2$$ (2.4-9, 10)

From these the two integration constants λ and C_2 are found to be

$$C_2 = -1; \quad 2\lambda^2 = \frac{1 - \kappa^2}{\ln(1/\kappa)}$$ (2.4-11, 12)

These expressions can be inserted into Eqs. 2.4-4 and 2.4-6 to give the momentum-flux distribution and the velocity distribution[1] as follows:

$$\tau_{rz} = \frac{(\mathscr{P}_0 - \mathscr{P}_L)R}{2L}\left[\left(\frac{r}{R}\right) - \frac{1 - \kappa^2}{2\ln(1/\kappa)}\left(\frac{R}{r}\right)\right]$$ (2.4-13)

$$v_z = \frac{(\mathscr{P}_0 - \mathscr{P}_L)R^2}{4\mu L}\left[1 - \left(\frac{r}{R}\right)^2 - \frac{1 - \kappa^2}{\ln(1/\kappa)}\ln\left(\frac{R}{r}\right)\right]$$ (2.4-14)

Note that when the annulus becomes very thin (i.e., κ only slightly less than unity), these results simplify to those for a plane slit (see Problem 2B.5). It is always a good idea to check "limiting cases" such as these whenever the opportunity presents itself.

The lower limit of $\kappa \to 0$ is not so simple, because the ratio $\ln(R/r)/\ln(1/\kappa)$ will always be important in a region close to the inner boundary. Hence Eq. 2.4-14 does not simplify to the parabolic distribution. However, Eq. 2.4-17 for the mass rate of flow does simplify to the Hagen–Poiseuille equation.

Once we have the momentum-flux and velocity distributions, it is straightforward to get other results of interest:

(i)   The *maximum velocity* is

$$v_{z,max} = v_z|_{r=\lambda R} = \frac{(\mathscr{P}_0 - \mathscr{P}_L)R^2}{4\mu L}[1 - \lambda^2(1 - \ln \lambda^2)]$$ (2.4-15)

where λ^2 is given in Eq. 2.4-12.

(ii)   The *average velocity* is given by

$$\langle v_z \rangle = \frac{\displaystyle\int_0^{2\pi}\int_{\kappa R}^R v_z r\, dr\, d\theta}{\displaystyle\int_0^{2\pi}\int_{\kappa R}^R r\, dr\, d\theta} = \frac{(\mathscr{P}_0 - \mathscr{P}_L)R^2}{8\mu L}\left[\frac{1 - \kappa^4}{1 - \kappa^2} - \frac{1 - \kappa^2}{\ln(1/\kappa)}\right]$$ (2.4-16)

(iii)   The *mass rate of flow* is $w = \pi R^2(1 - \kappa^2)\rho\langle v_z \rangle$, or

$$w = \frac{\pi(\mathscr{P}_0 - \mathscr{P}_L)R^4\rho}{8\mu L}\left[(1 - \kappa^4) - \frac{(1 - \kappa^2)^2}{\ln(1/\kappa)}\right]$$ (2.4-17)

---

[1] H. Lamb, *Hydrodynamics*, Cambridge University Press, 2nd edition (1895), p. 522.

(iv) The *force exerted by the fluid on the solid surfaces* is obtained by summing the forces acting on the inner and outer cylinders, as follows:

$$F_z = (2\pi\kappa RL)(-\tau_{rz}|_{r=\kappa R}) + (2\pi RL)(+\tau_{rz}|_{r=R})$$

$$= \pi R^2(1 - \kappa^2)(\mathscr{P}_0 - \mathscr{P}_L) \tag{2.4-18}$$

The reader should explain the choice of signs in front of the shear stresses above and also give an interpretation of the final result.

The equations derived above are valid only for laminar flow. The laminar–turbulent transition occurs in the neighborhood of Re = 2000, with the Reynolds number defined as Re = $2R(1 - \kappa)\langle v_z\rangle\rho/\mu$.

§2.5 FLOW OF TWO ADJACENT IMMISCIBLE FLUIDS[1]

Thus far we have considered flow situations with solid–fluid and liquid–gas boundaries. We now give one example of a flow problem with a liquid–liquid interface (see Fig. 2.5-1).

Two immiscible, incompressible liquids are flowing in the z direction in a horizontal thin slit of length L and width W under the influence of a horizontal pressure gradient $(p_0 - p_L)/L$. The fluid flow rates are adjusted so that the slit is half filled with fluid I (the more dense phase) and half filled with fluid II (the less dense phase). The fluids are flowing sufficiently slowly that no instabilities occur—that is, that the interface remains exactly planar. It is desired to find the momentum-flux and velocity distributions.

A differential momentum balance leads to the following differential equation for the momentum flux:

$$\frac{d\tau_{xz}}{dx} = \frac{p_0 - p_L}{L} \tag{2.5-1}$$

This equation is obtained for both phase I and phase II. Integration of Eq. 2.5-1 for the two regions gives

$$\tau_{xz}^{I} = \left(\frac{p_0 - p_L}{L}\right)x + C_1^{I} \tag{2.5-2}$$

$$\tau_{xz}^{II} = \left(\frac{p_0 - p_L}{L}\right)x + C_1^{II} \tag{2.5-3}$$

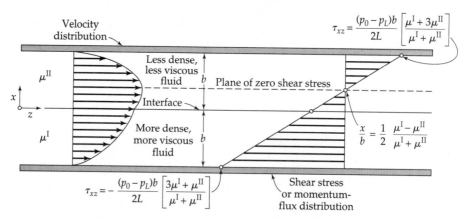

Fig. 2.5-1 Flow of two immiscible fluids between a pair of horizontal plates under the influence of a pressure gradient.

[1] The adjacent flow of gases and liquids in conduits has been reviewed by A. E. Dukler and M. Wicks, III, in Chapter 8 of *Modern Chemical Engineering*, Vol. 1, "Physical Operations," A. Acrivos (ed.), Reinhold, New York (1963).

We may immediately make use of one of the boundary conditions—namely, that the momentum flux τ_{xz} is continuous through the fluid–fluid interface:

B.C. 1:   at $x = 0$, $\tau_{xz}^{I} = \tau_{xz}^{II}$ (2.5-4)

This tells us that $C_1^{I} = C_1^{II}$; hence we drop the superscript and call both integration constants C_1.

When Newton's law of viscosity is substituted into Eqs. 2.5-2 and 2.5-3, we get

$$-\mu^{I}\frac{dv_z^{I}}{dx} = \left(\frac{p_0 - p_L}{L}\right)x + C_1 \tag{2.5-5}$$

$$-\mu^{II}\frac{dv_z^{II}}{dx} = \left(\frac{p_0 - p_L}{L}\right)x + C_1 \tag{2.5-6}$$

These two equations can be integrated to give

$$v_z^{I} = -\left(\frac{p_0 - p_L}{2\mu^{I}L}\right)x^2 - \frac{C_1}{\mu^{I}}x + C_2^{I} \tag{2.5-7}$$

$$v_z^{II} = -\left(\frac{p_0 - p_L}{2\mu^{II}L}\right)x^2 - \frac{C_1}{\mu^{II}}x + C_2^{II} \tag{2.5-8}$$

The three integration constants can be determined from the following no-slip boundary conditions:

B.C. 2:   at $x = 0$, $v_z^{I} = v_z^{II}$ (2.5-9)
B.C. 3:   at $x = -b$, $v_z^{I} = 0$ (2.5-10)
B.C. 4:   at $x = +b$, $v_z^{II} = 0$ (2.5-11)

When these three boundary conditions are applied, we get three simultaneous equations for the integration constants:

from B.C. 2: $C_2^{I} = C_2^{II}$ (2.5-12)

from B.C. 3: $0 = -\left(\frac{p_0 - p_L}{2\mu^{I}L}\right)b^2 + \frac{C_1}{\mu^{I}}b + C_2^{I}$ (2.5-13)

from B.C. 4: $0 = -\left(\frac{p_0 - p_L}{2\mu^{II}L}\right)b^2 - \frac{C_1}{\mu^{II}}b + C_2^{II}$ (2.5-14)

From these three equations we get

$$C_1 = -\frac{(p_0 - p_L)b}{2L}\left(\frac{\mu^{I} - \mu^{II}}{\mu^{I} + \mu^{II}}\right) \tag{2.5-15}$$

$$C_2^{I} = +\frac{(p_0 - p_L)b^2}{2\mu^{I}L}\left(\frac{2\mu^{I}}{\mu^{I} + \mu^{II}}\right) = C_2^{II} \tag{2.5-16}$$

The resulting momentum-flux and velocity profiles are

$$\boxed{\tau_{xz} = \frac{(p_0 - p_L)b}{L}\left[\left(\frac{x}{b}\right) - \frac{1}{2}\left(\frac{\mu^{I} - \mu^{II}}{\mu^{I} + \mu^{II}}\right)\right]} \tag{2.5-17}$$

$$\boxed{v_z^{I} = \frac{(p_0 - p_L)b^2}{2\mu^{I}L}\left[\left(\frac{2\mu^{I}}{\mu^{I} + \mu^{II}}\right) + \left(\frac{\mu^{I} - \mu^{II}}{\mu^{I} + \mu^{II}}\right)\left(\frac{x}{b}\right) - \left(\frac{x}{b}\right)^2\right]} \tag{2.5-18}$$

$$\boxed{v_z^{II} = \frac{(p_0 - p_L)b^2}{2\mu^{II}L}\left[\left(\frac{2\mu^{II}}{\mu^{I} + \mu^{II}}\right) + \left(\frac{\mu^{I} - \mu^{II}}{\mu^{I} + \mu^{II}}\right)\left(\frac{x}{b}\right) - \left(\frac{x}{b}\right)^2\right]} \tag{2.5-19}$$

These distributions are shown in Fig. 2.5-1. If both viscosities are the same, then the velocity distribution is parabolic, as one would expect for a pure fluid flowing between parallel plates (see Eq. 2B.3-2).

The *average velocity* in each layer can be obtained and the results are

$$\langle v_z^I \rangle = \frac{1}{b}\int_{-b}^{0} v_z^I dx = \frac{(p_0 - p_L)b^2}{12\mu^I L}\left(\frac{7\mu^I + \mu^{II}}{\mu^I + \mu^{II}}\right) \tag{2.5-20}$$

$$\langle v_z^{II} \rangle = \frac{1}{b}\int_{0}^{b} v_z^{II} dx = \frac{(p_0 - p_L)b^2}{12\mu^{II} L}\left(\frac{\mu^I + 7\mu^{II}}{\mu^I + \mu^{II}}\right) \tag{2.5-21}$$

From the velocity and momentum-flux distributions given above, one can also calculate the maximum velocity, the velocity at the interface, the plane of zero shear stress, and the drag on the walls of the slit.

§2.6   CREEPING FLOW AROUND A SPHERE[1,2,3,4]

In the preceding sections several elementary viscous flow problems have been solved. These have all dealt with rectilinear flows with only one nonvanishing velocity component. Since the flow around a sphere involves two nonvanishing velocity components, v_r and v_θ, it cannot be conveniently understood by the techniques explained at the beginning of this chapter. Nonetheless, a brief discussion of flow around a sphere is warranted here because of the importance of flow around submerged objects. In Chapter 4 we show how to obtain the velocity and pressure distributions. Here we only cite the results and show how they can be used to derive some important relations that we need in later discussions. The problem treated here, and also in Chapter 4, is concerned with "creeping flow"—that is, very slow flow. This type of flow is also referred to as "Stokes flow."

We consider here the flow of an incompressible fluid about a solid sphere of radius R and diameter D as shown in Fig. 2.6-1. The fluid, with density ρ and viscosity μ, ap-

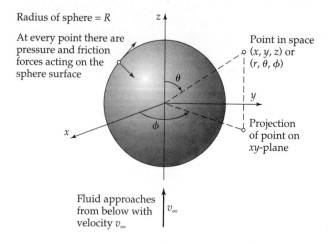

Radius of sphere = R

At every point there are pressure and friction forces acting on the sphere surface

Point in space (x, y, z) or (r, θ, ϕ)

Projection of point on xy-plane

Fluid approaches from below with velocity v_∞

Fig. 2.6-1 Sphere of radius R around which a fluid is flowing. The coordinates r, θ, and ϕ are shown. For more information on spherical coordinates, see Fig. A.8-2.

[1] G. G. Stokes, *Trans. Cambridge Phil. Soc.*, **9**, 8–106 (1851). For creeping flow around an object of arbitrary shape, see H. Brenner, *Chem. Engr. Sci.*, **19**, 703–727 (1964).

[2] L. D. Landau and E. M. Lifshitz, *Fluid Mechanics*, 2nd edition, Pergamon, London (1987), §20.

[3] G. K. Batchelor, *An Introduction to Fluid Dynamics*, Cambridge University Press (1967), §4.9.

[4] S. Kim and S. J. Karrila, *Microhydrodynamics: Principles and Selected Applications*, Butterworth-Heinemann, Boston (1991), §4.2.3; this book contains a thorough discussion of "creeping flow" problems.

proaches the fixed sphere vertically upward in the z direction with a uniform velocity v_∞. For this problem, "creeping flow" means that the Reynolds number $\mathrm{Re} = Dv_\infty\rho/\mu$ is less than about 0.1. This flow regime is characterized by the absence of eddy formation downstream from the sphere.

The velocity and pressure distributions for this creeping flow are found in Chapter 4 to be

$$v_r = v_\infty\left[1 - \frac{3}{2}\left(\frac{R}{r}\right) + \frac{1}{2}\left(\frac{R}{r}\right)^3\right]\cos\theta \tag{2.6-1}$$

$$v_\theta = v_\infty\left[-1 + \frac{3}{4}\left(\frac{R}{r}\right) + \frac{1}{4}\left(\frac{R}{r}\right)^3\right]\sin\theta \tag{2.6-2}$$

$$v_\phi = 0 \tag{2.6-3}$$

$$p = p_0 - \rho g z - \frac{3}{2}\frac{\mu v_\infty}{R}\left(\frac{R}{r}\right)^2\cos\theta \tag{2.6-4}$$

In the last equation the quantity p_0 is the pressure in the plane $z = 0$ far away from the sphere. The term $-\rho g z$ is the hydrostatic pressure resulting from the weight of the fluid, and the term containing v_∞ is the contribution of the fluid motion. Equations 2.6-1, 2, and 3 show that the fluid velocity is zero at the surface of the sphere. Furthermore, in the limit as $r \to \infty$, the fluid velocity is in the z direction with uniform magnitude v_∞; this follows from the fact that $v_z = v_r \cos\theta - v_\theta \sin\theta$, which can be derived by using Eq. A.6-33, and $v_x = v_y = 0$, which follows from Eqs. A.6-31 and 32.

The components of the stress tensor $\boldsymbol{\tau}$ in spherical coordinates may be obtained from the velocity distribution above by using Table B.1. They are

$$\tau_{rr} = -2\tau_{\theta\theta} = -2\tau_{\phi\phi} = \frac{3\mu v_\infty}{R}\left[-\left(\frac{R}{r}\right)^2 + \left(\frac{R}{r}\right)^4\right]\cos\theta \tag{2.6-5}$$

$$\tau_{r\theta} = \tau_{\theta r} = \frac{3}{2}\frac{\mu v_\infty}{R}\left(\frac{R}{r}\right)^4\sin\theta \tag{2.6-6}$$

and all other components are zero. Note that the normal stresses for this flow are nonzero, except at $r = R$.

Let us now determine the force exerted by the flowing fluid on the sphere. Because of the symmetry around the z-axis, the resultant force will be in the z direction. Therefore the force can be obtained by integrating the z-components of the normal and tangential forces over the sphere surface.

Integration of the Normal Force

At each point on the surface of the sphere the fluid exerts a force per unit area $-(p + \tau_{rr})|_{r=R}$ on the solid, acting normal to the surface. Since the fluid is in the region of greater r and the sphere in the region of lesser r, we have to affix a minus sign in accordance with the sign convention established in §1.2. The z-component of the force is $-(p + \tau_{rr})|_{r=R}(\cos \theta)$. We now multiply this by a differential element of surface $R^2 \sin \theta \, d\theta \, d\phi$ to get the force on the surface element (see Fig. A.8-2). Then we integrate over the surface of the sphere to get the resultant normal force in the z direction:

$$F^{(n)} = \int_0^{2\pi} \int_0^\pi (-(p + \tau_{rr})|_{r=R} \cos \theta) R^2 \sin \theta \, d\theta \, d\phi \tag{2.6-7}$$

According to Eq. 2.6-5, the normal stress τ_{rr} is zero[5] at $r = R$ and can be omitted in the integral in Eq. 2.6-7. The pressure distribution at the surface of the sphere is, according to Eq. 2.6-4,

$$p|_{r=R} = p_0 - \rho g R \cos \theta - \frac{3}{2} \frac{\mu v_\infty}{R} \cos \theta \tag{2.6-8}$$

When this is substituted into Eq. 2.6-7 and the integration performed, the term containing p_0 gives zero, the term containing the gravitational acceleration g gives the buoyant force, and the term containing the approach velocity v_∞ gives the "form drag" as shown below:

$$F^{(n)} = \tfrac{4}{3}\pi R^3 \rho g + 2\pi \mu R v_\infty \tag{2.6-9}$$

The buoyant force is the mass of displaced fluid ($\tfrac{4}{3}\pi R^3 \rho$) times the gravitational acceleration (g).

Integration of the Tangential Force

At each point on the solid surface there is also a shear stress acting tangentially. The force per unit area exerted in the $-\theta$ direction by the fluid (region of greater r) on the solid (region of lesser r) is $+\tau_{r\theta}|_{r=R}$. The z-component of this force per unit area is $(\tau_{r\theta}|_{r=R}) \sin \theta$. We now multiply this by the surface element $R^2 \sin \theta \, d\theta d\phi$ and integrate over the entire spherical surface. This gives the resultant force in the z direction:

$$F^{(t)} = \int_0^{2\pi} \int_0^\pi (\tau_{r\theta}|_{r=R} \sin \theta) R^2 \sin \theta \, d\theta \, d\phi \tag{2.6-10}$$

The shear stress distribution on the sphere surface, from Eq. 2.6-6, is

$$\tau_{r\theta}|_{r=R} = \frac{3}{2} \frac{\mu v_\infty}{R} \sin \theta \tag{2.6-11}$$

Substitution of this expression into the integral in Eq. 2.6-10 gives the "friction drag"

$$F^{(t)} = 4\pi \mu R v_\infty \tag{2.6-12}$$

Hence the total force F of the fluid on the sphere is given by the sum of Eqs. 2.6-9 and 2.6-12:

$$F = \underset{\substack{\text{buoyant} \\ \text{force}}}{\tfrac{4}{3}\pi R^3 \rho g} + \underset{\substack{\text{form} \\ \text{drag}}}{2\pi \mu R v_\infty} + \underset{\substack{\text{friction} \\ \text{drag}}}{4\pi \mu R v_\infty} \tag{2.6-13}$$

or

$$F = F_b + F_k = \underset{\substack{\text{buoyant} \\ \text{force}}}{\tfrac{4}{3}\pi R^3 \rho g} + \underset{\substack{\text{kinetic} \\ \text{force}}}{6\pi \mu R v_\infty} \tag{2.6-14}$$

---

[5] In Example 3.1-1 we show that, for incompressible, Newtonian fluids, all three of the normal stresses are zero at fixed solid surfaces in all flows.

The first term is the *buoyant force*, which would be present in a fluid at rest; it is the mass of the displaced fluid multiplied by the gravitational acceleration. The second term, the *kinetic force*, results from the motion of the fluid. The relation

$$F_k = 6\pi\mu R v_\infty \qquad (2.6\text{-}15)$$

is known as *Stokes' law*.[1] It is used in describing the motion of colloidal particles under an electric field, in the theory of sedimentation, and in the study of the motion of aerosol particles. Stokes' law is useful only up to a Reynolds number $\mathrm{Re} = Dv_\infty\rho/\mu$ of about 0.1. At Re = 1, Stokes' law predicts a force that is about 10% too low. The flow behavior for larger Reynolds numbers is discussed in Chapter 6.

This problem, which could not be solved by the shell balance method, emphasizes the need for a more general method for coping with flow problems in which the streamlines are not rectilinear. That is the subject of the following chapter.

<table>
<tr>
<td>

EXAMPLE 2.6-1

Determination of Viscosity from the Terminal Velocity of a Falling Sphere

</td>
<td>

Derive a relation that enables one to get the viscosity of a fluid by measuring the terminal velocity v_t of a small sphere of radius R in the fluid.

SOLUTION

If a small sphere is allowed to fall from rest in a viscous fluid, it will accelerate until it reaches a constant velocity—the *terminal velocity*. When this steady-state condition has been reached the sum of all the forces acting on the sphere must be zero. The force of gravity on the solid acts in the direction of fall, and the buoyant and kinetic forces act in the opposite direction:

$$\tfrac{4}{3}\pi R^3 \rho_s g = \tfrac{4}{3}\pi R^3 \rho g + 6\pi\mu R v_t \qquad (2.6\text{-}16)$$

Here ρ_s and ρ are the densities of the solid sphere and the fluid. Solving this equation for the terminal velocity gives

$$\mu = \tfrac{2}{9}R^2(\rho_s - \rho)g/v_t \qquad (2.6\text{-}17)$$

This result may be used only if the Reynolds number is less than about 0.1.

This experiment provides an apparently simple method for determining viscosity. However, it is difficult to keep a homogeneous sphere from rotating during its descent, and if it does rotate, then Eq. 2.6-17 cannot be used. Sometimes weighted spheres are used in order to preclude rotation; then the left side of Eq. 2.6-16 has to be replaced by m, the mass of the sphere, times the gravitational acceleration.

</td>
</tr>
</table>

QUESTIONS FOR DISCUSSION

1. Summarize the procedure used in the solution of viscous flow problems by the shell balance method. What kinds of problems can and cannot be solved by this method? How is the definition of the first derivative used in the method?
2. Which of the flow systems in this chapter can be used as a viscometer? List the difficulties that might be encountered in each.
3. How are the Reynolds numbers defined for films, tubes, and spheres? What are the dimensions of Re?
4. How can one modify the film thickness formula in §2.2 to describe a thin film falling down the interior wall of a cylinder? What restrictions might have to be placed on this modified formula?
5. How can the results in §2.3 be used to estimate the time required for a liquid to drain out of a vertical tube that is open at both ends?
6. Contrast the radial dependence of the shear stress for the laminar flow of a Newtonian liquid in a tube and in an annulus. In the latter, why does the function change sign?

7. Show that the Hagen–Poiseuille formula is dimensionally consistent.
8. What differences are there between the flow in a circular tube of radius R and the flow in the same tube with a thin wire placed along the axis?
9. Under what conditions would you expect the analysis in §2.5 to be inapplicable?
10. Is Stokes' law valid for droplets of oil falling in water? For air bubbles rising in benzene? For tiny particles falling in air, if the particle diameters are of the order of the mean free path of the molecules in the air?
11. Two immiscible liquids, A and B, are flowing in laminar flow between two parallel plates. Is it possible that the velocity profiles would be of the following form? Explain.

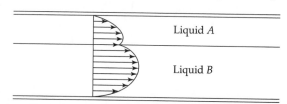

12. What is the terminal velocity of a spherical colloidal particle having an electric charge e in an electric field of strength $\mathscr{E}$? How is this used in the Millikan oil-drop experiment?

PROBLEMS

2A.1 Thickness of a falling film. Water at 20°C is flowing down a vertical wall with Re = 10. Calculate (a) the flow rate, in gallons per hour per foot of wall width, and (b) the film thickness in inches.

Answers: (a) 0.727 gal/hr · ft; (b) 0.00361 in.

2A.2 Determination of capillary radius by flow measurement. One method for determining the radius of a capillary tube is by measuring the rate of flow of a Newtonian liquid through the tube. Find the radius of a capillary from the following flow data:

Length of capillary tube	50.02 cm
Kinematic viscosity of liquid	$4.03 \times 10^{-5} \text{ m}^2/\text{s}$
Density of liquid	$0.9552 \times 10^3 \text{ kg/m}^3$
Pressure drop in the horizontal tube	$4.829 \times 10^5 \text{ Pa}$
Mass rate of flow through tube	$2.997 \times 10^{-3} \text{ kg/s}$

What difficulties may be encountered in this method? Suggest some other methods for determining the radii of capillary tubes.

2A.3 Volume flow rate through an annulus. A horizontal annulus, 27 ft in length, has an inner radius of 0.495 in. and an outer radius of 1.1 in. A 60% aqueous solution of sucrose ($C_{12}H_{22}O_{11}$) is to be pumped through the annulus at 20°C. At this temperature the solution density is 80.3 lb/ft³ and the viscosity is 136.8 lb$_m$/ft · hr. What is the volume flow rate when the impressed pressure difference is 5.39 psi?

Answer: 0.110 ft³/s

2A.4 Loss of catalyst particles in stack gas.

(a) Estimate the maximum diameter of microspherical catalyst particles that could be lost in the stack gas of a fluid cracking unit under the following conditions:

Gas velocity at axis of stack = 1.0 ft/s (vertically upward)
Gas viscosity = 0.026 cp
Gas density = 0.045 lb/ft³
Density of a catalyst particle = 1.2 g/cm³
Express the result in microns (1 micron = 10^{-6}m = 1μm).

(b) Is it permissible to use Stokes' law in (a)?

Answers: (a) 110 μm; Re = 0.93

2B.1 Different choice of coordinates for the falling film problem. Rederive the velocity profile and the average velocity in §2.2, by replacing x by a coordinate $\bar{x}$ measured away from the wall; that is, $\bar{x} = 0$ is the wall surface, and $\bar{x} = \delta$ is the liquid–gas interface. Show that the velocity distribution is then given by

$$v_z = (\rho g \delta^2/\mu)[(\bar{x}/\delta) - \tfrac{1}{2}(\bar{x}/\delta)^2] \cos \beta \qquad (2B.1\text{-}1)$$

and then use this to get the average velocity. Show how one can get Eq. 2B.1-1 from Eq. 2.2-18 by making a change of variable.

2B.2 Alternate procedure for solving flow problems. In this chapter we have used the following procedure: (i) derive an equation for the momentum flux, (ii) integrate this equation, (iii) insert Newton's law to get a first-order differential equation for the velocity, (iv) integrate the latter to get the velocity distribution. Another method is: (i) derive an equation for the momentum flux, (ii) insert Newton's law to get a second-order differential equation for the velocity profile, (iii) integrate the latter to get the velocity distribution. Apply this second method to the falling film problem by substituting Eq. 2.2-14 into Eq. 2.2-10 and continuing as directed until the velocity distribution has been obtained and the integration constants evaluated.

2B.3 Laminar flow in a narrow slit (see Fig. 2B.3).

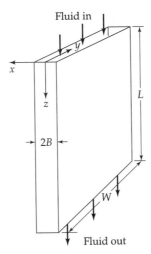

Fig. 2B.3 Flow through a slit, with $B \ll W \ll L$.

(a) A Newtonian fluid is in laminar flow in a narrow slit formed by two parallel walls a distance $2B$ apart. It is understood that $B \ll W$, so that "edge effects" are unimportant. Make a differential momentum balance, and obtain the following expressions for the momentum-flux and velocity distributions:

$$\tau_{xz} = \left(\frac{\mathcal{P}_0 - \mathcal{P}_L}{L}\right)x \qquad (2B.3\text{-}1)$$

$$v_z = \frac{(\mathcal{P}_0 - \mathcal{P}_L)B^2}{2\mu L}\left[1 - \left(\frac{x}{B}\right)^2\right] \qquad (2B.3\text{-}2)$$

In these expressions $\mathcal{P} = p + \rho gh = p - \rho gz$.
(b) What is the ratio of the average velocity to the maximum velocity for this flow?
(c) Obtain the slit analog of the Hagen–Poiseuille equation.
(d) Draw a meaningful sketch to show why the above analysis is inapplicable if $B = W$.
(e) How can the result in (b) be obtained from the results of §2.5?

Answers: **(b)** $\langle v_z \rangle / v_{z,\max} = \tfrac{2}{3}$

$$\text{(c) } w = \frac{2}{3}\frac{(\mathcal{P}_0 - \mathcal{P}_L)B^3 W\rho}{\mu L}$$

2B.4   Laminar slit flow with a moving wall ("plane Couette flow"). Extend Problem 2B.3 by allowing the wall at $x = B$ to move in the positive z direction at a steady speed v_0. Obtain (*a*) the shear-stress distribution and (*b*) the velocity distribution. Draw carefully labeled sketches of these functions.

Answers: $\tau_{xz} = \left(\dfrac{\mathcal{P}_0 - \mathcal{P}_L}{L}\right)x - \dfrac{\mu v_0}{2B}; \ v_z = \dfrac{(\mathcal{P}_0 - \mathcal{P}_L)B^2}{2\mu L}\left[1 - \left(\dfrac{x}{B}\right)^2\right] + \dfrac{v_0}{2}\left(1 + \dfrac{x}{B}\right)$

2B.5   Interrelation of slit and annulus formulas. When an annulus is very thin, it may, to a good approximation, be considered as a thin slit. Then the results of Problem 2B.3 can be taken over with suitable modifications. For example, the mass rate of flow in an annulus with outer wall of radius R and inner wall of radius $(1 - \varepsilon)R$, where ε is small, may be obtained from Problem 2B.3 by replacing $2B$ by εR, and W by $2\pi(1 - \frac{1}{2}\varepsilon)R$. In this way we get for the mass rate of flow:

$$w = \frac{\pi(\mathcal{P}_0 - \mathcal{P}_L)R^4\varepsilon^3\rho}{6\mu L}\,(1 - \tfrac{1}{2}\varepsilon) \tag{2B.5-1}$$

Show that this same result may be obtained from Eq. 2.4-17 by setting κ equal to $1 - \varepsilon$ everywhere in the formula and then expanding the expression for w in powers of ε. This requires using the Taylor series (see §C.2)

$$\ln(1 - \varepsilon) = -\varepsilon - \tfrac{1}{2}\varepsilon^2 - \tfrac{1}{3}\varepsilon^3 - \tfrac{1}{4}\varepsilon^4 - \cdots \tag{2B.5-2}$$

and then performing a long division. The first term in the resulting series will be Eq. 2B.5-1. *Caution:* In the derivation it is necessary to use the first *four* terms of the Taylor series in Eq. 2B.5-2.

2B.6   Flow of a film on the outside of a circular tube (see Fig. 2B.6). In a gas absorption experiment a viscous fluid flows upward through a small circular tube and then downward in laminar flow on the outside. Set up a momentum balance over a shell of thickness Δr in the film,

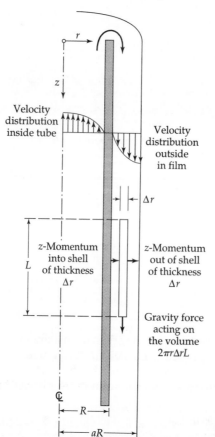

Fig. 2B.6 Velocity distribution and z-momentum balance for the flow of a falling film on the outside of a circular tube.

as shown in Fig. 2B.6. Note that the "momentum in" and "momentum out" arrows are always taken in the positive coordinate direction, even though in this problem the momentum is flowing through the cylindrical surfaces in the negative r direction.

(a) Show that the velocity distribution in the falling film (neglecting end effects) is

$$v_z = \frac{\rho g R^2}{4\mu}\left[1 - \left(\frac{r}{R}\right)^2 + 2a^2 \ln\left(\frac{r}{R}\right)\right] \tag{2B.6-1}$$

(b) Obtain an expression for the mass rate of flow in the film.

(c) Show that the result in (b) simplifies to Eq. 2.2-21 if the film thickness is very small.

2B.7 Annular flow with inner cylinder moving axially (see Fig. 2B.7). A cylindrical rod of radius κR moves axially with velocity $v_z = v_0$ along the axis of a cylindrical cavity of radius R as seen in the figure. The pressure at both ends of the cavity is the same, so that the fluid moves through the annular region solely because of the rod motion.

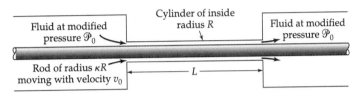

Fig. 2B.7 Annular flow with the inner cylinder moving axially.

(a) Find the velocity distribution in the narrow annular region.

(b) Find the mass rate of flow through the annular region.

(c) Obtain the viscous force acting on the rod over the length L.

(d) Show that the result in (c) can be written as a "plane slit" formula multiplied by a "curvature correction." Problems of this kind arise in studying the performance of wire-coating dies.[1]

Answers: (a) $\dfrac{v_z}{v_0} = \dfrac{\ln(r/R)}{\ln \kappa}$

(b) $w = \dfrac{\pi R^2 v_0 \rho}{2}\left[\dfrac{(1 - \kappa^2)}{\ln(1/\kappa)} - 2\kappa^2\right]$

(c) $F_z = -2\pi L\mu v_0/\ln(1/\kappa)$

(d) $F_z = \dfrac{-2\pi L\mu v_0}{\varepsilon}\left(1 - \tfrac{1}{2}\varepsilon - \tfrac{1}{12}\varepsilon^2 + \cdots\right)$ where $\varepsilon = 1 - \kappa$ (see Problem 2B.5)

2B.8 Analysis of a capillary flowmeter (see Fig. 2B.8). Determine the rate of flow (in lb/hr) through the capillary flow meter shown in the figure. The fluid flowing in the inclined tube is

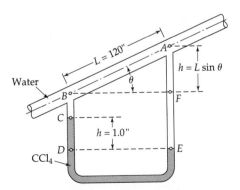

Fig. 2B.8 A capillary flow meter.

[1] J. B. Paton, P. H. Squires, W. H. Darnell, F. M. Cash, and J. F. Carley, *Processing of Thermoplastic Materials*, E. C. Bernhardt (ed.), Reinhold, New York (1959), Chapter 4.

water at 20°C, and the manometer fluid is carbon tetrachloride (CCl_4) with density 1.594 g/cm³. The capillary diameter is 0.010 in. *Note:* Measurements of H and L are sufficient to calculate the flow rate; θ need not be measured. Why?

2B.9   Low-density phenomena in compressible tube flow[2,3] (Fig. 2B.9). As the pressure is decreased in the system studied in Example 2.3-2, deviations from Eqs. 2.3-28 and 2.3-29 arise. The gas behaves as if it slips at the tube wall. It is conventional[2] to replace the customary "no-slip" boundary condition that $v_z = 0$ at the tube wall by

$$v_z = -\zeta \frac{dv_z}{dr}, \qquad \text{at } r = R \tag{2B.9-1}$$

in which ζ is the *slip coefficient*. Repeat the derivation in Example 2.3-2 using Eq. 2B.9-1 as the boundary condition. Also make use of the experimental fact that the slip coefficient varies inversely with the pressure $\zeta = \zeta_0/p$, in which ζ_0 is a constant. Show that the mass rate of flow is

$$w = \frac{\pi(p_0 - p_L)R^4 \rho_{\text{avg}}}{8\mu L}\left(1 + \frac{4\zeta_0}{R p_{\text{avg}}}\right) \tag{2B.9-2}$$

in which $p_{\text{avg}} = \frac{1}{2}(p_0 + p_L)$.

When the pressure is decreased further, a flow regime is reached in which the mean free path of the gas molecules is large with respect to the tube radius (*Knudsen flow*). In that regime[3]

$$w = \sqrt{\frac{2m}{\pi \kappa T}}\,(\tfrac{4}{3}\pi R^3)\left(\frac{p_0 - p_L}{L}\right) \tag{2B.9-3}$$

in which m is the molecular mass and κ is the Boltzmann constant. In the derivation of this result it is assumed that all collisions of the molecules with the solid surfaces are *diffuse* and not *specular*. The results in Eqs. 2.3-29, 2B.9-2, and 2B.9-3 are summarized in Fig. 2B.9.

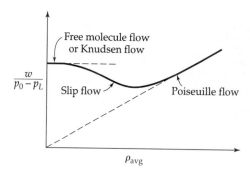

Fig. 2B.9 A comparison of the flow regimes in gas flow through a tube.

2B.10   Incompressible flow in a slightly tapered tube. An incompressible fluid flows through a tube of circular cross section, for which the tube radius changes linearly from R_0 at the tube entrance to a slightly smaller value R_L at the tube exit. Assume that the Hagen–Poiseuille equation is *approximately* valid over a differential length, dz, of the tube so that the mass flow rate is

$$w = \frac{\pi[R(z)]^4 \rho}{8\mu}\left(-\frac{d\mathscr{P}}{dz}\right) \tag{2B.10-1}$$

This is a differential equation for $\mathscr{P}$ as a function of z, but, when the explicit expression for $R(z)$ is inserted, it is not easily solved.

---

[2] E. H. Kennard, *Kinetic Theory of Gases*, McGraw-Hill, New York (1938), pp. 292–295, 300–306.

[3] M. Knudsen, *The Kinetic Theory of Gases*, Methuen, London, 3rd edition (1950). See also R. J. Silbey and R. A. Alberty, *Physical Chemistry*, Wiley, New York, 3rd edition (2001), §17.6.

(a) Write down the expression for R as a function of z.

(b) Change the independent variable in the above equation to R, so that the equation becomes

$$w = \frac{\pi R^4 \rho}{8\mu}\left(-\frac{d\mathcal{P}}{dR}\right)\left(\frac{R_L - R_0}{L}\right) \tag{2B.10-2}$$

(c) Integrate the equation, and then show that the solution can be rearranged to give

$$w = \frac{\pi(\mathcal{P}_0 - \mathcal{P}_L)R_0^4\rho}{8\mu L}\left[1 - \frac{1 + (R_L/R_0) + (R_L/R_0)^2 - 3(R_L/R_0)^3}{1 + (R_L/R_0) + (R_L/R_0)^2}\right] \tag{2B.10-3}$$

Interpret the result. The approximation used here that a flow between nonparallel surfaces can be regarded locally as flow between parallel surfaces is sometimes referred to as the *lubrication approximation* and is widely used in the theory of lubrication. By making a careful order-of-magnitude analysis, it can be shown that, for this problem, the lubrication approximation is valid as long as[4]

$$\frac{R_L}{R_0}\left(1 - \left(\frac{R_L}{R_0}\right)^2\right) \ll 1 \tag{2B.10-4}$$

2B.11 The cone-and-plate viscometer (see Fig. 2B.11). A cone-and-plate viscometer consists of a stationary flat plate and an inverted cone, whose apex just contacts the plate. The liquid whose viscosity is to be measured is placed in the gap between the cone and plate. The cone is rotated at a known angular velocity Ω, and the torque T_z required to turn the cone is measured. Find an expression for the viscosity of the fluid in terms of Ω, T_z, and the angle ψ_0 between the cone and plate. For commercial instruments ψ_0 is about 1 degree.

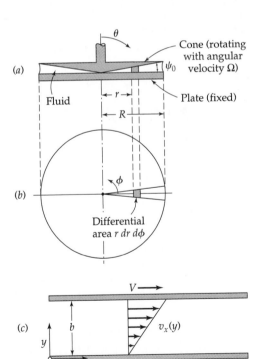

Fig. **2B.11** The cone-and-plate viscometer: (*a*) side view of the instrument; (*b*) top view of the cone–plate system, showing a differential element $r\,dr\,d\phi$; (*c*) an approximate velocity distribution within the differential region. To equate the systems in (*a*) and (*c*), we identify the following equivalences: $V = \Omega r$ and $b = r\sin\psi_0 \approx r\psi_0$.

[4] R. B. Bird, R. C. Armstrong, and O. Hassager, *Dynamics of Polymeric Liquids*, Vol. 1, Wiley-Interscience, New York, 2nd edition (1987), pp. 16–18.

(a) Assume that locally the velocity distribution in the gap can be very closely approximated by that for flow between parallel plates, the upper one moving with a constant speed. Verify that this leads to the *approximate* velocity distribution (in spherical coordinates)

$$\frac{v_\phi}{r} = \Omega\left(\frac{(\pi/2) - \theta}{\psi_0}\right) \tag{2B.11-1}$$

This approximation should be rather good, because ψ_0 is so small.

(b) From the velocity distribution in Eq. 2B.11-1 and Appendix B.1, show that a reasonable expression for the shear stress is

$$\tau_{\theta\phi} = \mu(\Omega/\psi_0) \tag{2B.11-2}$$

This result shows that the shear stress is uniform throughout the gap. It is this fact that makes the cone-and-plate viscometer quite attractive. The instrument is widely used, particularly in the polymer industry.

(c) Show that the torque required to turn the cone is given by

$$T_z = \tfrac{2}{3}\pi\mu\Omega R^3/\psi_0 \tag{2B.11-3}$$

This is the standard formula for calculating the viscosity from measurements of the torque and angular velocity for a cone–plate assembly with known R and ψ_0.

(d) For a cone-and-plate instrument with radius 10 cm and angle ψ_0 equal to 0.5 degree, what torque (in dyn · cm) is required to turn the cone at an angular velocity of 10 radians per minute if the fluid viscosity is 100 cp?

Answer: **(d)** 40,000 dyn · cm

2B.12 **Flow of a fluid in a network of tubes** (Fig. 2B.12). A fluid is flowing in laminar flow from A to B through a network of tubes, as depicted in the figure. Obtain an expression for the mass flow rate w of the fluid entering at A (or leaving at B) as a function of the modified pressure drop $\mathcal{P}_A - \mathcal{P}_B$. Neglect the disturbances at the various tube junctions.

$$\textit{Answer: } w = \frac{3\pi(\mathcal{P}_A - \mathcal{P}_B)R^4\rho}{20\mu L}$$

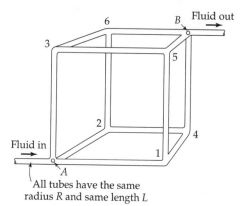

All tubes have the same radius R and same length L

Fig. 2B.12 Flow of a fluid in a network with branching.

2C.1 **Performance of an electric dust collector** (see Fig. 2C.1)[5].

(a) A dust precipitator consists of a pair of oppositely charged plates between which dust-laden gases flow. It is desired to establish a criterion for the minimum length of the precipitator in terms of the charge on the particle e, the electric field strength $\mathcal{E}$, the pressure difference

---

[5] The answer given in the first edition of this book was incorrect, as pointed out to us in 1970 by Nau Gab Lee of Seoul National University.

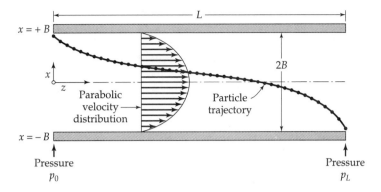

Fig. 2C.1 Particle trajectory in an electric dust collector. The particle that begins at $z = 0$ and ends at $x = +B$ may not necessarily travel the longest distance in the z direction.

$(p_0 - p_L)$, the particle mass m, and the gas viscosity μ. That is, for what length L will the smallest particle present (mass m) reach the bottom just before it has a chance to be swept out of the channel? Assume that the flow between the plates is laminar so that the velocity distribution is described by Eq. 2B.3-2. Assume also that the particle velocity in the z direction is the same as the fluid velocity in the z direction. Assume further that the Stokes drag on the sphere as well as the gravity force acting on the sphere as it is accelerated in the negative x direction can be neglected.

(b) Rework the problem neglecting acceleration in the x direction, but including the Stokes drag.

(c) Compare the usefulness of the solutions in (*a*) and (*b*), considering that stable aerosol particles have effective diameters of about 1–10 microns and densities of about 1 g/cm^3.

Answer: **(a)** $L_{\min} = [12(p_0 - p_L)^2 B^5 m / 25\mu^2 e\mathscr{E}]^{1/4}$

2C.2 **Residence time distribution in tube flow.** Define the *residence time function* $F(t)$ to be that fraction of the fluid flowing in a conduit which flows completely through the conduit in a time interval t. Also define the *mean residence time* t_m by the relation

$$t_m = \int_0^1 t\, dF \qquad (2C.2\text{-}1)$$

(a) An incompressible Newtonian liquid is flowing in a circular tube of length L and radius R, and the average flow velocity is $\langle v_z \rangle$. Show that

$$F(t) = 0 \qquad\qquad \text{for } t \leq (L/2\langle v_z \rangle)) \qquad (2C.2\text{-}2)$$
$$F(t) = 1 - (L/2\langle v_z \rangle t)^2 \qquad \text{for } t \geq (L/2\langle v_z \rangle)) \qquad (2C.2\text{-}3)$$

(b) Show that $t_m = (L/\langle v_z \rangle)$.

2C.3 **Velocity distribution in a tube.** You have received a manuscript to referee for a technical journal. The paper deals with heat transfer in tube flow. The authors state that, because they are concerned with nonisothermal flow, they must have a "general" expression for the velocity distribution, one that can be used even when the viscosity of the fluid is a function of temperature (and hence position). The authors state that a "general expression for the velocity distribution for flow in a tube" is

$$\frac{v_z}{\langle v_z \rangle} = \frac{\displaystyle\int_y^1 (\bar{y}/\mu)d\bar{y}}{\displaystyle\int_0^1 (\bar{y}^3/\mu)d\bar{y}} \qquad (2C.3\text{-}1)$$

in which $y = r/R$. The authors give no derivation, nor do they give a literature citation. As the referee you feel obliged to derive the formula and list any restrictions implied.

2C.4 **Falling-cylinder viscometer** (see Fig. 2C.4).[6] A falling-cylinder viscometer consists of a long vertical cylindrical container (radius R), capped at both ends, with a solid cylindrical slug (radius κR). The slug is equipped with fins so that its axis is coincident with that of the tube.

One can observe the rate of descent of the slug in the cylindrical container when the latter is filled with fluid. Find an equation that gives the viscosity of the fluid in terms of the terminal velocity v_0 of the slug and the various geometric quantities shown in the figure.

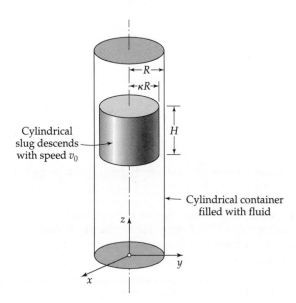

Cylindrical slug descends with speed v_0

H

Cylindrical container filled with fluid

Fig. 2C.4 A falling-cylinder viscometer with a tightly fitting solid cylinder moving vertically. The cylinder is usually equipped with fins to maintain centering within the tube. The fluid completely fills the tube, and the top and bottom are closed.

(a) Show that the velocity distribution in the annular slit is given by

$$\frac{v_z}{v_0} = -\frac{(1 - \xi^2) - (1 + \kappa^2)\ln(1/\xi)}{(1 - \kappa^2) - (1 + \kappa^2)\ln(1/\kappa)} \tag{2C.4-1}$$

in which $\xi = r/R$ is a dimensionless radial coordinate.

(b) Make a force balance on the cylindrical slug and obtain

$$\mu = \frac{(\rho_0 - \rho)g(\kappa R)^2}{2v_0}\left[\left(\ln\frac{1}{\kappa}\right) - \left(\frac{1 - \kappa^2}{1 + \kappa^2}\right)\right] \tag{2C.4-2}$$

in which ρ and ρ_0 are the densities of the fluid and the slug, respectively.

(c) Show that, for small slit widths, the result in (b) may be expanded in powers of $\varepsilon = 1 - \kappa$ to give

$$\mu = \frac{(\rho_0 - \rho)gR^2\varepsilon^3}{6v_0}(1 - \tfrac{1}{2}\varepsilon - \tfrac{13}{20}\varepsilon^2 + \cdots) \tag{2C.4-3}$$

See §C.2 for information on expansions in Taylor series.

2C.5 **Falling film on a conical surface** (see Fig. 2C.5).[7] A fluid flows upward through a circular tube and then downward on a conical surface. Find the film thickness as a function of the distance s down the cone.

[6] J. Lohrenz, G. W. Swift, and F. Kurata, *AIChE Journal*, **6**, 547–550 (1960) and **7**, 6S (1961); E. Ashare, R. B. Bird, and J. A. Lescarboura, *AIChE Journal*, **11**, 910–916 (1965).

[7] R. B. Bird, in *Selected Topics in Transport Phenomena*, CEP Symposium Series #58, **61**, 1–15 (1965).

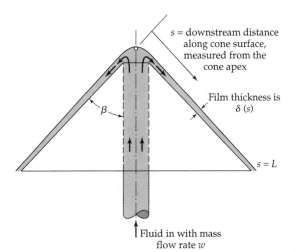

s = downstream distance
along cone surface,
measured from the
cone apex

Film thickness is
δ (s)

β

s = L

Fluid in with mass
flow rate w

Fig. 2C.5 A falling film on a conical
surface.

(a) Assume that the results of §2.2 apply *approximately* over any small region of the cone surface. Show that a mass balance on a ring of liquid contained between s and $s + \Delta s$ gives:

$$\frac{d}{ds}(s\delta\langle v\rangle) = 0 \quad \text{or} \quad \frac{d}{ds}(s\delta^3) = 0 \tag{2C.5-1}$$

(b) Integrate this equation and evaluate the constant of integration by equating the mass rate of flow w up the central tube to that flowing down the conical surface at $s = L$. Obtain the following expression for the film thickness:

$$\delta = \sqrt[3]{\frac{3\mu w}{\pi\rho^2 gL \sin 2\beta}\left(\frac{L}{s}\right)} \tag{2C.5-2}$$

2C.6 Rotating cone pump (see Fig. 2C.6). Find the mass rate of flow through this pump as a function of the gravitational acceleration, the impressed pressure difference, the angular velocity of the cone, the fluid viscosity and density, the cone angle, and other geometrical quantities labeled in the figure.

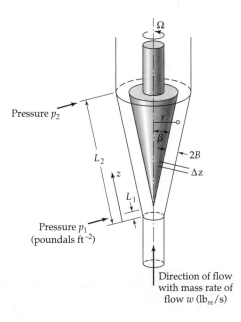

Ω

Pressure p_2

L_2

r

β

2B

Δz

z

L_1

Pressure p_1
(poundals ft^{-2})

Direction of flow
with mass rate of
flow w (lb$_m$/s)

Fig. 2C.6 A rotating-cone pump. The variable r
is the distance from the axis of rotation out to
the center of the slit.

(a) Begin by analyzing the system without the rotation of the cone. Assume that it is possible to apply the results of Problem 2B.3 locally. That is, adapt the solution for the mass flow rate from that problem by making the following replacements:

replace $(\mathcal{P}_0 - \mathcal{P}_L)/L$ by $\quad -d\mathcal{P}/dz$

replace W by $\qquad 2\pi r = 2\pi z \sin \beta$

thereby obtaining

$$w = \frac{2}{3}\left(-\frac{d\mathcal{P}}{dz}\right)\frac{B^3\rho \cdot 2\pi z \sin \beta}{\mu} \tag{2C.6-1}$$

The mass flow rate w is a constant over the range of z. Hence this equation can be integrated to give

$$(\mathcal{P}_1 - \mathcal{P}_2) = \frac{3}{4\pi}\frac{\mu w}{B^3\rho \sin \beta}\ln\frac{L_2}{L_1} \tag{2C.6-2}$$

(b) Next, modify the above result to account for the fact that the cone is rotating with angular velocity Ω. The mean centrifugal force per unit volume acting on the fluid in the slit will have a z-component *approximately* given by

$$(F_{\text{centrif.}})_z = K\rho\Omega^2 z \sin^2 \beta \tag{2C.6-3}$$

What is the value of K? Incorporate this as an additional force tending to drive the fluid through the channel. Show that this leads to the following expression for the mass rate of flow:

$$w = \frac{4\pi B^3\rho \sin \beta}{3\mu}\left[\frac{(\mathcal{P}_1 - \mathcal{P}_2) + (\frac{1}{2}K\rho\Omega^2 \sin^2 \beta)(L_2^2 - L_1^2)}{\ln (L_2/L_1)}\right] \tag{2C.6-4}$$

Here $\mathcal{P}_i = p_i + \rho g L_i \cos \beta$.

2C.7 A simple rate-of-climb indicator (see Fig. 2C.7). Under the proper circumstances the simple apparatus shown in the figure can be used to measure the rate of climb of an airplane. The gauge pressure inside the Bourdon element is taken as proportional to the rate of climb. For the purposes of this problem the apparatus may be assumed to have the following properties: (i) the capillary tube (of radius R and length L, with $R \ll L$) is of negligible volume but appreciable flow resistance; (ii) the Bourdon element has a constant volume V and offers negligible resistance to flow; and (iii) flow in the capillary is laminar and incompressible, and the volumetric flow rate depends only on the conditions at the ends of the capillary.

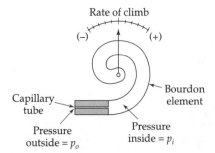

Fig. 2C.7 A rate-of-climb indicator.

(a) Develop an expression for the change of air pressure with altitude, neglecting temperature changes, and considering air to be an ideal gas of constant composition. (*Hint:* Write a shell balance in which the weight of gas is balanced against the static pressure.)

(b) By making a mass balance over the gauge, develop an approximate relation between gauge pressure $p_i - p_o$ and rate of climb v_z for a long continued constant-rate climb. Neglect change of air viscosity, and assume changes in air density to be small.

(c) Develop an approximate expression for the "relaxation time" t_{rel} of the indicator—that is, the time required for the gauge pressure to drop to $1/e$ of its initial value when the external pressure is suddenly changed from zero (relative to the interior of the gauge) to some different constant value, and maintained indefinitely at this new value.

(d) Discuss the usefulness of this type of indicator for small aircraft.

(e) Justify the plus and minus signs in the figure.

Answers: **(a)** $dp/dz = -\rho g = -(pM/RT)g$

(b) $p_i - p_o \approx v_z(8\mu L/\pi R^4)(MgV/R_gT)$, where R_g is the gas constant and M is the molecular weight.

(c) $t_0 = (128/\pi)(\mu VL/\pi D^4 \bar{p})$, where $\bar{p} = \frac{1}{2}(p_i + p_o)$

2D.1 **Rolling-ball viscometer.** An approximate analysis of the rolling-ball experiment has been given, in which the results of Problem 2B.3 are used.[8] Read the original paper and verify the results.

2D.2 **Drainage of liquids**[9] (see Fig. 2D.2). How much liquid clings to the inside surface of a large vessel when it is drained? As shown in the figure there is a thin film of liquid left behind on the wall as the liquid level in the vessel falls. The local film thickness is a function of both z (the distance down from the initial liquid level) and t (the elapsed time).

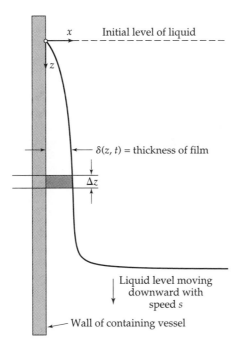

x — Initial level of liquid

$\delta(z, t)$ = thickness of film

Δz

Liquid level moving downward with speed s

Wall of containing vessel

Fig. 2D.2 Clinging of a viscous fluid to wall of vessel during draining.

[8] H. W. Lewis, *Anal. Chem.*, **25**, 507 (1953); R. B. Bird and R. M. Turian, *Ind. Eng. Chem. Fundamentals*, **3**, 87 (1964); J. Šesták and F. Ambros, *Rheol. Acta*, **12**, 70–76 (1973).

[9] J. J. van Rossum, *Appl. Sci. Research*, **A7**, 121–144 (1958); see also V. G. Levich, *Physicochemical Hydrodynamics*, Prentice-Hall, Englewood Cliffs, N.J. (1962), Chapter 12.

(a) Make an unsteady-state mass balance on a portion of the film between z and $z + \Delta z$ to get

$$\frac{\partial}{\partial z} \langle v_z \rangle \delta = -\frac{\partial \delta}{\partial t} \tag{2D.2-1}$$

(b) Use Eq. 2.2-18 and a quasi-steady-assumption to obtain the following first-order partial differential equation for $\delta(z, t)$:

$$\frac{\partial \delta}{\partial t} + \frac{\rho g}{\mu} \delta^2 \frac{\partial \delta}{\partial z} = 0 \tag{2D.2-2}$$

(c) Solve this equation to get

$$\delta(z, t) = \sqrt{\frac{\mu}{\rho g} \frac{z}{t}} \tag{2D.2-3}$$

What restrictions have to be placed on this result?

Chapter 3

The Equations of Change
for Isothermal Systems

In Chapter 2, velocity distributions were determined for several simple flow systems by the shell momentum balance method. The resulting velocity distributions were then used to get other quantities, such as the average velocity and drag force. The shell balance approach was used to acquaint the novice with the notion of a momentum balance. Even though we made no mention of it in Chapter 2, at several points we tacitly made use of the idea of a mass balance.

It is tedious to set up a shell balance for each problem that one encounters. What we need is a general mass balance and a general momentum balance that can be applied to any problem, including problems with nonrectilinear motion. That is the main point of this chapter. The two equations that we derive are called the *equation of continuity* (for the mass balance) and the *equation of motion* (for the momentum balance). These equations can be used as the starting point for studying all problems involving the isothermal flow of a pure fluid.

In Chapter 11 we enlarge our problem-solving capability by developing the equations needed for nonisothermal pure fluids by adding an equation for the temperature. In Chapter 19 we go even further and add equations of continuity for the concentrations of the individual species. Thus as we go from Chapter 3 to Chapter 11 and on to Chapter 19 we are able to analyze systems of increasing complexity, using the complete set of *equations of change*. It should be evident that Chapter 3 is a very important chapter—perhaps the most important chapter in the book—and it should be mastered thoroughly.

In §3.1 the equation of continuity is developed by making a mass balance over a small element of volume through which the fluid is flowing. Then the size of this element is allowed to go to zero (thereby treating the fluid as a continuum), and the desired partial differential equation is generated.

In §3.2 the equation of motion is developed by making a momentum balance over a small element of volume and letting the volume element become infinitesimally small. Here again a partial differential equation is generated. This equation of motion can be used, along with some help from the equation of continuity, to set up and solve all the problems given in Chapter 2 and many more complicated ones. It is thus a key equation in transport phenomena.

In §3.3 and §3.4 we digress briefly to introduce the equations of change for mechanical energy and angular momentum. These equations are obtained from the equation of motion and hence contain no new physical information. However, they provide a convenient starting point for several applications in this book—particularly the macroscopic balances in Chapter 7.

In §3.5 we introduce the "substantial derivative." This is the time derivative following the motion of the substance (i.e., the fluid). Because it is widely used in books on fluid dynamics and transport phenomena, we then show how the various equations of change can be rewritten in terms of the substantial derivatives.

In §3.6 we discuss the solution of flow problems by use of the equations of continuity and motion. Although these are partial differential equations, we can solve many problems by postulating the form of the solution and then discarding many terms in these equations. In this way one ends up with a simpler set of equations to solve. In this chapter we solve only problems in which the general equations reduce to one or more ordinary differential equations. In Chapter 4 we examine problems of greater complexity that require some ability to solve partial differential equations. Then in Chapter 5 the equations of continuity and motion are used as the starting point for discussing turbulent flow. Later, in Chapter 8, these same equations are applied to flows of polymeric liquids, which are non-Newtonian fluids.

Finally, §3.7 is devoted to writing the equations of continuity and motion in dimensionless form. This makes clear the origin of the Reynolds number, Re, often mentioned in Chapter 2, and why it plays a key role in fluid dynamics. This discussion lays the groundwork for scale-up and model studies. In Chapter 6 dimensionless numbers arise again in connection with experimental correlations of the drag force in complex systems.

At the end of §2.2, we emphasized the importance of experiments in fluid dynamics. We repeat those words of caution here and point out that photographs and other types of flow visualization have provided us with a much deeper understanding of flow problems than would be possible by theory alone.[1] Keep in mind that when one derives a flow field from the equations of change, it is not necessarily the only physically admissible solution.

Vector and tensor notations are occasionally used in this chapter, primarily for the purpose of abbreviating otherwise lengthy expressions. The beginning student will find that only an elementary knowledge of vector and tensor notation is needed for reading this chapter and for solving flow problems. The advanced student will find Appendix A helpful in getting a better understanding of vector and tensor manipulations. With regard to the notation, it should be kept in mind that we use *lightface italic* symbols for scalars, **boldface Roman** symbols for vectors, and **boldface Greek** symbols for tensors. Also dot-product operations enclosed in ( ) are scalars, and those enclosed in [ ] are vectors.

---

[1] We recommend particularly M. Van Dyke, *An Album of Fluid Motion*, Parabolic Press, Stanford (1982); H. Werlé, *Ann. Rev. Fluid Mech.*, **5**, 361–382 (1973); D. V. Boger and K. Walters, *Rheological Phenomena in Focus*, Elsevier, Amsterdam (1993).

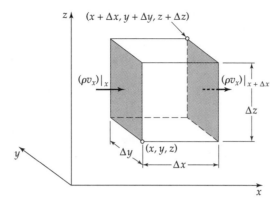

Fig. 3.1-1. Fixed volume element $\Delta x\, \Delta y\, \Delta z$ through which a fluid is flowing. The arrows indicate the mass flux in and out of the volume at the two shaded faces located at x and $x + \Delta x$.

§3.1   THE EQUATION OF CONTINUITY

This equation is developed by writing a mass balance over a volume element $\Delta x\, \Delta y\, \Delta z$, fixed in space, through which a fluid is flowing (see Fig. 3.1-1):

$$\begin{Bmatrix} \text{rate of} \\ \text{increase} \\ \text{of mass} \end{Bmatrix} = \begin{Bmatrix} \text{rate of} \\ \text{mass} \\ \text{in} \end{Bmatrix} - \begin{Bmatrix} \text{rate of} \\ \text{mass} \\ \text{out} \end{Bmatrix} \tag{3.1-1}$$

Now we have to translate this simple physical statement into mathematical language.

We begin by considering the two shaded faces, which are perpendicular to the x-axis. The rate of mass entering the volume element through the shaded face at x is $(\rho v_x)|_x \Delta y\, \Delta z$, and the rate of mass leaving through the shaded face at $x + \Delta x$ is $(\rho v_x)|_{x+\Delta x} \Delta y\, \Delta z$. Similar expressions can be written for the other two pairs of faces. The rate of increase of mass within the volume element is $\Delta x\, \Delta y\, \Delta z (\partial \rho / \partial t)$. The mass balance then becomes

$$\Delta x\, \Delta y\, \Delta z \frac{\partial \rho}{\partial t} = \Delta y\, \Delta z [(\rho v_x)|_x - (\rho v_x)|_{x+\Delta x}]$$

$$+ \Delta z\, \Delta x [(\rho v_y)|_y - (\rho v_y)|_{y+\Delta y}]$$

$$+ \Delta x\, \Delta y [(\rho v_z)|_z - (\rho v_z)|_{z+\Delta z}] \tag{3.1-2}$$

By dividing the entire equation by $\Delta x\, \Delta y\, \Delta z$ and taking the limit as Δx, Δy, and Δz go to zero, and then using the definitions of the partial derivatives, we get

$$\frac{\partial \rho}{\partial t} = -\left(\frac{\partial}{\partial x} \rho v_x + \frac{\partial}{\partial y} \rho v_y + \frac{\partial}{\partial z} \rho v_z \right) \tag{3.1-3}$$

This is the *equation of continuity*, which describes the time rate of change of the fluid density at a fixed point in space. This equation can be written more concisely by using vector notation as follows:

$$\frac{\partial \rho}{\partial t} = -(\nabla \cdot \rho \mathbf{v}) \tag{3.1-4}$$

| rate of increase of mass per unit volume | net rate of mass addition per unit volume by convection |

Here $(\nabla \cdot \rho\mathbf{v})$ is called the "divergence of $\rho\mathbf{v}$," sometimes written as "div $\rho\mathbf{v}$." The vector $\rho\mathbf{v}$ is the mass flux, and its divergence has a simple meaning: it is the net rate of mass efflux per unit volume. The derivation in Problem 3D.1 uses a volume element of arbitrary shape; it is not necessary to use a rectangular volume element as we have done here.

A very important special form of the equation of continuity is that for a fluid of constant density, for which Eq. 3.1-4 assumes the particularly simple form

(incompressible fluid) $\qquad\qquad$ $(\nabla \cdot \mathbf{v}) = 0$ $\qquad\qquad\qquad$ (3.1-5)

Of course, no fluid is truly incompressible, but frequently in engineering and biological applications, the assumption of constant density results in considerable simplification and very little error.[1,2]

EXAMPLE 3.1-1 *Normal Stresses at Solid Surfaces for Incompressible Newtonian Fluids*	Show that for any kind of flow pattern, the normal stresses are zero at fluid–solid boundaries, for Newtonian fluids with constant density. This is an important result that we shall use often. *SOLUTION* We visualize the flow of a fluid near some solid surface, which may or may not be flat. The flow may be quite general, with all three velocity components being functions of all three coordinates and time. At some point P on the surface we erect a Cartesian coordinate system with the origin at P. We now ask what the normal stress τ_{zz} is at P. According to Table B.1 or Eq. 1.2-6, $\tau_{zz} = -2\mu(\partial v_z/\partial z)$, because $(\nabla \cdot \mathbf{v}) = 0$ for incompressible fluids. Then at point P on the surface of the solid

$$\tau_{zz}\big|_{z=0} = -2\mu\left.\frac{\partial v_z}{\partial z}\right|_{z=0} = +2\mu\left(\frac{\partial v_x}{\partial x} + \frac{\partial v_y}{\partial y}\right)\bigg|_{z=0} = 0 \qquad\qquad (3.1\text{-}6)$$

First we replaced the derivative $\partial v_z/\partial z$ by using Eq. 3.1-3 with ρ constant. However, on the solid surface at $z = 0$, the velocity v_x is zero by the no-slip condition (see §2.1), and therefore the derivative $\partial v_x/\partial x$ on the surface must be zero. The same is true of $\partial v_y/\partial y$ on the surface. Therefore τ_{zz} is zero. It is also true that τ_{xx} and τ_{yy} are zero at the surface because of the vanishing of the derivatives at $z = 0$. (*Note:* The vanishing of the normal stresses on solid surfaces does not apply to polymeric fluids, which are viscoelastic. For compressible fluids, the normal stresses at solid surfaces are zero if the density is not changing with time, as is shown in Problem 3C.2.)

§3.2 THE EQUATION OF MOTION

To get the equation of motion we write a momentum balance over the volume element $\Delta x\,\Delta y\,\Delta z$ in Fig. 3.2-1 of the form

$$\begin{bmatrix} \text{rate of} \\ \text{increase} \\ \text{of momentum} \end{bmatrix} = \begin{bmatrix} \text{rate of} \\ \text{momentum} \\ \text{in} \end{bmatrix} - \begin{bmatrix} \text{rate of} \\ \text{momentum} \\ \text{out} \end{bmatrix} + \begin{bmatrix} \text{external} \\ \text{force on} \\ \text{the fluid} \end{bmatrix} \qquad (3.2\text{-}1)$$

---

[1] L. D. Landau and E. M. Lifshitz, *Fluid Mechanics*, Pergamon Press, Oxford (1987), p. 21, point out that, for steady, isentropic flows, commonly encountered in aerodynamics, the incompressibility assumption is valid when the fluid velocity is small compared to the velocity of sound (i.e., low Mach number).

[2] Equation 3.1-5 is the basis for Chapter 2 in G. K. Batchelor, *An Introduction to Fluid Dynamics*, Cambridge University Press (1967), which is a lengthy discussion of the kinematical consequences of the equation of continuity.

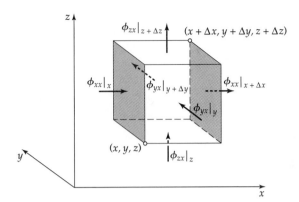

Fig. 3.2-1. Fixed volume element Δx $\Delta y\,\Delta z$, with six arrows indicating the directions of the fluxes of x-momentum through the surfaces by all mechanisms. The shaded faces are located at x and $x + \Delta x$.

Note that Eq. 3.2-1 is an extension of Eq. 2.1-1 to unsteady-state problems. Therefore we proceed in much the same way as in Chapter 2. However, in addition to including the unsteady-state term, we must allow the fluid to move through all six faces of the volume element. Remember that Eq. 3.2-1 is a vector equation with components in each of the three coordinate directions x, y, and z. We develop the x-component of each term in Eq. 3.2-1; the y- and z-components may be treated analogously.[1]

First, we consider the rates of flow of the x-component of momentum into and out of the volume element shown in Fig. 3.2-1. Momentum enters and leaves $\Delta x\,\Delta y\,\Delta z$ by two mechanisms: convective transport (see §1.7), and molecular transport (see §1.2).

The rate at which the x-component of momentum enters across the shaded face at x by all mechanisms—both convective and molecular—is $(\phi_{xx})|_x\,\Delta y\,\Delta z$ and the rate at which it leaves the shaded face at $x + \Delta x$ is $(\phi_{xx})|_{x+\Delta x}\,\Delta y\,\Delta z$. The rates at which x-momentum enters and leaves through the faces at y and $y + \Delta y$ are $(\phi_{yx})|_y\,\Delta z\,\Delta x$ and $(\phi_{yx})|_{y+\Delta y}\,\Delta z\,\Delta x$, respectively. Similarly, the rates at which x-momentum enters and leaves through the faces at z and $z + \Delta z$ are $(\phi_{zx})|_z\,\Delta x\,\Delta y$ and $(\phi_{zx})|_{z+\Delta z}\,\Delta x\,\Delta y$. When these contributions are added we get for the net rate of addition of x-momentum

$$\Delta y\,\Delta z(\phi_{xx}|_x - \phi_{xx}|_{x+\Delta x}) + \Delta z\,\Delta x(\phi_{yx}|_y - \phi_{yx}|_{y+\Delta y}) + \Delta x\,\Delta y(\phi_{zx}|_z - \phi_{zx}|_{z+\Delta z}) \quad (3.2\text{-}2)$$

across all three pairs of faces.

Next there is the external force (typically the gravitational force) acting on the fluid in the volume element. The x-component of this force is

$$\rho g_x \Delta x\,\Delta y\,\Delta z \quad (3.2\text{-}3)$$

Equations 3.2-2 and 3.2-3 give the x-components of the three terms on the right side of Eq. 3.2-1. The sum of these terms must then be equated to the rate of increase of x-momentum within the volume element: $\Delta x\,\Delta y\,\Delta z\,\partial(\rho v_x)/\partial t$. When this is done, we have the x-component of the momentum balance. When this equation is divided by $\Delta x\,\Delta y\,\Delta z$ and the limit is taken as Δx, Δy, and Δz go to zero, the following equation results:

$$\frac{\partial}{\partial t}\rho v_x = -\left(\frac{\partial}{\partial x}\phi_{xx} + \frac{\partial}{\partial y}\phi_{yx} + \frac{\partial}{\partial z}\phi_{zx}\right) + \rho g_x \quad (3.2\text{-}4)$$

---

[1] In this book all the equations of change are derived by applying the conservation laws to a region $\Delta x\,\Delta y\,\Delta z$ fixed in space. The same equations can be obtained by using an arbitrary region fixed in space or one moving along with the fluid. These derivations are described in Problem 3D.1. Advanced students should become familiar with these derivations.

Here we have made use of the definitions of the partial derivatives. Similar equations can be developed for the y- and z-components of the momentum balance:

$$\frac{\partial}{\partial t}\rho v_y = -\left(\frac{\partial}{\partial x}\phi_{xy} + \frac{\partial}{\partial y}\phi_{yy} + \frac{\partial}{\partial z}\phi_{zy}\right) + \rho g_y \tag{3.2-5}$$

$$\frac{\partial}{\partial t}\rho v_z = -\left(\frac{\partial}{\partial x}\phi_{xz} + \frac{\partial}{\partial y}\phi_{yz} + \frac{\partial}{\partial z}\phi_{zz}\right) + \rho g_z \tag{3.2-6}$$

By using vector-tensor notation, these three equations can be written as follows:

$$\frac{\partial}{\partial t}\rho v_i = -[\nabla \cdot \boldsymbol{\phi}]_i + \rho g_i \qquad i = x, y, z \tag{3.2-7}$$

That is, by letting i be successively x, y, and z, Eqs. 3.2-4, 5, and 6 can be reproduced. The quantities ρv_i are the Cartesian components of the vector $\rho\mathbf{v}$, which is the momentum per unit volume at a point in the fluid. Similarly, the quantities ρg_i are the components of the vector $\rho\mathbf{g}$, which is the external force per unit volume. The term $-[\nabla \cdot \boldsymbol{\phi}]_i$ is the ith component of the vector $-[\nabla \cdot \boldsymbol{\phi}]$.

When the ith component of Eq. 3.2-7 is multiplied by the unit vector in the ith direction and the three components are added together vectorially, we get

$$\frac{\partial}{\partial t}\rho\mathbf{v} = -[\nabla \cdot \boldsymbol{\phi}] + \rho\mathbf{g} \tag{3.2-8}$$

which is the differential statement of the law of conservation of momentum. It is the translation of Eq. 3.2-1 into mathematical symbols.

In Eq. 1.7-1 it was shown that the combined momentum flux tensor $\boldsymbol{\phi}$ is the sum of the convective momentum flux tensor $\rho\mathbf{vv}$ and the molecular momentum flux tensor $\boldsymbol{\pi}$, and that the latter can be written as the sum of $p\boldsymbol{\delta}$ and $\boldsymbol{\tau}$. When we insert $\boldsymbol{\phi} = \rho\mathbf{vv} + p\boldsymbol{\delta} + \boldsymbol{\tau}$ into Eq. 3.2-8, we get the following *equation of motion*:[2]

$\dfrac{\partial}{\partial t}\rho\mathbf{v}$	$=$	$-[\nabla \cdot \rho\mathbf{vv}]$	$-\nabla p$	$-[\nabla \cdot \boldsymbol{\tau}]$	$+\rho\mathbf{g}$	(3.2-9)
rate of increase of momentum per unit volume		rate of momentum addition by convection per unit volume	rate of momentum addition by molecular transport per unit volume		external force on fluid per unit volume	

In this equation ∇p is a vector called the "gradient of (the scalar) p" sometimes written as "grad p." The symbol $[\nabla \cdot \boldsymbol{\tau}]$ is a vector called the "divergence of (the tensor) $\boldsymbol{\tau}$" and $[\nabla \cdot \rho\mathbf{vv}]$ is a vector called the "divergence of (the dyadic product) $\rho\mathbf{vv}$."

In the next two sections we give some formal results that are based on the equation of motion. The equations of change for mechanical energy and angular momentum are not used for problem solving in this chapter, but will be referred to in Chapter 7. Beginners are advised to skim these sections on first reading and to refer to them later as the need arises.

---

[2] This equation is attributed to A.-L. Cauchy, *Ex. de math.*, **2**, 108–111 (1827). **(Baron) Augustin-Louis Cauchy** (1789–1857) (pronounced "Koh-shee" with the accent on the second syllable), originally trained as an engineer, made great contributions to theoretical physics and mathematics, including the calculus of complex variables.

§3.3   THE EQUATION OF MECHANICAL ENERGY

Mechanical energy is not conserved in a flow system, but that does not prevent us from developing an equation of change for this quantity. In fact, during the course of this book, we will obtain equations of change for a number of nonconserved quantities, such as internal energy, enthalpy, and entropy. The equation of change for mechanical energy, which involves only mechanical terms, may be derived from the equation of motion of §3.2. The resulting equation is referred to in many places in the text that follows.

We take the dot product of the velocity vector **v** with the equation of motion in Eq. 3.2-9 and then do some rather lengthy rearranging, making use of the equation of continuity in Eq. 3.1-4. We also split up each of the terms containing p and τ into two parts. The final result is the *equation of change for kinetic energy*:

$$\frac{\partial}{\partial t}\left(\tfrac{1}{2}\rho v^2\right) \quad = \quad -(\nabla \cdot \tfrac{1}{2}\rho v^2 \mathbf{v}) \quad -(\nabla \cdot p\mathbf{v}) \quad -p(-\nabla \cdot \mathbf{v})$$

rate of increase of kinetic energy per unit volume	rate of addition of kinetic energy by convection per unit volume	rate of work done by pressure of surroundings on the fluid	rate of *reversible* conversion of kinetic energy into internal energy

$$-(\nabla \cdot (\boldsymbol{\tau} \cdot \mathbf{v}]) \quad -(-\boldsymbol{\tau}{:}\nabla\mathbf{v}) \quad +\rho(\mathbf{v} \cdot \mathbf{g}) \qquad (3.3\text{-}1)^1$$

rate of work done by viscous forces on the fluid	rate of irreversible conversion from kinetic to internal energy	rate of work by external force on the fluid

At this point it is not clear why we have attributed the indicated physical significance to the terms $p(\nabla \cdot \mathbf{v})$ and $(\boldsymbol{\tau}{:}\nabla\mathbf{v})$. Their meaning cannot be properly appreciated until one has studied the energy balance in Chapter 11. There it will be seen how these same two terms appear with opposite sign in the equation of change for internal energy.

We now introduce the *potential energy*[2] (per unit mass) $\hat{\Phi}$, defined by $\mathbf{g} = -\nabla\hat{\Phi}$. Then the last term in Eq. 3.3-1 may be rewritten as $-\rho(\mathbf{v} \cdot \nabla\hat{\Phi}) = -(\nabla \cdot \rho\mathbf{v}\hat{\Phi}) + \hat{\Phi}(\nabla \cdot \rho\mathbf{v})$. The equation of continuity in Eq. 3.1-4 may now be used to replace $+\hat{\Phi}(\nabla \cdot \rho\mathbf{v})$ by $-\hat{\Phi}(\partial\rho/\partial t)$. The latter may be written as $-\partial(\rho\hat{\Phi})/\partial t$, if the potential energy is independent of the time. This is true for the gravitational field for systems that are located on the surface of the earth; then $\hat{\Phi} = gh$, where g is the (constant) gravitational acceleration and h is the elevation coordinate in the gravitational field.

With the introduction of the potential energy, Eq. 3.3-1 assumes the following form:

$$\frac{\partial}{\partial t}\left(\tfrac{1}{2}\rho v^2 + \rho\hat{\Phi}\right) = -\left(\nabla \cdot \left(\tfrac{1}{2}\rho v^2 + \rho\hat{\Phi}\right)\mathbf{v}\right)$$

$$-(\nabla \cdot p\mathbf{v}) - p(-\nabla \cdot \mathbf{v}) - (\nabla \cdot [\boldsymbol{\tau} \cdot \mathbf{v}]) - (-\boldsymbol{\tau}{:}\nabla\mathbf{v}) \qquad (3.3\text{-}2)$$

This is an *equation of change for kinetic-plus-potential energy*. Since Eqs. 3.3-1 and 3.3-2 contain only mechanical terms, they are both referred to as the *equation of change for mechanical energy*.

The term $p(\nabla \cdot \mathbf{v})$ may be either positive or negative depending on whether the fluid is undergoing *expansion* or *compression*. The resulting temperature changes can be rather large for gases in compressors, turbines, and shock tubes.

---

[1] The interpretation under the $(\boldsymbol{\tau}{:}\nabla\mathbf{v})$ term is correct only for Newtonian fluids; for viscoelastic fluids, such as polymers, this term may include reversible conversion to elastic energy.

[2] If $\mathbf{g} = -\boldsymbol{\delta}_z g$ is a vector of magnitude g in the negative z direction, then the potential energy per unit mass is $\hat{\Phi} = gz$, where z is the elevation in the gravitational field.

The term $(-\boldsymbol{\tau}:\nabla\mathbf{v})$ is always positive for *Newtonian* fluids,[3] because it may be written as a sum of squared terms:

$$(-\boldsymbol{\tau}:\nabla\mathbf{v}) = \tfrac{1}{2}\mu \sum_i \sum_j \left[\left(\frac{\partial v_i}{\partial x_j} + \frac{\partial v_j}{\partial x_i}\right) - \tfrac{2}{3}(\nabla\cdot\mathbf{v})\delta_{ij}\right]^2 + \kappa(\nabla\cdot\mathbf{v})^2$$

$$= \mu\Phi_v + \kappa\Psi_v \qquad (3.3\text{-}3)$$

which serves to define the two quantities Φ_v and Ψ_v. When the index i takes on the values 1, 2, 3, the velocity components v_i become v_x, v_y, v_z and the Cartesian coordinates x_i become x, y, z. The symbol δ_{ij} is the Kronecker delta, which is 0 if $i \neq j$ and 1 if $i = j$.

The quantity $(-\boldsymbol{\tau}:\nabla\mathbf{v})$ describes the degradation of mechanical energy into thermal energy that occurs in all flow systems (sometimes called the *viscous dissipation heating*).[4] This heating can produce considerable temperature rises in systems with large viscosity and large velocity gradients, as in lubrication, rapid extrusion, and high-speed flight. (Another example of conversion of mechanical energy into heat is the rubbing of two sticks together to start a fire, which scouts are presumably able to do.)

When we speak of "isothermal systems," we mean systems in which there are no externally imposed temperature gradients and no appreciable temperature change resulting from expansion, contraction, or viscous dissipation.

The most important use of Eq. 3.3-2 is for the development of the macroscopic mechanical energy balance (or engineering Bernoulli equation) in Section 7.8.

§3.4 THE EQUATION OF ANGULAR MOMENTUM

Another equation can be obtained from the equation of motion by forming the cross product of the position vector $\mathbf{r}$ (which has Cartesian components x, y, z) with Eq. 3.2-9. The equation of motion as derived in §3.2 does not contain the assumption that the stress (or momentum-flux) tensor $\boldsymbol{\tau}$ is symmetric. (Of course, the expressions given in §2.3 for the Newtonian fluid are symmetric; that is, $\tau_{ij} = \tau_{ji}$.)

When the cross product is formed, we get—after some vector-tensor manipulations—the following *equation of change for angular momentum*:

$$\frac{\partial}{\partial t}\rho[\mathbf{r}\times\mathbf{v}] = -[\nabla\cdot\rho\mathbf{v}[\mathbf{r}\times\mathbf{v}]] - [\nabla\cdot\{\mathbf{r}\times p\boldsymbol{\delta}\}^\dagger] - [\nabla\cdot\{\mathbf{r}\times\boldsymbol{\tau}^\dagger\}^\dagger] + [\mathbf{r}\times\rho\mathbf{g}] - [\boldsymbol{\varepsilon}:\boldsymbol{\tau}] \quad (3.4\text{-}1)$$

Here $\boldsymbol{\varepsilon}$ is a third-order tensor with components ε_{ijk} (the permutation symbol defined in §A.2). If the stress tensor $\boldsymbol{\tau}$ is symmetric, as for Newtonian fluids, the last term is zero. According to the kinetic theories of dilute gases, monatomic liquids, and polymers, the tensor $\boldsymbol{\tau}$ is symmetric, in the absence of electric and magnetic torques.[1] If, on the other hand, $\boldsymbol{\tau}$ is asymmetric, then the last term describes the rate of conversion of bulk angular momentum to internal angular momentum.

The assumption of a symmetric stress tensor, then, is equivalent to an assertion that there is no interconversion between bulk angular momentum and internal angular momentum and that the two forms of angular momentum are conserved separately. This

---

[3] An amusing consequence of the viscous dissipation for air is the study by H. K. Moffatt [*Nature*, **404**, 833–834 (2000)] of the way in which a spinning coin comes to rest on a table.

[4] G. G. Stokes, *Trans. Camb. Phil. Soc.*, **9**, 8–106 (1851), see pp. 57–59.

[1] J. S. Dahler and L. E. Scriven, *Nature*, **192**, 36–37 (1961); S. de Groot and P. Mazur, *Nonequilibrium Thermodynamics*, North Holland, Amsterdam (1962), Chapter XII. A literature review can be found in G. D. C. Kuiken, *Ind. Eng. Chem. Res.*, **34**, 3568–3572 (1995).

corresponds, in Eq. 0.3-8, to equating the cross-product terms and the internal angular momentum terms separately.

Eq. 3.4-1 will be referred to only in Chapter 7, where we indicate that the macroscopic angular momentum balance can be obtained from it.

§3.5 THE EQUATIONS OF CHANGE IN TERMS OF THE SUBSTANTIAL DERIVATIVE

Before proceeding we point out that several different time derivatives may be encountered in transport phenomena. We illustrate these by a homely example—namely, the observation of the concentration of fish in the Mississippi River. Because fish swim around, the fish concentration will in general be a function of position (x, y, z) and time (t).

The Partial Time Derivative $\partial/\partial t$

Suppose we stand on a bridge and observe the concentration of fish just below us as a function of time. We can then record the time rate of change of the fish concentration at a fixed location. The result is $(\partial c/\partial t)|_{x,y,z}$, the partial derivative of c with respect to t, at constant x, y, and z.

The Total Time Derivative d/dt

Now suppose that we jump into a motor boat and speed around on the river, sometimes going upstream, sometimes downstream, and sometimes across the current. All the time we are observing fish concentration. At any instant, the time rate of change of the observed fish concentration is

$$\frac{dc}{dt} = \left(\frac{\partial c}{\partial t}\right)_{x,y,z} + \frac{dx}{dt}\left(\frac{\partial c}{\partial x}\right)_{y,z,t} + \frac{dy}{dt}\left(\frac{\partial c}{\partial y}\right)_{z,x,t} + \frac{dz}{dt}\left(\frac{\partial c}{\partial z}\right)_{x,y,t} \tag{3.5-1}$$

in which dx/dt, dy/dt, and dz/dt are the components of the velocity of the boat.

The Substantial Time Derivative D/Dt

Next we climb into a canoe, and not feeling energetic, we just float along with the current, observing the fish concentration. In this situation the velocity of the observer is the same as the velocity $\mathbf{v}$ of the stream, which has components v_x, v_y, and v_z. If at any instant we report the time rate of change of fish concentration, we are then giving

$$\frac{Dc}{Dt} = \frac{\partial c}{\partial t} + v_x\frac{\partial c}{\partial x} + v_y\frac{\partial c}{\partial y} + v_z\frac{\partial c}{\partial z} \quad \text{or} \quad \frac{Dc}{Dt} = \frac{\partial c}{\partial t} + (\mathbf{v} \cdot \nabla c) \tag{3.5-2}$$

The special operator $D/Dt = \partial/\partial t + \mathbf{v} \cdot \nabla$ is called the *substantial derivative* (meaning that the time rate of change is reported as one moves with the "substance"). The terms *material derivative*, *hydrodynamic derivative*, and *derivative following the motion* are also used.

Now we need to know how to convert equations expressed in terms of $\partial/\partial t$ into equations written with D/Dt. For any scalar function $f(x, y, z, t)$ we can do the following manipulations:

$$\frac{\partial}{\partial t}(\rho f) + \left(\frac{\partial}{\partial x}\rho v_x f\right) + \left(\frac{\partial}{\partial y}\rho v_y f\right) + \left(\frac{\partial}{\partial z}\rho v_z f\right)$$

$$= \rho\left(\frac{\partial f}{\partial t} + v_x\frac{\partial f}{\partial x} + v_y\frac{\partial f}{\partial y} + v_z\frac{\partial f}{\partial z}\right) + f\left(\frac{\partial \rho}{\partial t} + \frac{\partial}{\partial x}\rho v_x + \frac{\partial}{\partial y}\rho v_y + \frac{\partial}{\partial z}\rho v_z\right)$$

$$= \rho\frac{Df}{Dt} \tag{3.5-3}$$

Table 3.5-1   The Equations of Change for Isothermal Systems in the D/Dt-Form[a]
Note: At the left are given the equation numbers for the $\partial/\partial t$ forms.

(3.1-4)	$\dfrac{D\rho}{Dt} = -\rho(\nabla \cdot \mathbf{v})$	(A)
(3.2-9)	$\rho \dfrac{D\mathbf{v}}{Dt} = -\nabla p - [\nabla \cdot \boldsymbol{\tau}] + \rho\mathbf{g}$	(B)
(3.3-1)	$\rho \dfrac{D}{Dt}\left(\tfrac{1}{2}v^2\right) = -(\mathbf{v} \cdot \nabla p) - (\mathbf{v} \cdot [\nabla \cdot \boldsymbol{\tau}]) + \rho(\mathbf{v} \cdot \mathbf{g})$	(C)
(3.4-1)	$\rho \dfrac{D}{Dt}[\mathbf{r} \times \mathbf{v}] = -[\nabla \cdot \{\mathbf{r} \times p\boldsymbol{\delta}\}^\dagger] - [\nabla \cdot \{\mathbf{r} \times \boldsymbol{\tau}\}^\dagger] + [\mathbf{r} \times \rho\mathbf{g}]$	(D)[a]

[a] Equations (A) through (C) are obtained from Eqs. 3.1-4, 3.2-9, and 3.3-1 with *no assumptions.* Equation (D) is written for symmetrical $\boldsymbol{\tau}$ only.

The quantity in the second parentheses in the second line is zero according to the equation of continuity. Consequently Eq. 3.5-3 can be written in vector form as

$$\frac{\partial}{\partial t}(\rho f) + (\nabla \cdot \rho \mathbf{v} f) = \rho \frac{Df}{Dt} \tag{3.5-4}$$

Similarly, for any vector function $\mathbf{f}(x, y, z, t)$,

$$\frac{\partial}{\partial t}(\rho \mathbf{f}) + [\nabla \cdot \rho \mathbf{v} \mathbf{f}] = \rho \frac{D\mathbf{f}}{Dt} \tag{3.5-5}$$

These equations can be used to rewrite the equations of change given in §§3.1 to 3.4 in terms of the substantial derivative as shown in Table 3.5-1.

Equation A in Table 3.5-1 tells how the density is decreasing or increasing as one moves along with the fluid, because of the compression $[(\nabla \cdot \mathbf{v}) < 0]$ or expansion of the fluid $[(\nabla \cdot \mathbf{v}) > 0]$. Equation B can be interpreted as (mass) × (acceleration) = the sum of the pressure forces, viscous forces, and the external force. In other words, Eq. 3.2-9 is equivalent to Newton's second law of motion applied to a small blob of fluid whose envelope moves locally with the fluid velocity $\mathbf{v}$ (see Problem 3D.1).

We now discuss briefly the three most common simplifications of the equation of motion.[1]

(i) For *constant ρ and μ*, insertion of the Newtonian expression for $\boldsymbol{\tau}$ from Eq. 1.2-7 into the equation of motion leads to the very famous *Navier–Stokes equation*, first developed from molecular arguments by Navier and from continuum arguments by Stokes:[2]

$$\rho \frac{D}{Dt}\mathbf{v} = -\nabla p + \mu\nabla^2\mathbf{v} + \rho\mathbf{g} \quad \text{or} \quad \rho\frac{D}{Dt}\mathbf{v} = -\nabla \mathscr{P} + \mu\nabla^2\mathbf{v} \tag{3.5-6, 7}$$

In the second form we have used the "modified pressure" $\mathscr{P} = p + \rho gh$ introduced in Chapter 2, where h is the elevation in the gravitational field and gh is the gravitational

[1] For discussions of the history of these and other famous fluid dynamics relations, see H. Rouse and S. Ince, *History of Hydraulics*, Iowa Institute of Hydraulics, Iowa City (1959).

[2] L. M. H. Navier, *Mémoires de l'Académie Royale des Sciences*, **6**, 389–440 (1827); G. G. Stokes, *Proc. Cambridge Phil. Soc*, **8**, 287–319 (1845). The name Navier is pronounced "Nah-vyay."

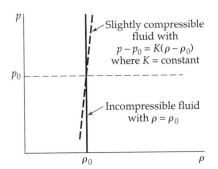

Fig. 3.5-1. The equation of state for a slightly compressible fluid and an incompressible fluid when T is constant.

potential energy per unit mass. Equation 3.5-6 is a standard starting point for describing isothermal flows of gases and liquids.

It must be kept in mind that, when constant ρ is assumed, the equation of state (at constant T) is a vertical line on a plot of p vs. ρ (see Fig. 3.5-1). Thus, the absolute pressure is no longer determinable from ρ and T, although pressure gradients and instantaneous differences remain determinate by Eq. 3.5-6 or Eq. 3.5-7. Absolute pressures are also obtainable if p is known at some point in the system.

(ii) When the *acceleration terms* in the Navier–Stokes equation are neglected—that is, when $\rho(D\mathbf{v}/Dt) = 0$—we get

$$0 = -\nabla p + \mu \nabla^2 \mathbf{v} + \rho \mathbf{g} \tag{3.5-8}$$

which is called the *Stokes flow equation*. It is sometimes called the *creeping flow equation*, because the term $\rho[\mathbf{v} \cdot \nabla \mathbf{v}]$, which is quadratic in the velocity, can be discarded when the flow is extremely slow. For some flows, such as the Hagen–Poiseuille tube flow, the term $\rho[\mathbf{v} \cdot \nabla \mathbf{v}]$ drops out, and a restriction to slow flow is not implied. The Stokes flow equation is important in lubrication theory, the study of particle motions in suspension, flow through porous media, and swimming of microbes. There is a vast literature on this subject.[3]

(iii) When *viscous forces are neglected*—that is, $[\nabla \cdot \boldsymbol{\tau}] = 0$—the equation of motion becomes

$$\rho \frac{D\mathbf{v}}{Dt} = -\nabla p + \rho \mathbf{g} \tag{3.5-9}$$

which is known as the *Euler equation* for "inviscid" fluids.[4] Of course, there are no truly "inviscid" fluids, but there are many flows in which the viscous forces are relatively unimportant. Examples are the flow around airplane wings (except near the solid boundary), flow of rivers around the upstream surfaces of bridge abutments, some problems in compressible gas dynamics, and flow of ocean currents.[5]

[3] J. Happel and H. Brenner, *Low Reynolds Number Hydrodynamics*, Martinus Nijhoff, The Hague (1983); S. Kim and S. J. Karrila, *Microhydrodynamics: Principles and Selected Applications*, Butterworth-Heinemann, Boston (1991).

[4] L. Euler, *Mém. Acad. Sci. Berlin*, **11**, 217–273, 274–315, 316–361 (1755). The Swiss-born mathematician **Leonhard Euler** (1707–1783) (pronounced "Oiler") taught in St. Petersburg, Basel, and Berlin and published extensively in many fields of mathematics and physics.

[5] See, for example, D. J. Acheson, *Elementary Fluid Mechanics*, Clarendon Press, Oxford (1990), Chapters 3–5; and G. K. Batchelor, *An Introduction to Fluid Dynamics*, Cambridge University Press (1967), Chapter 6.

The Bernoulli equation for steady flow of inviscid fluids is one of the most famous equations in classical fluid dynamics.[6] Show how it is obtained from the Euler equation of motion.

SOLUTION

Omit the time-derivative term in Eq. 3.5-9, and then use the vector identity $[\mathbf{v} \cdot \nabla \mathbf{v}] = \frac{1}{2}\nabla(\mathbf{v} \cdot \mathbf{v}) - [\mathbf{v} \times [\nabla \times \mathbf{v}]]$ (Eq. A.4-23) to rewrite the equation as

$$\rho\nabla\tfrac{1}{2}v^2 - \rho[\mathbf{v} \times [\nabla \times \mathbf{v}]] = -\nabla p - \rho g\nabla h \tag{3.5-10}$$

In writing the last term, we have expressed $\mathbf{g}$ as $-\nabla\hat{\Phi} = -g\nabla h$, where h is the elevation in the gravitational field.

Next we divide Eq. 3.5-10 by ρ and then form the dot product with the unit vector $\mathbf{s} = \mathbf{v}/|\mathbf{v}|$ in the flow direction. When this is done the term involving the curl of the velocity field can be shown to vanish (a nice exercise in vector analysis), and $(\mathbf{s} \cdot \nabla)$ can be replaced by d/ds, where s is the distance along a streamline. Thus we get

$$\frac{d}{ds}\left(\tfrac{1}{2}v^2\right) = -\frac{1}{\rho}\frac{d}{ds}p - g\frac{d}{ds}h \tag{3.5-11}$$

When this is integrated along a streamline from point 1 to point 2, we get

$$\tfrac{1}{2}(v_2^2 - v_1^2) + \int_{p_1}^{p_2}\frac{1}{\rho}\,dp + g(h_2 - h_1) = 0 \tag{3.5-12}$$

which is called the *Bernoulli equation*. It relates the velocity, pressure, and elevation of two points along a streamline in a fluid in steady-state flow. It is used in situations where it can be assumed that viscosity plays a rather minor role.

§3.6   USE OF THE EQUATIONS OF CHANGE TO SOLVE FLOW PROBLEMS

For most applications of the equation of motion, we have to insert the expression for $\boldsymbol{\tau}$ from Eq. 1.2-7 into Eq. 3.2-9 (or, equivalently, the components of $\boldsymbol{\tau}$ from Eq. 1.2-6 or Appendix B.1 into Eqs. 3.2-5, 3.2-6, and 3.2-7). Then to describe the flow of a Newtonian fluid at constant temperature, we need in general

The equation of continuity	Eq. 3.1-4
The equation of motion	Eq. 3.2-9
The components of $\boldsymbol{\tau}$	Eq. 1.2-6
The equation of state	$p = p(\rho)$
The equations for the viscosities	$\mu = \mu(\rho), \kappa = \kappa(\rho)$

These equations, along with the necessary boundary and initial conditions, determine completely the pressure, density, and velocity distributions in the fluid. They are seldom used in their complete form to solve fluid dynamics problems. Usually restricted forms are used for convenience, as in this chapter.

If it is appropriate to assume constant density and viscosity, then we use

| The equation of continuity | Eq. 3.1-4 and Table B.4 |
| The Navier-Stokes equation | Eq. 3.5-6 and Tables B.5, 6, 7 |

along with initial and boundary conditions. From these one determines the pressure and velocity distributions.

---

[6] **Daniel Bernoulli** (1700–1782) was one of the early researchers in fluid dynamics and also the kinetic theory of gases. His hydrodynamical ideas were summarized in D. Bernoulli, *Hydrodynamica sive de viribus et motibus fluidorum commentarii*, Argentorati (1738), however he did not actually give Eq. 3.5-12. The credit for the derivation of Eq. 3.5-12 goes to L. Euler, *Histoires de l'Académie de Berlin* (1755).

In Chapter 1 we gave the components of the stress tensor in Cartesian coordinates, and in this chapter we have derived the equations of continuity and motion in Cartesian coordinates. In Tables B.1, 4, 5, 6, and 7 we summarize these key equations in three much-used coordinate systems: Cartesian (x, y, z), cylindrical (r, θ, z), and spherical (r, θ, ϕ). Beginning students should not concern themselves with the derivation of these equations, but they should be very familiar with the tables in Appendix B and be able to use them for setting up fluid dynamics problems. Advanced students will want to go through the details of Appendix A and learn how to develop the expressions for the various ∇-operations, as is done in §§A.6 and A.7.

In this section we illustrate how to set up and solve some problems involving the steady, isothermal, laminar flow of Newtonian fluids. The relatively simple analytical solutions given here are not to be regarded as ends in themselves, but rather as a preparation for moving on to the analytical or numerical solution of more complex problems, the use of various approximate methods, or the use of dimensional analysis.

The complete solution of viscous flow problems, including proofs of uniqueness and criteria for stability, is a formidable task. Indeed, the attention of some of the world's best applied mathematicians has been devoted to the challenge of solving the equations of continuity and motion. The beginner may well feel inadequate when faced with these equations for the first time. All we attempt to do in the illustrative examples in this section is to solve a few problems for stable flows that are known to exist. In each case we begin by making some *postulates* about the form for the pressure and velocity distributions: that is, we *guess* how p and $\mathbf{v}$ should depend on position in the problem being studied. Then we discard all the terms in the equations of continuity and motion that are unnecessary according to the postulates made. For example, if one postulates that v_x is a function of y alone, terms like $\partial v_x/\partial x$ and $\partial^2 v_x/\partial z^2$ can be discarded. When all the unnecessary terms have been eliminated, one is frequently left with a small number of relatively simple equations; and if the problem is sufficiently simple, an analytical solution can be obtained.

It must be emphasized that in listing the postulates, one makes use of *intuition*. The latter is based on our daily experience with flow phenomena. Our intuition often tells us that a flow will be symmetrical about an axis, or that some component of the velocity is zero. Having used our intuition to make such postulates, we must remember that the final solution is correspondingly restricted. However, by starting with the equations of change, when we have finished the "discarding process" we do at least have a complete listing of all the assumptions used in the solution. In some instances it is possible to go back and remove some of the assumptions and get a better solution.

In several examples to be discussed, we will find one solution to the fluid dynamical equations. However, because the full equations are nonlinear, there may be other solutions to the problem. Thus a complete solution to a fluid dynamics problem requires the specification of the limits on the stable flow regimes as well as any ranges of unstable behavior. That is, we have to develop a "map" showing the various flow regimes that are possible. Usually analytical solutions can be obtained for only the simplest flow regimes; the remainder of the information is generally obtained by experiment or by very detailed numerical solutions. In other words, although we know the differential equations that govern the fluid motion, much is yet unknown about how to solve them. This is a challenging area of applied mathematics, well above the level of an introductory textbook.

When difficult problems are encountered, a search should be made through some of the advanced treatises on fluid dynamics.[1]

---

[1] R. Berker, *Handbuch der Physik*, Volume VIII-2, Springer, Berlin (1963), pp. 1–384; G. K. Batchelor, *An Introduction to Fluid Mechanics*, Cambridge University Press (1967); L. Landau and E. M. Lifshitz, *Fluid Mechanics*, Pergamon Press, Oxford, 2nd edition (1987); J. A. Schetz and A. E. Fuhs (eds.), *Handbook of Fluid Dynamics and Fluid Machinery*, Wiley-Interscience, New York (1996); R. W. Johnson (ed.), *The Handbook of Fluid Dynamics*, CRC Press, Boca Raton, Fla. (1998); C. Y. Wang, *Ann. Revs. Fluid Mech.*, **23**, 159–177 (1991).

We now turn to the illustrative examples. The first two are problems that were discussed in the preceding chapter; we rework these just to illustrate the use of the equations of change. Then we consider some other problems that would be difficult to set up by the shell balance method of Chapter 2.

EXAMPLE 3.6-1

Steady Flow in a Long Circular Tube

Rework the tube-flow problem of Example 2.3-1 using the equations of continuity and motion. This illustrates the use of the tabulated equations for constant viscosity and density in cylindrical coordinates, given in Appendix B.

SOLUTION

We postulate that $\mathbf{v} = \boldsymbol{\delta}_z v_z(r, z)$. This postulate implies that there is no radial flow ($v_r = 0$) and no tangential flow ($v_\theta = 0$), and that v_z does not depend on θ. Consequently, we can discard many terms from the tabulated equations of change, leaving

equation of continuity
$$\frac{\partial v_z}{\partial z} = 0 \tag{3.6-1}$$

r-equation of motion
$$0 = -\frac{\partial \mathcal{P}}{\partial r} \tag{3.6-2}$$

θ-equation of motion
$$0 = -\frac{\partial \mathcal{P}}{\partial \theta} \tag{3.6-3}$$

z-equation of motion
$$0 = -\frac{\partial \mathcal{P}}{\partial z} + \mu \frac{1}{r} \frac{\partial}{\partial r}\left(r \frac{\partial v_z}{\partial r}\right) \tag{3.6-4}$$

The first equation indicates that v_z depends only on r; hence the partial derivatives in the second term on the right side of Eq. 3.6-4 can be replaced by ordinary derivatives. By using the modified pressure $\mathcal{P} = p + \rho g h$ (where h is the height above some arbitrary datum plane), we avoid the necessity of calculating the components of $\mathbf{g}$ in cylindrical coordinates, and we obtain a solution valid for any orientation of the axis of the tube.

Equations 3.6-2 and 3.6-3 show that $\mathcal{P}$ is a function of z alone, and the partial derivative in the first term of Eq. 3.6-4 may be replaced by an ordinary derivative. The only way that we can have a function of r plus a function of z equal to zero is for each term individually to be a constant—say, C_0—so that Eq. 3.6-4 reduces to

$$\mu \frac{1}{r} \frac{d}{dr}\left(r \frac{dv_z}{dr}\right) = C_0 = \frac{d\mathcal{P}}{dz} \tag{3.6-5}$$

The $\mathcal{P}$ equation can be integrated at once. The v_z-equation can be integrated by merely "peeling off" one operation after another on the left side (do not "work out" the compound derivative there). This gives

$$\mathcal{P} = C_0 z + C_1 \tag{3.6-6}$$

$$v_z = \frac{C_0}{4\mu} r^2 + C_2 \ln r + C_3 \tag{3.6-7}$$

The four constants of integration can be found from the boundary conditions:

B.C. 1   at $z = 0$, $\mathcal{P} = \mathcal{P}_0$ (3.6-8)

B.C. 2   at $z = L$, $\mathcal{P} = \mathcal{P}_L$ (3.6-9)

B.C. 3   at $r = R$, $v_z = 0$ (3.6-10)

B.C. 4   at $r = 0$, $v_z = $ finite   (3.6-11)

The resulting solutions are:

$$\mathcal{P} = \mathcal{P}_0 - (\mathcal{P}_0 - \mathcal{P}_L)(z/L) \tag{3.6-12}$$

$$v_z = \frac{(\mathcal{P}_0 - \mathcal{P}_L)R^2}{4\mu L}\left[1 - \left(\frac{r}{R}\right)^2\right] \tag{3.6-13}$$

Equation 3.6-13 is the same as Eq. 2.3-18. The pressure profile in Eq. 3.6-12 was not obtained in Example 2.3-1, but was tacitly postulated; we could have done that here, too, but we chose to work with a minimal number of postulates.

As pointed out in Example 2.3-1, Eq. 3.6-13 is valid only in the laminar-flow regime, and at locations not too near the tube entrance and exit. For Reynolds numbers above about 2100, a turbulent-flow regime exists downstream of the entrance region, and Eq. 3.6-13 is no longer valid.

EXAMPLE 3.6-2

Falling Film with Variable Viscosity

Set up the problem in Example 2.2-2 by using the equations of Appendix B. This illustrates the use of the equation of motion in terms of τ.

SOLUTION

As in Example 2.2-2 we postulate a steady-state flow with constant density, but with viscosity depending on x. We postulate, as before, that the x- and y-components of the velocity are zero and that $v_z = v_z(x)$. With these postulates, the equation of continuity is identically satisfied. According to Table B.1, the only nonzero components of τ are $\tau_{xz} = \tau_{zx} = -\mu(dv_z/dx)$. The components of the equation of motion in terms of τ are, from Table B.5,

$$0 = -\frac{\partial p}{\partial x} + \rho g \sin \beta \tag{3.6-14}$$

$$0 = -\frac{\partial p}{\partial y} \tag{3.6-15}$$

$$0 = -\frac{\partial p}{\partial z} - \frac{d}{dx}\tau_{xz} + \rho g \cos \beta \tag{3.6-16}$$

where β is the angle shown in Fig. 2.2-2.

Integration of Eq. 3.6-14 gives

$$p = \rho g x \sin \beta + f(y, z) \tag{3.6-17}$$

in which $f(y, z)$ is an arbitrary function. Equation 3.6-15 shows that f cannot be a function of y. We next recognize that the pressure in the gas phase is very nearly constant at the prevailing atmospheric pressure p_{atm}. Therefore, at the gas–liquid interface $x = 0$, the pressure is also constant at the value p_{atm}. Consequently, f can be set equal to p_{atm} and we obtain finally

$$p = \rho g x \sin \beta + p_{atm} \tag{3.6-18}$$

Equation 3.5-16 then becomes

$$\frac{d}{dx}\tau_{xz} = \rho g \cos \beta \tag{3.6-19}$$

which is the same as Eq. 2.2-10. The remainder of the solution is the same as in §2.2.

EXAMPLE 3.6-3

Operation of a Couette Viscometer

We mentioned earlier that the measurement of pressure difference vs. mass flow rate through a cylindrical tube is the basis for the determination of viscosity in commercial capillary viscometers. The viscosity may also be determined by measuring the torque required to turn a solid object in contact with a fluid. The forerunner of all rotational viscometers is the Couette instrument, which is sketched in Fig. 3.6-1.

The fluid is placed in the cup, and the cup is then made to rotate with a constant angular velocity Ω_o (the subscript "o" stands for outer). The rotating viscous liquid causes the suspended bob to turn until the torque produced by the momentum transfer in the fluid equals the product of the torsion constant k_t and the angular displacement θ_b of the bob. The angular displacement can be measured by observing the deflection of a light beam reflected from a mirror mounted on the bob. The conditions of measurement are controlled so that there is a steady, tangential, laminar flow in the annular region between the two coaxial cylinders

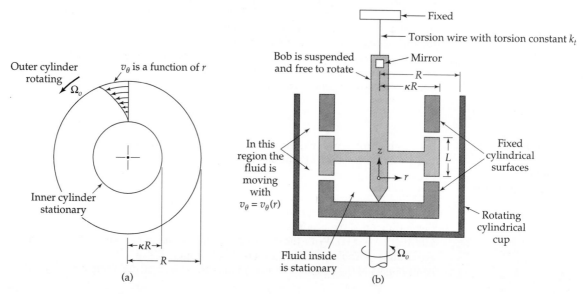

Fig. 3.6-1. (*a*)Tangential laminar flow of an incompressible fluid in the space between two cylinders; the outer one is moving with an angular velocity Ω_o. (*b*) A diagram of a Couette viscometer. One measures the angular velocity Ω_o of the cup and the deflection θ_B of the bob at steady-state operation. Equation 3.6-31 gives the viscosity μ in terms of Ω_o and the torque $T_z = k_t\theta_B$.

shown in the figure. Because of the arrangement used, end effects over the region including the bob height L are negligible.

To analyze this measurement, we apply the equations of continuity and motion for constant ρ and μ to the tangential flow in the annular region around the bob. Ultimately we want an expression for the viscosity in terms of (the z-component of) the torque T_z on the inner cylinder, the angular velocity Ω_o of the rotating cup, the bob height L, and the radii κR and R of the bob and cup, respectively.

SOLUTION

In the portion of the annulus under consideration the fluid moves in a circular pattern. Reasonable postulates for the velocity and pressure are: $v_\theta = v_\theta(r)$, $v_r = 0$, $v_z = 0$, and $p = p(r, z)$. We expect p to depend on z because of gravity and on r because of the centrifugal force.

For these postulates all the terms in the equation of continuity are zero, and the components of the equation of motion simplify to

r-component
$$-\rho\frac{v_\theta^2}{r} = -\frac{\partial p}{\partial r}\qquad(3.6\text{-}20)$$

θ-component
$$0 = \frac{d}{dr}\left(\frac{1}{r}\frac{d}{dr}(rv_\theta)\right)\qquad(3.6\text{-}21)$$

z-component
$$0 = -\frac{\partial p}{\partial z} - \rho g\qquad(3.6\text{-}22)$$

The second equation gives the velocity distribution. The third equation gives the effect of gravity on the pressure (the hydrostatic effect), and the first equation tells how the centrifugal force affects the pressure. For the problem at hand we need only the θ-component of the equation of motion.[2]

---

[2] See R. B. Bird, C. F. Curtiss, and W. E. Stewart, *Chem. Eng. Sci.*, **11**, 114–117 (1959) for a method of getting $p(r, z)$ for this system. The time-dependent buildup to the steady-state profiles is given by R. B. Bird and C. F. Curtiss, *Chem. Eng. Sci.*, **11**, 108–113 (1959).

A novice might have a compelling urge to perform the differentiations in Eq. 3.6-21 before solving the differential equation, but this should not be done. All one has to do is "peel off" one operation at a time—just the way you undress—as follows:

$$\frac{1}{r}\frac{d}{dr}(rv_\theta) = C_1 \tag{3.6-23}$$

$$\frac{d}{dr}(rv_\theta) = C_1 r \tag{3.6-24}$$

$$rv_\theta = \frac{1}{2}C_1 r^2 + C_2 \tag{3.6-25}$$

$$v_\theta = \frac{1}{2}C_1 r + \frac{C_2}{r} \tag{3.6-26}$$

The boundary conditions are that the fluid does not slip at the two cylindrical surfaces:

B.C. 1   at $r = \kappa R$, $v_\theta = 0$ (3.6-27)

B.C. 2   at $r = R$, $v_\theta = \Omega_o R$ (3.6-28)

These boundary conditions can be used to get the constants of integration, which are then inserted in Eq. 3.6-26. This gives

$$v_\theta = \Omega_o R \frac{\left(\dfrac{r}{\kappa R} - \dfrac{\kappa R}{r}\right)}{\left(\dfrac{1}{\kappa} - \kappa\right)} \tag{3.6-29}$$

By writing the result in this form, with similar terms in the numerator and denominator, it is clear that both boundary conditions are satisfied and that the equation is dimensionally consistent.

From the velocity distribution we can find the momentum flux by using Table B.2:

$$\tau_{r\theta} = -\mu r \frac{d}{dr}\left(\frac{v_\theta}{r}\right) = -2\mu\Omega_o\left(\frac{R}{r}\right)^2\left(\frac{\kappa^2}{1 - \kappa^2}\right) \tag{3.6-30}$$

The torque acting on the inner cylinder is then given by the product of the inward momentum flux $(-\tau_{r\theta})$, the surface of the cylinder, and the lever arm, as follows:

$$T_z = (-\tau_{r\theta})|_{r=\kappa R} \cdot 2\pi\kappa RL \cdot \kappa R = 4\pi\mu\Omega_o R^2 L\left(\frac{\kappa^2}{1 - \kappa^2}\right) \tag{3.6-31}$$

The torque is also given by $T_z = k_t\theta_b$. Therefore, measurement of the angular velocity of the cup and the angular deflection of the bob makes it possible to determine the viscosity. The same kind of analysis is available for other rotational viscometers.[3]

For any viscometer it is essential to know when turbulence will occur. The critical Reynolds number $(\Omega_o R^2 \rho/\mu)_{crit}$, above which the system becomes turbulent, is shown in Fig. 3.6-2 as a function of the radius ratio κ.

One might ask what happens if we hold the outer cylinder fixed and cause the inner cylinder to rotate with an angular velocity Ω_i (the subscript "i" stands for inner). Then the velocity distribution is

$$v_\theta = \Omega_i\kappa R \frac{\left(\dfrac{R}{r} - \dfrac{r}{R}\right)}{\left(\dfrac{1}{\kappa} - \kappa\right)} \tag{3.6-32}$$

This is obtained by making the same postulates (see before Eq. 3.6-20) and solving the same differential equation (Eq. 3.6-21), but with a different set of boundary conditions.

[3] J. R. VanWazer, J. W. Lyons, K. Y. Kim, and R. E. Colwell, *Viscosity and Flow Measurement*, Wiley, New York (1963); K. Walters, *Rheometry*, Chapman and Hall, London (1975).

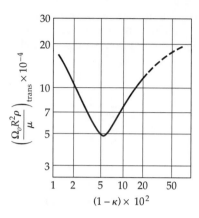

Fig. 3.6-2. Critical Reynolds number for the tangential flow in an annulus, with the outer cylinder rotating and the inner cylinder stationary [H. Schlichting, *Boundary Layer Theory*, McGraw-Hill, New York (1955), p. 357].

Equation 3.6-32 describes the flow accurately for small values of Ω_i. However, when Ω_i reaches a critical value ($\Omega_{i,\text{crit}} \approx 41.3(\mu/R^2(1 - \kappa)^{3/2}\rho)$ for $\kappa \approx 1$) the fluid develops a secondary flow, which is superimposed on the primary (tangential) flow and which is periodic in the axial direction. A very neat system of toroidal vortices, called *Taylor vortices*, is formed, as depicted in Figs. 3.6-3 and 3.6-4(b). The loci of the centers of these vortices are circles, whose centers are located on the common axis of the cylinders. This is still laminar motion—but certainly inconsistent with the postulates made at the beginning of the problem. When the angular velocity Ω_i is increased further, the loci of the centers of the vortices become traveling waves; that is, the flow becomes, in addition, periodic in the tangential direction [see Fig. 3.6-4(c)]. Furthermore, the angular velocity of the traveling waves is approximately $\frac{1}{3}\Omega_i$. When the angular velocity Ω_i is further increased, the flow becomes turbulent. Figure 3.6-5 shows the various flow regimes, with the inner and outer cylinders both rotating, determined for a specific apparatus and a

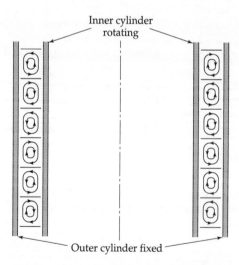

Fig. 3.6-3. Counter-rotating toroidal vortices, called *Taylor vortices*, observed in the annular space between two cylinders. The streamlines have the form of helices, with the axes wrapped around the common axis of the cylinders. This corresponds to Fig. 3.5-4(b).

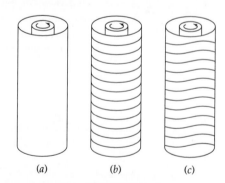

Fig. 3.6-4. Sketches showing the phenomena observed in the annular space between two cylinders: (a) purely tangential flow; (b) singly periodic flow (Taylor vortices); and (c) doubly periodic flow in which an undulatory motion is superposed on the Taylor vortices.

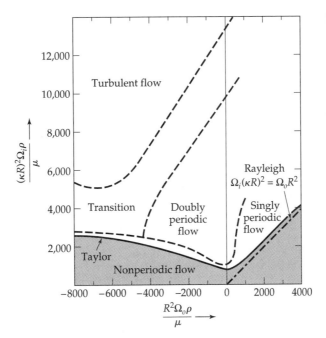

Fig. 3.6-5. Flow-regime diagram for the flow between two coaxial cylinders. The straight line labeled "Rayleigh" is Lord Rayleigh's analytic solution for an inviscid fluid. [See D. Coles, *J. Fluid. Mech.*, **21**, 385–425 (1965).]

specific fluid. This diagram demonstrates how complicated this apparently simple system is. Further details may be found elsewhere.[4,5]

The preceding discussion should serve as a stern warning that intuitive postulates may be misleading. Most of us would not think about postulating the singly and doubly periodic solutions just described. Nonetheless, this information *is* contained in the Navier–Stokes equations. However, since problems involving instability and transitions between several flow regimes are extremely complex, we are forced to use a combination of theory and experiment to describe them. Theory alone cannot yet give us all the answers, and carefully controlled experiments will be needed for years to come.

EXAMPLE 3.6-4

Shape of the Surface of a Rotating Liquid

A liquid of constant density and viscosity is in a cylindrical container of radius R as shown in Fig. 3.6-6. The container is caused to rotate about its own axis at an angular velocity Ω. The cylinder axis is vertical, so that $g_r = 0$, $g_\theta = 0$, and $g_z = -g$, in which g is the magnitude of the gravitational acceleration. Find the shape of the free surface of the liquid when steady state has been established.

---

[4] The initial work on this subject was done by **John William Strutt (Lord Rayleigh)** (1842–1919), who established the field of acoustics with his *Theory of Sound*, written on a houseboat on the Nile River. Some original references on Taylor instability are: J. W. Strutt (Lord Rayleigh), *Proc. Roy. Soc.*, **A93**, 148–154 (1916); G. I. Taylor, *Phil. Trans.*, **A223**, 289–343 (1923) and *Proc. Roy. Soc.* **A157**, 546–564 (1936); P. Schultz-Grunow and H. Hein, *Zeits. Flugwiss.*, **4**, 28–30 (1956); D. Coles, *J. Fluid. Mech.* **21**, 385–425 (1965). See also R. P. Feynman, R. B. Leighton, and M. Sands, *The Feynman Lectures in Physics*, Addison-Wesley, Reading, MA (1964), §41–6.

[5] Other references on Taylor instability, as well as instability in other flow systems, are: L. D. Landau and E. M. Lifshitz, *Fluid Mechanics*, Pergamon, Oxford, 2nd edition (1987), pp. 99–106; S. Chandrasekhar, *Hydrodynamic and Hydromagnetic Stability*, Oxford University Press (1961), pp. 272–342; H. Schlichting and K. Gersten, *Boundary-Layer Theory*, 8th edition (2000), Chapter 15; P. G. Drazin and W. H. Reid, *Hydrodynamic Stability*, Cambridge University Press (1981); M. Van Dyke, *An Album of Fluid Motion*, Parabolic Press, Stanford (1982).

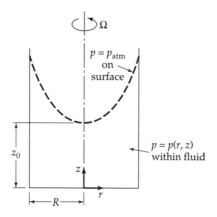

Fig. 3.6-6. Rotating liquid with a free surface, the shape of which is a paraboloid of revolution.

SOLUTION

Cylindrical coordinates are appropriate for this problem, and the equations of change are given in Tables B.2 and B.6. At steady state we postulate that v_r and v_z are both zero and that v_θ depends only on r. We also postulate that p depends on z because of the gravitational force and on r because of the centrifugal force but not on θ.

These postulates give $0 = 0$ for the equation of continuity, and the equation of motion gives:

r-component

$$-\rho \frac{v_\theta^2}{r} = -\frac{\partial p}{\partial r} \tag{3.6-33}$$

θ-component

$$0 = \mu \frac{d}{dr}\left(\frac{1}{r}\frac{d}{dr}(rv_\theta)\right) \tag{3.6-34}$$

z-component

$$0 = -\frac{\partial p}{\partial z} - \rho g \tag{3.6-35}$$

The θ-component of the equation of motion can be integrated to give

$$v_\theta = \frac{1}{2}C_1 r + \frac{C_2}{r} \tag{3.6-36}$$

in which C_1 and C_2 are constants of integration. Because v_θ cannot be infinite at $r = 0$, the constant C_2 must be zero. At $r = R$ the velocity v_θ is $R\Omega$. Hence $C_1 = 2\Omega$ and

$$v_\theta = \Omega r \tag{3.6-37}$$

This states that each element of the rotating liquid moves as an element of a rigid body (we could have actually postulated that the liquid would rotate as a rigid body and written down Eq. 3.6-37 directly). When the result in Eq. 3.6-37 is substituted into Eq. 3.6-33, we then have these two equations for the pressure gradients:

$$\frac{\partial p}{\partial r} = \rho\Omega^2 r \quad \text{and} \quad \frac{\partial p}{\partial z} = -\rho g \tag{3.6-38, 39}$$

Each of these equations can be integrated, as follows:

$$p = \tfrac{1}{2}\rho\Omega^2 r^2 + f_1(\theta, z) \quad \text{and} \quad p = -\rho g z + f_2(r, \theta) \tag{3.6-40, 41}$$

where f_1 and f_2 are arbitrary functions of integration. Since we have postulated that p does not depend on θ, we can choose $f_1 = -\rho g z + C$ and $f_2 = \tfrac{1}{2}\rho\Omega^2 r^2 + C$, where C is a constant, and satisfy Eqs. 3.6-38 and 39. Thus the solution to those equations has the form

$$p = -\rho g z + \tfrac{1}{2}\rho\Omega^2 r^2 + C \tag{3.6-42}$$

The constant C may be determined by requiring that $p = p_{\text{atm}}$ at $r = 0$ and $z = z_0$, the latter being the elevation of the liquid surface at $r = 0$. When C is obtained in this way, we get

$$p - p_{\text{atm}} = -\rho g(z - z_0) + \tfrac{1}{2}\rho\Omega^2 r^2 \tag{3.6-43}$$

This equation gives the pressure at all points within the liquid. Right at the liquid–air interface, $p = p_{atm}$, and with this substitution Eq. 3.6-43 gives the shape of the liquid–air interface:

$$z - z_0 = \left(\frac{\Omega^2}{2g}\right) r^2 \tag{3.6-44}$$

This is the equation for a parabola. The reader can verify that the free surface of a liquid in a rotating annular container obeys a similar relation.

EXAMPLE 3.6-5

Flow near a Slowly Rotating Sphere

A solid sphere of radius R is rotating slowly at a constant angular velocity Ω in a large body of quiescent fluid (see Fig. 3.6-7). Develop expressions for the pressure and velocity distributions in the fluid and for the torque T_z required to maintain the motion. It is assumed that the sphere rotates sufficiently slowly that it is appropriate to use the *creeping flow* version of the equation of motion in Eq. 3.5-8. This problem illustrates setting up and solving a problem in spherical coordinates.

SOLUTION

The equations of continuity and motion in spherical coordinates are given in Tables B.4 and B.6, respectively. We postulate that, for steady creeping flow, the velocity distribution will have the general form $\mathbf{v} = \boldsymbol{\delta}_\phi v_\phi(r, \theta)$, and that the modified pressure will be of the form $\mathcal{P} = \mathcal{P}(r, \theta)$. Since the solution is expected to be symmetric about the z-axis, there is no dependence on the angle ϕ.

With these postulates, the equation of continuity is exactly satisfied, and the components of the creeping flow equation of motion become

r-component
$$0 = -\frac{\partial \mathcal{P}}{\partial r} \tag{3.6-45}$$

θ-component
$$0 = -\frac{1}{r}\frac{\partial \mathcal{P}}{\partial \theta} \tag{3.6-46}$$

ϕ-component
$$0 = \frac{1}{r^2}\frac{\partial}{\partial r}\left(r^2 \frac{\partial v_\phi}{\partial r}\right) + \frac{1}{r^2}\frac{\partial}{\partial \theta}\left(\frac{1}{\sin\theta}\frac{\partial}{\partial \theta}(v_\phi \sin\theta)\right) \tag{3.6-47}$$

The boundary conditions may be summarized as

B.C. 1:   at $r = R$, $v_r = 0, v_\theta = 0, v_\phi = R\Omega \sin\theta$ (3.6-48)

B.C. 2:   as $r \to \infty$, $v_r \to 0, v_\theta \to 0, v_\phi \to 0$ (3.6-49)

B.C. 3:   as $r \to \infty$, $\mathcal{P} \to p_0$ (3.6-50)

where $\mathcal{P} = p + \rho gz$, and p_0 is the fluid pressure far from the sphere at $z = 0$.

Equation 3.6-47 is a partial differential equation for $v_\phi(r, \theta)$. To solve this, we try a solution of the form $v_\phi = f(r) \sin\theta$. This is just a guess, but it is consistent with the boundary condition in Eq. 3.6-48. When this trial form for the velocity distribution is inserted into Eq. 3.6-47 we get the following ordinary differential equation for $f(r)$:

$$\frac{d}{dr}\left(r^2 \frac{df}{dr}\right) - 2f = 0 \tag{3.6-51}$$

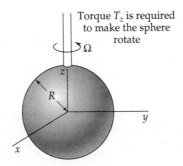

Torque T_z is required to make the sphere rotate

Fig. 3.6-7. A slowly rotating sphere in an infinite expanse of fluid. The primary flow is $v_\phi = \Omega R(R/r)^2 \sin\theta$.

This is an "equidimensional equation," which may be solved by assuming a trial solution $f = r^n$ (see Eq. C.1-14). Substitution of this trial solution into Eq. 3.6-51 gives $n = 1, -2$. The solution of Eq. 3.6-51 is then

$$f = C_1 r + \frac{C_2}{r^2} \tag{3.6-52}$$

so that

$$v_\phi(r, \theta) = \left(C_1 r + \frac{C_2}{r^2} \right) \sin \theta \tag{3.6-53}$$

Application of the boundary conditions shows that $C_1 = 0$ and $C_2 = \Omega R^3$. Therefore the final expression for the velocity distribution is

$$v_\phi = \Omega R \left(\frac{R}{r} \right)^2 \sin \theta \tag{3.6-54}$$

Next we evaluate the torque needed to maintain the rotation of the sphere. This will be the integral, over the sphere surface, of the tangential force $(\tau_{r\phi}|_{r=R}) R^2 \sin \theta d\theta d\phi$ exerted on the fluid by a solid surface element, multiplied by the lever arm $R \sin \theta$ for that element:

$$
\begin{aligned}
T_z &= \int_0^{2\pi} \int_0^\pi (\tau_{r\phi})|_{r=R} (R \sin \theta) R^2 \sin \theta d\theta d\phi \\
&= \int_0^{2\pi} \int_0^\pi (3\mu\Omega \sin \theta)(R \sin \theta) R^2 \sin \theta d\theta d\phi \\
&= 6\pi\mu\Omega R^3 \int_0^\pi \sin^3 \theta d\theta \\
&= 8\pi\mu\Omega R^3
\end{aligned}
\tag{3.6-55}
$$

In going from the first to the second line, we have used Table B.1, and in going from the second to the third line we have done the integration over the range of the ϕ variable. The integral in the third line is $\frac{4}{3}$.

As the angular velocity increases, deviations from the "primary flow" of Eq. 3.6-54 occur. Because of the centrifugal force effects, the fluid is pulled in toward the poles of the sphere and shoved outward from the equator as shown in Fig. 3.6-8. To describe this "secondary flow," one has to include the $[\mathbf{v} \cdot \nabla \mathbf{v}]$ term in the equation of motion. This can be done by the use of a stream-function method.[6]

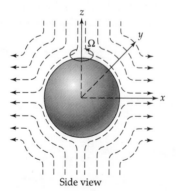

Side view

Fig. 3.6-8. Rough sketch showing the secondary flow which appears around a rotating sphere as the Reynolds number is increased.

[6] See, for example, the development by O. Hassager in R. B. Bird, R. C. Armstrong, and O. Hassager, *Dynamics of Polymeric Liquids*, Vol. 1., Wiley-Interscience, New York, 2nd edition (1987), pp. 31–33. See also L. Landau and E. M. Lifshitz, *Fluid Mechanics*, Pergamon, Oxford, 2nd edition (1987), p. 65; and L. G. Leal, *Laminar Flow and Convective Transport Processes*, Butterworth-Heinemann, Boston (1992), pp. 180–181.

§3.7   DIMENSIONAL ANALYSIS OF THE EQUATIONS OF CHANGE

Suppose that we have taken experimental data on, or made photographs of, the flow through some system that cannot be analyzed by solving the equations of change analytically. An example of such a system is the flow of a fluid through an orifice meter in a pipe (this consists of a disk with a centered hole in it, placed in the tube, with a pressure-sensing device upstream and downstream of the disk). Suppose now that we want to scale up (or down) the experimental system, in order to build a new one in which exactly the same flow patterns occur [but appropriately scaled up (or down)]. First of all, we need to have *geometric similarity*: that is, the ratios of all dimensions of the pipe and orifice plate in the original system and in the scaled-up (or scaled-down) system must be the same. In addition, we must have *dynamic similarity*: that is, the dimensionless groups (such as the Reynolds number) in the differential equations and boundary conditions must be the same. The study of dynamic similarity is best understood by writing the equations of change, along with boundary and initial conditions, in dimensionless form.[1,2]

For simplicity we restrict the discussion here to fluids of constant density and viscosity, for which the equations of change are Eqs. 3.1-5 and 3.5-7

$$(\nabla \cdot \mathbf{v}) = 0 \tag{3.7-1}$$

$$\rho \frac{D}{Dt} \mathbf{v} = -\nabla \mathscr{P} + \mu \nabla^2 \mathbf{v} \tag{3.7-2}$$

In most flow systems one can identify the following "scale factors": a characteristic length l_0, a characteristic velocity v_0, and a characteristic modified pressure $\mathscr{P}_0 = p_0 + \rho g h_0$ (for example, these might be a tube diameter, the average flow velocity, and the modified pressure at the tube exit). Then we can define dimensionless variables and differential operators as follows:

$$\check{x} = \frac{x}{l_0} \qquad \check{y} = \frac{y}{l_0} \qquad \check{z} = \frac{z}{l_0} \qquad \check{t} = \frac{v_0 t}{l_0} \tag{3.7-3}$$

$$\check{\mathbf{v}} = \frac{\mathbf{v}}{v_0} \qquad \check{\mathscr{P}} = \frac{\mathscr{P} - \mathscr{P}_0}{\rho v_0^2} \quad \text{or} \quad \check{\mathscr{P}} = \frac{\mathscr{P} - \mathscr{P}_0}{\mu v_0 / l_0} \tag{3.7-4}$$

$$\check{\nabla} = l_0 \nabla = \boldsymbol{\delta}_x (\partial/\partial \check{x}) + \boldsymbol{\delta}_y (\partial/\partial \check{y}) + \boldsymbol{\delta}_z (\partial/\partial \check{z}) \tag{3.7-5}$$

$$\check{\nabla}^2 = (\partial^2/\partial \check{x}^2) + (\partial^2/\partial \check{y}^2) + (\partial^2/\partial \check{z}^2) \tag{3.7-6}$$

$$D/D\check{t} = (l_0/v_0)(D/Dt) \tag{3.7-7}$$

We have suggested two choices for the dimensionless pressure, the first one being convenient for high Reynolds numbers and the second for low Reynolds numbers. When the equations of change in Eqs. 3.7-1 and 3.7-2 are rewritten in terms of the dimensionless quantities, they become

$$(\check{\nabla} \cdot \check{\mathbf{v}}) = 0 \tag{3.7-8}$$

$$\frac{D}{D\check{t}} \check{\mathbf{v}} = -\check{\nabla}\check{\mathscr{P}} + \left[\!\left[\frac{\mu}{l_0 v_0 \rho}\right]\!\right] \check{\nabla}^2 \check{\mathbf{v}} \tag{3.7-9a}$$

or

$$\frac{D}{D\check{t}} \check{\mathbf{v}} = -\left[\!\left[\frac{\mu}{l_0 v_0 \rho}\right]\!\right] \check{\nabla}\check{\mathscr{P}} + \left[\!\left[\frac{\mu}{l_0 v_0 \rho}\right]\!\right] \check{\nabla}^2 \check{\mathbf{v}} \tag{3.7-9b}$$

---

[1] G. Birkhoff, *Hydrodynamics,* Dover, New York (1955), Chapter IV. Our dimensional analysis procedure corresponds to Birkhoff's "complete inspectional analysis."

[2] R. W. Powell, *An Elementary Text in Hydraulics and Fluid Mechanics,* Macmillan, New York (1951), Chapter VIII; and H. Rouse and S. Ince, *History of Hydraulics,* Dover, New York (1963) have interesting historical material regarding the dimensionless groups and the persons for whom they were named.

In these dimensionless equations, the four scale factors l_0, v_0, ρ, and μ appear in one dimensionless group. The reciprocal of this group is named after a famous fluid dynamicist[3]

$$\mathrm{Re} = \left[\!\left[\frac{l_0 v_0 \rho}{\mu}\right]\!\right] = \textit{Reynolds number} \tag{3.7-10}$$

The magnitude of this dimensionless group gives an indication of the relative importance of inertial and viscous forces in the fluid system.

From the two forms of the equation of motion given in Eq. 3.7-9, we can gain some perspective on the special forms of the Navier–Stokes equation given in §3.5. Equation 3.7-9a gives the Euler equation of Eq. 3.5-9 when $\mathrm{Re} \to \infty$ and Eq. 3.7-9b gives the creeping flow equation of Eq. 3.5-8 when $\mathrm{Re} \ll 1$. The regions of applicability of these and other asymptotic forms of the equation of motion are considered further in §§4.3 and 4.4.

Additional dimensionless groups may arise in the initial and boundary conditions; two that appear in problems with fluid–fluid interfaces are

$$\mathrm{Fr} = \left[\!\left[\frac{v_0^2}{l_0 g}\right]\!\right] = \textit{Froude number} \tag{3.7-11}[4]$$

$$\mathrm{We} = \left[\!\left[\frac{\sigma}{l_0 v_0^2 \rho}\right]\!\right] = \textit{Weber number} \tag{3.7-12}[5]$$

The first of these contains the gravitational acceleration g, and the second contains the interfacial tension σ, which may enter into the boundary conditions, as described in Problem 3C.5. Still other groups may appear, such as ratios of lengths in the flow system (for example, the ratio of tube diameter to the diameter of the hole in an orifice meter).

EXAMPLE 3.7-1

Transverse Flow around a Circular Cylinder[6]

The flow of an incompressible Newtonian fluid past a circular cylinder is to be studied experimentally. We want to know how the flow patterns and pressure distribution depend on the cylinder diameter, length, the approach velocity, and the fluid density and viscosity. Show how to organize the work so that the number of experiments needed will be minimized.

SOLUTION

For the analysis we consider an idealized flow system: a cylinder of diameter D and length L, submerged in an otherwise unbounded fluid of constant density and viscosity. Initially the fluid and the cylinder are both at rest. At time $t = 0$, the cylinder is abruptly made to move with velocity v_∞ in the negative x direction. The subsequent fluid motion is analyzed by using coordinates fixed in the cylinder axis as shown in Fig. 3.7-1.

The differential equations describing the flow are the equation of continuity (Eq. 3.7-1) and the equation of motion (Eq. 3.7-2). The initial condition for $t = 0$ is:

I.C.   if $x^2 + y^2 > \frac{1}{4}D^2$ or if $|z| > \frac{1}{2}L$, $\mathbf{v} = \boldsymbol{\delta}_x v_\infty$ (3.7-13)

The boundary conditions for $t \geq 0$ and all z are:

B.C. 1   as $x^2 + y^2 + z^2 \to \infty$, $\mathbf{v} \to \boldsymbol{\delta}_x v_\infty$ (3.7-14)

B.C. 2   if $x^2 + y^2 \leq \frac{1}{4}D^2$ and $|z| \leq \frac{1}{2}L$, $\mathbf{v} = 0$ (3.7-15)

B.C. 3   as $x \to -\infty$ at $y = 0$, $\mathscr{P} \to \mathscr{P}_\infty$ (3.7-16)

---

[3] See fn. 1 in §2.2.

[4] **William Froude** (1810–1879) (rhymes with "food") studied at Oxford and worked as a civil engineer concerned with railways and steamships. The Froude number is sometimes defined as the square root of the group given in Eq. 3.7-11.

[5] **Moritz Weber** (1871–1951) (pronounced "Vayber") was a professor of naval architecture in Berlin; another dimensionless group involving the surface tension in the *capillary number*, defined as $\mathrm{Ca} = [\![\mu v_0/\sigma]\!]$.

[6] This example is adapted from R. P. Feynman, R. B. Leighton, and M. Sands, *The Feynman Lectures on Physics*, Vol. II, Addison-Wesley, Reading, Mass. (1964), §41-4.

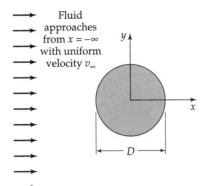

Fluid approaches from $x = -\infty$ with uniform velocity v_∞

Fig. 3.7-1. Transverse flow around a cylinder.

Now we rewrite the problem in terms of variables made dimensionless with the characteristic length D, velocity v_∞, and modified pressure $\mathcal{P}_\infty$. The resulting dimensionless equations of change are

$$(\breve{\nabla} \cdot \breve{v}) = 0, \quad \text{and} \quad \frac{\partial \breve{v}}{\partial \breve{t}} + [\breve{v} \cdot \breve{\nabla}\breve{v}] = -\breve{\nabla}\breve{\mathcal{P}} + \frac{1}{\text{Re}} \breve{\nabla}^2 \breve{v} \qquad (3.7\text{-}17, 18)$$

in which $\text{Re} = Dv_\infty\rho/\mu$. The corresponding initial and boundary conditions are:

I.C.   if $\breve{x}^2 + \breve{y}^2 > \frac{1}{4}$ or if $|\breve{z}| > \frac{1}{2}(L/D)$, $\breve{v} = \delta_x$ (3.7-19)

B.C. 1   as $\breve{x}^2 + \breve{y}^2 + \breve{z}^2 \to \infty$, $\breve{v} \to \delta_x$ (3.7-20)

B.C. 2   if $\breve{x}^2 + \breve{y}^2 \le \frac{1}{4}$ and $|\breve{z}| \le \frac{1}{2}(L/D)$, $\breve{v} = 0$ (3.7-21)

B.C. 3   as $\breve{x} \to -\infty$ at $y = 0$, $\breve{\mathcal{P}} \to 0$ (3.7-22)

If we were bright enough to be able to solve the dimensionless equations of change along with the dimensionless boundary conditions, the solutions would *have* to be of the following form:

$$\breve{v} = \breve{v}(\breve{x}, \breve{y}, \breve{z}, \breve{t}, \text{Re}, L/D) \quad \text{and} \quad \breve{\mathcal{P}} = \breve{\mathcal{P}}(\breve{x}, \breve{y}, \breve{z}, \breve{t}, \text{Re}, L/D) \qquad (3.7\text{-}23, 24)$$

That is, the dimensionless velocity and dimensionless modified pressure can depend only on the dimensionless parameters Re and L/D and the dimensionless independent variables $\breve{x}, \breve{y}, \breve{z}$, and $\breve{t}$.

This completes the dimensional analysis of the problem. We have not solved the flow problem, but have decided on a convenient set of dimensionless variables to restate the problem and suggest the form of the solution. The analysis shows that if we wish to catalog the flow patterns for flow past a cylinder, it will suffice to record them (e.g., photographically) for a series of Reynolds numbers $\text{Re} = Dv_\infty\rho/\mu$ and L/D values; thus, separate investigations into the roles of L, D, v_∞, ρ, and μ are unnecessary. Such a simplification saves a lot of time and expense. Similar comments apply to the tabulation of numerical results, in the event that one decides to make a numerical assault on the problem.[7,8]

---

[7] Analytical solutions of this problem at very small Re and infinite L/D are reviewed in L. Rosenhead (ed.), *Laminar Boundary Layers*, Oxford University Press (1963), Chapter IV. An important feature of this two-dimensional problem is the absence of a "creeping flow" solution. Thus the $[v \cdot \nabla v]$-term in the equation of motion has to be included, even in the limit as $\text{Re} \to 0$ (see Problem 3B.9). This is in sharp contrast to the situation for slow flow around a sphere (see §2.6 and §4.2) and around other finite, three-dimensional objects.

[8] For computer studies of the flow around a long cylinder, see F. H. Harlow and J. E. From, *Scientific American*, **212**, 104–110 (1965), and S. J. Sherwin and G. E. Karniadakis, *Comput. Math.*, **123**, 189–229 (1995).

Experiments involve some necessary departures from the above analysis: the stream is finite in size, and fluctuations of velocity are inevitably present at the initial state and in the upstream fluid. These fluctuations die out rapidly near the cylinder at Re < 1. For Re approaching 40 the damping of disturbances gets slower, and beyond this approximate limit unsteady flow is always observed.

The observed flow patterns at large $\breve{t}$ vary strongly with the Reynolds number as shown in Fig. 3.7-2. At Re << 1 the flow is orderly, as shown in (a). At Re of about 10, a pair of vortices appears behind the cylinder, as may be seen in (b). This type of flow persists up to about Re = 40, when there appear two "separation points," at which the streamlines separate from the solid surface. Furthermore the flow becomes permanently unsteady; vortices begin to "peel off" from the cylinder and move downstream. With further increase in Re, the vortices separate regularly from alternate sides of the cylinder, as shown in (c); such a regular array of vortices is known as a "von Kármán vortex street." At still higher Re there is a disorderly fluctuating motion (turbulence) in the wake of the cylinder, as shown in (d). Finally, at Re near 10^6, turbulence appears upstream of the separation point, and the wake abruptly narrows down as shown in (e). Clearly, the unsteady flows shown in the last three sketches would be very difficult to compute from the equations of change. It is much easier to observe them experimentally and correlate the results in terms of Eqs. 3.7-23 and 24.

Equations 3.7-23 and 24 can also be used for scale-up from a single experiment. Suppose that we wanted to predict the flow patterns around a cylinder of diameter $D_I = 5$ ft, around which air is to flow with an approach velocity $(v_\infty)_I = 30$ ft/s, by means of an ex-

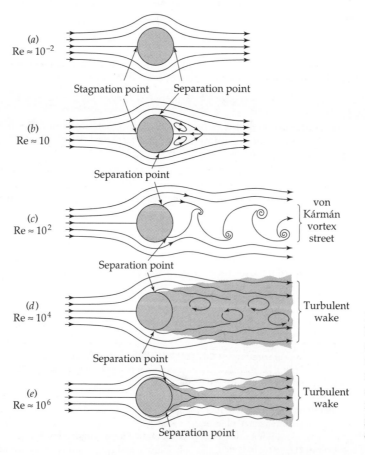

Fig. 3.7-2. The types of behavior for the flow around a cylinder, illustrating the various flow regimes that are observed as the Reynolds number increases. Regions of turbulent flow are shaded in gray.

periment on a scale model of diameter $D_{II} = 1$ ft. To have dynamic similarity, we must choose conditions such that $Re_{II} = Re_I$. Then if we use the same fluid in the small-scale experiment as in the large system, so that $\mu_{II}/\rho_{II} = \mu_I/\rho_I$, we find $(v_\infty)_{II} = 150$ ft/s as the required air velocity in the small-scale model. With the Reynolds numbers thus equalized, the flow patterns in the model and the full-scale system will look alike: that is, they are geometrically similar.

Furthermore, if Re is in the range of periodic vortex formation, the dimensionless time interval $t_v v_\infty/D$ between vortices will be the same in the two systems. Thus, the vortices will shed 25 times as fast in the model as in the full-scale system. The regularity of the vortex shedding at Reynolds numbers from about 10^2 to 10^4 is utilized commercially for precise flow metering in large pipelines.

EXAMPLE 3.7-2

Steady Flow in an Agitated Tank

It is desired to predict the flow behavior in a large, unbaffled tank of oil, shown in Fig. 3.7-3, as a function of the impeller rotation speed. We propose to do this by means of model experiments in a smaller, geometrically similar system. Determine the conditions necessary for the model studies to provide a direct means of prediction.

SOLUTION

We consider a tank of radius R, with a centered impeller of overall diameter D. At time $t = 0$, the system is stationary and contains liquid to a height H above the tank bottom. Immediately after time $t = 0$, the impeller begins rotating at a constant speed of N revolutions per minute. The drag of the atmosphere on the liquid surface is neglected. The impeller shape and initial position are described by the function $S_{imp}(r, \theta, z) = 0$.

The flow is governed by Eqs. 3.7-1 and 2, along with the initial condition

$$\text{at } t = 0, \text{ for } 0 \le r < R \text{ and } 0 < z < H, \qquad \mathbf{v} = 0 \qquad (3.7\text{-}25)$$

and the following boundary conditions for the liquid region:

tank bottom $\qquad$ at $z = 0$ and $0 \le r < R$, $\quad \mathbf{v} = 0$ $\qquad\qquad$ (3.7-26)

tank wall $\qquad$ at $r = R$, $\quad \mathbf{v} = 0$ $\qquad\qquad$ (3.7-27)

impeller surface $\qquad$ at $S_{imp}(r, \theta - 2\pi Nt, z) = 0$, $\quad \mathbf{v} = 2\pi Nr\boldsymbol{\delta}_\theta$ $\qquad$ (3.7-28)

gas–liquid interface $\qquad$ at $S_{int}(r, \theta, z, t) = 0$, $\quad (\mathbf{n} \cdot \mathbf{v}) = 0$ $\qquad$ (3.7-29)

$$\text{and} \quad \mathbf{n}p + [\mathbf{n} \cdot \boldsymbol{\tau}] = \mathbf{n}p_{atm} \qquad\qquad (3.7\text{-}30)$$

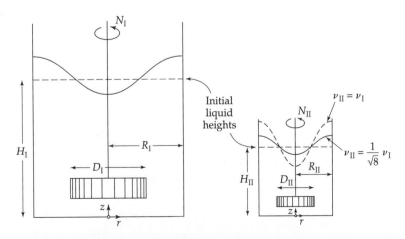

Fig. 3.7-3. Long-time average free-surface shapes, with $Re_I = Re_{II}$.

Equations 3.7-26 to 28 are the no-slip and impermeability conditions; the surface $S_{imp}(r, \theta - 2\pi Nt, z) = 0$ describes the location of the impeller after Nt rotations. Equation 3.7-29 is the condition of no mass flow through the gas–liquid interface, described by $S_{int}(r, \theta, z, t) = 0$, which has a local unit normal vector $\mathbf{n}$. Equation 3.7-30 is a force balance on an element of this interface (or a statement of the continuity of the normal component of the momentum flux tensor $\boldsymbol{\pi}$) in which the viscous contributions from the gas side are neglected. This interface is initially stationary in the plane $z = H$, and its motion thereafter is best obtained by measurement, though it is also predictable in principle by numerical solution of this equation system, which describes the initial conditions and subsequent acceleration $D\mathbf{v}/Dt$ of every fluid element.

Next we nondimensionalize the equations using the characteristic quantities $v_0 = ND$, $l_0 = D$, and $\mathcal{P}_0 = p_{atm}$ along with dimensionless polar coordinates $\check{r} = r/D$, θ, and $\check{z} = z/D$. Then the equations of continuity and motion appear as in Eqs. 3.7-8 and 9, with $Re = D^2 N\rho/\mu$. The initial condition takes the form

$$\text{at } \check{t} = 0, \text{ for } \check{r} = \left[\!\left[\frac{R}{D}\right]\!\right] \text{ and } 0 < \check{z} < \left[\!\left[\frac{H}{D}\right]\!\right], \qquad \check{\mathbf{v}} = 0 \qquad (3.7\text{-}31)$$

and the boundary conditions become:

tank bottom $\qquad$ at $\check{z} = 0$ and $0 < \check{r} < \left[\!\left[\dfrac{R}{D}\right]\!\right]$, $\qquad\qquad \check{\mathbf{v}} = 0 \qquad (3.7\text{-}32)$

tank wall $\qquad$ at $\check{r} = \left[\!\left[\dfrac{R}{D}\right]\!\right]$, $\qquad\qquad\qquad\qquad \check{\mathbf{v}} = 0 \qquad (3.7\text{-}33)$

impeller surface $\qquad$ at $\check{S}_{imp}(\check{r}, \theta - 2\pi\check{t}, \check{z}) = 0$, $\qquad \check{\mathbf{v}} = 2\pi\check{r}\boldsymbol{\delta}_\theta \qquad (3.7\text{-}34)$

gas–liquid interface $\qquad$ at $\check{S}_{int}(\check{r}, \theta, \check{z}, \check{t}) = 0$, $\qquad\qquad (\mathbf{n} \cdot \check{\mathbf{v}}) = 0 \qquad (3.7\text{-}35)$

$$\text{and} \qquad \mathbf{n}\check{\mathcal{P}} - \mathbf{n}\left[\!\left[\frac{g}{DN^2}\right]\!\right]\check{z} - \left[\!\left[\frac{\mu}{D^2 N\rho}\right]\!\right][\mathbf{n} \cdot \check{\dot{\gamma}}] = 0 \qquad (3.7\text{-}36)$$

In going from Eq. 3.7-30 to 3.7-36 we have used Newton's law of viscosity in the form of Eq. 1.2-7 (but with the last term omitted, as is appropriate for incompressible liquids). We have also used the abbreviation $\dot{\gamma} = \nabla\mathbf{v} + (\nabla\mathbf{v})^\dagger$ for the rate-of-deformation tensor, whose dimensionless Cartesian components are $\check{\dot{\gamma}} = \partial\check{v}_j/\partial\check{x}_i) + (\partial\check{v}_i/\partial\check{x}_j)$.

The quantities in double brackets are known dimensionless quantities. The function $\check{S}_{imp}(\check{r}, \theta - 2\pi\check{t}, \check{z})$ is known for a given impeller design. The unknown function $\check{S}_{int}(\check{r}, \theta, \check{z}, \check{t})$ is measurable photographically, or in principle is computable from the problem statement.

By inspection of the dimensionless equations, we find that the velocity and pressure profiles must have the form

$$\check{\mathbf{v}} = \check{\mathbf{v}}\left(\check{r}, \theta, \check{z}, \check{t}; \frac{R}{D}, \frac{H}{D}, Re, Fr\right) \qquad (3.7\text{-}37)$$

$$\check{\mathcal{P}} = \check{\mathcal{P}}\left(\check{r}, \theta, \check{z}, \check{t}; \frac{R}{D}, \frac{H}{D}, Re, Fr\right) \qquad (3.7\text{-}38)$$

for a given impeller shape and location. The corresponding locus of the free surface is given by

$$\check{S}_{int} = \check{S}_{int}\left(\check{r}, \theta, \check{z}, \check{t}; \frac{R}{D}, \frac{H}{D}, Re, Fr\right) = 0 \qquad (3.7\text{-}39)$$

in which $Re = D^2 N\rho/\mu$ and $Fr = DN^2/g$. For time-smoothed observations at large $\check{t}$, the dependence on $\check{t}$ will disappear, as will the dependence on θ for this axisymmetric tank geometry.

These results provide the necessary conditions for the proposed model experiment: the two systems must be (i) geometrically similar (same values of R/D and H/D, same impeller geometry and location), and (ii) operated at the same values of the Reynolds and Froude numbers. Condition (ii) requires that

$$\frac{D_I^2 N_I}{\nu_I} = \frac{D_{II}^2 N_{II}}{\nu_{II}} \qquad (3.7\text{-}40)$$

$$\frac{D_I N_I^2}{g_I} = \frac{D_{II} N_{II}^2}{g_{II}} \qquad (3.7\text{-}41)$$

in which the kinematic viscosity $\nu = \mu/\rho$ is used. Normally both tanks will operate in the same gravitational field $g_I = g_{II}$, so that Eq. 3.7-41 requires

$$\frac{N_{II}}{N_I} = \left(\frac{D_I}{D_{II}}\right)^{1/2} \tag{3.7-42}$$

Substitution of this into Eq. 3.7-40 gives the requirement

$$\frac{\nu_{II}}{\nu_I} = \left(\frac{D_{II}}{D_I}\right)^{3/2} \tag{3.7-43}$$

This is an important result—namely, that the smaller tank (II) requires a fluid of smaller kinematic viscosity to maintain dynamic similarity. For example, if we use a scale model with $D_{II} = \frac{1}{2}D_I$, then we need to use a fluid with kinematic viscosity $\nu_{II} = \nu_I/\sqrt{8}$ in the scaled-down experiment. Evidently the requirements for dynamic similarity are more stringent here than in the previous example, because of the additional dimensionless group Fr.

In many practical cases, Eq. 3.7-43 calls for unattainably low values of ν_{II}. Exact scale-up from a single model experiment is then not possible. In some circumstances, however, the effect of one or more dimensionless groups may be known to be small, or may be predictable from experience with similar systems; in such situations, approximate scale-up from a single experiment is still feasible.[9]

This example shows the importance of including the boundary conditions in a dimensional analysis. Here the Froude number appeared only in the free-surface boundary condition Eq. 3.7-36. Failure to use this condition would result in the omission of the restriction in Eq. 3.7-42, and one might improperly choose $\nu_{II} = \nu_I$. If one did this, with $\text{Re}_{II} = \text{Re}_I$, the Froude number in the smaller tank would be too large, and the vortex would be too deep, as shown by the dotted line in Fig. 3.7-3.

EXAMPLE 3.7-3	Show that the mean axial gradient of the modified pressure $\mathscr{P}$ for creeping flow of a fluid of constant ρ and μ through a tube of radius R, uniformly packed for a length $L >> D_p$ with solid particles of characteristic size $D_p << R$, is
Pressure Drop for Creeping Flow in a Packed Tube	

$$\frac{\Delta\langle\mathscr{P}\rangle}{L} = \frac{\mu\langle v_z\rangle}{D_p^2} K(\text{geom}) \tag{3.7-44}$$

Here $\langle\cdots\rangle$ denotes an average over a tube cross section within the packed length L, and the function $K(\text{geom})$ is a constant for a given bed geometry (i.e., a given shape and arrangement of the particles).

SOLUTION

We choose D_p as the characteristic length and $\langle v_z\rangle$ as the characteristic velocity. Then the interstitial fluid motion is determined by Eqs. 3.7-8 and 3.7-9b, with $\check{\mathbf{v}} = \mathbf{v}/\langle v_z\rangle$ and $\check{\mathscr{P}} = (\mathscr{P} - \mathscr{P}_0)D_p/\mu\langle v_z\rangle$, along with no-slip conditions on the solid surfaces and the modified pressure difference $\Delta\langle\mathscr{P}\rangle = \langle\mathscr{P}_0\rangle - \langle\mathscr{P}_L\rangle$. The solutions for $\check{\mathbf{v}}$ and $\check{\mathscr{P}}$ in creeping flow ($D_p\langle v_z\rangle\rho/\mu \to 0$) accordingly depend only on $\check{r}$, θ, and $\check{z}$ for a given particle arrangement and shape. Then the mean axial gradient

$$\frac{D_p}{L}\int_0^{L/D_p} \left(-\frac{d\langle\check{\mathscr{P}}\rangle}{d\check{z}}\right)d\check{z} = \frac{D_p}{L}(\check{\mathscr{P}}_0 - \check{\mathscr{P}}_L) \tag{3.7-45}$$

depends only on the bed geometry as long as R and L are large relative to D_p. Inserting the foregoing expression for $\check{\mathscr{P}}$, we immediately obtain Eq. 3.7-44.

---

[9] For an introduction to methods for scale-up with incomplete dynamic similarity, see R. W. Powell, *An Elementary Text in Hydraulics and Fluid Mechanics*, Macmillan, New York (1951).

QUESTIONS FOR DISCUSSION

1. What is the physical meaning of the term $\Delta x\, \Delta y (\rho v_z)|_z$ in Eq. 3.1-2? What is the physical meaning of $(\nabla \cdot \mathbf{v})$? of $(\nabla \cdot \rho \mathbf{v})$?

2. By making a mass balance over a volume element $(\Delta r)(r\Delta\theta)(\Delta z)$ derive the equation of continuity in cylindrical coordinates.

3. What is the physical meaning of the term $\Delta x\, \Delta y (\rho v_z v_x)|_z$ in Eq. 3.2-2? What is the physical meaning of $[\nabla \cdot \rho \mathbf{v v}]$?

4. What happens when f is set equal to unity in Eq. 3.5-4?

5. Equation B in Table 3.5-1 is *not* restricted to fluids with constant density, even though ρ is to the left of the substantial derivative. Explain.

6. In the tangential annular flow problem in Example 3.5-3, would you expect the velocity profiles relative to the inner cylinder to be the same in the following two situations: (i) the inner cylinder is fixed and the outer cylinder rotates with an angular velocity Ω; (ii) the outer cylinder is fixed and the inner cylinder rotates with an angular velocity $-\Omega$? Both flows are presumed to be laminar and stable.

7. Suppose that, in Example 3.6-4, there were two immiscible liquids in the rotating beaker. What would be the shape of the interface between the two liquid regions?

8. Would the system discussed in Example 3.6-5 be useful as a viscometer?

9. In Eq. 3.6-55, explain by means of a carefully drawn sketch the choice of limits in the integration and the meaning of each factor in the first integrand.

10. What factors would need to be taken into account in designing a mixing tank for use on the moon by using data from a similar tank on earth?

PROBLEMS

3A.1 Torque required to turn a friction bearing (Fig. 3A.1). Calculate the required torque in $\text{lb}_f \cdot \text{ft}$ and power consumption in horsepower to turn the shaft in the friction bearing shown in the figure. The length of the bearing surface on the shaft is 2 in, and the shaft is rotating at 200 rpm. The viscosity of the lubricant is 200 cp, and its density is 50 lb_m/ft^3. Neglect the effect of eccentricity.

Answers: 0.32 $\text{lb}_f \cdot \text{ft}$; 0.012 hp = 0.009 kW

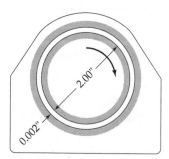

Fig. 3A.1. Friction bearing.

3A.2 Friction loss in bearings.[1] Each of two screws on a large motor-ship is driven by a 4000-hp engine. The shaft that connects the motor and the screw is 16 in. in diameter

---

[1] This problem was contributed by Prof. E. J. Crosby, University of Wisconsin.

and rests in a series of sleeve bearings that give a 0.005 in. clearance. The shaft rotates at 50 rpm, the lubricant has a viscosity of 5000 cp, and there are 20 bearings, each 1 ft in length. Estimate the fraction of engine power expended in rotating the shafts in their bearings. Neglect the effect of the eccentricity.

Answer: 0.115

3A.3 Effect of altitude on air pressure. When standing at the mouth of the Ontonagon River on the south shore of Lake Superior (602 ft above mean sea level), your portable barometer indicates a pressure of 750 mm Hg. Use the equation of motion to estimate the barometric pressure at the top of Government Peak (2023 ft above mean sea level) in the nearby Porcupine Mountains. Assume that the temperature at lake level is 70°F and that the temperature decreases with increasing altitude at a steady rate of 3°F per 1000 feet. The gravitational acceleration at the south shore of Lake Superior is about 32.19 ft/s^2, and its variation with altitude may be neglected in this problem.

Answer: 713 mm Hg = $9.49 \times 10^4\, \text{N}/\text{m}^2$

3A.4 Viscosity determination with a rotating-cylinder viscometer. It is desired to measure the viscosities of sucrose solutions of about 60% concentration by weight at about 20°C with a rotating-cylinder viscometer such as that shown in Fig. 3.5-1. This instrument has an inner cylinder 4.000 cm in diameter surrounded by a rotating concentric cylinder 4.500 cm in diameter. The length L is 4.00 cm. The viscosity of a 60% sucrose solution at 20°C is about 57 cp, and its density is about 1.29 g/cm³.

On the basis of past experience it seems possible that end effects will be important, and it is therefore decided to calibrate the viscometer by measurements on some known solutions of approximately the same viscosity as those of the unknown sucrose solutions.

Determine a reasonable value for the applied torque to be used in calibration if the torque measurements are reliable within 100 dyne/cm and the angular velocity can be measured within 0.5%. What will be the resultant angular velocity?

3A.5 Fabrication of a parabolic mirror. It is proposed to make a backing for a parabolic mirror, by rotating a pan of slow-hardening plastic resin at constant speed until it hardens. Calculate the rotational speed required to produce a mirror of focal length $f = 100$ cm. The focal length is one-half the radius of curvature at the axis, which in turn is given by

$$r_c = \left[1 + \left(\frac{dz}{dr}\right)^2\right]^{3/2}\left(\frac{d^2z}{dr^2}\right)^{-1} \qquad \text{(3A.5-1)}$$

Answer: 21.1 rpm

3A.6 Scale-up of an agitated tank. Experiments with a small-scale agitated tank are to be used to design a geometrically similar installation with linear dimensions 10 times as large. The fluid in the large tank will be a heavy oil with $\mu = 13.5$ cp and $\rho = 0.9$ g/cm³. The large tank is to have an impeller speed of 120 rpm.

(a) Determine the impeller speed for the small-scale model, in accordance with the criteria for scale-up given in Example 3.7-2.

(b) Determine the operating temperature for the model if water is to be used as the stirred fluid.

Answers: **(a)** 380 rpm, **(b)** $T = 60°C$

3A.7 Air entrainment in a draining tank (Fig. 3A.7). A molasses storage tank 60 ft in diameter is to be built with a draw-off line 1 ft in diameter, 4 ft from the sidewall of the

tank and extending vertically upward 1 ft from the tank bottom. It is known from experience that, as molasses is withdrawn from the tank, a vortex will form, and, as the liquid level drops, this vortex will ultimately reach the draw-off pipe, allowing air to be sucked into the molasses. This is to be avoided.

It is proposed to predict the minimum liquid level at which this entrainment can be avoided, at a draw-off rate of 800 gal/min, by a model study using a smaller tank. For convenience, water at 68°F is to be used for the fluid in the model study.

Determine the proper tank dimensions and operating conditions for the model if the density of the molasses is 1.286 g/cm³ and its viscosity is 56.7 cp. It may be assumed that, in either the full-size tank or the model, the vortex shape is dependent only on the amount of the liquid in the tank and the draw-off rate; that is, the vortex establishes itself very rapidly.

3B.1 Flow between coaxial cylinders and concentric spheres.

(a) The space between two coaxial cylinders is filled with an incompressible fluid at constant temperature. The radii of the inner and outer wetted surfaces are κR and R, respectively. The angular velocities of rotation of the inner and outer cylinders are Ω_i and Ω_o. Determine the velocity distribution in the fluid and the torques on the two cylinders needed to maintain the motion.

(b) Repeat part (a) for two concentric spheres.

Answers:

(a) $v_\theta = \dfrac{\kappa R}{1 - \kappa^2}\left[(\Omega_o - \Omega_i\kappa^2)\left(\dfrac{r}{\kappa R}\right) + (\Omega_i - \Omega_o)\left(\dfrac{\kappa R}{r}\right)\right]$

(b) $v_\phi = \dfrac{\kappa R}{1 - \kappa^3}\left[(\Omega_o - \Omega_i\kappa^3)\left(\dfrac{r}{\kappa R}\right) + (\Omega_i - \Omega_o)\left(\dfrac{\kappa R}{r}\right)^2\right]\sin\theta$

3B.2 Laminar flow in a triangular duct (Fig. 3B.2).[2] One type of compact heat exchanger is shown in Fig. 3B.2(a). In order to analyze the performance of such an apparatus, it is necessary to understand the flow in a duct whose cross section is an equilateral triangle. This is done most easily by installing a coordinate system as shown in Fig. 3B.2(b).

(a) Verify that the velocity distribution for the laminar flow of a Newtonian fluid in a duct of this type is given by

$$v_z = \frac{(\mathscr{P}_0 - \mathscr{P}_L)}{4\mu LH}(y - H)(3x^2 - y^2) \qquad \text{(3B.2-1)}$$

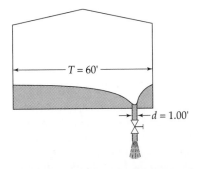

Fig. 3A.7. Draining of a molasses tank.

---

[2] An alternative formulation of the velocity profile is given by L. D. Landau and E. M. Lifshitz, *Fluid Mechanics*, Pergamon, Oxford, 2nd edition (1987), p. 54.

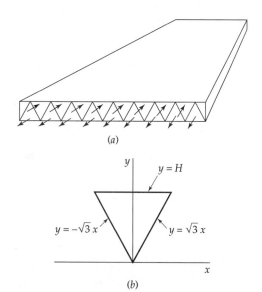

(a)

(b)

Fig. 3B.2. (a) Compact heat-exchanger element, showing channels of a triangular cross section; (b) coordinate system for an equilateral-triangular duct.

(b) From Eq. 3B.2-1 find the average velocity, maximum velocity, and mass flow rate.

Answers: **(b)** $\langle v_z \rangle = \dfrac{(\mathcal{P}_0 - \mathcal{P}_L)H^2}{60\mu L} = \dfrac{9}{20}v_{z,\text{max}};$

$$w = \frac{\sqrt{3}(\mathcal{P}_0 - \mathcal{P}_L)H^4 \rho}{180\mu L}$$

3B.3   Laminar flow in a square duct.

(a) A straight duct extends in the z direction for a length L and has a square cross section, bordered by the lines $x = \pm B$ and $y = \pm B$. A colleague has told you that the velocity distribution is given by

$$v_z = \frac{(\mathcal{P}_0 - \mathcal{P}_L)B^2}{4\mu L}\left[1 - \left(\frac{x}{B}\right)^2\right]\left[1 - \left(\frac{y}{B}\right)^2\right] \quad (3B.3\text{-}1)$$

Since this colleague has occasionally given you wrong advice in the past, you feel obliged to check the result. Does it satisfy the relevant boundary conditions and the relevant differential equation?

(b) According to the review article by Berker,[3] the mass rate of flow in a square duct is given by

$$w = \frac{0.563(\mathcal{P}_0 - \mathcal{P}_L)B^4 \rho}{\mu L} \quad (3B.3\text{-}2)$$

Compare the coefficient in this expression with the coefficient that one obtains from Eq. 3B.3-1.

---

[3] R. Berker, *Handbuch der Physik*, Vol. VIII/2, Springer, Berlin (1963); see pp. 67–77 for laminar flow in conduits of noncircular cross sections. See also W. E. Stewart, *AIChE Journal*, **8**, 425–428 (1962).

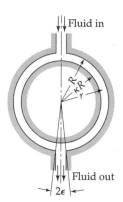

Fluid in

Fluid out

2ϵ

Fig. 3B.4. Creeping flow in the region between two stationary concentric spheres.

3B.4   Creeping flow between two concentric spheres

(Fig. 3B.4). A very viscous Newtonian fluid flows in the space between two concentric spheres, as shown in the figure. It is desired to find the rate of flow in the system as a function of the imposed pressure difference. Neglect end effects and postulate that v_θ depends only on r and θ with the other velocity components zero.

(a) Using the equation of continuity, show that $v_\theta \sin \theta = u(r)$, where $u(r)$ is a function of r to be determined.

(b) Write the θ-component of the equation of motion for this system, assuming the flow to be slow enough that the $[\mathbf{v} \cdot \nabla \mathbf{v}]$ term is negligible. Show that this gives

$$0 = -\frac{1}{r}\frac{\partial \mathcal{P}}{\partial \theta} + \mu\left[\frac{1}{\sin\theta}\frac{1}{r^2}\frac{d}{dr}\left(r^2\frac{du}{dr}\right)\right] \quad (3B.4\text{-}1)$$

(c) Separate this into two equations

$$\sin\theta\,\frac{d\mathcal{P}}{d\theta} = B; \qquad \frac{\mu}{r}\frac{d}{dr}\left(r^2\frac{du}{dr}\right) = B \quad (3B.4\text{-}2, 3)$$

where B is the separation constant, and solve the two equations to get

$$B = \frac{\mathcal{P}_2 - \mathcal{P}_1}{2\ln\,\cot\frac{1}{2}\varepsilon} \quad (3B.4\text{-}4)$$

$$u(r) = \frac{(\mathcal{P}_1 - \mathcal{P}_2)R}{4\mu\ln\,\cot(\varepsilon/2)}\left[\left(1 - \frac{r}{R}\right) + \kappa\left(1 - \frac{R}{r}\right)\right] \quad (3B.4\text{-}5)$$

where $\mathcal{P}_1$ and $\mathcal{P}_2$ are the values of the modified pressure at $\theta = \varepsilon$ and $\theta = \pi - \varepsilon$, respectively.

(d) Use the results above to get the mass rate of flow

$$w = \frac{\pi(\mathcal{P}_1 - \mathcal{P}_2)R^3(1 - \kappa)^3 \rho}{12\mu\ln\,\cot(\varepsilon/2)} \quad (3B.4\text{-}6)$$

3B.5   Parallel-disk viscometer (Fig. 3B.5).
A fluid, whose viscosity is to be measured, is placed in the gap of thickness B between the two disks of radius R. One measures the torque T_z required to turn the upper disk at an angular velocity Ω. Develop the formula for deducing the viscosity from these measurements. Assume creeping flow.

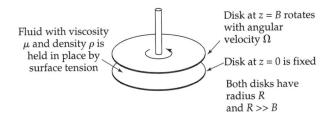

Fluid with viscosity μ and density ρ is held in place by surface tension

Disk at $z = B$ rotates with angular velocity Ω

Disk at $z = 0$ is fixed

Both disks have radius R and $R \gg B$

Fig. 3B.5. Parallel-disk viscometer.

(a) Postulate that for small values of Ω the velocity profiles have the form $v_r = 0$, $v_z = 0$, and $v_\theta = rf(z)$; why does this form for the tangential velocity seem reasonable? Postulate further that $\mathcal{P} = \mathcal{P}(r, z)$. Write down the resulting simplified equations of continuity and motion.

(b) From the θ-component of the equation of motion, obtain a differential equation for $f(z)$. Solve the equation for $f(z)$ and evaluate the constants of integration. This leads ultimately to the result $v_\theta = \Omega r(z/B)$. Could you have guessed this result?

(c) Show that the desired working equation for deducing the viscosity is $\mu = 2BT_z / \pi \Omega R^4$.

(d) Discuss the advantages and disadvantages of this instrument.

3B.6 Circulating axial flow in an annulus (Fig. 3B.6). A rod of radius κR moves upward with a constant velocity v_0 through a cylindrical container of inner radius R containing a Newtonian liquid. The liquid circulates in the cylinder, moving upward along the moving central rod and moving downward along the fixed container wall. Find the velocity distribution in the annular region, far from the end disturbances. Flows similar to this occur in the seals of some reciprocating machinery—for example, in the annular space between piston rings.

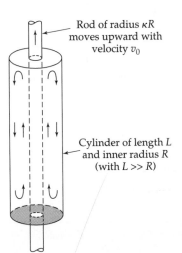

Rod of radius κR moves upward with velocity v_0

Cylinder of length L and inner radius R (with $L \gg R$)

Fig. 3B.6. Circulating flow produced by an axially moving rod in a closed annular region.

(a) First consider the problem where the annular region is quite narrow—that is, where κ is just slightly less than unity. In that case the annulus may be approximated by a thin plane slit and the curvature can be neglected. Show that in this limit, the velocity distribution is given by

$$\frac{v_z}{v_0} = 3\left(\frac{\xi - \kappa}{1 - \kappa}\right)^2 - 4\left(\frac{\xi - \kappa}{1 - \kappa}\right) + 1 \qquad (3B.6\text{-}1)$$

where $\xi = r/R$.

(b) Next work the problem without the thin-slit assumption. Show that the velocity distribution is given by

$$\frac{v_z}{v_0} = \frac{(1 - \xi^2)\left(1 - \dfrac{2\kappa^2}{1 - \kappa^2} \ln \dfrac{1}{\kappa}\right) - (1 - \kappa^2) \ln \dfrac{1}{\xi}}{(1 - \kappa^2) - (1 + \kappa^2) \ln \dfrac{1}{\kappa}} \qquad (3B.6\text{-}2)$$

3B.7 Momentum fluxes for creeping flow into a slot (Fig. 3.B-7). An incompressible Newtonian liquid is flowing very slowly into a thin slot of thickness $2B$ (in the y direction) and width W (in the z direction). The mass rate of flow in the slot is w. From the results of Problem 2B.3 it can be shown that the velocity distribution within the slot is

$$v_x = \frac{3w}{4BW\rho}\left[1 - \left(\frac{y}{B}\right)^2\right] \qquad v_y = 0 \quad v_z = 0 \qquad (3B.7\text{-}1)$$

at locations not too near the inlet. In the region outside the slot the components of the velocity for *creeping flow* are

$$v_x = -\frac{2w}{\pi W\rho} \frac{x^3}{(x^2 + y^2)^2} \qquad (3B.7\text{-}2)$$

$$v_y = -\frac{2w}{\pi W\rho} \frac{x^2 y}{(x^2 + y^2)^2} \qquad (3B.7\text{-}3)$$

$$v_z = 0 \qquad (3B.7\text{-}4)$$

Equations 3B.7-1 to 4 are only approximate in the region near the slot entry for both $x \geq 0$ and $x \leq 0$.

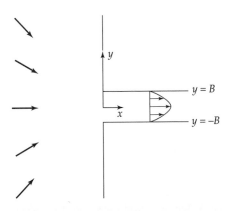

Fig. 3B.7. Flow of a liquid into a slot from a semi-infinite region $x < 0$.

(a) Find the components of the convective momentum flux $\rho\mathbf{vv}$ inside and outside the slot.

(b) Evaluate the xx-component of $\rho\mathbf{vv}$ at $x = -a, y = 0$.

(c) Evaluate the xy-component of $\rho\mathbf{vv}$ at $x = -a, y = +a$.

(d) Does the total flow of kinetic energy through the plane $x = -a$ equal the total flow of kinetic energy through the slot?

(e) Verify that the velocity distributions given in Eqs. 3B.7-1 to 4 satisfy the relation $(\nabla \cdot \mathbf{v}) = 0$.

(f) Find the normal stress τ_{xx} at the plane $y = 0$ and also on the solid surface at $x = 0$.

(g) Find the shear stress τ_{yx} on the solid surface at $x = 0$. Is this result surprising? Does sketching the velocity profile v_y vs. x at some plane $y = a$ assist in understanding the result?

3B.8   Velocity distribution for creeping flow toward a slot (Fig. 3B.7).[4] It is desired to get the velocity distribution given for the upstream region in the previous problem. We postulate that $v_\theta = 0$, $v_z = 0$, $v_r = v_r(r, \theta)$, and $\mathcal{P} = \mathcal{P}(r, \theta)$.

(a) Show that the equation of continuity in cylindrical coordinates gives $v_r = f(\theta)/r$, where $f(\theta)$ is a function of θ for which $df/d\theta = 0$ at $\theta = 0$, and $f = 0$ at $\theta = \pi/2$.

(b) Write the r- and θ-components of the creeping flow equation of motion, and insert the expression for $f(\theta)$ from (a).

(c) Differentiate the r-component of the equation of motion with respect to θ and the θ-component with respect to r. Show that this leads to

$$\frac{d^3f}{d\theta^3} + 4\frac{df}{d\theta} = 0 \tag{3B.8-1}$$

(d) Solve this differential equation and obtain an expression for $f(\theta)$ containing three integration constants.

(e) Evaluate the integration constants by using the two boundary conditions in (a) and the fact that the total mass-flow rate through any cylindrical surface must equal w. This gives

$$v_r = -\frac{2w}{\pi W\rho r}\cos^2\theta \tag{3B.8-2}$$

(f) Next from the equations of motion in (b) obtain $\mathcal{P}(r, \theta)$ as

$$\mathcal{P}(r, \theta) = \mathcal{P}_\infty - \frac{2\mu w}{\pi W\rho r^2}\cos 2\theta \tag{3B.8-3}$$

What is the physical meaning of $\mathcal{P}_\infty$?

(g) Show that the total normal stress exerted on the solid surface at $\theta = \pi/2$ is

$$(p + \tau_{\theta\theta})\big|_{\theta=\pi/2} = p_\infty + \frac{2\mu w}{\pi W\rho r^2} \tag{3B.8-4}$$

(h) Next evaluate $\tau_{\theta r}$ on the same solid surface.

(i) Show that the velocity profile obtained in Eq. 3B.8-2 is the equivalent to Eqs. 3B.7-2 and 3.

3B.9   Slow transverse flow around a cylinder (see Fig. 3.7-1). An incompressible Newtonian fluid approaches a stationary cylinder with a uniform, steady velocity v_∞ in the positive x direction. When the equations of change are solved for creeping flow, the following expressions[5] are found for the pressure and velocity in the immediate vicinity of the cylinder (they are *not* valid at large distances):

$$p(r, \theta) = p_\infty - C\mu\frac{v_\infty \cos\theta}{r} - \rho gr\sin\theta \tag{3B.9-1}$$

$$v_r = Cv_\infty\left[\frac{1}{2}\ln\left(\frac{r}{R}\right) - \frac{1}{4} + \frac{1}{4}\left(\frac{R}{r}\right)^2\right]\cos\theta \tag{3B.9-2}$$

$$v_\theta = -Cv_\infty\left[\frac{1}{2}\ln\left(\frac{r}{R}\right) + \frac{1}{4} - \frac{1}{4}\left(\frac{R}{r}\right)^2\right]\sin\theta \tag{3B.9-3}$$

Here p_∞ is the pressure far from the cylinder at $y = 0$ and

$$C = \frac{2}{\ln(7.4/\mathrm{Re})} \tag{3B.9-4}$$

with the Reynolds number defined as $\mathrm{Re} = 2Rv_\infty\rho/\mu$.

(a) Use these results to get the pressure p, the shear stress $\tau_{r\theta}$, and the normal stress τ_{rr} at the surface of the cylinder.

(b) Show that the x-component of the force per unit area exerted by the liquid on the cylinder is

$$-p|_{r=R}\cos\theta + \tau_{r\theta}|_{r=R}\sin\theta \tag{3B.9-5}$$

(c) Obtain the force $F_x = 2C\pi L\mu v_\infty$ exerted in the x direction on a length L of the cylinder.

3B.10   Radial flow between parallel disks (Fig. 3B.10). A part of a lubrication system consists of two circular disks between which a lubricant flows radially. The flow takes place because of a modified pressure difference $\mathcal{P}_1 - \mathcal{P}_2$ between the inner and outer radii r_1 and r_2, respectively.

(a) Write the equations of continuity and motion for this flow system, assuming steady-state, laminar, incompressible Newtonian flow. Consider only the region $r_1 \leq r \leq r_2$ and a flow that is radially directed.

---

[4]Adapted from R. B. Bird, R. C. Armstrong, and O. Hassager, *Dynamics of Polymeric Liquids*, Vol. 1, Wiley-Interscience, New York, 2nd edition (1987), pp. 42–43.

[5] See G. K. Batchelor, *An Introduction to Fluid Dynamics*, Cambridge University Press (1967), pp. 244–246, 261.

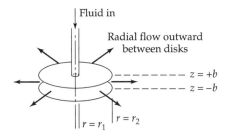

Fig. 3B.10. Outward radial flow in the space between two parallel, circular disks.

(b) Show how the equation of continuity enables one to simplify the equation of motion to give

$$-\rho \frac{\phi^2}{r^3} = -\frac{d\mathcal{P}}{dr} + \mu \frac{1}{r} \frac{d^2\phi}{dz^2} \qquad (3B.10\text{-}1)$$

in which $\phi = r v_r$ is a function of z only. Why is ϕ independent of r?

(c) It can be shown that no solution exists for Eq. 3B.10-1 unless the nonlinear term containing ϕ is omitted. Omission of this term corresponds to the "creeping flow assumption." Show that for creeping flow, Eq. 3B.10-1 can be integrated with respect to r to give

$$0 = (\mathcal{P}_1 - \mathcal{P}_2) + \left(\mu \ln \frac{r_2}{r_1} \right) \frac{d^2\phi}{dz^2} \qquad (3B.10\text{-}2)$$

(d) Show that further integration with respect to z gives

$$v_r(r, z) = \frac{(\mathcal{P}_1 - \mathcal{P}_2) b^2}{2\mu r \ln (r_2/r_1)} \left[1 - \left(\frac{z}{b} \right)^2 \right] \qquad (3B.10\text{-}3)$$

(e) Show that the mass flow rate is

$$w = \frac{4\pi (\mathcal{P}_1 - \mathcal{P}_2) b^3 \rho}{3\mu \ln (r_2/r_1)} \qquad (3B.10\text{-}4)$$

(f) Sketch the curves $\mathcal{P}(r)$ and $v_r(r, z)$.

3B.11 Radial flow between two coaxial cylinders. Consider an incompressible fluid, at constant temperature, flowing radially between two porous cylindrical shells with inner and outer radii κR and R.

(a) Show that the equation of continuity leads to $v_r = C/r$, where C is a constant.

(b) Simplify the components of the equation of motion to obtain the following expressions for the modified-pressure distribution:

$$\frac{d\mathcal{P}}{dr} = -\rho v_r \frac{dv_r}{dr} \quad \frac{d\mathcal{P}}{d\theta} = 0 \quad \frac{d\mathcal{P}}{dz} = 0 \quad (3B.11\text{-}1)$$

(c) Integrate the expression for $d\mathcal{P}/dr$ above to get

$$\mathcal{P}(r) - \mathcal{P}(R) = \tfrac{1}{2}\rho [v_r(R)]^2 \left[1 - \left(\frac{R}{r} \right)^2 \right] \quad (3B.11\text{-}2)$$

(d) Write out all the nonzero components of τ for this flow.

(e) Repeat the problem for concentric spheres.

3B.12 Pressure distribution in incompressible fluids. Penelope is staring at a beaker filled with a liquid, which for all practical purposes can be considered as incompressible; let its density be ρ_0. She tells you she is trying to understand how the pressure in the liquid varies with depth. She has taken the origin of coordinates at the liquid–air interface, with the positive z-axis pointing away from the liquid. She says to you:

"If I simplify the equation of motion for an incompressible liquid at rest, I get $0 = -dp/dz - \rho_0 g$. I can solve this and get $p = p_{atm} - \rho_0 g z$. That seems reasonable—the pressure increases with increasing depth.

"But, on the other hand, the equation of state for any fluid is $p = p(\rho, T)$, and if the system is at constant temperature, this just simplifies to $p = p(\rho)$. And, since the fluid is incompressible, $p = p(\rho_0)$, and p must be a constant throughout the fluid! How can that be?"

Clearly Penelope needs help. Provide a useful explanation.

3B.13 Flow of a fluid through a sudden contraction.

(a) An incompressible liquid flows through a sudden contraction from a pipe of diameter D_1 into a pipe of smaller diameter D_2. What does the Bernoulli equation predict for $\mathcal{P}_1 - \mathcal{P}_2$, the difference between the modified pressures upstream and downstream of the contraction? Does this result agree with experimental observations?

(b) Repeat the derivation for the isothermal horizontal flow of an ideal gas through a sudden contraction.

3B.14 Torricelli's equation for efflux from a tank (Fig. 3B.14). A large uncovered tank is filled with a liquid to a height h. Near the bottom of the tank, there is a hole that allows the fluid to exit to the atmosphere. Apply Bernoulli's equation to a streamline that extends from the surface of the liquid at the top to a point in the exit

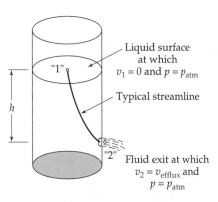

Fig. 3B.14. Fluid draining from a tank. Points "1" and "2" are on the same streamline.

stream just outside the vessel. Show that this leads to an efflux velocity $v_{\text{efflux}} = \sqrt{2gh}$. This is known as *Torricelli's equation*.

To get this result, one has to assume incompressibility (which is usually reasonable for most liquids), and that the height of the fluid surface is changing so slowly with time that the Bernoulli equation can be applied at any instant of time (the quasi-steady-state assumption).

3B.15   Shape of free surface in tangential annular flow.

(a) A liquid is in the annular space between two vertical cylinders of radii κR and R, and the liquid is open to the atmosphere at the top. Show that when the inner cylinder rotates with an angular velocity Ω_i, and the outer cylinder is fixed, the free liquid surface has the shape

$$z_R - z = \frac{1}{2g}\left(\frac{\kappa^2 R \Omega_i}{1 - \kappa^2}\right)^2 (\xi^{-2} + 4\ln\xi - \xi^2) \quad \text{(3B.15-1)}$$

in which z_R is the height of the liquid at the outer-cylinder wall, and $\xi = r/R$.

(b) Repeat (a) but with the inner cylinder fixed and the outer cylinder rotating with an angular velocity Ω_o. Show that the shape of the liquid surface is

$$z_R - z = \frac{1}{2g}\left(\frac{\kappa^2 R \Omega_o}{1 - \kappa^2}\right)^2 [(\xi^{-2} - 1) + 4\kappa^{-2}\ln\xi - \kappa^{-4}(\xi^2 - 1)]$$

$$\text{(3B.15-2)}$$

(c) Draw a sketch comparing these two liquid-surface shapes.

3B.16   Flow in a slit with uniform cross flow (Fig. 3B.16).

A fluid flows in the positive x-direction through a long flat duct of length L, width W, and thickness B, where $L \gg W \gg B$. The duct has porous walls at $y = 0$ and $y = B$, so that a constant cross flow can be maintained, with $v_y = v_0$, a constant, everywhere. Flows of this type are important in connection with separation processes using the sweep-diffusion effect. By carefully controlling the cross flow, one can concentrate the larger constituents (molecules, dust particles, etc.) near the upper wall.

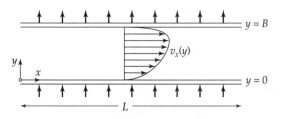

Fig. 3B.16. Flow in a slit of length L, width W, and thickness B. The walls at $y = 0$ and $y = B$ are porous, and there is a flow of the fluid in the y direction, with a uniform velocity $v_y = v_0$.

(a) Show that the velocity profile for the system is given by

$$v_x = \frac{(\mathcal{P}_0 - \mathcal{P}_L)B^2}{\mu L}\frac{1}{A}\left(\frac{y}{B} - \frac{e^{Ay/B} - 1}{e^A - 1}\right) \quad \text{(3B.16-1)}$$

in which $A = Bv_0\rho/\mu$.

(b) Show that the mass flow rate in the x direction is

$$w = \frac{(\mathcal{P}_0 - \mathcal{P}_L)B^3 W\rho}{\mu L}\frac{1}{A}\left(\frac{1}{2} - \frac{1}{A} + \frac{1}{e^A - 1}\right) \quad \text{(3B.16-2)}$$

(c) Verify that the above results simplify to those of Problem 2B.3 in the limit that there is no cross flow at all (that is, $A \to 0$).

(d) A colleague has also solved this problem, but taking a coordinate system with $y = 0$ at the midplane of the slit, with the porous walls located at $y = \pm b$. His answer to part (a) above is

$$\frac{v_x}{\langle v_x \rangle} = \frac{e^{\alpha\eta} - \eta\sinh\alpha - \cosh\alpha}{(1/\alpha)\sinh\alpha - \cosh\alpha} \quad \text{(3B.16-3)}$$

in which $\alpha = bv_0\rho/\mu$ and $\eta = y/b$. Is this result equivalent to Eq. 3B.16-1?

3C.1   Parallel-disk compression viscometer[6] (Fig. 3C.-1).

A fluid fills completely the region between two circular disks of radius R. The bottom disk is fixed, and the upper disk is made to approach the lower one very slowly with a constant speed v_0, starting from a height H_0 (and $H_0 \ll R$). The instantaneous height of the upper disk is $H(t)$. It is desired to find the force needed to maintain the speed v_0.

This problem is inherently a rather complicated unsteady-state flow problem. However, a useful approximate solution can be obtained by making two simplifications in

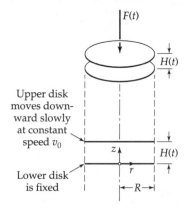

Fig. 3C.1. Squeezing flow in a parallel-disk compression viscometer.

[6] J. R. Van Wazer, J. W. Lyons, K. Y. Kim, and R. E. Colwell, *Viscosity and Flow Measurement*, Wiley-Interscience, New York (1963), pp. 292–295.

the equations of change: (i) we assume that the speed v_0 is so slow that all terms containing time derivatives can be omitted; this is the so-called "quasi-steady-state" assumption; (ii) we use the fact that $H_0 << R$ to neglect quite a few terms in the equations of change by order-of-magnitude arguments. Note that the rate of decrease of the fluid volume between the disks is $\pi R^2 v_0$, and that this must equal the rate of outflow from between the disks, which is $2\pi RH\langle v_r\rangle|_{r=R}$. Hence

$$\langle v_r\rangle|_{r=R} = \frac{Rv_0}{2H(t)} \qquad (3C.1\text{-}1)$$

We now argue that $v_r(r, z)$ will be of the order of magnitude of $\langle v_r\rangle|_{r=R}$ and that $v_z(r, z)$ is of the order of magnitude of v_0, so that

$$v_r \approx (R/H)v_0; \quad v_z \approx -v_0 \qquad (3C.1\text{-}2, 3)$$

and hence $|v_z| << v_r$. We may now estimate the order of magnitude of various derivatives as follows: as r goes from 0 to R, the radial velocity v_r goes from zero to approximately $(R/H)v_0$. By this kind of reasoning we get

$$\frac{\partial v_r}{\partial r} \approx \frac{(R/H)v_0 - 0}{R - 0} = \frac{v_0}{H} \qquad (3C.1\text{-}4)$$

$$\frac{\partial v_z}{\partial z} \approx \frac{(-v_0) - 0}{H - 0} = -\frac{v_0}{H}, \text{ etc.} \qquad (3C.1\text{-}5)$$

(a) By the above-outlined order-of-magnitude analysis, show that the continuity equation and the r-component of the equation of motion become (with g_z neglected)

continuity: $\quad \dfrac{1}{r}\dfrac{\partial}{\partial r}(rv_r) + \dfrac{\partial v_z}{\partial z} = 0 \qquad (3C.1\text{-}6)$

motion $\quad 0 = -\dfrac{dp}{dr} + \mu\dfrac{\partial^2 v_r}{\partial z^2} \qquad (3C.1\text{-}7)$

with the boundary conditions

B.C. 1: $\quad$ at $z = 0$, $\quad v_r = 0$, $\quad v_z = 0 \qquad (3C.1\text{-}8)$

B.C. 2: $\quad$ at $z = H(t)$, $\quad v_r = 0$, $\quad v_z = -v_0 \qquad (3C.1\text{-}9)$

B.C. 3: $\quad$ at $r = R$, $\quad p = p_{\text{atm}} \qquad (3C.1\text{-}10)$

(b) From Eqs. 3C.1-7 to 9 obtain

$$v_r = \frac{1}{2\mu}\left(\frac{dp}{dr}\right)z(z - H) \qquad (3C.1\text{-}11)$$

(c) Integrate Eq. 3C.1-6 with respect to z and substitute the result from Eq. 3.C.1-11 to get

$$v_0 = -\frac{H^3}{12\mu}\frac{1}{r}\frac{d}{dr}\left(r\frac{dp}{dr}\right) \qquad (3C.1\text{-}12)$$

(d) Solve Eq. 3C.1-12 to get the pressure distribution

$$p = p_{\text{atm}} + \frac{3\mu v_0 R^2}{H^3}\left[1 - \left(\frac{r}{R}\right)^2\right] \qquad (3C.1\text{-}13)$$

(e) Integrate $[(p + \tau_{zz}) - p_{\text{atm}}]$ over the moving-disk surface to find the total force needed to maintain the disk motion:

$$F(t) = \frac{3\pi\mu v_0 R^4}{2[H(t)]^3} \qquad (3C.1\text{-}14)$$

This result can be used to obtain the viscosity from the force and velocity measurements.

(f) Repeat the analysis for a viscometer that is operated in such a way that a centered, circular glob of liquid never completely fills the space between the two plates. Let the volume of the sample be V and obtain

$$F(t) = \frac{3\mu v_0 V^2}{2\pi[H(t)]^5} \qquad (3C.1\text{-}15)$$

(g) Repeat the analysis for a viscometer that is operated with constant applied force, F_0. The viscosity is then to be determined by measuring H as a function of time, and the upper-plate velocity is not a constant. Show that

$$\frac{1}{[H(t)]^2} = \frac{1}{H_0^2} + \frac{4F_0 t}{3\pi\mu R^4} \qquad (3C.1\text{-}16)$$

3C.2 Normal stresses at solid surfaces for compressible fluids. Extend example 3.1-1 to compressible fluids. Show that

$$\tau_{zz}|_{z=0} = (\tfrac{4}{3}\mu + \kappa)(\partial \ln \rho/\partial t)|_{z=0} \qquad (3C.2\text{-}1)$$

Discuss the physical significance of this result.

3C.3 Deformation of a fluid line (Fig. 3C.3). A fluid is contained in the annular space between two cylinders of radii κR and R. The inner cylinder is made to rotate with a constant angular velocity of Ω_i. Consider a line of fluid particles in the plane $z = 0$ extending from the inner cylinder to the outer cylinder and initially located at $\theta = 0$, normal to the two surfaces. How does this fluid line deform into a curve $\theta(r, t)$? What is the length, l, of the curve after N revolutions of the inner cylinder? Use Eq. 3.6-32.

Answer: $\dfrac{l}{R} = \displaystyle\int_\kappa^1 \sqrt{1 + \frac{16\pi^2 N^2}{[(1/\kappa)^2 - 1]^2\xi^4}}\, d\xi$

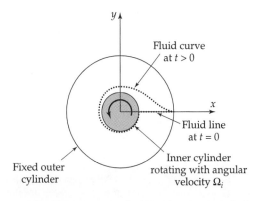

Fig. 3C.3. Deformation of a fluid line in Couette flow.

3C.4 Alternative methods of solving the Couette viscometer problem by use of angular momentum concepts (Fig. 3.6-1).

(a) By making a *shell angular-momentum balance* on a thin shell of thickness Δr, show that

$$\frac{d}{dr}(r^2 \tau_{r\theta}) = 0 \qquad (3C.4\text{-}1)$$

Next insert the appropriate expression for $\tau_{r\theta}$ in terms of the gradient of the tangential component of the velocity. Then solve the resulting differential equation with the boundary conditions to get Eq. 3.6-29.

(b) Show how to obtain Eq. 3C.4-1 from the *equation of change for angular momentum* given in Eq. 3.4-1.

3C.5 Two-phase interfacial boundary conditions. In §2.1, boundary conditions for solving viscous flow problems were given. At that point no mention was made of the role of interfacial tension. At the interface between two immiscible fluids, I and II, the following boundary condition should be used:[7]

$$\mathbf{n}^I(p^I - p^{II}) + [\mathbf{n}^I \cdot (\boldsymbol{\tau}^I - \boldsymbol{\tau}^{II})] = \mathbf{n}^I\left(\frac{1}{R_1} + \frac{1}{R_2}\right)\sigma \quad (3C.5\text{-}1)$$

This is essentially a momentum balance written for an interfacial element dS with no matter passing through it, and with no interfacial mass or viscosity. Here $\mathbf{n}^I$ is the unit vector normal to dS and pointing into phase I. The quantities R_1 and R_2 are the principal radii of curvature at dS, and each of these is positive if its center lies in phase I. The sum $(1/R_1) + (1/R_2)$ can also be expressed as $(\nabla \cdot \mathbf{n}^I)$. The quantity σ is the interfacial tension, assumed constant.

(a) Show that, for a spherical droplet of I at rest in a second medium II, *Laplace's equation*

$$p^I - p^{II} = \left(\frac{1}{R_1} + \frac{1}{R_2}\right)\sigma \qquad (3C.5\text{-}2)$$

relates the pressures inside and outside the droplet. Is the pressure in phase I greater than that in phase II, or the reverse? What is the relation between the pressures at a planar interface?

(b) Show that Eq. 3C.5-1 leads to the following dimensionless boundary condition

$$\mathbf{n}^I(\breve{\mathscr{P}}^I - \breve{\mathscr{P}}^{II}) + \mathbf{n}^I\left[\frac{\rho^{II} - \rho^I}{\rho^I}\right]\left[\frac{gl_0}{v_0^2}\right]\breve{h}$$

$$\left[\frac{\mu^I}{l_0 v_0 \rho^I}\right][\mathbf{n}^I \cdot \breve{\boldsymbol{\gamma}}^I] + \left[\frac{\mu^{II}}{l_0 v_0 \rho^{II}}\right][\mathbf{n}^I \cdot \breve{\boldsymbol{\gamma}}^{II}]$$

$$= \mathbf{n}^I\left(\frac{1}{\breve{R}_1} + \frac{1}{\breve{R}_2}\right)\left[\frac{\sigma}{l_0 v_0^2 \rho^I}\right] \qquad (3C.5\text{-}3)$$

---

[7] L. Landau and E. M. Lifshitz, *Fluid Mechanics*, Pergamon, Oxford, 2nd edition (1987), Eq. 61.13. More general formulas including the excess density and viscosity have been developed by L. E. Scriven, *Chem. Eng. Sci.*, **12**, 98–108 (1960).

in which $\breve{h} = (h - h_0)/l_0$ is the dimensionless elevation of dS, $\breve{\boldsymbol{\gamma}}^I$ and $\breve{\boldsymbol{\gamma}}^{II}$ are dimensionless rate-of-deformation tensors, and $\breve{R}_1 = R_1/l_0$ and $\breve{R}_2 = R_2/l_0$ are dimensionless radii of curvature. Furthermore

$$\breve{\mathscr{P}}^I = \frac{p^I - p_0 + \rho^I g(h - h_0)}{\rho^I v_0^2};$$

$$\breve{\mathscr{P}}^{II} = \frac{p^{II} - p_0 + \rho^{II} g(h - h_0)}{\rho^I v_0^2} \qquad (3C.5\text{-}4, 5)$$

In the above, the zero-subscripted quantities are the scale factors, valid in both phases. Identify the dimensionless groups that appear in Eq. 3C.5-3.

(c) Show how the result in **(b)** simplifies to Eq. 3.7-36 under the assumptions made in Example 3.7-2.

3D.1 Derivation of the equations of change by integral theorems (Fig. 3D.1).

(a) A fluid is flowing through some region of 3-dimensional space. Select an arbitrary "blob" of this fluid—that is, a region that is bounded by some surface $S(t)$ enclosing a volume $V(t)$, whose elements move with the local fluid velocity. Apply Newton's second law of motion to this system to get

$$\frac{d}{dt}\int_{V(t)} \rho \mathbf{v}\, dV = -\int_{S(t)} [\mathbf{n} \cdot \boldsymbol{\pi}]\, dS + \int_{V(t)} \rho \mathbf{g}\, dV \quad (3D.1\text{-}1)$$

in which the terms on the right account for the surface and volume forces acting on the system. Apply the Leibniz formula for differentiating an integral (see §A.5), recognizing that at all points on the surface of the blob, the surface velocity is identical to the fluid velocity. Next apply the Gauss theorem for a tensor (see §A.5) so that each term in the equation is a volume integral. Since the choice of the "blob" is arbitrary, all the integral signs may be removed, and the equation of motion in Eq. 3.2-9 is obtained.

(b) Derive the equation of motion by writing a momentum balance over an arbitrary region of volume V and surface S, fixed in space, through which a fluid is flowing. In doing this, just parallel the derivation given in §3.2 for a

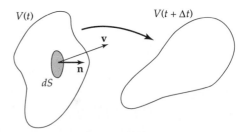

Fig. 3D.1. Moving "blob" of fluid to which Newton's second law of motion is applied. Every element of the fluid surface $dS(t)$ of the moving, deforming volume element $V(t)$ moves with the local, instantaneous fluid velocity $\mathbf{v}(t)$.

rectangular fluid element. The Gauss theorem for a tensor is needed to complete the derivation.

This problem shows that applying Newton's second law of motion to an arbitrary moving "blob" of fluid is equivalent to setting up a momentum balance over an arbitrary fixed region of space through which the fluid is moving. Both (a) and (b) give the same result as that obtained in §3.2.

(c) Derive the equation of continuity using a volume element of arbitrary shape, both moving and fixed, by the methods outlined in (a) and (b).

3D.2   The equation of change for vorticity.

(a) By taking the curl of the Navier–Stokes equation of motion (in either the D/Dt form or the $\partial/\partial t$ form), obtain an equation for the *vorticity*, $\mathbf{w} = [\nabla \times \mathbf{v}]$ of the fluid; this equation may be written in two ways:

$$\frac{D}{Dt}\mathbf{w} = \nu\nabla^2\mathbf{w} + [\mathbf{w} \cdot \nabla\mathbf{v}] \qquad (3D.2\text{-}1)$$

$$\frac{D}{Dt}\mathbf{w} = \nu\nabla^2\mathbf{w} + [\boldsymbol{\varepsilon}{:}[(\nabla\mathbf{v}) \cdot (\nabla\mathbf{v})]] \qquad (3D.2\text{-}2)$$

in which $\boldsymbol{\varepsilon}$ is a third-order tensor whose components are the permutation symbol ε_{ijk} (see §A.2) and $\nu = \mu/\rho$ is the kinematic viscosity.

(b) How do the equations in (a) simplify for two-dimensional flows?

3D.3   Alternate form of the equation of motion.[8] Show that, for an incompressible Newtonian fluid with constant viscosity, the equation of motion may be put into the form

$$4\nabla^2 p = \rho(\boldsymbol{\omega}{:}\boldsymbol{\omega}^\dagger - \dot{\boldsymbol{\gamma}}{:}\dot{\boldsymbol{\gamma}}) \qquad (3D.3\text{-}2)$$

where

$$\dot{\boldsymbol{\gamma}} = \nabla\mathbf{v} + (\nabla\mathbf{v})^\dagger \text{ and } \boldsymbol{\omega} = \nabla\mathbf{v} - (\nabla\mathbf{v})^\dagger \qquad (3D.3\text{-}2)$$

---

[8] P. G. Saffman, *Vortex Dynamics*, Cambridge University Press, corrected edition (1995).

Chapter 4

Velocity Distributions with More Than One Independent Variable

§4.1 Time-dependent flow of Newtonian fluids

§4.2° Solving flow problems using a stream function

§4.3° Flow of inviscid fluids by use of the velocity potential

§4.4° Flow near solid surfaces by boundary-layer theory

In Chapter 2 we saw that viscous flow problems with straight streamlines can be solved by shell momentum balances. In Chapter 3 we introduced the equations of continuity and motion, which provide a better way to set up problems. The method was illustrated in §3.6, but there we restricted ourselves to flow problems in which only ordinary differential equations had to be solved.

In this chapter we discuss several classes of problems that involve the solutions of partial differential equations: unsteady-state flow (§4.1), viscous flow in more than one direction (§4.2), the flow of inviscid fluids (§4.3), and viscous flow in boundary layers (§4.4). Since all these topics are treated extensively in fluid dynamics treatises, we provide here only an introduction to them and illustrate some widely used methods for problem solving.

In addition to the analytical methods given in this chapter, there is also a rapidly expanding literature on numerical methods.[1] The field of computational fluid dynamics is already playing an important role in the field of transport phenomena. The numerical and analytical methods play roles complementary to one another, with the numerical methods being indispensable for complicated practical problems.

§4.1 TIME-DEPENDENT FLOW OF NEWTONIAN FLUIDS

In §3.6 only steady-state problems were solved. However, in many situations the velocity depends on both position and time, and the flow is described by partial differential equations. In this section we illustrate three techniques that are much used in fluid dynamics, heat conduction, and diffusion (as well as in many other branches of physics and engineering). In each of these techniques the problem of solving a partial differential equation is converted into a problem of solving one or more ordinary differential equations.

---

[1] R. W. Johnson (ed.), *The Handbook of Fluid Dynamics*, CRC Press, Boca Raton, Fla. (1998);
C. Pozrikidis, *Introduction to Theoretical and Computational Fluid Dynamics*, Oxford University Press (1997).

The first example illustrates the *method of combination of variables* (or the *method of similarity solutions*). This method is useful only for semi-infinite regions, such that the initial condition and the boundary condition at infinity may be combined into a single new boundary condition.

The second example illustrates the *method of separation of variables*, in which the partial differential equation is split up into two or more ordinary differential equations. The solution is then an infinite sum of products of the solutions of the ordinary differential equations. These ordinary differential equations are usually discussed under the heading of "Sturm-Liouville" problems in intermediate-level mathematics textbooks.[1]

The third example demonstrates the *method of sinusoidal response*, which is useful in describing the way a system responds to external periodic disturbances.

The illustrative examples are chosen for their physical simplicity, so that the major focus can be on the mathematical methods. Since all the problems discussed here are linear in the velocity, Laplace transforms can also be used, and readers familiar with this subject are invited to solve the three examples in this section by that technique.

EXAMPLE 4.1-1

Flow near a Wall Suddenly Set in Motion

A semi-infinite body of liquid with constant density and viscosity is bounded below by a horizontal surface (the xz-plane). Initially the fluid and the solid are at rest. Then at time $t = 0$, the solid surface is set in motion in the positive x direction with velocity v_0 as shown in Fig. 4.1-1. Find the velocity v_x as a function of y and t. There is no pressure gradient or gravity force in the x direction, and the flow is presumed to be laminar.

SOLUTION

For this system $v_x = v_x(y, t)$, $v_y = 0$, and $v_z = 0$. Then from Table B.4 we find that the equation of continuity is satisfied directly, and from Table B.5 we get

$$\frac{\partial v_x}{\partial t} = \nu \frac{\partial^2 v_x}{\partial y^2} \tag{4.1-1}$$

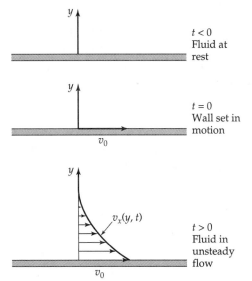

$t < 0$
Fluid at rest

$t = 0$
Wall set in motion

$v_x(y, t)$

$t > 0$
Fluid in unsteady flow

Fig. 4.1-1. Viscous flow of a fluid near a wall suddenly set in motion.

---

[1] See, for example, M. D. Greenberg, *Foundations of Applied Mathematics*, Prentice-Hall, Englewood Cliffs, N.J. (1978), §20.3.

in which $\nu = \mu/\rho$. The initial and boundary conditions are

I.C.:	at $t \leq 0$,	$v_x = 0$	for all y	(4.1-2)
B.C. 1:	at $y = 0$,	$v_x = v_0$	for all $t > 0$	(4.1-3)
B.C. 2:	at $y = \infty$,	$v_x = 0$	for all $t > 0$	(4.1-4)

Next we introduce a dimensionless velocity $\phi = v_x/v_0$, so that Eq. 4.4-1 becomes

$$\frac{\partial \phi}{\partial t} = \nu \frac{\partial^2 \phi}{\partial y^2} \tag{4.1-5}$$

with $\phi(y, 0) = 0$, $\phi(0, t) = 1$, and $\phi(\infty, t) = 0$. Since the initial and boundary conditions contain only pure numbers, the solution to Eq. 4.1-5 has to be of the form $\phi = \phi(y, t; \nu)$. However, since ϕ is a dimensionless function, the quantities y, t, and ν must always appear in a dimensionless combination. The only dimensionless combinations of these three quantities are $y/\sqrt{\nu t}$ or powers or multiples thereof. We therefore conclude that

$$\phi = \phi(\eta), \qquad \text{where } \eta = \frac{y}{\sqrt{4\nu t}} \tag{4.1-6}$$

This is the "method of combination of (independent) variables." The "4" is included so that the final result in Eq. 4.1-14 will look neater; we know to do this only after solving the problem without it. The form of the solution in Eq. 4.1-6 is possible essentially because there is no characteristic length or time in the physical system.

We now convert the derivatives in Eq. 4.1-5 into derivatives with respect to the "combined variable" η as follows:

$$\frac{\partial \phi}{\partial t} = \frac{d\phi}{d\eta} \frac{\partial \eta}{\partial t} = -\frac{1}{2} \frac{\eta}{t} \frac{d\phi}{d\eta} \tag{4.1-7}$$

$$\frac{\partial \phi}{\partial y} = \frac{d\phi}{d\eta} \frac{\partial \eta}{\partial y} = \frac{d\phi}{d\eta} \frac{1}{\sqrt{4\nu t}} \quad \text{and} \quad \frac{\partial^2 \phi}{\partial y^2} = \frac{d^2\phi}{d\eta^2} \frac{1}{4\nu t} \tag{4.1-8}$$

Substitution of these expressions into Eq. 4.1-5 then gives

$$\frac{d^2\phi}{d\eta^2} + 2\eta \frac{d\phi}{d\eta} = 0 \tag{4.1-9}$$

This is an ordinary differential equation of the type given in Eq. C.1-8, and the accompanying boundary conditions are

| B.C. 1: | at $\eta = 0$, | $\phi = 1$ | (4.1-10) |
| B.C. 2: | at $\eta = \infty$, | $\phi = 0$ | (4.1-11) |

The first of these boundary conditions is the same as Eq. 4.1-3, and the second includes Eqs. 4.1-2 and 4. If now we let $d\phi/d\eta = \psi$, we get a first-order separable equation for ψ, and it may be solved to give

$$\psi = \frac{d\phi}{d\eta} = C_1 \exp(-\eta^2) \tag{4.1-12}$$

A second integration then gives

$$\phi = C_1 \int_0^\eta \exp(-\bar{\eta}^2) \, d\bar{\eta} + C_2 \tag{4.1-13}$$

The choice of 0 for the lower limit of the integral is arbitrary; another choice would lead to a different value of C_2, which is still undetermined. Note that we have been careful to use an overbar for the variable of integration ($\bar{\eta}$) to distinguish it from the η in the upper limit.

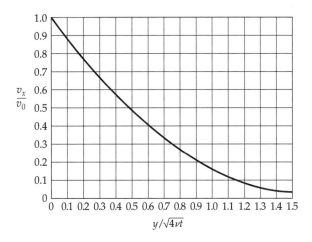

Fig. 4.1-2. Velocity distribution, in dimensionless form, for flow in the neighborhood of a wall suddenly set in motion.

Application of the two boundary conditions makes it possible to evaluate the two integration constants, and we get finally

$$\phi(\eta) = 1 - \frac{\int_0^\eta \exp(-\overline{\eta}^2)\, d\overline{\eta}}{\int_0^\infty \exp(-\overline{\eta}^2)\, d\overline{\eta}} = 1 - \frac{2}{\sqrt{\pi}} \int_0^\eta \exp(-\overline{\eta}^2)\, d\overline{\eta} = 1 - \operatorname{erf} \eta \qquad (4.1\text{-}14)$$

The ratio of integrals appearing here is called the *error function*, abbreviated erf η (see §C.6). It is a well-known function, available in mathematics handbooks and computer software programs. When Eq. 4.1-14 is rewritten in the original variables, it becomes

$$\frac{v_x(y, t)}{v_0} = 1 - \operatorname{erf} \frac{y}{\sqrt{4\nu t}} = \operatorname{erfc} \frac{y}{\sqrt{4\nu t}} \qquad (4.1\text{-}15)$$

in which erfc η is called the *complementary error function*. A plot of Eq. 4.1-15 is given in Fig. 4.1-2. Note that, by plotting the result in terms of dimensionless quantities, only one curve is needed.

The complementary error function erfc η is a monotone decreasing function that goes from 1 to 0 and drops to 0.01 when η is about 2.0. We can use this fact to define a "boundary-layer thickness" δ as that distance y for which v_x has dropped to a value of $0.01v_0$. This gives $\delta = 4\sqrt{\nu t}$ as a natural length scale for the diffusion of momentum. This distance is a measure of the extent to which momentum has "penetrated" into the body of the fluid. Note that this boundary-layer thickness is proportional to the square root of the elapsed time.

EXAMPLE 4.1-2

Unsteady Laminar Flow Between Two Parallel Plates

It is desired to re-solve the preceding illustrative example, but with a fixed wall at a distance b from the moving wall at $y = 0$. This flow system has a steady-state limit as $t \to \infty$, whereas the problem in Example 4.1-1 did not.

SOLUTION

As in Example 4.1-1, the equation for the x-component of the velocity is

$$\frac{\partial v_x}{\partial t} = \nu \frac{\partial^2 v_x}{\partial y^2} \qquad (4.1\text{-}16)$$

The boundary conditions are now

I.C.:	at $t \le 0$,	$v_x = 0$	for $0 \le y \le b$	(4.1-17)
B.C. 1:	at $y = 0$,	$v_x = v_0$	for all $t > 0$	(4.1-18)
B.C. 2:	at $y = b$,	$v_x = 0$	for all $t > 0$	(4.1-19)

It is convenient to introduce the following dimensionless variables:

$$\phi = \frac{v_x}{v_0}; \qquad \eta = \frac{y}{b}; \qquad \tau = \frac{\nu t}{b^2} \tag{4.1-20}$$

The choices for dimensionless velocity and position ensure that these variables will go from 0 to 1. The choice of the dimensionless time is made so that there will be no parameters occurring in the transformed partial differential equation:

$$\frac{\partial \phi}{\partial \tau} = \frac{\partial^2 \phi}{\partial \eta^2} \tag{4.1-21}$$

The initial condition is $\phi = 0$ at $\tau = 0$, and the boundary conditions are $\phi = 1$ at $\eta = 0$ and $\phi = 0$ at $\eta = 1$.

We know that at infinite time the system attains a steady-state velocity profile $\phi_\infty(\eta)$ so that at $\tau = \infty$ Eq. 4.1-21 becomes

$$0 = \frac{d^2\phi_\infty}{d\eta^2} \tag{4.1-22}$$

with $\phi_\infty = 1$ at $\eta = 0$, and $\phi_\infty = 0$ at $\eta = 1$. We then get

$$\phi_\infty = 1 - \eta \tag{4.1-23}$$

for the steady-state limiting profile.

We then can write

$$\phi(\eta, \tau) = \phi_\infty(\eta) - \phi_t(\eta, \tau) \tag{4.1-24}$$

where ϕ_t is the transient part of the solution, which fades out as time goes to infinity. Substitution of this expression into the original differential equation and boundary conditions then gives for ϕ_t

$$\frac{\partial \phi_t}{\partial \tau} = \frac{\partial^2 \phi_t}{\partial \eta^2} \tag{4.1-25}$$

with $\phi_t = \phi_\infty$ at $\tau = 0$, and $\phi_t = 0$ at $\eta = 0$ and 1.

To solve Eq. 4.1-25 we use the "method of separation of (dependent) variables," in which we assume a solution of the form

$$\phi_t = f(\eta)g(\tau) \tag{4.1-26}$$

Substitution of this trial solution into Eq. 4.1-25 and then division by the product fg gives

$$\frac{1}{g}\frac{dg}{d\tau} = \frac{1}{f}\frac{d^2f}{d\eta^2} \tag{4.1-27}$$

The left side is a function of τ alone, and the right side is a function of η alone. This means that both sides must equal a constant. We choose to designate the constant as $-c^2$ (we could equally well use c or $+c^2$, but experience tells us that these choices make the subsequent mathematics somewhat more complicated). Equation 4.1-27 can then be separated into two equations

$$\frac{dg}{d\tau} = -c^2 g \tag{4.1-28}$$

$$\frac{d^2f}{d\eta^2} + c^2 f = 0 \tag{4.1-29}$$

These equations have the following solutions (see Eqs. C.1-1 and 3):

$$g = Ae^{-c^2\tau} \tag{4.1-30}$$

$$f = B \sin c\eta + C \cos c\eta \tag{4.1-31}$$

in which A, B, and C are constants of integration.

We now apply the boundary and initial conditions in the following way:

B.C. 1: Because $\phi_t = 0$ at $\eta = 0$, the function f must be zero at $\eta = 0$. Hence C must be zero.

B.C. 2: Because $\phi_t = 0$ at $\eta = 1$, the function f must be zero at $\eta = 1$. This will be true if $B = 0$ or if $\sin c$ is zero. The first choice would lead to $f = 0$ for all η, and that would be physically unacceptable. Therefore we make the second choice, which leads to the requirement that $c = 0, \pm\pi, \pm2\pi, \pm3\pi, \cdots$. We label these various admissible values of c (called "eigenvalues") as c_n and write

$$c_n = n\pi, \qquad \text{with } n = 0, \pm1, \pm2, \pm3, \cdots \tag{4.1-32}$$

There are thus many admissible functions f_n (called "eigenfunctions") that satisfy Eq. 4.1-29 and the boundary conditions; namely,

$$f_n = B_n \sin n\pi\eta, \qquad \text{with } n = 0, \pm1, \pm2, \pm3, \cdots \tag{4.1-33}$$

The corresponding functions satisfying Eq. 4.1-28 are called g_n and are given by

$$g_n = A_n \exp(-n^2\pi^2\tau), \qquad \text{with } n = 0, \pm1, \pm2, \pm3, \cdots \tag{4.1-34}$$

I.C.: The combinations $f_n g_n$ satisfy the partial differential equation for ϕ_t in Eq. 4.1-25, and so will any superposition of such products. Therefore we write for the solution of Eq. 4.1-25

$$\phi_t = \sum_{n=-\infty}^{+\infty} D_n \exp(-n^2\pi^2\tau) \sin n\pi\eta \tag{4.1-35}$$

in which the expansion coefficients $D_n = A_n B_n$ have yet to be determined. In the sum, the term $n = 0$ does not contribute; also since $\sin(-n)\pi\eta = -\sin(+n)\pi\eta$, we may omit all the terms with negative values of n. Hence, Eq. 4.1-35 becomes

$$\phi_t = \sum_{n=1}^{\infty} D_n \exp(-n^2\pi^2\tau) \sin n\pi\eta \tag{4.1-36}$$

According to the initial condition, $\phi_t = 1 - \eta$ at $\tau = 0$, so that

$$1 - \eta = \sum_{n=1}^{\infty} D_n \sin n\pi\eta \tag{4.1-37}$$

We now have to determine all the D_n from this one equation! This is done by multiplying both sides of the equation by $\sin m\pi\eta$, where m is an integer, and then integrating over the physically pertinent range from $\eta = 0$ to $\eta = 1$, thus:

$$\int_0^1 (1 - \eta) \sin m\pi\eta\, d\eta = \sum_{n=1}^{\infty} D_n \int_0^1 \sin n\pi\eta \sin m\pi\eta\, d\eta \tag{4.1-38}$$

The left side gives $1/m\pi$; the integrals on the right side are zero when $n \neq m$ and $\frac{1}{2}$ when $n = m$. Hence the initial condition leads to

$$D_m = \frac{2}{m\pi} \tag{4.1-39}$$

The final expression for the dimensionless velocity profile is obtained from Eqs. 4.1-24, 36, and 39 as

$$\phi(\eta, \tau) = (1 - \eta) - \sum_{n=1}^{\infty} \left(\frac{2}{n\pi}\right) \exp(-n^2\pi^2\tau) \sin n\pi\eta \tag{4.1-40}$$

The solution thus consists of a steady-state-limit term minus a transient term, which fades out with increasing time.

Those readers who are encountering the method of separation of variables for the first time will have found the above sequence of steps rather long and complicated. However, no single step in the development is particularly difficult. The final solution in Eq. 4.1-40 looks rather involved because of the infinite sum. Actually, except for very small values of the time, only the first few terms in the series contribute appreciably.

Although we do not prove it here, the solution to this problem and that of the preceding problem are closely related.[2] In the limit of vanishingly small time, Eq. 4.1-40 becomes equivalent to Eq. 4.1-15. This is reasonable, since, at very small time, in this problem the fluid is in motion only very near the wall at $y = 0$, and the fluid cannot "feel" the presence of the wall at $y = b$. Since the solution and result in Example 4.1-1 are far simpler than those of this one, they are often used to represent the system if only small times are involved. This is, of course, an approximation, but a very useful one. It is often used in heat- and mass-transport problems as well.

<table>
<tr><td>**EXAMPLE 4.1-3**

Unsteady Laminar Flow near an Oscillating Plate</td><td>A semi-infinite body of liquid is bounded on one side by a plane surface (the xz-plane). Initially the fluid and solid are at rest. At time $t = 0$ the solid surface is made to oscillate sinusoidally in the x direction with amplitude X_0 and (circular) frequency ω. That is, the displacement X of the plane from its rest position is</td></tr>
</table>

$$X(t) = X_0 \sin \omega t \tag{4.1-41}$$

and the velocity of the fluid at $y = 0$ is then

$$v_x(0, t) = \frac{dX}{dt} = X_0 \omega \cos \omega t \tag{4.1-42}$$

We designate the amplitude of the velocity oscillation by $v_0 = X_0\omega$ and rewrite Eq. 4.1-42 as

$$v_x(0, t) = v_0 \cos \omega t = v_0 \Re\{e^{i\omega t}\} \tag{4.1-43}$$

where $\Re\{z\}$ means "the real part of z."

For oscillating systems we are generally not interested in the complete solution, but only the "periodic steady state" that exists after the initial "transients" have disappeared. In this state all the fluid particles in the system will be executing sinusoidal oscillations with frequency ω, but with phase and amplitude that are functions only of position. This "periodic steady state" solution may be obtained by an elementary technique that is widely used. Mathematically it is an asymptotic solution for $t \to \infty$.

SOLUTION   Once again the equation of motion is given by

$$\frac{\partial v_x}{\partial t} = \nu \frac{\partial^2 v_x}{\partial y^2} \tag{4.1-44}$$

and the initial and boundary conditions are given by

I.C.:	at $t \le 0$,	$v_x = 0$	for all y	(4.1-45)
B.C. 1:	at $y = 0$,	$v_x = v_0 \Re\{e^{i\omega t}\}$	for all $t > 0$	(4.1-46)
B.C. 2:	at $y = \infty$,	$v_x = 0$	for all $t > 0$	(4.1-47)

The initial condition will not be needed, since we are concerned only with the fluid response after the plate has been oscillating for a long time.

We postulate an oscillatory solution of the form

$$v_x(y, t) = \Re\{v°(y)e^{i\omega t}\} \tag{4.1-48}$$

---

[2] See H. S. Carslaw and J. C. Jaeger, *Conduction of Heat in Solids*, Oxford University Press, 2nd edition (1959), pp. 308–310, for a series solution that is particularly good for short times.

Here $v°$ is chosen to be a *complex* function of y, so that $v_x(y, t)$ will differ from $v_x(0, t)$ both in amplitude and phase. We substitute this trial solution into Eq. 4.1-44 and obtain

$$\Re\{v°i\omega e^{i\omega t}\} = \nu\Re\left\{\frac{d^2v°}{dy^2}e^{i\omega t}\right\} \tag{4.1-49}$$

Next we make use of the fact that, if $\Re\{z_1 w\} = \Re\{z_2 w\}$, where z_1 and z_2 are two complex quantities and w is an arbitrary complex quantity, then $z_1 = z_2$. Then Eq. 4.1-49 becomes

$$\frac{d^2v°}{dy^2} - \left(\frac{i\omega}{\nu}\right)v° = 0 \tag{4.1-50}$$

with the following boundary conditions:

B.C. 1:   at $y = 0$, $v° = v_0$ (4.1-51)

B.C. 2:   at $y = \infty$, $v° = 0$ (4.1-52)

Equation 4.1-50 is of the form of Eq. C.1-4 and has the solution

$$v° = C_1 e^{\sqrt{i\omega/\nu}\,y} + C_2 e^{-\sqrt{i\omega/\nu}\,y} \tag{4.1-53}$$

Since $\sqrt{i} = \pm(1/\sqrt{2})(1 + i)$, this equation can be rewritten as

$$v° = C_1 e^{\sqrt{\omega/2\nu}(1+i)y} + C_2 e^{-\sqrt{\omega/2\nu}(1+i)y} \tag{4.1-54}$$

The second boundary condition requires that $C_1 = 0$, and the first boundary condition gives $C_2 = v_0$. Therefore the solution to Eq. 4.1-50 is

$$v° = v_0 e^{-\sqrt{\omega/2\nu}(1+i)y} \tag{4.1-55}$$

From this result and Eq. 4.1-48, we get

$$v_x(y, t) = \Re\{v_0 e^{-\sqrt{\omega/2\nu}(1+i)y}e^{i\omega t}\}$$
$$= v_0 e^{-\sqrt{\omega/2\nu}\,y}\Re\{e^{-i(\sqrt{\omega/2\nu}\,y - \omega t)}\} \tag{4.1-56}$$

or finally

$$v_x(y, t) = v_0 e^{-\sqrt{\omega/2\nu}\,y}\cos(\omega t - \sqrt{\omega/2\nu}\,y) \tag{4.1-57}$$

In this expression, the exponential describes the *attenuation* of the oscillatory motion—that is, the decrease in the amplitude of the fluid oscillations with increasing distance from the plate. In the argument of the cosine, the quantity $-\sqrt{\omega/2\nu}\,y$ is called the *phase shift*; that is, it describes how much the fluid oscillations at a distance y from the wall are "out-of-step" with the oscillations of the wall itself.

Keep in mind that Eq. 4.1-57 is not the complete solution to the problem as stated in Eqs. 4.1-44 to 47, but only the "periodic-steady-state" solution. The complete solution is given in Problem 4D.1.

§4.2 SOLVING FLOW PROBLEMS USING A STREAM FUNCTION

Up to this point the examples and problems have been chosen so that there was only one nonvanishing component of the fluid velocity. Solutions of the complete Navier–Stokes equation for flow in two or three dimensions are more difficult to obtain. The basic procedure is, of course, similar: one solves simultaneously the equations of continuity and motion, along with the appropriate initial and boundary conditions, to obtain the pressure and velocity profiles.

However, having both velocity and pressure as dependent variables in the equation of motion presents more difficulty in multidimensional flow problems than in the simpler ones discussed previously. It is therefore frequently convenient to eliminate the pressure by taking the curl of the equation of motion, after making use of the vector identity $[\mathbf{v} \cdot \nabla \mathbf{v}] = \frac{1}{2}\nabla(\mathbf{v} \cdot \mathbf{v}) - [\mathbf{v} \times [\nabla \times \mathbf{v}]]$, which is given in Eq. A.4-23. For fluids of constant viscosity and density, this operation gives

$$\frac{\partial}{\partial t}[\nabla \times \mathbf{v}] - [\nabla \times [\mathbf{v} \times [\nabla \times \mathbf{v}]]] = \nu\nabla^2[\nabla \times \mathbf{v}] \tag{4.2-1}$$

This is the *equation of change for the vorticity* $[\nabla \times \mathbf{v}]$; two other ways of writing it are given in Problem 3D.2.

For viscous flow problems one can then solve the vorticity equation (a third-order vector equation) together with the equation of continuity and the relevant initial and boundary conditions to get the velocity distribution. Once that is known, the pressure distribution can be obtained from the Navier–Stokes equation in Eq. 3.5-6. This method of solving flow problems is sometimes convenient even for the one-dimensional flows previously discussed (see, for example, Problem 4B.4).

For planar or axisymmetric flows the vorticity equation can be reformulated by introducing the *stream function* ψ. To do this, we express the two nonvanishing components of the velocity as derivatives of ψ in such a way that the equation of continuity is automatically satisfied (see Table 4.2-1). The component of the vorticity equation corresponding to the direction in which there is no flow then becomes a fourth-order scalar equation for ψ. The two nonvanishing velocity components can then be obtained after the equation for the scalar ψ has been found. The most important problems that can be treated in this way are given in Table 4.1-1.[1]

The stream function itself is not without interest. Surfaces of constant ψ contain the *streamlines*,[2] which in steady-state flow are the paths of fluid elements. The volumetric rate of flow between the surfaces $\psi = \psi_1$ and $\psi = \psi_2$ is proportional to $\psi_2 - \psi_1$.

In this section we consider, as an example, the steady, creeping flow past a stationary sphere, which is described by the Stokes equation of Eq. 3.5-8, valid for Re $\ll$ 1 (see the discussion right after Eq. 3.7-9). For creeping flow the second term on the left side of Eq. 4.2-1 is set equal to zero. The equation is then linear, and therefore there are many methods available for solving the problem.[3] We use the stream function method based on Eq. 4.2-1.

EXAMPLE 4.2-1

Creeping Flow around a Sphere

Use Table 4.2-1 to set up the differential equation for the stream function for the flow of a Newtonian fluid around a stationary sphere of radius R at Re $\ll$ 1. Obtain the velocity and pressure distributions when the fluid approaches the sphere in the positive z direction, as in Fig. 2.6-1.

---

[1] For a technique applicable to more general flows, see J. M. Robertson, *Hydrodynamics in Theory and Application*, Prentice-Hall, Englewood Cliffs, N.J. (1965), p. 77; for examples of three-dimensional flows using two stream functions, see Problem 4D.5 and also J. P. Sørensen and W. E. Stewart, *Chem. Eng. Sci.*, **29**, 819–825 (1974). A. Lahbabi and H.-C. Chang, *Chem. Eng. Sci.*, **40**, 434–447 (1985) dealt with high-Re flow through cubic arrays of spheres, including steady-state solutions and transition to turbulence. W. E. Stewart and M. A. McClelland, *AIChE Journal*, **29**, 947–956 (1983) gave matched asymptotic solutions for forced convection in three-dimensional flows with viscous heating.

[2] See, for example, G. K. Batchelor, *An Introduction to Fluid Dynamics*, Cambridge University Press (1967), §2.2. Chapter 2 of this book is an extensive discussion of the kinematics of fluid motion.

[3] The solution given here follows that given by L. M. Milne-Thomson, *Theoretical Hydrodynamics*, Macmillan, New York, 3rd edition (1955), pp. 555–557. For other approaches, see H. Lamb, *Hydrodynamics*, Dover, New York (1945), §§337, 338. For a discussion of unsteady flow around a sphere, see R. Berker, in *Handbuch der Physik*, Volume VIII-2, Springer, Berlin (1963), §69; or H. Villat and J. Kravtchenko, *Leçons sur les Fluides Visqueux*, Gauthier-Villars, Paris (1943), Chapter VII. The problem of finding the forces and torques on objects of arbitrary shapes is discussed thoroughly by S. Kim and S. J. Karrila, *Microhydrodynamics: Principles and Selected Applications*, Butterworth-Heinemann, Boston (1991), Chapter II.

Table 4.2-1 Equations for the Stream Function[a]

Type of motion	Coordinate system	Velocity components	Differential equations for ψ which are equivalent to the Navier-Stokes equation[b]	Expressions for operators
Two-dimensional (planar)	Rectangular with $v_z = 0$ and no z-dependence	$v_x = -\dfrac{\partial \psi}{\partial y}$ $v_y = +\dfrac{\partial \psi}{\partial x}$	(A) $\dfrac{\partial}{\partial t}(\nabla^2\psi) + \dfrac{\partial(\psi, \nabla^2\psi)}{\partial(x,y)} = \nu\nabla^4\psi$	$\nabla^2 \equiv \dfrac{\partial^2}{\partial x^2} + \dfrac{\partial^2}{\partial y^2}$ $\nabla^4\psi = \nabla^2(\nabla^2\psi)$ $= \left(\dfrac{\partial^4}{\partial x^4} + 2\dfrac{\partial^4}{\partial x^2\partial y^2} + \dfrac{\partial^4}{\partial y^4}\right)\psi$
Two-dimensional (planar)	Cylindrical with $v_z = 0$ and no z-dependence	$v_r = -\dfrac{1}{r}\dfrac{\partial \psi}{\partial \theta}$ $v_\theta = +\dfrac{\partial \psi}{\partial r}$	(B) $\dfrac{\partial}{\partial t}(\nabla^2\psi) + \dfrac{1}{r}\dfrac{\partial(\psi, \nabla^2\psi)}{\partial(r,\theta)} = \nu\nabla^4\psi$	$\nabla^2 \equiv \dfrac{\partial^2}{\partial r^2} + \dfrac{1}{r}\dfrac{\partial}{\partial r} + \dfrac{1}{r^2}\dfrac{\partial^2}{\partial \theta^2}$
Axisymmetrical	Cylindrical with $v_\theta = 0$ and no θ-dependence	$v_z = \dfrac{1}{r}\dfrac{\partial \psi}{\partial r}$ $v_r = +\dfrac{1}{r}\dfrac{\partial \psi}{\partial z}$	(C) $\dfrac{\partial}{\partial t}(E^2\psi) - \dfrac{1}{r}\dfrac{\partial(\psi, E^2\psi)}{\partial(r,z)} - \dfrac{2}{r^2}\dfrac{\partial\psi}{\partial z}E^2\psi = \nu E^4\psi$	$E^2 \equiv \dfrac{\partial^2}{\partial r^2} - \dfrac{1}{r}\dfrac{\partial}{\partial r} + \dfrac{\partial^2}{\partial z^2}$ $E^4\psi = E^2(E^2\psi)$
Axisymmetrical	Spherical with $v_\phi = 0$ and no φ-dependence	$v_r = -\dfrac{1}{r^2\sin\theta}\dfrac{\partial \psi}{\partial \theta}$ $v_\theta = +\dfrac{1}{r\sin\theta}\dfrac{\partial \psi}{\partial r}$	(D) $\dfrac{\partial}{\partial t}(E^2\psi) + \dfrac{1}{r^2\sin\theta}\dfrac{\partial(\psi, E^2\psi)}{\partial(r,\theta)}$ $- \dfrac{2E^2\psi}{r^2\sin^2\theta}\left(\dfrac{\partial\psi}{\partial r}\cos\theta - \dfrac{1}{r}\dfrac{\partial\psi}{\partial\theta}\sin\theta\right) = \nu E^4\psi$	$E^2 \equiv \dfrac{\partial^2}{\partial r^2} + \dfrac{\sin\theta}{r^2}\dfrac{\partial}{\partial\theta}\left(\dfrac{1}{\sin\theta}\dfrac{\partial}{\partial\theta}\right)$

[a] Similar relations in general orthogonal coordinates may be found in S. Goldstein, *Modern Developments in Fluid Dynamics*, Dover, N.Y. (1965), pp. 114–115; in this reference, formulas are also given for axisymmetrical flows with a nonzero component of the velocity around the axis.

[b] Here the Jacobians are designated by

$$\frac{\partial(f,g)}{\partial(x,y)} = \begin{vmatrix} \partial f/\partial x & \partial f/\partial y \\ \partial g/\partial x & \partial g/\partial y \end{vmatrix}$$

SOLUTION

For steady, creeping flow, the entire left side of Eq. D of Table 4.2-1 may be set equal to zero, and the ψ equation for axisymmetric flow becomes

$$E^4 \psi = 0 \tag{4.2-2}$$

or, in spherical coordinates

$$\left[\frac{\partial^2}{\partial r^2} + \frac{\sin \theta}{r^2} \frac{\partial}{\partial \theta} \left(\frac{1}{\sin \theta} \frac{\partial}{\partial \theta} \right) \right]^2 \psi = 0 \tag{4.2-3}$$

This is to be solved with the following boundary conditions:

B.C. 1:   at $r = R$, $v_r = -\dfrac{1}{r^2 \sin \theta} \dfrac{\partial \psi}{\partial \theta} = 0$ (4.2-4)

B.C. 2:   at $r = R$, $v_\theta = +\dfrac{1}{r \sin \theta} \dfrac{\partial \psi}{\partial r} = 0$ (4.2-5)

B.C. 3:   as $r \to \infty$, $\psi \to -\tfrac{1}{2} v_\infty r^2 \sin^2 \theta$ (4.2-6)

The first two boundary conditions describe the no-slip condition at the sphere surface. The third implies that $v_z \to v_\infty$ far from the sphere (this can be seen by recognizing that $v_r = v_\infty \cos \theta$ and $v_\theta = -v_\infty \sin \theta$ far from the sphere).

We now postulate a solution of the form

$$\psi(r, \theta) = f(r) \sin^2 \theta \tag{4.2-7}$$

since it will at least satisfy the third boundary condition in Eq. 4.2-6. When it is substituted into Eq. 4.2-4, we get

$$\left(\frac{d^2}{dr^2} - \frac{2}{r^2} \right) \left(\frac{d^2}{dr^2} - \frac{2}{r^2} \right) f = 0 \tag{4.2-8}$$

The fact that the variable θ does not appear in this equation suggests that the postulate in Eq. 4.2-7 is satisfactory. Equation 4.2-8 is an "equidimensional" fourth-order equation (see Eq. C.1-14). When a trial solution of the form $f(r) = C r^n$ is substituted into this equation, we find that n may have the values $-1, 1, 2,$ and 4. Therefore $f(r)$ has the form

$$f(r) = C_1 r^{-1} + C_2 r + C_3 r^2 + C_4 r^4 \tag{4.2-9}$$

To satisfy the third boundary condition, C_4 must be zero, and C_3 has to be $-\tfrac{1}{2} v_\infty$. Hence the stream function is

$$\psi(r, \theta) = (C_1 r^{-1} + C_2 r - \tfrac{1}{2} v_\infty r^2) \sin^2 \theta \tag{4.2-10}$$

The velocity components are then obtained by using Table 4.2-1 as follows:

$$v_r = -\frac{1}{r^2 \sin \theta} \frac{\partial \psi}{\partial \theta} = \left(v_\infty - 2\frac{C_2}{r} - 2\frac{C_1}{r^3} \right) \cos \theta \tag{4.2-11}$$

$$v_\theta = +\frac{1}{r \sin \theta} \frac{\partial \psi}{\partial r} = \left(-v_\infty + \frac{C_2}{r} - \frac{C_1}{r^3} \right) \sin \theta \tag{4.2-12}$$

The first two boundary conditions now give $C_1 = -\tfrac{1}{4} v_\infty R^3$ and $C_2 = \tfrac{3}{4} v_\infty R$, so that

$$v_r = v_\infty \left(1 - \frac{3}{2} \left(\frac{R}{r} \right) + \frac{1}{2} \left(\frac{R}{r} \right)^3 \right) \cos \theta \tag{4.2-13}$$

$$v_\theta = -v_\infty \left(1 - \frac{3}{4} \left(\frac{R}{r} \right) - \frac{1}{4} \left(\frac{R}{r} \right)^3 \right) \sin \theta \tag{4.2-14}$$

These are the velocity components given in Eqs. 2.6-1 and 2 without proof.

To get the pressure distribution, we substitute these velocity components into the r- and θ-components of the Navier–Stokes equation (given in Table B.7). After some tedious manipulations we get

$$\frac{\partial \mathscr{P}}{\partial r} = 3\left(\frac{\mu v_\infty}{R^2}\right)\left(\frac{R}{r}\right)^3 \cos \theta \qquad (4.2\text{-}15)$$

$$\frac{\partial \mathscr{P}}{\partial \theta} = \frac{3}{2}\left(\frac{\mu v_\infty}{R}\right)\left(\frac{R}{r}\right)^2 \sin \theta \qquad (4.2\text{-}16)$$

These equations may be integrated (cf. Eqs. 3.6-38 to 41), and, when use is made of the boundary condition that as $r \to \infty$ the modified pressure $\mathscr{P}$ tends to p_0 (the pressure in the plane $z = 0$ far from the sphere), we get

$$p = p_0 - \rho g z - \frac{3}{2}\left(\frac{\mu v_\infty}{R}\right)\left(\frac{R}{r}\right)^2 \cos \theta \qquad (4.2\text{-}17)$$

This is the same as the pressure distribution given in Eq. 2.6-4.

In §2.6 we showed how one can integrate the pressure and velocity distributions over the sphere surface to get the drag force. That method for getting the force of the fluid on the solid is general. Here we evaluate the "kinetic force" F_k by equating the rate of doing work on the sphere (force × velocity) to the rate of viscous dissipation within the fluid, thus

$$F_k v_\infty = -\int_0^{2\pi} \int_0^{\pi} \int_R^{\infty} (\boldsymbol{\tau}\!:\!\nabla\mathbf{v}) r^2 dr \sin \theta d\theta d\phi \qquad (4.2\text{-}18)$$

Insertion of the function $(-\boldsymbol{\tau}\!:\!\nabla\mathbf{v})$ in spherical coordinates from Table B.7 gives

$$F_k v_\infty = \mu \int_0^{2\pi} \int_0^{\pi} \int_R^{\infty} \left[2\left(\frac{\partial v_r}{\partial r}\right)^2 + 2\left(\frac{1}{r}\frac{\partial v_\theta}{\partial \theta} + \frac{v_r}{r}\right)^2 + 2\left(\frac{v_r}{r} + \frac{v_\theta \cot \theta}{r}\right)^2\right.$$

$$\left. + \left(r\frac{\partial}{\partial r}\left(\frac{v_\theta}{r}\right) + \frac{1}{r}\frac{\partial v_r}{\partial \theta}\right)^2\right] r^2 dr \sin \theta d\theta d\phi \qquad (4.2\text{-}19)$$

Then the velocity profiles from Eqs. 4.2-13 and 14 are substituted into Eq. 4.2-19. When the indicated differentiations and integrations (lengthy!) are performed, one finally gets

$$F_k = 6\pi\mu v_\infty R \qquad (4.2\text{-}20)$$

which is *Stokes' law*.

As pointed out in §2.6, Stokes' law is restricted to Re < 0.1. The expression for the drag force can be improved by going back and including the $[\mathbf{v}\cdot\nabla\mathbf{v}]$ term. Then use of the *method of matched asymptotic expansions* leads to the following result[4]

$$F_k = 6\pi\mu v_\infty R[1 + \tfrac{3}{16}\,\text{Re} + \tfrac{9}{160}\,\text{Re}^2(\ln \tfrac{1}{2}\,\text{Re} + \gamma + \tfrac{5}{3}\ln 2 - \tfrac{323}{360}) + \tfrac{27}{640}\,\text{Re}^3\ln \tfrac{1}{2}\,\text{Re} + O\,(\text{Re}^3)]$$

$$(4.2\text{-}21)$$

where $\gamma = 0.5772$ is Euler's constant. This expression is good up to Re of about 1.

---

[4] I. Proudman and J. R. A. Pearson, *J. Fluid Mech.* **2**, 237–262 (1957); W. Chester and D. R. Breach, *J. Fluid. Mech.* **37**, 751–760 (1969).

§4.3   FLOW OF INVISCID FLUIDS BY USE OF THE VELOCITY POTENTIAL[1]

Of course, we know that inviscid fluids (i.e., fluids devoid of viscosity) do not actually exist. However, the Euler equation of motion of Eq. 3.5-9 has been found to be useful for describing the flows of low-viscosity fluids at Re >> 1 around streamlined objects and gives a reasonably good description of the velocity profile, except very near the object and beyond the line of separation.

Then the vorticity equation in Eq. 3D.2-1 may be simplified by omitting the term containing the kinematic viscosity. If, in addition, the flow is steady and two-dimensional, then the terms $\partial/\partial t$ and $[\mathbf{w} \cdot \nabla \mathbf{v}]$ vanish. This means that the vorticity $\mathbf{w} = [\nabla \times \mathbf{v}]$ is constant along a streamline. If the fluid approaching a submerged object has no vorticity far away from the object, then the flow will be such that $\mathbf{w} = [\nabla \times \mathbf{v}]$ will be zero throughout the entire flow field. That is, the flow will be *irrotational*.

To summarize, if we assume that ρ = constant and $[\nabla \times \mathbf{v}] = 0$, then we can expect to get a reasonably good description of the flow of low-viscosity fluids around submerged objects in two-dimensional flows. This type of flow is referred to as *potential flow*.

Of course we know that this flow description will be inadequate in the neighborhood of solid surfaces. Near these surfaces we make use of a different set of assumptions, and these lead to *boundary-layer theory*, which is discussed in §4.4. By solving the potential flow equations for the "far field" and the boundary-layer equations for the "near field" and then matching the solutions asymptotically for large Re, it is possible to develop an understanding of the entire flow field around a streamlined object.[2]

To describe potential flow we start with the equation of continuity for an incompressible fluid and the Euler equation for an inviscid fluid (Eq. 3.5-9):

(continuity) $$(\nabla \cdot \mathbf{v}) = 0 \tag{4.3-1}$$

(motion) $$\rho\left(\frac{\partial \mathbf{v}}{\partial t} + \nabla \tfrac{1}{2}v^2 - [\mathbf{v} \times [\nabla \times \mathbf{v}]]\right) = -\nabla \mathscr{P} \tag{4.3-2}$$

In the equation of motion we have made use of the vector identity $[\mathbf{v} \cdot \nabla \mathbf{v}] = \nabla \tfrac{1}{2}v^2 - [\mathbf{v} \times [\nabla \times \mathbf{v}]]$ (see Eq. A.4-23).

For the two-dimensional, irrotational flow the statement that $[\nabla \times \mathbf{v}] = 0$ is

(irrotational) $$\frac{\partial v_x}{\partial y} - \frac{\partial v_y}{\partial x} = 0 \tag{4.3-3}$$

and the equation of continuity is

(continuity) $$\frac{\partial v_x}{\partial x} + \frac{\partial v_y}{\partial y} = 0 \tag{4.3-4}$$

The equation of motion for steady, irrotational flow can be integrated to give

(motion) $$\tfrac{1}{2}\rho(v_x^2 + v_y^2) + \mathscr{P} = \text{constant} \tag{4.3-5}$$

That is, the sum of the pressure and the kinetic and potential energy per unit volume is constant throughout the entire flow field. This is the *Bernoulli equation* for incompressible, potential flow, and the constant is the same for all streamlines. (This has to be contrasted with Eq. 3.5-12, the Bernoulli equation for a compressible fluid in any kind of flow; there the sum of the three contributions is a different constant on each streamline.)

---

[1] R. H. Kirchhoff, Chapter 7 in *Handbook of Fluid Dynamics* (R. W. Johnson, ed.), CRC Press, Boca Raton, Fla. (1998).

[2] M. Van Dyke, *Perturbation Methods in Fluid Dynamics*, The Parabolic Press, Stanford, Cal. (1975).

We want to solve Eqs. 4.3-3 to 5 to obtain v_x, v_y, and $\mathcal{P}$ as functions of x and y. We have already seen in the previous section that the equation of continuity in two-dimensional flows can be satisfied by writing the components of the velocity in terms of a *stream function* $\psi(x, y)$. However, any vector that has a zero curl can also be written as the gradient of a scalar function (that is, $[\nabla \times \mathbf{v}] = 0$ implies that $\mathbf{v} = -\nabla\phi$). It is very convenient, then, to introduce a *velocity potential* $\phi(x, y)$. Instead of working with the velocity components v_x and v_y, we choose to work with $\psi(x, y)$ and $\phi(x, y)$. We then have the following relations:

(stream function) $$v_x = -\frac{\partial\psi}{\partial y} \qquad v_y = \frac{\partial\psi}{\partial x}$$ (4.3-6, 7)

(velocity potential) $$v_x = -\frac{\partial\phi}{\partial x} \qquad v_y = -\frac{\partial\phi}{\partial y}$$ (4.3-8, 9)

Now Eqs. 4.3-3 and 4.3-4 will automatically be satisfied. By equating the expressions for the velocity components we get

$$\frac{\partial\phi}{\partial x} = \frac{\partial\psi}{\partial y} \quad \text{and} \quad \frac{\partial\phi}{\partial y} = -\frac{\partial\psi}{\partial x}$$ (4.3-10, 11)

These are the *Cauchy–Riemann equations*, which are the relations that must be satisfied by the real and imaginary parts of any analytic function[3] $w(z) = \phi(x, y) + i\psi(x, y)$, where $z = x + iy$. The quantity $w(z)$ is called the *complex potential*. Differentiation of Eq. 4.3-10 with respect to x and Eq. 4.3-11 with respect to y and then adding gives $\nabla^2\phi = 0$. Differentiating with respect to the variables in reverse order and then substracting gives $\nabla^2\psi = 0$. That is, both $\phi(x, y)$ and $\psi(x, y)$ satisfy the two-dimensional Laplace equation.[4]

As a consequence of the preceding development, it appears that *any* analytic function $w(z)$ yields a pair of functions $\phi(x, y)$ and $\psi(x, y)$ that are the velocity potential and stream function for *some* flow problem. Furthermore, the curves $\phi(x, y) = $ constant and $\psi(x, y) = $ constant are then the *equipotential lines* and *streamlines* for the problem. The velocity components are then obtained from Eqs. 4.3-6 and 7 or Eqs. 4.3-8 and 9 or from

$$\frac{dw}{dz} = -v_x + iv_y$$ (4.3-12)

in which dw/dz is called the *complex velocity*. Once the velocity components are known, the modified pressure can then be found from Eq. 4.3-5.

Alternatively, the equipotential lines and streamlines can be obtained from the *inverse function* $z(w) = x(\phi, \psi) + iy(\phi, \psi)$, in which $z(w)$ is *any* analytic function of w. Between the functions $x(\phi, \psi)$ and $y(\phi, \psi)$ we can eliminate ψ and get

$$F(x, y, \phi) = 0$$ (4.3-13)

---

[3] Some knowledge of the analytic functions of a complex variable is assumed here. Helpful introductions to the subject can be found in V. L. Streeter, E. B. Wylie, and K. W. Bedford, *Fluid Mechanics*, McGraw-Hill, New York, 9th ed. (1998), Chapter 8, and in M. D. Greenberg, *Foundations of Applied Mathematics*, Prentice-Hall, Englewood Cliffs, N.J. (1978), Chapters 11 and 12.

[4] Even for three-dimensional flows the assumption of irrotational flow still permits the definition of a velocity potential. When $\mathbf{v} = -\nabla\phi$ is substituted into $(\nabla \cdot \mathbf{v}) = 0$, we get the three-dimensional Laplace equation $\nabla^2\phi = 0$. The solution of this equation is the subject of "potential theory," for which there is an enormous literature. See, for example, P. M. Morse and H. Feshbach, *Methods of Theoretical Physics*, McGraw-Hill, New York (1953), Chapter 11; and J. M. Robertson, *Hydrodynamics in Theory and Application*, Prentice-Hall, Englewood Cliffs, N.J. (1965), which emphasizes the engineering applications. There are many problems in flow through porous media, heat conduction, diffusion, and electrical conduction that are described by Laplace's equation.

Similar elimination of ϕ gives

$$G(x, y, \psi) = 0 \tag{4.3-14}$$

Setting ϕ = a constant in Eq. 4.3-13 gives the equations for the equipotential lines for *some* flow problem, and setting ψ = constant in Eq. 4.3-14 gives equations for the stream-lines. The velocity components can be obtained from

$$-\frac{dz}{dw} = \frac{v_x + iv_y}{v_x^2 + v_y^2} \tag{4.3-15}$$

Thus from any analytic function $w(z)$, or its inverse $z(w)$, we can construct a flow net with streamlines ψ = constant and equipotential lines ϕ = constant. The task of finding $w(z)$ or $z(w)$ to satisfy a given flow problem is, however, considerably more difficult. Some special methods are available[4,5] but it is frequently more expedient to consult a table of conformal mappings.[6]

In the next two illustrative examples we show how to use the complex potential $w(z)$ to describe the potential flow around a cylinder, and the inverse function $z(w)$ to solve the problem of the potential flow into a channel. In the third example we solve the flow in the neighborhood of a corner, which is treated further in §4.4 by the boundary-layer method. A few general comments should be kept in mind:

(a) The streamlines are everywhere perpendicular to the equipotential lines. This property, evident from Eqs. 4.3-10, 11, is useful for the approximate construction of flow nets.

(b) Streamlines and equipotential lines can be interchanged to get the solution of another flow problem. This follows from (a) and the fact that both ϕ and ψ are solutions to the two-dimensional Laplace equation.

(c) Any streamline may be replaced by a solid surface. This follows from the boundary condition that the normal component of the velocity of the fluid is zero at a solid surface. The tangential component is not restricted, since in potential flow the fluid is presumed to be able to slide freely along the surface (the complete-slip assumption).

EXAMPLE 4.3-1

Potential Flow around a Cylinder

(a) Show that the complex potential

$$w(z) = -v_\infty R\left(\frac{z}{R} + \frac{R}{z}\right) \tag{4.3-16}$$

describes the potential flow around a circular cylinder of radius R, when the approach velocity is v_∞ in the positive x direction.

(b) Find the components of the velocity vector.

(c) Find the pressure distribution on the cylinder surface, when the modified pressure far from the cylinder is $\mathcal{P}_\infty$.

SOLUTION

(a) To find the stream function and velocity potential, we write the complex potential in the form $w(z) = \phi(x, y) + i\psi(x, y)$:

$$w(z) = -v_\infty x\left(1 + \frac{R^2}{x^2 + y^2}\right) - iv_\infty y\left(1 - \frac{R^2}{x^2 + y^2}\right) \tag{4.3-17}$$

---

[5] J. Fuka, Chapter 21 in K. Rektorys, *Survey of Applicable Mathematics*, MIT Press, Cambridge, Mass. (1969).

[6] H. Kober, *Dictionary of Conformal Representations*, Dover, New York, 2nd edition (1957).

Hence the stream function is

$$\psi(x, y) = -v_\infty y\left(1 - \frac{R^2}{x^2 + y^2}\right) \tag{4.3-18}$$

To make a plot of the streamlines it is convenient to rewrite Eq. 4.3-18 in dimensionless form

$$\Psi(X, Y) = -Y\left(1 - \frac{1}{X^2 + Y^2}\right) \tag{4.3-19}$$

in which $\Psi = \psi/v_\infty R$, $X = x/R$, and $Y = y/R$.

In Fig. 4.3-1 the streamlines are plotted as the curves $\Psi = $ constant. The streamline $\Psi = 0$ gives a unit circle, which represents the surface of the cylinder. The streamline $\Psi = -\frac{3}{2}$ goes through the point $X = 0$, $Y = 2$, and so on.

(b) The velocity components are obtainable from the stream function by using Eqs. 4.3-6 and 7. They may also be obtained from the complex velocity according to Eq. 4.3-12, as follows:

$$\frac{dw}{dz} = -v_\infty\left(1 - \frac{R^2}{z^2}\right) = -v_\infty\left(1 - \frac{R^2}{r^2}e^{-2i\theta}\right)$$

$$= -v_\infty\left(1 - \frac{R^2}{r^2}(\cos 2\theta - i\sin 2\theta)\right) \tag{4.3-20}$$

Therefore the velocity components as function of position are

$$v_x = v_\infty\left(1 - \frac{R^2}{r^2}\cos 2\theta\right) \tag{4.3-21}$$

$$v_y = -v_\infty\left(\frac{R^2}{r^2}\sin 2\theta\right) \tag{4.3-22}$$

(c) On the surface of the cylinder, $r = R$, and

$$v^2 = v_x^2 + v_y^2$$
$$= v_\infty^2[(1 - \cos 2\theta)^2 + (\sin 2\theta)^2]$$
$$= 4v_\infty^2 \sin^2 \theta \tag{4.3-23}$$

When θ is zero or π, the fluid velocity is zero; such points are known as *stagnation points*. From Eq. 4.3-5 we know that

$$\tfrac{1}{2}\rho v^2 + \mathscr{P} = \tfrac{1}{2}\rho v_\infty^2 + \mathscr{P}_\infty \tag{4.3-24}$$

Then from the last two equations we get the pressure distribution on the surface of the cylinder

$$(\mathscr{P} - \mathscr{P}_\infty) = \tfrac{1}{2}\rho v_\infty^2(1 - 4\sin^2 \theta) \tag{4.3-25}$$

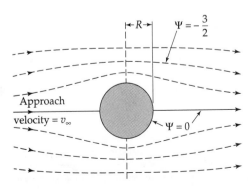

Fig. 4.3-1. The streamlines for the potential flow around a cylinder according to Eq. 4.3-19.

Note that the modified pressure distribution is symmetric about the x-axis; that is, for potential flow there is no form drag on the cylinder (*d'Alembert's paradox*).[7] Of course, we know now that this is not really a paradox, but simply the result of the fact that the inviscid fluid does not permit applying the no-slip boundary condition at the interface.

EXAMPLE 4.3-2

*Flow Into a
Rectangular Channel*

Show that the inverse function

$$z(w) = \frac{w}{v_\infty} + \frac{b}{\pi} \exp(\pi w / b v_\infty) \tag{4.3-26}$$

represents the potential flow into a rectangular channel of half-width b. Here v_∞ is the magnitude of the velocity far downstream from the entrance to the channel.

SOLUTION

First we introduce dimensionless distance variables

$$X = \frac{\pi x}{b} \qquad Y = \frac{\pi y}{b} \qquad Z = X + iY = \frac{\pi z}{b} \tag{4.3-27}$$

and the dimensionless quantities

$$\Phi = \frac{\pi \phi}{b v_\infty} \qquad \Psi = \frac{\pi \psi}{b v_\infty} \qquad W = \Phi + i\Psi = \frac{\pi w}{b v_\infty} \tag{4.3-28}$$

The inverse function of Eq. 4.3-26 may now be expressed in terms of dimensionless quantities and split up into real and imaginary parts

$$Z = W + e^W = (\Phi + e^\Phi \cos \Psi) + i(\Psi + e^\Phi \sin \Psi) \tag{4.3-29}$$

Therefore

$$X = \Phi + e^\Phi \cos \Psi \qquad Y = \Psi + e^\Phi \sin \Psi \tag{4.3-30, 31}$$

We can now set Ψ equal to a constant, and the streamline $Y = Y(X)$ is expressed parametrically in Φ. For example, the streamline $\Psi = 0$ is given by

$$X = \Phi + e^\Phi \qquad Y = 0 \tag{4.3-32, 33}$$

As Φ goes from $-\infty$ to $+\infty$, X also goes from $-\infty$ to $+\infty$; hence the X-axis is a streamline. Next, the streamline $\Psi = \pi$ is given by

$$X = \Phi - e^\Phi \qquad Y = \pi \tag{4.3-34, 35}$$

As Φ goes from $-\infty$ to $+\infty$, X goes from $-\infty$ to -1 and then back to $-\infty$; that is, the streamline doubles back on itself. We select this streamline to be one of the solid walls of the rectangular channel. Similarly, the streamline $\Psi = -\pi$ is the other wall. The streamlines $\Psi = C$, where $-\pi < C < \pi$, then give the flow pattern for the flow into the rectangular channel as shown in Fig. 4.3-2.

Next, from Eq. 4.3-29 the derivative $-dz/dw$ can be found:

$$-\frac{dz}{dw} = -\frac{1}{v_\infty} \frac{dZ}{dW} = -\frac{1}{v_\infty}(1 + e^W) = -\frac{1}{v_\infty}(1 + e^\Phi \cos \Psi + ie^\Phi \sin \Psi) \tag{4.3-36}$$

Comparison of this expression with Eq. 4.3-15 gives for the velocity components

$$\frac{v_x v_\infty}{v^2} = -(1 + e^\Phi \cos \Psi) \qquad \frac{v_y v_\infty}{v^2} = -(e^\Phi \sin \Psi) \tag{4.3-37}$$

These equations have to be used in conjunction with Eqs. 4.3-30 and 31 to eliminate Φ and Ψ in order to get the velocity components as functions of position.

---

[7] Hydrodynamic paradoxes are discussed in G. Birkhoff, *Hydrodynamics*, Dover, New York (1955).

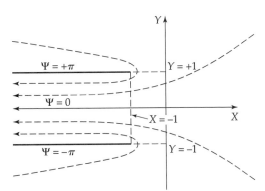

Fig. 4.3-2. The streamlines for the potential flow into a rectangular channel, as predicted from potential flow theory in Eqs. 4.3-30 and 31. A more realistic flow pattern is shown in Fig. 4.3-5.

EXAMPLE 4.3-3

Flow Near a Corner[8]

Figure 4.3-3 shows the potential flow in the neighborhood of two walls that meet at a corner at O. The flow in the neighborhood of this corner can be described by the complex potential

$$w(z) = -cz^\alpha \tag{4.3-38}$$

in which c is a constant. We can now consider two situations: (i) an "interior corner flow," with $\alpha > 1$; and (ii) an "exterior corner flow," with $\alpha < 1$.

(a) Find the velocity components.

(b) Obtain the tangential velocity at both parts of the wall.

(c) Describe how to get the streamlines.

(d) How can this result be applied to the flow around a wedge?

SOLUTION

(a) The velocity components are obtained from the complex velocity

$$\frac{dw}{dz} = -c\alpha z^{\alpha-1} = -c\alpha r^{\alpha-1}e^{i(\alpha-1)\theta} \tag{4.3-39}$$

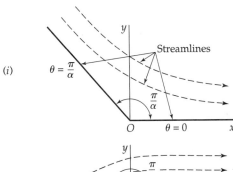

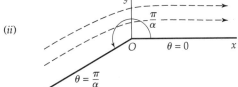

Fig. 4.3-3. Potential flow near a corner. On the left portion of the wall, $v_r = -cr^{\alpha-1}$, and on the right, $v_r = +cr^{\alpha-1}$. (i) Interior-corner flow, with $\alpha > 1$; and (ii) exterior-corner flow, with $\alpha < 1$.

[8] R. L. Panton, *Compressible Flow*, Wiley, New York, 2nd edition (1996).

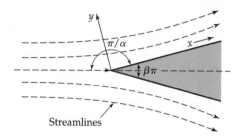

Fig. 4.3-4. Potential flow along a wedge. On the upper surface of the wedge, $v_x = cx^{\alpha-1} = cx^{\beta/(2-\beta)}$. The quantities α and β are related by $\beta = (2/\alpha)(\alpha - 1)$.

Streamlines

Hence from Eq. 4.3-12 we get

$$v_x = +c\alpha r^{\alpha-1} \cos(\alpha - 1)\theta \tag{4.3-40}$$

$$v_y = -c\alpha r^{\alpha-1} \sin(\alpha - 1)\theta \tag{4.3-41}$$

(b) The tangential velocity at the walls is

at $\theta = 0$: $\quad v_x = v_r = c\alpha r^{\alpha-1} = c\alpha x^{\alpha-1} \tag{4.3-42}$

at $\theta = \pi/\alpha$: $\quad v_r = v_x \cos\theta + v_y \sin\theta$

$$= +c\alpha r^{\alpha-1} \cos(\alpha - 1)\theta \cos\theta - c\alpha r^{\alpha-1} \sin(\alpha - 1)\theta \sin\theta$$

$$= c\alpha r^{\alpha-1} \cos\alpha\theta$$

$$= -c\alpha r^{\alpha-1} \tag{4.3-43}$$

Hence, in Case (i), the incoming fluid at the wall decelerates as it approaches the junction, and the departing fluid accelerates as it moves away from the junction. In Case (ii) the velocity components become infinite at the corner as $\alpha - 1$ is then negative.

(c) The complex potential can be decomposed into its real and imaginary parts

$$w = \phi + i\psi = -cr^{\alpha}(\cos\alpha\theta + i\sin\alpha\theta) \tag{4.3-44}$$

Hence the stream function is

$$\psi = -cr^{\alpha} \sin\alpha\theta \tag{4.3-45}$$

To get the streamlines, one selects various values for the stream function—say, $\psi_1, \psi_2, \psi_3 \cdots$ —and then for each value one plots r as a function of θ.

(d) Since for ideal flow any streamline may be replaced by a wall, and vice versa, the results found here for $\alpha > 0$ describe the inviscid flow over a wedge (see Fig. 4.3-4). We make use of this in Example 4.4-3.

A few words of warning are in order concerning the applicability of potential-flow theory to real systems:

a. For the flow around a cylinder, the streamlines shown in Fig. 4.3-1 do not conform to any of the flow regimes sketched in Fig. 3.7-2.

b. For the flow into a channel, the predicted flow pattern of Fig. 4.3-2 is unrealistic inside the channel and just upstream from the channel entrance. A much better approximation to the actual behavior is shown in Fig. 4.3-5.

Both of these failures of the elementary potential theory result from the phenomenon of *separation*: the departure of streamlines from a boundary surface.

Separation tends to occur at sharp corners of solid boundaries, as in channel flow, and on the downstream sides of bluff objects, as in the flow around a cylinder. Generally, separation is likely to occur in regions where the pressure increases in the direction

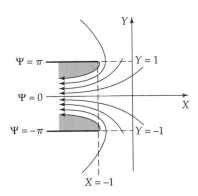

Fig. 4.3-5. Potential flow into a rectangular channel with separation, as calculated by H. von Helmholtz, *Phil. Mag.*, **36**, 337–345 (1868). The streamlines for $\Psi = \pm\pi$ separate from the inner surface of the channel. The velocity along this separated streamline is constant. Between the separated streamline and the wall is an empty region.

of flow. Potential-flow analyses are not useful in the separated region. They can, however, be used upstream of this region if the location of the *separation streamline* is known. Methods of making such calculations have been highly developed. Sometimes the position of the separation streamline can be estimated successfully from potential-flow theory. This is true for flow into a channel, and, in fact, Fig. 4.3-5 was obtained in this way.[9] For other systems, such as the flow around the cylinder, the separation point and separation streamline must be located by experiment. Even when the position of the separation streamline is not known, potential flow solutions may be valuable. For example, the flow field of Ex. 4.3-1 has been found useful for estimating aerosol impaction coefficients on cylinders.[10] This success is a result of the fact that most of the particle impacts occur near the forward stagnation point, where the flow is not affected very much by the position of the separation streamline. Valuable semiquantitative conclusions concerning heat- and mass-transfer behavior can also be made on the basis of potential flow calculations ignoring the separation phenomenon.

The techniques described in this section all assume that the velocity vector can be written as the gradient of a scalar function that satisfies Laplace's equation. The equation of motion plays a much less prominent role than for the viscous flows discussed previously, and its primary use is for the determination of the pressure distribution once the velocity profiles are found.

§4.4 FLOW NEAR SOLID SURFACES BY BOUNDARY-LAYER THEORY

The potential flow examples discussed in the previous section showed how to predict the flow field by means of a stream function and a velocity potential. The solutions for the velocity distribution thus obtained do not satisfy the usual "no-slip" boundary condition at the wall. Consequently, the potential flow solutions are of no value in describing the transport phenomena in the immediate neighborhood of the wall. Specifically, the viscous drag force cannot be obtained, and it is also not possible to get reliable descriptions of interphase heat- and mass-transfer at solid surfaces.

To describe the behavior near the wall, we use *boundary-layer theory*. For the description of a viscous flow, we obtain an approximate solution for the velocity components in a very thin boundary layer near the wall, taking the viscosity into account. Then we "match" this solution to the potential flow solution that describes the flow outside the

---

[9] H. von Helmholtz, *Phil Mag.* (4), **36**, 337–345 (1868). **Herman Ludwig Ferdinand von Helmholtz** (1821–1894) studied medicine and became an army doctor; he then served as professor of medicine and later as professor of physics in Berlin.

[10] W. E. Ranz, *Principles of Inertial Impaction*, Bulletin #66, Department of Engineering Research, Pennsylvania State University, University Park, Pa. (1956).

boundary layer. The success of the method depends on the thinness of the boundary layer, a condition that is met at high Reynolds number.

We consider the steady, two-dimensional flow of a fluid with constant ρ and μ around a submerged object, such as that shown in Fig. 4.4-1. We assert that the main changes in the velocity take place in a very thin region, the boundary layer, in which the curvature effects are not important. We can then set up a Cartesian coordinate system with x pointing downstream, and y perpendicular to the solid surface. The continuity equation and the Navier–Stokes equations then become:

$$\frac{\partial v_x}{\partial x} + \frac{\partial v_y}{\partial y} = 0 \tag{4.4-1}$$

$$\left(v_x \frac{\partial v_x}{\partial x} + v_y \frac{\partial v_x}{\partial y} \right) = -\frac{1}{\rho}\frac{\partial \mathcal{P}}{\partial x} + \nu\left(\frac{\partial^2 v_x}{\partial x^2} + \frac{\partial^2 v_x}{\partial y^2} \right) \tag{4.4-2}$$

$$\left(v_x \frac{\partial v_y}{\partial x} + v_y \frac{\partial v_y}{\partial y} \right) = -\frac{1}{\rho}\frac{\partial \mathcal{P}}{\partial y} + \nu\left(\frac{\partial^2 v_y}{\partial x^2} + \frac{\partial^2 v_y}{\partial y^2} \right) \tag{4.4-3}$$

Some of the terms in these equations can be discarded by order-of-magnitude arguments. We use three quantities as "yardsticks": the approach velocity v_∞, some linear dimension l_0 of the submerged body, and an average thickness δ_0 of the boundary layer. The presumption that $\delta_0 \ll l_0$ allows us to make a number of rough calculations of orders of magnitude.

Since v_x varies from zero at the solid surface to v_∞ at the outer edge of the boundary layer, we can say that

$$\frac{\partial v_x}{\partial y} = O\left(\frac{v_\infty}{\delta_0}\right) \tag{4.4-4}$$

where O means "order of magnitude of." Similarly, the maximum variation in v_x over the length l_0 of the surface will be v_∞, so that

$$\frac{\partial v_x}{\partial x} = O\left(\frac{v_\infty}{l_0}\right) \text{ and } \frac{\partial v_y}{\partial y} = O\left(\frac{v_\infty}{l_0}\right) \tag{4.4-5}$$

Here we have made use of the equation of continuity to get one more derivative (we are concerned here only with orders of magnitude and not the signs of the quantities). Integration of the second relation suggests that $v_y = O((\delta_0/l_0)v_\infty) \ll v_x$. The various terms in Eq. 4.4-2 may now be estimated as

$$v_x \frac{\partial v_x}{\partial x} = O\left(\frac{v_\infty^2}{l_0}\right); \; v_y \frac{\partial v_x}{\partial y} = O\left(\frac{v_\infty^2}{l_0}\right) \qquad \frac{\partial^2 v_x}{\partial x^2} = O\left(\frac{v_\infty}{l_0^2}\right) \qquad \frac{\partial^2 v_x}{\partial y^2} = O\left(\frac{v_\infty}{\delta_0^2}\right) \tag{4.4-6}$$

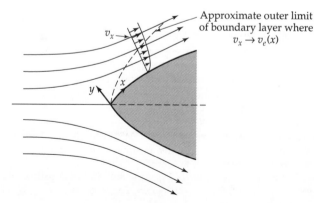

Approximate outer limit of boundary layer where $v_x \rightarrow v_e(x)$

Fig. 4.4-1. Coordinate system for the two-dimensional flow around a submerged object. The boundary-layer thickness is greatly exaggerated for purposes of illustration. Because the boundary layer is in fact quite thin, it is permissible to use rectangular coordinates locally along the curved surface.

This suggests that $\partial^2 v_x/\partial x^2 << \partial^2 v_x/\partial y^2$, so that the former may be safely neglected. In the boundary layer it is expected that the terms on the left side of Eq. 4.4-2 should be of the same order of magnitude as those on the right side, and therefore

$$\frac{v_\infty^2}{l_0} = O\!\left(\nu\,\frac{v_\infty}{\delta_0^2}\right) \quad \text{or} \quad \frac{\delta_0}{l_0} = O\!\left(\sqrt{\frac{\nu}{v_\infty l_0}}\right) = O\!\left(\frac{1}{\sqrt{\text{Re}}}\right) \tag{4.4-7}$$

The second of these relations shows that the boundary-layer thickness is small compared to the dimensions of the submerged object in high-Reynolds-number flows.

Similarly it can be shown, with the help of Eq. 4.4-7, that three of the derivatives in Eq. 4.4-3 are of the same order of magnitude:

$$v_x\frac{\partial v_y}{\partial x},\; v_y\frac{\partial v_y}{\partial y},\; \nu\frac{\partial^2 v_y}{\partial y^2} = O\!\left(\frac{v_\infty^2\delta_0}{l_0^2}\right) >> \nu\frac{\partial^2 v_y}{\partial x^2} \tag{4.4-8}$$

Comparison of this result with Eq. 4.4-6 shows that $\partial \mathscr{P}/\partial y << \partial \mathscr{P}/\partial x$. This means that the y-component of the equation of motion is not needed and that the modified pressure can be treated as a function of x alone.

As a result of these order-of-magnitude arguments, we are left with the *Prandtl boundary layer equations*:[1]

(continuity) $$\frac{\partial v_x}{\partial x} + \frac{\partial v_y}{\partial y} = 0 \tag{4.4-9}$$

(motion) $$v_x\frac{\partial v_x}{\partial x} + v_y\frac{\partial v_x}{\partial y} = -\frac{1}{\rho}\frac{d\mathscr{P}}{dx} + \nu\frac{\partial^2 v_x}{\partial y^2} \tag{4.4-10}$$

The modified pressure $\mathscr{P}(x)$ is presumed known from the solution of the corresponding potential-flow problem or from experimental measurements.

The usual boundary conditions for these equations are the no-slip condition ($v_x = 0$ at $y = 0$), the condition of no mass transfer from the wall ($v_y = 0$ at $y = 0$), and the statement that the velocity merges into the external (potential-flow) velocity at the outer edge of the boundary layer ($v_x(x, y) \to v_e(x)$). The function $v_e(x)$ is related to $\mathscr{P}(x)$ according to the potential-flow equation of motion in Eq. 4.3-5. Consequently the term $-(1/\rho)(d\mathscr{P}/dx)$ in Eq. 4.4-10 can be replaced by $v_e(dv_e/dx)$ for steady flow. Thus Eq. 4.4-10 may also be written as

$$v_x\frac{\partial v_x}{\partial x} + v_y\frac{\partial v_x}{\partial y} = v_e\frac{dv_e}{dx} + \nu\frac{\partial^2 v_x}{\partial y^2} \tag{4.4-11}$$

The equation of continuity may be solved for v_y by using the boundary condition that $v_y = 0$ at $y = 0$ (i.e., no mass transfer), and then this expression for v_y may be substituted into Eq. 4.4-11 to give

$$v_x\frac{\partial v_x}{\partial x} - \left(\int_0^y \frac{\partial v_x}{\partial x}\,dy\right)\frac{\partial v_x}{\partial y} = v_e\frac{dv_e}{dx} + \nu\frac{\partial^2 v_x}{\partial y^2} \tag{4.4-12}$$

This is a partial differential equation for the single dependent variable v_x.

---

[1] **Ludwig Prandtl** (1875–1953) (pronounced "Prahn-t'l"), who taught in Hannover and Göttingen and later served as the Director of the Kaiser Wilhelm Institute for Fluid Dynamics, was one of the people who shaped the future of his field at the beginning of the twentieth century; he made contributions to turbulent flow and heat transfer, but his development of the boundary-layer equations was his crowning achievement. L. Prandtl, *Verhandlungen des III Internationalen Mathematiker-Kongresses* (Heidelberg, 1904), Leipzig, pp. 484–491; L. Prandtl, *Gesammelte Abhandlungen*, **2**, Springer-Verlag, Berlin (1961), pp. 575–584. For an introductory discussion of matched asymptotic expressions, see D. J. Acheson, *Elementary Fluid Mechanics*," Oxford University Press (1990), pp. 269–271. An exhaustive discussion of the subject may be found in M. Van Dyke, *Perturbation Methods in Fluid Dynamics*, The Parabolic Press, Stanford, Cal. (1975).

This equation may now be multiplied by ρ and integrated from $y = 0$ to $y = \infty$ to give the *von Kármán momentum balance*[2]

$$\mu \frac{\partial v_x}{\partial y}\bigg|_{y=0} = \frac{d}{dx} \int_0^\infty \rho v_x (v_e - v_x) dy + \frac{dv_e}{dx} \int_0^\infty \rho (v_e - v_x) dy \qquad (4.4\text{-}13)$$

Here use has been made of the condition that $v_x(x, y) \rightarrow v_e(x)$ as $y \rightarrow \infty$. The quantity on the left side of Eq. 4.4-13 is the shear stress exerted by the fluid on the wall: $-\tau_{yx}|_{y=0}$.

The original Prandtl boundary-layer equations, Eqs. 4.4-9 and 10, have thus been transformed into Eq. 4.4-11, Eq. 4.4-12, and Eq. 4.4-13, and any of these may be taken as the starting point for solving two-dimensional boundary-layer problems. Equation 4.4-13, with assumed expressions for the velocity profile, is the basis of many "approximate boundary-layer solutions" (see Example 4.4-1). On the other hand, the analytical or numerical solutions of Eqs. 4.4-11 or 12 are called "exact boundary-layer solutions" (see Example 4.4-2).

The discussion here is for steady, laminar, two-dimensional flows of fluids with constant density and viscosity. Corresponding equations are available for unsteady flow, turbulent flow, variable fluid properties, and three-dimensional boundary layers.[3-6]

Although many exact and approximate boundary-layer solutions have been obtained and applications of the theory to streamlined objects have been quite successful, considerable work remains to be done on flows with adverse pressure gradients (i.e., positive $\partial \mathcal{P}/\partial x$) in Eq. 4.4-10, such as the flow on the downstream side of a blunt object. In such flows the streamlines usually separate from the surface before reaching the rear of the object (see Fig. 3.7-2). The boundary-layer approach described here is suitable for such flows only in the region upstream from the separation point.

EXAMPLE 4.4-1

Laminar Flow along a Flat Plate (Approximate Solution)

Use the von Kármán momentum balance to estimate the steady-state velocity profiles near a semi-infinite flat plate in a tangential stream with approach velocity v_∞ (see Fig. 4.4-2). For this system the potential-flow solution is $v_e = v_\infty$.

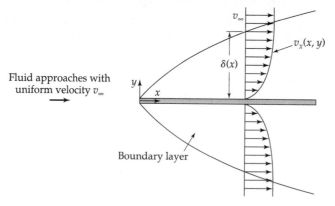

Fig. 4.4-2. Boundary-layer development near a flat plate of negligible thickness.

---

[2] Th. von Kármán, *Zeits. für angew. Math. u. Mech.*, **1**, 233–252 (1921). Hungarian-born **Theodor von Kármán** taught in Göttingen, Aachen, and California Institute of Technology; he contributed much to the theory of turbulence and aerodynamics.

[3] H. Schlichting and K. Gersten, *Boundary-Layer Theory*, Springer Verlag, Berlin, 8th edition (2000).

[4] L. Rosenhead, *Laminar Boundary Layers*, Oxford University Press, London (1963).

[5] K. Stewartson, *The Theory of Laminar Boundary Layers in Compressible Fluids*, Oxford University Press (1964).

[6] W. H. Dorrance, *Viscous Hypersonic Flow*, McGraw-Hill, New York (1962).

SOLUTION

We know intuitively what the velocity profile $v_x(y)$ looks like. Hence we can guess a form for $v_x(y)$ and substitute it directly into the von Kármán momentum balance. One reasonable choice is to let $v_x(y)$ by a function of y/δ, where $\delta(x)$ is the "thickness" of the boundary layer. The function is so chosen that $v_x = 0$ at $y = 0$ and $v_x = v_e$ at $y = \delta$. This is tantamount to assuming geometrical similarity of the velocity profiles for various values of x. When this assumed profile is substituted into the von Kármán momentum balance, an ordinary differential equation for the boundary-layer thickness $\delta(x)$ is obtained. When this equation has been solved, the $\delta(x)$ so obtained can then be used to get the velocity profile and other quantities of interest.

For the present problem a plausible guess for the velocity distribution, with a reasonable shape, is

$$\frac{v_x}{v_\infty} = \frac{3}{2}\frac{y}{\delta} - \frac{1}{2}\left(\frac{y}{\delta}\right)^3 \qquad \text{for } 0 \leq y \leq \delta(x) \qquad \text{(boundary-layer region)} \qquad (4.4\text{-}14)$$

$$\frac{v_x}{v_\infty} = 1 \qquad \text{for } y \geq \delta(x) \qquad \text{(potential flow region)} \qquad (4.4\text{-}15)$$

This is "reasonable" because this velocity profile satisfies the no-slip condition at $y = 0$, and $\partial v_x/\partial y = 0$ at the outer edge of the boundary layer. Substitution of this profile into the von Kármán integral balance in Eq. 4.4-13 gives

$$\frac{3}{2}\frac{\mu v_\infty}{\delta} = \frac{d}{dx}\left(\frac{39}{280}\rho v_\infty^2 \delta\right) \qquad (4.4\text{-}16)$$

This first-order, separable differential equation can now be integrated to give for the boundary-layer thickness

$$\delta(x) = \sqrt{\frac{280}{13}\frac{\nu x}{v_\infty}} = 4.64\sqrt{\frac{\nu x}{v_\infty}} \qquad (4.4\text{-}17)$$

Therefore, the boundary-layer thickness increases as the square root of the distance from the upstream end of the plate. The resulting approximate solution for the velocity distribution is then

$$\frac{v_x}{v_\infty} = \frac{3}{2}\left(y\sqrt{\frac{13}{280}\frac{v_\infty}{\nu x}}\right) - \frac{1}{2}\left(y\sqrt{\frac{13}{280}\frac{v_\infty}{\nu x}}\right)^3 \qquad (4.4\text{-}18)$$

From this result we can estimate the drag force on a plate of finite size wetted on both sides. For a plate of width W and length L, integration of the momentum flux over the two solid surfaces gives:

$$F_x = 2\int_0^W\int_0^L\left(+\mu\frac{\partial v_x}{\partial y}\right)\bigg|_{y=0} dx\,dz = 1.293\sqrt{\rho\mu LW^2 v_\infty^3} \qquad (4.4\text{-}19)$$

The exact solution, given in the next example, gives the same result, but with a numerical coefficient of 1.328. Both solutions predict the drag force within the scatter of the experimental data. However, the exact solution gives somewhat better agreement with the measured velocity profiles.[3] This additional accuracy is essential for stability calculations.

EXAMPLE 4.4-2

Laminar Flow along a Flat Plate (Exact Solution)[7]

Obtain the exact solution for the problem given in the previous example.

SOLUTION

This problem may be solved by using the definition of the stream function in Table 4.2-1. Inserting the expressions for the velocity components in the first row of entries, we get

$$\frac{\partial \psi}{\partial y}\frac{\partial^2 \psi}{\partial x\partial y} - \frac{\partial \psi}{\partial x}\frac{\partial^2 \psi}{\partial y^2} = -\nu\frac{\partial^3 \psi}{\partial y^3} \qquad (4.4\text{-}20)$$

---

[7] This problem was treated originally by H. Blasius, *Zeits. Math. Phys.*, **56**, 1–37 (1908).

The boundary conditions for this equation for $\psi(x, y)$ are

B.C. 1: at $y = 0$, $\dfrac{\partial \psi}{\partial x} = v_y = 0$ for $x \geq 0$ (4.4-21)

B.C. 2: at $y = 0$, $\dfrac{\partial \psi}{\partial y} = -v_x = 0$ for $x \geq 0$ (4.4-22)

B.C. 3: as $y \to \infty$, $\dfrac{\partial \psi}{\partial y} = -v_x \to -v_\infty$ for $x \geq 0$ (4.4-23)

B.C. 4: at $x = 0$, $\dfrac{\partial \psi}{\partial y} = -v_x = -v_\infty$ for $y > 0$ (4.4-24)

Inasmuch as there is no characteristic length appearing in the above relations, the method of combination of independent variables seems appropriate. By dimensional arguments similar to those used in Example 4.1-1, we write

$$\frac{v_x}{v_\infty} = \Pi(\eta), \qquad \text{where } \eta = y\sqrt{\frac{1}{2}\frac{v_\infty}{\nu x}} \tag{4.4-25}$$

The factor of 2 is included to avoid having any numerical factors occur in the differential equation in Eq. 4.4-27. The stream function that gives the velocity distribution in Eq. 4.4-25 is

$$\psi(x, y) = -\sqrt{2v_\infty \nu x}f(\eta), \qquad \text{where } f(\eta) = \int_0^\eta \Pi'(\bar{\eta})d\bar{\eta} \tag{4.4-26}$$

This expression for the stream function is consistent with Eq. 4.4-25 as may be seen by using the relation $v_x = -\partial \psi/\partial y$ (given in Table 4.2-1). Substitution of Eq. 4.4-26 into Eq. 4.4-20 gives

$$-ff'' = f''' \tag{4.4-27}$$

Substitution into the boundary conditions gives

B.C. 1 and 2: at $\eta = 0$, $f = 0$ and $f' = 0$ (4.4-28)
B.C. 3 and 4: as $\eta \to \infty$, $f' \to 1$ (4.4-29)

Thus the determination of the flow field is reduced to the solution of one third-order ordinary differential equation.

This equation, along with the boundary conditions given, can be solved by numerical integration, and accurate tables of the solution are available.[3,4] The problem was originally solved by Blasius[7] using analytic approximations that proved to be quite accurate. A plot of his solution is shown in Fig. 4.4-3 along with experimental data taken subsequently. The agreement between theory and experiment is remarkably good.

The drag force on a plate of width W and length L may be calculated from the dimensionless velocity gradient at the wall, $f''(0) = 0.4696 \ldots$ as follows:

$$F_x = 2\int_0^W \int_0^L \left(+\mu \frac{\partial v_x}{\partial y}\right)\bigg|_{y=0} dx \, dz$$

$$= 2\int_0^W \int_0^L \left(+\mu v_\infty \frac{df'}{d\eta}\frac{\partial \eta}{\partial y}\right)\bigg|_{y=0} dx \, dz$$

$$= 2\int_0^W \int_0^L \mu v_\infty f''(0)\sqrt{\frac{1}{2}\frac{v_\infty}{\nu x}} \, dx \, dz$$

$$= 1.328\sqrt{\rho \mu L W^2 v_\infty^3} \tag{4.4-30}$$

This result has also been confirmed experimentally.[3,4]

Because of the approximations made in Eq. 4.4-10, the solution is most accurate at large local Reynolds numbers; that is, $Re_x = xv_\infty/\nu \gg 1$. The excluded region of lower Reynolds numbers is small enough to ignore in most drag calculations. More complete

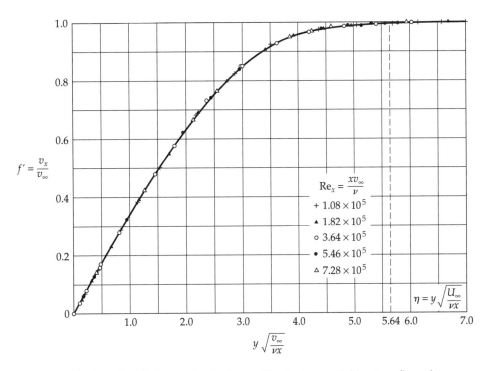

Fig. 4.4-3. Predicted and observed velocity profiles for tangential laminar flow along a flat plate. The solid line represents the solution of Eqs. 4.4-20 to 24, obtained by Blasius [see H. Schlichting, *Boundary-Layer Theory*, McGraw-Hill, New York, 7th edition (1979), p. 137].

analyses[8] indicate that Eq. 4.4-30 is accurate to within 3% for $Lv_\infty/\nu \geq 10^4$ and within 0.3% for $Lv_\infty/\nu \geq 10^6$.

The growth of the boundary layer with increasing x eventually leads to an unstable situation, and turbulent flow sets in. The transition is found to begin somewhere in the range of local Reynolds number of $\text{Re}_x = xv_\infty/\nu \geq 3 \times 10^5$ to 3×10^6, depending on the uniformity of the approaching stream.[8] Upstream of the transition region the flow remains laminar, and downstream it is turbulent.

EXAMPLE 4.4-3 *Flow near a Corner*	We now want to treat the boundary-layer problem analogous to Example 4.3-3, namely the flow near a corner (see Fig. 4.3-4). If $\alpha > 1$, the problem may also be interpreted as the flow along a wedge of included angle $\beta\pi$, with $\alpha = 2/(2 - \beta)$. For this system the external flow v_e is known from Eqs. 4.3-42 and 43, where we found that

$$v_e(x) = cx^{\beta/(2-\beta)} \qquad (4.4\text{-}31)$$

This was the expression that was found to be valid right at the wall (i.e., at $y = 0$). Here, it is assumed that the boundary layer is so thin that using the wall expression from ideal flow is adequate for the outer limit of the boundary-layer solution, at least for small values of x.

---

[8] Y. H. Kuo, *J. Math. Phys.*, **32**, 83–101 (1953); I. Imai, *J. Aero. Sci.*, **24**, 155–156 (1957).

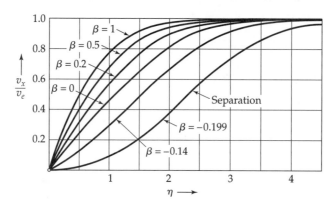

Fig. 4.4-4. Velocity profile for wedge flow with included angle $\beta\pi$. Negative values of β correspond to the flow around an "external corner" [see Fig. 4.3-4(ii)] with slip at the wall upstream of the corner.

SOLUTION

We now have to solve Eq. 4.4-11, using Eq. 4.4-31 for $v_e(x)$. When we introduce the stream function from the first row of Table 4.2-1, we obtain the following differential equation for ψ:

$$\frac{\partial\psi}{\partial y}\frac{\partial^2\psi}{\partial x\partial y} - \frac{\partial\psi}{\partial x}\frac{\partial^2\psi}{\partial y^2} = \left(\frac{c^2\beta}{2-\beta}\right)\frac{1}{x^{(2-3\beta)/(2-\beta)}} - \nu\frac{\partial^3\psi}{\partial y^3} \tag{4.4-32}$$

which corresponds to Eq. 4.4-20 with the term $v_e(dv_e/dx)$ added. It was discovered[9] that this equation can be reduced to a single ordinary differential equation by introducing a dimensionless stream function $f(\eta)$ by

$$\psi(x, y) = \sqrt{c\nu(2-\beta)}\,x^{1/(2-\beta)}f(\eta) \tag{4.4-33}$$

in which the independent variable is

$$\eta = \sqrt{\frac{c}{(2-\beta)\nu}}\frac{y}{x^{(1-\beta)/(2-\beta)}} \tag{4.4-34}$$

Then Eq. 4.4-32 becomes the *Falkner–Skan equation*[9]

$$f''' - ff'' - \beta(1 - f'^2) = 0 \tag{4.4-35}$$

This equation has been solved numerically with the appropriate boundary conditions, and the results are shown in Fig. 4.4-4.

It can be seen that for positive values of β, which corresponds to the systems shown in Fig. 4.3-4(a) and Fig. 4.3-5, the fluid is accelerating and the velocity profiles are stable. For negative values of β, down to $\beta = -0.199$, the flows are decelerating but stable, and no separation occurs. However, if $\beta > -0.199$, the velocity gradient at the wall becomes zero, and separation of the flow occurs. Therefore, for the interior corner flows and for wedge flows, there is no separation, but for the exterior corner flows, separation may occur.

QUESTIONS FOR DISCUSSION

1. For what types of problems is the method of combination of variables useful? The method of separation of variables?
2. Can the flow near a cylindrical rod of infinite length suddenly set in motion in the axial direction be described by the method in Example 4.1-1?

[9] V. M. Falkner and S. W. Skan, *Phil. Mag.*, **12**, 865–896 (1931); D. R. Hartree, *Proc. Camb. Phil. Soc.*, **33**, Part II, 223–239 (1937); H. Rouse (ed.), *Advanced Mechanics of Fluids*, Wiley, New York (1959), Chapter VII, Sec. D; H. Schlichting and K. Gersten, *Boundary-Layer Theory*, Springer-Verlag, Berlin (2000), pp. 169–173 (isothermal), 220–221 (nonisothermal); W. E. Stewart and R. Prober, *Int. J. Heat Mass Transfer*, **5**, 1149–1163 (1962); **6**, 221–229, 872 (1963), include wedge flow with heat and mass transfer.

3. What happens in Example 4.1-2 if one tries to solve Eq. 4.1-21 by the method of separation of variables without first recognizing that the solution can be written as the sum of a steady-state solution and a transient solution?

4. What happens if the separation constant after Eq. 4.1-27 is taken to be c or c^2 instead of $-c^2$?

5. Try solving the problem in Example 4.1-3 using trigonometric quantities in lieu of complex quantities.

6. How is the vorticity equation obtained and how may it be used?

7. How is the stream function defined, and why is it useful?

8. In what sense are the potential flow solutions and the boundary-layer flow solutions complementary?

9. List all approximate forms of the equations of change encountered thus far, and indicate their range of applicability.

PROBLEMS

4A.1 Time for attainment of steady state in tube flow.

(a) A heavy oil, with a kinematic viscosity of $3.45 \times 10^{-4}\,\mathrm{m^2/s}$, is at rest in a long vertical tube with a radius of 0.7 cm. The fluid is suddenly allowed to flow from the bottom of the tube by virtue of gravity. After what time will the velocity at the tube center be within 10% of its final value?

(b) What is the result if water at 68°F is used?

Note: The result shown in Fig. 4D.2 should be used.

Answers: (a) 6.4×10^{-2} s; (b) 0.22 s

4A.2 Velocity near a moving sphere. A sphere of radius R is falling in creeping flow with a terminal velocity v_∞ through a quiescent fluid of viscosity μ. At what horizontal distance from the sphere does the velocity of the fluid fall to 1% of the terminal velocity of the sphere?

Answer: About 37 diameters

4A.3 Construction of streamlines for the potential flow around a cylinder. Plot the streamlines for the flow around a cylinder using the information in Example 4.3-1 by the following procedure:

(a) Select a value of $\Psi = C$ (that is, select a streamline).

(b) Plot $Y = C + K$ (straight lines parallel to the X-axis) and $Y = K(X^2 + Y^2)$ (circles with radius $1/2K$, tangent to the X-axis at the origin).

(c) Plot the intersections of the lines and circles that have the same value of K.

(d) Join these points to get the streamline for $\Psi = C$.

Then select other values of C and repeat the process until the pattern of streamlines is clear.

4A.4 Comparison of exact and approximate profiles for flow along a flat plate. Compare the values of v_x/v_∞ obtained from Eq. 4.4-18 with those from Fig. 4.4-3, at the following values of $y\sqrt{v_\infty/\nu x}$: (a) 1.5, (b) 3.0, (c) 4.0. Express the results as the ratio of the approximate to the exact values.

Answers: (a) 0.96; (b) 0.99; (c) 1.01

4A.5 Numerical demonstration of the von Kármán momentum balance.

(a) Evaluate the integrals in Eq. 4.4-13 numerically for the Blasius velocity profile given in Fig. 4.4-3.

(b) Use the results of (a) to determine the magnitude of the wall shear stress $\tau_{yx}|_{y=0}$.

(c) Calculate the total drag force, F_x, for a plate of width W and length L, wetted on both sides. Compare your result with that obtained in Eq. 4.4-30.

Answers: (a) $\displaystyle\int_0^\infty \rho v_x(v_e - v_x)\,dy = 0.664\sqrt{\rho\mu v_\infty^3 x}$

$\displaystyle\int_0^\infty \rho(v_e - v_x)\,dy = 1.73\sqrt{\rho\mu v_\infty x}$

4A.6 Use of boundary-layer formulas. Air at 1 atm and 20°C flows tangentially on both sides of a thin, smooth flat plate of width $W = 10$ ft, and of length $L = 3$ ft in the direction of the flow. The velocity outside the boundary layer is constant at 20 ft/s.

(a) Compute the local Reynolds number $\mathrm{Re}_x = xv_\infty/\nu$ at the trailing edge.

(b) Assuming laminar flow, compute the approximate boundary-layer thickness, in inches, at the trailing edge. Use the results of Example 4.4-1.

(c) Assuming laminar flow, compute the total drag of the plate in lb_f. Use the results of Examples 4.4-1 and 2.

4A.7 Entrance flow in conduits.

(a) Estimate the entrance length for laminar flow in a circular tube. Assume that the boundary-layer thickness δ is given adequately by Eq. 4.4-17, with v_∞ of the flat-plate problem corresponding to v_{max} in the tube-flow problem. Assume further that the entrance length L_e can be taken to be the value of x at which $\delta = R$. Compare your result with the expression for L_e cited in §2.3—namely, $L_e = 0.035D\,\mathrm{Re}$.

(b) Rewrite the transition Reynolds number $xv_\infty/\nu \approx 3.5 \times 10^5$ (for the flat plate) by inserting δ from Eq. 4.4-17 in place of x as the characteristic length. Compare the quantity $\delta v_\infty/\nu$ thus obtained with the corresponding minimum transition Reynolds number for the flow through long smooth tubes.

(c) Use the method of (a) to estimate the entrance length in the flat duct shown in Fig. 4C.1. Compare the result with that given in Problem 4C.1(d).

4B.1 Flow of a fluid with a suddenly applied constant wall stress. In the system studied in Example 4.1-1, let the fluid be at rest before $t = 0$. At time $t = 0$ a constant force is applied to the fluid at the wall in the positive x direction, so that the shear stress τ_{yx} takes on a new constant value τ_0 at $y = 0$ for $t > 0$.

(a) Differentiate Eq. 4.1-1 with respect to y and multiply by $-\mu$ to obtain a partial differential equation for $\tau_{yx}(y, t)$.

(b) Write the boundary and initial conditions for this equation.

(c) Solve using the method in Example 4.1-1 to obtain

$$\frac{\tau_{yx}}{\tau_0} = 1 - \mathrm{erf}\,\frac{y}{\sqrt{4\nu t}} \tag{4B.1-1}$$

(d) Use the result in (c) to obtain the velocity profile. The following relation[1] will be helpful

$$\int_x^\infty (1 - \mathrm{erf}\,u)du = \frac{1}{\sqrt{\pi}}e^{-x^2} - x(1 - \mathrm{erf}\,x) \tag{4B.1-2}$$

4B.2 Flow near a wall suddenly set in motion (approximate solution) (Fig. 4B.2). Apply a procedure like that of Example 4.4-1 to get an approximate solution for Example 4.1.1.

(a) Integrate Eq. 4.4-1 over y to get

$$\int_0^\infty \frac{\partial v_x}{\partial t}\,dy = \nu\,\frac{\partial v_x}{\partial y}\bigg|_0^\infty \tag{4B.2-1}$$

Make use of the boundary conditions and the Leibniz rule for differentiating an integral (Eq. C.3-2) to rewrite Eq. 4B.2-1 in the form

$$\frac{d}{dt}\int_0^\infty \rho v_x\,dy = \tau_{yx}|_{y=0} \tag{4B.2-2}$$

Interpret this result physically.

---

[1] A useful summary of error functions and their properties can be found in H. S. Carslaw and J. C. Jaeger, *Conduction of Heat in Solids*, Oxford University Press, 2nd edition (1959), Appendix II.

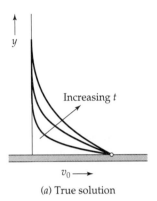
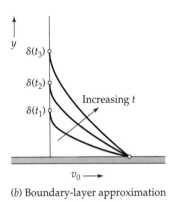

(a) True solution   (b) Boundary-layer approximation

Fig. 4B.2. Comparison of true and approximate velocity profiles near a wall suddenly set in motion with velocity v_0.

(b) We know roughly what the velocity profiles look like. We can make the following reasonable postulate for the profiles:

$$\frac{v_x}{v_\infty} = 1 - \frac{3}{2}\frac{y}{\delta(t)} + \frac{1}{2}\left(\frac{y}{\delta(t)}\right)^3 \qquad \text{for } 0 \le y \le \delta(t) \tag{4B.2-3}$$

$$\frac{v_x}{v_\infty} = 1 \qquad \text{for } y \ge \delta(t) \tag{4B.2-4}$$

Here $\delta(t)$ is a time-dependent boundary-layer thickness. Insert this approximate expression into Eq. 4B.2-2 to obtain

$$\delta\frac{d\delta}{dt} = 4\nu \tag{4B.2-5}$$

(c) Integrate Eq. 4B.2-5 with a suitable initial value of $\delta(t)$, and insert the result into Eq. 4B.2-3 to get the approximate velocity profiles.

(d) Compare the values of v_x/v_∞ obtained from (c) with those from Eq. 4.1-15 at $y/\sqrt{4\nu t} = 0.2, 0.5,$ and 1.0. Express the results as the ratio of the approximate value to the exact value.

Answer **(d)** 1.015, 1.026, 0.738

4B.3 Creeping flow around a spherical bubble. When a liquid flows around a gas bubble, circulation takes place within the bubble. This circulation lowers the interfacial shear stress, and, to a first approximation, we may assume that it is entirely eliminated. Repeat the development of Ex. 4.2-1 for such a gas bubble, assuming it is spherical.

(a) Show that B.C. 2 of Ex. 4.2-1 is replaced by

B.C. 2:   at $r = R$, $\dfrac{d}{dr}\left(\dfrac{1}{r^2}\dfrac{df}{dr}\right) + 2\dfrac{f}{r^4} = 0$ \qquad (4B.3-1)

and that the problem set-up is otherwise the same.

(b) Obtain the following velocity components:

$$v_r = v_\infty\left[1 - \left(\frac{R}{r}\right)\right]\cos\theta \tag{4B.3-2}$$

$$v_\theta = -v_\infty\left[1 - \frac{1}{2}\left(\frac{R}{r}\right)\right]\sin\theta \tag{4B.3-3}$$

(c) Next obtain the pressure distribution by using the equation of motion:

$$p = p_0 - \rho gh - \left(\frac{\mu v_\infty}{R}\right)\left(\frac{R}{r}\right)^2\cos\theta \tag{4B.3-4}$$

(d) Evaluate the total force of the fluid on the sphere to obtain

$$F_z = \tfrac{4}{3}\pi R^3 \rho g + 4\pi \mu R v_\infty \tag{4B.3-5}$$

This result may be obtained by the method of §2.6 or by integrating the z-component of $-[\mathbf{n} \cdot \boldsymbol{\pi}]$ over the sphere surface (**n** being the outwardly directed unit normal on the surface of the sphere).

4B.4 Use of the vorticity equation.

(a) Work Problem 2B.3 using the y-component of the vorticity equation (Eq. 3D.2-1) and the following boundary conditions: at $x = \pm B$, $v_z = 0$ and at $x = 0$, $v_z = v_{z,max}$. Show that this leads to

$$v_z = v_{z,max}[1 - (x/B)^2] \tag{4B.4-1}$$

Then obtain the pressure distribution from the z-component of the equation of motion.

(b) Work Problem 3B.6(b) using the vorticity equation, with the following boundary conditions: at $r = R$, $v_z = 0$ and at $r = \kappa R$, $v_z = v_0$. In addition an integral condition is needed to state that there is no net flow in the z direction. Find the pressure distribution in the system.

(c) Work the following problems using the vorticity equation: 2B.6, 2B.7, 3B.1, 3B.10, 3B.16.

4B.5 Steady potential flow around a stationary sphere.[2] In Example 4.2-1 we worked through the creeping flow around a sphere. We now wish to consider the flow of an incompressible, inviscid fluid in irrotational flow around a sphere. For such a problem, we know that the velocity potential must satisfy Laplace's equation (see text after Eq. 4.3-11).

(a) State the boundary conditions for the problem.

(b) Give reasons why the velocity potential ϕ can be postulated to be of the form $\phi(r, \theta) = f(r) \cos \theta$.

(c) Substitute the trial expression for the velocity potential in (b) into Laplace's equation for the velocity potential.

(d) Integrate the equation obtained in (c) and obtain the function $f(r)$ containing two constants of integration; determine these constants from the boundary conditions and find

$$\phi = -v_\infty R\left[\left(\frac{r}{R}\right) + \frac{1}{2}\left(\frac{R}{r}\right)^2\right] \cos \theta \tag{4B.5-1}$$

(e) Next show that

$$v_r = v_\infty\left[1 - \left(\frac{R}{r}\right)^3\right] \cos \theta \tag{4B.5-2}$$

$$v_\theta = -v_\infty\left[1 + \frac{1}{2}\left(\frac{R}{r}\right)^3\right] \sin \theta \tag{4B.5-3}$$

(f) Find the pressure distribution, and then show that at the sphere surface

$$\mathcal{P} - \mathcal{P}_\infty = \tfrac{1}{2}\rho v_\infty^2(1 - \tfrac{9}{4} \sin^2 \theta) \tag{4B.5-4}$$

4B.6 Potential flow near a stagnation point (Fig. 4B.6).

(a) Show that the complex potential $w = -v_0 z^2$ describes the flow near a plane stagnation point.

(b) Find the velocity components $v_x(x, y)$ and $v_y(x, y)$.

(c) Explain the physical significance of v_0.

---

[2] L. Landau and E. M. Lifshitz, *Fluid Mechanics*, Pergamon, Boston, 2nd edition (1987), pp. 21–26, contains a good collection of potential-flow problems.

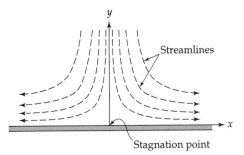

4B.7 Vortex flow.

(a) Show that the complex potential $w = (i\Gamma/2\pi) \ln z$ describes the flow in a vortex. Verify that the tangential velocity is given by $v_\theta = \Gamma/2\pi r$ and that $v_r = 0$. This type of flow is sometimes called a *free vortex*. Is this flow irrotational?

(b) Compare the functional dependence of v_θ on r in (a) with that which arose in Example 3.6-4. The latter kind of flow is sometimes called a *forced vortex*. Actual vortices, such as those that occur in a stirred tank, have a behavior intermediate between these two idealizations.

4B.8 The flow field about a line source. Consider the symmetric radial flow of an incompressible, inviscid fluid outward from an infinitely long uniform source, coincident with the z-axis of a cylindrical coordinate system. Fluid is being generated at a volumetric rate Γ per unit length of source.

(a) Show that the Laplace equation for the velocity potential for this system is

$$\frac{1}{r}\frac{d}{dr}\left(r\frac{d\phi}{dr}\right) = 0 \tag{4B.8-1}$$

(b) From this equation find the velocity potential, velocity, and pressure as functions of position:

$$\phi = -\frac{\Gamma}{2\pi}\ln r \qquad v_r = \frac{\Gamma}{2\pi r} \qquad \mathcal{P}_\infty - \mathcal{P} = \frac{\rho\Gamma^2}{8\pi^2 r^2} \tag{4B.8-2}$$

where $\mathcal{P}_\infty$ is the value of the modified pressure far away from the source.

(c) Discuss the applicability of the results in (b) to the flow field about a well drilled into a large body of porous rock.

(d) Sketch the flow net of streamlines and equipotential lines.

4B.9 Checking solutions to unsteady flow problems.

(a) Verify the solutions to the problems in Examples 4.1-1, 2, and 3 by showing that they satisfy the partial differential equations, initial conditions, and boundary conditions. To show that Eq. 4.1-15 satisfies the differential equation, one has to know how to differentiate an integral using the *Leibniz formula* given in §C.3.

(b) In Example 4.1-3 the initial condition is not satisfied by Eq. 4.1-57. Why?

4C.1 Laminar entrance flow in a slit[3] (Fig. 4C.1). Estimate the velocity distribution in the entrance region of the slit shown in the figure. The fluid enters at $x = 0$ with $v_y = 0$ and $v_x = \langle v_x \rangle$, where $\langle v_x \rangle$ is the average velocity inside the slit. Assume that the velocity distribution in the entrance region $0 < x < L_e$ is

$$\frac{v_x}{v_e} = 2\left(\frac{y}{\delta}\right) - \left(\frac{y}{\delta}\right)^2 \qquad \text{(boundary layer region, } 0 < y < \delta) \tag{4C.1-1}$$

$$\frac{v_x}{v_e} = 1 \qquad \text{(potential flow region, } \delta < y < B) \tag{4C.1-2}$$

in which δ and v_e are functions of x, yet to be determined.

---

[3] A numerical solution to this problem using the Navier–Stokes equation has been given by Y. L. Wang and P. A. Longwell, *AIChE Journal*, **10**, 323–329 (1964).

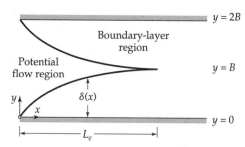

$y = 2B$ **Fig. 4C.1.** Entrance flow into a slit.

Boundary-layer region

Potential flow region

$y = B$

y

$\delta(x)$

x

$y = 0$

L_e

(a) Use the above two equations to get the mass flow rate w through an arbitrary cross section in the region $0 < x < L_e$. Then evaluate w from the inlet conditions and obtain

$$\frac{v_e(x)}{\langle v_x \rangle} = \frac{B}{B - \frac{1}{3}\delta(x)} \tag{4C.1-3}$$

(b) Next use Eqs. 4.4-13, 4C.1-1, and 4C.1-2 with ∞ replaced by B (why?) to obtain a differential equation for the quantity $\Delta = \delta/B$:

$$\frac{6\Delta + 7\Delta^2}{(3 - \Delta)^2} \frac{d\Delta}{dx} = 10\left(\frac{\nu}{\langle v_x \rangle B^2}\right) \tag{4C.1-4}$$

(c) Integrate this equation with a suitable initial condition to obtain the following relation between the boundary-layer thickness and the distance down the duct:

$$\frac{\nu x}{\langle v_x \rangle B^2} = \frac{1}{10}\left[7\Delta + 48 \ln\left(1 - \tfrac{1}{3}\Delta\right) + \frac{27\Delta}{3 - \Delta}\right] \tag{4C.1-5}$$

(d) Compute the entrance length L_e from Eq. 4C.1-5, where L_e is that value of x for which $\delta(x) = B$.

(e) Using potential flow theory, evaluate $\mathcal{P} - \mathcal{P}_0$ in the entrance region, where $\mathcal{P}_0$ is the value of the modified pressure at $x = 0$.

Answers: **(d)** $L_e = 0.104\langle v_x \rangle B^2/\nu$; **(e)** $\mathcal{P} - \mathcal{P}_0 = \frac{1}{2}\rho\langle v_x \rangle^2\left[1 - \left(\frac{3}{3 - \Delta}\right)^2\right]$

4C.2 Torsional oscillatory viscometer (Fig. 4C.2). In the torsional oscillatory viscometer, the fluid is placed between a "cup" and "bob" as shown in the figure. The cup is made to undergo small sinusoidal oscillations in the tangential direction. This motion causes the bob, suspended by a torsion wire, to oscillate with the same frequency, but with a different amplitude

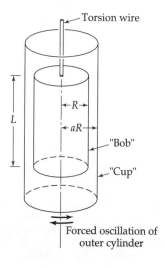

Torsion wire

R

L

aR

"Bob"

"Cup"

Forced oscillation of outer cylinder

Fig. 4C.2. Sketch of a torsional oscillatory viscometer.

and phase. The amplitude ratio (ratio of amplitude of output function to input function) and phase shift both depend on the viscosity of the fluid and hence can be used for determining the viscosity. It is assumed throughout that the oscillations are of *small* amplitude. Then the problem is a linear one, and it can be solved either by Laplace transform or by the method outlined in this problem.

(a) First, apply Newton's second law of motion to the cylindrical bob for the special case that the annular space is completely evacuated. Show that the natural frequency of the system is $\omega_0 = \sqrt{k/I}$, in which I is the moment of inertia of the bob, and k is the spring constant for the torsion wire.

(b) Next, apply Newton's second law when there is a fluid of viscosity μ in the annular space. Let θ_R be the angular displacement of the bob at time t, and v_θ be the tangential velocity of the fluid as a function of r and t. Show that the equation of motion of the bob is

(Bob)
$$I\frac{d^2\theta_R}{dt^2} = -k\theta_R + (2\pi RL)(R)\left(\mu r \frac{\partial}{\partial r}\left(\frac{v_\theta}{r}\right)\right)\Bigg|_{r=R} \qquad (4C.2\text{-}1)$$

If the system starts from rest, we have the initial conditions

I.C.: $\qquad\qquad$ at $t = 0$, $\qquad \theta_R = 0 \quad$ and $\quad \dfrac{d\theta_R}{dt} = 0 \qquad\qquad (4C.2\text{-}2)$

(c) Next, write the equation of motion for the fluid along with the relevant initial and boundary conditions:

(Fluid)
$$\rho\frac{\partial v_\theta}{\partial t} = \mu \frac{\partial}{\partial r}\left(\frac{1}{r}\frac{\partial}{\partial r}(rv_\theta)\right) \qquad (4C.2\text{-}3)$$

I.C.: $\qquad\qquad$ at $t = 0$, $\qquad v_\theta = 0 \qquad\qquad\qquad\qquad (4C.2\text{-}4)$

B.C. 1: $\qquad\qquad$ at $r = R$, $\qquad v_\theta = R\dfrac{d\theta_R}{dt} \qquad\qquad (4C.2\text{-}5)$

B.C. 2: $\qquad\qquad$ at $r = aR$, $\qquad v_\theta = aR\dfrac{d\theta_{aR}}{dt} \qquad\qquad (4C.2\text{-}6)$

The function $\theta_{aR}(t)$ is a specified sinusoidal function (the "input"). Draw a sketch showing θ_{aR} and θ_R as functions of time, and defining the *amplitude ratio* and the *phase shift*.

(d) Simplify the starting equations, Eqs. 4C.2-1 to 6, by making the assumption that a is only slightly greater than unity, so that the curvature may be neglected (the problem can be solved without making this assumption[4]). This suggests that a suitable dimensionless distance variable is $x = (r - R)/[(a - 1)R]$. Recast the entire problem in dimensionless quantities in such a way that $1/\omega_0 = \sqrt{I/k}$ is used as a characteristic time, and so that the viscosity appears in just one dimensionless group. The only choice turns out to be:

time: $\qquad\qquad\qquad\qquad \tau = \sqrt{\dfrac{k}{I}}\,t \qquad\qquad\qquad (4C.2\text{-}7)$

velocity: $\qquad\qquad\qquad \phi = \dfrac{2\pi R^3 L\rho(a - 1)}{kI}\,v_\theta \qquad\qquad (4C.2\text{-}8)$

viscosity: $\qquad\qquad\qquad M = \dfrac{\mu/\rho}{(a-1)^2 R^2}\sqrt{\dfrac{I}{k}} \qquad\qquad (4C.2\text{-}9)$

reciprocal of moment of inertia: $\quad A = \dfrac{2\pi R^4 L\rho(a - 1)}{I} \qquad\qquad (4C.2\text{-}10)$

---

[4] H. Markovitz, *J. Appl. Phys.*, **23**, 1070–1077 (1952) has solved the problem without assuming a small spacing between the cup and bob. The cup-and-bob instrument has been used by L. J. Wittenberg, D. Ofte, and C. F. Curtiss, *J. Chem. Phys.*, **48**, 3253–3260 (1968), to measure the viscosity of liquid plutonium alloys.

Show that the problem can now be restated as follows:

(bob)
$$\frac{d^2\theta_R}{d\tau^2} = -\theta_R + M\left(\frac{\partial\phi}{\partial x}\right)\Big|_{x=0} \qquad \text{at } t = 0, \theta_R = 0 \qquad (4C.2\text{-}11)$$

(fluid)
$$\frac{\partial\phi}{\partial\tau} = M\frac{\partial^2\phi}{\partial x^2} \qquad \begin{cases} \text{at } \tau = 0, & \phi = 0 \\ \text{at } x = 0, & \phi = A(d\theta_R/dt) \\ \text{at } x = 1, & \phi = A(d\theta_{aR}/dt) \end{cases} \qquad (4C.2\text{-}12)$$

From these two equations we want to get θ_R and ϕ as functions of x and τ, with M and A as parameters.

(e) Obtain the "sinusoidal steady-state" solution by taking the input function θ_{aR} (the displacement of the cup) to be of the form

$$\theta_{aR}(\tau) = \theta_{aR}^{\circ}\Re\{e^{i\bar{\omega}\tau}\} \qquad (\theta_{aR}^{\circ} \text{ is real}) \qquad (4C.2\text{-}13)$$

in which $\bar{\omega} = \omega/\omega_0 = \omega\sqrt{I/k}$ is a dimensionless frequency. Then postulate that the bob and fluid motions will also be sinusoidal, but with different amplitudes and phases:

$$\theta_R(\tau) = \Re\{\theta_R^{\circ}e^{i\bar{\omega}\tau}\} \qquad (\theta_R^{\circ} \text{ is complex}) \qquad (4C.2\text{-}14)$$
$$\phi(x, \tau) = \Re\{\phi^{\circ}(x)e^{i\bar{\omega}\tau}\} \qquad (\phi^{\circ}(x) \text{ is complex}) \qquad (4C.2\text{-}15)$$

Verify that the amplitude ratio is given by $|\theta_R^{\circ}|/\theta_{aR}^{\circ}$, where $|\cdots|$ indicates the absolute magnitude of a complex quantity. Further show that the phase angle α is given by $\tan\alpha = \Im\{\theta_R^{\circ}\}/\Re\{\theta_R^{\circ}\}$, where $\Re$ and $\Im$ stand for the real and imaginary parts, respectively.

(f) Substitute the postulated solutions of (e) into the equations in (d) to obtain equations for the complex amplitudes θ_R° and ϕ°.

(g) Solve the equation for $\phi^{\circ}(x)$ and verify that

$$\frac{d\phi^{\circ}}{dx}\Big|_{x=0} = -\frac{A(i\bar{\omega})^{3/2}}{\sqrt{M}}\left(\frac{\theta_R^{\circ}\cosh\sqrt{i\bar{\omega}/M} - \theta_{aR}^{\circ}}{\sinh\sqrt{i\bar{\omega}/M}}\right) \qquad (4C.2\text{-}16)$$

(h) Next, solve the θ_R° equation to obtain

$$\frac{\theta_R^{\circ}}{\theta_{aR}^{\circ}} = \frac{AMi\bar{\omega}}{(1 - \bar{\omega}^2)\dfrac{\sinh\sqrt{i\bar{\omega}/M}}{\sqrt{i\bar{\omega}/M}} + AMi\bar{\omega}\cosh\sqrt{i\bar{\omega}/M}} \qquad (4C.2\text{-}17)$$

from which the amplitude ratio $|\theta_R^{\circ}|/\theta_{aR}^{\circ}$ and phase shift α can be found.

(i) For high-viscosity fluids, we can seek a power series by expanding the hyperbolic functions in Eq. 4C.2-17 to get a power series in $1/M$. Show that this leads to

$$\frac{\theta_{aR}^{\circ}}{\theta_R^{\circ}} = 1 + \frac{i}{M}\left(\frac{\bar{\omega}^2 - 1}{A\bar{\omega}} + \frac{\bar{\omega}}{2}\right) - \frac{1}{M^2}\left(\frac{\bar{\omega}^2 - 1}{6A} + \frac{\bar{\omega}^2}{24}\right) + O\left(\frac{1}{M^3}\right) \qquad (4C.2\text{-}18)$$

From this, find the amplitude ratio and the phase angle.

(j) Plot $|\theta_R^{\circ}|/\theta_{aR}^{\circ}$ versus $\bar{\omega}$ for $\mu/\rho = 10 \text{ cm}^2/\text{s}$, $L = 25$ cm, $R = 5.5$ cm, $I = 2500 \text{ gm}/\text{cm}^2$, $k = 4 \times 10^6$ dyn cm. Where is the maximum in the curve?

4C.3   Darcy's equation for flow through porous media. For the flow of a fluid through a porous medium, the equations of continuity and motion may be replaced by

smoothed continuity equation
$$\varepsilon\frac{\partial\rho}{\partial t} = -(\nabla\cdot\rho\mathbf{v}_0) \qquad (4C.3\text{-}1)$$

Darcy's equation[5]
$$\mathbf{v}_0 = -\frac{\kappa}{\mu}(\nabla p - \rho\mathbf{g}) \qquad (4C.3\text{-}2)$$

---

[5] **Henry Philibert Gaspard Darcy** (1803–1858) studied in Paris and became famous for designing the municipal water-supply system in Dijon, the city of his birth. H. Darcy, *Les Fontaines Publiques de la Ville de Dijon*, Victor Dalmont, Paris (1856). For further discussions of "Darcy's law," see J. Happel and H. Brenner, *Low Reynolds Number Hydrodynamics*, Martinus Nihjoff, Dordrecht (1983); and H. Brenner and D. A. Edwards, *Macrotransport Processes*, Butterworth-Heinemann, Boston (1993).

in which ε, the *porosity*, is the ratio of pore volume to total volume, and κ is the *permeability* of the porous medium. The velocity $\mathbf{v}_0$ in these equations is the *superficial velocity*, which is defined as the volume rate of flow through a unit cross-sectional area of the solid plus fluid, averaged over a small region of space—small with respect to the macroscopic dimensions in the flow system, but large with respect to the pore size. The density and pressure are averaged over a region available to flow that is large with respect to the pore size. Equation 4C.3-2 was proposed empirically to describe the slow seepage of fluids through granular media.

When Eqs. 4C.3-1 and 2 are combined we get

$$\left(\frac{\varepsilon\mu}{\kappa}\right)\frac{\partial\rho}{\partial t} = (\nabla \cdot \rho(\nabla p - \rho\mathbf{g})) \tag{4C.3-3}$$

for constant viscosity and permeability. This equation and the equation of state describe the motion of a fluid in a porous medium. For most purposes we may write the *equation of state* as

$$\rho = \rho_0 p^m e^{\beta p} \tag{4C.3-4}$$

in which ρ_0 is the fluid density at unit pressure, and the following parameters have been given:[6]

1. Incompressible liquids $m = 0$ $\beta = 0$
2. Compressible liquids $m = 0$ $\beta \neq 0$
3. Isothermal expansion of gases $\beta = 0$ $m = 1$
4. Adiabatic expansion of gases $\beta = 0$ $m = C_V/C_p = 1/\gamma$

Show that Eqs. 4C.3-3 and 4 can be combined and simplified for these four categories to give (for gases it is customary to neglect the gravity terms since they are small compared with the pressure terms):

Case 1.

$$\nabla^2 \mathcal{P} = 0 \tag{4C.3-5}$$

Case 2.

$$\left(\frac{\varepsilon\mu\beta}{\kappa}\right)\frac{\partial\rho}{\partial t} = \nabla^2\rho - (\nabla \cdot \rho^2\beta\mathbf{g}) \tag{4C.3-6}$$

Case 3.

$$\left(\frac{2\varepsilon\mu\rho_0}{\kappa}\right)\frac{\partial\rho}{\partial t} = \nabla^2\rho^2 \tag{4C.3-7}$$

Case 4.

$$\left(\frac{(m+1)\varepsilon\mu\rho_0^{1/m}}{\kappa}\right)\frac{\partial\rho}{\partial t} = \nabla^2\rho^{(1+m)/m} \tag{4C.3-8}$$

Note that Case 1 leads to *Laplace's equation*, Case 2 without the gravity term leads to the *heat-conduction* or *diffusion equation*, and Cases 3 and 4 lead to nonlinear equations.[7]

4C.4 Radial flow through a porous medium (Fig. 4C.4). A fluid flows through a porous cylindrical shell with inner and outer radii R_1 and R_2, respectively. At these surfaces, the pressures are known to be p_1 and p_2, respectively. The length of the cylindrical shell is h.

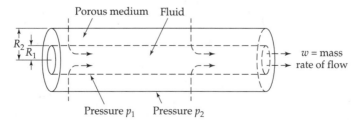

Fig. 4C.4. Radial flow through a porous medium.

---

[6] M. Muskat, *Flow of Homogeneous Fluids Through Porous Media*, McGraw-Hill (1937).

[7] For the boundary condition at a porous surface that bounds a moving fluid, see G. S. Beavers and D. D. Joseph, *J. Fluid Mech.*, **30**, 197–207 (1967) and G. S. Beavers, E. M. Sparrow, and B. A. Masha, *AIChE Journal*, **20**, 596–597 (1974).

(a) Find the pressure distribution, radial flow velocity, and mass rate of flow for an incompressible fluid.

(b) Rework (a) for a compressible liquid and for an ideal gas.

$$Answers: \text{(a)} \quad \frac{\mathcal{P} - \mathcal{P}_1}{\mathcal{P}_2 - \mathcal{P}_1} = \frac{\ln{(r/R_1)}}{\ln{(R_2/R_1)}} \qquad v_{0r} = -\frac{\kappa}{\mu r}\frac{\mathcal{P}_2 - \mathcal{P}_1}{\ln{(R_2/R_1)}} \qquad w = \frac{2\pi\kappa h(p_2 - p_1)\rho}{\mu \ln{(R_2/R_1)}}$$

4D.1 **Flow near an oscillating wall.**[8] Show, by using Laplace transforms, that the complete solution to the problem stated in Eqs. 4.1-44 to 47 is

$$\frac{v_x}{v_0} = e^{-\sqrt{\omega/2\nu}y}\cos{(\omega t - \sqrt{\omega/2\nu}y)} - \frac{1}{\pi}\int_0^\infty e^{-\bar{\omega}t}(\sin\sqrt{\bar{\omega}/\nu}y)\frac{\bar{\omega}}{\omega^2 + \bar{\omega}^2}d\bar{\omega} \qquad (4D.1-1)$$

4D.2 **Start-up of laminar flow in a circular tube** (Fig. 4D.2). A fluid of constant density and viscosity is contained in a very long pipe of length L and radius R. Initially the fluid is at rest. At time $t = 0$, a pressure gradient $(\mathcal{P}_0 - \mathcal{P}_L)/L$ is imposed on the system. Determine how the velocity profiles change with time.

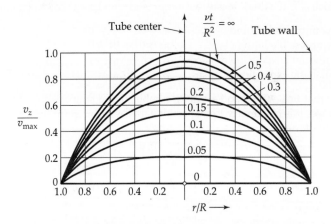

Fig. 4D.2. Velocity distribution for the unsteady flow resulting from a suddenly impressed pressure gradient in a circular tube [P. Szymanski, *J. Math. Pures Appl.*, Series 9, **11**, 67–107 (1932)].

(a) Show that the relevant equation of motion can be put into dimensionless form as follows:

$$\frac{\partial\phi}{\partial\tau} = 4 + \frac{1}{\xi}\frac{\partial}{\partial\xi}\left(\xi\frac{\partial\phi}{\partial\xi}\right) \qquad (4D.2-1)$$

in which $\xi = r/R$, $\tau = \mu t/\rho R^2$, and $\phi = [(\mathcal{P}_0 - \mathcal{P}_L)R^2/4\mu L]^{-1}v_z$.

(b) Show that the asymptotic solution for large time is $\phi_\infty = 1 - \xi^2$. Then define ϕ_t by $\phi(\xi, \tau) = \phi_\infty(\xi) - \phi_t(\xi, \tau)$, and solve the partial differential equation for ϕ_t by the method of separation of variables.

(c) Show that the final solution is

$$\phi(\xi, \tau) = (1 - \xi^2) - 8\sum_{n=1}^\infty \frac{J_0(\alpha_n\xi)}{\alpha_n^3 J_1(\alpha_n)}\exp(-\alpha_n^2\tau) \qquad (4D.2-2)$$

in which $J_n(\xi)$ is the nth order Bessel function of ξ, and the α_n are the roots of the equation $J_0(\alpha_n) = 0$. The result is plotted in Fig. 4D.2.

---

[8] H. S. Carslaw and J. C. Jaeger, *Conduction of Heat in Solids*, Oxford University Press, 2nd edition (1959), p. 319, Eq. (8), with $\varepsilon = \frac{1}{2}\pi$ and $\bar{\omega} = \kappa u^2$.

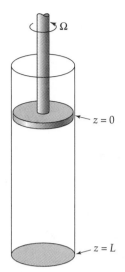

Fig. 4D.3. Rotating disk in a circular tube.

4D.3 Flows in the disk-and-tube system (Fig. 4D.3).[9]

(a) A fluid in a circular tube is caused to move tangentially by a tightly fitting rotating disk at the liquid surface at $z = 0$; the bottom of the tube is located at $z = L$. Find the steady-state velocity distribution $v_\theta(r, z)$, when the angular velocity of the disk is Ω. Assume that creeping flow prevails throughout, so that there is no secondary flow. Find the limit of the solution as $L \to \infty$.

(b) Repeat the problem for the unsteady flow. The fluid is at rest before $t = 0$, and the disk suddenly begins to rotate with an angular velocity Ω at $t = 0$. Find the velocity distribution $v_\theta(r, z, t)$ for a column of fluid of height L. Then find the solution for the limit as $L \to \infty$.

(c) If the disk is oscillating sinusoidally in the tangential direction with amplitude Ω_0, obtain the velocity distribution in the tube when the "oscillatory steady state" has been attained. Repeat the problem for a tube of infinite length.

4D.4 Unsteady annular flows.

(a) Obtain a solution to the Navier–Stokes equation for the start-up of *axial* annular flow by a sudden impressed pressure gradient. Check your result against the published solution.[10]

(b) Solve the Navier–Stokes equation for the unsteady *tangential* flow in an annulus. The fluid is at rest for $t < 0$. Starting at $t = 0$ the outer cylinder begins rotating with a constant angular velocity to cause laminar flow for $t > 0$. Compare your result with the published solution.[11]

4D.5 Stream functions for three-dimensional flow.

(a) Show that the velocity functions $\rho \mathbf{v} = [\nabla \times \mathbf{A}]$ and $\rho \mathbf{v} = [(\nabla \psi_1) \times (\nabla \psi_2)]$ both satisfy the equation of continuity identically for steady flow. The functions ψ_1, ψ_2, and $\mathbf{A}$ are arbitrary, except that their derivatives appearing in $(\nabla \cdot \rho \mathbf{v})$ must exist.

(b) Show that the expression $\mathbf{A}/\rho = \boldsymbol{\delta}_3 \psi/h_3$ reproduces the velocity components for the four incompressible flows of Table 4.2-1. Here h_3 is the scale factor for the third coordinate (see §A.7). (Read the general vector $\mathbf{v}$ of Eq. A.7-18 here as $\mathbf{A}$.)

(c) Show that the streamlines of $[(\nabla \psi_1) \times (\nabla \psi_2)]$ are given by the intersections of the surfaces ψ_1 = constant and ψ_2 = constant. Sketch such a pair of surfaces for the flow in Fig. 4.3-1.

(d) Use Stokes' theorem (Eq. A.5-4) to obtain an expression in terms of $\mathbf{A}$ for the mass flow rate through a surface S bounded by a closed curve C. Show that the vanishing of $\mathbf{v}$ on C does not imply the vanishing of $\mathbf{A}$ on C.

[9] W. Hort, *Z. tech. Phys.*, **10**, 213 (1920); C. T. Hill, J. D. Huppler, and R. B. Bird, *Chem. Engr. Sci.*, **21**, 815–817 (1966).

[10] W. Müller, *Zeits. für angew. Math. u. Mech.*, **16**, 227–228 (1936).

[11] R. B. Bird and C. F. Curtiss, *Chem. Engr. Sci*, **11**, 108–113 (1959).

Chapter 5

Velocity Distributions in Turbulent Flow

§5.1 Comparisons of laminar and turbulent flows

§5.2 Time-smoothed equations of change for incompressible fluids

§5.3 The time-smoothed velocity profile near a wall

§5.4 Empirical expressions for the turbulent momentum flux

§5.5 Turbulent flow in ducts

§5.6° Turbulent flow in jets

In the previous chapters we discussed laminar flow problems only. We have seen that the differential equations describing laminar flow are well understood and that, for a number of simple systems, the velocity distribution and various derived quantities can be obtained in a straightforward fashion. The limiting factor in applying the equations of change is the mathematical complexity that one encounters in problems for which there are several velocity components that are functions of several variables. Even there, with the rapid development of computational fluid dynamics, such problems are gradually yielding to numerical solution.

In this chapter we turn our attention to turbulent flow. Whereas laminar flow is orderly, turbulent flow is chaotic. It is this chaotic nature of turbulent flow that poses all sorts of difficulties. In fact, one might question whether or not the equations of change given in Chapter 3 are even capable of describing the violently fluctuating motions in turbulent flow. Since the sizes of the turbulent eddies are several orders of magnitude larger than the mean free path of the molecules of the fluid, the equations of change *are* applicable. Numerical solutions of these equations are obtainable and can be used for studying the details of the turbulence structure. For many purposes, however, we are not interested in having such detailed information, in view of the computational effort required. Therefore, in this chapter we shall concern ourselves primarily with methods that enable us to describe the time-smoothed velocity and pressure profiles.

In §5.1 we start by comparing the experimental results for laminar and turbulent flows in several flow systems. In this way we can get some qualitative ideas about the main differences between laminar and turbulent motions. These experiments help to define some of the challenges that face the fluid dynamicist.

In §5.2 we define several *time-smoothed* quantities, and show how these definitions can be used to time-average the equations of change over a short time interval. These equations describe the behavior of the time-smoothed velocity and pressure. The time-smoothed equation of motion, however, contains the *turbulent momentum flux*. This flux cannot be simply related to velocity gradients in the way that the momentum flux is given by Newton's law of viscosity in Chapter 1. At the present time the turbulent momentum flux is usually estimated experimentally or else modeled by some type of empiricism based on experimental measurements.

Fortunately, for turbulent flow near a solid surface, there are several rather general results that are very helpful in fluid dynamics and transport phenomena: the Taylor series development for the velocity near the wall; and the logarithmic and power law velocity profiles for regions further from the wall, the latter being obtained by dimensional reasoning. These expressions for the time-smoothed velocity distribution are given in §5.3.

In the following section, §5.4, we present a few of the empiricisms that have been proposed for the turbulent momentum flux. These empiricisms are of historical interest and have also been widely used in engineering calculations. When applied with proper judgment, these empirical expressions can be useful.

The remainder of the chapter is devoted to a discussion of two types of turbulent flows: flows in closed conduits (§5.5) and flows in jets (§5.6). These flows illustrate the two classes of flows that are usually discussed under the headings of *wall turbulence* and *free turbulence*.

In this brief introduction to turbulence we deal primarily with the description of the fully developed turbulent flow of an incompressible fluid. We do not consider the theoretical methods for predicting the inception of turbulence nor the experimental techniques devised for probing the structure of turbulent flow. We also give no discussion of the statistical theories of turbulence and the way in which the turbulent energy is distributed over the various modes of motion. For these and other interesting topics, the reader should consult some of the standard books on turbulence.[1-6] There is a growing literature on experimental and computational evidence for "coherent structures" (vortices) in turbulent flows.[7]

Turbulence is an important subject. In fact, most flows encountered in engineering are turbulent and not laminar! Although our understanding of turbulence is far from satisfactory, it is a subject that must be studied and appreciated. For the solution to industrial problems we cannot get neat analytical results, and, for the most part, such problems are attacked by using a combination of dimensional analysis and experimental data. This method is discussed in Chapter 6.

---

[1] S. Corrsin, "Turbulence: Experimental Methods," in *Handbuch der Physik*, Springer, Berlin (1963), Vol. VIII/2. **Stanley Corrsin** (1920–1986), a professor at The Johns Hopkins University, was an excellent experimentalist and teacher; he studied the interaction between chemical reactions and turbulence and the propagation of the double temperature correlations.

[2] A. A. Townsend, *The Structure of Turbulent Shear Flow*, Cambridge University Press, 2nd edition (1976); see also A. A. Townsend in *Handbook of Fluid Dynamics* (V. L. Streeter, ed.), McGraw-Hill (1961) for a readable survey.

[3] J. O. Hinze, *Turbulence*, McGraw-Hill, New York, 2nd edition (1975).

[4] H. Tennekes and J. L. Lumley, *A First Course in Turbulence*, MIT Press, Cambridge, Mass. (1972); Chapters 1 and 2 of this book provide an introduction to the physical interpretations of turbulent flow phenomena.

[5] M. Lesieur, *La Turbulence*, Presses Universitaires de Grenoble (1994); this book contains beautiful color photographs of turbulent flow systems.

[6] Several books that cover material beyond the scope of this text are: W. D. McComb, *The Physics of Fluid Turbulence*, Oxford University Press (1990); T. E. Faber, *Fluid Dynamics for Physicists*, Cambridge University Press (1995); U. Frisch, *Turbulence*, Cambridge University Press (1995).

[7] P. Holmes, J. L. Lumley, and G. Berkooz, *Turbulence, Coherent Structures, Dynamical Systems, and Symmetry*, Cambridge University Press (1996); F. Waleffe, *Phys. Rev. Lett.*, **81**, 4140–4148 (1998).

§5.1   COMPARISONS OF LAMINAR AND TURBULENT FLOWS

Before discussing any theoretical ideas about turbulence, it is important to summarize the differences between laminar and turbulent flows in several simple systems. Specifically we consider the flow in conduits of circular and triangular cross section, flow along a flat plate, and flows in jets. The first three of these were considered for laminar flow in §2.3, Problem 3B.2, and §4.4.

Circular Tubes

For the steady, fully developed, laminar flow in a circular tube of radius R we know that the velocity distribution and the average velocity are given by

$$\frac{v_z}{v_{z,\max}} = 1 - \left(\frac{r}{R}\right)^2 \quad \text{and} \quad \frac{\langle v_z \rangle}{v_{z,\max}} = \frac{1}{2} \qquad (\text{Re} < 2100) \qquad (5.1\text{-}1, 2)$$

and that the pressure drop and mass flow rate w are linearly related:

$$\mathcal{P}_0 - \mathcal{P}_L = \left(\frac{8\mu L}{\pi \rho R^4}\right) w \qquad (\text{Re} < 2100) \qquad (5.1\text{-}3)$$

For turbulent flow, on the other hand, the velocity is fluctuating with time chaotically at each point in the tube. We can measure a "time-smoothed velocity" at each point with, say, a Pitot tube. This type of instrument is not sensitive to rapid velocity fluctuations, but senses the velocity averaged over several seconds. The time-smoothed velocity (which is defined in the next section) will have a z-component represented by $\bar{v}_z$, and its shape and average value will be given very roughly by[1]

$$\frac{\bar{v}_z}{\bar{v}_{z,\max}} \approx \left(1 - \frac{r}{R}\right)^{1/7} \quad \text{and} \quad \frac{\langle \bar{v}_z \rangle}{\bar{v}_{z,\max}} \approx \frac{4}{5} \qquad (10^4 < \text{Re} < 10^5) \qquad (5.1\text{-}4, 5)$$

This $\frac{1}{7}$-power expression for the velocity distribution is too crude to give a realistic velocity derivative at the wall. The laminar and turbulent velocity profiles are compared in Fig. 5.1-1.

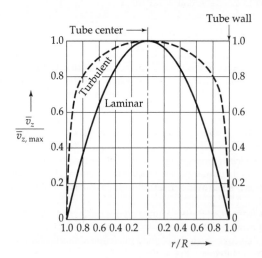

Fig. 5.1-1. Qualitative comparison of laminar and turbulent velocity profiles. For a more detailed description of the turbulent velocity distribution near the wall, see Fig. 5.5-3.

---

[1] H. Schlichting, *Boundary-Layer Theory*, McGraw-Hill, New York, 7th edition (1979), Chapter XX (tube flow), Chapters VII and XXI (flat plate flow), Chapters IX and XXIIV (jet flows).

Over the same range of Reynolds numbers the mass rate of flow and the pressure drop are no longer proportional but are related approximately by

$$\mathscr{P}_0 - \mathscr{P}_L \approx 0.0198\left(\frac{2}{\pi}\right)^{7/4}\left(\frac{\mu^{1/4}L}{\rho R^{19/4}}\right)w^{7/4} \qquad (10^4 < \mathrm{Re} < 10^5) \tag{5.1-6}$$

The stronger dependence of pressure drop on mass flow rate for turbulent flow results from the fact that more energy has to be supplied to maintain the violent eddy motion in the fluid.

The laminar–turbulent transition in circular pipes normally occurs at a *critical Reynolds number* of roughly 2100, although this number may be higher if extreme care is taken to eliminate vibrations in the system.[2] The transition from laminar flow to turbulent flow can be demonstrated by the simple experiment originally performed by Reynolds. One sets up a long transparent tube equipped with a device for injecting a small amount of dye into the stream along the tube axis. When the flow is laminar, the dye moves downstream as a straight, coherent filament. For turbulent flow, on the other hand, the dye spreads quickly over the entire cross section, similarly to the motion of particles in Fig. 2.0-1, because of the eddying motion (turbulent diffusion).

Noncircular Tubes

For developed laminar flow in the triangular duct shown in Fig. 3B.2(b), the fluid particles move rectilinearly in the z direction, parallel to the walls of the duct. By contrast, in turbulent flow there is superposed on the time-smoothed flow in the z direction (the *primary flow*) a time-smoothed motion in the xy-plane (the *secondary flow*). The secondary flow is much weaker than the primary flow and manifests itself as a set of six vortices arranged in a symmetric pattern around the duct axis (see Fig. 5.1-2). Other noncircular tubes also exhibit secondary flows.

Flat Plate

In §4.4 we found that for the laminar flow around a flat plate, wetted on both sides, the solution of the boundary layer equations gave the drag force expression

$$F = 1.328\sqrt{\rho\mu L W^2 v_\infty^3} \qquad \text{(laminar)}\ 0 < \mathrm{Re}_L < 5 \times 10^5 \tag{5.1-7}$$

in which $\mathrm{Re}_L = L v_\infty \rho / \mu$ is the Reynolds number for a plate of length L; the plate width is W, and the approach velocity of the fluid is v_∞.

Fig. 5.1-2. Sketch showing the secondary flow patterns for turbulent flow in a tube of triangular cross section [H. Schlichting, *Boundary-Layer Theory*, McGraw-Hill, New York, 7th edition (1979), p. 613].

---

[2] O. Reynolds, *Phil. Trans. Roy. Soc.*, **174**, Part III, 935–982 (1883). See also A. A. Draad and F. M. T. Nieuwstadt, *J. Fluid Mech.*, **361**, 297–308 (1998).

Table 5.1-1 Dependence of Jet Parameters on Distance z from Wall

	Laminar flow			Turbulent flow		
	Width of jet	Centerline velocity	Mass flow rate	Width of jet	Centerline velocity	Mass flow rate
Circular jet	z	z^{-1}	z	z	z^{-1}	z
Plane jet	$z^{2/3}$	$z^{-1/3}$	$z^{1/3}$	z	$z^{-1/2}$	$z^{1/2}$

For turbulent flow, on the other hand, the dependence on the geometrical and physical properties is quite different:[1]

$$F \approx 0.74 \sqrt[5]{\rho^4 \mu L^4 W^5 v_\infty^9} \qquad \text{(turbulent)} \ (5 \times 10^5 < \text{Re}_L < 10^7) \qquad (5.1\text{-}8)$$

Thus the force is proportional to the $\frac{3}{2}$-power of the approach velocity for laminar flow, but to the $\frac{9}{5}$-power for turbulent flow. The stronger dependence on the approach velocity reflects the extra energy needed to maintain the irregular eddy motions in the fluid.

Circular and Plane Jets

Next we examine the behavior of jets that emerge from a flat wall, which is taken to be the xy-plane (see Fig. 5.6-1). The fluid comes out from a circular tube or a long narrow slot, and flows into a large body of the same fluid. Various observations on the jets can be made: the width of the jet, the centerline velocity of the jet, and the mass flow rate through a cross section parallel to the xy-plane. All these properties can be measured as functions of the distance z from the wall. In Table 5.1-1 we summarize the properties of the circular and two-dimensional jets for laminar and turbulent flow.[1] It is curious that, for the circular jet, the jet width, centerline velocity, and mass flow rate have exactly the same dependence on z in both laminar and turbulent flow. We shall return to this point later in §5.6.

The above examples should make it clear that the gross features of laminar and turbulent flow are generally quite different. One of the many challenges in turbulence theory is to try to explain these differences.

§5.2 TIME-SMOOTHED EQUATIONS OF CHANGE FOR INCOMPRESSIBLE FLUIDS

We begin by considering a turbulent flow in a tube with a constant imposed pressure gradient. If at one point in the fluid we observe one component of the velocity as a function of time, we find that it is fluctuating in a chaotic fashion as shown in Fig. 5.2-1(a). The fluctuations are irregular deviations from a mean value. The actual velocity can be regarded as the sum of the mean value (designated by an overbar) and the fluctuation (designated by a prime). For example, for the z-component of the velocity we write

$$v_z = \bar{v}_z + v_z' \qquad (5.2\text{-}1)$$

which is sometimes called the *Reynolds decomposition*. The mean value is obtained from $v_z(t)$ by making a time average over a large number of fluctuations

$$\bar{v}_z = \frac{1}{t_0} \int_{t-\frac{1}{2}t_0}^{t+\frac{1}{2}t_0} v_z(s) \, ds \qquad (5.2\text{-}2)$$

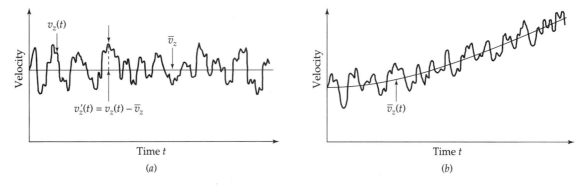

Fig. 5.2-1. Sketch showing the velocity component v_z as well as its time-smoothed value $\bar{v}_z$ and its fluctuation v_z' in turbulent flow (a) for "steadily driven turbulent flow" in which $\bar{v}_z$ does not depend on time, and (b) for a situation in which $\bar{v}_z$ does depend on time.

the period t_0 being long enough to give a smooth averaged function. For the system at hand, the quantity $\bar{v}_z$, which we call the *time-smoothed velocity*, is independent of time, but of course depends on position. When the time-smoothed velocity does not depend on time, we speak of *steadily driven turbulent flow*. The same comments we have made for velocity can also be made for pressure.

Next we consider turbulent flow in a tube with a time-dependent pressure gradient. For such a flow one can define time-smoothed quantities as above, but one has to understand that the period t_0 must be small with respect to the changes in the pressure gradient, but still large with respect to the periods of fluctuations. For such a situation the time-smoothed velocity and the actual velocity are illustrated in Fig. 5.2-1(b).[1]

According to the definition in Eq. 5.2-2, it is easy to verify that the following relations are true:

$$\overline{v_z'} = 0 \qquad \overline{\bar{v}_z} = \bar{v}_z \qquad \overline{\bar{v}_z v_z'} = 0 \qquad \overline{\frac{\partial}{\partial x} v_z} = \frac{\partial}{\partial x} \bar{v}_z \qquad \overline{\frac{\partial}{\partial t} v_z} = \frac{\partial}{\partial t} \bar{v}_z \qquad (5.2\text{-}3)$$

The quantity $\overline{v_z'^2}$ will not, however, be zero, and in fact the ratio $\sqrt{\overline{v_z'^2}}/\langle \bar{v}_z \rangle$ can be taken to be a measure of the magnitude of the turbulent fluctuations. This quantity, known as the *intensity of turbulence*, may have values from 1 to 10% in the main part of a turbulent stream and values of 25% or higher in the neighborhood of a solid wall. Hence, it must be emphasized that we are not necessarily dealing with tiny disturbances; sometimes the fluctuations are actually quite violent and large.

Quantities such as $\overline{v_x' v_y'}$ are also nonzero. The reason for this is that the local motions in the x and y directions are *correlated*. In other words, the fluctuations in the x direction are not independent of the fluctuations in the y direction. We shall see presently that these time-smoothed values of the products of fluctuating properties have an important role in turbulent momentum transfer. Later we shall find similar correlations arising in turbulent heat and mass transport.

---

[1] One can also define the "overbar" quantities in terms of an "ensemble average." For most purposes the results are equivalent or are assumed to be so. See, for example, A. A. Townsend, *The Structure of Turbulent Shear Flow*, Cambridge University Press, 2nd edition (1976). See also P. K. Kundu, *Fluid Mechanics*, Academic Press, New York (1990), p. 421, regarding the last of the formulas given in Eq. 5.2-3.

Having defined the time-smoothed quantities and discussed some of the properties of the fluctuating quantities, we can now move on to the time-smoothing of the equations of change. To keep the development as simple as possible, we consider here only the equations for a fluid of constant density and viscosity. We start by writing the equations of continuity and motion with $\mathbf{v}$ replaced by its equivalent $\bar{\mathbf{v}} + \mathbf{v}'$ and p by its equivalent $\bar{p} + p'$. The equation of continuity is then $(\nabla \cdot \mathbf{v}) = 0$, and we write the x-component of the equation of motion, Eq. 3.5-6, in the $\partial/\partial t$ form by using Eq. 3.5-5:

$$\frac{\partial}{\partial x}(\bar{v}_x + v'_x) + \frac{\partial}{\partial y}(\bar{v}_y + v'_y) + \frac{\partial}{\partial z}(\bar{v}_z + v'_z) = 0 \tag{5.2-4}$$

$$\frac{\partial}{\partial t}\rho(\bar{v}_x + v'_x) = -\frac{\partial}{\partial x}(\bar{p} + p') - \left(\frac{\partial}{\partial x}\rho(\bar{v}_x + v'_x)(\bar{v}_x + v'_x) + \frac{\partial}{\partial y}\rho(\bar{v}_y + v'_y)(\bar{v}_x + v'_x)\right.$$

$$\left. + \frac{\partial}{\partial z}\rho(\bar{v}_z + v'_z)(\bar{v}_x + v'_x)\right) + \mu\nabla^2(\bar{v}_x + v'_x) + \rho g_x \tag{5.2-5}$$

The y- and z-components of the equation of motion can be similarly written. We next time-smooth these equations, making use of the relations given in Eq. 5.2-3. This gives

$$\frac{\partial}{\partial x}\bar{v}_x + \frac{\partial}{\partial y}\bar{v}_y + \frac{\partial}{\partial z}\bar{v}_z = 0 \tag{5.2-6}$$

$$\frac{\partial}{\partial t}\rho\bar{v}_x = -\frac{\partial}{\partial x}\bar{p} - \left(\frac{\partial}{\partial x}\rho\bar{v}_x\bar{v}_x + \frac{\partial}{\partial y}\rho\bar{v}_y\bar{v}_x + \frac{\partial}{\partial z}\rho\bar{v}_z\bar{v}_x\right)$$

$$- \left(\frac{\partial}{\partial x}\rho\overline{v'_xv'_x} + \frac{\partial}{\partial y}\rho\overline{v'_yv'_x} + \frac{\partial}{\partial z}\rho\overline{v'_zv'_x}\right) + \mu\nabla^2\bar{v}_x + \rho g_x \tag{5.2-7}$$

with similar relations for the y- and z-components of the equation of motion. These are then the *time-smoothed equations of continuity and motion* for a fluid with constant density and viscosity. By comparing them with the corresponding equations in Eq. 3.1-5 and Eq. 3.5-6 (the latter rewritten in terms of $\partial/\partial t$), we conclude that

a. The equation of continuity is the same as we had previously, except that $\mathbf{v}$ is now replaced by $\bar{\mathbf{v}}$.

b. The equation of motion now has $\bar{\mathbf{v}}$ and $\bar{p}$ where we previously had $\mathbf{v}$ and p. In addition there appear the dashed-underlined terms, which describe the momentum transport associated with the turbulent fluctuations.

We may rewrite Eq. 5.2-7 by introducing the *turbulent momentum flux tensor* $\bar{\boldsymbol{\tau}}^{(t)}$ with components

$$\bar{\tau}^{(t)}_{xx} = \rho\overline{v'_xv'_x} \qquad \bar{\tau}^{(t)}_{xy} = \rho\overline{v'_xv'_y} \qquad \bar{\tau}^{(t)}_{xz} = \rho\overline{v'_xv'_z} \text{ and so on} \tag{5.2-8}$$

These quantities are usually referred to as the *Reynolds stresses*. We may also introduce a symbol $\bar{\boldsymbol{\tau}}^{(v)}$ for the time-smoothed viscous momentum flux. The components of this tensor have the same appearance as the expressions given in Appendices B.1 to B.3, except that the time-smoothed velocity components appear in them:

$$\bar{\tau}^{(v)}_{xx} = -2\mu\frac{\partial\bar{v}_x}{\partial x} \qquad \bar{\tau}^{(v)}_{xy} = -\mu\left(\frac{\partial\bar{v}_y}{\partial x} + \frac{\partial\bar{v}_x}{\partial y}\right) \text{ and so on} \tag{5.2-9}$$

This enables us then to write the equations of change in vector-tensor form as

$$(\nabla \cdot \bar{\mathbf{v}}) = 0 \quad \text{and} \quad (\nabla \cdot \mathbf{v}') = 0 \tag{5.2-10, 11}$$

$$\frac{\partial}{\partial t}\rho\bar{\mathbf{v}} = -\nabla\bar{p} - [\nabla \cdot \rho\bar{\mathbf{v}}\,\bar{\mathbf{v}}] - [\nabla \cdot (\bar{\boldsymbol{\tau}}^{(v)} + \bar{\boldsymbol{\tau}}^{(t)})] + \rho\mathbf{g} \tag{5.2-12}$$

Equation 5.2-11 is an extra equation obtained by subtracting Eq. 5.2-10 from the original equation of continuity.

The principal result of this section is that the equation of motion in terms of the stress tensor, summarized in Appendix Table B.5, can be adapted for time-smoothed turbulent flow by changing all v_i to $\bar{v}_i$ and p to $\bar{p}$ as well as τ_{ij} to $\bar{\tau}_{ij} = \bar{\tau}_{ij}^{(v)} + \bar{\tau}_{ij}^{(t)}$ in any of the coordinate systems given.

We have now arrived at the main stumbling block in the theory of turbulence. The Reynolds stresses $\bar{\tau}_{ij}^{(t)}$ above are not related to the velocity gradients in a simple way as are the time-smoothed viscous stresses $\bar{\tau}_{ij}^{(v)}$ in Eq. 5.2-9. They are, instead, complicated functions of the position and the turbulence intensity. To solve flow problems we must have experimental information about the Reynolds stresses or else resort to some empirical expression. In §5.4 we discuss some of the empiricisms that are available.

Actually one can also obtain equations of change for the Reynolds stresses (see Problem 5D.1). However, these equations contain quantities like $\overline{v_i' v_j' v_k'}$. Similarly, the equations of change for the $\overline{v_i' v_j' v_k'}$ contain the next higher-order correlation $\overline{v_i' v_j' v_k' v_l'}$, and so on. That is, there is a never-ending hierarchy of equations that must be solved. To solve flow problems one has to "truncate" this hierarchy by introducing empiricisms. If we use empiricisms for the Reynolds stresses, we then have a "first-order" theory. If we introduce empiricisms for the $\overline{v_i' v_j' v_k'}$, we then have a "second-order theory," and so on. The problem of introducing empiricisms to get a closed set of equations that can be solved for the velocity and pressure distributions is referred to as the "closure problem." The discussion in §5.4 deals with closure at the first order. At the second order the "k-ε empiricism" has been extensively studied and widely used in computational fluid mechanics.[2]

§5.3   THE TIME-SMOOTHED VELOCITY PROFILE NEAR A WALL

Before we discuss the various empirical expressions used for the Reynolds stresses, we present here several developments that do not depend on any empiricisms. We are concerned here with the fully developed, time-smoothed velocity distribution in the neighborhood of a wall. We discuss several results: a Taylor expansion of the velocity near the wall, and the universal logarithmic and power law velocity distributions a little further out from the wall.

The flow near a flat surface is depicted in Fig. 5.3-1. It is convenient to distinguish four regions of flow:

- the *viscous sublayer* very near the wall, in which viscosity plays a key role
- the *buffer layer* in which the transition occurs between the viscous and inertial sublayers
- the *inertial sublayer* at the beginning of the main turbulent stream, in which viscosity plays at most a minor role
- the *main turbulent stream*, in which the time-smoothed velocity distribution is nearly flat and viscosity is unimportant

It must be emphasized that this classification into regions is somewhat arbitrary.

---

[2] J. L. Lumley, *Adv. Appl. Mech.*, **18**, 123–176 (1978); C. G. Speziale, *Ann. Revs. Fluid Mech.*, **23**, 107–157 (1991); H. Schlichting and K. Gersten, *Boundary-Layer Theory*, Springer, Berlin, 8th edition (2000), pp. 560–563.

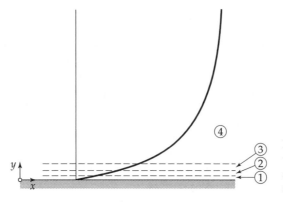

Fig. 5.3-1. Flow regions for describing turbulent flow near a wall: ① viscous sublayer, ② buffer layer, ③ inertial sublayer, ④ main turbulent stream.

The Logarithmic and Power Law Velocity Profiles in the Inertial Sublayer[1–4]

Let the time-smoothed shear stress acting on the wall $y = 0$ be called τ_0 (this is the same as $-\bar{\tau}_{yx}|_{y=0}$). Then the shear stress in the inertial sublayer will not be very different from the value τ_0. We now ask: On what quantities will the time-smoothed velocity gradient $d\bar{v}_x/dy$ depend? It should not depend on the viscosity, since, out beyond the buffer layer, momentum transport should depend primarily on the velocity fluctuations (loosely referred to as "eddy motion"). It may depend on the density ρ, the wall shear stress τ_0, and the distance y from the wall. The only combination of these three quantities that has the dimensions of a velocity gradient is $\sqrt{\tau_0/\rho}/y$. Hence we write

$$\frac{d\bar{v}_x}{dy} = \frac{1}{\kappa}\sqrt{\frac{\tau_0}{\rho}}\frac{1}{y} \tag{5.3-1}$$

in which κ is an arbitrary dimensionless constant, which must be determined experimentally. The quantity $\sqrt{\tau_0/\rho}$ has the dimensions of velocity; it is called the *friction velocity* and given the symbol v_*. When Eq. 5.3-1 is integrated we get

$$\bar{v}_x = \frac{v_*}{\kappa}\ln\, y + \lambda' \tag{5.3-2}$$

λ' being an integration constant. To use dimensionless groupings, we rewrite Eq. 5.3-2 as

$$\frac{\bar{v}_x}{v_*} = \frac{1}{\kappa}\ln\left(\frac{yv_*}{\nu}\right) + \lambda \tag{5.3-3}$$

in which λ is a constant simply related to λ'; the kinematic viscosity ν was included in order to construct the dimensionless argument of the logarithm. Experimentally it has been found that reasonable values of the constants[2] are $\kappa = 0.4$ and $\lambda = 5.5$, giving

$$\frac{\bar{v}_x}{v_*} = 2.5\ln\left(\frac{yv_*}{\nu}\right) + 5.5 \qquad \frac{yv_*}{\nu} > 30 \tag{5.3-4}$$

[1] L. Landau and E. M. Lifshitz, *Fluid Mechanics*, Pergamon, Oxford, 2nd edition (1987), pp. 172–178.

[2] H. Schlichting and K. Gersten, *Boundary-Layer Theory*, Springer-Verlag, Berlin, 8th edition (2000), §17.2.3.

[3] T. von Kármán, *Nachr. Ges. Wiss. Göttingen, Math-Phys. Klasse* (1930), pp. 58–76; L. Prandtl, *Ergeb. Aerodyn. Versuch.*, Series 4, Göttingen (1932).

[4] G. I. Barenblatt and A. J. Chorin, *Proc. Nat. Acad. Sci. USA*, **93**, 6749–6752 (1996) and *SIAM Rev.*, **40**, 265–291 (1981); G. I. Barenblatt, A. J. Chorin, and V. M. Prostokishin, *Proc. Nat. Acad. Sci. USA*, **94**, 773–776 (1997). See also G. I. Barenblatt, *Scaling, Self-Similarity, and Intermediate Asymptotics*, Cambridge University Press (1992), §10.2.

This is called the *von Kármán-Prandtl universal logarithmic velocity distribution;*[3] it is intended to apply only in the inertial sublayer. Later we shall see (in Fig. 5.5-3) that this function describes moderately well the experimental data somewhat beyond the inertial sublayer.

If Eq. 5.3-1 were correct, then the constants κ and λ would indeed be "universal constants," applicable at any Reynolds number. However, values of κ in the range 0.40 to 0.44 and values of λ in the range 5.0 to 6.3 can be found in the literature, depending on the range of Reynolds numbers. This suggests that the right side of Eq. 5.3-1 should be multiplied by some function of Reynolds number and that y could be raised to some power involving the Reynolds number. Theoretical arguments have been advanced[4] that Eq. 5.3-1 should be replaced by

$$\frac{d\bar{v}_x}{dy} = \frac{v_*}{y}\left(B_0 + \frac{B_1}{\ln \mathrm{Re}}\right)\left(\frac{yv_*}{\nu}\right)^{\beta_1/\ln \mathrm{Re}} \tag{5.3-5}$$

in which $B_0 = \frac{1}{2}\sqrt{3}$, $B_1 = \frac{15}{4}$, and $\beta_1 = \frac{3}{2}$. When Eq. 5.3-5 is integrated with respect to y, the *Barenblatt–Chorin universal velocity distribution* is obtained:

$$\frac{\bar{v}_x}{v_*} = \left(\frac{1}{\sqrt{3}}\ln \mathrm{Re} + \frac{5}{2}\right)\left(\frac{yv_*}{\nu}\right)^{3/(2\ln \mathrm{Re})} \tag{5.3-6}$$

Equation 5.3-6 describes regions ③ and ④ of Fig. 5.3-1 better than does Eq. 5.3-4.[4] Region ① is better described by Eq. 5.3-13.

Taylor-Series Development in the Viscous Sublayer

We start by writing a Taylor series for $\bar{v}_x$ as a function of y, thus

$$\bar{v}_x(y) = \bar{v}_x(0) + \left.\frac{\partial \bar{v}_x}{\partial y}\right|_{y=0} y + \frac{1}{2!}\left.\frac{\partial^2 \bar{v}_x}{\partial y^2}\right|_{y=0} y^2 + \frac{1}{3!}\left.\frac{\partial^3 \bar{v}_x}{\partial y^3}\right|_{y=0} y^3 + \cdots \tag{5.3-7}$$

To evaluate the terms in this series, we need the expression for the time-smoothed shear stress in the vicinity of a wall. For the special case of the steadily driven flow in a slit of thickness $2B$, the shear stress will be of the form $\bar{\tau}_{yx} = \bar{\tau}_{yx}^{(v)} + \bar{\tau}_{yx}^{(t)} = -\tau_0[1 - (y/B)]$. Then from Eqs. 5.2-8 and 9, we have

$$+\mu\frac{\partial \bar{v}_x}{\partial y} - \rho\overline{v_x'v_y'} = \tau_0\left(1 - \frac{y}{B}\right) \tag{5.3-8}$$

Now we examine one by one the terms that appear in Eq. 5.3-7:[5]

(i) The first term is zero by the no-slip condition.

(ii) The coefficient of the second term can be obtained from Eq. 5.3-8, recognizing that both v_x' and v_y' are zero at the wall so that

$$\left.\frac{\partial \bar{v}_x}{\partial y}\right|_{y=0} = \frac{\tau_0}{\mu} \tag{5.3-9}$$

(iii) The coefficient of the third term involves the second derivative, which may be obtained by differentiating Eq. 5.3-8 with respect to y and then setting $y = 0$, as follows,

$$\left.\frac{\partial^2 \bar{v}_x}{\partial y^2}\right|_{y=0} = \frac{\rho}{\mu}\left.\left(\overline{v_x'\frac{\partial v_y'}{\partial y} + v_y'\frac{\partial v_x'}{\partial y}}\right)\right|_{y=0} - \frac{\tau_0}{\mu B} = -\frac{\tau_0}{\mu B} \tag{5.3-10}$$

since both v_x' and v_y' are zero at the wall.

[5] A. A. Townsend, *The Structure of Turbulent Shear Flow*, Cambridge University Press, 2nd edition (1976), p. 163.

(iv) The coefficient of the fourth term involves the third derivative, which may be obtained from Eq. 5.3-8, and this is

$$
\left. \frac{\partial^3 \bar{v}_x}{\partial y^3} \right|_{y=0} = \frac{\rho}{\mu} \overline{\left(v_x' \frac{\partial^2 v_y'}{\partial y^2} + 2 \frac{\partial v_y'}{\partial y} \frac{\partial v_x'}{\partial y} + v_y' \frac{\partial^2 v_x'}{\partial y^2} \right)} \Bigg|_{y=0}
$$

$$
= -\frac{\rho}{\mu} \overline{\left(+2 \left(\frac{\partial v_x'}{\partial x} + \frac{\partial v_z'}{\partial z} \right) \frac{\partial v_x'}{\partial y} \right)} \Bigg|_{y=0} = 0 \qquad (5.3\text{-}11)
$$

Here Eq. 5.2-11 has been used.

There appears to be no reason to set the next coefficient equal to zero, so we find that the Taylor series, in dimensionless quantities, has the form

$$
\frac{\bar{v}_x}{v_*} = \frac{y v_*}{\nu} - \frac{1}{2} \left(\frac{\nu}{v_* B} \right) \left(\frac{y v_*}{\nu} \right)^2 + C \left(\frac{y v_*}{\nu} \right)^4 + \cdots \qquad (5.3\text{-}12)
$$

The coefficient C has been obtained experimentally,[6] and therefore we have the final result:

$$
\frac{\bar{v}_x}{v_*} = \frac{y v_*}{\nu} \left[1 - \frac{1}{2} \left(\frac{\nu}{v_* B} \right) \left(\frac{y v_*}{\nu} \right) - \frac{1}{4} \left(\frac{y v_*}{14.5 \nu} \right)^3 + \cdots \right] \qquad 0 < \frac{y v_*}{\nu} < 5 \qquad (5.3\text{-}13)
$$

The y^3 term in the brackets will turn out to be very important in connection with turbulent heat and mass transfer correlations in Chapters 13, 14, 21, and 22.

For the region $5 < y v_* / \nu < 30$ no simple analytical derivations are available, and empirical curve fits are sometimes used. One of these is shown in Fig. 5.5-3 for circular tubes.

§5.4 EMPIRICAL EXPRESSIONS FOR THE TURBULENT MOMENTUM FLUX

We now return to the problem of using the time-smoothed equations of change in Eqs. 5.2-11 and 12 to obtain the time-smoothed velocity and pressure distributions. As pointed out in the previous section, some information about the velocity distribution can be obtained without having a specific expression for the turbulent momentum flux $\bar{\tau}^{(t)}$. However, it has been popular among engineers to use various empiricisms for $\bar{\tau}^{(t)}$ that involve velocity gradients. We mention a few of these, and many more can be found in the turbulence literature.

The Eddy Viscosity of Boussinesq

By analogy with Newton's law of viscosity, Eq. 1.1-1, one may write for a turbulent shear flow[1]

$$
\bar{\tau}_{yx}^{(t)} = -\mu^{(t)} \frac{d\bar{v}_x}{dy} \qquad (5.4\text{-}1)
$$

[6] C. S. Lin, R. W. Moulton, and G. L. Putnam, *Ind. Eng. Chem.*, **45**, 636–640 (1953); the numerical coefficient was determined from mass transfer experiments in circular tubes. The importance of the y^4 term in heat and mass transfer was recognized earlier by E. V. Murphree, *Ind. Eng. Chem.*, **24**, 726–736 (1932). **Eger Vaughn Murphree** (1898–1962) was captain of the University of Kentucky football team in 1920 and became President of the Standard Oil Development Company.

[1] J. Boussinesq, *Mém. prés. par div. savants à l'acad. sci. de Paris*, **23**, #1, 1–680 (1877), **24**, #2, 1–64 (1877). **Joseph Valentin Boussinesq** (1842–1929), university professor in Lille, wrote a two-volume treatise on heat, and is famous for the "Boussinesq approximation" and the idea of "eddy viscosity."

in which $\mu^{(t)}$ is the *turbulent viscosity* (often called the *eddy viscosity*, and given the symbol ε). As one can see from Table 5.1-1, for at least one of the flows given there, the circular jet, one might expect Eq. 5.4-1 to be useful. Usually, however, $\mu^{(t)}$ is a strong function of position and the intensity of turbulence. In fact, for some systems[2] $\mu^{(t)}$ may even be negative in some regions. It must be emphasized that the viscosity μ is a property of the *fluid*, whereas the eddy viscosity $\mu^{(t)}$ is primarily a property of the *flow*.

For two kinds of turbulent flows (i.e., flows along surfaces and flows in jets and wakes), special expressions for $\mu^{(t)}$ are available:

(i) Wall turbulence:
$$\mu^{(t)} = \mu\left(\frac{yv_*}{14.5\nu}\right)^3 \qquad 0 < \frac{yv_*}{\nu} < 5 \tag{5.4-2}$$

This expression, derivable from Eq. 5.3-13, is valid only very near the wall. It is of considerable importance in the theory of turbulent heat and mass transfer at fluid–solid interfaces.[3]

(ii) Free turbulence:
$$\mu^{(t)} = \rho\kappa_0 b(\bar{v}_{z,\max} - \bar{v}_{z,\min}) \tag{5.4-3}$$

in which κ_0 is a dimensionless coefficient to be determined experimentally, b is the width of the mixing zone at a downstream distance z, and the quantity in parentheses represents the maximum difference in the z-component of the time-smoothed velocities at that distance z. Prandtl[4] found Eq. 5.4-3 to be a useful empiricism for jets and wakes.

The Mixing Length of Prandtl

By assuming that eddies move around in a fluid very much as molecules move around in a low-density gas (not a very good analogy) Prandtl[5] developed an expression for momentum transfer in a turbulent fluid. The "mixing length" l plays roughly the same role as the mean free path in kinetic theory (see §1.4). This kind of reasoning led Prandtl to the following relation:

$$\bar{\tau}_{yx}^{(t)} = -\rho l^2 \left|\frac{d\bar{v}_x}{dy}\right| \frac{d\bar{v}_x}{dy} \tag{5.4-4}$$

If the mixing length were a universal constant, Eq. 5.4-4 would be very attractive, but in fact l has been found to be a function of position. Prandtl proposed the following expressions for l:

(i) Wall turbulence: $l = \kappa_1 y$ (y = distance from wall)   (5.4-5)

(ii) Free turbulence: $l = \kappa_2 b$ (b = width of mixing zone)   (5.4-6)

in which κ_1 and κ_2 are constants. A result similar to Eq. 5.4-4 was obtained by Taylor[6] by his "vorticity transport theory" some years prior to Prandtl's proposal.

---

[2] J. O. Hinze, *Appl. Sci. Res.*, **22**, 163–175 (1970); V. Kruka and S. Eskinazi, *J. Fluid. Mech.*, **20**, 555–579 (1964).

[3] C. S. Lin, R. W. Moulton, and G. L. Putnam, *Ind. Eng. Chem.*, **45**, 636–640 (1953).

[4] L. Prandtl, *Zeits. f. angew. Math. u. Mech.*, **22**, 241–243 (1942).

[5] L. Prandtl, *Zeits. f. angew. Math. u. Mech.*, **5**, 136–139 (1925).

[6] G. I. Taylor, *Phil. Trans.* **A215**, 1–26 (1915), and *Proc. Roy. Soc.* (London), **A135**, 685–701 (1932).

The Modified van Driest Equation

There have been numerous attempts to devise empirical expressions that can describe the turbulent shear stress all the way from the wall to the main turbulent stream. Here we give a modification of the equation of van Driest.[7] This is a formula for the mixing length of Eq. 5.4-4:

$$l = 0.4y \frac{1 - \exp(-yv_*/26\nu)}{\sqrt{1 - \exp(-0.26yv_*/\nu)}} \tag{5.4-7}$$

This relation has been found to be useful for predicting heat and mass transfer rates in flow in tubes.

In the next two sections and in several problems at the end of the chapter, we illustrate the use of the above empiricisms. Keep in mind that these expressions for the Reynolds stresses are little more than crutches that can be used for the representation of experimental data or for solving problems that fall into rather special classes.

EXAMPLE 5.4-1	Obtain an expression for $\overline{\tau}_{yx}^{(t)} = \rho \overline{v'_x v'_y}$ as a function of y in the neighborhood of the wall.
Development of the Reynolds Stress Expression in the Vicinity of the Wall	**SOLUTION** **(a)** We start by making a Taylor series development of the three components of $\mathbf{v'}$:

$$v'_x(y) = v'_x(0) + \left.\frac{\partial v'_x}{\partial y}\right|_{y=0} y + \frac{1}{2!}\left.\frac{\partial^2 v'_x}{\partial y^2}\right|_{y=0} y^2 + \cdots \tag{5.4-8}$$

$$v'_y(y) = v'_y(0) + \left.\frac{\partial v'_y}{\partial y}\right|_{y=0} y + \frac{1}{2!}\left.\frac{\partial^2 v'_y}{\partial y^2}\right|_{y=0} y^2 + \cdots \tag{5.4-9}$$

$$v'_z(y) = v'_z(0) + \left.\frac{\partial v'_z}{\partial y}\right|_{y=0} y + \frac{1}{2!}\left.\frac{\partial^2 v'_z}{\partial y^2}\right|_{y=0} y^2 + \cdots \tag{5.4-10}$$

The first term in Eqs. 5.4-8 and 10 must be zero because of the no-slip condition; the first term in Eq. 5.4-9 is zero in the absence of mass transfer. Next we can write Eq. 5.2-11 at $y = 0$,

$$\left.\frac{\partial v'_x}{\partial x}\right|_{y=0} + \left.\frac{\partial v'_y}{\partial y}\right|_{y=0} + \left.\frac{\partial v'_z}{\partial z}\right|_{y=0} = 0 \tag{5.4-11}$$

The first and third terms in this equation are zero because of the no-slip condition. Therefore we have to conclude that the second term is zero as well. Hence all the dashed-underlined terms in Eqs. 5.4-8 to 10 are zero, and we may conclude that

$$\overline{\tau}_{yx}^{(t)} = \rho \overline{v'_x v'_y} = Ay^3 + By^4 + \cdots \tag{5.4-12}$$

This suggests—but does not prove[8]—that the lead term in the Reynolds stress near a wall should be proportional to y^3. Extensive studies of mass transfer rates in closed channels[9] have, however, established that $A \neq 0$.

---

[7] E. R. van Driest, *J. Aero. Sci.*, **23**, 1007–1011 and 1036 (1956). Van Driest's original equation did not have the square root divisor. This modification was made by O. T. Hanna, O. C. Sandall, and P. R. Mazet, *AIChE Journal*, **27**, 693–697 (1981) so that the turbulent viscosity would be proportional to y^3 as $y \to 0$, in accordance with Eq. 5.4-2.

[8] H. Reichardt, *Zeits. f. angew. Math. u. Mech.*, **31**, 208–219 (1951). See also J. O. Hinze, *Turbulence*, McGraw-Hill, New York, 2nd edition (1975), pp. 620–621.

[9] R. H. Notter and C. A. Sleicher, *Chem. Eng. Sci.*, **26**, 161–171 (1971); O. C. Sandall and O. T. Hanna, *AIChE Journal*, **25**, 190–192 (1979); D. W. Hubbard and E. N. Lightfoot, *Ind. Eng. Chem. Fundamentals*, **5**, 370–379 (1966).

(b) For the flow between parallel plates, we can use the expression found in Eq. 5.3-12 for the time-smoothed velocity profile to get the turbulent momentum flux:

$$\bar{\tau}_{yx}^{(t)} = \rho\overline{v_x'v_y'} = -\tau_0\left(1 - \frac{y}{B}\right) + \mu\frac{d\bar{v}_x}{dy}$$

$$= -\tau_0\left(1 - \frac{y}{B}\right) + \left(\tau_0 - \tau_0\frac{y}{B} + Ay^3 + \cdots\right) \tag{5.4-13}$$

where $A = 4C(v_*/\nu)^4$. This is in accord with Eq. 5.4-12.

§5.5   TURBULENT FLOW IN DUCTS

We start this section with a short discussion of experimental measurements for turbulent flow in rectangular ducts, in order to give some impressions about the Reynolds stresses. In Figs. 5.5-1 and 2 are shown some experimental measurements of the time-smoothed quantities $\overline{v_z'^2}$, $\overline{v_x'^2}$, and $\overline{v_x'v_z'}$ for the flow in the z direction in a rectangular duct.

In Fig. 5.5-1 note that quite close to the wall, $\sqrt{\overline{v_z'^2}}$ is about 13% of the time-smoothed centerline velocity $\bar{v}_{z,max}$, whereas $\sqrt{\overline{v_x'^2}}$ is about 5%. This means that, near the wall, the velocity fluctuations in the flow direction are appreciably greater than those in the transverse direction. Near the center of the duct, the two fluctuation amplitudes are nearly equal and we say that the turbulence is nearly *isotropic* there.

In Fig. 5.5-2 the turbulent shear stress $\bar{\tau}_{xz}^{(t)} = \rho\overline{v_x'v_z'}$ is compared with the total shear stress $\bar{\tau}_{xz} = \bar{\tau}_{xz}^{(t)} + \bar{\tau}_{xz}^{(v)}$ across the duct. It is evident that the turbulent contribution is the

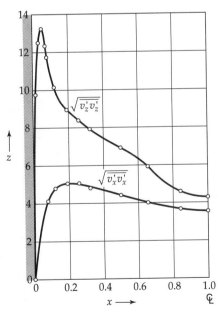

Fig. 5.5-1. Measurements of H. Reichardt [*Naturwissenschaften*, 404 (1938), *Zeits. f. angew. Math. u. Mech.*, **13**, 177–180 (1933), **18**, 358–361 (1938)] for the turbulent flow of air in a rectangular duct with $\bar{v}_{z,max} = 100$ cm/s. Here the quantities $\sqrt{\overline{v_x'v_x'}}$ and $\sqrt{\overline{v_z'v_z'}}$ are shown.

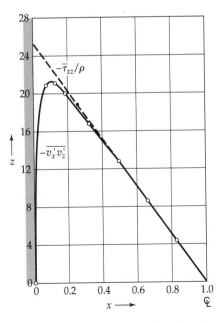

Fig. 5.5-2. Measurements of Reichardt (see Fig. 5.5-1) for the quantity $\overline{v_x'v_z'}$ in a rectangular duct. Note that this quantity differs from $\bar{\tau}_{xz}/\rho$ only near the duct wall.

more important over most of the cross section and that the viscous contribution is important only in the vicinity of the wall. This is further illustrated in Example 5.5-3. Analogous behavior is observed in tubes of circular cross section.

EXAMPLE 5.5-1 *Estimation of the Average Velocity in a Circular Tube*	Apply the results of §5.3 to obtain the average velocity for turbulent flow in a circular tube. **SOLUTION** We can use the velocity distribution in the caption to Fig. 5.5-3. To get the average velocity in the tube, one should integrate over four regions: the viscous sublayer ($y^+ < 5$), the buffer zone $5 < y^+ < 30$, the inertial sublayer, and the main turbulent stream, which is roughly parabolic in shape. One can certainly do this, but it has been found that integrating the logarithmic profile of Eq. 5.3-4 (or the power law profile of Eq. 5.3-6) over the entire cross section gives results that are roughly of the right form. For the *logarithmic profile* one gets

$$\frac{\langle \bar{v}_z \rangle}{v_*} = 2.5 \ln \left(\frac{Rv_*}{\nu} \right) + 1.75 \qquad (5.5\text{-}1)$$

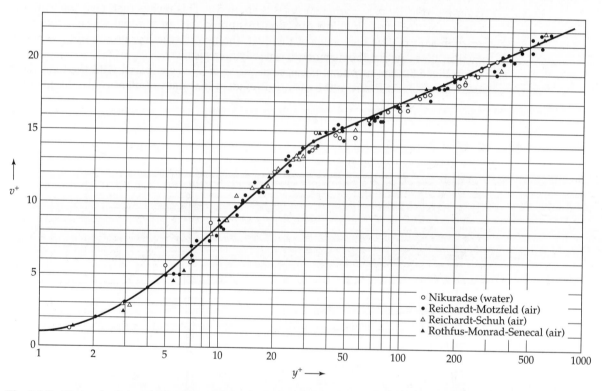

Fig. 5.5-3. Dimensionless velocity distribution for turbulent flow in circular tubes, presented as $v^+ = \bar{v}_z/v_*$ vs. $y^+ = yv_*\rho/\mu$, where $v_* = \sqrt{\tau_0/\rho}$ and τ_0 is the wall shear stress. The solid curves are those suggested by Lin, Moulton, and Putnam [*Ind. Eng. Chem.*, **45**, 636–640 (1953)]:

$$0 < y^+ < 5: \qquad v^+ = y^+[1 - \tfrac{1}{4}(y^+/14.5)^3]$$
$$5 < y^+ < 30: \qquad v^+ = 5 \ln(y^+ + 0.205) - 3.27$$
$$30 < y^+: \qquad v^+ = 2.5 \ln y^+ + 5.5$$

The experimental data are those of J. Nikuradse for water (○) [*VDI Forschungsheft*, **H356** (1932)]; Reichardt and Motzfeld for air (●); Reichardt and Schuh (△) for air [H. Reichardt, NACA Tech. Mem. 1047 (1943)]; and R. R. Rothfus, C. C. Monrad, and V. E. Senecal for air (■) [*Ind. Eng. Chem.*, **42**, 2511–2520 (1950)}.

If this is compared with experimental data on flow rate versus pressure drop, it is found that good agreement can be obtained by changing 2.5 to 2.45 and 1.75 to 2.0. This "fudging" of the constants would probably not be necessary if the integration over the cross section had been done by using the local expression for the velocity in the various layers. On the other hand, there is some virtue in having a simple logarithmic relation such as Eq. 5.5-1 to describe pressure drop vs. flow rate.

In a similar fashion the *power law profile* can be integrated over the entire cross section to give (see Ref. 4 of §5.3)

$$\frac{\langle \bar{v}_z \rangle}{v_*} = \frac{2}{(\alpha + 1)(\alpha + 2)} \left(\frac{1}{\sqrt{3}} \ln \text{Re} + \frac{5}{2} \right) \left(\frac{R v_*}{\nu} \right)^\alpha \tag{5.5-2}$$

in which $\alpha = 3/(2 \ln \text{Re})$. This relation is useful over the range $3.07 \times 10^3 < \text{Re} < 3.23 \times 10^6$.

EXAMPLE 5.5-2

Application of Prandtl's Mixing Length Formula to Turbulent Flow in a Circular Tube

Show how Eqs. 5.4-4 and 5 can be used to describe turbulent flow in a circular tube.

SOLUTION

Equation 5.2-12 gives for the steadily driven flow in a circular tube,

$$0 = \frac{\mathscr{P}_0 - \mathscr{P}_L}{L} - \frac{1}{r} \frac{d}{dr} (r \bar{\tau}_{rz}) \tag{5.5-3}$$

in which $\bar{\tau}_{rz} = \bar{\tau}_{rz}^{(v)} + \bar{\tau}_{rz}^{(t)}$. Over most of the tube the viscous contribution is quite small; here we neglect it entirely. Integration of Eq. 5.5-3 then gives

$$\bar{\tau}_{rz}^{(t)} = \frac{(\mathscr{P}_0 - \mathscr{P}_L)r}{2L} = \tau_0 \left(1 - \frac{y}{R} \right) \tag{5.5-4}$$

where τ_0 is the wall shear stress and $y = R - r$ is the distance from the tube wall.

According to the mixing length theory in Eq. 5.4-4, with the empirical expression in Eq. 5.4-5, we have for $d\bar{v}_z/dr$ negative

$$\bar{\tau}_{rz}^{(t)} = -\rho l^2 \left| \frac{d\bar{v}_z}{dr} \right| \frac{d\bar{v}_z}{dr} = + \rho (\kappa_1 y)^2 \left(\frac{d\bar{v}_z}{dy} \right)^2 \tag{5.5-5}$$

Substitution of this into Eq. 5.5-4 gives a differential equation for the time-smoothed velocity. If we follow Prandtl and extrapolate the inertial sublayer to the wall, then in Eq. 5.5-5 it is appropriate to replace $\bar{\tau}_{rz}^{(t)}$ by τ_0. When this is done, Eq. 5.5-5 can be integrated to give

$$\bar{v}_z = \frac{v_*}{\kappa_1} \ln y + \text{constant} \tag{5.5-6}$$

Thus a logarithmic profile is obtained and hence the results from Example 5.5-1 can be used; that is, one can apply Eq. 5.5-6 as a very rough approximation over the entire cross section of the tube.

EXAMPLE 5.5-3

Relative Magnitude of Viscosity and Eddy Viscosity

Determine the ratio $\mu^{(t)}/\mu$ at $y = R/2$ for water flowing at a steady rate in a long, smooth, round tube under the following conditions:

$$R = \text{tube radius} = 3 \text{ in.} = 7.62 \text{ cm}$$
$$\tau_0 = \text{wall shear stress} = 2.36 \times 10^{-5} \text{ lb}_f/\text{in.}^2 = 0.163 \text{ Pa}$$
$$\rho = \text{density} = 62.4 \text{ lb}_m/\text{ft}^3 = 1000 \text{ kg/m}^3$$
$$\nu = \text{kinematic viscosity} = 1.1 \times 10^{-5} \text{ ft}^2/\text{s} = 1.02 \times 10^{-7} \text{ m}^2/\text{s}$$

SOLUTION

The expression for the time-smoothed momentum flux is

$$\bar{\tau}_{rz} = -\mu \frac{d\bar{v}_z}{dr} - \mu^{(t)} \frac{d\bar{v}_z}{dr} \tag{5.5-7}$$

This result may be solved for $\mu^{(t)}/\mu$ and the result can be expressed in terms of dimensionless variables:

$$
\begin{aligned}
\frac{\mu^{(t)}}{\mu} &= \frac{1}{\mu}\frac{\bar{\tau}_{rz}}{d\bar{v}_z/dy} - 1 \\
&= \frac{1}{\mu}\frac{\tau_0[1 - (y/R)]}{d\bar{v}_z/dy} - 1 \\
&= \frac{[1 - (y/R)]}{dv^+/dy^+} - 1
\end{aligned}
\tag{5.5-8}
$$

where $y^+ = yv_*\rho/\mu$ and $v^+ = \bar{v}_z/v_*$. When $y = R/2$, the value of y^+ is

$$
y^+ = \frac{yv_*\rho}{\mu} = \frac{(R/2)\sqrt{\tau_0/\rho}\,\rho}{\mu} = 485
\tag{5.5-9}
$$

For this value of y^+, the logarithmic distribution in the caption of Fig. 5.5-3 gives

$$
\frac{dv^+}{dy^+} = \frac{2.5}{485} = 0.0052
\tag{5.5-10}
$$

Substituting this into Eq. 5.5-8 gives

$$
\frac{\mu^{(t)}}{\mu} = \frac{1/2}{0.0052} - 1 = 95
\tag{5.5-11}
$$

This result emphasizes that, far from the tube wall, molecular momentum transport is negligible in comparison with eddy transport.

§5.6   TURBULENT FLOW IN JETS

In the previous section we discussed the flow in ducts, such as circular tubes; such flows are examples of *wall turbulence*. Another main class of turbulent flows is *free turbulence*, and the main examples of these flows are jets and wakes. The time-smoothed velocity in these types of flows can be described adequately by using Prandtl's expression for the eddy viscosity in Fig. 5.4-3, or by using Prandtl's mixing length theory with the empiricism given in Eq. 5.4-6. The former method is simpler, and hence we use it in the following illustrative example.

EXAMPLE 5.6-1

Time-Smoothed Velocity Distribution in a Circular Wall Jet[1-4]

A jet of fluid emerges from a circular hole into a semi-infinite reservoir of the same fluid as depicted in Fig. 5.6-1. In the same figure we show roughly what we expect the profiles of the z-component of the velocity to look like. We would expect that for various values of z the profiles will be similar in shape, differing only by a scale factor for distance and velocity. We also can imagine that as the jet moves outward, it will create a net radial inflow so that some of the surrounding fluid will be dragged along. We want to find the time-smoothed velocity distribution in the jet and also the amount of fluid crossing each plane of constant z. Before working through the solution, it may be useful to review the information on jets in Table 5.1-1.

---

[1] H. Schlichting, *Boundary-Layer Theory*, McGraw-Hill, New York, 7th edition (1979), pp. 747–750.

[2] A. A. Townsend, *The Structure of Turbulent Shear Flow*, Cambridge University Press, 2nd edition (1976), Chapter 6.

[3] J. O. Hinze, *Turbulence*, McGraw-Hill, New York, 2nd edition (1975), Chapter 6.

[4] S. Goldstein, *Modern Developments in Fluid Dynamics*, Oxford University Press (1938), and Dover reprint (1965), pp. 592–597.

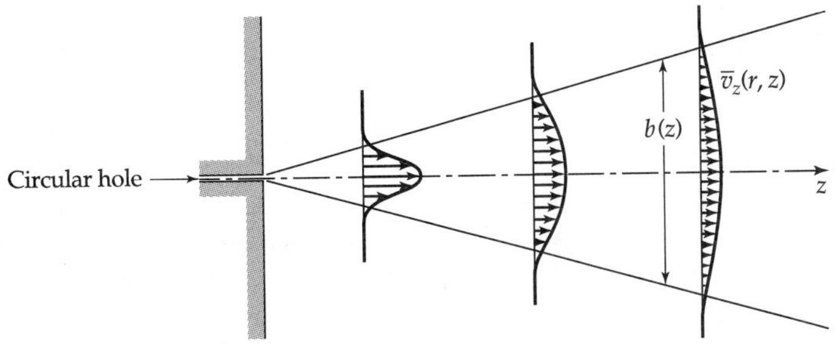

Fig. 5.6-1. Circular jet emerging from a plane wall.

SOLUTION

In order to use Eq. 5.4-3 it is necessary to know how b and $\bar{v}_{z,\max} - \bar{v}_{z,\min}$ vary with z for the circular jet. We know that the total rate of flow of z-momentum J will be the same for all values of z. We presume that the convective momentum flux is much greater than the viscous momentum flux. This permits us to postulate that the jet width b depends on J, on the density ρ and the kinematic viscosity ν of the fluid, and on the downstream distance z from the wall. The only combination of these variables that has the dimensions of length is $b \propto Jz/\rho\nu^2$, so that the jet width is proportional to z.

We next postulate that the velocity profiles are "similar," that is,

$$\frac{\bar{v}_z}{\bar{v}_{z,\max}} = f(\xi) \qquad \text{where } \xi = \frac{r}{b(z)} \tag{5.6-1}$$

which seems like a plausible proposal; here $\bar{v}_{z,\max}$ is the velocity along the centerline. When this is substituted into the expression for the rate of momentum flow in the jet (neglecting the contribution from $\bar{\tau}_{xx}$)

$$J = \int_0^{2\pi} \int_0^\infty \rho \bar{v}_z^2 r \, dr \, d\theta \tag{5.6-2}$$

we find that

$$J = 2\pi\rho b^2 \bar{v}_{z,\max}^2 \int_0^\infty f^2 \xi d\xi = \text{constant} \times \rho b^2 \bar{v}_{z,\max}^2 \tag{5.6-3}$$

Since J does not depend on z and since b is proportional to z, then $\bar{v}_{z,\max}$ has to be inversely proportional to z.

The $\bar{v}_{z,\min}$ in Eq. 5.4-3 occurs at the outer edge of the jet and is zero. Therefore because $b \propto z$ and $\bar{v}_{z,\max} \propto z^{-1}$, we find from Eq. 5.4-3 that $\mu^{(t)}$ is a constant. Thus we can use the equations of motion for laminar flow and replace the viscosity μ by the eddy viscosity $\mu^{(t)}$, or ν by $\nu^{(t)}$.

In the jet the main motion is in the z direction; that is $|\bar{v}_r| \ll |\bar{v}_z|$. Hence we can use a boundary layer approximation (see §4.4) for the time-smoothed equations of change and write

continuity:

$$\frac{1}{r}\frac{\partial}{\partial r}(r\bar{v}_r) + \frac{\partial \bar{v}_z}{\partial z} = 0 \tag{5.6-4}$$

motion:

$$\bar{v}_r \frac{\partial \bar{v}_z}{\partial r} + \bar{v}_z \frac{\partial \bar{v}_z}{\partial z} = \nu^{(t)} \frac{1}{r}\frac{\partial}{\partial r}\left(r \frac{\partial \bar{v}_z}{\partial r}\right) \tag{5.6-5}$$

These equations are to be solved with the following boundary conditions:

B.C. 1:   at $r = 0$, $\bar{v}_r = 0$ (5.6-6)

B.C. 2:   at $r = 0$, $\partial \bar{v}_z / \partial r = 0$ (5.6-7)

B.C. 3:   at $z = \infty$, $\bar{v}_z = 0$ (5.6-8)

The last boundary condition is automatically satisfied, inasmuch as we have already found that $\bar{v}_{z,\max}$ is inversely proportional to z. We now seek a solution to Eq. 5.6-5 of the form of Eq. 5.6-1 with $b = z$.

To avoid working with two dependent variables, we introduce the stream function as discussed in §4.2. For axially symmetric flow, the stream function is defined as follows:

$$\bar{v}_z = -\frac{1}{r}\frac{\partial \psi}{\partial r} \qquad \bar{v}_r = \frac{1}{r}\frac{\partial \psi}{\partial z} \tag{5.6-9, 10}$$

This definition ensures that the equation of continuity in Eq. 5.6-4 is satisfied. Since we know that $\bar{v}_z$ is $z^{-1} \times$ some function of ξ, we deduce from Eq. 5.6-9 that ψ must be proportional to z. Furthermore ψ must have dimensions of (velocity) $\times$ (length)^2, hence the stream function must have the form

$$\psi(r, z) = \nu^{(t)} z F(\xi) \tag{5.6-11}$$

in which F is a dimensionless function of $\xi = r/z$. From Eqs. 5.6-9 and 10 we then get

$$\bar{v}_z = -\frac{\nu^{(t)}}{z}\frac{F'}{\xi} \qquad \bar{v}_r = \frac{\nu^{(t)}}{z}\left(\frac{F}{\xi} - F'\right) \tag{5.6-12, 13}$$

The first two boundary conditions may now be rewritten as

B.C. 1:   at $\xi = 0$, $\dfrac{F}{\xi} - F' = 0$ $\qquad\qquad$ (5.6-14)

B.C. 2:   at $\xi = 0$, $\dfrac{F''}{\xi} - \dfrac{F'}{\xi^2} = 0$ $\qquad\qquad$ (5.6-15)

If we expand F in a Taylor series about $\xi = 0$,

$$F(\xi) = a + b\xi + c\xi^2 + d\xi^3 + e\xi^4 + \cdots \tag{5.6-16}$$

then the first boundary condition gives $a = 0$, and the second gives $b = d = 0$. We will use this result presently.

Substitution of the velocity expressions of Eqs. 5.6-12 and 13 into the equation of motion in Eq. 5.6-5 then gives a third-order differential equation for F,

$$\frac{d}{d\xi}\left(\frac{FF'}{\xi}\right) = \frac{d}{d\xi}\left(F'' - \frac{F'}{\xi}\right) \tag{5.6-17}$$

This may be integrated to give

$$\frac{FF'}{\xi} = F'' - \frac{F'}{\xi} + C_1 \tag{5.6-18}$$

in which the constant of integration must be zero; this can be seen by using the Taylor series in Eq. 5.6-16 along with the fact that a, b, and d are all zero.

Equation 5.6-18 was first solved by Schlichting.[5] First one changes the independent variable by setting $\xi = \ln \beta$. The resulting second-order differential equation contains only the dependent variable and its first two derivatives. Equations of this type can be solved by elementary methods. The first integration gives

$$\xi F' = 2F + \tfrac{1}{2}F^2 + C_2 \tag{5.6-19}$$

Once again, knowing the behavior of F near $\xi = 0$, we conclude that the second constant of integration is zero. Equation 5.6-19 is then a first-order separable equation, and it may be solved to give

$$F(\xi) = -\frac{(C_3\xi)^2}{1 + \tfrac{1}{4}(C_3\xi)^2} \tag{5.6-20}$$

---

[5] H. Schlichting, *Zeits. f. angew. Math. u. Mech.*, **13**, 260–263 (1933).

in which C_3 is the third constant of integration. Substitution of this into Eqs. 5.6-12 and 13 then gives

$$\bar{v}_z = \frac{\nu^{(t)}}{z} \frac{2C_3^2}{[1 + \frac{1}{4}(C_3 r/z)^2]^2} \tag{5.6-21}$$

$$\bar{v}_r = \frac{C_3 \nu^{(t)}}{z} \frac{(C_3 r/z) - \frac{1}{4}(C_3 r/z)^3}{[1 + \frac{1}{4}(C_3 r/z)^2]^2} \tag{5.6-22}$$

When the above expression for $\bar{v}_z$ is substituted into Eq. 5.6-2 for J, we get an expression for the third integration constant in terms of J:

$$C_3 = \sqrt{\frac{3}{16\pi}} \sqrt{\frac{J}{\rho}} \frac{1}{\nu^{(t)}} \tag{5.6-23}$$

The last three equations then give the time-smoothed velocity profiles in terms of J, ρ, and $\nu^{(t)}$.

A measurable quantity in jet flow is the radial position corresponding to an axial velocity one-half the centerline value; we call this half-width $b_{1/2}$. From Eq. 5.6-21 we then obtain

$$\frac{\bar{v}_z(b_{1/2}, z)}{\bar{v}_{z,max}(z)} = \frac{1}{2} = \frac{1}{[1 + \frac{1}{4}(C_3 b_{1/2}/z)^2]^2} \tag{5.6-24}$$

Experiments indicate[6] that $b_{1/2} = 0.0848z$. When this is inserted into Eq. 5.6-24, it is found that $C_3 = 15.1$. Using this value, we can get the turbulent viscosity $\nu^{(t)}$ as a function of J and ρ from Eq. 5.6-23.

Figure 5.6-2 gives a comparison of the above axial velocity profile with experimental data. The calculated curve obtained from the Prandtl mixing length theory is also shown.[7] Both methods appear to give reasonably good curve fits of the experimental profiles. The

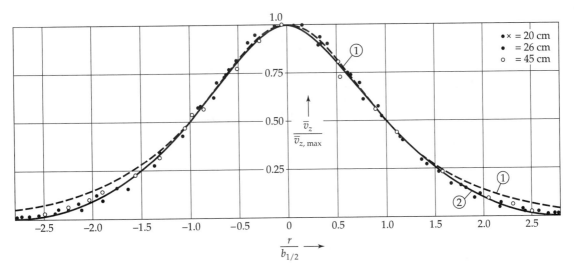

Fig. 5.6-2. Velocity distribution in a circular jet in turbulent flow [H. Schlichting, *Boundary-Layer Theory*, McGraw-Hill, New York, 7th edition (1979), Fig. 24.9]. The eddy viscosity calculation (curve 1) and the Prandtl mixing length calculation (curve 2) are compared with the measurements of H. Reichardt [*VDI Forschungsheft*, 414 (1942), 2nd edition (1951)]. Further measurements by others are cited by S. Corrsin ["Turbulence: Experimental Methods," in *Handbuch der Physik*, Vol. VIII/2, Springer, Berlin (1963)].

---

[6] H. Reichardt, *VDI Forschungsheft*, **414** (1942).
[7] W. Tollmien, *Zeits. f. angew. Math. u. Mech.*, **6**, 468–478 (1926).

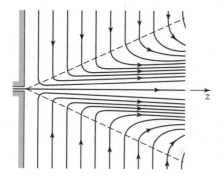

Fig. 5.6-3. Streamline pattern in a circular jet in turbulent flow [H. Schlichting, *Boundary-Layer Theory*, McGraw-Hill, New York, 7th edition (1979), Fig. 24.10].

eddy viscosity method seems to be somewhat better in the neighborhood of the maximum, whereas the mixing length results are better in the outer part of the jet.

Once the velocity profiles are known, the streamlines can be obtained. From the streamlines, shown in Fig. 5.6-3, it can be seen how the jet draws in fluid from the surrounding mass of fluid. Hence the mass of fluid carried by the jet increases with the distance from the source. This mass rate of flow is

$$ w = \int_0^{2\pi} \int_0^{\infty} \rho \bar{v}_z r \, dr \, d\theta = 8\pi\rho\nu^{(t)}z \tag{5.6-25} $$

This result corresponds to an entry in Table 5.1-1.

The two-dimensional jet issuing from a thin slot may be analyzed similarly. In that problem, however, the turbulent viscosity is a function of position.

QUESTIONS FOR DISCUSSION

1. Compare and contrast the procedures for solving laminar flow problems and turbulent flow problems.
2. Why must Eq. 5.1-4 *not* be used for evaluating the velocity gradient at the solid boundary?
3. What does the logarithmic profile of Eq. 5.3-4 give for the fluid velocity at the wall? Why does this not create a problem in Example 5.5-1 when the logarithmic profile is integrated over the cross section of the tube?
4. Discuss the physical interpretation of each term in Eq. 5.2-12.
5. Why is the absolute value sign used in Eq. 5.4-4? How is it eliminated in Eq. 5.5-5?
6. In Example 5.6-1, how do we know that the momentum flow through any plane of constant z is a constant? Can you imagine a modification of the jet problem in which that would not be the case?
7. Go through some of the volumes of *Ann. Revs. Fluid Mech.* and summarize the topics in turbulent flow that are found there.
8. In Eq. 5.3-1 why do we investigate the functional dependence of the velocity gradient rather than the velocity itself?
9. Why is turbulence such a difficult topic?

PROBLEMS

5A.1 Pressure drop needed for laminar-turbulent transition. A fluid with viscosity 18.3 cp and density 1.32 g/cm³ is flowing in a long horizontal tube of radius 1.05 in. (2.67 cm). For what pressure gradient will the flow become turbulent?

Answer: 26 psi/mi (1.1 × 10⁵ Pa/km)

5A.2 **Velocity distribution in turbulent pipe flow.** Water is flowing through a long, straight, level run of smooth 6.00 in. i.d. pipe, at a temperature of 68°F. The pressure gradient along the length of the pipe is 1.0 psi/mi.

(a) Determine the wall shear stress τ_0 in psi ($lb_f/in.^2$) and Pa.

(b) Assume the flow to be turbulent and determine the radial distances from the pipe wall at which $\bar{v}_z/\bar{v}_{z,max} = 0.0, 0.1, 0.2, 0.4, 0.7, 0.85, 1.0$.

(c) Plot the complete velocity profile, $\bar{v}_z/\bar{v}_{z,max}$ vs. $y = R - r$.

(d) Is the assumption of turbulent flow justified?

(e) What is the mass flow rate?

5B.1 **Average flow velocity in turbulent tube flow.**

(a) For the turbulent flow in smooth circular tubes, the function[1]

$$\frac{\bar{v}_z}{\bar{v}_{z,max}} = \left(1 - \frac{r}{R}\right)^{1/n} \tag{5B.1-1}$$

is sometimes useful for curve-fitting purposes: near $Re = 4 \times 10^3$, $n = 6$; near $Re = 1.1 \times 10^5$, $n = 7$; and near $Re = 3.2 \times 10^6$, $n = 10$. Show that the ratio of average to maximum velocity is

$$\frac{\langle \bar{v}_z \rangle}{\bar{v}_{z,max}} = \frac{2n^2}{(n+1)(2n+1)} \tag{5B.1-2}$$

and verify the result in Eq. 5.1-5.

(b) Sketch the logarithmic profile in Eq. 5.3-4 as a function of r when applied to a circular tube of radius R. Then show how this function may be integrated over the tube cross section to get Eq. 5.5-1. List all the assumptions that have been made to get this result.

5B.2 **Mass flow rate in a turbulent circular jet.**

(a) Verify that the velocity distributions in Eqs. 5.6-21 and 22 do indeed satisfy the differential equations and boundary conditions.

(b) Verify that Eq. 5.6-25 follows from Eq. 5.6-21.

5B.3 **The eddy viscosity expression in the viscous sublayer.** Verify that Eq. 5.4-2 for the eddy viscosity comes directly from the Taylor series expression in Eq. 5.3-13.

5C.1 **Two-dimensional turbulent jet.** A fluid jet issues forth from a slot perpendicular to the xy-plane and emerges in the z direction into a semi-infinite medium of the same fluid. The width of the slot in the y direction is W. Follow the pattern of Example 5.6-1 to find the time-smoothed velocity profiles in the system.

(a) Assume the similar profiles

$$\bar{v}_z/\bar{v}_{z,max} = f(\xi) \qquad \text{with } \xi = x/z \tag{5C.1-1}$$

Show that the momentum conservation statement leads to the fact that the centerline velocity must be proportional to $z^{-1/2}$.

(b) Introduce a stream function ψ such that $\bar{v}_z = -\partial\psi/\partial x$ and $\bar{v}_x = +\partial\psi/\partial z$. Show that the result in (a) along with dimensional considerations leads to the following form for ψ:

$$\psi = z^{1/2}\sqrt{J/\rho W}F(\xi) \tag{5C.1-2}$$

Here $F(\xi)$ is a dimensionless stream function, which will be determined from the equation of motion for the fluid.

---

[1] H. Schlichting, *Boundary-Layer Theory*, McGraw-Hill, New York, 7th edition (1979), pp. 596–600.

(c) Show that Eq. 5.4-2 and dimensional considerations lead to the following form for the turbulent kinematic viscosity:

$$\nu^{(t)} = \mu^{(t)}/\rho = \lambda\sqrt{J/\rho W}z^{1/2} \tag{5C.1-3}$$

Here λ is a dimensionless constant that has to be determined from experiments.

(d) Rewrite the equation of motion for the jet using the expression for the turbulent kinematic viscosity from (c) and the stream function from (b). Show that this leads to the following differential equation:

$$\tfrac{1}{2}F'^2 + \tfrac{1}{2}FF'' - \lambda F''' = 0 \tag{5C.1-4}$$

For the sake of convenience, introduce a new variable

$$\eta = \xi/4\lambda = x/4\lambda z \tag{5C.1-5}$$

and rewrite Eq. 5C.1-4.

(e) Next verify that the boundary conditions for Eq. 5C.1-4 are $F(0) = 0$, $F''(0) = 0$, and $F'(\infty) = 0$.

(f) Show that Eq. 5C.1-4 can be integrated to give

$$2FF' - F'' = \text{constant} \tag{5C.1-6}$$

and that the boundary conditions require that the constant be zero.

(g) Show that further integration leads to

$$F^2 - F' = C^2 \tag{5C.1-7}$$

where C is a constant of integration.

(h) Show that another integration leads to

$$F = -C \tanh C\eta \tag{5C.1-8}$$

and that the axial velocity can be found from this to be

$$\bar{v}_z = \frac{\sqrt{J/\rho W}C^2}{4\lambda\sqrt{z}}\,\text{sech}^2\,C\eta \tag{5C.1-9}$$

(i) Next show that putting the axial velocity into the expression for the total momentum of the jet leads to the value $C = \sqrt[3]{3\lambda}$ for the integration constant. Rewrite Eq. 5C.1-9 in terms of λ rather than C. The value of $\lambda = 0.0102$ gives good agreement with the experimental data.[2] The agreement is believed to be slightly better than that for the Prandtl mixing length empiricism.

(j) Show that the mass flow rate across any line $z = $ constant is given by

$$w = 2\sqrt[3]{3\lambda}\sqrt{\frac{J\rho z}{W}} \tag{5C.1-10}$$

5C.2 Axial turbulent flow in an annulus. An annulus is bounded by cylindrical walls at $r = aR$ and $r = R$ (where $a < 1$). Obtain expressions for the turbulent velocity profiles and the mass flow rate. Apply the logarithmic profile of Eq. 5.3-3 for the flow in the neighborhood of each wall. Assume that the location of the maximum in the velocity occurs on the same cylindrical surface $r = bR$ found for laminar annular flow:

$$b = \sqrt{\frac{1 - a^2}{2\ln(1/a)}} \tag{5C.2-1}$$

---

[2] H. Schlichting, *Boundary-Layer Theory*, McGraw-Hill, New York, 4th edition (1960), p. 607 and Fig. 23.7.

Measured velocity profiles suggest that this assumption for b is reasonable, at least for high Reynolds numbers.[3] Assume further that κ in Eq. 5.3-3 is the same for the inner and outer walls.

(a) Show that direct application of Eq. 5.3-3 leads immediately to the following velocity profiles[4] in the region $r < bR$ (designated by $<$) and $r > bR$ (designated by $>$):

$$\frac{\bar{v}_z^<}{v_*^<} = \frac{1}{\kappa} \ln\left(\frac{(r - aR)v_*^<}{\nu}\right) + \lambda^< \qquad \text{where } v_*^< = v_{**}\sqrt{\frac{b^2 - a^2}{a}} \qquad (5C.2-2)$$

$$\frac{\bar{v}_z^>}{v_*^>} = \frac{1}{\kappa} \ln\left(\frac{(R - r)v_*^>}{\nu}\right) + \lambda^> \qquad \text{where } v_*^> = v_{**}\sqrt{1 - b^2} \qquad (5C.2-3)$$

in which $v_{**} = \sqrt{(\mathcal{P}_0 - \mathcal{P}_L)R/2L\rho}$.

(b) Obtain a relation between the constants $\lambda^<$ and $\lambda^>$ by requiring that the velocity be continuous at $r = bR$.

(c) Use the results of (b) to show that the mass flow rate through the annulus is

$$w = \pi R^2 \rho v_{**}\left\{\left[\frac{(b^2 - a^2)^{3/2}}{\sqrt{a}} + (1 - b^2)^{3/2}\right]\left[\frac{1}{\kappa}\ln\frac{R(1 - b)\sqrt{1 - b^2}v_{**}}{\nu} + \lambda^>\right] - B\right\} \qquad (5C.2-4)$$

in which B is

$$B = \frac{(b^2 - a^2)^{3/2}}{\kappa\sqrt{a}}\left(\frac{a}{a + b} + \frac{1}{2}\right) + \frac{(1 - b^2)^{3/2}}{\kappa}\left(\frac{1}{1 + b} + \frac{1}{2}\right) \qquad (5C.2-5)$$

5C.3 Instability in a simple mechanical system (Fig. 5C.3).

(a) A disk is rotating with a constant angular velocity Ω. Above the center of the disk a sphere of mass m is suspended by a massless rod of length L. Because of the rotation of the disk, the sphere experiences a centrifugal force and the rod makes an angle θ with the vertical. By making a force balance on the sphere, show that

$$\cos\theta = \frac{g}{\Omega^2 L} \qquad (5C.3-1)$$

What happens when Ω goes to zero?

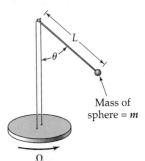

Mass of sphere = m

Ω

Fig. 5C.3. A simple mechanical system for illustrating concepts in stability.

[3] J. G. Knudsen and D. L. Katz, *Fluid Dynamics and Heat Transfer*, McGraw-Hill, New York (1958); R. R. Rothfus (1948), J. E. Walker (1957), and G. A. Whan (1956), Doctoral theses, Carnegie Institute of Technology (now Carnegie-Mellon University), Pittsburgh, Pa.

[4] W. Tiedt, *Berechnung des laminaren u. turbulenten Reibungswiderstandes konzentrischer u. exzentrischer Ringspalten*, Technischer Bericht Nr. 4, Inst. f. Hydraulik u. Hydraulogie, Technische Hochschule, Darmstadt (1968); D. M. Meter and R. B. Bird, *AIChE Journal*, **7**, 41–45 (1961) did the same analysis using the Prandtl mixing length theory.

(b) Show that, if Ω is below some threshold value Ω_{thr}, the angle θ is zero. Above the threshold value, show that there are two admissible values for θ. Explain by means of a carefully drawn sketch of θ vs. Ω. Above Ω_{thr} label the two curves *stable* and *unstable*.

(c) In (a) and (b) we considered only the steady-state operation of the system. Next show that the equation of motion for the sphere of mass m is

$$mL \frac{d^2\theta}{dt^2} = m\Omega^2 L \sin\theta \cos\theta - mg \sin\theta \tag{5C.3-2}$$

Show that for steady-state operation this leads to Eq. 5C.3-1. We now want to use this equation to make a small-amplitude stability analysis. Let $\theta = \theta_0 + \theta_1$, where θ_0 is a steady-state solution (independent of time) and θ_1 is a very small perturbation (dependent on time).

(d) Consider first the lower branch in (b), which is $\theta_0 = 0$. Then $\sin\theta = \sin\theta_1 \approx \theta_1$ and $\cos\theta = \cos\theta_1 \approx 1$, so that Eq. 5B.2-2 becomes

$$\frac{d^2\theta_1}{dt^2} = \left(\Omega^2 - \frac{g}{L}\right)\theta_1 \tag{5C.3-3}$$

We now try a small-amplitude oscillation of the form $\theta_1 = A\Re\{e^{-i\omega t}\}$ and find that

$$\omega_\pm = \pm i\sqrt{\Omega^2 - \frac{g}{L}} \tag{5C.3-4}$$

Now consider two cases: (i) If $\Omega^2 < g/L$, both ω_+ and ω_- are real, and hence θ_1 oscillates; this indicates that for $\Omega^2 < g/L$ the system is stable. (ii) If $\Omega^2 > g/L$, the root ω_+ is positive imaginary and $e^{-i\omega t}$ will increase indefinitely with time; this indicates that for $\Omega^2 > g/L$ the system is unstable with respect to infinitesimal perturbations.

(e) Next consider the upper branch in (b). Do an analysis similar to that in (d). Set up the equation for θ_1 and drop terms in the square of θ_1 (that is, linearize the equation). Once again try a solution of the form $\theta_1 = A\Re\{e^{-i\omega t}\}$. Show that for the upper branch the system is stable with respect to infinitesimal perturbations.

(f) Relate the above analysis, which is for a system with one degree of freedom, to the problem of laminar-turbulent transition for the flow of a Newtonian fluid in the flow between two counter-rotating cylinders. Read the discussion by Landau and Lifshitz[5] on this point.

5D.1 **Derivation of the equation of change for the Reynolds stresses.** At the end of §5.2 it was pointed out that there is an equation of change for the Reynolds stresses. This can be derived by (a) multiplying the ith component of the vector form of Eq. 5.2-5 by v_j' and time smoothing, (b) multiplying the jth component of the vector form of Eq. 5.2-5 by v_i' and time smoothing, and (c) adding the results of (a) and (b). Show that one finally gets

$$\rho \frac{D}{Dt}\overline{\mathbf{v}'\mathbf{v}'} = -\rho\overline{\{\mathbf{v}'\mathbf{v}'} \cdot \nabla\overline{\mathbf{v}}\} - \rho\overline{\{\mathbf{v}'\mathbf{v}'} \cdot \nabla\overline{\mathbf{v}}\}^\dagger - \rho\overline{\{\nabla \cdot \mathbf{v}'\mathbf{v}'\mathbf{v}'\}}$$

$$- \overline{\{\mathbf{v}'\nabla p'\}} - \overline{\{\mathbf{v}'\nabla p'\}}^\dagger + \mu\left\{\overline{\mathbf{v}'\nabla^2\mathbf{v}'} + \overline{\{\mathbf{v}'\nabla^2\mathbf{v}'\}}^\dagger\right\} \tag{5D.1-1}$$

Equations 5.2-10 and 11 will be needed in this development.

5D.2 **Kinetic energy of turbulence.** By taking the trace of Eq. 5D.1-1 obtain the following:

$$\frac{D}{Dt}(\tfrac{1}{2}\rho\overline{v'^2}) = -\rho(\overline{\mathbf{v}'\mathbf{v}'}:\nabla\overline{\mathbf{v}}) - (\nabla \cdot \tfrac{1}{2}\rho\overline{v'^2\mathbf{v}'}) - (\nabla \cdot \overline{p'\mathbf{v}'}) + \mu(\overline{\mathbf{v}' \cdot \nabla^2\mathbf{v}'}) \tag{5D.2-1}$$

Interpret the equation.[6]

---

[5] L. Landau and E. M. Lifshitz, *Fluid Mechanics*, Pergamon, Oxford, 2nd edition (1987), §§26–27.

[6] H. Tennekes and J. L. Lumley, *A First Course in Turbulence*, MIT Press, Cambridge, Mass. (1972), §3.2.

Chapter 6

Interphase Transport
in Isothermal Systems

In Chapters 2–4 we showed how laminar flow problems may be formulated and solved. In Chapter 5 we presented some methods for solving turbulent flow problems by dimensional arguments or by semiempirical relations between the momentum flux and the gradient of the time-smoothed velocity. In this chapter we show how flow problems can be solved by a combination of dimensional analysis and experimental data. The technique presented here has been widely used in chemical, mechanical, aeronautical, and civil engineering, and it is useful for solving many practical problems. It is a topic worth learning well.

Many engineering flow problems fall into one of two broad categories: flow in channels and flow around submerged objects. Examples of channel flow are the pumping of oil through pipes, the flow of water in open channels, and extrusion of plastics through dies. Examples of flow around submerged objects are the motion of air around an airplane wing, motion of fluid around particles undergoing sedimentation, and flow across tube banks in heat exchangers.

In channel flow the main object is usually to get a relationship between the volume rate of flow and the pressure drop and/or elevation change. In problems involving flow around submerged objects the desired information is generally the relation between the velocity of the approaching fluid and the drag force on the object. We have seen in the preceding chapters that, if one knows the velocity and pressure distributions in the system, then the desired relationships for these two cases may be obtained. The derivation of the Hagen–Poiseuille equation in §2.3 and the derivation of the Stokes equation in §2.6 and §4.2 illustrate the two categories we are discussing here.

For many systems the velocity and pressure profiles cannot be easily calculated, particularly if the flow is turbulent or the geometry is complicated. One such system is the flow through a packed column; another is the flow in a tube in the shape of a helical coil. For such systems we can take carefully chosen experimental data and then construct "correlations" of dimensionless variables that can be used to estimate the flow behavior in geometrically similar systems. This method is based on §3.7.

We start in §6.1 by defining the "friction factor," and then we show in §§6.2 and 6.3 how to construct friction factor charts for flow in circular tubes and flow around spheres. These are both systems we have already studied and, in fact, several results from earlier chapters are included in these charts. Finally in §6.4 we examine the flow in packed columns, to illustrate the treatment of a geometrically complicated system. The more complex problem of fluidized beds is not included in this chapter.[1]

§6.1 DEFINITION OF FRICTION FACTORS

We consider the steadily driven flow of a fluid of constant density in one of two systems: (*a*) the fluid flows in a straight conduit of uniform cross section; (*b*) the fluid flows around a submerged object that has an axis of symmetry (or two planes of symmetry) parallel to the direction of the approaching fluid. There will be a force $\mathbf{F}_{f \to s}$ exerted by the fluid on the solid surfaces. It is convenient to split this force into two parts: $\mathbf{F}_s$, the force that would be exerted by the fluid even if it were stationary; and $\mathbf{F}_k$, the additional force associated with the motion of the fluid (see §2.6 for the discussion of $\mathbf{F}_s$ and $\mathbf{F}_k$ for flow around spheres). In systems of type (*a*), $\mathbf{F}_k$ points in the same direction as the average velocity $\langle \mathbf{v} \rangle$ in the conduit, and in systems of type (*b*), $\mathbf{F}_k$ points in the same direction as the approach velocity $\mathbf{v}_\infty$.

For both types of systems we state that the magnitude of the force $\mathbf{F}_k$ is proportional to a characteristic area A and a characteristic kinetic energy K per unit volume; thus

$$F_k = AKf \qquad (6.1\text{-}1)^1$$

in which the proportionality constant f is called the *friction factor*. Note that Eq. 6.1-1 is *not* a law of fluid dynamics, but only a definition for f. This is a useful definition, because the dimensionless quantity f can be given as a relatively simple function of the Reynolds number and the system shape.

Clearly, for any given flow system, f is not defined until A and K are specified. Let us now see what the customary definitions are:

(a) For *flow in conduits*, A is usually taken to be the wetted surface, and K is taken to be $\frac{1}{2}\rho\langle v \rangle^2$. Specifically, for circular tubes of radius R and length L we define f by

$$F_k = (2\pi R L)(\tfrac{1}{2}\rho\langle v \rangle^2)f \qquad (6.1\text{-}2)$$

Generally, the quantity measured is not F_k, but rather the pressure difference $p_0 - p_L$ and the elevation difference $h_0 - h_L$. A force balance on the fluid between 0 and L in the direction of flow gives for fully developed flow

$$\begin{aligned} F_k &= [(p_0 - p_L) + \rho g(h_0 - h_L)]\pi R^2 \\ &= (\mathscr{P}_0 - \mathscr{P}_L)\pi R^2 \end{aligned} \qquad (6.1\text{-}3)$$

Elimination of F_k between the last two equations then gives

$$f = \frac{1}{4}\left(\frac{D}{L}\right)\left(\frac{\mathscr{P}_0 - \mathscr{P}_L}{\frac{1}{2}\rho\langle v \rangle^2}\right) \qquad (6.1\text{-}4)$$

---

[1] R. Jackson, *The Dynamics of Fluidized Beds*, Cambridge University Press (2000).

[1] For systems lacking symmetry, the fluid exerts both a force and a torque on the solid. For discussions of such systems see J. Happel and H. Brenner, *Low Reynolds Number Hydrodynamics*, Martinus Nijhoff, The Hague (1983), Chapter 5; H. Brenner, in *Adv. Chem. Engr.*, **6**, 287–438 (1966); S. Kim and S. J. Karrila, *Microhydrodynamics: Principles and Selected Applications*, Butterworth-Heinemann, Boston (1991), Chapter 5.

in which $D = 2R$ is the tube diameter. Equation 6.1-4 shows how to calculate f from experimental data. The quantity f is sometimes called the *Fanning friction factor*.[2]

(b) For *flow around submerged objects*, the characteristic area A is usually taken to be the area obtained by projecting the solid onto a plane perpendicular to the velocity of the approaching fluid; the quantity K is taken to be $\frac{1}{2}\rho v_\infty^2$, where v_∞ is the approach velocity of the fluid at a large distance from the object. For example, for flow around a sphere of radius R, we define f by the equation

$$F_k = (\pi R^2)(\tfrac{1}{2}\rho v_\infty^2)f \qquad (6.1\text{-}5)[3]$$

If it is not possible to measure F_k, then we can measure the terminal velocity of the sphere when it falls through the fluid (in that case, v_∞ has to be interpreted as the terminal velocity of the sphere). For the steady-state fall of a sphere in a fluid, the force F_k is just counterbalanced by the gravitational force on the sphere less the buoyant force (cf. Eq. 2.6-14):

$$F_k = \tfrac{4}{3}\pi R^3 \rho_{\text{sph}} g - \tfrac{4}{3}\pi R^3 \rho g \qquad (6.1\text{-}6)$$

Elimination of F_k between Eqs. 6.1-5 and 6.1-6 then gives

$$f = \frac{4}{3}\frac{gD}{v_\infty^2}\left(\frac{\rho_{\text{sph}} - \rho}{\rho}\right) \qquad (6.1\text{-}7)$$

This expression can be used to obtain f from terminal velocity data. The friction factor used in Eqs. 6.1-5 and 7 is sometimes called the *drag coefficient* and given the symbol c_D.

We have seen that the "drag coefficient" for submerged objects and the "friction factor" for channel flow are defined in the same general way. For this reason we prefer to use the same symbol and name for both of them.

§6.2 FRICTION FACTORS FOR FLOW IN TUBES

We now combine the definition of f in Eq. 6.1-2 with the dimensional analysis of §3.7 to show what f must depend on in this kind of system. We consider a "test section" of inner radius R and length L, shown in Fig. 6.2-1, carrying a fluid of constant density and viscosity at a steady mass flow rate. The pressures $\mathcal{P}_0$ and $\mathcal{P}_L$ at the ends of the test section are known.

---

[2] This friction factor definition is due to J. T. Fanning, *A Practical Treatise on Hydraulic and Water Supply Engineering*, Van Nostrand, New York, 1st edition (1877), 16th edition (1906); the name "Fanning" is used to avoid confusion with the "Moody friction factor," which is larger by a factor of 4 than the f used here [L. F. Moody, *Trans. ASME*, **66**, 671–684 (1944)].

If we use the "friction velocity" $v_* = \sqrt{\tau_0/\rho} = \sqrt{(\mathcal{P}_0 - \mathcal{P}_L)R/2L\rho}$, introduced in §5.3, then Eq. 6.1-4 assumes the form

$$f = 2(v_*/\langle v\rangle)^2 \qquad (6.1\text{-}4a)$$

John Thomas Fanning (1837–1911) studied architectural and civil engineering, served as an officer in the Civil War, and after the war became prominent in hydraulic engineering. The 14th edition of his book *A Practical Treatise on Hydraulic and Water-Supply Engineering* appeared in 1899.

[3] For the translational motion of a sphere in three dimensions, one can write *approximately*

$$\mathbf{F}_k = (\pi R^2)(\tfrac{1}{2}\rho v_\infty^2)f\mathbf{n} \qquad (6.1\text{-}5a)$$

where $\mathbf{n}$ is a unit vector in the direction of $\mathbf{v}_\infty$. See Problem 6C.1.

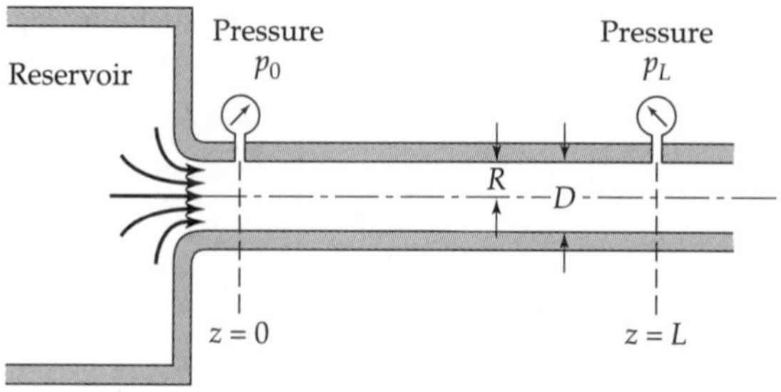

Fig. 6.2-1. Section of a circular pipe from $z = 0$ to $z = L$ for the discussion of dimensional analysis.

The system is either in steady laminar flow or steadily driven turbulent flow (i.e., turbulent flow with a steady total throughput). In either case the force in the z direction of the fluid on the inner wall of the test section is

$$F_k(t) = \int_0^L \int_0^{2\pi} \left(-\mu \frac{\partial v_z}{\partial r} \right) \Bigg|_{r=R} R \, d\theta \, dz \tag{6.2-1}$$

In turbulent flow the force may be a function of time, not only because of the turbulent fluctuations, but also because of occasional ripping off of the boundary layer from the wall, which results in some distances with long time scales. In laminar flow it is understood that the force will be independent of time.

Equating Eqs. 6.2-1 and 6.1-2, we get the following expression for the friction factor:

$$f(t) = \frac{\displaystyle\int_0^L \int_0^{2\pi} \left(-\mu \frac{\partial v_z}{\partial r} \right) \Bigg|_{r=R} R \, d\theta \, dz}{(2\pi R L)(\frac{1}{2}\rho\langle v_z \rangle^2)} \tag{6.2-2}$$

Next we introduce the dimensionless quantities from §3.7: $\check{r} = r/D$, $\check{z} = z/D$, $\check{v}_z = v_z/\langle v_z \rangle$, $\check{t} = \langle v_z \rangle t/D$, $\check{\mathscr{P}} = (\mathscr{P} - \mathscr{P}_0)/\rho\langle v_z \rangle^2$, and $\mathrm{Re} = D\langle v_z \rangle \rho/\mu$. Then Eq. 6.2-2 may be rewritten as

$$f(\check{t}) = \frac{1}{\pi} \frac{D}{L} \frac{1}{\mathrm{Re}} \int_0^{L/D} \int_0^{2\pi} \left(-\frac{\partial \check{v}_z}{\partial \check{r}} \right) \Bigg|_{\check{r}=1/2} d\theta \, d\check{z} \tag{6.2-3}$$

This relation is valid for laminar or turbulent flow in smooth circular tubes. We see that for flow systems in which the drag depends on viscous forces alone (i.e., no "form drag") the product of $f\mathrm{Re}$ is essentially a dimensionless velocity gradient averaged over the surface.

Recall now that, in principle, $\partial \check{v}_z/\partial \check{r}$ can be evaluated from Eqs. 3.7-8 and 9 along with the boundary conditions[1]

B.C. 1:	at $\check{r} = \frac{1}{2}$,	$\check{\mathbf{v}} = 0 \quad$ for $z > 0$	(6.2-4)
B.C. 2:	at $\check{z} = 0$,	$\check{\mathbf{v}} = \boldsymbol{\delta}_z$	(6.2-5)
B.C. 3:	at $\check{r} = 0$ and $\check{z} = 0$,	$\check{\mathscr{P}} = 0$	(6.2-6)

---

[1] Here we follow the customary practice of neglecting the $(\partial^2/\partial \check{z}^2)\mathbf{v}$ terms of Eq. 3.7-9, on the basis of order-of-magnitude arguments such as those given in §4.4. With those terms suppressed, we do not need an outlet boundary condition on $\mathbf{v}$.

and appropriate initial conditions. The uniform inlet velocity profile in Eq. 6.2-5 is accurate except very near the wall, for a well-designed nozzle and upstream system. If Eqs. 3.7-8 and 9 could be solved with these boundary and initial conditions to get $\check{\mathbf{v}}$ and $\check{\mathscr{P}}$, the solutions would necessarily be of the form

$$\check{\mathbf{v}} = \check{\mathbf{v}}(\check{r}, \theta, \check{z}, \check{t}; \mathrm{Re}) \tag{6.2-7}$$

$$\check{\mathscr{P}} = \check{\mathscr{P}}(\check{r}, \theta, \check{z}, \check{t}; \mathrm{Re}) \tag{6.2-8}$$

That is, the functional dependence of $\check{\mathbf{v}}$ and $\check{\mathscr{P}}$ must, in general, include all the dimensionless variables and the one dimensionless group appearing in the differential equations. No additional dimensionless groups enter via the preceding boundary conditions. As a consequence, $\partial \check{v}_z / \partial \check{r}$ must likewise depend on $\check{r}, \theta, \check{z}, \check{t}$, and Re. When $\partial \check{v}_z / \partial \check{r}$ is evaluated at $\check{r} = \frac{1}{2}$ and then integrated over $\check{z}$ and θ in Eq. 6.2-3, the result depends only on $\check{t}$, Re, and L/D (the latter appearing in the upper limit in the integration over $\check{z}$). Therefore we are led to the conclusion that $f(\check{t}) = f(\mathrm{Re}, L/D, \check{t})$, which, when time averaged, becomes

$$f = f(\mathrm{Re}, L/D) \tag{6.2-9}$$

when the time average is performed over an interval long enough to include any long-time turbulent disturbances. The measured friction factor then depends only on the Reynolds number and the length-to-diameter ratio.

The dependence of f on L/D arises from the development of the time-average velocity distribution from its flat entry shape toward more rounded profiles at downstream z values. This development occurs within an entrance region, of length $L_e \cong 0.03D \, \mathrm{Re}$ for laminar flow or $L_e \approx 60D$ for turbulent flow, beyond which the shape of the velocity distribution is "fully developed." In the transportation of fluids, the entrance length is usually a small fraction of the total; then Eq. 6.2-9 reduces to the long-tube form

$$f = f(\mathrm{Re}) \tag{6.2-10}$$

and f can be evaluated experimentally from Eq. 6.1-4, which was written for fully developed flow at the inlet and outlet.

Equations 6.2-9 and 10 are useful results, since they provide a guide for the systematic presentation of data on flow rate versus pressure difference for laminar and turbulent flow in circular tubes. For *long* tubes we need only a single curve of f plotted versus the single combination $D\langle \bar{v}_z \rangle \rho / \mu$. Think how much simpler this is than plotting pressure drop versus the flow rate for separate values of D, L, ρ, and μ, which is what the uninitiated might do.

There is much experimental information for pressure drop versus flow rate in tubes, and hence f can be calculated from the experimental data by Eq. 6.1-4. Then f can be plotted versus Re for smooth tubes to obtain the *solid* curves shown in Fig. 6.2-2. These solid curves describe the laminar and turbulent behavior for fluids flowing in *long, smooth, circular* tubes.

Note that the *laminar* curve on the friction factor chart is merely a plot of the *Hagen–Poiseuille* equation in Eq. 2.3-21. This can be seen by substituting the expression for $(\mathscr{P}_0 - \mathscr{P}_L)$ from Eq. 2.3-21 into Eq. 6.1-4 and using the relation $w = \rho \langle \bar{v}_z \rangle \pi R^2$; this gives

$$f = \frac{16}{\mathrm{Re}} \left\{ \begin{array}{ll} \mathrm{Re} < 2100 & \text{stable} \\ \mathrm{Re} > 2100 & \text{usually unstable} \end{array} \right\} \tag{6.2-11}$$

in which $\mathrm{Re} = D\langle \bar{v}_z \rangle \rho / \mu$; this is exactly the laminar line in Fig. 6.2-2.

Analogous *turbulent* curves have been constructed by using *experimental data*. Some analytical curve-fit expressions are also available. For example, Eq. 5.1-6 can be put into the form

$$f = \frac{0.0791}{\mathrm{Re}^{1/4}} \qquad 2.1 \times 10^3 < \mathrm{Re} < 10^5 \tag{6.2-12}$$

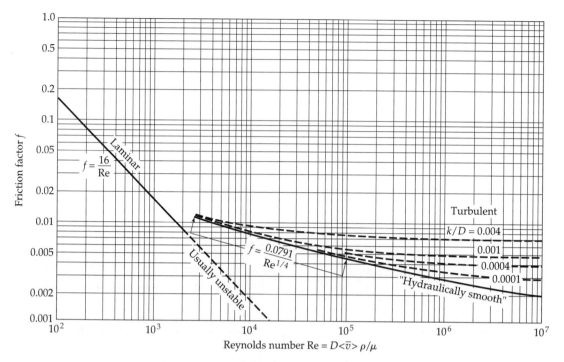

Fig. 6.2-2. Friction factor for tube flow (see definition of f in Eqs. 6.1-2 and 6.1-3. [Curves of L. F. Moody, *Trans. ASME*, **66**, 671–684 (1944) as presented in W. L. McCabe and J. C. Smith, *Unit Operations of Chemical Engineering*, McGraw-Hill, New York (1954).]

which is known as the *Blasius formula*.[2] Equation 5.5-1 (with 2.5 replaced by 2.45 and 1.75 by 2.00) is equivalent to

$$\frac{1}{\sqrt{f}} = 4.0 \; \log_{10} \mathrm{Re}\sqrt{f} - 0.4 \qquad 2.3 \times 10^3 < \mathrm{Re} < 4 \times 10^6 \tag{6.2-13}$$

which is known as the *Prandtl formula*.[3] Finally, corresponding to Eq. 5.5-2, we have

$$f = \frac{2}{\Psi^{2/(\alpha+1)}} \qquad \text{where} \qquad \Psi = \frac{e^{3/2}(\sqrt{3} + 5\alpha)}{2^{\alpha}\alpha(\alpha + 1)(\alpha + 2)} \tag{6.2-14}$$

and $\alpha = 3/(2 \ln \mathrm{Re})$. This has been found to represent the experimental data well for $3.07 \times 10^3 < \mathrm{Re} < 3.23 \times 10^6$. Equation 6.2-14 is called the *Barenblatt formula*.[4]

A further relation, which includes the dashed curves for rough pipes in Fig. 6.2-2, is the empirical *Haaland equation*[5]

$$\frac{1}{\sqrt{f}} = -3.6 \; \log_{10}\left[\frac{6.9}{\mathrm{Re}} + \left(\frac{k/D}{3.7}\right)^{10/9}\right] \qquad \begin{cases} 4 \times 10^4 < \mathrm{Re} < 10^8 \\ 0 < k/D < 0.05 \end{cases} \tag{6.2-15}$$

[2] H. Blasius, *Forschungsarbeiten des Ver. Deutsch. Ing.*, no. 131 (1913).

[3] L. Prandtl, *Essentials of Fluid Dynamics*, Hafner, New York (1952), p. 165.

[4] G. I. Barenblatt, *Scaling, Self-Similarity, and Intermediate Asymptotics*, Cambridge University Press (1996), §10.2.

[5] S. E. Haaland, *Trans. ASME, JFE*, **105**, 89–90 (1983). For other empiricisms see D. J. Zigrang and N. D. Sylvester, *AIChE Journal*, **28**, 514–515 (1982).

This equation is stated[5] to be accurate within 1.5%. As can be seen in Fig. 6.2-2, the frictional resistance to flow increases with the height, k, of the protuberances. Of course, k has to enter into the correlation in a dimensionless fashion and hence appears via the ratio k/D.

For *turbulent flow* in *noncircular tubes* it is common to use the following empiricism: First we define a "mean hydraulic radius" R_h as follows:

$$R_h = S/Z \tag{6.2-16}$$

in which S is the cross section of the conduit and Z is the wetted perimeter. Then we can use Eq. 6.1-4 and Fig. 6.2-2, with the diameter D of the circular pipe replaced by $4R_h$. That is, we calculate pressure differences by replacing Eq. 6.1-4 by

$$f = \left(\frac{R_h}{L}\right)\left(\frac{\mathcal{P}_0 - \mathcal{P}_L}{\frac{1}{2}\rho\langle v_z\rangle^2}\right) \tag{6.2-17}$$

and getting f from Fig. 6.2-2 with a Reynolds number defined as

$$\text{Re}_h = \frac{4R_h\langle v_z\rangle\rho}{\mu} \tag{6.2-18}$$

For laminar flows in noncircular passages, this method is less satisfactory.

EXAMPLE 6.2-1

Pressure Drop Required for a Given Flow Rate

What pressure gradient is required to cause diethylaniline, $C_6H_5N(C_2H_5)_2$, to flow in a horizontal, smooth, circular tube of inside diameter $D = 3$ cm at a mass rate of 1028 g/s at 20°C? At this temperature the density of diethylaniline is $\rho = 0.935$ g/cm^3 and its viscosity is $\mu = 1.95$ cp.

SOLUTION

The Reynolds number for the flow is

$$\text{Re} = \frac{D\langle v_z\rangle\rho}{\mu} = \frac{Dw}{(\pi D^2/4)\mu} = \frac{4w}{\pi D\mu}$$

$$= \frac{4(1028 \text{ g/s})}{\pi(3 \text{ cm})(1.95 \times 10^{-2} \text{ g/cm} \cdot \text{s})} = 2.24 \times 10^4 \tag{6.2-19}$$

From Fig. 6.2-2, we find that for this Reynolds number the friction factor f has a value of 0.0063 for smooth tubes. Hence the pressure gradient required to maintain the flow is (according to Eq. 6.1-4)

$$\frac{p_0 - p_L}{L} = \left(\frac{4}{D}\right)\left(\frac{1}{2}\rho\langle v_z\rangle^2\right)f = \frac{2}{D}\rho\left(\frac{4w}{\pi D^2\rho}\right)^2 f$$

$$= \frac{32w^2 f}{\pi^2 D^5\rho} = \frac{(32)(1028)^2(0.0063)}{\pi^2(3.0)^5(0.935)}$$

$$= 95(\text{dyne/cm}^2)/\text{cm} = 0.071(\text{mm Hg})/\text{cm} \tag{6.2-20}$$

EXAMPLE 6.2-2

Flow Rate for a Given Pressure Drop

Determine the flow rate, in pounds per hour, of water at 68°F through a 1000-ft length of horizontal 8-in. schedule 40 steel pipe (internal diameter 7.981 in.) under a pressure difference of 3.00 psi. For such a pipe use Fig. 6.2-2 and assume that $k/D = 2.3 \times 10^{-4}$.

SOLUTION

We want to use Eq. 6.1-4 and Fig. 6.2-2 to solve for $\langle v_z\rangle$ when $p_0 - p_L$ is known. However, the quantity $\langle v_z\rangle$ appears explicitly on the left side of the equation and implicitly on the right side in f, which depends on $\text{Re} = D\langle v_z\rangle\rho/\mu$. Clearly a trial-and-error solution can be found.

However, if one has to make more than a few calculations of $\langle v_z \rangle$, it is advantageous to develop a systematic approach; we suggest two methods here. Because experimental data are often presented in graphical form, it is important for engineering students to use their originality in devising special methods such as those described here.

Method A. Figure 6.2-2 may be used to construct a plot[6] of Re versus the group $\text{Re}\sqrt{f}$, which does not contain $\langle v_z \rangle$:

$$\text{Re}\sqrt{f} = \frac{D\langle v_z \rangle \rho}{\mu} \sqrt{\frac{(p_0 - p_L)D}{2L\rho\langle \bar{v}_z \rangle^2}} = \frac{D\rho}{\mu} \sqrt{\frac{(p_0 - p_L)D}{2L\rho}} \tag{6.2-21}$$

The quantity $\text{Re}\sqrt{f}$ can be computed for this problem, and a value of the Reynolds number can be read from the Re versus $\text{Re}\sqrt{f}$ plot. From Re the average velocity and flow rate can then be calculated.

Method B. Figure 6.2-2 may also be used directly without any replotting, by devising a scheme that is equivalent to the graphical solution of two simultaneous equations. The two equations are

$$f = f(\text{Re}, k/D) \qquad \text{curve given in Fig. 6.2-2} \tag{6.2-22}$$

$$f = \frac{(\text{Re}\sqrt{f})^2}{\text{Re}^2} \qquad \text{straight line of slope } -2 \text{ on log-log plot} \tag{6.2-23}$$

The procedure is then to compute $\text{Re}\sqrt{f}$ according to Eq. 6.2-21 and then to plot Eq. 6.2-23 on the log-log plot of f versus Re in Fig. 6.2-2. The intersection point gives the Reynolds number of the flow, from which $\langle \bar{v}_z \rangle$ can then be computed.

For the problem at hand, we have

$$p_0 - p_L = (3.00 \text{ lb}_f/\text{in.}^2)(32.17 \text{ (lb}_m\text{ft/s}^2)/\text{lb}_f)(144 \text{ in.}^2/\text{ft}^2)$$

$$= 1.39 \times 10^4 \text{ lb}_m/\text{ft} \cdot \text{s}^2$$

$$D = (7.981 \text{ in.})(\tfrac{1}{12} \text{ ft/in.}) = 0.665 \text{ ft}$$

$$L = 1000 \text{ ft}$$

$$\rho = 62.3 \text{ lb}_m/\text{ft}^3$$

$$\mu = (1.03 \text{ cp})(6.72 \times 10^{-4}(\text{lb}_m/\text{ft} \cdot \text{s})/\text{cp})$$

$$= 6.93 \times 10^{-4} \text{ lb}_m/\text{ft} \cdot \text{s}$$

Then according to Eq. 6.2-21,

$$\text{Re}\sqrt{f} = \frac{D\rho}{\mu} \sqrt{\frac{(p_0 - p_L)D}{2L\rho}} = \frac{(0.665)(62.3)}{(6.93 \times 10^{-4})} \sqrt{\frac{(1.39 \times 10^4)(0.665)}{2(1000)(62.3)}}$$

$$= 1.63 \times 10^4 \qquad \text{(dimensionless)} \tag{6.2-24}$$

The line of Eq. 6.2-23 for this value of $\text{Re}\sqrt{f}$ passes through $f = 1.0$ at $\text{Re} = 1.63 \times 10^4$ and through $f = 0.01$ at $\text{Re} = 1.63 \times 10^5$. Extension of the straight line through these points to the curve of Fig. 6.2-2 for $k/D = 0.00023$ gives the solution to the two simultaneous equations:

$$\text{Re} = \frac{D\langle v_z \rangle \rho}{\mu} = \frac{4w}{\pi D \mu} = 2.4 \times 10^5 \tag{6.2-25}$$

Solving for w then gives

$$w = (\pi/4)D\mu\,\text{Re}$$

$$= (0.7854)(0.665)(6.93 \times 10^{-4})(3600)(2.4 \times 10^5)$$

$$= 3.12 \times 10^5 \text{ lb}_m/\text{hr} = 39 \text{ kg/s} \tag{6.2-26}$$

---

[6] A related plot was proposed by T. von Kármán, *Nachr. Ges. Wiss. Göttingen, Fachgruppen*, **I, 5**, 58–76 (1930).

§6.3   FRICTION FACTORS FOR FLOW AROUND SPHERES

In this section we use the definition of the friction factor in Eq. 6.1-5 along with the dimensional analysis of §3.7 to determine the behavior of f for a stationary sphere in an infinite stream of fluid approaching with a uniform, steady velocity v_∞. We have already studied the flow around a sphere in §2.6 and §4.2 for Re < 0.1 (the "creeping flow" region). At Reynolds numbers above about 1 there is a significant unsteady eddy motion in the wake of the sphere. Therefore, it will be necessary to do a time average over a time interval long with respect to this eddy motion.

Recall from §2.6 that the total force acting in the z direction on the sphere can be written as the sum of a contribution from the normal stresses (F_n) and one from the tangential stresses (F_t). One part of the normal-stress contribution is the force that would be present even if the fluid were stationary, F_s. Thus the "kinetic force," associated with the fluid motion, is

$$F_k = (F_n - F_s) + F_t = F_{\text{form}} + F_{\text{friction}} \tag{6.3-1}$$

The forces associated with the form drag and the friction drag are then obtained from

$$F_{\text{form}}(t) = \int_0^{2\pi} \int_0^{\pi} (-\mathscr{P}|_{r=R} \cos\theta)R^2 \sin\theta \, d\theta \, d\phi \tag{6.3-2}$$

$$F_{\text{friction}}(t) = \int_0^{2\pi} \int_0^{\pi} \left(-\mu \left[r\frac{\partial}{\partial r}\left(\frac{v_\theta}{r}\right) + \frac{1}{r}\frac{\partial v_r}{\partial \theta} \right]\bigg|_{r=R} \sin\theta \right)R^2 \sin\theta \, d\theta \, d\phi \tag{6.3-3}$$

Since v_r is zero everywhere on the sphere surface, the term containing $\partial v_r/\partial\theta$ is zero.

If now we split f into two parts as follows

$$f = f_{\text{form}} + f_{\text{friction}} \tag{6.3-4}$$

then, from the definition in Eq. 6.1-5, we get

$$f_{\text{form}}(\check{t}) = \frac{2}{\pi} \int_0^{2\pi} \int_0^{\pi} (-\check{\mathscr{P}}|_{\check{r}=1} \cos\theta) \sin\theta \, d\theta \, d\phi \tag{6.3-5}$$

$$f_{\text{friction}}(\check{t}) = -\frac{4}{\pi}\frac{1}{\text{Re}} \int_0^{2\pi} \int_0^{\pi} \left[\check{r}\frac{\partial}{\partial\check{r}}\left(\frac{\check{v}_\theta}{\check{r}}\right) \right]\bigg|_{\check{r}=1} \sin^2\theta \, d\theta \, d\phi \tag{6.3-6}$$

The friction factor is expressed here in terms of dimensionless variables

$$\check{\mathscr{P}} = \frac{\mathscr{P}}{\rho v_\infty^2} \qquad \check{v}_\theta = \frac{v_\theta}{v_\infty} \qquad \check{r} = \frac{r}{R} \qquad \check{t} = \frac{v_\infty t}{R} \tag{6.3-7}$$

and a Reynolds number defined as

$$\text{Re} = \frac{Dv_\infty\rho}{\mu} = \frac{2Rv_\infty\rho}{\mu} \tag{6.3-8}$$

To evaluate $f(\check{t})$ one would have to know $\check{\mathscr{P}}$ and $\check{v}_\theta$ as functions of $\check{r}$, θ, ϕ, and $\check{t}$.

We know that for incompressible flow these distributions can *in principle* be obtained from the solution of Eqs. 3.7-8 and 9 along with the boundary conditions

B.C. 1:   at $\check{r} = 1$, $\check{v}_r = 0$ and $\check{v}_\theta = 0$ $\qquad\qquad$ (6.3-9)

B.C. 2:   at $\check{r} = \infty$, $\check{v}_z = 1$ $\qquad\qquad\qquad\qquad\qquad$ (6.3-10)

B.C. 3:   at $\check{r} = \infty$, $\check{\mathscr{P}} = 0$ $\qquad\qquad\qquad\qquad\qquad$ (6.3-11)

and some appropriate initial condition on $\check{\mathbf{v}}$. Because no additional dimensionless groups enter via the boundary and initial conditions, we know that the dimensionless pressure and velocity profiles will have the following form:

$$\check{\mathscr{P}} = \check{\mathscr{P}}(\check{r}, \theta, \phi, \check{t}; \text{Re}) \qquad \check{\mathbf{v}} = \check{\mathbf{v}}(\check{r}, \theta, \phi, \check{t}; \text{Re}) \tag{6.3-12}$$

When these expressions are substituted into Eqs. 6.3-5 and 6, it is then evident that the friction factor in Eq. 6.3-4 must have the form $f(\check{t}) = f(\mathrm{Re}, \check{t})$, which, when time averaged over the turbulent fluctuations, simplifies to

$$f = f(\mathrm{Re}) \tag{6.3-13}$$

by using arguments similar to those in §6.2. Hence from the definition of the friction factor and the dimensionless form of the equations of change and the boundary conditions, we find that f must be a function of Re alone.

Many experimental measurements of the drag force on spheres are available, and when these are plotted in dimensionless form, Fig. 6.3-1 results. For this system there is no sharp transition from an unstable laminar flow curve to a stable turbulent flow curve as for long tubes at a Reynolds number of about 2100 (see Fig. 6.2-2). Instead, as the approach velocity increases, f varies smoothly and moderately up to Reynolds numbers of the order of 10^5. The kink in the curve at about $\mathrm{Re} = 2 \times 10^5$ is associated with the shift of the boundary layer separation zone from in front of the equator to in back of the equator of the sphere.[1]

We have juxtaposed the discussions of tube flow and flow around a sphere to emphasize the fact that various flow systems behave quite differently. Several points of difference between the two systems are:

Flow in Tubes	**Flow Around Spheres**
• Rather well defined laminar–turbulent transition at about Re = 2100	• No well defined laminar–turbulent transition
• The only contribution to f is the friction drag (if the tubes are smooth)	• Contributions to f from both friction and form drag
• No boundary layer separation	• There is a kink in the f vs. Re curve associated with a shift in the separation zone

The general shape of the curves in Figs. 6.2-2 and 6.3-1 should be carefully remembered.

For the *creeping flow region*, we already know that the drag force is given by *Stokes' law*, which is a consequence of solving the continuity equation and the Navier–Stokes equation of motion without the $\rho D\mathbf{v}/Dt$ term. Stokes' law can be rearranged into the form of Eq. 6.1-5 to get

$$F_k = (\pi R^2)(\tfrac{1}{2}\rho v_\infty^2)\left(\frac{24}{Dv_\infty\rho/\mu}\right) \tag{6.3-14}$$

Hence for *creeping flow* around a sphere

$$f = \frac{24}{\mathrm{Re}} \qquad \text{for Re} < 0.1 \tag{6.3-15}$$

and this is the straight-line asymptote as $\mathrm{Re} \to 0$ of the friction factor curve in Fig. 6.3-1.

For higher values of the Reynolds number, Eq. 4.2-21 can describe f accurately up to about Re = 1. However, the empirical expression[2]

$$f = \left(\sqrt{\frac{24}{\mathrm{Re}}} + 0.5407\right)^2 \qquad \text{for Re} < 6000 \tag{6.3-16}$$

[1] R. K. Adair, *The Physics of Baseball*, Harper and Row, New York (1990).
[2] F. F. Abraham, *Physics of Fluids*, **13**, 2194 (1970); M. Van Dyke, *Physics of Fluids*, **14**, 1038–1039 (1971).

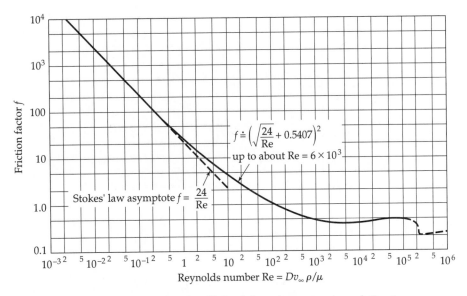

Fig. 6.3-1. Friction factor (or drag coefficient) for spheres moving relative to a fluid with a velocity v_∞. The definition of f is given in Eq. 6.1-5. [Curve taken from C. E. Lapple, "Dust and Mist Collection," in *Chemical Engineers' Handbook*, (J. H. Perry, ed.), McGraw-Hill, New York, 3rd edition (1950), p. 1018.]

is both simple and useful. It is important to remember that

$$f \approx 0.44 \qquad \text{for } 5 \times 10^2 < \text{Re} < 1 \times 10^5 \tag{6.3-17}$$

which covers a remarkable range of Reynolds numbers. Eq. 6.3-17 is sometimes called *Newton's resistance law*; it is handy for a seat-of-the-pants calculation. According to this, the drag force is proportional to the square of the approach velocity of the fluid.

Many extensions of Fig. 6.3-1 have been made, but a systematic study is beyond the scope of this text. Among the effects that have been investigated are wall effects[3] (see Prob. 6C.2), fall of droplets with internal circulation,[4] hindered settling (i.e., fall of clusters of particles[5] that interfere with one another), unsteady flow,[6] and the fall of nonspherical particles.[7]

EXAMPLE 6.3-1 *Determination of the Diameter of a Falling Sphere*	Glass spheres of density $\rho_{sph} = 2.62$ g/cm³ are to be allowed to fall through liquid CCl₄ at 20°C in an experiment for studying human reaction times in making time observations with stopwatches and more elaborate devices. At this temperature the relevant properties of CCl₄ are $\rho = 1.59$ g/cm³ and $\mu = 9.58$ millipoises. What diameter should the spheres be to have a terminal velocity of about 65 cm/s?

[3] J. R. Strom and R. C. Kintner, *AIChE Journal*, **4**, 153–156 (1958).

[4] L. Landau and E. M. Lifshitz, *Fluid Mechanics*, Pergamon, Oxford, 2nd edition (1987), pp. 65–66; S. Hu and R. C. Kintner, *AIChE Journal*, **1**, 42–48 (1955).

[5] C. E. Lapple, *Fluid and Particle Mechanics*, University of Delaware Press, Newark, Del. (1951), Chapter 13; R. F. Probstein, *Physicochemical Hydrodynamics*, Wiley, New York, 2nd edition (1994), §5.4.

[6] R. R. Hughes and E. R. Gilliland, *Chem. Eng. Prog.*, **48**, 497–504 (1952); L. Landau and E. M. Lifshitz, *Fluid Mechanics*, Pergamon, Oxford, 2nd edition (1987), pp. 90–91.

[7] E. S. Pettyjohn and E. B. Christiansen, *Chem. Eng. Prog.*, **44**, 147 (1948); H. A. Becker, *Can. J. Chem. Eng.*, **37**, 885–891 (1959); S. Kim and S. J. Karrila, *Microhydrodynamics: Principles and Selected Applications*, Butterworth-Heinemann, Boston (1991), Chapter 5.

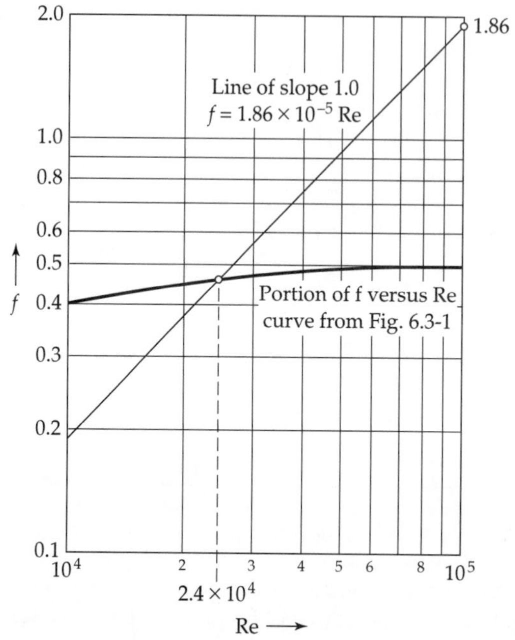

Fig. 6.3-2. Graphical procedure used in Example 6.3-1.

SOLUTION

To find the sphere diameter, we have to solve Eq. 6.1-7 for D. However, in this equation one has to know D in order to get f; and f is given by the solid curve in Fig. 6.3-1. A trial-and-error procedure can be used, taking $f = 0.44$ as a first guess.

Alternatively, we can solve Eq. 6.1-7 for f and then note that f/Re is a quantity independent of D:

$$\frac{f}{\text{Re}} = \frac{4}{3} \frac{g\mu}{\rho v_\infty^3} \left(\frac{\rho_{\text{sph}} - \rho}{\rho} \right) \tag{6.3-18}$$

The quantity on the right side can be calculated with the information above, and we call it C. Hence we have two simultaneous equations to solve:

$$f = C\,\text{Re} \qquad \text{from Eq. 6.3-18} \tag{6.3-19}$$
$$f = f(\text{Re}) \qquad \text{from Fig. 6.3-1} \tag{6.3-20}$$

Equation 6.3-19 is a straight line with slope of unity on the log-log plot of f versus Re.

For the problem at hand we have

$$C = \frac{4}{3} \frac{(980)(9.58 \times 10^{-3})}{(1.59)(65)^3} \left(\frac{2.62 - 1.59}{1.59} \right) = 1.86 \times 10^{-5} \tag{6.3-21}$$

Hence at $\text{Re} = 10^5$, according to Eq. 6.3-19, $f = 1.86$. The line of slope 1 passing through $f = 1.86$ at $\text{Re} = 10^5$ is shown in Fig. 6.3-2. This line intersects the curve of Eq. 6.3-20 (i.e., the curve of Fig. 6.3-1) at $\text{Re} = Dv_\infty \rho / \mu = 2.4 \times 10^4$. The sphere diameter is then found to be

$$D = \frac{\text{Re}\,\mu}{\rho v_\infty} = \frac{(2.4 \times 10^4)(9.58 \times 10^{-3})}{(1.59)(65)} = 2.2 \text{ cm} \tag{6.3-22}$$

§6.4 FRICTION FACTORS FOR PACKED COLUMNS

In the preceding two sections we have discussed the friction factor correlations for two simple flow systems of rather wide interest. Friction factor charts are available for a number of other systems, such as transverse flow past a cylinder, flow across tube banks, flow near baffles, and flow near rotating disks. These and many more are summarized in various reference works.[1] One complex system of considerable interest in chemical engineering is the packed column, widely used for catalytic reactors and for separation processes.

There have been two main approaches for developing friction factor expressions for packed columns. In one method the packed column is visualized as a bundle of tangled tubes of weird cross section; the theory is then developed by applying the previous results for single straight tubes to the collection of crooked tubes. In the second method the packed column is regarded as a collection of submerged objects, and the pressure drop is obtained by summing up the resistances of the submerged particles.[2] The tube bundle theories have been somewhat more successful, and we discuss them here. Figure 6.4-1(a) depicts a packed column, and Fig. 6.4-1(b) illustrates the tube bundle model.

A variety of materials may be used for the packing in columns: spheres, cylinders, Berl saddles, and so on. It is assumed throughout the following discussion that the packing is statistically uniform, so that there is no "channeling" (in actual practice, channeling frequently occurs, and then the development given here does not apply). It is further assumed that the diameter of the packing particles is small in comparison to the diameter of the column in which the packing is contained, and that the column diameter is uniform.

We define the friction factor for the packed column analogously to Eq. 6.1-4:

$$f = \frac{1}{4}\left(\frac{D_p}{L}\right)\left(\frac{\mathscr{P}_0 - \mathscr{P}_L}{\frac{1}{2}\rho v_0^2}\right) \tag{6.4-1}$$

in which L is the length of the packed column, D_p is the effective particle diameter (defined presently), and v_0 is the *superficial velocity*; this is the volume flow rate divided by the empty column cross section, $v_0 = w/\rho S$.

The pressure drop through a representative tube in the tube bundle model is given by Eq. 6.2-17

$$\mathscr{P}_0 - \mathscr{P}_L = \frac{1}{2}\rho\langle v\rangle^2\left(\frac{L}{R_h}\right)f_{\text{tube}} \tag{6.4-2}$$

 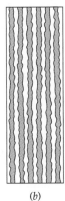

(a)   (b)

Fig. 6.4-1. (a) A cylindrical tube packed with spheres; (b) a "tube bundle" model for the packed column in (a).

[1] P. C. Carman, *Flow of Gases through Porous Media*, Butterworths, London (1956); J. G. Richardson, Section 16 in *Handbook of Fluid Dynamics* (V. L. Streeter, ed.), McGraw-Hill, New York (1961); M. Kaviany, Chapter 21 in *The Handbook of Fluid Dynamics* (R. W. Johnson, ed.), CRC Press, Boca Raton, Fla. (1998).

[2] W. E. Ranz, *Chem. Eng. Prog.*, **48**, 274–253 (1952); H. C. Brinkman, *Appl. Sci. Research.*, **A1**, 27–34, 81–86, 333–346 (1949). **Henri Coenraad Brinkman** (1908–1961) did research on viscous dissipation heating, flow in porous media, and plasma physics; he taught at the University of Bandung, Indonesia, from 1949 to 1954, where he wrote *The Application of Spinor Invariants to Atomic Physics*.

in which the friction factor for a single tube, f_{tube}, is a function of the Reynolds number $\text{Re}_h = 4R_h\langle v\rangle\rho/\mu$. When this pressure difference is substituted into Eq. 6.4-1, we get

$$f = \frac{1}{4}\frac{D_p}{R_h}\frac{\langle v\rangle^2}{v_0^2}f_{\text{tube}} = \frac{1}{4\epsilon^2}\frac{D_p}{R_h}f_{\text{tube}} \tag{6.4-3}$$

In the second expression, we have introduced the *void fraction*, ε, the fraction of space in the column not occupied by the packing. Then $v_0 = \langle v\rangle\varepsilon$, which results from the definition of the superficial velocity. We now need an expression for R_h.

The hydraulic radius can be expressed in terms of the void fraction ε and the wetted surface a per unit volume of bed as follows:

$$R_h = \left(\frac{\text{cross section available for flow}}{\text{wetted perimeter}}\right)$$

$$= \left(\frac{\text{volume available for flow}}{\text{total wetted surface}}\right)$$

$$= \frac{\left(\dfrac{\text{volume of voids}}{\text{volume of bed}}\right)}{\left(\dfrac{\text{wetted surface}}{\text{volume of bed}}\right)} = \frac{\varepsilon}{a} \tag{6.4-4}$$

The quantity a is related to the "specific surface" a_v (total particle surface per volume of particles) by

$$a_v = \frac{a}{1-\varepsilon} \tag{6.4-5}$$

The quantity a_v is in turn used to define the mean particle diameter D_p as follows:

$$D_p = \frac{6}{a_v} \tag{6.4-6}$$

This definition is chosen because, for spheres of uniform diameter, D_p is exactly the diameter of a sphere. From the last three expressions we find that the hydraulic radius is $R_h = D_p\varepsilon/6(1-\varepsilon)$. When this is substituted into Eq. 6.4-3, we get

$$f = \frac{3}{2}\left(\frac{1-\varepsilon}{\varepsilon^3}\right)f_{\text{tube}} \tag{6.4-7}$$

We now adapt this result to laminar and turbulent flows by inserting appropriate expressions for f_{tube}.

(a) For *laminar flow* in tubes, $f_{\text{tube}} = 16/\text{Re}_h$. This is exact for circular tubes only. To account for the noncylindrical surfaces and tortuous fluid paths encountered in typical packed-column operations, it has been found that replacing 16 by 100/3 allows the tube bundle model to describe the packed-column data. When this modified expression for the tube friction factor is used, Eq. 6.4-7 becomes

$$f = \frac{(1-\varepsilon)^2}{\varepsilon^3}\frac{75}{(D_pG_0/\mu)} \tag{6.4-8}$$

in which $G_0 = \rho v_0$ is the mass flux through the system. When this expression for f is substituted into Eq. 6.4-1 we get

$$\frac{\mathscr{P}_0 - \mathscr{P}_L}{L} = 150\left(\frac{\mu v_0}{D_p^2}\right)\frac{(1-\varepsilon)^2}{\varepsilon^3} \tag{6.4-9}$$

which is the *Blake–Kozeny equation*.[3] Equations 6.4-8 and 9 are generally good for $(D_p G_0/\mu(1 - \varepsilon)) < 10$ and for void fractions less than $\varepsilon = 0.5$.

(b) For *highly turbulent flow* a treatment similar to the above can be given. We begin again with the expression for the friction factor definition for flow in a circular tube. This time, however, we note that, for highly turbulent flow in tubes with any appreciable roughness, the friction factor is a function of the roughness only, and is independent of the Reynolds number. If we assume that the tubes in all packed columns have similar roughness characteristics, then the value of f_{tube} may be taken to be the same constant for all systems. Taking $f_{\text{tube}} = 7/12$ proves to be an acceptable choice. When this is inserted into Eq. 6.4-7, we get

$$f = \frac{7}{8}\left(\frac{1 - \varepsilon}{\varepsilon^3}\right) \tag{6.4-10}$$

When this is substituted into Eq. 6.4-1, we get

$$\frac{\mathscr{P}_0 - \mathscr{P}_L}{L} = \frac{7}{4}\left(\frac{\rho v_0^2}{D_p}\right)\frac{1 - \varepsilon}{\varepsilon^3} \tag{6.4-11}$$

which is the *Burke–Plummer*[4] equation, valid for $(D_p G_0/\mu(1 - \varepsilon)) > 1000$. Note that the dependence on the void fraction is different from that for laminar flow.

(c) For the *transition region*, we may superpose the pressure drop expressions for (*a*) and (*b*) above to get

$$\frac{\mathscr{P}_0 - \mathscr{P}_L}{L} = 150\left(\frac{\mu v_0}{D_p^2}\right)\frac{(1 - \varepsilon)^2}{\varepsilon^3} + \frac{7}{4}\left(\frac{\rho v_0^2}{D_p}\right)\frac{1 - \varepsilon}{\varepsilon^3} \tag{6.4-12}$$

For very small v_0, this simplifies to the Blake–Kozeny equation, and for very large v_0, to the Burke–Plummer equation. Such empirical superpositions of asymptotes often lead to satisfactory results. Equation 6.4-12 may be rearranged to form dimensionless groups:

$$\left(\frac{(\mathscr{P}_0 - \mathscr{P}_L)\rho}{G_0^2}\right)\left(\frac{D_p}{L}\right)\left(\frac{\varepsilon^3}{1 - \varepsilon}\right) = 150\left(\frac{1 - \varepsilon}{D_p G_0/\mu}\right) + \frac{7}{4} \tag{6.4-13}$$

This is the *Ergun equation*,[5] which is shown in Fig. 6.4-2 along with the Blake–Kozeny and Burke–Plummer equations and experimental data. It has been applied with success to gas flow through packed columns by using the density $\bar{\rho}$ of the gas at the arithmetic average of the end pressures. Note that G_0 is constant through the column, whereas v_0 changes through the column for a compressible fluid. For large pressure drops, however, it seems more appropriate to apply Eq. 6.4-12 locally by expressing the pressure gradient in differential form.

The Ergun equation is but one of many[6] that have been proposed for describing packed columns. For example, the *Tallmadge equation*[7]

$$\left(\frac{(\mathscr{P}_0 - \mathscr{P}_L)\rho}{G_0^2}\right)\left(\frac{D_p}{L}\right)\left(\frac{\varepsilon^3}{1 - \varepsilon}\right) = 150\left(\frac{1 - \varepsilon}{D_p G_0/\mu}\right) + 4.2\left(\frac{1 - \varepsilon}{D_p G_0/\mu}\right)^{1/6} \tag{6.4-14}$$

is reported to give good agreement with experimental data over the range $0.1 < (D_p G_0/\mu(1 - \varepsilon)) < 10^5$.

---

[3] F. C. Blake, *Trans. Amer. Inst. Chem. Engrs.*, **14**, 415–421 (1922); J. Kozeny, *Sitzungsber. Akad. Wiss. Wien*, Abt. II*a*, **136**, 271–306 (1927).

[4] S. P. Burke and W. B. Plummer, *Ind. Eng. Chem.*, **20**, 1196–1200 (1928).

[5] S. Ergun, *Chem. Engr. Prog.*, **48**, 89–94 (1952).

[6] I. F. Macdonald, M. S. El-Sayed, K. Mow, and F. A. Dullien, *Ind. Eng. Chem. Fundam.*, **18**, 199–208 (1979).

[7] J. A. Tallmadge, *AIChE Journal*, **16**, 1092–1093 (1970).

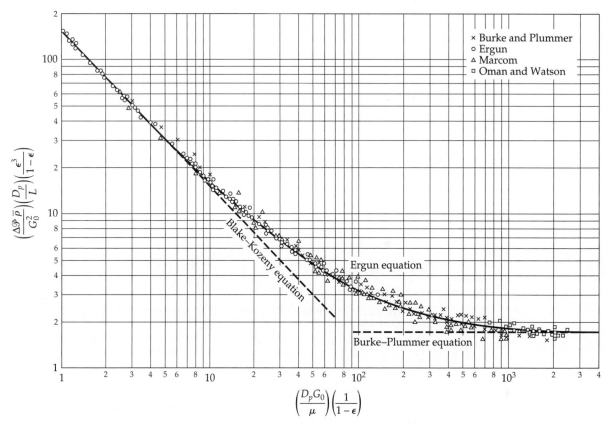

Fig. 6.4-2. The Ergun equation for flow in packed beds, and the two related asymptotes, the Blake–Kozeny equation and the Burke–Plummer equation [S. Ergun, *Chem. Eng. Prog.*, **48**, 89–94 (1952)].

The above discussion of packed beds illustrates how one can often combine solutions of elementary problems to create useful models for complex systems. The constants appearing in the models are then determined from experimental data. As better data become available the modeling can be improved.

QUESTIONS FOR DISCUSSION

1. How are graphs of friction factors versus Reynolds numbers generated from experimental data, and why are they useful?
2. Compare and contrast the friction factor curves for flow in tubes and flow around spheres. Why do they have different shapes?
3. In Fig. 6.2-2, why does the f versus Re curve for turbulent flow lie above the curve for laminar flow rather than below?
4. Discuss the caveat after Eq. 6.2-18. Will the use of the mean hydraulic radius for laminar flow predict a pressure drop that is too high or too low for a given flow rate?
5. Can friction factor correlations be used for unsteady flows?
6. What is the connection, if any, between the Blake–Kozeny equation (Eq. 6.4-9) and Darcy's law (Eq. 4C.3-2)?

7. Discuss the flow of water through a 1/2-in. rubber garden hose that is attached to a house faucet with a pressure of 70 psig available.

8. Why was Eq. 6.4-12 rewritten in the form of Eq. 6.4-13?

9. A baseball announcer says: "Because of the high humidity today, the baseball cannot go as far through the heavy humid air as it would on a dry day." Comment critically on this statement.

PROBLEMS

6A.1 Pressure drop required for a pipe with fittings. What pressure drop is needed for pumping water at 20°C through a pipe of 25 cm diameter and 1234 m length at a rate of 1.97 m³/s? The pipe is at the same elevation throughout and contains four standard radius 90° elbows and two 45° elbows. The resistance of a standard radius 90° elbow is roughly equivalent to that offered by a pipe whose length is 32 diameters; a 45° elbow, 15 diameters. (An alternative method for calculating losses in fittings is given in §7.5.)

Answer: 4.7×10^3 psi = 33 MPa

6A.2 Pressure difference required for flow in pipe with elevation change (Fig. 6A.2). Water at 68°F is to be pumped through 95 ft of standard 3-in. pipe (internal diameter 3.068 in.) into an overhead reservoir.

(a) What pressure is required at the outlet of the pump to

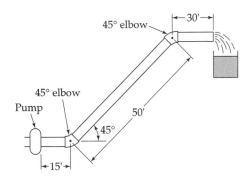

Fig. 6A.2. Pipe flow system.

supply water to the overhead reservoir at a rate of 18 gal/min? At 68°F the viscosity of water is 1.002 cp and the density is 0.9982 g/ml.

(b) What percentage of the pressure drop is needed for overcoming the pipe friction?

Answer: **(a)** 15.2 psig

6A.3 Flow rate for a given pressure drop. How many gal/hr of water at 68°F can be delivered through a 1320-ft length of smooth 6.00-in. i.d. pipe under a pressure difference of 0.25 psi? Assume that the pipe is "hydraulically smooth."

(a) Solve by Method A of Example 6.2-2.

(b) Solve by Method B of Example 6.2-2.

Answer: 68 U.S. gal/min

6A.4 Motion of a sphere in a liquid. A hollow sphere, 5.00 mm in diameter, with a mass of 0.0500 g, is released in a column of liquid and attains a terminal velocity of 0.500 cm/s. The liquid density is 0.900 g/cm³. The local gravitational acceleration is 980.7 cm/sec². The sphere is far enough from the containing walls so that their effect can be neglected.

(a) Compute the drag force on the sphere in dynes.

(b) Compute the friction factor.

(c) Determine the viscosity of the liquid.

Answers: **(a)** 8.7 dynes; **(b)** $f = 396$; **(c)** 3.7g/cm·s

6A.5 Sphere diameter for a given terminal velocity.

(a) Explain how to find the sphere diameter D corresponding to given values of v_∞, ρ, ρ_s, μ, and g by making a direct construction on Fig. 6.3-1.

(b) Rework Problem 2A.4 by using Fig. 6.3-1.

(c) Rework (b) when the gas velocity is 10 ft/s.

6A.6 Estimation of void fraction of a packed column. A tube of 146 sq. in. cross section and 73 in. height is packed with spherical particles of diameter 2 mm. When a pressure difference of 158 psi is maintained across the column, a 60% aqueous sucrose solution at 20°C flows through the bed at a rate of 244 lb/min. At this temperature, the viscosity of the solution is 56.5 cp and its density is 1.2865 g/cm³. What is the void fraction of the bed? Discuss the usefulness of this method of obtaining the void fraction.

Answer: 0.30

6A.7 Estimation of pressure drops in annular flow. For flow in an annulus formed by cylindrical surfaces of diameters D and κD (with $\kappa < 1$) the friction factors for laminar and turbulent flow are

Laminar
$$f = \frac{16}{\text{Re}_\kappa} \tag{6A.7-1}$$

Turbulent
$$\sqrt{\frac{1}{f}} = G \log_{10}(\text{Re}_\kappa \sqrt{f}) - H \tag{6A.7-2}$$

in which the Reynolds number is defined by

$$\mathrm{Re}_\kappa = K\frac{D(1-\kappa)\langle \bar{v}_z\rangle \rho}{\mu} \qquad (6A.7\text{-}3)$$

The values of G, H, and K are given as:[1]

κ	G	H	K
0.00	4.000	0.400	1.000
0.05	3.747	0.293	0.7419
0.10	3.736	0.239	0.7161
0.15	3.738	0.208	0.7021
0.20	3.746	0.186	0.6930
0.30	3.771	0.154	0.6820
0.40	3.801	0.131	0.6757
0.50	3.833	0.111	0.6719
0.60	3.866	0.093	0.6695
0.70	3.900	0.076	0.6681
0.80	3.933	0.060	0.6672
0.90	3.967	0.046	0.6668
1.00	4.000	0.031	0.6667

Equation 6A.7-2 is based on Problem 5C.2 and reproduces the experimental data within about 3% up to Reynolds numbers of 20,000.

(a) Verify that, for developed laminar flow, Eqs. 6A.7-1 and 3 with the tabulated K values are consistent with Eq. 2.4-16.

(b) An annular duct is formed from cylindrical surfaces of diameters 6 in. and 15 in. It is desired to pump water at 60°F at a rate of 1500 cu ft per second. How much pressure drop is required per unit length of conduit, if the annulus is horizontal? Use Eq. 6A.7-2.

(c) Repeat (b) using the "mean hydraulic radius" empiricism.

6A.8 Force on a water tower in a gale. A water tower has a spherical storage tank 40 ft in diameter. In a 100-mph gale what is the force of the wind on the spherical tank at 0°C? Take the density of air to be 1.29 g/liter or 0.08 lb/ft^3 and the viscosity to be 0.017 cp.

Answer: $1.7 \times 10^4 \mathrm{lb}_f$

6A.9 Flow of gas through a packed column. A horizontal tube with diameter 4 in. and length 5.5 ft is packed with glass spheres of diameter 1/16 in., and the void fraction is 0.41. Carbon dioxide is to be pumped through the tube at 300K, at which temperature its viscosity is known to be 1.495×10^{-4} g/cm · s. What will be the mass flow rate through the column when the inlet and outlet pressures are 25 atm and 3 atm, respectively?

Answer: 480 g/s

6A.10 Determination of pipe diameter. What size of circular pipe is needed to produce a flow rate of 250 firkins per fortnight when there is a pressure drop of 3×10^5 scruples per square barleycorn? The pipe is horizontal. (The authors are indebted to Professor R. S. Kirk of the University of Massachusetts, who introduced them to these units.)

6B.1 Effect of error in friction factor calculations. In a calculation using the Blasius formula for turbulent flow in pipes, the Reynolds number used was too low by 4%. Calculate the resulting error in the friction factor.

Answer: Too high by 1%

6B.2 Friction factor for flow along a flat plate.[2]

(a) An expression for the drag force on a flat plate, wetted on both sides, is given in Eq. 4.4-30. This equation was derived by using *laminar* boundary layer theory and is known to be in good agreement with experimental data. Define a friction factor and Reynolds number, and obtain the *f* versus Re relation.

(b) For *turbulent* flow, an approximate boundary layer treatment based on the 1/7 power velocity distribution gives

$$F_k = 0.072\rho v_\infty^2 WL(Lv_\infty\rho/\mu)^{-1/5} \qquad (6B.2\text{-}1)$$

When 0.072 is replaced by 0.074, this relation describes the drag force within experimental error for $5 \times 10^5 < Lv_\infty\rho/\mu < 2 \times 10^7$. Express the corresponding friction factor as a function of the Reynolds number.

6B.3 Friction factor for laminar flow in a slit. Use the results of Problem 2B.3 to show that for the laminar flow in a thin slit of thickness 2B the friction factor is f = 12/Re, if the Reynolds number is defined as Re = $2B\langle v_z\rangle \rho/\mu$. Compare this result for *f* with what one would get from the mean hydraulic radius empiricism.

6B.4 Friction factor for a rotating disk.[3] A thin circular disk of radius R is immersed in a large body of fluid with density ρ and viscosity μ. If a torque T_z is required to make the disk rotate at an angular velocity Ω, then a friction factor *f* may be defined analogously to Eq. 6.1-1 as follows,

$$T_z/R = AKf \qquad (6B.4\text{-}1)$$

where reasonable definitions for K and A are $K = \frac{1}{2}\rho(\Omega R)^2$ and $A = 2(\pi R^2)$. An appropriate choice for the Reynolds number for the system is Re = $R^2\Omega\rho/\mu$.

For *laminar* flow, an exact boundary layer development gives

$$T_z = 0.616\pi\rho R^4\sqrt{\mu\Omega^3/\rho} \qquad (6B.4\text{-}2)$$

---

[1] D. M. Meter and R. B. Bird, *AIChE Journal*, **7**, 41–45 (1961).

[2] H. Schlichting, *Boundary-Layer Theory*, McGraw-Hill, New York, 7th edition (1979), Chapter XXI.

[3] T. von Kármán, *Zeits. für angew. Math. u. Mech.*, **1**, 233–252 (1921).

For *turbulent* flow, an approximate boundary layer treatment based on the 1/7 power velocity distribution leads to

$$T_z = 0.073\rho\Omega^2 R^5 \sqrt[5]{\mu/R^2\Omega\rho} \qquad (6B.4\text{-}3)$$

Express these results as relations between f and Re.

6B.5 Turbulent flow in horizontal pipes. A fluid is flowing with a mass flow rate w in a smooth horizontal pipe of length L and diameter D as the result of a pressure difference $p_0 - p_L$. The flow is known to be turbulent.

The pipe is to be replaced by one of diameter $D/2$ but with the same length. The same fluid is to be pumped at the same mass flow rate w. What pressure difference will be needed?

(a) Use Eq. 6.2-12 as a suitable equation for the friction factor.

(b) How can this problem be solved using Fig. 6.2-2 if Eq. 6.2-12 is not appropriate?

Answer: **(a)** A pressure difference 27 times greater will be needed.

6B.6 Inadequacy of mean hydraulic radius for laminar flow.

(a) For laminar flow in an annulus with radii κR and R, use Eqs. 6.2-17 and 18 to get an expression for the average velocity in terms of the pressure difference analogous to the exact expression given in Eq. 2.4-16.

(b) What is the percentage of error in the result in (a) for $\kappa = \frac{1}{2}$?

Answer: 49%

6B.7 Falling sphere in Newton's drag-law region. A sphere initially at rest at $z = 0$ falls under the influence of gravity. Conditions are such that, after a negligible interval, the sphere falls with a resisting force proportional to the square of the velocity.

(a) Find the distance z that the sphere falls as a function of t.

(b) What is the terminal velocity of the sphere? Assume that the density of the fluid is much less than the density of the sphere.

Answer: **(a)** The distance is $z = (1/c^2 g) \ln \cosh cgt$, where $c^2 = \frac{3}{8}(0.44)(\rho/\rho_{sph})(1/gR)$; **(b)** $1/c$

6B.8 Design of an experiment to verify the f vs. Re chart for spheres. It is desired to design an experiment to test the friction factor chart in Fig. 6.3-1 for flow around a sphere. Specifically, we want to test the plotted value $f = 1$ at Re = 100. This is to be done by dropping bronze spheres ($\rho_{sph} = 8$ g/cm^3) in water ($\rho = 1$ g/cm^3, $\mu = 10^{-2}$ g/cm · s). What sphere diameter must be used?

(a) Derive a formula that gives the required diameter as a function of f, Re, g, μ, ρ, and ρ_{sph} for terminal velocity conditions.

(b) Insert numerical values and find the value of the sphere diameter.

Answers: **(a)** $D = \sqrt[3]{\dfrac{3f \, \text{Re}^2 \, \mu^2}{4(\rho_{sph} - \rho)\rho g}}$; **(b)** $D = 0.048$ cm

6B.9 Friction factor for flow past an infinite cylinder.[4] The flow past a long cylinder is very different from the flow past a sphere, and the method introduced in §4.2 cannot be used to describe this system. It is found that, when the fluid approaches with a velocity v_∞, the kinetic force acting on a length L of the cylinder is

$$F_k = \frac{4\pi\mu v_\infty L}{\ln (7.4/\text{Re})} \qquad (6B.9\text{-}1)$$

The Reynolds number is defined here as Re $= D v_\infty \rho / \mu$. Equation 6B.9-1 is valid only up to about Re $= 1$. In this range of Re, what is the formula for the friction factor as a function of the Reynolds number?

6C.1 Two-dimensional particle trajectories. A sphere of radius R is fired horizontally (in the x direction) at high velocity in still air above level ground. As it leaves the propelling device, an identical sphere is dropped from the same height above the ground (in the y direction).

(a) Develop differential equations from which the particle trajectories can be computed, and that will permit comparison of the behavior of the two spheres. Include the effects of fluid friction, and make the assumption that steady-state friction factors may be used (this is a "quasi-steady-state assumption").

(b) Which sphere will reach the ground first?

(c) Would the answer to (b) have been the same if the sphere Reynolds numbers had been in the Stokes' law region?

Answers: **(a)** $\dfrac{dv_x}{dt} = -\dfrac{3}{8}\dfrac{v_x}{R}\sqrt{v_x^2 + v_y^2}\, f \dfrac{\rho_{air}}{\rho_{sph}}$,

$\dfrac{dv_y}{dt} = -\dfrac{3}{8}\dfrac{v_y}{R}\sqrt{v_x^2 + v_y^2}\, f \dfrac{\rho_{air}}{\rho_{sph}} + \left(1 - \dfrac{\rho_{air}}{\rho_{sph}}\right)g$,

in which $f = f(\text{Re})$ as given by Fig. 5.3-1, with

$$\text{Re} = \frac{2R\sqrt{v_x^2 + v_y^2}\,\rho_{air}}{\mu_{air}}$$

6C.2 Wall effects for a sphere falling in a cylinder.[5-7]

(a) Experiments on friction factors of spheres are generally performed in cylindrical tubes. Show by dimensional analysis that, for such an arrangement, the friction factor for the sphere will have the following dependence:

$$f = f(\text{Re}, R/R_{cyl}) \qquad (6C.2\text{-}1)$$

Here Re $= 2Rv_\infty \rho/\mu$, in which R is the sphere radius, v_∞ is the terminal velocity of the sphere, and R_{cyl} is the inside

---

[4] G. K. Batchelor, *An Introduction to Fluid Dynamics,* Cambridge University Press (1967), pp. 244–246, 257–261. For flow past finite cylinders, see J. Happel and H. Brenner, *Low Reynolds Number Hydrodynamics*, Martinus Nijhoff, The Hague (1983), pp. 227–230.

radius of the cylinder. For the *creeping flow* region, it has been found empirically that the dependence of f on R/R_{cyl} may be described by the *Ladenburg–Faxén correction*,[5] so that

$$f = \frac{24}{Re}\left(1 + 2.1\frac{R}{R_{cyl}}\right) \qquad (6C.2\text{-}2)$$

Wall effects for falling droplets have also been studied.[6]

(b) Design an experiment to check the graph for spheres in Fig. 6.3-1. Select sphere sizes, cylinder dimensions, and appropriate materials for the experiment.

6C.3   Power input to an agitated tank (Fig. 6C.3). Show by dimensional analysis that the power, P, imparted by a rotating impeller to an incompressible fluid in an agitated tank may be correlated, for any specific tank and impeller shape, by the expression

$$\frac{P}{\rho N^3 D^5} = \Phi\left(\frac{D^2 N\rho}{\mu}, \frac{DN^2}{g}, Nt\right) \qquad (6C.3\text{-}1)$$

Here N is the rate of rotation of the impeller, D is the impeller diameter, t is the time since the start of the operation, and Φ is a function whose form has to be determined experimentally.

For the commonly used geometry shown in the figure, the power is given by the sum of two integrals representing the contributions of friction drag of the cylindrical tank

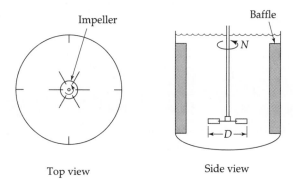

Fig. 6C.3. Agitated tank with a six-bladed impeller and four vertical baffles.

body and bottom and the form drag of the radial baffles, respectively:

$$P = NT_z = N\left(\int_S R(\partial v_\theta/\partial n)_{surf}\,dS + \int_A R p_{surf}\,dA\right) \qquad (6C.3\text{-}2)$$

Here T_z is the torque required to turn the impeller, S is the total surface area of the tank, A is the surface area of the baffles, (considered positive on the "upstream" side and negative on the "downstream side"), R is the radial distance to any surface element dS or dA from the impeller axis of rotation, and n is the distance measured normally into the fluid from any element of tank surface dS.

The desired solution may now be obtained by dimensional analysis of the equations of motion and continuity by rewriting the integrals above in dimensionless form. Here it is convenient to use D, DN, and $\rho N^2 D^2$ for the characteristic length, velocity, and pressure, respectively.

6D.1   Friction factor for a bubble in a clean liquid.[7,8] When a gas bubble moves through a liquid, the bulk of the liquid behaves as if it were in potential flow; that is, the flow field in the liquid phase is very nearly given by Eqs. 4B.5-2 and 3.

The drag force is closely related to the energy dissipation in the liquid phase (see Eq. 4.2-18)

$$F_k v_\infty = E_v \qquad (6D.1\text{-}1)$$

Show that for irrotational flow the general expression for the energy dissipation can be transformed into the following surface integral:

$$E_v = \mu \int (\mathbf{n} \cdot \nabla v^2)\, dS \qquad (6D.1\text{-}2)$$

Next show that insertion of the potential flow velocity profiles into Eq. 6D.1-2, and use of Eq. 6D.1-1 leads to

$$f = \frac{48}{Re} \qquad (6D.1\text{-}3)$$

A somewhat improved calculation that takes into account the dissipation in the boundary layer and in the turbulent wake leads to the following result:[9]

$$f = \frac{48}{Re}\left(1 - \frac{2.2}{\sqrt{Re}}\right) \qquad (6D.1\text{-}4)$$

This result seems to hold rather well up to a Reynolds number of about 200.

[5] R. Ladenburg, *Ann. Physik* (4), **23**, 447 (1907); H. Faxén, dissertation, Uppsala (1921). For extensive discussions of wall effects for falling spheres, see J. Happel and H. Brenner, *Low Reynolds Number Hydrodynamics*, Martinus Nijhoff, The Hague (1983).
[6] J. R. Strom and R. C. Kintner, *AIChE Journal*, **4**, 153–156 (1958).

[7] L. Landau and E. M. Lifshitz, *Fluid Mechanics*, Pergamon, Oxford (1987), pp. 182–183.
[8] G. K. Batchelor, *An Introduction to Fluid Dynamics*, Cambridge University Press, (1967), pp. 367–370.
[9] D. W. Moore, *J. Fluid Mech.*, **16**, 161–176 (1963).

Chapter 7

Macroscopic Balances for Isothermal Flow Systems

In the first four sections of Chapter 3 the *equations of change* for isothermal systems were presented. These equations were obtained by writing conservation laws over a "microscopic system"—namely, a small element of volume through which the fluid is flowing. In this way partial differential equations were obtained for the changes in mass, momentum, angular momentum, and mechanical energy in the system. The microscopic system has no solid bounding surfaces, and the interactions of the fluid with solid surfaces in specific flow systems are accounted for by boundary conditions on the differential equations.

In this chapter we write similar conservation laws for "macroscopic systems"—that is, large pieces of equipment or parts thereof. A sample macroscopic system is shown in Fig. 7.0-1. The balance statements for such a system are called the *macroscopic balances*; for

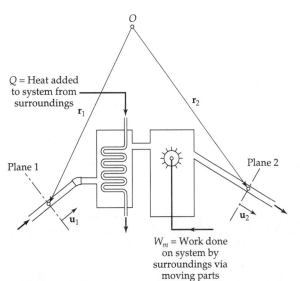

Fig. 7.0-1. Macroscopic flow system with fluid entering at plane 1 and leaving at plane 2. It may be necessary to add heat at a rate Q to maintain the system temperature constant. The rate of doing work *on* the system *by* the surroundings by means of moving surfaces is W_m. The symbols $\mathbf{u}_1$ and $\mathbf{u}_2$ denote *unit vectors* in the direction of flow at planes 1 and 2. The quantities $\mathbf{r}_1$ and $\mathbf{r}_2$ are position vectors giving the location of the centers of the inlet and outlet planes with respect to some designated origin of coordinates.

unsteady-state systems, these are ordinary differential equations, and for steady-state systems, they are algebraic equations. The macroscopic balances contain terms that account for the interactions of the fluid with the solid surfaces. The fluid can exert forces and torques on the surfaces of the system, and the surroundings can do work W_m on the fluid by means of moving surfaces.

The macroscopic balances can be obtained from the equations of change by integrating the latter over the entire volume of the flow system:[1,2]

$$\int_{V(t)} (\text{eq. of continuity}) \, dV = \text{macroscopic mass balance}$$

$$\int_{V(t)} (\text{eq. of motion}) \, dV = \text{macroscopic momentum balance}$$

$$\int_{V(t)} (\text{eq. of angular momentum}) \, dV = \text{macroscopic angular momentum balance}$$

$$\int_{V(t)} (\text{eq. of mechanical energy}) \, dV = \text{macroscopic mechanical energy balance}$$

The first three of these macroscopic balances can be obtained either by writing the conservation laws directly for the macroscopic system or by doing the indicated integrations. However, to get the macroscopic mechanical energy balance, the corresponding equation of change must be integrated over the macroscopic system.

In §§7.1 to 7.3 we set up the macroscopic mass, momentum, and angular momentum balances by writing the conservation laws. In §7.4 we present the macroscopic mechanical energy balance, postponing the detailed derivation until §7.8. In the macroscopic mechanical energy balance, there is a term called the "friction loss," and we devote §7.5 to estimation methods for this quantity. Then in §7.6 and §7.7 we show how the set of macroscopic balances can be used to solve flow problems.

The macroscopic balances have been widely used in many branches of engineering. They provide global descriptions of large systems without much regard for the details of the fluid dynamics inside the systems. Often they are useful for making an initial appraisal of an engineering problem and for making order-of-magnitude estimates of various quantities. Sometimes they are used to derive approximate relations, which can then be modified with the help of experimental data to compensate for terms that have been omitted or about which there is insufficient information.

In using the macroscopic balances one often has to decide which terms can be omitted, or one has to estimate some of the terms. This requires (i) intuition, based on experience with similar systems, (ii) some experimental data on the system, (iii) flow visualization studies, or (iv) order-of-magnitude estimates. This will be clear when we come to specific examples.

The macroscopic balances make use of nearly all the topics covered thus far; therefore Chapter 7 provides a good opportunity for reviewing the preceding chapters.

§7.1   THE MACROSCOPIC MASS BALANCE

In the system shown in Fig. 7.0-1 the fluid enters the system at plane 1 with cross section S_1 and leaves at plane 2 with cross section S_2. The average velocity is $\langle v_1 \rangle$ at the entry plane and $\langle v_2 \rangle$ at the exit plane. In this and the following sections, we introduce two assumptions that are not very restrictive: (i) at the planes 1 and 2 the time-smoothed veloc-

---

[1] R. B. Bird, *Chem. Eng. Sci.*, **6**, 123–131 (1957); *Chem. Eng. Educ.*, **27**(2), 102–109 (Spring 1993).
[2] J. C. Slattery and R. A. Gaggioli, *Chem. Eng. Sci.*, **17**, 893–895 (1962).

ity is perpendicular to the relevant cross section, and (ii) at planes 1 and 2 the density and other physical properties are uniform over the cross section.

The law of conservation of mass for this system is then

$$\frac{d}{dt} m_{tot} = \rho_1 \langle v_1 \rangle S_1 - \rho_2 \langle v_2 \rangle S_2 \tag{7.1-1}$$

$$\underbrace{\phantom{\frac{d}{dt}m}}_{\substack{\text{rate of} \\ \text{increase} \\ \text{of mass}}} \quad \underbrace{}_{\substack{\text{rate of} \\ \text{mass in} \\ \text{at plane 1}}} \quad \underbrace{}_{\substack{\text{rate of} \\ \text{mass out} \\ \text{at plane 2}}}$$

Here $m_{tot} = \int \rho\, dV$ is the total mass of fluid contained in the system between planes 1 and 2. We now introduce the symbol $w = \rho \langle v \rangle S$ for the mass rate of flow, and the notation $\Delta w = w_2 - w_1$ (exit value minus entrance value). Then the *unsteady-state macroscopic mass balance* becomes

$$\boxed{\frac{d}{dt} m_{tot} = -\Delta w} \tag{7.1-2}$$

If the total mass of fluid does not change with time, then we get the *steady-state macroscopic mass balance*

$$\Delta w = 0 \tag{7.1-3}$$

which is just the statement that the rate of mass entering equals the rate of mass leaving.

For the macroscopic mass balance we use the term "steady state" to mean that the time derivative on the left side of Eq. 7.1-2 is zero. Within the system, because of the possibility for moving parts, flow instabilities, and turbulence, there may well be regions of unsteady flow.

EXAMPLE 7.1-1

Draining of a Spherical Tank

A spherical tank of radius R and its drainpipe of length L and diameter D are completely filled with a heavy oil. At time $t = 0$ the valve at the bottom of the drainpipe is opened. How long will it take to drain the tank? There is an air vent at the very top of the spherical tank. Ignore the amount of oil that clings to the inner surface of the tank, and assume that the flow in the drainpipe is laminar.

SOLUTION

We label three planes as in Fig. 7.1-1, and we let the instantaneous liquid level above plane 2 be $h(t)$. Then, at any time t the total mass of liquid in the sphere is

$$m_{tot} = \pi R h^2 \left(1 - \frac{1}{3} \frac{h}{R} \right) \rho \tag{7.1-4}$$

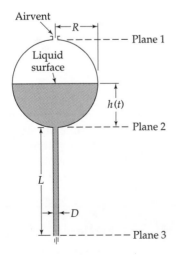

Fig. 7.1-1. Spherical tank with drainpipe.

which can be obtained by using integral calculus. Since no fluid crosses plane 1 we know that $w_1 = 0$. The outlet mass flow rate w_2, as determined from the Hagen–Poiseuille formula, is

$$w_2 = \frac{\pi(\mathcal{P}_2 - \mathcal{P}_3)D^4\rho}{128\mu L} = \frac{\pi(\rho g h + \rho g L)D^4\rho}{128\mu L} \tag{7.1-5}$$

The Hagen–Poiseuille formula was derived for steady-state flow, but we use it here since the volume of liquid in the tank is changing slowly with time; this is an example of a "quasi-steady-state" approximation. When these expressions for m_{tot} and w_2 are substituted into Eq. 7.1-2, we get, after some rearrangement,

$$-\frac{(2R - h)h}{h + L}\frac{dh}{dt} = \frac{\rho g D^4}{128\mu L} \tag{7.1-6}$$

We now abbreviate the constant on the right side of the equation as A. The equation is easier to integrate if we make the change of variable $H = h + L$ so that

$$\frac{[H - (2R + L)](H - L)}{H}\frac{dH}{dt} = A \tag{7.1-7}$$

We now integrate this equation between $t = 0$ (when $h = 2R$ or $H = 2R + L$), and $t = t_{efflux}$ (when $h = 0$ or $H = L$). This gives for the efflux time

$$t_{efflux} = \frac{L^2}{A}\left[2\frac{R}{L}\left(1 + \frac{R}{L}\right) - \left(1 + 2\frac{R}{L}\right)\ln\left(1 + 2\frac{R}{L}\right)\right] \tag{7.1-8}$$

in which A is given by the right side of Eq. 7.1-6. Note that we have obtained this result without any detailed analysis of the fluid motion within the sphere.

§7.2   THE MACROSCOPIC MOMENTUM BALANCE

We now apply the law of conservation of momentum to the system in Fig. 7.0-1, using the same two assumptions mentioned in the previous section, plus two additional assumptions: (iii) the forces associated with the stress tensor $\boldsymbol{\tau}$ are neglected at planes 1 and 2, since they are generally small compared to the pressure forces at the entry and exit planes, and (iv) the pressure does not vary over the cross section at the entry and exit planes.

Since momentum is a vector quantity, each term in the balance must be a vector. We use unit vectors $\mathbf{u}_1$ and $\mathbf{u}_2$ to represent the direction of flow at planes 1 and 2. The law of conservation of momentum then reads

$$\frac{d}{dt}\mathbf{P}_{tot} = \rho_1\langle v_1^2\rangle S_1\mathbf{u}_1 - \rho_2\langle v_2^2\rangle S_2\mathbf{u}_2 + p_1 S_1\mathbf{u}_1 - p_2 S_2\mathbf{u}_2 + \mathbf{F}_{s\to f} + m_{tot}\mathbf{g} \tag{7.2-1}$$

rate of increase of momentum	rate of momentum in at plane 1	rate of momentum out at plane 2	pressure force on fluid at plane 1	pressure force on fluid at plane 2	force of solid surface on fluid	force of gravity on fluid

Here $\mathbf{P}_{tot} = \int \rho \mathbf{v}\, dV$ is the total momentum in the system. The equation states that the total momentum within the system changes because of the convection of momentum into and out of the system, and because of the various forces acting on the system: the pressure forces at the ends of the system, the force of the solid surfaces acting on the fluid in the system, and the force of gravity acting on the fluid within the walls of the system. The subscript "$s \to f$" serves as a reminder of the direction of the force.

By introducing the symbols for the mass rate of flow and the Δ symbol we finally get for the *unsteady-state macroscopic momentum balance*

$$\boxed{\frac{d}{dt}\mathbf{P}_{tot} = -\Delta\left(\frac{\langle v^2\rangle}{\langle v\rangle}w + pS\right)\mathbf{u} + \mathbf{F}_{s\to f} + m_{tot}\mathbf{g}} \tag{7.2-2}$$

If the total amount of momentum in the system does not change with time, then we get the *steady-state macroscopic momentum balance*

$$\mathbf{F}_{f \to s} = -\Delta\left(\frac{\langle v^2\rangle}{\langle v\rangle} w + pS\right)\mathbf{u} + m_{\text{tot}}\mathbf{g} \tag{7.2-3}$$

Once again we emphasize that this is a vector equation. It is useful for computing the force of the fluid on the solid surfaces, $\mathbf{F}_{f \to s}$, such as the force on a pipe bend or a turbine blade. Actually we have already used a simplified version of the above equation in Eq. 6.1-3.

Notes regarding turbulent flow: (i) For turbulent flow it is customary to replace $\langle v\rangle$ by $\langle \bar{v}\rangle$ and $\langle v^2\rangle$ by $\langle \bar{v}^2\rangle$; in the latter we are neglecting the term $\langle \overline{v'^2}\rangle$, which is generally small with respect to $\langle \bar{v}^2\rangle$. (ii) Then we further replace $\langle \bar{v}^2\rangle/\langle \bar{v}\rangle$ by $\langle \bar{v}\rangle$. The error in doing this is quite small; for the empirical $\frac{1}{7}$ power law velocity profile given in Eq. 5.1-4, $\langle \bar{v}^2\rangle/\langle \bar{v}\rangle = \frac{50}{49}\langle \bar{v}\rangle$, so that the error is about 2%. (iii) When we make this assumption we will normally drop the angular brackets and overbars to simplify the notation. That is, we will let $\langle \bar{v}_1\rangle \equiv v_1$ and $\langle \bar{v}_1^2\rangle \equiv v_1^2$, with similar simplifications for quantities at plane 2.

<table>
<tr><td>

EXAMPLE 7.2-1

Force Exerted by a Jet (Part a)

</td><td>

A turbulent jet of water emerges from a tube of radius $R_1 = 2.5$ cm with a speed $v_1 = 6$ m/s, as shown in Fig. 7.2-1. The jet impinges on a disk-and-rod assembly of mass $m = 5.5$ kg, which is free to move vertically. The friction between the rod and the sleeve will be neglected. Find the height h at which the disk will "float" as a result of the jet.[1] Assume that the water is incompressible.

</td></tr>
<tr><td>

SOLUTION

</td><td>

To solve this problem one has to imagine how the jet behaves. In Fig. 7.2-1(a) we make the assumption that the jet has a constant radius, R_1, between the tube exit and the disk, whereas in Fig. 7.2-1(b) we assume that the jet spreads slightly. In this example, we make the first assumption, and in Example 7.4-1 we account for the jet spreading.

We apply the z-component of the steady-state momentum balance between planes 1 and 2. The pressure terms can be omitted, since the pressure is atmospheric at both planes. The z component of the fluid velocity at plane 2 is zero. The momentum balance then becomes

</td></tr>
</table>

$$mg = v_1(\rho v_1 \pi R_1^2) - (\pi R_1^2 h)\rho g \tag{7.2-4}$$

When this is solved for h, we get (in SI units)

$$h = \frac{v_1^2}{g} - \frac{m}{\rho\pi R_1^2} = \frac{(6)^2}{(9.807)} - \frac{5.5}{\pi(0.025)^2(1000)} = 0.87 \text{ m} \tag{7.2-5}$$

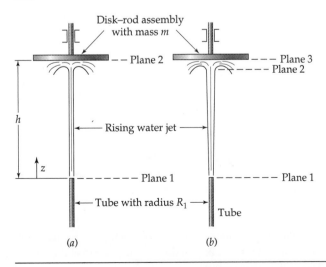

Disk–rod assembly with mass m

Plane 2

Plane 3
Plane 2

h

Rising water jet

z

Plane 1

Plane 1

Tube with radius R_1

Tube

(a)

(b)

Fig. 7.2-1. Sketches corresponding to the two solutions to the jet-and-disk problem. In (*a*) the water jet is assumed to have a uniform radius R_1. In (*b*) allowance is made for the spreading of the liquid jet.

---

[1] K. Federhofer, *Aufgaben aus der Hydromechanik*, Springer-Verlag, Vienna (1954), pp. 36 and 172.

§7.3   THE MACROSCOPIC ANGULAR MOMENTUM BALANCE

The development of the macroscopic angular momentum balance parallels that for the (linear) momentum balance in the previous section. All we have to do is to replace "momentum" by "angular momentum" and "force" by "torque."

To describe the angular momentum and torque we have to select an origin of coordinates with respect to which these quantities are evaluated. The origin is designated by "O" in Fig. 7.0-1, and the locations of the midpoints of planes 1 and 2 with respect to this origin are given by the position vectors $\mathbf{r}_1$ and $\mathbf{r}_2$.

Once again we make assumptions (i)–(iv) introduced in §§7.1 and 7.2. With these assumptions the rate of entry of angular momentum at plane 1, which is $\int [\mathbf{r} \times \rho \mathbf{v}](\mathbf{v} \cdot \mathbf{u}) dS$ evaluated at that plane, becomes $\rho_1 \langle v_1^2 \rangle S_1 [\mathbf{r}_1 \times \mathbf{u}_1]$, with a similar expression for the rate at which angular momentum leaves the system at 2.

The *unsteady-state macroscopic angular momentum balance* may now be written as

$$\frac{d}{dt} \mathbf{L}_{\text{tot}} = \rho_1 \langle v_1^2 \rangle S_1 [\mathbf{r}_1 \times \mathbf{u}_1] - \rho_2 \langle v_2^2 \rangle S_2 [\mathbf{r}_2 \times \mathbf{u}_2]$$

$$\begin{array}{cccc}
\text{rate of} & \text{rate of angular} & \text{rate of angular} \\
\text{increase of} & \text{momentum} & \text{momentum} \\
\text{angular} & \text{in at plane 1} & \text{out at plane 2} \\
\text{momentum}
\end{array}$$

$$+ \; p_1 S_1 [\mathbf{r}_1 \times \mathbf{u}_1] - p_2 S_2 [\mathbf{r}_2 \times \mathbf{u}_2] + \mathbf{T}_{s \to f} + \mathbf{T}_{\text{ext}} \qquad (7.3\text{-}1)$$

$$\begin{array}{cccc}
\text{torque due to} & \text{torque due to} & \text{torque} & \text{external} \\
\text{pressure on} & \text{pressure on} & \text{of solid} & \text{torque} \\
\text{fluid at} & \text{fluid at} & \text{surface} & \text{on fluid} \\
\text{plane 1} & \text{plane 2} & \text{on fluid}
\end{array}$$

Here $\mathbf{L}_{\text{tot}} = \int \rho [\mathbf{r} \times \mathbf{v}] dV$ is the total angular momentum within the system, and $\mathbf{T}_{\text{ext}} = \int [\mathbf{r} \times \rho \mathbf{g}] dV$ is the torque on the fluid in the system resulting from the gravitational force. This equation can also be written as

$$\boxed{\frac{d}{dt} \mathbf{L}_{\text{tot}} = -\Delta \left(\frac{\langle v^2 \rangle}{\langle v \rangle} w + pS \right)[\mathbf{r} \times \mathbf{u}] + \mathbf{T}_{s \to f} + \mathbf{T}_{\text{ext}}} \qquad (7.3\text{-}2)$$

Finally, the *steady-state macroscopic angular momentum balance* is

$$\mathbf{T}_{f \to s} = -\Delta \left(\frac{\langle v^2 \rangle}{\langle v \rangle} w + pS \right)[\mathbf{r} \times \mathbf{u}] + \mathbf{T}_{\text{ext}} \qquad (7.3\text{-}3)$$

This gives the torque exerted by the fluid on the solid surfaces.

EXAMPLE 7.3-1

Torque on a Mixing Vessel

A mixing vessel, shown in Fig. 7.3-1, is being operated at steady state. The fluid enters tangentially at plane 1 in turbulent flow with a velocity v_1 and leaves through the vertical pipe with a velocity v_2. Since the tank is baffled there is no swirling motion of the fluid in the vertical exit pipe. Find the torque exerted on the mixing vessel.

SOLUTION

The origin of the coordinate system is taken to be on the tank axis in a plane passing through the axis of the entrance pipe and parallel to the tank top. Then the vector $[\mathbf{r}_1 \times \mathbf{u}_1]$ is a vector pointing in the z direction with magnitude R. Furthermore $[\mathbf{r}_2 \times \mathbf{u}_2] = 0$, since the two vectors are collinear. For this problem Eq. 7.3-3 gives

$$\mathbf{T}_{f \to s} = (\rho v_1^2 S_1 + p_1 S_1) R \boldsymbol{\delta}_z \qquad (7.3\text{-}4)$$

Thus the torque is just "force × lever arm," as would be expected. If the torque is sufficiently large, the equipment must be suitably braced to withstand the torque produced by the fluid motion and the inlet pressure.

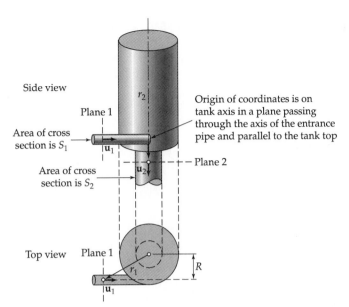

Fig. 7.3-1. Torque on a tank, showing side view and top view.

Side view

Plane 1

Area of cross section is S_1

r_2

Origin of coordinates is on tank axis in a plane passing through the axis of the entrance pipe and parallel to the tank top

Plane 2

Area of cross section is S_2

Top view   Plane 1

r_1

R

§7.4   THE MACROSCOPIC MECHANICAL ENERGY BALANCE

Equations 7.1-2, 7.2-2, and 7.3-2 have been set up by applying the laws of conservation of mass, (linear) momentum, and angular momentum over the macroscopic system in Fig. 7.0-1. The three macroscopic balances thus obtained correspond to the equations of change in Eqs. 3.1-4, 3.2-9, and 3.4-1, and, in fact, they are very similar in structure. These three macroscopic balances can also be obtained by integrating the three equations of change over the volume of the flow system.

Next we want to set up the macroscopic mechanical energy balance, which corresponds to the equation of mechanical energy in Eq. 3.3-2. There is no way to do this directly as we have done in the preceding three sections, since there is no conservation law for mechanical energy. In this instance we *must* integrate the equation of change of mechanical energy over the volume of the flow system. The result, which has made use of the same assumptions (i–iv) used above, is the *unsteady-state macroscopic mechanical energy balance* (sometimes called the *engineering Bernoulli equation*). The equation is derived in §7.8; here we state the result and discuss its meaning:

$$\frac{d}{dt}(K_{tot} + \Phi_{tot}) = \left(\tfrac{1}{2}\rho_1\langle v_1^3\rangle + \rho_1\hat{\Phi}_1\langle v_1\rangle\right)S_1 - \left(\tfrac{1}{2}\rho_2\langle v_2^3\rangle + \rho_2\hat{\Phi}_2\langle v_2\rangle\right)S_2$$

rate of increase of kinetic and potential energy in system

rate at which kinetic and potential energy enter system at plane 1

rate at which kinetic and potential energy leave system at plane 2

$$+ \left(p_1\langle v_1\rangle S_1 - p_2\langle v_2\rangle S_2\right) + W_m + \int_{V(t)} p(\boldsymbol{\nabla}\cdot\mathbf{v})\,dV + \int_{V(t)} (\boldsymbol{\tau}{:}\boldsymbol{\nabla}\mathbf{v})\,dV \qquad (7.4\text{-}1)$$

net rate at which the surroundings do work on the fluid at planes 1 and 2 by the pressure

rate of doing work on fluid by moving surfaces

rate at which mechanical energy increases or decreases because of expansion or compression of fluid

rate at which mechanical energy decreases because of viscous dissipation[1]

---

[1] This interpretation of the term is valid only for Newtonian fluids; polymeric liquids have elasticity and the interpretation given above no longer holds.

Here $K_{tot} = \int \frac{1}{2}\rho v^2 \, dV$ and $\Phi_{tot} = \int \rho\hat{\Phi} \, dV$ are the total kinetic and potential energies within the system. According to Eq. 7.4-1, the total mechanical energy (i.e., kinetic plus potential) changes because of a difference in the rates of addition and removal of mechanical energy, because of work done on the fluid by the surroundings, and because of compressibility effects and viscous dissipation. Note that, at the system entrance (plane 1), the force $p_1 S_1$ multiplied by the velocity $\langle v_1 \rangle$ gives the rate at which the surroundings do work on the fluid. Furthermore, W_m is the work done by the surroundings on the fluid by means of moving surfaces.

The macroscopic mechanical energy balance may now be written more compactly as

$$\frac{d}{dt}(K_{tot} + \Phi_{tot}) = -\Delta\left(\frac{1}{2}\frac{\langle v^3 \rangle}{\langle v \rangle} + \hat{\Phi} + \frac{p}{\rho}\right)w + W_m - E_c - E_v \tag{7.4-2}$$

in which the terms E_c and E_v are defined as follows:

$$E_c = -\int_{V(t)} p(\nabla \cdot \mathbf{v}) \, dV \quad \text{and} \quad E_v = -\int_{V(t)} (\boldsymbol{\tau}:\nabla\mathbf{v}) \, dV \tag{7.4-3, 4}$$

The *compression term* E_c is positive in compression and negative in expansion; it is zero when the fluid is assumed to be incompressible. The term E_v is the *viscous dissipation* (or *friction loss) term*, which is always positive for Newtonian liquids, as can be seen from Eq. 3.3-3. (For polymeric fluids, which are viscoelastic, E_v is not necessarily positive; these fluids are discussed in the next chapter.)

If the total kinetic plus potential energy in the system is not changing with time, we get

$$\Delta\left(\frac{1}{2}\frac{\langle v^3 \rangle}{\langle v \rangle} + gh + \frac{p}{\rho}\right)w = W_m - E_c - E_v \tag{7.4-5}$$

which is the *steady-state macroscopic mechanical energy balance*. Here h is the height above some arbitrarily chosen datum plane.

Next, if we assume that it is possible to draw a representative streamline through the system, we may combine the $\Delta(p/\rho)$ and E_c terms to get the following *approximate* relation (see §7.8)

$$\Delta\left(\frac{p}{\rho}\right)w + E_c \approx w \int_1^2 \frac{1}{\rho} \, dp \tag{7.4-6}$$

Then, after dividing Eq. 7.4-5 by $w_1 = w_2 = w$, we get

$$\Delta\left(\frac{1}{2}\frac{\langle v^3 \rangle}{\langle v \rangle}\right) + g\Delta h + \int_1^2 \frac{1}{\rho} \, dp = \hat{W}_m - \hat{E}_v \tag{7.4-7}$$

Here $\hat{W}_m = W_m/w$ and $\hat{E}_v = E_v/w$. Equation 7.4-7 is the version of the steady-state mechanical energy balance that is most often used. For isothermal systems, the integral term can be calculated as long as an expression for density as a function of pressure is available.

Equation 7.4-7 should now be compared with Eq. 3.5-12, which is the "classical" Bernoulli equation for an inviscid fluid. If, to the right side of Eq. 3.5-12, we just add the work $\hat{W}_m$ done by the surroundings and subtract the viscous dissipation term $\hat{E}_v$, and reinterpret the velocities as appropriate averages over the cross sections, then we get Eq. 7.4-7. This provides a "plausibility argument" for Eq. 7.4-7 and still preserves the fundamental idea that the macroscopic mechanical energy balance is derived from the equation of motion (that is, from the law of conservation of momentum). The full derivation of the macroscopic mechanical energy balance is given in §7.8 for those who are interested.

Notes for turbulent flow: (i) For turbulent flows we replace $\langle v^3 \rangle$ by $\langle \bar{v}^3 \rangle$, and ignore the contribution from the turbulent fluctuations. (ii) It is common practice to replace the quotient $\langle \bar{v}^3 \rangle / \langle \bar{v} \rangle$ by $\langle \bar{v} \rangle^2$. For the empirical $\frac{1}{7}$ power law velocity profile given in Eq. 5.1-4, it can be shown that $\langle \bar{v}^3 \rangle / \langle \bar{v} \rangle = \frac{43200}{40817} \langle \bar{v} \rangle^2$, so that the error amounts to about 6%. (iii) We further omit the brackets and overbars to simplify the notation in turbulent flow.

EXAMPLE 7.4-1

Force Exerted by a Jet (Part b)

Continue the problem in Example 7.2-1 by accounting for the spreading of the jet as it moves upward.

SOLUTION

We now permit the jet diameter to increase with increasing z as shown in Fig. 7.2-1(b). It is convenient to work with three planes and to make balances between pairs of planes. The separation between planes 2 and 3 is taken to be quite small.

A mass balance between planes 1 and 2 gives

$$w_1 = w_2 \tag{7.4-8}$$

Next we apply the mechanical energy balance of Eq. 7.4-5 or 7.4-7 between the same two planes. The pressures at planes 1 and 2 are both atmospheric, and there is no work done by moving parts W_m. We assume that the viscous dissipation term E_v can be neglected. If z is measured upward from the tube exit, then $g\Delta h = g(h_2 - h_1) \approx g(h - 0)$, since planes 2 and 3 are so close together. Thus the mechanical energy balance gives

$$\tfrac{1}{2}(v_2^2 - v_1^2) + gh = 0 \tag{7.4-9}$$

We now apply the z-momentum balance between planes 2 and 3. Since the region is very small, we neglect the last term in Eq. 7.2-3. Both planes are at atmospheric pressure, so the pressure terms do not contribute. The fluid velocity is zero at plane 3, so there are only two terms left in the momentum balance

$$mg = v_2 w_2 \tag{7.4-10}$$

From the above three equations we get

$$
\begin{aligned}
h &= \frac{v_1^2}{2g}\left(1 - \frac{v_2^2}{v_1^2}\right) && \text{from Eq. 7.4–9} \\[2mm]
&= \frac{v_1^2}{2g}\left(1 - \frac{(mg/w_2)^2}{v_1^2}\right) && \text{from Eq. 7.4–10} \\[2mm]
&= \frac{v_1^2}{2g}\left(1 - \left(\frac{mg}{v_1 w_1}\right)^2\right) && \text{from Eq. 7.4–8}
\end{aligned}
\tag{7.4-11}
$$

in which mg and $v_1 w_1 = \pi R_1^2 \rho v_1^2$ are known. When the numerical values are substituted into Eq. 7.4-10, we get $h = 0.77$ m. This is probably a better result than the value of 0.87 m obtained in Example 7.2-1, since it accounts for the spreading of the jet. We have not, however, considered the clinging of the water to the disk, which gives the disk–rod assembly a somewhat greater effective mass. In addition, the frictional resistance of the rod in the sleeve has been neglected. It is necessary to run an experiment to assess the validity of Eq. 7.4-10.

§7.5   ESTIMATION OF THE VISCOUS LOSS

This section is devoted to methods for estimating the viscous loss (or friction loss), E_v, which appears in the macroscopic mechanical energy balance. The general expression for E_v is given in Eq. 7.4-4. For incompressible Newtonian fluids, Eq. 3.3-3 may be used to rewrite E_v as

$$E_v = \int \mu \Phi_v \, dV \tag{7.5-1}$$

which shows that it is the integral of the local rate of viscous dissipation over the volume of the entire flow system.

We now want to examine E_v from the point of view of dimensional analysis. The quantity Φ_v is a sum of squares of velocity gradients; hence it has dimensions of $(v_0/l_0)^2$, where v_0 and l_0 are a characteristic velocity and length, respectively. We can therefore write

$$E_v = (\rho v_0^3 l_0^2)(\mu/l_0 v_0 \rho) \int \check{\Phi}_v \, d\check{V} \tag{7.5-2}$$

where $\check{\Phi}_v = (l_0/v_0)^2 \Phi_v$ and $d\check{V} = l_0^{-3} dV$ are dimensionless quantities. If we make use of the dimensional arguments of §§3.7 and 6.2, we see that the integral in Eq. 7.5-2 depends only on the various dimensionless groups in the equations of change and on various geometrical factors that enter into the boundary conditions. Hence, if the only significant dimensionless group is a Reynolds number, $\text{Re} = l_0 v_0 \rho/\mu$, then Eq. 7.5-2 must have the general form

$$E_v = (\rho v_0^3 l_0^2) \times \left(\begin{array}{c} \text{a dimensionless function of Re} \\ \text{and various geometrical ratios} \end{array} \right) \tag{7.5-3}$$

In *steady-state flow* we prefer to work with the quantity $\hat{E}_v = E_v/w$, in which $w = \rho\langle v \rangle S$ is the mass rate of flow passing through *any* cross section of the flow system. If we select the reference velocity v_0 to be $\langle v \rangle$ and the reference length l_0 to be $\sqrt{S}$, then

$$\hat{E}_v = \tfrac{1}{2}\langle v \rangle^2 e_v \tag{7.5-4}$$

in which e_v, the *friction loss factor*, is a function of a Reynolds number and relevant dimensionless geometrical ratios. The factor $\tfrac{1}{2}$ has been introduced in keeping with the form of several related equations. We now want to summarize what is known about the friction loss factor for the various parts of a piping system.

For a straight conduit the friction loss factor is closely related to the friction factor. We consider only the steady flow of a fluid of constant density in a straight conduit of arbitrary, but constant, cross section S and length L. If the fluid is flowing in the z direction under the influence of a pressure gradient and gravity, then Eqs. 7.2-2 and 7.4-7 become

(z-momentum) $$F_{f \to s} = (p_1 - p_2)S + (\rho S L)g_z \tag{7.5-5}$$

(mechanical energy) $$\hat{E}_v = \frac{1}{\rho}(p_1 - p_2) + Lg_z \tag{7.5-6}$$

Multiplication of the second of these by ρS and subtracting gives

$$\hat{E}_v = \frac{F_{f \to s}}{\rho S} \tag{7.5-7}$$

If, in addition, the flow is *turbulent* then the expression for $F_{f \to s}$ in terms of the mean hydraulic radius R_h may be used (see Eqs. 6.2-16 to 18) so that

$$\hat{E}_v = \tfrac{1}{2}\langle v \rangle^2 \frac{L}{R_h} f \tag{7.5-8}$$

in which f is the friction factor discussed in Chapter 6. Since this equation is of the form of Eq. 7.5-4, we get a simple relation between the friction loss factor and the friction factor

$$e_v = \frac{L}{R_h} f \tag{7.5-9}$$

for turbulent flow in sections of straight pipe with uniform cross section. For a similar treatment for conduits of variable cross section, see Problem 7B.2.

Table 7.5-1   Brief Summary of Friction Loss Factors for Use with Eq. 7.5-10 (Approximate Values for Turbulent Flow)[a]

Disturbances	e_v
Sudden changes in cross-sectional area[b]	
Rounded entrance to pipe	0.05
Sudden contraction	$0.45(1 - \beta)$
Sudden expansion[c]	$\left(\dfrac{1}{\beta} - 1\right)^2$
Orifice (sharp-edged)	$2.71(1 - \beta)(1 - \beta^2)\dfrac{1}{\beta^2}$
Fittings and valves	
90° elbows (rounded)	0.4–0.9
90° elbows (square)	1.3–1.9
45° elbows	0.3–0.4
Globe valve (open)	6–10
Gate valve (open)	0.2

[a] Taken from H. Kramers, *Physische Transportverschijnselen*, Technische Hogeschool Delft, Holland (1958), pp. 53–54.

[b] Here $\beta = $ (smaller cross-sectional area)/(larger cross-sectional area).

[c] See derivation from the macroscopic balances in Example 7.6-1. If $\beta = 0$, then $\hat{E}_v = \frac{1}{2}\langle v \rangle^2$, where $\langle v \rangle$ is the velocity *upstream* from the enlargement.

Most flow systems contain various "obstacles," such as fittings, sudden changes in diameter, valves, or flow measuring devices. These also contribute to the friction loss $\hat{E}_v$. Such additional resistances may be written in the form of Eq. 7.5-4, with e_v determined by one of two methods: (a) simultaneous solution of the macroscopic balances, or (b) experimental measurement. Some rough values of e_v are tabulated in Table 7.5-1 for the convention that $\langle v \rangle$ is the average velocity *downstream* from the disturbance. These e_v values are for *turbulent flow* for which the Reynolds number dependence is not too important.

Now we are in a position to rewrite Eq. 7.4-7 in the *approximate* form frequently used for *turbulent flow* calculations in a system composed of various kinds of piping and additional resistances:

$$\tfrac{1}{2}(v_2^2 - v_1^2) + g(z_2 - z_1) + \int_{p_1}^{p_2} \frac{1}{\rho}\,dp = \hat{W}_m - \underbrace{\sum_i \left(\tfrac{1}{2}v^2 \frac{L}{R_h} f\right)_i}_{\substack{\text{sum over all} \\ \text{sections of} \\ \text{straight conduits}}} - \underbrace{\sum_i \left(\tfrac{1}{2}v^2 e_v\right)_i}_{\substack{\text{sum over all} \\ \text{fittings, valves,} \\ \text{meters, etc.}}} \qquad (7.5\text{-}10)$$

Here R_h is the mean hydraulic radius defined in Eq. 6.2-16, f is the friction factor defined in Eq. 6.1-4, and e_v is the friction loss factor given in Table 7.5-1. Note that the v_1 and v_2 in the first term refer to the velocities at planes 1 and 2; the v in the first sum is the average velocity in the ith pipe segment; and the v in the second sum is the average velocity *downstream* from the ith fitting, valve, or other obstacle.

EXAMPLE 7.5-1

Power Requirement for Pipeline Flow

What is the required power output from the pump at steady state in the system shown in Fig. 7.5-1? Water at 68°F ($\rho = 62.4$ lb$_m$/ft^3; $\mu = 1.0$ cp) is to be delivered to the upper tank at a rate of 12 ft^3/min. All of the piping is 4-in. internal diameter smooth circular pipe.

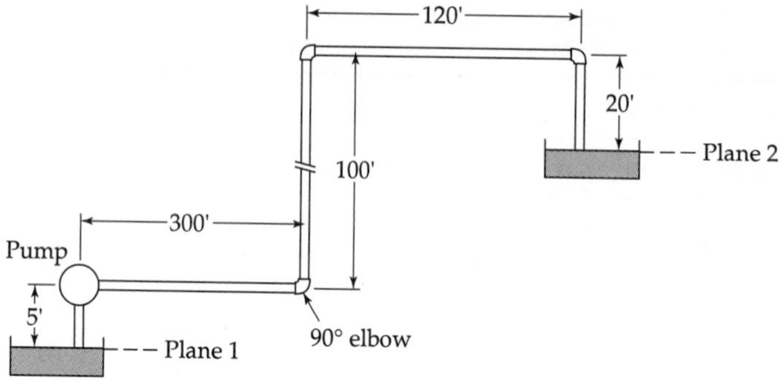

Fig. 7.5-1. Pipeline flow with friction losses because of fittings. Planes 1 and 2 are just under the surface of the liquid.

SOLUTION

The average velocity in the pipe is

$$\langle v \rangle = \frac{w/\rho}{\pi R^2} = \frac{(12/60)}{\pi (1/6)^2} = 2.30 \text{ ft/s} \tag{7.5-11}$$

and the Reynolds number is

$$\text{Re} = \frac{D\langle v \rangle \rho}{\mu} = \frac{(1/3)(2.30)(62.4)}{(1.0)(6.72 \times 10^{-4})} = 7.11 \times 10^4 \tag{7.5-12}$$

Hence the flow is *turbulent*.

The contribution to $\hat{E}_v$ from the various lengths of pipe will be

$$\sum_i \left(\tfrac{1}{2} v^2 \frac{L}{R_h} f \right)_i = \frac{2v^2 f}{D} \sum_i L_i$$

$$= \frac{2(2.30)^2 (0.0049)}{(1/3)} (5 + 300 + 100 + 120 + 20)$$

$$= (0.156)(545) = 85 \text{ ft}^2/\text{s}^2 \tag{7.5-13}$$

The contribution to $\hat{E}_v$ from the sudden contraction, the three 90° elbows, and the sudden expansion (see Table 7.5-1) will be

$$\sum_i (\tfrac{1}{2} v^2 e_v)_i = \tfrac{1}{2}(2.30)^2 (0.45 + 3(\tfrac{1}{2}) + 1) = 8 \text{ ft}^2/\text{s}^2 \tag{7.5-14}$$

Then from Eq. 7.5-10 we get

$$0 + (32.2)(105 - 20) + 0 = \hat{W}_m - 85 - 8 \tag{7.5-15}$$

Solving for $\hat{W}_m$ we get

$$\hat{W}_m = 2740 + 85 - 8 \approx 2830 \text{ ft}^2/\text{s}^2 \tag{7.5-16}$$

This is the work (per unit mass of fluid) done *on* the fluid *in* the pump. Hence the pump does 2830 ft²/s² or 2830/32.2 = 88 ft lb$_f$/lb$_m$ of work on the fluid passing through the system. The mass rate of flow is

$$w = (12/60)(62.4) = 12.5 \text{ lb}_m/\text{s} \tag{7.5-17}$$

Consequently

$$W_m = w\hat{W}_m = (12.5)(88) = 1100 \text{ ft lb}_f/\text{s} = 2 \text{ hp} = 1.5 \text{ kW} \tag{7.5-18}$$

which is the power delivered by the pump.

Table 7.6-1   Steady-State Macroscopic Balances for Turbulent Flow in Isothermal Systems

Mass:	$\Sigma w_1 - \Sigma w_2 = 0$	(A)
Momentum:	$\Sigma(v_1 w_1 + p_1 S_1)\mathbf{u}_1 - \Sigma(v_2 w_2 + p_2 S_2)\mathbf{u}_2 + m_{tot}\mathbf{g} = \mathbf{F}_{f \to s}$	(B)
Angular momentum:	$\Sigma(v_1 w_1 + p_1 S_1)[\mathbf{r}_1 \times \mathbf{u}_1] - \Sigma(v_2 w_2 + p_2 S_2)[\mathbf{r}_2 \times \mathbf{u}_2] + \mathbf{T}_{ext} = \mathbf{T}_{f \to s}$	(C)
Mechanical energy:	$\Sigma\left(\tfrac{1}{2}v_1^2 + gh_1 + \dfrac{p_1}{\rho_1}\right)w_1 - \Sigma\left(\tfrac{1}{2}v_2^2 + gh_2 + \dfrac{p_2}{\rho_2}\right)w_2 = -W_m + E_c + E_v$	(D)

Notes:
(a) All formulas here assume flat velocity profiles.
(b) $\Sigma w_1 = w_{1a} + w_{1b} + w_{1c} + \ldots$, where $w_{1a} = \rho_{1a}v_{1a}S_{1a}$, etc.
(c) h_1 and h_2 are elevations above an arbitrary datum plane.
(d) All equations are written for compressible flow; for incompressible flow, $E_c = 0$.

§7.6   USE OF THE MACROSCOPIC BALANCES FOR STEADY-STATE PROBLEMS

In §3.6 we saw how to set up the differential equations to calculate the velocity and pressure profiles for isothermal flow systems by simplifying the equations of change. In this section we show how to use the set of steady-state macroscopic balances to obtain the algebraic equations for describing large systems.

For each problem we start with the four macroscopic balances. By keeping track of the discarded or approximated terms, we automatically have a complete listing of the assumptions inherent in the final result. All of the examples given here are for isothermal, incompressible flow. The incompressibility assumption means that the velocity of the fluid must be less than the velocity of sound in the fluid and the pressure changes must be small enough that the resulting density changes can be neglected.

The steady-state macroscopic balances may be easily generalized for systems with multiple inlet streams (called 1a, 1b, 1c, . . .) and multiple outlet streams (called 2a, 2b, 2c, . . .). These balances are summarized in Table 7.6-1 for turbulent flow (where the velocity profiles are regarded as flat).

EXAMPLE 7.6-1

Pressure Rise and Friction Loss in a Sudden Enlargement

An incompressible fluid flows from a small circular tube into a large tube in turbulent flow, as shown in Fig. 7.6-1. The cross-sectional areas of the tubes are S_1 and S_2. Obtain an expression for the pressure change between planes 1 and 2 and for the friction loss associated with the sudden enlargement in cross section. Let $\beta = S_1/S_2$, which is less than unity.

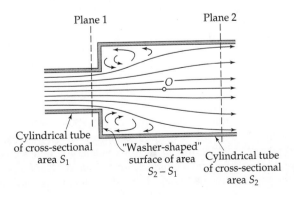

Plane 1   Plane 2

Cylindrical tube of cross-sectional area S_1

"Washer-shaped" surface of area $S_2 - S_1$

Cylindrical tube of cross-sectional area S_2

Fig. 7.6-1. Flow through a sudden enlargement.

SOLUTION

(a) Mass balance. For steady flow the mass balance gives

$$w_1 = w_2 \quad \text{or} \quad \rho_1 v_1 S_1 = \rho_2 v_2 S_2 \tag{7.6-1}$$

For a fluid of constant density, this gives

$$\frac{v_1}{v_2} = \frac{1}{\beta} \tag{7.6-2}$$

(b) Momentum balance. The downstream component of the momentum balance is

$$\mathbf{F}_{f \to s} = (v_1 w_1 - v_2 w_2) + (p_1 S_1 - p_2 S_2) \tag{7.6-3}$$

The force $\mathbf{F}_{f \to s}$ is composed of two parts: the viscous force on the cylindrical surfaces parallel to the direction of flow, and the pressure force on the washer-shaped surface just to the right of plane 1 and perpendicular to the flow axis. The former contribution we neglect (by intuition) and the latter we take to be $p_1(S_2 - S_1)$ by assuming that the pressure on the washer-shaped surface is the same as that at plane 1. We then get, by using Eq. 7.6-1,

$$-p_1(S_2 - S_1) = \rho v_2 S_2 (v_1 - v_2) + (p_1 S_1 - p_2 S_2) \tag{7.6-4}$$

Solving for the pressure difference gives

$$p_2 - p_1 = \rho v_2 (v_1 - v_2) \tag{7.6-5}$$

or, in terms of the downstream velocity,

$$p_2 - p_1 = \rho v_2^2 \left(\frac{1}{\beta} - 1 \right) \tag{7.6-6}$$

Note that the momentum balance predicts (correctly) a *rise* in pressure.

(c) Angular momentum balance. This balance is not needed. If we take the origin of coordinates on the axis of the system at the center of gravity of the fluid located between planes 1 and 2, then $[\mathbf{r}_1 \times \mathbf{u}_1]$ and $[\mathbf{r}_2 \times \mathbf{u}_2]$ are both zero, and there are no torques on the fluid system.

(d) Mechanical energy balance. There is no compressive loss, no work done via moving parts, and no elevation change, so that

$$\hat{E}_v = \tfrac{1}{2}(v_1^2 - v_2^2) + \frac{1}{\rho}(p_1 - p_2) \tag{7.6-7}$$

Insertion of Eq. 7.6-6 for the pressure rise then gives, after some rearrangement,

$$\hat{E}_v = \tfrac{1}{2} v_2^2 \left(\frac{1}{\beta} - 1 \right)^2 \tag{7.6-8}$$

which is an entry in Table 7.5-1.

This example has shown how to use the macroscopic balances to estimate the friction loss factor for a simple resistance in a flow system. Because of the assumptions mentioned after Eq. 7.6-3, the results in Eqs. 7.6-6 and 8 are approximate. If great accuracy is needed, a correction factor based on experimental data should be introduced.

EXAMPLE 7.6-2

Performance of a Liquid–Liquid Ejector

A diagram of a liquid–liquid ejector is shown in Fig. 7.6-2. It is desired to analyze the mixing of the two streams, both of the same fluid, by means of the macroscopic balances. At plane 1 the two fluid streams merge. Stream 1a has a velocity v_0 and a cross-sectional area $\tfrac{1}{3}S_1$, and stream 1b has a velocity $\tfrac{1}{2}v_0$ and a cross-sectional area $\tfrac{2}{3}S_1$. Plane 2 is chosen far enough downstream that the two streams have mixed and the velocity is almost uniform at v_2. The flow is

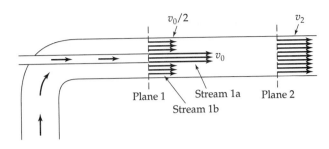

Fig. 7.6-2. Flow in a liquid–liquid ejector pump.

turbulent and the velocity profiles at planes 1 and 2 are assumed to be flat. In the following analysis $\mathbf{F}_{f \to s}$ is neglected, since it is felt to be less important than the other terms in the momentum balance.

SOLUTION

(a) Mass balance. At steady state, Eq. (A) of Table 7.6-1 gives

$$w_{1a} + w_{1b} = w_2 \tag{7.6-9}$$

or

$$\rho v_0 (\tfrac{1}{3} S_1) + \rho (\tfrac{1}{2} v_0)(\tfrac{2}{3} S_1) = \rho v_2 S_2 \tag{7.6-10}$$

Hence, since $S_1 = S_2$, this equation gives

$$v_2 = \tfrac{2}{3} v_0 \tag{7.6-11}$$

for the velocity of the exit stream. We also note, for later use, that $w_{1a} = w_{1b} = \tfrac{1}{2} w_2$.

(b) Momentum balance. From Eq. (B) of Table 7.6-1 the component of the momentum balance in the flow direction is

$$(v_{1a} w_{1a} + v_{1b} w_{1b} + p_1 S_1) - (v_2 w_2 + p_2 S_2) = 0 \tag{7.6-12}$$

or using the relation at the end of (a)

$$(p_2 - p_1) S_2 = (\tfrac{1}{2}(v_{1a} + v_{1b}) - v_2) w_2$$
$$= (\tfrac{1}{2}(v_0 + \tfrac{1}{2} v_0) - \tfrac{2}{3} v_0)(\rho(\tfrac{2}{3} v_0) S_2) \tag{7.6-13}$$

from which

$$p_2 - p_1 = \tfrac{1}{18} \rho v_0^2 \tag{7.6-14}$$

This is the expression for the pressure rise resulting from the mixing of the two streams.

(c) Angular momentum balance. This balance is not needed.

(d) Mechanical energy balance. Equation (D) of Table 7.6-1 gives

$$(\tfrac{1}{2} v_{1a}^2 w_{1a} + \tfrac{1}{2} v_{1b}^2 w_{1b}) - \left(\tfrac{1}{2} v_2^2 + \frac{p_2 - p_1}{\rho} \right) w_2 = E_v \tag{7.6-15}$$

or, using the relation at the end of (a), we get

$$(\tfrac{1}{2} v_{1a}^2 (\tfrac{1}{2} w_2) + \tfrac{1}{2} (\tfrac{1}{2} v_0)^2 (\tfrac{1}{2} w_2)) - (\tfrac{1}{2}(\tfrac{2}{3} v_0)^2 + \tfrac{1}{18} v_0^2) w_2 = E_v \tag{7.6-16}$$

Hence

$$\hat{E}_v = \frac{E_v}{w_2} = \frac{5}{144} v_0^2 \tag{7.6-17}$$

is the energy dissipation per unit mass. The preceding analysis gives fairly good results for liquid–liquid ejector pumps. In gas–gas ejectors, however, the density varies significantly and it is necessary to include the macroscopic total energy balance as well as an equation of state in the analysis. This is discussed in Example 15.3-2.

EXAMPLE 7.6-3

Thrust on a Pipe Bend

Water at 95°C is flowing at a rate of 2.0 ft³/s through a 60° bend, in which there is a contraction from 4 to 3 in. internal diameter (see Fig. 7.6-3). Compute the force exerted on the bend if the pressure at the downstream end is 1.1 atm. The density and viscosity of water at the conditions of the system are 0.962 g/cm³ and 0.299 cp, respectively.

SOLUTION

The Reynolds number for the flow in the 3-in. pipe is

$$\mathrm{Re} = \frac{D\langle v\rangle\rho}{\mu} = \frac{4w}{\pi D\mu}$$

$$= \frac{4(2.0 \times (12 \times 2.54)^3)(0.962)}{\pi(3 \times 2.54)(0.00299)} = 3 \times 10^6 \tag{7.6-18}$$

At this Reynolds number the flow is highly turbulent, and the assumption of flat velocity profiles is reasonable.

(a) Mass balance. For steady-state flow, $w_1 = w_2$. If the density is constant throughout,

$$\frac{v_1}{v_2} = \frac{S_2}{S_1} \equiv \beta \tag{7.6-19}$$

in which β is the ratio of the smaller to the larger cross section.

(b) Mechanical energy balance. For steady, incompressible flow, Eq. (d) of Table 7.6-1 becomes, for this problem,

$$\tfrac{1}{2}(v_2^2 - v_1^2) + g(h_2 - h_1) + \frac{1}{\rho}(p_2 - p_1) + \hat{E}_v = 0 \tag{7.6-20}$$

According to Table 7.5-1 and Eq. 7.5-4, we can take the friction loss as approximately $\tfrac{2}{5}(\tfrac{1}{2}v_2^2) = \tfrac{1}{5}v_2^2$. Inserting this into Eq. 7.6-20 and using the mass balance we get

$$p_1 - p_2 = \rho v_2^2(\tfrac{1}{2} - \tfrac{1}{2}\beta^2 + \tfrac{1}{5}) + \rho g(h_2 - h_1) \tag{7.6-21}$$

This is the pressure drop through the bend in terms of the known velocity v_2 and the known geometrical factor β.

(c) Momentum balance. We now have to consider both the x- and y-components of the momentum balance. The inlet and outlet unit vectors will have x- and y-components given by $u_{1x} = 1$, $u_{1y} = 0$, $u_{2x} = \cos\theta$, and $u_{2y} = \sin\theta$.

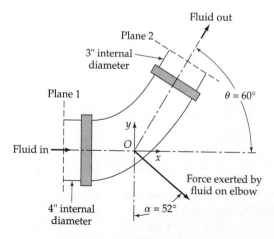

Fig. 7.6-3. Reaction force at a reducing bend in a pipe.

The x-component of the momentum balance then gives

$$F_x = (v_1w_1 + p_1S_1) - (v_2w_2 + p_2S_2) \cos \theta \tag{7.6-22}$$

where F_x is the x-component of $\mathbf{F}_{f \to s}$. Introducing the specific expressions for w_1 and w_2, we get

$$\begin{aligned} F_x &= v_1(\rho v_1 S_1) - v_2(\rho v_2 S_2) \cos \theta + p_1 S_1 - p_2 S_2 \cos \theta \\ &= \rho v_2^2 S_2(\beta - \cos \theta) + (p_1 - p_2)S_1 + p_2(S_1 - S_2 \cos \theta) \end{aligned} \tag{7.6-23}$$

Substituting into this the expression for $p_1 - p_2$ from Eq. 7.6-21 gives

$$\begin{aligned} F_x &= \rho v_2^2 S_2(\beta - \cos \theta) + \rho v_2^2 S_2 \beta^{-1}(\tfrac{7}{10} - \tfrac{1}{2}\beta^2) \\ &\quad + \rho g(h_2 - h_1)S_2\beta^{-1} + p_2 S_2(\beta^{-1} - \cos \theta) \\ &= w^2(\rho S_2)^{-1}(\tfrac{7}{10}\beta^{-1} - \cos \theta + \tfrac{1}{2}\beta) \\ &\quad + \rho g(h_2 - h_1)S_2\beta^{-1} + p_2 S_2(\beta^{-1} - \cos \theta) \end{aligned} \tag{7.6-24}$$

The y-component of the momentum balance is

$$F_y = -(v_2w_2 + p_2S_2) \sin \theta - m_{\text{tot}}g \tag{7.6-25}$$

or

$$F_y = -w^2(\rho S_2)^{-1} \sin \theta - p_2 S_2 \sin \theta - \pi R^2 L \rho g \tag{7.6-26}$$

in which R and L are the radius and length of a roughly equivalent cylinder.

We now have the components of the reaction force in terms of known quantities. The numerical values needed are

$\rho = 60 \text{ lb}_m/\text{ft}^3$ 　　　　　　　　　$S_2 = \tfrac{1}{64}\pi = 0.049 \text{ ft}^2$

$w = (2.0)(60) = 120 \text{ lb}_m/\text{s}$ 　　　$\beta = S_2/S_1 = 3^2/4^2 = 0.562$

$\cos \theta = \tfrac{1}{2}$ 　　　　　　　　　　　$R \approx \tfrac{1}{8} \text{ ft}$

$\sin \theta = \tfrac{1}{2}\sqrt{3}$ 　　　　　　　　　$L \approx \tfrac{5}{6} \text{ ft}$

$p_2 = 16.2 \text{ lb}_f/\text{in.}^2$ 　　　　　　　$h_2 - h_1 \approx \tfrac{1}{2} \text{ ft}$

With these values we then get

$$\begin{aligned} F_x &= \frac{(120)^2}{2(0.049)(32.2)}\left(\frac{7}{10}\frac{1}{0.562} - \frac{1}{2} + \frac{0.562}{2}\right) + (60)(\tfrac{1}{2})(0.049)\left(\frac{1}{0.562}\right) \\ &\quad + (16.2)(0.049)(144)\left(\frac{1}{0.562} - \frac{1}{2}\right) \text{ lb}_f \\ &= (152)(1.24 - 0.50 + 0.28) + 2.6 + (144)(1.78 - 0.50) \\ &= 155 + 2.6 + 146 = 304 \text{ lb}_f = 1352\text{N} \end{aligned} \tag{7.6-27}$$

$$\begin{aligned} F_y &= -\frac{(120)^2}{2(0.049)(32.2)}\left(\tfrac{1}{2}\sqrt{3}\right) - (16.2)(0.049)(144)\left(\tfrac{1}{2}\sqrt{3}\right) - \pi(\tfrac{1}{8})^2(\tfrac{5}{6})(60) \text{ lb}_f \\ &= -132 - 99 - 2.5 = -234 \text{ lb}_f = -1041 \text{ N} \end{aligned} \tag{7.6-28}$$

Hence the magnitude of the force is

$$|\mathbf{F}| = \sqrt{F_x^2 + F_y^2} = \sqrt{304^2 + 234^2} = 384 \text{ lb}_f = 1708 \text{ N} \tag{7.6-29}$$

The angle that this force makes with the vertical is

$$\alpha = \arctan(F_x/F_y) = \arctan 1.30 = 52° \tag{7.6-30}$$

In looking back over the calculation, we see that all the effects we have included are important, with the possible exception of the gravity terms of 2.6 lb$_f$ in F_x and 2.5 lb$_f$ in F_y.

EXAMPLE 7.6-4

The Impinging Jet

A rectangular incompressible fluid jet of thickness b_1 emerges from a slot of width c, hits a flat plate and splits into two streams of thicknesses b_{2a} and b_{2b} as shown in Fig. 7.6-4. The emerging turbulent jet stream has a velocity v_1 and a mass flow rate w_1. Find the velocities and mass rates of flow in the two streams on the plate.[1]

SOLUTION

We neglect viscous dissipation and gravity, and assume that the velocity profiles of all three streams are flat and that their pressures are essentially equal. The macroscopic balances then give

Mass balance

$$w_1 = w_{2a} + w_{2b} \tag{7.6-31}$$

Momentum balance (in the direction parallel to the plate)

$$v_1 w_1 \cos \theta = v_{2a} w_{2a} - v_{2b} w_{2b} \tag{7.6-32}$$

Mechanical energy balance

$$\tfrac{1}{2} v_1^2 w_1 = \tfrac{1}{2} v_{2a}^2 w_{2a} + \tfrac{1}{2} v_{2b}^2 w_{2b} \tag{7.6-33}$$

Angular momentum balance (put the origin of coordinates on the centerline of the jet and at an altitude of $\tfrac{1}{2}b_1$; this is done so that there will be no angular momentum of the incoming jet)

$$0 = (v_{2a} w_{2a}) \cdot \tfrac{1}{2}(b_1 - b_{2a}) - (v_{2b} w_{2b}) \cdot \tfrac{1}{2}(b_1 - b_{2b}) \tag{7.6-34}$$

This last equation can be rewritten to eliminate the b's in favor of the w's. Since $w_1 = \rho v_1 b_1 c$ and $w_{2a} = \rho v_{2a} b_{2a} c$, we can replace $b_1 - b_{2a}$ by $(w_1/\rho v_1 c) - (w_{2a}/\rho v_{2a} c)$ and replace $b_1 - b_{2b}$ correspondingly. Then the angular momentum balance becomes

$$(v_{2a} w_{2a}) \left(\frac{w_1}{v_1} - \frac{w_{2a}}{v_{2a}} \right) = (v_{2b} w_{2b}) \left(\frac{w_1}{v_1} - \frac{w_{2b}}{v_{2b}} \right) \tag{7.6-35}$$

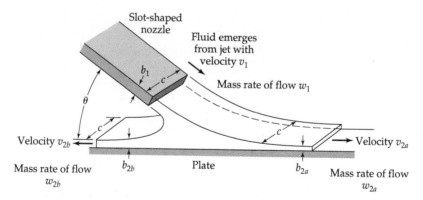

Fig. 7.6-4. Jet impinging on a wall and splitting into two streams. The point O, which is the origin of coordinates for the angular momentum balance, is taken to be the intersection of the centerline of the incoming jet and a plane that is at an elevation $\tfrac{1}{2}b_1$.

---

[1] For alternative solutions to this problem, see G. K. Batchelor, *An Introduction to Fluid Dynamics*, Cambridge University Press (1967), pp. 392–394, and S. Whitaker, *Introduction to Fluid Dynamics*, Prentice-Hall, Englewood Cliffs, N.J. (1968), p. 260. An application of the *compressible* impinging jet problem has been given by J. V. Foa, U.S. Patent 3,361,336 (Jan. 2, 1968). There, use is made of the fact that if the slot-shaped nozzle moves to the left in Fig. 7.6-4 (i.e., left with respect to the plate), then, for a compressible fluid, the right stream will be cooler than the jet and the left stream will be warmer.

or

$$w_{2a}^2 - w_{2b}^2 = \frac{w_1}{v_1}(v_{2a}w_{2a} - v_{2b}w_{2b}) \tag{7.6-36}$$

Now Eqs. 7.6-31, 32, 33, and 36 are four equations with four unknowns. When these are solved we find that

$$v_{2a} = v_1 \qquad w_{2a} = \tfrac{1}{2}w_1(1 + \cos\theta) \tag{7.6-37, 38}$$

$$v_{2b} = v_1 \qquad w_{2b} = \tfrac{1}{2}w_1(1 - \cos\theta) \tag{7.6-39, 40}$$

Hence the velocities of all three streams are equal. The same result is obtained by applying the classical Bernoulli equation for the flow of an inviscid fluid (see Example 3.5-1).

EXAMPLE 7.6-5

Isothermal Flow of a Liquid Through an Orifice

A common method for determining the mass rate of flow through a pipe is to measure the pressure drop across some "obstacle" in the pipe. An example of this is the orifice, which is a thin plate with a hole in the middle. There are pressure taps at planes 1 and 2, upstream and downstream of the orifice plate. Fig. 7.6-5(a) shows the orifice meter, the pressure taps, and the general behavior of the velocity profiles as observed experimentally. The velocity profile at plane 1

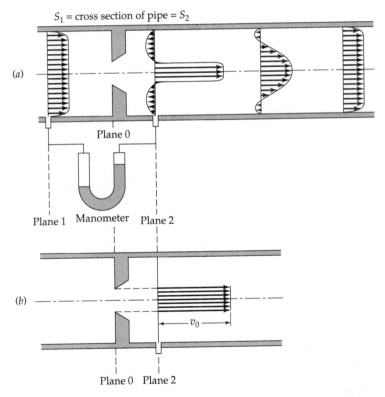

Fig. 7.6-5. (a)A sharp-edged orifice, showing the approximate velocity profiles at several planes near the orifice plate. The fluid jet emerging from the hole is somewhat smaller than the hole itself. In highly turbulent flow this jet necks down to a minimum cross section at the *vena contracta*. The extent of this necking down can be given by the *contraction coefficient*, $C_c = (S_{\text{vena contracta}}/S_0)$. According to inviscid flow theory, $C_c = \pi/(\pi + 2) = 0.611$ if $S_0/S_1 = 0$ [H. Lamb, *Hydrodynamics*, Dover, New York (1945), p. 99]. Note that there is some back flow near the wall. (b) Approximate velocity profile at plane 2 used to estimate $\langle v_2^3 \rangle / \langle v_2 \rangle$.

will be assumed to be flat. In Fig. 7.6-5(b) we show an approximate velocity profile at plane 2, which we use in the application of the macroscopic balances. The standard orifice meter equation is obtained by applying the macroscopic mass and mechanical energy balances.

SOLUTION

(a) *Mass balance.* For a fluid of constant density with a system for which $S_1 = S_2 = S$, the mass balance in Eq. 7.1-1 gives

$$\langle v_1 \rangle = \langle v_2 \rangle \tag{7.6-41}$$

With the assumed velocity profiles this becomes

$$v_1 = \frac{S_0}{S} v_0 \tag{7.6-42}$$

and the volume rate of flow is $w = \rho v_1 S$.

(b) *Mechanical energy balance.* For a constant-density fluid in a flow system with no elevation change and no moving parts, Eq. 7.4-5 gives

$$\frac{1}{2} \frac{\langle v_2^3 \rangle}{\langle v_2 \rangle} - \frac{1}{2} \frac{\langle v_1^3 \rangle}{\langle v_1 \rangle} + \frac{p_2 - p_1}{\rho} + \hat{E}_v = 0 \tag{7.6-43}$$

The viscous loss $\hat{E}_v$ is neglected, even though it is certainly not equal to zero. With the assumed velocity profiles, Eq. 7.6-43 then becomes

$$\tfrac{1}{2}(v_0^2 - v_1^2) + \frac{p_2 - p_1}{\rho} = 0 \tag{7.6-44}$$

When Eqs. 7.6-42 and 44 are combined to eliminate v_0, we can solve for v_1 to get

$$v_1 = \sqrt{\frac{2(p_1 - p_2)}{\rho} \frac{1}{(S/S_0)^2 - 1}} \tag{7.6-45}$$

We can now multiply by ρS to get the volume rate of flow. Then to account for the errors introduced by neglecting $\hat{E}_v$ and by the assumptions regarding the velocity profiles we include a *discharge coefficient*, C_d, and obtain

$$w = C_d S_0 \sqrt{\frac{2\rho(p_1 - p_2)}{1 - (S_0/S)^2}} \tag{7.6-46}$$

Experimental discharge coefficients have been correlated as a function of S_0/S and the Reynolds number.[2] For Reynolds numbers greater than 10^4, C_d approaches about 0.61 for all practical values of S_0/S.

This example has illustrated the use of the macroscopic balances to get the general form of the result, which is then modified by introducing a multiplicative function of dimensionless groups to correct for errors introduced by unwarranted assumptions. This combination of macroscopic balances and dimensional considerations is often used and can be quite useful.

§7.7   USE OF THE MACROSCOPIC BALANCES FOR UNSTEADY-STATE PROBLEMS

In the preceding section we have illustrated the use of the macroscopic balances for solving steady-state problems. In this section we turn our attention to unsteady-state problems. We give two examples to illustrate the use of the time-dependent macroscopic balance equations.

---

[2] G. L. Tuve and R. E. Sprenkle, *Instruments*, **6**, 202–205, 225, 232–234 (1935); see also R. H. Perry and C. H. Chilton, *Chemical Engineers' Handbook*, McGraw-Hill, New York, 5th edition (1973), Fig. 5-18; *Fluid Meters: Their Theory and Applications*, 6th edition, American Society of Mechanical Engineers, New York (1971), pp. 58–65; Measurement of Fluid Flow Using Small Bore Precision Orifice Meters, American Society of Mechanical Engineers, MFC-14-M, New York (1995).

EXAMPLE 7.7-1

Acceleration Effects in Unsteady Flow from a Cylindrical Tank

SOLUTION

An open cylinder of height H and radius R is initially entirely filled with a liquid. At time $t = 0$ the liquid is allowed to drain out through a small hole of radius R_0 at the bottom of the tank (see Fig. 7.7-1).

(a) Find the efflux time by using the unsteady-state mass balance and by assuming Torricelli's equation (see Problem 3B.14) to describe the relation between efflux velocity and the instantaneous height of the liquid.

(b) Find the efflux time using the unsteady-state mass and mechanical energy balances.

(a) We apply Eq. 7.1-2 to the system in Fig. 7.7-1, taking plane 1 to be at the top of the tank (so that $w_1 = 0$). If the instantaneous liquid height is $h(t)$, then

$$\frac{d}{dt}(\pi R^2 h \rho) = -\rho v_2(\pi R_0^2) \tag{7.7-1}$$

Here we have assumed that the velocity profile at plane 2 is flat. According to Torricelli's equation $v_2 = \sqrt{2gh}$, so that Eq. 7.7-1 becomes

$$\frac{dh}{dt} = -\left(\frac{R_0}{R}\right)^2 \sqrt{2gh} \tag{7.7-2}$$

When this is integrated from $t = 0$ to $t = t_{efflux}$ we get

$$t_{efflux} = \sqrt{\frac{2NH}{g}} \tag{7.7-3}$$

in which $N = (R/R_0)^4 \gg 1$. This is effectively a quasi-steady-state solution, since we have used the unsteady-state mass balance along with Torricelli's equation, which was derived for a steady-state flow.

(b) We now use Eq. 7.7-1 and the mechanical energy balance in Eq. 7.4-2. In the latter, the terms W_m and E_c are identically zero, and we assume that E_v is negligibly small, since the velocity gradients in the system will be small. We take the datum plane for the potential energy to be at the bottom of the tank, so that $\hat{\Phi}_2 = gz_2 = 0$; at plane 1 no liquid is entering, and therefore the potential energy term is not needed there. Since the top of the tank is open to the atmosphere and the tank is discharging into the atmosphere, the pressure contributions cancel one another.

To get the total kinetic energy in the system at any time t, we have to know the velocity of every fluid element in the tank. At every point in the tank, we assume that the fluid is moving downward at the same velocity, namely $v_2(R_0/R)^2$ so that the kinetic energy per unit volume is everywhere $\frac{1}{2}\rho v_2^2(R_0/R)^4$.

To get the total potential energy in the system at any time t, we have to integrate the potential energy per unit volume ρgz over the volume of fluid from 0 to h. This gives $\pi R^2 \rho g(\frac{1}{2}h^2)$.

Therefore the mechanical energy balance in Eq. 7.4-2 becomes

$$\frac{d}{dt}[(\pi R^2 h)(\tfrac{1}{2}\rho v_2^2)(R_0/R)^4 + \pi R^2 \rho g(\tfrac{1}{2}h^2)] = -\tfrac{1}{2}v_2^2(\rho v_2 \pi R_0^2) \tag{7.7-4}$$

From the unsteady-state mass balance, $v_2 = -(R/R_0)^2(dh/dt)$. When this is inserted into Eq. 7.7-4 we get (after dividing by dh/dt)

$$2h\frac{d^2h}{dt^2} - (N-1)\left(\frac{dh}{dt}\right)^2 + 2gh = 0 \tag{7.7-5}$$

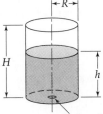

Outlet of radius R_0 **Fig. 7.7-1.** Flow out of a cylindrical tank.

This is to be solved with the two initial conditions:

I.C. 1:   at $t = 0$, $h = H$ (7.7-6)

I.C. 2:   at $t = 0$, $\dfrac{dh}{dt} = \sqrt{2gH}(R_0/R)^2$ (7.7-7)

The second of these is Torricelli's equation at the initial instant of time.

The second-order differential equation for h can be converted to a first-order equation for the function $u(h)$ by making the change of variable $(dh/dt)^2 = u$. This gives

$$h\frac{du}{dh} - (N - 1)u + 2gh = 0 \tag{7.7-8}$$

The solution to this first-order equation can be verified to be[1]

$$u = Ch^{N-1} + 2gh/(N - 2) \tag{7.7-9}$$

The second initial condition then gives $C = -4g/[N(N - 2)H^{N-2}]$ for the integration constant; since $N \gg 1$, we need not concern ourselves with the special case that $N = 2$. We can next take the square root of Eq. 7.7-9 and introduce a dimensionless liquid height $\eta = h/H$; this gives

$$\frac{d\eta}{dt} = \pm\sqrt{\frac{2g}{(N - 2)H}}\sqrt{\eta - \frac{2}{N}\,\eta^{N-1}} \tag{7.7-10}$$

in which the minus sign must be chosen on physical grounds. This separable, first-order equation can be integrated from $t = 0$ to $t = t_{\text{efflux}}$ to give

$$t_{\text{efflux}} = \sqrt{\frac{(N - 2)H}{2g}}\int_0^1 \frac{d\eta}{\sqrt{\eta - (2/N)\eta^{N-1}}} \equiv \sqrt{\frac{2NH}{g}}\,\phi(N) \tag{7.7-11}$$

The function $\phi(N)$ gives the deviation from the quasi-steady-state solution obtained in Eq. 7.7-3. This function can be evaluated as follows:

$$\phi(N) = \frac{1}{2}\sqrt{\frac{N - 2}{N}}\int_0^1 \frac{d\eta}{\sqrt{\eta - (2/N)\eta^{N-1}}}$$

$$= \frac{1}{2}\sqrt{\frac{N - 2}{N}}\int_0^1 \frac{1}{\sqrt{\eta}}\left(1 - \frac{2}{N}\,\eta^{N-2}\right)^{-1/2}d\eta$$

$$= \frac{1}{2}\sqrt{\frac{N - 2}{N}}\int_0^1 \frac{1}{\sqrt{\eta}}\left(1 + \frac{1}{2}\left(\frac{2}{N}\,\eta^{N-2}\right) + \frac{3}{8}\left(\frac{2}{N}\,\eta^{N-2}\right)^2 + \cdots\right)d\eta \tag{7.7-12}$$

The integrations can now be performed. When the result is expanded in inverse powers of N, one finds that

$$\phi(N) = 1 - \frac{1}{N} + O\!\left(\frac{1}{N^3}\right) \tag{7.7-13}$$

Since $N = (R/R_0)^4$ is a very large number, it is evident that the factor $\phi(N)$ differs only very slightly from unity.

It is instructive now to return to Eq. 7.7-4 and omit the term describing the change in total kinetic energy with time. If this is done, one obtains exactly the expression for efflux time in Eq. 7.7-3 (or Eq. 7.7-11, with $\phi(N) = 1$. We can therefore conclude that in this type of problem, the change in kinetic energy with time can safely be neglected.

---

[1] See E. Kamke, *Differentialgleichungen: Lösungsmethoden und Lösungen*, Chelsea Publishing Company, New York (1948), p. 311, #1.94; G. M. Murphy, *Ordinary Differential Equations and Their Solutions*, Van Nostrand, Princeton, N.J. (1960), p. 236, #157.

EXAMPLE 7.7-2

Manometer Oscillations[2]

The liquid in a U-tube manometer, initially at rest, is set in motion by suddenly imposing a pressure difference $p_a - p_b$. Determine the differential equation for the motion of the manometer fluid, assuming incompressible flow and constant temperature. Obtain an expression for the tube radius for which critical damping occurs. Neglect the motion of the gas above the manometer liquid. The notation is summarized in Fig. 7.7-2.

SOLUTION

We designate the manometric liquid as the system to which we apply the macroscopic balances. In that case, there are no planes 1 and 2 through which liquid enters or exits. The free liquid surfaces are capable of performing work on the surroundings, W_m, and hence play the role of the moving mechanical parts in §7.4. We apply the mechanical energy balance of Eq. 7.4-2, with E_c set equal to zero (since the manometer liquid is regarded as incompressible). Because of the choice of the system, both w_1 and w_2 are zero, so that the only terms on the right side are $-W_m$ and $-E_v$.

To evaluate dK_{tot}/dt and E_v it is necessary to make some kind of assumption about the velocity profile. Here we take the velocity profile to be parabolic:

$$v(r, t) = 2\langle v \rangle \left[1 - \left(\frac{r}{R} \right)^2 \right] \tag{7.7-14}$$

in which $\langle v \rangle = dh/dt$ is a function of time, defined to be positive when the flow is from left to right.

The kinetic energy term may then be evaluated as follows:

$$\frac{dK_{tot}}{dt} = \frac{d}{dt} \int_0^L \int_0^{2\pi} \int_0^R (\tfrac{1}{2}\rho v^2) r \, dr \, d\theta \, dl$$

$$= 2\pi L(\tfrac{1}{2}\rho) \frac{d}{dt} \int_0^R v^2 r \, dr$$

$$= 2\pi L R^2 (\tfrac{1}{2}\rho) \frac{d}{dt} \int_0^1 (2\langle v \rangle (1 - \xi^2))^2 \xi \, d\xi$$

$$= \frac{4}{3} \rho L S \langle v \rangle \frac{d}{dt} \langle v \rangle \tag{7.7-15}$$

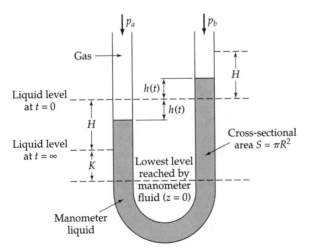

Fig. 7.7-2. Damped oscillations of a manometer fluid.

---

[2] For a summary of experimental and theoretical work on manometer oscillations, see J. C. Biery, *AIChE Journal*, **9**, 606–614 (1963); **10**, 551–557 (1964); **15**, 631–634 (1969). Biery's experimental data show that the assumption made in Eq. 7.7-14 is not very good.

Here l is a coordinate running along the axis of the manometer tube, and L is the distance along this axis from one manometer interface to the other—that is, the total length of the manometer fluid. The dimensionless coordinate ξ is r/R, and S is the cross-sectional area of the tube.

The change of potential energy with time is given by

$$\frac{d\Phi_{tot}}{dt} = \frac{d}{dt}\int_0^L \int_0^{2\pi} \int_0^R (\rho g z) r\, dr\, d\theta\, dl$$

$$= \frac{d}{dt}\left[\begin{pmatrix}\text{integral over portion}\\ \text{below } z = 0, \text{ which}\\ \text{is constant}\end{pmatrix} + \rho g S \int_0^{K+H-h} z\, dz + \rho g S \int_0^{K+H+h} z\, dz\right]$$

$$= 2\rho g S h \frac{dh}{dt} \tag{7.7-16}$$

The viscous loss term can also be evaluated as follows:

$$E_v = -\int_0^L \int_0^{2\pi} \int_0^R (\boldsymbol{\tau}:\nabla\mathbf{v}) r\, dr\, d\theta\, dl$$

$$= 2\pi L \mu \int_0^R \left(\frac{dv}{dr}\right)^2 r\, dr$$

$$= 8\pi L \mu \langle v\rangle^2 \int_0^1 (-2\xi)^2 \xi\, d\xi$$

$$= 8 L S \mu \langle v\rangle^2 / R^2 \tag{7.7-17}$$

Furthermore, the net work done by the surroundings on the system is

$$W_m = (p_a - p_b) S \langle v\rangle \tag{7.7-18}$$

Substitution of the above terms into the mechanical energy balance and letting $\langle v\rangle = dh/dt$ then gives the differential equation for $h(t)$ as

$$\frac{d^2h}{dt^2} + \left(\frac{6\mu}{\rho R^2}\right)\frac{dh}{dt} + \left(\frac{3g}{2L}\right)h = \frac{3}{4}\left(\frac{p_a - p_b}{\rho L}\right) \tag{7.7-19}$$

which is to be solved with the initial conditions that $h = 0$ and $dh/dt = 0$ at $t = 0$. This second-order, linear, nonhomogeneous equation can be rendered homogeneous by introducing a new variable k defined by

$$k = 2h - \frac{p_a - p_b}{\rho L} \tag{7.7-20}$$

Then the equation for the motion of the manometer liquid is

$$\frac{d^2k}{dt^2} + \left(\frac{6\mu}{\rho R^2}\right)\frac{dk}{dt} + \left(\frac{3g}{2L}\right)k = 0 \tag{7.7-21}$$

This equation also arises in describing the motion of a mass connected to a spring and dash-pot as well as the current in an RLC circuit (see Eq. C.1-7).

We now try a solution of the form $k = e^{mt}$. Substituting this trial function into Eq. 7.7-21 shows that there are two admissible values for m:

$$m_{\pm} = \tfrac{1}{2}[-(6\mu/\rho R^2) \pm \sqrt{(6\mu/\rho R^2)^2 - (6g/L)}] \tag{7.7-22}$$

and the solution is

$$k = C_+ e^{m_+ t} + C_- e^{m_- t} \qquad \text{when } m_+ \neq m_- \tag{7.7-23}$$

$$k = C_1 e^{mt} + C_2 t e^{mt} \qquad \text{when } m_+ = m_- = m \tag{7.7-24}$$

with the constants being determined by the initial conditions.

The type of motion that the manometer liquid exhibits depends on the value of the discriminant in Eq. 7.7-22:

(a) If $(6\mu/\rho R^2)^2 > (6g/L)$, the system is *overdamped*, and the liquid moves slowly to its final position.

(b) If $(6\mu/\rho R^2)^2 < (6g/L)$, the system is *underdamped*, and the liquid oscillates about its final position, the oscillations becoming smaller and smaller.

(c) If $(6\mu/\rho R^2)^2 = (6g/L)$, the system is *critically damped*, and the liquid moves to its final position in the most rapid monotone fashion.

The tube radius for critical damping is then

$$R_{cr} = \left(\frac{6\mu^2 L}{\rho^2 g}\right)^{1/4} \tag{7.7-25}$$

If the tube radius R is greater than R_{cr}, an oscillatory motion occurs.

§7.8 DERIVATION OF THE MACROSCOPIC MECHANICAL ENERGY BALANCE[1]

In Eq. 7.4-2 the macroscopic mechanical energy balance was presented without proof. In this section we show how the equation is obtained by integrating the equation of change for mechanical energy (Eq. 3.3-2) over the entire volume of the flow system of Fig. 7.0-1. We begin by doing the formal integration:

$$\int_{V(t)} \frac{\partial}{\partial t} (\tfrac{1}{2}\rho v^2 + \rho\hat{\Phi}) \, dV = -\int_{V(t)} (\nabla \cdot (\tfrac{1}{2}\rho v^2 + \rho\hat{\Phi})\mathbf{v}) \, dV - \int_{V(t)} (\nabla \cdot p\mathbf{v}) \, dV - \int_{V(t)} (\nabla \cdot [\boldsymbol{\tau} \cdot \mathbf{v}]) \, dV$$

$$+ \int_{V(t)} p(\nabla \cdot \mathbf{v}) \, dV + \int_{V(t)} (\boldsymbol{\tau}{:}\nabla\mathbf{v}) \, dV \tag{7.8-1}$$

Next we apply the 3-dimensional Leibniz formula (Eq. A.5-5) to the left side and the Gauss divergence theorem (Eq. A.5-2) to terms 1, 2, and 3 on the right side.

$$\frac{d}{dt} \int_{V(t)} (\tfrac{1}{2}\rho v^2 + \rho\hat{\Phi}) \, dV = -\int_{S(t)} (\mathbf{n} \cdot (\tfrac{1}{2}\rho v^2 + \rho\hat{\Phi})(\mathbf{v} - \mathbf{v}_S)) \, dS - \int_{S(t)} (\mathbf{n} \cdot p\mathbf{v}) \, dS$$

$$- \int_{S(t)} (\mathbf{n} \cdot [\boldsymbol{\tau} \cdot \mathbf{v}]) \, dS + \int_{V(t)} p(\nabla \cdot \mathbf{v}) \, dV + \int_{V(t)} (\boldsymbol{\tau}{:}\nabla\mathbf{v}) \, dV \tag{7.8-2}$$

The term containing $\mathbf{v}_S$, the velocity of the surface of the system, arises from the application of the Leibniz formula. The surface $S(t)$ consists of four parts:

- the fixed surface S_f (on which both $\mathbf{v}$ and $\mathbf{v}_S$ are zero)
- the moving surfaces S_m (on which $\mathbf{v} = \mathbf{v}_S$ with both nonzero)
- the cross section of the entry port S_1 (where $\mathbf{v}_S = 0$)
- the cross section of the exit port S_2 (where $\mathbf{v}_S = 0$)

Presently each of the surface integrals will be split into four parts corresponding to these four surfaces.

We now interpret the terms in Eq. 7.8-2 and, in the process, introduce several assumptions; these assumptions have already been mentioned in §§7.1 to 7.4, but now the reasons for them will be made clear.

---

[1] R. B. Bird, *Korean J. Chem. Eng.*, **15**, 105–123 (1998), §3.

The term on the left side can be interpreted as the time rate of change of the total kinetic and potential energy ($K_{tot} + \Phi_{tot}$) within the "control volume," whose shape and volume are changing with time.

We next examine one by one the five terms on the right side:

Term 1 (including the minus sign) contributes only at the entry and exit ports and gives the rates of influx and efflux of kinetic and potential energy:

$$\text{Term 1} = (\tfrac{1}{2}\rho_1\langle v_1^3\rangle S_1 + \rho_1\hat{\Phi}_1\langle v_1\rangle S_1) - (\tfrac{1}{2}\rho_2\langle v_2^3\rangle S_2 + \rho_2\hat{\Phi}_2\langle v_2\rangle S_2) \tag{7.8-3}$$

The angular brackets indicate an average over the cross section. To get this result we have to assume that the fluid density and potential energy per unit mass are constant over the cross section, and that the fluid is flowing parallel to the tube walls at the entry and exit ports. The first term in Eq. 7.8-3 is positive, since at plane 1, $(-\mathbf{n} \cdot \mathbf{v}) = (\mathbf{u}_1 \cdot (\mathbf{u}_1 v_1)) = v_1$, and the second term is negative, since at plane 2, $(-\mathbf{n} \cdot \mathbf{v}) = (-\mathbf{u}_2 \cdot (\mathbf{u}_2 v_2)) = -v_2$.

Term 2 (including the minus sign) gives no contribution on S_f since $\mathbf{v}$ is zero there. On each surface element dS of S_m there is a force $-\mathbf{n}p\,dS$ acting on a surface moving with a velocity $\mathbf{v}$, and the dot product of these quantities gives the rate at which the surroundings do work on the fluid through the moving surface element dS. We use the symbol $W_m^{(p)}$ to indicate the sum of all these surface terms. Furthermore, the integrals over the stationary surfaces S_1 and S_2 give the work required to push the fluid into the system at plane 1 minus the work required to push the fluid out of the system at plane 2. Therefore term 2 finally gives

$$\text{Term 2} = p_1\langle v_1\rangle S_1 - p_2\langle v_2\rangle S_2 + W_m^{(p)} \tag{7.8-4}$$

Here we have assumed that the pressure does not vary over the cross section at the entry and exit ports.

Term 3 (including the minus sign) gives no contribution on S_f since $\mathbf{v}$ is zero there. The integral over S_m can be interpreted as the rate at which the surroundings do work on the fluid by means of the viscous forces, and this integral is designated as $W_m^{(\tau)}$. At the entry and exit ports it is conventional to neglect the work terms associated with the viscous forces, since they are generally quite small compared with the pressure contributions. Therefore we get

$$\text{Term 3} = W_m^{(\tau)} \tag{7.8-5}$$

We now introduce the symbol $W_m = W_m^{(p)} + W_m^{(\tau)}$ to represent the total rate at which the surroundings do work on the fluid within the system through the agency of the moving surfaces.

Terms 4 and 5 cannot be further simplified, and hence we define

$$\text{Term 4} = +\int_{V(t)} p(\boldsymbol{\nabla} \cdot \mathbf{v})\,dV = -E_c \tag{7.8-6}$$

$$\text{Term 5} = +\int_{V(t)} (\boldsymbol{\tau}{:}\boldsymbol{\nabla}\mathbf{v})\,dV = -E_v \tag{7.8-7}$$

For Newtonian fluids the viscous loss E_v is the rate at which mechanical energy is *irreversibly* degraded into thermal energy because of the viscosity of the fluid and is always a positive quantity (see Eq. 3.3-3). We have already discussed methods for estimating E_v in §7.5. (For viscoelastic fluids, which we discuss in Chapter 8, E_v has to be interpreted differently and may even be negative.) The compression term E_c is the rate at which mechanical energy is *reversibly* changed into thermal energy because of the compressiblity of the fluid; it may be either positive or negative. If the fluid is being regarded as incompressible, then E_c is zero.

When all the contributions are inserted into Eq. 7.8-2 we finally obtain the macroscopic mechanical energy balance:

$$\frac{d}{dt}(K_{tot} + \Phi_{tot}) = (\tfrac{1}{2}\rho_1\langle v_1^3\rangle S_1 + \rho_1\hat{\Phi}_1\langle v_1\rangle S_1 + p_1\langle v_1\rangle S_1) - (\tfrac{1}{2}\rho_2\langle v_2^3\rangle S_2$$
$$+ \rho_2\hat{\Phi}_2\langle v_2\rangle S_2 + p_2\langle v_2\rangle S_2) + W_m - E_c - E_v \qquad (7.8\text{-}8)$$

If, now, we introduce the symbols $w_1 = \rho_1\langle v_1\rangle S_1$ and $w_2 = \rho_2\langle v_2\rangle S_2$ for the mass rates of flow in and out, then Eq. 7.8-8 can be rewritten in the form of Eq. 7.4-2. Several assumptions have been made in this development, but normally they are not serious. If the situation warrants, one can go back and include the neglected effects.

It should be noted that the above derivation of the mechanical energy balance does not require that the system be isothermal. Therefore the results in Eqs. 7.4-2 and 7.8-8 are valid for nonisothermal systems.

To get the mechanical energy balance in the form of Eq. 7.4-7 we have to develop an *approximate* expression for E_c. We imagine that there is a representative streamline running through the system, and we introduce a coordinate s along the streamline. We assume that pressure, density, and velocity do not vary over the cross section. We further imagine that at each position along the streamline, there is a cross section $S(s)$ perpendicular to the s-coordinate, so that we can write $dV = S(s)ds$. If there are moving parts in the system and if the system geometry is complex, it may not be possible to do this.

We start by using the fact that $(\nabla \cdot \rho\mathbf{v}) = 0$ at steady state so that

$$E_c = -\int_V p(\nabla \cdot \mathbf{v})\, dV = +\int_V \frac{p}{\rho}(\mathbf{v} \cdot \nabla\rho)\, dV \qquad (7.8\text{-}9)$$

Then we use the assumption that the pressure and density are constant over the cross section to write approximately

$$E_c \approx \int_1^2 \frac{p}{\rho}\left(v\frac{d\rho}{ds}\right)S(s)ds \qquad (7.8\text{-}10)$$

Even though ρ, v, and S are functions of the streamline coordinate s, their product, $w = \rho vS$, is a constant for steady-state operation and hence may be taken outside the integral. This gives

$$E_c \approx w\int_1^2 \frac{p}{\rho^2}\frac{d\rho}{ds}\, ds = -w\int_1^2 p\frac{d}{ds}\left(\frac{1}{\rho}\right)ds \qquad (7.8\text{-}11)$$

Then an integration by parts can be performed:

$$E_c \approx -w\left[\frac{p}{\rho}\Big|_1^2 - \int_1^2 \frac{1}{\rho}\frac{dp}{ds}\, ds\right] = -w\Delta\left(\frac{p}{\rho}\right) + w\int_1^2 \frac{1}{\rho}\, dp \qquad (7.8\text{-}12)$$

When this result is put into Eq. 7.4-5, the approximate relation in Eq. 7.4-7 is obtained. Because of the questionable nature of the assumptions made (the existence of a representative streamline and the constancy of p and ρ over a cross section), it seems preferable to use Eq. 7.4-5 rather than Eq. 7.4-7. Also, Eq. 7.4-5 is easily generalized to systems with multiple inlet and outlet ports, whereas Eq. 7.4-7 is not; the generalization is given in Eq. (D) of Table 7.6-1.

QUESTIONS FOR DISCUSSION

1. Discuss the origin, meaning, and use of the macroscopic balances, and explain what assumptions have been made in deriving them.
2. How does one decide which macroscopic balances to use for a given problem? What auxiliary information might one need in order to solve problems with the macroscopic balances?
3. Are friction factors and friction loss factors related? If so, how?

4. Discuss the viscous loss E_v and the compression term E_c, with regard to physical interpretation, sign, and methods of estimation.

5. How is the macroscopic mechanical energy balance related to the Bernoulli equation for inviscid fluids? How is it derived?

6. What happens in Example 7.3-1 if one makes a different choice for the origin of the coordinate system?

7. In Example 7.5-1 what would be the error in the final result if the estimation of the viscous loss E_v were off by a factor of 2? Under what circumstances would such an error be more serious?

8. In Example 7.5-1 what would happen if 5 ft were replaced by 50 ft?

9. In Example 7.6-3, how would the results be affected if the outlet pressure were 11 atm instead of 1.1 atm?

10. List all the assumptions that are inherent in the equations given in Table 7.6-1.

PROBLEMS

7A.1 Pressure rise in a sudden enlargement (Fig. 7.6-1). An aqueous salt solution is flowing through a sudden enlargement at a rate of 450 U.S. gal/min $= 0.0384$ m^3/s. The inside diameter of the smaller pipe is 5 in. and that of the large pipe is 9 in. What is the pressure rise in pounds per square inch if the density of the solution is 63 lb$_m$/ft^3? Is the flow in the smaller pipe laminar or turbulent?

Answer: 0.157 psi $= 1.08 \times 10^3$ N/m^2

7A.2 Pumping a hydrochloric acid solution (Fig. 7A.2). A dilute HCl solution of constant density and viscosity ($\rho = 62.4$ lb$_m$/ft^3, $\mu = 1$ cp) is to be pumped from tank 1 to tank 2 with no overall change in elevation. The pressures in the gas spaces of the two tanks are $p_1 = 1$ atm and $p_2 = 4$ atm. The pipe radius is 2 in. and the Reynolds number is 7.11×10^4. The average velocity in the pipe is to be 2.30 ft/s. What power must be delivered by the pump?

Answer: 2.4 hp $= 1.8$ kW

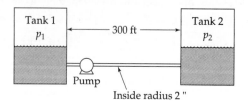

Fig. 7A.2. Pumping of a hydrochloric acid solution.

7A.3 Compressible gas flow in a cylindrical pipe. Gaseous nitrogen is in isothermal turbulent flow at 25°C through a straight length of horizontal pipe with 3-in. inside diameter at a rate of 0.28 lb$_m$/s. The absolute pressures at the inlet and outlet are 2 atm and 1 atm, respectively. Evaluate $\hat{E}_v$, assuming ideal gas behavior and radially uniform velocity distribution.

Answer: 26.3 Btu/lb$_m$ $= 6.12 \times 10^4$ J/kg

7A.4 Incompressible flow in an annulus. Water at 60°F is being delivered from a pump through a coaxial annular conduit 20.3 ft long at a rate of 241 U.S. gal/min. The inner and outer radii of the annular space are 3 in. and 7 in. The inlet is 5 ft lower than the outlet. Determine the power output required from the pump. Use the mean hydraulic radius empiricism to solve the problem. Assume that the pressures at the pump inlet and the annular outlet are the same.

Answer: 0.31 hp $= 0.23$ kW

7A.5 Force on a U-bend (Fig. 7A.5). Water at 68°F ($\rho = 62.4$ lb$_m$/ft^3, $\mu = 1$ cp) is flowing in turbulent flow in a U-shaped pipe bend at 3 ft^3/s. What is the horizontal force exerted by the water on the U-bend?

Answer: 903 lb$_f$

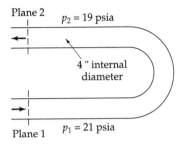

Plane 2 $p_2 = 19$ psia

4 " internal diameter

Plane 1 $p_1 = 21$ psia

Fig. 7A.5. Flow in a U-bend; both arms of the bend are at the same elevation.

7A.6 Flow-rate calculation (Fig. 7A.6). For the system shown in the figure, calculate the volume flow rate of water at 68°F.

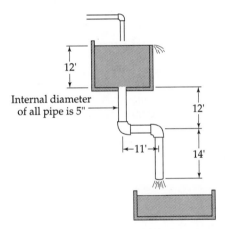

12'

Internal diameter of all pipe is 5"

12'

←11'→

14'

Fig. 7A.6. Flow from a constant-head tank.

7A.7 Evaluation of various velocity averages from Pitot tube data. Following are some experimental data[1] for a Pitot tube traverse for the flow of water in a pipe of internal radius 3.06 in.:

Position	Distance from tube center (in.)	Local velocity (ft/s)	Position	Distance from tube center (in.)	Local velocity (ft/s)
1	2.80	7.85	6	0.72	11.70
2	2.17	10.39	7	1.43	11.47
3	1.43	11.31	8	2.17	11.10
4	0.72	11.66	9	2.80	9.26
5	0.00	11.79			

Plot these data and find out whether the flow is laminar or turbulent. Then use Simpson's rule for numerical integration to compute $\langle v \rangle / v_{max}$, $\langle v^2 \rangle / v_{max}^2$, and $\langle v^3 \rangle / v_{max}^3$. Are these results consistent with the values of 50/49 (given just before Example 7.2-1) and 43200/40817 (given just before Example 7.4-1)?

---

[1] B. Bird, C. E. thesis, University of Wisconsin (1915).

7B.1   Velocity averages from the $\frac{1}{7}$ power law. Evaluate the velocity ratios in Problem 7A.7 according to the velocity distribution in Eq. 5.1-4.

7B.2   Relation between force and viscous loss for flow in conduits of variable cross section. Equation 7.5-6 gives the relation $F_{f \to s} = \rho S \hat{E}_v$ between the drag force and viscous loss for straight conduits of arbitrary, but constant, cross section. Here we consider a straight horizontal channel whose cross section varies gradually with the downstream distance. We restrict ourselves to axisymmetrical channels, so that the drag force is axially directed.

If the cross section and pressure at the entrance are S_1 and p_1, and those at the exit are S_2 and p_2, then prove that the relation analogous to Eq. 7.5-7 is

$$F_{f \to s} = \rho S_m \hat{E}_v + p_m(S_1 - S_2) \tag{7B.2-1}$$

where

$$\frac{1}{S_m} = \frac{1}{2}\left(\frac{1}{S_1} + \frac{1}{S_2}\right) \tag{7B.2-2}$$

$$p_m = \frac{p_1 S_1 + p_2 S_2}{S_1 + S_2} \tag{7B.2-3}$$

Interpret the results.

7B.3   Flow through a sudden enlargement (Fig. 7.6-1). A fluid is flowing through a sudden enlargement, in which the initial and final diameters are D_1 and D_2 respectively. At what ratio D_2/D_1 will the pressure rise $p_2 - p_1$ be a maximum for a given value of v_1?
Answer: $D_2/D_1 = \sqrt{2}$

7B.4   Flow between two tanks (Fig. 7B.4). *Case I:* A fluid flows between two tanks A and B because $p_A > p_B$. The tanks are at the same elevation and there is no pump in the line. The connecting line has a cross-sectional area S_I and the mass rate of flow is w for a pressure drop of $(p_A - p_B)_I$.

Case II: It is desired to replace the connecting line by two lines, each with cross section $S_{II} = \frac{1}{2}S_I$. What pressure difference $(p_A - p_B)_{II}$ is needed to give the same total mass flow rate as in Case I? Assume turbulent flow and use the Blasius formula (Eq. 6.2-12) for the friction factor. Neglect entrance and exit losses.
Answer: $(p_A - p_B)_{II}/(p_A - p_B)_I = 2^{5/8}$

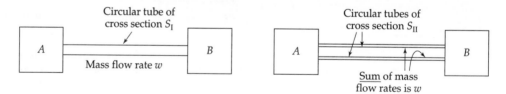

Fig. 7B.4. Flow between two tanks.

7B.5   Revised design of an air duct (Fig. 7B.5). A straight, horizontal air duct was to be installed in a factory. The duct was supposed to be 4 ft × 4 ft in cross section. Because of an obstruction, the duct may be only 2 ft high, but it may have any width. How wide should the duct be to have the same terminal pressures and same volume rate of flow? Assume that the flow is turbulent and that the Blasius formula (Eq. 6.2-12) is satisfactory for this calculation. Air can be regarded as incompressible in this situation.

(a) Write the simplified versions of the mechanical energy balance for ducts I and II.

(b) Equate the pressure drops for the two ducts and obtain an equation relating the widths and heights of the two ducts.

(c) Solve the equation in (b) numerically to find the width that should be used for duct II.
Answer: (c) 9.2 ft

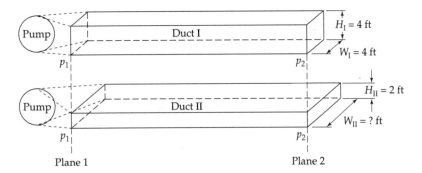

Fig. 7B.5. Installation of an air duct.

7B.6 **Multiple discharge into a common conduit**[2] (Fig. 7B.6). Extend Example 7.6-1 to an incompressible fluid discharging from several tubes into a larger tube with a net increase in cross section. Such systems are important in heat exchangers of certain types, for which the expansion and contraction losses account for an appreciable fraction of the overall pressure drop. The flows in the small tubes and the large tube may be laminar or turbulent. Analyze this system by means of the macroscopic mass, momentum, and mechanical energy balances.

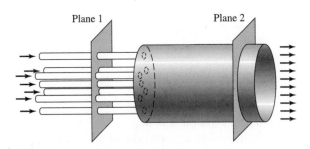

Fig. 7B.6. Multiple discharge into a common conduit. The total cross sectional area at plane 1 available for flow is S_1 and that at plane 2 is S_2.

7B.7 **Inventory variations in a gas reservoir.** A natural gas reservoir is to be supplied from a pipeline at a steady rate of w_1 lb$_m$/hr. During a 24-hour period, the fuel demand from the reservoir, w_2, varies approximately as follows,

$$w_2 = A + B \cos \omega t \tag{7B.7-1}$$

where ωt is a dimensionless time measured from the time of peak demand (approximately 6 A.M.).

(a) Determine the maximum, minimum, and average values of w_2 for a 24-hour period in terms of A and B.

(b) Determine the required value of w_1 in terms of A and B.

(c) Let $m_{tot} = m_{tot}^0$ at $t = 0$, and integrate the unsteady mass balance with this initial condition to obtain m_{tot} as a function of time.

(d) If $A = 5000$ lb$_m$/hr, $B = 2000$ lb$_m$/hr, and $\rho = 0.044$ lb$_m$/ft^3 in the reservoir, determine the absolute minimum reservoir capacity in cubic feet to meet the demand without interruption. At what time of day must the reservoir be full to permit such operation?

(e) Determine the minimum reservoir capacity in cubic feet required to permit maintaining at least a three-day reserve at all times.

Answer: 3.47×10^5 ft^3; 8.53×10^6 ft^3

[2] W. M. Kays, *Trans. ASME*, **72**, 1067–1074 (1950).

7B.8 Change in liquid height with time (Fig. 7.1-1).

(a) Derive Eq. 7.1-4 by using integral calculus.

(b) In Example 7.1-1, obtain the expression for the liquid height h as a function of time t.

(c) Make a graph of Eq. 7.1-8 using dimensionless quantities. Is this useful?

7B.9 Draining of a cylindrical tank with exit pipe (Fig. 7B.9).

(a) Rework Example 7.1-1, but with a cylindrical tank instead of a spherical tank. Use the quasi-steady-state approach; that is, use the unsteady-state mass balance along with the Hagen–Poiseuille equation for the laminar flow in the pipe.

(b) Rework the problem for turbulent flow in the pipe.

Answer: (a) $t_{\text{efflux}} = \dfrac{128\mu LR^2}{\rho g D^4} \ln\left(1 + \dfrac{H}{L}\right)$

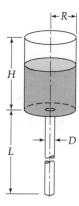

Fig. 7B.9. A cylindrical tank with a long pipe attached. The fluid surface and pipe exit are open to the atmosphere.

7B.10 Efflux time for draining a conical tank (Fig. 7B.10). A conical tank, with dimensions given in the figure, is initially filled with a liquid. The liquid is allowed to drain out by gravity. Determine the efflux time. In parts (a)–(c) take the liquid in the cone to be the "system."

(a) First use an unsteady macroscopic mass balance to show that the exit velocity is

$$v_2 = -\frac{z^2}{z_2^2}\frac{dz}{dt} \qquad (7B.10\text{-}1)$$

(b) Write the unsteady-state mechanical energy balance for the system. Discard the viscous loss term and the term containing the time derivative of the kinetic energy, and give reasons for doing so. Show that Eq. 7B.10-1 then leads to

$$v_2 = \sqrt{2g(z - z_2)} \qquad (7B.10\text{-}2)$$

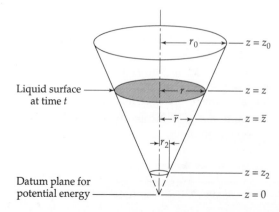

Fig. 7B.10. A conical container from which a fluid is allowed to drain. The quantity r is the radius of the liquid surface at height z, and $\bar{r}$ is the radius of the cone at some arbitrary height $\bar{z}$.

(c) Combine the results of (a) and (b). Solve the resulting differential equation with an appropriate initial condition to get the liquid level z as a function of t. From this get the efflux time

$$t_{\text{efflux}} = \frac{1}{5}\left(\frac{z_0}{z_2}\right)^2 \sqrt{\frac{2z_0}{g}}$$ (7B.10-3)

List all the assumptions that have been made and discuss how serious they are. How could these assumptions be avoided?

(d) Rework part (b) by choosing plane 1 to be stationary and slightly below the liquid surface at time t. It is understood that the liquid surface does not go below plane 1 during the differential time interval dt over which the unsteady mechanical energy balance is made. With this choice of plane 1 the derivative $d\Phi_{\text{tot}}/dt$ is zero and there is no work term W_m. Furthermore the conditions at plane 1 are very nearly those at the liquid surface. Then with the pseudo-steady-state approximation that the derivative dK_{tot}/dt is approximately zero and the neglect of the viscous loss term, the mechanical energy balance, with $w_1 = w_2$, takes the form

$$0 = \tfrac{1}{2}(v_1^2 - v_2^2) + g(h_1 - h_2)$$ (7B.10-4)

7B.11 Disintegration of wood chips (Fig. 7B.11). In the manufacture of paper pulp the cellulose fibers of wood chips are freed from the lignin binder by heating in alkaline solutions under pressure in large cylindrical tanks called digesters. At the end of the "cooking" period, a small port in one end of the digester is opened, and the slurry of softened wood chips is allowed to blow against an impact plate to complete the breakup of the chips and the separation of the fibers. Estimate the velocity of the discharging stream and the additional force on the impact plate shortly after the discharge begins. Frictional effects inside the digester, and the small kinetic energy of the fluid inside the tank, may be neglected. (*Note:* See Problem 7B.10 for two different methods for selecting the entrance and exit planes.)

Answer: 2810 lb_m/s (or 1275 kg/s); 10,900 lb_f (or 48,500 N)

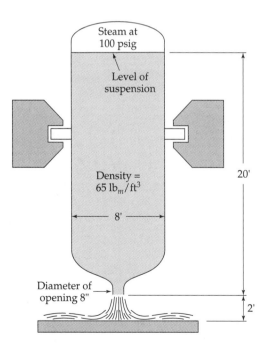

Fig. 7B.11. Pulp digester.

7B.12 Criterion for vapor-free flow in a pipeline. To ensure that a pipeline is completely liquid-filled, it is necessary that $p > p_{\text{vap}}$ at every point. Apply this criterion to the system in Fig. 7.5-1, by using mechanical energy balances over appropriate portions of the system.

7C.1   **End corrections in tube viscometers** (Fig. 7C.1).[3] In analyzing tube-flow viscometric data to determine viscosity, one compares pressure drop versus flow rate data with the theoretical expression (the Hagen–Poiseuille equation of Eq. 2.3-21). The latter assumes that the flow is fully developed in the region between the two planes at which the pressure is measured. In an apparatus such as that shown in the figure, the pressure is known at the tube exit (2) and also above the fluid in the reservoir (1). However, in the entrance region of the tube, the velocity profiles are not yet fully developed. Hence the theoretical expression relating the pressure drop to the flow rate is not valid.

There is, however, a method in which the Hagen–Poiseuille equation can be used, by making flow measurements in two tubes of different lengths, L_A and L_B; the shorter of the two tubes must be long enough so that the velocity profiles are fully developed at the exit. Then the end section of the long tube, of length $L_B - L_A$, will be a region of fully developed flow. If we knew the value of $\mathscr{P}_0 - \mathscr{P}_4$ for this region, then we could apply the Hagen–Poiseuille equation.

Show that proper combination of the mechanical energy balances, written for the systems 1–2, 3–4, and 0–4 gives the following expression for $\mathscr{P}_0 - \mathscr{P}_4$ when each viscometer has the *same flow rate*.

$$\frac{\mathscr{P}_0 - \mathscr{P}_4}{L_B - L_A} = \frac{p_B - p_A}{L_B - L_A} + \rho g\left(1 + \frac{l_B - l_A}{L_B - L_A}\right) \tag{7C.1-1}$$

where $\mathscr{P}_0 = p_0 + \rho g z_0$. Explain carefully how you would use Eq. 7C.1-1 to analyze experimental measurements. Is Eq. 7C.1-1 valid for ducts with noncircular, uniform cross section?

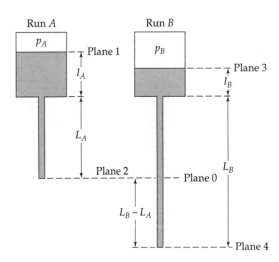

Fig. 7C.1. Two tube viscometers with the same flow rate and the same exit pressure. The pressures p_A and p_B are maintained by an inert gas.

7D.1   **Derivation of the macroscopic balances from the equations of change.** Derive the macroscopic mass and momentum balances by integrating the equations of continuity and motion over the flow system of Fig. 7.0-1. Follow the procedure given in §7.8 for the macroscopic mechanical energy balance, using the Gauss divergence theorem and the Leibniz formula.

---

[3] A. G. Fredrickson, PhD Thesis, University of Wisconsin (1959); *Principles and Applications of Rheology*, Prentice-Hall, Englewood Cliffs, N.J. (1964), §9.2.

Chapter 8

Polymeric Liquids

§8.1   **Examples of the behavior of polymeric liquids**

§8.2   **Rheometry and material functions**

§8.3   **Non-Newtonian viscosity and the generalized Newtonian models**

§8.4°   **Elasticity and the linear viscoelastic models**

§8.5•   **The corotational derivatives and the nonlinear viscoelastic models**

§8.6•   **Molecular theories for polymeric liquids**

In the first seven chapters we have considered only *Newtonian fluids*. The relations between stresses and velocity gradients are described by Eq. 1.1-2 for simple shear flow and by Eq. 1.2-6 (or Eq. 1.2-7) for arbitrary time-dependent flows. For the Newtonian fluid, two material parameters are needed—the two coefficients of viscosity μ and κ—which depend on temperature, pressure, and composition, but not on the velocity gradients. All gases and all liquids composed of "small" molecules (up to molecular weights of about 5000) are accurately described by the Newtonian fluid model.

There are many fluids that are not described by Eq. 1.2-6, and these are called *non-Newtonian fluids*. These structurally complex fluids include polymer solutions, polymer melts, soap solutions, suspensions, emulsions, pastes, and some biological fluids. In this chapter we focus on polymeric liquids.

Because they contain high-molecular-weight molecules with many internal degrees of freedom, polymer solutions and molten polymers have behavior qualitatively different from that of Newtonian fluids. Their viscosities depend strongly on the velocity gradients, and in addition they may display pronounced "elastic effects." Also in the steady simple shear flow between two parallel plates, there are nonzero and unequal normal stresses (τ_{xx}, τ_{yy}, and τ_{zz}) that do not arise in Newtonian fluids. In §8.1 we describe some experiments that emphasize the differences between Newtonian and polymeric fluids.

In dealing with Newtonian fluids the science of the measurement of viscosity is called *viscometry*, and in earlier chapters we have seen examples of simple flow systems that can be used as *viscometers* (the circular tube, the cone–plate system, and coaxial cylinders). To characterize non-Newtonian fluids we have to measure not only the viscosity, but the normal stresses and the viscoelastic responses as well. The science of measurement of these properties is called *rheometry*, and the instruments are called *rheometers*. We treat this subject briefly in §8.2. The science of *rheology* includes all aspects of the study of deformation and flow of non-Hookean solids and non-Newtonian liquids.

After the first two sections, which deal with experimental facts, we turn to the presentation of various non-Newtonian "models" (that is, empirical expressions for the stress tensor) that are commonly used for describing polymeric liquids. In §8.3 we start with the *generalized Newtonian models*, which are relatively simple, but which can describe only the non-Newtonian viscosity (and not the viscoelastic effects). Then in §8.4 we give examples of *linear viscoelastic models*, which can describe the viscoelastic responses, but only in flows with exceedingly small displacement gradients. Next in §8.5 we give several *nonlinear viscoelastic models*, and these are intended to be applicable in all flow situations. As we go from elementary to more complicated models, we enlarge the set of observed phenomena that we can describe (but also the mathematical difficulties). Finally in §8.6 there is a brief discussion about the kinetic theory approach to polymer fluid dynamics.

Polymeric liquids are encountered in the fabrication of plastic objects, and as additives to lubricants, foodstuffs, and inks. They represent a vast and important class of liquids, and many scientists and engineers must deal with them. Polymer fluid dynamics, heat transfer, and diffusion form a rapidly growing part of the subject of transport phenomena, and there are many textbooks,[1] treatises,[2] and journals devoted to the subject. The subject has also been approached from the kinetic theory standpoint, and molecular theories of the subject have contributed much to our understanding of the mechanical, thermal, and diffusional behavior of these fluids.[3] Finally, for those interested in the history of the subject, the reader is referred to the book by Tanner and Walters.[4]

§8.1   EXAMPLES OF THE BEHAVIOR OF POLYMERIC LIQUIDS

In this section we discuss several experiments that contrast the flow behavior of Newtonian and polymeric fluids.[1]

Steady-State Laminar Flow in Circular Tubes

Even for the steady-state, axial, laminar flow in circular tubes, there is an important difference between the behavior of Newtonian liquids and that of polymeric liquids. For Newtonian liquids the velocity distribution, average velocity, and pressure drop are given by Eqs. 2.3-18, 2.3-20, and 2.3-21, respectively.

For *polymeric liquids*, experimental data suggest that the following equations are reasonable:

$$\frac{v_z}{v_{z,\max}} \approx 1 - \left(\frac{r}{R}\right)^{(1/n)+1} \quad \text{and} \quad \frac{\langle v_z \rangle}{v_{z,\max}} \approx \frac{(1/n)+1}{(1/n)+3} \qquad (8.1\text{-}1, 2)$$

where n is a positive parameter characterizing the fluid, usually with a value less than unity. That is, the velocity profile is more blunt than it is for the Newtonian fluid, for which $n = 1$. It is further found experimentally that

$$\mathcal{P}_0 - \mathcal{P}_L \sim w^n \qquad (8.1\text{-}3)$$

The pressure drop thus increases much less rapidly with the mass flow rate than for Newtonian fluids, for which the relation is linear.

---

[1] A. S. Lodge, *Elastic Liquids*, Academic Press, New York (1964); R. B. Bird, R. C. Armstrong, and O. Hassager, *Dynamics of Polymeric Liquids, Vol. 1., Fluid Mechanics*, Wiley-Interscience, New York, 2nd edition (1987); R. I. Tanner, *Engineering Rheology*, Clarendon Press, Oxford (1985).

[2] H. A. Barnes, J. F. Hutton, and K. Walters, *An Introduction to Rheology*, Elsevier, Amsterdam (1989); H. Giesekus, *Phänomenologische Rheologie: Eine Einführung*, Springer Verlag, Berlin (1994). Books emphasizing the engineering aspects of the subject include Z. Tadmor and C. G. Gogos, *Principles of Polymer Processing*, Wiley, New York (1979), D. G. Baird and D. I. Collias, *Polymer Processing: Principles and Design*, Butterworth-Heinemann, Boston (1995), J. Dealy and K. Wissbrun, *Melt Rheology and its Role in Plastics Processing*, Van Nostrand Reinhold, New York (1990).

[3] R. B. Bird, C. F. Curtiss, R. C. Armstrong, and O. Hassager, *Dynamics of Polymeric Liquids, Vol. 2, Kinetic Theory*, Wiley-Interscience, New York, 2nd edition (1987); C. F. Curtiss and R. B. Bird, *Adv. Polymer Sci*, **125**, 1–101 (1996) and *J. Chem. Phys.* **111**, 10362–10370 (1999).

[4] R. I. Tanner and K. Walters, *Rheology: An Historical Perspective*, Elsevier, Amsterdam (1998).

[1] More details about these and other experiments can be found in R. B. Bird, R. C. Armstrong, and O. Hassager, *Dynamics of Polymeric Liquids, Vol. 1, Fluid Dynamics*, Wiley-Interscience, New York, 2nd edition (1987), Chapter 2. See also A. S. Lodge, *Elastic Liquids*, Academic Press, New York (1964), Chapter 10.

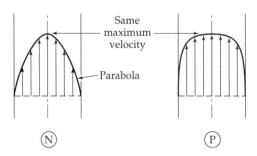

Fig. 8.1-1. Laminar flow in a circular tube. The symbols Ⓝ (Newtonian liquid) and Ⓟ (Polymeric liquid) are used in this and the next six figures.

In Fig. 8.1-1 we show typical velocity profiles for laminar flow of Newtonian and polymeric fluids for the same maximum velocity. This simple experiment suggests that the polymeric fluids have a viscosity that depends on the velocity gradient. This point will be elaborated on in §8.3.

For laminar flow in tubes of noncircular cross section, polymeric liquids exhibit secondary flows superposed on the axial motion. Recall that for turbulent Newtonian flows secondary flows are also observed—in Fig. 5.1-2 it is shown that the fluid moves toward the corners of the conduit and then back in toward the center. For laminar flow of polymeric fluids, the secondary flows go in the opposite direction—from the corners of the conduit and then back toward the walls.[2] In turbulent flows the secondary flows result from inertial effects, whereas in the flow of polymers the secondary flows are associated with the "normal stresses."

Recoil after Cessation of Steady-State Flow in a Circular Tube

We start with a fluid at rest in a circular tube and, with a syringe, we "draw" a dye line radially in the fluid as shown in Fig. 8.1-2. Then we pump the fluid and watch the dye deform.[3]

For a Newtonian fluid the dye line deforms into a continuously stretching parabola. If the pump is turned off, the dye parabola stops moving. After some time diffusion occurs and the parabola begins to get fuzzy, of course.

For a *polymeric liquid* the dye line deforms into a curve that is more blunt than a parabola (see Eq. 8.1-1). If the pump is stopped and the fluid is not axially constrained, the fluid will begin to "recoil" and will retreat from this maximum stretched shape; that

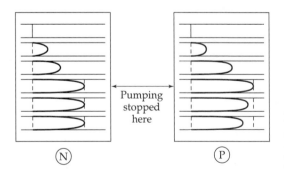

Fig. 8.1-2. Constrained recoil after cessation of flow in a circular tube, observed in polymeric liquids, but not in Newtonian liquids.

---

[2] B. Gervang and P. S. Larsen, *J. Non-Newtonian Fluid Mech.*, **39**, 217–237 (1991).

[3] For the details of this experiment see N. N. Kapoor, M.S. thesis, University of Minnesota, Minneapolis (1964), as well as A. G. Fredrickson, *Principles and Applications of Rheology*, Prentice-Hall, Englewood Cliffs, N.J. (1964), p. 120.

is, the fluid snaps back somewhat like a rubber band. However, whereas a rubber band returns to its original shape, the fluid retreats only part way toward its original configuration.

If we permit ourselves an anthropomorphism, we can say that a rubber band has "perfect memory," since it returns to its initial unstressed state. The polymeric fluid, on the other hand, has a "fading memory," since it gradually "forgets" its original state. That is, as it recoils, its memory becomes weaker and weaker.

Fluid recoil is a manifestation of *elasticity*, and any complete description of polymeric fluids must be able to incorporate the idea of elasticity into the expression for the stress tensor. The theory must also include the notion of fading memory.

"Normal Stress" Effects

Other striking differences in the behavior of Newtonian and polymeric liquids appear in the "normal stress" effects. The reason for this nomenclature will be given in the next section.

A rotating rod in a beaker of a Newtonian fluid causes the fluid to undergo a tangential motion. At steady state, the fluid surface is lower near the rotating rod. Intuitively we know that this comes about because the centrifugal force causes the fluid to move radially toward the beaker wall. For a *polymeric liquid*, on the other hand, the fluid moves toward the rotating rod, and, at steady state, the fluid surface is as shown in Fig. 8.1-3. This phenomenon is called the *Weissenberg rod-climbing effect*.[4] Evidently some kinds of forces are induced that cause the polymeric liquid to behave in a way that is qualitatively different from that of a Newtonian liquid.

In a closely related experiment, we can put a rotating disk on the surface of a fluid in a cylindrical container as shown in Fig. 8.1-4. If the fluid is Newtonian, the rotating disk causes the fluid to move in a tangential direction (the "primary flow"), but, in addition, the fluid moves slowly outward toward the cylinder wall because of the centrifugal force, then moves downward, and then back up along the cylinder axis. This superposed radial and axial flow is weaker than the primary flow and is termed a "secondary flow." For a *polymeric liquid*, the fluid also develops a primary tangential flow with a weak ra-

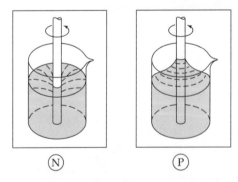

Fig. 8.1-3. The free surface of a liquid near a rotating rod. The polymeric liquid shows the Weissenberg rod-climbing effect.

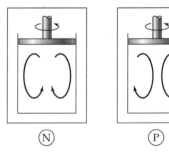

Fig. 8.1-4. The secondary flows in a cylindrical container with a rotating disk at the liquid surface have the opposite directions for Newtonian and polymeric fluids.

---

[4] This phenomenon was first described by F. H. Garner and A. H. Nissan, *Nature*, **158**, 634–635 (1946) and by R. J. Russel, Ph.D. thesis, Imperial College, University of London (1946), p. 58. The experiment was analyzed by K. Weissenberg, *Nature*, **159**, 310–311 (1947).

 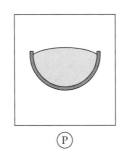

Ⓝ   Ⓟ

Fig. 8.1-5. Flow down a tilted semicylindrical trough. The convexity of the polymeric liquid surface is somewhat exaggerated here.

dial and axial secondary flow, but the latter goes in a direction opposite to that seen in the Newtonian fluid.[5]

In another experiment we can let a liquid flow down a tilted, semi-cylindrical trough as shown in Fig. 8.1-5. If the fluid is Newtonian, the liquid surface is flat, except for the meniscus effects at the outer edges. For most *polymeric liquids*, however, the liquid surface is found to be slightly convex. The effect is small but reproducible.[6]

Some Other Experiments

The operation of a simple siphon is familiar to everyone. We know from experience that, if the fluid is Newtonian, the removal of the siphon tube from the liquid means that the siphoning action ceases. However, as may be seen in Fig. 8.1-6, for polymeric liquids the siphoning can continue even when the siphon is lifted several centimeters above the liquid surface. This is called the *tubeless siphon* effect. One can also just lift some of the fluid up over the edge of the beaker and then the fluid will flow upward along the inside of the beaker and then down the outside until the beaker is nearly empty.[7]

In another experiment a long cylindrical rod, with its axis in the z direction, is made to oscillate back and forth in the x direction with the axis parallel to the z axis (see Fig.

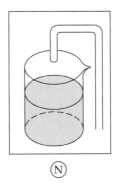 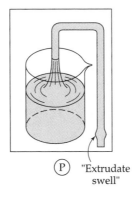

Ⓝ   Ⓟ   "Extrudate
                   swell"

Fig. 8.1-6. Siphoning continues to occur when the tube is raised above the surface of a polymeric liquid, but not so for a Newtonian liquid. Note the swelling of the polymeric liquid as it leaves the siphon tube.

---

[5] C. T. Hill, J. D. Huppler, and R. B. Bird, *Chem. Eng. Sci.* **21**, 815–817 (1966); C. T. Hill, *Trans. Soc. Rheol.*, **16**, 213–245 (1972). Theoretical analyses have been given by J. M. Kramer and M. W. Johnson, Jr., *Trans. Soc. Rheol.* **16**, 197–212 (1972), and by J. P. Nirschl and W. E. Stewart, *J. Non-Newtonian Fluid Mech.*, **16**, 233–250 (1984).

[6] This experiment was first done by R. I. Tanner, *Trans. Soc. Rheol.*, **14**, 483–507 (1970), prompted by a suggestion by A. S. Wineman and A. C. Pipkin, *Acta Mech.* **2**, 104–115 (1966). See also R. I. Tanner, *Engineering Rheology*, Oxford University Press (1985), 102–105.

[7] D. F. James, *Nature*, **212**, 754–756 (1966).

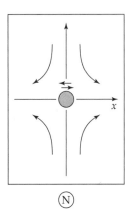 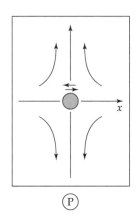

⒩   ⒫

Fig. 8.1-7. The "acoustical streaming" near a laterally oscillating rod, showing that the induced secondary flow goes in the opposite directions for Newtonian and polymeric fluids.

8.1-7). In a Newtonian fluid, a secondary flow is induced whereby the fluid moves toward the cylinder from above and below (i.e., from the $+y$ and $-y$ directions, and moves away to the left and right (i.e., toward the $-x$ and $+x$ direction). For the *polymeric liquid*, however, the induced secondary motion is in the opposite direction: the fluid moves inward from the left and right along the x axis and outward in the up and down directions along the y axis.[8]

The preceding examples are only a few of many interesting experiments that have been performed.[9] The polymeric behavior can be illustrated easily and inexpensively with a 0.5% aqueous solution of polyethylene oxide.

There are also some fascinating effects that occur when even tiny quantities of polymers are present. The most striking of these is the phenomenon of *drag reduction*.[10] With only parts per million of some polymers ("drag-reducing agents"), the friction loss in turbulent pipe flow may be lowered dramatically—by 30–50%. Such polymeric drag-reducing agents are used by fire departments to increase the flow of water, and by oil companies to lower the costs for pumping crude oil over long distances.

For discussions of other phenomena that arise in polymeric fluids, the reader should consult the summary articles in *Annual Review of Fluid Mechanics*.[11]

§8.2   RHEOMETRY AND MATERIAL FUNCTIONS

The experiments described in §8.1 make it abundantly clear that polymeric liquids do not obey Newton's law of viscosity. In this section we discuss several simple, controllable flows in which the stress components can be measured. From these experiments one can measure a number of *material functions* that describe the mechanical response of complex fluids. Whereas incompressible Newtonian fluids are described by only one material constant (the viscosity), one can measure many different material functions for non-Newtonian liquids. Here we show how a few of the more commonly used material

---

[8] C. F. Chang and W. R. Schowalter, *J. Non-Newtonian Fluid Mech.*, **6**, 47–67 (1979).

[9] The book by D. V. Boger and K. Walters, *Rheological Phenomena in Focus*, Elsevier, Amsterdam (1993), contains many photographs of fluid behavior in a variety of non-Newtonian flow systems.

[10] This is sometimes called the *Toms phenomenon*, since it was perhaps first reported in B. A. Toms, *Proc. Int. Congress on Rheology*, North-Holland, Amsterdam (1949). The phenomenon has also been studied in connection with the drag-reducing nature of fish slime [T. L. Daniel, *Biol. Bull.*, **160**, 376–382 (1981)], which is thought to explain, at least in part, "Gray's paradox"—the fact that fish seem to be able to swim faster than energy considerations permit.

[11] For example, M. M. Denn, *Ann. Rev. Fluid Mech.*, **22**, 13–34 (1990); E. S. G. Shaqfeh, *Ann. Rev. Fluid Mech.*, **28**, 129–185 (1996); G. G. Fuller, *Ann. Rev. Fluid Mech.*, **22**, 387–417 (1992).

functions are defined and measured. Information about the actual measurement equipment and other material functions can be found elsewhere.[1,2] It is assumed throughout this chapter that the polymeric liquids can be regarded as incompressible.

Steady Simple Shear Flow

We consider now the steady shear flow between a pair of parallel plates, where the velocity profile is given by $v_x = \dot{\gamma}y$, the other velocity components being zero (see Fig. 8.2-1). The quantity $\dot{\gamma}$, here taken to be positive, is called the "shear rate." For a Newtonian fluid the shear stress τ_{yx} is given by Eq. 1.1-2, and the normal stresses (τ_{xx}, τ_{yy}, and τ_{zz}) are all zero.

For incompressible non-Newtonian fluids, the normal stresses are nonzero and unequal. For these fluids it is conventional to define three material functions as follows:

$$\tau_{yx} = -\eta \frac{dv_x}{dy} \tag{8.2-1}$$

$$\tau_{xx} - \tau_{yy} = -\Psi_1 \left(\frac{dv_x}{dy}\right)^2 \tag{8.2-2}$$

$$\tau_{yy} - \tau_{zz} = -\Psi_2 \left(\frac{dv_x}{dy}\right)^2 \tag{8.2-3}$$

in which η is the non-Newtonian viscosity, Ψ_1 is the first normal stress coefficient, and Ψ_2 is the second normal stress coefficient. These three quantities—η, Ψ_1, Ψ_2—are all functions of the shear rate $\dot{\gamma}$. For many polymeric liquids η may decrease by a factor of as much as 10^4 as the shear rate increases. Similarly, the normal stress coefficients may decrease by a factor of as much as 10^7 over the usual range of shear rates. For polymeric fluids made up of flexible macromolecules, the functions $\eta(\dot{\gamma})$ and $\Psi_1(\dot{\gamma})$ have been found experimentally to be positive, whereas $\Psi_2(\dot{\gamma})$ is almost always negative. It can be shown that for positive $\Psi_1(\dot{\gamma})$ the fluid behaves as though it were under tension in the flow (or x) direction, and that the negative $\Psi_2(\dot{\gamma})$ means that the fluid is under tension in the transverse (or z) direction. For the Newtonian fluid $\eta = \mu$, $\Psi_1 = 0$, and $\Psi_2 = 0$.

The strongly shear-rate–dependent non-Newtonian viscosity is connected with the behavior given in Eqs. 8.1-1 to 3, as is shown in the next section. The positive Ψ_1 is primarily responsible for the Weissenberg rod-climbing effect. Because of the tangential flow, there is a tension in the tangential direction, and this tension pulls the fluid toward the rotating rod, overcoming the centrifugal force. The secondary flows in the disk-and-cylinder experiment (Fig. 8.1-4) can also be explained qualitatively in terms of the positive Ψ_1. Also, the negative Ψ_2 can be shown to explain the convex surface shape in the tilted-trough experiment (Fig. 8.1-5).

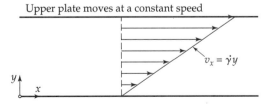

Upper plate moves at a constant speed

$v_x = \dot{\gamma}y$

y

x

Fig. 8.2-1. Steady simple shear flow between parallel plates, with shear rate $\dot{\gamma}$. For Newtonian fluids in this flow, $\tau_{xx} = \tau_{yy} = \tau_{zz} = 0$, but for polymeric fluids the normal stresses are in general nonzero and unequal.

[1] J. R. Van Wazer, J. W. Lyons, K. Y. Kim, and R. E. Colwell, *Viscosity and Flow Measurement*, Interscience (Wiley), New York (1963).

[2] K. Walters, *Rheometry*, Wiley, New York (1975).

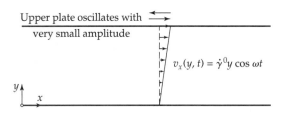

Upper plate oscillates with
very small amplitude

$v_x(y, t) = \dot{\gamma}^0 y \cos \omega t$

Fig. 8.2-2. Small-amplitude oscillatory motion. For small plate spacing and highly viscous fluids, the velocity profile may be *assumed* to be linear.

Many ingenious devices have been developed to measure the three material functions for steady shearing flow, and the theories needed for the use of the instruments are explained in detail elsewhere.[2] See Problem 8C.1 for the use of the cone-and-plate instrument for measuring the material functions.

Small-Amplitude Oscillatory Motion

A standard method for measuring the elastic response of a fluid is the small-amplitude oscillatory shear experiment, depicted in Fig. 8.2-2. Here the top plate moves back and forth in sinusoidal fashion, and with a tiny amplitude. If the plate spacing is extremely small and the fluid has a very high viscosity, then the velocity profile will be nearly linear, so that $v_x(y, t) = \dot{\gamma}^0 y \cos \omega t$, in which $\dot{\gamma}^0$, a real quantity, gives the amplitude of the shear rate excursion.

The shear stress required to maintain the oscillatory motion will also be periodic in time and, in general, of the form

$$\tau_{yx} = -\eta' \dot{\gamma}^0 \cos \omega t - \eta'' \dot{\gamma}^0 \sin \omega t \qquad (8.2\text{-}4)$$

in which η' and η'' are the components of the *complex viscosity*, $\eta^* = \eta' - i\eta''$, which is a function of the frequency. The first (in-phase) term is the "viscous response," and the second (out-of-phase) term is the "elastic response." Polymer chemists use the curves of $\eta'(\omega)$ and $\eta''(\omega)$ (or the storage and loss moduli, $G' = \eta''\omega$ and $G'' = \eta'\omega$) for "characterizing" polymers, since much is known about the connection between the shapes of these curves and the chemical structure.[3] For the Newtonian fluid, $\eta' = \mu$ and $\eta'' = 0$.

Steady-State Elongational Flow

A third experiment that can be performed involves the stretching of the fluid, in which the velocity distribution is given by $v_z = \dot{\varepsilon}z$, $v_x = -\frac{1}{2}\dot{\varepsilon}x$, and $v_y = -\frac{1}{2}\dot{\varepsilon}y$ (see Fig. 8.2-3), where the positive quantity $\dot{\varepsilon}$ is called the "elongation rate." Then the relation

$$\tau_{zz} - \tau_{xx} = -\bar{\eta} \frac{dv_z}{dz} \qquad (8.2\text{-}5)$$

defines the *elongational viscosity* $\bar{\eta}$, which depends on $\dot{\varepsilon}$. When $\dot{\varepsilon}$ is negative, the flow is referred to as *biaxial stretching*. For the Newtonian fluid it can be shown that $\bar{\eta} = 3\mu$, and this is sometimes called the "Trouton viscosity."

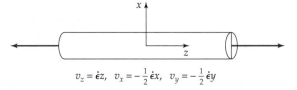

$v_z = \dot{\varepsilon}z, \quad v_x = -\frac{1}{2}\dot{\varepsilon}x, \quad v_y = -\frac{1}{2}\dot{\varepsilon}y$

Fig. 8.2-3. Steady elongational flow with elongation rate $\dot{\varepsilon} = dv_z/dz$.

[3] J. D. Ferry, *Viscoelastic Properties of Polymers*, Wiley, New York, 3rd edition (1980).

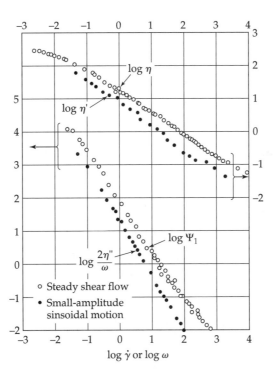

Fig. 8.2-4. The material functions $\eta(\dot{\gamma})$, $\Psi_1(\dot{\gamma})$, $\eta'(\omega)$, and $\eta''(\omega)$ for a 1.5% polyacrylamide solution in a 50/50 mixture of water and glycerin. The quantities η, η', and η'' are given in Pa · s, and Ψ_1 in Pa · s². Both $\dot{\gamma}$ and ω are given in s⁻¹. The data are from J. D. Huppler, E. Ashare, and L. Holmes, *Trans. Soc. Rheol.*, **11**, 159–179 (1967), as replotted by J. M. Wiest. The oscillatory normal stresses have also been studied experimentally and theoretically (see M. C. Williams and R. B. Bird, *Ind. Eng. Chem. Fundam.*, **3**, 42–48 (1964); M. C. Williams, *J. Chem. Phys.*, **42**, 2988–2989 (1965); E. B. Christiansen and W. R. Leppard, *Trans. Soc. Rheol.*, **18**, 65–86 (1974), in which the ordinate of Fig. 15 should be multiplied by 39.27.

The elongational viscosity $\bar{\eta}$ cannot be measured for all fluids, since a steady-state elongational flow cannot always be attained.[4]

The three experiments described above are only a few of the rheometric tests that can be performed. Other tests include stress relaxation after cessation of flow, stress growth at the inception of flow, recoil, and creep—each of which can be performed in shear, elongation, and other types of flow. Each experiment results in the definition of one or more material functions. These can be used for fluid characterization and also for determining the empirical constants in the models described in §§8.3 to 8.5.

Some sample material functions are displayed in Figs. 8.2-4 to 8.2-6. Since there is a wide range of complex fluids, as regards chemical structure and constitution,

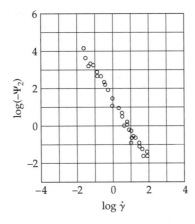

Fig. 8.2-5. Dependence of the second normal stress coefficient on shear rate for a 2.5% solution of polyacrylamide in a 50/50 mixture of water and glycerin. The quantity Ψ_2 is given in Pa · s², and ω is in s⁻¹. The data of E. B. Christiansen and W. R. Leppard, *Trans. Soc. Rheol.*, **18**, 65–86 (1974), have been replotted by J. M. Wiest.

---

[4] C. J. S. Petrie, *Elongational Flows*, Pitman, London (1979); J. Meissner, *Chem. Engr. Commun.*, **33**, 159–180 (1985).

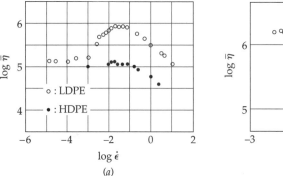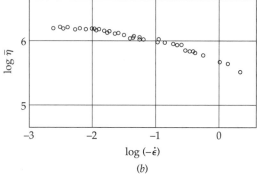

Fig. 8.2-6. (*a*) Elongational viscosity for uniaxial stretching of low- and high-density polyethylene. [From H. Münstedt and H. M. Laun, *Rheol. Acta*, **20**, 211–221 (1981).] (*b*) Elongational viscosity for biaxial stretching of low-density polyethylene, deduced from flow-birefringence data. [From J. A. van Aken and H. Janeschitz-Kriegl, *Rheol. Acta*, **20**, 419–432 (1981).] In both graphs the quantity $\bar{\eta}$ is given in Pa · s and $\dot{\varepsilon}$ is in s^{-1}.

there are many types of mechanical responses in these various experiments. More complete discussions of the data obtained in rheometric experiments are given elsewhere.[5]

§8.3 NON-NEWTONIAN VISCOSITY AND THE GENERALIZED NEWTONIAN MODELS

This is the first of three sections devoted to empirical stress tensor expressions for non-Newtonian fluids. One might say, very roughly, that these three sections satisfy three different groups of people:

§8.3 The generalized Newtonian models are primarily used to describe steady-state shear flows and have been widely used by *engineers* for designing flow systems.

§8.4 The linear viscoelastic models are primarily used to describe unsteady-state flows in systems with very small displacement gradients and have been used mainly by *chemists* interested in understanding polymer structure.

§8.5 The nonlinear viscoelastic models represent an attempt to describe all types of flow (including the two listed above) and have been developed largely by *physicists* and *applied mathematicians* interested in finding an all-inclusive theory.

Actually the three classes of models are interrelated, and each is important for understanding the subject of non-Newtonian flow. In the following discussion of non-Newtonian models, we assume throughout that the fluids are incompressible.

The *generalized Newtonian models*[1] discussed here are the simplest of the three types of models to be discussed. However, they can describe only the non-Newtonian viscosity, and none of the normal stress effects, time-dependent effects, or elastic effects. Nonethe-

---

[5] R. B. Bird, R. C. Armstrong, and O. Hassager, *Dynamics of Polymeric Liquids, Vol. 1, Fluid Mechanics*, Wiley-Interscience, 2nd edition (1987).

[1] K. Hohenemser and W. Prager, *Zeits. f. Math. u. Mech.*, **12**, 216–226 (1932); J. G. Oldroyd, *Proc. Camb. Phil. Soc.*, **45**, 595–611 (1949), and **47**, 410–418 (1950). **James Gardner Oldroyd** (1921–1982), a professor at the University of Liverpool, made many contributions to the theory of non-Newtonian fluids, in particular his ideas on the construction of constitutive equations and the principles of continuum mechanics.

less, in many processes in the polymer industry, such as pipe flow with heat transfer, distributor design, extrusion, and injection molding, the non-Newtonian viscosity and its enormous variation with shear rate are central to describing the flows of interest.

For incompressible Newtonian fluids the expression for the stress tensor is given by Eq. 1.2-7 with the last term omitted:

$$\boldsymbol{\tau} = -\mu(\nabla\mathbf{v} + (\nabla\mathbf{v})^\dagger) \equiv -\mu\dot{\boldsymbol{\gamma}} \tag{8.3-1}$$

in which we have introduced the symbol $\dot{\boldsymbol{\gamma}} = \nabla\mathbf{v} + (\nabla\mathbf{v})^\dagger$, the *rate-of-strain tensor* (or *rate-of-deformation tensor*). The generalized Newtonian fluid model is obtained by simply replacing the constant viscosity μ by the non-Newtonian viscosity η, a function of the shear rate, which in general can be written as the "magnitude of the rate-of-strain tensor" $\dot{\gamma} = \sqrt{\frac{1}{2}(\dot{\boldsymbol{\gamma}}:\dot{\boldsymbol{\gamma}})}$; it is understood that when the square root is taken, the sign must be so chosen that $\dot{\gamma}$ is a positive quantity. Then the generalized Newtonian fluid model is

$$\boldsymbol{\tau} = -\eta(\nabla\mathbf{v} + (\nabla\mathbf{v})^\dagger) \equiv -\eta\dot{\boldsymbol{\gamma}} \quad \text{with } \eta = \eta(\dot{\gamma}) \tag{8.3-2}$$

The components of the rate-of-strain tensor $\dot{\boldsymbol{\gamma}}$ can be obtained in Cartesian, cylindrical, and spherical coordinates from the right sides of the equations in Table B.1 by omitting the $(\nabla \cdot \mathbf{v})$ terms as well as the factor $(-\mu)$ in the remaining terms.

We now have to give an empiricism for the non-Newtonian viscosity function $\eta(\dot{\gamma})$. Dozens of such expressions have been proposed, but we mention only two here:

(a) The simplest empiricism for $\eta(\dot{\gamma})$ is the two-parameter *power law* expression:[2]

$$\eta = m\dot{\gamma}^{n-1} \tag{8.3-3}$$

in which m and n are constants characterizing the fluid. This simple relation describes the non-Newtonian viscosity curve over the linear portion of the log-log plot of the viscosity versus shear rate for many materials (see, for example, the viscosity data in Fig. 8.2-4). The parameter m has units of Pa $\cdot$ s^n, and $n - 1$ is the slope of the log η vs. log $\dot{\gamma}$ plot. Some sample values of power law parameters are given in Table 8.3-1.

Although the power law model was proposed as an empirical expression, it will be seen in Eq. 8.6-11 that a simple molecular theory leads to a power law expression for high shear rates, with $n = \frac{1}{3}$.

Table 8.3-1 Power Law Parameters for Aqueous Solutions[a]

Solution	Temperature (K)	m(Pa $\cdot$ s^n)	n(—)
2.0% hydroxyethylcellulose	293	93.5	0.189
	313	59.7	0.223
	333	38.5	0.254
0.5% hydroxyethylcellulose	293	0.84	0.509
	313	0.30	0.595
	333	0.136	0.645
1.0% polyethylene oxide	293	0.994	0.532
	313	0.706	0.544
	333	0.486	0.599

[a] R. M. Turian, Ph.D. Thesis, University of Wisconsin, Madison (1964), pp. 142–148.

[2] W. Ostwald, *Kolloid-Zeitschrift*, **36**, 99–117 (1925); A. de Waele, *Oil Color Chem. Assoc. J.*, **6**, 33–88 (1923).

Table 8.3-2   Parameters in the Carreau Model for Some Solutions of Linear Polystyrene in 1-Chloronaphthalene[a]

Properties of solution		Parameters in Eq. 8.3-4 (η_∞ is taken to be zero)		
$\overline{M}_w$ (g/mol)	c (g/ml)	η_0 (Pa · s)	λ (s)	n (- - -)
3.9×10^5	0.45	8080	1.109	0.304
3.9×10^5	0.30	135	3.61×10^{-2}	0.305
1.1×10^5	0.52	1180	9.24×10^{-2}	0.441
1.1×10^5	0.45	166	1.73×10^{-2}	0.538
3.7×10^4	0.62	3930	1×10^{-1}	0.217

[a] Values of the parameters are taken from K. Yasuda, R. C. Armstrong, and R. E. Cohen, *Rheol. Acta*, **20**, 163–178 (1981).

(b) A better curve fit for most data can be obtained by using the four-parameter *Carreau equation*,[3] which is

$$\frac{\eta - \eta_\infty}{\eta_0 - \eta_\infty} = [1 + (\lambda \dot{\gamma})^2]^{(n-1)/2} \tag{8.3-4}$$

in which η_0 is the zero shear rate viscosity, η_∞ is the infinite shear rate viscosity, λ is a parameter with units of time, and n is a dimensionless parameter. Some sample parameters for the Carreau model are given in Table 8.3-2.

We now give some examples of how to use the power law model. These are extensions of problems discussed in Chapters 2 and 3 for Newtonian fluids.[4]

EXAMPLE 8.3-1

Laminar Flow of an Incompressible Power Law Fluid in a Circular Tube[4,5]

Derive the expression for the mass flow rate of a polymer liquid, described by the power law model. The fluid is flowing in a long circular tube of radius R and length L, as a result of a pressure difference, gravity, or both.

SOLUTION

Equation 2.3-13 gives the shear stress distribution for any fluid in developed steady flow in a circular tube. Into this expression we have to insert the shear stress for the power law fluid (instead of using Eq. 2.3-14). This expression may be obtained from Eqs. 8.3-2 and 3 above.

$$\tau_{rz} = -m\dot{\gamma}^{n-1}\frac{dv_z}{dr} \tag{8.3-5}$$

Since v_z is postulated to be a function of r alone, from Eq. B.1-13 we find that $\dot{\gamma} = \sqrt{\tfrac{1}{2}(\dot{\gamma}:\dot{\gamma})} = \sqrt{(dv_z/dr)^2}$. We have to choose the sign for the square root so that $\dot{\gamma}$ will be positive. Since dv_z/dr is negative in tube flow, we have to choose the minus sign, so that

$$\tau_{rz} = -m\left(-\frac{dv_z}{dr}\right)^{n-1}\frac{dv_z}{dr} = m\left(-\frac{dv_z}{dr}\right)^n \tag{8.3-6}$$

---

[3] P. J. Carreau, Ph.D. thesis, University of Wisconsin, Madison (1968). See also K. Yasuda, R. C. Armstrong, and R. E. Cohen, *Rheol. Acta*, **20**, 163–178 (1981).

[4] For additional examples, including nonisothermal flows, see R. B. Bird, R. C. Armstrong, and O. Hassager, *Dynamics of Polymeric Liquids, Vol. 1. Fluid Mechanics*, Wiley-Interscience, New York, 2nd edition (1998), Chapter 4.

[5] M. Reiner, *Deformation, Strain and Flow*, Interscience, New York, 2nd edition (1960), pp. 243–245.

Combining Eq. 8.3-6 and 2.3-13 then gives the following differential equation for the velocity:

$$m\left(-\frac{dv_z}{dr}\right)^n = \left(\frac{\mathcal{P}_0 - \mathcal{P}_L}{2L}\right)r \tag{8.3-7}$$

After taking the nth root the equation may be integrated, and when the no-slip boundary condition at $r = R$ is used, we get

$$v_z = \left(\frac{(\mathcal{P}_0 - \mathcal{P}_L)R}{2mL}\right)^{1/n} \frac{R}{(1/n) + 1}\left[1 - \left(\frac{r}{R}\right)^{(1/n)+1}\right] \tag{8.3-8}$$

for the velocity distribution (see Eq. 8.1-1). When this is integrated over the cross section of the circular tube we get

$$w = \frac{\pi R^3 \rho}{(1/n) + 3}\left(\frac{(\mathcal{P}_0 - \mathcal{P}_L)R}{2mL}\right)^{1/n} \tag{8.3-9}$$

which simplifies to the Hagen–Poiseuille law for Newtonian fluids (Eq. 2.3-21) when $n = 1$ and $m = \mu$. Equation 8.3-9 can be used along with data on pressure drop versus flow rate to determine the power law parameters m and n.

EXAMPLE 8.3-2

Flow of a Power Law Fluid in a Narrow Slit[4]

The flow of a Newtonian fluid in a narrow slit is solved in Problem 2B.3. Find the velocity distribution and the mass flow rate for a power law fluid flowing in the slit.

SOLUTION

The expression for the shear stress τ_{xz} as a function of position x in Eq. 2B.3-1 can be taken over here, since it does not depend on the type of fluid. The power law formula for τ_{xz} from Eq. 8.3-3 is

$$\tau_{xz} = m\left(-\frac{dv_z}{dx}\right)^n \qquad \text{for } 0 \leq x \leq B \tag{8.3-10}$$

$$\tau_{xz} = -m\left(\frac{dv_z}{dx}\right)^n \qquad \text{for } -B \leq x \leq 0 \tag{8.3-11}$$

To get the velocity distribution for $0 \leq x \leq B$, we substitute τ_{xz} from Eq. 8.3-10 into Eq. 2B.3-1 to get:

$$m\left(-\frac{dv_z}{dx}\right)^n = \frac{(\mathcal{P}_0 - \mathcal{P}_L)x}{L} \qquad 0 \leq x \leq B \tag{8.3-12}$$

Integrating and using the no-slip boundary condition at $x = B$ gives

$$v_z = \left(\frac{(\mathcal{P}_0 - \mathcal{P}_L)B}{mL}\right)^{1/n} \frac{B}{(1/n) + 1}\left[1 - \left(\frac{x}{B}\right)^{(1/n)+1}\right] \qquad 0 \leq x \leq B \tag{8.3-13}$$

Since we expect the velocity profile to be symmetric about the midplane $x = 0$, we can get the mass rate of flow as follows:

$$w = \int_0^W \int_{-B}^B \rho v_z dx\, dy = 2\int_0^W \int_0^B \rho v_z dx\, dy$$

$$= 2\left(\frac{(\mathcal{P}_0 - \mathcal{P}_L)B}{mL}\right)^{1/n} \frac{WB^2\rho}{(1/n) + 1}\int_0^1\left[1 - \left(\frac{x}{B}\right)^{(1/n)+1}\right]d\left(\frac{x}{B}\right)$$

$$= \frac{2WB^2\rho}{(1/n) + 2}\left(\frac{(\mathcal{P}_0 - \mathcal{P}_L)B}{mL}\right)^{1/n} \tag{8.3-14}$$

When $n = 1$ and $m = \mu$, the Newtonian result in Problem 2B.3 is recovered. Experimental data on pressure drop and mass flow rate through a narrow slit can be used with Eq. 8.3-14 to determine the power law parameters.

EXAMPLE 8.3-3	Rework Example 3.6-3 for a power law fluid.

Tangential Annular Flow of a Power Law Fluid[4,5]

SOLUTION

Equations 3.6-20 and 3.6-22 remain unchanged for a non-Newtonian fluid, but in lieu of Eq. 3.6-21 we write the θ-component of the equation of motion in terms of the shear stress by using Table B.5:

$$0 = -\frac{1}{r^2}\frac{d}{dr}(r^2\tau_{r\theta}) \tag{8.3-15}$$

For the postulated velocity profile, we get for the power law model (with the help of Table B.1)

$$\tau_{r\theta} = -\eta r \frac{d}{dr}\left(\frac{v_\theta}{r}\right)$$

$$= -m\left(r\frac{d}{dr}\left(\frac{v_\theta}{r}\right)\right)^{n-1} r\frac{d}{dr}\left(\frac{v_\theta}{r}\right)$$

$$= -m\left(r\frac{d}{dr}\left(\frac{v_\theta}{r}\right)\right)^{n} \tag{8.3-16}$$

Combining Eqs. 8.3-15 and 16 we get

$$\frac{d}{dr}\left(r^2 m\left(r\frac{d}{dr}\left(\frac{v_\theta}{r}\right)\right)^{n}\right) = 0 \tag{8.3-17}$$

Integration gives

$$r^2\left(r\frac{d}{dr}\left(\frac{v_\theta}{r}\right)\right)^{n} = C_1 \tag{8.3-18}$$

Dividing by r^2 and taking the nth root gives a first-order differential equation for the angular velocity

$$\frac{d}{dr}\left(\frac{v_\theta}{r}\right) = \frac{1}{r}\left(\frac{C_1}{r^2}\right)^{1/n} \tag{8.3-19}$$

This may be integrated with the boundary conditions in Eqs. 3.6-27 and 28 to give

$$\frac{v_\theta}{\Omega_o r} = \frac{1 - (\kappa R/r)^{2/n}}{1 - \kappa^{2/n}} \tag{8.3-20}$$

The (z-component of the) torque needed on the outer cylinder to maintain the motion is then

$$T_z = (-\tau_{r\theta})|_{r=R} \cdot 2\pi RL \cdot R$$

$$= m\left(r\frac{d}{dr}\left(\frac{v_\theta}{r}\right)\right)^{n}\bigg|_{r=R} \cdot 2\pi RL \cdot R \tag{8.3-21}$$

Combining Eqs. 8.3-20 and 21 then gives

$$T_z = 2\pi m\Omega_o(\kappa R)^2 L\left(\frac{(2/n)}{1 - \kappa^{2/n}}\right)^{n} \tag{8.3-22}$$

The Newtonian result can be recovered by setting $n = 1$ and $m = \mu$. Equation 8.3-22 can be used along with torque versus angular velocity data to determine the power law parameters m and n.

§8.4   ELASTICITY AND THE LINEAR VISCOELASTIC MODELS

Just after Eq. 1.2-3, in the discussion about generalizing Newton's "law of viscosity," we specifically excluded time derivatives and time integrals in the construction of a linear expression for the stress tensor in terms of the velocity gradients. In this section, we allow for the inclusion of time derivatives or time integrals, but still require a linear relation between $\boldsymbol{\tau}$ and $\dot{\boldsymbol{\gamma}}$. This leads to *linear viscoelastic* models.

We start by writing Newton's expression for the stress tensor for an incompressible viscous liquid along with Hooke's analogous expression for the stress tensor for an incompressible elastic solid:[1]

Newton:
$$\boldsymbol{\tau} = -\mu(\nabla\mathbf{v} + (\nabla\mathbf{v})^\dagger) \equiv -\mu\dot{\boldsymbol{\gamma}} \tag{8.4-1}$$

Hooke:
$$\boldsymbol{\tau} = -G(\nabla\mathbf{u} + (\nabla\mathbf{u})^\dagger) \equiv -G\boldsymbol{\gamma} \tag{8.4-2}$$

In the second of these expressions G is the elastic modulus, and $\mathbf{u}$ is the "displacement vector," which gives the distance and direction that a point in the solid has moved from its initial position as a result of the applied stresses. The quantity $\boldsymbol{\gamma}$ is called the "infinitesimal strain tensor." The rate-of-strain tensor and the infinitesimal strain tensor are related by $\dot{\boldsymbol{\gamma}} = \partial\boldsymbol{\gamma}/\partial t$. The Hookean solid has a perfect memory; when imposed stresses are removed, the solid returns to its initial configuration. Hooke's law is valid only for very small displacement gradients, $\nabla\mathbf{u}$. Now we want to combine the ideas embodied in Eqs. 8.4-1 and 2 to describe viscoelastic fluids.

The Maxwell Model

The simplest equation for describing a fluid that is both viscous and elastic is the following *Maxwell model*:[2]

$$\boldsymbol{\tau} + \lambda_1 \frac{\partial}{\partial t}\boldsymbol{\tau} = -\eta_0\dot{\boldsymbol{\gamma}} \tag{8.4-3}$$

Here λ_1 is a time constant (the *relaxation time*) and η_0 is the *zero shear rate viscosity*. When the stress tensor changes imperceptibly with time, then Eq. 8.4-3 has the form of Eq. 8.4-1 for a Newtonian liquid. When there are very rapid changes in the stress tensor with time, then the first term on the left side of Eq. 8.4-3 can be omitted, and when the equation is integrated with respect to time, we get an equation of the form of Eq. 8.4-2 for the Hookean solid. In that sense, Eq. 8.4-3 incorporates both viscosity and elasticity.

A simple experiment that illustrates the behavior of a viscoelastic liquid involves "silly putty." This material flows easily when squeezed slowly between the palms of the hands, and this indicates that it is a viscous fluid. However, when it is rolled into a ball, the ball will bounce when dropped onto a hard surface. During the impact the stresses change rapidly, and the material behaves as an elastic solid.

The Jeffreys Model

The Maxwell model of Eq. 8.4-3 is a linear relation between the stresses and the velocity gradients, involving a time derivative of the stresses. One could also include a time derivative of the velocity gradients and still have a linear relation:

$$\boldsymbol{\tau} + \lambda_1 \frac{\partial}{\partial t}\boldsymbol{\tau} = -\eta_0\left(\dot{\boldsymbol{\gamma}} + \lambda_2 \frac{\partial}{\partial t}\dot{\boldsymbol{\gamma}}\right) \tag{8.4-4}$$

---

[1] R. Hooke, *Lectures de Potentia Restitutiva* (1678).
[2] This relation was proposed by J. C. Maxwell, *Phil. Trans. Roy. Soc.*, **A157**, 49–88 (1867), to investigate the possibility that gases might be viscoelastic.

This *Jeffreys model*[3] contains three constants: the zero shear rate viscosity and two time constants (the constant λ_2 is called the *retardation time*).

One could clearly add terms containing second, third, and higher derivatives of the stress and rate-of-strain tensors with appropriate multiplicative constants, to get a still more general linear relation among the stress and rate-of-strain tensors. This gives greater flexibility in fitting experimental data.

The Generalized Maxwell Model

Another way of generalizing Maxwell's original idea is to "superpose" equations of the form of Eq. 8.4-3 and write the *generalized Maxwell model* as

$$\boldsymbol{\tau}(t) = \sum_{k=1}^{\infty} \boldsymbol{\tau}_k(t) \qquad \text{where } \boldsymbol{\tau}_k + \lambda_k \frac{\partial}{\partial t} \boldsymbol{\tau}_k = -\eta_k \dot{\boldsymbol{\gamma}} \qquad (8.4\text{-}5, 6)$$

in which there are many relaxation times λ_k (with $\lambda_1 \geq \lambda_2 \geq \lambda_3 \ldots$) and many constants η_k with dimensions of viscosity. Much is known about the constants in this model from polymer molecular theories and the extensive experiments that have been done on polymeric liquids.[4]

The total number of parameters can be reduced to three by using the following empirical expressions:[5]

$$\eta_k = \eta_0 \frac{\lambda_k}{\sum_j \lambda_j} \quad \text{and} \quad \lambda_k = \frac{\lambda}{k^\alpha} \qquad (8.4\text{-}7, 8)$$

in which η_0 is the zero shear rate viscosity, λ is a time constant, and α is a dimensionless constant (usually between 1.5 and 4).

Since Eq. 8.4-6 is a linear differential equation, it can be integrated analytically, with the condition that the fluid is at rest at $t = -\infty$. Then when the various $\boldsymbol{\tau}_k$ are summed according to Eq. 8.4-5, we get the integral form of the generalized Maxwell model:

$$\boldsymbol{\tau}(t) = -\int_{-\infty}^{t} \left\{ \sum_{k=1}^{\infty} \frac{\eta_k}{\lambda_k} \exp[-(t - t')/\lambda_k] \right\} \dot{\boldsymbol{\gamma}}(t') dt' = -\int_{-\infty}^{t} G(t - t') \, \dot{\boldsymbol{\gamma}}(t') dt' \qquad (8.4\text{-}9)$$

In this form, the "fading memory" idea is clearly present: the stress at time t depends on the velocity gradients at all past times t', but, because of the exponentials in the integrand, greatest weight is given to times t' that are near t; that is, the fluid "memory" is better for recent times than for more remote times in the past. The quantity within braces { } is called the *relaxation modulus* of the fluid and is denoted by $G(t - t')$. The integral ex-

---

[3] This model was suggested by H. Jeffreys, *The Earth*, Cambridge University Press, 1st edition (1924), and 2nd edition (1929), p. 265, to describe the propagation of waves in the earth's mantle. The parameters in this model have been related to the structure of suspensions and emulsions by H. Fröhlich and R. Sack, *Proc. Roy. Soc.*, **A185**, 415–430 (1946) and by J. G. Oldroyd, *Proc. Roy. Soc.*, **A218**, 122–132 (1953), respectively. Another interpretation of Eq. 8.4-4 is to regard it as the sum of a Newtonian solvent contribution (*s*) and a polymer contribution (*p*), the latter being described by a Maxwell model:

$$\boldsymbol{\tau}_s = -\eta_s \dot{\boldsymbol{\gamma}}; \qquad \boldsymbol{\tau}_p + \lambda_1 \frac{\partial}{\partial t} \boldsymbol{\tau}_p = -\eta_p \dot{\boldsymbol{\gamma}} \qquad (8.4\text{-}4a, b)$$

so that $\boldsymbol{\tau} = \boldsymbol{\tau}_s + \boldsymbol{\tau}_p$. Then if Eqs. 8.4-4a, 8.4-4b, and λ_1 times the time derivative of Eq. 8.4-4a are added, we get the Jeffreys model of Eq. 8.4-4, with $\eta_0 = \eta_s + \eta_p$ and $\lambda_2 = (\eta_s/(\eta_s + \eta_p))\lambda_1$.

[4] J. D. Ferry, *Viscoelastic Properties of Polymers*, Wiley, New York, 3rd edition (1980). See also N. W. Tschoegl, *The Phenomenological Theory of Linear Viscoelastic Behavior*, Springer-Verlag, Berlin (1989); and R. B. Bird, R. C. Armstrong, and O. Hassager, *Dynamics of Polymeric Liquids, Vol. 1, Fluid Mechanics*, Wiley-Interscience, New York, 2nd edition (1987), Chapter 5.

[5] T. W. Spriggs, *Chem. Eng. Sci.*, **20**, 931–940 (1965).

pression in Eq. 8.4-9 is sometimes more convenient for solving linear viscoelastic problems than are the differential equations in Eqs. 8.4-5 and 6.

The Maxwell, Jeffreys, and generalized Maxwell models are all examples of linear viscoelastic models, and their use is restricted to motions with very small displacement gradients. Polymeric liquids have many internal degrees of freedom and therefore many relaxation times are needed to describe their linear response. For this reason, the generalized Maxwell model has been widely used for interpreting experimental data on linear viscoelasticity. By fitting Eq. 8.4-9 to experimental data one can determine the relaxation function $G(t - t')$. One can then relate the shapes of the relaxation functions to the molecular structure of the polymer. In this way a sort of "mechanical spectroscopy" is developed, which can be used to investigate structure via linear viscoelastic measurements (such as the complex viscosity).

Models describing flows with very small displacement gradients might seem to have only limited interest to engineers. However, an important reason for studying them is that some background in linear viscoelasticity helps us in the study of nonlinear viscoelasticity, where flows with large displacement gradients are discussed.

EXAMPLE 8.4-1

Small-Amplitude Oscillatory Motion

Obtain an expression for the components of the complex viscosity by using the generalized Maxwell model. The system is described in Fig. 8.2-2.

SOLUTION

We use the yx-component of Eq. 8.4-9, and for this problem the yx-component of the rate-of-strain tensor is

$$\dot{\gamma}_{yx}(t) = \frac{\partial v_x}{\partial y} = \dot{\gamma}^0 \cos \omega t \tag{8.4-10}$$

where ω is the angular frequency. When this is substituted into Eq. 8.4-9, with the relaxation modulus (in braces) expressed as $G(t - t')$, we get

$$\tau_{yx} = -\int_{-\infty}^{t} G(t - t')\dot{\gamma}^0 \cos \omega t' dt'$$

$$= -\dot{\gamma}^0 \int_{0}^{\infty} G(s) \cos \omega(t - s) ds$$

$$= -\dot{\gamma}^0 \left[\int_{0}^{\infty} G(s) \cos \omega s \, ds\right] \cos \omega t - \dot{\gamma}^0 \left[\int_{0}^{\infty} G(s) \sin \omega s \, ds\right] \sin \omega t \tag{8.4-11}$$

in which $s = t - t'$. When this equation is compared with Eq. 8.2-4, we obtain

$$\eta'(\omega) = \int_{0}^{\infty} G(s) \cos \omega s \, ds \tag{8.4-12}$$

$$\eta''(\omega) = \int_{0}^{\infty} G(s) \sin \omega s \, ds \tag{8.4-13}$$

for the components of the complex viscosity $\eta^* = \eta' - i\eta''$. When the generalized Maxwell expression for the relaxation modulus is introduced and the integrals are evaluated, we find that

$$\eta'(\omega) = \sum_{k=1}^{\infty} \frac{\eta_k}{1 + (\lambda_k \omega)^2} \tag{8.4-14}$$

$$\eta''(\omega) = \sum_{k=1}^{\infty} \frac{\eta_k \lambda_k \omega}{1 + (\lambda_k \omega)^2} \tag{8.4-15}$$

If the empiricisms in Eqs. 8.4-7 and 8 are used, it can be shown that both η' and η'' decrease as $1/\omega^{1-(1/\alpha)}$ at very high frequencies (see Fig. 8.2-4).

EXAMPLE 8.4-2

Unsteady Viscoelastic Flow Near an Oscillating Plate

Extend Example 4.1-3 to viscoelastic fluids, using the Maxwell model, and obtain the attenuation and phase shift in the "periodic steady state."

SOLUTION

For the postulated shearing flow, the equation of motion, written in terms of the stress tensor component gives

$$\rho \frac{\partial v_x}{\partial t} = -\frac{\partial}{\partial y} \tau_{yx} \tag{8.4-16}$$

The Maxwell model in integral form is like Eq. 8.4-9, but with a single exponential:

$$\tau_{yx}(y, t) = -\int_{-\infty}^{t} \left\{ \frac{\eta_0}{\lambda_1} \exp[-(t - t')/\lambda_1] \right\} \frac{\partial v_x(y, t')}{\partial y} dt' \tag{8.4-17}$$

Combining these two equations, we get

$$\rho \frac{\partial v_x}{\partial t} = \int_{-\infty}^{t} \left\{ \frac{\eta_0}{\lambda_1} \exp[-(t - t')/\lambda_1] \right\} \frac{\partial^2 v_x(y, t')}{\partial y^2} dt' \tag{8.4-18}$$

As in Example 4.1-3 we postulate a solution of the form

$$v_x(y, t) = \Re\{v^0(y)e^{i\omega t}\} \tag{8.4-19}$$

where $v^0(y)$ is complex. Substituting this into Eq. 8.4-19, we get

$$\rho \Re\{i\omega v^0 e^{i\omega t}\} = \int_{-\infty}^{t} \left\{ \frac{\eta_0}{\lambda_1} \exp[-(t - t')/\lambda_1] \right\} \Re\left\{ \frac{d^2 v^0}{dy^2} e^{i\omega t'} \right\} dt'$$

$$= \Re\left\{ \frac{d^2 v^0}{dy^2} e^{i\omega t} \int_{0}^{\infty} \frac{\eta_0}{\lambda_1} e^{-s/\lambda_1} e^{-i\omega s} ds \right\}$$

$$= \Re\left\{ \frac{d^2 v^0}{dy^2} e^{i\omega t} \left[\frac{\eta_0}{1 + i\lambda_1 \omega} \right] \right\} \tag{8.4-20}$$

Removing the real operator then gives an equation for $v^0(y)$

$$\frac{d^2 v^0}{dy^2} - \left[\frac{i\rho\omega(1 + i\lambda_1\omega)}{\eta_0} \right] v^0 = 0 \tag{8.4-21}$$

Then if the complex quantity in the brackets [ ] is set equal to $(\alpha + i\beta)^2$, the solution to the differential equation is

$$v^0 = v_0 e^{-(\alpha + i\beta)y} \tag{8.4-22}$$

Multiplying this by $e^{i\omega t}$ and taking the real part gives

$$v_x(y, t) = v_0 e^{-\alpha y} \cos(\omega t - \beta y) \tag{8.4-23}$$

This result has the same form as that in Eq. 4.1-57, but the quantities α and β depend on frequency:

$$\alpha(\omega) = \sqrt{\frac{\rho\omega}{2\eta_0}} \left[\sqrt{1 + (\lambda_1\omega)^2} - \lambda_1\omega \right]^{+1/2} \tag{8.4-24}$$

$$\beta(\omega) = \sqrt{\frac{\rho\omega}{2\eta_0}} \left[\sqrt{1 + (\lambda_1\omega)^2} - \lambda_1\omega \right]^{-1/2} \tag{8.4-25}$$

That is, with increasing frequency, α decreases and β increases, because of the fluid elasticity. This result shows how elasticity affects the transmission of shear waves near an oscillating surface.

Note that there is an important difference between the problems in the last two examples. In Example 8.4-1 the velocity profile is prescribed, and we have derived an expression for the shear stress required to maintain the motion; the equation of motion was not used. In Example 8.4-2 no assumption was made about the velocity distribution, and we derived the velocity distribution by using the equation of motion.

§8.5   THE COROTATIONAL DERIVATIVES AND THE NONLINEAR VISCOELASTIC MODELS

In the previous section it was shown that the inclusion of time derivatives (or time integrals) in the stress tensor expression allows for the description of elastic effects. The linear viscoelastic models can describe the complex viscosity and the transmission of small-amplitude shearing waves. It can also be shown that the linear models can describe elastic recoil, although the results are restricted to flows with negligible displacement gradients (and hence of little practical interest).

In this section we introduce the hypothesis[1,2] that the relation between the stress tensor and the kinematic tensors at a fluid particle should be independent of the instantaneous orientation of that particle in space. This seems like a reasonable hypothesis; if you measure the stress–strain relation in a rubber band, it should not matter whether you are stretching the rubber band in the north–south direction or the east–west direction, or even rotating as you take data (provided, of course, that you do not rotate so rapidly that centrifugal forces interfere with the measurements).

One way to implement the above hypothesis is to introduce at each fluid particle a corotating coordinate frame. This orthogonal frame rotates with the local instantaneous angular velocity as it moves along with the fluid particle through space (see Fig. 8.5-1). In the corotating coordinate system we can now write down some kind of relation

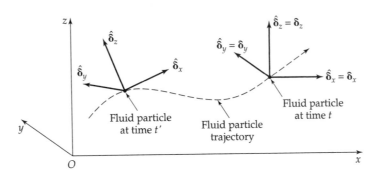

Fig. 8.5-1. Fixed coordinate frame with origin at O, and the corotating frame with unit vectors $\hat{\boldsymbol{\delta}}_1, \hat{\boldsymbol{\delta}}_2, \hat{\boldsymbol{\delta}}_3$ that move with a fluid particle and rotate with the local, instantaneous angular velocity $\frac{1}{2}[\nabla \times \mathbf{v}]$ of the fluid.

[1] G. Jaumann, *Grundlagen der Bewegungslehre*, Leipzig (1905); *Sitzungsberichte Akad. Wiss. Wien*, IIa, **120**, 385–530 (1911); S. Zaremba, *Bull. Int. Acad. Sci., Cracovie*, 594-614, 614–621 (1903). **Gustaf Andreas Johannes Jaumann** (1863–1924) (pronounced "Yow-mahn") who taught at the German university in Brünn (now Brno), for whom the "Jaumann derivative" is named, was an important contributor to the field of continuum mechanics at the beginning of the twentieth century; he was the first to give the equation of change for entropy, including the "entropy flux" and the "rate of entropy production" (see §24.1).
[2] J. G. Oldroyd, *Proc. Roy. Soc.*, **A245**, 278–297 (1958). For an extension of the corotational idea, see L. E. Wedgewood, *Rheol. Acta*, **38**, 91–99 (1999).

between the stress tensor and the rate-of-strain tensor; for example, we can write the Jeffreys model and then add some additional nonlinear terms for good measure:

$$\hat{\boldsymbol{\tau}} + \lambda_1 \frac{\partial}{\partial t}\hat{\boldsymbol{\tau}} + \tfrac{1}{2}\mu_0(\mathrm{tr}\,\hat{\boldsymbol{\tau}})\hat{\dot{\boldsymbol{\gamma}}} - \tfrac{1}{2}\mu_1\{\hat{\dot{\boldsymbol{\gamma}}} \cdot \hat{\boldsymbol{\tau}} + \hat{\boldsymbol{\tau}} \cdot \hat{\dot{\boldsymbol{\gamma}}}\} = -\eta_0\left(\hat{\dot{\boldsymbol{\gamma}}} + \lambda_2 \frac{\partial}{\partial t}\hat{\dot{\boldsymbol{\gamma}}} - \mu_2\{\hat{\dot{\boldsymbol{\gamma}}} \cdot \hat{\dot{\boldsymbol{\gamma}}}\}\right) \quad (8.5\text{-}1)$$

in which the circumflexes ($\,\hat{}\,$) on the tensors indicate that their components are those with respect to the corotating coordinate frame. In Eq. 8.5-1 the constants λ_1, λ_2, μ_0, μ_1, and μ_2 all have dimensions of time.

Since the equations of continuity and motion are written for the usual xyz-coordinate frame, fixed in space, it seems reasonable to transform Eq. 8.5-1 from the $\hat{x}\hat{y}\hat{z}$ frame into the xyz frame. This is a purely mathematical problem, which was worked out long ago,[1] and the solution is well known. It can be shown that the partial time derivatives $\partial/\partial t$, $\partial^2/\partial t^2, \cdots$ are changed into *corotational* (or *Jaumann*[1-4]) *time derivatives* $\mathcal{D}/\mathcal{D}t$, $\mathcal{D}^2/\mathcal{D}t^2, \cdots$. The corotational time derivative of a second-order tensor is defined as

$$\frac{\mathcal{D}}{\mathcal{D}t}\boldsymbol{\alpha} = \frac{D}{Dt}\boldsymbol{\alpha} + \frac{1}{2}\{\boldsymbol{\omega} \cdot \boldsymbol{\alpha} - \boldsymbol{\alpha} \cdot \boldsymbol{\omega}\} \quad (8.5\text{-}2)$$

in which $\boldsymbol{\omega} = \nabla\mathbf{v} - (\nabla\mathbf{v})^\dagger$ is the *vorticity tensor*, and D/Dt is the substantial time derivative defined in §3.5. The tensor dot products appearing in Eq. 8.5-1, with components in the $\hat{x}\hat{y}\hat{z}$ frame, transform into the corresponding dot products, with the components given in the xyz frame.

When transformed into the xyz frame, Eq. 8.5-1 becomes

$$\boldsymbol{\tau} + \lambda_1 \frac{\mathcal{D}}{\mathcal{D}t}\boldsymbol{\tau} + \tfrac{1}{2}\mu_0(\mathrm{tr}\,\boldsymbol{\tau})\dot{\boldsymbol{\gamma}} - \tfrac{1}{2}\mu_1\{\boldsymbol{\tau} \cdot \dot{\boldsymbol{\gamma}} + \dot{\boldsymbol{\gamma}} \cdot \boldsymbol{\tau}\} = -\eta_0\left(\dot{\boldsymbol{\gamma}} + \lambda_2 \frac{\mathcal{D}}{\mathcal{D}t}\dot{\boldsymbol{\gamma}} - \mu_2\{\dot{\boldsymbol{\gamma}} \cdot \dot{\boldsymbol{\gamma}}\}\right) \quad (8.5\text{-}3)$$

which is the *Oldroyd 6-constant model*. This model, then, has no dependence on the local instantaneous orientation of the fluid particles in space. It should be emphasized that Eq. 8.5-3 is an empirical model; the use of the corotating frame guarantees only that the instantaneous local rotation of the fluid has been "subtracted off."

With proper choice of these parameters most of the observed phenomena in polymer fluid dynamics can be described *qualitatively*. As a result this model has been widely used in exploratory fluid dynamics calculations. A 3-constant simplification of Eq. 8.5-3 with $\mu_1 = \lambda_1$, $\mu_2 = \lambda_2$, and $\mu_0 = 0$ is called the *Oldroyd-B model*. In Example 8.5-1 we show what Eq. 8.5-3 gives for the material functions defined in §8.2.

Another nonlinear viscoelastic model is the 3-constant *Giesekus model*,[5] which contains a term that is quadratic in the stress components:

$$\boldsymbol{\tau} + \lambda\left(\frac{\mathcal{D}}{\mathcal{D}t}\boldsymbol{\tau} - \tfrac{1}{2}\{\boldsymbol{\tau} \cdot \dot{\boldsymbol{\gamma}} + \dot{\boldsymbol{\gamma}} \cdot \boldsymbol{\tau}\}\right) - \alpha\frac{\lambda}{\eta_0}\{\boldsymbol{\tau} \cdot \boldsymbol{\tau}\} = -\eta_0\dot{\boldsymbol{\gamma}} \quad (8.5\text{-}4)$$

Here λ is a time constant, η_0 is the zero shear rate viscosity, and α is a dimensionless parameter. This model gives reasonable shapes for most material functions, and the analytical expressions for them are summarized in Table 8.5-1. Because of the $\{\boldsymbol{\tau} \cdot \boldsymbol{\tau}\}$ term, they

---

[3] J. D. Goddard and C. Miller, *Rheol. Acta*, **5**, 177–184 (1966).

[4] R. B. Bird, R. C. Armstrong, and O. Hassager, *Dynamics of Polymeric Liquids, Vol. 1, Fluid Mechanics*, Wiley, New York, 1st edition (1977), Chapters 7 and 8; the corotational models are not discussed in the second edition of this book, where emphasis is placed on the use of "convected coordinates" and the "codeforming" frame. For differential models, either the corotating or codeforming frame can be used, but the former is simpler conceptually and mathematically.

[5] H. Giesekus, *J. Non-Newtonian Fluid Mech.*, **11**, 69–109 (1982); **12**, 367–374; *Rheol. Acta*, **21**, 366–375 (1982). See also R. B. Bird and J. M. Wiest, *J. Rheol.*, **29**, 519–532 (1985), and R. B. Bird, R. C. Armstrong, and O. Hassager, *Dynamics of Polymeric Liquids, Vol. 1, Fluid Dynamics*, Wiley-Interscience, New York, 2nd edition (1987), §7.3(c).

Table 8.5-1   Material Functions for the Giesekus Model

Steady shear flow:

$$\frac{\eta}{\eta_0} = \frac{(1-f)^2}{1+(1-2\alpha)f} \qquad \text{(A)}$$

$$\frac{\Psi_1}{2\eta_0\lambda} = \frac{f(1-\alpha f)}{\alpha(1-f)}\frac{1}{(\lambda\dot\gamma)^2} \qquad \text{(B)}$$

$$\frac{\Psi_2}{\eta_0\lambda} = -f\frac{1}{(\lambda\dot\gamma)^2} \qquad \text{(C)}$$

where

$$f = \frac{1-\chi}{1+(1-2\alpha)\chi} \quad \text{and} \quad \chi^2 = \frac{[1+16\alpha(1-\alpha)(\lambda\dot\gamma)^2]^{1/2}-1}{8\alpha(1-\alpha)(\lambda\dot\gamma)^2} \qquad \text{(D, E)}$$

Small-amplitude oscillatory shear flow:

$$\frac{\eta'}{\eta_0} = \frac{1}{1+(\lambda\omega)^2} \quad \text{and} \quad \frac{\eta''}{\eta_0} = \frac{\lambda\omega}{1+(\lambda\omega)^2} \qquad \text{(F, G)}$$

Steady elongational flow:

$$\frac{\bar\eta}{3\eta_0} = \frac{1}{6\alpha}\left[3 + \frac{1}{\lambda\dot\varepsilon}\left(\sqrt{1-4(1-2\alpha)\lambda\dot\varepsilon + 4(\lambda\dot\varepsilon)^2} - \sqrt{1+2(1-2\alpha)\lambda\dot\varepsilon + (\lambda\dot\varepsilon)^2}\right)\right] \qquad \text{(H)}$$

are not particularly simple. Superpositions of Giesekus models can be made to describe the shapes of the measured material functions almost quantitatively.[6] The model has been used widely for fluid dynamics calculations.

EXAMPLE 8.5-1	Obtain the material functions for steady shear flow, small amplitude oscillatory motion, and steady uniaxial elongational flow. Make use of the fact that in shear flows, the stress tensor components τ_{xz} and τ_{yz} are zero, and that in elongational flow, the off-diagonal elements of the stress tensor are zero (these results are obtained by symmetry arguments[7]).
Material Functions for the Oldroyd 6-Constant Model[2,4]	

SOLUTION

(a) First we simplify Eq. 8.5-3 for *unsteady shear flow*, with the velocity distribution $v_x(y, t) = \dot\gamma(t)y$. By writing out the components of the equation we get

$$\left(1 + \lambda_1\frac{\partial}{\partial t}\right)\tau_{xx} - (\lambda_1 + \mu_1)\tau_{yx}\dot\gamma = +\eta_0(\lambda_2 + \mu_2)\dot\gamma^2 \qquad \text{(8.5-5)}$$

$$\left(1 + \lambda_1\frac{\partial}{\partial t}\right)\tau_{yy} + (\lambda_1 - \mu_1)\tau_{yx}\dot\gamma = -\eta_0(\lambda_2 - \mu_2)\dot\gamma^2 \qquad \text{(8.5-6)}$$

$$\left(1 + \lambda_1\frac{\partial}{\partial t}\right)\tau_{zz} = 0 \qquad \text{(8.5-7)}$$

$$\left(1 + \lambda_1\frac{\partial}{\partial t}\right)\tau_{yx} + \tfrac{1}{2}(\lambda_1 - \mu_1 + \mu_0)\tau_{xx}\dot\gamma - \tfrac{1}{2}(\lambda_1 + \mu_1 - \mu_0)\tau_{yy}\dot\gamma + \tfrac{1}{2}\mu_0\tau_{zz}\dot\gamma = -\eta_0\left(1 + \lambda_2\frac{\partial}{\partial t}\right)\dot\gamma \qquad \text{(8.5-8)}$$

---

[6] W. R. Burghardt, J.-M. Li, B. Khomami, and B. Yang, *J. Rheol.*, **147**, 149–165 (1999).
[7] See, for example, R. B. Bird, R. C. Armstrong, and O. Hassager, *Dynamics of Polymeric Liquids*, Vol. 1, *Fluid Dynamics*, Wiley-Interscience, New York, 2nd edition (1987), §3.2.

(b) For *steady-state shear flow*, Eqs. 8.5-7 gives $\tau_{zz} = 0$, and the other three equations give a set of simultaneous algebraic equations that can be solved to get the remaining stress tensor components. Then with the definitions of the material functions in §8.2, we can obtain

$$\frac{\eta}{\eta_0} = \frac{1 + [\lambda_1\lambda_2 + (\mu_0 - \mu_1)\mu_2]\dot\gamma^2}{1 + [\lambda_1^2 + (\mu_0 - \mu_1)\mu_1]\dot\gamma^2} \equiv \frac{1 + \sigma_2\dot\gamma^2}{1 + \sigma_1\dot\gamma^2} \tag{8.5-9}$$

$$\frac{\Psi_1}{2\eta_0\lambda_1} = \frac{1 + \sigma_2\dot\gamma^2}{1 + \sigma_1\dot\gamma^2} - \frac{\lambda_2}{\lambda_1} \tag{8.5-10}$$

$$\frac{\Psi_2}{\eta_0\lambda_1} = -\left(1 - \frac{\mu_1}{\lambda_1}\right)\frac{1 + \sigma_2\dot\gamma^2}{1 + \sigma_1\dot\gamma^2} + \left(1 - \frac{\mu_2}{\lambda_2}\right)\frac{\lambda_2}{\lambda_1} \tag{8.5-11}$$

The model thus gives a shear-rate-dependent viscosity as well as shear-rate-dependent normal-stress coefficients. (For the Oldroyd-B model the viscosity and normal-stress coefficients are independent of the shear rate.) For most polymers the non-Newtonian viscosity decreases with the shear rate, and for such fluids we conclude that $0 < \sigma_2 < \sigma_1$. Moreover, since measured values of $|\tau_{yx}|$ always increase monotonically with shear rate, we also require that $\sigma_2 > \frac{1}{9}\sigma_1$. Although the model gives shear-rate-dependent viscosity and normal stresses, the shapes of the curves are not in satisfactory agreement with experimental data over a wide range of shear rates.

If $\mu_1 < \lambda_1$ and $\mu_2 < \lambda_2$, the second normal-stress coefficient has the opposite sign of the first normal-stress coefficient, in agreement with the data for most polymeric liquids. Since the second normal-stress coefficient is much smaller than the first for many fluids and in some flows plays a negligible role, setting $\mu_1 = \lambda_1$ and $\mu_2 = \lambda_2$ may be reasonable, thereby reducing the number of parameters from 6 to 4.

This discussion shows how to evaluate a proposed empirical model by comparing the model predictions with experimental data obtained in rheometric experiments. We have also seen that the experimental data may necessitate restrictions on the parameters. Clearly this is a tremendous task, but it is not unlike the problem that the thermodynamicist faces in developing empirical equations of state for mixtures, for example. The rheologist, however, is dealing with tensor equations, whereas the thermodynamicist is concerned only with scalar equations.

(c) For *small-amplitude oscillatory motion* the nonlinear terms in Eqs. 8.5-5 to 8 may be omitted, and the material functions are the same as those obtained from the Jeffreys model of linear viscoelasticity:

$$\frac{\eta'}{\eta_0} = \frac{1 + \lambda_1\lambda_2\omega^2}{1 + \lambda_1^2\omega^2} \quad \text{and} \quad \frac{\eta''}{\eta_0} = \frac{(\lambda_1 - \lambda_2)\omega}{1 + \lambda_1^2\omega^2} \tag{8.5-12, 13}$$

For η' to be a monotone decreasing function of the frequency and for η'' to be positive (as seen in all experiments), we have to require that $\lambda_2 < \lambda_1$. Here again, the model gives qualitatively correct results, but the shapes of the curves are not correct.

(d) For the *steady elongational flow* defined in §8.2, the Oldroyd 6-constant model gives

$$\frac{\bar\eta}{3\eta_0} = \frac{1 - \mu_2\dot\varepsilon + \mu_2(3\mu_0 - 2\mu_1)\dot\varepsilon^2}{1 - \mu_1\dot\varepsilon + \mu_1(3\mu_0 - 2\mu_1)\dot\varepsilon^2} \tag{8.5-14}$$

Since, for most polymers, the slope of the elongational viscosity versus elongation rate curve is positive at $\dot\varepsilon = 0$, we must require that $\mu_1 > \mu_2$. Equation 8.5-14 predicts that the elongational viscosity may become infinite at some finite value of the elongation rate; this may possibly present a problem in fiber-stretching calculations.

Note that the time constants λ_1 and λ_2 do not appear in the expression for elongational viscosity, whereas the constants μ_0, μ_1, and μ_2 do not enter into the components of the complex viscosity in Eqs. 8.5-14 and 15. This emphasizes the fact that a wide range of rheometric experiments is necessary for determining the parameters in an empirical expression for the stress tensor. To put it in another way, various experiments emphasize different parts of the model.

§8.6 MOLECULAR THEORIES FOR POLYMERIC LIQUIDS[1,2,3]

It should be evident from the previous section that proposing and testing empirical expressions for the stress tensor in nonlinear viscoelasticity is a formidable task. Recall that, in turbulence, seeking empirical expressions for the Reynolds stress tensor is equally daunting. However, in nonlinear viscoelasticity we have the advantage that we can narrow the search for stress tensor expressions considerably by using molecular theory. Although the kinetic theory of polymers is considerably more complicated than the kinetic theory of gases, it nonetheless guides us in suggesting possible forms for the stress tensor. However, the constants appearing in the molecular expressions must still be determined from rheometric measurements.

The kinetic theories for polymers can be divided roughly into two classes: *network theories* and *single-molecule theories*:

a. The network theories[3] were originally developed for describing the mechanical properties of rubber. One imagines that the polymer molecules in the rubber are joined chemically during vulcanization. The theories have been extended to describe molten polymers and concentrated solutions by postulating an ever-changing network in which the junction points are temporary, formed by adjacent strands that move together for a while and then gradually pull apart (see Fig. 8.6-1). It is necessary in the theory to make some empirical statements about the rates of formation and rupturing of the junctions.

b. The single-molecule theories[1] were originally designed for describing the polymer molecules in a very dilute solution, where polymer–polymer interactions are infrequent. The molecule is usually represented by means of some kind of "bead spring" model, a series of small spheres connected by linear or nonlinear springs in such a way as to represent the molecular architecture; the bead spring model is then allowed to move about in the solvent, with the beads experiencing a Stokes' law drag force by the solvent as well as being buffeted about by Brownian motion (see Fig. 8.6-2a). Then from the kinetic theory one obtains the "distribution function" for the orientations of the molecules (modeled as bead spring structures); once this function is known, various macroscopic properties can be calculated. The same kind of theory may be applied to concentrated solutions and molten polymers by examining the motion of a single bead spring model in the "mean force field" exerted by the surrounding molecules. That is,

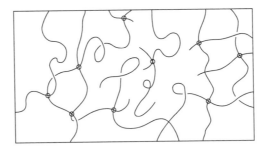

Fig. 8.6-1. Portion of a polymer network formed by "temporary junctions," indicated here by circles.

[1] R. B. Bird, C. F. Curtiss, R. C. Armstrong, and O. Hassager, *Dynamics of Polymeric Liquids, Vol. 2, Kinetic Theory*, Wiley-Interscience, New York, 2nd edition (1987).

[2] M. Doi and S. F. Edwards, *The Theory of Polymer Dynamics*, Clarendon Press, Oxford (1986); J. D. Schieber, "Polymer Dynamics," in *Encyclopedia of Applied Physics*, Vol. 14, VCH Publishers, Inc. (1996), pp. 415–443. R. B. Bird and H. C. Öttinger, *Ann. Rev. Phys. Chem.*, **43**, 371–406 (1992).

[3] A. S. Lodge, *Elastic Liquids*, Academic Press, New York (1964); *Body Tensor Fields in Continuum Mechanics*, Academic Press, New York (1974); *Understanding Elastomer Molecular Network Theory*, Bannatek Press, Madison, Wis. (1999).

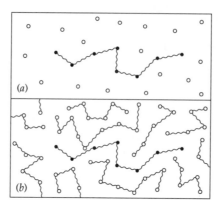

Fig. 8.6-2. Single-molecule bead spring models for (a) a dilute polymer solution, and (b) an undiluted polymer (a polymer "melt" with no solvent). In the dilute solution, the polymer molecule can move about in all directions through the solvent. In the undiluted polymer, a typical polymer molecule (black beads) is constrained by the surrounding molecules and tends to execute snakelike motion ("reptation") by sliding back and forth along its backbone direction.

because of the proximity of the surrounding molecules, it is easier for the "beads" of the model to move in the direction of the polymer chain backbone than perpendicular to it. In other words, the polymer finds itself executing a sort of snakelike motion, called "reptation" (see Fig. 8.6-2b).

As an illustration of the kinetic theory approach we discuss the results for a simple system: a dilute solution of a polymer, modeled as an elastic dumbbell consisting of two beads connected by a spring. We take the spring to be nonlinear and finitely extensible, with the force in the connecting spring being given by[4]

$$\mathbf{F}^{(c)} = \frac{H\mathbf{Q}}{1 - (Q/Q_0)^2} \tag{8.6-1}$$

in which H is a spring constant, $\mathbf{Q}$ is the end-to-end vector of the dumbbell representing the stretching and orientation of the dumbbell, and Q_0 is the maximum elongation of the spring. The friction coefficient for the motion of the beads through the solvent is given by Stokes' law as $\zeta = 6\pi\eta_s a$, where a is the bead radius and η_s is the solvent viscosity. Although this model is greatly oversimplified, it does embody the key physical ideas of molecular orientation, molecular stretching, and finite extensibility.

When the details of the kinetic theory are worked out, one gets the following expression for the stress tensor, written as the sum of a Newtonian solvent and a polymer contribution (see fn. 3 in §8.4):[5]

$$\boldsymbol{\tau} = \boldsymbol{\tau}_s + \boldsymbol{\tau}_p \tag{8.6-2}$$

Here

$$\boldsymbol{\tau}_s = -\eta_s \dot{\boldsymbol{\gamma}} \tag{8.6-3}$$

$$Z\boldsymbol{\tau}_p + \lambda_H \left(\frac{\mathscr{D}}{\mathscr{D}t} \boldsymbol{\tau}_p - \tfrac{1}{2} \{ \boldsymbol{\tau}_p \cdot \dot{\boldsymbol{\gamma}} + \dot{\boldsymbol{\gamma}} \cdot \boldsymbol{\tau}_p \} \right) - \lambda_H (\boldsymbol{\tau}_p - n\kappa T\boldsymbol{\delta}) \frac{D \ln Z}{Dt} = -n\kappa T\lambda_H \dot{\boldsymbol{\gamma}} \tag{8.6-4}$$

where n is the number density of polymer molecules (i.e., dumbbells), $\lambda_H = \zeta/4H$ is a time constant (typically between 0.01 and 10 seconds), $Z = 1 + (3/b)[1 - (\mathrm{tr}\, \boldsymbol{\tau}_p/3n\kappa T)]$, and $b = HQ_0^2/\kappa T$ is the finite extensibility parameter, usually between 10 and 100. The

[4] H. R. Warner, Jr., *Ind. Eng. Chem. Fundamentals*, **11**, 379–387 (1972); R. L. Christiansen and R. B. Bird, *J. Non-Newtonian Fluid Mech.*, **3**, 161–177 (1977/1978).

[5] R. I. Tanner, *Trans. Soc. Rheol.*, **19**, 37–65 (1975); R. B. Bird, P. J. Dotson, and N. L. Johnson, *J. Non-Newtonian Fluid Mech.*, **7**, 213–235 (1980)—in the last publication, Eqs. 58–85 are in error.

molecular theory has thus resulted in a model with four adjustable constants: η_s, λ_H, n, and b, which can be determined from rheometric experiments. Thus the molecular theory suggests the form of the stress tensor expression, and the rheometric data are used to determine the values of the parameters. The model described by Eqs. 8.6-2, 3, and 4 is called the FENE-P model (finitely extensible nonlinear elastic model, in the Peterlin approximation) in which $(Q/Q_0)^2$ in Eq. 8.6-1 is replaced by $\langle Q^2 \rangle / Q_0^2$.

This model is more difficult to work with than the Oldroyd 6-constant model, because it is nonlinear in the stresses. However, it gives better shapes for some of the material functions. Also, since we are dealing here with a molecular model, we can get information about the molecular stretching and orientation after a flow problem has been solved. For example, it can be shown that the average molecular stretching is given by $\langle Q^2 \rangle / Q_0^2 = 1 - Z^{-1}$ where the angular brackets indicate a statistical average.

The following examples illustrate how one obtains the material functions for the model and compares the results with experimental data. If the model is acceptable, then it must be combined with the equations of continuity and motion to solve interesting flow problems. This requires large-scale computing.

EXAMPLE 8.6-1

Material Functions for the FENE-P Model

Obtain the material functions for the steady-state shear flow and the steady-state elongational flow of a polymer described by the FENE-P model.

SOLUTION

(a) For steady-state shear flow the model gives the following equations for the nonvanishing components of the polymer contribution to the stress tensor:

$$Z\tau_{p,xx} = 2\tau_{p,yx}\lambda_H\dot{\gamma} \tag{8.6-5}$$

$$Z\tau_{p,yx} = -n\kappa T\lambda_H\dot{\gamma} \tag{8.6-6}$$

Here the quantity Z is given by

$$Z = 1 + (3/b)[1 - (\tau_{p,xx}/3n\kappa T)] \tag{8.6-7}$$

These equations can be combined to give a cubic equation for the dimensionless shear stress contribution $T_{yx} = \tau_{p,yx}/3n\kappa T$

$$T_{yx}^3 + 3pT_{yx} + 2q = 0 \tag{8.6-8}$$

in which $p = (b/54) + (1/18)$ and $q = (b/108)\lambda_H\dot{\gamma}$. This cubic equation may be solved to give[6]

$$T_{yx} = -2p^{1/2}\sinh(\tfrac{1}{3}\operatorname{arcsinh} qp^{-3/2}) \tag{8.6-9}$$

The non-Newtonian viscosity based on this function is shown in Fig. 8.6-3 along with some experimental data for some polymethyl-methacrylate solutions. From Eq. 8.6-9 we find for the limiting values of the viscosity

For $\dot{\gamma} = 0$:

$$\eta - \eta_s = n\kappa T\lambda_H\left(\frac{b}{b+3}\right) \tag{8.6-10}$$

For $\dot{\gamma} \to \infty$:

$$\eta - \eta_s \sim n\kappa T\lambda_H\left(\frac{b}{2}\frac{1}{\lambda_H^2\dot{\gamma}^2}\right)^{1/3} \tag{8.6-11}$$

Hence, at high shear rates one obtains a power law behavior (Eq. 8.3-3) with $n = \tfrac{1}{3}$. This can be taken as a molecular justification for use of the power law model.

---

[6] K. Rektorys, *Survey of Applicable Mathematics*, MIT Press, Cambridge, MA (1969), pp. 78–79.

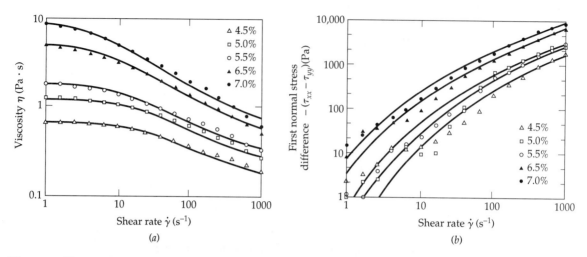

Fig. 8.6-3. Viscosity and first-normal-stress difference data for polymethylmethacrylate solutions from D. D. Joseph, G. S. Beavers, A. Cers, C. Dewald, A. Hoger, and P. T. Than, *J. Rheol.*, **28**, 325–345 (1984), along with the FENE-P curves for the following constants, determined by L. E. Wedgewood:

Polymer concentration [%]	η_0 [Pa·s]	λ_H [s]	a [Pa]	b [- - -]
4.5	0.13	0.157	3.58	47.9
5.0	0.19	0.192	5.94	38.3
5.5	0.25	0.302	5.98	30.6
6.5	0.38	0.447	11.8	25.0
7.0	0.45	0.553	19.1	16.0

The quantity $a = n\kappa T$ was taken to be a parameter determined from the rheometric data.

From Eq. 8.6-5 one finds that Ψ_1 is given by $\Psi_1 = 2(\eta - \eta_s)^2/n\kappa T$; a comparison of this result with experimental data is shown in Fig. 8.6-3. The second normal stress coefficient Ψ_2 for this model is zero. As pointed out above, once we have solved the flow problem, we can also get the molecular stretching from the quantity Z. In Fig. 8.6-4 we show how the molecules are stretched, on the average, as a function of the shear rate.

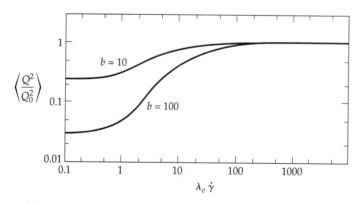

Fig. 8.6-4. Molecular stretching as a function of shear rate $\dot{\gamma}$ in steady shear flow, according to the FENE-P dumbbell model. The experimentally accessible time constant $\lambda_e = [\eta_0]\eta_s M/RT$, where $[\eta_0]$ is the zero shear rate intrinsic viscosity, is related to λ_H by $\lambda_e = \lambda_H b/(b + 3)$. [From R. B. Bird, P. J. Dotson, and N. L. Johnson, *J. Non-Newtonian Fluid Mech.*, **7**, 213–235 (1980).]

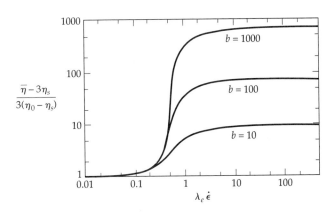

Fig. 8.6-5. Steady elongational viscosity $\bar{\eta}$ as a function of the elongation rate $\dot{\varepsilon}$ according to the FENE-P dumbbell model. The time constant is given by $\lambda_e = \lambda_H b/(b+3)$. [From R. B. Bird, P. J. Dotson, and N. L. Johnson, *J. Non-Newtonian Fluid Mech.*, **7**, 213–235 (1980).]

(b) For steady-state elongational flow we get

$$Z\tau_{p,xx} + \tau_{p,xx}\lambda_H\dot{\varepsilon} = +n\kappa T\lambda_H\dot{\varepsilon} \tag{8.6-12}$$

$$Z\tau_{p,yy} + \tau_{p,yy}\lambda_H\dot{\varepsilon} = +n\kappa T\lambda_H\dot{\varepsilon} \tag{8.6-13}$$

$$Z\tau_{p,zz} - 2\tau_{p,zz}\lambda_H\dot{\varepsilon} = -2n\kappa T\lambda_H\dot{\varepsilon} \tag{8.6-14}$$

$$Z = 1 + \frac{3}{b}\left(1 - \frac{\tau_{p,xx} + \tau_{p,yy} + \tau_{p,zz}}{3n\kappa T}\right) \tag{8.6-15}$$

This set of equations leads to a cubic equation for $\tau_{p,xx} - \tau_{p,zz}$, from which the elongational viscosity can be obtained (see Fig. 8.6-5). Limited experimental data on polymer solutions indicate that the shapes of the curves are probably approximately correct.

The limiting expressions for the elongational viscosity are

For $\dot{\varepsilon} = 0$:
$$\bar{\eta} - 3\eta_s = 3n\kappa T\lambda_H\left(\frac{b}{b+3}\right) \tag{8.6-16}$$

For $\dot{\varepsilon} \to \infty$:
$$\bar{\eta} - 3\eta_s = 2n\kappa T\lambda_H b \tag{8.6-17}$$

Having found the stresses in the system, we can then get the average stretching of the molecules as a function of the elongation rate; this is shown in Fig. 8.6-6.

It is worth noting that for a typical value of b—say, 50—the elongational viscosity can increase by a factor of about 30 as the elongation rate increases, thereby having a profound effect on flows in which there is a strong elongational component.[7]

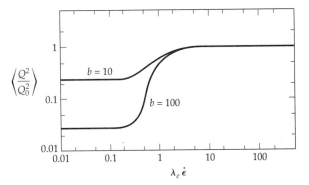

Fig. 8.6-6. Molecular stretching as a function of the elongation rate $\dot{\varepsilon}$ in steady elongational flow, as predicted by the FENE-P dumbbell model. The time constant is given by $\lambda_e = \lambda_H b/(b+3)$. [From R. B. Bird, P. J. Dotson, and N. L. Johnson, *J. Non-Newtonian Fluid Mech.*, **7**, 213–235 (1980).]

---

[7] The FENE-P and Giesekus models have been used successfully to describe the details of turbulent drag reduction, which is closely related to elongational viscosity, by R. Sureshkumar, A. N. Beris and R. A. Handler, *Phys. Fluids*, **9**, 743–755 (1997), and C. D. Dimitropoulos, R. Sureshkumar, and A. N. Beris, *J. Non-Newtonian Fluid Mechanics*, **79**, 433–468 (1998).

QUESTIONS FOR DISCUSSION

1. Compare the behavior of Newtonian liquids and polymeric liquids in the various experiments discussed in §§8.1 and 8.2.

2. Why do we deal only with differences in normal stresses for incompressible liquids (see Eqs. 8.2-2 and 3)?

3. In Fig. 8.2-2 the postulated velocity profile is linear in y. What would you expect the velocity distribution to look like if the gap between the plates were not small and the fluid had a very low viscosity?

4. How is the parameter n in Eq. 8.3-3 related to the parameter n in Eq. 8.3-4? How is it related to the slope of the non-Newtonian velocity curve from the dumbbell kinetic theory model in §8.6?

5. What limitations have to be placed on use of the generalized Newtonian models and the linear viscoelastic models?

6. Compare and contrast Examples 8.4-1 and 2 regarding the geometry of the flow system and the assumptions regarding the velocity profiles.

7. To what extent does the Oldroyd model in Eq. 8.5-3 include a generalized Newtonian model and a linear viscoelastic model? Can the Oldroyd model describe effects that are not described by these other models?

8. Why is it necessary to put restrictions on the parameters in the Oldroyd model? What is the relation between these restrictions and the subject of rheometry?

9. What advantages do molecular expressions for the stress tensor have over the empirical expressions?

10. For what kinds of industrial problems would you use the various kinds of models described in this chapter?

11. Why may the power law model be unsatisfactory for describing the axial flow in an annulus?

PROBLEMS

8A.1 Flow of a polyisoprene solution in a pipe. A 13.5% (by weight) solution of polyisoprene in isopentane has the following power law parameters at 323 K: $n = 0.2$ and $m = 5 \times 10^3$ Pa · s^n. It is being pumped (in laminar flow) through a horizontal pipe that has a length of 10.2 m and an internal diameter of 1.3 cm. It is desired to use another pipe with a length of 30.6 m with the same mass flow rate and the same pressure drop. What should the pipe radius be?

8A.2 Pumping of a polyethylene oxide solution. A 1% aqueous solution of polyethylene oxide at 333 K has power law parameters $n = 0.6$ and $m = 0.50$ Pa · s^n. The solution is being pumped between two tanks, with the first tank at pressure p_1 and the second at pressure p_2. The pipe carrying the solution has a length of 14.7m and an internal diameter of 0.27 m.

It has been decided to replace the single pipe by a pair of pipes of the same length, but with smaller diameter. What diameter should these pipes have so that the mass flow rate will be the same as in the single pipe?

8B.1 Flow of a polymeric film. Work the problem in §2.2 for the power law fluid. Show that the result simplifies properly to the Newtonian result.

8B.2 Power law flow in a narrow slit. In Example 8.3-2 show how to derive the velocity distribution for the region $-B \le x \le 0$. Is it possible to combine this result with that in Eq. 8.3-13 into one equation?

8B.3 Non-Newtonian flow in an annulus. Rework Problem 2B.7 for the annular flow of a power law fluid with the flow being driven by the axial motion of the inner cylinder.

(a) Show that the velocity distribution for the fluid is

$$\frac{v_z}{v_0} = \frac{(r/R)^{1-(1/n)} - 1}{\kappa^{1-(1/n)} - 1} \tag{8B.3-1}$$

(b) Verify that the result in (a) simplifies to the Newtonian result when n goes to unity.

(c) Show that the mass flow rate in the annular region is given by

$$w = \frac{2\pi R^2 v_0 \rho}{\kappa^{1-(1/n)} - 1}\left(\frac{1 - \kappa^{3-(1/n)}}{3 - (1/n)} - \frac{1 - \kappa^2}{2}\right) \qquad \text{(for } n \neq \tfrac{1}{3}\text{)} \qquad \text{(8B.3-2)}$$

(d) What is the mass flow rate for fluids with $n = \tfrac{1}{3}$?

(e) Simplify Eq. 8B.3-2 for the Newtonian fluid.

8B.4 Flow of a polymeric liquid in a tapered tube. Work Problem 2B.10 for a power law fluid, using the lubrication approximation.

8B.5 Slit flow of a Bingham fluid.[1] For thick suspensions and pastes it is found that no flow occurs until a certain critical stress, the *yield stress*, is reached, and then the fluid flows in such a way that part of the stream is in "plug flow." The simplest model of a fluid with a yield value is the *Bingham model*:

$$\begin{cases} \eta = \infty & \text{when } \tau \leq \tau_0 \\ \eta = \mu_0 + \dfrac{\tau_0}{\dot{\gamma}} & \text{when } \tau \geq \tau_0 \end{cases} \qquad \text{(8B.5-1)}$$

in which τ_0 is the yield stress, the stress below which no flow occurs, and μ_0 is a parameter with units of viscosity. The quantity $\tau = \sqrt{\tfrac{1}{2}(\boldsymbol{\tau}:\boldsymbol{\tau})}$ is the magnitude of the stress tensor.

Find the mass flow rate in a slit for the Bingham fluid (see Problem 2B.3 and Example 8.3-2). The expression for the shear stress τ_{xz} as a function of position x in Eq. 2B.3-1 can be taken over here, since it does not depend on the type of fluid. We see that $|\tau_{xz}|$ is just equal to the yield stress τ_0 at $x = \pm x_0$, where x_0 is defined by

$$\tau_0 = \frac{\mathcal{P}_0 - \mathcal{P}_L}{L} x_0 \qquad \text{(8B.5-2)}$$

(a) Show that the upper equation of Eq. 8B.5-1 requires that $dv_z/dx = 0$ for $|x| \leq x_0$, since $\tau_{xz} = -\eta dv_z/dx$ and τ_{xz} is finite; this is then the "plug-flow" region. Then show that, since for x positive, $\dot{\gamma} = -dv_z/dx$, and for x negative, $\dot{\gamma} = +dv_z/dx$, the lower equation of Eq. 8B.5-1 requires that

$$\tau_{xz} = \begin{cases} -\mu_0(dv_z/dx) + \tau_0 & \text{for } +x_0 \leq x \leq +B \\ -\mu_0(dv_z/dx) - \tau_0 & \text{for } -B \leq x \leq -x_0 \end{cases} \qquad \text{(8B.5-3)}$$

(b) To get the velocity distribution for $+x_0 \leq x \leq +B$, substitute the upper relation from Eq. 8B.5-3 into Eq. 2B.3-1 and get the differential equation for v_z. Show that this may be integrated with the boundary condition that the velocity is zero at $x = B$ to give

$$v_z = \frac{(\mathcal{P}_0 - \mathcal{P}_L)B^2}{2\mu_0 L}\left[1 - \left(\frac{x}{B}\right)^2\right] - \frac{\tau_0 B}{\mu_0}\left(1 - \frac{x}{B}\right) \qquad \text{for } +x_0 \leq x \leq +B \qquad \text{(8B.5-4)}$$

What is the velocity in the range $|x| \leq x_0$? Draw a sketch of $v_z(x)$.

(c) The mass flow rate can then be obtained from

$$w = W\rho \int_{-B}^{+B} v_z dx = 2W\rho \int_0^B v_z dx = 2W\rho \int_{x_0}^B x\left(-\frac{dv_z}{dx}\right)dx \qquad \text{(8B.5-5)}$$

---

[1] E. C. Bingham, *Fluidity and Plasticity*, McGraw-Hill, New York (1922), pp. 215–218. See R. B. Bird, G. C. Dai, and B. J. Yarusso, *Reviews in Chemical Engineering*, **1**, 1–70 (1982) for a review of models with a yield stress.

The integration by parts allows the integration to be done more easily. Show that the final result is

$$w = \frac{2}{3}\frac{(\mathscr{P}_0 - \mathscr{P}_L)WB^3\rho}{\mu_0 L}\left[1 - \frac{3}{2}\left(\frac{\tau_0 L}{(\mathscr{P}_0 - \mathscr{P}_L)B}\right) + \frac{1}{2}\left(\frac{\tau_0 L}{(\mathscr{P}_0 - \mathscr{P}_L)B}\right)^3\right] \tag{8B.5-6}$$

Verify that, when the yield stress goes to zero, this result simplifies to the Newtonian fluid result in Problem 2B.3.

8B.6 Derivation of the Buckingham–Reiner equation.[2] Rework Example 8.3-1 for the Bingham model. First find the velocity distribution. Then show that the mass rate of flow is given by

$$w = \frac{\pi(\mathscr{P}_0 - \mathscr{P}_L)R^4\rho}{8\mu_0 L}\left[1 - \frac{4}{3}\left(\frac{\tau_0}{\tau_R}\right) + \frac{1}{3}\left(\frac{\tau_0}{\tau_R}\right)^4\right] \tag{8B.6-1}$$

in which $\tau_R = (\mathscr{P}_0 - \mathscr{P}_L)R/2L$ is the shear stress at the tube wall. This expression is valid only when $\tau_R \geq \tau_0$.

8B.7 The complex-viscosity components for the Jeffreys fluid.

(a) Work Example 8.4-1 for the Jeffreys model of Eq. 8.4-4, and show that the results are Eqs. 8.5-12 and 13. How are these results related to Eqs. (F) and (G) of Table 8.5-1?

(b) Obtain the complex-viscosity components for the Jeffreys model by using the superposition suggested in fn. 3 of §8.4.

8B.8 Stress relaxation after cessation of shear flow. A viscoelastic fluid is in steady-state flow between a pair of parallel plates, with $v_x = \dot{\gamma}y$. If the flow is suddenly stopped (i.e., $\dot{\gamma}$ becomes zero), the stresses do not go to zero as would be the case for a Newtonian fluid. Explore this *stress relaxation* phenomenon using a 3-constant Oldroyd model (Eq. 8.5-3 with $\lambda_2 = \mu_2 = \mu_1 = \mu_0 = 0$).

(a) Show that in steady-state flow

$$\tau_{yx} = -\eta_0\dot{\gamma}\frac{1}{1 + (\lambda_1\dot{\gamma})^2} \tag{8B.8-1}$$

To what extent does this expression agree with the experimental data in Fig. 8.2-4?

(b) By using Example 8.5-1 (part a) show that, if the flow is stopped at $t = 0$, the shear stress for $t \geq 0$ will be

$$\tau_{yx} = -\eta_0\dot{\gamma}\frac{1}{1 + (\lambda_1\dot{\gamma})^2}e^{-t/\lambda_1} \tag{8B.8-2}$$

This shows why λ_1 is called the "relaxation time." This relaxation of stresses after the fluid motion has stopped is characteristic of viscoelastic materials.

(c) What is the normal stress τ_{xx} during steady shear flow and after cessation of the flow?

8B.9 Draining of a tank with an exit pipe (Fig. 7B.9). Rework Problem 7B.9(a) for the power law fluid.

8B.10 The Giesekus model.

(a) Use the results in Table 8.5-1 to get the limiting values for the non-Newtonian viscosity and the normal stress differences as the shear rate goes to zero.

(b) Find the limiting expressions for the non-Newtonian viscosity and the two normal-stress coefficients in the limit as the shear rate becomes infinitely large.

(c) What is the steady-state elongational viscosity in the limit that the elongation rate tends to zero? Show that the elongational viscosity has a finite limit as the elongation rate goes to infinity.

---

[2] E. Buckingham, *Proc. ASTM*, **21**, 1154–1161 (1921); M. Reiner, *Deformation and Flow*, Lewis, London (1949).

8C.1 The cone-and-plate viscometer (Fig. 2B.11).[3] Review the Newtonian analysis of the cone-and-plate instrument in Problem 2B.11 and then do the following:

(a) Show that the shear rate $\dot{\gamma}$ is uniform throughout the gap and equal to $\dot{\gamma} = -\dot{\gamma}_{\theta\phi} = \Omega/\psi_0$. Because of the uniformity of $\dot{\gamma}$, the components of the stress tensor are also constant throughout the gap.

(b) Show that the non-Newtonian viscosity is then obtained from measurements of the torque T_z and rotation speed Ω by using

$$\eta(\dot{\gamma}) = \frac{3T_z\psi_0}{2\pi R^3 \Omega} \qquad (8C.1\text{-}1)$$

(c) Show that for the cone-and-plate system the radial component of the equation of motion is

$$0 = -\frac{\partial p}{\partial r} - \frac{1}{r^2}\frac{\partial}{\partial r}(r^2\tau_{rr}) + \frac{\tau_{\theta\theta} - \tau_{\phi\phi}}{r} \qquad (8C.1\text{-}2)$$

if the centrifugal force term $-\rho v_\phi^2/r$ can be neglected. Rearrange this to get

$$0 = -\partial\pi_{rr}/\partial \ln r + (\tau_{\phi\phi} - \tau_{\theta\theta}) + 2(\tau_{\theta\theta} - \tau_{rr}) \qquad (8C.1\text{-}3)$$

Then introduce the normal stress coefficients, and use the result of (a) to replace $\partial\pi_{rr}/\partial \ln r$ by $\partial\pi_{\theta\theta}/\partial \ln r$, to get

$$\partial\pi_{\theta\theta}/\partial \ln r = -(\Psi_1 + 2\Psi_2)\dot{\gamma}^2 \qquad (8C.1\text{-}4)$$

Integrate this from r to R and use the boundary condition $\pi_{rr}(R) = p_a$ to get

$$\pi_{\theta\theta}(r) = \pi_{\theta\theta}(R) - (\Psi_1 + 2\Psi_2)\dot{\gamma}^2 \ln(r/R)$$
$$= p_a - \Psi_2\dot{\gamma}^2 - (\Psi_1 + 2\Psi_2)\dot{\gamma}^2 \ln(r/R) \qquad (8C.1\text{-}5)$$

in which p_a is the atmospheric pressure acting on the fluid at the rim of the cone-and-plate instrument.

(d) Show that the total thrust in the z direction exerted by the fluid on the cone is

$$F_z = \int_0^{2\pi}\int_0^R [\pi_{\theta\theta}(r) - p_a]\, r\, dr\, d\theta = \tfrac{1}{2}\pi R^2\Psi_1\dot{\gamma}^2 \qquad (8C.1\text{-}6)$$

From this one can obtain the first normal-stress coefficient by measuring the force that the fluid exerts.

(e) Suggest a method for measuring the second normal-stress coefficient using results in part (c) if small pressure transducers are flush-mounted in the plate at several different radial locations.

8C.2 Squeezing flow between parallel disks (Fig. 3C.1).[4] Rework Problem 3C.1(g) for the power law fluid. This device can be useful for determining the power law parameters for materials that are highly viscous. Show that the power law analog of Eq. 3C.1-16 is

$$\frac{1}{H^{(n+1)/n}} = \frac{1}{H_0^{(n+1)/n}} + \frac{(n+1)}{2n+1}\left(\frac{n+3}{2\pi m R^{n+3}}\right)^{1/n} F_0^{1/n} t \qquad (8C.2\text{-}1)$$

[3] R. B. Bird, R. C. Armstrong, and O. Hassager, *Dynamics of Polymeric Liquids, Vol. 1, Fluid Mechanics,* Wiley-Interscience, New York, 2nd Edition (1987), pp. 521–524.

[4] P. J. Leider, *Ind. Eng. Chem. Fundam.*, **13**, 342–346 (1974); R. J. Grimm, *AIChE Journal*, **24**, 427–439 (1978).

8C.3   **Verification of Giesekus viscosity function.**[5]

(a) To check the shear-flow entries in Table 8.5-1, introduce dimensionless stress tensor components $T_{ij} = (\lambda/\eta_0)\tau_{ij}$ and a dimensionless shear rate $\dot{\Gamma} = \lambda\dot{\gamma}$, and then show that for steady-state shear flow Eq. 8.5-4 becomes

$$T_{xx} - 2\dot{\Gamma}T_{yx} - \alpha(T_{xx}^2 + T_{yx}^2) = 0 \tag{8C.3-1}$$

$$T_{yy} - \alpha(T_{yx}^2 + T_{yy}^2) = 0 \tag{8C.3-2}$$

$$T_{yx} - \dot{\Gamma}T_{yy} - \alpha T_{yx}(T_{xx} + T_{yy}) = -\dot{\Gamma} \tag{8C.3-3}$$

There is also a fourth equation, which leads to $T_{zz} = 0$.

(b) Rewrite these equations in terms of the dimensionless normal-stress differences $N_1 = T_{xx} - T_{yy}$ and $N_2 = T_{yy} - T_{zz}$, and T_{yx}.

(c) It is difficult to solve the equations in (b) to get the dimensionless shear stress and normal-stress differences in terms of the dimensionless shear rate. Instead, solve for N_1, T_{yx}, and $\dot{\Gamma}$ as functions of N_2:

$$T_{yx}^2 = \frac{N_2(1 - \alpha N_2)}{\alpha} \tag{8C.3-4}$$

$$N_1 = -\frac{2N_2(1 - \alpha N_2)}{\alpha(1 - N_2)} \tag{8C.3-5}$$

$$\dot{\Gamma}^2 = \frac{N_2(1 - \alpha N_2)[1 + (1 - 2\alpha)N_2]^2}{\alpha(1 - N_2)^4} \tag{8C.3-6}$$

(d) Solve the last equation for N_2 as a function of $\dot{\Gamma}$ to get

$$N_2 = f(\chi) = (1 - \chi)/[1 + (1 - 2\alpha)\chi] \tag{8C.3-7}$$

where

$$\chi^2 = \frac{\sqrt{1 + 16\alpha(1 - \alpha)\dot{\Gamma}^2} - 1}{8\alpha(1 - \alpha)\dot{\Gamma}^2} = 1 - 4\alpha(1 - \alpha)\dot{\Gamma}^2 + \cdots \tag{8C.3-8}$$

Then get the expression for the non-Newtonian viscosity and plot the curve of $\eta(\dot{\gamma})$.

8C.4   **Tube Flow for the Oldroyd 6-Constant Model.** Find the mass flow rate for the steady flow in a long circular tube[6] using Eq. 8.5-3.

8C.5   **Chain Models with Rigid-Rod Connectors.** Read and discuss the following publications: M. Gottlieb, *Computers in Chemistry*, **1**, 155–160 (1977); O. Hassager, *J. Chem. Phys.*, **60**, 2111–2124 (1974); X. J. Fan and T. W. Liu, *J. Non-Newtonian Fluid Mech.*, **19**, 303–321 (1986); T. W. Liu, *J. Chem. Phys.*, **90**, 5826–5842 (1989); H. H. Saab, R. B. Bird, and C. F. Curtiss, *J. Chem. Phys.*, **77**, 4758–4766 (1982); J. D. Schieber, *J. Chem. Phys.*, **87**, 4917–4927, 4928–4936 (1987). Why are rodlike connectors more difficult to handle than springs? What kinds of problems can be solved by computer simulations?

---

[5] H. Giesekus, *J. Non-Newtonian Fluid Mech.*, **11**, 69–109 (1982).
[6] M. C. Williams and R. B. Bird, *AIChE Journal*, **8**, 378–382 (1962).

Part Two

Energy Transport

Chapter 9

Thermal Conductivity and the Mechanisms of Energy Transport

It is common knowledge that some materials such as metals conduct heat readily, whereas others such as wood act as thermal insulators. The physical property that describes the rate at which heat is conducted is the thermal conductivity k.

Heat conduction in fluids can be thought of as *molecular energy transport*, inasmuch as the basic mechanism is the motion of the constituent molecules. Energy can also be transported by the bulk motion of a fluid, and this is referred to as *convective energy transport*; this form of transport depends on the density ρ of the fluid. Another mechanism is that of *diffusive energy transport*, which occurs in mixtures that are interdiffusing. In addition, energy can be transmitted by means of *radiative energy transport*, which is quite distinct in that this form of transport does not require a material medium as do conduction and convection. This chapter introduces the first two mechanisms, conduction and convection. Radiation is treated separately in Chapter 16, and the subject of diffusive heat transport arises in §19.3 and again in §24.2.

We begin in §9.1 with the definition of the thermal conductivity k by Fourier's law for the heat flux vector $\mathbf{q}$. In §9.2 we summarize the temperature and pressure dependence of k for fluids by means of the principle of corresponding states. Then in the next four sections we present information about thermal conductivities of gases, liquids, solids, and solid composites, giving theoretical results when available.

Since in Chapters 10 and 11 we will be setting up problems by using the law of conservation of energy, we need to know not only how *heat* moves into and out of a system but also how *work* is done on or by a system by means of molecular mechanisms. The nature of the molecular work terms is discussed in §9.8. Finally, by combining the conductive heat flux, the convective energy flux, and the work flux we can create a *combined energy flux vector* $\mathbf{e}$, which is useful in setting up energy balances.

§9.1   FOURIER'S LAW OF HEAT CONDUCTION (MOLECULAR ENERGY TRANSPORT)

Consider a slab of solid material of area A located between two large parallel plates a distance Y apart. We imagine that initially (for time $t < 0$) the solid material is at a temperature T_0 throughout. At $t = 0$ the lower plate is suddenly brought to a slightly higher temperature T_1 and maintained at that temperature. As time proceeds, the temperature profile in the slab changes, and ultimately a linear steady-state temperature distribution is attained (as shown in Fig. 9.1-1). When this steady-state condition has been reached, a constant rate of heat flow Q through the slab is required to maintain the temperature difference $\Delta T = T_1 - T_0$. It is found then that for sufficiently small values of ΔT the following relation holds:

$$\frac{Q}{A} = k \frac{\Delta T}{Y} \tag{9.1-1}$$

That is, the rate of heat flow per unit area is proportional to the temperature decrease over the distance Y. The constant of proportionality k is the *thermal conductivity* of the slab. Equation 9.1-1 is also valid if a liquid or gas is placed between the two plates, provided that suitable precautions are taken to eliminate convection and radiation.

In subsequent chapters it is better to work with the above equation in differential form. That is, we use the limiting form of Eq. 9.1-1 as the slab thickness approaches zero. The local rate of heat flow per unit area (heat flux) in the positive y direction is designated by q_y. In this notation Eq. 9.1-1 becomes

$$q_y = -k \frac{dT}{dy} \tag{9.1-2}$$

This equation, which serves to define k, is the one-dimensional form of *Fourier's law of heat conduction*.[1,2] It states that the heat flux by conduction is proportional to the tempera-

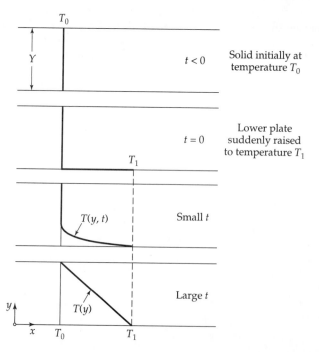

Fig. 9.1-1. Development of the steady-state temperature profile for a solid slab between two parallel plates. See Fig. 1.1-1 for the analogous situation for momentum transport.

ture gradient, or, to put it pictorially, "heat slides downhill on the temperature versus distance graph." Actually Eq. 9.1-2 is not really a "law" of nature, but rather a suggestion, which has proven to be a very useful empiricism. However, it does have a theoretical basis, as discussed in Appendix D.

If the temperature varies in all three directions, then we can write an equation like Eq. 9.1-2 for each of the coordinate directions:

$$q_x = -k \frac{\partial T}{\partial x} \qquad q_y = -k \frac{\partial T}{\partial y} \qquad q_z = -k \frac{\partial T}{\partial z} \qquad (9.1\text{-}3, 4, 5)$$

If each of these equations is multiplied by the appropriate unit vector and the equations are then added, we get

$$\boxed{\mathbf{q} = -k\nabla T} \qquad (9.1\text{-}6)$$

which is the three-dimensional form of Fourier's law. This equation describes the molecular transport of heat in isotropic media. By "isotropic" we mean that the material has no preferred direction, so that heat is conducted with the same thermal conductivity k in all directions.

Some solids, such as single noncubic crystals, fibrous materials, and laminates, are anisotropic.[3] For such substances one has to replace Eq. 9.1-6 by

$$\mathbf{q} = -[\boldsymbol{\kappa} \cdot \nabla T] \qquad (9.1\text{-}7)$$

in which $\boldsymbol{\kappa}$ is a symmetric second-order tensor called the *thermal conductivity tensor*. Thus, the heat flux vector does not point in the same direction as the temperature gradient. For polymeric liquids in the shearing flow $v_x(y, t)$, the thermal conductivity may increase above the equilibrium value by 20% in the x direction and decrease by 10% in the z direction. Anisotropic heat conduction in packed beds is discussed briefly in §9.6.

---

[1] J. B. Fourier, *Théorie analytique de la chaleur*, *Œuvres de Fourier*, Gauthier-Villars et Fils, Paris (1822). **(Baron) Jean-Baptiste-Joseph Fourier** (pronounced "Foo-ree-ay") (1768–1830) was not only a brilliant mathematician and the originator of the Fourier series and the Fourier transform, but also famous as an Egyptologist and a political figure (he was prefect of the province of Isère).

[2] Some authors prefer to write Eq. 9.1-2 in the form

$$q_y = -J_e k \frac{dT}{dy} \qquad (9.1\text{-}2a)$$

in which J_e is the "mechanical equivalent of heat," which displays explicitly the conversion of thermal units into mechanical units. For example, in the c.g.s. system one would use the following units: $q_y \, [=]$ erg/cm$^2 \cdot$ s, $k \, [=]$ cal/cm $\cdot$ s $\cdot$ C, $T \, [=]$ C, $y \, [=]$ cm, and $J_e \, [=]$ erg/cal. We will not use Eq. 9.1-2a in this book.

[3] Although polymeric liquids at rest are isotropic, kinetic theory suggests that when they are flowing the heat conduction is anisotropic [see B. H. A. A. van den Brule, *Rheol. Acta*, **28**, 257–266 (1989); and C. F. Curtiss and R. B. Bird, *Advances in Polymer Science*, **25**, 1–101 (1996)]. Experimental measurements for shear and elongational flows have been reported by D. C. Venerus, J. D. Schieber, H. Iddir, J. D. Guzman, and A. W. Broerman, *Phys. Rev. Letters*, **82**, 366–369 (1999); A. W. Broerman, D. C. Venerus, and J. D. Schieber, *J. Chem. Phys.*, **111**, 6965–6969 (1999); H. Iddir, D. C. Venerus, and J. D. Schieber, *AIChE Journal*, **46**, 610–615 (2000). For oriented polymer solids, enhanced thermal conductivity in the direction of orientation has been measured by B. Poulaert, J.-C. Chielens, C. Vandenhaende, J.-P. Issi, and R. Legras, *Polymer Comm.*, **31**, 148–151 (1989). In connection with the bead spring models of polymer thermal conductivity, it has been shown by R. B. Bird, C. F. Curtiss, and K. J. Beers [*Rheol. Acta*, **36**, 269–276 (1997)] that the predicted thermal conductivity is exceedingly sensitive to the form of the potential energy used for describing the springs.

Another possible generalization of Eq. 9.1-6 is to include a term containing the time derivative of **q** multiplied by a time constant, by analogy with the Maxwell model of linear viscoelasticity in Eq. 8.4-3. There seems to be little experimental evidence that such a generalization is warranted.[4]

The reader will have noticed that Eq. 9.1-2 for heat conduction and Eq. 1.1-2 for viscous flow are quite similar. In both equations the flux is proportional to the negative of the gradient of a macroscopic variable, and the coefficient of proportionality is a physical property characteristic of the material and dependent on the temperature and pressure. For the situations in which there is three-dimensional transport, we find that Eq. 9.1-6 for heat conduction and Eq. 1.2-7 for viscous flow differ in appearance. This difference arises because energy is a scalar, whereas momentum is a vector, and the heat flux **q** is a vector with three components, whereas the momentum flux $\boldsymbol{\tau}$ is a second-order tensor with nine components. We can anticipate that the transport of energy and momentum will in general not be mathematically analogous except in certain geometrically simple situations.

In addition to the thermal conductivity k, defined by Eq. 9.1-2, a quantity known as the *thermal diffusivity* α is widely used. It is defined as

$$\alpha = \frac{k}{\rho \hat{C}_p} \tag{9.1-8}$$

Here $\hat{C}_p$ is the heat capacity at constant pressure; the circumflex ($\wedge$) over the symbol indicates a quantity "per unit mass." Occasionally we will need to use the symbol $\tilde{C}_p$ in which the tilde ($\sim$) over the symbol stands for a quantity "per mole."

The thermal diffusivity α has the same dimensions as the kinematic viscosity ν—namely, (length)^2/time. When the assumption of constant physical properties is made, the quantities ν and α occur in similar ways in the equations of change for momentum and energy transport. Their ratio ν/α indicates the relative ease of momentum and energy transport in flow systems. This dimensionless ratio

$$\text{Pr} = \frac{\nu}{\alpha} = \frac{\hat{C}_p \mu}{k} \tag{9.1-9}$$

is called the *Prandtl number*.[5] Another dimensionless group that we will encounter in subsequent chapters is the *Péclet number*,[6] Pé = RePr.

The units that are commonly used for thermal conductivity and related quantities are given in Table 9.1-1. Other units, as well as the interrelations among the various systems, may be found in Appendix F.

Thermal conductivity can vary all the way from about 0.01 W/m · K for gases to about 1000 W/m · K for pure metals. Some experimental values of the thermal con-

---

[4] The linear theory of thermoviscoelasticity does predict relaxation effects in heat conduction, as discussed by R. M. Christensen, *Theory of Viscoelasticity*, Academic Press, 2nd edition (1982). The effect has also been found from a kinetic theory treatment of the energy equation by R. B. Bird and C. F. Curtiss, *J. Non-Newtonian Fluid Mechanics*, **79**, 255–259 (1998).

[5] This dimensionless group, named for Ludwig Prandtl, involves only the physical properties of the fluid.

[6] **Jean-Claude-Eugène Péclet** (pronounced "Pay-clay" with the second syllable accented) (1793–1857) authored several books including one on heat conduction.

Table 9.1-1   Summary of Units for Quantities in Eqs. 9.1-2 and 9

	SI	c.g.s.	British
q_y	W/m^2	$cal/cm^2 \cdot s$	$Btu/hr \cdot ft^2$
T	K	C	F
y	m	cm	ft
k	$W/m \cdot K$	$cal/cm \cdot s \cdot C$	$Btu/hr \cdot ft \cdot F$
$\hat{C}_p$	$J/K \cdot kg$	$cal/C \cdot g$	$Btu/F \cdot lb_m$
α	m^2/s	cm^2/s	ft^2/s
μ	$Pa \cdot s$	$g/cm \cdot s$	$lb_m/ft \cdot hr$
Pr	—	—	—

Note: The watt (W) is the same as J/s, the joule (J) is the same as $N \cdot m$, the newton (N) is $kg \cdot m/s^2$, and the Pascal (Pa) is N/m^2. For more information on interconversion of units, see Appendix F.

ductivity of gases, liquids, liquid metals, and solids are given in Tables 9.1-2, 9.1-3, 9.1-4, and 9.1-5. In making calculations, experimental values should be used when possible. In the absence of experimental data, one can make estimates by using the methods outlined in the next several sections or by consulting various engineering handbooks.[7]

Table 9.1-2   Thermal Conductivities, Heat Capacities, and Prandtl Numbers of Some Common Gases at 1 atm Pressure[a]

Gas	Temperature T (K)	Thermal conductivity k (W/m $\cdot$ K)	Heat capacity $\hat{C}_p$ (J/kg $\cdot$ K)	Prandtl number Pr (—)
H_2	100	0.06799	11,192	0.682
	200	0.1282	13,667	0.724
	300	0.1779	14,316	0.720
O_2	100	0.00904	910	0.764
	200	0.01833	911	0.734
	300	0.02657	920	0.716
NO	200	0.01778	1015	0.781
	300	0.02590	997	0.742
CO_2	200	0.00950	734	0.783
	300	0.01665	846	0.758
CH_4	100	0.01063	2073	0.741
	200	0.02184	2087	0.721
	300	0.03427	2227	0.701

[a] Taken from J. O. Hirschfelder, C. F. Curtiss, and R. B. Bird, *Molecular Theory of Gases and Liquids*, Wiley, New York, 2nd corrected printing (1964), Table 8.4-10. The k values are measured, the $\hat{C}_p$ values are calculated from spectroscopic data, and μ is calculated from Eq. 1.4-18. The values of $\hat{C}_p$ for H_2 represent a 3:1 ortho-para mixture.

---

[7] For example, W. M. Rohsenow, J. P. Hartnett, and Y. I. Cho, eds., *Handbook of Heat Transfer*, McGraw-Hill, New York (1998); Landolt-Börnstein, *Zahlenwerte und Funktionen*, Vol. II, 5, Springer (1968–1969).

Table 9.1-3   Thermal Conductivities, Heat Capacities, and Prandtl Numbers for Some Nonmetallic Liquids at Their Saturation Pressures[a]

Liquid	Temperature T (K)	Thermal conductivity k (W/m · K)	Viscosity $\mu \times 10^4$ (Pa · s)	Heat capacity $\hat{C}_p \times 10^{-3}$ (J/kg · K)	Prandtl number Pr (—)
1-Pentene	200	0.1461	6.193	1.948	8.26
	250	0.1307	3.074	2.070	4.87
	300	0.1153	1.907	2.251	3.72
CCl_4	250	0.1092	20.32	0.8617	16.0
	300	0.09929	8.828	0.8967	7.97
	350	0.08935	4.813	0.9518	5.13
$(C_2H_5)_2O$	250	0.1478	3.819	2.197	5.68
	300	0.1274	2.213	2.379	4.13
	350	0.1071	1.387	2.721	3.53
C_2H_5OH	250	0.1808	30.51	2.120	35.8
	300	0.1676	10.40	2.454	15.2
	350	0.1544	4.486	2.984	8.67
Glycerol	300	0.2920	7949	2.418	6580
	350	0.2977	365.7	2.679	329
	400	0.3034	64.13	2.940	62.2
H_2O	300	0.6089	8.768	4.183	6.02
	350	0.6622	3.712	4.193	2.35
	400	0.6848	2.165	4.262	1.35

[a] The entries in this table were prepared from functions provided by T. E. Daubert, R. P. Danner, H. M. Sibul, C. C. Stebbins, J. L. Oscarson, R. L. Rowley, W. V. Wilding, M. E. Adams, T. L. Marshall, and N. A. Zundel, *DIPPR® Data Compilation of Pure Compound Properties*, Design Institute for Physical Property Data®, AIChE, New York, NY (2000).

EXAMPLE 9.1-1

Measurement of Thermal Conductivity

A plastic panel of area $A = 1$ ft^2 and thickness $Y = 0.252$ in. was found to conduct heat at a rate of 3.0 W at steady state with temperatures $T_0 = 24.00°C$ and $T_1 = 26.00°C$ imposed on the two main surfaces. What is the thermal conductivity of the plastic in cal/cm · s · K at 25°C?

SOLUTION

First convert units with the aid of Appendix F:

$$A = 144 \text{ in.}^2 \times (2.54)^2 = 929 \text{ cm}^2$$

$$Y = 0.252 \text{ in.} \times 2.54 = 0.640 \text{ cm}$$

$$Q = 3.0 \text{ W} \times 0.23901 = 0.717 \text{ cal/s}$$

$$\Delta T = 26.00 - 24.00 = 2.00 \text{K}$$

Substitution into Eq. 9.1-1 then gives

$$k = \frac{QY}{A\Delta T} = \frac{0.717 \times 0.640}{929 \times 2} = 2.47 \times 10^{-4} \text{ cal/cm} \cdot \text{s} \cdot \text{K} \tag{9.1-20}$$

For ΔT as small as 2 degrees C, it is reasonable to assume that the value of k applies at the average temperature, which in this case is 25°C. See Problem 10B.12 and 10C.1 for methods of accounting for the variation of k with temperature.

Table 9.1-4 Thermal Conductivities, Heat Capacities, and Prandtl Numbers of Some Liquid Metals at Atmospheric Pressure[a]

Metal	Temperature T (K)	Thermal conductivity k (W/m · K)	Heat capacity $\hat{C}_p$ (J/kg · K)	Prandtl number[c] Pr (—)
Hg	273.2	8.20	140.2	0.0288
	373.2	10.50	137.2	0.0162
	473.2	12.34	156.9	0.0116
Pb	644.2	15.9	15.9	0.024
	755.2	15.5	15.5	0.017
	977.2	15.1	14.6[b]	0.013[b]
Bi	589.2	16.3	14.4	0.0142
	811.2	15.5	15.4	0.0110
	1033.2	15.5	16.4	0.0083
Na	366.2	86.2	13.8	0.011
	644.2	72.8	13.0	0.0051
	977.2	59.8	12.6	0.0037
K	422.2	45.2	795	0.0066
	700.2	39.3	753	0.0034
	977.2	33.1	753	0.0029
Na-K alloy[c]	366.2	25.5	1130	0.026
	644.2	27.6	1054	0.0091
	977.2	28.9	1042	0.0058

[a] Data taken from *Liquid Metals Handbook*, 2nd edition, U.S. Government Printing Office, Washington, D.C. (1952), and from E. R. G. Eckert and R. M. Drake, Jr., *Heat and Mass Transfer*, McGraw-Hill, New York, 2nd edition (1959), Appendix A.
[b] Based on an extrapolated heat capacity.
[c] 56% Na by weight, 44% K by weight.

Table 9.1-5 Experimental Values of Thermal Conductivities of Some Solids[a]

Substance	Temperature T (K)	Thermal conductivity k (W/m · K)
Aluminum	373.2	205.9
	573.2	268
	873.2	423
Cadmium	273.2	93.0
	373.2	90.4
Copper	291.2	384.1
	373.2	379.9
Steel	291.2	46.9
	373.2	44.8
Tin	273.2	63.93
	373.2	59.8
Brick (common red)	—	0.63
Concrete (stone)	—	0.92
Earth's crust (average)	—	1.7
Glass (soda)	473.2	0.71
Graphite	—	5.0
Sand (dry)	—	0.389
Wood (fir)		
parallel to axis	—	0.126
normal to axis	—	0.038

[a] Data taken from the *Reactor Handbook*, Vol. 2, Atomic Energy Commission AECD-3646, U.S. Government Printing Office, Washington, D.C. (May 1955), pp. 1766 *et seq.*

§9.2 TEMPERATURE AND PRESSURE DEPENDENCE OF THERMAL CONDUCTIVITY

When thermal conductivity data for a particular compound cannot be found, one can make an estimate by using the corresponding-states chart in Fig. 9.2-1, which is based on thermal conductivity data for several monatomic substances. This chart, which is similar to that for viscosity shown in Fig. 1.3-1, is a plot of the reduced thermal conductivity $k_r = k/k_c$, which is the thermal conductivity at pressure p and temperature T divided by the thermal conductivity at the critical point. This quantity is plotted as a function of the reduced temperature $T_r = T/T_c$ and the reduced pressure $p_r = p/p_c$. Figure 9.2-1 is based on a limited amount of experimental data for monatomic substances, but may be used

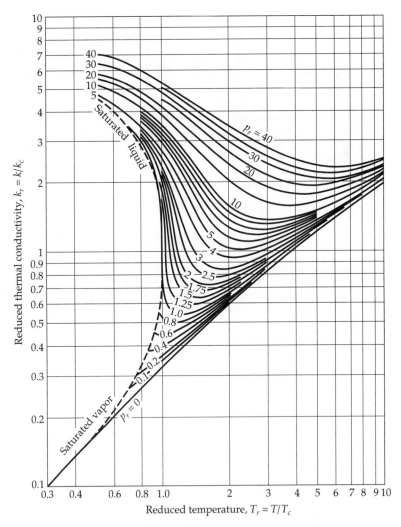

Fig. 9.2-1. Reduced thermal conductivity for monatomic substances as a function of the reduced temperature and pressure [E. J. Owens and G. Thodos, *AIChE Journal*, **3**, 454–461 (1957)]. A large-scale version of this chart may be found in O. A. Hougen, K. M. Watson, and R. A. Ragatz, *Chemical Process Principles Charts*, 2nd edition, Wiley, New York (1960).

for rough estimates for polyatomic materials. It should *not* be used in the neighborhood of the critical point.[1]

It can be seen that the thermal conductivity of a gas approaches a limiting function of T at low pressures; for most gases this limit is reached at about 1 atm pressure. The thermal conductivities of *gases* at low density *increase* with increasing temperature, whereas the thermal conductivities of most *liquids decrease* with increasing temperature. The correlation is less reliable in the liquid region; polar or associated liquids, such as water, may exhibit a maximum in the curve of k versus T. The main virtue of the corresponding-states chart is that one gets a global view of the behavior of the thermal conductivity of gases and liquids.

The quantity k_c may be estimated in one of two ways: (i) given k at a known temperature and pressure, preferably close to the conditions at which k is to be estimated, one can read k_r from the chart and compute $k_c = k/k_r$; or (ii) one can estimate a value of k in the low-density region by the methods given in §9.3 and then proceed as in (i). Values of k_c obtained by method (i) are given in Appendix E.

For mixtures, one might estimate the thermal conductivity by methods analogous to those described in §1.3. Very little is known about the accuracy of pseudocritical procedures as applied to thermal conductivity, largely because there are so few data on mixtures at elevated pressures.

EXAMPLE 9.2-1

Effect of Pressure on Thermal Conductivity

Estimate the thermal conductivity of ethane at 153°F and 191.9 atm from the experimental value[2] $k = 0.0159$ Btu/hr · ft · F at 1 atm and 153°F.

SOLUTION

Since a measured value of k is known, we use method (i). First we calculate p_r and T_r at the condition of the measured value:

$$T_r = \frac{153 + 460}{(1.8)(305.4)} = 1.115 \qquad p_r = \frac{1}{48.2} = 0.021 \tag{9.2-1}$$

From Fig. 9.2-1 we read $k_r = 0.36$. Hence k_c is

$$k_c = \frac{k}{k_r} = \frac{0.0159}{0.36} = 0.0442 \text{ Btu/hr · ft · F} \tag{9.2-2}$$

At 153°F ($T_r = 1.115$) and 191.9 atm ($p_r = 3.98$), we read from the chart $k_r = 2.07$. The predicted thermal conductivity is then

$$k = k_r k_c = (2.07)(0.0422) = 0.0914 \text{ Btu/hr · ft · F} \tag{9.2-3}$$

An observed value of 0.0453 Btu/hr · ft · F has been reported.[2] The poor agreement shows that one should not rely heavily on this correlation for polyatomic substances nor for conditions near the critical point.

---

[1] In the vicinity of the critical point, where the thermal conductivity diverges, it is customary to write $k = k^b + \Delta k$, where k^b is the "background" contribution and Δk is the "critical enhancement" contribution. The k_c being used in the corresponding states correlation is the background contribution. For the behavior of transport properties near the critical point, see J. V. Sengers and J. Luettmer Strathmann, in *Transport Properties of Fluids* (J. H. Dymond, J. Millat, and C. A. Nieto de Castro, eds.), Cambridge University Press (1995); E. P. Sakonidou, H. R. van den Berg, C. A. ten Seldam, and J. V. Sengers, *J. Chem. Phys.*, **105**, 10535–10555 (1996) and **109**, 717–736 (1998).

[2] J. M. Lenoir, W. A. Junk, and E. W. Comings, *Chem. Eng. Progr.*, **49**, 539–542 (1949).

§9.3 THEORY OF THERMAL CONDUCTIVITY OF GASES AT LOW DENSITY

The thermal conductivities of dilute *monatomic* gases are well understood and can be described by the kinetic theory of gases at low density. Although detailed theories for *polyatomic* gases have been developed,[1] it is customary to use some simple approximate theories. Here, as in §1.5, we give a simplified mean free path derivation for monatomic gases, and then summarize the result of the Chapman–Enskog kinetic theory of gases.

We use the model of rigid, nonattracting spheres of mass m and diameter d. The gas as a whole is at rest ($\mathbf{v} = 0$), but the molecular motions must be accounted for.

As in §1.5, we use the following results for a rigid-sphere gas:

$$\bar{u} = \sqrt{\frac{8\kappa T}{\pi m}} = \text{mean molecular speed} \tag{9.3-1}$$

$$Z = \tfrac{1}{4}n\bar{u} = \text{wall collision frequency per unit area} \tag{9.3-2}$$

$$\lambda = \frac{1}{\sqrt{2}\pi d^2 n} = \text{mean free path} \tag{9.3-3}$$

The molecules reaching any plane in the gas have had, on an average, their last collision at a distance a from the plane, where

$$a = \tfrac{2}{3}\lambda \tag{9.3-4}$$

In these equations κ is the Boltzmann constant, n is the number of molecules per unit volume, and m is the mass of a molecule.

The only form of energy that can be exchanged in a collision between two smooth rigid spheres is translational energy. The mean translational energy per molecule under equilibrium conditions is

$$\tfrac{1}{2}m\overline{u^2} = \tfrac{3}{2}\kappa T \tag{9.3-5}$$

as shown in Prob. 1C.1. For such a gas, the molar heat capacity at constant volume is

$$\tilde{C}_V = \left(\frac{\partial \tilde{U}}{\partial T}\right)_V = \tilde{N}\frac{d}{dT}(\tfrac{1}{2}m\overline{u^2}) = \tfrac{3}{2}R \tag{9.3-6}$$

in which R is the gas constant. Equation 9.3-6 is satisfactory for monatomic gases up to temperatures of several thousand degrees.

To determine the thermal conductivity, we examine the behavior of the gas under a temperature gradient dT/dy (see Fig. 9.3-1). We assume that Eqs. 9.3-1 to 6 remain valid in this nonequilibrium situation, except that $\tfrac{1}{2}m\overline{u^2}$ in Eq. 9.3-5 is taken as the average kinetic energy for molecules that had their last collision in a region of temperature T. The heat flux q_y across any plane of constant y is found by summing the kinetic energies of the molecules that cross the plane per unit time in the positive y direction and subtracting the kinetic energies of the equal number that cross in the negative y direction:

$$\begin{aligned} q_y &= Z(\tfrac{1}{2}m\overline{u^2}\big|_{y-a} - \tfrac{1}{2}m\overline{u^2}\big|_{y+a}) \\ &= \tfrac{3}{2}\kappa Z(T\big|_{y-a} - T\big|_{y+a}) \end{aligned} \tag{9.3-7}$$

[1] C. S. Wang Chang, G. E. Uhlenbeck, and J. de Boer, *Studies in Statistical Mechanics*, Wiley-Interscience, New York, Vol. II (1964), pp. 241–265; E. A. Mason and L. Monchick, *J. Chem. Phys.*, **35**, 1676–1697 (1961) and **36**, 1622–1639, 2746–2757 (1962); L. Monchick, A. N. G. Pereira, and E. A. Mason, *J. Chem. Phys.*, **42**, 3241–3256 (1965). For an introduction to the kinetic theory of the transport properties, see R. S. Berry, S. A. Rice, and J. Ross, *Physical Chemistry*, 2nd edition (2000), Chapter 28.

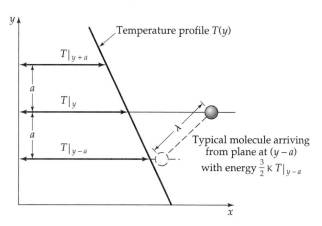

Fig. 9.3-1. Molecular transport of (kinetic) energy from plane at $(y - a)$ to plane at y.

Temperature profile $T(y)$

$T|_{y+a}$

$T|_y$

$T|_{y-a}$

Typical molecule arriving from plane at $(y - a)$ with energy $\frac{3}{2}\kappa T|_{y-a}$

Equation 9.3-7 is based on the assumption that all molecules have velocities representative of the region of their last collision and that the temperature profile $T(y)$ is linear for a distance of several mean free paths. In view of the latter assumption we may write

$$T|_{y-a} = T|_y - \tfrac{2}{3}\lambda \frac{dT}{dy} \tag{9.3-8}$$

$$T|_{y+a} = T|_y + \tfrac{2}{3}\lambda \frac{dT}{dy} \tag{9.3-9}$$

By combining the last three equations we get

$$q_y = -\tfrac{1}{2}n\kappa\bar{u}\,\lambda\,\frac{dT}{dy} \tag{9.3-10}$$

This corresponds to Fourier's law of heat conduction (Eq. 9.1-2) with the thermal conductivity given by

$$k = \tfrac{1}{2}n\kappa\bar{u}\lambda = \tfrac{1}{3}\rho\hat{C}_V\bar{u}\lambda \qquad \text{(monatomic gas)} \tag{9.3-11}$$

in which $\rho = nm$ is the gas density, and $\hat{C}_V = \tfrac{3}{2}\kappa/m$ (from Eq. 9.3-6).

Substitution of the expressions for $\bar{u}$ and λ from Eqs. 9.3-1 and 3 then gives

$$k = \frac{\sqrt{m\kappa T/\pi}}{\pi d^2}\frac{\kappa}{m} = \frac{2}{3\pi}\frac{\sqrt{\pi m\kappa T}}{\pi d^2}\hat{C}_V \qquad \text{(monatomic gas)} \tag{9.3-12}$$

which is the thermal conductivity of a dilute gas composed of rigid spheres of diameter d. This equation predicts that k is independent of pressure. Figure 9.2-1 indicates that this prediction is in good agreement with experimental data up to about 10 atm for most gases. The predicted temperature dependence is too weak, as was the case for viscosity.

For a more accurate treatment of the monatomic gas, we turn again to the rigorous Chapman–Enskog treatment discussed in §1.5. The Chapman–Enskog formula[2] for the thermal conductivity of a monatomic gas at low density and temperature T is

$$k = \frac{25}{32}\frac{\sqrt{\pi m\kappa T}}{\pi\sigma^2\Omega_k}\hat{C}_V \quad \text{or} \quad k = 1.9891 \times 10^{-4}\frac{\sqrt{T/M}}{\sigma^2\Omega_k} \qquad \text{(monatomic gas)} \tag{9.3-13}$$

In the second form of this equation, $k\,[=]\,\text{cal/cm}\cdot\text{s}\cdot\text{K}$, $T\,[=]\,\text{K}$, $\sigma\,[=]\,\text{Å}$, and the "collision integral" for thermal conductivity, Ω_k, is identical to that for viscosity, Ω_μ in §1.4.

---

[2] J. O. Hirschfelder, C. F. Curtiss, and R. B. Bird, *Molecular Theory of Gases and Liquids*, Wiley, New York, 2nd corrected printing (1964), p. 534.

Values of $\Omega_k = \Omega_\mu$ are given for the Lennard-Jones intermolecular potential in Table E.2 as a function of the dimensionless temperature $\kappa T/\varepsilon$. Equation 9.3-13, together with Table E.2, has been found to be remarkably accurate for predicting thermal conductivities of *monatomic* gases when the parameters σ and ε deduced from viscosity measurements are used (that is, the values given in Table E.1).

Equation 9.3-13 is very similar to the corresponding viscosity formula, Eq. 1.4-14. From these two equations we can then get

$$k = \frac{15}{4}\frac{R}{M}\mu = \frac{5}{2}\hat{C}_V\mu \qquad \text{(monatomic gas)} \tag{9.3-14}$$

The simplified rigid-sphere theory (see Eqs. 1.4-8 and 9.3-11) gives $k = \hat{C}_V\mu$ and is thus in error by a factor 2.5. This is not surprising in view of the many approximations that were made in the simple treatment.

So far we have discussed only *monatomic* gases. We know from the discussion in §0.3 that, in binary collisions between diatomic molecules, there may be interchanges between kinetic and internal (i.e., vibrational and rotational) energy. Such interchanges are not taken into account in the Chapman–Enskog theory for monatomic gases. It can therefore be anticipated that the Chapman–Enskog theory will not be adequate for describing the thermal conductivity of polyatomic molecules.

A simple semiempirical method of accounting for the energy exchange in polyatomic gases was developed by Eucken.[3] His equation for thermal conductivity of a polyatomic gas at low density is

$$k = \left(\hat{C}_p + \frac{5}{4}\frac{R}{M}\right)\mu \qquad \text{(polyatomic gas)} \tag{9.3-15}$$

This *Eucken formula* includes the monatomic formula (Eq. 9.3-14) as a special case, because $\hat{C}_p = \frac{5}{2}(R/M)$ for monatomic gases. Hirschfelder[4] obtained a formula similar to that of Eucken by using multicomponent-mixture theory (see Example 19.4-4). Other theories, correlations, and empirical formulas are also available.[5,6]

Equation 9.3-15 provides a simple method for estimating the Prandtl number, defined in Eq. 9.1-8:

$$\text{Pr} = \frac{\hat{C}_p\mu}{k} = \frac{\tilde{C}_p}{\tilde{C}_p + \frac{5}{4}R} \qquad \text{(polyatomic gas)} \tag{9.3-16}$$

This equation is fairly satisfactory for nonpolar polyatomic gases at low density, as can be seen in Table 9.3-1; it is less accurate for polar molecules.

The thermal conductivities for gas mixtures at low density may be estimated by a method[7] analogous to that previously given for viscosity (see Eqs. 1.4-15 and 16):

$$k_{\text{mix}} = \sum_{\alpha=1}^{N} \frac{x_\alpha k_\alpha}{\Sigma_\beta x_\beta \Phi_{\alpha\beta}} \tag{9.3-17}$$

The x_α are the mole fractions, and the k_α are the thermal conductivities of the pure chemical species. The coefficients $\Phi_{\alpha\beta}$ are identical to those appearing in the viscosity equation

---

[3] A. Eucken, *Physik. Z.*, **14**, 324–333 (1913); "Eucken" is pronounced "Oy-ken."

[4] J. O. Hirschfelder, *J. Chem. Phys.*, **26**, 274–281, 282–285 (1957).

[5] J. H. Ferziger and H. G. Kaper, *Mathematical Theory of Transport Processes in Gases*, North-Holland, Amsterdam (1972).

[6] R. C. Reid, J. M. Prausnitz, and B. E. Poling, *The Properties of Gases and Liquids*, McGraw-Hill, New York, 4th edition (1987).

[7] E. A. Mason and S. C. Saxena, *Physics of Fluids*, **1**, 361–369 (1958). Their method is an approximation to a more accurate method given by J. O. Hirschfelder, *J. Chem. Phys.*, **26**, 274–281, 282–285 (1957). With Professor Mason's approval we have omitted here an empirical factor 1.065 in his Φ_{ij} expression for $i \neq j$ to establish self-consistency for mixtures of identical species.

Table 9.3-1 Predicted and Observed Values of the Prandtl Number for Gases at Atmospheric Pressure[a]

Gas	$T(K)$	$\hat{C}_p \mu / k$ from Eq. 9.3-16	$\hat{C}_p \mu / k$ from observed values of $\hat{C}_p$, μ, and k
Ne[b]	273.2	0.667	0.66
Ar[b]	273.2	0.667	0.67
H$_2$	90.6	0.68	0.68
	273.2	0.73	0.70
	673.2	0.74	0.65
N$_2$	273.2	0.74	0.73
O$_2$	273.2	0.74	0.74
Air	273.2	0.74	0.73
CO	273.2	0.74	0.76
NO	273.2	0.74	0.77
Cl$_2$	273.2	0.76	0.76
H$_2$O	373.2	0.77	0.94
	673.2	0.78	0.90
CO$_2$	273.2	0.78	0.78
SO$_2$	273.2	0.79	0.86
NH$_3$	273.2	0.77	0.85
C$_2$H$_4$	273.2	0.80	0.80
C$_2$H$_6$	273.2	0.83	0.77
CHCl$_3$	273.2	0.86	0.78
CCl$_4$	273.2	0.89	0.81

[a] Calculated from values given by M. Jakob, *Heat Transfer*, Wiley, New York (1949), pp. 75–76.
[b] J. O. Hirschfelder, C. F. Curtiss, and R. B. Bird, *Molecular Theory of Gases and Liquids*, Wiley, New York, corrected printing (1964), p. 16.

(see Eq. 1.4-16). All values of k_α in Eq. 9.3-17 and μ_α in Eq. 1.4-16 are low-density values at the given temperature. If viscosity data are not available, they may be estimated from k and $\hat{C}_p$ via Eq. 9.3-15. Comparisons with experimental data[7] indicate an average deviation of about 4% for mixtures containing nonpolar polyatomic gases, including O$_2$, N$_2$, CO, C$_2$H$_2$, and CH$_4$.

EXAMPLE 9.3-1

Computation of the Thermal Conductivity of a Monatomic Gas at Low Density

Compute the thermal conductivity of Ne at 1 atm and 373.2K.

SOLUTION

From Table E.1 the Lennard-Jones constants for neon are $\sigma = 2.789$ Å and $\varepsilon/\kappa = 35.7$K, and its molecular weight M is 20.183. Then, at 373.2K, we have $\kappa T/\varepsilon = 373.2/35.7 = 10.45$. From Table E.2 we find that $\Omega_k = \Omega_\mu = 0.821$. Substitution into Eq. 9.3-13 gives

$$k = (1.9891 \times 10^{-4}) \frac{\sqrt{T/M}}{\sigma^2 \Omega_k}$$

$$= (1.9891 \times 10^{-4}) \frac{\sqrt{(373.2)/(20.183)}}{(2.789)^2 (0.821)}$$

$$= 1.338 \times 10^{-4} \text{ cal/cm} \cdot \text{s} \cdot \text{K} \tag{9.3-18}$$

A measured value of 1.35×10^{-4} cal/cm · s · K has been reported[8] at 1 atm and 373.2K.

---

[8] W. G. Kannuluik and E. H. Carman, *Proc. Phys. Soc.* (London), **65B**, 701–704 (1952).

EXAMPLE 9.3-2

Estimation of the Thermal Conductivity of a Polyatomic Gas at Low Density

Estimate the thermal conductivity of molecular oxygen at 300K and low pressure.

SOLUTION

The molecular weight of O_2 is 32.0000; its molar heat capacity $\tilde{C}_p$ at 300°K and low pressure is 7.019 cal/g-mole · K. From Table E.1 we find the Lennard-Jones parameters for molecular oxygen to be $\sigma = 3.433$ Å and $\varepsilon/\kappa = 113$K. At 300K, then, $\kappa T/\varepsilon = 300/113 = 2.655$. From Table E.2, we find $\Omega_\mu = 1.074$. The viscosity, from Eq. 1.4-14, is

$$\mu = (2.6693 \times 10^{-5}) \frac{\sqrt{MT}}{\sigma^2 \Omega_\mu}$$

$$= (2.6693 \times 10^{-5}) \frac{\sqrt{(32.00)(300)}}{(3.433)^2 (1074)}$$

$$= 2.065 \times 10^{-5} \, \text{g/cm} \cdot \text{s} \tag{9.3-19}$$

Then, from Eq. 9.3-15, the Eucken approximation to the thermal conductivity is

$$k = (\tilde{C}_p + \tfrac{5}{4}R)(\mu/M)$$

$$= (7.019 + 2.484)((2.065 \times 10^{-4})/(32.000))$$

$$= 6.14 \times 10^{-5} \, \text{cal/cm} \cdot \text{s} \cdot \text{K} = 0.0257 \, \text{W/m} \cdot \text{K} \tag{9.3-20}$$

This compares favorably with the experimental value of 0.02657 W/m · K in Table 9.1-2.

EXAMPLE 9.3-3

Prediction of the Thermal Conductivity of a Gas Mixture at Low Density

Predict the thermal conductivity of the following gas mixture at 1 atm and 293K from the given data on the pure components at the same pressure and temperature:

Species	α	Mole fraction x_α	Molecular weight M_α	$\mu_\alpha \times 10^7$ (g/cm · s)	$k_\alpha \times 10^7$ (cal/cm · s · K)
CO_2	1	0.133	44.010	1462	383
O_2	2	0.039	32.000	2031	612
N_2	3	0.828	28.016	1754	627

SOLUTION

Use Eq. 9.3-17. We note that the $\Phi_{\alpha\beta}$ for this gas mixture at these conditions have already been computed in the viscosity calculation in Example 1.4-2. In that example we evaluated the following summations, which also appear in Eq. 9.3-17:

$\alpha \rightarrow$	1	2	3
$\sum_{\beta=1}^{3} x_\beta \Phi_{\alpha\beta}$	0.763	1.057	1.049

Substitution in Eq. 9.3-17 gives

$$k_{\text{mix}} = \sum_{\alpha=1}^{N} \frac{x_\alpha k_\alpha}{\sum_\beta x_\beta \Phi_{\alpha\beta}}$$

$$= \frac{(0.133)(383)(10^{-7})}{0.763} + \frac{(0.039)(612)(10^{-7})}{1.057} + \frac{(0.828)(627)(10^{-7})}{1.049}$$

$$= 584 \times (10^{-7}) \, \text{cal/cm} \cdot \text{s} \cdot \text{K} \tag{9.3-22}$$

No data are available for comparison at these conditions.

§9.4 THEORY OF THERMAL CONDUCTIVITY OF LIQUIDS

A very detailed kinetic theory for the thermal conductivity of monatomic liquids was developed a half-century ago,[1] but it has not yet been possible to implement it for practical calculations. As a result we have to use rough theories or empirical estimation methods.[2]

We choose to discuss here Bridgman's simple theory[3] of energy transport in pure liquids. He assumed that the molecules are arranged in a cubic lattice, with a center-to-center spacing given by $(\tilde{V}/\tilde{N})^{1/3}$, in which $\tilde{V}/\tilde{N}$ is the volume per molecule. He further assumed energy to be transferred from one lattice plane to the next at the sonic velocity v_s for the given fluid. The development is based on a reinterpretation of Eq. 9.3-11 of the rigid-sphere gas theory:

$$k = \tfrac{1}{3}\rho\hat{C}_V\bar{u}\lambda = \rho\hat{C}_V\overline{|u_y|}a \tag{9.4-1}$$

The heat capacity at constant volume of a monatomic liquid is about the same as for a solid at high temperature, which is given by the Dulong and Petit formula[4] $\hat{C}_V = 3(\kappa/m)$. The mean molecular speed in the y direction, $\overline{|u_y|}$, is replaced by the sonic velocity v_s. The distance a that the energy travels between two successive collisions is taken to be the lattice spacing $(\tilde{V}/\tilde{N})^{1/3}$. Making these substitutions in Eq. 9.4-1 gives

$$k = 3(\tilde{N}/\tilde{V})^{2/3}\kappa v_s \tag{9.4-2}$$

which is *Bridgman's equation*. Experimental data show good agreement with Eq. 9.4-2, even for polyatomic liquids, but the numerical coefficient is somewhat too high. Better agreement is obtained if the coefficient is changed to 2.80:

$$k = 2.80(\tilde{N}/\tilde{V})^{2/3}\kappa v_s \tag{9.4-3}[5]$$

This equation is limited to densities well above the critical density, because of the tacit assumption that each molecule oscillates in a "cage" formed by its nearest neighbors. The success of this equation for polyatomic fluids seems to imply that the energy transfer in collisions of polyatomic molecules is incomplete, since the heat capacity used here, $\hat{C}_V = 3(\kappa/m)$, is less than the heat capacities of polyatomic liquids.

The velocity of low-frequency sound is given (see Problem 11C.1) by

$$v_s = \sqrt{\frac{C_p}{C_V}\left(\frac{\partial p}{\partial\rho}\right)_T} \tag{9.4-4}$$

The quantity $(\partial p/\partial\rho)_T$ may be obtained from isothermal compressibility measurements or from an equation of state, and (C_p/C_V) is very nearly unity for liquids, except near the critical point.

---

[1] J. H. Irving and J. G. Kirkwood, *J. Chem. Phys.*, **18**, 817–829 (1950). This theory has been extended to polymeric liquids by C. F. Curtiss and R. B. Bird, *J. Chem. Phys.*, **107**, 5254–5267 (1997).

[2] R. C. Reid, J. M. Prausnitz, and B. E. Poling, *The Properties of Gases and Liquids*, McGraw-Hill, New York (1987); L. Riedel, *Chemie-Ing.-Techn.*, **27**, 209–213 (1955).

[3] P. W. Bridgman, *Proc. Am. Acad. Arts and Sci.*, **59**, 141–169 (1923). Bridgman's equation is often misquoted, because he gave it in terms of a little-known gas constant equal to $\tfrac{3}{2}\kappa$.

[4] This empirical equation has been justified, and extended, by A. Einstein [*Ann. Phys.* [4], **22**, 180–190 (1907)] and P. Debye [*Ann. Phys.*, [4] **39**, 789–839 (1912)].

[5] Equation 9.4-3 is in approximate agreement with a formula derived by R. E. Powell, W. E. Roseveare, and H. Eyring, *Ind. Eng. Chem.*, **33**, 430–435 (1941).

EXAMPLE 9.4-1

Prediction of the Thermal Conductivity of a Liquid

The density of liquid CCl_4 at 20°C and 1 atm is 1.595 g/cm^3, and its isothermal compressibility $(1/\rho)(\partial\rho/\partial p)_T$ is 90.7×10^{-6} atm^{-1}. What is its thermal conductivity?

SOLUTION

First compute

$$\left(\frac{\partial p}{\partial \rho}\right)_T = \frac{1}{\rho(1/\rho)(\partial\rho/\partial p)_T} = \frac{1}{(1.595)(90.7 \times 10^{-6})} = 6.91 \times 10^3 \text{ atm} \cdot \text{cm}^3/\text{g}$$

$$= 7.00 \times 10^9 \text{ cm}^2/\text{s}^2 \text{ (using Appendix F)} \tag{9.4-5}$$

Assuming that $C_p/C_V = 1.0$, we get from Eq. 9.4-4

$$v_s = \sqrt{(1.0)(7.00 \times 10^9)} = 8.37 \times 10^4 \text{ cm/s} \tag{9.4-6}$$

The molar volume is $\tilde{V} = M/\rho = 153.84/1.595 = 96.5$ cm^3/g-mole. Substitution of these values in Eq. 9.4-3 gives

$$k = 2.80(\tilde{N}/\tilde{V})^{2/3} \kappa v_s$$

$$= 2.80\left(\frac{6.023 \times 10^{23}}{0.965 \times 10^2}\right)^{2/3} (1.3805 \times 10^{-16})(8.37 \times 10^4)$$

$$= 1.10 \times 10^4 \text{ (cm}^{-2}\text{)(erg/K)(cm/s)}$$

$$= 0.110 \text{ W/m} \cdot \text{K} \tag{9.4-7}$$

The experimental value as interpolated from Table 9.1-3 is 0.101 W/m · K.

§9.5 THERMAL CONDUCTIVITY OF SOLIDS

Thermal conductivities of solids have to be measured experimentally, since they depend on many factors that are difficult to measure or predict.[1] In crystalline materials, the phase and crystallite size are important; in amorphous solids the degree of molecular orientation has a considerable effect. In porous solids, the thermal conductivity is strongly dependent on the void fraction, the pore size, and the fluid contained in the pores. A detailed discussion of thermal conductivity of solids has been given by Jakob.[2]

In general, metals are better heat conductors than nonmetals, and crystalline materials conduct heat more readily than amorphous materials. Dry porous solids are very poor heat conductors and are therefore excellent for thermal insulation. The conductivities of most pure metals decrease with increasing temperature, whereas the conductivities of nonmetals increase; alloys show intermediate behavior. Perhaps the most useful of the rules of thumb is that thermal and electrical conductivity go hand in hand.

For pure metals, as opposed to alloys, the thermal conductivity k and the electrical conductivity k_e are related approximately[3] as follows:

$$\frac{k}{k_e T} = L = \text{constant} \tag{9.5-1}$$

This is the *Wiedemann–Franz–Lorenz equation*; this equation can also be explained theoretically (see Problem 9A.6). The "Lorenz number" L is about 22 to 29 $\times 10^{-9}$ volt^2/K^2 for

---

[1] A. Goldsmith, T. E. Waterman, and H. J. Hirschhorn, eds., *Handbook of Thermophysical Properties of Solids*, Macmillan, New York (1961).

[2] M. Jakob, *Heat Transfer*, Vol. 1, Wiley, New York (1949), Chapter 6. See also W. H. Rohsenow, J. P. Hartnett, and Y. I. Cho, eds., Handbook of *Heat Transfer*, McGraw-Hill, New York (1998).

[3] G. Wiedemann and R. Franz, *Ann. Phys. u. Chemie*, **89**, 497–531 (1853); L. Lorenz, *Poggendorff's Annalen*, **147**, 429–452 (1872).

pure metals at 0°C and changes but little with temperatures above 0°C, increases of 10–20% per 1000°C being typical. At very low temperatures (-269.4°C for mercury) metals become superconductors of electricity but not of heat, and L thus varies strongly with temperature near the superconducting region. Equation 9.5-1 is of limited use for alloys, since L varies strongly with composition and, in some cases, with temperature.

The success of Eq. 9.5-1 for pure metals is due to the fact that free electrons are the major heat carriers in pure metals. The equation is not suitable for nonmetals, in which the concentration of free electrons is so low that energy transport by molecular motion predominates.

§9.6   EFFECTIVE THERMAL CONDUCTIVITY OF COMPOSITE SOLIDS

Up to this point we have discussed homogeneous materials. Now we turn our attention briefly to the thermal conductivity of two-phase solids—one solid phase dispersed in a second solid phase, or solids containing pores, such as granular materials, sintered metals, and plastic foams. A complete description of the heat transport through such materials is clearly extremely complicated. However, for steady conduction these materials can be regarded as homogeneous materials with an *effective thermal conductivity k_{eff}*, and the temperature and heat flux components are reinterpreted as the analogous quantities averaged over a volume that is large with respect to the scale of the heterogeneity but small with respect to the overall dimensions of the heat conduction system.

The first major contribution to the estimation of the conductivity of heterogeneous solids was by Maxwell.[1] He considered a material made of spheres of thermal conductivity k_1 embedded in a continuous solid phase with thermal conductivity k_0. The volume fraction ϕ of embedded spheres is taken to be sufficiently small that the spheres do not "interact" thermally; that is, one needs to consider only the thermal conduction in a large medium containing only one embedded sphere. Then by means of a surprisingly simple derivation, Maxwell showed that for *small volume fraction ϕ*

$$\frac{k_{eff}}{k_0} = 1 + \frac{3\phi}{\left(\dfrac{k_1 + 2k_0}{k_1 - k_0}\right) - \phi} \tag{9.6-1}$$

(see Problems 11B.8 and 11C.5).

For *large volume fraction ϕ*, Rayleigh[2] showed that, if the spheres are located at the intersections of a cubic lattice, the thermal conductivity of the composite is given by

$$\frac{k_{eff}}{k_0} = 1 + \frac{3\phi}{\left(\dfrac{k_1 + 2k_0}{k_1 - k_0}\right) - \phi + 1.569\left(\dfrac{k_1 - k_0}{3k_1 - 4k_0}\right)\phi^{10/3} + \cdots} \tag{9.6-2}$$

Comparison of this result with Eq. 9.6-1 shows that the interaction between the spheres is small, even at $\phi = \frac{1}{6}\pi$, the maximum possible value of ϕ for the cubic lattice arrangement. Therefore the simpler result of Maxwell is often used, and the effects of nonuniform sphere distribution are usually neglected.

---

[1] Maxwell's derivation was for electrical conductivity, but the same arguments apply for thermal conductivity. See J. C. Maxwell, *A Treatise on Electricity and Magnetism*, Oxford University Press, 3rd edition (1891, reprinted 1998), Vol. 1, §314; H. S. Carslaw and J. C. Jaeger, *Conduction of Heat in Solids*, Clarendon Press, Oxford, 2nd edition (1959), p. 428.

[2] J. W. Strutt (Lord Rayleigh), *Phil. Mag.* (5), **34**, 431–502 (1892).

For *nonspherical inclusions*, however, Eq. 9.6-1 does require modification. Thus for square arrays of long cylinders parallel to the z axis, Rayleigh[2] showed that the zz component of the thermal conductivity tensor κ is

$$\frac{\kappa_{\text{eff},zz}}{k_0} = 1 + \left(\frac{k_1 - k_0}{k_0}\right)\phi \tag{9.6-3}$$

and the other two components are

$$\frac{\kappa_{\text{eff},xx}}{k_0} = \frac{\kappa_{\text{eff},yy}}{k_0} = 1 + \cfrac{2\phi}{\left(\cfrac{k_1 + k_0}{k_1 - k_0}\right) - \phi + \left(\cfrac{k_1 - k_0}{k_1 + k_0}\right)(0.30584\phi^4 + 0.013363\phi^8 + \cdots)} \tag{9.6-4}$$

That is, the composite solid containing aligned embedded cylinders is anisotropic. The effective thermal conductivity tensor has been computed up to $O(\phi^2)$ for a medium containing spheroidal inclusions.[3]

For *complex nonspherical inclusions*, often encountered in practice, no exact treatment is possible, but some approximate relations are available.[4,5,6] For simple unconsolidated granular beds the following expression has proven successful:

$$\frac{k_{\text{eff}}}{k_0} = \frac{(1 - \phi) + \alpha\phi(k_1/k_0)}{(1 - \phi) + \alpha\phi} \tag{9.6-5}$$

in which

$$\alpha = \frac{1}{3}\sum_{k=1}^{3}\left[1 + \left(\frac{k_1}{k_0} - 1\right)g_k\right]^{-1} \tag{9.6-6}$$

The g_k are "shape factors" for the granules of the medium,[7] and they must satisfy $g_1 + g_2 + g_3 = 1$. For spheres $g_1 = g_2 = g_3 = \frac{1}{3}$, and Eq. 9.6-5 reduces to Eq. 9.6-1. For unconsolidated soils,[5] $g_1 = g_2 = \frac{1}{8}$ and $g_3 = \frac{3}{4}$. The structure of consolidated porous beds—for example, sandstones—is considerably more complex. Some success is claimed for predicting the effective conductivity of such substances,[4,6,8] but the generality of the methods is not yet known.

For *solids containing gas pockets*,[9] thermal radiation (see Chapter 16) may be important. The special case of parallel planar fissures perpendicular to the direction of heat conduction is particularly important for high-temperature insulation. For such systems it may be shown that

$$\frac{k_{\text{eff}}}{k_0} = \cfrac{1}{1 - \phi + \left(\cfrac{k_1}{k_0\phi} + \cfrac{4\sigma T^3 L}{k_0}\right)^{-1}} \tag{9.6-7}$$

where σ is the Stefan–Boltzmann constant, k_1 is the thermal conductivity of the gas, and L is the total thickness of the material in the direction of the heat conduction. A modification of this equation for fissures of other shapes and orientations is available.[7]

---

[3] S.-Y. Lu and S. Kim, *AIChE Journal*, **36**, 927–938 (1990).

[4] V. I. Odelevskii, *J. Tech. Phys.* (USSR), **24**, 667 and 697 (1954); F. Euler, *J. Appl. Phys.*, **28**, 1342–1346 (1957).

[5] D. A. de Vries, *Mededelingen van de Landbouwhogeschool te Wageningen*, (1952); see also Ref. 6 and D. A. de Vries, Chapter 7 in *Physics of Plant Environment*, W. R. van Wijk, ed., Wiley, New York (1963).

[6] W. Woodside and J. H. Messmer, *J. Appl. Phys.*, **32**, 1688–1699, 1699–1706 (1961).

[7] A. L. Loeb, *J. Amer. Ceramic Soc.*, **37**, 96–99 (1954).

[8] Sh. N. Plyat, *Soviet Physics JETP*, **2**, 2588–2589 (1957).

[9] M. Jakob, *Heat Transfer*, Wiley, New York (1959), Vol. 1, §6.5.

For *gas-filled granular beds*[6,9] a different type of complication arises. Since the thermal conductivities of gases are much lower than those of solids, most of the gas-phase heat conduction is concentrated near the points of contact of adjacent solid particles. As a result, the distances over which the heat is conducted through the gas may approach the mean free path of the gas molecules. When this is true, the conditions for the developments of §9.3 are violated, and the thermal conductivity of the gas decreases. Very effective insulators can thus be prepared from partially evacuated beds of fine powders.

Cylindrical ducts filled with granular materials through which a fluid is flowing (in the z direction) are of considerable importance in separation processes and chemical reactors. In such systems the effective thermal conductivities in the radial and axial directions are quite different and are designated[10] by $\kappa_{\mathrm{eff},rr}$ and $\kappa_{\mathrm{eff},zz}$. Conduction, convection, and radiation all contribute to the flow of heat through the porous medium.[11] For highly turbulent flow, the energy is transported primarily by the tortuous flow of the fluid in the interstices of the granular material; this gives rise to a highly anisotropic thermal conductivity. For a bed of uniform spheres, the radial and axial components are approximately

$$\kappa_{\mathrm{eff},rr} = \tfrac{1}{10}\rho\hat{C}_p v_0 D_p; \qquad \kappa_{\mathrm{eff},zz} = \tfrac{1}{2}\rho\hat{C}_p v_0 D_p \qquad (9.6\text{-}8, 9)$$

in which v_0 is the "superficial velocity" defined in §4.3 and §6.4, and D_p is the diameter of the spherical particles. These simplified relations hold for $\mathrm{Re} = D_p v_0 \rho/\mu$ greater than 200. The behavior at lower Reynolds numbers is discussed in several references.[12] Also, the behavior of the effective thermal conductivity tensor as a function of the Péclet number has been studied in considerable detail.[13]

§9.7   CONVECTIVE TRANSPORT OF ENERGY

In §9.1 we gave Fourier's law of heat conduction, which accounts for the energy transported through a medium by virtue of the molecular motions.

Energy may also be transported by the bulk motion of the fluid. In Fig. 9.7-1 we show three mutually perpendicular elements of area dS at the point P, where the fluid

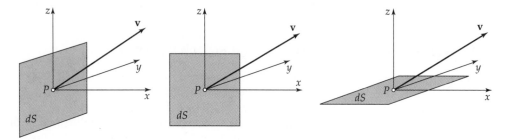

Fig. 9.7-1. Three mutually perpendicular surface elements of area dS across which energy is being transported by convection by the fluid moving with the velocity $\mathbf{v}$. The volume rate of flow across the face perpendicular to the x-axis is $v_x dS$, and the rate of flow of energy across dS is then $(\tfrac{1}{2}\rho v^2 + \rho\hat{U})v_x dS$. Similar expressions can be written for the surface elements perpendicular to the y- and z-axes.

---

[10] See Eq. 9.1-7 for the modification of Fourier's law for anisotropic materials. The subscripts rr and zz emphasize that these quantities are components of a second-order symmetrical tensor.

[11] W. B. Argo and J. M. Smith, *Chem. Engr. Progress*, **49**, 443–451 (1953).

[12] J. Beek, *Adv. Chem. Engr.*, **3**, 203–271 (1962); H. Kramers and K. R. Westerterp, *Elements of Chemical Reactor Design and Operation*, Academic Press, New York (1963), §III.9; O. Levenspiel and K. B. Bischoff, *Adv. Chem. Engr.*, **4**, 95–198 (1963).

[13] D. L. Koch and J. F. Brady, *J. Fluid Mech.*, **154**, 399–427 (1985).

velocity is **v**. The volume rate of flow across the surface element dS perpendicular to the x-axis is $v_x dS$. The rate at which energy is being swept across the same surface element is then

$$(\tfrac{1}{2}\rho v^2 + \rho \hat{U})v_x \, dS \tag{9.7-1}$$

in which $\tfrac{1}{2}\rho v^2 = \tfrac{1}{2}\rho(v_x^2 + v_y^2 + v_z^2)$ is the kinetic energy per unit volume, and $\rho \hat{U}$ is the internal energy per unit volume.

The definition of the internal energy in a nonequilibrium situation requires some care. From the *continuum* point of view, the internal energy at position **r** and time t is assumed to be the same function of the local, instantaneous density and temperature that one would have at equilibrium. From the *molecular* point of view, the internal energy consists of the sum of the kinetic energies of all the constituent atoms (relative to the flow velocity **v**), the intramolecular potential energies, and the intermolecular energies, within a small region about the point **r** at time t.

Recall that, in the discussion of molecular collisions in §0.3, we found it convenient to regard the energy of a colliding pair of molecules to be the sum of the kinetic energies referred to the center of mass of the molecule plus the intramolecular potential energy of the molecule. Here also we split the energy of the fluid (regarded as a continuum) into kinetic energy associated with the bulk fluid motion and the internal energy associated with the kinetic energy of the molecules with respect to the flow velocity and the intra- and intermolecular potential energies.

We can write expressions similar to Eq. 9.7-1 for the rate at which energy is being swept through the surface elements perpendicular to the y- and z-axes. If we now multiply each of the three expressions by the corresponding unit vector and add, we then get, after division by dS,

$$(\tfrac{1}{2}\rho v^2 + \rho \hat{U})\boldsymbol{\delta}_x v_x + (\tfrac{1}{2}\rho v^2 + \rho \hat{U})\boldsymbol{\delta}_y v_y + (\tfrac{1}{2}\rho v^2 + \rho \hat{U})\boldsymbol{\delta}_z v_z = (\tfrac{1}{2}\rho v^2 + \rho \hat{U})\mathbf{v} \tag{9.7-2}$$

and this quantity is called the *convective energy flux vector*. To get the convective energy flux across a unit surface whose normal unit vector is **n**, we form the dot product $(\mathbf{n} \cdot (\tfrac{1}{2}\rho v^2 + \rho \hat{U})\mathbf{v})$. It is understood that this is the flux from the negative side of the surface to the positive side. Compare this with the convective momentum flux in Fig. 1.7-2.

§9.8   WORK ASSOCIATED WITH MOLECULAR MOTIONS

Presently we will be concerned with applying the law of conservation of energy to "shells" (as in the shell balances in Chapter 10) or to small elements of volume fixed in space (to develop the equation of change for energy in §11.1). The law of conservation of energy for an open flow system is an extension of the first law of classical thermodynamics (for a closed system at rest). In the latter we state that the change in internal energy is equal to the amount of heat added to the system plus the amount of work done on the system. For flow systems we shall need to account for the heat added to the system (by molecular motions and by bulk fluid motion) and also for the work done on the system by the molecular motions. Therefore it is appropriate that we develop here the expression for the rate of work done by the molecular motions.

First we recall that, when a force **F** acts on a body and causes it to move through a distance $d\mathbf{r}$, the work done is $dW = (\mathbf{F} \cdot d\mathbf{r})$. Then the rate of doing work is $dW/dt = (\mathbf{F} \cdot d\mathbf{r}/dt) = (\mathbf{F} \cdot \mathbf{v})$—that is, the dot product of the force times the velocity. We now apply this formula to the three perpendicular planes at a point P in space shown in Fig. 9.8-1.

First we consider the surface element perpendicular to the x-axis. The fluid on the minus side of the surface exerts a force $\boldsymbol{\pi}_x dS$ on the fluid that is on the plus side (see

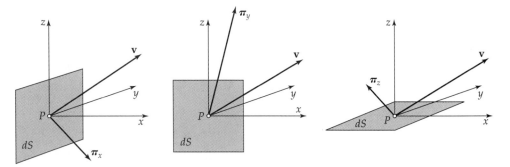

Fig. 9.8-1. Three mutually perpendicular surface elements of area dS at point P along with the stress vectors $\boldsymbol{\pi}_x, \boldsymbol{\pi}_y, \boldsymbol{\pi}_z$ acting on these surfaces. In the first figure, the rate at which work is done by the fluid on the minus side of dS on the fluid on the plus side of dS is then $(\boldsymbol{\pi}_x \cdot \mathbf{v})dS = [\boldsymbol{\pi} \cdot \mathbf{v}]_x dS$. Similar expressions hold for the surface elements perpendicular to the other two coordinate axes.

Table 1.2-1). Since the fluid is moving with a velocity $\mathbf{v}$, the rate at which work is done by the minus fluid on the plus fluid is $(\boldsymbol{\pi}_x \cdot \mathbf{v})dS$. Similar expressions may be written for the work done across the other two surface elements. When written out in component form, these rate of work expressions, per unit area, become

$$(\boldsymbol{\pi}_x \cdot \mathbf{v}) = \pi_{xx}v_x + \pi_{xy}v_y + \pi_{xz}v_z \equiv [\boldsymbol{\pi} \cdot \mathbf{v}]_x \tag{9.8-1}$$

$$(\boldsymbol{\pi}_y \cdot \mathbf{v}) = \pi_{yx}v_x + \pi_{yy}v_y + \pi_{yz}v_z \equiv [\boldsymbol{\pi} \cdot \mathbf{v}]_y \tag{9.8-2}$$

$$(\boldsymbol{\pi}_z \cdot \mathbf{v}) = \pi_{zx}v_x + \pi_{zy}v_y + \pi_{zz}v_z \equiv [\boldsymbol{\pi} \cdot \mathbf{v}]_z \tag{9.8-3}$$

When these scalar components are multiplied by the unit vectors and added, we get the "rate of doing work vector per unit area," and we can call this, for short, the *work flux*:

$$[\boldsymbol{\pi} \cdot \mathbf{v}] = \boldsymbol{\delta}_x(\boldsymbol{\pi}_x \cdot \mathbf{v}) + \boldsymbol{\delta}_y(\boldsymbol{\pi}_y \cdot \mathbf{v}) + \boldsymbol{\delta}_z(\boldsymbol{\pi}_z \cdot \mathbf{v}) \tag{9.8-4}$$

Furthermore, the rate of doing work across a unit area of surface with orientation given by the unit vector $\mathbf{n}$ is $(\mathbf{n} \cdot [\boldsymbol{\pi} \cdot \mathbf{v}])$.

Equations 9.8-1 to 9.8-4 are easily written for cylindrical coordinates by replacing x, y, z by r, θ, z and, for spherical coordinates by replacing x, y, z by r, θ, ϕ.

We now define, for later use, the *combined energy flux vector* $\mathbf{e}$ as follows:

$$\mathbf{e} = (\tfrac{1}{2}\rho v^2 + \rho \hat{U})\mathbf{v} + [\boldsymbol{\pi} \cdot \mathbf{v}] + \mathbf{q} \tag{9.8-5}$$

The $\mathbf{e}$ vector is the sum of (a) the convective energy flux, (b) the rate of doing work (per unit area) by molecular mechanisms, and (c) the rate of transporting heat (per unit area) by molecular mechanisms. All the terms in Eq. 9.8-5 have the same sign convention, so that e_x is the energy transport in the positive x direction per unit area per unit time.

The total molecular stress tensor $\boldsymbol{\pi}$ can now be split into two parts: $\boldsymbol{\pi} = p\boldsymbol{\delta} + \boldsymbol{\tau}$ so that $[\boldsymbol{\pi} \cdot \mathbf{v}] = p\mathbf{v} + [\boldsymbol{\tau} \cdot \mathbf{v}]$. The term $p\mathbf{v}$ can then be combined with the internal energy term $\rho \hat{U}\mathbf{v}$ to give an enthalpy term $\rho \hat{U}\mathbf{v} + p\mathbf{v} = \rho(\hat{U} + (p/\rho))\mathbf{v} = \rho(\hat{U} + p\hat{V})\mathbf{v} = \rho\hat{H}\mathbf{v}$, so that

$$\mathbf{e} = (\tfrac{1}{2}\rho v^2 + \rho \hat{H})\mathbf{v} + [\boldsymbol{\tau} \cdot \mathbf{v}] + \mathbf{q} \tag{9.8-6}$$

We shall usually use the $\mathbf{e}$ vector in this form. For a surface element dS of orientation $\mathbf{n}$, the quantity $(\mathbf{n} \cdot \mathbf{e})$ gives the convective energy flux, the heat flux, and the work flux across the surface element dS from the negative side to the positive side of dS.

Table 9.8-1   Summary of Notation for Energy Fluxes

Symbol	Meaning	Reference
$(\frac{1}{2}\rho v^2 + \rho\hat{U})\mathbf{v}$	convective energy flux vector	Eq. 9.7-2
$\mathbf{q}$	molecular heat flux vector	Eq. 9.1-6
$[\boldsymbol{\pi} \cdot \mathbf{v}]$	molecular work flux vector	Eq. 9.8-4
$\mathbf{e} = \mathbf{q} + [\boldsymbol{\pi} \cdot \mathbf{v}] + (\frac{1}{2}\rho v^2 + \rho\hat{U})\mathbf{v}$ $\phantom{\mathbf{e}} = \mathbf{q} + [\boldsymbol{\tau} \cdot \mathbf{v}] + (\frac{1}{2}\rho v^2 + \rho\hat{H})\mathbf{v}$	combined energy flux vector	Eq. 9.8-5, 6

In Table 9.8-1 we summarize the notation for the various energy flux vectors introduced in this section. All of them have the same sign convention.

To evaluate the enthalpy in Eq. 9.8-6, we make use of the standard equilibrium thermodynamics formula

$$d\hat{H} = \left(\frac{\partial\hat{H}}{\partial T}\right)_p dT + \left(\frac{\partial\hat{H}}{\partial p}\right)_T dp = \hat{C}_p dT + \left[\hat{V} - T\left(\frac{\partial\hat{V}}{\partial T}\right)_p\right] dp \tag{9.8-7}$$

When this is integrated from some reference state $p°$, $T°$ to the state p, T, we then get[1]

$$\hat{H} - \hat{H}° = \int_{T°}^{T} \hat{C}_p dT + \int_{p°}^{p}\left[\hat{V} - T\left(\frac{\partial\hat{V}}{\partial T}\right)_p\right] dp \tag{9.8-8}$$

in which $\hat{H}°$ is the enthalpy per unit mass at the reference state. The integral over p is zero for an ideal gas and $(1/\rho)(p - p°)$ for fluids of constant density. The integral over T becomes $\hat{C}_p(T - T°)$ if the heat capacity can be regarded as constant over the relevant temperature range. It is assumed that Eq. 9.8-7 is valid in nonequilibrium systems, where p and T are the *local* values of the pressure and temperature.

QUESTIONS FOR DISCUSSION

1. Define and give the dimensions of thermal conductivity k, thermal diffusivity α, heat capacity $\hat{C}_p$, heat flux $\mathbf{q}$, and combined energy flux $\mathbf{e}$. For the dimensions use m = mass, l = length, T = temperature, and t = time.
2. Compare the orders of magnitude of the thermal conductivities of gases, liquids, and solids.
3. In what way are Newton's law of viscosity and Fourier's law of heat conduction similar? Dissimilar?
4. Are gas viscosities and thermal conductivities related? If so, how?
5. Compare the temperature dependence of the thermal conductivities of gases, liquids, and solids.
6. Compare the orders of magnitudes of Prandtl numbers for gases and liquids.
7. Are the thermal conductivities of gaseous Ne[20] and Ne[22] the same?
8. Is the relation $\tilde{C}_p - \tilde{C}_V = R$ true only for ideal gases, or is it also true for liquids? If it is not true for liquids, what formula should be used?
9. What is the kinetic energy flux in the axial direction for the laminar Poiseuille flow of a Newtonian liquid in a circular tube?
10. What is $[\boldsymbol{\pi} \cdot \mathbf{v}] = p\mathbf{v} + [\boldsymbol{\tau} \cdot \mathbf{v}]$ for Poiseuille flow?

---

[1] See, for example, R. J. Silbey and R. A. Alberty, *Physical Chemistry*, Wiley, 3rd edition (2001), §2.11.

PROBLEMS

9A.1 Prediction of thermal conductivities of gases at low density.

(a) Compute the thermal conductivity of argon at 100°C and atmospheric pressure, using the Chapman–Enskog theory and the Lennard-Jones constants derived from viscosity data. Compare your result with the observed value[1] of 506×10^{-7} cal/cm $\cdot$ s $\cdot$ K.

(b) Compute the thermal conductivities of NO and CH_4 at 300K and atmospheric pressure from the following data for these conditions:

	$\mu \times 10^7$ (g/cm $\cdot$ s)	$\tilde{C}_p$ (cal/g-mole $\cdot$ K)
NO	1929	7.15
CH_4	1116	8.55

Compare your results with the experimental values given in Table 9.1-2.

9A.2 Computation of the Prandtl numbers for gases at low density.

(a) By using the Eucken formula and experimental heat capacity data, estimate the Prandtl number at 1 atm and 300K for each of the gases listed in the table.

(b) For the same gases, compute the Prandtl number directly by substituting the following values of the physical properties into the defining formula $\text{Pr} = \hat{C}_p\mu/k$, and compare the values with the results obtained in (a). All properties are given at low pressure and 300K.

Gas[a]	$\hat{C}_p \times 10^{-3}$ J/kg $\cdot$ K	$\mu \times 10^5$ Pa $\cdot$ s	k W/m $\cdot$ K
He	5.193	1.995	0.1546
Ar	0.5204	2.278	0.01784
H_2	14.28	0.8944	0.1789
Air	1.001	1.854	0.02614
CO_2	0.8484	1.506	0.01661
H_2O	1.864	1.041	0.02250

[a] The entries in this table were prepared from functions provided by T. E. Daubert, R. P.Danner, H. M. Sibul, C. C. Stebbins, J. L. Oscarson, R. L. Rowley, W. V. Wilding, M. E. Adams, T. L. Marshall, and N. A. Zundel, *DIPPR ® Data Compilation of Pure Compound Properties*, Design Institute for Physical Property Data®, AIChE, New York (2000).

9A.3. Estimation of the thermal conductivity of a dense gas. Predict the thermal conductivity of methane at 110.4 atm and 127°F by the following methods:

(a) Use Fig. 9.2-1. Obtain the necessary critical properties from Appendix E.

(b) Use the Eucken formula to get the thermal conductivity at 127°F and low pressure. Then apply a pressure correction by using Fig. 9.2-1. The experimental value[2] is 0.0282 Btu/hr $\cdot$ ft $\cdot$ F.

Answer: **(a)** 0.0294 Btu/hr $\cdot$ ft $\cdot$ F.

9A.4. Prediction of the thermal conductivity of a gas mixture. Calculate the thermal conductivity of a mixture containing 20 mole % CO_2 and 80 mole % H_2 at 1 atm and 300K. Use the data of Problem 9A.2 for your calculations.

Answer: 0.1204 W/m $\cdot$ K

9A.5. Estimation of the thermal conductivity of a pure liquid. Predict the thermal conductivity of liquid H_2O at 40°C and 40 megabars pressure (1 megabar $= 10^6$ dyn/cm^2). The isothermal compressibility, $(1/\rho)(\partial\rho/\partial p)_T$, is 38×10^{-6} megabar^{-1} and the density is 0.9938 g/cm^3. Assume that $\tilde{C}_p = \tilde{C}_V$.

Answer: 0.375 Btu/hr $\cdot$ ft $\cdot$ F

9A.6. Calculation of the Lorenz number.

(a) Application of kinetic theory to the "electron gas" in a metal[3] gives for the Lorenz number

$$L = \frac{\pi^2}{3}\left(\frac{\kappa}{e}\right)^2 \qquad (9A.6\text{-}1)$$

in which κ is the Boltzmann constant and e is the charge on the electron. Compute L in the units given under Eq. 9.5-1.

(b) The electrical resistivity, $1/k_e$, of copper at 20°C is 1.72×10^{-6} ohm $\cdot$ cm. Estimate its thermal conductivity in W/m $\cdot$ K using Eq. 9A.6-1, and compare your result with the experimental value given in Table 9.1-4.

Answers: **(a)** 2.44×10^{-8} volt^2/K^2; **(b)** 416 W/m $\cdot$ K

9A.7. Corroboration of the Wiedemann–Franz–Lorenz law. Given the following experimental data at 20°C for pure metals, compute the corresponding values of the Lorenz number, L, defined in Eq. 9.5-1.

---

[1] W. G. Kannuluik and E. H. Carman, *Proc. Phys. Soc. (London)*, **65B**, 701–704 (1952).

[2] J. M. Lenoir, W. A. Junk, and E. W. Comings, *Chem. Engr. Prog.*, **49**, 539–542 (1953).

[3] J. E. Mayer and M. G. Mayer, *Statistical Mechanics*, Wiley, New York (1946), p. 412; P. Drude, *Ann. Phys.*, **1**, 566–613 (1900).

Metal	$(1/k_e)$ (ohm $\cdot$ cm)	k (cal/cm $\cdot$ s $\cdot$ K)
Na	4.6×10^{-6}	0.317
Ni	6.9×10^{-6}	0.140
Cu	1.69×10^{-6}	0.92
Al	2.62×10^{-6}	0.50

9A.8. Thermal conductivity and Prandtl number of a polyatomic gas.

(a) Estimate the thermal conductivity of CH_4 at 1500K and 1.37 atm. The molar heat capacity at constant pressure[4] at 1500K is 20.71 cal/g-mole $\cdot$ K.

(b) What is the Prandtl number at the same pressure and temperature?

Answers: **(a)** 5.06×10^{-4} cal/cm $\cdot$ s $\cdot$ K; **(b)** 0.89

9A.9. Thermal conductivity of gaseous chlorine. Use Eq. 9.3-15 to calculate the thermal conductivity of gaseous chlorine. To do this you will need to use Eq. 1.4-14 to estimate the viscosity, and will also need the following values of the heat capacity:

T (K)	200	300	400	500	600
$\tilde{C}_p$ (cal/g-mole $\cdot$ K)	(8.06)	8.12	8.44	8.62	8.74

Check to see how well the calculated values agree with the following experimental thermal conductivity data[5]

T (K)	p (mm Hg)	$k \times 10^5$ cal/cm $\cdot$ s $\cdot$ K
198	50	1.31 ± 0.03
275	220	1.90 ± 0.02
276	120	1.93 ± 0.01
	220	1.92 ± 0.01
363	100	2.62 ± 0.02
	200	2.61 ± 0.02
395	210	3.04 ± 0.02
453	150	3.53 ± 0.03
	250	3.42 ± 0.02
495	250	3.72 ± 0.07
553	100	4.14 ± 0.04
583	170	4.43 ± 0.04
	210	4.45 ± 0.08
676	150	5.07 ± 0.10
	250	4.90 ± 0.03

9A.10. Thermal conductivity of chlorine–air mixtures. Using Eq. 9.3-17, predict thermal conductivities of chlorine–air mixtures at 297K and 1 atm for the following mole fractions of chlorine: 0.25, 0.50, 0.75. Air may be considered a single substance, and the following data may be assumed:

Substance[a]	μ (Pa $\cdot$ s)	k (W/m $\cdot$ K)	$\hat{C}_p$ (J/kg $\cdot$ K)
Air	1.854×10^{-5}	2.614×10^{-2}	1.001×10^3
Chlorine	1.351×10^{-5}	8.960×10^{-3}	4.798×10^2

[a] The entries in this table were prepared from functions provided by T. E. Daubert, R. P. Danner, H. M. Sibul, C. C. Stebbins, J. L. Oscarson, R. L. Rowley, W. V. Wilding, M. E. Adams, T. L. Marshall, and N. A. Zundel, *DIPPR ® Data Compilation of Pure Compound Properties*, Design Institute for Physical Property Data®, AIChE, New York (2000).

9A.11. Thermal conductivity of quartz sand. A typical sample of quartz sand has the following properties at 20°C:

Component	Volume fraction ϕ_i	k cal/cm $\cdot$ s $\cdot$ K
$i = 1$: Silica	0.510	20.4×10^{-3}
$i = 2$: Feldspar	0.063	7.0×10^{-3}

Continuous phase ($i = 0$) is one of the following:

(i) Water	0.427	1.42×10^{-3}
(ii) Air	0.427	0.0615×10^{-3}

Estimate the thermal conductivity of the sand (i) when it is water saturated, and (ii) when it is completely dry.

(a) Use the following generalization of Eqs. 9.6-5 and 6:

$$\frac{k_{\text{eff}}}{k_0} = \frac{\sum_{i=0}^{N} \alpha_i (k_i/k_0) \phi_i}{\sum_{i=0}^{N} \alpha_i \phi_i} \quad (9A.11\text{-}1)$$

$$\alpha_i = \frac{1}{3} \sum_{j=1}^{3} \left[1 + \left(\frac{k_i}{k_0} - 1 \right) g_j \right]^{-1} \quad (9A.11\text{-}2)$$

Here N is the number of solid phases. Compare the prediction for spheres ($g_1 = g_2 = g_3 = \frac{1}{3}$) with the recommendation of de Vries ($g_1 = g_2 = \frac{1}{8}; g_3 = \frac{3}{4}$). The latter g_i values closely approximate the fitted ones[6] for the present sample. The right-hand member of Eq. 9A.11-1 is to be multiplied by 1.25 for completely dry sand.[6]

(b) Use Eq. 9.6-1 with $k_1 = 18.9 \times 10^{-3}$ cal/cm $\cdot$ s $\cdot$ K, which is the volume-average thermal conductivity of the two solids. Observed values, accurate within about 3%, are

[4] O. A. Hougen, K. M. Watson, and R. A. Ragatz, *Chemical Process Principles*, Vol. 1, Wiley, New York (1954), p. 253.

[5] Interpolated from data of E. U. Frank, *Z. Elektrochem.*, **55**, 636 (1951), as reported in *Nouveau Traité de Chimie Minerale*, P. Pascal, ed., Masson et Cᶦᵉ, Paris (1960), pp. 158–159.

[6] The behavior of partially wetted soil has been treated by D. A. de Vries, Chapter 7 in *Physics and Plant Environment*, W. R. van Wijk, ed., Wiley, New York (1963).

6.2 and 0.58 × 10^{-3} cal/cm · s · K for wet and dry sand, respectively.[6]

Answers in cal/cm · s · K for wet and dry sand respectively: **(a)** Eq. 9A.11-1 gives k_{eff} = 6.3 × 10^{-3} and 0.38 × 10^{-3} with $g_1 = g_2 = g_3 = \frac{1}{3}$, vs. 6.2 × 10^{-3} and 0.54 × 10^{-3} with $g_1 = g_2 = \frac{1}{8}$ and $g_3 = \frac{3}{4}$. **(b)** Eq. 9.6-1 gives k_{eff} = 5.1 × 10^{-3} and 0.30 × 10^{-3}.

9A.12. Calculation of molecular diameters from transport properties.

(a) Determine the molecular diameter d for argon from Eq. 1.4-9 and the experimental viscosity given in Problem 9A.2.

(b) Repeat part (a), but using Eq. 9.3-12 and the measured thermal conductivity in Problem 9A.2. Compare this result with the value obtained in (a).

(c) Calculate and compare the values of the Lennard-Jones collision diameter σ from the same experimental data used in (a) and (b), using ε/κ from Table E.1.

(d) What can be concluded from the above calculations?

Answer: **(a)** 2.95 Å; **(b)** 1.86 Å; **(c)** 3.415 Å from Eq. 1.4-14, 3.409 Å from Eq. 9.3-13

9C.1. Enskog theory for dense gases.
Enskog[7] developed a kinetic theory for the transport properties of dense gases. He showed that for molecules that are idealized as rigid spheres of diameter σ_0

$$\frac{\mu}{\mu^\circ} \frac{\tilde{V}}{b_0} = \frac{1}{y} + 0.8 + 0.761y \qquad (9C.1-1)$$

$$\frac{k}{k^\circ} \frac{\tilde{V}}{b_0} = \frac{1}{y} + 1.2 + 0.755y \qquad (9C.1-2)$$

Here μ° and k° are the low-pressure properties (computed, for example, from Eqs. 1.4-14 and 9.3-13), $\tilde{V}$ is the molar volume, and $b_0 = \frac{2}{3}\pi\tilde{N}\sigma_0^3$, where $\tilde{N}$ is Avogadro's number. The quantity y is related to the equation of state of a gas of rigid spheres:

$$y = \frac{p\tilde{V}}{RT} - 1 = \left(\frac{b_0}{\tilde{V}}\right) + 0.6250\left(\frac{b_0}{\tilde{V}}\right)^2 + 0.2869\left(\frac{b_0}{\tilde{V}}\right)^3 + \cdots$$
$$(9C.1-3)$$

These three equations give the density corrections to the viscosity and thermal conductivity of a hypothetical gas made up of rigid spheres.

Enskog further suggested that for real gases, (i) y can be given empirically by

$$y = \frac{\tilde{V}}{R}\left(\frac{\partial p}{\partial T}\right)_{\tilde{V}} - 1 \qquad (9C.1-4)$$

where experimental p-$\tilde{V}$-T data are used, and (ii) b_0 can be determined by fitting the minimum in the curve of $(\mu/\mu^\circ)\tilde{V}$ versus y.

(a) A useful way to summarize the equation of state is to use the corresponding-states presentation[8] of $Z = Z(p_r, T_r)$, where $Z = p\tilde{V}/RT$, $p_r = p/p_c$, and $T_r = T/T_c$. Show that the quantity y defined by Eq. 9C.1-4 can be computed as a function of the reduced pressure and temperature from

$$y = Z \frac{1 + (\partial\ln Z/\partial\ln T_r)_{p_r}}{1 - (\partial\ln Z/\partial\ln p_r)_{T_r}} - 1 \qquad (9C.1-5)$$

(b) Show how Eqs. 9C.1-1, 2, and 5, together with the Hougen–Watson Z-chart and the Uyehara–Watson μ/μ_c chart in Fig. 1.3-1, can be used to develop a chart of k/k_c as a function of p_r and T_r. What would be the limitations of the resulting chart? Such a procedure (but using specific p-$\tilde{V}$-T data instead of the Hougen–Watson Z-chart) was used by Comings and Nathan.[9]

(c) How might one use the Redlich and Kwong[10] equation of state

$$\left(p + \frac{a}{\sqrt{T}\tilde{V}(\tilde{V} + b)}\right)(\tilde{V} - b) = RT \qquad (9C.1-6)$$

for the same purpose? The quantities a and b are constants characteristic of each gas.

---

[7] D. Enskog, *Kungliga Svenska Vetenskapsakademiens Handlingar*, **62**, No. 4 (1922), in German. See also J. O. Hirschfelder, C. F. Curtiss, and R. B. Bird, *Molecular Theory of Gases and Liquids*, 2nd printing with corrections (1964), pp. 647–652.

[8] O. A. Hougen and K. M. Watson, *Chemical Process Principles*, Vol. II, Wiley, New York (1947), p. 489.

[9] E. W. Comings and M. F. Nathan, *Ind. Eng. Chem.*, **39**, 964–970 (1947).

[10] O. Redlich and J. N. S. Kwong, *Chem. Rev.*, **44**, 233–244 (1949).

Chapter 10

Shell Energy Balances and Temperature Distributions in Solids and Laminar Flow

In Chapter 2 we saw how certain simple viscous flow problems are solved by a two-step procedure: (i) a momentum balance is made over a thin slab or shell perpendicular to the direction of momentum transport, which leads to a first-order differential equation that gives the momentum flux distribution; (ii) then into the expression for the momentum flux we insert Newton's law of viscosity, which leads to a first-order differential equation for the fluid velocity as a function of position. The integration constants that appear are evaluated by using the boundary conditions, which specify the velocity or momentum flux at the bounding surfaces.

In this chapter we show how a number of heat conduction problems are solved by an analogous procedure: (i) an energy balance is made over a thin slab or shell perpendicular to the direction of the heat flow, and this balance leads to a first-order differential equation from which the heat flux distribution is obtained; (ii) then into this expression for the heat flux, we substitute Fourier's law of heat conduction, which gives a first-order differential equation for the temperature as a function of position. The integration constants are then determined by use of boundary conditions for the temperature or heat flux at the bounding surfaces.

It should be clear from the similar wording of the preceding two paragraphs that the mathematical methods used in this chapter are the same as those introduced in Chapter 2—only the notation and terminology are different. However, we will encounter here a number of physical phenomena that have no counterpart in Chapter 2.

After a brief introduction to the shell energy balance in §10.1, we give an analysis of the heat conduction in a series of uncomplicated systems. Although these examples are somewhat idealized, the results find application in numerous standard engineering calculations. The problems were chosen to introduce the beginner to a number of important physical concepts associated with the heat transfer field. In addition, they serve to show how to use a variety of boundary conditions and to illustrate problem solving in Cartesian, cylindrical, and spherical coordinates. In §§10.2–10.5 we consider four kinds of heat sources: electrical, nuclear, viscous, and chemical. In §§10.6 and 10.7 we cover two topics with widespread applications—namely, heat flow through composite walls and heat loss from fins. Finally, in §§10.8 and 10.9, we analyze two limiting cases of heat transfer in moving fluids: forced convection and free convection. The study of these topics paves the way for the general equations in Chapter 11.

§10.1   SHELL ENERGY BALANCES; BOUNDARY CONDITIONS

The problems discussed in this chapter are set up by means of shell energy balances. We select a slab (or shell), the surfaces of which are normal to the direction of heat conduction, and then we write for this system a statement of the law of conservation of energy. For *steady-state* (i.e., time-independent) systems, we write:

$$
\begin{Bmatrix} \text{rate of} \\ \text{energy in} \\ \text{by convective} \\ \text{transport} \end{Bmatrix} - \begin{Bmatrix} \text{rate of} \\ \text{energy out} \\ \text{by convective} \\ \text{transport} \end{Bmatrix} + \begin{Bmatrix} \text{rate of} \\ \text{energy in} \\ \text{by molecular} \\ \text{transport} \end{Bmatrix} - \begin{Bmatrix} \text{rate of} \\ \text{energy out} \\ \text{by molecular} \\ \text{transport} \end{Bmatrix} +
$$

$$
\begin{Bmatrix} \text{rate of} \\ \text{work done} \\ \text{on system} \\ \text{by molecular} \\ \text{transport} \end{Bmatrix} - \begin{Bmatrix} \text{rate of} \\ \text{work done} \\ \text{by system} \\ \text{by molecular} \\ \text{transport} \end{Bmatrix} + \begin{Bmatrix} \text{rate of} \\ \text{work done} \\ \text{on system} \\ \text{by external} \\ \text{forces} \end{Bmatrix} + \begin{Bmatrix} \text{rate of} \\ \text{energy} \\ \text{production} \end{Bmatrix} = 0 \quad (10.1\text{-}1)
$$

The *convective transport* of energy was discussed in §9.7, and the *molecular transport* (heat conduction) in §9.1. The *molecular work terms* were explained in §9.8. These three terms can be added to give the "combined energy flux" **e**, as shown in Eq. 9.8-6. In setting up problems here (and in the next chapter) we will use the **e** vector along with the expression for the enthalpy in Eq. 9.8-8. Note that in nonflow systems (for which **v** is zero) the **e** vector simplifies to the **q** vector, which is given by Fourier's law.

The *energy production* term in Eq. 10.1-1 includes (i) the degradation of electrical energy into heat, (ii) the heat produced by slowing down of neutrons and nuclear fragments liberated in the fission process, (iii) the heat produced by viscous dissipation, and (iv) the heat produced in chemical reactions. The chemical reaction heat source will be discussed further in Chapter 19. Equation 10.1-1 is a statement of the first law of thermodynamics, written for an "open" system at steady-state conditions. In Chapter 11 this same statement—extended to unsteady-state systems—will be written as an equation of change.

After Eq. 10.1-1 has been written for a thin slab or shell of material, the thickness of the slab or shell is allowed to approach zero. This procedure leads ultimately to an expression for the temperature distribution containing constants of integration, which we evaluate by use of boundary conditions. The commonest types of boundary conditions are:

    a. The temperature may be specified at a surface.

    b. The heat flux normal to a surface may be given (this is equivalent to specifying the normal component of the temperature gradient).

    c. At interfaces the continuity of temperature and of the heat flux normal to the interface are required.

d. At a solid–fluid interface, the normal heat flux component may be related to the difference between the solid surface temperature T_0 and the "bulk" fluid temperature T_b:

$$q = h(T_0 - T_b) \tag{10.1-2}$$

This relation is referred to as *Newton's law of cooling*. It is not really a "law" but rather the defining equation for h, which is called the *heat transfer coefficient*. Chapter 14 deals with methods for estimating heat-transfer coefficients.

All four types of boundary conditions are encountered in this chapter. Still other kinds of boundary conditions are possible, and they will be introduced as needed.

§10.2 HEAT CONDUCTION WITH AN ELECTRICAL HEAT SOURCE

The first system we consider is an electric wire of circular cross section with radius R and electrical conductivity k_e ohm^{-1} cm^{-1}. Through this wire there is an electric current with current density I amp/cm^2. The transmission of an electric current is an irreversible process, and some electrical energy is converted into heat (thermal energy). The rate of heat production per unit volume is given by the expression

$$S_e = \frac{I^2}{k_e} \tag{10.2-1}$$

The quantity S_e is the heat source resulting from electrical dissipation. We assume here that the temperature rise in the wire is not so large that the temperature dependence of either the thermal or electrical conductivity need be considered. The surface of the wire is maintained at temperature T_0. We now show how to find the radial temperature distribution within the wire.

For the energy balance we take the system to be a cylindrical shell of thickness Δr and length L (see Fig. 10.2-1). Since $\mathbf{v} = 0$ in this system, the only contributions to the energy balance are

Rate of heat in
across cylindrical $$(2\pi r L)q_r|_r = (2\pi r L q_r)|_r \tag{10.2-2}$$
surface at r

Rate of heat out
across cylindrical $$(2\pi (r + \Delta r)L)(q_r|_{r+\Delta r}) = (2\pi r L q_r)|_{r+\Delta r} \tag{10.2-3}$$
surface at $r + \Delta r$

Rate of thermal
energy production by $$(2\pi r \Delta r L)S_e \tag{10.2-4}$$
electrical dissipation

The notation q_r means "heat flux in the r direction," and $(\cdots)|_{r+\Delta r}$ means "evaluated at $r + \Delta r$." Note that we take "in" and "out" to be in the positive r direction.

We now substitute these quantities into the energy balance of Eq. 9.1-1. Division by $2\pi L \Delta r$ and taking the limit as Δr goes to zero gives

$$\lim_{\Delta r \to 0} \frac{(rq_r)|_{r+\Delta r} - (rq_r)|_r}{\Delta r} = S_e r \tag{10.2-5}$$

The expression on the left side is the first derivative of rq_r with respect to r, so that Eq. 10.2-5 becomes

$$\frac{d}{dr}(rq_r) = S_e r \tag{10.2-6}$$

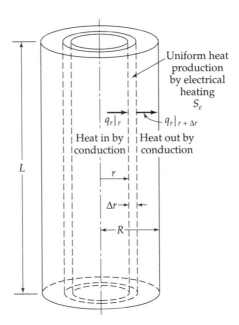

Fig. 10.2-1. An electrically heated wire, showing the cylindrical shell over which the energy balance is made.

This is a first-order differential equation for the energy flux, and it may be integrated to give

$$q_r = \frac{S_e r}{2} + \frac{C_1}{r} \tag{10.2-7}$$

The integration constant C_1 must be zero because of the boundary condition that

B.C. 1:   at $r = 0$, q_r is not infinite   (10.2-8)

Hence the final expression for the heat flux distribution is

$$\boxed{q_r = \frac{S_e r}{2}} \tag{10.2-9}$$

This states that the heat flux increases linearly with r.

 We now substitute Fourier's law in the form $q_r = -k(dT/dr)$ (see Eq. B.2-4) into Eq. 10.2-9 to obtain

$$-k\frac{dT}{dr} = \frac{S_e r}{2} \tag{10.2-10}$$

When k is assumed to be constant, this first-order differential equation can be integrated to give

$$T = -\frac{S_e r^2}{4k} + C_2 \tag{10.2-11}$$

The integration constant is determined from

B.C. 2:   at $r = R$, $T = T_0$ (10.2-12)

Hence $C_2 = (S_e R^2 / 4k) + T_0$ and Eq. 10.2-11 becomes

$$\boxed{T - T_0 = \frac{S_e R^2}{4k}\left[1 - \left(\frac{r}{R}\right)^2\right]} \tag{10.2-13}$$

Equation 10.2-13 gives the temperature rise as a parabolic function of the distance r from the wire axis.

Once the temperature and heat flux distributions are known, various information about the system may be obtained:

(i) *Maximum temperature rise* (at $r = 0$)

$$T_{max} - T_0 = \frac{S_e R^2}{4k}$$

(10.2-14)

(ii) *Average temperature rise*

$$\langle T \rangle - T_0 = \frac{\int_0^{2\pi} \int_0^R (T(r) - T_0) r \, dr \, d\theta}{\int_0^{2\pi} \int_0^R r \, dr \, d\theta} = \frac{S_e R^2}{8k}$$

(10.2-15)

Thus the temperature rise, averaged over the cross section, is half the maximum temperature rise.

(iii) *Heat outflow at the surface* (for a length L of wire)

$$Q|_{r=R} = 2\pi RL \cdot q_r|_{r=R} = 2\pi RL \cdot \frac{S_e R}{2} = \pi R^2 L \cdot S_e$$

(10.2-16)

This result is not surprising, since, at steady state, all the heat produced by electrical dissipation in the volume $\pi R^2 L$ must leave through the surface $r = R$.

The reader, while going through this development, may well have had the feeling of *déja vu*. There is, after all, a pronounced similarity between the heated wire problem and the viscous flow in a circular tube. Only the notation is different:

	Tube flow	Heated wire
First integration gives	$\tau_{rz}(r)$	$q_r(r)$
Second integration gives	$v_z(r)$	$T(r) - T_0$
Boundary condition at $r = 0$	τ_{rz} = finite	q_r = finite
Boundary condition at $r = R$	$v_z = 0$	$T - T_0 = 0$
Transport property	μ	k
Source term	$(\mathscr{P}_0 - \mathscr{P}_L)/L$	S_e
Assumptions	μ = constant	k, k_e = constant

That is, when the quantities are properly chosen, the differential equations *and* the boundary conditions for the two problems are identical, and the physical processes are said to be "analogous." Not all problems in momentum transfer have analogs in energy and mass transport. However, when such analogies can be found, they may be useful in taking over known results from one field and applying them in another. For example, the reader should have no trouble in finding a heat conduction analog for the viscous flow in a liquid film on an inclined plane.

There are many examples of heat conduction problems in the electrical industry.[1] The minimizing of temperature rises inside electrical machinery prolongs insulation life. One example is the use of internally liquid-cooled stator conductors in very large (500,000 kw) AC generators.

---

[1] M. Jakob, *Heat Transfer*, Vol. 1, Wiley, New York (1949), Chapter 10, pp. 167–199.

To illustrate further problems in electrical heating, we give two examples concerning the temperature rise in wires: the first indicates the order of magnitude of the heating effect, and the second shows how to handle different boundary conditions. In addition, in Problem 10C.2 we show how to take into account the temperature dependence of the thermal and electrical conductivities.

EXAMPLE 10.2-1

Voltage Required for a Given Temperature Rise in a Wire Heated by an Electric Current

A copper wire has a radius of 2 mm and a length of 5 m. For what voltage drop would the temperature rise at the wire axis be 10°C, if the surface temperature of the wire is 20°C?

SOLUTION

Combining Eq. 10.2-14 and 10.2-1 gives

$$T_{max} - T_0 = \frac{I^2 R^2}{4kk_e}$$ (10.2-17)

The current density is related to the voltage drop E over a length L by

$$I = k_e \frac{E}{L}$$ (10.2-18)

Hence

$$T_{max} - T_0 = \frac{E^2 R^2}{4L^2}\left(\frac{k_e}{k}\right)$$ (10.2-19)

from which

$$E = 2\frac{L}{R}\sqrt{\frac{k}{k_e T_0}}\sqrt{T_0(T_{max} - T_0)}$$ (10.2-20)

For copper, the Lorenz number of §9.5 is $k/k_e T_0 = 2.23 \times 10^{-8}$ volt^2/K^2. Therefore, the voltage drop needed to cause a 10°C temperature rise is

$$E = 2\left(\frac{5000 \text{ mm}}{2 \text{ mm}}\right)\sqrt{2.23 \times 10^{-8}\frac{\text{volt}}{\text{K}}}\sqrt{(293)(10)}\text{K}$$

$$= (5000)(1.49 \times 10^{-4})(54.1) = 40 \text{ volts}$$ (10.2-21)

EXAMPLE 10.2.2

Heated Wire with Specified Heat Transfer Coefficient and Ambient Air Temperature

Repeat the analysis in §10.2, assuming that T_0 is not known, but that instead the heat flux at the wall is given by Newton's "law of cooling" (Eq. 10.1-2). Assume that the heat transfer coefficient h and the ambient air temperature T_{air} are known.

SOLUTION I

The solution proceeds as before through Eq. 10.2-11, but the second integration constant is determined from Eq. 10.1-2:

B.C. 2′: at $r = R$, $\quad -k\frac{dT}{dr} = h(T - T_{air})$ (10.2-22)

Substituting Eq. 10.2-11 into Eq. 10.2-22 gives $C_2 = (S_e R/2h) + (S_e R^2/4k) + T_{air}$, and the temperature profile is then

$$T - T_{air} = \frac{S_e R^2}{4k}\left[1 - \left(\frac{r}{R}\right)^2\right] + \frac{S_e R}{2h}$$ (10.2-23)

From this the surface temperature of the wire is found to be $T_{air} + S_e R/2h$.

SOLUTION II

Another method makes use of the result obtained previously in Eq. 10.2-13. Although T_0 is not known in the present problem, we can nonetheless use the result. From Eqs. 10.1-2 and 10.2-16 we can get the temperature difference

$$T_0 - T_{\text{air}} = \frac{\pi R^2 L S_e}{h(2\pi RL)} = \frac{S_e R}{2h} \tag{10.2-24}$$

Substraction of Eq. 10.2-24 from Eq. 10.2-13 enables us to eliminate the unknown T_0 and gives Eq. 10.2-23.

§10.3 HEAT CONDUCTION WITH A NUCLEAR HEAT SOURCE

We consider a spherical nuclear fuel element as shown in Fig. 10.3-1. It consists of a sphere of fissionable material with radius $R^{(F)}$, surrounded by a spherical shell of aluminum "cladding" with outer radius $R^{(C)}$. Inside the fuel element, fission fragments are produced that have very high kinetic energies. Collisions between these fragments and the atoms of the fissionable material provide the major source of thermal energy in the reactor. Such a volume source of thermal energy resulting from nuclear fission we call S_n (cal/cm^3 · s). This source will not be uniform throughout the sphere of fissionable material; it will be the smallest at the center of the sphere. For the purpose of this problem, we assume that the source can be approximated by a simple parabolic function

$$S_n = S_{n0}\left[1 + b\left(\frac{r}{R^{(F)}}\right)^2\right] \tag{10.3-1}$$

Here S_{n0} is the volume rate of heat production at the center of the sphere, and b is a dimensionless positive constant.

We select as the system a spherical shell of thickness Δr within the sphere of fissionable material. Since the system is not in motion, the energy balance will consist only of heat conduction terms and a source term. The various contributions to the energy balance are:

Rate of heat in
by conduction
at r

$$q_r^{(F)}\big|_r \cdot 4\pi r^2 = (4\pi r^2 q_r^{(F)})\big|_r \tag{10.3-2}$$

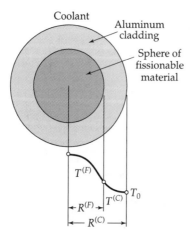

Coolant

Aluminum cladding

Sphere of fissionable material

$T^{(F)}$

$R^{(F)}$ $T^{(C)}$ T_0

$R^{(C)}$

Fig. 10.3-1. A spherical nuclear fuel assembly, showing the temperature distribution within the system.

Rate of heat out by conduction at $r + \Delta r$	$q_r^{(F)}\big|_{r+\Delta r} \cdot 4\pi(r + \Delta r)^2 = (4\pi r^2 q_r^{(F)})\big|_{r+\Delta r}$	(10.3-3)
Rate of thermal energy produced by nuclear fission	$S_n \cdot 4\pi r^2 \, \Delta r$	(10.3-4)

Substitution of these terms into the energy balance of Eq. 10.1-1 gives, after dividing by $4\pi \, \Delta r$ and taking the limit as $\Delta r \to 0$

$$\lim_{\Delta r \to 0} \frac{(r^2 q_r^{(F)})\big|_{r+\Delta r} - (r^2 q_r^{(F)})\big|_r}{\Delta r} = S_n r^2 \tag{10.3-5}$$

Taking the limit and introducing the expression in Eq. 10.3-1 leads to

$$\frac{d}{dr}(r^2 q_r^{(F)}) = S_{n0}\left[1 + b\left(\frac{r}{R^{(F)}}\right)^2\right]r^2 \tag{10.3-6}$$

The differential equation for the heat flux $q_r^{(C)}$ in the cladding is of the same form as Eq. 10.3-6, except that there is no significant source term:

$$\frac{d}{dr}(r^2 q_r^{(C)}) = 0 \tag{10.3-7}$$

Integration of these two equations gives

$$q_r^{(F)} = S_{n0}\left(\frac{r}{3} + \frac{b}{R^{(F)2}}\frac{r^3}{5}\right) + \frac{C_1^{(F)}}{r^2} \tag{10.3-8}$$

$$q_r^{(C)} = +\frac{C_1^{(C)}}{r^2} \tag{10.3-9}$$

in which $C_1^{(F)}$ and $C_1^{(C)}$ are integration constants. These are evaluated by means of the boundary conditions:

B.C. 1:	at $r = 0$, $\quad q_r^{(F)}$ is not infinite	(10.3-10)
B.C. 2:	at $r = R^{(F)}$, $\quad q_r^{(F)} = q_r^{(C)}$	(10.3-11)

Evaluation of the constants then leads to

$$\boxed{q_r^{(F)} = S_{n0}\left(\frac{r}{3} + \frac{b}{R^{(F)2}}\frac{r^3}{5}\right)} \tag{10.3-12}$$

$$\boxed{q_r^{(C)} = S_{n0}\left(\frac{1}{3} + \frac{b}{5}\right)\frac{R^{(F)3}}{r^2}} \tag{10.3-13}$$

These are the heat flux distributions in the fissionable sphere and in the spherical-shell cladding.

Into these distributions we now substitute Fourier's law of heat conduction (Eq. B.2-7):

$$-k^{(F)}\frac{dT^{(F)}}{dr} = S_{n0}\left(\frac{r}{3} + \frac{b}{R^{(F)2}}\frac{r^3}{5}\right) \tag{10.3-14}$$

$$-k^{(C)}\frac{dT^{(C)}}{dr} = S_{n0}\left(\frac{1}{3} + \frac{b}{5}\right)\frac{R^{(F)3}}{r^2} \tag{10.3-15}$$

These equations may be integrated for constant $k^{(F)}$ and $k^{(C)}$ to give

$$T^{(F)} = -\frac{S_{n0}}{k^{(F)}}\left(\frac{r^2}{6} + \frac{b}{R^{(F)2}}\frac{r^4}{20}\right) + C_2^{(F)} \tag{10.3-16}$$

$$T^{(C)} = +\frac{S_{n0}}{k^{(C)}}\left(\frac{1}{3} + \frac{b}{5}\right)\frac{R^{(F)3}}{r} + C_2^{(C)} \tag{10.3-17}$$

The integration constants can be determined from the boundary conditions

B.C. 3:   at $r = R^{(F)}$, $T^{(F)} = T^{(C)}$ (10.3-18)

B.C. 4:   at $r = R^{(C)}$, $T^{(C)} = T_0$ (10.3-19)

where T_0 is the known temperature at the outside of the cladding. The final expressions for the temperature profiles are

$$T^{(F)} = \frac{S_{n0}R^{(F)2}}{6k^{(F)}}\left\{\left[1 - \left(\frac{r}{R^{(F)}}\right)^2\right] + \frac{3}{10}b\left[1 - \left(\frac{r}{R^{(F)}}\right)^4\right]\right\}$$
$$+ \frac{S_{n0}R^{(F)2}}{3k^{(C)}}\left(1 + \frac{3}{5}b\right)\left(1 - \frac{R^{(F)}}{R^{(C)}}\right) + T_0 \tag{10.3-20}$$

$$T^{(C)} = \frac{S_{n0}R^{(F)2}}{3k^{(C)}}\left(1 + \frac{3}{5}b\right)\left(\frac{R^{(F)}}{r} - \frac{R^{(F)}}{R^{(C)}}\right) + T_0 \tag{10.3-21}$$

To find the maximum temperature in the sphere of fissionable material, all we have to do is set r equal to zero in Eq. 10.3-20. This is a quantity one might well want to know when making estimates of thermal deterioration.

This problem has illustrated two points: (i) how to handle a position-dependent source term, and (ii) the application of the continuity of temperature and normal heat flux at the boundary between two solid materials.

§10.4   HEAT CONDUCTION WITH A VISCOUS HEAT SOURCE

Next we consider the flow of an incompressible Newtonian fluid between two coaxial cylinders as shown in Fig. 10.4-1. The surfaces of the inner and outer cylinders are maintained at $T = T_0$ and $T = T_b$, respectively. We can expect that T will be a function of r alone.

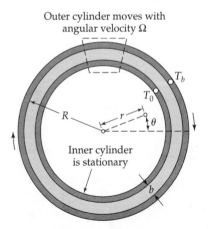

Outer cylinder moves with
angular velocity Ω

T_b

T_0

R r θ

Inner cylinder
is stationary

b

Fig. 10.4-1. Flow between cylinders with viscous heat generation. That part of the system enclosed within the dotted lines is shown in modified form in Fig. 10.4-2.

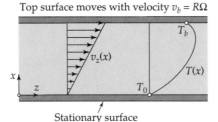

Top surface moves with velocity $v_b = R\Omega$

Fig. 10.4-2. Modification of a portion of the flow system in Fig. 10.4-1, in which the curvature of the bounding surfaces is neglected.

Stationary surface

As the outer cylinder rotates, each cylindrical shell of fluid "rubs" against an adjacent shell of fluid. This friction between adjacent layers of the fluid produces heat; that is, the mechanical energy is degraded into thermal energy. The volume heat source resulting from this "viscous dissipation," which can be designated by S_v, appears automatically in the shell balance when we use the combined energy flux vector **e** defined at the end of Chapter 9, as we shall see presently.

If the slit width b is small with respect to the radius R of the outer cylinder, then the problem can be solved approximately by using the somewhat simplified system depicted in Fig. 10.4-2. That is, we ignore curvature effects and solve the problem in Cartesian coordinates. The velocity distribution is then $v_z = v_b(x/b)$, where $v_b = \Omega R$.

We now make an energy balance over a shell of thickness Δx, width W, and length L. Since the fluid is in motion, we use the combined energy flux vector **e** as written in Eq. 9.8-6. The balance then reads

$$WLe_x|_x - WLe_x|_{x+\Delta x} = 0 \tag{10.4-1}$$

Dividing by $WL\,\Delta x$ and letting the shell thickness Δx go to zero then gives

$$\frac{de_x}{dx} = 0 \tag{10.4-2}$$

This equation may be integrated to give

$$e_x = C_1 \tag{10.4-3}$$

Since we do not know any boundary conditions for e_x, we cannot evaluate the integration constant at this point.

We now insert the expression for e_x from Eq. 9.8-6. Since the velocity component in the x direction is zero, the term $(\frac{1}{2}\rho v^2 + \rho \hat{U})\mathbf{v}$ can be discarded. The x-component of **q** is $-k(dT/dx)$ according to Fourier's law. The x-component of $[\boldsymbol{\tau} \cdot \mathbf{v}]$ is, as shown in Eq. 9.8-1, $\tau_{xx}v_x + \tau_{xy}v_y + \tau_{xz}v_z$. Since the only nonzero component of the velocity is v_z and since $\tau_{xz} = -\mu(dv_z/dx)$ according to Newton's law of viscosity, the x-component of $[\boldsymbol{\tau} \cdot \mathbf{v}]$ is $-\mu v_z(dv_z/dx)$. We conclude, then, that Eq. 10.4-3 becomes

$$-k\frac{dT}{dx} - \mu v_z \frac{dv_z}{dx} = C_1 \tag{10.4-4}$$

When the linear velocity profile $v_z = v_b(x/b)$ is inserted, we get

$$-k\frac{dT}{dx} - \mu x \left(\frac{v_b}{b}\right)^2 = C_1 \tag{10.4-5}$$

in which $\mu(v_b/b)^2$ can be identified as the rate of viscous heat production per unit volume S_v.

When Eq. 10.4-5 is integrated we get

$$T = -\left(\frac{\mu}{k}\right)\left(\frac{v_b}{b}\right)^2 \frac{x^2}{2} - \frac{C_1}{k}x + C_2 \tag{10.4-6}$$

The two integration constants are determined from the boundary conditions

B.C. 1:   at $x = 0$, $T = T_0$ (10.4-7)

B.C. 2:   at $x = b$, $T = T_b$ (10.4-8)

This yields finally, for $T_b \neq T_0$

$$\boxed{\left(\frac{T - T_0}{T_b - T_0}\right) = \frac{1}{2} \operatorname{Br} \frac{x}{b}\left(1 - \frac{x}{b}\right) + \frac{x}{b}}$$ (10.4-9)

Here $\operatorname{Br} = \mu v_b^2/k(T_b - T_0)$ is the dimensionless *Brinkman number*,[1] which is a measure of the importance of the viscous dissipation term. If $T_b = T_0$, then Eq. 10.4-9 can be written as

$$\frac{T - T_0}{T_0} = \frac{1}{2}\frac{\mu v_b^2}{kT_0}\frac{x}{b}\left(1 - \frac{x}{b}\right)$$ (10.4-10)

and the maximum temperature is at $x/b = \frac{1}{2}$.

If the temperature rise is appreciable, the temperature dependence of the viscosity has to be taken into account. This is discussed in Problem 10C.1.

The viscous heating term $S_v = \mu(v_b/b)^2$ may be understood by the following arguments. For the system in Fig. 10.4-2, the rate at which work is done is the force acting on the upper plate times the velocity with which it moves, or $(-\tau_{xz}WL)(v_b)$. The rate of energy addition per unit volume is then obtained by dividing this quantity by WLb, which gives $(-\tau_{xz}v_b/b) = \mu(v_b/b)^2$. This energy all appears as heat and is hence S_v.

In most flow problems viscous heating is not important. However if there are large velocity gradients, then it cannot be neglected. Examples of situations where viscous heating must be accounted for include: (i) flow of a lubricant between rapidly moving parts, (ii) flow of molten polymers through dies in high-speed extrusion, (iii) flow of highly viscous fluids in high-speed viscometers, and (iv) flow of air in the boundary layer near an earth satellite or rocket during reentry into the earth's atmosphere. The first two of these are further complicated because many lubricants and molten plastics are non-Newtonian fluids. Viscous heating for non-Newtonian fluids is illustrated in Problem 10B.5.

§10.5   HEAT CONDUCTION WITH A CHEMICAL HEAT SOURCE

A chemical reaction is being carried out in a tubular, fixed-bed flow reactor with inner radius R as shown in Fig. 10.5-1. The reactor extends from $z = -\infty$ to $z = +\infty$ and is divided into three zones:

        Zone I: Entrance zone packed with noncatalytic spheres

        Zone II: Reaction zone packed with catalyst spheres, extending from $z = 0$ to $z = L$

        Zone III: Exit zone packed with noncatalytic spheres

It is assumed that the fluid proceeds through the reactor tube in "plug flow"—that is, with axial velocity uniform at a superficial value $v_0 = w/\pi R^2\rho$ (see text below Eq. 6.4-1 for the definition of "superficial velocity"). The density, mass flow rate, and superficial

---

[1] H. C. Brinkman, *Appl. Sci. Research*, **A2**, 120–124 (1951), solved the viscous dissipation heating problem for the Poiseuille flow in a circular tube. Other dimensionless groups that may be used for characterizing viscous heating have been summarized by R. B. Bird, R. C. Armstrong, and O. Hassager, *Dynamics of Polymeric Liquids*, Vol. 1, 2nd edition, Wiley, New York (1987), pp. 207–208.

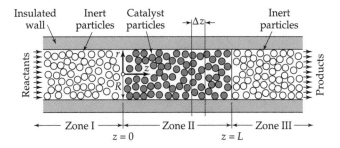

Fig. 10.5-1. Fixed-bed axial-flow reactor. Reactants enter at $z = -\infty$ and leave at $z = +\infty$. The reaction zone extends from $z = 0$ to $z = L$.

velocity are all treated as independent of r and z. In addition, the reactor wall is assumed to be well insulated, so that the temperature can be considered essentially independent of r. It is desired to find the steady-state axial temperature distribution $T(z)$ when the fluid enters at $z = -\infty$ with a uniform temperature T_1.

When a chemical reaction occurs, thermal energy is produced or consumed when the reactant molecules rearrange to form the products. The volume rate of thermal energy production by chemical reaction, S_c, is in general a complicated function of pressure, temperature, composition, and catalyst activity. For simplicity, we represent S_c here as a function of temperature only: $S_c = S_{c1}F(\Theta)$, where $\Theta = (T - T_0)/(T_1 - T_0)$. Here T is the local temperature in the catalyst bed (assumed equal for catalyst and fluid), and S_{c1} and T_0 are empirical constants for the given reactor inlet conditions.

For the shell balance we select a disk of radius R and thickness Δz in the catalyst zone (see Fig. 10.5-1), and we choose Δz to be much larger than the catalyst particle dimensions. In setting up the energy balance, we use the combined energy flux vector **e** inasmuch as we are dealing with a flow system. Then, at steady state, the energy balance is

$$\pi R^2 e_z|_z - \pi R^2 e_z|_{z+\Delta z} + (\pi R^2 \, \Delta z)S_c = 0 \tag{10.5-1}$$

Next we divide by $\pi R^2 \, \Delta z$ and take the limit as Δz goes to zero. Strictly speaking, this operation is not "legal," since we are not dealing with a continuum but rather with a granular structure. Nevertheless, we perform this limiting process with the understanding that the resulting equation describes, not point values, but rather average values of e_z and S_c for reactor cross sections of constant z. This gives

$$\frac{de_z}{dz} = S_c \tag{10.5-2}$$

Now we substitute the z-component of Eq. 9.8-6 into this equation to get

$$\frac{d}{dz}\left((\tfrac{1}{2}\rho v^2 + \rho \hat{H})v_z + \tau_{zz}v_z + q_z\right) = S_c \tag{10.5-3}$$

We now use Fourier's law for q_z, Eq. 1.2-6 for τ_{zz}, and the enthalpy expression in Eq. 9.8-8 (with the assumption that the heat capacity is constant) to get

$$\frac{d}{dz}\left(\tfrac{1}{2}\rho v_z^2 v_z + \rho \hat{C}_p(T - T^\circ)v_z + (p - p^\circ)v_z + \rho \hat{H}^\circ v_z - 2\mu v_z \frac{dv_z}{dz} - \kappa_{\text{eff},zz}\frac{dT}{dz}\right) = S_c \tag{10.5-4}$$

in which the effective thermal conductivity in the z direction $\kappa_{\text{eff},zz}$ has been used (see Eq. 9.6-9). The first, fourth and fifth terms on the left side may be discarded, since the velocity is not changing with z. The third term may be discarded if the pressure does not change significantly in the axial direction. Then in the second term we replace v_z by the superficial velocity v_0, because the latter is the effective fluid velocity in the reactor. Then Eq. 10.5-4 becomes

$$\rho \hat{C}_p v_0 \frac{dT}{dz} = \kappa_{\text{eff},zz} \frac{d^2 T}{dz^2} + S_c \tag{10.5-5}$$

This is the differential equation for the temperature in zone II. The same equation applies in zones I and III with the source term set equal to zero. The differential equations for the temperature are then

Zone I $\qquad\qquad (z < 0) \qquad \rho \hat{C}_p v_0 \dfrac{dT^{\text{I}}}{dz} = \kappa_{\text{eff},zz} \dfrac{d^2 T^{\text{I}}}{dz^2}$ $\hspace{2cm}$ (10.5-6)

Zone II $\qquad\qquad (0 < z < L) \qquad \rho \hat{C}_p v_0 \dfrac{dT^{\text{II}}}{dz} = \kappa_{\text{eff},zz} \dfrac{d^2 T^{\text{II}}}{dz^2} + S_{c1} F(\Theta)$ $\hspace{1cm}$ (10.5-7)

Zone III $\qquad\qquad (z > L) \qquad \rho \hat{C}_p v_0 \dfrac{dT^{\text{III}}}{dz} = \kappa_{\text{eff},zz} \dfrac{d^2 T^{\text{III}}}{dz^2}$ $\hspace{2cm}$ (10.5-8)

Here we have assumed that we can use the same value of the effective thermal conductivity in all three zones. These three second-order differential equations are subject to the following six boundary conditions:

B.C. 1: $\qquad\qquad$ at $z = -\infty$, $\qquad T^{\text{I}} = T_1$ $\hspace{5cm}$ (10.5-9)

B.C. 2: $\qquad\qquad$ at $z = 0$, $\qquad T^{\text{I}} = T^{\text{II}}$ $\hspace{5cm}$ (10.5-10)

B.C. 3: $\qquad\qquad$ at $z = 0$, $\qquad \kappa_{\text{eff},zz} \dfrac{dT^{\text{I}}}{dz} = \kappa_{\text{eff},zz} \dfrac{dT^{\text{II}}}{dz}$ $\hspace{2.5cm}$ (10.5-11)

B.C. 4: $\qquad\qquad$ at $z = L$, $\qquad T^{\text{II}} = T^{\text{III}}$ $\hspace{5cm}$ (10.5-12)

B.C. 5: $\qquad\qquad$ at $z = L$, $\qquad \kappa_{\text{eff},zz} \dfrac{dT^{\text{II}}}{dz} = \kappa_{\text{eff},zz} \dfrac{dT^{\text{III}}}{dz}$ $\hspace{2.3cm}$ (10.5-13)

B.C. 6: $\qquad\qquad$ at $z = \infty$, $\qquad T^{\text{III}} = \text{finite}$ $\hspace{4.5cm}$ (10.5-14)

Equations 10.5-10 to 13 express the continuity of temperature and heat flux at the boundaries between the zones. Equations 10.5-9 and 14 specify requirements at the two ends of the system.

The solution of Eqs. 10.5-6 to 14 is considered here for arbitrary $F(\Theta)$. In many cases of practical interest, the convective heat transport is far more important than the axial conductive heat transport. Therefore, here we drop the conductive terms entirely (those containing $\kappa_{\text{eff},zz}$). This treatment of the problem still contains the salient features of the solution in the limit of large $\text{Pé} = \text{RePr}$ (see Problem 10B.18 for a fuller treatment).

If we introduce a dimensionless axial coordinate $Z = z/L$ and a dimensionless chemical heat source $N = S_{c1} L / \rho \hat{C}_p v_0 (T_1 - T_0)$, then Eqs. 10.5-6 to 8 become

Zone I $\qquad\qquad (Z < 0) \qquad \dfrac{d\Theta^{\text{I}}}{dZ} = 0$ $\hspace{5cm}$ (10.5-15)

Zone II $\qquad\qquad (0 < Z < 1) \qquad \dfrac{d\Theta^{\text{II}}}{dZ} = NF(\Theta)$ $\hspace{3.5cm}$ (10.5-16)

Zone III $\qquad\qquad (Z > 1) \qquad \dfrac{d\Theta^{\text{III}}}{dZ} = 0$ $\hspace{5cm}$ (10.5-17)

for which we need three boundary conditions:

B.C. 1: $\qquad\qquad$ at $Z = -\infty$, $\qquad \Theta^{\text{I}} = 1$ $\hspace{4.5cm}$ (10.5-18)

B.C. 2: $\qquad\qquad$ at $Z = 0$, $\qquad \Theta^{\text{I}} = \Theta^{\text{II}}$ $\hspace{4.5cm}$ (10.5-19)

B.C. 3: $\qquad\qquad$ at $Z = 1$, $\qquad \Theta^{\text{II}} = \Theta^{\text{III}}$ $\hspace{4.3cm}$ (10.5-20)

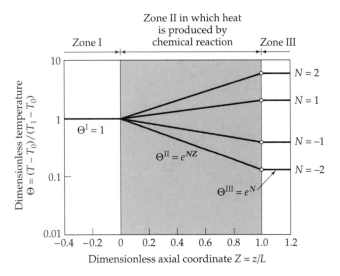

Fig. 10.5-2. Predicted temperature profiles in a fixed-bed axial-flow reactor when the heat production varies linearly with the temperature and when there is negligible axial diffusion.

The above first-order, separable differential equations, with boundary conditions, are easily solved to get

Zone I

$$\Theta^{\mathrm{I}} = 1$$

(10.5-21)

Zone II

$$\int_{\Theta^{\mathrm{I}}}^{\Theta^{\mathrm{II}}} \frac{1}{F(\Theta)}\, d\Theta = NZ$$

(10.5-22)

Zone III

$$\Theta^{\mathrm{III}} = \Theta^{\mathrm{II}}\big|_{Z=1}$$

(10.5-23)

These results are shown in Fig. 10.5-2 for a simple choice for the source function—namely, $F(\Theta) = \Theta$—which is reasonable for small changes in temperature, if the reaction rate is insensitive to concentration.

Here in this section we ended up discarding the axial conduction terms. In Problem 10B.18, these terms are not discarded, and then the solution shows that there is some preheating (or precooling) in region I.

§10.6 HEAT CONDUCTION THROUGH COMPOSITE WALLS

In industrial heat transfer problems one is often concerned with conduction through walls made up of layers of various materials, each with its own characteristic thermal conductivity. In this section we show how the various resistances to heat transfer are combined into a total resistance.

In Fig. 10.6-1 we show a composite wall made up of three materials of different thicknesses, $x_1 - x_0$, $x_2 - x_1$, and $x_3 - x_2$, and different thermal conductivities k_{01}, k_{12}, and k_{23}. At $x = x_0$, substance 01 is in contact with a fluid with ambient temperature T_a, and at $x = x_3$, substance 23 is in contact with a fluid at temperature T_b. The heat transfer at the boundaries $x = x_0$ and $x = x_3$ is given by Newton's "law of cooling" with heat transfer

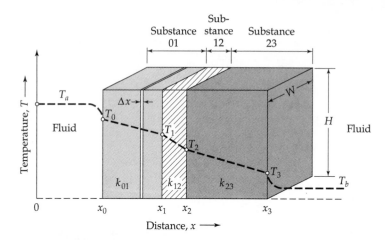

Fig. 10.6-1. Heat conduction through a composite wall, located between two fluid streams at temperatures T_a and T_b.

coefficients h_0 and h_3, respectively. The anticipated temperature profile is sketched in Fig. 10.6-1.

First we set up the energy balance for the problem. Since we are dealing with heat conduction in a solid, the terms containing velocity in the **e** vector can be discarded, and the only relevant contribution is the **q** vector, describing heat conduction. We first write the energy balance for a slab of volume $WH\,\Delta x$

Region 01:
$$q_x|_x WH - q_x|_{x+\Delta x} WH = 0 \tag{10.6-1}$$

which states that the heat entering at x must be equal to the heat leaving at $x + \Delta x$, since no heat is produced within the region. After division by $WH\,\Delta x$ and taking the limit as $\Delta x \to 0$, we get

Region 01:
$$\frac{dq_x}{dx} = 0 \tag{10.6-2}$$

Integration of this equation gives

Region 01:
$$q_x = q_0 \quad \text{(a constant)} \tag{10.6-3}$$

The constant of integration, q_0, is the heat flux at the plane $x = x_0$. The development in Eqs. 10.6-1, 2, and 3 can be repeated for regions 12 and 23 with continuity conditions on q_x at interfaces, so that the heat flux is constant and the same for all three slabs:

Regions 01, 12, 23:
$$q_x = q_0 \tag{10.6-4}$$

with the same constant for each of the regions. We may now introduce a Fourier's law for each of the three regions and get

Region 01:
$$-k_{01}\frac{dT}{dx} = q_0 \tag{10.6-5}$$

Region 12:
$$-k_{12}\frac{dT}{dx} = q_0 \tag{10.6-6}$$

Region 23:
$$-k_{23}\frac{dT}{dx} = q_0 \tag{10.6-7}$$

We now assume that k_{01}, k_{12}, and k_{23} are constants. Then we integrate each equation over the entire thickness of the relevant slab of material to get

Region 01:
$$T_0 - T_1 = q_0\left(\frac{x_1 - x_0}{k_{01}}\right) \tag{10.6-8}$$

Region 12:
$$T_1 - T_2 = q_0\left(\frac{x_2 - x_1}{k_{12}}\right) \tag{10.6-9}$$

Region 23:
$$T_2 - T_3 = q_0\left(\frac{x_3 - x_2}{k_{23}}\right) \tag{10.6-10}$$

In addition we have the two statements regarding the heat transfer at the surfaces according to Newton's law of cooling:

At surface 0:
$$T_a - T_0 = \frac{q_0}{h_0} \tag{10.6-11}$$

At surface 3:
$$T_3 - T_b = \frac{q_0}{h_3} \tag{10.6-12}$$

Addition of these last five equations then gives

$$T_a - T_b = q_0\left(\frac{1}{h_0} + \frac{x_1 - x_0}{k_{01}} + \frac{x_2 - x_1}{k_{12}} + \frac{x_3 - x_2}{k_{23}} + \frac{1}{h_3}\right) \tag{10.6-13}$$

or

$$q_0 = \frac{T_a - T_b}{\left(\dfrac{1}{h_0} + \displaystyle\sum_{j=1}^{3}\dfrac{x_j - x_{j-1}}{k_{j-1,j}} + \dfrac{1}{h_3}\right)} \tag{10.6-14}$$

Sometimes this result is rewritten in a form reminiscent of Newton's law of cooling, either in terms of the heat flux q_0 (J/m$^2 \cdot$ s) or the heat flow Q_0 (J/s):

$$q_0 = U(T_a - T_b) \quad \text{or} \quad Q_0 = U(WH)(T_a - T_b) \tag{10.6-15}$$

The quantity U, called the "overall heat transfer coefficient," is given then by the following famous formula for the "additivity of resistances":

$$\frac{1}{U} = \frac{1}{h_0} + \sum_{j=1}^{n}\frac{x_j - x_{j-1}}{k_{j-1,j}} + \frac{1}{h_n} \tag{10.6-16}$$

Here we have generalized the formula to a system with n slabs of material. Equations 10.6-15 and 16 are useful for calculating the heat transfer rate through a composite wall separating two fluid streams, when the heat transfer coefficients and thermal conductivities are known. The estimation of heat transfer coefficients is discussed in Chapter 14.

In the above development it has been tacitly assumed that the solid slabs are contiguous with no intervening "air spaces." If the solid surfaces touch each other only at several points, the resistance to heat transfer will be appreciably increased.

EXAMPLE 10.6-1

Composite Cylindrical Walls

Develop a formula for the overall heat transfer coefficient for the composite cylindrical pipe wall shown in Fig. 10.6-2.

SOLUTION

An energy balance on a shell of volume $2\pi r L\, \Delta r$ for region 01 is

Region 01:
$$q_r|_r \cdot 2\pi r L - q_r|_{r+\Delta r} \cdot 2\pi(r + \Delta r)L = 0 \tag{10.6-17}$$

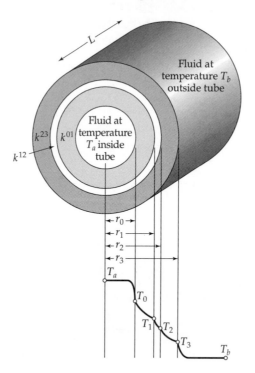

Fig. 10.6-2. Heat conduction through a laminated tube with a fluid at temperature T_a inside the tube and temperature T_b outside.

which can also be written as

Region 01:
$$(2\pi r L q_r)|_r - (2\pi r L q_r)|_{r+\Delta r} = 0 \tag{10.6-18}$$

Dividing by $2\pi L\,\Delta r$ and taking the limit as Δr goes to zero gives

Region 01:
$$\frac{d}{dr}(rq_r) = 0 \tag{10.6-19}$$

Integration of this equation gives

$$rq_r = r_0 q_0 \tag{10.6-20}$$

in which r_0 is the inner radius of region 01, and q_0 is the heat flux there. In regions 12 and 23, rq_r is equal to the same constant. Application of Fourier's law to the three regions gives

Region 01:
$$-k_{01}r\frac{dT}{dr} = r_0 q_0 \tag{10.6-21}$$

Region 12:
$$-k_{12}r\frac{dT}{dr} = r_0 q_0 \tag{10.6-22}$$

Region 23:
$$-k_{23}r\frac{dT}{dr} = r_0 q_0 \tag{10.6-23}$$

If we assume that the thermal conductivities in the three annular regions are constants, then each of the above three equations can be integrated across its region to give

Region 10:
$$T_0 - T_1 = r_0 q_0 \frac{\ln(r_1/r_0)}{k_{01}} \tag{10.6-24}$$

Region 12:
$$T_1 - T_2 = r_0 q_0 \frac{\ln(r_2/r_1)}{k_{12}} \tag{10.6-25}$$

Region 23:
$$T_2 - T_3 = r_0 q_0 \frac{\ln (r_3/r_2)}{k_{23}} \tag{10.6-26}$$

At the two fluid–solid interfaces we can write Newton's law of cooling:

Surface 0:
$$T_a - T_0 = \frac{q_0}{h_0} \tag{10.6-27}$$

Surface 3:
$$T_3 - T_b = \frac{q_3}{h_3} = \frac{q_0}{h_3} \frac{r_0}{r_3} \tag{10.6-28}$$

Addition of the preceding five equations gives an equation for $T_a - T_b$. Then the equation is solved for q_0 to give

$$Q_0 = 2\pi L r_0 q_0 = \frac{2\pi L(T_a - T_b)}{\left(\dfrac{1}{r_0 h_0} + \dfrac{\ln (r_1/r_0)}{k_{01}} + \dfrac{\ln (r_2/r_1)}{k_{12}} + \dfrac{\ln (r_3/r_2)}{k_{23}} + \dfrac{1}{r_3 h_3} \right)} \tag{10.6-29}$$

We now define an "overall heat transfer coefficient based on the inner surface" U_0 by

$$Q_0 = 2\pi L r_0 q_0 = U_0(2\pi L r_0)(T_a - T_b) \tag{10.6-30}$$

Combination of the last two equations gives, on generalizing to a system with n annular layers,

$$\frac{1}{r_0 U_0} = \left(\frac{1}{r_0 h_0} + \sum_{j=1}^{n} \frac{\ln (r_j/r_{j-1})}{k_{j-1,j}} + \frac{1}{r_n h_n} \right) \tag{10.6-31}$$

The subscript "0" on U_0 indicates that the overall heat transfer coefficient is referred to the radius r_0.

§10.7   HEAT CONDUCTION IN A COOLING FIN[1]

Another simple, but practical application of heat conduction is the calculation of the efficiency of a cooling fin. Fins are used to increase the area available for heat transfer between metal walls and poorly conducting fluids such as gases. A simple rectangular fin is shown in Fig. 10.7-1. The wall temperature is T_w and the ambient air temperature is T_a.

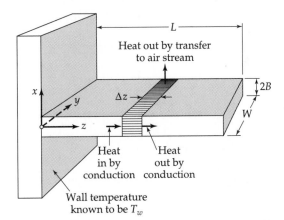

Heat out by transfer to air stream

Heat in by conduction

Heat out by conduction

Wall temperature known to be T_w

Fig. 10.7-1. A simple cooling fin with $B \ll L$ and $B \ll W$.

---

[1] For further information on fins, see M. Jakob, *Heat Transfer*, Vol. I, Wiley, New York (1949), Chapter 11; and H. D. Baehr and K. Stephan, *Heat and Mass Transfer*, Springer, Berlin (1998), §2.2.3.

A reasonably good description of the system may be obtained by approximating the true physical situation by a simplified model:

True situation	Model
1. T is a function of x, y, and z, but the dependence on z is most important.	1. T is a function of z alone.
2. A small quantity of heat is lost from the fin at the end (area $2BW$) and at the edges (area $(2BL + 2BL)$.	2. No heat is lost from the end or from the edges.
3. The heat transfer coefficient is a function of position.	3. The heat flux at the surface is given by $q_z = h(T - T_a)$, where h is constant and T depends on z.

The energy balance is made over a segment Δz of the bar. Since the bar is stationary, the terms containing $\mathbf{v}$ in the combined energy flux vector $\mathbf{e}$ may be discarded, and the only contribution to the energy flux is $\mathbf{q}$. Therefore the energy balance is

$$2BWq_z|_z - 2BWq_z|_{z+\Delta z} - h(2W\Delta z)(T - T_a) = 0 \tag{10.7-1}$$

Division by $2BW\,\Delta z$ and taking the limit as Δz approaches zero gives

$$-\frac{dq_z}{dz} = \frac{h}{B}(T - T_a) \tag{10.7-2}$$

We now insert Fourier's law ($q_z = -kdT/dz$), in which k is the thermal conductivity of the metal. If we assume that k is constant, we then get

$$\frac{d^2T}{dz^2} = \frac{h}{kB}(T - T_a) \tag{10.7-3}$$

This equation is to be solved with the boundary conditions

B.C. 1:   at $z = 0$, $T = T_w$ (10.7-4)

B.C. 2:   at $z = L$, $\dfrac{dT}{dz} = 0$ (10.7-5)

We now introduce the following dimensionless quantities:

$$\Theta = \frac{T - T_a}{T_w - T_a} = \text{dimensionless temperature} \tag{10.7-6}$$

$$\zeta = \frac{z}{L} \qquad = \text{dimensionless distance} \tag{10.7-7}$$

$$N^2 = \frac{hL^2}{kB} \qquad = \text{dimensionless heat transfer coefficient}^2 \tag{10.7-8}$$

The problem then takes the form

$$\frac{d^2\Theta}{d\zeta^2} = N^2\Theta \qquad \text{with } \Theta|_{\zeta=0} = 1 \quad \text{and} \quad \frac{d\Theta}{d\zeta}\bigg|_{\zeta=1} = 0 \qquad \text{(10.7-9, 10, 11)}$$

---

[2] The quantity N^2 may be rewritten as $N^2 = (hL/k)(L/B) = \text{Bi}(L/B)$, where Bi is called the *Biot number*, named after **Jean Baptiste Biot** (1774–1862) (pronounced "Bee-oh"). Professor of physics at the Collège de France, he received the Rumford Medal for his development of a simple, nondestructive test to determine sugar concentration.

Equation 10.7-9 may be integrated to give hyperbolic functions (see Eq. C.1-4 and §C.5). When the two integration constants have been determined, we get

$$\Theta = \cosh N\zeta - (\tanh N) \sinh N\zeta \tag{10.7-12}$$

This may be rearranged to give

$$\boxed{\Theta = \frac{\cosh N(1 - \zeta)}{\cosh N}} \tag{10.7-13}$$

This result is reasonable only if the heat lost at the end and at the edges is negligible.

The "effectiveness" of the fin surface is defined[3] by

$$\eta = \frac{\text{actual rate of heat loss from the fin}}{\text{rate of heat loss from an isothermal fin at } T_w} \tag{10.7-14}$$

For the problem being considered here η is then

$$\eta = \frac{\displaystyle\int_0^W \int_0^L h(T - T_a)\,dz\,dy}{\displaystyle\int_0^W \int_0^L h(T_w - T_a)\,dz\,dy} = \frac{\displaystyle\int_0^1 \Theta\,d\zeta}{\displaystyle\int_0^1 d\zeta} \tag{10.7-15}$$

or

$$\eta = \frac{1}{\cosh N}\left(-\frac{1}{N}\sinh N(1 - \zeta)\right)\Bigg|_0^1 = \frac{\tanh N}{N} \tag{10.7-16}$$

in which N is the dimensionless quantity defined in Eq. 10.7-8.

EXAMPLE 10.7-1

Error in Thermocouple Measurement

In Fig. 10.7-2 a thermocouple is shown in a cylindrical well inserted into a gas stream. Estimate the true temperature of the gas stream if

$T_1 = 500°F$ = temperature indicated by thermocouple

$T_w = 350°F$ = wall temperature

$h = 120 \text{ Btu/hr} \cdot \text{ft}^2 \cdot \text{F}$ = heat transfer coefficient

$k = 60 \text{ Btu/hr} \cdot \text{ft}^3 \cdot \text{F}$ = thermal conductivity of well wall

$B = 0.08 \text{ in.}$ = thickness of well wall

$L = 0.2 \text{ ft}$ = length of well

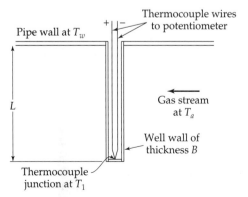

Pipe wall at T_w

Thermocouple wires to potentiometer

L

Gas stream at T_a

Well wall of thickness B

Thermocouple junction at T_1

Fig. 10.7-2. A thermocouple in a cylindrical well.

---

[3] M. Jakob, *Heat Transfer*, Vol. I, Wiley, New York (1949), p. 235.

SOLUTION

The thermocouple well wall of thickness B is in contact with the gas stream on one side only, and the tube thickness is small compared with the diameter. Hence the temperature distribution along this wall will be about the same as that along a bar of thickness $2B$, in contact with the gas stream on both sides. According to Eq. 10.7-13, the temperature at the end of the well (that registered by the thermocouple) satisfies

$$\frac{T_1 - T_a}{T_w - T_a} = \frac{\cosh 0}{\cosh N} = \frac{1}{\cosh\sqrt{hL^2/kB}}$$

$$= \frac{1}{\cosh\sqrt{(120)(0.2)^2/(60)(\frac{1}{12} \cdot 0.08)}}$$

$$= \frac{1}{\cosh(2\sqrt{3})} = \frac{1}{16.0} \tag{10.7-17}$$

Hence the actual ambient gas temperature is obtained by solving this equation for T_a:

$$\frac{500 - T_a}{350 - T_a} = \frac{1}{16.0} \tag{10.7-18}$$

and the result is

$$T_a = 510°F \tag{10.7-19}$$

Therefore, the reading is 10 F° too low.

This example has focused on one kind of error that can occur in thermometry. Frequently a simple analysis, such as the foregoing, can be used to estimate the measurement errors.[4]

§10.8   FORCED CONVECTION

In the preceding sections the emphasis has been placed on heat conduction in solids. In this and the following section we study two limiting types of heat transport in fluids: *forced convection* and *free convection* (also called *natural convection*). The main differences between these two modes of convection are shown in Fig. 10.8-1. Most industrial heat transfer problems are usually put into either one or the other of these two limiting categories. In some problems, however, both effects must be taken into account, and then we speak of *mixed convection* (see §14.6 for some empiricisms for handling this situation).

In this section we consider forced convection in a circular tube, a limiting case of which is simple enough to be solved analytically.[1,2] A viscous fluid with physical properties ($\mu, k, \rho, \hat{C}_p$) assumed constant is in laminar flow in a circular tube of radius R. For $z < 0$ the fluid temperature is uniform at the inlet temperature T_1. For $z > 0$ there is a constant radial heat flux $q_r = -q_0$ at the wall. Such a situation exists, for example, when a pipe is wrapped uniformly with an electrical heating coil, in which case q_0 is positive. If the pipe is being chilled, then q_0 has to be taken as negative.

As indicated in Fig. 10.8-1, the first step in solving a forced convection heat transfer problem is the calculation of the velocity profiles in the system. We have seen in §2.3

---

[4] For further discussion, see M. Jakob, *Heat Transfer*, Vol. II, Wiley, New York (1949), Chapter 33, pp. 147–201.

[1] A. Eagle and R. M. Ferguson, *Proc. Roy. Soc. (London)*, **A127**, 540–566 (1930).

[2] S. Goldstein, *Modern Developments in Fluid Dynamics*, Oxford University Press (1938), Dover Edition (1965), Vol. II, p. 622.

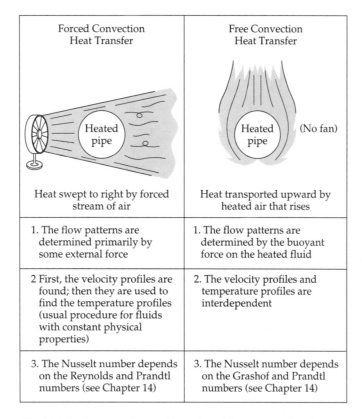

Forced Convection Heat Transfer	Free Convection Heat Transfer
Heat swept to right by forced stream of air	Heat transported upward by heated air that rises
1. The flow patterns are determined primarily by some external force	1. The flow patterns are determined by the buoyant force on the heated fluid
2 First, the velocity profiles are found; then they are used to find the temperature profiles (usual procedure for fluids with constant physical properties)	2. The velocity profiles and temperature profiles are interdependent
3. The Nusselt number depends on the Reynolds and Prandtl numbers (see Chapter 14)	3. The Nusselt number depends on the Grashof and Prandtl numbers (see Chapter 14)

Fig. 10.8-1. A comparison of forced and free convection in non-isothermal systems.

how this may be done for tube flow by using the shell balance method. We know that the velocity distribution so obtained is $v_r = 0$, $v_\theta = 0$, and

$$v_z = \frac{(\mathcal{P}_0 - \mathcal{P}_L)R^2}{4\mu L}\left[1 - \left(\frac{r}{R}\right)^2\right] = v_{z,\max}\left[1 - \left(\frac{r}{R}\right)^2\right] \tag{10.8-1}$$

This parabolic distribution is valid sufficiently far downstream from the inlet that the entrance length has been exceeded.

In this problem, heat is being transported in both the r and the z directions. Therefore, for the energy balance we use a "washer-shaped" system, which is formed by the intersection of an annular region of thickness Δr with a slab of thickness Δz (see Fig. 10.8-2). In this problem, we are dealing with a flowing fluid, and therefore all terms in the **e** vector will be retained. The various contributions to Eq. 10.1-1 are

Total energy in at r	$e_r\|_r \cdot 2\pi r\Delta z = (2\pi r e_r)\|_r \, \Delta z$	(10.8-2)
Total energy out at $r + \Delta r$	$e_r\|_{r+\Delta r} \cdot 2\pi(r + \Delta r)\Delta z = (2\pi r e_r)\|_{r+\Delta r} \, \Delta z$	(10.8-3)
Total energy in at z	$e_z\|_z \cdot 2\pi r\Delta r$	(10.8-4)
Total energy out at $z + \Delta z$	$e_z\|_{z+\Delta z} \cdot 2\pi r\Delta r$	(10.8-5)
Work done on fluid by gravity	$\rho v_z g_z \cdot 2\pi r\Delta r\Delta z$	(10.8-6)

The last contribution is the rate at which work is done on the fluid within the ring by gravity—that is, the force per unit volume ρg_z times the volume $2\pi r \, \Delta r \, \Delta z$ multiplied by the downward velocity of the fluid.

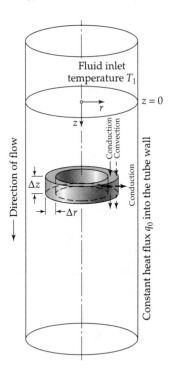

Fig. 10.8-2. Heating of a fluid in laminar flow through a circular tube, showing the annular ring over which the energy balance is made.

The energy balance is obtained by summing these contributions and equating the sum to zero. Then we divide by $2\pi\,\Delta r\,\Delta z$ to get

$$\frac{(re_r)|_r - (re_r)|_{r+\Delta r}}{\Delta r} + r\frac{e_z|_z - e_z|_{z+\Delta z}}{\Delta z} + \rho v_z g_z r = 0 \tag{10.8-7}$$

In the limit as Δr and Δz go to zero, we find

$$-\frac{1}{r}\frac{\partial}{\partial r}(re_r) - \frac{\partial e_z}{\partial z} + \rho v_z g = 0 \tag{10.8-8}$$

The subscript z in g_z has been omitted, since the gravity vector is acting in the $+z$ direction.

Next we use Eqs. 9.8-6 and 9.8-8 to write out the expressions for the r- and z-components of the combined energy flux vector, using the fact that the only nonzero component of **v** is the axial component v_z:

$$e_r = \tau_{rz}v_z + q_r = -\left(\mu\frac{\partial v_z}{\partial r}\right)v_z - k\frac{\partial T}{\partial r} \tag{10.8-9}$$

$$e_z = (\tfrac{1}{2}\rho v_z^2)v_z + \rho\hat{H}v_z + \tau_{zz}v_z + q_z$$

$$= (\tfrac{1}{2}\rho v_z^2)v_z + \rho\hat{H}^\circ v_z + (p - p^\circ)v_z + \rho\hat{C}_p(T - T^\circ)v_z - \left(2\mu\frac{\partial v_z}{\partial z}\right)v_z - k\frac{\partial T}{\partial z} \tag{10.8-10}$$

Substituting these flux expressions into Eq. 10.8-8 and using the fact that v_z depends only on r gives, after some rearrangement,

$$\rho\hat{C}_p v_z\frac{\partial T}{\partial z} = k\left[\frac{1}{r}\frac{\partial}{\partial r}\left(r\frac{\partial T}{\partial r}\right) + \frac{\partial^2 T}{\partial z^2}\right] + \mu\left(\frac{\partial v_z}{\partial r}\right)^2 + v_z\left[-\frac{\partial p}{\partial z} + \mu\frac{1}{r}\frac{\partial}{\partial r}\left(r\frac{\partial v_z}{\partial r}\right) + \rho g\right] \tag{10.8-11}$$

The second bracket is exactly zero, as can be seen from Eq. 3.6-4, which is the z-component term containing the viscosity is the viscous heating, which we shall neglect in this discussion. The last term in the first bracket, corresponding to heat conduction in the axial direction, will be omitted, since we know from experience that it is usually small in comparison with the heat convection in the axial direction. Therefore, the equation that we want to solve here is

$$\rho \hat{C}_p v_{z,max} \left[1 - \left(\frac{r}{R} \right)^2 \right] \frac{\partial T}{\partial z} = k \left[\frac{1}{r} \frac{\partial}{\partial r} \left(r \frac{\partial T}{\partial r} \right) \right]$$ (10.8-12)

This partial differential equation, when solved, describes the temperature in the fluid as a function of r and z. The boundary conditions are

B.C. 1:   at $r = 0$, T = finite   (10.8-13)

B.C. 2:   at $r = R$, $k \dfrac{\partial T}{\partial r} = q_0$ (constant)   (10.8-14)

B.C. 3:   at $z = 0$, $T = T_1$ (10.8-15)

We now put the problem statement into dimensionless form. The choice of the dimensionless quantities is arbitrary. We choose

$$\Theta = \frac{T - T_1}{q_0 R/k} \qquad \xi = \frac{r}{R} \qquad \zeta = \frac{z}{\rho \hat{C}_p v_{z,max} R^2/k}$$ (10.8-16, 17, 18)

Generally one tries to select dimensionless quantities so as to minimize the number of parameters in the final problem formulation. In this problem the choice of $\xi = r/R$ is a natural one, because of the appearance of r/R in the differential equation. The choice for the dimensionless temperature is suggested by the second and third boundary conditions. Having specified these two dimensionless variables, the choice of dimensionless axial coordinate follows naturally.

The resulting problem statement, in dimensionless form, is now

$$(1 - \xi^2) \frac{\partial \Theta}{\partial \zeta} = \frac{1}{\xi} \frac{\partial}{\partial \xi} \left(\xi \frac{\partial \Theta}{\partial \xi} \right)$$ (10.8-19)

with boundary conditions

B.C. 1:   at $\xi = 0$, Θ = finite   (10.8-20)

B.C. 2:   at $\xi = 1$, $\dfrac{\partial \Theta}{\partial \xi} = 1$ (10.8-21)

B.C. 3:   at $\zeta = 0$, $\Theta = 0$ (10.8-22)

The partial differential equation in Eq. 10.8-19 has been solved for these boundary conditions,[3] but in this section we do not give the complete solution.

It is, however, instructive to obtain the asymptotic solution to Eq. 10.8-19 for large ζ. After the fluid is sufficiently far downstream from the beginning of the heated section, one expects that the constant heat flux through the wall will result in a rise of the fluid temperature that is linear in ζ. One further expects that the shape of the temperature profiles as a function of ξ will ultimately not undergo further change with increasing ζ (see Fig. 10.8-3). Hence a solution of the following form seems reasonable for large ζ:

$$\Theta(\xi, \zeta) = C_0 \zeta + \Psi(\xi)$$ (10.8-23)

in which C_0 is a constant to be determined presently.

---

[3] R. Siegel, E. M. Sparrow, and T. M. Hallman, *Appl. Sci. Research*, **A7**, 386–392 (1958). See Example 12.2-1 for the complete solution and Example 12.2-2 for the asymptotic solution for small ζ.

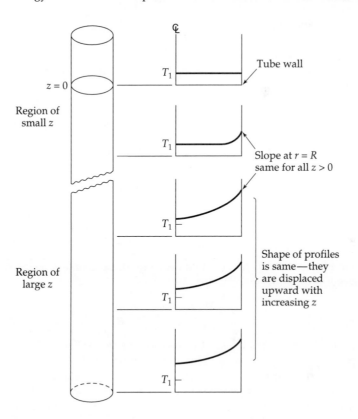

Fig. 10.8-3. Sketch showing how one expects the temperature $T(r, z)$ to look for the system shown in Fig. 10.8-2 when the fluid is heated by means of a heating coil wrapped uniformly around the tube (corresponding to q_0 positive).

The function in Eq. 10.8-23 is clearly not the complete solution to the problem; it does allow the partial differential equation and boundary conditions 1 and 2 to be satisfied, but clearly does not satisfy boundary condition 3. Hence we replace the latter by an integral condition (see Fig. 10.8-4),

Condition 4:
$$2\pi R z q_0 = \int_0^{2\pi} \int_0^R \rho \hat{C}_p (T - T_1) v_z r \, dr \, d\theta \tag{10.8-24}$$

or, in dimensionless form,

$$\zeta = \int_0^1 \Theta(\xi, \zeta)(1 - \xi^2)\xi \, d\xi \tag{10.8-25}$$

This condition states that the energy entering through the walls over a distance ζ is the same as the difference between the energy leaving through the cross section at ζ and that entering at $\zeta = 0$.

Substitution of the postulated function of Eq. 10.8-23 into Eq. 10.8-19 leads to the following ordinary differential equation for Ψ (see Eq. C.1-11):

$$\frac{1}{\xi}\frac{d}{d\xi}\left(\xi \frac{d\Psi}{d\xi}\right) = C_0(1 - \xi^2) \tag{10.8-26}$$

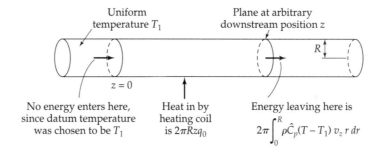

Fig. 10.8-4. Energy balance used for boundary condition 4 given in Eq. 10.8-24.

This equation may be integrated twice with respect to ξ and the result substituted into Eq. 10.8-23 to give

$$\Theta(\xi, \zeta) = C_0\zeta + C_0\left(\frac{\xi^2}{4} - \frac{\xi^4}{16}\right) + C_1 \ln\, \xi + C_2 \tag{10.8-27}$$

The three constants are determined from the conditions 1, 2, and 4 above:

B.C. 1:	$C_1 = 0$	(10.8-28)
B.C. 2:	$C_0 = 4$	(10.8-29)
Condition 4:	$C_2 = -\frac{7}{24}$	(10.8-30)

Substitution of these values into Eq. 10.8-27 gives finally

$$\boxed{\Theta(\xi, \zeta) = 4\zeta + \xi^2 - \tfrac{1}{4}\xi^4 - \tfrac{7}{24}} \tag{10.8-31}$$

This result gives the dimensionless temperature as a function of the dimensionless radial and axial coordinates. It is exact in the limit as $\zeta \rightarrow \infty$; for $\zeta > 0.1$, it predicts the local value of Θ to within about 2%.

Once the temperature distribution is known, one can get various derived quantities. There are two kinds of average temperatures commonly used in connection with the flow of fluids with constant ρ and $\hat{C}_p$:

$$\langle T \rangle = \frac{\displaystyle\int_0^{2\pi} \int_0^R T(r, z)r\, dr\, d\theta}{\displaystyle\int_0^{2\pi} \int_0^R r\, dr\, d\theta} = T_1 + (4\zeta + \tfrac{7}{24})\frac{q_0 R}{k} \tag{10.8-32}$$

$$T_b = \frac{\langle v_z T \rangle}{\langle v_z \rangle} = \frac{\displaystyle\int_0^{2\pi} \int_0^R v_z(r)T(r, z)r\, dr\, d\theta}{\displaystyle\int_0^{2\pi} \int_0^R v_z(r)r\, dr\, d\theta} = T_1 + (4\zeta)\frac{q_0 R}{k} \tag{10.8-33}$$

Both averages are functions of z. The quantity $\langle T \rangle$ is the arithmetic average of the temperatures over the cross section at z. The "bulk temperature" T_b is the temperature one would obtain if the tube were chopped off at z and if the fluid issuing forth were collected in a container and thoroughly mixed. This average temperature is sometimes referred to as the "cup-mixing temperature" or the "flow-average temperature."

Now let us evaluate the local heat transfer driving force, $T_0 - T_b$, which is the difference between the wall and bulk temperatures at a distance z down the tube:

$$T_0 - T_b = \frac{11}{24}\frac{q_0 R}{k} = \frac{11}{48}\frac{q_0 D}{k} \tag{10.8-34}$$

where D is the tube diameter. We may now rearrange this result in the form of a dimensionless wall heat flux

$$\frac{q_0 D}{k(T_0 - T_b)} = \frac{48}{11} \tag{10.8-35}$$

which, in Chapter 14, will be identified as a *Nusselt number*.

Before leaving this section, we point out that the dimensionless axial coordinate ζ introduced above may be rewritten in the following way:

$$\zeta = \left[\frac{\mu}{D\langle v_z\rangle\rho}\right]\left[\frac{k}{\hat{C}_p\mu}\right]\left[\frac{z}{R}\right] = \frac{1}{\mathrm{RePr}}\left[\frac{z}{R}\right] = \frac{1}{\mathrm{P\acute{e}}}\left[\frac{z}{R}\right] \tag{10.8-36}$$

Here D is the tube diameter, Re is the Reynolds number used in Part I, and Pr and Pé are the Prandtl and Péclet numbers introduced in Chapter 9. We shall find in Chapter 11 that the Reynolds and Prandtl numbers can be expected to appear in forced convection problems. This point will be reinforced in Chapter 14 in connection with correlations for heat transfer coefficients.

§10.9 FREE CONVECTION

In §10.8 we gave an example of forced convection. In this section we turn our attention to an elementary free convection problem—namely, the flow between two parallel walls maintained at different temperatures (see Fig. 10.9-1).

A fluid with density ρ and viscosity μ is located between two vertical walls a distance $2B$ apart. The heated wall at $y = -B$ is maintained at temperature T_2, and the cooled wall at $y = +B$ is maintained at temperature T_1. It is assumed that the temperature difference is sufficiently small that terms containing $(\Delta T)^2$ can be neglected.

Because of the temperature gradient in the system, the fluid near the hot wall rises and that near the cold wall descends. The system is closed at the top and bottom, so that the fluid is continuously circulating between the plates. The mass rate of flow of the

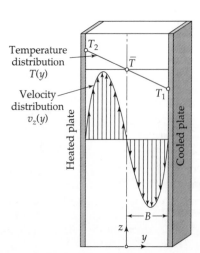

Temperature distribution $T(y)$

Velocity distribution $v_z(y)$

Fig. 10.9-1. Laminar free convection flow between two vertical plates at two different temperatures. The velocity is a cubic function of the coordinate y.

fluid in the upward-moving stream is the same as that in the downward-moving stream. The plates are presumed to be very tall, so that end effects near the top and bottom can be disregarded. Then for all practical purposes the temperature is a function of y alone.

An energy balance can now be made over a thin slab of fluid of thickness Δy, using the y-component of the combined energy flux vector $\mathbf{e}$ as given in Eq. 9.8-6. The term containing the kinetic energy and enthalpy can be disregarded, since the y-component of the $\mathbf{v}$ vector is zero. The y-component of the term $[\boldsymbol{\tau} \cdot \mathbf{v}]$ is $\tau_{yz}v_z = -\mu(dv_z/dy)v_z$, which would lead to the viscous heating contribution discussed in §10.4. However, in the very slow flows encountered in free convection, this term will be extremely small and can be neglected. The energy balance then leads to the equation

$$-\frac{dq_y}{dy} = 0 \quad \text{or} \quad k\frac{d^2T}{dy^2} = 0 \tag{10.9-1}$$

for constant k. The temperature equation is to be solved with the boundary conditions:

B.C. 1:   at $y = -B$, $T = T_2$ (10.9-2)

B.C. 2:   at $y = +B$, $T = T_1$ (10.9-3)

The solution to this problem is

$$\boxed{T = \overline{T} - \frac{1}{2}\Delta T \frac{y}{B}} \tag{10.9-4}$$

in which $\Delta T = T_2 - T_1$ is the difference of the wall temperatures, and $\overline{T} = \frac{1}{2}(T_1 + T_2)$ is their arithmetic mean.

By making a momentum balance over the same slab of thickness Δy, one arrives at a differential equation for the velocity distribution

$$\mu\frac{d^2v_z}{dy^2} = \frac{dp}{dz} + \rho g \tag{10.9-5}$$

Here the viscosity has been assumed constant (see Problem 10B.11 for a solution with temperature-dependent viscosity).

The phenomenon of free convection results from the fact that when the fluid is heated, the density (usually) decreases and the fluid rises. The mathematical description of the system must take this essential feature of the phenomenon into account. Because the temperature difference $\Delta T = T_2 - T_1$ is taken to be small in this problem, it can be expected that the density changes in the system will be small. This suggests that we should expand ρ in a Taylor series about the temperature $\overline{T} = \frac{1}{2}(T_1 + T_2)$ thus:

$$\rho = \rho|_{T=\overline{T}} + \frac{d\rho}{dT}\bigg|_{T=\overline{T}}(T - \overline{T}) + \cdots$$

$$= \overline{\rho} - \overline{\rho}\overline{\beta}(T - \overline{T}) + \cdots \tag{10.9-6}$$

Here $\overline{\rho}$ and $\overline{\beta}$ are the density and coefficient of volume expansion evaluated at the temperature $\overline{T}$. The coefficient of volume expansion is defined as

$$\beta = \frac{1}{V}\left(\frac{\partial V}{\partial T}\right)_p = \frac{1}{(1/\rho)}\left(\frac{\partial(1/\rho)}{\partial T}\right)_p = -\frac{1}{\rho}\left(\frac{\partial \rho}{\partial T}\right)_p \tag{10.9-7}$$

We now introduce the "Taylor-made" equation of state of Eq. 10.9-6 (keeping two terms only) into the equation of motion in Eq. 10.9-5 to get

$$\mu\frac{d^2v_z}{dy^2} = \frac{dp}{dz} + \overline{\rho}g - \overline{\rho}g\overline{\beta}(T - \overline{T}) \tag{10.9-8}$$

This equation describes the balance among the viscous force, the pressure force, the gravity force, and the buoyant force $-\bar{\rho}g\beta(T - \bar{T})$ (all per unit volume). Into this we now substitute the temperature distribution given in Eq. 10.9-4 to get the differential equation

$$\mu \frac{d^2 v_z}{dy^2} = \left(\frac{dp}{dz} + \bar{\rho}g\right) + \tfrac{1}{2}\bar{\rho}g\beta\Delta T \frac{y}{B} \tag{10.9-9}$$

which is to be solved with the boundary conditions

B.C. 1:   at $y = -B$, $v_z = 0$ (10.9-10)

B.C. 2:   at $y = +B$, $v_z = 0$ (10.9-11)

The solution is

$$v_z = \frac{(\bar{\rho}g\bar{\beta}\Delta T)B^2}{12\mu}\left[\left(\frac{y}{B}\right)^3 - \left(\frac{y}{B}\right)\right] + \frac{B^2}{2\mu}\left(\frac{dp}{dz} + \bar{\rho}g\right)\left[\left(\frac{y}{B}\right)^2 - 1\right] \tag{10.9-12}$$

We now require that the net mass flow in the z direction be zero, that is,

$$\int_{-B}^{+B} \rho v_z dy = 0 \tag{10.9-13}$$

Substitution of v_z from Eq. 10.9-12 and ρ from Eqs. 10.9-6 and 4 into this integral leads to the conclusion that

$$\frac{dp}{dz} = -\bar{\rho}g \tag{10.9-14}$$

when terms containing the square of the small quantity ΔT are neglected. Equation 10.9-14 states that the pressure gradient in the system is due solely to the weight of the fluid, and the usual hydrostatic pressure distribution prevails. Therefore the second term on the right side of Eq. 10.9-12 drops out and the final expression for the velocity distribution is

$$\boxed{v_z = \frac{(\bar{\rho}g\bar{\beta}\Delta T)B^2}{12\mu}\left[\left(\frac{y}{B}\right)^3 - \left(\frac{y}{B}\right)\right]} \tag{10.9-15}$$

The average velocity in the upward-moving stream is

$$\langle v_z \rangle = \frac{(\bar{\rho}g\bar{\beta}\Delta T)B^2}{48\mu} \tag{10.9-16}$$

The motion of the fluid is thus a direct result of the buoyant force term in Eq. 10.9-8, associated with the temperature gradient in the system. The velocity distribution of Eq. 10.9-15 is shown in Fig. 10.9-1. It is this sort of velocity distribution that occurs in the air space in a double-pane window or in double-wall panels in buildings. It is also this kind of flow that occurs in the operation of a Clusius–Dickel column used for separating isotopes or organic liquid mixtures by the combined effects of thermal diffusion and free convection.[1]

---

[1] Thermal diffusion is the diffusion resulting from a temperature gradient. For a lucid discussion of the Clusius–Dickel column see K. E. Grew and T. L. Ibbs, *Thermal Diffusion in Gases*, Cambridge University Press (1952), pp. 94–106.

The velocity distribution in Eq. 10.9-15 may be rewritten using a dimensionless velocity $\breve{v}_z = B v_z \bar{\rho}/\mu$ and a dimensionless coordinate $\breve{y} = y/B$ thus:

$$\breve{v}_z = \tfrac{1}{12} \text{Gr}(\breve{y}^3 - \breve{y}) \tag{10.9-17}$$

Here Gr is the dimensionless *Grashof number*,[2] defined by

$$\text{Gr} = \left[\frac{(\bar{\rho}^2 g \bar{\beta} \Delta T) B^3}{\mu^2} \right] = \left[\frac{\bar{\rho} g B^3 \, \Delta \rho}{\mu^2} \right] \tag{10.9-18}$$

where $\Delta\rho = \rho_1 - \rho_2$. The second form of the Grashof number is obtained from the first form by using Eq. 10.9-6. The Grashof number is the characteristic group occurring in analyses of free convection, as is shown by dimensional analysis in Chapter 11. It arises in heat transfer coefficient correlations in Chapter 14.

QUESTIONS FOR DISCUSSION

1. Verify that the Brinkman, Biot, Prandtl, and Grashof numbers are dimensionless.
2. To what problem in electrical circuits is the addition of thermal resistances analogous?
3. What is the coefficient of volume expansion for an ideal gas? What is the corresponding expression for the Grashof number?
4. What might be some consequences of large temperature gradients produced by viscous heating in viscometry, lubrication, and plastics extrusion?
5. In §10.8 would there be any advantage to choosing the dimensionless temperature and dimensionless axial coordinate to be $\Theta = (T - T_1)/T_1$ and $\zeta = z/R$?
6. What would happen in §9.9 if the fluid were water and $\bar{T}$ were 4°C?
7. Is there any advantage to solving Eq. 9.7-9 in terms of hyperbolic functions rather than exponential functions?
8. In going from Eq. 10.8-11 to Eq. 10.8-12 the axial conduction term was neglected with respect to the axial convection term. To justify this, put in some reasonable numerical values to estimate the relative sizes of the terms.
9. How serious is it to neglect the dependence of viscosity on temperature in solving forced convection problems? Viscous dissipation heating problems?
10. At steady state the temperature profiles in a laminated system appear thus:
    Which material has the higher thermal conductivity?

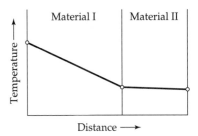

11. Show that Eq. 10.6-4 can be obtained directly by rewriting Eq. 10.6-1 with $x + \Delta x$ replaced by x_0. Similarly, one gets Eq. 10.6-20 from Eq. 10.6-17, with $r + \Delta r$ replaced by r_0.

---

[2] Named for **Franz Grashof** (1826–1893) (pronounced "Grahss-hoff). He was professor of applied mechanics in Karlsruhe and one of the founders of the Verein Deutscher Ingenieure in 1856.

PROBLEMS

10A.1. Heat loss from an insulated pipe. A standard schedule 40, 2-in. steel pipe (inside diameter 2.067 in. and wall thickness 0.154 in.) carrying steam is insulated with 2 in. of 85% magnesia covered in turn with 2 in. of cork. Estimate the heat loss per hour per foot of pipe if the inner surface of the pipe is at 250°F and the outer surface of the cork is at 90°F. The thermal conductivities (in Btu/hr · ft · F) of the substances concerned are: steel, 26.1; 85% magnesia, 0.04; cork, 0.03.

Answer: 24 Btu/hr · ft

10A.2. Heat loss from a rectangular fin. Calculate the heat loss from a rectangular fin (see Fig. 10.7-1) for the following conditions:

Air temperature	350°F
Wall temperature	500°F
Thermal conductivity of fin	60 Btu/hr · ft · F
Thermal conductivity of air	0.0022 Btu/hr · ft · F
Heat transfer coefficient	120 Btu/hr · ft² · F
Length of fin	0.2 ft
Width of fin	1.0 ft
Thickness of fin	0.16 in.

Answer: 2074 Btu/hr

10A.3. Maximum temperature in a lubricant. An oil is acting as a lubricant for a pair of cylindrical surfaces such as those shown in Fig. 10.4-1. The angular velocity of the outer cylinder is 7908 rpm. The outer cylinder has a radius of 5.06 cm, and the clearance between the cylinders is 0.027 cm. What is the maximum temperature in the oil if both wall temperatures are known to be 158°F? The physical properties of the oil are assumed constant at the following values:

Viscosity	92.3 cp
Density	1.22 g/cm³
Thermal conductivity	0.0055 cal/s · cm · C

Answer: 174°F

10A.4. Current-carrying capacity of wire. A copper wire of 0.040 in. diameter is insulated uniformly with plastic to an outer diameter of 0.12 in. and is exposed to surroundings at 100°F. The heat transfer coefficient from the outer surface of the plastic to the surroundings is 1.5 Btu/hr · ft² · F. What is the maximum steady current, in amperes, that this wire can carry without heating any part of the plastic above its operating limit of 200°F? The thermal and electrical conductivities may be assumed constant at the values given here:

	k (Btu/hr · ft · F)	k_e (ohm^{-1} cm^{-1})
Copper	220	5.1×10^5
Plastic	0.20	0.0

Answer: 13.4 amp

10A.5. Free convection velocity.

(a) Verify the expression for the average velocity in the upward-moving stream in Eq. 10.9-16.

(b) Evaluate $\bar{\beta}$ for the conditions given below.

(c) What is the average velocity in the upward-moving stream in the system described in Fig. 10.9-1 for air flowing under these conditions?

Pressure	1 atm
Temperature of the heated wall	100°C
Temperature of the cooled wall	20°C
Spacing between the walls	0.6 cm

Answer: 2.3 cm/s

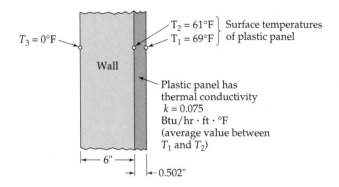

$T_3 = 0°F$

$T_2 = 61°F$ ⎱ Surface temperatures
$T_1 = 69°F$ ⎰ of plastic panel

Wall

Plastic panel has
thermal conductivity
$k = 0.075$
Btu/hr · ft · °F
(average value between
T_1 and T_2)

6"

0.502"

Fig. 10A.6. Determination of the thermal resistance of
a wall.

10A.6. **Insulating power of a wall** (Fig. 10A.6). The "insulating power" of a wall can be measured
by means of the arrangement shown in the figure. One places a plastic panel against the wall.
In the panel two thermocouples are mounted flush with the panel surfaces. The thermal conductivity and thickness of the plastic panel are known. From the measured steady-state temperatures shown in the figure, calculate:

(a) The steady-state heat flux through the wall (and panel).

(b) The "thermal resistance" (wall thickness divided by thermal conductivity).

Answers: **(a)** 14.3 Btu/hr · ft²; **(b)** 4.2 ft² · hr · F/Btu

10A.7. **Viscous heating in a ball-point pen.** You are asked to decide whether the apparent decrease
in viscosity in ball-point pen inks during writing results from "shear thinning" (decrease in
viscosity because of non-Newtonian effects) or "temperature thinning" (decrease in viscosity
because of temperature rise caused by viscous heating). If the temperature rise is less than 1K,
then "temperature thinning" will not be important. Estimate the temperature rise using Eq.
10.4-9 and the following estimated data:

Clearance between ball and holding cavity	5×10^{-5} in.
Diameter of ball	1 mm
Viscosity of ink	10^4 cp
Speed of writing	100 in./min
Thermal conductivity of ink (rough guess)	5×10^{-4} cal/s · cm · C

10A.8. **Temperature rise in an electrical wire.**

(a) A copper wire, 5 mm in diameter and 15 ft long, has a voltage drop of 0.6 volts. Find the
maximum temperature in the wire if the ambient air temperature is 25°C and the heat transfer
coefficient h is 5.7 Btu/hr · ft² · F.

(b) Compare the temperature drops across the wire and the surrounding air.

10B.1. **Heat conduction from a sphere to a stagnant fluid.** A heated sphere of radius R is suspended in a large, motionless body of fluid. It is desired to study the heat conduction in the
fluid surrounding the sphere in the absence of convection.

(a) Set up the differential equation describing the temperature T in the surrounding fluid as a
function of r, the distance from the center of the sphere. The thermal conductivity k of the
fluid is considered constant.

(b) Integrate the differential equation and use these boundary conditions to determine the integration constants: at $r = R$, $T = T_R$; and at $r = \infty$, $T = T_\infty$.

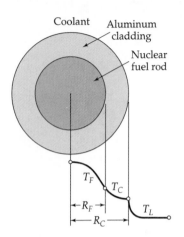

Fig. 10B.3. Temperature distribution in a cylindrical fuel-rod assembly.

(c) From the temperature profile, obtain an expression for the heat flux at the surface. Equate this result to the heat flux given by "Newton's law of cooling" and show that a dimensionless heat transfer coefficient (known as the *Nusselt number*) is given by

$$\text{Nu} = \frac{hD}{k} = 2 \tag{10B.1-1}$$

in which D is the sphere diameter. This well-known result provides the limiting value of Nu for heat transfer from spheres at low Reynolds and Grashof numbers (see §14.4).

(d) In what respect are the Biot number and the Nusselt number different?

10B.2. Viscous heating in slit flow. Find the temperature profile for the viscous heating problem shown in Fig. 10.4-2, when given the following boundary conditions: at $x = 0$, $T = T_0$; at $x = b$, $q_x = 0$.

Answer: $\dfrac{T - T_0}{\mu v_b^2 / k} = \left(\dfrac{x}{b}\right) - \dfrac{1}{2}\left(\dfrac{x}{b}\right)^2$

10B.3 Heat conduction in a nuclear fuel rod assembly (Fig. 10B.3). Consider a long cylindrical nuclear fuel rod, surrounded by an annular layer of aluminum cladding. Within the fuel rod heat is produced by fission; this heat source depends on position approximately as

$$S_n = S_{n0}\left[1 + b\left(\frac{r}{R_F}\right)^2\right] \tag{10B.3-1}$$

Here S_{n0} and b are known constants, and r is the radial coordinate measured from the axis of the cylindrical fuel rod. Calculate the maximum temperature in the fuel rod if the outer surface of the cladding is in contact with a liquid coolant at temperature T_L. The heat transfer coefficient at the cladding–coolant interface is h_L, and the thermal conductivities of the fuel rod and cladding are k_F and k_C.

Answer: $T_{F,\text{max}} - T_L = \dfrac{S_{n0}R_F^2}{4k_F}\left(1 + \dfrac{b}{4}\right) + \dfrac{S_{n0}R_F^2}{2k_C}\left(1 + \dfrac{b}{2}\right)\left(\dfrac{k_C}{R_C h_L} + \ln\dfrac{R_C}{R_F}\right)$

10B.4. Heat conduction in an annulus (Fig. 10B.4).

(a) Heat is flowing through an annular wall of inside radius r_0 and outside radius r_1. The thermal conductivity varies linearly with temperature from k_0 at T_0 to k_1 at T_1. Develop an expression for the heat flow through the wall.

(b) Show how the expression in (a) can be simplified when $(r_1 - r_0)/r_0$ is very small. Interpret the result physically.

Answer: **(a)** $Q = 2\pi L(T_0 - T_1)\left(\dfrac{k_0 + k_1}{2}\right)\left(\ln\dfrac{r_1}{r_0}\right)^{-1}$; **(b)** $Q = 2\pi r_0 L\left(\dfrac{k_0 + k_1}{2}\right)\left(\dfrac{T_0 - T_1}{r_1 - r_0}\right)$

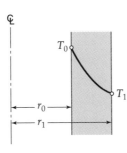

Fig. 10B.4. Temperature profile in an annular wall.

10B.5. Viscous heat generation in a polymer melt. Rework the problem discussed in §10.4 for a molten polymer, whose viscosity can be adequately described by the power law model (see Chapter 8). Show that the temperature distribution is the same as that in Eq. 10.4-9 but with the Brinkman number replaced by

$$\mathrm{Br}_n = \left[\frac{m v_b^{n+1}}{b^{n-1} k(T_b - T_0)}\right] \tag{10B.5-1}$$

10B.6. Insulation thickness for a furnace wall (Fig. 10B.6). A furnace wall consists of three layers: (i) a layer of heat-resistant or refractory brick, (ii) a layer of insulating brick, and (iii) a steel plate, 0.25 in. thick, for mechanical protection. Calculate the thickness of each layer of brick to give minimum total wall thickness if the heat loss through the wall is to be 5000 Btu/ft² · hr, assuming that the layers are in excellent thermal contact. The following information is available:

Material	Maximum allowable temperature	Thermal conductivity (Btu/hr · ft · F) at 100°F	at 2000°F
Refractory brick	2600°F	1.8	3.6
Insulating brick	2000°F	0.9	1.8
Steel	—	26.1	—

Answer: Refractory brick, 0.39 ft; insulating brick, 0.51 ft.

10B.7. Forced-convection heat transfer in flow between parallel plates (Fig. 10B.7). A viscous fluid with temperature-independent physical properties is in fully developed laminar flow between two flat surfaces placed a distance $2B$ apart. For $z < 0$ the fluid temperature is uniform at $T = T_1$. For $z > 0$ heat is added at a constant, uniform flux q_0 at both walls. Find the temperature distribution $T(x, z)$ for large z.

(a) Make a shell energy balance to obtain the differential equation for $T(x, z)$. Then discard the viscous dissipation term and the axial heat conduction term.

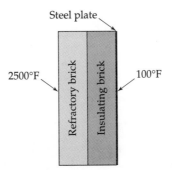

Fig. 10B.6. A composite furnace wall.

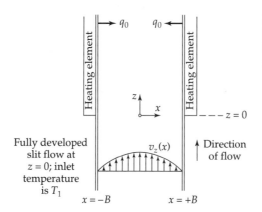

Fig. 10B.7. Laminar, incompressible flow between parallel plates, both of which are being heated by a uniform wall heat flux q_0 starting at $z = 0$.

(b) Recast the problem in terms of the dimensionless quantities

$$\Theta = \frac{T - T_1}{q_0 B / k} \qquad \sigma = \frac{x}{B} \qquad \zeta = \frac{kz}{\rho \hat{C}_p v_{z,\max} B^2} \tag{10B.7-1, 2, 3}$$

(c) Obtain the asymptotic solution for large z:

$$\Theta(\sigma, \zeta) = \tfrac{3}{2}\zeta + \tfrac{3}{4}\sigma^2 - \tfrac{1}{8}\sigma^4 - \tfrac{39}{280} \tag{10B.7-4}$$

10B.8. **Electrical heating of a pipe** (Fig. 10B.8). In the manufacture of glass-coated steel pipes, it is common practice first to heat the pipe to the melting range of glass and then to contact the hot pipe surface with glass granules. These granules melt and wet the pipe surface to form a tightly adhering nonporous coat. In one method of preheating the pipe, an electric current is passed along the pipe, with the result that the pipe is heated (as in §10.2). For the purpose of this problem make the following assumptions:

(i) The electrical conductivity of the pipe k_e is constant over the temperature range of interest. The local rate of electrical heat production S_e is then uniform throughout the pipe wall.

(ii) The top and bottom of the pipe are capped in such a way that heat losses through them are negligible.

(iii) Heat loss from the outer surface of the pipe to the surroundings is given by Newton's law of cooling: $q_r = h(T_1 - T_a)$. Here h is a suitable heat transfer coefficient.

How much electrical power is needed to maintain the inner pipe surface at some desired temperature, T_κ, for known k, T_a, h, and pipe dimensions?

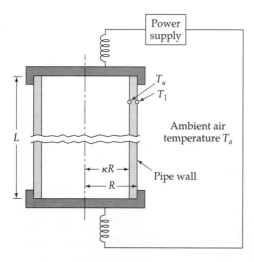

Fig. 10B.8. Electrical heating of a pipe.

Answer: $P = \dfrac{\pi R^2 (1 - \kappa^2) L (T_\kappa - T_a)}{\dfrac{(1 - \kappa^2)R}{2h} - \dfrac{(\kappa R)^2}{4k}\left[\left(1 - \dfrac{1}{\kappa^2}\right) - 2 \ln \kappa\right]}$

10B.9. **Plug flow with forced-convection heat transfer.** Very thick slurries and pastes sometimes move in channels almost as a solid plug. Thus, one can approximate the velocity by a constant value v_0 over the conduit cross section.

(a) Rework the problem of §10.8 for plug flow in a *circular tube* of radius R. Show that the temperature distribution analogous to Eq. 10.8-31 is

$$\Theta(\xi, \zeta) = 2\zeta + \tfrac{1}{2}\xi^2 - \tfrac{1}{4} \tag{10B.9-1}$$

in which $\zeta = kz / \rho \hat{C}_p v_0 R^2$, and Θ and ξ are defined as in §10.8.

(b) Show that for plug flow in a *plane slit* of width $2B$ the temperature distribution analogous to Eq. 10B.7-4 is

$$\Theta(\xi, \zeta) = \zeta + \tfrac{1}{2}\sigma^2 - \tfrac{1}{6} \tag{10B.9-2}$$

in which $\zeta = kz / \rho \hat{C}_p v_0 B^2$, and Θ and σ are defined as in Problem 10B.7.

10B.10. **Free convection in an annulus of finite height** (Fig. 10B.10). A fluid is contained in a vertical annulus closed at the top and bottom. The inner wall of radius κR is maintained at the temperature T_κ, and the outer wall of radius R is kept at temperature T_1. Using the assumptions and approach of §10.9, obtain the velocity distribution produced by free convection.

(a) First derive the temperature distribution

$$\frac{T_1 - T}{T_1 - T_\kappa} = \frac{\ln \xi}{\ln \kappa} \tag{10B.10-1}$$

in which $\xi = r/R$.

(b) Then show that the equation of motion is

$$\frac{1}{\xi}\frac{d}{d\xi}\left(\xi \frac{dv_z}{d\xi}\right) = A + B \ln \xi \tag{10B.10-2}$$

in which $A = (R^2/\mu)(dp/dz + \rho_1 g)$ and $B = ((\rho_1 g \beta_1 \Delta T)R^2/\mu \ln \kappa)$ where $\Delta T = T_1 - T_\kappa$.

(c) Integrate the equation of motion (see Eq. C.1-11) and apply the boundary conditions to evaluate the constants of integration. Then show that A can be evaluated by the requirement of no net mass flow through any plane z = constant, with the final result that

$$v_z = \frac{\rho_1 g \beta_1 \Delta T R^2}{16\mu}\left[\frac{(1 - \kappa^2)(1 - 3\kappa^2) - 4\kappa^4 \ln \kappa}{(1 - \kappa^2)^2 + (1 - \kappa^4)\ln \kappa}\left((1 - \xi^2) - (1 - \kappa^2)\frac{\ln \xi}{\ln \kappa}\right) + 4(\xi^2 - \kappa^2)\frac{\ln \xi}{\ln \kappa}\right] \tag{10B.10-3}$$

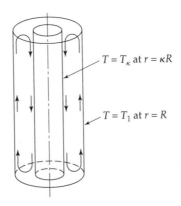

$T = T_\kappa$ at $r = \kappa R$

$T = T_1$ at $r = R$

Fig. 10B.10. Free convection pattern in an annular space with $T_1 > T_\kappa$.

10B.11. Free convection with temperature-dependent viscosity. Rework the problem in §10.9, taking into account the variation of viscosity with temperature. Assume that the "fluidity" (reciprocal of viscosity) is the following linear function of the temperature

$$\frac{1}{\mu} = \frac{1}{\bar{\mu}}[1 + \bar{\beta}_\mu(T - \bar{T})] \tag{10B.11-1}$$

Use the $\breve{y}$, $\breve{v}_z$, and Gr defined in §10.9 (but with $\bar{\mu}$ instead of μ) and in addition

$$b_T = \tfrac{1}{2}\bar{\beta}\Delta T, \quad b_\mu = \tfrac{1}{2}\bar{\beta}_\mu \Delta T \quad \text{and} \quad P = \frac{\bar{\rho}B^3}{\bar{\mu}^2}\left(\frac{dp}{dz} + \bar{\rho}g\right) \tag{10B.11-2, 3}$$

and show that the differential equation for the velocity distribution is

$$\frac{d}{d\breve{y}}\left(\frac{1}{1 - b_\mu\breve{y}}\frac{d\breve{v}_z}{d\breve{y}}\right) = P + \tfrac{1}{2}\text{Gr}\,\breve{y} \tag{10B.11-4}$$

Follow the procedure in §10.9, discarding terms containing the third and higher powers of ΔT. Show that this leads to $P = \tfrac{1}{30}\text{Gr}b_T + \tfrac{1}{15}\text{Gr}b_\mu$ and finally:

$$\breve{v}_z = \tfrac{1}{12}\text{Gr}(\breve{y}^3 - \breve{y}) + \tfrac{1}{60}\text{Gr}b_T(\breve{y}^2 - 1) - \tfrac{1}{80}\text{Gr}b_\mu(\breve{y}^2 - 1)(5\breve{y}^2 - 1) \tag{10B.11-5}$$

Sketch the result to show how the velocity profile becomes skewed because of the temperature-dependent viscosity.

10B.12. Heat conduction with temperature-dependent thermal conductivity (Fig. 10B.12). The curved surfaces and the end surfaces (both shaded in the figure) of the solid in the shape of a half-cylindrical shell are insulated. The surface $\theta = 0$, of area $(r_2 - r_1)L$, is maintained at temperature T_0, and the surface at $\theta = \pi$, also of area $(r_2 - r_1)L$, is kept at temperature T_π.

The thermal conductivity of the solid varies linearly with temperature from k_0 at $T = T_0$ to k_π at $T = T_\pi$.

(a) Find the steady-state temperature distribution.

(b) Find the total heat flow through the surface at $\theta = 0$.

10B.13. Flow reactor with exponentially temperature-dependent source. Formulate the function $F(\Theta)$ of Eq. 10.5-7 for a zero-order reaction with the temperature dependence

$$S_c = Ke^{-E/RT} \tag{10B.13-1}$$

in which K and E are constants, and R is the gas constant. Then insert $F(\Theta)$ into Eqs. 10.5-15 through 20 and solve for the dimensionless temperature profile with $k_{z,\text{eff}}$ neglected.

10B.14. Evaporation loss from an oxygen tank.

(a) Liquefied gases are sometimes stored in well-insulated spherical containers vented to the atmosphere. Develop an expression for the steady-state heat transfer rate through the walls of such a container, with the radii of the inner and outer walls being r_0 and r_1 respectively and

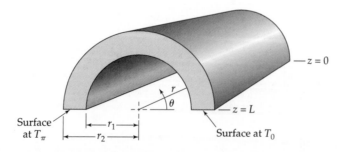

Fig. 10B.12. Tangential heat conduction in an annular shell.

the temperatures at the inner and outer walls being T_0 and T_1. The thermal conductivity of the insulation varies linearly with temperature from k_0 at T_0 to k_1 at T_1.

(b) Estimate the rate of evaporation of liquid oxygen from a spherical container of 6 ft inside diameter covered with a 1-ft-thick annular evacuated jacket filled with particulate insulation. The following information is available:

Temperature at inner surface of insulation $\qquad$ $-183°C$
Temperature at outer surface of insulation $\qquad$ $0°C$
Boiling point of O_2 $\qquad$ $-183°C$
Heat of vaporization of O_2 $\qquad$ 1636 cal/g-mol
Thermal conductivity of insulation at $0°C$ $\qquad$ 9.0×10^{-4} Btu/hr $\cdot$ ft $\cdot$ F
Thermal conductivity of insulation at $-183°C$ $\qquad$ 7.2×10^{-4} Btu/hr $\cdot$ ft $\cdot$ F

Answers: **(a)** $Q_0 = 4\pi r_0 r_1 \left(\dfrac{k_0 + k_1}{2} \right) \left(\dfrac{T_0 - T_1}{r_1 - r_0} \right)$; **(b)** 0.198 kg/hr

10B.15. Radial temperature gradients in an annular chemical reactor. A catalytic reaction is being carried out at constant pressure in a packed bed between coaxial cylindrical walls with inner radius r_0 and outer radius r_1. Such a configuration occurs when temperatures are measured with a centered thermowell, and is in addition useful for controlling temperature gradients if a thin annulus is used. The entire inner wall is at uniform temperature T_0, and it can be assumed that there is no heat transfer through this surface. The reaction releases heat at a uniform volumetric rate S_c throughout the reactor. The effective thermal conductivity of the reactor contents is to be treated as a constant throughout.

(a) By a shell energy balance, derive a second-order differential equation that describes the temperature profiles, assuming that the temperature gradients in the axial direction can be neglected. What boundary conditions must be used?

(b) Rewrite the differential equation and boundary conditions in terms of the dimensionless radial coordinate and dimensionless temperature defined as

$$\xi = \frac{r}{r_0}; \qquad \Theta = \frac{T - T_0}{S_c r_0^2 / 4k_{\text{eff}}} \qquad \text{(10B.15-1)}$$

Explain why these are logical choices.

(c) Integrate the dimensionless differential equation to get the radial temperature profile. To what viscous flow problem is this conduction problem analogous?

(d) Develop expressions for the temperature at the outer wall and for the volumetric average temperature of the catalyst bed.

(e) Calculate the outer wall temperature when $r_0 = 0.45$ in., $r_1 = 0.50$ in., $k_{\text{eff}} = 0.3$ Btu/hr $\cdot$ ft $\cdot$ F, $T_0 = 900°F$, and $S_c = 4800$ cal/hr $\cdot$ cm^3.

(f) How would the results of part (*e*) be affected if the inner and outer radii were doubled?

Answer: (e) $888°F$

10B.16. Temperature distribution in a hot-wire anemometer. A hot-wire anemometer is essentially a fine wire, usually made of platinum, which is heated electrically and exposed to a flowing fluid. Its temperature, which is a function of the fluid temperature, fluid velocity, and the rate of heating, may be determined by measuring its electrical resistance. It is used for measuring velocities and velocity fluctuations in flow systems. In this problem we analyze the temperature distribution in the wire element.

We consider a wire of diameter D and length $2L$ supported at its ends ($z = -L$ and $z = +L$) and mounted perpendicular to an air stream. An electric current of density I amp/cm^2 flows through the wire, and the heat thus generated is partially lost by convection to the air stream (see Eq. 10.1-2) and partially by conduction toward the ends of the wire. Because of their size and their high electrical and thermal conductivity, the supports are not appreciably heated by the current, but remain at the temperature T_L, which is the same as that of the approaching air stream. Heat loss by radiation is to be neglected.

(a) Derive an equation for the steady-state temperature distribution in the wire, assuming that T depends on z alone; that is, the radial temperature variation in the wire is neglected. Further, assume uniform thermal and electrical conductivities in the wire, and a uniform heat transfer coefficient from the wire to the air stream.

(b) Sketch the temperature profile obtained in (a).

(c) Compute the current, in amperes, required to heat a platinum wire to a midpoint temperature of 50°C under the following conditions:

$$T_L = 20°C \qquad\qquad h = 100 \text{ Btu/hr} \cdot \text{ft}^2 \cdot \text{F}$$
$$D = 0.127 \text{ mm} \qquad\qquad k = 40.2 \text{ Btu/hr} \cdot \text{ft} \cdot \text{F}$$
$$L = 0.5 \text{ cm} \qquad\qquad k_e = 1.00 \times 10^5 \text{ ohm}^{-1} \text{cm}^{-1}$$

Answers: (a) $T - T_L = \dfrac{D I^2}{4 h k_e}\left(1 - \dfrac{\cosh\sqrt{4h/kDz}}{\cosh\sqrt{4h/kDL}}\right)$; (c) 1.01 amp

10B.17. Non-Newtonian flow with forced-convection heat transfer.[1] For estimating the effect of non-Newtonian viscosity on heat transfer in ducts, the power law model of Chapter 8 gives velocity profiles that show rather well the deviation from parabolic shape.

(a) Rework the problem of §10.8 (heat transfer in a *circular tube*) for the power law model given in Eqs. 8.3-2, 3. Show that the final temperature profile is

$$\Theta = \frac{2(s+3)}{(s+1)}\zeta + \frac{(s+3)}{2(s+1)}\xi^2 - \frac{2}{(s+1)(s+3)}\xi^{s+3} - \frac{(s+3)^3 - 8}{4(s+1)(s+3)(s+5)} \qquad (10B.17\text{-}1)$$

in which $s = 1/n$.

(b) Rework Problem 10B.7 (heat transfer in a *plane slit*) for the power law model. Obtain the dimensionless temperature profile:

$$\Theta = \frac{(s+2)}{(s+1)}\left[\zeta + \frac{1}{2}\sigma^2 - \frac{1}{(s+2)(s+3)}|\sigma|^{s+3} - \frac{(s+2)(s+3)(2s+5) - 6}{6(s+3)(s+4)(2s+5)}\right] \qquad (10B.17\text{-}2)$$

Note that these results contain the Newtonian results ($s = 1$) and the plug flow results ($s = \infty$). See Problem 10D.2 for a generalization of this approach.

10B.18. Reactor temperature profiles with axial heat flux[2] (Fig. 10B.18).

(a) Show that for a heat source that depends linearly on the temperature, Eqs. 10.5-6 to 14 have the solutions (for $m_+ \neq m_-$)

$$\Theta^I = 1 + \frac{m_+ m_-(\exp m_+ - \exp m_-)}{m_+^2 \exp m_+ - m_-^2 \exp m_-} \exp[(m_+ + m_-)Z] \qquad (10B.18\text{-}1)$$

$$\Theta^{II} = \frac{m_+ (\exp m_+)(\exp m_- Z) - m_- (\exp m_-)(\exp m_+ Z)}{m_+^2 \exp m_+ - m_-^2 \exp m_-}(m_+ + m_-) \qquad (10B.18\text{-}2)$$

$$\Theta^{III} = \frac{m_+^2 - m_-^2}{m_+^2 \exp m_+ - m_-^2 \exp m_-} \exp(m_+ + m_-) \qquad (10B.18\text{-}3)$$

Here $m_\pm = \frac{1}{2}B(1 \pm \sqrt{1 - (4N/B)}$, in which $B = \rho v_0 \hat{C}_p L/\kappa_{\text{eff},zz}$. Some profiles calculated from these equations are shown in Fig. 10B.18.

---

[1] R. B. Bird, *Chem.-Ing. Technik*, **31**, 569–572 (1959).

[2] Taken from the corresponding results of G. Damköhler, *Z. Elektrochem.*, **43**, 1–8, 9–13 (1937), and J. F. Wehner and R. H. Wilhelm, *Chem. Engr. Sci.*, **6**, 89–93 (1956); **8**, 309 (1958), for isothermal flow reactors with longitudinal diffusion and first-order reaction. **Gerhard Damköhler** (1908–1944) achieved fame for his work on chemical reactions in flowing, diffusing systems; a key publication was in *Der Chemie-Ingenieur*, Leipzig (1937), pp. 359–485. **Richard Herman Wilhelm** (1909–1968), chairman of the Chemical Engineering Department at Princeton University, was well known for his work on fixed-bed catalytic reactors, fluidized transport, and the "parametric pumping" separation process.

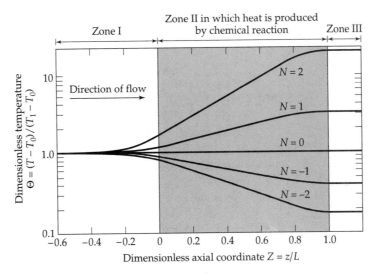

Fig. 10B.18. Predicted temperature profiles in a fixed-bed axial-flow reactor for $B = 8$ and various values of N.

(b) Show that, in the limit as B goes to infinity, the above solution agrees with that in Eqs. 10.5-21, 22, and 23.

(c) Make numerical comparisons of the results in Eq. 10.5-22 and Fig. 10B.18 for $N = 2$ at $Z = 0.0$, 0.5, 0.9, and 1.0.

(d) Assuming the applicability of Eq. 9.6-9, show that the results in Fig. 10B.18 correspond to a catalyst bed length L of 4 particle diameters. Since the ratio L/D_p is seldom less than 100 in industrial reactors, it follows that the neglect of $\kappa_{\text{eff},zz}$ is a reasonable assumption in steady-state design calculations.

10C.1. **Heating of an electric wire with temperature-dependent electrical and thermal conductivity.**[3] Find the temperature distribution in an electrically heated wire when the thermal and electrical conductivities vary with temperature as follows:

$$\frac{k}{k_0} = 1 - \alpha_1\Theta - \alpha_2\Theta^2 + \cdots \tag{10C.1-1}$$

$$\frac{k_e}{k_{e0}} = 1 - \beta_1\Theta - \beta_2\Theta^2 + \cdots \tag{10C.1-2}$$

Here k_0 and k_{e0} are the values of the conductivities at temperature T_0, and $\Theta = (T - T_0)/T_0$ is a dimensionless temperature rise. The coefficients α_i and β_i are constants. Such series expansions are useful over moderate temperature ranges.

(a) Because of the temperature gradient in the wire, the electrical conductivity is a function of position, $k_e(r)$. Therefore, the current density is also a function of r: $I(r) = k_e(r) \cdot (E/L)$, and the electrical heat source also is position dependent: $S_e(r) = k_e(r) \cdot (E/L)^2$. The equation for the temperature distribution is then

$$-\frac{1}{r}\frac{d}{dr}\left(rk(r)\frac{dT}{dr}\right) = k_e(r)\left(\frac{E}{L}\right)^2 \tag{10C.1-3}$$

---

[3] The solution given here was suggested by L. J. F. Broer (personal communication, 20 August 1958).

Now introduce the dimensionless quantities $\xi = r/R$ and $B = k_{e0}R^2E^2/k_0L^2T_0$ and show that Eq. 10C.1-3 then becomes

$$-\frac{1}{\xi}\frac{d}{d\xi}\left(\frac{k}{k_0}\xi\frac{d\Theta}{d\xi}\right) = B\frac{k_e}{k_{e0}} \tag{10C.1-4}$$

When the power series expressions for the conductivities are inserted into this equation we get

$$-\frac{1}{\xi}\frac{d}{d\xi}\left((1 - \alpha_1\Theta - \alpha_2\Theta^2 + \cdots)\xi\frac{d\Theta}{d\xi}\right) = B(1 - \beta_1\Theta - \beta_2\Theta^2 + \cdots) \tag{10C.1-5}$$

This is the equation that is to be solved for the dimensionless temperature distribution.

(b) Begin by noting that if all the α_i and β_i were zero (that is, both conductivities constant), then Eq. 10C.1-5 would simplify to

$$-\frac{1}{\xi}\frac{d}{d\xi}\left(\xi\frac{d\Theta}{d\xi}\right) = B \tag{10C.1-6}$$

When this is solved with the boundary conditions that $\Theta = $ finite at $\xi = 0$, and $\Theta = 0$ at $\xi = 1$, we get

$$\Theta = \tfrac{1}{4}B(1 - \xi^2) \tag{10C.1-7}$$

This is Eq. 10.2-13 in dimensionless notation.

Note that Eq. 10C.-5 will have the solution in Eq. 10C.1-7 for small values of B—that is, for weak heat sources. For stronger heat sources, postulate that the temperature distribution can be expressed as a power series in the dimensionless heat source strength B:

$$\Theta = \tfrac{1}{4}B(1 - \xi^2)(1 + B\Theta_1 + B^2\Theta_2 + \cdots) \tag{10C.1-8}$$

Here the Θ_n are functions of ξ but not of B. Substitute Eq. 10C.1-8 into Eq. 10C.1-5, and equate the coefficients of like powers of B to get a set of ordinary differential equations for the Θ_n, with $n = 1, 2, 3, \ldots$. These may be solved with the boundary conditions that $\Theta_n = $ finite at $\xi = 0$ and $\Theta_n = 0$ at $\xi = 1$. In this way obtain

$$\Theta = \tfrac{1}{4}B(1 - \xi^2)[1 + B(\tfrac{1}{8}\alpha_1(1 - \xi^2) - \tfrac{1}{16}\beta_1(3 - \xi^2)) + O(B^2)] \tag{10C.1-9}$$

where $O(B^2)$ means "terms of the order of B^2 and higher."

(c) For materials that are described by the Wiedemann–Franz–Lorenz law (see §9.5), the ratio k/k_eT is a constant (independent of temperature). Hence

$$\frac{k}{k_eT} = \frac{k_0}{k_{e0}T_0} \tag{10C.1-10}$$

Combine this with Eqs. 10C.1-1 and 2 to get

$$1 - \alpha_1\Theta - \alpha_2\Theta^2 + \cdots = (1 - \beta_1\Theta - \beta_2\Theta^2 + \cdots)(1 + \Theta) \tag{10C.1-11}$$

Equate coefficients of equal powers of the dimensionless temperature to get relations among the α_i and the β_i: $\alpha_1 = \beta_1 - 1$, $\alpha_2 = \beta_1 + \beta_2$, and so on. Use these relations to get

$$\Theta = \tfrac{1}{4}B(1 - \xi^2)[1 - \tfrac{1}{16}B((\beta_1 + 2) + (\beta_1 - 2)\xi^2) + O(B^2)] \tag{10C.1-12}$$

10C.2. Viscous heating with temperature-dependent viscosity and thermal conductivity (Figs. 10.4-1 and 2). Consider the flow situation shown in Fig. 10.4-2. Both the stationary surface and the moving surface are maintained at a constant temperature T_0. The temperature dependences of k and μ are given by

$$\frac{k}{k_0} = 1 + \alpha_1\Theta + \alpha_2\Theta^2 + \cdots \tag{10C.2-1}$$

$$\frac{\mu_0}{\mu} = \frac{\varphi}{\varphi_0} = 1 + \beta_1\Theta + \beta_2\Theta^2 + \cdots \tag{10C.2-2}$$

in which the α_i and β_i are constants, $\varphi = 1/\mu$ is the fluidity, and the subscript "0" means "evaluated at $T = T_0$." The dimensionless temperature is defined as $\Theta = (T - T_0)/T_0$.

(a) Show that the differential equations describing the viscous flow and heat conduction may be written in the forms

$$\frac{d}{d\xi}\left(\frac{\mu}{\mu_0}\frac{d\phi}{d\xi}\right) = 0 \tag{10C.2-3}$$

$$\frac{d}{d\xi}\left(\frac{k}{k_0}\frac{d\Theta}{d\xi}\right) + \text{Br}\,\frac{\mu}{\mu_0}\left(\frac{d\phi}{d\xi}\right)^2 = 0 \tag{10C.2-4}$$

in which $\phi = v_z/v_b$, $\xi = x/b$, and $\text{Br} = \mu_0 v_b^2/k_0 T_0$ (the Brinkman number).

(b) The equation for the dimensionless velocity distribution may be integrated once to give $d\phi/d\xi = C_1 \cdot (\varphi/\varphi_0)$, in which C_1 is an integration constant. This expression is then substituted into the energy equation to get

$$\frac{d}{d\xi}\left((1 + \alpha_1\Theta + \alpha_2\Theta^2 + \cdots)\frac{d\Theta}{d\xi}\right) + \text{Br}C_1^2(1 + \beta_1\Theta + \beta_2\Theta^2 + \cdots) = 0 \tag{10C.2-5}$$

Obtain the first two terms of a solution in the form

$$\Theta(\xi; \text{Br}) = \text{Br}\Theta_1(\xi) + \text{Br}^2\Theta_2(\xi) + \cdots \tag{10C.2-6}$$

$$\phi(\xi; \text{Br}) = \phi_0 + \text{Br}\phi_1(\xi) + \text{Br}^2\phi_2(\xi) + \cdots \tag{10C.2-7}$$

It is further suggested that the constant of integration C_1 also be expanded as a power series in the Brinkman number, thus

$$C_1(\text{Br}) = C_{10} + \text{Br}C_{11} + \text{Br}^2C_{12} + \cdots \tag{10C.2-8}$$

(c) Repeat the problem, changing the boundary condition at $y = b$ to $q_x = 0$ (instead of specifying the temperature).[4]

Answers: **(b)** $\phi = \xi - \frac{1}{12}\text{Br}\beta_1(\xi - 3\xi^2 + 2\xi^3) + \cdots$

$\quad\quad\quad \Theta = \frac{1}{2}\text{Br}(\xi - \xi^2) - \frac{1}{8}\text{Br}^2\alpha_1(\xi^2 - 2\xi^3 + \xi^4) - \frac{1}{24}\text{Br}^2\beta_1(\xi - 2\xi^2 + 2\xi^3 - \xi^4) + \cdots$

$\quad\quad$ **(c)** $\phi = \xi - \frac{1}{6}\text{Br}\beta_1(2\xi - 3\xi^2 + \xi^3) + \cdots$

$\quad\quad\quad \Theta = \text{Br}(\xi - \frac{1}{2}\xi^2) - \frac{1}{8}\text{Br}^2\alpha_1(4\xi^2 - 4\xi^3 + \xi^4) + \frac{1}{24}\text{Br}^2\beta_1(-8\xi + 8\xi^2 - 4\xi^3 + \xi^4) + \cdots$

10C.3. Viscous heating in a cone-and-plate viscometer.[5] In Eq. 2B.11-3 there is an expression for the torque $\mathcal{T}$ required to maintain an angular velocity Ω in a cone-and-plate viscometer with included angle ψ_0 (see Fig. 2B.11). It is desired to obtain a correction factor to account for the change in torque caused by the change in viscosity resulting from viscous heating. This effect can be a disturbing factor in viscometric measurements, causing errors as large as 20%.

(a) Adapt the result of Problem 10C.2 to the cone-and-plate system as was done in Problem 2B.11(a). The boundary condition of zero heat flux at the cone surface seems to be more realistic than the assumption that the cone and plate temperatures are the same, inasmuch as the plate is thermostatted and the cone is not.

(b) Show that this leads to the following modification of Eq. 2B.11-3:

$$T_z = \frac{2\pi\mu_0\Omega R^3}{3\psi_0}\left(1 - \tfrac{1}{5}\overline{\text{Br}}\beta_1 + \tfrac{1}{35}\overline{\text{Br}}^2(3\beta_1^2 + \alpha_1\beta_1 - 2\beta_2) + \cdots\right) \tag{10C.3-1}$$

where $\overline{\text{Br}} = \mu_0\Omega^2R^2/k_0T_0$ is the Brinkman number. The symbol μ_0 stands for the viscosity at the temperature T_0.

---

[4] R. M. Turian and R. B. Bird, *Chem. Eng. Sci.*, **18**, 689–696 (1963).

[5] R. M. Turian, *Chem. Eng. Sci.*, **20**, 771–781 (1965); the viscous heating correction for non-Newtonian fluids is discussed in this publication (see also R. B. Bird, R. C. Armstrong, and O. Hassager, *Dynamics of Polymeric Liquids*, Vol. 1, 2nd edition, Wiley-Interscience, New York (1987), pp. 223–227.

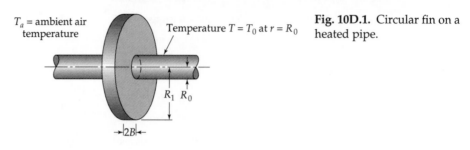

T_a = ambient air temperature

Temperature $T = T_0$ at $r = R_0$

Fig. 10D.1. Circular fin on a heated pipe.

R_1 R_0

$2B$

10D.1. Heat loss from a circular fin (Fig. 10D.1).

(a) Obtain the temperature profile $T(r)$ for a circular fin of thickness $2B$ on a pipe with outside wall temperature T_0. Make the same assumptions that were made in the study of the rectangular fin in §10.7.

(b) Derive an expression for the total heat loss from the fin.

10D.2. Duct flow with constant wall heat flux and arbitrary velocity distribution.

(a) Rework the problem in §10.8 for an arbitrary fully developed, axisymmetric flow velocity distribution $v_z/v_{z,\max} = \phi(\xi)$, where $\xi = r/R$. verify that the temperature distribution is given by

$$\Theta = C_0 \zeta + C_0 \int_0^\xi \frac{I(\bar{\xi})}{\bar{\xi}} \, d\bar{\xi} + C_1 \ln \xi + C_2 \tag{10D.2-1}$$

in which

$$I(\bar{\xi}) = \int_0^{\bar{\xi}} \phi \bar{\bar{\xi}} \, d\bar{\bar{\xi}} \tag{10D.2-2}$$

Show that $C_1 = 0$ and $C_0 = [I(1)]^{-1}$. Then show that the remaining constant is

$$C_2 = -[I(1)]^{-2} \int_0^1 \phi \xi \left[\int_0^\xi \bar{\xi}^{-1} I(\bar{\xi}) d\bar{\xi} \right] d\xi \tag{10D.2-3}$$

Verify that the above equations lead to Eqs. 10.8-27 to 30 when the velocity profile is parabolic.

These results can be used to compute the temperature profiles for the fully developed tube flow of any kind of material as long as a reasonable estimation can be made for the velocity distribution. As special cases, one can get results for Newtonian flow, plug flow, non-Newtonian flow, and even, with some modifications, turbulent flow (see §13.4).[6]

(b) Show that the dimensionless temperature difference driving force $\Theta_0 - \Theta_b$ is

$$\Theta_0 - \Theta_b = [I(1)]^{-2} \int_0^1 \xi^{-1} [I(\xi)]^2 \, d\xi \tag{10D.2-4}$$

(c) Verify that the dimensionless wall heat flux is

$$\frac{q_w D}{k(T_0 - T_b)} = \frac{2}{\Theta_0 - \Theta_b} \tag{10D.2-5}$$

and that, for the laminar flow of Newtonian fluids, this quantity has the value $\frac{48}{11}$.

(d) What is the physical interpretation of $I(1)$?

[6] R. N. Lyon, *Chem. Engr. Prog.*, **47**, 75–59 (1951); note that the definition of $\phi(\xi)$ used here is different from that in Tables 14.2-1 and 2.

Chapter 11

The Equations of Change for Nonisothermal Systems

§11.1 The energy equation

§11.2 Special forms of the energy equation

§11.3 The Boussinesq equation of motion for forced and free convection

§11.4 Use of the equations of change to solve steady-state problems

§11.5 Dimensional analysis of the equations of change for nonisothermal systems

In Chapter 10 we introduced the shell energy balance method for solving relatively simple, steady-state heat flow problems. We obtained the temperature profiles, as well as some derived properties such as average temperature and energy fluxes. In this chapter we generalize the shell energy balance and obtain the *equation of energy*, a partial differential equation that describes the transport of energy in a homogeneous fluid or solid.

This chapter is also closely related to Chapter 3, where we introduced the equation of continuity (conservation of mass) and the equation of motion (conservation of momentum). The addition of the equation of energy (conservation of energy) allows us to extend our problem-solving ability to include nonisothermal systems.

We begin in §11.1 by deriving the equation of change for the *total energy*. As in Chapter 10, we use the combined energy flux vector **e** in applying the law of conservation of energy. In §11.2 we subtract the *mechanical energy* equation (given in §3.3) from the total energy equation to get an equation of change for the *internal energy*. From the latter we can get an equation of change for the *temperature*, and it is this kind of energy equation that is most commonly used.

Although our main concern in this chapter will be with the various energy equations just mentioned, we find it useful to discuss in §11.3 an approximate equation of motion that is convenient for solving problems involving free convection.

In §11.4 we summarize the equations of change encountered up to this point. Then we proceed to illustrate the use of these equations in a series of examples, in which we begin with the general equations and discard terms that are not needed. In this way we have a standard procedure for setting up and solving problems.

Finally, in §11.5 we extend the dimensional analysis discussion of §3.7 and show how additional dimensionless groups arise in heat transfer problems.

§11.1 THE ENERGY EQUATION

The equation of change for energy is obtained by applying the law of conservation of energy to a small element of volume $\Delta x\, \Delta y\, \Delta z$ (see Fig. 3.1-1) and then allowing the dimensions of the volume element to become vanishingly small. The law of conservation of energy is an extension of the first law of classical thermodynamics, which concerns the difference in internal energies of two equilibrium states of a closed system because of the heat added to the system and the work done on the system (that is, the familiar $\Delta U = Q + W$).[1]

Here we are interested in a stationary volume element, fixed in space, through which a fluid is flowing. Both kinetic energy and internal energy may be entering and leaving the system by convective transport. Heat may enter and leave the system by heat conduction as well. As we saw in Chapter 9, heat conduction is fundamentally a molecular process. Work may be done on the moving fluid by the stresses, and this, too, is a molecular process. This term includes the work done by pressure forces and by viscous forces. In addition, work may be done on the system by virtue of the external forces, such as gravity.

We can summarize the preceding paragraph by writing the conservation of energy in words as follows:

$$\left\{ \begin{array}{c} \text{rate of} \\ \text{increase of} \\ \text{kinetic and} \\ \text{internal} \\ \text{energy} \end{array} \right\} = \left\{ \begin{array}{c} \text{net rate of kinetic} \\ \text{and internal} \\ \text{energy addition} \\ \text{by convective} \\ \text{transport} \end{array} \right\} + \left\{ \begin{array}{c} \text{net rate of heat} \\ \text{addition by} \\ \text{molecular} \\ \text{transport} \\ \text{(conduction)} \end{array} \right\} +$$

$$\left\{ \begin{array}{c} \text{rate of work} \\ \text{done on system} \\ \text{by molecular} \\ \text{mechanisms} \\ \text{(i.e., by stresses)} \end{array} \right\} + \left\{ \begin{array}{c} \text{rate of work} \\ \text{done on system} \\ \text{by external} \\ \text{forces} \\ \text{(e.g., by gravity)} \end{array} \right\} \qquad (11.1\text{-}1)$$

In developing the energy equation we will use the **e** vector of Eq. 9.8-5 or 6, which includes the first three brackets on the right side of Eq. 11.1-1. Several comments need to be made before proceeding:

(i) By *kinetic energy* we mean that energy associated with the observable motion of the fluid, which is $\frac{1}{2}\rho v^2 \equiv \frac{1}{2}\rho(\mathbf{v} \cdot \mathbf{v})$, per unit volume. Here **v** is the fluid velocity vector.

(ii) By *internal energy* we mean the kinetic energies of the constituent molecules calculated in a frame moving with the velocity **v**, plus the energies associated with the vibrational and rotational motions of the molecules and also the energies of interaction among all the molecules. It is *assumed* that the internal energy U for a flowing fluid is the same function of temperature and density as that for a fluid at equilibrium. Keep in mind that a similar assumption is made for the thermodynamic pressure $p(\rho, T)$ for a flowing fluid.

(iii) The *potential energy* does not appear in Eq. 11.1-1, since we prefer instead to consider the work done on the system by gravity. At the end of this section, however, we show how to express this work in terms of the potential energy.

(iv) In Eq. 10.1-1 various *source terms* were included in the shell energy balance. In §10.4 the viscous heat source S_v appeared automatically, because the mechanical energy terms in **e** were properly accounted for; the same situation prevails here, and the viscous heating term $-(\tau:\nabla\mathbf{v})$ will appear automatically in Eq. 11.2-1. The chemical, electrical, and nuclear source terms (S_c, S_e, and S_n) do not appear automatically, since chemical reactions, electrical effects, and nuclear

---

[1] R. J. Silbey and R. A. Alberty, *Physical Chemistry*, Wiley, New York, 3rd edition (2001), §2.3.

disintegrations have not been included in the energy balance. In Chapter 19, where the energy equation for mixtures with chemical reactions is considered, the chemical heat source S_c appears naturally, as does a "diffusive source term," $\Sigma_\alpha(\mathbf{j}_\alpha \cdot \mathbf{g}_\alpha)$.

We now translate Eq. 11.1-1 into mathematical terms. The rate of increase of kinetic and internal energy within the volume element $\Delta x\,\Delta y\,\Delta z$ is

$$\Delta x\,\Delta y\,\Delta z\,\frac{\partial}{\partial t}(\tfrac{1}{2}\rho v^2 + \rho\hat{U}) \tag{11.1-2}$$

Here $\hat{U}$ is the internal energy per unit mass (sometimes called the "specific internal energy"). The product $\rho\hat{U}$ is the internal energy per unit volume, and $\tfrac{1}{2}\rho v^2 = \tfrac{1}{2}\rho(v_x^2 + v_y^2 + v_z^2)$ is the kinetic energy per unit volume.

Next we have to know how much energy enters and leaves across the faces of the volume element $\Delta x\,\Delta y\,\Delta z$.

$$\Delta y\,\Delta z(e_x|_x - e_x|_{x+\Delta x}) + \Delta x\,\Delta z(e_y|_y - e_y|_{y+\Delta y}) + \Delta x\,\Delta y(e_z|_z - e_z|_{z+\Delta z}) \tag{11.1-3}$$

Keep in mind that the $\mathbf{e}$ vector includes the convective transport of kinetic and internal energy, the heat conduction, and the work associated with molecular processes.

The rate at which work is done on the fluid by the external force is the dot product of the fluid velocity $\mathbf{v}$ and the force acting on the fluid ($\rho\,\Delta x\,\Delta y\,\Delta z)\mathbf{g}$, or

$$\rho\,\Delta x\,\Delta y\,\Delta z(v_x g_x + v_y g_y + v_z g_z) \tag{11.1-4}$$

We now insert these various contributions into Eq. 11.1-1 and then divide by $\Delta x\,\Delta y\,\Delta z$. When Δx, Δy, and Δz are allowed to go to zero, we get

$$\frac{\partial}{\partial t}(\tfrac{1}{2}\rho v^2 + \rho\hat{U}) = -\left(\frac{\partial e_x}{\partial x} + \frac{\partial e_y}{\partial y} + \frac{\partial e_z}{\partial z}\right) + \rho(v_x g_x + v_y g_y + v_z g_z) \tag{11.1-5}$$

This equation may be written more compactly in vector notation as

$$\frac{\partial}{\partial t}(\tfrac{1}{2}\rho v^2 + \rho\hat{U}) = -(\nabla\cdot\mathbf{e}) + \rho(\mathbf{v}\cdot\mathbf{g}) \tag{11.1-6}$$

Next we insert the expression for the $\mathbf{e}$ vector from Eq. 9.8-5 to get the *equation of energy*:

$$
\boxed{
\begin{array}{l}
\dfrac{\partial}{\partial t}(\tfrac{1}{2}\rho v^2 + \rho\hat{U}) = -(\nabla\cdot(\tfrac{1}{2}\rho v^2 + \rho\hat{U})\mathbf{v}) - (\nabla\cdot\mathbf{q}) \\[4pt]
\quad\text{rate of increase of}\qquad \text{rate of energy addition}\qquad \text{rate of energy addition}\\
\quad\text{energy per unit}\qquad\quad \text{per unit volume by}\qquad\quad \text{per unit volume by}\\
\quad\text{volume}\qquad\qquad\qquad \text{convective transport}\qquad \text{heat conduction}\\[8pt]
\quad -\,(\nabla\cdot p\mathbf{v})\quad -\,(\nabla\cdot[\boldsymbol{\tau}\cdot\mathbf{v}])\quad +\,\rho(\mathbf{v}\cdot\mathbf{g})\\
\quad\text{rate of work}\qquad \text{rate of work done}\qquad \text{rate of work done}\\
\quad\text{done on fluid per}\quad \text{on fluid per unit}\qquad \text{on fluid per unit}\\
\quad\text{unit volume by}\qquad \text{volume by viscous}\qquad \text{volume by external}\\
\quad\text{pressure forces}\qquad \text{forces}\qquad\qquad\qquad \text{forces}
\end{array}}
\tag{11.1-7}
$$

This equation does not include nuclear, radiative, electromagnetic, or chemical forms of energy. For viscoelastic fluids, the next-to-last term has to be reinterpreted by replacing "viscous" by "viscoelastic."

Equation 11.1-7 is the main result of this section, and it provides the basis for the remainder of the chapter. The equation can be written in another form to include the potential energy per unit mass, $\hat{\Phi}$, which has been defined earlier by $\mathbf{g} = -\nabla\hat{\Phi}$ (see §3.3). For moderate elevation changes, this gives $\hat{\Phi} = gh$, where h is a coordinate in the direction opposed to the gravitational field. For terrestrial problems, where the gravitational field is independent of time, we can write

$$\rho(\mathbf{v} \cdot \mathbf{g}) = -(\rho\mathbf{v} \cdot \nabla\hat{\Phi}) \tag{11.1-8}$$

$$= -(\nabla \cdot \rho\mathbf{v}\hat{\Phi}) + \hat{\Phi}(\nabla \cdot \rho\mathbf{v}) \qquad \text{Use vector identity in Eq. A.4-19}$$

$$= -(\nabla \cdot \rho\mathbf{v}\hat{\Phi}) - \hat{\Phi}\frac{\partial\rho}{\partial t} \qquad \text{Use Eq. 3.1-4}$$

$$= -(\nabla \cdot \rho\mathbf{v}\hat{\Phi}) - \frac{\partial}{\partial t}(\rho\hat{\Phi}) \qquad \text{Use } \hat{\Phi} \text{ independent of } t$$

When this result is inserted into Eq. 11.1-7 we get

$$\frac{\partial}{\partial t}\left(\tfrac{1}{2}\rho v^2 + \rho\hat{U} + \rho\hat{\Phi}\right) = -\left(\nabla \cdot \left(\tfrac{1}{2}\rho v^2 + \rho\hat{U} + \rho\hat{\Phi}\right)\mathbf{v}\right)$$

$$- (\nabla \cdot \mathbf{q}) - (\nabla \cdot p\mathbf{v}) - (\nabla \cdot [\boldsymbol{\tau} \cdot \mathbf{v}]) \tag{11.1-9}$$

Sometimes it is convenient to have the energy equation in this form.

§11.2   SPECIAL FORMS OF THE ENERGY EQUATION

The most useful form of the energy equation is one in which the temperature appears. The object of this section is to arrive at such an equation, which can be used for prediction of temperature profiles.

First we subtract the mechanical energy equation in Eq. 3.3-1 from the energy equation in 11.1-7. This leads to the following *equation of change for internal energy*:

$$
\frac{\partial}{\partial t}\rho\hat{U} = \quad -(\nabla \cdot \rho\hat{U}\mathbf{v}) \quad - (\nabla \cdot \mathbf{q})
$$

| rate of increase in internal energy per unit volume | net rate of addition of internal energy by convective transport, per unit volume | rate of internal energy addition by heat conduction, per unit volume |

$$\tag{11.2-1}$$

$$
- p(\nabla \cdot \mathbf{v}) \quad - (\boldsymbol{\tau}:\nabla\mathbf{v})
$$

| *reversible* rate of internal energy increase per unit volume by compression | *irreversible* rate of internal energy increase per unit volume by viscous dissipation |

It is now of interest to compare the mechanical energy equation of Eq. 3.3-1 and the internal energy equation of Eq. 11.2-1. Note that the terms $p(\nabla \cdot \mathbf{v})$ and $(\boldsymbol{\tau}:\nabla\mathbf{v})$ appear in both equations—but with opposite signs. Therefore, these terms describe the interconversion of mechanical and thermal energy. The term $p(\nabla \cdot \mathbf{v})$ can be either positive or negative, depending on whether the fluid is expanding or contracting; therefore it represents a *reversible* mode of interchange. On the other hand, for Newtonian fluids, the quantity $-(\boldsymbol{\tau}:\nabla\mathbf{v})$ is always positive (see Eq. 3.3-3) and therefore represents an *irreversible* degradation of mechanical into internal energy. For viscoelastic fluids, discussed in Chapter 8, the quantity $-(\boldsymbol{\tau}:\nabla\mathbf{v})$ does not have to be positive, since some energy may be stored as elastic energy.

We pointed out in §3.5 that the equations of change can be written somewhat more compactly by using the substantial derivative (see Table 3.5-1). Equation 11.2-1 can be put in the substantial derivative form by using Eq. 3.5-4. This gives, with no further assumptions

$$\rho \frac{D\hat{U}}{Dt} = -(\nabla \cdot \mathbf{q}) - p(\nabla \cdot \mathbf{v}) - (\boldsymbol{\tau}:\nabla\mathbf{v}) \tag{11.2-2}$$

Next it is convenient to switch from internal energy to enthalpy, as we did at the very end of §9.8. That is, in Eq. 11.2-2 we set $\hat{U} = \hat{H} - p\hat{V} = \hat{H} - (p/\rho)$, making the standard assumption that thermodynamic formulas derived from equilibrium thermodynamics may be applied locally for nonequilibrium systems. When we substitute this formula into Eq. 11.2-2 and use the equation of continuity (Eq. A of Table 3.5-1), we get

$$\rho \frac{D\hat{H}}{Dt} = -(\nabla \cdot \mathbf{q}) - (\boldsymbol{\tau}:\nabla\mathbf{v}) + \frac{Dp}{Dt} \tag{11.2-3}$$

Next we may use Eq. 9.8-7, which presumes that the enthalpy is a function of p and T (this restricts the subsequent development to *Newtonian fluids*). Then we may get an expression for the change in the enthalpy in an element of fluid moving with the fluid velocity, which is

$$\begin{aligned}
\rho \frac{D\hat{H}}{Dt} &= \rho \hat{C}_p \frac{DT}{Dt} + \rho \left[\hat{V} - T\left(\frac{\partial \hat{V}}{\partial T}\right)_p \right] \frac{Dp}{Dt} \\
&= \rho \hat{C}_p \frac{DT}{Dt} + \rho \left[\frac{1}{\rho} - T\left(\frac{\partial(1/\rho)}{\partial T}\right)_p \right] \frac{Dp}{Dt} \\
&= \rho \hat{C}_p \frac{DT}{Dt} + \left[1 + \left(\frac{\partial \ln \rho}{\partial \ln T}\right)_p \right] \frac{Dp}{Dt}
\end{aligned} \tag{11.2-4}$$

Equating the right sides of Eqs. 11.2-3 and 11.2-4 gives

$$\boxed{\rho \hat{C}_p \frac{DT}{Dt} = -(\nabla \cdot \mathbf{q}) - (\boldsymbol{\tau}:\nabla\mathbf{v}) - \left(\frac{\partial \ln \rho}{\partial \ln T}\right)_p \frac{Dp}{Dt}} \tag{11.2-5}$$

This is the *equation of change for temperature*, in terms of the heat flux vector $\mathbf{q}$ and the viscous momentum flux tensor $\boldsymbol{\tau}$. To use this equation we need expressions for these fluxes:

(i) When Fourier's law of Eq. 9.1-4 is used, the term $-(\nabla \cdot \mathbf{q})$ becomes $+(\nabla \cdot k\nabla T)$, or, if the thermal conductivity is assumed constant, $+k\nabla^2 T$.

(ii) When Newton's law of Eq. 1.2-7 is used, the term $-(\boldsymbol{\tau}:\nabla\mathbf{v})$ becomes $\mu\Phi_v + \kappa\Psi_v$, the quantity given explicitly in Eq. 3.3-3.

We do not perform the substitutions here, because the equation of change for temperature is almost never used in its complete generality.

We now discuss several special *restricted* versions of the equation of change for temperature. In all of these we use Fourier's law with constant k, and we omit the viscous dissipation term, since it is important only in flows with enormous velocity gradients:

(i) For an *ideal gas*, $(\partial \ln \rho/\partial \ln T)_p = -1$, and

$$\rho \hat{C}_p \frac{DT}{Dt} = k\nabla^2 T + \frac{Dp}{Dt} \tag{11.2-6}$$

Or, if use is made of the relation $\tilde{C}_p - \tilde{C}_V = R$, the equation of state in the form $pM = \rho RT$, and the equation of continuity as written in Eq. A of Table 3.5-1, we get

$$\rho \hat{C}_V \frac{DT}{Dt} = k\nabla^2 T - p(\nabla \cdot \mathbf{v}) \tag{11.2-7}$$

(ii) For a *fluid flowing in a constant pressure system*, $Dp/Dt = 0$, and

$$\rho \hat{C}_p \frac{DT}{Dt} = k \nabla^2 T \tag{11.2-8}$$

(iii) For a *fluid with constant density*,[1] $(\partial \ln \rho / \partial \ln T)_p = 0$, and

$$\rho \hat{C}_p \frac{DT}{Dt} = k \nabla^2 T \tag{11.2-9}$$

(iv) For a *stationary solid*, $\mathbf{v}$ is zero and

$$\rho \hat{C}_p \frac{\partial T}{\partial t} = k \nabla^2 T \tag{11.2-10}$$

These last five equations are the ones most frequently encountered in textbooks and research publications. Of course, one can always go back to Eq. 11.2-5 and develop less restrictive equations when needed. Also, one can add chemical, electrical, and nuclear source terms on an ad hoc basis, as was done in Chapter 10.

Equation 11.2-10 is the heat conduction equation for solids, and much has been written about this famous equation developed first by Fourier.[2] The famous reference work by Carslaw and Jaeger deserves special mention. It contains hundreds of solutions of this equation for a wide variety of boundary and initial conditions.[3]

§11.3 THE BOUSSINESQ EQUATION OF MOTION FOR FORCED AND FREE CONVECTION

The equation of motion given in Eq. 3.2-9 (or Eq. B of Table 3.5-1) is valid for both isothermal and nonisothermal flow. In nonisothermal flow, the fluid density and viscosity depend in general on temperature as well as on pressure. The variation in the density is particularly important because it gives rise to buoyant forces, and thus to free convection, as we have already seen in §10.9.

The buoyant force appears automatically when an equation of state is inserted into the equation of motion. For example, we can use the simplified equation of state introduced in Eq. 10.9-6 (this is called the *Boussinesq approximation*)[1]

$$\rho(T) = \bar{\rho} - \bar{\rho}\bar{\beta}(T - \bar{T}) \tag{11.3-1}$$

in which $\bar{\beta}$ is $-(1/\rho)(\partial\rho/\partial T)_p$ evaluated at $T = \bar{T}$. This equation is obtained by writing the Taylor series for ρ as a function of T, considering the pressure p to be constant, and keeping only the first two terms of the series. When Eq. 11.3-1 is substituted into the $\rho\mathbf{g}$ term (but not into the $\rho(D\mathbf{v}/Dt)$ term) of Eq. B of Table 3.5-1, we get the *Boussinesq equation*:

$$\boxed{\rho \frac{D\mathbf{v}}{Dt} = (-\nabla p + \bar{\rho}\mathbf{g}) - [\nabla \cdot \boldsymbol{\tau}] - \bar{\rho}\mathbf{g}\bar{\beta}(T - \bar{T})} \tag{11.3-2}$$

---

[1] The assumption of constant density is made here, instead of the less stringent assumption that $(\partial \ln \rho / \partial \ln T)_p = 0$, since Eq. 11.2-9 is customarily used along with Eq. 3.1-5 (equation of continuity for constant density) and Eq. 3.5-6 (equation of motion for constant density and viscosity). Note that the hypothetical equation of state ρ = constant has to be supplemented by the statement that $(\partial p/\partial T)_p$ = finite, in order to permit the evaluation of certain thermodynamic derivatives. For example, the relation

$$\hat{C}_p - \hat{C}_V = -\frac{1}{\rho}\left(\frac{\partial \ln \rho}{\partial \ln T}\right)_p\left(\frac{\partial p}{\partial T}\right)_\rho \tag{11.2-9a}$$

leads to the result that $\hat{C}_p = \hat{C}_V$ for the "incompressible fluid" thus defined.

[2] J. B. Fourier, *Théorie analytique de la chaleur, Œuvres de Fourier*, Gauthier-Villars et Fils, Paris (1822).

[3] H. S. Carslaw and J. C. Jaeger, *Conduction of Heat in Solids*, Oxford University Press, 2nd edition (1959).

[1] J. Boussinesq, *Théorie Analytique de Chaleur*, Vol. 2, Gauthier-Villars, Paris (1903).

This form of the equation of motion is very useful for heat transfer analyses. It describes the limiting cases of forced convection and free convection (see Fig. 10.8-1), and the region between these extremes as well. In *forced convection* the buoyancy term $-\bar{\rho}\mathbf{g}\bar{\beta}(T - \bar{T})$ is neglected. In *free convection* (or *natural convection*) the term $(-\nabla p + \bar{\rho}\mathbf{g})$ is small, and omitting it is usually appropriate, particularly for vertical, rectilinear flow and for the flow near submerged objects in large bodies of fluid. Setting $(-\nabla p + \bar{\rho}\mathbf{g})$ equal to zero is equivalent to assuming that the pressure distribution is just that for a fluid at rest.

It is also customary to replace ρ on the left side of Eq. 11.3-2 by $\bar{\rho}$. This substitution has been successful for free convection at moderate temperature differences. Under these conditions the fluid motion is slow, and the acceleration term $D\mathbf{v}/Dt$ is small compared to $\mathbf{g}$.

However, in systems where the acceleration term is large with respect to $\mathbf{g}$, one must also use Eq. 11.3-1 for the density on the left side of the equation of motion. This is particularly true, for example, in gas turbines and near hypersonic missiles, where the term $(\rho - \bar{\rho})D\mathbf{v}/Dt$ may be at least as important as $\bar{\rho}\mathbf{g}$.

§11.4   USE OF THE EQUATIONS OF CHANGE TO SOLVE STEADY-STATE PROBLEMS

In §§3.1 to 3.4 and in §§11.1 to 11.3 we have derived various equations of change for a pure fluid or solid. It seems appropriate here to present a summary of these equations for future reference. Such a summary is given in Table 11.4-1, with most of the equations given in both the $\partial/\partial t$ form and the D/Dt form. Reference is also made to the first place where each equation has been presented.

Although Table 11.4-1 is a useful summary, for problem solving we use the equations written out explicitly in the several commonly used coordinate systems. This has been done in Appendix B, and readers should thoroughly familiarize themselves with the tables there.

In general, to describe the nonisothermal flow of a Newtonian fluid one needs

- the equation of continuity
- the equation of motion (containing μ and κ)
- the equation of energy (containing μ, κ, and k)
- the thermal equation of state ($p = p(\rho, T)$)
- the caloric equation of state ($\hat{C}_p = \hat{C}_p(\rho, T)$)

as well as expressions for the density and temperature dependence of the viscosity, dilatational viscosity, and thermal conductivity. In addition one needs the boundary and initial conditions. The entire set of equations can then—in principle—be solved to get the pressure, density, velocity, and temperature as functions of position and time. If one wishes to solve such a detailed problem, numerical methods generally have to be used.

Often one may be content with a restricted solution, for making an order-of-magnitude analysis of a problem, or for investigating limiting cases prior to doing a complete numerical solution. This is done by making some standard assumptions:

   (i)   *Assumption of constant physical properties.* If it can be assumed that all physical properties are constant, then the equations become considerably simpler, and in some cases analytical solutions can be found.

   (ii) *Assumption of zero fluxes.* Setting τ and $\mathbf{q}$ equal to zero may be useful for (a) adiabatic flow processes in systems designed to minimize frictional effects (such as Venturi meters and turbines), and (b) high-speed flows around streamlined objects. The solutions obtained would be of no use for describing the situation near fluid–solid boundaries, but may be adequate for analysis of phenomena far from the solid boundaries.

Table 11.4-1 Equations of Change for Pure Fluids in Terms of the Fluxes

Eq.	Special form	In terms of D/Dt		Comments
Cont.	—	$\dfrac{D\rho}{Dt} = -\rho(\nabla \cdot \mathbf{v})$	Table 3.5-1 (A)	For ρ = constant, simplifies to $(\nabla \cdot \mathbf{v}) = 0$
Motion	General	$\rho\dfrac{D\mathbf{v}}{Dt} = -\nabla p - [\nabla \cdot \boldsymbol{\tau}] + \rho\mathbf{g}$	Table 3.5-1 (B)	For $\boldsymbol{\tau} = 0$ this becomes Euler's equation
	Approximate	$\rho\dfrac{D\mathbf{v}}{Dt} = -\nabla p - [\nabla \cdot \boldsymbol{\tau}] + \bar{\rho}\mathbf{g} - \bar{\rho}\bar{g}\bar{\beta}(T - \bar{T})$	11.3-2 (C)	Displays buoyancy term
Energy	In terms of $\hat{K} + \hat{U} + \hat{\Phi}$	$\rho\dfrac{D(\hat{K} + \hat{U} + \hat{\Phi})}{Dt} = -(\nabla \cdot \mathbf{q}) - (\nabla \cdot p\mathbf{v}) - (\nabla \cdot [\boldsymbol{\tau} \cdot \mathbf{v}])$	— (D)	Exact only for Φ time independent
	In terms of $\hat{K} + \hat{U}$	$\rho\dfrac{D(\hat{K} + \hat{U})}{Dt} = -(\nabla \cdot \mathbf{q}) - (\nabla \cdot p\mathbf{v}) - (\nabla \cdot [\boldsymbol{\tau} \cdot \mathbf{v}]) + \rho(\mathbf{v} \cdot \mathbf{g})$	— (E)	
	In terms of $\hat{K} = \tfrac{1}{2}v^2$	$\rho\dfrac{D\hat{K}}{Dt} = -(\mathbf{v} \cdot \nabla p) - (\mathbf{v} \cdot [\nabla \cdot \boldsymbol{\tau}]) + \rho(\mathbf{v} \cdot \mathbf{g})$	Table 3.5-1 (F)	From equation of motion
	In terms of $\hat{U}$	$\rho\dfrac{D\hat{U}}{Dt} = -(\nabla \cdot \mathbf{q}) - p(\nabla \cdot \mathbf{v}) - (\boldsymbol{\tau}:\nabla\mathbf{v})$	11.2-2 (G)	Term containing $(\nabla \cdot \mathbf{v})$ is zero for constant ρ
	In terms of $\hat{H}$	$\rho\dfrac{D\hat{H}}{Dt} = -(\nabla \cdot \mathbf{q}) - (\boldsymbol{\tau}:\nabla\mathbf{v}) + \dfrac{Dp}{Dt}$	11.2-3 (H)	$\hat{H} = \hat{U} + (p/\rho)$
	In terms of $\hat{C}_v$ and T	$\rho\hat{C}_v\dfrac{DT}{Dt} = -(\nabla \cdot \mathbf{q}) - T\left(\dfrac{\partial p}{\partial T}\right)_\rho (\nabla \cdot \mathbf{v}) - (\boldsymbol{\tau}:\nabla\mathbf{v})$	— (I)	For an ideal gas $T(\partial p/\partial T)_\rho = p$
	In terms of $\hat{C}_p$ and T	$\rho\hat{C}_p\dfrac{DT}{Dt} = -(\nabla \cdot \mathbf{q}) - \left(\dfrac{\partial \ln \rho}{\partial \ln T}\right)_p \dfrac{Dp}{Dt} - (\boldsymbol{\tau}:\nabla\mathbf{v})$	11.2-5 (J)	For an ideal gas $(\partial \ln \rho/\partial \ln T)_p = -1$

Cont.	—	$\dfrac{\partial}{\partial t}\rho = -(\nabla\cdot\rho\mathbf{v})$	3.1-4 (K)	For ρ = constant, simplifies to $(\nabla\cdot\mathbf{v})=0$
Motion	General	$\dfrac{\partial}{\partial t}\rho\mathbf{v} = -[\nabla\cdot\rho\mathbf{v}\mathbf{v}] - \nabla p - [\nabla\cdot\boldsymbol{\tau}] + \rho\mathbf{g}$	3.2-9 (L)	For $\boldsymbol{\tau}=0$ this becomes Euler's equation
	Approximate	$\dfrac{\partial}{\partial t}\rho\mathbf{v} = -[\nabla\cdot\rho\mathbf{v}\mathbf{v}] - \nabla p - [\nabla\cdot\boldsymbol{\tau}] + \bar{\rho}\mathbf{g} - \bar{\rho}\bar{\mathbf{g}}\bar{\beta}(T - \bar{T})$	— (M)	Displays buoyancy term
Energy	In terms of $\hat{K}+\hat{U}+\hat{\Phi}$	$\dfrac{\partial}{\partial t}\rho(\hat{K}+\hat{U}+\hat{\Phi}) = -(\nabla\cdot\rho(\hat{K}+\hat{H}+\hat{\Phi})\mathbf{v}) - (\nabla\cdot\mathbf{q}) - (\nabla\cdot[\boldsymbol{\tau}\cdot\mathbf{v}])$	11.1-9 (N)	Exact only for Φ time independent
	In terms of $\hat{K}+\hat{\Phi}$	$\dfrac{\partial}{\partial t}\rho(\hat{K}+\hat{\Phi}) = -(\nabla\cdot\rho(\hat{K}+\hat{\Phi})\mathbf{v}) - (\mathbf{v}\cdot\nabla p) - (\mathbf{v}\cdot[\nabla\cdot\boldsymbol{\tau}])$	3.3-2 (O)	Exact only for Φ time independent. From equation of motion
	In terms of $\hat{K}+\hat{U}$	$\dfrac{\partial}{\partial t}\rho(\hat{K}+\hat{U}) = -(\nabla\cdot\rho(\hat{K}+\hat{H})\mathbf{v}) - (\nabla\cdot\mathbf{q}) - (\nabla\cdot[\boldsymbol{\tau}\cdot\mathbf{v}]) + \rho(\mathbf{v}\cdot\mathbf{g})$	11.1-7 (P)	
	In terms of $\hat{K}=\tfrac{1}{2}v^2$	$\dfrac{\partial}{\partial t}\rho\hat{K} = -(\nabla\cdot\rho\hat{K}\mathbf{v}) - (\mathbf{v}\cdot\nabla p) - (\mathbf{v}\cdot[\nabla\cdot\boldsymbol{\tau}]) + \rho(\mathbf{v}\cdot\mathbf{g})$	3.3-1 (Q)	From equation of motion
	In terms of $\hat{U}$	$\dfrac{\partial}{\partial t}\rho\hat{U} = -(\nabla\cdot\rho\hat{U}\mathbf{v}) - (\nabla\cdot\mathbf{q}) - p(\nabla\cdot\mathbf{v}) - (\boldsymbol{\tau}:\nabla\mathbf{v})$	11.2-1 (R)	Term containing $(\nabla\cdot\mathbf{v})$ is zero for constant ρ
	In terms of $\hat{H}$	$\dfrac{\partial}{\partial t}\rho\hat{H} = -(\nabla\cdot\rho\hat{H}\mathbf{v}) - (\nabla\cdot\mathbf{q}) - (\boldsymbol{\tau}:\nabla\mathbf{v}) + \dfrac{Dp}{Dt}$	— (S)	$\hat{H} = \hat{U} + (p/\rho)$
Entropy	—	$\dfrac{\partial}{\partial t}\rho\hat{S} = -(\nabla\cdot\rho\hat{S}\,\mathbf{v}) - \left(\nabla\cdot\dfrac{\mathbf{q}}{T}\right) - \dfrac{1}{T^2}(\mathbf{q}\cdot\nabla T) - \dfrac{1}{T}(\boldsymbol{\tau}:\nabla\mathbf{v})$	11D.1-1 (T)	Last two terms describe entropy production

To illustrate the solution of problems in which the energy equation plays a significant role, we solve a series of (idealized) problems. We restrict ourselves here to steady-state flow problems and consider unsteady-state problems in Chapter 12. In each problem we start by listing the postulates that lead us to simplified versions of the equations of change.

EXAMPLE 11.4-1

Steady-State Forced-Convection Heat Transfer in Laminar Flow in a Circular Tube

Show how to set up the equations for the problem considered in §10.8—namely, that of finding the fluid temperature profiles for the fully developed laminar flow in a tube.

SOLUTION

We assume constant physical properties, and we postulate a solution of the following form: $\mathbf{v} = \boldsymbol{\delta}_z v_z(r)$, $\mathcal{P} = \mathcal{P}(z)$, and $T = T(r, z)$. Then the equations of change, as given in Appendix B, may be simplified to

Continuity:
$$0 = 0 \tag{11.4-1}$$

Motion:
$$0 = -\frac{d\mathcal{P}}{dz} + \mu\left[\frac{1}{r}\frac{d}{dr}\left(r\frac{dv_z}{dr}\right)\right] \tag{11.4-2}$$

Energy:
$$\rho\hat{C}_p v_z \frac{\partial T}{\partial z} = k\left[\frac{1}{r}\frac{\partial}{\partial r}\left(r\frac{\partial T}{\partial r}\right) + \frac{\partial^2 T}{\partial z^2}\right] + \mu\left(\frac{dv_z}{dr}\right)^2 \tag{11.4-3}$$

The equation of continuity is automatically satisfied as a result of the postulates. The equation of motion, when solved as in Example 3.6-1, gives the velocity distribution (the parabolic velocity profile). This expression is then substituted into the convective heat transport term on the left side of Eq. 11.4-3 and into the viscous dissipation heating term on the right side.

Next, as in §10.8, we make two assumptions: (i) in the z direction, heat conduction is much smaller than heat convection, so that the term $\partial^2 T/\partial z^2$ can be neglected, and (ii) the flow is not sufficiently fast that viscous heating is significant, and hence the term $\mu(\partial v_z/\partial r)^2$ can be omitted. When these assumptions are made, Eq. 11.4-3 becomes the same as Eq. 10.8-12. From that point on, the asymptotic solution, valid for large z only, proceeds as in §10.8. Note that we have gone through three types of restrictive processes: (i) *postulates*, in which a tentative guess is made as to the form of the solution; (ii) *assumptions*, in which we eliminate some physical phenomena or effects by discarding terms or assuming physical properties to be constant; and (iii) an *asymptotic solution*, in which we obtain only a portion of the entire mathematical solution. It is important to distinguish among these various kinds of restrictions.

EXAMPLE 11.4-2

Tangential Flow in an Annulus with Viscous Heat Generation

Determine the temperature distribution in an incompressible liquid confined between two coaxial cylinders, the outer one of which is rotating at a steady angular velocity Ω_o (see §10.4 and Example 3.6-3). Use the nomenclature of Example 3.6-3, and consider the radius ratio κ to be fairly small so that the curvature of the fluid streamlines must be taken into account.

The temperatures of the inner and outer surfaces of the annular region are maintained at T_κ and T_1, respectively, with $T_\kappa \neq T_1$. Assume steady laminar flow, and neglect the temperature dependence of the physical properties.

This is an example of a forced convection problem: The equations of continuity and motion are solved to get the velocity distribution, and then the energy equation is solved to get the temperature distribution. This problem is of interest in connection with heat effects in coaxial cylinder viscometers[1] and in lubrication systems.

---

[1] J. R. Van Wazer, J. W. Lyons, K. Y. Kim, and R. E. Colwell, *Viscosity and Flow Measurement*, Wiley, New York (1963), pp. 82–85.

SOLUTION

We begin by postulating that $\mathbf{v} = \boldsymbol{\delta}_\theta v_\theta(r)$, that $\mathcal{P} = \mathcal{P}(r, z)$, and that $T = T(r)$. Then the simplification of the equations of change leads to Eqs. 3.6-20, 21, and 22 (the r-, θ-, and z-components of the equation of motion), and the energy equation

$$0 = k\frac{1}{r}\frac{d}{dr}\left(r\frac{dT}{dr}\right) + \mu\left[r\frac{d}{dr}\left(\frac{v_\theta}{r}\right)\right]^2 \tag{11.4-4}$$

When the solution to the θ-component of the equation of motion, given in Eq. 3.6-29, is substituted into the energy equation, we get

$$0 = k\frac{1}{r}\frac{d}{dr}\left(r\frac{dT}{dr}\right) + \frac{4\mu\Omega_o^2\kappa^4 R^4}{(1 - \kappa^2)^2}\frac{1}{r^4} \tag{11.4-5}$$

This is the differential equation for the temperature distribution. It may be rewritten in terms of dimensionless quantities by putting

$$\xi = \frac{r}{R} \qquad \Theta = \frac{T - T_\kappa}{T_1 - T_\kappa} \qquad N = \frac{\mu\Omega_o^2 R^2}{k(T_1 - T_\kappa)} \cdot \frac{\kappa^4}{(1 - \kappa^2)^2} \tag{11.4-6, 7, 8}$$

The parameter N is closely related to the Brinkman number of §10.4. Equation 11.4-5 now becomes

$$\frac{1}{\xi}\frac{d}{d\xi}\left(\xi\frac{d\Theta}{d\xi}\right) = -4N\frac{1}{\xi^4} \tag{11.4-9}$$

This is of the form of Eq. C.1-11 and has the solution

$$\Theta = -N\frac{1}{\xi^2} + C_1 \ln \xi + C_2 \tag{11.4-10}$$

The integration constants are found from the boundary conditions

B.C. 1:   at $\xi = \kappa$, $\Theta = 0$ (11.4-11)

B.C. 2:   at $\xi = 1$, $\Theta = 1$ (11.4-12)

Determination of the constants then leads to

$$\Theta = \left(1 - \frac{\ln \xi}{\ln \kappa}\right) + N\left[\left(1 - \frac{1}{\xi^2}\right) - \left(1 - \frac{1}{\kappa^2}\right)\frac{\ln \xi}{\ln \kappa}\right] \tag{11.4-13}$$

When $N = 0$, we obtain the temperature distribution for a motionless cylindrical shell of thickness $R(1 - \kappa)$ with inner and outer temperatures T_κ and T_1. If N is large enough, there will be a maximum in the temperature distribution, located at

$$\xi = \sqrt{\frac{2 \ln (1/\kappa)}{(1/\kappa^2) - 1 + (1/N)}} \tag{11.4-14}$$

with the temperature at this point greater than either T_κ or T_1.

Although this example provides an illustration of the use of the tabulated equations of change in cylindrical coordinates, in most viscometric and lubrication applications the clearance between the cylinders is so small that numerical values computed from Eq. 11.4-13 will not differ substantially from those computed from Eq. 10.4-9.

EXAMPLE 11.4-3

Steady Flow in a Nonisothermal Film

A liquid is flowing downward in steady laminar flow along an inclined plane surface, as shown in Figs. 2.2-1 to 3. The free liquid surface is maintained at temperature T_0, and the solid surface at $x = \delta$ is maintained at T_δ. At these temperatures the liquid viscosity has values μ_0 and μ_δ, respectively, and the liquid density and thermal conductivity may be assumed constant. Find the velocity distribution in this nonisothermal flow system, neglecting end effects and recognizing that viscous heating is unimportant in this flow. Assume that the temperature dependence of viscosity may be expressed by an equation of the form $\mu = Ae^{B/T}$, with A and B being empirical constants; this is suggested by the Eyring theory given in §1.5.

We first solve the energy equation to get the temperature profile, and then use the latter to find the dependence of viscosity on position. Then the equation of motion can be solved to get the velocity profile.

SOLUTION

We postulate that $T = T(x)$ and that $\mathbf{v} = \boldsymbol{\delta}_z v_z(x)$. Then the energy equation simplifies to

$$\frac{d^2T}{dx^2} = 0 \tag{11.4-15}$$

This can be integrated between the known terminal temperatures to give

$$\frac{T - T_0}{T_\delta - T_0} = \frac{x}{\delta} \tag{11.4-16}$$

The dependence of viscosity on temperature may be written as

$$\frac{\mu(T)}{\mu_0} = \exp\left[B\left(\frac{1}{T} - \frac{1}{T_0}\right)\right] \tag{11.4-17}$$

in which B is a constant, to be determined from experimental data for viscosity versus temperature. To get the dependence of viscosity on position, we combine the last two equations to get

$$\frac{\mu(x)}{\mu_0} = \exp\left[B\frac{T_0 - T_\delta}{T_0 T}\left(\frac{x}{\delta}\right)\right] \cong \exp\left[B\frac{T_0 - T_\delta}{T_0 T_\delta}\left(\frac{x}{\delta}\right)\right] \tag{11.4-18}$$

The second expression is a good approximation if the temperature does not change greatly through the film. When this equation is combined with Eq. 11.4-17, written for $T = T_\delta$, we then get

$$\frac{\mu(x)}{\mu_0} = \exp\left[\left(\ln \frac{\mu_\delta}{\mu_0}\right)\left(\frac{x}{\delta}\right)\right] = \left(\frac{\mu_\delta}{\mu_0}\right)^{x/\delta} \tag{11.4-19}$$

This is the same as the expression used in Example 2.2-2, if we set α equal to $-\ln(\mu_\delta/\mu_0)$. Therefore we may take over the result from Example 2.2-2 and write the velocity profile as

$$v_z = \left(\frac{\rho g \cos \beta}{\mu_0}\right)\left(\frac{\delta}{\ln (\mu_\delta/\mu_0)}\right)^2\left[\frac{1 + (x/\delta) \ln (\mu_\delta/\mu_0)}{(\mu_\delta/\mu_0)^{x/\delta}} - \frac{1 + \ln (\mu_\delta/\mu_0)}{(\mu_\delta/\mu_0)}\right] \tag{11.4-20}$$

This completes the analysis of the problem begun in Example 2.2-2, by providing the appropriate value of the constant α.

EXAMPLE 11.4-4

Transpiration Cooling[2]

A system with two concentric porous spherical shells of radii κR and R is shown in Fig. 11.4-1. The inner surface of the outer shell is at temperature T_1, and the outer surface of the inner shell is at a lower temperature T_κ. Dry air at T_κ is blown outward radially from the inner shell into the intervening space and then through the outer shell. Develop an expression for the required rate of heat removal from the inner sphere as a function of the mass rate of flow of the gas. Assume steady laminar flow and low gas velocity.

In this example the equations of continuity and energy are solved to get the temperature distribution. The equation of motion gives information about the pressure distribution in the system.

---

[2] M. Jakob, *Heat Transfer*, Vol. 2, Wiley, New York (1957), pp. 394–415.

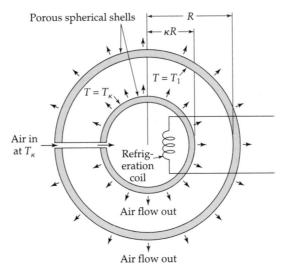

Porous spherical shells

$T = T_\kappa$

$T = T_1$

Air in at T_κ

Refrig-eration coil

Air flow out

Air flow out

Fig. 11.4-1. Transpiration cooling. The inner sphere is being cooled by means of a refrigeration coil to maintain its temperature at T_κ. When air is blown outward, as shown, less refrigeration is required.

SOLUTION

We postulate that for this system $\mathbf{v} = \boldsymbol{\delta}_r v_r(r)$, $T = T(r)$, and $\mathscr{P} = \mathscr{P}(r)$. The *equation of continuity* in spherical coordinates then becomes

$$\frac{1}{r^2}\frac{d}{dr}(r^2 \rho v_r) = 0 \tag{11.4-21}$$

This equation can be integrated to give

$$r^2 \rho v_r = \text{const.} = \frac{w_r}{4\pi} \tag{11.4-22}$$

Here w_r is the radial mass flow rate of the gas.

The r-component of the *equation of motion* in spherical coordinates is, from Eq. B.6-7,

$$\rho v_r \frac{dv_r}{dr} = -\frac{d\mathscr{P}}{dr} + \mu\left(\frac{1}{r^2}\frac{d^2}{dr^2}(r^2 v_r)\right) \tag{11.4-23}$$

The viscosity term drops out because of Eq. 11.4-21. Integration of Eq. 11.4-23 then gives

$$\mathscr{P}(r) - \mathscr{P}(R) = \frac{w_r^2}{32\pi^2 \rho R^4}\left[1 - \left(\frac{R}{r}\right)^4\right] \tag{11.4-24}$$

Hence the modified pressure $\mathscr{P}$ increases with r, but only very slightly for the low gas velocity assumed here.

The *energy equation* in terms of the temperature, in spherical coordinates, is, according to Eq. B.9-3,

$$\rho \hat{C}_p v_r \frac{dT}{dr} = k\frac{1}{r^2}\frac{d}{dr}\left(r^2\frac{dT}{dr}\right) \tag{11.4-25}$$

Here we have used Eq. 11.2-8, for which we assume that the thermal conductivity is constant, the pressure is constant, and there is no viscous dissipation—all reasonable assumptions for the problem at hand.

When Eq. 11.4-22 for the velocity distribution is used for v_r in Eq. 11.4-25, we obtain the following differential equation for the temperature distribution $T(r)$ in the gas between the two shells:

$$\frac{dT}{dr} = \frac{4\pi k}{w_r \hat{C}_p}\frac{d}{dr}\left(r^2\frac{dT}{dr}\right) \tag{11.4-26}$$

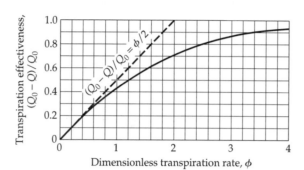

Fig. 11.4-2. The effect of transpiration cooling.

We make the change of variable $u = r^2(dT/dr)$ and obtain a first-order, separable differential equation for $u(r)$. This may be integrated, and when the boundary conditions are applied, we get

$$\frac{T - T_1}{T_\kappa - T_1} = \frac{e^{-R_0/r} - e^{-R_0/R}}{e^{-R_0/\kappa R} - e^{-R_0/R}} \tag{11.4-27}$$

in which $R_0 = w_r \hat{C}_p/4\pi k$ is a constant with units of length.

The rate of heat flow toward the inner sphere is

$$Q = -4\pi\kappa^2 R^2 q_r|_{r=\kappa R} \tag{11.4-28}$$

and this is the required rate of heat removal by the refrigerant. Insertion of Fourier's law for the r-component of the heat flux gives

$$Q = +4\pi\kappa^2 R^2 k \left.\frac{dT}{dr}\right|_{r=\kappa R} \tag{11.4-29}$$

Next we evaluate the temperature gradient at the surface with the aid of Eq. 11.4-27 to obtain the expression for the heat removal rate.

$$Q = \frac{4\pi R_0 k(T_1 - T_\kappa)}{\exp[(R_0/\kappa R)(1-\kappa)] - 1} \tag{11.4-30}$$

In the limit that the mass flow rate of the gas is zero, so that $R_0 = 0$, the heat removal rate becomes

$$Q_0 = \frac{4\pi\kappa Rk(T_1 - T_\kappa)}{1 - \kappa} \tag{11.4-31}$$

The fractional reduction in heat removal as a result of the transpiration of the gas is then

$$\frac{Q_0 - Q}{Q_0} = 1 - \frac{\phi}{e^\phi - 1} \tag{11.4-32}$$

Here $\phi = R_0(1 - \kappa)/\kappa R = w_r \hat{C}_p(1 - \kappa)/4\pi\kappa Rk$ is the "dimensionless transpiration rate." Equation 11.4-32 is shown graphically in Fig. 11.4-2. For small values of ϕ, the quantity $(Q_0 - Q)/Q_0$ approaches the asymptote $\frac{1}{2}\phi$.

EXAMPLE 11.4-5

Free-Convection Heat Transfer from a Vertical Plate

A flat plate of height H and width W (with $W \gg H$) heated to a temperature T_0 is suspended in a large body of fluid, which is at ambient temperature T_1. In the neighborhood of the heated plate the fluid rises because of the buoyant force (see Fig. 11.4-3). From the equations of change, deduce the dependence of the heat loss on the system variables. The physical properties of the fluid are considered constant, except that the change in density with temperature will be accounted for by the Boussinesq approximation.

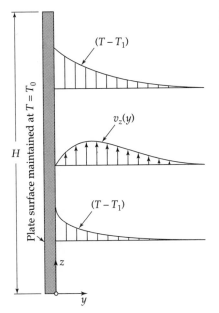

Fig. 11.4-3. The temperature and velocity profiles in the neighborhood of a vertical heated plate.

SOLUTION

We postulate that $\mathbf{v} = \boldsymbol{\delta}_y v_y(y, z) + \boldsymbol{\delta}_z v_z(y, z)$ and that $T = T(y, z)$. We assume that the heated fluid moves almost directly upward, so that $v_y << v_z$. Then the x- and y-components of Eq. 11.3-2 give $p = p(z)$, so that the pressure is given to a very good approximation by $-dp/dz - \bar{\rho}g = 0$, which is the hydrostatic pressure distribution. The remaining equations of change are

Continuity
$$\frac{\partial v_y}{\partial y} + \frac{\partial v_z}{\partial z} = 0 \qquad (11.4\text{-}33)$$

Motion
$$\bar{\rho}\left(v_y \frac{\partial}{\partial y} + v_z \frac{\partial}{\partial z}\right)v_z = \mu\left(\frac{\partial^2}{\partial y^2} + \frac{\partial^2}{\partial z^2}\right)v_z + \bar{\rho}g\bar{\beta}(T - T_1) \qquad (11.4\text{-}34)$$

Energy
$$\bar{\rho}\hat{C}_p\left(v_y \frac{\partial}{\partial y} + v_z \frac{\partial}{\partial z}\right)(T - T_1) = k\left(\frac{\partial^2}{\partial y^2} + \frac{\partial^2}{\partial z^2}\right)(T - T_1) \qquad (11.4\text{-}35)$$

in which $\bar{\rho}$ and $\bar{\beta}$ are evaluated at the ambient temperature T_1. The dashed-underlined terms will be omitted on the ground that momentum and energy transport by molecular processes in the z direction is small compared with the corresponding convective terms on the left side of the equations. These omissions should give a satisfactory description of the system except for a small region around the bottom of the plate. With this simplification, the following boundary conditions suffice to analyze the system up to $z = H$:

B.C. 1: at $y = 0$, $v_y = v_z = 0$ and $T = T_0$ (11.4-36)

B.C. 2: as $y \to \pm\infty$, $v_z \to 0$ and $T \to T_1$ (11.4-37)

B.C. 3: at $z = 0$, $v_z = 0$ (11.4-38)

Note that the temperature rise appears in the equation of motion and that the velocity distribution appears in the energy equation. Thus these equations are "coupled." Analytic solutions of such coupled, nonlinear differential equations are very difficult, and we content ourselves here with a dimensional analysis approach.

To do this we introduce the following dimensionless variables:

$$\Theta = \frac{T - T_1}{T_0 - T_1} = \text{dimensionless temperature} \qquad (11.4\text{-}39)$$

$$\zeta = \frac{z}{H} = \text{dimensionless vertical coordinate} \qquad (11.4\text{-}40)$$

$$\eta = \left(\frac{B}{\mu\alpha H}\right)^{1/4} y \;=\; \text{dimensionless horizontal coordinate} \tag{11.4-41}$$

$$\phi_z = \left(\frac{\mu}{\alpha BH}\right)^{1/2} v_z \;=\; \text{dimensionless vertical velocity} \tag{11.4-42}$$

$$\phi_y = \left(\frac{\mu H}{\alpha^3 B}\right)^{1/4} v_y \;=\; \text{dimensionless horizontal velocity} \tag{11.4-43}$$

in which $\alpha = k/\rho\hat{C}_p$ and $B = \bar{\rho}g\bar{\beta}(T_0 - T_1)$.

When the equations of change, without the dashed-underlined terms, are written in terms of these dimensionless variables, we get

Continuity
$$\frac{\partial \phi_y}{\partial \eta} + \frac{\partial \phi_z}{\partial \zeta} = 0 \tag{11.4-44}$$

Motion
$$\frac{1}{\text{Pr}}\left(\phi_y \frac{\partial}{\partial \eta} + \phi_z \frac{\partial}{\partial \zeta}\right)\phi_z = \frac{\partial^2 \phi_z}{\partial \eta^2} + \Theta \tag{11.4-45}$$

Energy
$$\left(\phi_y \frac{\partial}{\partial \eta} + \phi_z \frac{\partial}{\partial \zeta}\right)\Theta = \frac{\partial^2 \Theta}{\partial \eta^2} \tag{11.4-46}$$

The preceding boundary conditions then become

B.C. 1:   at $\eta = 0$, $\phi_y = \phi_z = 0$, $\Theta = 1$ (11.4-47)

B.C. 2:   as $\eta \to \infty$, $\phi_z \to 0$, $\Theta \to 0$ (11.4-48)

B.C. 3:   at $\zeta = 0$, $\phi_z = 0$ (11.4-49)

One can see immediately from these equations and boundary conditions that the dimensionless velocity components ϕ_y and ϕ_z and the dimensionless temperature Θ will depend on η and ζ and also on the Prandtl number, Pr. Since the flow is usually very slow in free convection, the terms in which Pr appears will generally be rather small; setting them equal to zero would correspond to the "creeping flow assumption." Hence we expect that the dependence of the solution on the Prandtl number will be weak.

The average heat flux from one side of the plate may be written as

$$q_{\text{avg}} = \frac{1}{H}\int_0^H \left(-k\frac{\partial T}{\partial y}\right)\bigg|_{y=0} dz \tag{11.4-50}$$

The integral may now be written in terms of the dimensionless quantities

$$
\begin{aligned}
q_{\text{avg}} &= k(T_0 - T_1)\left(\frac{B}{\mu\alpha H}\right)^{1/4} \cdot \int_0^1 \left(-\frac{\partial \Theta}{\partial \eta}\right)\bigg|_{\eta=0} d\zeta \\[2mm]
&= k(T_0 - T_1)\left(\frac{B}{\mu\alpha H}\right)^{1/4} \cdot C \\[2mm]
&= C \cdot \frac{k}{H}(T_0 - T_1)\left(\left(\frac{\hat{C}_p \mu}{k}\right)\left(\frac{\bar{\rho}^2 g\bar{\beta}(T_0 - T_1)H^3}{\mu^2}\right)\right)^{1/4} \\[2mm]
&= C \cdot \frac{k}{H}(T_0 - T_1)(\text{GrPr})^{1/4}
\end{aligned}
\tag{11.4-51}
$$

in which the grouping Ra = GrPr is referred to as the *Rayleigh number*. Because Θ is a function of η, ζ, and Pr, the derivative $\partial\Theta/\partial\eta$ is also a function of η, ζ, and Pr. Then $\partial\Theta/\partial\eta$, evaluated at $\eta = 0$, depends only on ζ and Pr. The definite integral over ζ is thus a function of Pr. From the remarks made earlier, we can infer that this function, called C, will be only a weak function of the Prandtl number—that is, nearly a constant.

The preceding analysis shows that, even without solving the partial differential equations, we can predict that the average heat flux is proportional to the $\frac{5}{4}$-power of the temperature difference $(T_0 - T_1)$ and inversely proportional to the $\frac{1}{4}$-power of H. Both predictions have been confirmed by experiment. The only thing we could not do was to find C as a function of Pr.

To determine that function, we have to make experimental measurements or solve Eqs. 11.4-44 to 49. In 1881, Lorenz[3] obtained an approximate solution to these equations and found $C = 0.548$. Later, more refined calculations[4] gave the following dependence of C on Pr:

Pr	0.73 (air)	1	10	100	1000	∞
C	0.518	0.535	0.620	0.653	0.665	0.670

These values of C are nearly in exact agreement with the best experimental measurements in the laminar flow range (i.e., for GrPr $< 10^9$).[5]

EXAMPLE 11.4-6

Adiabatic Frictionless Processes in an Ideal Gas

Develop equations for the relationship of local pressure to density or temperature in a stream of ideal gas in which the momentum flux τ and the heat flux $\mathbf{q}$ are negligible.

SOLUTION

With τ and $\mathbf{q}$ neglected, the equation of energy [Eq. (J) in Table 11.4-1] may be rewritten as

$$\rho \hat{C}_p \frac{DT}{Dt} = \left(\frac{\partial \ln \hat{V}}{\partial \ln T}\right)_p \frac{Dp}{Dt} \tag{11.4-52}$$

For an ideal gas, $p\hat{V} = RT/M$, where M is the molecular weight of the gas, and Eq. 11.4-52 becomes

$$\rho \hat{C}_p \frac{DT}{Dt} = \frac{Dp}{Dt} \tag{11.4-53}$$

Dividing this equation by p and assuming the molar heat capacity $\tilde{C}_p = M\hat{C}_p$ to be constant, we can again use the ideal gas law to get

$$\frac{D}{Dt}\left(\frac{\tilde{C}_p}{R} \ln T - \ln p\right) = 0 \tag{11.4-54}$$

Hence the quantity in parentheses is a constant along the path of a fluid element, as is its antilogarithm, so that we have

$$T^{\tilde{C}_p/R} p^{-1} = \text{constant} \tag{11.4-55}$$

This relation applies to all thermodynamic states p, T that a fluid element encounters as it moves along with the fluid.

Introducing the definition $\gamma = \hat{C}_p/\hat{C}_V$ and the ideal gas relations $\tilde{C}_p - \tilde{C}_V = R$ and $p = \rho RT/M$, one obtains the related expressions

$$p^{(\gamma-1)/\gamma} T^{-1} = \text{constant} \tag{11.4-56}$$

and

$$p\rho^{-\gamma} = \text{constant} \tag{11.4-57}$$

These last three equations find frequent use in the study of frictionless adiabatic processes in ideal gas dynamics. Equation 11.4-57 is a famous relation well worth remembering.

---

[3] L. Lorenz, *Wiedemann's Ann. der Physik u. Chemie*, **13**, 422–447, 582–606 (1881). See also U. Grigull, *Die Grundgesetze der Wärmeübertragung*, Springer-Verlag, Berlin, 3rd edition (1955), pp. 263–269.

[4] See S. Whitaker, *Fundamental Principles of Heat Transfer*, Krieger, Malabar Fla. (1977), §5.11. The limiting case of Pr → ∞ has been worked out numerically by E. J. LeFevre [Heat Div. Paper 113, Dept. Sci. and Ind. Res., Mech. Engr. Lab. (Great Britain), Aug. 1956] and it was found that

$$\left.\frac{\partial \Theta}{\partial \eta}\right|_{\eta=0} = \frac{0.5028}{\zeta^{1/4}} \qquad \left.\frac{\partial \phi_z}{\partial \eta}\right|_{\eta=0} = \frac{1.16}{\zeta^{1/4}} \tag{11.4-51a, b}$$

Equation 11.4-51a corresponds to the value $C = 0.670$ above. This result has been verified experimentally by C. R. Wilke, C. W. Tobias, and M. Eisenberg, *J. Electrochem. Soc.*, **100**, 513–523 (1953), for the analogous mass transfer problem.

[5] For an analysis of free convection in three-dimensional creeping flow, see W. E. Stewart, *Int. J. Heat and Mass Transfer*, **14**, 1013–1031 (1971).

When the momentum flux τ and the heat flux $\mathbf{q}$ are zero, there is no change in entropy following an element of fluid (see Eq. 11D.1-3). Hence the derivative $d \ln p/d \ln T = \gamma/(\gamma - 1)$ following the fluid motion has to be understood to mean $(\partial \ln p/\partial \ln T)_S = \gamma/(\gamma - 1)$. This equation is a standard formula from equilibrium thermodynamics.

EXAMPLE 11.4-7

One-Dimensional Compressible Flow: Velocity, Temperature, and Pressure Profiles in a Stationary Shock Wave

We consider here the adiabatic expansion[6-10] of an ideal gas through a convergent–divergent nozzle under such conditions that a stationary shock wave is formed. The gas enters the nozzle from a reservoir, where the pressure is p_0, and discharges to the atmosphere, where the pressure is p_a. In the absence of a shock wave, the flow through a well-designed nozzle is virtually frictionless (hence *isentropic* for the adiabatic situation being considered). If, in addition, p_a/p_0 is sufficiently small, it is known that the flow is essentially sonic at the throat (the region of minimum cross section) and is supersonic in the divergent portion of the nozzle. Under these conditions the pressure will continually *decrease*, and the velocity will *increase* in the direction of the flow, as indicated by the curves in Fig. 11.4-4.

However, for any nozzle design there is a range of p_a/p_0 for which such an isentropic flow produces a pressure less than p_a at the exit. Then the isentropic flow becomes unstable. The simplest of many possibilities is a stationary normal shock wave, shown schematically in the Fig. 11.4-4 as a pair of closely spaced parallel lines. Here the velocity falls off very rapidly

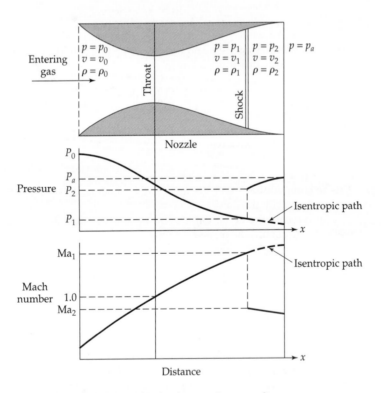

Fig. 11.4-4. Formation of a shock wave in a nozzle.

[6] H. W. Liepmann and A. Roshko, *Elements of Gas Dynamics*, Wiley, New York (1957), §§5.4 and 13.13.

[7] J. O. Hirschfelder, C. F. Curtiss, and R. B. Bird, *Molecular Theory of Gases and Liquids*, Wiley, New York, 2nd corrected printing (1964), pp. 791–797.

[8] M. Morduchow and P. A. Libby, *J. Aeronautical Sci.*, **16**, 674–684 (1948).

[9] R. von Mises, *J. Aeronautical Sci.*, **17**, 551–554 (1950).

[10] G. S. S. Ludford, *J. Aeronautical Sci.*, **18**, 830–834 (1951).

to a subsonic value, while both the pressure and the density rise. These changes take place in an extremely thin region, which may therefore be considered locally one-dimensional and laminar, and they are accompanied by a very substantial dissipation of mechanical energy. Viscous dissipation and heat conduction effects are thus concentrated in an extremely small region of the nozzle, and it is the purpose of the example to explore the fluid behavior there. For simplicity the shock wave will be considered normal to the fluid streamlines; in practice, much more complicated shapes are often observed. The velocity, pressure, and temperature just upstream of the shock can be calculated and will be considered as known for the purposes of this example.

Use the three equations of change to determine the conditions under which a shock wave is possible and to find the velocity, temperature, and pressure distributions in such a shock wave. Assume steady, one-dimensional flow of an ideal gas, neglect the dilatational viscosity κ, and ignore changes of μ, k, and $\hat{C}_p$ with temperature and pressure.

SOLUTION

The equations of change in the neighborhood of the stationary shock wave may be simplified to

Continuity:
$$\frac{d}{dx}\rho v_x = 0 \tag{11.4-58}$$

Motion:
$$\rho v_x \frac{dv_x}{dx} = -\frac{dp}{dx} + \frac{4}{3}\frac{d}{dx}\left(\mu\frac{dv_x}{dx}\right) \tag{11.4-59}$$

Energy:
$$\rho\hat{C}_p v_x \frac{dT}{dx} = \frac{d}{dx}\left(k\frac{dT}{dx}\right) + v_x\frac{dp}{dx} + \frac{4}{3}\mu\left(\frac{dv_x}{dx}\right)^2 \tag{11.4-60}$$

The energy equation is in the form of Eq. J of Table 11.4-1, written for an ideal gas in a steady-state situation.

The equation of *continuity* may be integrated to give

$$\rho v_x = \rho_1 v_1 \tag{11.4-61}$$

in which ρ_1 and v_1 are quantities evaluated a short distance upstream from the shock.

In the *energy* equation we eliminate ρv_x by use of Eq. 11.4-61 and dp/dx by using the equation of motion to get (after some rearrangement)

$$\rho_1\hat{C}_p v_1 \frac{dT}{dx} = \frac{d}{dx}\left(k\frac{dT}{dx}\right) - \rho_1 v_1 \frac{d}{dx}\left(\frac{1}{2}v_x^2\right) + \frac{4}{3}\mu\frac{d}{dx}\left(v_x\frac{dv_x}{dx}\right) \tag{11.4-62}$$

We next move the second term on the right side over to the left side and divide the entire equation by $\rho_1 v_1$. Then each term is integrated with respect to x to give

$$\hat{C}_p T + \tfrac{1}{2}v_x^2 = \frac{k}{\rho_1\hat{C}_p v_1}\frac{d}{dx}(\hat{C}_p T + (\tfrac{4}{3}\text{Pr})\tfrac{1}{2}v_x^2) + C_{\text{I}} \tag{11.4-63}$$

in which C_{I} is a constant of integration and $\text{Pr} = \hat{C}_p\mu/k$. For most gases Pr is between 0.65 and 0.85, with an average value close to 0.75. Therefore, to simplify the problem we set Pr equal to $\frac{3}{4}$. Then Eq. 11.4-63 becomes a first-order, linear ordinary differential equation, for which the solution is

$$\hat{C}_p T + \tfrac{1}{2}v_x^2 = C_{\text{I}} + C_{\text{II}}\exp[(\rho_1\hat{C}_p v_1/k)x] \tag{11.4-64}$$

Since $\hat{C}_p T + \tfrac{1}{2}v_x^2$ cannot increase without limit in the positive x direction, the second integration constant C_{II} must be zero. The first integration constant is evaluated from the upstream conditions, so that

$$\hat{C}_p T + \tfrac{1}{2}v_x^2 = \hat{C}_p T_1 + \tfrac{1}{2}v_1^2 \tag{11.4-65}$$

Of course, if we had not chosen Pr to be $\frac{3}{4}$, a numerical integration of Eq. 11.4-63 would have been required.

Next we substitute the integrated continuity equation into the equation of *motion* and integrate once to obtain

$$\rho_1 v_1 v_x = -p + \frac{4}{3}\mu\frac{dv_x}{dx} + C_{III} \tag{11.4-66}$$

Evaluation of the constant C_{III} from upstream conditions, where $dv_x/dx = 0$, gives $C_{III} = \rho_1 v_1^2 + p_1 = \rho_1[v_1^2 + (RT_1/M)]$. We now multiply both sides by v_x and divide by $\rho_1 v_1$. Then, with the help of the ideal gas law, $p = \rho RT/M$, and Eqs. 11.4-61 and 65, we may eliminate p from Eq. 11.4-60 to obtain a relation containing only v_x and x as variables:

$$\frac{4}{3}\frac{\mu}{\rho_1 v_1}v_x\frac{dv_x}{dx} = \frac{\gamma+1}{2\gamma}v_x^2 + \frac{\gamma-1}{\gamma}C_I - \frac{C_{III}}{\rho_1 v_1}v_x \tag{11.4-67}$$

This equation can, after considerable rearrangement, be rewritten in terms of dimensionless variables:

$$\phi\frac{d\phi}{d\xi} = \beta\mathrm{Ma}_1(\phi - 1)(\phi - \alpha) \tag{11.4-68}$$

The relevant dimensionless quantities are

$$\phi = \frac{v_x}{v_1} = \text{dimensionless velocity} \tag{11.4-69}$$

$$\xi = \frac{x}{\lambda} = \text{dimensionless coordinate} \tag{11.4-70}$$

$$\mathrm{Ma}_1 = \frac{v_1}{\sqrt{\gamma RT_1/M}} = \text{Mach number at the upstream condition} \tag{11.4-71}$$

$$\alpha = \frac{\gamma-1}{\gamma+1} + \frac{2}{\gamma+1}\frac{1}{\mathrm{Ma}_1^2} \tag{11.4-72}$$

$$\beta = \tfrac{9}{8}(\gamma+1)\sqrt{\pi/8\gamma} \tag{11.4-73}$$

The reference length λ is the mean free path defined in Eq. 1.4-3 (with d^2 eliminated by use of Eq. 1.4-9):

$$\lambda = \frac{3\mu_1}{\rho_1}\sqrt{\frac{\pi M}{8RT_1}} \tag{11.4-74}$$

We may integrate Eq. 11.4-68 to obtain

$$\frac{1-\phi}{(\phi-\alpha)^\alpha} = \exp[\beta\mathrm{Ma}_1(1-\alpha)(\xi-\xi_0)] \qquad (\alpha < \phi < 1) \tag{11.4-75}$$

This equation describes the dimensionless velocity distribution $\phi(\xi)$ containing an integration constant $\xi_0 = x_0/\lambda$, which specifies the position of the shock wave in the nozzle; here ξ_0 is considered to be known. It can be seen from the plot of Eq. 11.4-85 in Fig. 11.4-5 that shock waves

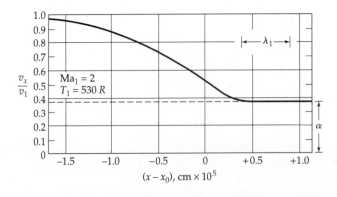

Fig. 11.4-5. Velocity distribution in a stationary shock wave.

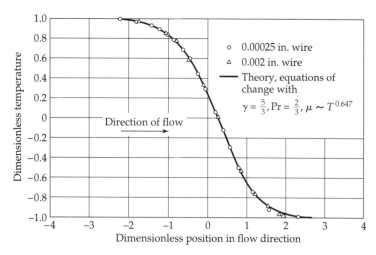

Fig. 11.4-6. Semi-log plot of the temperature profile through a shock wave, for helium with $Ma_1 = 1.82$. The experimental values were measured with a resistance-wire thermometer. [Adapted from H. W. Liepmann and A. Roshko, *Elements of Gas Dynamics*, Wiley, New York (1957), p. 333]

are indeed very thin. The temperature and pressure distributions may be determined from Eq. 11.4-75 and Eqs. 11.4-65 and 66. Since ϕ must approach unity as $\xi \to -\infty$, the constant α is less than 1. This can be true only if $Ma_1 > 1$—that is, if the upstream flow is supersonic. It can also be seen that for very large positive ξ, the dimensionless velocity ϕ approaches α. The Mach number Ma_1 is defined as the ratio of v_1 to the velocity of sound at T_1 (see Problem 11C.1).

In the above development we chose the Prandtl number Pr to be $\frac{3}{4}$, but the solution has been extended[8] to include other values of Pr as well as the temperature variation of the viscosity.

The tendency of a gas in supersonic flow to revert spontaneously to subsonic flow is important in wind tunnels and in the design of high-velocity systems—for example, in turbines and rocket engines. Note that the changes taking place in shock waves are irreversible and that, since the velocity gradients are so very steep, a considerable amount of mechanical energy is dissipated.

In view of the thinness of the predicted shock wave, one may question the applicability of the analysis given here, based on the continuum equations of change. Therefore it is desirable to compare the theory with experiment. In Fig. 11.4-6 experimental temperature measurements for a shock wave in helium are compared with the theory for $\gamma = \frac{5}{3}$, $Pr = \frac{2}{3}$, and $\mu \sim T^{0.647}$. We can see that the agreement is excellent. Nevertheless we should recognize that this is a simple system, inasmuch as helium is monatomic, and therefore internal degrees of freedom are not involved. The corresponding analysis for a diatomic or polyatomic gas would need to consider the exchange of energy between translational and internal degrees of freedom, which typically requires hundreds of collisions, broadening the shock wave considerably. Further discussion of this matter can be found in Chapter 11 of Ref. 7.

§11.5   DIMENSIONAL ANALYSIS OF THE EQUATIONS OF CHANGE FOR NONISOTHERMAL SYSTEMS

Now that we have shown how to use the equations of change for nonisothermal systems to solve some representative heat transport problems, we discuss the dimensional analysis of these equations.

Just as the dimensional analysis discussion in §3.7 provided an introduction for the discussion of friction factors in Chapter 6, the material in this section provides the background needed for the discussion of heat transfer coefficient correlations in Chapter 14. As in Chapter 3, we write the equations of change and boundary conditions in dimensionless form. In this way we find some dimensionless parameters that can be used to characterize nonisothermal flow systems.

We shall see, however, that the analysis of nonisothermal systems leads us to a larger number of dimensionless groups than we had in Chapter 3. As a result, greater reliance has to be placed on judicious simplifications of the equations of change and on carefully chosen physical models. Examples of the latter are the Boussinesq equation of motion for free convection (§11.3) and the laminar boundary layer equations (§12.4).

As in §3.7, for the sake of simplicity we restrict ourselves to a fluid with constant μ, k, and $\hat{C}_p$. The density is taken to be $\rho = \bar{\rho} - \bar{\rho}\bar{\beta}(T - \bar{T})$ in the $\rho\mathbf{g}$ term in the equation of motion, and $\rho = \bar{\rho}$ everywhere else (the "Boussinesq approximation"). The equations of change then become with $p + \bar{\rho}gh$ expressed as $\mathscr{P}$,

Continuity:
$$(\nabla \cdot \mathbf{v}) = 0 \tag{11.5-1}$$

Motion:
$$\bar{\rho}\frac{D\mathbf{v}}{Dt} = -\nabla\mathscr{P} + \mu\nabla^2\mathbf{v} + \bar{\rho}\mathbf{g}\bar{\beta}(T - \bar{T}) \tag{11.5-2}$$

Energy:
$$\bar{\rho}\hat{C}_p\frac{DT}{Dt} = k\nabla^2T + \mu\Phi_v \tag{11.5-3}$$

We now introduce quantities made dimensionless with the characteristic quantities (subscript 0 or 1) as follows:

$$\check{x} = \frac{x}{l_0} \qquad \check{y} = \frac{y}{l_0} \qquad \check{z} = \frac{z}{l_0} \qquad \check{t} = \frac{v_0 t}{l_0} \tag{11.5-4}$$

$$\check{\mathbf{v}} = \frac{\mathbf{v}}{v_0} \qquad \check{\mathscr{P}} = \frac{\mathscr{P} - \mathscr{P}_0}{\bar{\rho}v_0^2} \qquad \check{T} = \frac{T - T_0}{T_1 - T_0} \tag{11.5-5}$$

$$\check{\Phi}_v = \left(\frac{l_0}{v_0}\right)^2\Phi_v \qquad \check{\nabla} = l_0\nabla \qquad \frac{D}{D\check{t}} = \left(\frac{l_0}{v_0}\right)\frac{D}{Dt} \tag{11.5-6}$$

Here l_0, v_0, and $\mathscr{P}_0$ are the reference quantities introduced in §3.7, and T_0 and T_1 are temperatures appearing in the boundary conditions. In Eq. 11.5-2 the value $\bar{T}$ is the temperature around which the density ρ was expanded.

In terms of these dimensionless variables, the equations of change in Eqs. 11.5-1 to 3 take the forms

Continuity:
$$(\check{\nabla} \cdot \check{\mathbf{v}}) = 0 \tag{11.5-7}$$

Motion:
$$\frac{D\check{\mathbf{v}}}{D\check{t}} = -\check{\nabla}\check{\mathscr{P}} + \left[\frac{\mu}{l_0v_0\bar{\rho}}\right]\check{\nabla}^2\check{\mathbf{v}} - \left[\frac{gl_0\bar{\beta}(T_1 - T_0)}{v_0^2}\right]\left(\frac{\mathbf{g}}{g}\right)(\check{T} - \check{\bar{T}}) \tag{11.5-8}$$

Energy:
$$\frac{D\check{T}}{D\check{t}} = \left[\frac{k}{l_0v_0\bar{\rho}\hat{C}_p}\right]\check{\nabla}^2\check{T} + \left[\frac{\mu v_0}{l_0\bar{\rho}\hat{C}_p(T_1 - T_0)}\right]\check{\Phi}_v \tag{11.5-9}$$

The characteristic velocity can be chosen in several ways, and the consequences of the choices are summarized in Table 11.5-1. The dimensionless groups appearing in Eqs. 11.5-8 and 9, along with some combinations of these groups, are summarized in Table 11.5-2. Further dimensionless groups may arise in the boundary conditions or in the equation of state. The Froude and Weber numbers have already been introduced in §3.7, and the Mach number in Ex. 11.4-7.

We already saw in Chapter 10 how several dimensionless groups appeared in the solution of nonisothermal problems. Here we have seen that the same groupings appear naturally when the equations of change are made dimensionless. These dimensionless groups are used widely in correlations of heat transfer coefficients.

Table 11.5-1   Dimensionless Groups in Equations 11.5-7, 8, and 9

Special cases →	Forced convection	Intermediate	Free convection (A)	Free convection (B)
Choice for v_0 →	v_0	v_0	ν/l_0	α/l_0
$\left[\dfrac{\mu}{l_0 v_0 \bar{\rho}}\right]$	$\dfrac{1}{\mathrm{Re}}$	$\dfrac{1}{\mathrm{Re}}$	1	Pr
$\left[\dfrac{g l_0 \bar{\beta}(T_1 - T_0)}{v_0^2}\right]$	Neglect	$\dfrac{\mathrm{Gr}}{\mathrm{Re}^2}$	Gr	GrPr^2
$\left[\dfrac{k}{l_0 v_0 \bar{\rho}\hat{C}_p}\right]$	$\dfrac{1}{\mathrm{RePr}}$	$\dfrac{1}{\mathrm{RePr}}$	$\dfrac{1}{\mathrm{Pr}}$	1
$\left[\dfrac{\mu v_0}{l_0 \bar{\rho}\hat{C}_p(T_1 - T_0)}\right]$	$\dfrac{\mathrm{Br}}{\mathrm{RePr}}$	$\dfrac{\mathrm{Br}}{\mathrm{RePr}}$	Neglect	Neglect

Notes:

[a] For forced convection and forced-plus-free ("intermediate") convection, v_0 is generally taken to be the approach velocity (for flow around submerged objects) or an average velocity in the system (for flow in conduits).

[b] For free convection there are two standard choices for v_0, labeled as A and B. In §10.9, Case A arises naturally. Case B proves convenient if the assumption of creeping flow is appropriate, so that $D\check{v}/D\check{t}$ can be neglected (see Example 11.5-2). Then a new dimensionless pressure difference $\bar{\mathscr{P}} = \mathrm{Pr}\check{\mathscr{P}}$, different from $\check{\mathscr{P}}$ in Eq. 3.7-4, can be introduced, so that when the equation of motion is divided by Pr, the only dimensionless group appearing in the equation is GrPr. Note that in Case B, no dimensionless groups appear in the equation of energy.

It is sometimes useful to think of the dimensionless groups as ratios of various forces or effects in the system, as shown in Table 11.5-3. For example, the inertial term in the equation of motion is $\rho[\mathbf{v}\cdot\nabla\mathbf{v}]$ and the viscous term is $\mu\nabla^2\mathbf{v}$. To get "typical" values of these terms, replace the variables by the characteristic "yardsticks" used in constructing dimensionless variables. Hence replace $\rho[\mathbf{v}\cdot\nabla\mathbf{v}]$ by $\rho v_0^2/l_0$, and replace $\mu\nabla^2\mathbf{v}$ by $\mu v_0/l_0^2$ to get rough orders of magnitude. The ratio of these two terms then gives the Reynolds number, as shown in the table. The other dimensionless groups are obtained in similar fashion.

Table 11.5-2   Dimensionless Groups Used in Nonisothermal Systems

$\mathrm{Re} = [\![l_0 v_0 \rho/\mu]\!] = [\![l_0 v_0/\nu]\!]$	= *Reynolds number*
$\mathrm{Pr} = [\![\hat{C}_p\mu/k]\!] = [\![\nu/\alpha]\!]$	= *Prandtl number*
$\mathrm{Gr} = [\![g\beta(T_1 - T_0)l_0^3/\nu^2]\!]$	= *Grashof number*
$\mathrm{Br} = [\![\mu v_0^2/k(T_1 - T_0)]\!]$	= *Brinkman number*
$\mathrm{P\acute{e}} = \mathrm{RePr}$	= *Péclet number*
$\mathrm{Ra} = \mathrm{GrPr}$	= *Rayleigh number*
$\mathrm{Ec} = \mathrm{Br/Pr}$	= *Eckert number*

Table 11.5-3   Physical Interpretation of Dimensionless Groups

$$\text{Re} = \frac{\rho v_0^2 / l_0}{\mu v_0 / l_0^2} = \frac{\text{inertial force}}{\text{viscous force}}$$

$$\text{Fr} = \frac{\rho v_0^2 / l_0}{\rho g} = \frac{\text{inertial force}}{\text{gravity force}}$$

$$\frac{\text{Gr}}{\text{Re}^2} = \frac{\rho g \beta (T_1 - T_0)}{\rho v_0^2 / l_0} = \frac{\text{buoyant force}}{\text{inertial force}}$$

$$\text{Pé} = \text{RePr} = \frac{\rho \hat{C}_p v_0 (T_1 - T_0) / l_0}{k(T_1 - T_0) / l_0^2} = \frac{\text{heat transport by convection}}{\text{heat transport by conduction}}$$

$$\text{Br} = \frac{\mu (v_0 / l_0)^2}{k(T_1 - T_0) / l_0^2} = \frac{\text{heat production by viscous dissipation}}{\text{heat transport by conduction}}$$

A low value for the Reynolds number means that viscous forces are large in comparison with inertial forces. A low value of the Brinkman number indicates that the heat produced by viscous dissipation can be transported away quickly by heat conduction. When Gr/Re^2 is large, the buoyant force is important in determining the flow pattern.

Since dimensional analysis is an art requiring judgment and experience, we give three illustrative examples. In the first two we analyze forced and free convection in simple geometries. In the third we discuss scale-up problems in a relatively complex piece of equipment.

EXAMPLE 11.5-1

Temperature Distribution about a Long Cylinder

It is desired to predict the temperature distribution in a gas flowing about a long, internally cooled cylinder (system I) from experimental measurements on a one-quarter scale model (system II). If possible the same fluid should be used in the model as in the full-scale system. The system, shown in Fig. 11.5-1, is the same as that in Example 3.7-1 except that it is now nonisothermal. The fluid approaching the cylinder has a speed v_∞ and a temperature T_∞, and the cylinder surface is maintained at T_0, for example, by the boiling of a refrigerant contained within it.

Show by means of dimensional analysis how suitable experimental conditions can be chosen for the model studies. Perform the dimensional analysis for the "intermediate case" in Table 11.5-1.

SOLUTION

The two systems, I and II, are geometrically similar. To ensure dynamical similarity, as pointed out in §3.7, the dimensionless differential equations and boundary conditions must be the same, and the dimensionless groups appearing in them must have the same numerical values.

Here we choose the characteristic length to be the diameter D of the cylinder, the characteristic velocity to be the approach velocity v_∞ of the fluid, the characteristic pressure to be the pressure at $x = -\infty$ and $y = 0$, and the characteristic temperatures to be the temperature T_∞ of the approaching fluid and the temperature T_0 of the cylinder wall. These characteristic quantities will carry a label I or II corresponding to the system being described.

Both systems are described by the dimensionless differential equations given in Eqs. 11.5-7 to 9, and by boundary conditions

B.C. 1	as $\check{x}^2 + \check{y}^2 \to \infty$,	$\check{\mathbf{v}} \to \boldsymbol{\delta}_x$,	$\check{T} \to 1$	(11.5-10)
B.C. 2	at $\check{x}^2 + \check{y}^2 = \frac{1}{4}$,	$\check{\mathbf{v}} = 0$,	$\check{T} = 0$	(11.5-11)
B.C. 3	at $\check{x} \to -\infty$ and $\check{y} = 0$,		$\check{\mathscr{P}} \to 0$	(11.5-12)

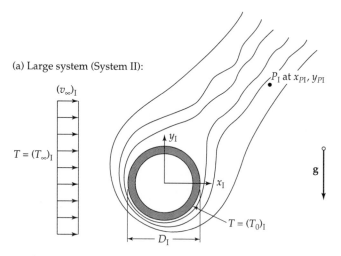

(a) Large system (System II):

(b) Small system (System II):

Fig. 11.5-1. Temperature profiles about long heated cylinders. The contour lines in the two figures represent surfaces of constant temperature.

in which $\check{T} = (T - T_0)/(T_\infty - T_0)$. For this simple geometry, the boundary conditions contain no dimensionless groups. Therefore, the requirement that the differential equations and boundary conditions in dimensionless form be identical is that the following dimensionless groups be equal in the two systems: $Re = Dv_\infty \rho/\mu$, $Pr = \hat{C}_p \mu/k$, $Br = \mu v_\infty^2/k(T_\infty - T_0)$, and $Gr = \rho^2 g \beta(T_\infty - T_0)D^3/\mu^2$. In the latter group we use the ideal gas expression $\beta = 1/T$.

To obtain the necessary equality for the four governing dimensionless groups, we may use different values of the four disposable parameters in the two systems: the approach velocity v_∞, the fluid temperature T_∞, the approach pressure $\mathcal{P}_\infty$, and the cylinder temperature T_0.

The similarity requirements are then (for $D_I = 4D_{II}$):

Equality of Pr

$$\frac{\nu_I}{\nu_{II}} = \frac{\alpha_I}{\alpha_{II}}$$

(11.5-13)

Equality of Re

$$\frac{\nu_I}{\nu_{II}} = 4\frac{v_{\infty I}}{v_{\infty II}}$$

(11.5-14)

Equality of Gr

$$\left(\frac{\nu_I}{\nu_{II}}\right)^2 = 64\frac{T_{\infty II}}{T_{\infty I}}\frac{(T_\infty - T_0)_I}{(T_\infty - T_0)_{II}}$$

(11.5-15)

Equality of Br

$$\left(\frac{Pr_I}{Pr_{II}}\right)\left(\frac{v_{\infty I}}{v_{\infty II}}\right)^2 = \frac{\hat{C}_{pI}}{\hat{C}_{pII}}\frac{(T_\infty - T_0)_I}{(T_\infty - T_0)_{II}}$$

(11.5-16)

Here $\nu = \mu/\rho$ is the kinematic viscosity and $\alpha = k/\rho\hat{C}_p$ is the thermal diffusivity.

The simplest way to satisfy Eq. 11.5-13 is to use the same fluid at the same approach pressure $\mathcal{P}_\infty$ and temperature T_∞ in the two systems. If that is done, Eq. 11.5-14 requires that the approach velocity in the small model (II) be four times that used in the full-scale system (I). If the fluid velocity is moderately large and the temperature differences small, the equality of Pr and Re in the two systems provides a sufficient approximation to dynamic similarity. This is the limiting case of forced convection with negligible viscous dissipation.

If, however, the temperature differences $T_\infty - T_0$ are large, free-convection effects may be appreciable. Under these conditions, according to Eq. 11.5-15, temperature differences in the model must be 64 times those in the large system to ensure similarity.

From Eq. 11.5-16 it may be seen that such a ratio of temperature differences will not permit equality of the Brinkman number. For the latter a ratio of 16 would be needed. This conflict will not normally arise, however, as free-convection and viscous heating effects are seldom important simultaneously. Free-convection effects arise in low-velocity systems, whereas viscous heating occurs to a significant degree only when velocity gradients are very large.

EXAMPLE 11.5-2 *Free Convection in a Horizontal Fluid Layer; Formation of Bénard Cells*	We wish to investigate the free-convection motion in the system shown in Fig. 11.5-2. It consists of a thin layer of fluid between two horizontal parallel plates, the lower one at temperature T_0, and the upper one at T_1, with $T_1 < T_0$. In the absence of fluid motion, the conductive heat flux will be the same for all z, and a nearly uniform temperature gradient will be established at steady state. This temperature gradient will in turn cause a density gradient. If the density decreases with increasing z, the system will clearly be stable, but if it increases a potentially unstable situation occurs. It appears possible in this latter case that any chance disturbance may cause the more dense fluid to move downward and displace the lighter fluid beneath it. If the temperatures of the top and bottom surfaces are maintained constant, the result may be a continuing free-convection motion. This motion will, however, be opposed by viscous forces and may, therefore, occur only if the temperature difference tending to cause it is greater than some critical minimum value. Determine by means of dimensional analysis the functional dependence of this fluid motion and the conditions under which it may be expected to arise.

SOLUTION

The system is described by Eqs. 11.5-1 to 3 along with the following boundary conditions:

B.C. 1: at $z = 0$, $\mathbf{v} = 0$ $T = T_0$ (11.5-17)

B.C. 2: at $z = h$, $\mathbf{v} = 0$ $T = T_1$ (11.5-18)

B.C. 3: at $r = R$, $\mathbf{v} = 0$ $\partial T/\partial r = 0$ (11.5-19)

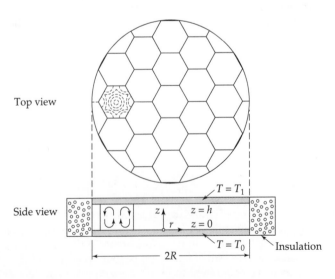

Top view

Side view

$T = T_1$

$z = h$

$z = 0$

$T = T_0$

Insulation

2R

Fig. 11.5-2. Bénard cells formed in the region between two horizontal parallel plates, with the bottom plate at a higher temperature than the upper one. If the Rayleigh number exceeds a certain critical value, the system becomes unstable and hexagonal Bénard cells are produced.

We now restate the problem in dimensionless form, using $l_0 = h$. We use the dimensionless quantities listed under Case B in Table 11.5-1, and we select the reference temperature $\overline{T}$ to be $\frac{1}{2}(T_0 + T_1)$, so that

Continuity:

$$(\breve{\nabla} \cdot \breve{\mathbf{v}}) = 0 \tag{11.5-20}$$

Motion:

$$\frac{D\breve{\mathbf{v}}}{D\breve{t}} = -\breve{\nabla}\breve{\mathscr{P}} + \mathrm{Pr}\breve{\nabla}^2\breve{\mathbf{v}} - \mathrm{GrPr}^2\left(\frac{\mathbf{g}}{g}\right)(\breve{T} - \tfrac{1}{2}) \tag{11.5-21}$$

Energy:

$$\frac{D\breve{T}}{D\breve{t}} = \breve{\nabla}^2\breve{T} \tag{11.5-22}$$

with dimensionless boundary conditions

B.C. 1:   at $\breve{z} = 0$, $\breve{\mathbf{v}} = 0$ $\breve{T} = 0$ (11.5-23)

B.C. 2:   at $\breve{z} = 1$, $\breve{\mathbf{v}} = 0$ $\breve{T} = 1$ (11.5-24)

B.C. 3:   at $\breve{r} = R/h$, $\breve{\mathbf{v}} = 0$ $\partial\breve{T}/\partial\breve{r} = 0$ (11.5-25)

If the above dimensionless equations could be solved along with the dimensionless boundary conditions, we would find that the velocity and temperature profiles would depend only on Gr, Pr, and R/h. Furthermore, the larger the ratio R/h is, the less prominent its effect will be, and in the limit of extremely large horizontal plates, the system behavior will depend solely on Gr and Pr.

If we consider only steady creeping flows, then the term $D\breve{\mathbf{v}}/D\breve{t}$ may be set equal to zero. Then we define a new dimensionless pressure difference as $\tilde{\mathscr{P}} = \mathrm{Pr}\breve{\mathscr{P}}$. With the left side of Eq. 11.5-21 equal to zero, we may now divide by Pr and the resulting equation contains only one dimensionless group, namely the Rayleigh number[1] $\mathrm{Ra} = \mathrm{GrPr} = \rho^2 g\beta(T_1 - T_0)h^3\hat{C}_p/\mu k$, whose value will determine the behavior of the system. This illustrates how one may reduce the number of dimensionless groups that are needed to describe a nonisothermal flow system.

The preceding analysis suggests that there may be a critical value of the Rayleigh number, and when this critical value is exceeded, fluid motion will occur. This suggestion has been amply confirmed experimentally[2,3] and the critical Rayleigh number has been found to be 1700 ± 51 for $R/h \gg 1$. For Rayleigh numbers below the critical value, the fluid is stationary, as evidenced by the observation that the heat flux across the liquid layer is the same as that predicted for conduction through a static fluid: $q_z = k(T_0 - T_1)/h$. As soon as the critical Rayleigh number is exceeded, however, the heat flux rises sharply, because of convective energy transport. An increase of the thermal conductivity reduces the Rayleigh number, thus moving Ra toward its stable range.

The assumption of creeping flow is a reasonable one for this system and is asymptotically correct when $\mathrm{Pr} \to \infty$. It is also very convenient, inasmuch as it allows analytic solutions of the relevant equations of change.[4] One such solution, which agrees well with experiment, is sketched qualitatively in Fig. 11.5-2. This flow pattern is cellular and hexagonal, with upflow at the center of each hexagon and downflow at the periphery. The units of this fascinating pattern are called *Bénard cells.*[5] The analytic solution also confirms the existence of a critical Rayleigh number. For the boundary conditions of this problem and very large R/h it has been calculated[4] to be 1708, which is in excellent agreement with the experimental result cited above.

[1] The Rayleigh number is named after Lord Rayleigh (J. W. Strutt), *Phil. Mag.*, (6) **32**, 529–546 (1916).

[2] P. L. Silveston, *Forsch. Ingenieur-Wesen*, **24**, 29–32, 59–69 (1958).

[3] S. Chandrasekhar, *Hydrodynamic and Hydromagnetic Instability*, Oxford University Press (1961); T. E. Faber, *Fluid Dynamics for Physicists*, Cambridge University Press (1995), §8.7.

[4] A. Pellew and R. V. Southwell, *Proc. Roy. Soc.*, **A176**, 312–343 (1940).

[5] H. Bénard, *Revue générale des sciences pures et appliquées*, **11**, 1261–1271, 1309–1328 (1900); *Annales de Chimie et de Physique*, **23**, 62–144 (1901).

Similar behavior is observed for other boundary conditions. If the upper plate of Fig. 11.5-2 is replaced by a liquid–gas interface, so that the surface shear stress in the liquid is negligible, cellular convection is predicted theoretically[3] for Rayleigh numbers above about 1101. A spectacular example of this type of instability occurs in the occasional spring "turnover" of water in northern lakes. If the lake water is cooled to near freezing during the winter, an adverse density gradient will occur as the surface waters warm toward 4°C, the temperature of maximum density for water.

In shallow liquid layers with free surfaces, instabilities can also arise from surface-tension gradients. The resulting surface stresses produce cellular convection superficially similar to that resulting from temperature gradients, and the two effects may be easily confused. Indeed, it appears that the steady flows first seen by Bénard, and ascribed to buoyancy effects, may actually have been produced by surface-tension gradients.[6]

EXAMPLE 11.5-3 *Surface Temperature of an Electrical Heating Coil*	An electrical heating coil of diameter D is being designed to keep a large tank of liquid above its freezing point. It is desired to predict the temperature that will be reached at the coil surface as a function of the heating rate Q and the tank surface temperature T_0. This prediction is to be made on the basis of experiments with a smaller, geometrically similar apparatus filled with the same liquid. Outline a suitable experimental procedure for making the desired prediction. Temperature dependence of the physical properties, other than the density, may be neglected. The entire heating coil surface may be assumed to be at a uniform temperature T_1.
SOLUTION	This is a free-convection problem, and we use the column labeled A in Table 11.5-1 for the dimensionless groups. From the equations of change and the boundary conditions, we know that the dimensionless temperature $\check{T} = (T - T_0)/(T_1 - T_0)$ must be a function of the dimensionless coordinates and depend on the dimensionless groups Pr and Gr. The total energy input rate through the coil surface is

$$Q = -k \int_S \left.\frac{\partial T}{\partial r}\right|_S dS \tag{11.5-26}$$

Here r is the coordinate measured outward from and normal to the coil surface, S is the surface area of the coil, and the temperature gradient is that of the fluid immediately adjacent to the coil surface. In dimensionless form this relation is

$$\frac{Q}{k(T_1 - T_0)D} = -\int_{\check{S}} \left.\frac{\partial \check{T}}{\partial \check{r}}\right|_{\check{S}} d\check{S} = \psi(\text{Pr}, \text{Gr}) \tag{11.5-27}$$

in which ψ is a function of $\text{Pr} = \hat{C}_p\mu/k$ and $\text{Gr} = \rho^2 g\beta(T_1 - T_0)D^3/\mu^2$. Since the large-scale and small-scale systems are geometrically similar, the dimensionless function $\check{S}$ describing the surface of integration will be the same for both systems and hence does not need to be included in the function ψ. Similarly, if we write the boundary conditions for temperature, velocity, and pressure at the coil and tank surfaces, we will obtain only size ratios that will be identical in the two systems.

We now note that the desired quantity $(T_1 - T_0)$ appears on both sides of Eq. 11.5-27. If we multiply both sides of the equation by the Grashof number, then $(T_1 - T_0)$ appears only on the right side:

$$\frac{Q\rho^2 g\beta D^2}{k\mu^2} = \text{Gr} \cdot \psi(\text{Pr}, \text{Gr}) \tag{11.5-28}$$

[6] C. V. Sternling and L. E. Scriven, *AIChE Journal*, **5**, 514–523 (1959); L. E. Scriven and C. V. Sternling, *J. Fluid Mech.*, **19**, 321–340 (1964).

In principle, we may solve Eq. 11.5-28 for Gr and obtain an expression for $(T_1 - T_0)$. Since we are neglecting the temperature dependence of physical properties, we may consider the Prandtl number constant for the given fluid and write

$$T_1 - T_0 = \frac{\mu^2}{\rho^2 g \beta D^3} \cdot \phi\left(\frac{Q\rho^2 g \beta D^2}{k\mu^2}\right) \qquad (11.5\text{-}29)$$

Here ϕ is an experimentally determinable function of the group $Q\rho^2 g \beta D^2 / k\mu^2$. We may then construct a plot of Eq. 11.5-29 from the experimental measurements of T_1, T_0, and D for the small-scale system, and the known physical properties of the fluid. This plot may then be used to predict the behavior of the large-scale system.

Since we have neglected the temperature dependence of the fluid properties, we may go even further. If we maintain the ratio of the Q values in the two systems equal to the inverse square of the ratio of the diameters, then the corresponding ratio of the values of $(T_1 - T_0)$ will be equal to the inverse cube of the ratio of the diameters.

QUESTIONS FOR DISCUSSION

1. Define energy, potential energy, kinetic energy, and internal energy. What common units are used for these?
2. How does one assign the physical meaning to the individual terms in Eqs. 11.1-7 and 11.2-1?
3. In getting Eq. 11.2-7 we used the relation $\tilde{C}_p - \tilde{C}_V = R$, which is valid for ideal gases. What is the corresponding equation for nonideal gases and liquids?
4. Summarize all the steps required in obtaining the equation of change for the temperature.
5. Compare and contrast forced convection and free convection, with regard to methods of problem solving, dimensional analysis, and occurrence in industrial and meteorological problems.
6. If a rocket nose cone were made of a porous material and a volatile liquid were forced slowly through the pores during reentry into the atmosphere, how would the cone surface temperature be affected and why?
7. What is Archimedes' principle, and how is it related to the term $\bar{\rho} \mathbf{g} \beta(T - \bar{T})$ in Eq. 11.3-2?
8. Would you expect to see Bénard cells while heating a shallow pan of water on a stove?
9. When, if ever, can the equation of energy be completely and exactly solved without detailed knowledge of the velocity profiles of the system?
10. When, if ever, can the equation of motion be completely solved for a nonisothermal system without detailed knowledge of the temperature profiles of the system?

PROBLEMS

11A.1. Temperature in a friction bearing. Calculate the maximum temperature in the friction bearing of Problem 3A.1, assuming the thermal conductivity of the lubricant to be 4.0×10^{-4} cal/s $\cdot$ cm $\cdot$ C, the metal temperature 200°C, and the rate of rotation 4000 rpm.

Answer: About 217°C (from both Eq. 11.4-13 and Eq. 10.4-9)

11A.2. Viscosity variation and velocity gradients in a nonisothermal film. Water is falling down a vertical wall in a film 0.1 mm thick. The water temperature is 100°C at the free liquid surface and 80°C at the wall surface.

(a) Show that the maximum fractional deviation between viscosities predicted by Eqs. 11.4-17 and 18 occurs when $T = \sqrt{T_0 T_\delta}$.

(b) Calculate the maximum fractional deviation for the conditions given.

Answer: (b) 0.5%

11A.3. Transpiration cooling.

(a) Calculate the temperature distribution between the two shells of Example 11.4-4 for radial mass flow rates of zero and 10^{-5} g/s for the following conditions:

$$R = 500 \text{ microns} \qquad T_R = 300°C$$
$$\kappa R = 100 \text{ microns} \qquad T_\kappa = 100°C$$
$$k = 6.13 \times 10^{-5} \text{ cal/cm} \cdot s \cdot C$$
$$\hat{C}_p = 0.25 \text{ cal/g} \cdot C$$

(b) Compare the rates of heat conduction to the surface at κR in the presence and absence of convection.

11A.4. Free-convection heat loss from a vertical surface. A small heating panel consists essentially of a flat, vertical, rectangular surface 30 cm high and 50 cm wide. Estimate the total rate of heat loss from one side of this panel by free convection, if the panel surface is at 150°F, and the surrounding air is at 70°F and 1 atm. Use the value $C = 0.548$ of Lorenz in Eq. 11.4-51 and the value of C recommended by Whitaker, and compare the results of the two calculations.

Answer: 8.1 cal/sec by Lorenz expression

11A.5. Velocity, temperature, and pressure changes in a shock wave. Air at 1 atm and 70°F is flowing at an upstream Mach number of 2 across a stationary shock wave. Calculate the following quantities, assuming that γ is constant at 1.4 and that $\hat{C}_p = 0.24$ Btu/lb$_m \cdot$ F:

(a) The initial velocity of the air.

(b) The velocity, temperature, and pressure downstream from the shock wave.

(c) The changes of internal and kinetic energy across the shock wave.

Answer: (a) 2250 ft/s
(b) 844 ft/s; 888 R; 4.48 atm
(c) $\Delta \hat{U} = +61.4$ Btu/lb$_m$; $\Delta \hat{K} - 86.9$ Btu/lb$_m$

11A.6. Adiabatic frictionless compression of an ideal gas. Calculate the temperature attained by compressing air, initially at 100°F and 1 atm, to 0.1 of its initial volume. It is assumed that $\gamma = 1.40$ and that the compression is frictionless and adiabatic. Discuss the result in relation to the operation of an internal combustion engine.

Answer: 950°F

11A.7. Effect of free convection on the insulating value of a horizontal air space. Two large parallel horizontal metal plates are separated by a 2.5 cm air gap, with the air at an average temperature of 100°C. How much hotter may the lower plate be (than the upper plate) without causing the onset of the cellular free convection discussed in Example 11.5-2? How much may this temperature difference be increased if a very thin metal sheet is placed midway between the two plates?

Answers: Approximately 3 and 48°C, respectively.

11B.1. Adiabatic frictionless processes in an ideal gas.

(a) Note that a gas that obeys the ideal gas law may deviate appreciably from $\tilde{C}_p$ = constant. Hence, rework Example 11.4-6 using a molar heat capacity expression of the form

$$\tilde{C}_p = a + bT + cT^2 \tag{11B.1-1}$$

(b) Determine the final pressure, p_2, required if methane (CH_4) is to be heated from 300K and 1 atm to 800K by adiabatic frictionless compression. The recommended empirical constants[1]

---

[1] O. A. Hougen, K. M. Watson, and R. A. Ragatz, *Chemical Process Principles*, Part I, 2nd edition, Wiley, New York (1958), p. 255. See also Part II, pp. 646–653, for a fuller discussion of isentropic process calculations.

for methane are: $a = 2.322$ cal/g-mole $\cdot$ K, $b = 38.04 \times 10^{-3}$ cal/g-mole $\cdot$ K^2, and $c = -10.97 \times 10^{-6}$ cal/g-mole $\cdot$ K^3.

Answers: **(a)** $pT^{-a/R} \exp[-(b/R)T - (c/2R)T^2] = $ constant;

        (b) 270 atm

11B.2. **Viscous heating in laminar tube flow (asymptotic solutions).**

(a) Show that for fully developed laminar Newtonian flow in a circular tube of radius R, the energy equation becomes

$$\rho \hat{C}_p v_{z,\text{max}} \left[1 - \left(\frac{r}{R} \right)^2 \right] \frac{\partial T}{\partial z} = k \frac{1}{r} \frac{\partial}{\partial r} \left(r \frac{\partial T}{\partial r} \right) + \frac{4\mu v_{z,\text{max}}^2}{R^2} \left(\frac{r}{R} \right)^2 \tag{11B.2-1}$$

if the viscous dissipation terms are not neglected. Here $v_{z,\text{max}}$ is the maximum velocity in the tube. What restrictions have to be placed on any solutions of Eq. 11B.2-1?

(b) For the *isothermal wall* problem ($T = T_0$ at $r = R$ for $z > 0$ and at $z = 0$ for all r), find the asymptotic expression for $T(r)$ at large z. Do this by recognizing that $\partial T / \partial z$ will be zero at large z. Solve Eq. 11B.2-1 and obtain

$$T - T_0 = \frac{\mu v_{z,\text{max}}^2}{4k} \left[1 - \left(\frac{r}{R} \right)^4 \right] \tag{11B.2-2}$$

(c) For the *adiabatic wall* problem ($q_r = 0$ at $r = R$ for all z) an asymptotic expression for large z may be found as follows: Multiply by $r\,dr$ and then integrate from $r = 0$ to $r = R$. Then integrate the resulting equation over z to get

$$T_b - T_1 = (4\mu v_{z,\text{max}} / \rho \hat{C}_p R^2) z \tag{11B.2-3}$$

in which T_1 is the inlet temperature at $z = 0$. Postulate now that an asymptotic temperature profile at large z is of the form

$$T - T_1 = (4\mu v_{z,\text{max}} / \rho \hat{C}_p R^2) z + f(r) \tag{11B.2-4}$$

Substitute this into Eq. 11B.2-1 and integrate the resulting equation for $f(r)$ to obtain

$$T - T_1 = \frac{4\mu v_{z,\text{max}}}{\rho \hat{C}_p R^2} z + \frac{\mu v_{z,\text{max}}^2}{k} \left[\left(\frac{r}{R} \right)^2 - \frac{1}{2} \left(\frac{r}{R} \right)^4 - \frac{1}{4} \right] \tag{11B.2-5}$$

after determining the integration constant by an energy balance over the tube from 0 to z. Keep in mind that Eqs. 11B.2-2 and 5 are valid solutions only for large z. The complete solutions for small z are discussed in Problem 11D.2.

11B.3. **Velocity distribution in a nonisothermal film.** Show that Eq. 11.4-20 meets the following requirements:

(a) At $x = \delta$, $v_z = 0$.

(b) At $x = 0$, $\partial v_z / \partial x = 0$.

(c) $\lim\limits_{\mu_\delta \to \mu_0} v_z(x) = (\rho g \delta^2 \cos \beta / 2\mu_0)[1 - (x/\delta)^2]$

11B.4. **Heat conduction in a spherical shell** (Fig. 11B.4). A spherical shell has inner and outer radii R_1 and R_2. A hole is made in the shell at the north pole by cutting out the conical segment in the region $0 \leq \theta \leq \theta_1$. A similar hole is made at the south pole by removing the portion $(\pi - \theta_1) \leq \theta \leq \pi$. The surface $\theta = \theta_1$ is kept at temperature $T = T_1$, and the surface at $\theta = \pi - \theta_1$ is held at $T = T_2$. Find the steady-state temperature distribution, using the heat conduction equation.

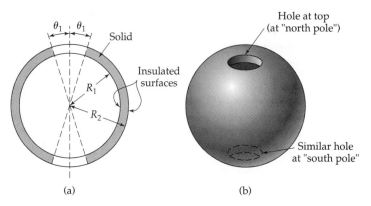

Fig. 11B.4. Heat conduction in a spherical shell: (*a*) cross section containing the *z*-axis; (*b*) view of the sphere from above.

11B.5. **Axial heat conduction in a wire**[2] (Fig. 11B.5). A wire of constant density ρ moves downward with uniform speed v into a liquid metal bath at temperature T_0. It is desired to find the temperature profile $T(z)$. Assume that $T = T_\infty$ at $z = \infty$, and that resistance to radial heat conduction is negligible. Assume further that the wire temperature is $T = T_0$ at $z = 0$.

(a) First solve the problem for constant physical properties $\hat{C}_p$ and k. Obtain

$$\Theta = \frac{T - T_\infty}{T_0 - T_\infty} = \exp\left(-\frac{\rho \hat{C}_p v z}{k}\right) \tag{11B.5-1}$$

(b) Next solve the problem when $\hat{C}_p$ and k are known functions of the dimensionless temperature Θ: $k = k_\infty K(\Theta)$ and $\hat{C}_p = \hat{C}_{p\infty} L(\Theta)$. Obtain the temperature profile,

$$-\left(\frac{\rho \hat{C}_{p\infty} v z}{k_\infty}\right) = \int_1^\Theta \frac{K(\overline{\Theta}) d\overline{\Theta}}{\int_0^{\overline{\Theta}} L(\overline{\overline{\Theta}}) d\overline{\overline{\Theta}}} \tag{11B.5-2}$$

(c) Verify that the solution in (b) satisfies the differential equation from which it was derived.

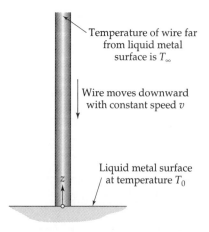

Temperature of wire far from liquid metal surface is T_∞

Wire moves downward with constant speed v

Liquid metal surface at temperature T_0

Fig. 11B.5. Wire moving into a liquid metal bath.

---

[2] Suggested by Prof. G. L. Borman, Mechanical Engineering Department, University of Wisconsin.

11B.6. Transpiration cooling in a planar system. Two large flat porous horizontal plates are separated by a relatively small distance L. The upper plate at $y = L$ is at temperature T_L, and the lower one at $y = 0$ is to be maintained at a lower temperature T_0. To reduce the amount of heat that must be removed from the lower plate, an ideal gas at T_0 is blown upward through both plates at a steady rate. Develop an expression for the temperature distribution and the amount of heat q_0 that must be removed from the cold plate per unit area as a function of the fluid properties and gas flow rate. Use the abbreviation $\phi = \rho \hat{C}_p v_y L / k$.

Answer: $\dfrac{T - T_L}{T_0 - T_L} = \dfrac{e^{\phi y/L} - e^\phi}{1 - e^\phi}$; $q_0 = \dfrac{k(T_L - T_0)}{L}\left(\dfrac{\phi}{e^\phi - 1}\right)$

11B.7. Reduction of evaporation losses by transpiration (Fig. 11B.7). It is proposed to reduce the rate of evaporation of liquefied oxygen in small containers by taking advantage of transpiration. To do this, the liquid is to be stored in a spherical container surrounded by a spherical shell of a porous insulating material as shown in the figure. A thin space is to be left between the container and insulation, and the opening in the insulation is to be stoppered. In operation, the evaporating oxygen is to leave the container proper, move through the gas space, and then flow uniformly out through the porous insulation.

Calculate the rate of heat gain and evaporation loss from a tank 1 ft in diameter covered with a shell of insulation 6 in. thick under the following conditions with and without transpiration.

Temperature of liquid oxygen	$-297°F$
Temperature of outer surface of insulation	$30°F$
Effective thermal conductivity of insulation	0.02 Btu/hr $\cdot$ ft $\cdot$ F
Heat of evaporation of oxygen	91.7 Btu/lb
Average $\hat{C}_p$ of O_2 flowing through insulation	0.22 Btu/lb $\cdot$ F

Neglect the thermal resistance of the liquid oxygen, container wall, and gas space, and neglect heat losses through the stopper. Assume the particles of insulation to be in local thermal equilibrium with the gas.

Answers: 82 Btu/hr without transpiration; 61 Btu/hr with transpiration

11B.8. Temperature distribution in an embedded sphere. A sphere of radius R and thermal conductivity k_1 is embedded in an infinite solid of thermal conductivity k_0. The center of the sphere is located at the origin of coordinates, and there is a constant temperature gradient A in the positive z direction far from the sphere. The temperature at the center of the sphere is $T°$.

The steady-state temperature distributions in the sphere T_1 and in the surrounding medium T_0 have been shown to be:[3]

$$T_1(r,\theta) - T° = \left[\frac{3k_0}{k_1 + 2k_0}\right]Ar\cos\theta \qquad\qquad r \le R \qquad\qquad (11B.8\text{-}1)$$

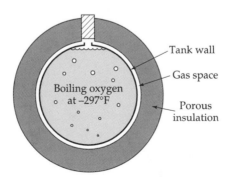

Fig. 11B.7. Use of transpiration to reduce the evaporation rate.

[3] L. D. Landau and E. M. Lifshitz, *Fluid Mechanics*, 2nd edition, Pergamon Press, Oxford (1987), p. 199.

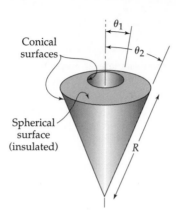

Fig. 11B.9. Body formed from the intersection of two cones and a sphere.

Conical surfaces

Spherical surface (insulated)

$$T_0(r,\theta) - T^\circ = \left[1 - \frac{k_1 - k_0}{k_1 + 2k_0}\left(\frac{R}{r}\right)^3\right]Ar\cos\theta \quad r \ge R \qquad (11B.8\text{-}2)$$

(a) What are the partial differential equations that must be satisfied by Eqs. 11B.8-1 and 2?

(b) Write down the boundary conditions that apply at $r = R$.

(c) Show that T_1 and T_0 satisfy their respective partial differential equations in (a).

(d) Show that Eqs. 11B.8-1 and 2 satisfy the boundary conditions in (b).

11B.9. **Heat flow in a solid bounded by two conical surfaces** (Fig. 11B.9). A solid object has the shape depicted in the figure. The conical surfaces $\theta_1 = $ constant and $\theta_2 = $ constant are held at temperatures T_1 and T_2, respectively. The spherical surface at $r = R$ is insulated. For steady-state heat conduction, find

(a) The partial differential equation that $T(\theta)$ must satisfy.

(b) The solution to the differential equation in (a) containing two constants of integration.

(c) Expressions for the constants of integration.

(d) The expression for the θ-component of the heat flux vector.

(e) The total heat flow (cal/sec) across the conical surface at $\theta = \theta_1$.

Answer: **(e)** $Q = \dfrac{2\pi Rk(T_1 - T_2)}{\ln\left(\dfrac{\tan\frac{1}{2}\theta_2}{\tan\frac{1}{2}\theta_1}\right)}$

11B.10. **Freezing of a spherical drop** (Fig. 11B.10). To evaluate the performance of an atomizing nozzle, it is proposed to atomize a nonvolatile liquid wax into a stream of cool air. The atomized wax particles are expected to solidify in the air, from which they may later be collected and

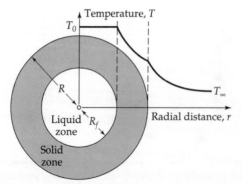

Fig. 11B.10. Temperature profile in the freezing of a spherical drop.

examined. The wax droplets leave the atomizer only slightly above their melting point. Estimate the time t_f required for a drop of radius R to freeze completely, if the drop is initially at its melting point T_0 and the surrounding air is at T_∞. Heat is lost from the drop to the surrounding air according to Newton's law of cooling, with a constant heat-transfer coefficient h. Assume that there is no volume change in the solidification process. Solve the problem by using a quasi-steady-state method.

(a) First solve the steady-state heat conduction problem in the solid phase in the region between $r = R_f$ (the liquid–solid interface) and $r = R$ (the solid–air interface). Let k be the thermal conductivity of the solid phase. Then find the radial heat flow Q across the spherical surface at $r = R$.

(b) Then write an unsteady–state energy balance, by equating the heat liberation at $r = R_f(t)$ resulting from the freezing of the liquid to the heat flow Q across the spherical surface at $r = R$. Integrate the resulting separable, first-order differential equation between the limits 0 and R, to obtain the time that it takes for the drop to solidify. Let $\Delta \hat{H}_f$ be the latent heat of freezing (per unit mass).

$Answers:$ **(a)** $Q = \dfrac{h \cdot 4\pi R^2 (T_0 - T_\infty)}{[1 - (hR/k)] + (hR^2/kR_f)}$; **(b)** $t_f = \dfrac{\rho \Delta \hat{H}_f R}{h(T_0 - T_\infty)} \left[\dfrac{1}{3} + \dfrac{1}{6} \dfrac{hR}{k} \right]$

11B.11. **Temperature rise in a spherical catalyst pellet** (Fig. 11B.11). A catalyst pellet has a radius R and a thermal conductivity k (which may be assumed constant). Because of the chemical reaction occurring within the porous pellet, heat is generated at a rate of S_c cal/cm$^3 \cdot$ s. Heat is lost at the outer surface of the pellet to a gas stream at constant temperature T_g by convective heat transfer with heat transfer coefficient h. Find the steady-state temperature profile, assuming that S_c is constant throughout.

(a) Set up the differential equation by making a shell energy balance.

(b) Set up the differential equation by simplifying the appropriate form of the energy equation.

(c) Integrate the differential equation to get the temperature profile. Sketch the function $T(r)$.

(d) What is the limiting form of $T(r)$ when $h \to \infty$?

(e) What is the maximum temperature in the system?

(f) Where in the derivation would one modify the procedure to account for variable k and variable S_c?

11B.12. **Stability of an exothermic reaction system.**[3] Consider a porous slab of thickness $2B$, width W, and length L, with $B \ll W$ and $B \ll L$. Within the slab an exothermic reaction occurs, with a temperature-dependent rate of heat production $S_c(T) = S_{c0} \exp A(T - T_0)$.

(a) Use the energy equation to obtain a differential equation for the temperature in the slab. Assume constant physical properties, and postulate a steady-state solution $T(x)$.

(b) Write the differential equation and boundary conditions in terms of these dimensionless quantities: $\xi = x/B$, $\Theta = A(T - T_0)$, and $\lambda = S_{c0}AB^2/k$; here A is a constant.

(c) Integrate the differential equation (*hint:* first multiply by $2d\Theta/d\xi$) to obtain

$$\left(\frac{d\Theta}{d\xi} \right)^2 = 2\lambda (\exp \Theta_0 - \exp \Theta) \tag{11B.12-1}$$

in which Θ_0 is an auxiliary constant representing the value of Θ at $\xi = 0$.

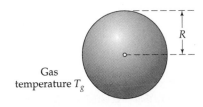

Gas temperature T_g

Fig. 11B.11. Sphere with internal heat generation.

(d) Integrate the result of (c) and make use of the boundary conditions to obtain the relation between the slab thickness and midplane temperature

$$\exp(-\tfrac{1}{2}\Theta_0)\,\text{arccosh}(\exp(\tfrac{1}{2}\Theta_0)) = \sqrt{\tfrac{1}{2}\lambda} \tag{11B.12-2}$$

(e) Calculate λ at $\Theta_0 = 0.5, 1.0, 1.2, 1.4,$ and 2.0; graph these results to find the maximum value of λ for steady-state conditions. If this value of λ is exceeded, the system will explode.

11B.13. Laminar annular flow with constant wall heat flux. Repeat the development of §10.8 for flow in an annulus of inner and outer radii κR and R, respectively, starting with the equations of change. Heat is added to the fluid through the inner cylinder wall at a rate q_0 (heat per unit per unit time), and the outer cylinder wall is thermally insulated.

11B.14. Unsteady-state heating of a sphere. A sphere of radius R and thermal diffusivity α is initially at a uniform temperature T_0. For $t > 0$ the sphere is immersed in a well-stirred water bath maintained at a temperature $T_1 > T_0$. The temperature within the sphere is then a function of the radial coordinate r and the time t. The solution to the heat conduction equation is given by:[4]

$$\frac{T - T_0}{T_1 - T_0} = 1 + 2\sum_{n=1}^{\infty} (-1)^n \left(\frac{R}{n\pi r}\right) \sin\left(\frac{n\pi r}{R}\right) \exp\left(-\alpha n^2 \pi^2 t / R^2\right) \tag{11B.14-1}$$

It is desired to verify that this equation satisfies the differential equation, the boundary conditions, and the initial condition.

(a) Write down the differential equation describing the problem.

(b) Show that Eq. 11B.14-1 for $T(r, t)$ satisfies the differential equation in (a).

(c) Show that the boundary condition at $r = R$ is satisfied.

(d) Show that T is finite at $r = 0$.

(e) To show that Eq. 11B.14-1 satisfies the initial condition, set $t = 0$ and $T = T_0$ and obtain the following:

$$-1 = 2\sum_{n=1}^{\infty} (-1)^n \left(\frac{R}{n\pi r}\right) \sin\left(\frac{n\pi r}{R}\right) \tag{11B.14-2}$$

To show that this is true, multiply both sides by $(r/R)\sin(m\pi r/R)$, where m is any integer from 1 to ∞, and integrate from $r = 0$ to $r = R$. In the integration all terms with $m \neq n$ vanish on the right side. The term with $m = n$, when integrated, equals the integral on the left side.

11B.15. Dimensionless variables for free convection.[5] The dimensionless variables in Eqs. 11.4-39 to 43 can be obtained by simple arguments. The form of Θ is dictated by the boundary conditions and that of ζ is suggested by the geometry. The remaining dimensionless variables may be found as follows:

(a) Set $\eta = y/y_0$, $\phi_z = v_z/v_{z0}$, and $\phi_y = v_y/v_{y0}$, the subscript-zero quantities being constants. Then the differential equations in Eqs. 11.4-33 to 35 become

$$\frac{\partial \phi_y}{\partial \eta} + \left[\frac{v_{z0}y_0}{v_{y0}H}\right]\frac{\partial \phi_z}{\partial \zeta} = 0 \tag{11B.15-1}$$

$$\phi_y \frac{\partial \phi_y}{\partial \eta} + \left[\frac{v_{z0}y_0}{v_{y0}H}\right]\phi_z \frac{\partial \phi_z}{\partial \zeta} = \left[\frac{\mu}{\rho y_0 v_{y0}}\right]\frac{\partial^2 \phi_z}{\partial \eta^2} + \left[\frac{y_0 g \beta(T_0 - T_1)}{v_{y0}v_{z0}}\right]\Theta \tag{11B.15-2}$$

$$\phi_y \frac{\partial \Theta}{\partial \eta} + \left[\frac{v_{z0}y_0}{v_{y0}H}\right]\phi_z \frac{\partial \Theta}{\partial \zeta} = \left[\frac{k}{\rho \hat{C}_p y_0 v_{y0}}\right]\frac{\partial^2 \Theta}{\partial \eta^2} \tag{11B.15-3}$$

with the boundary conditions given in Eqs. 11.4-47 to 49.

---

[4] H. S. Carslaw and J. C. Jaeger, *Conduction of Heat in Solids*, 2nd edition, Oxford University Press (1959), p. 233, Eq. (4).

[5] The procedure used here is similar to that suggested by J. D. Hellums and S. W. Churchill, *AIChE Journal*, **10**, 110–114 (1964).

(b) Choose appropriate values of v_{z0}, v_{y0}, and y_0 to convert the equations in (a) into Eqs. 11.4-44 to 46, and show that the definitions in Eqs. 11.4-41 to 43 follow directly.

(c) Why is the choice of variables developed in (b) preferable to that obtained by setting the dimensionless groups in Eqs. 11B.15-1 and 2 equal to unity?

11C.1. The speed of propagation of sound waves. Sound waves are harmonic compression waves of very small amplitude traveling through a compressible fluid. The velocity of propagation of such waves may be estimated by assuming that the momentum flux tensor $\boldsymbol{\tau}$ and the heat flux vector $\mathbf{q}$ are zero and that the velocity $\mathbf{v}$ of the fluid is small.[6] The neglect of $\boldsymbol{\tau}$ and $\mathbf{q}$ is equivalent to assuming that the entropy is constant following the motion of a given fluid element (see Problem 11D.1).

(a) Use equilibrium thermodynamics to show that

$$\left(\frac{\partial p}{\partial \rho}\right)_S = \gamma \left(\frac{\partial p}{\partial \rho}\right)_T \tag{11C.1-1}$$

in which $\gamma = C_p/C_V$.

(b) When sound is being propagated through a fluid, there are slight perturbations in the pressure, density, and velocity from the rest state: $p = p_0 + p'$, $\rho = \rho_0 + \rho'$, and $\mathbf{v} = \mathbf{v}_0 + \mathbf{v}'$, the subscript-zero quantities being constants associated with the rest state (with $\mathbf{v}_0$ being zero), and the primed quantities being very small. Show that when these quantities are substituted into the equation of continuity and the equation of motion (with the $\boldsymbol{\tau}$ and $\mathbf{g}$ terms omitted) and products of the small primed quantities are omitted, we get

Equation of continuity
$$\frac{\partial \rho}{\partial t} = -\rho_0 (\nabla \cdot \mathbf{v}) \tag{11C.1-2}$$

Equation of motion
$$\rho_0 \frac{\partial \mathbf{v}}{\partial t} = -\nabla p \tag{11C.1-3}$$

(c) Next use the result in (a) to rewrite the equation of motion as

$$\rho_0 \frac{\partial \mathbf{v}}{\partial t} = -v_s^2 \nabla \rho \tag{11C.1-4}$$

in which $v_s^2 = \gamma (\partial p/\partial \rho)_T$.

(d) Show how Eqs. 11C.1-2 and 4 can be combined to give

$$\frac{\partial^2 \rho}{\partial t^2} = v_s^2 \nabla^2 \rho \tag{11C.1-5}$$

(e) Show that a solution of Eq. 11C.1-5 is

$$\rho = \rho_0 \left[1 + A \, \sin \left(\frac{2\pi}{\lambda} (z - v_s t) \right) \right] \tag{11C.1-6}$$

This solution represents a harmonic wave of wavelength λ and amplitude $\rho_0 A$ traveling in the z direction at a speed v_s. More general solutions may be constructed by a superposition of waves of different wavelengths and directions.

11C.2. Free convection in a slot. A fluid of constant viscosity, with density given by Eq. 11.3-1, is confined in a rectangular slot. The slot has vertical walls at $x = \pm B$, $y = \pm W$, and a top and bottom at $z = \pm H$, with $H \gg W \gg B$. The walls are nonisothermal, with temperature distribution $T_w = \overline{T} + Ay$, so that the fluid circulates by free convection. The velocity profiles are to be predicted, for steady laminar flow conditions and small deviations from the mean density $\overline{\rho}$.

---

[6] See L. Landau and E. M. Lifshitz, *Fluid Mechanics*, 2nd edition, Pergamon, Oxford (1987), Chapter VIII; R. J. Silbey and R. A. Alberty, *Physical Chemistry*, 3rd edition, Wiley, New York (2001), §17.4.

(a) Simplify the equations of continuity, motion, and energy according to the postulates: $\mathbf{v} = \boldsymbol{\delta}_z v_z(x, y)$, $\partial^2 v_z / \partial y^2 << \partial^2 v_z / \partial x^2$, and $T = T(y)$. These postulates are reasonable for slow flow, except near the edges $y = \pm W$ and $z = \pm H$.

(b) List the boundary conditions to be used with the problem as simplified in (a).

(c) Solve for the temperature, pressure, and velocity profiles.

(d) When making diffusion measurements in closed chambers, free convection can be a serious source of error, and temperature gradients must be avoided. By way of illustration, compute the maximum tolerable temperature gradient, A, for an experiment with water at 20°C in a chamber with $B = 0.1$ mm, $W = 2.0$ mm, and $H = 2$ cm, if the maximum permissible convective movement is 0.1% of H in a one-hour experiment.

Answers: (c) $v_z(x, y) = \dfrac{\overline{\rho g} \overline{\beta} A}{2\mu} (B^2 - x^2)y$; (d) 2.7×10^{-3} K/cm

11C.3. **Tangential annular flow of a highly viscous liquid.** Show that Eq. 11.4-13 for flow in an annular region reduces to Eq. 10.4-9 for plane slit flow in the limit as κ approaches unity. Comparisons of this kind are often useful for checking results.

The right side of Eq. 11.4-13 is indeterminate at $\kappa = 1$, but its limit as $\kappa \to 1$ can be obtained by expanding in powers of $\varepsilon = 1 - \kappa$. To do this, set $\kappa = 1 - \varepsilon$ and $\xi = 1 - \varepsilon[1 - (x/b)]$; then the range $\kappa \leq \xi \leq 1$ in Problem 11.4-2 corresponds to the range $0 \leq x \leq b$ in §10.4. After making the substitutions, expand the right side of Eq. 11.4-13 in powers of ε (neglecting terms beyond ε^2) and show that Eq. 10.4-9 is obtained.

11C.4. **Heat conduction with variable thermal conductivity.**

(a) For steady-state heat conduction in solids, Eq. 11.2-5 becomes $(\nabla \cdot \mathbf{q}) = 0$, and insertion of Fourier's law gives $(\nabla \cdot k\nabla T) = 0$. Show that the function $F = \int k\, dT + $ const. satisfies the Laplace equation $\nabla^2 F = 0$, provided that k depends only on T.

(b) Use the result in (a) to solve Problem 10B.12 (part a), using an arbitrary function $k(T)$.

11C.5. **Effective thermal conductivity of a solid with spherical inclusions** (Fig. 11C.5). Derive Eq. 9.6-1 for the effective thermal conductivity of a two-phase system by starting with Eqs. 11B.8-1 and 2. We construct two systems both contained within a spherical region of radius R': (a) the "true" system, a medium with thermal conductivity k_0, in which there are embedded n tiny spheres of thermal conductivity k_1 and radius R; and (b) an "equivalent" system, which is a continuum, with an effective thermal conductivity k_{eff}. Both of these systems are placed in a temperature gradient A, and both are surrounded by a medium with thermal conductivity k_0.

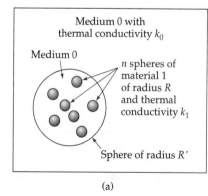
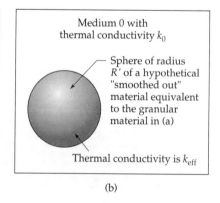

Fig. 11C.5. Thought experiment used by Maxwell to get the thermal conductivity of a composite solid: (*a*) the "true" discrete system, and (*b*) the "equivalent" continuum system.

(a) For the "true" system we know that at a large distance L from the system (i.e., $L >> R'$), the temperature field will be given by a slight modification of Eq. 11B.8-2, provided that the tiny occluded spheres are very "dilute" in the true system:

$$T_0(r, \theta) - T^\circ = \left[1 - n \frac{k_1 - k_0}{k_1 + 2k_0} \left(\frac{R}{r} \right)^3 \right] Ar \cos \theta \qquad (11C.5\text{-}1)$$

Explain carefully how this result is obtained.

(b) Next, for the "equivalent system," we can write from Eq. 11B.8-2

$$T_0(r, \theta) - T^\circ = \left[1 - \frac{k_{\text{eff}} - k_0}{k_{\text{eff}} + 2k_0} \left(\frac{R'}{r} \right)^3 \right] Ar \cos \theta \qquad (11C.5\text{-}2)$$

(c) Next derive the relation $nR^3 = \phi R'^3$, in which ϕ is the volume fraction of the occlusions in the "true system."

(d) Equate the right sides of Eqs. 11C.5-1 and 2 to get Maxwell's equation[7] in Eq. 9.6-1.

11C.6. Interfacial boundary conditions. Consider a nonisothermal interfacial surface $S(t)$ between pure phases I and II in a nonisothermal system. The phases may consist of two immiscible fluids (so that no material crosses $S(t)$), or two different pure phases of a single substance (between which mass may be interchanged by condensation, evaporation, freezing, or melting). Let $\mathbf{n}^{\text{I}}$ be the local unit normal to $S(t)$ directed into phase I. A superscript I or II will be used for values along S in each phase, and a superscript s for values in the interface itself. The usual interfacial boundary conditions on tangential velocity v_t and temperature T on S are

$$v_t^{\text{I}} = v_t^{\text{II}} \qquad \text{(no slip)} \qquad (11C.6\text{-}1)$$

$$T^{\text{I}} = T^{\text{II}} \qquad \text{(continuity of temperature)} \qquad (11C.6\text{-}2)$$

In addition, the following simplified conservation equations are suggested[8] for surfactant-free interfaces:

Interfacial mass balance

$$(\mathbf{n}^{\text{I}} \cdot \{\rho^{\text{I}}(\mathbf{v}^{\text{I}} - \mathbf{v}^s) - \rho^{\text{II}}(\mathbf{v}^{\text{II}} - \mathbf{v}^s)\}) = 0 \qquad (11C.6\text{-}3)$$

Interfacial momentum balance

$$\mathbf{n}^{\text{I}} \left[(p^{\text{I}} - p^{\text{II}}) + (\rho^{\text{I}} v^{\text{I2}} - \rho^{\text{II}} v^{\text{II2}}) + \sigma \left(\frac{1}{R_1} + \frac{1}{R_2} \right) \right] + [\mathbf{n}^{\text{I}} \cdot \{\tau^{\text{I}} - \tau^{\text{II}}\}] = -\nabla^s \sigma \qquad (11C.6\text{-}4)$$

Interfacial internal energy balance

$$(\mathbf{n}^{\text{I}} \cdot \rho^{\text{I}} \{\mathbf{v}^{\text{I}} - \mathbf{v}^s\})[(\hat{H}^{\text{I}} - \hat{H}^{\text{II}}) + \tfrac{1}{2}(v^{\text{I2}} - v^{\text{II2}})] + (\mathbf{n}^{\text{I}} \cdot \{\mathbf{q}^{\text{I}} - \mathbf{q}^{\text{II}}\}) = \sigma(\nabla^s \cdot \mathbf{v}^s) \qquad (11C.6\text{-}5)$$

The momentum balance of Eq. 3C.5-1 has been extended here to include the surface gradient $\nabla^s \sigma$ of the interfacial tension; the resulting tangential force gives rise to a variety of interfacial flow phenomena, known as *Marangoni effects*.[9,10] Equation 11C.6-5 is obtained in the manner of §11.2, from total and mechanical energy balances on S, neglecting interfacial excess energy $\hat{U}^s$, heat flux $\mathbf{q}^s$, and viscous dissipation $(\tau^s:\nabla^s \mathbf{v}^s)$; fuller results are given elsewhere.[8]

---

[7] J. C. Maxwell, *A Treatise on Electricity and Magnetism*, Vol. 1, Oxford University Press (1891, reprinted 1998), §314.

[8] J. C. Slattery, *Advanced Transport Phenomena*, Cambridge University Press (1999), pp. 58, 435; more complete conditions are given in Ref. 10.

[9] C. G. M. Marangoni, *Ann. Phys.* (*Poggendorf*), **3**, 337–354 (1871); C. V. Sternling and L. E. Scriven, *AIChE Journal*, **5**, 514–523 (1959).

[10] D. A. Edwards, H. Brenner, and D. T. Wasan, *Interfacial Transport Processes and Rheology*, Butterworth-Heinemann, Stoneham, Mass. (1991).

(a) Verify the dimensional consistency of each interfacial balance equation.

(b) Under what conditions are $\mathbf{v}^{\mathrm{I}}$ and $\mathbf{v}^{\mathrm{II}}$ equal?

(c) Show how the balance equations simplify when phases I and II are two pure immiscible liquids.

(d) Show how the balance equations simplify when one phase is a solid.

11C.7.   Effect of surface-tension gradients on a falling film.

(a) Repeat the determination of the shear-stress and velocity distributions of Example 2.1-1 in the presence of a small temperature gradient dT/dz in the direction of flow. Assume that this temperature gradient produces a constant surface-tension gradient $d\sigma/dz = A$ but has no other effect on system physical properties. Note that this surface-tension gradient will produce a shear stress at the free surface of the film (see Problem 11C.6) and, hence, will require a nonzero velocity gradient there. Once again, postulate a stable, nonrippling, laminar film.

(b) Calculate the film thickness as a function of the net downward flow rate and discuss the physical significance of the result.

Answer: **(a)** $\tau_{xz} = \rho g x \cos \beta + A$; $v_z = \dfrac{\rho g \delta^2 \cos \beta}{2\mu} \left[1 - \left(\dfrac{x}{\delta} \right)^2 \right] + \dfrac{A\delta}{\mu} \left(1 - \dfrac{x}{\delta} \right)$

11D.1.   Equation of change for entropy. This problem is an introduction to the thermodynamics of irreversible processes. A treatment of multicomponent mixtures is given in §§24.1 and 2.

(a) Write an entropy balance for the fixed volume element $\Delta x\, \Delta y\, \Delta z$. Let $\mathbf{s}$ be the *entropy flux vector*, measured with respect to the fluid velocity vector $\mathbf{v}$. Further, let the *rate of entropy production* per unit volume be designated by g_S. Show that when the volume element $\Delta x\, \Delta y\, \Delta z$ is allowed to become vanishingly small, one finally obtains an *equation of change for entropy* in either of the following two forms:[11]

$$\frac{\partial}{\partial t} \rho \hat{S} = -(\nabla \cdot \rho \hat{S} \mathbf{v}) - (\nabla \cdot \mathbf{s}) + g_S \tag{11D.1-1}$$

$$\rho \frac{D\hat{S}}{Dt} = -(\nabla \cdot \mathbf{s}) + g_S \tag{11D.1-2}$$

in which $\hat{S}$ is the entropy per unit mass.

(b) If one assumes that the thermodynamic quantities can be defined locally in a nonequilibrium situation, then $\hat{U}$ can be related to $\hat{S}$ and $\hat{V}$ according to the thermodynamic relation $d\hat{U} = T d\hat{S} - p d\hat{V}$. Combine this relation with Eq. 11.2-2 to get

$$\rho \frac{D\hat{S}}{Dt} = -\frac{1}{T}(\nabla \cdot \mathbf{q}) - \frac{1}{T}(\boldsymbol{\tau}{:}\nabla \mathbf{v}) \tag{11D.1-3}$$

(c) The local entropy flux is equal to the local energy flux divided by the local temperature[12–15]; that is, $\mathbf{s} = \mathbf{q}/T$. Once this relation between $\mathbf{s}$ and $\mathbf{q}$ is recognized, we can compare Eqs. 11D.1-2 and 3 to get the following expression for the rate of entropy production per unit volume:

$$g_S = -\frac{1}{T^2}(\mathbf{q} \cdot \nabla T) - \frac{1}{T}(\boldsymbol{\tau}{:}\nabla \mathbf{v}) \tag{11D.1-4}$$

---

[11] G. A. J. Jaumann, *Sitzungsber. der Math.-Naturwiss. Klasse der Kaiserlichen Akad. der Wissenschaften* (*Wien*), **102**, Abt. IIa, 385–530 (1911).

[12] **Carl Henry Eckart** (1902–1973), vice-chancellor of the University of California at San Diego (1965–1969), made fundamental contributions to quantum mechanics, geophysical hydrodynamics, and the thermodynamics of irreversible processes; his key contributions to transport phenomena are in C. H. Eckart, *Phys. Rev.*, **58**, 267–268, 269–275 (1940).

[13] C. F. Curtiss and J. O. Hirschfelder, *J. Chem. Phys.*, **18**, 171–173 (1950).

[14] J. G. Kirkwood and B. L. Crawford, Jr., *J. Phys. Chem.* **56**, 1048–1051 (1952).

[15] S. R. de Groot and P. Mazur, *Non-Equilibrium Thermodynamics*, North-Holland, Amsterdam (1962).

The first term on the right side is the rate of entropy production associated with heat transport, and the second is the rate of entropy production resulting from momentum transport. Equation 11D.1-4 is the starting point for the thermodynamic study of the irreversible processes in a pure fluid.

(d) What conclusions can be drawn when Newton's law of viscosity and Fourier's law of heat conduction are inserted into Eq. 11D.1-4?

11D.2. **Viscous heating in laminar tube flow.**

(a) Continue the analysis begun in Problem 11B.2—namely, that of finding the temperature profiles in a Newtonian fluid flowing in a circular tube at a speed sufficiently high that viscous heating effects are important. Assume that the velocity profile at the inlet ($z = 0$) is fully developed, and that the inlet temperature is uniform over the cross section. Assume all physical properties to be constant.

(b) Repeat the analysis for a power law non-Newtonian viscosity.[16]

11D.3. **Derivation of the energy equation using integral theorems.** In §11.1 the energy equation is derived by accounting for the energy changes occurring in a small rectangular volume element $\Delta x\, \Delta y\, \Delta z$.

(a) Repeat the derivation using an arbitrary volume element V with a fixed boundary S by following the procedure outlined in Problem 3D.1. Begin by writing the law of conservation of energy as

$$\frac{d}{dt} \int_V (\rho\hat{U} + \tfrac{1}{2}\rho v^2)\, dV = -\int_S (\mathbf{n} \cdot \mathbf{e})\, dS + \int_V (\mathbf{v} \cdot \mathbf{g})\, dV \tag{11D.3-1}$$

Then use the Gauss divergence theorem to convert the surface integral into a volume integral, and obtain Eq. 11.1-6.

(b) Do the analogous derivation for a moving "blob" of fluid.

---

[16] R. B. Bird, *Soc. Plastics Engrs. Journal*, **11**, 35–40 (1955).

Chapter 12

Temperature Distributions with More Than One Independent Variable

§12.1   Unsteady heat conduction in solids

§12.2°   Steady heat conduction in laminar, incompressible flow

§12.3°   Steady potential flow of heat in solids

§12.4°   Boundary layer theory for nonisothermal flow

In Chapter 10 we saw how simple heat flow problems can be solved by means of shell energy balances. In Chapter 11 we developed the energy equation for flow systems, which describes the heat transport processes in more complex situations. To illustrate the usefulness of the energy equation, we gave in §11.4 a series of examples, most of which required no knowledge of solving partial differential equations.

In this chapter we turn to several classes of heat transport problems that involve more than one dependent variable, either two spatial variables, or one space variable and the time variable. The types of problems and the mathematical methods parallel those given in Chapter 4.

§12.1   UNSTEADY HEAT CONDUCTION IN SOLIDS

For solids, the energy equation of Eq. 11.2-5, when combined with Fourier's law of heat conduction, becomes

$$\rho \hat{C}_p \frac{\partial T}{\partial t} = (\nabla \cdot k \nabla T) \tag{12.1-1}$$

If the thermal conductivity can be assumed to be independent of the temperature and position, then Eq. 12.1-1 becomes

$$\frac{\partial T}{\partial t} = \alpha \nabla^2 T \tag{12.1-2}$$

in which $\alpha = k/\rho \hat{C}_p$ is the thermal diffusivity of the solid. Many solutions to this equation have been worked out. The treatise of Carslaw and Jaeger[1] contains a thorough dis-

---

[1] H. S. Carslaw and J. C. Jaeger, *Conduction of Heat in Solids*, 2nd edition, Oxford University Press (1959).

cussion of solution methods as well as a very comprehensive tabulation of solutions for a wide variety of boundary and initial conditions. Many frequently encountered heat conduction problems may be solved just by looking up the solution in this impressive reference work.

In this section we illustrate four important methods for solving unsteady heat conduction problems: the method of combination of variables, the method of separation of variables, the method of sinusoidal response, and the method of Laplace transform. The first three of these were also used in §4.1.

EXAMPLE 12.1-1

Heating a Semi-Infinite Slab

A solid material occupying the space from $y = 0$ to $y = \infty$ is initially at temperature T_0. At time $t = 0$, the surface at $y = 0$ is suddenly raised to temperature T_1 and maintained at that temperature for $t > 0$. Find the time-dependent temperature profiles $T(y, t)$.

SOLUTION

For this problem, Eq. 12.1-2 becomes

$$\frac{\partial \Theta}{\partial t} = \alpha \frac{\partial^2 \Theta}{\partial y^2} \tag{12.1-3}$$

Here a dimensionless temperature difference $\Theta = (T - T_0)/(T_1 - T_0)$ has been introduced. The initial and boundary conditions are then

I.C.:	at $t \le 0$,	$\Theta = 0$	for all y	(12.1-4)
B.C. 1:	at $y = 0$,	$\Theta = 1$	for all $t > 0$	(12.1-5)
B.C. 2:	at $y = \infty$,	$\Theta = 0$	for all $t > 0$	(12.1-6)

This problem is mathematically analogous to that formulated in Eqs. 4.1-1 to 4. Hence the solution in Eq. 4.1-15 can be taken over directly by appropriate changes in notation:

$$\Theta = 1 - \frac{2}{\sqrt{\pi}} \int_0^{y/\sqrt{4\alpha t}} \exp(-\eta^2)\, d\eta \tag{12.1-7}$$

or

$$\frac{T - T_0}{T_1 - T_0} = 1 - \operatorname{erf} \frac{y}{\sqrt{4\alpha t}} \tag{12.1-8}$$

The solution shown in Fig. 4.1-2 describes the temperature profiles when the ordinate is labeled $(T - T_0)/(T_1 - T_0)$ and the abscissa $y/\sqrt{4\alpha t}$.

Since the error function reaches a value of 0.99 when the argument is about 2, the *thermal penetration thickness* δ_T is

$$\delta_T = 4\sqrt{\alpha t} \tag{12.1-9}$$

That is, for distances $y > \delta_T$, the temperature has changed by less than 1% of the difference $T_1 - T_0$. If it is necessary to calculate the temperature in a slab of finite thickness, the solution in Eq. 12.1-8 will be a good approximation when δ_T is small with respect to the slab thickness. However, when δ_T is of the order of magnitude of the slab thickness or greater, then the series solution of Example 12.1-2 has to be used.

The wall heat flux can be calculated from Eq. 12.1-8 as follows:

$$q_y\big|_{y=0} = -k \frac{\partial T}{\partial y}\bigg|_{y=0} = \frac{k}{\sqrt{\pi \alpha t}} (T_1 - T_0) \tag{12.1-10}$$

Hence, the wall heat flux varies as $t^{-1/2}$, whereas the penetration thickness varies as $t^{1/2}$.

EXAMPLE 12.1-2

Heating of a Finite Slab

A solid slab occupying the space between $y = -b$ and $y = +b$ is initially at temperature T_0. At time $t = 0$ the surfaces at $y = \pm b$ are suddenly raised to T_1 and maintained there. Find $T(y, t)$.

SOLUTION

For this problem we define the following dimensionless variables:

Dimensionless temperature
$$\Theta = \frac{T_1 - T}{T_1 - T_0} \tag{12.1-11}$$

Dimensionless coordinate
$$\eta = \frac{y}{b} \tag{12.1-12}$$

Dimensionless time
$$\tau = \frac{\alpha t}{b^2} \tag{12.1-13}$$

With these dimensionless variables, the differential equation and boundary conditions are

$$\frac{\partial \Theta}{\partial \tau} = \frac{\partial^2 \Theta}{\partial \eta^2} \tag{12.1-14}$$

I.C.:   at $\tau = 0$, $\Theta = 1$ (12.1-15)

B.C. 1 and 2: at $\eta = \pm 1$, $\Theta = 0$ for $\tau > 0$ (12.1-16)

Note that no parameters appear when the problem is restated thus.

We can solve this problem by the method of separation of variables. We start by postulating that a solution of the following product form can be obtained:

$$\Theta(\eta, \tau) = f(\eta)g(\tau) \tag{12.1-17}$$

Substitution of this trial function into Eq. 12.1-14 and subsequent division by the product $f(\eta)g(\tau)$ gives

$$\frac{1}{g}\frac{dg}{d\tau} = \frac{1}{f}\frac{d^2f}{d\eta^2} \tag{12.1-18}$$

The left side is a function of τ alone, and the right side is a function of η alone. This can be true only if both sides equal a constant, which we call $-c^2$. If the constant is called $+c^2$, $+c$, or $-c$, the same final result is obtained, but the solution is a bit messier. Equation 12.1-18 can then be separated into two ordinary differential equations

$$\frac{dg}{d\tau} = -c^2 g \tag{12.1-19}$$

$$\frac{d^2f}{d\eta^2} = -c^2 f \tag{12.1-20}$$

These equations are of the form of Eq. C.1-1 and 3 and may be integrated to give

$$g = A \exp(-c^2\tau) \tag{12.1-21}$$

$$f = B \sin c\eta + C \cos c\eta \tag{12.1-22}$$

in which A, B, and C are constants of integration.

Because of the symmetry about the xz-plane, we must have $\Theta(\eta, \tau) = \Theta(-\eta, \tau)$, and thus $f(\eta) = f(-\eta)$. Since the sine function does not have this kind of behavior, we have to require that B be zero. Use of either of the two boundary conditions gives

$$C \cos c = 0 \tag{12.1-23}$$

Clearly C cannot be zero, because that choice leads to a physically inadmissible solution. However, the equality can be satisfied by many different choices of c, which we call c_n:

$$c_n = (n + \tfrac{1}{2})\pi \qquad n = 0, \pm 1, \pm 2, \pm 3 \ldots, \pm\infty \tag{12.1-24}$$

Hence Eq. 12.1-14 can be satisfied by

$$\Theta_n = A_n C_n \exp[-(n + \tfrac{1}{2})^2 \pi^2 \tau] \, \cos(n + \tfrac{1}{2})\pi\eta \qquad (12.1\text{-}25)$$

The subscripts n remind us that A and C may be different for each value of n. Because of the linearity of the differential equation, we may now superpose all the solutions of the form of Eq. 12.1-25. In doing this we note that the exponentials and cosines for n have the same values as those for $-(n + 1)$, so that the terms with negative indices combine with those with positive indices. The superposition then gives

$$\Theta = \sum_{n=0}^{\infty} D_n \exp[-(n + \tfrac{1}{2})^2 \pi^2 \tau] \, \cos(n + \tfrac{1}{2})\pi\eta \qquad (12.1\text{-}26)$$

in which $D_n = A_n C_n + A_{-(n+1)} C_{-(n+1)}$.

The D_n are now determined by using the initial condition, which gives

$$1 = \sum_{n=0}^{\infty} D_n \cos(n + \tfrac{1}{2})\pi\eta \qquad (12.1\text{-}27)$$

Multiplication by $\cos(m + \tfrac{1}{2})\pi\eta$ and integration from $\eta = -1$ to $\eta = +1$ gives

$$\int_{-1}^{+1} \cos(m + \tfrac{1}{2})\pi\eta \, d\eta = \sum_{n=0}^{\infty} D_n \int_{-1}^{+1} \cos(m + \tfrac{1}{2})\pi\eta \, \cos(n + \tfrac{1}{2})\pi\eta \, d\eta \qquad (12.1\text{-}28)$$

When the integrations are performed, all integrals on the right side are identically zero, except for the term in which $n = m$. Hence we get

$$\frac{\sin(m + \tfrac{1}{2})\pi\eta}{(m + \tfrac{1}{2})\pi}\bigg|_{\eta=-1}^{\eta=+1} = D_m \frac{\tfrac{1}{2}(m + \tfrac{1}{2})\pi\eta + \tfrac{1}{4}\sin 2(m + \tfrac{1}{2})\pi\eta}{(m + \tfrac{1}{2})\pi}\bigg|_{\eta=-1}^{\eta=+1} \qquad (12.1\text{-}29)$$

After inserting the limits, we may solve for D_m to get

$$D_m = \frac{2(-1)^m}{(m + \tfrac{1}{2})\pi} \qquad (12.1\text{-}30)$$

Substitution of this expression into Eq. 12.1-26 gives the temperature profiles, which we now rewrite in terms of the original variables[2]

$$\frac{T_1 - T}{T_1 - T_0} = 2 \sum_{n=0}^{\infty} \frac{(-1)^n}{(n + \tfrac{1}{2})\pi} \exp[-(n + \tfrac{1}{2})^2 \pi^2 \alpha t / b^2] \, \cos(n + \tfrac{1}{2})\frac{\pi y}{b} \qquad (12.1\text{-}31)$$

The solutions to many unsteady-state heat conduction problems come out as infinite series, such as that just obtained here. These series converge rapidly for large values[2] of the dimensionless time, $\alpha t / b^2$. For very short times the convergence is very slow, and in the limit as $\alpha t / b^2$ approaches zero, the solution in Eq. 12.1-31 may be shown to approach that given in Eq. 12.1-8 (see Problem 12D.1). Although Eq. 12.1-31 is unwieldy for some practical calculations, a graphical presentation, such as that in Fig. 12.1-1, is easy to use (see Problem 12A.3). From the figure it is clear that when the dimensionless time $\tau = \alpha t / b^2$ is 0.1, the heat has "penetrated" measurably to the center plane of the slab, and that at $\tau = 1.0$ the heating is 90% complete at the center plane.

Results analogous to Fig. 12.1-1 are given for infinite cylinders and for spheres in Figs. 12.1-2 and 3. These charts can also be used to build up the solutions for the analogous heat conduction problems in rectangular parallelepipeds and cylinders of finite length (see Prob. 12C.1).

---

[2] H. S. Carslaw and J. C. Jaeger, *Conduction of Heat in Solids*, 2nd edition, Oxford University Press (1959), p. 97, Eq. (8); the alternate solution in Eq. (9) converges rapidly for small times.

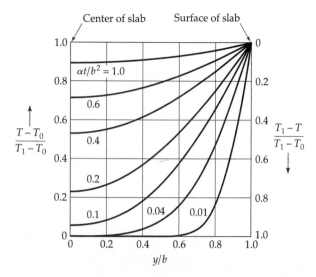

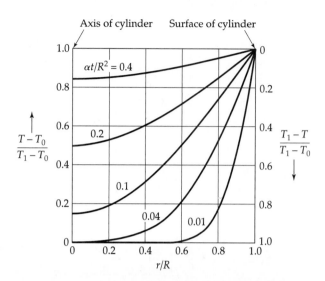

Fig. 12.1-1. Temperature profiles for unsteady-state heat conduction in a slab of finite thickness $2b$. The initial temperature of the slab is T_0, and T_1 is the temperature imposed at the slab surfaces for time $t > 0$. [H. S. Carslaw and J. C. Jaeger, *Conduction of Heat in Solids*, 2nd edition, Oxford University Press (1959), p. 101.]

Fig. 12.1-2. Temperature profiles for unsteady-state heat conduction in a cylinder of radius R. The initial temperature of the cylinder is T_0, and T_1 is the temperature imposed at the cylinder surface for time $t > 0$. [H. S. Carslaw and J. C. Jaeger, *Conduction of Heat in Solids*, 2nd edition, Oxford University Press (1959), p. 200.]

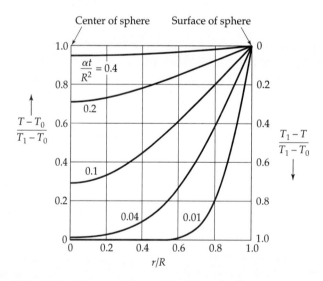

Fig. 12.1-3. Temperature profiles for unsteady-state heat conduction in a sphere of radius R. The initial temperature of the sphere is T_0, and T_1 is the temperature imposed at the sphere surface for time $t > 0$. [H. S. Carslaw and J. C. Jaeger, *Conduction of Heat in Solids*, 2nd edition, Oxford University Press (1959), p. 234.]

EXAMPLE 12.1-3

Unsteady Heat Conduction near a Wall with Sinusoidal Heat Flux

A solid body occupying the space from $y = 0$ to $y = \infty$ is initially at temperature T_0. Beginning at time $t = 0$, a periodic heat flux given by

$$q_y = q_0 \cos \omega t = q_0 \Re\{e^{i\omega t}\} \tag{12.1-32}$$

is imposed at $y = 0$. Here q_0 is the amplitude of the heat flux oscillations, and ω is the (circular) frequency. It is desired to find the temperature in this system, $T(y, t)$, in the "periodic steady state" (see Problem 4.1-3).

SOLUTION

For one-dimensional heat conduction, Eq. 12.1-2 is

$$\frac{\partial T}{\partial t} = \alpha \frac{\partial^2 T}{\partial y^2} \tag{12.1-33}$$

Multiplying by $-k$ and operating on the entire equation with $\partial/\partial y$ gives

$$\frac{\partial}{\partial t}\left(-k\frac{\partial T}{\partial y}\right) = \alpha \frac{\partial^2}{\partial y^2}\left(-k\frac{\partial T}{\partial y}\right) \tag{12.2-34}$$

or, by making use of $q_y = -k(\partial T/\partial y)$,

$$\frac{\partial q_y}{\partial t} = \alpha \frac{\partial^2 q_y}{\partial y^2} \tag{12.1-35}$$

Hence q_y satisfies the same differential equation as T. The boundary conditions are

B.C. 1:   at $y = 0$, $q_y = q_0\Re\{e^{i\omega t}\}$ (12.1-36)

B.C. 2:   at $y = \infty$, $q_y = 0$ (12.1-37)

This problem is formally exactly the same as that given in Eqs. 4.1-44, 46, and 47. Hence the solution in Eq. 4.1-57 may be taken over with appropriate notational changes:

$$q_y(y, t) = q_0 e^{-\sqrt{\omega/2\alpha}\,y} \cos\left(\omega t - \sqrt{\frac{\omega}{2\alpha}}\,y\right) \tag{12.1-38}$$

Then by integrating Fourier's law

$$-k\int_T^{T_0} dT = \int_y^\infty q_y(y, t)\, dy \tag{12.1-39}$$

Substitution of the heat flux distribution into the right side of this equation gives after integration

$$T - T_0 = \frac{q_0}{k}\sqrt{\frac{\alpha}{\omega}}\, e^{-\sqrt{\omega/2\alpha}\,y} \cos\left(\omega t - \sqrt{\frac{\omega}{2\alpha}}\,y - \frac{\pi}{4}\right) \tag{12.1-40}$$

Thus, at the surface $y = 0$, the temperature oscillations lag behind the heat flux oscillations by $\pi/4$.

This problem illustrates a standard procedure for obtaining the "periodic steady state" in heat conduction systems. It also shows how one can use the heat conduction equation in terms of the heat flux, when boundary conditions on the heat flux are known.

EXAMPLE 12.1-4

Cooling of a Sphere in Contact with a Well-Stirred Fluid

A homogeneous solid sphere of radius R, initially at a uniform temperature T_1, is suddenly immersed at time $t = 0$ in a volume V_f of well-stirred fluid of temperature T_0 in an insulated tank. It is desired to find the thermal diffusivity $\alpha_s = k_s/\rho_s \hat{C}_{ps}$ of the solid by observing the change of the fluid temperature T_f with time. We use the following dimensionless variables:

$$\Theta_s(\xi, \tau) = \frac{T_1 - T_s}{T_1 - T_0} = \text{dimensionless solid temperature} \tag{12.1-41}$$

$$\Theta_f(\tau) = \frac{T_1 - T_f}{T_1 - T_0} = \text{dimensionless fluid temperature} \tag{12.1-42}$$

$$\xi = \frac{r}{R} = \text{dimensionless radial coordinate} \tag{12.1-43}$$

$$\tau = \frac{\alpha_s t}{R^2} = \text{dimensionless time} \tag{12.1-44}$$

SOLUTION   The reader may verify that the problem stated in dimensionless variables is

Solid		Fluid	
$\dfrac{\partial \Theta_s}{\partial \tau} = \dfrac{1}{\xi^2} \dfrac{\partial}{\partial \xi}\left(\xi^2 \dfrac{\partial \Theta_s}{\partial \xi}\right)$	(12.1-45)	$\dfrac{d\Theta_f}{d\tau} = -\dfrac{3}{B} \dfrac{\partial \Theta_s}{\partial \xi}\Big|_{\xi=1}$	(12.1-49)
At $\tau = 0$, $\Theta_s = 0$	(12.1-46)	At $\tau = 0$, $\Theta_f = 1$	(12.1-50)
At $\xi = 1$, $\Theta_s = \Theta_f$	(12.1-47)		
At $\xi = 0$, $\Theta_s = $ finite	(12.1-48)		

in which $B = \rho_f \hat{C}_{pf} V_f / \rho_s \hat{C}_{ps} V_s$, the V's representing the volume of the fluid and of the solid.

Linear problems with complicated boundary conditions and/or coupling between equations are often solved readily by the Laplace transform method. We now take the Laplace transform of the preceding equations and their boundary conditions to get:

Solid		Fluid	
$p\overline{\Theta}_s = \dfrac{1}{\xi^2} \dfrac{d}{d\xi}\left(\xi^2 \dfrac{d\overline{\Theta}_s}{d\xi}\right)$	(12.1-51)	$p\overline{\Theta}_f - 1 = -\dfrac{3}{B} \dfrac{d\overline{\Theta}_s}{d\xi}\Big|_{\xi=1}$	(12.1-54)
At $\xi = 1$, $\overline{\Theta}_s = \overline{\Theta}_f$	(12.1-52)		
At $\xi = 0$, $\overline{\Theta}_s = $ finite	(12.1-53)		

Here p is the transform variable.[3] The solution to Eq. 12.1-51 is

$$\overline{\Theta}_s = \frac{C_1}{\xi} \sinh\sqrt{p}\xi + \frac{C_2}{\xi} \cosh\sqrt{p}\xi \tag{12.1-55}$$

Because of the boundary condition at $\xi = 0$, we must set C_2 equal to zero. Substitution of this result into Eq. 12.1-54 then gives

$$\overline{\Theta}_f = \frac{1}{p} + 3\frac{C_1}{Bp}(\sinh\sqrt{p} - \sqrt{p}\cosh\sqrt{p}) \tag{12.1-56}$$

Next, we insert these last two results into the boundary condition at $\xi = 1$, in order to determine C_1. This gives us for $\overline{\Theta}_f$:

$$\overline{\Theta}_f = \frac{1}{p} + 3\left(\frac{1 - (1/\sqrt{p})\tanh\sqrt{p}}{(3 - Bp)\sqrt{p}\tanh\sqrt{p} - 3p}\right) \tag{12.1-57}$$

We now divide the numerator and denominator within the parentheses by p, and take the inverse Laplace transform to get

$$\Theta_f = 1 + 3\mathcal{L}^{-1}\left\{\frac{(1/p) - (1/p^{3/2})\tanh\sqrt{p}}{(3 - Bp)(1/\sqrt{p})\tanh\sqrt{p} - 3}\right\} \equiv 1 + 3\mathcal{L}^{-1}\left\{\frac{N(p)}{D(p)}\right\} \tag{12.1-58}$$

---

[3] We use the definition $\mathcal{L}\{f(t)\} = \bar{f}(p) = \displaystyle\int_0^\infty f(t)e^{-pt}\,dt$.

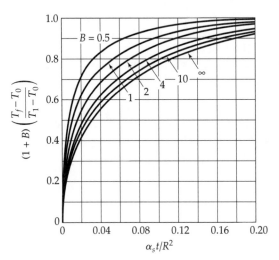

Fig. 12.1-4. Variation of the fluid temperature with time after a sphere of radius R at temperature T_1 is placed in a well-stirred fluid initially at a temperature T_0. The dimensionless parameter B is defined in the text following Eq. 12.1-50. [H. S. Carslaw and J. C. Jaeger, *Conduction of Heat in Solids*, 2nd edition, Oxford University Press (1959), p. 241.]

It can be shown that $D(p)$ has a single root at $p = 0$, and roots at $\sqrt{p_k} = ib_k$ (with $k = 1, 2, 3, \ldots, \infty$), where the b_k are the nonzero roots of $\tan b_k = 3b_k/(3 + Bb_k^2)$. The Heaviside partial fractions expansion theorem[4] may now be used with

$$\frac{N(0)}{D'(0)} = -\frac{1/3}{1 + B} \qquad \frac{N(p_k)}{D'(p_k)} = \frac{2B}{9(1 + B) + B^2 b_k^2} \qquad \text{(12.1-59, 60)}$$

to get

$$\Theta_f = \frac{B}{1 + B} + 6B \sum_{k=1}^{\infty} \frac{\exp(-b_k^2 \tau)}{9(1 + B) + B^2 b_k^2} \qquad \text{(12.1-61)}$$

Equation 12.1-61 is shown graphically in Fig. 12.1-4. In this result the only place where the thermal diffusivity of the solid α_s appears is in the dimensionless time $\tau = \alpha_s t/R^2$, so that the temperature rise of the fluid can be used to determine experimentally the thermal diffusivity of the solid. Note that the Laplace transform technique allows us to get the temperature history of the fluid without obtaining the temperature profiles in the solid.

§12.2   STEADY HEAT CONDUCTION IN LAMINAR, INCOMPRESSIBLE FLOW

In the preceding discussion of heat conduction in solids, we needed to use only the energy equation. For problems involving flowing fluids, however, all three equations of change are needed. Here we restrict the discussion to steady flow of incompressible, Newtonian fluids with constant fluid properties, for which the relevant equations of change are:

Continuity	$(\nabla \cdot \mathbf{v}) = 0$	(12.2-1)
Motion	$\rho[\mathbf{v} \cdot \nabla \mathbf{v}] = \mu \nabla^2 \mathbf{v} - \nabla \mathscr{P}$	(12.2-2)
Energy	$\rho \hat{C}_p (\mathbf{v} \cdot \nabla T) = k \nabla^2 T + \mu \Phi_v$	(12.2-3)

---

[4] A. Erdélyi, W. Magnus, F. Oberhettinger, and F. G. Tricomi, *Tables of Integral Transforms*, Vol. 1, McGraw-Hill, New York (1954), p. 232, Eq. 20; see also C. R. Wylie and L. C. Barrett, *Advanced Engineering Mathematics*, McGraw-Hill, New York, 6th Edition (1995), §10.9.

In Eq. 12.2-3, Φ_v is the dissipation function given in Eq. 3.3-3. To get the temperature profiles for forced convection, a two-step procedure is used: first Eqs. 12.2-1 and 2 are solved to obtain the velocity distribution $\mathbf{v}(\mathbf{r}, t)$; then the expression for $\mathbf{v}$ is substituted into Eq. 12.2-3, which may in turn be solved to get the temperature distribution $T(\mathbf{r}, t)$.

Many analytical solutions of Eqs. 12.2-1 to 3 are available for commonly encountered situations.[1-7] One of the oldest forced-convection problems is the *Graetz–Nusselt problem*,[8] describing the temperature profiles in tube flow where the wall temperature undergoes a sudden step change at some position along the tube (see Problems 12D.2, 3, and 4). Analogous solutions have been obtained for arbitrary variations of wall temperature and wall flux.[9] The Graetz–Nusselt problem has also been extended to non-Newtonian fluids.[10] Solutions have also been developed for a large class of laminar heat exchanger problems,[11] in which the wall boundary condition is provided by the continuity of heat flux across the surfaces separating the two streams. A further problem of interest is duct flow with significant viscous heating effects (the *Brinkman problem*[12]).

In this section we extend the discussion of the problem treated in §10.8—namely, the determination of temperature profiles for laminar flow of an incompressible fluid in a circular tube. In that section we set up the problem and found the asymptotic solution for distances far downstream from the beginning of the heated zone. Here, we give the complete solution to the partial differential equation as well as the asymptotic solution for short distances. That is, the system shown in Fig. 10.8-2 is discussed from three viewpoints in this book:

    a. Complete solution of the partial differential equation by the method of separation of variables (Example 12.2-1).

    b. Asymptotic solution for short distances down the tube by the method of combination of variables (Example 12.2-2).

    c. Asymptotic solution for large distances down the tube (§10.8).

---

[1] M. Jakob, *Heat Transfer*, Vol. I, Wiley, New York (1949), pp. 451–464.

[2] H. Gröber, S. Erk, and U. Grigull, *Die Grundgesetze der Wärmeübertragung*, Springer, Berlin (1961), Part II.

[3] R. K. Shah and A. L. London, *Laminar Flow Forced Convection in Ducts*, Academic Press, New York (1978).

[4] L. C. Burmeister, *Convective Heat Transfer*, Wiley-Interscience, New York (1983).

[5] L. D. Landau and E. M. Lifshitz, *Fluid Mechanics*, Pergamon, Oxford (1987), Chapter 5.

[6] L. G. Leal, *Laminar Flow and Convective Transport Processes*, Butterworth-Heinemann (1992), Chapters 8 and 9.

[7] W. M. Deen, *Analysis of Transport Phenomena*, Oxford University Press (1998), Chapters 9 and 10.

[8] L. Graetz, *Ann. Phys. (N.F.)*, **18**, 79–94 (1883), **25**, 337–357 (1885); W. Nusselt, *Zeits. Ver. deutch. Ing.*, **54**, 1154–1158 (1910). For the "extended Graetz problem," which includes axial conduction, see E. Papoutsakis, D. Ramkrishna, and H. C. Lim, *Appl. Sci. Res.*, **36**, 13–34 (1980).

[9] E. N. Lightfoot, C. Massot, and F. Irani, *Chem. Eng. Progress Symp. Series*, Vol. 61, No. 58 (1965), pp. 28–60.

[10] R. B. Bird, R. C. Armstrong, and O. Hassager, *Dynamics of Polymeric Liquids*, Wiley-Interscience (1987), 2nd edition, Vol. 1, §4.4.

[11] R. J. Nunge and W. N. Gill, *AIChE Journal*, **12**, 279–289 (1966).

[12] H. C. Brinkman, *Appl. Sci. Research*, **A2**, 120–124 (1951); R. B. Bird, *SPE Journal*, **11**, 35–40 (1955); H. L. Toor, *Ind. Eng. Chem.*, **48**, 922–926 (1956).

EXAMPLE 12.2-1

Laminar Tube Flow with Constant Heat Flux at the Wall

Solve Eq. 10.8-19 with the boundary conditions given in Eqs. 10.8-20, 21, and 22.

SOLUTION

The complete solution for the temperature is postulated to be of the following form:

$$\Theta(\xi, \zeta) = \Theta_\infty(\xi, \zeta) - \Theta_d(\xi, \zeta) \tag{12.2-4}$$

in which $\Theta_\infty(\xi, \zeta)$ is the asymptotic solution given in Eq. 10.8-31, and $\Theta_d(\xi, \zeta)$ is a function that will be damped out exponentially with ζ. By substituting the expression for $\Theta(\xi, \zeta)$ in Eq. 12.2-4 into Eq. 10.8-19, it may be shown that the function $\Theta_d(\xi, \zeta)$ must satisfy Eq. 10.8-19 and also the following boundary conditions:

B.C. 1:   at $\xi = 0$, $\dfrac{\partial \Theta_d}{\partial \xi} = 0$ (12.2-5)

B.C. 2:   at $\xi = 1$, $\dfrac{\partial \Theta_d}{\partial \xi} = 0$ (12.2-6)

B.C. 3:   at $\zeta = 0$, $\Theta_d = \Theta_\infty(\xi, 0)$ (12.2-7)

We anticipate that a solution to the equation for $\Theta_d(\xi, \zeta)$ will be factorable,

$$\Theta_d(\xi, \zeta) = X(\xi) Z(\zeta) \tag{12.2-8}$$

Then Eq. 10.8-19 can be separated into two ordinary differential equations

$$\frac{dZ}{d\zeta} = -c^2 Z \tag{12.2-9}$$

$$\frac{1}{\xi} \frac{d}{d\xi} \left(\xi \frac{dX}{d\xi} \right) + c^2 (1 - \xi^2) X = 0 \tag{12.2-10}$$

in which $-c^2$ is the separation constant. Since the boundary conditions on X are $dX/d\xi = 0$ at $\xi = 0, 1$, we have a Sturm–Liouville problem.[13] Therefore we know there will be an infinite number of eigenvalues c_k and eigenfunctions X_k, and that the final solution must be of the form:

$$\Theta(\xi, \zeta) = \Theta_\infty(\xi, \zeta) - \sum_{k=1}^{\infty} B_k \exp(-c_k^2 \zeta) X_k(\xi) \tag{12.2-11}$$

where

$$B_k = \frac{\displaystyle\int_0^1 \Theta_\infty(\xi, 0)[X_k(\xi)](1 - \xi^2)\xi \, d\xi}{\displaystyle\int_0^1 [X_k(\xi)]^2 (1 - \xi^2)\xi \, d\xi} \tag{12.2-12}$$

The problem is thus reduced to finding the eigenfunctions $X_k(\xi)$ by solving Eq. 12.2-10, and then getting the eigenvalues c_k by applying the boundary condition at $\xi = 1$. This has been done for k up to 7 for this problem.[14]

---

[13] M. D. Greenberg, *Advanced Engineering Mathematics*, Prentice-Hall, Upper Saddle River, N.J., Second Edition (1998), §17.7.

[14] R. Siegel, E. M. Sparrow, and T. M. Hallman, *Appl. Sci. Research*, **A7**, 386–392 (1958).

EXAMPLE 12.2-2

Laminar Tube Flow with Constant Heat Flux at the Wall: Asymptotic Solution for the Entrance Region

Note that the sum in Eq. 12.2-11 converges rapidly for large z but slowly for small z. Develop an expression for $T(r, z)$ that is useful for small values.

SOLUTION

For small z the heat addition affects only a very thin region near the wall, so that the following three approximations lead to results that are accurate in the limit as $z \to 0$:

a. Curvature effects may be neglected and the problem treated as though the wall were flat; call the distance from the wall $y = R - r$.

b. The fluid may be regarded as extending from the (flat) heat transfer surface ($y = 0$) to $y = \infty$.

c. The velocity profile may be regarded as linear, with a slope given by the slope of the parabolic velocity profile at the wall: $v_z(y) = v_0 y/R$, in which $v_0 = (\mathcal{P}_0 - \mathcal{P}_L)R^2/2\mu L$.

This is the way the system would appear to a tiny "observer" who is located within the very thin shell of heated fluid. To this observer, the wall would seem flat, the fluid would appear to be of infinite extent, and the velocity profile would seem to be linear.

The energy equation then becomes, in the region just slightly beyond $z = 0$,

$$v_0 \frac{y}{R} \frac{\partial T}{\partial z} = \alpha \frac{\partial^2 T}{\partial y^2} \tag{12.2-13}$$

Actually it is easier to work with the corresponding equation for the heat flux in the y direction ($q_y = -k\, \partial T/\partial y$). This equation is obtained by dividing Eq. 12.2-13 by y and differentiating with respect to y:

$$v_0 \frac{1}{R} \frac{\partial q_y}{\partial z} = \alpha \frac{\partial}{\partial y}\left(\frac{1}{y}\frac{\partial q_y}{\partial y}\right) \tag{12.2-14}$$

It is more convenient to work with dimensionless variables defined as

$$\psi = \frac{q_y}{q_0} \qquad \eta = \frac{y}{R} \qquad \lambda = \frac{\alpha z}{v_0 R^2} \tag{12.2-15}$$

Then Eq. 12.2-14 becomes

$$\frac{\partial \psi}{\partial \lambda} = \frac{\partial}{\partial \eta}\left(\frac{1}{\eta}\frac{\partial \psi}{\partial \eta}\right) \tag{12.2-16}$$

with these boundary conditions:

B.C. 1:	at $\lambda = 0$,	$\psi = 0$	(12.2-17)
B.C. 2:	at $\eta = 0$,	$\psi = 1$	(12.2-18)
B.C. 3:	as $\eta \to \infty$,	$\psi \to 0$	(12.2-19)

This problem can be solved by the method of combination of variables (see Examples 4.1-1 and 12.1-1) by using the new independent variable $\chi = \eta/\sqrt[3]{9\lambda}$. Then Eq. 12.2-16 becomes

$$\chi \frac{d^2\psi}{d\chi^2} + (3\chi^3 - 1)\frac{d\psi}{d\chi} = 0 \tag{12.2-20}$$

The boundary conditions are: at $\chi = 0$, $\psi = 1$, and as $\chi \to \infty$, $\psi \to 0$. The solution of Eq. 12.2-20 is found by first letting $d\psi/d\chi = p$, and getting a first-order equation for p. The equation for p can be solved and then ψ is obtained as

$$\psi(\chi) = \frac{\displaystyle\int_\chi^\infty \bar{\chi}\exp(-\bar{\chi}^3)\,d\bar{\chi}}{\displaystyle\int_0^\infty \bar{\chi}\exp(-\bar{\chi}^3)\,d\bar{\chi}} = \frac{3}{\Gamma(\tfrac{2}{3})}\int_\chi^\infty \bar{\chi}\exp(-\bar{\chi}^3)\,d\bar{\chi} \tag{12.2-21}$$

The temperature profile may then be obtained by integrating the heat flux:

$$\int_T^{T_1} dT = -\frac{1}{k} \int_y^\infty q_y dy \tag{12.2-22}$$

or, in dimensionless form,

$$\Theta(\eta, \lambda) = \frac{T - T_1}{q_0 R / k} = \sqrt[3]{9\lambda} \int_\chi^\infty \psi d\bar\chi \tag{12.2-23}$$

Then the expression for ψ is inserted into the integral, and the order of integration in the double integral can be reversed (see Problem 12D.7). The result is

$$\Theta(\eta, \lambda) = \sqrt[3]{9\lambda} \left[\frac{\exp(-\chi^3)}{\Gamma(\frac{2}{3})} - \chi \left(1 - \frac{\Gamma(\frac{2}{3}, \chi^3)}{\Gamma(\frac{2}{3})} \right) \right] \tag{12.2-24}$$

Here $\Gamma(\frac{2}{3})$ is the (complete) gamma function, and $\Gamma(\frac{2}{3}, \chi^3)$ is an incomplete gamma function.[15] To compare this result with that in Example 12.2-1, we note that $\eta = 1 - \xi$ and $\lambda = \frac{1}{2}\zeta$. The dimensionless temperature is defined identically in §10.8, in Example 12.2-1, and here.

§12.3   STEADY POTENTIAL FLOW OF HEAT IN SOLIDS

The steady flow of heat in solids of constant thermal conductivity is described by

Fourier's law $\mathbf{q} = -k\nabla T$ \hfill (12.3-1)

Heat conduction equation $\nabla^2 T = 0$ \hfill (12.3-2)

These equations are exactly analogous to the expression for the velocity in terms of the velocity potential ($\mathbf{v} = -\nabla\phi$), and the Laplace equation for the velocity potential ($\nabla^2\phi = 0$), which we encountered in §4.3. Steady heat conduction problems can therefore be solved by application of potential theory.

For two-dimensional heat conduction in solids with constant thermal conductivity, the temperature satisfies the two-dimensional Laplace equation:

$$\frac{\partial^2 T}{\partial x^2} + \frac{\partial^2 T}{\partial y^2} = 0 \tag{12.3-3}$$

We now use the fact that *any* analytic function $w(z) = f(x, y) + ig(x, y)$ provides two scalar functions f and g, which are solutions of Eq. 12.3-3. Curves of f = constant may be interpreted as lines of heat flow, and curves of g = constant are the corresponding isothermals for *some* heat flow problems. These two sets of curves are orthogonal—that is, they intersect at right angles. Furthermore, the components of the heat flux vector at any point are given by

$$ik\frac{dw}{dz} = q_x - iq_y \tag{12.3-4}$$

Given an analytic function, it is easy to find heat flow problems that are described by it. But the inverse process of finding an analytic function suitable for a given heat flow problem is generally very difficult. Some methods for this are available, but they are outside the scope of this textbook.[1,2]

---

[15] M. Abramowitz and I. A. Stegun, eds., *Handbook of Mathematical Functions*, Dover, New York, 9th Printing (1973), pp. 255 et seq.

[1] H. S. Carslaw and J. C. Jaeger, *Conduction of Heat in Solids*, 2nd edition, Oxford University Press (1959), Chapter XVI.

[2] M. D. Greenberg, *Advanced Engineering Mathematics*, Prentice-Hall, Upper Saddle River, N.J., 2nd Edition (1998), Chapter 22.

For every complex function $w(z)$, two heat flow nets are obtained by interchanging the lines of constant f and the lines of constant g. Furthermore, two additional nets are obtained by working with the inverse function $z(w)$ as illustrated in Chapter 4 for ideal fluid flow.

Note that potential fluid flow and potential heat flow are mathematically similar, the two-dimensional flow nets in both cases being described by analytic functions. Physically, however, there are certain important differences. The fluid flow nets described in §4.3 are for a fluid with no viscosity (a fictitious fluid!), and therefore one cannot use them to calculate the drag forces at surfaces. On the other hand, the heat flow nets described here are for solids that have a finite thermal conductivity, and therefore the results can be used to calculate the heat flow at all surfaces. Moreover, both the velocity components (in Cartesian coordinates!) of §4.3 and the temperature profiles of this section satisfy the Laplace equation. Further information about analogous physical processes described by the Laplace equation is available in books on partial differential equations.[3]

Here we give just one example to provide a glimpse of the use of analytic functions; further examples may be found in the references cited.

EXAMPLE 12.3-1	Consider a wall of thickness b extending from 0 to ∞ in the y direction, and from $-\infty$ to $+\infty$ in the direction perpendicular to the x and y directions (see Fig. 12.3-1). The surfaces at $x = \pm\frac{1}{2}b$

Temperature Distribution in a Wall

are held at temperature T_0, whereas the bottom of the wall at the surface $y = 0$ is maintained at temperature T_1. Show that the imaginary part of the function[4]

$$w(z) = \frac{1}{\pi} \ln\left(\frac{(\sin \pi z/b) - 1}{(\sin \pi z/b) + 1}\right) \tag{12.3-5}$$

gives the steady temperature distribution $\Theta(x, y) \doteq (T - T_0)/(T_1 - T_0)$.

SOLUTION

The imaginary part of $w(z)$ in Eq. 12.3-5 is

$$\Theta(x, y) = \frac{2}{\pi} \arctan\left(\frac{\cos \pi x/b}{\sinh \pi y/b}\right) \tag{12.3-6}$$

in which the arctangent is in the range from 0 to $\frac{\pi}{2}$. When $x = \pm\frac{1}{2}b$, Eq. 12.3-6 gives $\Theta = 0$, and when $y = 0$, it gives $\Theta = (2/\pi) \arctan \infty = 1$.

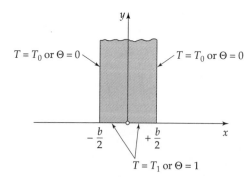

$T = T_0$ or $\Theta = 0$ $T = T_0$ or $\Theta = 0$

$-\dfrac{b}{2}$ $+\dfrac{b}{2}$

$T = T_1$ or $\Theta = 1$

Fig. 12.3-1. Steady two-dimensional temperature distribution in a wall.

---

[3] I. N. Sneddon, *Elements of Partial Differential Equations*, Dover, New York (1996), Chapter 4.

[4] R. V. Churchill, *Introduction to Complex Variables and Applications*, McGraw-Hill, New York (1948), Chapter IX. See also C. R. Wylie and L. C. Barrett, *Advanced Engineering Mathematics*, McGraw-Hill, New York, 6th Edition (1995), Chapter 20.

From Eq. 12.3-6 the heat flux through the base of the wall may be obtained:

$$q_y\big|_{y=0} = -k\frac{\partial T}{\partial y}\bigg|_{y=0} = \frac{2k\sec\pi x/b}{b}(T_1 - T_0) \tag{12.3-7}$$

§12.4   BOUNDARY LAYER THEORY FOR NONISOTHERMAL FLOW[1,2,3]

In §4.4 the use of boundary layer approximations for steady, laminar flow of incompressible fluids at constant temperature was discussed. We saw that, in the neighborhood of a solid surface, the equations of continuity and motion could be simplified, and that these equations may be solved to get "exact boundary layer solutions" and that an integrated form of these equations (the von Kármán momentum balance) enables one to get "approximate boundary layer solutions." In this section we extend the previous development by including the boundary layer equation for energy transport, so that the temperature profiles near solid surfaces can be obtained.

As in §4.4 we consider the steady two-dimensional flow around a submerged object such as that shown in Fig. 4.4-1. In the vicinity of the solid surface the equations of change may be written (omitting the bars over ρ and β) as:

Continuity
$$\frac{\partial v_x}{\partial x} + \frac{\partial v_y}{\partial y} = 0 \tag{12.4-1}$$

Motion
$$\rho\left(v_x\frac{\partial v_x}{\partial x} + v_y\frac{\partial v_x}{\partial y}\right) = \rho v_e\frac{dv_e}{dx} + \mu\frac{\partial^2 v_x}{\partial y^2} + \rho g_x\beta(T - T_\infty) \tag{12.4-2}$$

Energy
$$\rho\hat{C}_p\left(v_x\frac{\partial T}{\partial x} + v_y\frac{\partial T}{\partial y}\right) = k\frac{\partial^2 T}{\partial y^2} + \mu\left(\frac{\partial v_x}{\partial y}\right)^2 \tag{12.4-3}$$

Here ρ, μ, k, and $\hat{C}_p$ are regarded as constants, and $\mu(\partial v_x/\partial y)^2$ is the viscous heating effect, which is henceforth disregarded. Solutions of these equations are asymptotically accurate for small momentum diffusivity $\nu = \mu/\rho$ in Eq. 12.4-2, and for small thermal diffusivity $\alpha = k/\rho\hat{C}_p$ in Eq. 12.4-3.

Equation 12.4-1 is the same as Eq. 4.4-1. Equation 12.4-2 differs from Eq. 4.4-2 because of the inclusion of the buoyant force term (see §11.3), which can be significant even when fractional changes in density are small. Equation 12.4-3 is obtained from Eq. 11.2-9 by neglecting the heat conduction in the x direction. More complete forms of the boundary layer equations may be found elsewhere.[2,3]

The usual boundary conditions for Eqs. 12.4-1 and 2 are that $v_x = v_y = 0$ at the solid surface, and that the velocity merges into the potential flow at the outer edge of the *velocity boundary layer*, so that $v_x \to v_e(x)$. For Eq. 12.4-3 the temperature T is specified to be T_0 at the solid surface and T_∞ at the outer edge of the *thermal boundary layer*. That is, the velocity and temperature are different from $v_e(x)$ and T_∞ only in thin layers near the solid surface. However, the velocity and temperature boundary layers will be of different thicknesses corresponding to the relative ease of the diffusion of momentum and heat. Since $Pr = \nu/\alpha$, for $Pr > 1$ the temperature boundary layer usually lies inside the veloc-

---

[1] H. Schlichting, *Boundary-Layer Theory*, 7th edition, McGraw-Hill, New York (1979), Chapter 12.

[2] K. Stewartson, *The Theory of Laminar Boundary Layers in Compressible Fluids*, Oxford University Press (1964).

[3] E. R. G. Eckert and R. M. Drake, Jr., *Analysis of Heat and Mass Transfer*, McGraw-Hill, New York, (1972), Chapters 6 and 7.

ity boundary layer, whereas for $Pr < 1$ the relative thicknesses are just reversed (keep in mind that for gases Pr is about $\frac{3}{4}$, whereas for ordinary liquids $Pr > 1$ and for liquid metals $Pr << 1$).

In §4.4 we showed that the boundary layer equation of motion could be integrated formally from $y = 0$ to $y = \infty$, if use is made of the equation of continuity. In a similar fashion the integration of Eqs. 12.4-1 to 3 can be performed to give

Momentum
$$\mu \frac{\partial v_x}{\partial y}\bigg|_{y=0} = \frac{d}{dx} \int_0^\infty \rho v_x (v_e - v_x) dy + \frac{dv_e}{dx} \int_0^\infty \rho(v_e - v_x) dy$$

$$+ \int_0^\infty \rho g_x \beta (T - T_\infty) dy \tag{12.4-4}$$

Energy
$$k \frac{\partial T}{\partial y}\bigg|_{y=0} = \frac{d}{dx} \int_0^\infty \rho \hat{C}_p v_x (T_\infty - T) dy \tag{12.4-5}$$

Equations 12.4-4 and 5 are the *von Kármán momentum and energy balances*, valid for forced-convection and free-convection systems. The no-slip condition $v_y = 0$ at $y = 0$ has been used here, as in Eq. 4.4-4; nonzero velocities at $y = 0$ occur in mass transfer systems and will be considered in Chapter 20.

As mentioned in §4.4, there are two approaches for solving boundary layer problems: analytical or numerical solutions of Equations 12.4-1 to 3 are called "exact boundary layer solutions," whereas solutions obtained from Eqs. 12.4-4 and 5, with reasonable guesses for the velocity and temperature profiles, are called "approximate boundary layer solutions." Often considerable physical insight can be obtained by the second method, and with relatively little effort. Example 12.4-1 illustrates this method.

Extensive use has been made of the boundary layer equations to establish correlations of momentum- and heat-transfer rates, as we shall see in Chapter 14. Although in this section we do not treat free convection, in Chapter 14 many useful results are given along with the appropriate literature citations.

EXAMPLE 12.4-1

Heat Transfer in Laminar Forced Convection along a Heated Flat Plate (von Kármán Integral Method)

Obtain the temperature profiles near a flat plate, along which a Newtonian fluid is flowing, as shown in Fig. 12.4-1. The wetted surface of the plate is maintained at temperature T_0 and the temperature of the approaching fluid is T_∞.

SOLUTION

In order to use the von Kármán balances we first postulate reasonable forms for the velocity and temperature profiles. The following polynomial form gives 0 at the wall and 1 at the outer limit of the boundary layer, with a slope of zero at the outer limit:

$$\begin{cases} \dfrac{v_x}{v_\infty} = 2\left(\dfrac{y}{\delta}\right) - 2\left(\dfrac{y}{\delta}\right)^3 + \left(\dfrac{y}{\delta}\right)^4 & y \le \delta(x) \\[2mm] \dfrac{v_x}{v_\infty} = 1 & y \ge \delta(x) \end{cases} \tag{12.4-6, 7}$$

$$\begin{cases} \dfrac{T_0 - T}{T_0 - T_\infty} = 2\left(\dfrac{y}{\delta_T}\right) - 2\left(\dfrac{y}{\delta_T}\right)^3 + \left(\dfrac{y}{\delta_T}\right)^4 & y \le \delta_T(x) \\[2mm] \dfrac{T_0 - T}{T_0 - T_\infty} = 1 & y \ge \delta_T(x) \end{cases} \tag{12.4-8, 9}$$

That is, we assume that the dimensionless velocity and temperature profiles have the same form within their respective boundary layers. We further *assume* that the boundary layer thicknesses $\delta(x)$ and $\delta_T(x)$ have a constant ratio, so that $\Delta = \delta_T(x)/\delta(x)$ is independent of x. Two possibilities have to be considered: $\Delta \le 1$ and $\Delta \ge 1$. We consider here $\Delta \le 1$ and relegate the other case to Problem 12D.8.

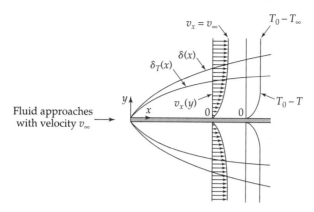

Fig. 12.4-1. Boundary layer development for the flow along a heated flat plate, showing the thermal boundary layer for $\Delta = \delta_T(x)/\delta(x) < 1$. The surface of the plate is at temperature T_0, and the approaching fluid is at T_∞.

The use of Eqs. 12.4-4 and 5 is now straightforward but tedious. Substitution of Eqs. 12.4-6 through 9 into the integrals gives (with v_e set equal to v_∞ here)

$$\int_0^\infty \rho v_x(v_\infty - v_x)dy = \rho v_\infty^2 \delta(x) \int_0^1 (2\eta - 2\eta^3 + \eta^4)(1 - 2\eta + 2\eta^3 - \eta^4)d\eta = \tfrac{37}{315}\rho v_\infty^2 \delta(x) \quad (12.4\text{-}10)$$

$$\int_0^\infty \rho \hat{C}_p v_x(T_\infty - T)dy = \rho \hat{C}_p v_\infty(T_\infty - T_0)\delta_T(x) \int_0^\infty (2\eta_T\Delta - 2\eta_T^3\Delta^3 + \eta_T^4\Delta^4)$$

$$\cdot (1 - 2\eta_T + 2\eta_T^3 - \eta_T^4)d\eta_T$$

$$= (\tfrac{2}{15}\Delta - \tfrac{3}{140}\Delta^3 + \tfrac{1}{180}\Delta^4)\rho \hat{C}_p v_\infty(T_\infty - T_0)\delta_T(x) \quad (12.4\text{-}11)$$

In these integrals $\eta = y/\delta(x)$ and $\eta_T = y/\delta_T(x) = y/\Delta\delta(x)$. Next, substitution of these integrals into Eqs. 12.4-4 and 5 gives differential equations for the boundary layer thicknesses. These first-order separable differential equations are easily integrated, and we get

$$\delta(x) = \sqrt{\frac{1260}{37}\left(\frac{\nu x}{v_\infty}\right)} \quad (12.4\text{-}12)$$

$$\delta_T(x) = \sqrt{\frac{4}{\tfrac{2}{15}\Delta - \tfrac{3}{140}\Delta^3 + \tfrac{1}{180}\Delta^4}\left(\frac{\alpha x}{v_\infty}\right)} \quad (12.4\text{-}13)$$

The boundary layer thicknesses are now determined, except for the evaluation of Δ in Eq. 12.4-13. The ratio of Eq. 12.4-12 to Eq. 12.4-13 gives an equation for Δ as a function of the Prandtl number:

$$\tfrac{2}{15}\Delta^3 - \tfrac{3}{140}\Delta^5 + \tfrac{1}{180}\Delta^6 = \tfrac{37}{315}\text{Pr}^{-1} \qquad \Delta \le 1 \quad (12.4\text{-}14)$$

When this sixth-order algebraic equation is solved for Δ as a function of Pr, it is found that the solution may be curve-fitted by the simple relation[4]

$$\Delta = \text{Pr}^{-1/3} \qquad \Delta < 1 \quad (12.4\text{-}15)$$

within about 5%.

---

[4] H. Schlichting, *Boundary-Layer Theory*, 7th edition, McGraw-Hill, New York (1979), pp. 292–308.

Table 12.4-1   Comparison of Boundary Layer Heat Transfer Calculations for Flow along a Flat Plate

Method	Value of numerical coefficient in expression for heat transfer rate in Eq. 12.4-17
Von Kármán method with profiles of Eqs. 12.4-9 to 12	$\sqrt{148/315} = 0.685$
Exact solution of Eqs. 12.4-1 to 3 by Pohlhausen	0.657 at Pr = 0.6
	0.664 at Pr = 1.0
	0.670 at Pr = 2.0
Curve fit of exact calculations (Pohlhausen)	0.664
Asymptotic solution of Eqs. 12.4-1 to 3 for Pr $\gg$ 1	0.677

The temperature profile is then finally given (for $\Delta \leq 1$) by

$$\frac{T_0 - T}{T_0 - T_\infty} = 2\left(\frac{y}{\Delta\delta}\right) - 2\left(\frac{y}{\Delta\delta}\right)^3 + \left(\frac{y}{\Delta\delta}\right)^4 \tag{12.4-16}$$

in which $\Delta \approx \mathrm{Pr}^{-1/3}$ and $\delta(x) = \sqrt{(1260/37)(\nu x/v_\infty)}$. The assumption of laminar flow made here is valid for $x < x_{crit}$, where $x_{crit}v_\infty\rho/\mu$ is usually greater than 10^5.

Finally, the rate of heat loss from both sides of a heated plate of width W and length L can be obtained from Eqs. 12.4-5, 11, 12, 15, and 16:

$$Q = 2\int_0^W \int_0^L q_y|_{y=0}dx\, dz$$

$$= 2\int_0^W \int_0^\infty \rho\hat{C}_p v_x(T - T_\infty)|_{x=L}dy\, dz$$

$$= 2W\rho\hat{C}_p v_\infty(T_0 - T_\infty)(\tfrac{2}{15}\Delta - \tfrac{3}{140}\Delta^3 + \tfrac{1}{180}\Delta^4)\delta_T(L)$$

$$\approx \sqrt{\tfrac{148}{315}}(2WL)(T_0 - T_\infty)\left(\frac{k}{L}\right)\mathrm{Pr}^{1/3}\mathrm{Re}_L^{1/2} \tag{12.4-17}$$

in which $\mathrm{Re}_L = Lv_\infty\rho/\mu$. Thus the boundary layer approach allows one to obtain the dependence of the rate of heat loss Q on the dimensions of the plate, the flow conditions, and the thermal properties of the fluid.

Eq. 12.4-17 is in good agreement with more detailed solutions based on Eqs. 12.4-1 to 3. The asymptotic solution for Q at large Prandtl numbers, given in the next example,[5] has the same form except that the numerical coefficient $\sqrt{148/315} = 0.685$ is replaced by 0.677. The exact solution for Q at finite Prandtl numbers, obtained numerically,[6] has the same form except that the coefficient is replaced by a slowly varying function $C(\mathrm{Pr})$, shown in Table 12.4-1. The value $C = 0.664$ is exact at Pr = 1 and good within $\pm 2\%$ for Pr > 0.6.

The proportionality of Q to $\mathrm{Pr}^{1/3}$, found here, is asymptotically correct in the limit as $\mathrm{Pr} \to \infty$, not only for the flat plate but also for all geometries that permit a laminar, nonseparating boundary layer, as illustrated in the next example. Deviations from $Q \sim \mathrm{Pr}^{1/3}$ occur at finite Prandtl numbers for flow along a flat plate and even more so for flows near other-shaped objects and near rotating surfaces. These deviations arise from nonlinearity of the velocity profiles within the thermal boundary layer. Asymptotic expansions for the Pr dependence of Q have been presented by Merk and others.[7]

[5] M. J. Lighthill, *Proc. Roy. Soc.*, **A202**, 359–377 (1950).
[6] E. Pohlhausen, *Zeits. f. angew. Math. u. Mech.*, **1**, 115–121 (1921).
[7] H. J. Merk, *J. Fluid Mech.*, **5**, 460–480 (1959).

EXAMPLE 12.4-2

Heat Transfer in Laminar Forced Convection along a Heated Flat Plate (Asymptotic Solution for Large Prandtl Numbers)[5]

In the preceding example we used the von Kármán boundary layer integral expressions. Now we repeat the same problem but obtain an exact solution of the boundary layer equations in the limit that the Prandtl number is large—that is, for liquids (see §9.1). In this limit, the outer edge of the thermal boundary layer is well inside the velocity boundary layer. Therefore it can safely be assumed that v_x varies linearly with y throughout the entire thermal boundary layer.

SOLUTION

By combining the boundary layer equations of continuity and energy (Eqs. 12.4-1 and 3) we get

$$v_x \frac{\partial T}{\partial x} + \left(-\int_0^y \frac{\partial v_x}{\partial x} \, dy \right) \frac{\partial T}{\partial y} = \alpha \frac{\partial^2 T}{\partial y^2} \tag{12.4-18}$$

in which $\alpha = k/\rho \hat{C}_p$. The leading term of a Taylor expansion for the velocity distribution near the wall is

$$\frac{v_x}{v_\infty} = c \frac{y}{\sqrt{\nu x / v_\infty}} \tag{12.4-19}$$

in which the constant $c = 0.4696/\sqrt{2} = 0.332$ can be inferred from Eq. 4.4-30.

Substitution of this velocity expression into Eq. 12.4-18 gives

$$\left(c \frac{y v_\infty}{\sqrt{\nu x / v_\infty}} \right) \frac{\partial T}{\partial x} + \left(\frac{c}{4} \frac{y^2 v_\infty / x}{\sqrt{\nu x / v_\infty}} \right) \frac{\partial T}{\partial y} = \alpha \frac{\partial^2 T}{\partial y^2} \tag{12.4-20}$$

This has to be solved with the boundary conditions that $T = T_0$ at $y = 0$, and $T = T_\infty$ at $x = 0$.

This equation can be solved by the method of combination of variables. The choice of the dimensionless variables

$$\Pi(\eta) = \frac{T_0 - T}{T_0 - T_\infty} \quad \text{and} \quad \eta = \left(\frac{c v_\infty^{3/2}}{12 \alpha \nu^{1/2}} \right)^{1/3} \frac{y}{x^{1/2}} \tag{12.4-21, 22}$$

makes it possible to rewrite Eq. 12.4-20 (see Eq. C.1-9) as

$$\frac{d^2 \Pi}{d\eta^2} + 3\eta^2 \frac{d\Pi}{d\eta} = 0 \tag{12.4-23}$$

Integration of this equation with the boundary conditions that $\Pi = 0$ at $\eta = 0$ and $\Pi \to 1$ as $\eta \to \infty$ gives

$$\Pi(\eta) = \frac{\int_0^\eta \exp(-\bar{\eta}^3) \, d\bar{\eta}}{\int_0^\infty \exp(-\bar{\eta}^3) \, d\bar{\eta}} = \frac{\int_0^\eta \exp(-\bar{\eta}^3) \, d\bar{\eta}}{\Gamma(\frac{4}{3})} \tag{12.4-24}$$

for the dimensionless temperature distribution. See §C.4 for a discussion of the gamma function $\Gamma(n)$.

For the rate of heat loss from both sides of a heated plate of width W and length L, we get

$$Q = 2 \int_0^W \int_0^L q_y|_{y=0} \, dx \, dz$$

$$= 2W \int_0^L \left(-k \frac{\partial T}{\partial y} \right)\bigg|_{y=0} dx$$

$$= (2WL)(T_0 - T_\infty)\left(\frac{k}{L} \right) \int_0^L \left(+k \frac{d\Pi}{d\eta} \right)\bigg|_{\eta=0} \left(\frac{c v_\infty^{3/2}}{12 \alpha \nu^{1/2}} \right)^{1/3} \frac{dx}{x^{1/2}}$$

$$= (2WL)(T_0 - T_\infty)\left(\frac{k}{L} \right)\left[\frac{2}{\Gamma(\frac{4}{3})} \left(\frac{c}{12} \right)^{1/3} \right] \mathrm{Pr}^{1/3} \mathrm{Re}_L^{1/2} \tag{12.4-25}$$

which is the same result as that in Eq. 12.4-17 aside from a numerical constant. The quantity within brackets equals 0.677, the asymptotic value that appears in Table 12.4-1.

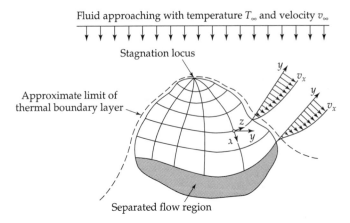

Fluid approaching with temperature T_∞ and velocity v_∞

Fig. 12.4-2. Heat transfer from a three-dimensional surface. The asymptotic analysis applies upstream of the separated and turbulent flow regions. These regions are illustrated for cylinders in Fig. 3.7-2.

EXAMPLE 12.4-3

Forced Convection in Steady Three-Dimensional Flow at High Prandtl Numbers[8,9]

The technique introduced in the preceding example has been extended to flow around objects of arbitrary shape. Consider the steady flow of a fluid over a stationary object as shown in Fig. 12.4-2. The fluid approaches at a uniform temperature T_∞, and the solid surface is maintained at a uniform temperature T_0. The temperature distribution and heat transfer rate are to be found for the region of laminar flow, which extends downstream from the stagnation locus to the place where turbulence or flow separation begins. The velocity profiles are considered to be known.

The thermal boundary layer is considered to be very thin. This implies that the isotherms nearly coincide with the solid surface, so that the heat flux **q** is nearly normal to the surface. It also implies that the complete velocity profiles are not needed here. We need to know the state of the motion only near the solid surface.

To capitalize on these simplifications, we choose the coordinates in a special way (see Fig. 12.4-2). We define y as the distance from the surface into the fluid just as in Fig. 12.4-1. We further define x and z as the coordinates of the nearest point on the surface, measured parallel and perpendicular to the tangential motion next to the surface. We express elements of arc in the x and z directions as $h_x dx$ and $h_z dz$, where h_x and h_z are position-dependent "scale factors" discussed in §A.7. Since we are interested here in the region of small y, the scale factors are treated as functions only of x and z evaluated at $y = 0$, with $h_y = 1$.

With this choice of coordinates, the velocity components for small y become

$$v_x = \beta(x, z)y \tag{12.4-26}$$

$$v_y = \left(-\frac{1}{2h_x h_z} \frac{\partial}{\partial x} (h_z \beta) \right) y^2 \tag{12.4-27}$$

$$v_z = 0 \tag{12.4-28}$$

Here $\beta(x, z)$ is the local value of $\partial v_x / \partial y$ on the surface; it is positive in the nonseparated region, but may vanish at points of stagnation or separation. These equations are obtained by writing Taylor series for v_x and v_z, retaining terms through the first degree in y, and then inte-

[8] W. E. Stewart, *AIChE Journal*, **9**, 528–535 (1963).
[9] For related two-dimensional analyses, see M. J. Lighthill, *Proc. Roy. Soc.*, **A202**, 359–377 (1950); V. G. Levich, *Physico-Chemical Hydrodynamics*, Chapter 2, Prentice-Hall, Englewood Cliffs, N.J. (1962); A. Acrivos, *Physics of Fluids*, **3**, 657–658 (1960).

grating the continuity equation with the boundary condition $v_y = 0$ at the surface to obtain v_y. These results are valid for Newtonian or non-Newtonian flow with temperature-independent density and viscosity.[10]

By a procedure analogous to that used in Example 12.4-2, one obtains a result similar to that given in Eq. 12.4-24. The only difference is that η is defined more generally as $\eta = y/\delta_T$, where δ_T is the thermal boundary layer thickness given by

$$\delta_T = \frac{1}{\sqrt{h_z\beta}}\left(9\alpha \int_{x_1(z)}^{x} \sqrt{h_z\beta}h_x h_z d\bar{x}\right)^{1/3} \tag{12.4-29}$$

and $x_1(z)$ is the upstream limit of the heat transfer region. From Eqs. 12.4-24 and 25 the local surface heat flux q_0 and the total heat flow for a heated region of the form $x_1(z) < x < x_2(z)$, $z_1 < z < z_2$ are

$$q_0 = \frac{k(T_0 - T_\infty)}{\Gamma(\tfrac{4}{3})\delta_T} \tag{12.4-30}$$

$$Q = \frac{3^{1/3}k(T_0 - T_\infty)}{2\alpha^{1/3}\Gamma(\tfrac{4}{3})}\int_{z_1}^{z_2}\left(\int_{x_1(z)}^{x_2(z)} \sqrt{h_z\beta}h_x h_z d\bar{x}\right)^{2/3} dz \tag{12.4-31}$$

This last result shows how Q depends on the fluid properties, the velocity profiles, and the geometry of the system. We see that Q is proportional to the temperature difference, to $k/\alpha^{1/3} = k^{2/3}\rho^{1/3}\hat{C}_p^{1/3}$, and to the $\frac{1}{3}$-power of a mean velocity gradient over the surface.

Show how the above results can be used to obtain the heat transfer rate from a heated sphere of radius R with a viscous fluid streaming past it in creeping flow[11] (see Example 4.2-1 and Fig. 2.6-1).

SOLUTION

The boundary-layer coordinates x, y, and z may be identified here with $\pi - \theta$, $r - R$, and ϕ of Fig. 2.6-1. Then stagnation occurs at $\theta = \pi$, and separation occurs at $\theta = 0$. The scale factors are $h_x = R$, and $h_z = R \sin\theta$. The interfacial velocity gradient β is

$$\beta = -\frac{\partial v_\theta}{\partial r}\bigg|_{r=R} = \frac{3}{2}\frac{v_\infty}{R}\sin\theta \tag{12.4-32}$$

Insertion of the above into Eqs. 12.4-29 and 31 gives the following results for forced convection heat transfer from an isothermal sphere of diameter D:

$$\delta_T = \frac{1}{\sqrt{\tfrac{3}{2}v_\infty \sin^2\theta}}\left(-9\alpha\int_\pi^0 \sqrt{\tfrac{3}{2}v_\infty \sin^2\theta}\, R^2 \sin\theta\, d\theta\right)^{1/3}$$

$$= (\tfrac{3}{4})^{1/3}D(\text{Re Pr})^{-1/3}\frac{(\pi - \theta + \tfrac{1}{2}\sin 2\theta)^{1/3}}{\sin\theta} \tag{12.4-33}$$

$$Q = \frac{3^{1/3}k(T_0 - T_\infty)}{2\alpha^{1/3}\Gamma(\tfrac{4}{3})}\int_0^{2\pi}\left(-\int_\pi^0 \sqrt{\tfrac{3}{2}v_\infty \sin^2\theta}\, R^2 \sin\theta\, d\theta\right)^{2/3} d\phi$$

$$= (\pi D^2)(T_0 - T_\infty)\left(\frac{k}{D}\right)\left[\frac{(3\pi)^{2/3}}{2^{7/3}\Gamma(\tfrac{4}{3})}\right](\text{Re Pr})^{1/3} \tag{12.4-34}$$

The constant in brackets is 0.991.

The behavior predicted by Eq. 12.4-33 is sketched in Fig. 12.4-3. The boundary layer thickness increases steadily from a small value at the stagnation point to an infinite value at separation, where the boundary layer becomes a wake extending downstream. The analysis here is most accurate for the forward part of the sphere, where δ_T is small; fortunately, that is

[10] Temperature-dependent properties have been included by Acrivos, *loc. cit.*

[11] The solution to this problem was first obtained by V. G. Levich, *loc. cit.* It has been extended to somewhat higher Reynolds numbers by A. Acrivos and T. D. Taylor, *Phys. Fluids*, **5**, 387–394 (1962).

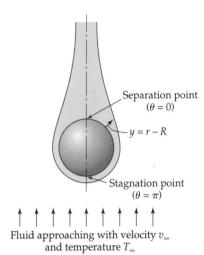

Fig. 12.4-3. Forced-convection heat transfer from a sphere in creeping flow. The shaded region shows the thermal boundary layer (defined by $\Pi_T \leq 0.99$ or $y \leq 1.5\delta_T$) for Pé $=$ RePr ≈ 200.

Separation point
$(\theta = 0)$

$y = r - R$

Stagnation point
$(\theta = \pi)$

Fluid approaching with velocity v_∞
and temperature T_∞

also the region where most of the heat transfer occurs. The result for Q is good within about 5% for RePr > 100; this limits its use primarily to fluids with Pr > 100, since creeping flow is obtained only at Re of the order of 1 or less.[12]

Results of the same form as Eq. 12.4-34 are obtained for creeping flow in other geometries, including packed beds.[8,13]

It should be emphasized that the asymptotic solutions are particularly important: they are relatively easy to obtain, and for many applications they are sufficiently accurate. We will see in Chapter 14 that some of the standard heat transfer correlations are based on asymptotic solutions of the type discussed here.

QUESTIONS FOR DISCUSSION

1. How does Eq. 12.1-2 have to be modified if there is a heat source within the solid?
2. Show how Eq. 12.1-10 is obtained from Eq. 12.1-8. What is the viscous flow analog of this equation?
3. What kinds of heat conduction problems can be solved by Laplace transform and which cannot?
4. In Example 12.1-3 the heat flux and the temperature both satisfy the "heat conduction equation." Is this always true?
5. Draw a carefully labeled sketch of the results in Eqs. 12.1-38 and 40 showing what is meant by the statement that the "temperature oscillations lag behind the heat flux oscillations by $\pi/4$."
6. Verify that Eq. 12.1-40 satisfies the boundary conditions. Does it have to satisfy an initial condition? If so, what is it?
7. In Ex. 12.2-1, would the method of separation of variables work if applied directly to the function $\Theta(\xi, \zeta)$ rather than to $\Theta_d(\xi, \zeta)$?
8. In Example 12.2-2, how does the wall temperature depend on the downstream coordinate z?
9. By means of a carefully labeled diagram, show what is meant by the two cases $\Delta \leq 1$ and $\Delta \geq 1$ in §12.4. Which case applies to dilute polyatomic gases? Organic liquids? Molten metals?
10. Summarize the situations in which the four mathematical methods in §12.1 are applicable.

---

[12] A review of analyses for a wide range of Pé $=$ RePr is given by S. K. Friedlander, *AIChE Journal*, **7**, 347–348 (1961).

[13] J. P. Sørensen and W. E. Stewart, *Chem. Eng. Sci.*, **29**, 833–837 (1974).

PROBLEMS   **12A.1.** **Unsteady-state heat conduction in an iron sphere.** An iron sphere of 1-in. diameter has the following physical properties: $k = 30$ Btu/hr · ft · F, $\hat{C}_p = 0.12$ Btu/lb$_m$ · F. and $\rho = 436$ lb$_m$/ft^3. Initially the sphere is at a temperature of 70°F.

(a) What is the thermal diffusivity of the sphere?

(b) If the sphere is suddenly plunged into a large body of fluid of temperature 270°F, how much time is needed for the center of the sphere to attain a temperature of 128°F?

(c) A sphere of the same size and same initial temperature, but made of another material, requires twice as long for its center to reach 128°F. What is its thermal diffusivity?

(d) The chart used in the solution of (b) and (c) was prepared from the solution to a partial differential equation. What is that differential equation?

Answers: **(a)** 0.574 ft^2/hr;   **(b)** 1.1 sec;   **(c)** 0.287 ft^2/hr

12A.2 **Comparison of the two slab solutions for short times.** What error is made by using Eq. 12.1-8 (based on the semi-infinite slab) instead of Eq. 12.1-31 (based on the slab of finite thickness), when $\alpha t/b^2 = 0.01$ and for a position 0.9 of the way from the midplane to the slab surface? Use the graphically presented solutions for making the comparison.

Answer: 4%

12A.3 **Bonding with a thermosetting adhesive**[1] (Fig. 12A.3). It is desired to bond together two sheets of a solid material, each of thickness 0.77 cm. This is done by using a thin layer of thermosetting material, which fuses and forms a good bond at 160°C. The two plates are inserted in a press, with both platens of the press maintained at a constant temperature of 220°C. How long will the sheets have to be held in the press, if they are initially at 20°C? The solid sheets have a thermal diffusivity of 4.2×10^{-3} cm^2/s.

Answer: 85 s

12A.4. **Quenching of a steel billet.** A cylindrical steel billet 1 ft in diameter and 3 ft long, initially at 1000°F, is quenched in oil. Assume that the surface of the billet is at 200°F during the quenching process. The steel has the following properties, which may be assumed to be independent of the temperature: $k = 25$ Btu/hr · ft · F, $\rho = 7.7$ g/cm^3, and $\tilde{C}_p = 0.12$ cal/g · C.

Estimate the temperature of the hottest point in the billet after five minutes of quenching. Neglect end effects; that is, make the calculation for a cylinder of the given diameter but of infinite length. See Problem 12C.1 for the method for taking end effects into account.

Answer: 750°F

12A.5. **Measurement of thermal diffusivity from amplitude of temperature oscillations.**

(a) It is desired to use the results of Example 12.1-3 to measure the thermal diffusivity $\alpha = k/\rho\hat{C}_p$ of a solid material. This may be done by measuring the amplitudes A_1 and A_2 at two

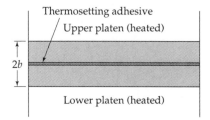

Thermosetting adhesive
Upper platen (heated)

$2b$

Lower platen (heated)

Fig. 12A.3. Two sheets of solid material with a thin layer of adhesive in between.

---

[1] This problem is based on Example 10 of J. M. McKelvey, Chapter 2 of *Processing of Thermoplastic Materials* (E. C. Bernhardt, ed.), Reinhold, New York (1959), p. 93.

points at distances y_1 and y_2 from the periodically heated surface. Show that the thermal diffusivity may then be estimated from the formula

$$\alpha = \frac{\omega}{2}\left(\frac{y_2 - y_1}{\ln(A_1/A_2)}\right)^2 \tag{12A.4-1}$$

(b) Calculate the thermal diffusivity α when the sinusoidal surface heat flux has a frequency 0.0030 cycles/s, if $y_2 - y_1 = 6.19$ cm and the amplitude ratio A_1/A_2 is 6.05.

Answer: $\alpha = 0.111$ cm^2/s

12A.6. Forced convection from a sphere in creeping flow. A sphere of diameter D, whose surface is maintained at a temperature T_0, is located in a fluid stream approaching with a velocity v_∞ and temperature T_∞. The flow around the sphere is in the "creeping flow" regime—that is, with the Reynolds number less than about 0.1. The heat loss from the sphere is described by Eq. 12.4-34.

(a) Verify that the equation is dimensionally correct.

(b) Estimate the rate of heat transfer, Q, for the flow around a sphere of diameter 1 mm. The fluid is an oil at $T_\infty = 50°C$ moving at a velocity 1.0 cm/sec with respect to the sphere, the surface of which is at a temperature of 100°C. The oil has the following properties: $\rho = 0.9$ g/cm^3, $\hat{C}_p = 0.45$ cal/g · K, $k = 3.0 \times 10^{-4}$ cal/s · cm · K, and $\mu = 150$ cp.

12B.1. Measurement of thermal diffusivity in an unsteady-state experiment. A solid slab, 1.90 cm thick, is brought to thermal equilibrium in a constant-temperature bath at 20.0°C. At a given instant ($t = 0$) the slab is clamped tightly between two thermostatted copper plates, the surfaces of which are carefully maintained at 40.0°C. The midplane temperature of the slab is sensed as a function of time by means of a thermocouple. The experimental data are:

t (sec)	0	120	240	360	480	600
T (C)	20.0	24.4	30.5	34.2	36.5	37.8

Determine the thermal diffusivity and thermal conductivity of the slab, given that $\rho = 1.50$ g/cm^3 and $\hat{C}_p = 0.365$ cal/g · C.

Answer: $\alpha = 1.50 \times 10^{-3}$ cm^2/s; $k = 8.2 \times 10^{-4}$ cal/s · cm · C or 0.20 Btu/hr · ft · F

12B.2. Two-dimensional forced convection with a line heat source. A fluid at temperature T_∞ flows in the x direction across a long, infinitesimally thin wire, which is heated electrically at a rate Q/L (energy per unit time per unit length). The wire thus acts as a line heat source. It is assumed that the wire does not disturb the flow appreciably. The fluid properties (density, thermal conductivity, and heat capacity) are assumed constant and the flow is assumed uniform. Furthermore, radiant heat transfer from the wire is neglected.

(a) Simplify the energy equation to the appropriate form, by neglecting the heat conduction in the x direction with respect to the heat transport by convection. Verify that the following conditions on the temperature are reasonable:

$$T \to T_\infty \quad \text{as } y \to \infty \quad \text{for all } x \tag{12B.2-1}$$

$$T = T_\infty \quad \text{at } x < 0 \quad \text{for all } y \tag{12B.2-2}$$

$$\int_{-\infty}^{+\infty} \rho\hat{C}_p(T - T_\infty)|_x v_x dy = Q/L \quad \text{for all } x > 0 \tag{12B.2-3}$$

(b) Postulate a solution of the form (for $x > 0$)

$$T(x, y) - T_\infty = f(x)g(\eta) \quad \text{where } \eta = y/\delta(x) \tag{12B.2-4}$$

Show by means of Eq. 12B.2-3 that $f(x) = C_1/\delta(x)$. Then insert Eq. 12B.2-4 into the energy equation and obtain

$$-\left[\frac{v_x\delta}{\alpha}\frac{d\delta}{dx}\right]\frac{d}{d\eta}(\eta g) = \frac{d^2g}{d\eta^2} \tag{12B.2-5}$$

(c) Set the quantity in brackets in Eq. 12B.2-5 equal to 2 (why?), and then solve to get $\delta(x)$.

(d) Then solve the equation for $g(\eta)$.

(e) Finally, evaluate the constant C_1, and thereby complete the derivation of the temperature distribution.

12B.3. **Heating of a wall (constant wall heat flux).** A very thick solid wall is initially at the temperature T_0. At time $t = 0$, a constant heat flux q_0 is applied to one surface of the wall (at $y = 0$), and this heat flux is maintained. Find the time-dependent temperature profiles $T(y, t)$ for small times. Since the wall is very thick it can be safely assumed that the two wall surfaces are an infinite distance apart in obtaining the temperature profiles.

(a) Follow the procedure used in going from Eq. 12.1-33 to Eq. 12.1-35, and then write the appropriate boundary and initial conditions. Show that the analytical solution of the problem is

$$T(y, t) - T_0 = \frac{q_0}{k} \left(\sqrt{\frac{4\alpha t}{\pi}} \exp\left(-y^2/4\alpha t\right) - \frac{2y}{\sqrt{\pi}} \int_{y/\sqrt{4\alpha t}}^{\infty} \exp(-u^2)\, du \right) \tag{12B.3-1}$$

(b) Verify that the solution is correct by substituting it into the one-dimensional heat conduction equation for the temperature (see Eq. 12.1-33). Also show that the boundary and initial conditions are satisfied.

12B.4. **Heat transfer from a wall to a falling film (short contact time limit)[2]** (Fig. 12B.4). A cold liquid film flowing down a vertical solid wall, as shown in the figure, has a considerable cooling effect on the solid surface. Estimate the rate of heat transfer from the wall to the fluid for such short contact times that the fluid temperature changes appreciably only in the immediate vicinity of the wall.

(a) Show that the velocity distribution in the falling film, given in §2.2, may be written as $v_z = v_{z,max}[2(y/\delta) - (y/\delta)^2]$, in which $v_{z,max} = \rho g \delta^2/2\mu$. Then show that in the vicinity of the wall the velocity is a linear function of y given by

$$v_z \approx \frac{\rho g \delta}{\mu} y \tag{12B.4-1}$$

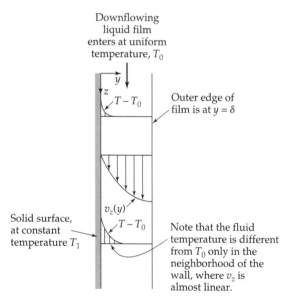

Downflowing
liquid film
enters at uniform
temperature, T_0

$T - T_0$

Outer edge of
film is at $y = \delta$

$v_z(y)$

Solid surface,
at constant
temperature T_1

$T - T_0$

Note that the fluid
temperature is different
from T_0 only in the
neighborhood of the
wall, where v_z is
almost linear.

Fig. 12B.4. Heat transfer to a film falling down a vertical wall.

[2] R. L. Pigford, *Chemical Engineering Progress Symposium Series*, **51**, No. 17, 79–92 (1955). **Robert Lamar Pigford** (1917–1988), who taught at both the University of Delaware and the University of California in Berkeley, researched many aspects of diffusion and mass transfer; he was the founding editor of *Industrial and Engineering Chemistry Fundamentals*.

(b) Show that the energy equation for this situation reduces to

$$\rho \hat{C}_p v_z \frac{\partial T}{\partial z} = k \frac{\partial^2 T}{\partial y^2} \tag{12B.4-2}$$

List all the simplifying assumptions needed to get this result. Combine the preceding two equations to obtain

$$y \frac{\partial T}{\partial z} = \beta \frac{\partial^2 T}{\partial y^2} \tag{12B.4-3}$$

in which $\beta = \mu k / \rho^2 \hat{C}_p g \delta$.

(c) Show that for short contact times we may write as boundary conditions

B.C. 1:	$T = T_0$	for $z = 0$	and $y > 0$	(12B.4-4)
B.C. 2:	$T = T_0$	for $y = \infty$	and z finite	(12B.4-5)
B.C. 3:	$T = T_1$	for $y = 0$	and $z > 0$	(12B.4-6)

Note that the true boundary condition at $y = \delta$ is replaced by a fictitious boundary condition at $y = \infty$. This is possible because the heat is penetrating just a very short distance into the fluid.

(d) Use the dimensionless variables $\Theta(\eta) = (T - T_0)/(T_1 - T_0)$ and $\eta = y/\sqrt[3]{9\beta z}$ to rewrite the differential equation as (see Eq. C.1-9):

$$\frac{d^2\Theta}{d\eta^2} + 3\eta^2 \frac{d\Theta}{d\eta} = 0 \tag{12B.4-7}$$

Show that the boundary conditions are $\Theta = 0$ for $\eta = \infty$ and $\Theta = 1$ at $\eta = 0$.

(e) In Eq. 12B.4-7, set $d\Theta/d\eta = p$ and obtain an equation for $p(\eta)$. Solve that equation to get $d\Theta/d\eta = p(\eta) = C_1 \exp(-\eta^3)$. Show that a second integration and application of the boundary conditions give

$$\Theta = \frac{\int_\eta^\infty \exp(-\bar{\eta}^3) d\bar{\eta}}{\int_0^\infty \exp(-\bar{\eta}^3) d\bar{\eta}} = \frac{1}{\Gamma(\frac{4}{3})} \int_\eta^\infty \exp(-\bar{\eta}^3) d\bar{\eta} \tag{12B.4-8}$$

(f) Show that the average heat flux to the fluid is

$$q_{avg}|_{y=0} = \frac{3}{2} \frac{(9\beta L)^{-1/3}}{\Gamma(\frac{4}{3})} k(T_1 - T_0) \tag{12B.4-9}$$

where use is made of the Leibniz formula in §C.3.

12B.5. Temperature in a slab with heat production. The slab of thermal conductivity k in Example 12.1-2 is initially at a temperature T_0. For time $t > 0$ there is a uniform volume production of heat S_0 within the slab.

(a) Obtain an expression for the dimensionless temperature $k(T - T_0)/S_0 b^2$ as a function of the dimensionless coordinate $\eta = y/b$ and the dimensionless time by looking up the solution in the book by Carslaw and Jaeger.

(b) What is the maximum temperature reached at the center of the slab?

(c) How much time elapses before 90% of the temperature rise occurs?

Answer: **(c)** $t \approx b^2/\alpha$

12B.6. Forced convection in slow flow across a cylinder (Fig. 12B.6). A long cylinder of radius R is suspended in an infinite fluid of constant properties ρ, μ, $\hat{C}_p$, and k. The fluid approaches with

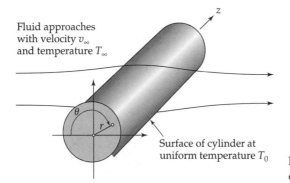

Fluid approaches
with velocity v_∞
and temperature T_∞

Surface of cylinder at
uniform temperature T_0

Fig. 12B.6. Heat transfer from a long
cylinder of radius R.

temperature T_∞ and velocity v_∞. The cylindrical surface is maintained at temperature T_0. For
this system the velocity distribution has been determined by Lamb[3] in the limit of Re $\ll 1$.
His result for the region close to the cylinder is

$$\psi = -\frac{v_\infty R \sin \theta}{2S}\left[\frac{r}{R}\left(2 \ln \frac{r}{R} - 1\right) + \frac{R}{r}\right] \tag{12B.6-1}$$

in which ψ is the first polar-coordinate stream function in Table 4.2-1. The dimensionless
quantity S is given by $S = \frac{1}{2} - \gamma + \ln(8/\text{Re})$, where $\gamma = 0.5772 \cdots$ is "Euler's constant," and
$\text{Re} = Dv_\infty\rho/\mu$.

(a) For this system, determine the interfacial velocity gradient β defined in Example 12.4-3.

(b) Determine the rate of heat loss Q from a length L of the cylinder using the method of Example 12.4-3. Note that

$$\int_0^\pi \sqrt{\sin \theta}d\theta = B(\tfrac{3}{4}, \tfrac{1}{2}) = 2.3963 \ldots \tag{12B.6-2}$$

where $B(m, n) = \Gamma(m)\Gamma(n)/\Gamma(m + n)$ is the "beta function."

(c) Determine δ_T/R at $\theta = 0, \frac{1}{2}\pi$, and π.

Answers: **(a)** $\beta = \dfrac{2v_\infty \sin \theta}{RS}$

(b) $Q = C(\pi DL)(T_0 - T_\infty)\left(\dfrac{k}{D}\right)\left(\dfrac{\text{Re Pr}}{S}\right)^{1/3}$ (Evaluate the constant C)

(c) $\dfrac{\delta_T}{R} = \left(\dfrac{9S}{\text{Re Pr}}\right)^{1/3} f(\theta); f = (\tfrac{2}{3})^{1/3}, 1.1981, \infty$

12B.7. Timetable for roasting turkey

(a) A homogeneous solid body of arbitrary shape is initially at temperature T_0 throughout. At
$t = 0$ it is immersed in a fluid medium of temperature T_1. Let L be a characteristic length in
the solid. Show that dimensional analysis predicts that

$$\Theta = \Theta(\xi, \eta, \zeta, \tau, \text{and geometrical ratios}) \tag{12B.7-1}$$

where $\Theta = (T - T_0)/(T_1 - T_0)$, $\xi = x/L$, $\eta = y/L$, $\zeta = z/L$, and $\tau = \alpha t/L^2$. Relate this result to
the graphs given in §12.1.

---

[3] H. Lamb, *Phil. Mag.*, (6) **21**, 112–110 (1911). For a survey of more detailed analyses, see
L. Rosenhead (ed.), *Laminar Boundary Layers*, Oxford University Press, London (1963), Chapter 4.

(b) A typical timetable for roasting turkey at 350°F is[4]

Mass of turkey (lb_m)	Time required per unit mass (min/lb_m)
6–10	20–25
10–16	18–20
18–25	15–18

Compare this empirically determined cooking schedule with the results of part (a), for geometrically similar turkeys at an initial temperature T_0, cooked with a given surface temperature T_1 to the same dimensionless temperature distribution $\Theta = \Theta(\xi, \eta, \zeta)$.

12B.8. Use of asymptotic boundary layer solution. Use the results of Ex. 12.4-2 to obtain δ_T and q_0 for the system in Problem 12D.4. By comparing δ_T with D, estimate the range of applicability of the solution obtained in Problem 12D.4.

12B.9. Non-Newtonian heat transfer with constant wall heat flux (asymptotic solution for small axial distances). Rework Example 12.2-2 for a fluid whose non-Newtonian behavior is described adequately by the power law model. Show that the solution given in Eq. 12.2-2 may be taken over for the power law model simply by an appropriate modification in the definition of v_0.

12C.1. Product solutions for unsteady heat conduction in solids.

(a) In Example 12.1-2 the unsteady state heat conduction equation is solved for a slab of thickness $2b$. Show that the solution to Eq. 12.1-2 for the analogous problem for a rectangular block of finite dimensions $2a$, $2b$, and $2c$ may be written as the product of the solutions for three slabs of corresponding dimensions:

$$\frac{T_1 - T(x, y, z, t)}{T_1 - T_0} = \Theta\left(\frac{x}{a}, \frac{\alpha t}{a^2}\right)\Theta\left(\frac{y}{b}, \frac{\alpha t}{b^2}\right)\Theta\left(\frac{z}{c}, \frac{\alpha t}{c^2}\right) \tag{12C.1-1}$$

in which $\Theta(y/b, \alpha t/b^2)$ is the right side of Eq. 12.1-31.

(b) Prove a similar result for cylinders of finite length; then rework Problem 12A.4 without the assumption that the cylinder is infinitely long.

12C.2. Heating of a semi-infinite slab with variable thermal conductivity. Rework Example 12.1-1 for a solid whose thermal conductivity varies with temperature as follows:

$$\frac{k}{k_0} = 1 + \beta\left(\frac{T - T_0}{T_1 - T_0}\right) \tag{12C.2-1}$$

in which k_0 is the thermal conductivity at temperature T_0, and β is a constant. Use the following approximate procedure:

(a) Let $\Theta = (T - T_0)/(T_1 - T_0)$ and $\eta = y/\delta(t)$, where $\delta(t)$ is a boundary layer thickness that changes with time. Then assume that

$$\Theta(y, t) = \Phi(\eta) \tag{12C.2-2}$$

in which the function $\Phi(\eta)$ gives the shapes of the "similar" profiles. This is tantamount to assuming that the temperature profiles have the same shape for all values of β, which, of course, is not really true.

(b) Substitute the above approximate profiles into the heat conduction equation and obtain the following differential equation for the boundary layer thickness:

$$M\delta\frac{d\delta}{dt} = \alpha_0 N \tag{12C.2-3}$$

---

[4] *Woman's Home Companion Cook Book*, Garden City Publishing Co., (1946), courtesy of Jean Stewart.

in which $\alpha_0 = k_0 / \rho \hat{C}_p$ and

$$M = \int_0^1 \Phi(\eta) d\eta \quad \text{and} \quad N = (1 + \beta\Phi)(d\Phi/d\eta)|_0^1 \qquad (12C.2\text{-}4, 5)$$

Then solve for the function $\delta(t)$.

(c) Now let $\Phi(\eta) = 1 - \frac{3}{2}\eta + \frac{1}{2}\eta^3$. Why is this a felicitous choice? Then find the time-dependent temperature distribution $T(y, t)$ as well as the heat flux at $y = 0$.

12C.3. **Heat conduction with phase change (the *Neumann–Stefan problem*)** (Fig. 12C.3)[5]. A liquid, contained in a long cylinder, is initially at temperature T_1. For time $t \geq 0$, the bottom of the container is maintained at a temperature T_0, which is below the melting point T_m. We want to estimate the movement of the solid–liquid interface, $Z(t)$, during the freezing process.

For the sake of simplicity, we assume here that the physical properties ρ, k, and $\hat{C}_p$ are constants and the same in both the solid and liquid phases. Let $\Delta\hat{H}_f$ be the heat of fusion per gram, and use the abbreviation $\Lambda = \Delta\hat{H}_f / \hat{C}_p(T_1 - T_0)$.

(a) Write the equation for heat conduction for the liquid (L) and solid (S) regions; state the boundary and initial conditions.

(b) Assume solutions of the form:

$$\Theta_S \equiv \frac{T_S - T_0}{T_1 - T_0} = C_1 + C_2 \operatorname{erf} \frac{z}{\sqrt{4\alpha t}} \qquad (12C.3\text{-}1)$$

$$\Theta_L \equiv \frac{T_L - T_0}{T_1 - T_0} = C_3 + C_4 \operatorname{erf} \frac{z}{\sqrt{4\alpha t}} \qquad (12C.3\text{-}2)$$

(c) Use the boundary condition at $z = 0$ to show that $C_1 = 0$, and the condition at $z = \infty$ to show that $C_3 = 1 - C_4$. Then use the fact that $T_S = T_L = T_m$ at $z = Z(t)$ to conclude that $Z(t) = \lambda\sqrt{4\alpha t}$, where λ is some (as yet undetermined) constant. Then get C_3 and C_4 in terms of λ. Use the remaining boundary condition to get λ in terms of Λ and $\Theta_m = (T_m - T_0)/(T_1 - T_0)$:

$$\sqrt{\pi}\Lambda\lambda \exp\lambda^2 = \frac{\Theta_m}{\operatorname{erf}\lambda} - \frac{1 - \Theta_m}{1 - \operatorname{erf}\lambda} \qquad (12C.3\text{-}3)$$

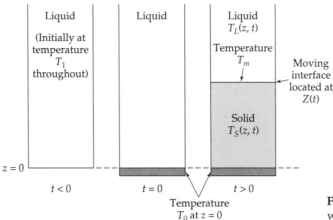

Fig. 12C.3. Heat conduction with solidification.

---

[5] For literature references and related problems, see H. S. Carslaw and J. C. Jaeger, *Conduction of Heat in Solids*, 2nd edition, Oxford University Press (1959), Chapter XI; on pp. 283–286 the problem considered here is worked out for the situation that the physical properties of the liquid and solid phases are different. See also S. G. Bankoff, *Advances in Chemical Engineering*, Vol. 5, Academic Press, New York (1964), pp. 75–150; J. Crank, *Free and Moving Boundary Problems*, Oxford University Press (1984); J. M. Hill, *One-Dimensional Stefan Problems*, Longmans (1987).

What is the final expression for $Z(t)$? (*Note:* In this problem it has been assumed that a phase change occurs instantaneously and that no supercooling of the liquid phase occurs. It turns out that in the freezing of many liquids, this assumption is untenable. That is, to describe the solidification process correctly, one has to take into account the kinetics of the crystallization process.[6])

12C.4. **Viscous heating in oscillatory flow.**[7] Viscous heating can be a disturbing factor in viscosity measurements. Here we see how viscous heating can affect the measurement of viscosity in an oscillating-plate system.

A Newtonian fluid is located in the region between two parallel plates separated by a distance b. Both plates are maintained at a temperature T_0. The lower plate (at $z = 0$) is made to oscillate sinusoidally in the z direction with a velocity amplitude v_0 and a circular frequency ω. Estimate the temperature rise resulting from viscous heating. Consider only the high-frequency limit.

(a) Show that the velocity distribution is given by

$$\frac{v_z(x, t)}{v_0} = \frac{\left[\begin{pmatrix} \sinh a(1 - \xi) \cos a(1 - \xi) \sinh a \cos a \\ + \sin a(1 - \xi) \cosh a(1 - \xi) \sin a \cosh a \end{pmatrix} \cos \omega t \\ + \begin{pmatrix} - \sin a(1 - \xi) \cosh a(1 - \xi) \sinh a \cos a \\ + \sinh a(1 - \xi) \cos a(1 - \xi) \sin a \cosh a \end{pmatrix} \sin \omega t \right]}{\sinh^2 a \cos^2 a + \cosh^2 a \sin^2 a} \tag{12C.4-1}$$

where $a = \sqrt{\rho \omega b^2 / 2\mu}$ and $\xi = x/b$.

(b) Next calculate the dissipation function Φ_v for the velocity profile in Eq. 12C.4-1. Then obtain a time-averaged dissipation function $\overline{\Phi}_v$, by averaging over one cycle. Use the formulas

$$\overline{\cos^2 \omega t} = \overline{\sin^2 \omega t} = \tfrac{1}{2} \quad \text{and} \quad \overline{\sin \omega t \cos \omega t} = 0 \tag{12C.4-2}$$

which may be verified. Then simplify the result for high frequencies (i.e., for large values of a) to obtain

$$\overline{\Phi}_v \text{ (large } \omega) = a^2 \left(\frac{v_0}{b} \right)^2 e^{-2a\xi} \tag{12C.4-3}$$

(c) Next take a time average of the heat conduction equation to obtain

$$0 = k \frac{d^2 \overline{T}}{dx^2} + \mu \overline{\Phi}_v \tag{12C.4-4}$$

in which $\overline{T}$ is the temperature averaged over one cycle. Solve this to get

$$\overline{T} - T_0 = \left(\frac{\mu v_0^2}{4k} \right)[(1 - e^{-2a\xi}) - (1 - e^{-2a})\xi] \tag{12C.4-5}$$

This shows how the temperature in the slit depends on position. From this function, the maximum temperature rise can be calculated. For reasonably high frequencies, $\overline{T} - T_0 \approx \mu v_0^2 / 4k$.

12C.5. **Solar heat penetration.** Many desert animals protect themselves from excessive diurnal temperature fluctuations by burrowing sufficiently far underground that they can maintain

---

[6] H. Janeschitz-Kriegl, *Plastics and Rubber Processing and Applications*, **4**, 145–158 (1984); H. Janeschitz-Kriegl, in *One-Hundred Years of Chemical Engineering* (N. A. Peppas, ed.), Kluwer Academic Publishers, Dordrecht (Netherlands) (1989), pp. 111–124; H. Janeschitz-Kriegl, E. Ratajski, and G. Eder, *Ind. Eng. Chem. Res.*, **34**, 3481–3487 (1995); G. Astarita and J. M. Kenny, *Chem. Eng. Comm.*, **53**, 69–84 (1987).

[7] R. B. Bird, *Chem. Eng. Prog. Symposium Series*, Vol. 61, No. 58 (1965), pp. 13–14; see also F. Ding, A. J. Giacomin, R. B. Bird, and C-B Kweon, J. *Non-Newtonian Fluid Mech.*, **86**, 359–374 (1999).

themselves at a reasonably steady temperature. Let the temperature in the ground be $T(y, t)$, where y is the depth below the surface of the earth and t is the time, measured from the time of maximum temperature T_0. Further, let the temperature far beneath the surface be T_∞, and let the surface temperature be given by

$$T(0, t) - T_\infty = 0 \qquad\qquad \text{for } t < 0$$
$$T(0, t) - T_\infty = (T_0 - T_\infty) \cos \omega t \qquad \text{for } t \geq 0 \qquad\qquad (12C.5\text{-}1)$$

Here $\omega = 2\pi/t_{per}$, in which t_{per} is the time for one full cycle of the oscillating temperature—namely, 24 hours. Then it can be shown that the temperature at any depth is given by

$$\frac{T(y, t) - T_\infty}{T_0 - T_\infty} = e^{-\sqrt{\omega/2\alpha}\, y} \cos (\omega t - \sqrt{\omega/2\alpha}\, y)$$
$$- \frac{1}{\pi} \int_0^\infty e^{-\bar{\omega} t}(\sin \sqrt{\bar{\omega}/\alpha}\, y)\, \frac{\bar{\omega}}{\omega^2 + \bar{\omega}^2}\, d\bar{\omega} \qquad\qquad (12C.5\text{-}2)$$

This equation is the heat conduction analog of Eq. 4D.1-1, which describes the response of the velocity profiles near an oscillating plate. The first term describes the "periodic steady state" and the second the "transient" behavior. Assume the following properties for the soil:[8] $\rho = 1515 \text{ kg/m}^3$, $k = 0.027 \text{ W/m} \cdot \text{K}$, and $\hat{C}_p = 800 \text{ J/kg} \cdot \text{K}$.

(a) Assume that the heating of the earth's surface is exactly sinusoidal, and find the amplitude of the temperature variation beneath the surface at a distance y. To do this, use only the periodic steady state term in Eq. 12C.5-2. Show that at a depth of 10 cm, this amplitude has the value of 0.0172.

(b) Discuss the importance of the transient term in Eq. 12C.5-2. Estimate the size of this contribution.

(c) Next consider an arbitrary formal expression for the daily surface temperature, given as a Fourier series of the form

$$\frac{T(0, t) - T_\infty}{T_0 - T_\infty} = \sum_{n=0}^{\infty} (a_n \cos n\omega t + b_n \sin n\omega t) \qquad\qquad (12C.5\text{-}3)$$

How many terms in this series are used to solve part (a)?

12C.6. Heat transfer in a falling non-Newtonian film. Repeat Problem 12B.4 for a polymeric fluid that is reasonably well described by the power law model of Eq. 8.3-3.

12D.1. Unsteady-state heating of a slab (Laplace transform method).

(a) Re-solve the problem in Example 12.1-2 by using the Laplace transform, and obtain the result in Eq. 12.1-31.

(b) Note that the series in Eq. 12.1-31 does not converge rapidly at short times. By inverting the Laplace transform in a way different from that in (a), obtain a different series that is rapidly convergent for small times.[9]

(c) Show how the first term in the series in (b) is related to the "short contact time" solution of Example 12.1-1.

12D.2. The Graetz-Nusselt problem (Table 12D.2).

(a) A fluid (Newtonian or generalized Newtonian) is in laminar flow in a circular tube of radius R. In the inlet region $z < 0$, the fluid temperature is uniform at T_1. In the region $z > 0$, the wall temperature is maintained at T_0. Assume that all physical properties are constant and

---

[8] W. M. Rohsenow, J. P. Hartnett, and Y. I. Cho, eds., *Handbook of Heat Transfer*, 3rd edition, McGraw-Hill (1998), p. 2.68.

[9] H. S. Carslaw and J. C. Jaeger, *Conduction of Heat in Solids*, 2nd edition, Oxford University Press (1959), pp. 308–310.

that viscous dissipation and axial heat conduction effects are negligible. Use the following dimensionless variables:

$$\Theta = \frac{T - T_0}{T_1 - T_0} \qquad \phi = \frac{v_z}{\langle v_z \rangle} \qquad \xi = \frac{r}{R} \qquad \zeta = \frac{\alpha z}{\langle v_z \rangle R^2} \tag{12D.2-1}$$

Show that the temperature profiles in this system are

$$\Theta = \sum_{i=1}^{\infty} A_i X_i(\xi) \exp(-\beta_i^2 \zeta) \tag{12D.2-2}$$

in which X_i and β_i are the eigenfunctions and eigenvalues obtained from the solution to the following equation:

$$\frac{1}{\xi} \frac{d}{d\xi} \left(\xi \frac{dX_i}{d\xi} \right) + \beta_i^2 \phi X_i = 0 \tag{12D.2-3}$$

with boundary conditions $X =$ finite at $\xi = 0$ and $X = 0$ at $\xi = 1$. Show further that

$$A_i = \frac{\displaystyle\int_0^1 X_i \phi \xi \, d\xi}{\displaystyle\int_0^1 X_i^2 \xi \, d\xi} \tag{12D.2-4}$$

(b) Solve Eq. 12D.2-3 for the Newtonian fluid by obtaining a power series solution for X_i. Calculate the lowest eigenvalue by solving an algebraic equation. Check your result against that given in Table 12D.2.

(c) From the work involved in (b) in computing β_i^2 it can be inferred that the computation of the higher eigenvalues is quite tedious. For eigenvalues higher than the second or third the Wenzel–Kramers–Brillouin (WKB) method[10] can be used; the higher the eigenvalue, the more accurate the WKB method is. Read about this method, and verify that for the Newtonian fluid

$$\beta_i^2 = \tfrac{1}{2}(4i - \tfrac{4}{3})^2 \tag{12D.2-5}$$

A similar formula has been derived for the power law model.[11]

Table 12D.2   Eigenvalues β_i^2 for the Graetz-Nusselt Problem for Newtonian Fluids[a]

i	By direct calculation[b]	By WKB method[c]	By Stodola and Vianello method[d]
1	3.67	3.56	3.661[c]
2	22.30	22.22	—
3	56.95	56.88	—
4	107.6	107.55	—

[a] The β_i^2 here correspond to $\tfrac{1}{2}\lambda_i^2$ in W. M. Rohsenow, J. P. Hartnett, and Y. I. Cho, *Handbook of Heat Transfer*, McGraw-Hill (New York), Table 5.3 on p. 510.
[b] Values taken from K. Yamagata, *Memoirs of the Faculty of Engineering*, Kyûshû University, Volume VIII, No. 6, Fukuoka, Japan (1940).
[c] Computed from Eq. 12D.2-5.
[d] For the particular trial function in part (d) of the problem.

[10] J. Heading, *An Introduction to Phase-Integral Methods*, Wiley, New York (1962); J. R. Sellars, M. Tribus, and J. S. Klein, *Trans. ASME*, **78**, 441–448 (1956).
[11] I. R. Whiteman and W. B. Drake, *Trans. ASME*, **80**, 728–732 (1958).

(d) Obtain the lowest eigenvalue by the method of Stodola and Vianello. Use Eqs. 71a and 72b on p. 203 of Hildebrand's book,[12] with $\phi = 2(1 - \xi^2)$ for Newtonian flow and $X_1 = 1 - \xi^2$ as a simple, but suitable, trial function. Show that this leads quickly to the value $\beta_1^2 = 3.661$.

12D.3. **The Graetz-Nusselt problem (asymptotic solution for large z).** Note that, in the limit of very large z, only one term ($i = 1$) is needed in Eq. 12D.2-2. It is desired to use this result to compute the heat flux at the wall, q_0, at large z and to express the result as

$$q_0 = (\text{a function of system and fluid properties}) \times (T_b - T_0) \qquad (12D.3\text{-}1)$$

where T_b is the "bulk fluid temperature" defined in Eq. 10.8-33.

(a) First verify that

$$q_0 = -\frac{k}{R} \frac{\partial\Theta/\partial\xi|_{\xi=1}}{\Theta_b} (T_b - T_0) \qquad (12D.3\text{-}2)$$

Here Θ is the same as in Problem 12D.2, and $\Theta_b = (T_b - T_0)/(T_1 - T_0)$.

(b) Show that for large z, Eq. 12D.3-2 and Eq. 12D.2-2 both give

$$q_0 = \frac{k}{2R} \beta_1^2 (T_b - T_0) \qquad (12D.3\text{-}3)$$

Hence for large z, all one needs to know is the first eigenvalue; the eigenfunctions need not be calculated. This shows how useful the method of Stodola and Vianello[12] is for computing the limiting value of a heat flux.

12D.4. **The Graetz-Nusselt problem (asymptotic solution for small z).**

(a) Apply the method of Example 12.2-2 to the solution of the problem discussed in Problem 12D.2. Consider a Newtonian fluid and use the following dimensionless quantities:

$$\Theta = \frac{T_1 - T}{T_1 - T_0} \qquad \zeta = \frac{z}{R} \qquad \sigma = \frac{R - r}{R} = \frac{s}{R} \qquad N = \frac{4\langle v_z\rangle R}{\alpha} \qquad (12D.4\text{-}1)$$

Show that the method of combination of variables gives

$$\Theta = \frac{1}{\Gamma(\frac{4}{3})} \int_\eta^\infty \exp(-\overline{\eta}^3) d\overline{\eta} \qquad (12D.4\text{-}2)$$

in which $\eta = (N\sigma^3/9\zeta)^{1/3}$.

(b) Show that the wall flux is

$$q_r|_{r=R} = \frac{k}{R} \left[\frac{1}{9^{1/3}\Gamma(\frac{4}{3})} \left(\text{Re Pr} \frac{D}{z} \right)^{1/3} \right] (T_1 - T_0) \qquad (12D.4\text{-}3)$$

The quantity $(\text{Re Pr } D/z) = (4/\pi)(w\hat{C}_p/kz)$ appears frequently; the grouping $\text{Gz} = (w\hat{C}_p/kz)$ is called the *Graetz number*. Compare this result with that in Eq. 12D.3-3, with regard to the dependence on the dimensionless groups.

(c) How may the results be written so that they are valid for any generalized Newtonian model?

12D.5. **The Graetz problem for flow between parallel plates.** Work through Problems 12D.2, 3, and 4 for flow between parallel plates (or flow in a thin rectangular duct).

12D.6. **The constant wall heat flux problem for parallel plates.** Apply the methods used in §10.8, Example 12.2-1, and Ex. 12.2-2 to the flow between parallel plates.

---

[12] F. B. Hildebrand, *Advanced Calculus for Applications*, Prentice-Hall, Englewood Cliffs, N.J. (1963), §5.5.

12D.7. Asymptotic solution for small z for laminar tube flow with constant heat flux. Fill in the missing steps between Eq. 12.2-23 and Eq. 12.2-24. Insertion of the expression for ψ into Eq. 12.2-23 gives

$$\Theta = \sqrt[3]{9\lambda} \int_{\chi}^{\infty} \left[\frac{3}{\Gamma(\frac{2}{3})} \int_{\bar{\chi}}^{\infty} \bar{\bar{\chi}} \exp(-\bar{\bar{\chi}}^3) \, d\bar{\bar{\chi}} \right] d\bar{\chi} \tag{12D.7-1}$$

Why do we introduce the symbols $\bar{\chi}$ and $\bar{\bar{\chi}}$? Next, exchange the order of integration to get

$$\Theta = \sqrt[3]{9\lambda} \int_{\chi}^{\infty} \left[\frac{3}{\Gamma(\frac{2}{3})} \int_{\chi}^{\bar{\bar{\chi}}} \bar{\chi} \exp(-\bar{\bar{\chi}}^3) d\bar{\chi} \right] d\bar{\bar{\chi}} \tag{12D.7-2}$$

Then perform the integration over $\bar{\chi}$ and obtain

$$\Theta(\eta, \lambda) = \frac{\sqrt[3]{9\lambda}}{\Gamma(\frac{2}{3})} \left[\exp(-\chi^3) - 3\chi \left(\int_0^{\infty} \chi \exp(-\chi^3) \, d\chi - \int_0^{\chi} \bar{\chi} \exp(-\bar{\bar{\chi}}^3) \, d\bar{\chi} \right) \right] \tag{12D.7-3}$$

Then use the definitions $\Gamma(a) = \int_0^{\infty} t^{a-1} e^{-t} dt$ and $\Gamma(a, x) = \int_x^{\infty} t^{a-1} e^{-t} dt$ for the complete and incomplete gamma functions.

12D.8. Forced conduction heat transfer from a flat plate (thermal boundary layer extends beyond the momentum boundary layer). Show that the result analogous to Eq. 12.4-14 for $\Delta > 1$ is[13]

$$\frac{3}{10} \Delta^2 - \frac{3}{10} \Delta + \frac{2}{15} - \frac{3}{140} \frac{1}{\Delta^2} + \frac{1}{180\Delta^3} = \frac{37}{315} \frac{1}{Pr} \tag{12D.8-1}$$

---

[13] H. Schlichting, *Boundary-Layer Theory*, 7th edition, McGraw-Hill, New York (1979), p. 306.

Chapter 13

Temperature Distributions in Turbulent Flow

In Chapters 10 to 12 we have shown how to obtain temperature distributions in solids and in fluids in laminar motion. The procedure has involved solving the equations of change with appropriate boundary and initial conditions.

We now turn to the problem of finding temperature profiles in turbulent flow. This discussion is quite similar to that given in Chapter 5. We begin by time-smoothing the equations of change. In the time-smoothed energy equation there appears a turbulent heat flux $\overline{\mathbf{q}}^{(t)}$, which is expressed in terms of the correlation of velocity and temperature fluctuations. There are several rather useful empiricisms for $\overline{\mathbf{q}}^{(t)}$, which enable one to predict time-smoothed temperature distributions in wall turbulence and in free turbulence. We use heat transfer in tube flow to illustrate the method.

The most apparent influence of turbulence on heat transport is the enhanced transport perpendicular to the main flow. If heat is injected into a fluid flowing in *laminar* flow in the z direction, then the movement of heat in the x and y directions is solely by conduction and proceeds very slowly. On the other hand, if the flow is *turbulent*, the heat "spreads out" in the x and y directions extremely rapidly. This rapid dispersion of heat is a characteristic feature of turbulent flow. This mixing process is worked out in some detail here for flow in tubes and in circular jets.

Although it has been conventional to study turbulent heat transport via the time-smoothed energy equation, it is also possible to analyze the heat flux at a wall by use of a Fourier transform technique without time-smoothing. This is set forth in the last section.

§13.1 TIME-SMOOTHED EQUATIONS OF CHANGE FOR INCOMPRESSIBLE NONISOTHERMAL FLOW

In §5.2 we introduced the notions of time-smoothed quantities and turbulent fluctuations. In this chapter we shall be primarily concerned with the temperature profiles. We introduce the time-smoothed temperature $\overline{T}$ and temperature fluctuation T', and write analogously to Eq. 5.2-1

$$T = \overline{T} + T' \tag{13.1-1}$$

Clearly T' averages to zero so that $\overline{T'} = 0$, but quantities like $\overline{v_x'T'}$, $\overline{v_y'T'}$, and $\overline{v_z'T'}$ will not be zero because of the "correlation" between the velocity and temperature fluctuations at any point.

For a nonisothermal pure fluid we need three equations of change, and we want to discuss here their time-smoothed forms. The time-smoothed equations of continuity and motion for a fluid with constant density and viscosity were given in Eqs. 5.2-10 and 12, and need not be repeated here. For a fluid with constant μ, ρ, $\hat{C}_p$, and k, Eq. 11.2-5, when put in the $\partial/\partial t$ form by using Eq. 3.5-4, and with Newton's and Fourier's law included, becomes

$$\frac{\partial}{\partial t} \rho \hat{C}_p T = -\left(\frac{\partial}{\partial x} \rho \hat{C}_p v_x T + \frac{\partial}{\partial y} \rho \hat{C}_p v_y T + \frac{\partial}{\partial z} \rho \hat{C}_p v_z T \right) + k\left(\frac{\partial^2 T}{\partial x^2} + \frac{\partial^2 T}{\partial y^2} + \frac{\partial^2 T}{\partial z^2} \right)$$
$$+ \mu\left[2\left(\frac{\partial v_x}{\partial x}\right)^2 + \left(\frac{\partial v_x}{\partial y}\right)^2 + 2\left(\frac{\partial v_x}{\partial y}\right)\left(\frac{\partial v_y}{\partial x}\right) + \cdots \right] \tag{13.1-2}$$

in which only a few sample terms in the viscous dissipation term $-(\boldsymbol{\tau}:\nabla\mathbf{v}) = \mu\Phi_v$ have been written (see Eq. B.7-1 for the complete expression).

In Eq. 13.1-2 we replace T by $T = \overline{T} + T'$, v_x by $\overline{v}_x + v_x'$, and so on. Then the equation is time-smoothed to give

$$\frac{\partial}{\partial t} \rho \hat{C}_p \overline{T} = -\left(\frac{\partial}{\partial x} \rho \hat{C}_p \overline{v}_x \overline{T} + \frac{\partial}{\partial y} \rho \hat{C}_p \overline{v}_y \overline{T} + \frac{\partial}{\partial z} \rho \hat{C}_p \overline{v}_z \overline{T} \right)$$
$$- \left(\frac{\partial}{\partial x} \rho \hat{C}_p \overline{v_x'T'} + \frac{\partial}{\partial y} \rho \hat{C}_p \overline{v_y'T'} + \frac{\partial}{\partial z} \rho \hat{C}_p \overline{v_z'T'} \right)$$
$$+ k\left(\frac{\partial^2 \overline{T}}{\partial x^2} + \frac{\partial^2 \overline{T}}{\partial y^2} + \frac{\partial^2 \overline{T}}{\partial z^2} \right)$$
$$+ \mu\left[2\left(\frac{\partial \overline{v}_x}{\partial x}\right)^2 + \left(\frac{\partial \overline{v}_x}{\partial y}\right)^2 + 2\left(\frac{\partial \overline{v}_x}{\partial y}\right)\left(\frac{\partial \overline{v}_y}{\partial x}\right) + \cdots \right]$$
$$+ \mu\left[2\overline{\left(\frac{\partial v_x'}{\partial x}\right)\left(\frac{\partial v_x'}{\partial x}\right)} + \overline{\left(\frac{\partial v_x'}{\partial y}\right)\left(\frac{\partial v_x'}{\partial y}\right)} + 2\overline{\left(\frac{\partial v_x'}{\partial y}\right)\left(\frac{\partial v_y'}{\partial x}\right)} + \cdots \right] \tag{13.1-3}$$

Comparison of this equation with the preceding one shows that the time-smoothed equation has the same form as the original equation, except for the appearance of the terms indicated by dashed underlines, which are concerned with the turbulent fluctuations. We are thus led to the definition of the turbulent heat flux $\overline{\mathbf{q}}^{(t)}$ with components

$$\overline{q}_x^{(t)} = \rho \hat{C}_p \overline{v_x'T'} \qquad \overline{q}_y^{(t)} = \rho \hat{C}_p \overline{v_y'T'} \qquad \overline{q}_z^{(t)} = \rho \hat{C}_p \overline{v_z'T'} \tag{13.1-4}$$

and the turbulent energy dissipation function $\overline{\Phi}_v^{(t)}$:

$$\overline{\Phi}_v^{(t)} = \sum_{i=1}^{3} \sum_{j=1}^{3} \left(\overline{\left(\frac{\partial v_i'}{\partial x_j}\right)\left(\frac{\partial v_i'}{\partial x_j}\right)} + \overline{\left(\frac{\partial v_i'}{\partial x_j}\right)\left(\frac{\partial v_j'}{\partial x_i}\right)} \right) \tag{13.1-5}$$

The similarity between the components of $\overline{\mathbf{q}}^{(t)}$ in Eq. 13.1-4 and those of $\overline{\boldsymbol{\tau}}^{(t)}$ in Eq. 5.2-8 should be noted. In Eq. 13.1-5, v_1', v_2', and v_3' are synonymous with v_x', v_y', and v_z', and x_1, x_2, and x_3 have the same meaning as x, y, and z.

To summarize, we list all three time-smoothed equations of change for turbulent flows of pure fluids with constant μ, ρ, $\hat{C}_p$, and k in their D/Dt form (the first two were given in Eqs. 5.2-10 and 12):

Continuity $$(\nabla \cdot \overline{\mathbf{v}}) = 0 \tag{13.1-6}$$

Motion $$\rho \frac{D\overline{\mathbf{v}}}{Dt} = -\nabla\overline{p} - [\nabla \cdot (\overline{\boldsymbol{\tau}}^{(v)} + \overline{\boldsymbol{\tau}}^{(t)})] + \rho\mathbf{g} \tag{13.1-7}$$

Energy $$\rho\hat{C}_p \frac{D\overline{T}}{Dt} = -(\nabla \cdot (\overline{\mathbf{q}}^{(v)} + \overline{\mathbf{q}}^{(t)})) + \mu(\overline{\Phi}_v^{(v)} + \overline{\Phi}_v^{(t)}) \tag{13.1-8}$$

in which it is understood that $D/Dt = \partial/\partial t + \overline{\mathbf{v}} \cdot \nabla$. Here $\overline{\mathbf{q}}^{(v)} = -k\nabla\overline{T}$, and $\overline{\Phi}_v^{(v)}$ is the viscous dissipation function of Eq. B.7-1, but with all the v_i replaced by $\overline{v}_i$.

In discussing turbulent heat flow problems, it has been customary to drop the viscous dissipation terms. Then, one sets up a turbulent heat transfer problem as for laminar flow, except that $\boldsymbol{\tau}$ and $\mathbf{q}$ are replaced by $\overline{\boldsymbol{\tau}}^{(v)} + \overline{\boldsymbol{\tau}}^{(t)}$ and $\overline{\mathbf{q}}^{(v)} + \overline{\mathbf{q}}^{(t)}$, respectively, and time-smoothed $\overline{p}$, $\overline{\mathbf{v}}$, and $\overline{T}$ are used in the remaining terms.

§13.2   THE TIME-SMOOTHED TEMPERATURE PROFILE NEAR A WALL[1]

Before giving empiricisms for $\overline{\mathbf{q}}^{(t)}$ in the next section, we present a short discussion of some results that do not depend on any empiricism.

We consider the turbulent flow along a flat wall as shown in Fig. 13.2-1, and we inquire as to the temperature in the inertial sublayer. We pattern the development after that for Eq. 5.3-1. We let the heat flux into the fluid at $y = 0$ be $q_0 = \overline{q}_y|_{y=0}$ and we postulate that the heat flux in the inertial sublayer will not be very different from that at the wall.

We seek to relate q_0 to the time-smoothed temperature gradient in the inertial sublayer. Because transport in this region is dominated by turbulent convection, the viscosity μ and the thermal conductivity k will not play an important role. Therefore the only parameters on which $d\overline{T}/dy$ can depend are q_0, $v_* = \sqrt{\tau_0/\rho}$, ρ, $\hat{C}_p$, and y. We must further use the fact that the linearity of the energy equation implies that $d\overline{T}/dy$ must be proportional to q_0. The only combination that satisfies these requirements is

$$-\frac{d\overline{T}}{dy} = \frac{\beta q_0}{\kappa\rho\hat{C}_p v_* y} \tag{13.2-1}$$

in which κ is the dimensionless constant in Eq. 5.3-1, and β is an additional constant (which turns out[1] to be the turbulent Prandtl number $\mathrm{Pr}^{(t)} = \nu^{(t)}/\alpha^{(t)}$).

When Eq. 13.2-1 is integrated we get

$$T_0 - \overline{T} = \frac{\beta q_0}{\kappa\rho\hat{C}_p v_*} \ln y + C \tag{13.2-2}$$

where T_0 is the wall temperature and C is a constant of integration. The constant is to be determined by matching the logarithmic expression with the expression for $\overline{T}(y)$ that

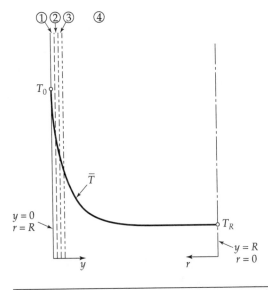

Fig. 13.2-1. Temperature profile in a tube with turbulent flow. The regions are (1) viscous sublayer, (2) buffer layer, (3) inertial sublayer, and (4) main turbulent stream.

[1] L. Landau and E. M. Lifshitz, *Fluid Mechanics*, 2nd edition, Pergamon Press, New York (1987), §54.

holds at the junction with the viscous sublayer. The latter expression will involve both μ and k; hence C will necessarily contain μ and k, and will therefore include the dimensionless group $\mathrm{Pr} = \hat{C}_p \mu / k$. If, in addition, we introduce the dimensionless coordinate $y v_* / \nu$, then Eq. 13.2-2 can be rewritten as

$$T_0 - \overline{T} = \frac{\beta q_0}{\kappa \rho \hat{C}_p v_*} \left[\ln \left(\frac{y v_*}{\nu} \right) + f(\mathrm{Pr}) \right] \qquad \text{for } \frac{y v_*}{\nu} > 1 \qquad (13.2\text{-}3)$$

in which $f(\mathrm{Pr})$ is a function representing the thermal resistance between the wall and the inertial sublayer. Landau and Lifshitz (see Ref. 1 on page 409) estimate, from a mixing-length argument (see Eq. 13.3-3), that, for large Prandtl numbers, $f(\mathrm{Pr}) = \text{constant} \cdot \mathrm{Pr}^{3/4}$; however, Example 13.3-1 implies that the function $f(\mathrm{Pr}) = \text{constant} \cdot \mathrm{Pr}^{2/3}$ is better. Keep in mind that Eq. 13.2-3 can be expected to be valid only in the inertial sublayer and that it should not be used in the immediate neighborhood of the wall.

§13.3 EMPIRICAL EXPRESSIONS FOR THE TURBULENT HEAT FLUX

In §13.1 we saw that the time-smoothing of the energy equation gives rise to a turbulent heat flux $\overline{\mathbf{q}}^{(t)}$. In order to solve the energy equation for the time-smoothed temperature profiles, it is customary to postulate a relation between $\overline{\mathbf{q}}^{(t)}$ and the time-smoothed temperature gradient. We summarize here two of the most popular empirical expressions; more of these can be found in the heat transfer literature.

Eddy Thermal Conductivity

By analogy with the Fourier law of heat conduction we may write

$$\overline{q}_y^{(t)} = -k^{(t)} \frac{d\overline{T}}{dy} \tag{13.3-1}$$

in which the quantity $k^{(t)}$ is called the *turbulent thermal conductivity* or the *eddy thermal conductivity*. This quantity is not a physical property of the fluid, but depends on position, direction, and the nature of the turbulent flow.

The eddy kinematic viscosity $\nu^{(t)} = \mu^{(t)}/\rho$ and the eddy thermal diffusivity $\alpha^{(t)} = k^{(t)}/\rho \hat{C}_p$ have the same dimensions. Their ratio is a dimensionless group

$$\mathrm{Pr}^{(t)} = \frac{\nu^{(t)}}{\alpha^{(t)}} \tag{13.3-2}$$

called the *turbulent Prandtl number*. This dimensionless quantity is of the order of unity, values in the literature varying from 0.5 to 1.0. For gas flow in conduits, $\mathrm{Pr}^{(t)}$ ranges from 0.7 to 0.9 (for circular tubes the value 0.85 has been recommended[1]), whereas for flow in jets and wakes the value is more nearly 0.5. The assumption that $\mathrm{Pr}^{(t)} = 1$ is called the *Reynolds analogy*.

The Mixing-Length Expression of Prandtl and Taylor

According to Prandtl's mixing-length theory, momentum and energy are transferred in turbulent flow by the same mechanism. Hence, by analogy with Eq. 5.4-4, one obtains

$$\overline{q}_y^{(t)} = -\rho \hat{C}_p l^2 \left| \frac{d\overline{v}_x}{dy} \right| \frac{d\overline{T}}{dy} \tag{13.3-3}$$

---

[1] W. M. Kays and M. E. Crawford, *Convective Heat and Mass Transfer*, 3rd edition, McGraw-Hill, New York (1993), pp. 259–266.

where l is the Prandtl mixing length introduced in Eq. 5.4-4. Note that this expression predicts that $\mathrm{Pr}^{(t)} = 1$. The Taylor vorticity transport theory[2] gives $\mathrm{Pr}^{(t)} = \frac{1}{2}$.

EXAMPLE 13.3-1

An Approximate Relation for the Wall Heat Flux for Turbulent Flow in a Tube

Use the Reynolds analogy ($\nu^{(t)} = \alpha^{(t)}$), along with Eq. 5.4-2 for the eddy viscosity, to estimate the wall heat flux q_0 for the turbulent flow in a tube of diameter $D = 2R$. Express the result in terms of the temperature-difference driving force $T_0 - \overline{T}_R$, where T_0 is the temperature at the wall ($y = 0$) and $\overline{T}_R$ is the time-smoothed temperature at the tube axis ($y = R$).

SOLUTION

The time-smoothed radial heat flux in a tube is given by the sum of $\overline{q}_r^{(v)}$ and $\overline{q}_r^{(t)}$:

$$\overline{q}_r = -(k + k^{(t)})\frac{d\overline{T}}{dr} = -(1 + \frac{\alpha^{(t)}}{\alpha})k\frac{d\overline{T}}{dr}$$

$$= +(1 + \frac{\nu^{(t)}}{\alpha})k\frac{d\overline{T}}{dy} \tag{13.3-4}$$

Here we have used Eq. 13.3-1 and the Reynolds analogy, and we have switched to the coordinate y, which is the distance from the wall. We now use the empirical expression of Eq. 5.4-2, which applies across the viscous sublayer next to the wall:

$$\overline{q}_y = -\left[1 + \mathrm{Pr}\left(\frac{yv_*}{14.5\nu}\right)^3\right]k\frac{d\overline{T}}{dy} \qquad \text{for } \frac{yv_*}{\nu} < 5 \tag{13.3-5}$$

where $\overline{q}_r = -\overline{q}_y$ has been used.

If now we approximate the heat flux $\overline{q}_y$ in Eq. 13.3-5 by its wall value q_0, then integration from $y = 0$ to $y = R$ gives

$$q_0 \int_0^R \frac{dy}{1 + \mathrm{Pr}(yv_*/14.5\nu)^3} = k(T_0 - \overline{T}_R) \tag{13.3-6}$$

For very large Prandtl numbers, the upper limit R in the integral can be replaced by ∞, since the integrand is decreasing rapidly with increasing y. Then when the integration on the left side is performed and the result is put into dimensionless form, we get

$$\frac{q_0 D}{k(T_0 - \overline{T}_R)} = \frac{3\sqrt{3}}{2\pi(14.5)}\left(\frac{v_*}{\langle v_z \rangle}\right)\mathrm{Re}\,\mathrm{Pr}^{1/3} = \frac{1}{17.5}\sqrt{\frac{f}{2}}\,\mathrm{Re}\,\mathrm{Pr}^{1/3} \tag{13.3-7}$$

in which Eq. 6.1-4a has been used to eliminate v_* in favor of the friction factor.

The above development is only approximate. We have not taken into account the change of the bulk temperature as the fluid moves axially through the tube, nor have we taken into account the change in the heat flux throughout the tube. Furthermore, the result is restricted to very high Pr, because of the extension of the integration to $y = \infty$. Another derivation is given in the next section, which is free from these assumptions. However, we will see that at large Prandtl numbers the result in Eq. 13.4-20 simplifies to that in Eq. 13.3-7 but with a different numerical constant.

---

§13.4   TEMPERATURE DISTRIBUTION FOR TURBULENT FLOW IN TUBES

In §10.8 we showed how to get the asymptotic behavior of the temperature profiles for large z in a fluid in laminar flow in a circular tube. We repeat that problem here, but for a fluid in fully developed turbulent flow. The fluid enters the tube of radius R at an inlet temperature T_1. For $z > 0$ the fluid is heated because of a uniform radial heat flux q_0 at the wall (see Fig. 13.4-1).

---

[2] G. I. Taylor, *Proc. Roy. Soc.* (London), **A135**, 685–702 (1932); *Phil. Trans.*, **A215**, 1–26 (1915).

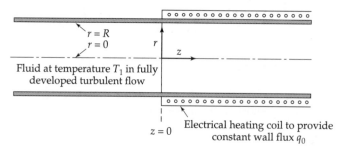

Fig. 13.4-1. System used for heating a liquid in fully developed turbulent flow with constant heat flux for $z > 0$.

We start from the energy equation, Eq. 13.1-8, written in cylindrical coordinates

$$\rho \hat{C}_p \bar{v}_z \frac{\partial \bar{T}}{\partial z} = -\frac{1}{r} \frac{\partial}{\partial r} (r(\bar{q}_r^{(v)} + \bar{q}_r^{(t)})) \tag{13.4-1}$$

Then insertion of the expression for the radial heat flux from Eq. 13.3-4 gives

$$\bar{v}_z \frac{\partial \bar{T}}{\partial z} = \frac{1}{r} \frac{\partial}{\partial r} \left(r(\alpha + \alpha^{(t)}) \frac{\partial \bar{T}}{\partial r} \right) \tag{13.4-2}$$

This is to be solved with the boundary conditions

B.C. 1:   at $r = 0$, $\bar{T} = $ finite   (13.4-3)

B.C. 2:   at $r = R$, $+k \dfrac{\partial \bar{T}}{\partial r} = q_0$ (a constant)   (13.4-4)

B.C. 3:   at $z = 0$, $\bar{T} = T_1$ (13.4-5)

We now use the same dimensionless variables as already given in Eqs. 10.8-16 to 18 (with $\bar{T}$ in place of T in the definition of the dimensionless temperature). Then Eq. 13.4-2 in dimensionless form is

$$\phi \frac{\partial \Theta}{\partial \zeta} = \frac{1}{\xi} \frac{\partial}{\partial \xi} \left(\xi \left(1 + \frac{\alpha^{(t)}}{\alpha} \right) \frac{\partial \Theta}{\partial \xi} \right) \tag{13.4-6}$$

in which $\phi(\xi) = \bar{v}_z/v_{max}$ is the dimensionless turbulent velocity profile. This equation is to be solved with the dimensionless boundary conditions

B.C. 1:   at $\xi = 0$, $\Theta = $ finite   (13.4-7)

B.C. 2:   at $\xi = 1$, $+\dfrac{\partial \Theta}{\partial \xi} = 1$ (13.4-8)

B.C. 3:   at $\zeta = 0$, $\Theta = 0$ (13.4-9)

The complete solution to this problem has been given,[1] but we content ourselves here with the solution for large z.

We begin by assuming an asymptotic solution of the form of Eq. 10.8-23

$$\Theta(\xi, \zeta) = C_0 \zeta + \Psi(\xi) \tag{13.4-10}$$

which must satisfy the differential equation, together with B.C. 1 and 2 and Condition 4 in Eq. 10.8-24 (with T and $v_z = v_{max}(1 - \xi^2)$ replaced by $\bar{T}$ and $v_z = v_{max}\phi(\xi)$). The resulting equation for Ψ is

$$\frac{1}{\xi} \frac{d}{d\xi} \left(\xi \left(1 + \frac{\alpha^{(t)}}{\alpha} \right) \frac{d\Psi}{d\xi} \right) = C_0 \phi \tag{13.4-11}$$

[1] R. H. Notter and C. A. Sleicher, *Chem. Eng. Sci.*, **27**, 2073–2093 (1972).

Integrating this equation twice and then constructing the function Θ using Eq. 13.4-10, we get

$$\Theta = C_0\zeta + C_0\int_0^\xi \frac{I(\bar\xi)}{\bar\xi[1+(\alpha^{(t)}/\alpha)]}\,d\bar\xi + C_1\int_0^\xi \frac{1}{\bar\xi[1+(\alpha^{(t)}/\alpha)]}\,d\bar\xi + C_2 \qquad (13.4\text{-}12)$$

in which it is understood that $\alpha^{(t)}$ is a function of $\bar\xi$, and $I(\bar\xi)$ is shorthand for the integral

$$I(\bar\xi) = \int_0^{\bar\xi} \phi\bar{\bar\xi}\,d\bar{\bar\xi} \qquad (13.4\text{-}13)$$

The constant of integration C_1 is set equal to zero in order to satisfy B.C. 1. The constant C_0 is found by applying B.C. 2, which gives

$$C_0 = \left(\int_0^1 \phi\xi d\xi\right)^{-1} = [I(1)]^{-1} \qquad (13.4\text{-}14)$$

The remaining constant, C_2, can, if desired, be obtained from Condition 4, but we shall not need it here (see Problem 13D.1).

We next get an expression for the dimensionless temperature difference $\Theta_0 - \Theta_b$, the "driving force" for the heat transfer at the tube wall:

$$\Theta_0 - \Theta_b = C_0\int_0^1 \frac{I(\xi)}{\xi[1+(\alpha^{(t)}/\alpha)]}\,d\xi - \frac{C_0}{I(1)}\int_0^1 \phi\xi\left[\int_0^\xi \frac{I(\bar\xi)}{\bar\xi[1+(\alpha^{(t)}/\alpha)]}\,d\bar\xi\right]d\xi$$

$$= C_0\int_0^1 \frac{I(\xi)}{\xi[1+(\alpha^{(t)}/\alpha)]}\,d\xi - \frac{C_0}{I(1)}\int_0^1 \frac{I(\bar\xi)}{\bar\xi[1+(\alpha^{(t)}/\alpha)]}\left[\int_{\bar\xi}^1 \phi\xi d\xi\right]d\bar\xi \qquad (13.4\text{-}15)$$

In the second line, the order of integration of the double integral has been reversed. The inner integral in the second term on the right is just $I(1) - I(\bar\xi)$, and the portion containing $I(1)$ exactly cancels the first term in Eq. 13.4-15. Hence when Eq. 13.4-14 is used, we get

$$\Theta_0 - \Theta_b = \int_0^1 \frac{[I(\xi)/I(1)]^2}{\xi[1+(\alpha^{(t)}/\alpha)]}\,d\xi \qquad (13.4\text{-}16)$$

But the quantity $I(1)$ appearing in Eq. 13.4-16 has a simple interpretation:

$$I(1) = \int_0^1 \phi\,\xi\,d\xi = \left(\int_0^R \bar{v}_z r\,dr\right)\frac{1}{\bar{v}_{z,max}R^2} = \frac{1}{2}\frac{\langle\bar{v}_z\rangle}{\bar{v}_{z,max}} \qquad (13.4\text{-}17)$$

Finally, we want to get the dimensionless wall heat flux,

$$\frac{q_0 D}{k(T_0 - T_b)} = \frac{2}{\Theta_0 - \Theta_b} \qquad (13.4\text{-}18)$$

the reciprocal of which is[2]

$$\frac{k(T_0 - T_b)}{q_0 D} = 2\left(\frac{\bar{v}_{z,max}}{\langle\bar{v}_z\rangle}\right)^2\int_0^1 \frac{[I(\xi)]^2}{\xi[1+(\nu^{(t)}/\nu)(Pr/Pr^{(t)})]}\,d\xi \qquad (13.4\text{-}19)$$

To use this result, it is necessary to have an expression for the time-smoothed velocity distribution $\bar{v}_z$ (which appears in $I(\xi)$), the turbulent kinematic viscosity $\nu^{(t)}$ as a function of position, and a postulate for the turbulent Prandtl number $Pr^{(t)}$.

---

[2] Equation 13.4-19 was first developed by R. N. Lyon, *Chem. Eng. Prog.*, **47**, 75–79 (1950) in a paper on liquid–metal heat transfer. The left side of Eq. 13.4-19 is the reciprocal of the Nusselt number, $Nu = hD/k$, which is a dimensionless heat transfer coefficient. This nomenclature is discussed in the next chapter.

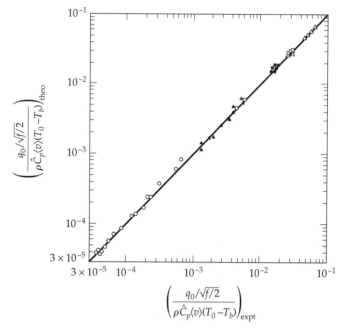

Fig. 13.4-2. Comparison of the expression in Eq. 13.4-20 for the wall heat flux in fully developed turbulent flow with the experimental data of R. G. Deissler and C. S. Eian, *NACA Tech. Note #2629* (1952); R. W. Allen and E. R. G. Eckert, *J. Heat Transfer, Trans. ASME, Ser. C.*, **86**, 301–310 (1964); J. A. Malina and E. M. Sparrow, *Chem. Eng. Sci*, **19**, 953–962 (1964); W. L. Friend and A. B. Metzner, *AIChE Journal*, **4**, 393–402 (1958); P. Harriott and R. M. Hamilton, *Chem. Eng. Sci.*, **20**, 1073–1078 (1965). The data of Harriott and Hamilton are for the analogous mass transfer experiment, for which Eq. 13.4-20 also applies.

Extensive calculations based on Eq. 13.4-19 were performed by Sandall, Hanna, and Mazet.[3] These authors took the turbulent Prandtl number to be unity. They divided the region of integration into two parts, one near the wall and the other for the turbulent core. In the "wall region" they used the modified van Driest equation of Eq. 5.4-7 for the mixing length, and in the "core region" they used a logarithmic velocity distribution. Their final result[3] is given as

$$\frac{q_0 D}{k(T_0 - T_b)} = \frac{\text{Re Pr}\sqrt{f/2}}{12.48 \, \text{Pr}^{2/3} - 7.853 \, \text{Pr}^{1/3} + 3.613 \ln \text{Pr} + 5.8 + 2.78 \ln (\frac{1}{45} \text{Re}\sqrt{f/8})} \tag{13.4-20}$$

In obtaining this result, Eq. 6.1-4a has been used.

Equation 13.4-20 agrees with the available data on heat transfer (and mass transfer) within 3.6 and 8.1% over the range $0.73 < \text{Pr} < 590$, depending on the sets of data studied. The analogous mass transfer expression, containing $\text{Sc} = \mu/\rho \mathscr{D}_{AB}$ instead of Pr, was reported[3] to agree with the mass transfer data within 8% over the range $452 < \text{Sc} < 97600$. The agreement of the theory with the heat transfer and mass transfer data, shown in Fig. 13.4-2, is quite convincing.

[3] O. C. Sandall, O. T. Hanna, and P. R. Mazet, *Canad. J. Chem. Eng.*, **58**, 443–447 (1980). See also O. T. Hanna and O. C. Sandall, *AIChE Journal*, **18**, 527–533 (1972).

§13.5   TEMPERATURE DISTRIBUTION FOR TURBULENT FLOW IN JETS[1]

In §5.6 we derived an expression for the velocity distribution in a circular fluid jet discharging into an infinite expanse of the same fluid (see Fig. 5.6-1). Here we wish to extend this problem by considering an incoming jet with temperature T_0 higher than that of the surrounding fluid T_1. The problem then is to find the time-smoothed temperature distribution $\overline{T}(r, z)$ in a steadily driven jet. We expect that this distribution will be monotone decreasing in both the r and z directions.

We start by assuming that viscous dissipation is negligible, and we neglect the contribution $\overline{\mathbf{q}}^{(v)}$ to the heat flux as well as the axial contribution to $\overline{\mathbf{q}}^{(t)}$. Then Eq. 13.1-8 takes the time-averaged form

$$\rho \hat{C}_p \left(\overline{v}_r \frac{\partial \overline{T}}{\partial r} + \overline{v}_z \frac{\partial \overline{T}}{\partial z} \right) = -\frac{1}{r} \frac{\partial}{\partial r} (r \overline{q}_r^{(t)}) \tag{13.5-1}$$

Then we express the turbulent heat flux in terms of the turbulent thermal conductivity introduced in Eq. 13.3-1:

$$\overline{q}_r^{(t)} = -k^{(t)} \frac{\partial \overline{T}}{\partial r} = -\rho \hat{C}_p \alpha^{(t)} \frac{\partial \overline{T}}{\partial r} = \rho \hat{C}_p \frac{\nu^{(t)}}{\mathrm{Pr}^{(t)}} \frac{\partial \overline{T}}{\partial r} \tag{13.5-2}$$

When Eq. 13.5-1 is written in terms of a dimensionless temperature function

$$\Theta(\xi, \zeta) = \frac{\overline{T} - T_1}{T_0 - T_1} \tag{13.5-3}$$

it becomes

$$\left(\overline{v}_r \frac{\partial \Theta}{\partial r} + \overline{v}_z \frac{\partial \Theta}{\partial z} \right) = \frac{\nu^{(t)}}{\mathrm{Pr}^{(t)}} \frac{1}{r} \frac{\partial}{\partial r} \left(r \frac{\partial \Theta}{\partial r} \right) \tag{13.5-4}$$

Here it has been assumed that the turbulent Prandtl number and the turbulent kinematic viscosity are constants (see the discussion after Eq. 5.6-3). This equation is to be solved with the boundary conditions:

B.C. 1:   at $z = 0$, $\Theta = 1$ (13.5-5)

B.C. 2:   at $r = 0$, Θ is finite   (13.5-6)

B.C. 3:   at $r = \infty$, $\Theta = 0$ (13.5-7)

Next we introduce the expressions for the time-smoothed velocity components $\overline{v}_r$ and $\overline{v}_z$ in terms of a stream function $F(\xi)$, as given in Eqs. 5.6-12 and 13, and a trial expression for the dimensionless time-smoothed temperature function:

$$\Theta(\xi, \zeta) = \frac{1}{\zeta} f(\xi) \tag{13.5-8}$$

Here $\xi = r/z$ and $\zeta = (\rho \nu^{(t)} / w) z$, where w is the total mass flow rate in the jet. The proposal in Eq. 13.5-8 is motivated by the expression for $\overline{v}_z$ that was found in Eq. 5.6-21.

When these expressions for the velocity components and the dimensionless temperature are substituted into Eq. 13.5-1, some terms cancel and others can be combined, and as a result, the following rather simple equation is obtained:

$$\mathrm{Pr}^{(t)} \frac{1}{\xi} \frac{d}{d\xi} (Ff) = \frac{1}{\xi} \frac{d}{d\xi} \left(\xi \frac{df}{d\xi} \right) \tag{13.5-9}$$

---

[1] J. O. Hinze, *Turbulence*, 2nd edition, McGraw-Hill, New York (1975), pp. 531–546.

This equation can be integrated once to give

$$\text{Pr}^{(t)} Ff = \xi \frac{df}{d\xi} + C \tag{13.5-10}$$

The constant of integration may be set equal to zero, since, according to Eq. 5.6-20, $F = 0$ at $\xi = 0$. A second integration from 0 to ξ then gives

$$\ln \frac{f(\xi)}{f(0)} = \text{Pr}^{(t)} \int_0^\xi \frac{F}{\xi} d\xi = -\text{Pr}^{(t)} \int_0^\xi \frac{C_3^2 \xi}{1 + \frac{1}{4}(C_3 \xi)^2} d\xi$$

$$= -\text{Pr}^{(t)} \ln (1 + \tfrac{1}{4}(C_3\xi)^2)^2 \tag{13.5-11}$$

or

$$\frac{f(\xi)}{f(0)} = (1 + \tfrac{1}{4}(C_3\xi)^2)^{-2\text{Pr}^{(t)}} \tag{13.5-12}$$

Finally, comparison of Eqs. 13.5-12 and 13.5-8 with Eq. 5.6-21 shows that the shapes of the time-smoothed temperature and axial velocity profiles are closely related,

$$\frac{\Theta}{\Theta_{\max}} = \left(\frac{\bar{v}_z}{\bar{v}_{z,\max}} \right)^{\text{Pr}^{(t)}} \tag{13.5-13}$$

an equation attributed to Reichardt.[2] This theory provides a moderately satisfactory explanation for the shapes of the temperature profiles.[1] The turbulent Prandtl (or Schmidt) number deduced from temperature (or concentration) measurements in circular jets is about 0.7.

The quantity C_3 appearing in Eq. 13.5-12 was given explicitly in Eq. 5.6-23 as $C_3 = \sqrt{3/16\pi}\sqrt{J/\rho}(1/\nu^{(t)})$, where J is the rate of momentum flow in the jet, defined in Eq. 5.6-2. Similarly, an expression for the quantity $f(0)$ in Eq. 13.5-12 can be found by equating the energy in the incoming jet to the energy crossing any plane downstream:

$$w\hat{C}_p(T_0 - T_1) = \int_0^{2\pi} \int_0^\infty \rho \hat{C}_p \bar{v}_z (\bar{T} - T_1) r \, dr \, d\theta \tag{13.5-14}$$

Insertion of the expressions for the velocity and temperature profiles and integrating then gives

$$\frac{1}{f(0)} = 4\pi C_3^2 \int_0^\infty (1 + \tfrac{1}{4}(C_3\xi)^2)^{-(2+2\text{Pr}^{(t)})} \xi \, d\xi = \frac{8}{C_3} \frac{1}{1 + 2\,\text{Pr}^{(t)}} \tag{13.5-15}$$

Combining Eqs. 13.5-3, 13.5-8, 5.6-23, 13.5-12, and 13.5-15 then gives the complete expression for the temperature profiles $\bar{T}(r, z)$ in the circular turbulent jet, in terms of the total momentum of the jet, the turbulent viscosity, the turbulent Prandtl number, and the fluid density.

13.6   FOURIER ANALYSIS OF ENERGY TRANSPORT IN TUBE FLOW AT LARGE PRANDTL NUMBERS

In the preceding two sections we analyzed energy transport in turbulent systems by use of time-smoothed equations of change. Empirical expressions were then required to describe the turbulent fluxes in terms of time-smoothed profiles, using eddy transport coef-

---

[2] H. Reichardt, *Zeits. f. angew. Math. u. Mech.*, **24**, 268–272 (1944).

ficients estimated from experiments. In this section we analyze a turbulent energy transport problem without time-smoothing—that is, by direct use of the energy equation with fluctuating velocity and temperature fields. The Fourier transform[1] is well suited for such problems, and the "method of dominant balance"[2] gives useful information without detailed computations.

The specific question considered here is the influence of the thermal diffusivity, $\alpha = k/\rho \hat{C}_p$, on the expected distribution and fluctuations of the fluid temperature in turbulent forced convection near a wall.[3] This topic was discussed in Example 13.3-1 by an approximate procedure.

Let us consider a fluid with constant ρ, $\hat{C}_p$, and k in turbulent flow through a tube of inner radius $R = \frac{1}{2}D$. The flow enters at $z = -\infty$ with uniform temperature T_1 and exits at $z = L$. The tube wall is adiabatic for $z < 0$, and isothermal at T_0 for $0 \leq z \leq L$. Heat conduction in the z direction is neglected. The temperature distribution $T(r, \theta, z, t)$ is to be analyzed in the long-time limit, in the thin thermal boundary layer that forms for $z > 0$ when the *molecular* thermal diffusivity α is small (as in a Newtonian fluid when the Prandtl number, $\mathrm{Pr} = \hat{C}_p \mu / k = \mu/\rho\alpha$, is large). A stretching function $\kappa(\alpha)$ will be derived for the average thickness of the thermal boundary layer without introducing an *eddy* thermal diffusivity $\alpha^{(t)}$.

In the limit as $\alpha \to 0$, the thermal boundary layer lies entirely within the viscous sublayer, where the velocity components are given by truncated Taylor expansions in the distance $y = R - r$ from the wall (compare these expansions with those in Eqs. 5.4-8 to 10)

$$v_\theta = \beta_\theta y + \mathrm{O}(y^2) \tag{13.6-1}$$

$$v_z = \beta_z y + \mathrm{O}(y^2) \tag{13.6-2}$$

$$v_r = -\left(\frac{1}{R}\frac{\partial\beta_\theta}{\partial\theta} + \frac{\partial\beta_z}{\partial z}\right)\frac{y^2}{2} + \mathrm{O}(y^3) \tag{13.6-3}$$

Here the coefficients β_θ and β_z are treated as given functions of θ, z, and t. These velocity expressions satisfy the no-slip conditions and the wall-impermeability condition at $y = 0$ and the continuity equation at small y, and are consistent with the equation of motion to the indicated orders in y. The energy equation can then be written as

$$\frac{\partial T}{\partial t} + \left(\frac{\beta_\theta}{R}\frac{\partial T}{\partial\theta} + \beta_z\frac{\partial T}{\partial z}\right)y - \left(\frac{1}{R}\frac{\partial\beta_\theta}{\partial\theta} + \frac{\partial\beta_z}{\partial z}\right)\frac{y^2}{2}\frac{\partial T}{\partial y} = \alpha\frac{\partial^2 T}{\partial y^2} \tag{13.6-4}$$

with the usual boundary layer approximation for $\nabla^2 T$, and with the following boundary conditions on $T(y, \theta, z, t)$:

Inlet condition:	at $z = 0$,	$T(y, \theta, 0, t) = T_1$	for $0 < y \leq R$	(13.6-5)
Wall condition:	at $y = 0$,	$T(0, \theta, z, t) = T_0$	for $0 \leq z \leq L$	(13.6-6)

The initial temperature distribution $T(y, \theta, z, 0)$ is not needed, since its effect disappears in the long-time limit.

To obtain results asymptotically valid for $\alpha \to 0$, we introduce a stretched coordinate $Y = y/\kappa(\alpha)$, which is the distance from the wall relative to the average boundary layer thickness $\kappa(\alpha)$. The range of Y is from 0 at $y = 0$ to ∞ at $y = R$ in the limit as $\alpha \to 0$.

---

[1] R. N. Bracewell, *The Fourier Transform and its Applications*, 2nd edition, McGraw-Hill, New York (1978).

[2] This method is well presented in C. M. Bender and S. A. Orzag, *Advanced Mathematical Methods for Scientists and Engineers*, McGraw-Hill, New York (1978), pp. 435–437.

[3] W. E. Stewart, *AIChE Journal*, **33**, 2008–2016 (1987); errata, *ibid.*, **34**, 1030 (1988); W. E. Stewart and D. G. O'Sullivan, *AIChE Journal* (to be submitted).

Use of κY in place of y, and introduction of the dimensionless temperature function $\Theta(Y, \theta, z, t) = (T - T_1)/(T_0 - T_1)$, enable us to rewrite Eq. 13.6-4 as

$$\frac{\partial \Theta}{\partial t} + \left(\frac{\beta_\theta}{R} \frac{\partial \Theta}{\partial \theta} + \beta_z \frac{\partial \Theta}{\partial z} \right) \kappa Y - \left(\frac{1}{R} \frac{\partial \beta_\theta}{\partial \theta} + \frac{\partial \beta_z}{\partial z} \right) \frac{\kappa Y^2}{2} \frac{\partial \Theta}{\partial y} = \frac{\alpha}{\kappa^2} \frac{\partial^2 \Theta}{\partial Y^2} \tag{13.6-7}$$

with boundary conditions as follows:

Inlet condition:	at $z = 0$,	$\Theta(Y, \theta, 0, t) = 0$	for $Y > 0$	(13.6-8)
Wall condition:	at $Y = 0$,	$\Theta(0, \theta, z, t) = 1$	for $0 \leq z \leq L$	(13.6-9)

Equation 13.6-7 contains an unbounded derivative $\partial \Theta / \partial t$ with a coefficient 1 independent of α. Thus a change of variables is needed to analyze the influence of the parameter α in this problem. For this purpose we turn to the Fourier transform, a standard tool for analyzing noisy processes.

We choose the following definition[1] for the Fourier transform of a function $g(t)$ into the domain of frequency ν at a particular position Y, θ, z:

$$\mathcal{F}\{g(t)\} = \int_{-\infty}^{\infty} e^{-2\pi i \nu t} g(t) dt = \tilde{g}(\nu) \tag{13.6-10}$$

The corresponding transforms for the t-derivative and for products of functions of t are

$$\int_{-\infty}^{\infty} e^{-2\pi i \nu t} \frac{\partial}{\partial t} g(t) dt = 2\pi i \nu \tilde{g}(\nu) \tag{13.6-11}$$

$$\int_{-\infty}^{\infty} e^{-2\pi i \nu t} g(t) h(t) dt = \int_{-\infty}^{\infty} \tilde{g}(\nu) \tilde{h}(\nu - \nu_1) d\nu_1 = \tilde{g} * \tilde{h} \tag{13.6-12}$$

and the latter integral is known as the *convolution* of the transforms $\tilde{g}$ and $\tilde{h}$.

Before taking the Fourier transforms of Eqs. 13.6-7 to 9, we express each included function $g(t)$ as a time average $\bar{g}$ plus a fluctuating function $g'(t)$ and expand each product of such functions. The resulting expressions have the following Fourier transforms:

$$\mathcal{F}\{\bar{g} + g'\} = \delta(\nu)\bar{g} + \tilde{g}'(\nu) \tag{13.6-13}$$

$$\mathcal{F}\{(\bar{g} + g')(\bar{h} + h')\} = \mathcal{F}\{\bar{g}\bar{h} + \bar{g}h' + g'\bar{h} + g'h'\}$$

$$= \delta(\nu)\bar{g}\bar{h} + \bar{g}\tilde{h}' + \tilde{g}'\bar{h} + \tilde{g}' * \tilde{h}' \tag{13.6-14}$$

Here $\delta(\nu)$ is the Dirac delta function, obtained as the Fourier transform of the function $g(t) = 1$ in the long-duration limit. The leading term in the last line is a real-valued impulse at $\nu = 0$, coming from the time-independent product $\bar{g}\bar{h}$. The next two terms are complex-valued functions of the frequency ν. The convolution term $\tilde{g}' * \tilde{h}'$ may contain complex-valued functions of ν, along with a real-valued impulse $\delta(\nu)\overline{g'h'}$ coming from time-independent products of simple harmonic oscillations present in g' and h'.

Taking the Fourier transform of Eq. 13.6-7 by the method just given and noting that $\partial \overline{\Theta} / \partial t$ is identically zero, we obtain the differential equation

$$2\pi i \nu \tilde{\Theta}' + \left(\delta(\nu) \frac{\overline{\beta_\theta}}{R} \frac{\partial \overline{\Theta}}{\partial \theta} + \frac{\overline{\beta_\theta}}{R} \frac{\partial \tilde{\Theta}'}{\partial \theta} + \frac{\tilde{\beta}_\theta'}{R} \frac{\partial \overline{\Theta}}{\partial \theta} + \frac{\tilde{\beta}_\theta'}{R} * \frac{\partial \tilde{\Theta}'}{\partial \theta} \right) \kappa Y$$

$$+ \left(\delta(\nu)\overline{\beta_z} \frac{\partial \overline{\Theta}}{\partial z} + \overline{\beta_z} \frac{\partial \tilde{\Theta}'}{\partial z} + \tilde{\beta}_z' \frac{\partial \overline{\Theta}}{\partial z} + \tilde{\beta}_z' * \frac{\partial \tilde{\Theta}'}{\partial z} \right) \kappa Y$$

$$- \left(\frac{\delta(\nu)}{R} \frac{\partial \overline{\beta_\theta}}{\partial \theta} \frac{\partial \overline{\Theta}}{\partial Y} + \frac{1}{R} \frac{\partial \overline{\beta_\theta}}{\partial \theta} \frac{\partial \tilde{\Theta}'}{\partial Y} + \frac{1}{R} \frac{\partial \tilde{\beta}_\theta'}{\partial \theta} \frac{\partial \overline{\Theta}}{\partial Y} + \frac{1}{R} \frac{\partial \tilde{\beta}_\theta'}{\partial \theta} * \frac{\partial \tilde{\Theta}'}{\partial Y} \right) \frac{\kappa Y^2}{2}$$

$$- \left(\delta(\nu) \frac{\partial \overline{\beta_z}}{\partial z} \frac{\partial \overline{\Theta}}{\partial Y} + \frac{\partial \overline{\beta_z}}{\partial z} \frac{\partial \tilde{\Theta}'}{\partial Y} + \frac{\partial \tilde{\beta}_z'}{\partial z} \frac{\partial \overline{\Theta}}{\partial Y} + \frac{\partial \tilde{\beta}_z'}{\partial z} * \frac{\partial \tilde{\Theta}'}{\partial Y} \right) \frac{\kappa Y^2}{2}$$

$$= \left(\delta(\nu) \frac{\partial^2 \overline{\Theta}}{\partial Y^2} + \frac{\partial^2 \tilde{\Theta}'}{\partial Y^2} \right) \frac{\alpha}{\kappa^2} \tag{13.6-15}$$

for the Fourier-transformed temperature $\tilde{\Theta}(Y, \theta, z, \nu)$. The transformed boundary conditions are

Inlet condition:	at $z = 0$,	$\tilde{\Theta}(Y, \theta, z, \nu) = 0$	for $Y > 0$

(13.6-16)

Wall condition:	at $Y = 0$,	$\tilde{\Theta}(Y, \theta, z, \nu) = \delta(\nu)$	for $0 \leq z \leq L$

(13.6-17)

Here again, the unit impulse function $\delta(\nu)$ appears as the Fourier transform of the function $g(t) = 1$ in the long-duration limit.

Two types of contributions appear in Eq. 13.6-15: real-valued zero-frequency impulses $\delta(\nu)$ from functions and products independent of t, and complex-valued functions of ν from time-dependent product terms. We consider these two types of contributions separately here, thus decoupling Eq. 13.6-15 into two equations.

We begin with the zero-frequency impulse terms. In addition to the explicit $\delta(\nu)$ terms of Eq. 13.6-15, implicit impulses arise in the convolution terms from synchronous oscillations of velocity and temperature, giving rise to the turbulent energy flux $\overline{\mathbf{q}}^{(t)} = \rho \hat{C}_p \overline{\mathbf{v}'T'}$ discussed in §13.2. The coefficients of all the impulse terms must be proportional functions of α, in order that the dominant terms at each point remain balanced (i.e., of comparable size) as $\alpha \to 0$. Therefore, the coefficient κ of the convective impulse terms, including those from synchronous fluctuations, must be proportional to the coefficient α / κ^2 of the conductive impulse term, giving $\kappa \propto \alpha^{1/3}$, or

$$\kappa = \text{Pr}^{-1/3} D \tag{13.6-18}$$

for the dependence of the average thermal boundary layer thickness on the Prandtl number.

The remaining terms in Eq. 13.6-15 describe the turbulent temperature fluctuations. They include the accumulation term $2\pi i \nu \tilde{\Theta}'$ and the remaining convection and conduction terms. The coefficients of all these terms (including $2\pi i \nu$ in the leading term) must be proportional functions of α in order that these terms likewise remain balanced as $\alpha \to 0$. This reasoning confirms Eq. 13.6-18 and gives the further relation $\nu \propto \kappa$, or

$$\frac{D\Delta\nu}{\langle v_z \rangle} \propto \kappa = \text{Pr}^{-1/3} D \tag{13.6-19}$$

for the frequency bandwidth $\Delta\nu$ of the temperature fluctuations. Consequently, the stretched frequency $\text{Pr}^{1/3}\nu$ and stretched time $\text{Pr}^{-1/3}t$ are natural variables for reporting Fourier analyses of turbulent forced convection. Shaw and Hanratty[4] reported turbulence spectra for their mass transfer experiments analogously, in terms of a stretched frequency variable proportional to $\text{Sc}^{1/3}\nu$ (here $\text{Sc} = \mu/\rho\mathscr{D}_{AB}$ is the Schmidt number, the mass transfer analog of the Prandtl number, which contains the binary diffusivity $\mathscr{D}_{AB}$, to be introduced in Chapter 16).

Thus far we have considered only the leading term of a Taylor expansion in κ for each term in the energy equations. More accurate results are obtainable by continuing the Taylor expansions to higher powers of κ, and thus of $\text{Pr}^{-1/3}D$. The resulting formal solution is a perturbation expansion

$$\tilde{\Theta} = \tilde{\Theta}_0(Y, \theta, z, \text{Pr}^{1/3}\nu) + \kappa \tilde{\Theta}_1(Y, \theta, z, \text{Pr}^{1/3}\nu) + \cdots \tag{13.6-20}$$

for the distribution of the fluctuating temperature over position and frequency in a given velocity field.

The expansion for $\overline{T}$ (the long-time average of the temperature) corresponding to Eq. 13.6-20 is obtained from the zero-frequency part of $\tilde{\Theta}$,

$$\overline{\Theta} = \overline{\Theta}_0(Y, \theta, z) + \kappa \overline{\Theta}_1(Y, \theta, z) + \cdots \tag{13.6-21}$$

---

[4] D. A. Shaw and T. J. Hanratty, *AIChE Journal*, **23**, 160–169 (1977); D. A. Shaw and T. J. Hanratty, *AIChE Journal*, **23**, 28–37 (1977).

From this we can calculate the local time-averaged heat flux at the wall:

$$q_0 = -k \left. \frac{\partial \overline{T}}{\partial y} \right|_{y=0} = -\frac{k(T_0 - T_1)}{\Pr^{-1/3}D} \left. \frac{\partial \overline{\Theta}}{\partial Y} \right|_{Y=0} \tag{13.6-22}$$

and the local Nusselt number is then

$$\mathrm{Nu}_{\mathrm{loc}} = \frac{q_0 D}{k(T_0 - T_1)} = \Pr^{1/3}\left(-\frac{\partial \overline{\Theta}}{\partial Y}\right)\bigg|_{Y=0} \tag{13.6-23}$$

Then the mean Nusselt number over the wall surface for heat transfer, and the analogous quantity for mass transfer, are

$$\mathrm{Nu}_m = \Pr^{1/3}\left\langle \left(-\frac{\partial \overline{\Theta}}{\partial Y}\right)\bigg|_{Y=0}\right\rangle = a_1 \Pr^{1/3} + a_2 \Pr^0 + \cdots \tag{13.6-24}$$

$$\mathrm{Sh}_m = \mathrm{Sc}^{1/3}\left\langle \left(-\frac{\partial \overline{\Theta}_A}{\partial Y}\right)\bigg|_{Y=0}\right\rangle = a_1 \mathrm{Sc}^{1/3} + a_2 \mathrm{Sc}^0 + \cdots \tag{13.6-25}$$

In this last equation Sh_m, $\overline{\Theta}_A$, and Sc are the mass transfer analogs of Nu_m, $\overline{\Theta}$, and Pr. We give the mass transfer expression here (rather than wait until Part III) because electrochemical mass transfer experiments give better precision than heat transfer experiments and the available range of Schmidt numbers is much greater than that of Prandtl numbers.

If the expansions in Eq. 13.6-24 and 25 are truncated to one term, we are led to $\mathrm{Nu}_m \propto \Pr^{1/3}$ and $\mathrm{Sh}_m \propto \mathrm{Sc}^{1/3}$. These expressions are essential ingredients in the famous Chilton–Colburn relations[5] (see Eqs. 14.3-18 and 19, and Eqs. 22.3-22 to 24). The first term in Eq. 13.6-24 or 25 also corresponds to the high Prandtl (or Schmidt) number asymptote of Eq. 13.4-20.[6]

With the development of electrochemical methods of measuring mass transfer at surfaces, it has become possible to investigate the second term in Eq. 13.6-25. In Fig. 13.6-1 are shown the data of Shaw and Hanratty, who measured the diffusion-limited current to a wall electrode for values of the Schmidt number $\mathrm{Sc} = \mu/\rho \mathcal{D}_{AB}$ from 693 to 37,200. These data are fitted[3] very well by the expression

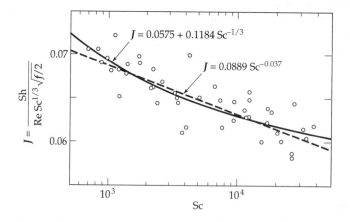

Fig. 13.6-1. Turbulent mass-transfer data of D. A. Shaw and T. J. Hanratty [*AIChE Journal*, **28**, 23–37, 160–169 (1977)] compared to a curve based on Eq. 13.6-25 (solid curve). Shown also is a simple power law function obtained by Shaw and Hanratty.

[5] T. H. Chilton and A. P. Colburn, *Ind. Eng. Chem.*, **26**, 1183–1187 (1934). **Thomas Hamilton Chilton** (1899–1972) had his entire professional career at the E. I. du Pont de Nemours Company, Inc., in Wilmington, Delaware; he was President of AIChE in 1951. After "retiring" he was a guest professor at a dozen or so universities.

[6] See also O. C. Sandall and O. T. Hanna, *AIChE Journal*, **25**, 290–192 (1979).

$$\frac{\text{Sh}}{\text{ReSc}^{1/3}\sqrt{f/2}} = 0.0575 + 0.1184\text{Sc}^{-1/3} \qquad (13.6\text{-}26)$$

in which $f(\text{Re})$ is the friction factor defined in Chapter 6. Equation 13.6-26 combines the observed Re number dependence of the Sherwood number with the two leading terms of Eq. 13.6-25 (that is, the coefficients $a_1, a_2, \ldots$ are proportional to $\text{Re}\sqrt{f/2}$). Equation 13.6-26 lends itself to clear physical interpretation: The leading term corresponds to a diffusional boundary layer so thin that the tangential velocity is linear in y and the wall curvature can be neglected, whereas the second term accounts for wall curvature and the y^2 terms in the tangential velocity expansions of Eqs. 13.6-1 and 2). In higher approximations, special terms can be expected to arise from edge effects as noted by Newman[7] and Stewart.[3]

QUESTIONS FOR DISCUSSION

1. Compare turbulent thermal conductivity and turbulent viscosity as to definition, order of magnitude, and dependence on physical properties and the nature of the flow.
2. What is the "Reynolds analogy," and what is its significance?
3. Is there any connection between Eq. 13.2-3 and Eq. 13.4-12, after the integration constants in the latter have been evaluated?
4. Is the analogy between Fourier's law of heat conduction and Eq. 13.3-1 a valid one?
5. What is the physical significance of the fact that the turbulent Prandtl number is of the order of unity?

PROBLEMS

13B.1. Wall heat flux for turbulent flow in tubes (approximate). Work through Example 13.3-1, and fill in the missing steps. In particular, verify the integration in going from Eq. 13.3-6 to Eq. 13.3-7.

13B.2. Wall heat flux for turbulent flow in tubes.

(a) Summarize the assumptions in §13.4.

(b) Work through the mathematical details of that section, taking particular care with the steps connecting Eq. 13.4-12 and Eq. 13.4-16.

(c) When is it not necessary to find the constant C_2 in Eq. 13.4-12?

13C.1. Wall heat flux for turbulent flow between two parallel plates.

(a) Work through the development in §13.4, and then perform a similar derivation for turbulent flow in a thin slit shown in Fig. 2B.3. Show that the analog of Eq. 13.4-19 is

$$\frac{k(T_0 - T_b)}{q_0 B} = \left(\frac{\bar{v}_{z,\max}}{\langle \bar{v}_z \rangle}\right)^2 \int_0^1 \frac{[J(\xi)]^2}{[1 + (\nu^{(t)}/\nu)(\text{Pr}/\text{Pr}^{(t)})]}\, d\xi \quad (13\text{C.1-1})$$

in which $\xi = x/B$ and $J(\xi) = \int_0^{\xi} \phi(\bar{\xi}) d\bar{\xi}$.

(b) Show how the result in (a) simplifies for laminar flow of Newtonian fluids, and for "plug flow" (flat velocity profiles).

Answer: **(b)** $\frac{35}{17}, 3$

13D.1. The temperature profile for turbulent flow in tubes. To calculate the temperature distribution for turbulent flow in circular tubes from Eq. 13.4-12, it is necessary to know C_2.

(a) Show how to get C_2 by applying B.C. 4 as was done in §10.8. The result is

$$C_2 = \int_0^1 \frac{[I(\xi)/I(1)]^2 - [I(\xi)/I(1)]}{\xi[1 + (\alpha^{(t)}/\alpha)]}\, d\xi \quad (13\text{D.1-1})$$

(b) Verify that Eq. 13D.1-1 gives $C_2 = \frac{7}{24}$ for a Newtonian fluid.

---

[7] J. S. Newman, *Electroanalytical Chemistry*, **6**, 187–352 (1973).

Chapter 14

Interphase Transport in Nonisothermal Systems

§14.1   Definitions of heat transfer coefficients

§14.2   Analytical calculations of heat transfer coefficients for forced convection through tubes and slits

§14.3   Heat transfer coefficients for forced convection in tubes

§14.4   Heat transfer coefficients for forced convection around submerged objects

§14.5   Heat transfer coefficients for forced convection through packed beds

§14.6^{O} Heat transfer coefficients for free and mixed convection

§14.7^{O} Heat transfer coefficients for condensation of pure vapors on solid surfaces

In Chapter 10 we saw how shell energy balances may be set up for various simple problems and how these balances lead to differential equations from which the temperature profiles may be calculated. We also saw in Chapter 11 that the energy balance over an arbitrary differential fluid element leads to a partial differential equation—the energy equation—which may be used to set up more complex problems. Then in Chapter 13 we saw that the time-smoothed energy equation, together with empirical expressions for the turbulent heat flux, provides a useful basis for summarizing and extrapolating temperature profile measurements in turbulent systems. Hence, at this point the reader should have a fairly good appreciation for the meaning of the equations of change for nonisothermal flow and their range of applicability.

It should be apparent that all of the problems discussed have pertained to systems of rather simple geometry and furthermore that most of these problems have contained assumptions, such as temperature-independent viscosity and constant fluid density. For some purposes, these solutions may be adequate, especially for order-of-magnitude estimates. Furthermore, the study of simple systems provides the stepping stones to the discussion of more complex problems.

In this chapter we turn to some of the problems in which it is convenient or necessary to use a less detailed analysis. In such problems the usual engineering approach is to formulate energy balances over pieces of equipment, or parts thereof, as described in Chapter 15. In the macroscopic energy balance thus obtained, there are usually terms that require estimating the heat that is transferred through the system boundaries. This requires knowing the *heat transfer coefficient* for describing the interphase transport. Usually the heat transfer coefficient is given, for the flow system of interest, as an empirical correlation of the *Nusselt number*[1] (a dimensionless wall heat flux or heat transfer coefficient) as a function of the relevant dimensionless quantities, such as the Reynolds and Prandtl numbers.

This situation is not unlike that in Chapter 6, where we learned how to use dimensionless correlations of the friction factor to solve momentum transfer problems. However, for nonisothermal problems the number of dimensionless groups is larger, the types of boundary conditions are more numerous, and the temperature dependence of the physical properties is often important. In addition, the phenomena of free convection, condensation, and boiling are encountered in nonisothermal systems.

We have purposely limited ourselves here to a small number of heat transfer formulas and correlations—just enough to introduce the reader to the subject without attempting to be encyclopedic. Many treatises and handbooks treat the subject in much greater depth.[2,3,4,5,6]

§14.1 DEFINITIONS OF HEAT TRANSFER COEFFICIENTS

Let us consider a flow system with the fluid flowing either in a conduit or around a solid object. Suppose that the solid surface is warmer than the fluid, so that heat is being transferred from the solid to the fluid. Then the rate of heat flow across the solid–fluid interface would be expected to depend on the area of the interface and on the temperature drop between the fluid and the solid. It is customary to define a proportionality factor h (the *heat transfer coefficient*) by

$$Q = hA \, \Delta T \tag{14.1-1}$$

in which Q is the heat flow into the fluid (J/hr or Btu/hr), A is a characteristic area, and ΔT is a characteristic temperature difference. Equation 14.1-1 can also be used when the fluid is cooled. Equation 14.1-1, in slightly different form, has been encountered in Eq. 10.1-2. Note that h is not defined until the area A and the temperature difference ΔT have been specified. We now consider the usual definitions for h for two types of flow geometry.

As an example of *flow in conduits*, we consider a fluid flowing through a circular tube of diameter D (see Fig. 14.1-1), in which there is a heated wall section of length L and varying inside surface temperature $T_0(z)$, going from T_{01} to T_{02}. Suppose that the bulk temperature T_b of the fluid (defined in Eq. 10.8-33 for fluids with constant ρ and $\hat{C}_p$) increases from T_{b1} to T_{b2} in the heated section. Then there are three conventional definitions of heat transfer coefficients for the fluid in the heated section:

$$Q = h_1(\pi DL)(T_{01} - T_{b1}) \equiv h_1(\pi DL)\Delta T_1 \tag{14.1-2}$$

$$Q = h_a(\pi DL)\left(\frac{(T_{01} - T_{b1}) + (T_{02} - T_{b2})}{2}\right) \equiv h_a(\pi DL)\Delta T_a \tag{14.1-3}$$

$$Q = h_{\ln}(\pi DL)\left(\frac{(T_{01} - T_{b1}) - (T_{02} - T_{b2})}{\ln(T_{01} - T_{b1}) - \ln(T_{02} - T_{b2})}\right) \equiv h_{\ln}(\pi DL)\Delta T_{\ln} \tag{14.1-4}$$

---

[1] This dimensionless group is named for **Ernst Kraft Wilhelm Nusselt** (1882–1957), the German engineer who was the first major figure in the field of convective heat and mass transfer. See, for example, W. Nusselt, *Zeits. d. Ver. deutsch. Ing.*, **53**, 1750–1755 (1909), *Forschungsarb. a. d. Geb. d. Ingenieurwes.*, No. 80, 1–38, Berlin (1910), and *Gesundheits-Ing.*, **38**, 477–482, 490–496 (1915).

[2] M. Jakob, *Heat Transfer*, Vol. 1 (1949) and Vol. 2 (1957), Wiley, New York.

[3] W. M. Kays and M. E. Crawford, *Convective Heat and Mass Transfer*, 3rd edition, McGraw-Hill, New York (1993).

[4] H. D. Baehr and K. Stephan, *Heat and Mass Transfer*, Springer, Berlin (1998).

[5] W. M. Rohsenow, J. P. Hartnett, and Y. I. Cho (eds.), *Handbook of Heat Transfer*, McGraw-Hill, New York (1998).

[6] H. Gröber, S. Erk, and U. Grigull, *Die Grundgesetze der Wärmeübertragung*, Springer, Berlin, 3rd edition (1961).

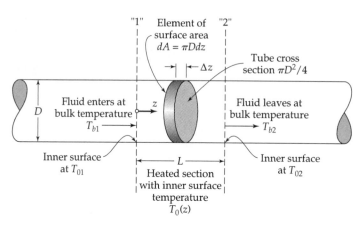

Fig. 14.1-1. Heat transfer in a circular tube.

That is, h_1 is based on the temperature difference ΔT_1 at the inlet, h_a is based on the *arithmetic mean* ΔT_a of the terminal temperature differences, and $h_{\ln}$ is based on the corresponding *logarithmic mean* temperature difference $\Delta T_{\ln}$. For most calculations $h_{\ln}$ is preferable, because it is less dependent on L/D than the other two, although it is not always used.[1] In using heat transfer correlations from treatises and handbooks, one must be careful to note the definitions of the heat transfer coefficients.

If the wall temperature distribution is initially unknown, or if the fluid properties change appreciably along the pipe, it is difficult to predict the heat transfer coefficients defined above. Under these conditions, it is customary to rewrite Eq. 14.1-2 in the differential form:

$$dQ = h_{\mathrm{loc}}(\pi D\, dz)(T_0 - T_b) \equiv h_{\mathrm{loc}}(\pi D\, dz)\Delta T_{\mathrm{loc}} \tag{14.1-5}$$

Here dQ is the heat added to the fluid over a distance dz along the pipe, ΔT_{loc} is the local temperature difference (at position z), and h_{loc} is the *local heat transfer coefficient*. This equation is widely used in engineering design. Actually, the definition of h_{loc} and ΔT_{loc} is not complete without specifying the shape of the element of area. In Eq. 14.1-5 we have set $dA = \pi D\, dz$, which means that h_{loc} and ΔT_{loc} are the mean values for the shaded area dA in Fig. 14.1-1.

As an example of *flow around submerged objects,* consider a fluid flowing around a sphere of radius R, whose surface temperature is maintained at a uniform value T_0. Suppose that the fluid approaches the sphere with a uniform temperature T_∞. Then we may define a *mean heat transfer coefficient,* h_m, for the entire surface of the sphere by the relation

$$Q = h_m(4\pi R^2)(T_0 - T_\infty) \tag{14.1-6}$$

The characteristic area is here taken to be the heat transfer surface (as in Eqs. 14.1-2 to 5), whereas in Eq. 6.1-5 we used the sphere cross section.

A local coefficient can also be defined for submerged objects by analogy with Eq. 14.1-5:

$$dQ = h_{\mathrm{loc}}(dA)(T_0 - T_\infty) \tag{14.1-7}$$

This coefficient is more informative than h_m because it predicts how the heat flux is distributed over the surface. However, most experimentalists report only h_m, which is easier to measure.

---

[1] If $\Delta T_2/\Delta T_1$ is between 0.5 and 2.0, then ΔT_a may be substituted for $\Delta T_{\ln}$, and h_a for $h_{\ln}$, with a maximum error of 4%. This degree of accuracy is acceptable in most heat transfer calculations.

Table 14.1-1 Typical Orders of Magnitude for Heat Transfer Coefficients[a]

System	h (W/m$^2 \cdot$ K) or (kcal/m$^2 \cdot$ hr $\cdot$ C)	h (Btu/ft$^2 \cdot$ hr $\cdot$ F)
Free convection		
Gases	3–20	1–4
Liquids	100–600	20–120
Boiling water	1000–20,000	200–4000
Forced convection		
Gases	10–100	2–20
Liquids	50–500	10–100
Water	500–10,000	100–2000
Condensing vapors	1000–100,000	200–20,000

[a] Taken from H. Gröber, S. Erk, and U. Grigull, *Wärmeübertragung*, Springer, Berlin, 3rd edition (1955), p. 158. When given h in kcal/m$^2 \cdot$ hr $\cdot$ C, multiply by 0.204 to get h in Btu/ft$^2 \cdot$ hr $\cdot$ F, and by 1.162 to get h in W/m$^2 \cdot$ K. For additional conversion factors, see Appendix F.

Let us emphasize that the definitions of A and ΔT must be made clear before h is defined. Keep in mind, also, that h is not a constant characteristic of the fluid medium. On the contrary, the heat transfer coefficient depends in a complicated way on many variables, including the fluid properties (k, μ, ρ, $\hat{C}_p$), the system geometry, and the flow velocity. The remainder of this chapter is devoted to predicting the dependence of h on these quantities. Usually this is done by using experimental data and dimensional analysis to develop correlations. It is also possible, for some very simple systems, to calculate the heat transfer coefficient directly from the equations of change. Some typical ranges of h are given in Table 14.1-1.

We saw in §10.6 that, in the calculation of heat transfer rates between two fluid streams separated by one or more solid layers, it is convenient to use an *overall heat transfer coefficient*, U_0, which expresses the combined effect of the series of resistances through which the heat flows. We give here a definition of U_0 and show how to calculate it in the special case of heat exchange between two coaxial streams with bulk temperatures T_h ("hot") and T_c ("cold"), separated by a cylindrical tube of inside diameter D_0 and outside diameter D_1:

$$dQ = U_0(\pi D_0 dz)(T_h - T_c) \tag{14.1-8}$$

$$\frac{1}{D_0 U_0} = \left(\frac{1}{D_0 h_0} + \frac{\ln (D_1/D_0)}{2k_{01}} + \frac{1}{D_1 h_1} \right)_{\text{loc}} \tag{14.1-9}$$

Note that U_0 is defined as a local coefficient. This is the definition implied in most design procedures (see Example 15.4-1).

Equations 14.1-8 and 9 are, of course, restricted to thermal resistances connected in *series*. In some situations there may be appreciable *parallel* heat flux at one or both surfaces by radiation, and Eqs. 14.1-8 and 9 will require special modification (see Example 16.5-2).

To illustrate the physical significance of heat transfer coefficients and illustrate one method of measuring them, we conclude this section with an analysis of a hypothetical set of heat transfer data.

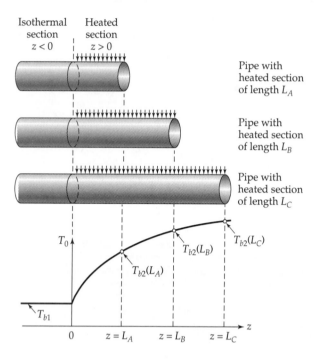

EXAMPLE 14.1-1

Calculation of Heat Transfer Coefficients from Experimental Data

A series of simulated steady-state experiments on the heating of air in tubes is shown in Fig. 14.1-2. In the first experiment, air at $T_{b1} = 200.0°F$ is flowing in a 0.5-in. i.d. tube with fully developed laminar velocity profile in the isothermal pipe section for $z < 0$. At $z = 0$ the wall temperature is suddenly increased to $T_0 = 212.0°F$ and maintained at that value for the remaining tube length L_A. At $z = L_A$ the fluid flows into a mixing chamber in which the cup-mixing (or "bulk") temperature T_{b2} is measured. Similar experiments are done with tubes of different lengths, L_B, L_C, and so on, with the following results:

Experiment	A	B	C	D	E	F	G
L (in.)	1.5	3.0	6.0	12.0	24.0	48.0	96.0
T_{b2} (°F)	201.4	202.2	203.1	204.6	206.6	209.0	211.0

In all experiments, the air flow rate w is 3.0 lb_m/hr. Calculate h_1, h_a, h_{ln}, and the exit value of h_{loc} as functions of the L/D ratio.

SOLUTION

First we make a steady-state energy balance over a length L of the tube, by stating that the heat in through the walls plus the energy entering at $z = 0$ by convection equals the energy leaving the tube at $z = L$. The axial energy flux at the tube entry and exit may be calculated from Eq. 9.8-6. For fully developed flow, changes in the kinetic energy flux $\frac{1}{2}\rho v^2 \mathbf{v}$ and the work term $[\tau \cdot \mathbf{v}]$ will be negligible relative to changes in the enthalpy flux. We also assume that $q_z << \rho \hat{H} v_z$, so that the axial heat conduction term may be neglected. Hence the only contribution to the energy flux entering and leaving with the flow will be the term containing the enthalpy, which can be computed with the help of Eq. 9.8-8 and the assumptions that the heat capacity and density of the fluid are constant throughout. Therefore the steady-state energy balance becomes simply "rate of energy flow in = rate of energy flow out," or

$$Q + w\hat{C}_p T_{b1} = w\hat{C}_p T_{b2} \tag{14.1-10}$$

Using Eq. 14.1-2 to evaluate Q and rearranging gives

$$w\hat{C}_p(T_{b2} - T_{b1}) = h_1(\pi DL)(T_0 - T_{b1}) \tag{14.1-11}$$

from which

$$h_1 = \frac{w\hat{C}_p}{\pi D^2} \frac{(T_{b2} - T_{b1})}{(T_0 - T_{b1})} \left(\frac{D}{L}\right)$$

(14.1-12)

This gives us the formula for calculating h_1 from the data given above.

Analogously, use of Eqs. 14.1-3 and 14.1-4 gives

$$h_a = \frac{w\hat{C}_p}{\pi D^2} \frac{(T_{b2} - T_{b1})}{(T_0 - T_b)_a} \left(\frac{D}{L}\right)$$

(14.1-13)

$$h_{\ln} = \frac{w\hat{C}_p}{\pi D^2} \frac{(T_{b2} - T_{b1})}{(T_0 - T_b)_{\ln}} \left(\frac{D}{L}\right)$$

(14.1-14)

for obtaining h_a and $h_{\ln}$ from the data.

To evaluate h_{loc}, we have to use the preceding data to construct a continuous curve $T_b(z)$, as in Fig. 14.1-2, to represent the change in bulk temperature with z in the longest (96-in.) tube. Then Eq. 14.1-10 becomes

$$Q(z) + w\hat{C}_p T_{b1} = w\hat{C}_p T_b(z)$$

(14.1-15)

By differentiating this expression with respect to z and combining the result with Eq. 14.1-5, we get

$$w\hat{C}_p \frac{dT_b}{dz} = h_{\text{loc}} \pi D(T_0 - T_b)$$

(14.1-16)

or

$$h_{\text{loc}} = \frac{w\hat{C}_p}{\pi D} \frac{1}{(T_0 - T_b)} \frac{dT_b}{dz}$$

(14.1-17)

Since T_0 is constant, this becomes

$$h_{\text{loc}} = -\frac{w\hat{C}_p}{\pi D^2} \frac{d \ln (T_0 - T_b)}{d(z/L)} \left(\frac{D}{L}\right)$$

(14.1-18)

The derivative in this equation is conveniently determined from a plot of $\ln(T_0 - T_b)$ versus z/L. Because a differentiation is involved, it is difficult to determine h_{loc} precisely.

The calculated results are shown in Fig. 14.1-3. Note that all of the coefficients decrease with increasing L/D, but that h_{loc} and $h_{\ln}$ vary less than the others. They approach a common asymptote (see Problem 14B.5 and Fig. 14.1-3). Somewhat similar behavior is observed in turbulent flow with constant wall temperature, except that h_{loc} approaches the asymptote much more rapidly (see Fig. 14.3-2).

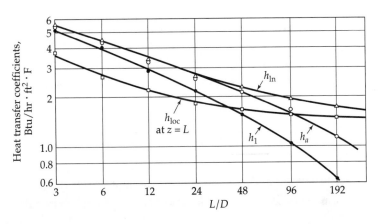

Fig. 14.1-3. Heat transfer coefficients calculated in Example 14.1-1.

§14.2 ANALYTICAL CALCULATIONS OF HEAT TRANSFER COEFFICIENTS FOR FORCED CONVECTION THROUGH TUBES AND SLITS

Recall from Chapter 6, where we defined and discussed friction factors, that for some very simple laminar flow systems we could obtain analytical formulas for the (dimensionless) friction factor as a function of the (dimensionless) Reynolds number. We would like to do the same for the heat transfer coefficient, h, which, however, is not dimensionless. Nonetheless we can construct with it a dimensionless quantity, $\mathrm{Nu} = hD/k$, the *Nusselt number*, using the fluid thermal conductivity k and a characteristic length D that must be specified for each flow system. Two other related dimensionless groups are commonly used: the *Stanton number*, $\mathrm{St} = \mathrm{Nu}/\mathrm{RePr}$, and the *Chilton–Colburn j-factor* for heat transfer, $j_H = \mathrm{Nu}/\mathrm{RePr}^{1/3}$. Each of these dimensionless groups may be "decorated" with subscript 1, a, ln, or m, corresponding to the subscript on the Nusselt number.

By way of illustration, let us return to §10.8 where we discussed the heating of a fluid in laminar flow in a tube, with all the fluid properties being considered constant. From Eq. 10.8-33 and Eq. 10.8-31 we can get the difference between the wall temperature and the bulk temperature:

$$T_0 - T_b = \left(4\zeta + \frac{11}{24}\right)\left(\frac{q_0 R}{k}\right) - 4\zeta\left(\frac{q_0 R}{k}\right)$$

$$= \frac{11}{24}\left(\frac{q_0 R}{k}\right) = \frac{11}{48}\left(\frac{q_0 D}{k}\right) \tag{14.2-1}$$

in which R and D are the radius and diameter of the tube. Solving for the wall flux we get

$$q_0 = \frac{48}{11}\left(\frac{k}{D}\right)(T_0 - T_b) \tag{14.2-2}$$

Then making use of the definition of the local heat transfer coefficient h_{loc}—namely, that $q_0 = h_{\mathrm{loc}}(T_0 - T_b)$—we find that

$$h_{\mathrm{loc}} = \frac{48}{11}\left(\frac{k}{D}\right) \quad \text{or} \quad \mathrm{Nu}_{\mathrm{loc}} = \frac{hD}{k} = \frac{48}{11} \tag{14.2-3}$$

This result is the entry in Eq. (L) of Table 14.2-1—namely, for the laminar flow of a constant-property fluid with a constant wall heat flux, for very large z. The other entries in Table 14.2-1 and 2 may be obtained in a similar way.[1] Some Nusselt numbers for Newtonian fluids with constant physical properties are shown in Fig. 14.2-1.[2]

---

[1] These tables are taken from R. B. Bird, R. C. Armstrong, and O. Hassager, *Dynamics of Polymeric Liquids, Vol. 1, Fluid Mechanics*, 1st edition, Wiley, New York, (1987), pp. 212–213. They are based, in turn, on W. J. Beek and R. Eggink, *De Ingenieur*, **74**, (35) Ch. 81–Ch. 89 (1962) and J. M. Valstar and W. J. Beek, *De Ingenieur*, **75**, (1), Ch. 1–Ch. 7 (1963).

[2] The correspondence between the entries of Tables 14.2-1 and 2 and problems in this book is as follows ($\odot$ = circular tube, $\parallel$ = plane slit):

Eq. (C)	Problem 12D.4 $\odot$; 12D.5 $\parallel$	Laminar Newtonian
Eq. (F)	Problem 12D.3 $\odot$; 12D.5 $\parallel$	Laminar Newtonian
Eq. (G)	Problem 10B.9(a) $\odot$; 10B.9(b) $\parallel$	Plug flow
Eq. (I)	Problem 12D.7 $\odot$; 12D.6 $\parallel$	Laminar Newtonian
Eq. (K)	Problem 10D.2 $\odot$	Laminar non-Newtonian
Eq. (L)	Problem 12D.6 $\parallel$	Laminar Newtonian

Equations analogous to Eqs. (K) in Tables 14.2-1 and 2 are given for turbulent flow in Eqs. 13.4-19 and 13C.1-1.

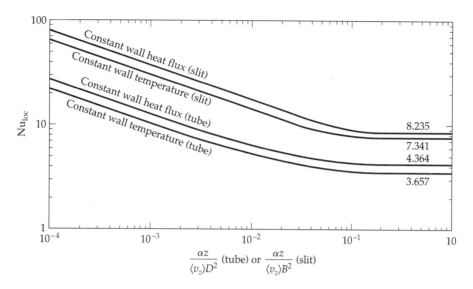

Fig. 14.2-1. The Nusselt number for fully developed, laminar flow of Newtonian fluids with constant physical properties: $Nu_{loc} = h_{loc}D/k$ for circular tubes of diameter D, and $Nu_{loc} = 4h_{loc}B/k$ for slits of half-width B. The limiting expressions are given in Tables 14.2-1 and 14.2-2.

For *turbulent flow* in a circular tube with constant heat flux, the Nusselt number can be obtained from Eq. 13.4-20 (which in turn originated with Eq. (K) of Table 14.2-1):[3]

$$Nu_{loc} = \frac{RePr\sqrt{f/2}}{12.48Pr^{2/3} - 7.853Pr^{1/3} + 3.613 \ln Pr + 5.8 + 2.78 \ln(\frac{1}{45}Re\sqrt{f/8})} \quad (14.2-4)$$

This is valid only for $\alpha z/\langle v_z \rangle D^2 \gg 1$, for fluids with constant physical properties, and for tubes with no roughness. It has been applied successfully over the Prandtl-number range $0.7 < Pr < 590$. Note that, for very large Prandtl numbers, Eq. 14.2-4 gives

$$Nu_{loc} = 0.0566 \, RePr^{1/3} \sqrt{f} \quad (14.2-5)$$

The $Pr^{1/3}$ dependence agrees exactly with the large Pr limit in §13.6 and Eq. 13.3-7. For turbulent flow there is little difference between Nu for constant wall temperature and for constant wall heat flux.

For the turbulent flow of *liquid metals*, for which the Prandtl numbers are generally much less than unity, there are two results of importance. Notter and Sleicher[4] solved the energy equation numerically, using a realistic turbulent velocity profile, and obtained the rates of heat transfer through the wall. The final results were curve-fitted to simple analytical expressions for two cases:

Constant wall temperature: $Nu_{loc} = 4.8 + 0.0156 \, Re^{0.85} \, Pr^{0.93}$ (14.2-6)

Constant wall heat flux: $Nu_{loc} = 6.3 + 0.0167 \, Re^{0.85} \, Pr^{0.93}$ (14.2-7)

These equations are limited to $L/D > 60$ and constant physical properties. Equation 14.2-7 is displayed in Fig. 14.2-2.

---

[3] O. C. Sandall, O. T. Hanna, and P. R. Mazet, *Canad. J. Chem. Eng.*, **58**, 443–447 (1980).
[4] R. H. Notter and C. A. Sleicher, *Chem. Eng. Sci*, **27**, 2073–2093 (1972).

Table 14.2-1 Asymptotic Results for Local Nusselt Numbers (Tube Flow)[a,b]; $Nu_{loc} = h_{loc}D/k$

All values are local Nu numbers	Constant wall temperature		Constant wall heat flux	
Thermal entrance region[c] $\dfrac{\langle v_z\rangle D^2}{\alpha z} >> 1$	Plug flow	(A) $Nu = \dfrac{1}{\sqrt{\pi}}\left(\dfrac{\langle v_z\rangle D^2}{\alpha z}\right)^{1/2}$	Plug flow	(G) $Nu = \dfrac{\sqrt{\pi}}{2}\left(\dfrac{\langle v_z\rangle D^2}{\alpha z}\right)^{1/2}$
	Laminar non-Newtonian flow	(B) $Nu = \dfrac{2}{9^{1/3}\Gamma(\frac{4}{3})}\left[\dfrac{\langle v_z\rangle D^2}{\alpha z}\left(-\dfrac{1}{4}\dfrac{d\phi}{d\xi}\Big|_{\xi=1}\right)\right]^{1/3}$	Laminar non-Newtonian flow	(H) $Nu = \dfrac{2\Gamma(\frac{2}{3})}{9^{1/3}}\left[\dfrac{\langle v_z\rangle D^2}{\alpha z}\left(-\dfrac{1}{4}\dfrac{d\phi}{d\xi}\Big|_{\xi=1}\right)\right]^{1/3}$
	Laminar Newtonian flow	(C) $Nu = \dfrac{2}{9^{1/3}\Gamma(\frac{4}{3})}\left(\dfrac{\langle v_z\rangle D^2}{\alpha z}\right)^{1/3}$	Laminar Newtonian flow	(I) $Nu = \dfrac{2\Gamma(\frac{2}{3})}{9^{1/3}}\left(\dfrac{\langle v_z\rangle D^2}{\alpha z}\right)^{1/3}$
Thermally fully developed flow $\dfrac{\langle v_z\rangle D^2}{\alpha z} << 1$	Plug flow	(D) $Nu = 5.772$	Plug flow	(J) $Nu = 8$
	Laminar non-Newtonian flow	(E) $Nu = \beta_1^2$, where β_1 is the *lowest* eigenvalue of $\dfrac{1}{\xi}\dfrac{d}{d\xi}\left(\xi\dfrac{dX_n}{d\xi}\right)+\beta_n^2\phi(\xi)X_n = 0$; $X_n'(0)=0,\ X_n(1)=0$	Laminar non-Newtonian flow	(K) $Nu = \left[2\int_0^1\dfrac{1}{\xi}\left[\int_0^\xi \xi'\phi(\xi')d\xi'\right]^2 d\xi\right]^{-1}$
	Laminar Newtonian flow	(F) $Nu = 3.657$	Laminar Newtonian flow	(L) $Nu = \dfrac{48}{11} = 4.364$

[a] Note: $\phi(\xi) = v_z/\langle v_z\rangle$, where $\xi = r/R$ and $R = D/2$; for Newtonian fluids $\langle v_z\rangle D^2/\alpha z = RePr(D/z)$ with $Re = D\langle v_z\rangle\rho/\mu$. Here $\alpha = k/\rho\hat{C}_p$.

[b] W. J. Beek and R. Eggink, *De Ingenieur*, **74**, No. 35, Ch. 81–89 (1962); erratum, **75**, No. 1, Ch. 7 (1963).

[c] The grouping $\langle v_z\rangle D^2/\alpha z$ is sometimes written as $Gz \cdot (L/z)$ where $Gz = \langle v_z\rangle D^2/\alpha L$ is called the Graetz number; here L is the length of the pipe past $z = 0$. Thus the thermal entry region corresponds to large Graetz number.

430

Table 14.2-2 Asymptotic Results for Local Nusselt Numbers (Thin-Slit Flow)[a,b]; $Nu_{loc} = 4h_{loc}B/k$

All values are local Nu numbers[c]		Constant wall temperature		Constant wall heat flux	
Thermal entrance region[c] $\dfrac{\langle v_z\rangle B^2}{\alpha z} >> 1$	Plug flow (A)	$Nu = \dfrac{4}{\sqrt{\pi}}\left(\dfrac{\langle v_z\rangle B^2}{\alpha z}\right)^{1/2}$	Plug flow (G)	$Nu = 2\sqrt{\pi}\left(\dfrac{\langle v_z\rangle B^2}{\alpha z}\right)^{1/2}$	
	Laminar non-Newtonian flow (B)	$Nu = \dfrac{4}{9^{1/3}\Gamma(\frac{4}{3})}\left[\dfrac{\langle v_z\rangle B^2}{\alpha z}\left(-\dfrac{d\phi}{d\sigma}\Big|_{\sigma=1}\right)\right]^{1/3}$	Laminar non-Newtonian flow (H)	$Nu = \dfrac{4\Gamma(\frac{2}{3})}{9^{1/3}}\left[\dfrac{\langle v_z\rangle B^2}{\alpha z}\left(-\dfrac{d\phi}{d\sigma}\Big|_{\sigma=1}\right)\right]^{1/3}$	
	Laminar Newtonian flow (C)	$Nu = \dfrac{4}{3^{1/3}\Gamma(\frac{4}{3})}\left(\dfrac{\langle v_z\rangle B^2}{\alpha z}\right)^{1/3}$	Laminar Newtonian flow (I)	$Nu = \dfrac{4\Gamma(\frac{2}{3})}{3^{1/3}}\left(\dfrac{\langle v_z\rangle B^2}{\alpha z}\right)^{1/3}$	
Thermally fully developed flow $\dfrac{\langle v_z\rangle B^2}{\alpha z} << 1$	Plug flow (D)	$Nu = \pi^2 = 9.870$	Plug flow (J)	$Nu = 12$	
	Laminar non-Newtonian flow (E)	$Nu = 4\beta_1^2$, where β_1 is the *lowest* eigenvalue of $\dfrac{d^2X_n}{d\sigma^2} + \beta_n^2\phi(\sigma)X_n = 0$; $X_n(\pm 1) = 0$	Laminar non-Newtonian flow (K)	$Nu = \left[\dfrac{1}{4}\int_0^1\left[\int_0^\sigma \phi(\sigma')d\sigma'\right]^2 d\sigma\right]^{-1}$	
	Laminar Newtonian flow (F)	$Nu = 7.541$	Laminar Newtonian flow (L)	$Nu = \dfrac{140}{17} = 8.235$	

[a] Note: $\phi(\sigma) = v_z/\langle v_z\rangle$, where $\sigma = y/B$; for Newtonian fluids $\langle v_z\rangle D^2/\alpha z = 4\,RePr(B/z)$ with $Re = 4B\langle v_z\rangle\rho/\mu$. Here $\alpha = k/\rho\hat{C}_p$.

[b] J. M. Valstar and W. J. Beek, *De Ingenieur*, **75**, No. 1, Ch. 1-7 (1963).

[c] The grouping $\langle v_z\rangle B^2/\alpha z$ is sometimes written as $Gz \cdot (L/z)$ where $Gz = \langle v_z\rangle B^2/\alpha L$ is called the Graetz number; here L is the length of the slit past $z = 0$. Thus the thermal entry region corresponds to large Graetz number.

431

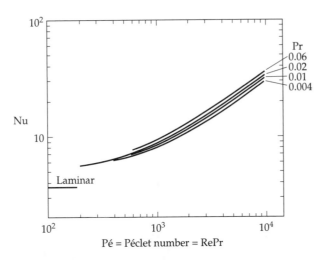

Fig. 14.2-2. Nusselt numbers for turbulent flow of liquid metals in circular tubes, based on the theoretical calculations of R. H. Notter and C. A. Sleicher, *Chem. Eng. Sci.*, **27**, 2073–2093 (1972).

It has been emphasized that all the results of this section are limited to fluids with constant physical properties. When there are large temperature differences in the system, it is necessary to take into account the temperature dependence of the viscosity, density, heat capacity, and thermal conductivity. Usually this is done by means of an empiricism—namely, by evaluating the physical properties at some appropriate average temperature. Throughout this chapter, unless explicitly stated otherwise, it is understood that all physical properties are to be calculated at the film temperature T_f defined as follows:[5]

a. For *tubes, slits, and other ducts,*

$$T_f = \tfrac{1}{2}(T_{0a} + T_{ba}) \tag{14.2-8}$$

in which T_{0a} is the arithmetic average of the surface temperatures at the two ends, $T_{0a} = \tfrac{1}{2}(T_{01} + T_{02})$, and T_{ba} is the arithmetic average of the inlet and outlet bulk temperatures, $T_{ba} = \tfrac{1}{2}(T_{b1} + T_{b2})$.

It is also recommended that the Reynolds number be written as $\mathrm{Re} = D\langle\rho v\rangle/\mu = Dw/S\mu$, in order to account for viscosity, velocity, and density changes over the cross section of area S.

b. For *submerged objects* with uniform surface temperature T_0 in a stream of liquid approaching with uniform temperature T_∞,

$$T_f = \tfrac{1}{2}(T_0 + T_\infty) \tag{14.2-9}$$

For flow systems involving more complicated geometries, it is preferable to use experimental correlations of the heat transfer coefficients. In the following sections we show how such correlations can be established by a combination of dimensional analysis and experimental data.

[5] W. J. M. Douglas and S. W. Churchill, *Chem. Eng. Prog. Symposium Series*, No. 18, **52**, 23–28 (1956); E. R. G. Eckert, *Recent Advances in Heat and Mass Transfer*, McGraw-Hill, New York (1961), pp. 51–81, Eq. (20); more detailed reference states have been proposed by W. E. Stewart, R. Kilgour, and K.-T. Liu, University of Wisconsin–Madison Mathematics Research Center Report #1310 (June 1973).

§14.3 HEAT TRANSFER COEFFICIENTS FOR FORCED CONVECTION IN TUBES

In the previous section we have shown that Nusselt numbers for some laminar flows can be computed from first principles. In this section we show how dimensional analysis leads us to a general form for the dependence of the Nusselt number on various dimensionless groups, and that this form includes not only the results of the preceding section, but turbulent flows as well. Then we present a dimensionless plot of Nusselt numbers that was obtained by correlating experimental data.

First we extend the dimensional analysis given in §11.5 to obtain a general form for correlations of heat transfer coefficients in forced convection. Consider the steadily driven laminar or turbulent flow of a Newtonian fluid through a straight tube of inner radius R, as shown in Fig. 14.3-1. The fluid enters the tube at $z = 0$ with velocity uniform out to very near the wall, and with a uniform inlet temperature T_1 ($= T_{b1}$). The tube wall is insulated except in the region $0 \leq z \leq L$, where a uniform inner-surface temperature T_0 is maintained by heat from vapor condensing on the outer surface. For the moment, we assume constant physical properties ρ, μ, k, and $\hat{C}_p$. Later we will extend the empiricism given in §14.2 to provide a fuller allowance for the temperature dependence of these properties.

We follow the same procedure used in §6.2 for friction factors. We start by writing the expression for the instantaneous heat flow from the tube wall into the fluid in the system described above,

$$Q(t) = \int_0^L \int_0^{2\pi} \left(+k\frac{\partial T}{\partial r} \right)\bigg|_{r=R} R\, d\theta\, dz \tag{14.3-1}$$

which is valid for laminar or turbulent flow (in laminar flow, Q would, of course, be independent of time). The $+$ sign appears here because the heat is added to the system in the negative r direction.

Equating the expressions for Q given in Eqs. 14.1-2 and 14.3-1 and solving for h_1, we get

$$h_1(t) = \frac{1}{\pi DL(T_0 - T_{b1})} \int_0^L \int_0^{2\pi} \left(+k\frac{\partial T}{\partial r} \right)\bigg|_{r=R} R\, d\theta\, dz \tag{14.3-2}$$

Next we introduce the dimensionless quantities $\check{r} = r/D$, $\check{z} = z/D$, and $\check{T} = (T - T_0)/(T_{b1} - T_0)$, and multiply by D/k to get an expression for the Nusselt number $\mathrm{Nu}_1 = h_1 D/k$:

$$\mathrm{Nu}_1(t) = \frac{1}{2\pi L/D} \int_0^{L/D} \int_0^{2\pi} \left(-\frac{\partial \check{T}}{\partial \check{r}} \right)\bigg|_{\check{r}=1/2} d\theta\, d\check{z} \tag{14.3-3}$$

Thus the (instantaneous) Nusselt number is basically a *dimensionless temperature gradient averaged over the heat transfer surface.*

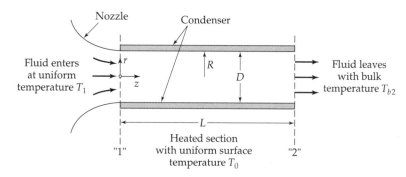

Fig. 14.3-1. Heat transfer in the entrance region of a tube.

The dimensionless temperature gradient appearing in Eq. 14.3-3 could, in principle, be evaluated by differentiating the expression for $\check{T}$ obtained by solving Eqs. 11.5-7, 8, and 9 with the boundary conditions

$$\text{at } \check{z} = 0, \qquad\qquad \check{\mathbf{v}} = \boldsymbol{\delta}_z \quad \text{for } 0 \leq \check{r} < \tfrac{1}{2} \tag{14.3-4}$$

$$\text{at } \check{r} = \tfrac{1}{2}, \qquad\qquad \check{\mathbf{v}} = 0 \quad\quad \text{for } \check{z} \geq 0 \tag{14.3-5}$$

$$\text{at } \check{r} = 0 \text{ and } \check{z} = 0, \quad \check{\mathscr{P}} = 0 \tag{14.3-6}$$

$$\text{at } \check{z} = 0, \qquad\qquad \check{T} = 1 \quad\quad \text{for } 0 \leq \check{r} \leq \tfrac{1}{2} \tag{14.3-7}$$

$$\text{at } \check{r} = \tfrac{1}{2} \qquad\qquad \check{T} = 0 \quad\quad \text{for } 0 \leq \check{z} \leq L/D \tag{14.3-8}$$

where $\check{\mathbf{v}} = \mathbf{v}/\langle v_z \rangle_1$ and $\check{\mathscr{P}} = (\mathscr{P} - \mathscr{P}_1)/\rho \langle v_z \rangle_1^2$. As in §6.2, we have neglected the $\partial^2/\partial \check{z}^2$ terms of the equations of change on the basis of order-of-magnitude reasoning similar to that in §4.4. With those terms suppressed, upstream transport of heat and momentum are excluded, so that the solutions upstream of plane 2 in Fig. 14.3-1 do not depend on L/D.

From Eqs. 11.5-7, 8, and 9 and these boundary conditions, we conclude that the dimensionless instantaneous temperature distribution must be of the following form:

$$\check{T} = \check{T}(\check{r}, \theta, \check{z}, \check{t}; \text{Re}, \text{Pr}, \text{Br}) \qquad \text{for } 0 \leq \check{z} \leq L/D \tag{14.3-9}$$

Substitution of this relation into Eq. 14.3-3 leads to the conclusion that $\text{Nu}_1(\check{t}) = \text{Nu}_1(\text{Re}, \text{Pr}, \text{Br}, L/D, \check{t})$. When time-averaged over an interval long enough to include all the turbulent disturbances, this becomes

$$\text{Nu}_1 = \text{Nu}_1(\text{Re}, \text{Pr}, \text{Br}, L/D) \tag{14.3-10}$$

A similar relation is valid when the flow at plane 1 is fully developed.

If, as is often the case, the viscous dissipation heating is small, the Brinkman number can be omitted. Then Eq. 14.3-10 simplifies to

$$\text{Nu}_1 = \text{Nu}_1(\text{Re}, \text{Pr}, L/D) \tag{14.3-11}$$

Therefore, dimensional analysis tells us that, for forced-convection heat transfer in circular tubes with constant wall temperature, experimental values of the heat transfer coefficient h_1 can be correlated by giving Nu_1 as a function of the Reynolds number, the Prandtl number, and the geometric ratio L/D. This should be compared with the similar, but simpler, situation with the friction factor (Eqs. 6.2-9 and 10).

The same reasoning leads us to similar expressions for the other heat transfer coefficients we have defined. It can be shown (see Problem 14.B-4) that

$$\text{Nu}_a = \text{Nu}_a(\text{Re}, \text{Pr}, L/D) \tag{14.3-12}$$

$$\text{Nu}_{\ln} = \text{Nu}_{\ln}(\text{Re}, \text{Pr}, L/D) \tag{14.3-13}$$

$$\text{Nu}_{\text{loc}} = \text{Nu}_{\text{loc}}(\text{Re}, \text{Pr}, z/D) \tag{14.3-14}$$

in which $\text{Nu}_a = h_a D/k$, $\text{Nu}_{\ln} = h_{\ln} D/k$, and $\text{Nu}_{\text{loc}} = h_{\text{loc}} D/k$. That is, to each of the heat transfer coefficients, there is a corresponding Nusselt number. These Nusselt numbers are, of course, interrelated (see Problem 14.B-5). These general functional forms for the Nusselt numbers have a firm scientific basis, since they involve only the dimensional analysis of the equations of change and boundary conditions.

Thus far we have assumed that the physical properties are constants over the temperature range encountered in the flow system. At the end of §14.2 we indicated that evaluating the physical properties at the film temperature is a suitable empiricism. However, for very large temperature differences, the viscosity variations may result in such a large distortion of the velocity profiles that it is necessary to account for this by introducing an additional dimensionless group, μ_b/μ_0, where μ_b is the viscosity at the arithmetic average bulk temperature and μ_0 is the viscosity at the arithmetic average wall temperature.[1] Then we may write

$$\text{Nu} = \text{Nu}(\text{Re}, \text{Pr}, L/D, \mu_b/\mu_0) \tag{14.3-15}$$

This type of correlation seems to have first been presented by Sieder and Tate.[2] If, in addition, the density varies significantly, then some free convection may occur. This effect can be accounted for in correlations by including the Grashof number along with the other dimensionless groups. This point is pursued further in §14.6.

Let us now pause to reflect on the significance of the above discussion for constructing heat transfer correlations. The heat transfer coefficient h depends on *eight* physical quantities (D, $\langle v \rangle$, ρ, μ_0, μ_b, $\hat{C}_p$, k, L). However, Eq. 14.3-15 tells us that this dependence can be expressed more concisely by giving Nu as a function of only *four* dimensionless groups (Re, Pr, L/D, μ_b/μ_0). Thus, instead of taking data on h for 5 values of each of the eight individual physical quantities (5^8 tests), we can measure h for 5 values of the dimensionless groups (5^4 tests)—a rather dramatic saving of time and effort.

A good global view of heat transfer in circular tubes with nearly constant wall temperature can be obtained from the Sieder and Tate[2] correlation shown in Fig. 14.3-2. This is of the form of Eq. 14.3-15. It has been found empirically[2,3] that transition to turbulence usually begins at about Re = 2100, even when the viscosity varies appreciably in the radial direction.

For *highly turbulent flow*, the curves for $L/D > 10$ converge to a single curve. For Re > 20,000 this curve is described by the equation

$$\text{Nu}_{\ln} = 0.026\, \text{Re}^{0.8}\, \text{Pr}^{1/3}\left(\frac{\mu_b}{\mu_0}\right)^{0.14} \tag{14.3-16}$$

This equation reproduces available experimental data within about ±20% in the ranges $10^4 < \text{Re} < 10^5$ and $0.6 < \text{Pr} < 100$.

For *laminar flow*, the descending lines at the left are given by the equation

$$\text{Nu}_{\ln} = 1.86\left(\text{RePr}\,\frac{D}{L}\right)^{1/3}\left(\frac{\mu_b}{\mu_0}\right)^{0.14} \tag{14.3-17}$$

---

[1] One can arrive at the viscosity ratio by inserting into the equations of change a temperature-dependent viscosity, described, for example, by a Taylor expansion about the wall temperature:

$$\mu = \mu_0 + \left.\frac{\partial \mu}{\partial T}\right|_{T=T_0}(T - T_0) + \cdots \tag{14.3-15a}$$

When the series is truncated and the differential quotient is approximated by a difference quotient, we get

$$\mu \cong \mu_0 + \left(\frac{\mu_b - \mu_0}{T_b - T_0}\right)(T - T_0) \tag{14.3-15b}$$

or, with some rearrangement,

$$\frac{\mu}{\mu_0} \cong 1 + \left(\frac{\mu_b}{\mu_0} - 1\right)\left(\frac{T - T_0}{T_b - T_0}\right) \tag{14.3-15c}$$

Thus, the viscosity ratio appears in the equation of motion and hence in the dimensionless correlation.

[2] E. N. Sieder and G. E. Tate, *Ind. Eng. Chem.*, **28**, 1429–1435 (1936).

[3] A. P. Colburn, *Trans. AIChE*, **29**, 174–210 (1933). **Alan Philip Colburn** (1904–1955), provost at the University of Delaware (1950–1955), made important contributions to the fields of heat and mass transfer, including the "Chilton–Colburn relations."

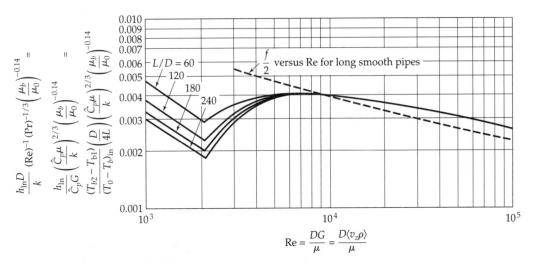

Fig. 14.3-2. Heat transfer coefficients for fully developed flow in smooth tubes. The lines for laminar flow should not be used in the range $RePrD/L < 10$, which corresponds to $(T_0 - T_b)_2/(T_0 - T_b)_1 < 0.2$. The laminar curves are based on data for $RePrD/L \gg 10$ and nearly constant wall temperature; under these conditions h_a and h_{ln} are indistinguishable. We recommend using h_{ln}, as opposed to the h_a suggested by Sieder and Tate, because this choice is conservative in the usual heat-exchanger design calculations [E. N. Sieder and G. E. Tate, *Ind. Eng. Chem.*, **28**, 1429–1435 (1936)].

which is based on Eq. (C) of Table 14.2-1[4] and Problem 12D.4. The numerical coefficient in Eq. (C) has been multiplied by a factor of $\frac{3}{2}$ to convert from h_{loc} to h_{ln}, and then further modified empirically to account for the deviations due to variable physical properties. This illustrates how a satisfactory empirical correlation can be obtained by modifying the result of an analytical derivation. Equation 14.3-17 is good within about 20% for $RePr$ $D/L > 10$, but at lower values of $RePr D/L$ it underestimates h_{ln} considerably. The occurrence of $Pr^{1/3}$ in Eqs. 14.3-16 and 17 is consistent with the large Prandtl number asymptote found in §§13.6 and 12.4.

The *transition region*, roughly $2100 < Re < 8000$ in Fig. 14.3-2, is not well understood and is usually avoided in design if possible. The curves in this region are supported by experimental measurements[2] but are less reliable than the rest of the plot.

The general characteristics of the curves in Fig. 14.3-2 deserve careful study. Note that for a heated section of given L and D and a fluid of given physical properties, the ordinate is proportional to the dimensionless temperature rise of the fluid passing through—that is, $(T_{b2} - T_{b1})/(T_0 - T_b)_{ln}$. Under these conditions, as the flow rate (or Reynolds number) is increased, the exit fluid temperature will first decrease until Re reaches about 2100, then increase until Re reaches about 8000, and then finally decrease again. The influence of L/D on h_{ln} is marked in laminar flow but becomes insignificant for $Re > 8000$ with $L/D > 60$.

---

[4] Equation (C) is an asymptotic solution of the *Graetz problem*, one of the classic problems of heat convection: L. Graetz, *Ann. d. Physik*, **13**, 79–94 (1883), **25**, 337–357 (1885); see J. Lévêque, *Ann. Mines* (Series 12), **13**, 201–299, 305–362, 381–415 (1928) for the asymptote in Eq. (C). An extensive summary can be found in M. A. Ebadian and Z. F. Dong, Chapter 5 of *Handbook of Heat Transfer*, 3rd edition, (W. M. Rohsenow, J. P. Hartnett, and Y. I. Cho, eds.), McGraw-Hill, New York (1998).

Note also that Fig. 14.3-2 somewhat resembles the friction-factor plot in Fig. 6.2-2, although the physical situation is quite different. In the highly turbulent range (Re > 10,000) the heat transfer ordinate agrees approximately with $f/2$ for the long smooth pipes under consideration. This was first pointed out by Colburn,[3] who proposed the following empirical analogy for long, smooth tubes:

$$j_{H,\ln} \approx \tfrac{1}{2} f \qquad (\text{Re} > 10{,}000) \tag{14.3-18}$$

in which

$$j_{H,\ln} = \frac{\text{Nu}_{\ln}}{\text{RePr}^{1/3}} = \frac{h_{\ln}}{\langle \rho v \rangle \hat{C}_p} \left(\frac{\hat{C}_p \mu}{k} \right)^{2/3} = \frac{h_{\ln} S}{w \hat{C}_p} \left(\frac{\hat{C}_p \mu}{k} \right)^{2/3} \tag{14.3-19}$$

where S is the area of the tube cross section, w is the mass rate of flow through the tube, and $f/2$ is obtainable from Fig. 6.2-2 using $\text{Re} = Dw/S\mu = 4w/\pi D\mu$. Clearly the analogy of Eq. 14.3-18 is not valid below $\text{Re} = 10{,}000$. For rough tubes with fully developed turbulent flow the analogy breaks down completely, because f is affected more by roughness than j_H is.

One additional remark about the use of Fig. 14.3-2 has to do with the application to conduits of noncircular cross section. For *highly turbulent* flow, one may use the mean hydraulic radius of Eq. 6.2-16. To apply that empiricism, D is replaced by $4R_h$ everywhere in the Reynolds and Nusselt numbers.

EXAMPLE 14.3-1

Design of a Tubular Heater

Air at 70°F and 1 atm is to be pumped through a straight 2-in. i.d. tube at a rate of 70 lb_m/hr. A section of the tube is to be heated to an inside wall temperature of 250°F to raise the air temperature to 230°F. What heated length is required?

SOLUTION

The arithmetic average bulk temperature is $T_{ba} = 150°F$, and the film temperature is $T_f = \tfrac{1}{2}(150 + 250) = 200°F$. At this temperature the properties of air are $\mu = 0.052 \ \text{lb}_m/\text{ft} \cdot \text{hr}$, $\hat{C}_p = 0.242$ Btu/$\text{lb}_m \cdot$F, $k = 0.0180$ Btu/hr $\cdot$ ft $\cdot$ F, and $\text{Pr} = \hat{C}_p \mu / k = 0.70$. The viscosities of air at 150°F and 250°F are 0.049 and 0.055 $\text{lb}_m/\text{ft} \cdot$ hr, respectively, so that the viscosity ratio is $\mu_b/\mu_0 = 0.049/0.055 = 0.89$.

The Reynolds number, evaluated at the film temperature, 200°F, is then

$$\text{Re} = \frac{Dw}{S\mu} = \frac{4w}{\pi D\mu} = \frac{4(70)}{\pi(2/12)(0.052)} = 1.02 \times 10^4 \tag{14.3-20}$$

From Fig. 14.3-1 we obtain

$$\frac{(T_{b2} - T_{b1})}{(T_0 - T_b)_{\ln}} \frac{D}{4L} \text{Pr}^{2/3} \left(\frac{\mu_b}{\mu_0} \right)^{-0.14} = 0.0039 \tag{14.3-21}$$

When this is solved for L/D we get

$$\begin{aligned}
\frac{L}{D} &= \frac{1}{4(0.0039)} \frac{(T_{b2} - T_{b1})}{(T_0 - T_b)_{\ln}} \text{Pr}^{2/3} \left(\frac{\mu_b}{\mu_0} \right)^{-0.14} \\
&= \frac{1}{4(0.0039)} \frac{(230 - 160)}{72.2} (0.70)^{2/3} (0.89)^{-0.14} \\
&= \frac{1}{4(0.0039)} \frac{160}{72.8} (0.788)(1.02) = 113
\end{aligned} \tag{14.3-22}$$

Hence the required length is

$$L = 113D = (113)(2/12) = 19 \text{ ft} \tag{14.3-23}$$

If Re_b had been much smaller, it would have been necessary to estimate L/D before reading Fig. 14.3-2, thus initiating a trial-and-error process.

Note that in this problem we did not have to calculate h. Numerical evaluation of h is necessary, however, in more complicated problems such as heat exchange between two fluids with an intervening wall.

§14.4 HEAT TRANSFER COEFFICIENTS FOR FORCED CONVECTION AROUND SUBMERGED OBJECTS

Another topic of industrial importance is the transfer of heat to or from an object around which a fluid is flowing. The object may be relatively simple, such as a single cylinder or sphere, or it may be more complex, such as a "tube bundle" made up of a set of cylindrical tubes with a stream of gas or liquid flowing between them. We examine here only a few selected correlations for simple systems: the flat plate, the sphere, and the cylinder. Many additional correlations may be found in the references cited in the introduction to the chapter.

Flow Along a Flat Plate

We first examine the flow along a flat plate, oriented parallel to the flow, with its surface maintained at T_0 and the approaching stream having a uniform temperature T_∞ and a uniform velocity v_∞. The heat transfer coefficient $h_{loc} = q_0/(T_0 - T_\infty)$ and the friction factor $f_{loc} = \tau_0/\frac{1}{2}\rho v_\infty^2$ are shown in Fig. 14.1-1. For the laminar region, which normally exists near the leading edge of the plate, the following theoretical expressions are obtained (see Eq. 4.4-30 as well as Eqs. 12.4-12, 12.4-15, and 12.4-16):

$$\tfrac{1}{2}f_{loc} = +\frac{\mu(\partial v_x/\partial y)|_{y=0}}{\rho v_\infty^2} = f''(0)\sqrt{\frac{\mu}{2xv_\infty\rho}} = 0.332\,\mathrm{Re}_x^{-1/2} \tag{14.4-1}$$

$$\mathrm{Nu}_{loc} = \frac{h_{loc}x}{k} = \frac{x}{(T_\infty - T_0)}\frac{\partial T}{\partial y}\bigg|_{y=0} = 2\sqrt{\frac{37}{1260}}\,\mathrm{Re}_x^{1/2}\mathrm{Pr}^{1/3} \tag{14.4-2}$$

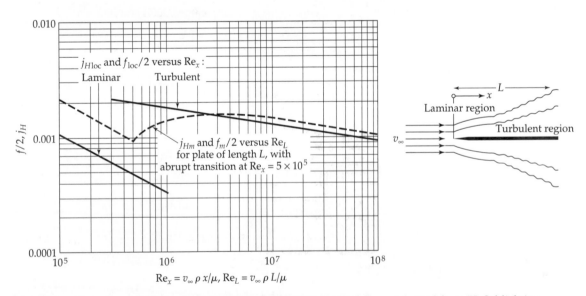

Fig. 14.4-1. Transfer coefficients for a smooth flat plate in tangential flow. Adapted from H. Schlichting, *Boundary-Layer Theory*, McGraw-Hill, New York (1955), pp. 438–439.

As shown in Table 12.4-1, a more accurate value of the numerical coefficient in Eq. 14.4-2 is that of Pohlhausen—namely, 0.332. If we use this value, then Eq. 14.4-2 gives

$$j_{H,\text{loc}} = \frac{\text{Nu}_{\text{loc}}}{\text{Re}\,\text{Pr}^{1/3}} = \frac{h_{\text{loc}}}{\rho \hat{C}_p v_\infty} \left(\frac{\hat{C}_p \mu}{k} \right)^{2/3} = 0.332\,\text{Re}_x^{-1/2} \tag{14.4-3}$$

Since the numerical coefficient in Eq. 14.4-3 is the same as that in Eq. 14.4-1, we then get

$$j_{H,\text{loc}} = \tfrac{1}{2} f_{\text{loc}} = 0.332\,\text{Re}_x^{-1/2} \tag{14.4-4}$$

for the Colburn analogy between heat transfer and fluid friction. This was to be expected, because there is no "form drag" in this flow geometry.

Equation 14.4-4 was derived for fluids with constant physical properties.[1] When the physical properties are evaluated at the film temperature $T_f = \tfrac{1}{2}(T_0 + T_\infty)$, Eq. 14.4-3 is known to work well for gases.[2] The analogy of Eq. 14.4-4 is accurate within 2% for $\text{Pr} > 0.6$, but becomes inaccurate at lower Prandtl numbers.

For highly turbulent flows, the Colburn analogy still holds with fair accuracy, with f_{loc} given by the empirical curve in Fig. 14.4-1. The transition between laminar and turbulent flow resembles that for pipes in Fig. 14.3-1, but the limits of the transition region are harder to predict. For smooth, sharp-edged flat plates in an isothermal flow the transition usually begins at a Reynolds number $\text{Re}_x = x v_\infty \rho / \mu$ of 100,000 to 300,000 and is almost complete at a 50% higher Reynolds number.

Flow Around a Sphere

In Problem 10B.1 it is shown that the Nusselt number for a sphere in a stationary fluid is 2. For the sphere with constant surface temperature T_0 in a flowing fluid approaching with a uniform velocity v_∞, the mean Nusselt number is given by the following empiricism[3]

$$\text{Nu}_m = 2 + 0.60\,\text{Re}^{1/2}\,\text{Pr}^{1/3} \tag{14.4-5}$$

This result is useful for predicting the heat transfer to or from droplets or bubbles.

Another correlation that has proven successful[4] is

$$\text{Nu}_m = 2 + (0.4\,\text{Re}^{1/2} + 0.06\text{Re}^{2/3})\text{Pr}^{0.4}\left(\frac{\mu_\infty}{\mu_0} \right)^{1/4} \tag{14.4-6}$$

in which the physical properties appearing in Nu_m, Re, and Pr are evaluated at the approaching stream temperature. This correlation is recommended for $3.5 < \text{Re} < 7.6 \times 10^4$, $0.71 < \text{Pr} < 380$, and $1.0 < \mu_\infty/\mu_0 < 3.2$. In contrast to Eq. 14.4-5, it is not valid in the limit that $\text{Pr} \to \infty$.

---

[1] The result in Eq. 14.4-1 was first obtained by H. Blasius, *Z. Math. Phys.*, **56**, 1–37 (1908), and that in Eq. 14.4-3 by E. Pohlhausen, *Z. angew. Math. Mech.*, **1**, 115–121 (1921).

[2] E. R. G. Eckert, *Trans. ASME*, **56**, 1273–1283 (1956). This article also includes high-velocity flows, for which compressibility and viscous dissipation become important.

[3] W. E. Ranz and W. R. Marshall, Jr., *Chem. Eng. Prog.*, **48**, 141–146, 173–180 (1952). N. Frössling, *Gerlands Beitr. Geophys.*, **52**, 170–216 (1938), first gave a correlation of this form, with a coefficient of 0.552 in lieu of 0.60 in the last term.

[4] S. Whitaker, *Fundamental Principles of Heat Transfer*, Krieger Publishing Co., Malabar, Fla. (1977), pp. 340–342; *AIChE Journal*, **18**, 361–371 (1972).

Flow Around a Cylinder

A cylinder in a stationary fluid of infinite extent does not admit a steady-state solution. Therefore the Nusselt number for a cylinder does not have the same form as that for a sphere. Whitaker recommends for the mean Nusselt number[4]

$$\mathrm{Nu}_m = (0.4\,\mathrm{Re}^{1/2} + 0.06\,\mathrm{Re}^{2/3})\mathrm{Pr}^{0.4}\left(\frac{\mu_\infty}{\mu_0}\right)^{1/4} \tag{14.4-7}$$

in the range $1.0 < \mathrm{Re} < 1.0 \times 10^5$, $0.67 < \mathrm{Pr} < 300$, and $0.25 < \mu_\infty/\mu_0 < 5.2$. Here, as in Eq. 14.4-6, the values of viscosity and thermal conductivity in Re and Pr are those at the approaching stream temperature. Similar results are available for banks of cylinders, which are used in certain types of heat exchangers.[4]

Another correlation,[5] based on a curve-fit of McAdams' compilation of heat transfer coefficient data,[6] and on the low-Re asymptote in Problem 12B.6, is

$$\mathrm{Nu}_m = (0.376\,\mathrm{Re}^{1/2} + 0.057\,\mathrm{Re}^{2/3})\mathrm{Pr}^{1/3} + 0.92\left[\ln\left(\frac{7.4055}{\mathrm{Re}}\right) + 4.18\,\mathrm{Re}\right]^{-1/3}\mathrm{Re}^{1/3}\mathrm{Pr}^{1/3}$$

$$\tag{14.4-8}$$

This correlation has the proper behavior in the limit that $\mathrm{Pr} \to \infty$, and also behaves properly for small values of the Reynolds number. This result can be used for analyzing the steady-state performance of hot-wire anemometers, which typically operate at low Reynolds numbers.

Flow Around Other Objects

We learn from the preceding three discussions that, for the flow around objects of shapes other than those described above, a fairly good guess for the heat transfer coefficients can be obtained by using the relation

$$\mathrm{Nu}_m - \mathrm{Nu}_{m,0} = 0.6\,\mathrm{Re}^{1/2}\,\mathrm{Pr}^{1/3} \tag{14.4-9}$$

in which $\mathrm{Nu}_{m,0}$ is the mean Nusselt number at zero Reynolds number. This generalization, which is shown in Fig. 14.4-2, is often useful in estimating the heat transfer from irregularly shaped objects.

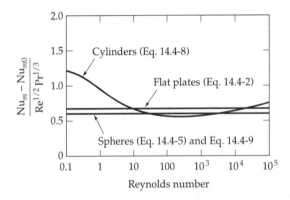

Fig. 14.4-2. Graph comparing the Nusselt numbers for flow around flat plates, spheres, and cylinders with Eq. 14.4-9.

---

[5] W. E. Stewart (to be published).

[6] W. H. McAdams, *Heat Transmission*, 3rd edition, McGraw-Hill, New York (1954), p. 259.

§14.5   HEAT TRANSFER COEFFICIENTS FOR FORCED CONVECTION THROUGH PACKED BEDS

Heat transfer coefficients between particles and fluid in packed beds are important in the design of fixed-bed catalytic reactors, absorbers, driers, and pebble-bed heat exchangers. The velocity profiles in packed beds exhibit a strong maximum near the wall, attributable partly to the higher void fraction there and partly to the more ordered interstitial passages along this smooth boundary. The resulting segregation of the flow into a fast outer stream and a slower interior one, which mix at the exit of the bed, leads to complicated behavior of mean Nusselt numbers in deep packed beds,[1] unless the tube-to-particle diameter ratio D_t/D_p is very large or close to unity. Experiments with wide, shallow beds show simpler behavior and are used in the following discussion.

We define h_{loc} for a representative volume $S\,dz$ of particles and fluid by the following modification of Eq. 14.1-5:

$$dQ = h_{loc}(aS\,dz)(T_0 - T_b) \tag{14.5-1}$$

Here a is the outer surface area of particles per unit bed volume, as in §6.4. Equations 6.4-5 and 6 give the effective particle size D_p as $6/a_v = 6(1 - \varepsilon)/a$ for a packed bed with void fraction ε.

Extensive data on forced convection for the flow of gases[2] and liquids[3] through shallow packed beds have been critically analyzed[4] to obtain the following local heat transfer correlation,

$$j_H = 2.19\,\mathrm{Re}^{-2/3} + 0.78\,\mathrm{Re}^{-0.381} \tag{14.5-2}$$

and an identical formula for the mass transfer function j_D defined in §22.3. Here the Chilton–Colburn j_H factor and the Reynolds number are defined by

$$j_H = \frac{h_{loc}}{\hat{C}_p G_0}\left(\frac{\hat{C}_p \mu}{k}\right)^{2/3} \tag{14.5-3}$$

$$\mathrm{Re} = \frac{D_p G_0}{(1 - \varepsilon)\mu\psi} = \frac{6 G_0}{a\mu\psi} \tag{14.5-4}$$

In this equation the physical properties are all evaluated at the film temperature $T_f = \frac{1}{2}(T_0 - T_b)$, and $G_0 = w/S$ is the superficial mass flux introduced in §6.4. The quantity ψ is a particle-shape factor, with a defined value of 1 for spheres and a fitted value[4] of 0.92 for cylindrical pellets. A related shape factor was used by Gamson[5] in Re and j_H; the present factor ψ is used in Re only.

For small Re, Eq. 14.5-2 yields the asymptote

$$j_H = 2.19\,\mathrm{Re}^{-2/3} \tag{14.5-5}$$

or

$$\mathrm{Nu}_{loc} = \frac{h_{loc} D_p}{k(1 - \varepsilon)\psi} = 2.19(\mathrm{RePr})^{1/3} \tag{14.5-6}$$

---

[1] H. Martin, *Chem. Eng. Sci.*, **33**, 913–919 (1978).

[2] B. W. Gamson, G. Thodos, and O. A. Hougen, *Trans. AIChE*, **39**, 1–35 (1943); C. R. Wilke and O. A. Hougen, *Trans. AIChE*, **41**, 445–451 (1945).

[3] L. K. McCune and R. H. Wilhelm, *Ind. Eng. Chem.*, **41**, 1124–1134 (1949); J. E. Williamson, K. E. Bazaire, and C. J. Geankoplis, *Ind. Eng. Chem. Fund.*, **2**, 126–129 (1963); E. J. Wilson and C. J. Geankoplis, *Ind. Eng. Chem. Fund.*, **5**, 9–14 (1966).

[4] W. E. Stewart, to be submitted.

[5] B. W. Gamson, *Chem. Eng. Prog.*, **47**, 19–28 (1951).

consistent with boundary layer theory[6] for creeping flow with RePr $\gg$ 1. The latter restriction gives Nu $\gg$ 1 corresponding to a thin thermal boundary layer relative to $D_p/(1 - \varepsilon)\psi$. This asymptote represents the creeping-flow mass-transfer data for liquids[3] very well.

The exponent $\frac{2}{3}$ in Eq. 14.5-3 is a high-Pr asymptote given by boundary layer theory for steady laminar flows[6] and for steadily driven turbulent flows.[7] This dependence is consistent with the cited data over the full range Pr > 0.6 and the corresponding range of the dimensionless group Sc for mass transfer.

§14.6 HEAT TRANSFER COEFFICIENTS FOR FREE AND MIXED CONVECTION[1]

Here we build on Example 11.4-5 to summarize the behavior of some important systems in the presence of appreciable buoyant forces, first by rephrasing the results obtained there in terms of Nusselt numbers and then by extension to other situations: (1) small buoyant forces, where the thin-boundary-layer assumption of Example 11.4-5 may not be valid; (2) very large buoyant forces, where turbulence can occur in the boundary layer, and (3) mixed forced and free convection. We shall confine ourselves to heat transfer between solid bodies and a large quiescent volume of surrounding fluid, and to the constant-temperature boundary conditions of Example 11.4-5. Discussions of other situations, including transient behavior and duct and cavity flows, are available elsewhere.[1]

In Example 11.4-5 we saw that for the free convection near a vertical flat plate, the principal dimensionless group is GrPr, which is often called the *Rayleigh number*, Ra. If we define the area mean Nusselt number as $\mathrm{Nu}_m = hH/k = q_{\mathrm{avg}}H/k(T_0 - T_1)$, then Eq. 11.4-51 may be written as

$$\mathrm{Nu}_m = C(\mathrm{GrPr})^{1/4} \tag{14.6-1}$$

where C was found to be a weak function of Pr. The heat transfer behavior at moderate values of Ra = GrPr is governed, for many shapes of solids, by laminar boundary layers of the type described in Example 11.4-5, and the results of those discussions are normally used directly.

However, at small values of GrPr direct heat conduction to the surroundings may invalidate the boundary layer result, and at sufficiently high values of GrPr the mechanism of heat transfer shifts toward random local eruptions or plumes of fluid, producing turbulence within the boundary layer. Then the Nusselt number becomes independent of the system size. The case of combined forced and free convection (normally referred to as *mixed convection*) is more complex: one must now consider Pr, Gr, and Re as independent variables, and also whether the forced and free convection effects are in the same or different directions. Only the former seems to be at all well understood. The description of the behavior is further complicated by lack of abrupt transitions between the various flow regimes.

---

[6] W. E. Stewart, *AIChE Journal*, **9**, 528–535 (1963); R. Pfeffer, *Ind. Eng. Chem. Fund.*, **3**, 380–383 (1964); J. P. Sørensen and W. E. Stewart, *Chem. Eng. Sci.*, **29**, 833–837 (1974). See also Example 12.4-3.

[7] W. E. Stewart, *AIChE Journal*, **33**, 2008–2016 (1987); corrigenda **34**, 1030 (1988).

[1] G. D. Raithby and K. G. T. Hollands, Chapter 4 in W. M. Rohsenow, J. P. Hartnett, and Y. I. Cho, eds., *Handbook of Heat Transfer*, 3rd edition, McGraw-Hill, New York (1998).

It has been shown, however, that simple and reliable predictions of heat transfer rates (expressed as area mean Nusselt numbers Nu_m) may be obtained for this wide variety of flow regimes by empirical combinations of asymptotic expressions:

  a. Nu_m^{cond}, for conduction in the absence of buoyant forces or forced convection

  b. Nu_m^{lam}, for thin laminar boundary layers, as in Example 11.4-5

  c. Nu_m^{turb}, for turbulent boundary layers

  d. Nu_m^{forced}, for pure forced convection

These are dealt with in the following subsections.

No Buoyant Forces

The limiting Nusselt number for vanishingly small free and forced convection is obtained by solving the heat conduction equation (the Laplace equation, $\nabla^2 T = 0$) for constant, uniform temperature over the solid surface and a different constant temperature at infinity. The mean Nusselt number then has the general form

$$Nu_m^{cond} = K(\text{shape}) \qquad (14.6\text{-}2)$$

With K equal to zero for all objects with at least one infinite dimension (e.g., infinitely long cylinders or infinitely wide plates). For finite bodies K is nonzero, and an important case is that of the sphere for which, according to Problem 10B.1,

$$Nu_m^{cond} = 2 \qquad (14.6\text{-}3)$$

with the characteristic length taken to be the sphere diameter. Oblate ellipsoids of revolution and circular disks are discussed in Problem 14D.1.

Thin Laminar Boundary Layers

For thin laminar boundary layers, the isothermal vertical flat plate is a representative system, conforming to Eq. 14.6-1. This equation may be generalized to

$$Nu_m^{lam} = C(\text{Pr, shape})(\text{GrPr})^{1/4} \qquad (14.6\text{-}4)$$

Moreover, the function of Pr and shape can be factored into the product

$$C = C_1(\text{shape})C_2(\text{Pr}) \qquad (14.6\text{-}5)$$

with[2]

$$C_2 \approx \frac{0.671}{[1 + (0.492/\text{Pr})^{9/16}]^{4/9}} \qquad (14.6\text{-}6)$$

Representative values[1,3] of C_1 and C_2 are given in Tables 14.6-1 and 2, respectively. Shape factors for a wide variety of other shapes are available.[3,4] For heated horizontal flat surfaces facing downward and cooled horizontal flat surfaces facing upward, the following correlation[5] is recommended:

$$Nu_m^{lam} = \frac{0.527}{[1 + (1.9/\text{Pr})^{9/10}]^{2/9}} (\text{GrPr})^{1/5} \qquad (14.6\text{-}7)$$

---

[2] S. W. Churchill and R. Usagi, *AIChE Journal*, **23**, 1121–1128 (1972).

[3] W. E. Stewart, *Int. J. Heat and Mass Transfer*, **14**, 1013–1031 (1971).

[4] A. Acrivos, *AIChE Journal*, **6**, 584–590 (1960).

[5] T. Fujii, M. Honda, and I. Morioka, *Int. J. Heat and Mass Transfer*, **15**, 755–767 (1972).

Table 14.6-1   The Factor C_1 in Eq. 14.6-5, and the D in the Nusselt Number, for Several Representative Shapes[a]

Shape →	Vertical plate	Horizontal plate[a]	Horizontal cylinder	Sphere
C_1	1.0	0.835	0.772	0.878
"D" in Nu	Height H	Width W	Diameter D	Diameter D

[a] For a hot upper surface and an insulated lower one, or the reverse for cold surfaces.

Table 14.6-2   The Factor C_2 as a Function of the Prandtl Number

	Hg	Gases		Water				Oils	
Pr	0.022	0.71	1.0	2.0	4.0	6.0	50	100	2000
C_2	0.287	0.515	0.534	0.568	0.595	0.608	0.650	0.656	0.668

For the vertical plate with a constant-heat-flux boundary condition, the recommended power on GrPr is also 1/5.

Laminar free-convection heat fluxes tend to be small, and a conduction correction is often necessary for accurate predictions. The conduction limit is determined by solving the equation $\nabla^2 T = 0$ for the given geometry, and this leads to the calculation of a "conduction Nusselt number," $\mathrm{Nu}_m^{\mathrm{cond}}$. Then the combined Nusselt number, $\mathrm{Nu}_m^{\mathrm{comb}}$, is estimated by combining the two contributing Nusselt numbers by an equation of the form[1]

$$\mathrm{Nu}_m^{\mathrm{comb}} \cong [(\mathrm{Nu}_m^{\mathrm{lam}})^n + (\mathrm{Nu}_m^{\mathrm{cond}})^n]^{1/n} \tag{14.6-8}$$

Optimum values of n are shape-dependent, but 1.07 is a suggested rough estimate in the absence of specific information.

Turbulent Boundary Layers

The effects of turbulence increase gradually, and it is common practice to combine the laminar and turbulent contributions as follows:[1]

$$\mathrm{Nu}_m^{\mathrm{free}} = [(\mathrm{Nu}_m^{\mathrm{comb}})^m + (\mathrm{Nu}_m^{\mathrm{turb}})^m]^{1/m} \tag{14.6-9}$$

Thus for the vertical isothermal flat plate, one writes[1]

$$\mathrm{Nu}_m^{\mathrm{turb}} = \frac{C_3(\mathrm{GrPr})^{1/3}}{1 + (1.4 \times 10^9/\mathrm{Gr})} \tag{14.6-10}$$

with

$$C_3 = \frac{0.13\mathrm{Pr}^{0.22}}{(1 + 0.61\mathrm{Pr}^{0.81})^{0.42}} \tag{14.6-11}$$

and $m = 6$. The values of m in Eq. 14.6-9 are heavily geometry-dependent.

Mixed Free and Forced Convection

Finally, one must deal with the problem of simultaneous free and forced convection, and this is again done through the use of an empirical combining rule:[6]

$$\text{Nu}_m^{\text{total}} = [(\text{Nu}_m^{\text{free}})^3 + (\text{Nu}_m^{\text{forced}})^3]^{1/3} \tag{14.6-12}$$

This rule appears to hold reasonably well for all geometries and situations, provided only that the forced and free convection have the *same* primary flow direction.

EXAMPLE 14.6-1

Heat Loss by Free Convection from a Horizontal Pipe

Estimate the rate of heat loss by free convection from a unit length of a long horizontal pipe, 6 in. in outside diameter, if the outer surface temperature is 100°F and the surrounding air is at 1 atm and 80°F.

SOLUTION

The properties of air at 1 atm and a film temperature $T_f = 90°F = 550°R$ are

$$\mu = 0.0190 \text{ cp} = 0.0460 \text{ lb}_m/\text{ft} \cdot \text{hr}$$
$$\rho = 0.0723 \text{ lb}_m/\text{ft}^3$$
$$\hat{C}_p = 0.241 \text{ Btu/lb}_m \cdot \text{R}$$
$$k = 0.0152 \text{ Btu/hr} \cdot \text{ft} \cdot \text{R}$$
$$\beta = 1/T_f = (1/550)\text{R}^{-1}$$

Other relevant values are $D = 0.5$ ft, $\Delta T = 20°R$, and $g = 4.17 \times 10^8$ ft/hr². From these data we obtain

$$\text{GrPr} = \left(\frac{(0.5)^3(0.0723)^2(4.17 \times 10^8)(20/550)}{(0.0460)^2}\right)\left(\frac{(0.241)(0.0460)}{0.0152}\right)$$
$$= (4.68 \times 10^6)(0.729) = 3.4 \times 10^6 \tag{14.6-13}$$

Then from Eqs. 14.6-4 to 6 and Table 14.6-1 we get

$$\text{Nu}_m = 0.772\left(\frac{0.671}{[1 + (0.492/0.729)^{9/16}]^{4/9}}\right)(3.4 \times 10^6)^{1/4}$$
$$= 0.772\left(\frac{0.671}{1.30}\right)(42.9) = 17.1 \tag{14.6-14}$$

The heat transfer coefficient is then

$$h_m = \text{Nu}_m \frac{k}{D} = 17.1\left(\frac{0.0152}{0.5}\right) = 0.52 \text{ Btu/hr} \cdot \text{ft}^2 \cdot \text{F} \tag{14.6-15}$$

The rate of heat loss per unit length of the pipe is

$$\frac{Q}{L} = \frac{h_m A \Delta T}{L} = h_m \pi D \Delta T$$
$$= (0.52)(3.1416)(0.5)(20) = 16 \text{ Btu/hr} \cdot \text{ft} \tag{14.6-16}$$

This is the heat loss by convection only. The radiation loss for the same problem is obtained in Example 16.5-2.

---

[6] E. Ruckenstein, *Adv. in Chem. Eng.*, **13**, 11–112 (1987) E. Ruckenstein and R. Rajagopalan, *Chem. Eng. Communications*, **4**, 15–29 (1980).

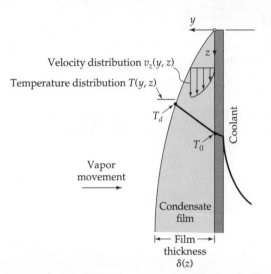

Fig. 14.7-1. Film condensation on a vertical surface (interfacial temperature discontinuity exaggerated).

§14.7 HEAT TRANSFER COEFFICIENTS FOR CONDENSATION OF PURE VAPORS ON SOLID SURFACES

The condensation of a pure vapor on a solid surface is a particularly complicated heat transfer process, because it involves two flowing fluid phases: the vapor and the condensate. Condensation occurs industrially in many types of equipment; for simplicity, we consider here only the common cases of condensation of a slowly moving vapor on the outside of horizontal tubes, vertical tubes, and vertical flat walls.

The condensation process on a vertical wall is illustrated schematically in Fig. 14.7-1. Vapor flows over the condensing surface and is moved toward it by the small pressure gradient near the liquid surface.[1] Some of the molecules from the vapor phase strike the liquid surface and bounce off; others penetrate the surface and give up their latent heat of condensation. The heat thus released must then move through the condensate to the wall, thence to the coolant on the other side of the wall. At the same time, the condensate must drain from the surface by gravity flow.

The condensate on the wall is normally the sole important resistance to heat transfer on the condensing wall. If the solid surface is clean, the condensate will usually form a continuous film over the surface, but if traces of certain impurities are present, (such as fatty acids in a steam condenser), the condensate will form in droplets. "Dropwise condensation"[2] gives much higher rates of heat transfer than "film condensation," but is difficult to maintain, so that it is common practice to assume film condensation in condenser design. The correlations that follow apply only to film condensation.

The usual definition of h_m for condensation of a pure vapor on a solid surface of area A and uniform temperature T_0 is

$$Q = h_m A(T_d - T_0) = w\Delta\hat{H}_{vap} \tag{14.7-1}$$

in which Q is the rate of heat flow into the solid surface, and T_d is the *dew point* of the vapor approaching the wall surface—that is, the temperature at which the vapor would

---

[1] Note that there occur small but abrupt changes in pressure and temperature at an interface. These discontinuities are essential to the condensation process, but are generally of negligible magnitude in engineering calculations for pure fluids. For mixtures, they may be important. See R. W. Schrage, *Interphase Mass Transfer*, Columbia University Press (1953).

[2] Dropwise condensation and boiling are discussed at length by J. G. Collier and J. R. Thome, *Convective Boiling and Condensation*, 3rd edition, Oxford University Press (1996).

condense if cooled slowly at the prevailing pressure. This temperature is very nearly that of the liquid at the liquid–gas interface. Therefore h_m may be regarded as a heat transfer coefficient for the liquid film.

Expressions for h_m have been derived[3] for *laminar nonrippling* condensate flow by approximate solution of the equations of energy and motion for a falling liquid film (see Problem 14C.1). For film condensation on a horizontal tube of diameter D, length L, and constant surface temperature T_0, the result of Nusselt[3] may be written as

$$h_m = 0.954\left(\frac{k^3\rho^2 gL}{\mu w}\right)^{1/3} \tag{14.7-2}$$

Here w/L is the mass rate of condensation per unit length of tube, and it is understood that all the physical properties of the condensate are to be calculated at the film temperature, $T_f = \frac{1}{2}(T_d + T_0)$.

For moderate temperature differences, Eq. 14.7-2 may be rewritten with the aid of an energy balance on the condensate to give

$$h_m = 0.725\left(\frac{k^3\rho^2 g\Delta\hat{H}_{\text{vap}}}{\mu D(T_d - T_0)}\right)^{1/4} \tag{14.7-3}$$

Equations 14.7-2 and 3 have been confirmed experimentally within $\pm10\%$ for single horizontal tubes. They also seem to give satisfactory results for bundles of horizontal tubes,[4] in spite of the complications introduced by condensate dripping from tube to tube.

For film condensation on *vertical tubes* or *vertical walls* of height L, the theoretical results corresponding to Eqs. 14.7-2 and 3 are

$$h_m = \frac{4}{3}\left(\frac{k^3\rho^2 g}{3\mu\Gamma}\right)^{1/3} \tag{14.7-4}$$

and

$$h_m = \frac{2\sqrt{2}}{3}\left(\frac{k^3\rho^2 g\Delta\hat{H}_{\text{vap}}}{\mu L(T_d - T_0)}\right)^{1/4} \tag{14.7-5}$$

respectively. The quantity Γ in Eq. 14.7-4 is the total rate of condensate flow from the bottom of the condensing surface per unit width of that surface. For a vertical tube, $\Gamma = w/\pi D$, where w is the total mass rate of condensation on the tube. For *short vertical tubes* ($L < 0.5$ ft), the experimental values of h_m confirm the theory well, but the measured values for *long vertical tubes* ($L > 8$ ft) may exceed the theory for a given $T_d - T_0$ by as much as 70%. This discrepancy is attributed to ripples that attain greatest amplitude on long vertical tubes.[5]

We now turn to the empirical expressions for *turbulent* condensate flow. Turbulent flow begins, on *vertical tubes or walls*, at a Reynolds number $\text{Re} = \Gamma/\mu$ of about 350. For higher Reynolds numbers, the following empirical formula has been proposed:[6]

$$h_m = 0.003\left(\frac{k^3\rho^2 g(T_d - T_0)L}{\mu^3\Delta\hat{H}_{\text{vap}}}\right)^{1/2} \tag{14.7-6}$$

This equation is equivalent, for small $T_d - T_0$, to the formula

$$h_m = 0.021\left(\frac{k^3\rho^2 g\Gamma}{\mu^3}\right)^{1/3} \tag{14.7-7}$$

[3] W. Nusselt, Z. *Ver. deutsch. Ing.*, **60**, 541–546, 596–575 (1916).

[4] B. E. Short and H. E. Brown, *Proc. General Disc. Heat Transfer*, London (1951), pp. 27–31. See also D. Butterworth, in *Handbook of Heat Exchanger Design* (G. F. Hewitt, ed.), Oxford University Press, London (1977), pp. 426–462.

[5] W. H. McAdams, *Heat Transmission*, 3rd edition, McGraw-Hill, New York (1954) p. 333.

[6] U. Grigull, *Forsch. Ingenieurwesen*, **13**, 49–57 (1942); Z. *Ver. dtsch. Ing.*, **86**, 444–445 (1942).

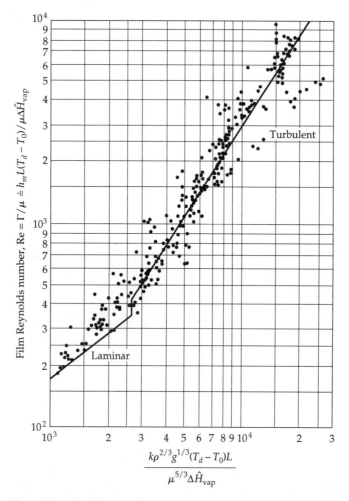

Fig. 14.7-2. Correlation of heat transfer data for film condensation of pure vapors on vertical surfaces. [H. Gröber, S. Erk, and U. Grigull, *Die Grundgesetze der Wärmeübertragung*, 3rd edition, Springer-Verlag, Berlin (1955), p. 296.]

Equations 14.7-4 to 7 are summarized in Fig. 14.7-2, for convenience of making calculations and to show the extent of agreement with the experimental data. Somewhat better agreement could have been obtained by using a family of lines in the turbulent range to represent the effect of Prandtl number. However, in view of the scattering of the data, a single line is adequate.

Turbulent condensate flow is very difficult to obtain on horizontal tubes, unless the tube diameters are very large or high temperature differences are encountered. Equations 14.7-2 and 3 are believed to be satisfactory up to the estimated transition Reynolds number, $Re = w_T/L\mu$, of about 1000, where w_T is the *total condensate flow* leaving a given tube, including the condensate from the tubes above.[7]

The inverse process of vaporization of a pure fluid is considerably more complicated than condensation. We do not attempt to discuss heat transfer to boiling liquids here, but refer the reader to some reviews.[2,8]

[7] W. H. McAdams, *Heat Transmission*, 3rd edition, McGraw-Hill, New York (1954), pp. 338–339.

[8] H. D. Baehr and K Stephan, *Heat and Mass Transfer*, Springer, Berlin (1998), Chapter 4.

EXAMPLE 14.7-1

Condensation of Steam on a Vertical Surface

A boiling liquid flowing in a vertical tube is being heated by condensation of steam on the outside of the tube. The steam-heated tube section is 10 ft high and 2 in. in outside diameter. If saturated steam is used, what steam temperature is required to supply 92,000 Btu/hr of heat to the tube at a tube-surface temperature of 200°F? Assume film condensation.

SOLUTION

The fluid properties depend on the unknown temperature T_d. We make a *guess* of $T_d = T_0 = 200°F$. Then the physical properties at the film temperature (also 200°F) are

$$\Delta \hat{H}_{vap} = 978 \text{ Btu/lb}_m$$
$$k = 0.393 \text{ Btu/hr} \cdot \text{ft} \cdot \text{F}$$
$$\rho = 60.1 \text{ lb}_m/\text{ft}^3$$
$$\mu = 0.738 \text{ lb}_m/\text{ft} \cdot \text{hr}$$

Assuming that the steam gives up only latent heat (the assumption $T_d = T_0 = 200°F$ implies this), an energy balance around the tube gives

$$Q = w\Delta\hat{H}_{vap} = \pi D\Gamma\Delta\hat{H}_{vap} \tag{14.7-8}$$

in which Q is the heat flow into the tube wall. The film Reynolds number is

$$\frac{\Gamma}{\mu} = \frac{Q}{\pi D\mu\Delta\hat{H}_{vap}} = \frac{92,000}{\pi(2/12)(0.738)(978)} = 244 \tag{14.7-9}$$

Reading Fig. 14.7-2 at this value of the ordinate, we find that the flow is laminar. Equation 14.7-2 is applicable, but it is more convenient to use the line based on this equation in Fig. 14.7-2, which gives

$$\frac{k\rho^{2/3}g^{1/3}(T_d - T_0)L}{\mu^{5/3}\Delta\hat{H}_{vap}} = 1700 \tag{14.7-10}$$

from which

$$T_d - T_0 = 1700 \frac{\mu^{5/3}\Delta\hat{H}_{vap}}{k\rho^{2/3}g^{1/3}L}$$

$$= 1700 \frac{(0.738)^{5/3}(978)}{(0.393)(60.1)^{2/3}(4.17 \times 10^8)^{1/3}(10)}$$

$$= 22°F \tag{14.7-11}$$

Therefore, the first approximation to the steam temperature is $T_d = 222°F$. This result is close enough; evaluation of the physical properties in accordance with this result gives $T_d = 220$ as a second approximation. It is apparent from Fig. 14.7-2 that this result represents an upper limit. On account of rippling, the temperature drop through the condensate film may be as little as half that predicted here.

QUESTIONS FOR DISCUSSION

1. Define the heat transfer coefficient, the Nusselt number, the Stanton number, and the Chilton-Colburn j_H. How can each of these be "decorated" to indicate the type of temperature-difference driving force that is being used?
2. What are the characteristic dimensionless groups that arise in the correlations for Nusselt numbers for forced convection? For free convection? For mixed convection?
3. To what extent can Nusselt numbers be calculated a priori from analytical solutions?
4. Explain how one develops an experimental correlation for Nusselt numbers as a function of the relevant dimensionless groups.
5. To what extent can empirical correlations be developed in which the Nusselt number is given as the product of the relevant dimensionless groups, each raised to a characteristic power?

6. In addition to the Nusselt number, we have met up with the Reynolds number Re, the Prandtl number Pr, the Grashof number Gr, the Péclet number Pé, and the Rayleigh number Ra. Define each of these and explain their meaning and usefulness.
7. Discuss the concept of wind-chill temperature.

PROBLEMS

14A.1. Average heat transfer coefficients (Fig. 14A.1). Ten thousand pounds per hour of an oil with a heat capacity of 0.6 Btu/lb$_m$ · F are being heated from 100°F to 200°F in the simple heat exchanger shown in the accompanying figure. The oil is flowing through the tubes, which are copper, 1 in. in outside diameter, with 0.065-in. walls. The combined length of the tubes is 300 ft. The required heat is supplied by condensation of saturated steam at 15.0 psia on the outside of the tubes. Calculate h_1, h_a, and h_{ln} for the oil, assuming that the inside surfaces of the tubes are at the saturation temperature of the steam, 213°F.

Answers: 78, 139, 190 Btu/hr · ft^2 · F

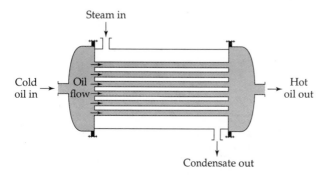

Steam in

Cold oil in → Oil flow → Hot oil out

Condensate out

Fig. 14A.1. A single-pass "shell-and-tube" heat exchanger.

14A.2. Heat transfer in laminar tube flow. One hundred pounds per hour of oil at 100°F are flowing through a 1-in. i.d. copper tube, 20 ft long. The inside surface of the tube is maintained at 215°F by condensing steam on the outside surface. Fully developed flow may be assumed through the length of the tube, and the physical properties of the oil may be considered constant at the following values: $\rho = 55$ lb$_m$/ft^3, $\hat{C}_p = 0.49$ Btu/lb$_m$ · F, $\mu = 1.42$ lb$_m$/hr · ft, $k = 0.0825$ Btu/hr · ft · F.

(a) Calculate Pr.
(b) Calculate Re.
(c) Calculate the exit temperature of the oil.

Answers: (a) 8.44; (b) 1075; (c) 155°F

14A.3. Effect of flow rate on exit temperature from a heat exchanger.

(a) Repeat parts (b) and (c) of Problem 14A.2 for oil flow rates of 200, 400, 800, 1600, and 3200 lb$_m$/hr.

(b) Calculate the total heat flow through the tube wall for each of the oil flow rates in (a).

14A.4. Local heat transfer coefficient for turbulent forced convection in a tube. Water is flowing in a 2-in. i.d. tube at a mass flow rate $w = 15,000$ lb$_m$/hr. The inner wall temperature at some point along the tube is 160°F, and the bulk fluid temperature at that point is 60°F. What is the local heat flux q_r at the pipe wall? Assume that h_{loc} has attained a constant asymptotic value.

Answer: -7.8×10^4 Btu/hr · ft^2

14A.5. Heat transfer from condensing vapors.

(a) The outer surface of a vertical tube 1 in. in outside diameter and 1 ft long is maintained at 190°F. If this tube is surrounded by saturated steam at 1 atm, what will be the total rate of heat transfer through the tube wall?

(b) What would the rate of heat transfer be if the tube were horizontal?

Answers: (a) 8400 Btu/hr; (b) 12,000 Btu/hr

14A.6. Forced-convection heat transfer from an isolated sphere.

(a) A solid sphere 1 in. in diameter is placed in an otherwise undisturbed air stream, which approaches at a velocity of 100 ft/s, a pressure of 1 atm, and a temperature of 100°F. The sphere surface is maintained at 200°F by means of an imbedded electric heating coil. What must be the rate of electrical heating in cal/s to maintain the stated conditions? Neglect radiation, and use Eq. 14.4-5.

(b) Repeat the problem in (a), but use Eq. 14.4-6.

Answer: (a) 12.9W = 3.1cal/s; (b) 16.8W = 4.0 cal/s

14A.7. Free convection heat transfer from an isolated sphere. If the sphere of Problem 14A.6 is suspended in still air at 1 atm pressure and 100°F ambient air temperature, and if the sphere surface is again maintained at 200°F, what rate of electrical heating would be needed? Neglect radiation.

Answer: 0.80W = 0.20 cal/s

14A.8. Heat loss by free convection from a horizontal pipe immersed in a liquid. Estimate the rate of heat loss by free convection from a unit length of a long horizontal pipe, 6 in. in outside diameter, if the outer surface temperature is 100°F and the surrounding water is at 80°F. Compare the result with that obtained in Example 14.6-1, in which air is the surrounding medium. The properties of water at a film temperature of 90°F (or 32.3°C) are $\mu =$

0.7632 cp, $\hat{C}_p = 0.9986$ cal/g·C and $k = 0.363$ Btu/hr·ft·F. Also, the density of water in the neighborhood of 90°F is

T(C)	30.3	31.3	32.3	33.3	34.3
ρ(g/cm³)	0.99558	0.99528	0.99496	0.99463	0.99430

Answer: $Q/L = 1930$ Btu/hr·ft

14A.9. The ice-fisherman on Lake Mendota. Compare the rates of heat loss of an ice-fisherman, when he is fishing in calm weather (wind velocity zero) and when the wind velocity is 20 mph out of the north. The ambient air temperature is $-10°F$. Assume that a bundled-up ice-fisherman can be approximated as a sphere 3 ft in diameter.

14B.1. Limiting local Nusselt number for plug flow with constant heat flux.

(a) Equation 10B.9-1 gives the asymptotic temperature distribution for heating a fluid of constant physical properties in plug flow in a long tube with constant heat flux at the wall. Use this temperature profile to show that the limiting Nusselt number for these conditions is Nu = 8.

(b) The asymptotic temperature distribution for the analogous problem for plug flow in a plane slit is given in Eq. 10B.9-2. Use this to show that the limiting Nusselt number is Nu = 12.

14B.2. Local overall heat transfer coefficient. In Problem 14A.1 the thermal resistances of the condensed steam film and wall were neglected. Justify this neglect by calculating the actual inner-surface temperature of the tubes at that cross section in the exchanger at which the oil bulk temperature is 150°F. You may assume that for the oil h_{loc} is constant throughout the exchanger at 190 Btu/hr·ft²·F. The tubes are horizontal.

14B.3. The hot-wire anemometer.[1] A hot-wire anemometer is essentially a fine wire, usually made of platinum, which is heated electrically and inserted into a flowing fluid. The wire temperature, which is a function of the fluid temperature, fluid velocity, and the rate of heating, may be determined by measuring its electrical resistance.

(a) A straight cylindrical wire 0.5 in. long and 0.01 in. in diameter is exposed to a stream of air at 70°F flowing past the wire at 100 ft/s. What must the rate of energy input be in watts to maintain the wire surface at 600°F? Neglect radiation as well as heat conduction along the wire.

(b) It has been reported[2] that for a given fluid and wire at given fluid and wire temperatures (hence a given wire resistance)

$$I^2 = B\sqrt{v_\infty} + C \qquad (14B.3-1)$$

in which I is the current required to maintain the desired temperature, v_∞ is the velocity of the approaching fluid, and C is a constant. How well does this equation agree with the predictions of Eq. 14.4-7 or Eq. 14.4-8 for the fluid and wire of (a) over a fluid velocity range of 100 to 300 ft/s? What is the significance of the constant C in Eq. 14B.3-1?

14B.4. Dimensional analysis. Consider the flow system described in the first paragraph of §14.3, for which dimensional analysis has already given the dimensionless velocity profile (Eq. 6.2-7) and temperature profile (Eq. 14.3-9).

(a) Use Eqs. 6.2-7 and 14.3-9 and the definition of cup-mixing temperature to get the time-averaged expression.

$$\frac{T_{b2} - T_{b1}}{T_0 - T_{b1}} = \text{a function of Re, Pr, } L/D \qquad (14B.4-1)$$

(b) Use the result just obtained and the definitions of the heat transfer coefficients to derive Eqs. 14.3-12, 13, and 14.

14B.5. Relation between h_{loc} and h_{ln}. In many industrial tubular heat exchangers (see Example 15.4-2) the tube-surface temperature T_0 varies linearly with the bulk fluid temperature T_b. For this common situation h_{loc} and h_{ln} may be simply interrelated.

(a) Starting with Eq. 14.1-5, show that

$$h_{loc}(\pi D dz)(T_b - T_0) = -(\tfrac{1}{4}\pi D^2)(\rho \hat{C}_p \langle v \rangle dT_b) \qquad (14B.5-1)$$

and therefore that

$$\int_0^L h_{loc} dz = \tfrac{1}{4}\rho \hat{C}_p D\langle v \rangle \frac{T_b(L) - T_b(0)}{(T_0 - T_b)_{ln}} \qquad (14B.5-2)$$

(b) Combine the result in (a) with Eq. 14.1-4 to show that

$$h_{ln} = \frac{1}{L}\int_0^L h_{loc} dz \qquad (14B.5-3)$$

in which L is the total tube length, and therefore that (if $(\partial h_{loc}/\partial L)_z = 0$, which is equivalent to the statement that axial heat conduction is neglected)

$$h_{loc}|_{z=L} = h_{ln} + L\frac{dh_{ln}}{dL} \qquad (14B.5-4)$$

14B.6. Heat loss by free convection from a pipe. In Example 14.6-1, would the heat loss be higher or lower if the pipe-surface temperature were 200°F and the air temperature were 180°F?

14C.1. The Nusselt expression for film condensation heat transfer coefficients (Fig. 14.7-1). Consider a laminar film of condensate flowing down a vertical wall, and assume that this liquid film constitutes the sole heat transfer resistance on the vapor side of the wall. Further assume that (i) the shear stress between liquid and vapor may be neglected; (ii) the physical properties in the film may be evaluated at the arithmetic mean of vapor and cooling-surface temperatures and that the cooling-surface temperature may be assumed constant; (iii) acceleration of fluid elements in the film may be neglected compared to the

---

[1] See, for example, G. Comte-Bellot, Chapter 34 in *The Handbook of Fluid Dynamics* (R. W. Johnson, ed.), CRC Press, Boca Raton, Fla. (1999).

[2] L. V. King, *Phil. Trans. Roy. Soc.* (London), **A214**, 373–432 (1914).

gravitational and viscous forces; (iv) sensible heat changes, $C_p dT$, in the condensate film are unimportant compared to the latent heat transferred through it; and (v) the heat flux is very nearly normal to the wall surface.

(a) Recall from §2.2 that the average velocity of a film of constant thickness δ is $\langle v_z \rangle = \rho g \delta^2 / 3\mu$. Assume that this relation is valid for any value of z.

(b) Write the energy equation for the film, neglecting film curvature and convection. Show that the heat flux through the film toward the cold surface is

$$-q_y = k\left(\frac{T_d - T_0}{\delta}\right) \tag{14C.1-1}$$

(c) As the film proceeds down the wall, it picks up additional material by the condensation process. In this process, heat is liberated to the extent of $\Delta \hat{H}_{vap}$ per unit mass of material that undergoes the change in state. Show that equating the heat liberation by condensation with the heat flowing through the film in a segment dz of the film leads to

$$\rho \Delta \hat{H}_{vap} d(\langle v_z \rangle \delta) = k\left(\frac{T_d - T_0}{\delta}\right) dz \tag{14C.1-2}$$

(d) Insert the expression for the average velocity from (a) into Eq. 14C.1-2 and integrate from $z = 0$ to $z = L$ to obtain

$$\delta(L) = \left(\frac{4k(T_d - T_0)\mu L}{\rho^2 g \Delta \hat{H}_{vap}}\right)^{1/4} \tag{14C.1-3}$$

(e) Use the definition of the heat transfer coefficient and the result in (d) to obtain Eq. 14.7-5.

(f) Show that Eqs. 14.7-4 and 5 are equivalent for the conditions of this problem.

14C.2.   Heat transfer correlations for agitated tanks (Fig. 14C.2). A liquid of essentially constant physical properties is being continuously heated by passage through an agitated tank, as shown in the accompanying figure. Heat is supplied by condensation of steam on the outer wall of the tank. The thermal resistance of the condensate film and the tank wall may be considered small compared to that of the fluid in the tank, and the unjacketed portion of the tank

may be assumed to be well insulated. The rate of liquid flow through the tank has a negligible effect on the flow pattern in the tank.

Develop a general form of dimensionless heat transfer correlation for the tank corresponding to the correlation for tube flow in §14.3. Choose the following reference quantities: reference length, D, the impeller diameter; reference velocity, ND, where N is the rate of shaft rotation in revolutions per unit time; reference pressure, $\rho N^2 D^2$, where ρ is the fluid density.

14D.1.   Heat transfer from an oblate ellipsoid of revolution. Systems of this sort are best described in oblate ellipsoidal coordinates (ξ, η, ψ)[1] for which

ξ = constant describes oblate ellipsoids $(0 \leq \xi < \infty)$

η = constant describes hyperboloids of revolution $(0 \leq \eta \leq \pi)$

ψ = constant describes half planes $(0 \leq \psi < 2\pi)$

Note that $\xi = \xi_0$ can describe oblate ellipsoids, with $\xi_0 = 0$ being a limiting case of the two-sided disk, and the limit as $\xi_0 \to \infty$ being a sphere. In this problem we investigate the corresponding two limiting values of the Nusselt number.

(a) First use Eq. A.7-13 to get the scale factors from the relation between oblate ellipsoidal coordinates and Cartesian coordinates:

$$x = a \cosh \xi \sin \eta \cos \psi \tag{14D.1-1}$$
$$y = a \cosh \xi \sin \eta \sin \psi \tag{14D.1-2}$$
$$z = a \sinh \xi \cos \eta \tag{14D.1-3}$$

in which a is one-half the distance between the foci. Show that

$$h_\xi = h_\eta = a\sqrt{\cosh^2 \xi - \sin^2 \eta} \tag{14D.1-4}$$
$$h_\psi = a \cosh \xi \sin \eta \tag{14D.1-5}$$

Equations A.7-13 and 14 can then be used to get any of the ∇-operations that are needed.

(b) Next obtain the temperature profile outside of an oblate ellipsoid with surface temperature T_0, which is embedded in an infinite medium with the temperature T_∞ far from the ellipsoid. Let $\Theta = (T - T_0)/(T_\infty - T_0)$ be a dimensionless temperature, and show that Laplace's equation describing the heat conduction exterior to the ellipsoid is

$$\frac{1}{a^2(\cosh^2 \xi - \sin^2 \eta)}\left[\frac{\partial}{\partial \xi}\left(\cosh \xi \frac{\partial \Theta}{\partial \xi}\right) + \cdots\right] = 0 \tag{14D.1-6}$$

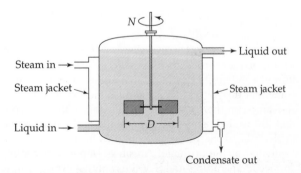

Fig. 14C.2. Continuous heating of a liquid in an agitated tank.

---

[1] For a discussion of oblate ellipsoidal coordinates, see P. Moon and D. E. Spencer, *Field Theory Handbook*, Springer, Berlin (1961), pp. 31–34. See also J. Happel and H. Brenner, *Low Reynolds Number Hydrodynamics*, Prentice-Hall, Englewood Cliffs, N.J. (1965), pp. 512–516; note that their scale factors are the reciprocals of those defined in this book.

The terms involving derivatives with respect to η and ψ have been omitted because they are not needed. Show that this equation may be solved with the boundary conditions that $\Theta(\xi_0) = 0$ and $\Theta(\infty) = 1$ to obtain

$$\Theta = 1 - \frac{\frac{1}{2}\pi - \arctan(\sinh \xi)}{\frac{1}{2}\pi - \arctan(\sinh \xi_0)} \qquad (14D.1\text{-}7)$$

(c) Next, specialize this result for the two-sided disk (that is, the limiting case that $\xi_0 = 0$), and show that the normal temperature gradient at the surface is

$$(\mathbf{n} \cdot \nabla\Theta)\big|_{\text{surf}} = \left(\mathbf{n}_\xi \cdot \mathbf{n}_\xi \frac{1}{h_\xi} \frac{\partial \Theta}{\partial \xi}\right)\bigg|_{\xi=0} = \frac{2}{\pi} \frac{1}{R \cos \eta} \qquad (14D.1\text{-}8)$$

where a has been expressed as R, the disk radius. Show further that the total heat loss through both sides of the disk is

$$Q = -2k\int(\mathbf{n} \cdot \nabla T)\, dS$$
$$= +2k(T_0 - T_\infty)\int(\mathbf{n} \cdot \nabla\Theta)\, dS$$
$$= 2k(T_0 - T_\infty) \int_0^{2\pi} \int_0^{\pi/2} \left(\frac{2}{\pi R \cos \eta}\right) R^2 \cos \eta \, \sin \eta \, d\eta \, d\psi$$
$$= 8kR(T_0 - T_\infty) \qquad (14D.1\text{-}9)$$

and that the Nusselt number is given by $\mathrm{Nu} = 16/\pi = 5.09$. Since $\mathrm{Nu} = 2$ for the analogous sphere problem, we see that the Nusselt number for any oblate ellipsoid must lie somewhere between 2 and 5.09.

(d) By dimensional analysis show that, without doing any detailed derivation (such as the above), one can predict that the heat loss from the ellipsoid must be proportional to the linear dimension a rather than to the surface area. Is this result limited to ellipsoids? Discuss.

Chapter 15

Macroscopic Balances for Nonisothermal Systems

§15.1 The macroscopic energy balance

§15.2 The macroscopic mechanical energy balance

§15.3 Use of the macroscopic balances to solve steady-state problems with flat velocity profiles

§15.4 The *d*-forms of the macroscopic balances

§15.5° Use of the macroscopic balances to solve unsteady-state problems and problems with nonflat velocity profiles

In Chapter 7 we discussed the macroscopic mass, momentum, angular momentum, and mechanical energy balances. The treatment there was restricted to systems at constant temperature. Actually this restriction is somewhat artificial, since in real flow systems mechanical energy is always being converted into thermal energy by viscous dissipation. What we really assumed in Chapter 7 is that any heat so produced is either too small to change the fluid properties or is immediately conducted away through the walls of the system containing the fluid. In this chapter we extend the previous results to describe the overall behavior of nonisothermal macroscopic flow systems.

For a nonisothermal system there are five macroscopic balances that describe the relations between the inlet and outlet conditions of the stream. They may be derived by integrating the equations of change over the macroscopic system:

$$\int_{V(t)} (\text{eq. of continuity}) \, dV = \text{macroscopic mass balance}$$

$$\int_{V(t)} (\text{eq. of motion}) \, dV = \text{macroscopic momentum balance}$$

$$\int_{V(t)} (\text{eq. of angular momentum}) \, dV = \text{macroscopic angular momentum balance}$$

$$\int_{V(t)} (\text{eq. of mechanical energy}) \, dV = \text{macroscopic mechanical energy balance}$$

$$\int_{V(t)} (\text{eq. of (total) energy}) \, dV = \text{macroscopic (total) energy balance}$$

The first four of these were discussed in Chapter 7, and their derivations suggest that they can be applied to nonisothermal systems just as well as to isothermal systems. In this chapter we add the fifth balance—namely, that for the total energy. This is derived in §15.1, not by performing the integration above, but rather by applying the law of conservation of total energy directly to the system shown in Fig. 7.0-1. Then in §15.2 we revisit the mechanical energy balance and examine it in the light of the discussion of the (total) energy balance. Next in §15.3 we give the simplified versions of the macroscopic balances for steady-state systems and illustrate their use.

In §15.4 we give the differential forms (*d*-forms) of the steady-state balances. In these forms, the entry and exit planes 1 and 2 are taken to be only a differential distance apart. The "*d*-forms" are frequently useful for problems involving flow in conduits in which the velocity, temperature, and pressure are continually changing in the flow direction.

Finally, in §15.5 we present several illustrations of unsteady-state problems that can be solved by the macroscopic balances.

This chapter will make use of nearly all the topics we have covered so far and provides an excellent opportunity to review the preceding chapters. Once again we take this opportunity to remind the reader that in using the macroscopic balances, it may be necessary to omit some terms and to estimate the values of others. This requires good intuition or some extra experimental data.

§15.1   THE MACROSCOPIC ENERGY BALANCE

We consider the system sketched in Fig. 7.0-1 and make the same assumptions that were made in Chapter 7 with regard to quantities at the entrance and exit planes:

(i) The time-smoothed velocity is perpendicular to the relevant cross section.

(ii) The density and other physical properties are uniform over the cross section.

(iii) The forces associated with the stress tensor $\boldsymbol{\tau}$ are neglected.

(iv) The pressure does not vary over the cross section.

To these we add (likewise at the entry and exit planes):

(v) The energy transport by conduction $\mathbf{q}$ is small compared to the convective energy transport and can be neglected.

(vi) The work associated with $[\boldsymbol{\tau} \cdot \mathbf{v}]$ can be neglected relative to $p\mathbf{v}$.

We now apply the statement of conservation of energy to the fluid in the macroscopic flow system. In doing this, we make use of the concept of potential energy to account for the work done against the external forces (this corresponds to using Eq. 11.1-9, rather than Eq. 11.1-7, as the equation of change for energy).

The statement of the law of conservation of energy then takes the form:

$$\frac{d}{dt}(U_{\text{tot}} + K_{\text{tot}} + \Phi_{\text{tot}}) = (\rho_1 \hat{U}_1 \langle v_1 \rangle + \tfrac{1}{2}\rho_1 \langle v_1^3 \rangle + \rho_1 \hat{\Phi}_1 \langle v_1 \rangle)S_1$$

rate of increase of internal, kinetic, and potential energy in the system

rate at which internal, kinetic, and potential energy enter the system at plane 1 by flow

$$- (\rho_2 \hat{U}_2 \langle v_2 \rangle + \tfrac{1}{2}\rho_2 \langle v_2^3 \rangle + \rho_2 \hat{\Phi}_2 \langle v_2 \rangle)S_2 \qquad (15.1\text{-}1)$$

rate at which internal, kinetic, and potential energy leave the system at plane 2 by flow

$+ Q$ $+ W_m$ $+ (p_1 \langle v_1 \rangle S_1 - p_2 \langle v_2 \rangle S_2)$

rate at which heat is added to the system across boundary

rate at which work is done on the system by the surroundings by means of the moving surfaces

rate at which work is done on the system by the surroundings at planes 1 and 2

Here $U_{\text{tot}} = \int \rho \hat{U} dV$, $K_{\text{tot}} = \int \tfrac{1}{2}\rho v^2 dV$, and $\Phi_{\text{tot}} = \int \rho \hat{\Phi} dV$ are the total internal, kinetic, and potential energy in the system, the integrations being performed over the entire volume of the system.

This equation may be written in a more compact form by introducing the mass rates of flow $w_1 = \rho_1 \langle v_1 \rangle S_1$ and $w_2 = \rho_2 \langle v_2 \rangle S_2$, and the total energy $E_{tot} = U_{tot} + K_{tot} + \Phi_{tot}$. We thus get for the *unsteady state macroscopic energy balance*

$$\frac{d}{dt} E_{tot} = -\Delta\left[\left(\hat{U} + p\hat{V} + \frac{1}{2}\frac{\langle v^3 \rangle}{\langle v \rangle} + \hat{\Phi}\right)w\right] + Q + W_m \qquad (15.1\text{-}2)$$

It is clear, from the derivation of Eq. 15.1-1, that the "work done on the system by the surroundings" consists of two parts: (1) the work done by the moving surfaces W_m, and (2) the work done at the ends of the system (planes 1 and 2), which appears as $-\Delta(p\hat{V}w)$ in Eq. 15.1-2. Although we have combined the pV terms with the internal, kinetic, and potential energy terms in Eq. 15.1-2, it is inappropriate to say that "pV energy enters and leaves the system" at the inlet and outlet. The pV terms originate as work terms and should be thought of as such.

We now consider the situation where the system is operating at steady state so that the total energy E_{tot} is constant, and the mass rates of flow in and out are equal ($w_1 = w_2 = w$). Then it is convenient to introduce the symbols $\hat{Q} = Q/w$ (the heat addition per unit mass of flowing fluid) and $\hat{W}_m = W_m/w$ (the work done on a unit mass of flowing fluid). Then the *steady state macroscopic energy balance* is

$$\Delta\left(\hat{H} + \frac{1}{2}\frac{\langle v^3 \rangle}{\langle v \rangle} + gh\right) = \hat{Q} + \hat{W}_m \qquad (15.1\text{-}3)$$

Here we have written $\hat{\Phi}_1 = gh_1$ and $\hat{\Phi}_2 = gh_2$, where h_1 and h_2 are heights above an arbitrarily chosen datum plane (see the discussion just before Eq. 3.3-2). Similarly, $\hat{H}_1 = \hat{U}_1 + p_1\hat{V}_1$ and $\hat{H}_2 = \hat{U}_2 + p_2\hat{V}_2$ are enthalpies per unit mass measured with respect to an arbitrarily specified reference state. The explicit formula for the enthalpy is given in Eq. 9.8-8.

For many problems in the chemical industry the kinetic energy, potential energy, and work terms are negligible compared with the thermal terms in Eq. 15.1-3, and the energy balance simplifies to $\hat{H}_2 - \hat{H}_1 = \hat{Q}$, often called an "enthalpy balance." However this relation should not be construed as a conservation equation for enthalpy.

§15.2 THE MACROSCOPIC MECHANICAL ENERGY BALANCE

The macroscopic mechanical energy balance, given in §7.4 and derived in §7.8, is repeated here for comparison with Eqs. 15.1-2 and 3. The *unsteady-state macroscopic mechanical energy balance*, as given in Eq. 7.4-2, is

$$\frac{d}{dt}(K_{tot} + \Phi_{tot}) = -\Delta\left(\frac{1}{2}\frac{\langle v^3 \rangle}{\langle v \rangle} + \hat{\Phi} + \frac{p}{\rho}\right)w + W_m - E_c - E_v \qquad (15.2\text{-}1)$$

where E_c and E_v are defined in Eqs. 7.4-3 and 4. An approximate form of the *steady-state macroscopic mechanical balance*, as given in Eq. 7.4-7, is

$$\Delta\left(\frac{1}{2}\frac{\langle v^3 \rangle}{\langle v \rangle}\right) + g\Delta h + \int_1^2 \frac{1}{\rho}\,dp = \hat{W}_m - \hat{E}_v \qquad (15.2\text{-}2)$$

The details of the approximation introduced here are explained in Eqs. 7.8-9 to 12.

The integral in Eq. 15.2-2 must be evaluated along a "representative streamline" in the system. To do this, one must know the equation of state $\rho = \rho(p, T)$ and also how T changes with p along the streamline. In Fig. 15.2-1 the surface $\hat{V} = \hat{V}(p, T)$ for an ideal gas is shown. In the pT-plane there is shown a curve beginning at p_1, T_1 (the inlet stream conditions) and ending at p_2, T_2 (the outlet stream conditions). The curve in the pT-plane indicates the succession of states through which the gas passes in going from the initial

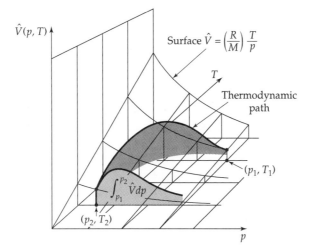

Fig. 15.2-1. Graphical representation of the integral in Eq. 15.2-2. The ruled area is $\int_{p_1}^{p_2} \hat{V} dp = \int_{p_1}^{p_2} (1/\rho) dp$. Note that the value of this integral is negative here, because we are integrating from right to left.

state to the final state. The integral $\int_1^2 (1/\rho)\, dp$ is then the projection of the shaded area in Fig. 15.2-1 onto the $p\hat{V}$-plane. It is evident that the value of this integral changes as the "thermodynamic path" of the process from plane 1 to 2 is altered. If one knows the path and the equation of state then one can compute $\int_1^2 (1/\rho)\, dp$.

In several special situations, it is not difficult to evaluate the integral:

a. For *isothermal systems*, the integral is evaluated by prescribing the isothermal equation of state—that is, by giving a relation for ρ as a function of p. For example, for ideal gases $\rho = pM/RT$ and

$$\int_1^2 \frac{1}{\rho}\, dp = \frac{RT}{M} \int_{p_1}^{p_2} \frac{1}{p}\, dp = \frac{RT}{M} \ln \frac{p_2}{p_1} \qquad \text{(ideal gases)} \qquad (15.2\text{-}3)$$

b. For *incompressible liquids*, ρ is constant so that

$$\int_1^2 \frac{1}{\rho}\, dp = \frac{1}{\rho} (p_2 - p_1) \qquad \text{(incompressible liquids)} \qquad (15.2\text{-}4)$$

c. For frictionless *adiabatic flow of ideal gases* with constant heat capacity, p and ρ are related by the expression $p\rho^{-\gamma} = \text{constant}$, in which $\gamma = \hat{C}_p/\hat{C}_V$ as shown in Example 11.4-6. Then the integral becomes

$$\int_1^2 \frac{1}{\rho}\, dp = \frac{p_1^{1/\gamma}}{\rho_1} \int_{p_1}^{p_2} \frac{1}{p^{1/\gamma}}\, dp = \frac{p_1}{\rho_1} \frac{\gamma}{\gamma - 1} \left[\left(\frac{p_2}{p_1} \right)^{(\gamma-1)/\gamma} - 1 \right]$$

$$= \frac{p_1}{\rho_1} \frac{\gamma}{\gamma - 1} \left[\left(\frac{\rho_2}{\rho_1} \right)^{\gamma-1} - 1 \right] \qquad (15.2\text{-}5)$$

Hence for this special case of nonisothermal flow, the integration can be performed analytically.

We now conclude with several comments involving both the mechanical energy balance and the total energy balance. We emphasized in §7.8 that Eq. 7.4-2 (same as Eq. 15.2-1) is derived by taking the dot product of **v** with the equation of motion and then integrating the result over the volume of the flow system. Since we start with the equation of motion—which is a statement of the law of conservation of linear momentum—the mechanical energy balance contains information different from that of the (total) energy balance, which is a statement of the law of conservation of energy. Therefore, in general, both balances are needed for problem solving. The mechanical energy balance is not "an alternative form" of the energy balance.

In fact, if we subtract the mechanical energy balance in Eq. 15.2-1 from the total energy balance in Eq. 15.1-2 we get the *macroscopic balance for the internal energy*

$$\frac{dU_{tot}}{dt} = -\Delta \hat{U} w + Q + E_c + E_v \tag{15.2-6}$$

This states that the total internal energy in the system changes because of the difference in the amount of internal energy entering and leaving the system by fluid flow, because of the heat entering (or leaving) the system through walls of the system, because of the heat produced (or consumed) within the fluid by compression (or expansion), and because of the heat produced in the system because of viscous dissipation heating. Equation 15.2-6 cannot be written a priori, since there is no conservation law for internal energy. It can, however, be obtained by integrating Eq. 11.2-1 over the entire flow system.

§15.3 USE OF THE MACROSCOPIC BALANCES TO SOLVE STEADY-STATE PROBLEMS WITH FLAT VELOCITY PROFILES

The most important applications of the macroscopic balances are to steady-state problems. Furthermore, it is usually assumed that the flow is turbulent so that the variation of the velocity over the cross section can be safely neglected (see "Notes" after Eqs. 7.2-3 and 7.4-7). The five macroscopic balances, with these additional restrictions, are summarized in Table 15.3-1. They have been generalized to multiple inlet and outlet ports to accommodate a larger set of problems.

Table 15.3-1   Steady-State Macroscopic Balances for Turbulent Flow in Nonisothermal Systems

Mass:	$\sum w_1 - \sum w_2 = 0$	(A)
Momentum:	$\sum (v_1 w_1 + p_1 S_1) \mathbf{u}_1 - \sum (v_2 w_2 + p_2 S_2) \mathbf{u}_2 + m_{tot} \mathbf{g} = \mathbf{F}_{f \to s}$	(B)
Angular momentum:	$\sum (v_1 w_1 + p_1 S_1)[\mathbf{r}_1 \times \mathbf{u}_1] - \sum (v_2 w_2 + p_2 S_2)[\mathbf{r}_2 \times \mathbf{u}_2] + \mathbf{T}_{ext} = \mathbf{T}_{f \to s}$	(C)
Mechanical energy:	$\sum \left(\frac{1}{2} v_1^2 + g h_1 + \frac{p_1}{\rho_1} \right) w_1 - \sum \left(\frac{1}{2} v_2^2 + g h_2 + \frac{p_2}{\rho_2} \right) w_2 = -W_m + E_c + E_v$	(D)
(Total) energy:	$\sum (\frac{1}{2} v_1^2 + g h_1 + \hat{H}_1) w_1 - \sum (\frac{1}{2} v_2^2 + g h_2 + \hat{H}_2) w_2 = -W_m - Q$	(E)

Notes:

[a] All formulas here imply flat velocity profiles.

[b] $\sum w_1 = w_{1a} + w_{1b} + w_{1c} + \cdots$, where $w_{1a} = \rho_{1a} v_{1a} S_{1a}$, and so on.

[c] h_1 and h_2 are elevations above an arbitrary datum plane.

[d] $\hat{H}_1$ and $\hat{H}_2$ are enthalpies per unit mass relative to some arbitrarily chosen reference state (see Eq. 9.8-8).

[e] All equations are written for compressible flow; for incompressible flow, $E_c = 0$. The quantities E_c and E_v are defined in Eqs. 7.3-3 and 4.

[f] $\mathbf{u}_1$ and $\mathbf{u}_2$ are unit vectors in the direction of flow.

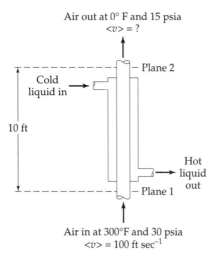

Air out at 0° F and 15 psia
$<v> = ?$

— Plane 2

Cold liquid in →

10 ft

Hot liquid out

— Plane 1

Air in at 300°F and 30 psia
$<v> = 100 \text{ ft sec}^{-1}$

Fig. 15.3-1. The cooling of air in a countercurrent heat exchanger.

EXAMPLE 15.3-1

The Cooling of an Ideal Gas

Two hundred pounds per hour of dry air enter the inner tube of the heat exchanger shown in Fig. 15.3-1 at 300°F and 30 psia, with a velocity of 100 ft/sec. The air leaves the exchanger at 0°F and 15 psia, at 10 ft above the exchanger entrance. Calculate the rate of energy removal across the tube wall. Assume turbulent flow and ideal gas behavior, and use the following expression for the heat capacity of air:

$$\tilde{C}_p = 6.39 + (9.8 \times 10^{-4})T - (8.18 \times 10^{-8})T^2 \tag{15.3-1}$$

where $\tilde{C}_p$ is in Btu/(lb-mole · R) and T is in degrees R.

SOLUTION

For this system, the macroscopic energy balance, Eq. 15.1-3, becomes

$$(\hat{H}_2 - \hat{H}_1) + \tfrac{1}{2}(v_2^2 - v_1^2) + g(h_2 - h_1) = \hat{Q} \tag{15.3-2}$$

The enthalpy difference may be obtained from Eq. 9.8-8, and the velocity may be obtained as a function of temperature and pressure with the aid of the macroscopic mass balance $\rho_1 v_1 = \rho_2 v_2$ and the ideal gas law $p = \rho RT/M$. Hence Eq. 15.3-2 becomes

$$\frac{1}{M}\int_{T_1}^{T_2} \tilde{C}_p dT + \frac{1}{2} v_1^2 \left[\left(\frac{p_1 T_2}{p_2 T_1} \right)^2 - 1 \right] + g(h_2 - h_1) = \hat{Q} \tag{15.3-3}$$

The explicit expression for $\tilde{C}_p$ in Eq. 15.3-1 may then be inserted into Eq. 15.3-3 and the integration performed. Next substitution of the numerical values gives the heat removal per pound of fluid passing through the heat exchanger:

$$-\hat{Q} = \tfrac{1}{29}[(6.39)(300) + \tfrac{1}{2}(9.8 \times 10^{-4})(5.78 - 2.12)(10^5)$$
$$-\tfrac{1}{3}(8.18 \times 10^{-8})(4.39 - 0.97)(10^8)]$$
$$+\frac{1}{2}\left(\frac{10^4}{(32.2)(778)}\right)[1 - (1.21)^2] - \left(\frac{10}{778}\right)$$
$$= 72.0 - 0.093 - 0.0128$$
$$= 71.9 \text{ Btu/hr} \tag{15.3-4}$$

The rate of heat removal is then

$$-\hat{Q}w = 14{,}380 \text{ Btu/hr} \tag{15.3-5}$$

Note, in Eq. 15.3-4, that the kinetic and potential energy contributions are negligible in comparison with the enthalpy change.

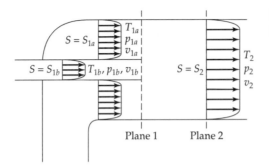

Fig. 15.3-2. The mixing of two ideal gas streams.

Plane 1   Plane 2

EXAMPLE 15.3-2

Mixing of Two Ideal Gas Streams

Two steady, turbulent streams of the same ideal gas flowing at different velocities, temperatures, and pressures are mixed as shown in Fig. 15.3-2. Calculate the velocity, temperature, and pressure of the resulting stream.

SOLUTION

The fluid behavior in this example is more complex than that for the incompressible, isothermal situation discussed in Example 7.6-2, because here changes in density and temperature may be important. We need to use the steady-state macroscopic energy balance, Eq. 15.2-3, and the ideal gas equation of state, in addition to the mass and momentum balances. With these exceptions, we proceed as in Example 7.6-2.

We choose the inlet planes (1a and 1b) to be cross sections at which the fluids first begin to mix. The outlet plane (2) is taken far enough downstream that complete mixing has occurred. As in Example 7.6-2 we assume flat velocity profiles, negligible shear stresses on the pipe wall, and no changes in the potential energy. In addition, we neglect the changes in the heat capacity of the fluid and assume adiabatic operation. We now write the following equations for this system with two entry ports and one exit port:

Mass: $$w_1 = w_{1a} + w_{1b} = w_2 \tag{15.3-6}$$

Momentum: $$v_2 w_2 + p_2 S_2 = v_{1a} w_{1a} + p_{1a} S_{1a} + v_{1b} w_{1b} + p_{1b} S_{1b} \tag{15.3-7}$$

Energy: $$w_2[\hat{C}_p(T_2 - T_{ref}) + \tfrac{1}{2}v_2^2] = w_{1a}[\hat{C}_p(T_{1a} - T_{ref}) + \tfrac{1}{2}v_{1a}^2] + w_{1b}[\hat{C}_p(T_{1b} - T_{ref}) + \tfrac{1}{2}v_{1b}^2] \tag{15.3-8}$$

Equation of state: $$p_2 = \rho_2 R T_2 / M \tag{15.3-9}$$

In this set of equations we know all the quantities at 1a and 1b, and the four unknowns are p_2, T_2, ρ_2, and v_2. T_{ref} is the reference temperature for the enthalpy. By multiplying Eq. 15.3-6 by $\hat{C}_p T_{ref}$ and adding the result to Eq. 15.3-8 we get

$$w_2[\hat{C}_p T_2 + \tfrac{1}{2}v_2^2] = w_{1a}[\hat{C}_p T_{1a} + \tfrac{1}{2}v_{1a}^2] + w_{1b}[\hat{C}_p T_{1b} + \tfrac{1}{2}v_{1b}^2] \tag{15.3-10}$$

The right sides of Eqs. 15.3-6, 7, and 10 contain known quantities and we designate them by w, P, and E, respectively. Note that w, P, and E are not independent, because the pressure, temperature, and density of each inlet stream must be related by the equation of state.

We now solve Eq. 15.3-7 for v_2 and eliminate p_2 by using the ideal gas law. In addition we write w_2 as $\rho_2 v_2 S_2$. This gives

$$v_2 + \frac{RT_2}{Mv_2} = \frac{P}{w} \tag{15.3-11}$$

This equation can be solved for T_2, which is inserted into Eq. 15.3-10 to give

$$v_2^2 - \left[2\left(\frac{\gamma}{\gamma+1}\right)\frac{P}{w}\right]v_2 + 2\left(\frac{\gamma-1}{\gamma+1}\right)\frac{E}{w} = 0 \tag{15.3-12}$$

in which $\gamma = C_p/C_V$, a quantity which varies from about 1.1 to 1.667 for gases. Here we have used the fact that $\tilde{C}_p/R = \gamma/(\gamma - 1)$ for an ideal gas. When Eq. 15.3-12 is solved for v_2 we get

$$v_2 = \left(\frac{\gamma}{\gamma + 1}\right)\frac{P}{w}\left[1 \pm \sqrt{1 - 2\left(\frac{\gamma^2 - 1}{\gamma^2}\right)\frac{wE}{P^2}}\right] \tag{15.3-13}$$

On physical grounds, the radicand cannot be negative. It can be shown (see Problem 15B.4) that, when the radicand is zero, the velocity of the final stream is sonic. Therefore, in general one of the solutions for v_2 is supersonic and one is subsonic. Only the lower (subsonic) solution can be obtained in the turbulent mixing process under consideration, since supersonic duct flow is unstable. The transition from supersonic to subsonic duct flow is illustrated in Example 11.4-7.

Once the velocity v_2 is known, the pressure and temperature may be calculated from Eqs. 15.3-7 and 11. The mechanical energy balance can be used to get $(E_c + E_v)$.

§15.4   THE *d*-FORMS OF THE MACROSCOPIC BALANCES

The estimation of E_v in the mechanical energy balance and Q in the total energy balance often presents some difficulties in nonisothermal systems.

For example, for E_v, consider the following two nonisothermal situations:

a. For liquids, the average flow velocity in a tube of constant cross section is nearly constant. However, the viscosity may change markedly in the direction of the flow because of the temperature changes, so that f in Eq. 7.5-9 changes with distance. Hence Eq. 7.5-9 cannot be applied to the entire pipe.

b. For gases, the viscosity does not change much with pressure, so that the local Reynolds number and local friction factor are nearly constant for ducts of constant cross section. However, the average velocity may change considerably along the duct as a result of the change in density with temperature. Hence Eq. 7.5-9 cannot be applied to the entire duct.

Similarly for pipe flow with the wall temperature changing with distance, it may be necessary to use local heat transfer coefficients. For such a situation, we can write Eq. 15.1-3 on an incremental basis and generate a differential equation. Or the cross sectional area of the conduit may be changing with downstream distance, and this situation also results in a need for handling the problem on an incremental basis.

It is therefore useful to rewrite the steady-state macroscopic mechanical energy balance and the total energy balance by taking planes 1 and 2 to be a differential distance dl apart. We then obtain what we call the "*d*-forms" of the balances:

The *d*-Form of the Mechanical Energy Balance

If we take planes 1 and 2 to be a differential distance apart, then we may write Eq. 15.2-2 in the following differential form (assuming flat velocity profiles):

$$d(\tfrac{1}{2}v^2) + g\,dh + \frac{1}{\rho}\,dp = d\hat{W} - d\hat{E}_v \tag{15.4-1}$$

Then using Eq. 7.5-9 for a differential length dl, we write

$$v\,dv + g\,dh + \frac{1}{\rho}\,dp = d\hat{W} - \tfrac{1}{2}v^2\frac{f}{R_h}\,dl \tag{15.4-2}$$

in which f is the local friction factor, and R_h is the local value of the mean hydraulic radius. In most applications we omit the $d\hat{W}$ term, since work is usually done at isolated points along the flow path. The term $d\hat{W}$ would be needed in tubes with extensible walls, magnetically driven flows, or systems with transport by rotating screws.

The d-Form of the Total Energy Balance

If we write Eq. 15.1-3 in differential form, we have (with flat velocity profiles)

$$d(\tfrac{1}{2}v^2) + g\,dh + d\hat{H} = d\hat{Q} + d\hat{W} \tag{15.4-3}$$

Then using Eq. 9.8-7 for $d\hat{H}$ and Eq. 14.1-8 for $d\hat{Q}$ we get

$$v\,dv + g\,dh + \hat{C}_p dT + \left[\hat{V} - T\left(\frac{\partial \hat{V}}{\partial T}\right)_p\right]dp = \frac{U_{loc}Z\Delta T}{w}\,dl + d\hat{W} \tag{15.4-4}$$

in which U_{loc} is the local overall heat transfer coefficient, Z is the corresponding local conduit perimeter, and ΔT is the local temperature difference between the fluids inside and outside of the conduit.

The examples that follow illustrate applications of Eqs. 15.4-2 and 15.4-4.

EXAMPLE 15.4-1

Parallel- or Counter-Flow Heat Exchangers

It is desired to describe the performance of the simple double-pipe heat exchanger shown in Fig. 15.4-1 in terms of the heat transfer coefficients of the two streams and the thermal resistance of the pipe wall. The exchanger consists essentially of two coaxial pipes with one fluid stream flowing through the inner pipe and another in the annular space; heat is transferred across the wall of the inner pipe. Both streams may flow in the same direction, as indicated in the figure, but normally it is more efficient to use counter flow—that is, to reverse the direction of one stream so that either w_h or w_c is negative. Steady-state turbulent flow may be assumed, and the heat losses to the surroundings may be neglected. Assume further that the local overall heat transfer coefficient is constant along the exchanger.

SOLUTION

(a) Macroscopic energy balance for each stream as a whole. We designate quantities referring to the hot stream with a subscript h and the cold stream with subscript c. The steady-state energy balance in Eq. 15.1-3 becomes, for negligible changes in kinetic and potential energy,

$$w_h(\hat{H}_{h2} - \hat{H}_{h1}) = Q_h \tag{15.4-5}$$

$$w_c(\hat{H}_{c2} - \hat{H}_{c1}) = Q_c \tag{15.4-6}$$

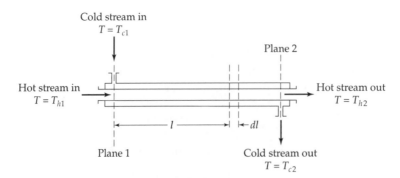

Fig. 15.4-1. A double-pipe heat exchanger.

Because there is no heat loss to the surroundings, $Q_h = -Q_c$. For incompressible liquids with a pressure drop that is not too large, or for ideal gases, Eq. 9.8-8 gives for constant $\hat{C}_p$ the relation $\Delta\hat{H} = \hat{C}_p\Delta T$. Hence Eqs. 15.4-5 and 6 can be rewritten as

$$w_h\hat{C}_{ph}(T_{h2} - T_{h1}) = Q_h \tag{15.4-7}$$

$$w_c\hat{C}_{pc}(T_{c2} - T_{c1}) = Q_c = -Q_h \tag{15.4-8}$$

(b) d-form of the macroscopic energy balance. Application of Eq. 15.4-4 to the hot stream gives

$$\hat{C}_{ph}dT_h = \frac{U_0(2\pi r_0)(T_c - T_h)}{w_h}dl \tag{15.4-9}$$

where r_0 is the outside radius of the inner tube, and U_0 is the overall heat transfer coefficient based on the radius r_0 (see Eq. 14.1-8).

Rearrangement of Eq. 15.4-9 gives

$$\frac{dT_h}{T_c - T_h} = U_0\frac{(2\pi r_0)dl}{w_h\hat{C}_{ph}} \tag{15.4-10}$$

The corresponding equation for the cold stream is

$$-\frac{dT_c}{T_c - T_h} = U_0\frac{(2\pi r_0)dl}{w_c\hat{C}_{pc}} \tag{15.4-11}$$

Adding Eqs. 15.4-10 and 11 gives a differential equation for the temperature difference of the two fluids as a function of l:

$$-\frac{d(T_h - T_c)}{T_h - T_c} = U_0\left(\frac{1}{w_h\hat{C}_{ph}} + \frac{1}{w_c\hat{C}_{pc}}\right)(2\pi r_0)dl \tag{15.4-12}$$

By assuming that U_0 is independent of l and integrating from plane 1 to plane 2, we get

$$\ln\left(\frac{T_{h1} - T_{c1}}{T_{h2} - T_{c2}}\right) = U_0\left(\frac{1}{w_h\hat{C}_{ph}} + \frac{1}{w_c\hat{C}_{pc}}\right)(2\pi r_0)L \tag{15.4-13}$$

This expression relates the terminal temperatures to the stream rates and exchanger dimensions, and it can thus be used to describe the performance of the exchanger. However, it is conventional to rearrange Eq. 15.4-13 by taking advantage of the steady-state energy balances in Eq. 15.4-7 and 8. We solve each of these equations for $w\hat{C}_p$ and substitute the results into Eq. 15.4-13 to obtain

$$Q_c = U_0(2\pi r_0 L)\left(\frac{(T_{h2} - T_{c2}) - (T_{h1} - T_{c1})}{\ln[(T_{h2} - T_{c2})/(T_{h1} - T_{c1})]}\right) \tag{15.4-14}$$

or

$$Q_c = U_0 A_0 (T_h - T_c)_{\ln} \tag{15.4-15}$$

Here A_0 is the total outer surface of the inner tube, and $(T_h - T_c)_{\ln}$ is the "logarithmic mean temperature difference" between the two streams. Equations 15.4-14 and 15 describe the rate of heat exchange between the two streams and find wide application in engineering practice. Note that the stream rates do not appear explicitly in these equations, which are valid for both parallel-flow and counter-flow exchangers (see Problem 15A.1).

From Eqs. 15.4-10 and 11 we can also get the stream temperatures as functions of l if desired. Considerable care must be used in applying the results of this example to laminar flow, for which the variation of the overall heat transfer coefficient may be quite large. An example of a problem with variable U_0 is Problem 15B.1.

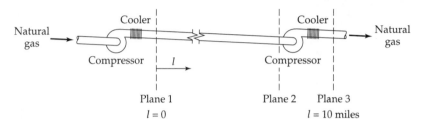

Fig. 15.4-2. Pumping a compressible fluid through a pipeline.

EXAMPLE 15.4-2

Power Requirement for Pumping a Compressible Fluid through a Long Pipe

A natural gas, which may be considered to be pure methane, is to be pumped through a long, smooth pipeline with a 2-ft inside diameter. The gas enters the line at 100 psia with a velocity of 40 ft/s and at the ambient temperature of 70°F. Pumping stations are provided every 10 miles along the line, and at each of these stations the gas is recompressed and cooled to its original temperature and pressure (see Fig. 15.4-2). Estimate the power that must be expended on the gas at each pumping station, assuming ideal gas behavior, flat velocity profiles, and negligible changes in elevation.

SOLUTION

We find it convenient to consider the pipe and compressor separately. First we apply Eq. 15.4-2 to a length dl of the pipe. We then integrate this equation between planes 1 and 2 to obtain the unknown pressure p_2. Once this is known, we may apply Eq. 15.2-2 to the system between planes 2 and 3 to obtain the work done by the pump.

(a) Flow through the pipe. For this portion of the system, Eq. 15.4-2 simplifies to

$$v\,dv + \frac{1}{\rho}\,dp + \frac{2v^2 f}{D}\,dl = 0 \tag{15.4-16}$$

where D is the pipe diameter. Since the pipe is quite long, we assume that the fluid is isothermal at 70°F. We may then eliminate both v and ρ from Eq. 15.4-16 by use of the assumed equation of state, $p = \rho RT/M$, and the macroscopic mass balance, which may be written $\rho v = \rho_1 v_1$. With ρ and v written in terms of the pressure, Eq. 15.4-16 becomes

$$-\frac{1}{p}\,dp + \frac{RT_1}{M(p_1 v_1)^2}\,p\,dp + \frac{2f}{D}\,dl = 0 \tag{15.4-17}$$

We pointed out in §1.3 that the viscosity of ideal gases is independent of the pressure. From this it follows that the Reynolds number of the gas, Re $= Dw/S\mu$, and hence the friction factor f, must be constants. We may then integrate Eq. 15.4-17 to obtain

$$-\ln\frac{p_2}{p_1} + \frac{1}{2}\left[\left(\frac{p_2}{p_1}\right)^2 - 1\right]\frac{RT_1}{Mv_1^2} + \frac{2fL}{D} = 0 \tag{15.4-18}$$

This equation gives p_2 in terms of quantities that are already known, except for f, which is easily calculated: the kinematic viscosity of methane at 100 psi and 70 F is about 2.61×10^{-5} ft²/s, and therefore Re $= Dv/\nu = (200$ ft$)(40$ ft/s$)/(2.61$ ft²/s$) = 3.07 \times 10^6$. The friction factor can then be estimated to be 0.0025 (see Fig. 6.2-2).

Substituting numerical values into Eq. 15.4-18, we get

$$-\ln\frac{p_2}{p_1} + \frac{1}{2}\left[\left(\frac{p_2}{p_1}\right)^2 - 1\right]\frac{(1545)(530)(32.2)}{(16.04)(40)^2} + \frac{(2)(0.0025)(52,800)}{(2.00)} = 0 \tag{15.4-19}$$

or

$$-\ln \frac{p_2}{p_1} + 513\left[\left(\frac{p_2}{p_1}\right)^2 - 1\right] + 132 = 0 \qquad (15.4\text{-}20)$$

By solving this equation with $p_1 = 100$ psia, we obtain $p_2 = 86$ psia.

(b) Flow through the compressor. We are now ready to apply the mechanical energy balance to the compressor. We start by putting Eq. 15.2-2 into the form

$$\hat{W}_m = \tfrac{1}{2}(v_3^2 - v_2^2) + \int_{p_2}^{p_3} \frac{1}{\rho}\, dp + \hat{E}_v \qquad (15.4\text{-}21)$$

To evaluate the integral in this equation, we assume that the compression is adiabatic and further that $\hat{E}_v$ between planes 2 and 3 can be neglected. We may use Eq. 15.2-5 to rewrite Eq. 15.2-21 as

$$\hat{W}_m = \tfrac{1}{2}(v_3^2 - v_2^2) + \frac{p_2^{1/\gamma}}{\rho_2} \int_{p_2}^{p_3} p^{-1/\gamma}\, dp$$

$$= \frac{v_1^2}{2}\left[1 - \left(\frac{p_1}{p_2}\right)^2\right] + \frac{RT_2}{M}\frac{\gamma}{\gamma - 1}\left[\left(\frac{p_1}{p_2}\right)^{(\gamma-1)/\gamma} - 1\right] \qquad (15.4\text{-}22)$$

in which $\hat{W}_m$ is the energy required of the compressor. By substituting numerical values into Eq. 15.4-22, we get

$$\hat{W}_m = \frac{(40)^2}{2(32.2)}[1 - (1.163)^2] + \frac{(1545)(530)}{16}\frac{1.3}{0.3}(1.163^{0.3/1.3} - 1)$$

$$= -9 + 7834 = 7825 \text{ ft lb}_f/\text{lb}_m \qquad (15.4\text{-}23)$$

The power required to compress the fluid is

$$w\hat{W}_m = \left(\frac{\pi D^2}{4}\right)\left(\frac{p_1 M}{RT_1}\right)v_1\hat{W}_m$$

$$= \frac{\pi(100)(16.04)(40)}{(10.73)(530)}(7825) \text{ ft lb}_f/\text{s}$$

$$= 277{,}000 \text{ ft lb}_f/\text{s} = 504 \text{ hp} \qquad (15.4\text{-}24)$$

The power required would be virtually the same if the flow in the pipeline were adiabatic (see Problem 15A.2).

The assumptions used here—assuming the compression to be adiabatic and neglecting the viscous dissipation—are conventional in the design of compressor–cooler combinations. Note that the energy required to run the compressor is greater than the calculated work, $\hat{W}_m$, by (i) $\hat{E}_v$ between planes 2 and 3, (ii) mechanical losses in the compressor itself, and (iii) errors in the assumed p–ρ path. Normally the energy required at the pump shaft is at least 15 to 20% greater than $\hat{W}_m$.

§15.5   USE OF THE MACROSCOPIC BALANCES TO SOLVE UNSTEADY-STATE PROBLEMS AND PROBLEMS WITH NONFLAT VELOCITY PROFILES

In Table 15.5-1 we summarize all five macroscopic balances for unsteady state and nonflat velocity profiles, and for systems with multiple entry and exit ports. One practically never needs to use these balances in this degree of completeness, but it is convenient to have the entire set of equations collected in one place. We illustrate their use in the examples that follow.

Table 15.5-1 Unsteady-State Macroscopic Balances for Flow in Nonisothermal Systems

Mass:
$$\frac{d}{dt}m_{tot} = \sum w_1 - \sum w_2 = \sum \rho_1 \langle v_1 \rangle S_1 - \sum \rho_2 \langle v_2 \rangle S_2 \tag{A}$$

Momentum:
$$\frac{d}{dt}\mathbf{P}_{tot} = \sum \left(\frac{\langle v_1^2 \rangle}{\langle v_1 \rangle}w_1 + p_1 S_1\right)\mathbf{u}_1 - \sum \left(\frac{\langle v_2^2 \rangle}{\langle v_2 \rangle}w_2 + p_2 S_2\right)\mathbf{u}_2 + m_{tot}\mathbf{g} - \mathbf{F}_{f \to s} \tag{B}$$

Angular momentum:
$$\frac{d}{dt}\mathbf{L}_{tot} = \sum \left(\frac{\langle v_1^2 \rangle}{\langle v_1 \rangle}w_1 + p_1 S_1\right)[\mathbf{r}_1 \times \mathbf{u}_1] - \sum \left(\frac{\langle v_2^2 \rangle}{\langle v_2 \rangle}w_2 + p_2 S_2\right)[\mathbf{r}_2 \times \mathbf{u}_2] + \mathbf{T}_{ext} - \mathbf{T}_{f \to s} \tag{C}$$

Mechanical energy:
$$\frac{d}{dt}(K_{tot} + \Phi_{tot}) = \sum \left(\frac{1}{2}\frac{\langle v_1^3 \rangle}{\langle v_1 \rangle} + gh_1 + \frac{p_1}{\rho_1}\right)w_1 - \sum \left(\frac{1}{2}\frac{\langle v_2^3 \rangle}{\langle v_2 \rangle} + gh_2 + \frac{p_2}{\rho_2}\right)w_2 + W_m - E_c - E_v \tag{D}$$

(Total) energy:
$$\frac{d}{dt}(K_{tot} + \Phi_{tot} + U_{tot}) = \sum \left(\frac{1}{2}\frac{\langle v_1^3 \rangle}{\langle v_1 \rangle} + gh_1 + \hat{H}_1\right)w_1 - \sum \left(\frac{1}{2}\frac{\langle v_2^3 \rangle}{\langle v_2 \rangle} + gh_2 + \hat{H}_2\right)w_2 + W_m + Q \tag{E}$$

Notes:

[a] $\sum w_1 = w_{1a} + w_{1b} + w_{1c} + \cdots$, where $w_{1a} = \rho_{1a}v_{1a}S_{1a}$, and so on.

[b] h_1 and h_2 are elevations above an arbitrary datum plane.

[c] $\hat{H}_1$ and $\hat{H}_2$ are enthalpies per unit mass relative to some arbitrarily chosen reference state; the formula for $\hat{H}$ is given in Eq. 9.8-8.

[d] All equations are written for compressible flow; for incompressible flow, $E_c = 0$. The quantities E_c and E_v are defined in Eqs. 7.3-3 and 4.

[e] $\mathbf{u}_1$ and $\mathbf{u}_2$ are unit vectors in the direction of flow.

EXAMPLE 15.5-1

Heating of a Liquid in an Agitated Tank[1]

A cylindrical tank capable of holding 1000 ft³ of liquid is equipped with an agitator having sufficient power to keep the liquid contents at a uniform temperature (see Fig. 15.5-1). Heat is transferred to the contents by means of a coil arranged in such a way that the area available for heat transfer is proportional to the quantity of liquid in the tank. This heating coil consists of 10 turns, 4 ft in diameter, of 1-in. o.d. tubing. Water at 20°C is fed into this tank at a rate of 20 lb/min, starting with no water in the tank at time $t = 0$. Steam at 105°C flows through the

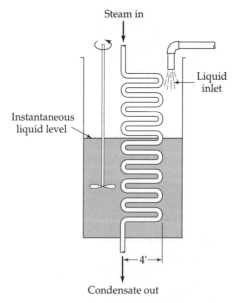

Steam in

Liquid inlet

Instantaneous liquid level

—4'—

Condensate out

Fig. 15.5-1. Heating of a liquid in a tank with a variable liquid level.

---

[1] This problem is taken in modified form from W. R. Marshall, Jr., and R. L. Pigford, *Applications of Differential Equations to Chemical Engineering Problems*, University of Delaware Press, Newark, Del. (1947), pp. 16–18.

heating coil, and the overall heat transfer coefficient is 100 Btu/hr · ft² · F. What is the temperature of the water when the tank is filled?

SOLUTION

We shall make the following assumptions:

a. The steam temperature is uniform throughout the coil.

b. The density and heat capacity do not change very much with temperature.

c. The fluid is approximately incompressible so that $\hat{C}_p \approx \hat{C}_V$.

d. The agitator maintains uniform temperature throughout the liquid.

e. The heat transfer coefficient is independent of position and time.

f. The walls of the tank are perfectly insulated so that no heat loss occurs.

We select the fluid within the tank as the system to be considered, and we make a time-dependent energy balance over this system. Such a balance is provided by Eq. (E) of Table 15.5-1. On the left side of the equation the time rates of change of kinetic and potential energies can be neglected relative to that of the internal energy. On the right side we can normally omit the work term, and the kinetic and potential energy terms can be discarded, since they will be small compared with the other terms. Inasmuch as there is no outlet stream, we can set w_2 equal to zero. Hence for this system the total energy balance simplifies to

$$\frac{d}{dt} U_{\text{tot}} = w_1 \hat{H}_1 + Q \tag{15.5-1}$$

This states that the internal energy of the system increases because of the enthalpy added by the incoming fluid, and because of the addition of heat through the steam coil.

Since U_{tot} and $\hat{H}_1$ cannot be given absolutely, we now select the inlet temperature T_1 as the thermal datum plane. Then $\hat{H}_1 = 0$ and $U_{\text{tot}} = \rho \hat{C}_V V (T - T_1) \approx \rho \hat{C}_p V (T - T_1)$, where T and V are the instantaneous temperature and volume of the liquid. Furthermore, the rate of heat addition to the liquid Q is given by $Q = U_0 A (T_s - T)$, in which T_s is the steam temperature, and A is the instantaneous heat transfer area. Hence Eq. 15.5-1 becomes

$$\rho \hat{C}_p \frac{d}{dt} V (T - T_1) = U_0 A (T_s - T) \tag{15.5-2}$$

The expressions for $V(t)$ and $A(t)$ are

$$V(t) = \frac{w_1}{\rho} t \qquad A(t) = \frac{V}{V_0} A_0 = \frac{w_1 t}{\rho V_0} A_0 \tag{15.5-3}$$

in which V_0 and A_0 are the volume and heat transfer area when the tank is full. Hence the energy balance equation becomes

$$w_1 \hat{C}_p t \frac{d}{dt} (T - T_1) + w_1 \hat{C}_p (T - T_1) = \frac{w_1 t}{\rho V_0} U_0 A_0 (T_s - T) \tag{15.5-4}$$

which is to be solved with the initial condition that $T = T_1$ at $t = 0$.

The equation is more easily solved in dimensionless form. We divide both sides by $w_1 \hat{C}_p (T_s - T_1)$ to get

$$t \frac{d}{dt} \left(\frac{T - T_1}{T_s - T_1} \right) + \left(\frac{T - T_1}{T_s - T_1} \right) = \frac{U_0 A_0 t}{\rho \hat{C}_p V_0} \left(\frac{T_s - T}{T_s - T_1} \right) \tag{15.5-5}$$

This equation suggests that suitable definitions of dimensionless temperature and time are

$$\Theta = \left(\frac{T - T_1}{T_s - T_1} \right) \quad \text{and} \quad \tau = \frac{U_0 A_0 t}{\rho \hat{C}_p V_0} \tag{15.5-6, 7}$$

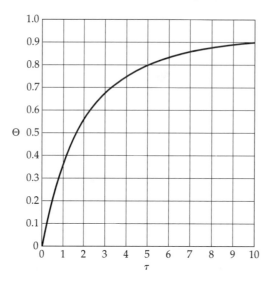

Fig. 15.5-2. Plot of dimensionless temperature, $\Theta = (T - T_1)/(T_s - T_1)$, versus dimensionless time, $\tau = (U_0 A_0/\rho \hat{C}_p V_0)t$, according to Eq. 15.5-10. [W. R. Marshall and R. L. Pigford, *Application of Differential Equations to Chemical Engineering*, University of Delaware Press, Newark, Del. (1947), p. 18.]

Then the equation in Eq. 15.5-5 becomes after some rearranging

$$\frac{d\Theta}{d\tau} + \left(1 + \frac{1}{\tau}\right)\Theta = 1 \tag{15.5-8}$$

and the initial condition requires that $\Theta = 0$ at $\tau = 0$.

This is a first-order linear differential equation whose solution is (see Eq. C.1-2)

$$\Theta = 1 - \frac{1 - Ce^{-\tau}}{\tau} \tag{15.5-9}$$

The constant of integration, C, can be obtained from the initial condition after first multiplying Eq. 15.5-9 by τ. In that way it is found that $C = 1$, so that the final solution is

$$\Theta = 1 - \frac{1 - e^{-\tau}}{\tau} \tag{15.5-10}$$

This function is shown in Fig. 15.5-2.

Finally, the temperature T_0 of the liquid in the tank, when it has been filled, is given by Eq. 15.5-10 when $t = \rho V_0/w_1$ (from Eq. 15.5-3) or $\tau = U_0 A_0/w_1 \hat{C}_p$ (from Eq. 15.5-7). Therefore, in terms of the original variables,

$$\frac{T_0 - T_1}{T_s - T_1} = 1 - \frac{1 - \exp(-U_0 A_0/w_1 \hat{C}_p)}{U_0 A_0/w_1 \hat{C}_p} \tag{15.5-11}$$

Thus it can be seen that the final liquid temperature is determined entirely by the dimensionless group $U_0 A_0/w_1 \hat{C}_p$ which, for this problem, has the value of 2.74. Knowing this we can find from Eq. 15.5-11 that $(T_0 - T_1)/(T_s - T_1) = 0.659$, whence $T_0 = 76°C$.

EXAMPLE 15.5-2

Operation of a Simple Temperature Controller

A well-insulated agitated tank is shown in Fig. 15.5-3. Liquid enters at a temperature $T_1(t)$, which may vary with time. It is desired to control the temperature, $T_2(t)$, of the fluid leaving the tank. It is presumed that the stirring is sufficiently thorough that the temperature in the tank is uniform and equal to the exit temperature. The volume of the liquid in the tank, V, and the mass rate of liquid flow, w, are both constant.

To accomplish the desired control, a metallic electric heating coil of surface area A is placed in the tank, and a temperature-sensing element is placed in the exit steam to measure $T_2(t)$. These devices are connected to a temperature controller that supplies energy to the heating coil at a rate $Q_e = b(T_{max} - T_2)$, in which T_{max} is the maximum temperature for which the controller is designed to operate, and b is a known parameter. It may be assumed that the

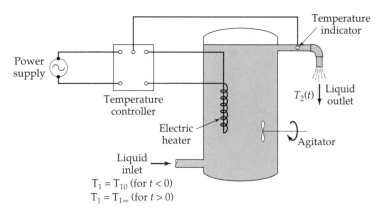

Fig. 15.5-3. An agitated tank with a temperature controller.

liquid temperature $T_2(t)$ is always less than T_{max} in normal operation. The heating coil supplies energy to the liquid in the tank at a rate $Q = UA(T_c - T_2)$, where U is the overall heat transfer coefficient between the coil and the liquid, and T_c is the instantaneous coil temperature, considered to be uniform.

Up to time $t = 0$, the system has been operating at steady state with liquid inlet temperature $T_1 = T_{10}$ and exit temperature $T_2 = T_{20}$. At time $t = 0$, the inlet stream temperature is suddenly increased to $T_1 = T_{1\infty}$ and held there. As a consequence of this disturbance, the tank temperature will begin to rise, and the temperature indicator in the outlet stream will signal the controller to decrease the power supplied to the heating coil. Ultimately, the liquid temperature in the tank will attain a new steady-state value $T_{2\infty}$. It is desired to describe the behavior of the liquid temperature $T_2(t)$. A qualitative sketch showing the various temperatures is given in Fig. 15.5-4.

SOLUTION   We first write the unsteady-state macroscopic energy balances [Eq. (E) of Table 15.5-1] for the liquid in the tank and for the heating coil:

(liquid)
$$\rho \hat{C}_p V \frac{dT_2}{dt} = w \hat{C}_p (T_1 - T_2) + UA(T_c - T_2) \tag{15.5-12}$$

(coil)
$$\rho_c \hat{C}_{pc} V_c \frac{dT_c}{dt} = b(T_{max} - T_2) - UA(T_c - T_2) \tag{15.5-13}$$

Note that in applying the macroscopic energy balance to the liquid, we have neglected kinetic and potential energy changes as well as the power input to the agitator.

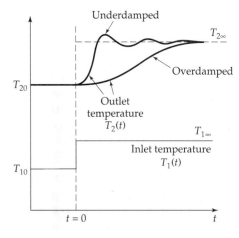

Fig. 15.5-4. Inlet and outlet temperatures as functions of time.

(a) *Steady-state behavior for t < 0.* When the time derivatives in Eqs. 15.5-12 and 13 are set equal to zero and the equations added, we get for $t < 0$, where $T_1 = T_{10}$:

$$T_{20} = \frac{w\hat{C}_p T_{10} + bT_{max}}{w\hat{C}_p + b} \qquad (15.5\text{-}14)$$

Then from Eq. 15.5-13 we can get the initial temperature of the coil

$$T_{c0} = T_{20}\left(1 - \frac{b}{UA}\right) + \frac{bT_{max}}{UA} \qquad (15.5\text{-}15)$$

(b) *Steady-state behavior for t → ∞.* When similar operations are performed with $T_1 = T_{1\infty}$, we get

$$T_{2\infty} = \frac{w\hat{C}_p T_{1\infty} + bT_{max}}{w\hat{C}_p + b} \qquad (15.5\text{-}16)$$

and

$$T_{c\infty} = T_{2\infty}\left(1 - \frac{b}{UA}\right) + \frac{bT_{max}}{UA} \qquad (15.5\text{-}17)$$

for the final temperature of the coil.

(c) *Unsteady state behavior for t > 0.* It is convenient to define dimensionless variables using the steady-state quantities for $t < 0$ and $t \to \infty$:

$$\Theta_2 = \frac{T_2 - T_{2\infty}}{T_{20} - T_{2\infty}} = \text{dimensionless liquid temperature} \qquad (15.5\text{-}18)$$

$$\Theta_c = \frac{T_c - T_{c\infty}}{T_{c0} - T_{c\infty}} = \text{dimensionless coil temperature} \qquad (15.5\text{-}19)$$

$$\tau = \frac{UAt}{\rho\hat{C}_p V} = \text{dimensionless time} \qquad (15.5\text{-}20)$$

In addition we define three dimensionless parameters:

$$R = \rho\hat{C}_p V / \rho_c \hat{C}_{pc} V_c = \text{ratio of thermal capacities} \qquad (15.5\text{-}21)$$

$$F = w\hat{C}_p / UA = \text{flow-rate parameter} \qquad (15.5\text{-}22)$$

$$b/UA = \text{controller parameter} \qquad (15.5\text{-}23)$$

In terms of these quantities, the unsteady-state balances in Eqs. 15.5-12 and 13 become (after considerable manipulation):

$$\frac{d\Theta_2}{d\tau} = -(1 + F)\Theta_2 + (1 - B)\Theta_c \qquad (15.5\text{-}24)$$

$$\frac{d\Theta_c}{d\tau} = R(\Theta_2 - \Theta_c) \qquad (15.5\text{-}25)$$

elimination of Θ_c between this pair of equations gives a single second-order linear ordinary differential equation for the exit liquid temperature as a function of time:

$$\frac{d^2\Theta_2}{d\tau^2} + (1 + R + F)\frac{d\Theta_2}{d\tau} + R(B + F)\Theta_2 = 0 \qquad (15.5\text{-}26)$$

This equation has the same form as that obtained for the damped manometer in Eq. 7.7-21 (see also Eq. C.1-7). The general solution is then of the form of Eq. 7.7-23 or 24:

$$\Theta_2 = C_+ \exp(m_+\tau) + C_- \exp(m_-\tau) \qquad (m_+ \neq m_-) \qquad (15.5\text{-}27)$$

$$\Theta_2 = C_1 \exp m\tau + C_2\tau \exp m\tau \qquad (m_+ = m_- = m) \qquad (15.5\text{-}28)$$

where

$$m_\pm = \tfrac{1}{2}[-(1 + R + F) \pm \sqrt{(1 + R + F)^2 - 4R(B + F)}] \tag{15.5-29}$$

Thus by analogy with Example 7.7-2, the fluid exit temperature may approach its final value as a monotone increasing function (overdamped or critically damped) or with oscillations (underdamped). The system parameters appear in the dimensionless time variable, as well as in the parameters B, F, and R. Therefore, numerical calculations are needed to determine whether in a particular system the temperature will oscillate or not.

EXAMPLE 15.5-3

Flow of Compressible Fluids Through Head Meters

Extend the development of Example 7.6-5 to the steady flow of compressible fluids through orifice meters and Venturi tubes.

SOLUTION

We begin, as in Example 7.6-5, by writing the steady-state mass and mechanical energy balances between reference planes 1 and 2 of the two flow meters shown in Fig. 15.5-5. For compressible fluids, these may be expressed as

$$w = \rho_1 \langle v_1 \rangle S_1 = \rho_2 \langle v_2 \rangle S_2 \tag{15.5-30}$$

$$\frac{\langle v_2 \rangle^2}{2\alpha_2} - \frac{\langle v_1 \rangle^2}{2\alpha_1} + \int_1^2 \frac{1}{\rho} \, dp + \tfrac{1}{2}\langle v_2 \rangle^2 e_v = 0 \tag{15.5-31}$$

in which the quantities $\alpha_i = \langle v_i \rangle^3 / \langle v_i^3 \rangle$ are included to allow for the replacement of the average of the cube by the cube of the average.

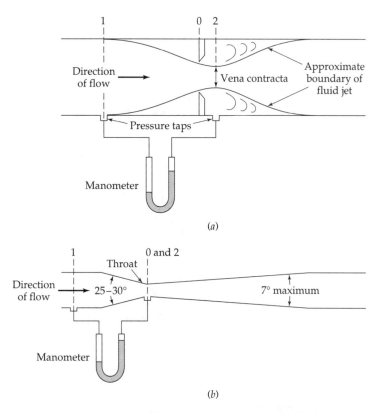

Fig. 15.5-5. Measurement of mass flow rate by use of (*a*) an orifice meter, and (*b*) a Venturi tube.

We next eliminate $\langle v_1 \rangle$ and $\langle v_2 \rangle$ from the above two equations to get an expression for the mass flow rate:

$$w = \rho_2 S_2 \sqrt{\frac{-2\alpha_2 \int_1^2 (1/\rho)dp}{1 - (\alpha_2/\alpha_1)(\rho_2 S_2/\rho_1 S_1)^2 + \alpha_2 e_v}} \tag{15.5-32}$$

We now repeat the assumptions of Example 7.6-5: (i) $e_v = 0$, (ii) $\alpha_1 = 1$, and (iii) $\alpha_2 = (S_0/S_2)^2$. Then Eq. 15.5-32 becomes

$$w = C_d \rho_2 S_0 \sqrt{\frac{-2 \int_1^2 (1/\rho)dp}{1 - (\rho_2 S_0/\rho_1 S_1)^2}} \tag{15.5-33}$$

The empirical "discharge coefficient," C_d, is included in this equation to permit correction of this expression for errors introduced by the three assumptions and must be determined experimentally.

For *Venturi meters*, it is convenient to put plane 2 at the point of minimum cross section of the meter so that $S_2 = S_0$. Then α_2 is very nearly unity, and it has been found experimentally that C_d is almost the same for compressible and incompressible fluids—that is, about 0.98 for well designed Venturi meters. For *orifice meters*, the degree of contraction of a compressible fluid stream at plane 2 is somewhat less than for incompressible fluids, especially at high flow rates, and a different discharge coefficient[2] is required.

In order to use Eq. 15.5-33, the fluid density must be known as a function of pressure. That is, one must know both the path of the expansion and the equation of state of the fluid. In most cases the assumption of frictionless adiabatic behavior appears to be acceptable. For ideal gases, one may write $p\rho^{-\gamma} = \text{constant}$, where $\gamma = C_p/C_V$ (see Eq. 15.2-5). Then Eq. 15.5-33 becomes

$$w = C_d \rho_2 S_0 \sqrt{\frac{2(p_1/\rho_1)[\gamma/(\gamma-1)][1 - (p_2/p_1)^{(\gamma-1)/\gamma}]}{1 - (S_0/S_1)^2(p_2/p_1)^{2/\gamma}}} \tag{15.5-34}$$

This formula expresses the mass flow rate as a function of measurable quantities and the discharge coefficient. Values of the latter may be found in engineering handbooks.[2]

EXAMPLE 15.5-4

Free Batch Expansion of a Compressible Fluid

A compressible gas, initially at $T = T_0$, $p = p_0$, and $\rho = \rho_0$, is discharged from a large stationary insulated tank through a small convergent nozzle, as shown in Fig. 15.5-6. Show how the fractional remaining mass of fluid in the tank, ρ/ρ_0, may be determined as a function of time. Develop working equations, assuming that the gas is ideal.

SOLUTION

For convenience, we divide the tank into two parts, separated by the surface 1 as shown in the figure. We assume that surface 1 is near enough to the tank exit that essentially all of the fluid mass is to left of it, but far enough from the exit that the fluid velocity through the surface 1 is negligible. We further assume that the average fluid properties to the left of 1 are identical with those at surface 1. We now consider the behavior of these two parts of the system separately.

(a) *The bulk of the fluid in the tank.* For the region to the left of surface 1, the unsteady state mass balance in Eq. (A) of Table 15.5-1 is

$$\frac{d}{dt}(\rho_1 V) = -w_1 \tag{15.5-35}$$

---

[2] R. H. Perry, D. W. Green, and J. O. Maloney, *Chemical Engineers' Handbook*, 7th Edition, McGraw-Hill, New York (1997); see also, Chapter 15 of *Handbook of Fluid Dynamics and Fluid Machinery* (J. A. Schertz and A. E. Fuhs, eds.), Wiley, New York (1996).

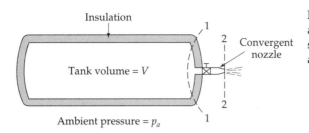

Fig. 15.5-6. Free batch expansion of a compressible fluid. The sketch shows the locations of surfaces 1 and 2.

For the same region, the energy balance of Eq. (E) of Table 15.5-1 becomes

$$\frac{d}{dt}\left(\rho_1 V(\hat{U}_1 + \hat{\Phi}_1)\right) = -w_1\left(\hat{U} + \frac{p_1}{\rho_1} + \hat{\Phi}_1\right) \tag{15.5-36}$$

in which V is the total volume in the system being considered, and w_1 is the mass rate of flow of gas leaving the system. In writing this equation, we have neglected the kinetic energy of the fluid.

Substituting the mass balance into both sides of the energy equation gives

$$\rho_1\left(\frac{d\hat{U}_1}{dt} + \frac{d\hat{\Phi}_1}{dt}\right) = \frac{p_1}{\rho_1}\frac{d\rho_1}{dt} \tag{15.5-37}$$

For a stationary system under the influence of no external forces other than gravity, $d\hat{\Phi}_1/dt = 0$, so that Eq. 15.5-37 becomes

$$\frac{d\hat{U}_1}{d\rho_1} = \frac{p_1}{\rho_1^2} \tag{15.5-38}$$

This equation may be combined with the thermal and caloric equations of state for the fluid in order to obtain $p_1(\rho_1)$ and $T_1(\rho_1)$. We find, thus, that the condition of the fluid in the tank depends only on the degree to which the tank has been emptied and not on the rate of discharge. For the special case of an ideal gas with constant $\hat{C}_V$, for which $d\hat{U} = \hat{C}_V dT$ and $p = \rho RT/M$, we may integrate Eq. 15.5-38 to obtain

$$p_1\rho_1^{-\gamma} = p_0\rho_0^{-\gamma} \tag{15.5-39}$$

in which $\gamma = C_p/C_V$. This result also follows from Eq. 11.4-57.

(b) *Discharge of the gas through the nozzle.* For the sake of simplicity we assume here that the flow between surfaces 1 and 2 is both frictionless and adiabatic. Also, since w_1 is not far different from w_2, it is also appropriate to consider at any one instant that the flow is quasi-steady-state. Then we can use the macroscopic mechanical energy balance in the form of Eq. 15.2-2 with the second, fourth, and fifth terms omitted. That is,

$$\tfrac{1}{2}v_2^2 + \int_1^2 \frac{1}{\rho}\,dp = 0 \tag{15.5-40}$$

Since we are dealing with an ideal gas, we may use the result in Eq. 15.5-34 to get the instantaneous discharge rate. Since in this problem the ratio S_2/S_1 is very small and its square is even smaller, we can replace the denominator under the square root sign in Eq. 15.5-34 by unity. Then the ρ_2 outside the square root sign is moved inside and use is made of Eq. 15.5-39. This gives

$$w_2 = -V\frac{d\rho_1}{dt} = S_2\sqrt{2p_1\rho_1[\gamma/(\gamma-1)][(p_2/p_0)^{2/\gamma} - (p_2/p_0)^{(\gamma+1)/\gamma}]} \tag{15.5-41}$$

in which S_2 is the cross-sectional area of the nozzle opening.

Now we use Eq. 15.5-39 to eliminate p_1 from Eq. 15.5-41. Then we have a first-order differential equation for ρ_1, which may be integrated to give

$$t = \frac{V/S_2}{\sqrt{2(p_0/\rho_0)[\gamma/(\gamma - 1)]}} \int_{\rho_1/\rho_0}^{1} \frac{d(\rho_1/\rho_0)}{\sqrt{(p_2/p_0)^{2/\gamma}(\rho_1/\rho_0)^{\gamma-1} - (p_2/\rho_0)^{(\gamma+1)/\gamma}}} \qquad (15.5\text{-}42)$$

From this equation we can obtain the time required to discharge any given fraction of the original gas.

At low flow rates the pressure p_2 at the nozzle opening is equal to the ambient pressure. However, examination of Eq. 15.5-41 shows that, as the ambient pressure is reduced, the calculated mass rate of flow reaches a maximum at a critical pressure ratio

$$r \equiv \left(\frac{p_2}{p_1}\right)_{\text{crit}} = \left(\frac{2}{\gamma + 1}\right)^{\gamma/(\gamma-1)} \qquad (15.5\text{-}43)$$

For air ($\gamma = 1.4$), this critical pressure ratio is 0.53. If the ambient pressure is further reduced, the pressure just inside the nozzle will remain at the value of p_2 calculated from Eq. 15.5-43, and the mass rate of flow will become independent of ambient pressure p_a. Under these conditions, the discharge rate is

$$w_{\text{max}} = S_2 \sqrt{p_1 \rho_1 \gamma \left(\frac{2}{\gamma + 1}\right)^{(\gamma+1)/(\gamma-1)}} \qquad (15.5\text{-}44)$$

Then, for $p_a/p_1 < r$, we may write Eq. 15.5-42 more simply:

$$t = \frac{V/S_2}{\sqrt{(p_0/\rho_0)\gamma(2/(\gamma + 1))^{(\gamma+1)/(\gamma-1)}}} \int_{\rho_1/\rho_0}^{1} \frac{dx}{x^{(\gamma+1)/2}} \qquad (15.5\text{-}45)$$

or

$$t = \frac{V/S_2}{\sqrt{(\gamma RT_0/M)(2/(\gamma + 1))^{(\gamma+1)/(\gamma-1)}}} \left(\frac{2}{\gamma - 1}\right)\left[\left(\frac{\rho_1}{\rho_0}\right)^{(1-\gamma)/2} - 1\right] \qquad (p_a/p_1 < r) \qquad (15.5\text{-}46)$$

If p_a/p_1 is initially less than r, both Eqs. 15.5-46 and 42 will be useful for calculating the total discharge time.

QUESTIONS FOR DISCUSSION

1. Give the physical significance of each term in the five macroscopic balances.
2. How are the equations of change related to the macroscopic balances?
3. Does each of the four terms within the parentheses in Eq. 15.1-2 represent a form of energy? Explain.
4. How is the macroscopic (total) energy balance related to the first law of thermodynamics, $\Delta U = Q + W$?
5. Explain how the averages $\langle v \rangle$ and $\langle v^3 \rangle$ arise in Eq. 15.1-1.
6. What is the physical significance of E_c and E_v? What sign do they have? How are they related to the velocity distribution? How can they be estimated?
7. How is the macroscopic balance for internal energy derived?
8. What information can be obtained from Eq. 15.2-2 about a fluid at rest?

PROBLEMS   **15A.1. Heat transfer in double-pipe heat exchangers.**

(a) Hot oil entering the heat exchanger in Example 15.4-1 at surface 2 is to be cooled by water entering at surface 1. That is, the exchanger is being operated in *countercurrent* flow. Compute the required exchanger area A, if the heat transfer coefficient U is 200 Btu/hr $\cdot$ ft^2 $\cdot$ F and the fluid streams have the following properties:

	Mass flow rate (lb_m/hr)	Heat capacity ($\text{Btu}/\text{lb}_m \cdot$ F)	Temperature entering (°F)	leaving (°F)
Oil	10,000	0.60	200	100
Water	5,000	1.00	60	—

(b) Repeat the calculation of part (a) if $U_1 = 50$ and $U_2 = 350$ Btu/hr $\cdot$ ft^2 $\cdot$ F. Assume that U varies linearly with the water temperature, and use the results of Problem 15B.1.

(c) What is the minimum amount of water that can be used in (a) and (b) to obtain the desired temperature change for the oil? What is the minimum amount of water that can be used in parallel flow?

(d) Calculate the required heat exchanger area for parallel flow operation, if the mass rate of flow of water is 15,500 lb_m/hr and U is constant at 200 Btu/hr $\cdot$ ft^2 $\cdot$F.

Answers: **(a)** 104 ft^2; **(b)** 122 ft^2; **(c)** 4290 lb_m/hr, 15,000 lb_m/hr; **(d)** about 101 ft^2

15A.2. Adiabatic flow of natural gas in a pipeline. Recalculate the power requirement $w\hat{W}$ in Example 15.4-2 if the flow in the pipeline were adiabatic rather than isothermal.

(a) Use the result of Problem 15B.3(d) to determine the density of the gas at plane 2.

(b) Use your answer to (a), along with the result of Problem 15B.3(e), to obtain p_2.

(c) Calculate the power requirement, as in Example 15.4-2.

Answers: **(a)** 0.243 lb_m/ft^3; **(b)** 86 psia; **(c)** 504 hp

15A.3. Mixing of two ideal-gas streams.

(a) Calculate the resulting velocity, temperature, and pressure when the following two air streams are mixed in an apparatus such as that described in Example 15.3-2. The heat capacity $\hat{C}_p$ of air may be considered constant at 6.97 Btu/lb-mole $\cdot$ F. The properties of the two streams are:

	w (lb_m/hr)	v (ft/s)	T (°F)	p (atm)
Stream 1a:	1000	1000	80	1.00
Stream 1b:	10,000	100	80	1.00

Answer: **(a)** 11,000 lb_m/hr;   about 110 ft/s;   86.5 °F;   1.00 atm

(b) What would the calculated velocity be, if the fluid density were treated as constant?

(c) Estimate $\hat{E}_v$ for this operation, basing your calculation on the results of part (b).

Answers: **(b)** 109 ft/s; **(c)** 1.4×10^3 ft lb_f/lb_m

15A.4. Flow through a Venturi tube. A Venturi tube, with a throat 3 in. in diameter, is placed in a circular pipe 1 ft in diameter carrying dry air. The discharge coefficient C_d of the meter is 0.98. Calculate the mass flow rate of air in the pipe if the air enters the Venturi at 70°F and 1 atm and the throat pressure is 0.75 atm.

(a) Assume adiabatic frictionless flow and $\gamma = 1.4$.

(b) Assume isothermal flow.

(c) Assume incompressible flow at the entering density.

Answers: **(a)** 2.07 lb_m/s; **(b)** 1.96 lb_m/s; **(c)** 2.43 lb_m/s

15A.5. Free batch expansion of a compressible fluid. A tank with volume $V = 10$ ft^3 (see Fig. 15.5-6) is filled with air ($\gamma = 1.4$) at $T_0 = 300$K and $p_0 = 100$ atm. At time $t = 0$ the valve is opened, allowing the air to expand to the ambient pressure of 1 atm through a convergent nozzle, with a throat cross section $S_2 = 0.1$ ft^2.

(a) Calculate the pressure and temperature at the throat of the nozzle, just after the start of the discharge.

(b) Calculate the pressure and temperature within the tank when p_2 attains its final value of 1 atm.

(c) How long will it take for the system to attain the state described in (b)?

15A.6. Heating of air in a tube. A horizontal tube of 20 ft length is heated by means of an electrical heating element wrapped uniformly around it. Dry air enters at 5°F and 40 psia at a velocity 75 ft/s and 185 lb_m/hr. The heating element provides heat at a rate of 800 Btu/hr per foot of tube. At what temperature will the air leave the tube, if the exit pressure is 15 psia? Assume turbulent flow and ideal gas behavior. For air in the range of interest the heat capacity at constant pressure in Btu/lb-mole · F is

$$\tilde{C}_p = 6.39 + (9.8 \times 10^{-4})T - (8.18 \times 10^{-8})T^2 \tag{15A.6-1}$$

where T is expressed in degrees Rankine.

Answer: $T_2 = 354°F$

15A.7. Operation of a simple double-pipe heat exchanger. A cold-water stream, 5400 lb_m/hr at 70°F, is to be heated by 8100 lb_m/hr of hot water at 200°F in a simple double-pipe heat exchanger. The cold water is to flow through the inner pipe, and the hot water through the annular space between the pipes. Two 20-ft lengths of heat exchanger are available, and also all the necessary fittings.

(a) By means of a sketch, show the way in which the two double-pipe heat exchangers should be connected in order to get the most effective heat transfer.

(b) Calculate the exit temperature of the cold stream for the arrangement decided on in (a) for the following situation:

(i) The heat-transfer coefficient for the annulus, based on the heat transfer area of the inner surface of the inner pipe is 2000 Btu/hr · ft² · F.

(ii) The inner pipe has the following properties: total length, 40 ft; inside diameter 0.0875 ft; heat transfer surface per foot, 0.2745 ft²; capacity at average velocity of 1 ft/s is 1345 lb_m/hr.

(iii) The average properties of the water in the inner pipe are:

$$\mu = 0.45 \text{ cp} = 1.09 \text{ } lb_m/\text{hr} \cdot \text{ft}$$
$$\hat{C}_p = 1.00 \text{ Btu}/lb_m \cdot \text{F}$$
$$k = 0.376 \text{ Btu}/\text{hr} \cdot \text{ft} \cdot \text{F}$$
$$\rho = 61.5 \text{ } lb_m/\text{ft}^3$$

(iv) The combined resistance of the pipe wall and encrustations is 0.001 hr · ft² · F/Btu based on the inner pipe surface area.

(c) Sketch the temperature profile in the exchanger.

Answer: **(b)** 136°F

15B.1. Performance of a double-pipe heat exchanger with variable overall heat transfer coefficient. Develop an expression for the amount of heat transferred in an exchanger of the type discussed in Example 15.4-1, if the overall heat transfer coefficient U varies linearly with the temperature of either stream.

(a) Since $T_h - T_c$ is a linear function of both T_h and T_c, show that

$$\frac{U - U_1}{U_2 - U_1} = \frac{\Delta T - \Delta T_1}{\Delta T_2 - \Delta T_1} \tag{15B.1-1}$$

in which $\Delta T = T_h - T_c$, and the subscripts 1 and 2 refer to the conditions at control surfaces 1 and 2.

(b) Substitute the result in (a) for $T_h - T_c$ into Eq. 15.4-12, and integrate the equation thus obtained over the length of the exchanger. Use this result to show that[1]

$$Q_c = A \frac{U_1 \Delta T_2 - U_2 \Delta T_1}{\ln(U_1 \Delta T_2 / U_2 \Delta T_1)} \tag{15B.1-2}$$

15B.2. Pressure drop in turbulent flow in a slightly converging tube (Fig. 15B.2). Consider the turbulent flow of an incompressible fluid in a circular tube with a diameter that varies linearly with distance according to the relation

$$D = D_1 + (D_2 - D_1)\frac{z}{L} \tag{15B.2-1}$$

At $z = 0$, the velocity is v_1 and may be assumed to be constant over the cross section. The Reynolds number for the flow is such that f is given approximately by the Blasius formula of Eq. 6.2-13,

$$f = \frac{0.0791}{\text{Re}^{1/4}} \tag{15B.2-2}$$

Obtain the pressure drop $p_1 - p_2$ in terms of v_1, D_1, D_2, ρ, L, and $\nu = \mu/\rho$.

(a) Integrate the d-form of the mechanical energy balance to get

$$\frac{1}{\rho}(p_1 - p_2) = \frac{1}{2}(v_2^2 - v_1^2) + 2\int_0^L \frac{v^2 f}{D}\, dz \tag{15B.2-3}$$

and then eliminate v_2 from the equation.

(b) Show that both v and f are functions of D:

$$v = v_1 \left(\frac{D_1}{D}\right)^2; \qquad f = \frac{0.0791}{(D_1 v_1/\nu)^{1/4}}\left(\frac{D}{D_1}\right)^{1/4} \tag{15B.2-4}$$

Of course, D is a function of z according to Eq. 15B.2-1.

(c) Make a change of variable in the integral in Eq. 15B.2-3 and show that

$$\int_0^L \frac{v^2 f}{D}\, dz = \frac{L}{D_2 - D_1}\int_{D_1}^{D_2} \frac{v^2 f}{D}\, dD \tag{15B.2-5}$$

(d) Combine the results of (b) and (c) to get finally

$$\frac{1}{\rho}(p_1 - p_2) = \frac{1}{2}v_1^2\left[\left(\frac{D_1}{D_2}\right)^4 - 1\right] + \frac{2Lv_1^2}{D_1 - D_2}\frac{\frac{4}{15}(0.0791)}{(D_1 v_1/\nu)^{1/4}}\left[\left(\frac{D_1}{D_2}\right)^{15/4} - 1\right] \tag{15B.2-6}$$

(e) Show that this result simplifies properly for $D_1 = D_2$.

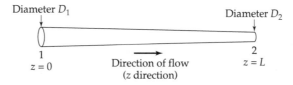

Diameter D_1 Diameter D_2

1   2
$z = 0$ Direction of flow $z = L$
            (z direction)

Fig. 15B.2. Turbulent flow in a horizontal, slightly tapered tube (D_1 is slightly greater than D_2).

---

[1] A. P. Colburn, *Ind. Eng. Chem.*, **25**, 873 (1933).

15B.3. Steady flow of ideal gases in ducts of constant cross section.

(a) Show that, for the horizontal flow of any fluid in a circular duct of uniform diameter D, the d-form of the mechanical energy balance, Eq. 15.4-1, may be written as

$$v \, dv + \frac{1}{\rho} dp + \frac{1}{2}v^2 de_v = 0 \tag{15B.3-1}$$

in which $de_v = (4f/D)dL$. Assume flat velocity profiles.

(b) Show that Eq. 15B.3-1 may be rewritten as

$$v \, dv + d\left(\frac{p}{\rho}\right) + \left(\frac{p}{\rho^2}\right)d\rho + \frac{1}{2}v^2 de_v = 0 \tag{15B.3-2}$$

Show further that, when use is made of the d-form of the mass balance, Eq. 15B.3-2 becomes for *isothermal flow* of an ideal gas

$$de_v = \frac{2RT}{M}\frac{dv}{v^3} - 2\frac{dv}{v} \tag{15B.3-3}$$

(c) Integrate Eq. 15B.3-3 between any two pipe cross sections 1 and 2 enclosing a total pipe length L. Make use of the ideal gas equation of state and the macroscopic mass balance to show that $v_2/v_1 = \rho_1/\rho_2 = p_1/p_2$, so that the "mass velocity" G can be put in the form

$$G \equiv \rho_1 v_1 = \sqrt{\frac{\rho_1 p_1(1-r)}{e_v - \ln r}} \qquad \text{(\textit{isothermal} flow of ideal gases)} \tag{15B.3-4}$$

in which $r = (p_2/p_1)^2$. Show that, for any given value of e_v and conditions at section 1, the quantity G reaches its maximum possible value at a critical value of r defined by $\ln r_c + (1 - r_c)/r_c = e_v$. See also Problem 15B.4.

(d) Show that, for the *adiabatic flow* of an ideal gas with constant $\hat{C}_p$ in a horizontal duct of constant cross section, the d-form of the total energy balance (Eq. 15.4-4) simplifies to

$$p\hat{V} + \left(\frac{\gamma - 1}{\gamma}\right)\frac{1}{2}v^2 = \text{constant} \tag{15B.3-5}$$

where $\gamma = C_p/C_V$. Combine this result with Eq. 15B.3-2 to get

$$\frac{\gamma + 1}{\gamma}\frac{dv}{v} - 2\left(\frac{p_1}{\rho_1} + \left(\frac{\gamma - 1}{\gamma}\right)\frac{1}{2}v_1^2\right)\frac{dv}{v^3} = -de_v \tag{15B.3-6}$$

Integrate this equation between sections 1 and 2 enclosing the resistance e_v, assuming γ constant. Rearrange the result with the aid of the macroscopic mass balance to obtain the following relation for the mass flux G.

$$G \equiv \rho_1 v_1 = \sqrt{\frac{\rho_1 p_1}{e_v - [(\gamma + 1)/2\gamma] \ln s \dfrac{}{1-s} - \dfrac{\gamma - 1}{2\gamma}}} \qquad \text{(\textit{adiabatic} flow of ideal gases)} \tag{15B.3-7}$$

in which $s = (\rho_2/\rho_1)^2$.

(e) Show by use of the macroscopic energy and mass balances that for horizontal adiabatic flow of ideal gases with constant γ,

$$\frac{p_2}{p_1} = \frac{\rho_2}{\rho_1}\left[1 + \frac{[1 - (\rho_1/\rho_2)^2]G^2}{\rho_1 p_1}\left(\frac{\gamma - 1}{2\gamma}\right)\right] \tag{15B.3-8}$$

This equation can be combined with Eq. 15B.3-7 to show that, as for isothermal flow, there is a critical pressure ratio p_2/p_1 corresponding to the maximum possible mass flow rate.

15B.4. The Mach number in the mixing of two fluid streams.

(a) Show that when the radicand in Eq. 15.3-13 is zero, the Mach number of the final stream is unity. Note that the Mach number, Ma, which is the ratio of the local fluid velocity to the velocity of sound at the local conditions, may be written for an ideal gas as $v/v_s = v/\sqrt{\gamma RT/M}$ (see Problem 11C.1).

(b) Show how the results of Example 15.3-2 may be used to predict the behavior of a gas passing through a sudden enlargement of duct cross section.

15B.5. Limiting discharge rates for Venturi meters.

(a) Starting with Eq. 15.5-34 (for *adiabatic flow*), show that as the throat pressure in a Venturi meter is reduced, the mass rate of flow reaches a maximum when the ratio $r = p_2/p_1$ of throat pressure to entrance pressure is defined by the expression

$$\frac{\gamma + 1}{r^{2/\gamma}} - \frac{2}{r^{(\gamma+1)/\gamma}} - \frac{\gamma - 1}{(S_1/S_0)^2} = 0 \tag{15B.5-1}$$

(b) Show that for $S_1 \gg S_0$ the mass flow rate under these limiting conditions is

$$w = C_d p_1 S_0 \sqrt{\frac{\gamma M}{RT_1} \left(\frac{2}{\gamma + 1}\right)^{(\gamma+1)/(\gamma-1)}} \tag{15B.5-2}$$

(c) Obtain results analogous to Eqs. 15B.5-1 and 2 for *isothermal flow*.

15B.6. Flow of a compressible fluid through a convergent–divergent nozzle (Fig. 15B.6). In many applications, such as steam turbines or rockets, hot compressed gases are expanded through nozzles of the kind shown in the accompanying figure in order to convert the gas enthalpy into kinetic energy. This operation is in many ways similar to the flow of gases through orifices. Here, however, the purpose of the expansion is to produce power—for example, by the impingement of the fast-moving fluid on a turbine blade, or by direct thrust, as in a rocket engine.

To explain the behavior of such a system and to justify the general shape of the nozzle described, follow the path of expansion of an ideal gas. Assume that the gas is initially in a very large reservoir at essentially zero velocity and that it expands through an adiabatic frictionless nozzle to zero pressure. Further assume flat velocity profiles, and neglect changes in elevation.

(a) Show, by writing the macroscopic mechanical energy balance or the total energy balance between planes 1 and 2, that

$$\tfrac{1}{2}v_2^2 = \frac{RT_1}{M} \frac{\gamma}{\gamma - 1} \left[1 - \left(\frac{p_2}{p_1}\right)^{(\gamma-1)/\gamma}\right] \tag{15B.6-1}$$

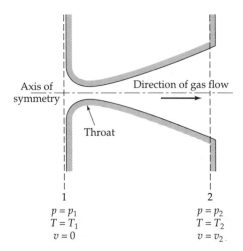

Fig. 15B.6. Schematic cross section of a convergent–divergent nozzle.

(b) Show, by use of the ideal gas law, the steady-state macroscopic mass balance, and Eq. 15B.6-1, that the cross section S of the expanding stream goes through a minimum at a critical pressure

$$p_{2,\text{crit}} = p_1 \left(\frac{2}{\gamma + 1} \right)^{\gamma/(\gamma-1)} \tag{15B.6-2}$$

(c) Show that the Mach number, $\text{Ma} = v_2/v_s(T_2)$, of the fluid at this minimum cross section is unity (v_s for low-frequency sound waves is derived in Problem 11C.1). How does the result of part (a) above compare with that in Problem 15B.5?

(d) Calculate fluid velocity v, fluid temperature T, and stream cross section S as a function of the local pressure p for the discharge of 10 lb-moles of air per second from 560°R and 10 atm to zero pressure. Discuss the significance of your results.

Answer:

p, atm	10	9	8	7	6	5.28	5	4	3	2	1	0
v, ft sec^{-1}	0	449	645	807	956	1058	1099	1245	1398	1574	1798	2591
T, °R	560	543	525	506	484	466	459	431	397	353	290	0
S, ft^2	∞	0.977	0.739	0.650	0.613	0.606	0.607	0.628	0.688	0.816	1.171	∞

15B.7. **Transient thermal behavior of a chromatographic device** (Fig. 15B.7). You are a consultant to an industrial concern that is experimenting, among other things, with transient thermal phenomena in gas chromatography. One of the employees first shows you some reprints of a well-known researcher and says that he is trying to apply some of the researcher's new approaches, but that he is currently stuck on a heat transfer problem. Although the problem is only ancillary to the main study, it must nonetheless be understood in connection with his interpretation of the data and the application of the new theories.

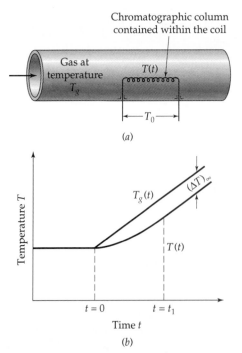

Fig. 15B.7. (*a*) Chromatographic device; (*b*) temperature response of the chromatographic system.

A very tiny chromatographic column is contained within a coil, which is in turn inserted into a pipe through which a gas is blown to control the temperature (see Fig. 15B.7a). The gas temperature will be called $T_g(t)$. The temperature at the ends of the coil (outside the pipe) is T_0, which is not very much different from the initial value of T_g. The actual temperature within the chromatographic column (i.e., within the coil) will be called $T(t)$. Initially the gas and the coil are both at the temperature T_{g0}. Then beginning at time $t = 0$, the gas temperature is increased linearly according to the equation

$$T_g(t) = T_{g0}\left(1 + \frac{t}{t_0}\right) \tag{15B.7-1}$$

where t_0 is a known constant with dimensions of time.

You are told that, by inserting thermocouples into the column itself, the people in the lab have obtained temperature curves that look like those in Fig. 15B.7(b). The $T(t)$ curve seems to become parallel to the $T_g(t)$ curves for large t. You are asked to explain the above pair of curves by means of some kind of theory. Specifically you are asked to find out the following:

(a) At any time t, what will $T_g - T$ be?

(b) What will the limiting value of $T_g - T$ be when $t \to \infty$? Call this quantity $(\Delta T)_\infty$.

(c) What time interval t_1 is required for $T_g - T$ to come within, say, 1% of $(\Delta T)_\infty$?

(d) What assumptions had to be made to model the system?

(e) What physical constants, physical properties, and so on, have to be known in order to make a comparison between the measured and theoretical values of $(\Delta T)_\infty$?

Devise the simplest possible theory to account for the temperature curves and to answer the above five questions.

15B.8. Continuous heating of a slurry in an agitated tank (Fig. 15B.8). A slurry is being heated by pumping it through a well-stirred heating tank. The inlet temperature of the slurry is T_i and the temperature of the outer surface of the steam coil is T_s. Use the following symbols:

$$V = \text{volume of the slurry in the tank}$$
$$\rho, \hat{C}_p = \text{density and heat capacity of the slurry}$$
$$w = \text{mass rate of flow of slurry through the tank}$$
$$U = \text{overall heat transfer coefficient of heating coil}$$
$$A = \text{total heat transfer area of the coil}$$

Assume that the stirring is sufficiently thorough that the fluid temperature in the tank is uniform and the same as the outlet fluid temperature.

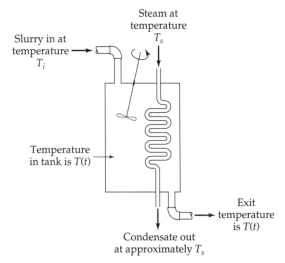

Steam at temperature T_s

Slurry in at temperature T_i

Temperature in tank is $T(t)$

Exit temperature is $T(t)$

Condensate out at approximately T_s

Fig. 15B.8. Heating of a slurry in an agitated tank.

(a) By means of an energy balance, show that the slurry temperature $T(t)$ is described by the differential equation

$$\frac{dT}{dt} = \left(\frac{UA}{\rho \hat{C}_p V}\right)(T_s - T) - \left(\frac{w}{\rho V}\right)(T - T_i) \tag{15B.8-1}$$

The variable t is the time since the start of heating.

(b) Rewrite this differential equation in terms of the dimensionless variables

$$\tau = \frac{wt}{\rho V} \qquad \Theta = \frac{T - T_\infty}{T_i - T_\infty} \tag{15B.8-2, 3}$$

where

$$T_\infty = \frac{(UA/w\hat{C}_p)T_s + T_i}{(UA/w\hat{C}_p) + 1} \tag{15B.8-4}$$

What is the physical significance of τ, Θ, and T_∞?

(c) Solve the dimensionless equation obtained in (b) for the initial condition that $T = T_i$ at $t = 0$.

(d) Check the solution to see that the differential equation and initial condition are satisfied. How does the system behave at large time? Is this limiting behavior in agreement with your intuition?

(e) How is the temperature at infinite time affected by the flow rate? Is this reasonable?

Answer: **(c)** $\dfrac{T - T_\infty}{T_i - T_\infty} = \exp\left[-\left(\dfrac{UA}{\rho \hat{C}_p V} + \dfrac{w}{\rho V}\right)t\right]$

15C.1. Parallel–counterflow heat exchangers (Fig. 15C.1). In the heat exchanger shown in the accompanying figure, the "tube fluid" (fluid A) enters and leaves at the same end of the heat exchanger, whereas the "shell fluid" (fluid B) always moves in the same direction. Thus there are both parallel flow and counterflow in the same apparatus. This flow arrangement is one of the simplest examples of "mixed flow," often used in practice to reduce exchanger length.[2]

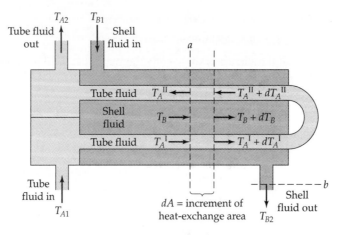

Fig. 15C.1. A parallel–counterflow heat exchanger.

---

[2] See D. Q. Kern, *Process Heat Transfer*, McGraw-Hill, New York (1950), pp. 127–189; J. H. Perry, *Chemical Engineers' Handbook*, 3rd edition, McGraw-Hill, New York, (1950), pp. 464–465; W. M. Rohsenow, J. P. Hartnett, and Y. I. Cho, *Handbook of Heat Transfer*, 3rd edition, McGraw-Hill, New York (1998), Chapter 17; S. Whitaker, *Fundamentals of Heat Transfer*, corrected edition, Krieger Publishing Company, Malabar, Fla., (1983), Chapter 11.

The behavior of this kind of equipment may be simply analyzed by making the following assumptions:

(i) Steady-state conditions exist.

(ii) The overall heat transfer coefficient U and the heat capacities of the two fluids are constants.

(iii) The shell-fluid temperature T_B is constant over any cross section perpendicular to the flow direction.

(iv) There is an equal amount of heating area in each tube fluid "pass"—that is, for streams I and II in the figure.

(a) Show by an energy balance over the portion of the system between planes a and b that

$$T_B - T_{B2} = R(T_A^{II} - T_A^I) \qquad \text{where } R = |w_A\hat{C}_{pA}/w_B\hat{C}_{pB}| \qquad (15C.1-1)$$

(b) Show that over a differential section of the exchanger, including a *total* heat exchange surface dA,

$$\frac{dT_A^I}{d\alpha} = \frac{1}{2}(T_B - T_A^I) \qquad (15C.1-2)$$

$$\frac{dT_A^{II}}{d\alpha} = \frac{1}{2}(T_A^{II} - T_B) \qquad (15C.1-3)$$

$$\frac{1}{R}\frac{dT_B}{d\alpha} = -\left[T_B - \frac{1}{2}(T_A^I + T_A^{II})\right] \qquad (15C.1-4)$$

in which $d\alpha = (U/w_A\hat{C}_{pA})dA$, and w_A and $\hat{C}_{pA}$ are defined as in Example 15.4-1.

(c) Show that when T_A^I and T_A^{II} are eliminated between these three equations, a differential equation for the shell fluid can be obtained:

$$\frac{d^2\Theta}{d\alpha^2} + R\frac{d\Theta}{d\alpha} - \frac{1}{4}\Theta = 0 \qquad (15C.1-5)$$

in which $\Theta(\alpha) = (T_B - T_{B2})/(T_{B1} - T_{B2})$. Solve this equation (see Eq. C.1-7) with the boundary conditions

B.C. 1:   at $\alpha = 0$, $\Theta = 1$ (15C.1-6)

B.C. 2:   at $\alpha = (UA_T/w_A\hat{C}_{pA})$, $\Theta = 0$ (15C.1-7)

in which A_T is the total heat-exchange surface of the exchanger.

(d) Use the result of part (c) to obtain an expression for $dT_B/d\alpha$. Eliminate $dT_B/d\alpha$ from this expression with the aid of Eq. 15C.1-4 and evaluate the resulting equation at $\alpha = 0$ to obtain the following relation for the performance of the exchanger:

$$\alpha_T = \frac{UA_T}{w_A\hat{C}_{pA}} = \frac{1}{\sqrt{R^2+1}}\ln\left[\frac{2 - \Psi(R + 1 - \sqrt{R^2+1})}{2 - \Psi(R + 1 + \sqrt{R^2+1})}\right] \qquad (15C.1-8)$$

in which $\Psi = (T_{A2} - T_{A1})/(T_{B1} - T_{A1})$.

(e) Use this result to obtain the following expression for the rate of heat transfer in the exchanger:

$$Q = UA(\Delta T)_{\ln} \cdot Y \qquad (15C.1-9)$$

in which

$$(\Delta T)_{\ln} = \frac{(T_{B1} - T_{A2}) - (T_{B2} - T_{A1})}{\ln[(T_{B1} - T_{A2})/(T_{B2} - T_{A1})]} \qquad (15C.1-10)$$

$$Y = \frac{\sqrt{R^2+1}\,\ln[(1 - \Psi)/(1 - R\Psi)]}{(R - 1)\ln\left[\dfrac{2 - \Psi(R + 1 - \sqrt{R^2+1})}{2 - \Psi(R + 1 + \sqrt{R^2+1})}\right]} \qquad (15C.1-11)$$

The quantity Y represents the ratio of the heat transferred in the "1–2 parallel–counterflow exchanger" shown to that transferred in a true counterflow exchanger of the same area and terminal fluid temperatures. Values of $Y(R, \Psi)$ are given graphically in Perry's handbook.[2] It may be seen that $Y(R, \Psi)$ is always less than unity.

15C.2. Discharge of air from a large tank. It is desired to withdraw 5 lb_m/s from a large storage tank through an equivalent length of 55 ft of new steel pipe 2.067 in. in diameter. The air undergoes a sudden contraction on entering the pipe, and the accompanying contraction loss is not included in the equivalent length of the pipe. Can the desired flow rate be obtained if the air in the tank is at 150 psig and 70°F and the pressure at the downstream end of the pipe is 50 psig?

The effect of the sudden contraction may be estimated with reasonable accuracy by considering the entrance to consist of an ideal nozzle converging to a cross section equal to that of the pipe, followed by a section of pipe with $e_v = 0.5$ (see Table 7.5-1). The behavior of the nozzle can be determined from Eq. 15.5-34 by assuming the cross sectional area S_1 to be infinite and C_d to be unity.

Answer: Yes. The calculated discharge rate is about 6 lb_m/s if *isothermal* flow is assumed (see Problem 15B.3) and about 6.3 lb_m/s for *adiabatic* flow. The actual rate should be between these limits for an ambient temperature of 70°F.

15C.3. Stagnation temperature (Fig. 15C.3). A "total temperature probe," as shown in the figure, is inserted in a steady stream of an ideal gas at a temperature T_1 and moving with a velocity v_1. Part of the moving gas enters the open end of the probe and is decelerated to nearly zero velocity before slowly leaking out of the bleed holes. This deceleration results in a temperature rise, which is measured by the thermocouple. Since the deceleration is rapid, it is nearly adiabatic.

(a) Develop an expression for the temperature registered by the thermocouple in terms of T_1 and v_1 by using the steady-state macroscopic energy balance, Eq. 15.1-3. Use as your system a representative stream of fluid entering the probe. Draw reference plane 1 far enough upstream that conditions may be assumed unaffected by the probe, and reference plane 2 in the probe itself. Assume zero velocity at plane 2, neglect radiation, and neglect conduction of heat from the fluid as it passes between the reference planes.

(b) What is the function of the bleed holes?

Answer: **(a)** $T_2 - T_1 = v_1^2/2\hat{C}_p$. Temperature rises within about 2% of those given by this expression and may be obtained with well-designed probes.

15D.1. The macroscopic entropy balance.

(a) Show that integration of the equation of change for entropy (Eq. 11D.1-3) over the flow system of Fig. 7.0-1 leads to

$$\frac{d}{dt} S_{\text{tot}} = -\Delta\left(\hat{S} + \frac{q}{\rho v T}\right)w + g_{S,\text{tot}} + Q_S \qquad (15\text{D}.1\text{-}1)$$

in which

$$S_{\text{tot}} = \int_V \rho \hat{S} dV \qquad (15\text{D}.1\text{-}2)$$

$$g_{S,\text{tot}} = -\int_V \frac{1}{T}((\mathbf{q} \cdot \nabla \ln T) + (\boldsymbol{\tau}:\nabla\mathbf{v}))dV \qquad (15\text{D}.1\text{-}3)$$

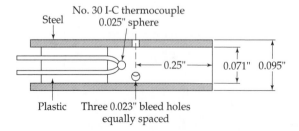

No. 30 I-C thermocouple
Steel   0.025" sphere
0.25" — — 0.071" 0.095"
Plastic   Three 0.023" bleed holes
equally spaced

Fig. 15C.3. A "total temperature probe." [H. C. Hottel and A. Kalitinsky, *J. Appl. Mech.*, **12**, A25 (1945).]

(b) Give a term-by-term interpretation of the equations in (a).

(c) Is the term in $g_{s,\text{tot}}$ involving the stress tensor the same as the energy dissipation by viscous heating?

15D.2. Derivation of the macroscopic energy balance. Show how to integrate Eq. (N) of Table 11.4-1 over the entire volume V of a flow system, which, because of moving parts, may be a function of time. With the help of the Gauss divergence theorem and the Leibniz formula for differentiating an integral, show that this gives the macroscopic total energy balance Eq. 15.1-2. What assumptions are made in the derivation? How is W_m to be interpreted? (*Hint:* Some suggestions on solving this problem may be obtained by studying the derivation of the macroscopic mechanical energy balance in §7.8.)

15D.3. Operation of a heat-exchange device (Fig. 15D.3). A hot fluid enters the circular tube of radius R_1 at position $z = 0$ and moves in the positive z direction to $z = L$, where it leaves the tube and flows back along the outside of that tube in the annular space. Heat is exchanged between the fluid in the tube and that in the annulus. Also heat is lost from the annulus to the air outside, which is at the ambient air temperature T_a (a constant). Assume that the density and heat capacity are constant. Use the following notation:

> U_1 = overall heat transfer coefficient between the fluid in the tube and the fluid in the annular space
>
> U_2 = overall heat transfer coefficient between the fluid in the annulus and the air at temperature T_a
>
> $T_1(z)$ = temperature of the fluid in the tube
>
> $T_2(z)$ = temperature of the fluid in the annular space
>
> w = mass flow rate through the system (a constant)

If the fluid enters at the inlet temperature T_i, what will be the outlet temperature T_o? It is suggested that the following dimensionless quantities be used: $\Theta_1 = (T_1 - T_a)/(T_i - T_a)$, $N_1 = 2\pi R_1 U_1 L/w \hat{C}_p$, and $\zeta = z/L$.

15D.4. Discharge of a gas from a moving tank (Fig. 15.5-6). Equation 15.5-38 in Example 15.5-4 was obtained by setting $d\hat{\Phi}/dt$ equal to zero, a procedure justified only because the tank was said to be stationary. It is nevertheless true that Eq. 15.5-38 is correct for moving tanks as well. This statement can be proved as follows:

(a) Consider a tank such as that pictured in Fig. 15.5-6, but moving at a velocity $\mathbf{v}$ that is much larger than the relative velocity of fluid and tank in the region to the left of surface 1. Show that for this region of the tank the macroscopic momentum balance becomes

$$-\left(\mathbf{F}_{f\to s} + \mathbf{u}_2 \int_{S_1} p_1 dS\right) = m_{\text{tot}}\left(\frac{d\mathbf{v}}{dt} - \mathbf{g}\right) \tag{15D.4-1}$$

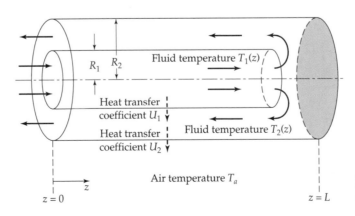

Air temperature T_a

Fig. 15D.3. A heat-exchange device.

in which the fluid velocity is assumed to be uniform and equal to $\mathbf{v}$. Then take the dot product of both sides of Eq. 15D.4-1 with $\mathbf{v}$ to obtain

$$W_m = m_{\text{tot}}\left(\frac{d\hat{K}}{dt} + \frac{d\hat{\Phi}}{dt}\right) \tag{15D.4-2}$$

where $\partial\hat{\Phi}/\partial t$ is neglected.

(b) Substitute this result into the macroscopic energy balance, and continue as in Example 15.5-4.

15D.5. **The classical Bernoulli equation.** Below Eq. 15.2-5 we have emphasized that the mechanical energy balance and the total energy balance contain different information, since the first is a consequence of conservation of momentum, whereas the second is a consequence of conservation of energy.

For the steady-state flow of a compressible fluid with zero transport properties, both balances lead to the classical Bernoulli equation. The derivation based on the equation of motion was given in Example 3.5-1. Make a similar derivation for the steady state energy equation, assuming zero transport properties, that is, for isentropic flow.[3]

---

[3] R. B. Bird and M. D. Graham, in *Handbook of Fluid Dynamics* (R. W. Johnson, ed.), CRC Press, Boca Raton, Fla. (1998), p. 3-13.

Chapter 16

Energy Transport by Radiation

We concluded Part I of this book with a chapter about fluids that cannot be described by Newton's law of viscosity, but that require various kinds of nonlinear and time-dependent expressions. We now end Part II with a brief discussion of radiative energy transport, which cannot be described by Fourier's law.

In Chapters 9 to 15 the transport of energy by conduction and by convection has been discussed. Both modes of transport rely on the presence of a material medium. For *heat conduction* to occur, there must be temperature inequalities between neighboring points. For *heat convection* to occur, there must be a fluid that is free to move and transport energy with it. In this chapter, we turn our attention to a third mechanism for energy transport—namely, *radiation*. Radiation is basically an electromagnetic mechanism, which allows energy to be transported with the speed of light through regions of space that are devoid of matter. The rate of energy transport between two "black" bodies in a vacuum is proportional to the difference of the fourth powers of their absolute temperatures. This mechanism is qualitatively very different from the three transport mechanisms considered elsewhere in this book: momentum transport in Newtonian fluids, proportional to the velocity gradient; energy transport by heat conduction, proportional to a temperature gradient; and mass transport by diffusion, proportional to a concentration gradient. Because of the uniqueness of radiation as a means of transport and because of the importance of radiant heat transfer in industrial calculations, we have devoted a separate chapter to this subject.

A thorough understanding of the physics of radiative transport requires the use of several different disciplines:[1,2] electromagnetic theory is needed to describe the essentially wavelike nature of radiation, in particular the energy and pressure associated with electromagnetic waves; thermodynamics is useful for obtaining some relations among

---

[1] M. Planck, *Theory of Heat*, Macmillan, London (1932), Parts III and IV. Nobel Laureate **Max Karl Ernst Ludwig Planck** (1858–1947) was the first to hypothesize the quantization of energy and thereby introduce a new fundamental constant *h* (Planck's constant); his name is also associated with the "Fokker–Planck" equation of stochastic dynamics.

[2] W. Heitler, *Quantum Theory of Radiation*, 2nd edition, Oxford University Press (1944).

the "bulk properties" of an enclosure containing radiation; quantum mechanics is necessary in order to describe in detail the atomic and molecular processes that occur when radiation is produced within matter and when it is absorbed by matter; and statistical mechanics is needed to describe the way in which the energy of radiation is distributed over the wavelength spectrum. All we can do in this elementary discussion is define the key quantities and set forth the results of theory and experiment. We then show how some of these results can be used to compute the rate of heat transfer by radiant processes in simple systems.

In §16.1 and §16.2 we introduce the basic concepts and definitions. Then in §16.3 some of the principal physical results concerning black-body radiation are given. In the following section, §16.4, the rate of heat exchange between two black bodies is discussed. This section introduces no new physical principles, the basic problems being those of geometry. Next, §16.5 is devoted to an extension of the preceding section to nonblack surfaces. Finally, in the last section, there is a brief discussion of radiation processes in absorbing media.[3]

§16.1   THE SPECTRUM OF ELECTROMAGNETIC RADIATION

When a solid body is heated—for example, by an electric coil—the surface of the solid emits radiation of wavelength primarily in the range 0.1 to 10 microns. Such radiation is usually referred to as *thermal radiation*. A quantitative description of the atomic and molecular mechanisms by which the radiation is produced is given by quantum mechanics and is outside the scope of this discussion. A qualitative description, however, is possible: When energy is supplied to a solid body, some of the constituent molecules and atoms are raised to "excited states." There is a tendency for the atoms or molecules to return spontaneously to lower energy states. When this occurs, energy is emitted in the form of electromagnetic radiation. Because the emitted radiation results from changes in the electronic, vibrational, and rotational states of the atoms and molecules, the radiation will be distributed over a range of wavelengths.

Actually, thermal radiation represents only a small part of the total spectrum of electromagnetic radiation. Figure 16.1-1 shows roughly the kinds of mechanisms that are responsible for the various parts of the radiation spectrum. The various kinds of radiation are distinguished from one another only by the range of wavelengths they include. In a vacuum, all these forms of radiant energy travel with the speed of light c. The wavelength λ, characterizing an electromagnetic wave, is then related to its frequency ν by the equation

$$\lambda = \frac{c}{\nu} \tag{16.1-1}$$

in which $c = 2.998 \times 10^8$ m/s. In the visible part of the spectrum, the various wavelengths are associated with the "color" of the light.

For some purposes, it is convenient to think of electromagnetic radiation from a corpuscular point of view. Then we associate with an electromagnetic wave of frequency ν a *photon*, which is a particle with charge zero and mass zero with an energy given by

$$\varepsilon = h\nu \tag{16.1-2}$$

---

[3] For additional information on radiative heat transfer and engineering applications, see the comprehensive textbook by R. Siegel and J. R. Howell, *Thermal Radiation Heat Transfer*, 3rd edition, Hemisphere Publishing Co., New York (1992). See also J. R. Howell and M. P. Mengöç, in *Handbook of Heat Transfer*, 3rd edition, (W. M. Rohsenow, J. P. Hartnett, and Y. I. Cho, eds.), McGraw-Hill, New York (1998), Chapter 7.

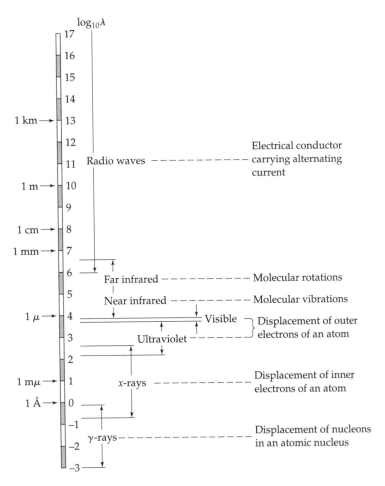

Fig. 16.1-1. The spectrum of electromagnetic radiation, showing roughly the mechanisms by which various wavelengths of radiation are produced (1 Å = Ångström unit = 10^{-8} cm = 0.1 nm; 1 μ = 1 micron = 10^{-6} m).

Here $h = 6.626 \times 10^{-34}$ J·s is Planck's constant. From these two equations and the information from Fig. 16.1-1, we see that decreasing the wavelength of electromagnetic radiation corresponds to increasing the energy of the corresponding photons. This fact ties in with the various mechanisms that produce the radiation. For example, relatively small energies are released when a molecule decreases its speed of rotation, and the associated radiation is in the infrared. On the other hand, relatively large energies are released when an atomic nucleus goes from a high energy state to a lower one, and the associated radiation is either gamma- or x-radiation. The foregoing statements also make it seem reasonable that the radiant energy emitted from heated objects will tend toward shorter wavelengths (higher energy photons) as the temperature of the body is raised.

Thus far we have sketched the phenomenon of the *emission* of radiant energy or photons when a molecular or atomic system goes from a high to a low energy state. The reverse process, known as absorption, occurs when the addition of radiant energy to a molecular or atomic system causes the system to go from a low to a high energy state. The latter process is then what occurs when radiant energy impinges on a solid surface and causes its temperature to rise.

§16.2 ABSORPTION AND EMISSION AT SOLID SURFACES

Having introduced the concepts of absorption and emission in terms of the atomic picture, we now proceed to the discussion of the same processes from a macroscopic viewpoint. We restrict the discussion here to opaque solids.

Radiation impinging on the surface of an opaque solid is either absorbed or reflected. The fraction of the incident radiation that is absorbed is called the *absorptivity* and is given the symbol a. Also the fraction of the incident radiation with frequency ν that is absorbed is designated by a_ν. That is, a and a_ν are defined as

$$a = \frac{q^{(a)}}{q^{(i)}} \qquad a_\nu = \frac{q_\nu^{(a)}}{q_\nu^{(i)}} \qquad\qquad (16.2\text{-}1, 2)$$

in which $q_\nu^{(a)}d\nu$ and $q_\nu^{(i)}d\nu$ are the absorbed and incident radiation per unit area per unit time in the frequency range ν to $\nu + d\nu$. For any *real body*, a_ν will be less than unity and will vary considerably with the frequency. A hypothetical body for which a_ν is a constant, less than unity, over the entire frequency range and at all temperatures is called a *gray body*. That is, a gray body always absorbs the same fraction of the incident radiation of all frequencies. A limiting case of the gray body is that for which $a_\nu = 1$ for all frequencies and all temperatures. This limiting behavior defines a *black body*.

All solid surfaces emit radiant energy. The total radiant energy emitted per unit area per unit time is designated by $q^{(e)}$, and that emitted in the frequency range ν to $\nu + d\nu$ is called $q_\nu^{(e)}d\nu$. The corresponding rates of energy emission from a black body are given the symbols $q_b^{(e)}$ and $q_{b\nu}^{(e)}d\nu$. In terms of these quantities, the *emissivity* for the total radiant-energy emission as well as that for a given frequency are defined as

$$e = \frac{q^{(e)}}{q_b^{(e)}} \qquad e_\nu = \frac{q_\nu^{(e)}}{q_{b\nu}^{(e)}} \qquad\qquad (16.2\text{-}3, 4)$$

The emissivity is also a quantity less than unity for real, nonfluorescing surfaces and is equal to unity for black bodies. At any given temperature the radiant energy emitted by a black body represents an upper limit to the radiant energy emitted by real, nonfluorescing surfaces.

We now consider the radiation within an evacuated enclosure or "cavity" with isothermal walls. We imagine that the entire system is at equilibrium. Under this condition, there is no net flux of energy across the interfaces between the solid and the cavity. We now show that the radiation in such a cavity is independent of the nature of the walls and dependent solely on the temperature of the walls of the cavity. We connect two cavities, the walls of which are at the same temperature, but are made of two different materials, as shown in Fig. 16.2-1. If the radiation intensities in the two cavities were different, there would be a net transport of radiant energy from one cavity to the other. Because such a flux would violate the second law of thermodynamics, the radiation intensities in the two cavities must be equal, regardless of the compositions of the cavity surfaces. Furthermore, it can be shown that the radiation is uniform and unpolarized throughout the cavity. This *cavity radiation* plays an important role in the development

Material 1   Material 2

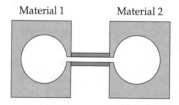

Fig. 16.2-1. Thought experiment for proof that cavity radiation is independent of the wall materials.

of Planck's law. We designate the intensity of the radiation as $q^{(cav)}$. This is the radiant energy that would impinge on a solid surface of unit area placed anywhere within the cavity.

We now perform two additional thought experiments. In the first, we put into a cavity a small black body at the same temperature as the walls of the cavity. There will be no net interchange of energy between the black body and the cavity walls. Hence the energy impinging on the black-body surface must equal the energy emitted by the black body:

$$q^{(cav)} = q_b^{(e)} \qquad (16.2\text{-}5)$$

From this result, we draw the important conclusion that the radiation emitted by a black body is the same as the equilibrium radiation intensity within a cavity at the same temperature.

In the second thought experiment, we put a small nonblack body into the cavity, once again specifying that its temperature be the same as that of the cavity walls. There is no net heat exchange between the nonblack body and the cavity walls. Hence we can state that the energy absorbed by the nonblack body will be the same as that radiating from it:

$$a q^{(cav)} = q^{(e)} \qquad (16.2\text{-}6)$$

Comparison of Eqs. 16.2-5 and 6 leads to the result

$$a = \frac{q^{(e)}}{q_b^{(e)}} \qquad (16.2\text{-}7)$$

The definition of the emissivity e in Eq. 16.2-3 allows us to conclude that

$$\boxed{e = a} \qquad (16.2\text{-}8)$$

This is *Kirchhoff's law*,[1] which states that at a given temperature the emissivity and absorptivity of any solid surface are the same when the radiation is in equilibrium with the solid surface. It can be shown that Eq. 16.2-8 is also valid for each wavelength separately:

$$\boxed{e_\nu = a_\nu} \qquad (16.2\text{-}9)$$

Values of the total emissivity e for some solids are given in Table 16.2-1. Actually, e depends also on the frequency and on the angle of emission, but the averaged values given there have found widespread use. The tabulated values are, with a few exceptions, for emission normal to the surface, but they may be used for hemispheric emissivity, particularly for rough surfaces. Unoxidized, clean, metallic surfaces have very low emissivities, whereas most nonmetals and metallic oxides have emissivities above 0.8 at room temperature or higher. Note that emissivity increases with increasing temperature for nearly all materials.

We have indicated that the radiant energy emitted by a black body is an upper limit to the radiant energy emitted by real surfaces and that this energy is a function of the temperature. It has been shown experimentally that the total emitted energy flux from a *black* surface is

$$\boxed{q_b^{(e)} = \sigma T^4} \qquad (16.2\text{-}10)$$

---

[1] G. Kirchhoff, *Monatsber. d. preuss. Akad. d. Wissenschaften*, p. 783 (1859); *Poggendorffs Annalen*, **109**, 275–301 (1860). **Gustav Robert Kirchhoff** (1824–1887) published his famous laws for electrical circuits while still a graduate student; he taught at Breslau, Heidelberg, and Berlin.

Table 16.2-1   The Total Emissivities of Various Surfaces for Perpendicular Emission[a]

	$T(°R)$	e	$T(°R)$	e
Aluminum				
Highly polished, 98.3% pure	900	0.039	1530	0.057
Oxidized at 1110°F	850	0.11	1570	0.19
Al-coated roofing	560	0.216		
Copper				
Highly polished, electrolytic	636	0.018		
Oxidized at 1110°F	850	0.57	1570	0.57
Iron				
Highly polished, electrolytic	810	0.052	900	0.064
Completely rusted	527	0.685		
Cast iron, polished	852	0.21		
Cast iron, oxidized at 1100°F	850	0.64	1570	0.78
Asbestos paper	560	0.93	1160	0.945
Brick				
Red, rough	530	0.93		
Silica, unglazed, rough	2292	0.80		
Silica, glazed, rough	2472	0.85		
Lampblack, 0.003 in. or thicker	560	0.945	1160	0.945
Paints				
Black shiny lacquer on iron	536	0.875		
White lacquer	560	0.80	660	0.95
Oil paints, 16 colors	672	0.92–0.96		
Aluminum paints, varying age and lacquer content	672	0.27–0.67		
Refractories, 40 different				
Poor radiators	1570	0.65–0.70	2290	0.75
Good radiators	1570	0.80–0.85	2290	0.85–0.90
Water, liquid, thick layer[b]	492	0.95	672	0.963

[a] Selected values from the table compiled by H. C. Hottel for W. H. McAdams, *Heat Transmission*, 3rd edition, McGraw-Hill, New York (1954), pp. 472–479.

[b] Calculated from spectroscopic data.

in which T is the absolute temperature. This is known as the *Stefan–Boltzmann law*.[2] The Stefan–Boltzmann constant σ has been found to have the value of 0.1712×10^{-8} Btu/hr · ft^2 · R or 1.355×10^{-12} cal/s · cm^2 · K. In the next section we indicate two routes by which this important formula has been obtained theoretically. For *nonblack* surfaces at temperature T the emitted energy flux is

$$\boxed{q^{(e)} = e\sigma T^4} \tag{16.2-11}$$

---

[2] J. Stefan, *Sitzber. Akad. Wiss. Wien*, **79**, part 2, 391–428 (1879); L. Boltzmann, *Ann. Phys.* (*Wied. Ann.*), Ser. 2, **22**, 291–294 (1884). Slovenian-born **Josef Stefan** (1835–1893), rector of the University of Vienna (1876–1877), in addition to being known for the law of radiation that bears his name, also contributed to the theory of multicomponent diffusion and to the problem of heat conduction with phase change. **Ludwig Eduard Boltzmann** (1844–1906), who held professorships in Vienna, Graz, Munich, and Leipzig, developed the basic differential equation for gas kinetic theory (see Appendix D) and the fundamental entropy-probability relation, $S = \kappa \ln W$, which is engraved on his tombstone in Vienna; κ is called the Boltzmann constant.

in which e must be evaluated at temperature T. The use of Eqs. 16.2-10 and 11 to calculate radiant heat transfer rates between heated surfaces is discussed in §§16.4 and 5.

We have mentioned that the Stefan–Boltzmann constant has been experimentally determined. This implies that we have a true black body at our disposal. Solids with perfectly black surfaces do not exist. However, we can get an excellent approximation to a black surface by piercing a very small hole in the wall of an isothermal cavity. The hole itself is then very nearly a black surface. The extent to which this is a good approximation may be seen from the following relation, which gives the effective emissivity of the hole, e_{hole}, in a rough-walled enclosure in terms of the actual emissivity e of the cavity walls and the fraction f of the total internal cavity area that is cut away by the hole:

$$e_{\text{hole}} \cong \frac{e}{e + f(1 - e)} \tag{16.2-12}$$

If $e = 0.8$ and $f = 0.001$, then $e_{\text{hole}} = 0.99975$. Therefore, 99.975% of the radiation that falls on the hole will be absorbed. The radiation that emerges from the hole will then be very nearly black-body radiation.

§16.3   PLANCK'S DISTRIBUTION LAW, WIEN'S DISPLACEMENT LAW, AND THE STEFAN–BOLTZMANN LAW[1,2,3]

The Stefan–Boltzmann law may be deduced from thermodynamics, provided that certain results of the theory of electromagnetic fields are known. Specifically, it can be shown that for cavity radiation the energy density (that is, the energy per unit volume) within the cavity is

$$u^{(r)} = \frac{4}{c} q_b^{(e)} \tag{16.3-1}$$

Since the radiant energy emitted by a black body depends on temperature alone, the energy density $u^{(r)}$ must also be a function of temperature only. It can further be shown that the electromagnetic radiation exerts a pressure $p^{(r)}$ on the walls of the cavity given by

$$p^{(r)} = \tfrac{1}{3} u^{(r)} \tag{16.3-2}$$

The preceding results for cavity radiation can also be obtained by considering the cavity to be filled with a gas made up of photons, each endowed with an energy $h\nu$ and momentum $h\nu/c$. We now apply the thermodynamic formula

$$\left(\frac{\partial U}{\partial V} \right)_T = T \left(\frac{\partial p}{\partial T} \right)_V - p \tag{16.3-3}$$

to the photon gas or radiation in the cavity. Insertion of $U^{(r)} = V u^{(r)}$ and $p^{(r)} = \tfrac{1}{3} u^{(r)}$ into this relation gives the following ordinary differential equation for $u^{(r)}(T)$:

$$u^{(r)} = \tfrac{1}{3} T \frac{du^{(r)}}{dT} - \tfrac{1}{3} u^{(r)} \tag{16.3-4}$$

[1] J. de Boer, Chapter VII in *Leerboek der Natuurkunde*, 3rd edition, (R. Kronig, ed.), Scheltema and Holkema, Amsterdam (1951).

[2] H. B. Callen, *Thermodynamics and an Introduction to Thermostatistics*, 2nd edition, Wiley, New York (1985), pp. 78–79.

[3] M. Planck, *Vorlesungen über die Theorie der Wärmestrahlung*, 5th edition, Barth, Leipzig (1923); *Ann. Phys.*, **4**, 553–563, 564–566 (1901).

This equation can be integrated to give

$$u^{(r)} = bT^4 \qquad (16.3\text{-}5)$$

in which b is a constant of integration. Combination of this result with Eq. 16.3-1 gives the radiant energy emitted from the surface of a black body per unit area per unit time:

$$q_b^{(e)} = \frac{c}{4} u^{(r)} = \frac{cb}{4} T^4 = \sigma T^4 \qquad (16.3\text{-}6)$$

This is the Stefan–Boltzmann law. Note that the thermodynamic development does not predict the numerical value of σ.

The second way of deducing the Stefan–Boltzmann law is by integrating the *Planck distribution law*. This famous equation gives the radiated energy flux $q_{b\lambda}^{(e)}$ from a black surface in the wavelength range λ to $\lambda + d\lambda$:

$$q_{b\lambda}^{(e)} = \frac{2\pi c^2 h}{\lambda^5} \frac{1}{e^{ch/\lambda\kappa T} - 1} \qquad (16.3\text{-}7)$$

Here h is Planck's constant. The result can be derived by applying quantum statistics to a photon gas in a cavity, the photons obeying Bose–Einstein statistics.[4,5] The Planck distribution, which is shown in Fig. 16.3-1, correctly predicts the entire energy versus wavelength curve and the shift of the maximum toward shorter wavelengths at higher temperatures. When Eq. 16.3-7 is integrated over all wavelengths, we get

$$q_b^{(e)} = \int_0^\infty q_{b\lambda}^{(e)} d\lambda$$

$$= 2\pi c^2 h \int_0^\infty \frac{\lambda^{-5}}{e^{ch/\lambda\kappa T} - 1} d\lambda$$

$$= \frac{2\pi\kappa^4 T^4}{c^2 h^3} \int_0^\infty \frac{x^3}{e^x - 1} dx$$

$$= \frac{2\pi\kappa^4 T^4}{c^2 h^3} \left(6 \sum_{n=1}^\infty \frac{1}{n^4} \right)$$

$$= \frac{2\pi\kappa^4 T^4}{c^2 h^3} \left(\frac{\pi^4}{15} \right) \qquad (16.3\text{-}8)$$

In the above integration we changed the variable of integration from λ to $x = ch/\lambda\kappa T$. Then the integration over x was performed by expanding $1/(e^x - 1)$ in a Taylor series in e^x (see §C.2) and integrating term by term. The quantum statistical approach thus gives the details of the spectral distribution of the radiation and also the expression for the Stefan–Boltzmann constant,

$$\sigma = \frac{2}{15} \frac{\pi^5 \kappa^4}{c^2 h^3} \qquad (16.3\text{-}9)$$

having the value 1.355×10^{-12} cal/s $\cdot$ cm^2 $\cdot$ K, which is confirmed within experimental uncertainty by direct radiation measurements. Equation 16.3-9 is an amazing formula, interrelating as it does the σ from radiation, the κ from statistical mechanics, the speed of light c from electromagnetism, and the h from quantum mechanics.

In addition to obtaining the Stefan–Boltzmann law from the Planck distribution, we can get an important relation pertaining to the maximum in the Planck distribution. First

---

[4] J. E. Mayer and M. G. Mayer, *Statistical Mechanics*, Wiley, New York (1940), pp. 363–374.

[5] L. D. Landau and E. M. Lifshitz, *Statistical Physics*, 3rd edition, Part 1, Pergamon, Oxford (1980), §63.

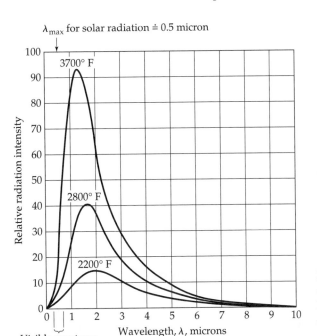

Fig. 16.3-1. The spectrum of equilibrium radiation as given by Planck's law. [M. Planck, *Verh. der deutschen physik. Gesell.*, **2**, 202, 237 (1900); *Ann. der Physik*, **4**, 553–563, 564–566 (1901).]

we rewrite Eq. 16.3-7 in terms of x and then set $dq_{b\lambda}^{(e)}/dx = 0$. This gives the following equation for x_{max}, which is the value of x for which the Planck distribution shows a maximum:

$$x_{max} = 5(1 - e^{-x_{max}}) \tag{16.3-10}$$

The solution to this equation is found numerically to be $x_{max} = 4.9651\ldots$ Hence at a given temperature T

$$\lambda_{max}T = \frac{ch}{\kappa x_{max}} \tag{16.3-11}$$

Inserting the values of the universal constants and the value for x_{max}, we then get

$$\lambda_{max}T = 0.2884 \text{ cm K} \tag{16.3-12}$$

This result, originally found experimentally,[6] is known as *Wien's displacement law*. It is useful primarily for estimating the temperature of remote objects. The law predicts, in agreement with experience, that the apparent color of radiation shifts from red (long wavelengths) toward blue (short wavelengths) as the temperature increases.

Finally, we may reinterpret some of our previous remarks in terms of the Planck distribution law. In Fig. 16.3-2 we have sketched three curves: the Planck distribution law for a hypothetical black body, the distribution curve for a hypothetical gray body, and a distribution curve for some real body. It is thus clear that when we use the total emissivity values, such as those in Table 16.2-1, we are just accounting empirically for the deviations from Planck's law over the entire spectrum.

We should not leave the subject of the Planck distribution without pointing out that Eq. 16.3-7 was presented at the October 1900 meeting of the German Physical Society as

---

[6] W. Wien, *Sitzungsber. d. kglch. preuss. Akad. d. Wissenschaften*, (VI), p. 55–62 (1893).

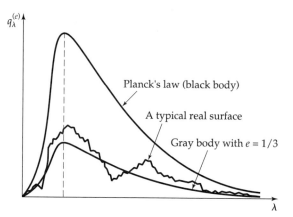

Fig. 16.3-2. Comparison of the emitted radiation from black, gray, and real surfaces.

an *empiricism* that fitted the available data.[7] However, before the end of the year,[8] Planck succeeded in *deriving* the equation, but at the expense of introducing the radical notion of the quantization of energy, an idea that was met with little enthusiasm. Planck himself had misgivings, as clearly stated in his textbook.[9] In a letter in 1931, he wrote: "... what I did can be described as an act of desperation. ... I had been wrestling unsuccessfully for six years... with the problem of equilibrium between radiation and matter, and I knew that the problem was of fundamental importance ..." Then Planck went on to say that he was "ready to sacrifice every one of my previous convictions about physical laws" except for the first and second laws of thermodynamics.[10] Planck's radical proposal ushered in a new and exciting era of physics, and quantum mechanics penetrated into chemistry and other fields in the twentieth century.

EXAMPLE 16.3-1

Temperature and Radiant-Energy Emission of the Sun

For approximate calculations, the sun may be considered a black body, emitting radiation with a maximum intensity at $\lambda = 0.5$ microns (5000 Å). With this information, estimate **(a)** the surface temperature of the sun, and **(b)** the emitted heat flux at the sun's surface.

SOLUTION

(a) From Wien's displacement law, Eq. 16.3-12,

$$T = \frac{0.2884}{\lambda_{\max}} = \frac{0.2884 \text{ cm K}}{0.5 \times 10^{-4} \text{ cm}} = 5760 \text{K} = 10{,}400 \text{ R} \tag{16.3-13}$$

(b) From the Stefan–Boltzmann law, Eq. 16.2-10,

$$q_b^{(e)} = \sigma T^4 = (0.1712 \times 10^{-8})(10{,}400)^4$$
$$= 2.0 \times 10^7 \text{ Btu/hr} \cdot \text{ft}^2 \tag{16.3-14}$$

---

[7] O. Lummer and E. Pringsheim, *Wied. Ann.*, **63**, 396 (1897); *Ann. der Physik*, **3**, 159 (1900).

[8] M. Planck, *Verhandl. d. deutsch. physik. Ges.*, **2**, 202 and 237 (1900); *Ann. Phys.*, **4**, 553–563, 564–566 (1901).

[9] M. Planck, *The Theory of Heat Radiation*, Dover, New York (1991), English translation of *Vorlesungen über die Theorie der Wärmestrahlung* (1913), p. 154.

[10] A. Hermann, *The Genesis of Quantum Theory*, MIT Press (1971), pp. 23–24.

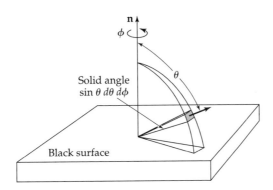

Fig. 16.4-1. Radiation at an angle θ from the normal to the surface into a solid angle $\sin\theta\,d\theta\,d\phi$.

§16.4 DIRECT RADIATION BETWEEN BLACK BODIES IN VACUO AT DIFFERENT TEMPERATURES

In the preceding sections we have given the Stefan–Boltzmann law, which describes the total radiant-energy emission from a perfectly black surface. In this section we discuss the radiant-energy transfer between two black bodies of arbitrary geometry and orientation. Hence we need to know how the radiant energy emanating from a black body is distributed with respect to angle. Because black-body radiation is isotropic, the following relation, known as *Lambert's cosine law*,[1] can be deduced:

$$q_{b\theta}^{(e)} = \frac{q_b^{(e)}}{\pi}\,\cos\,\theta = \frac{\sigma T^4}{\pi}\,\cos\,\theta \tag{16.4-1}$$

in which $q_{b\theta}^{(e)}$ is the energy emitted per unit area per unit time per unit solid angle in a direction θ (see Fig. 16.4-1). The energy emitted through the shaded solid angle is then $q_{b\theta}^{(e)}\sin\theta\,d\theta\,d\phi$ per unit area of black solid surface. Integration of the foregoing expression for $q_{b\theta}^{(e)}$ over the entire hemisphere gives the known total energy emission:

$$\int_0^{2\pi}\int_0^{\pi/2} q_{b\theta}^{(e)}\,\sin\,\theta\,d\theta\,d\phi = \frac{\sigma T^4}{\pi}\int_0^{2\pi}\int_0^{\pi/2}\cos\,\theta\,\sin\,\theta\,d\theta\,d\phi$$

$$= \sigma T^4 = q_b^{(e)} \tag{16.4-2}$$

This justifies the inclusion of the factor of $1/\pi$ in Eq. 16.4-1.

We are now in a position to get the net heat transfer rate from body 1 to body 2, where these are black bodies of any shape and orientation (see Fig. 16.4-2). We do this by getting the net heat transfer rate between a pair of surface elements dA_1 and dA_2 that can "see" each other, and then integrating over all such possible pairs of areas. The elements dA_1 and dA_2 are joined by a straight line of length r_{12}, which makes an angle θ_1 with the normal to dA_1 and an angle θ_2 with the normal to dA_2.

We start by writing an expression for the energy radiated from dA_1 into a solid angle $\sin\theta_1\,d\theta_1\,d\phi_1$ about r_{12}. We choose this solid angle large enough that dA_2 will lie entirely within the "beam" (see Fig. 16.4-2). According to Lambert's cosine law, the energy radiated per unit time will be

$$\left(\frac{\sigma T_1^4}{\pi}\,\cos\,\theta_1\right)dA_1\,\sin\,\theta_1\,d\theta_1\,d\phi_1 \tag{16.4-3}$$

---

[1] H. Lambert, *Photometria*, Augsburg (1760).

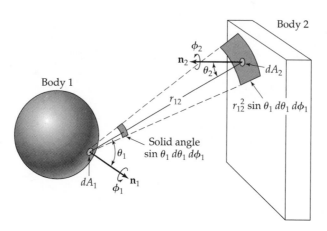

Fig. 16.4-2. Radiant interchange between two black bodies.

Of the energy leaving dA_1 at an angle θ_1, only the fraction given by the following ratio will be intercepted by dA_2:

$$\frac{\left(\begin{array}{l}\text{area of } dA_2 \text{ projected onto a}\\ \text{plane perpendicular to } r_{12}\end{array}\right)}{\left(\begin{array}{l}\text{area formed by the intersection}\\ \text{of the solid angle } \sin\theta_1\, d\theta_1\, d\phi_1\\ \text{with a sphere of radius } r_{12} \text{ with}\\ \text{center at } dA_1\end{array}\right)} = \frac{dA_2 \cos\theta_2}{r_{12}^2 \sin\theta_1\, d\theta_1\, d\phi_1} \tag{16.4-4}$$

Multiplication of these last two expressions then gives

$$dQ_{\underset{12}{\rightarrow}} = \frac{\sigma T_1^4}{\pi} \frac{\cos\theta_1 \cos\theta_2}{r_{12}^2} dA_1 dA_2 \tag{16.4-5}$$

This is the radiant energy emitted by dA_1 and intercepted by dA_2 per unit time. In a similar way we can write

$$dQ_{\underset{21}{\rightarrow}} = \frac{\sigma T_2^4}{\pi} \frac{\cos\theta_1 \cos\theta_2}{r_{12}^2} dA_1 dA_2 \tag{16.4-6}$$

which is the radiant energy emitted by dA_2 that is intercepted by dA_1 per unit time. The net rate of energy transport from dA_1 to dA_2 is then

$$dQ_{12} = dQ_{\underset{12}{\rightarrow}} - dQ_{\underset{21}{\rightarrow}}$$

$$= \frac{\sigma}{\pi}(T_1^4 - T_2^4)\frac{\cos\theta_1 \cos\theta_2}{r_{12}^2} dA_1 dA_2 \tag{16.4-7}$$

Therefore, the net rate of energy transfer from an isothermal black body 1 to another isothermal black body 2 is

$$Q_{12} = \frac{\sigma}{\pi}(T_1^4 - T_2^4)\int\int \frac{\cos\theta_1 \cos\theta_2}{r_{12}^2} dA_1 dA_2 \tag{16.4-8}$$

Here it is understood that the integration is restricted to those pairs of areas dA_1 and dA_2 that are in full view of each other. This result is conventionally written in the form

$$Q_{12} = A_1 F_{12}\sigma(T_1^4 - T_2^4) = A_2 F_{21}\sigma(T_1^4 - T_2^4) \tag{16.4-9}$$

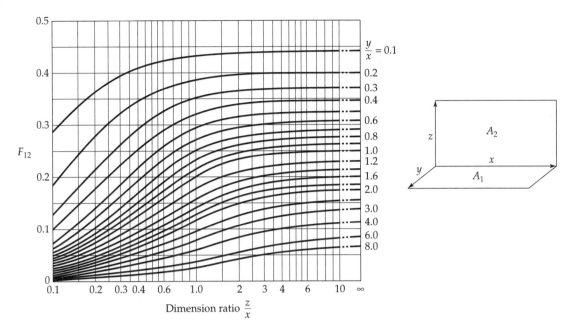

Fig. 16.4-3. View factors for direct radiation between adjacent rectangles in perpendicular planes [H. C. Hottel, Chapter 3 in W. H. McAdams, *Heat Transmission*, McGraw-Hill, New York (1954), p. 68].

where A_1 and A_2 are usually chosen to be the total areas of bodies 1 and 2. The dimensionless quantities F_{12} and F_{21}, called *view factors* (or *angle factors* or *configuration factors*), are given by

$$F_{12} = \frac{1}{\pi A_1} \int \int \frac{\cos \theta_1 \cos \theta_2}{r_{12}^2} dA_1 dA_2 \qquad (16.4\text{-}10)$$

$$F_{21} = \frac{1}{\pi A_2} \int \int \frac{\cos \theta_1 \cos \theta_2}{r_{12}^2} dA_1 dA_2 \qquad (16.4\text{-}11)$$

and the two view factors are related by $A_1 F_{12} = A_2 F_{21}$. The view factor F_{12} represents the fraction of radiation leaving body 1 that its directly intercepted by body 2.

The actual calculation of view factors is a difficult problem, except for some very simple situations. In Fig. 16.4-3 and Fig. 16.4-4 some view factors for direct radiation are shown.[2,3,4] When such charts are available, the calculations of energy interchanges by Eq. 16.4-9 are easy.

In the above development, we have assumed that Lambert's law and the Stefan–Boltzmann law may be used to describe the nonequilibrium transport process, in spite of the fact that they are strictly valid only for radiative equilibrium. The errors thus introduced do not seem to have been studied thoroughly, but apparently the resulting formulas give a good quantitative description.

---

[2] H. C. Hottel and A. F. Sarofim, *Radiative Transfer*, McGraw-Hill, New York (1967).

[3] H.C. Hottel, Chapter 4 in W. H. McAdams, *Heat Transmission*, McGraw-Hill, New York (1954).

[4] R. Siegel and J. R. Howell, *Thermal Radiation Heat Transfer*, 3rd edition, Hemisphere Publishing Co., New York (1992).

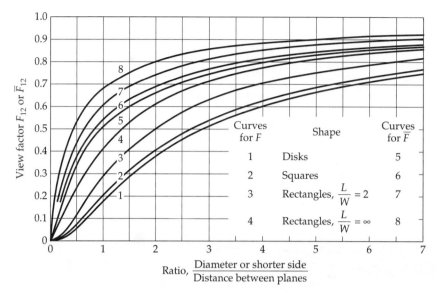

Fig. 16.4-4. View factors for direct radiation between opposed identical shapes in parallel planes. [H. C. Hottel, Chapter 3 in W. H. McAdams *Heat Transmission*, McGraw-Hill, New York (1954), Third Edition, p. 69.]

Thus far we have concerned ourselves with the radiative interactions between two black bodies. We now wish to consider a set of black surfaces 1, 2, . . . , n, which form the walls of a complete enclosure. The surfaces are maintained at temperatures $T_1, T_2, \ldots, T_n$, respectively. The net heat flow from any surface i to the enclosure surfaces is

$$Q_{ie} = \sigma A_i \sum_{j=1}^{n} F_{ij}(T_i^4 - T_j^4) \qquad i = 1, 2, \ldots, n \qquad (16.4\text{-}12)$$

or

$$Q_{ie} = \sigma A_i \left(T_i^4 - \sum_{j=1}^{n} F_{ij} T_j^4 \right) \qquad i = 1, 2, \ldots, n \qquad (16.4\text{-}13)$$

In writing the second form, we have used the relations

$$\sum_{j=1}^{n} F_{ij} = 1 \qquad i = 1, 2, \ldots, n \qquad (16.4\text{-}14)$$

The sums in Eqs. 16.4-13 and 14 include the term F_{ii}, which is zero for any object that intercepts none of its own rays. The set of n equations given in Eq. 16.4-12 (or Eq. 16.4-13) may be solved to get the temperatures or heat flows according to the data available.

A simultaneous solution of Eqs. 16.4-13 and 14 of special interest is that for which $Q_3 = Q_4 = \cdots = Q_n = 0$. Surfaces such as 3, 4, . . . , n are here called "adiabatic." In this situation one can eliminate the temperatures of all surfaces except 1 and 2 from the heat flow calculation and obtain an exact solution for the net heat flow from surface 1 to surface 2:

$$Q_{12} = A_1 \overline{F}_{12} \sigma (T_1^4 - T_2^4) = A_2 \overline{F}_{21} \sigma (T_1^4 - T_2^4) \qquad (16.4\text{-}15)$$

Values of $\overline{F}_{12}$ for use in this equation are given in Fig. 16.4-4. These values apply only when the adiabatic walls are formed from line elements perpendicular to surfaces 1 and 2.

The use of these view factors F and $\overline{F}$ greatly simplifies the calculations for blackbody radiation, when the temperatures of surfaces 1 and 2 are known to be uniform. The

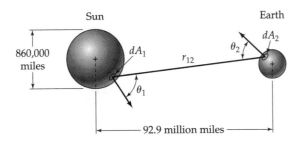

Fig. 16.4-5. Estimation of the solar constant.

reader wishing further information on radiative heat exchange in enclosures is referred to the literature.[4]

EXAMPLE 16.4-1

Estimation of the Solar Constant

The radiant heat flux entering the earth's atmosphere from the sun has been termed the "solar constant" and is important in solar energy utilization as well as in meteorology. Designate the sun as body 1 and the earth as body 2, and use the following data to calculate the solar constant: $D_1 = 8.60 \times 10^5$ miles; $r_{12} = 9.29 \times 10^7$ miles; $q_{b1}^{(e)} = 2.0 \times 10^7$ Btu/hr · ft² (from Example 16.3-1).

SOLUTION

In the terminology of Eq. 16.4-5 and Fig. 16.4-5,

$$\text{solar constant} = \frac{dQ_{\vec{12}}}{\cos\theta_2 dA_2} = \frac{\sigma T_1^4}{\pi r_{12}^2}\int \cos\theta_1 dA_1$$

$$= \frac{\sigma T_1^4}{\pi r_{12}^2}\left(\frac{\pi D_1^2}{4}\right) = \frac{q_{b1}^{(e)}}{4}\left(\frac{D_1}{r_{12}}\right)^2$$

$$= \frac{2.0 \times 10^7}{4}\left(\frac{8.60 \times 10^5}{9.29 \times 10^7}\right)^2$$

$$= 430 \text{ Btu/hr} \cdot \text{ft}^2 \tag{16.4-16}$$

This is in satisfactory agreement with other estimates that have been made. The treatment of r_{12}^2 as a constant in the integrand is permissible here because the distance r_{12} varies by less than 0.5% over the visible surface of the sun. The remaining integral, $\int \cos\theta_1 dA_1$, is the projected area of the sun as seen from the earth, or very nearly $\pi D_1^2/4$.

EXAMPLE 16.4-2

Radiant Heat Transfer Between Disks

Two black disks of diameter 2 ft are placed directly opposite one another at a distance of 4 ft. Disk 1 is maintained at 2000°R, and disk 2 at 1000°R. Calculate the heat flow between the two disks **(a)** when no other surfaces are present, and **(b)** when the two disks are connected by an adiabatic right-cylindrical black surface.

SOLUTION

(a) From Eq. 16.4-9 and curve 1 of Fig. 16.4-4,

$$Q_{12} = A_1 F_{12}\sigma(T_1^4 - T_2^4)$$
$$= \pi(0.06)(0.1712 \times 10^{-8})[(2000)^4 - (1000)^4]$$
$$= 4.83 \times 10^3 \text{ Btu/hr} \tag{16.4-17}$$

(b) From Eq. 16.4-15 and curve 5 of Fig. 16.4-4,

$$Q_{12} = A_1 \overline{F}_{12}\sigma(T_1^4 - T_2^4)$$
$$= \pi(0.34)(0.1712 \times 10^{-8})[(2000)^4 - (1000)^4]$$
$$= 27.4 \times 10^3 \text{ Btu/hr} \tag{16.4-18}$$

§16.5   RADIATION BETWEEN NONBLACK BODIES AT DIFFERENT TEMPERATURES

In principle, radiation between nonblack surfaces can be treated by differential analysis of emitted rays and their successive reflected components. For nearly black surfaces this is feasible, as only one or two reflections need be considered. For highly reflecting surfaces, however, the analysis is complicated, and the distributions of emitted and reflected rays with respect to angle and wavelength are not usually known with enough accuracy to justify a detailed calculation.

A reasonably accurate treatment is possible for a small convex surface in a large, nearly isothermal enclosure (i.e., a "cavity"), such as a steam pipe in a room with walls at constant temperature. The rate of energy emission from a nonblack surface 1 to the surrounding enclosure 2 is given by

$$Q_{1\bar{2}} = e_1 A_1 \sigma T_1^4 \tag{16.5-1}$$

and the rate of energy absorption from the surroundings by surface 1 is

$$Q_{\overrightarrow{21}} = a_1 A_1 \sigma T_2^4 \tag{16.5-2}$$

Here we have made use of the fact that the radiation impinging on surface 1 is very nearly cavity radiation or black-body radiation corresponding to temperature T_2. Since A_1 is convex, it intercepts none of its own rays; hence F_{12} has been set equal to unity. The net radiation rate from A_1 to the surroundings is therefore

$$Q_{12} = \sigma A_1 (e_1 T_1^4 - a_1 T_2^4) \tag{16.5-3}$$

In Eq. 16.5-3, e_1 is the value of the emissivity of surface 1 at T_1. The absorptivity a_1 is usually *estimated* as the value of e at T_2.

Next we consider an enclosure formed by n gray, opaque, diffuse-reflecting surfaces $A_1, A_2, A_3, \ldots, A_n$ at temperatures $T_1, T_2, T_3, \ldots, T_n$. Following Oppenheim[1] we define the *radiosity* J_i for each surface A_i as the sum of the fluxes of reflected and emitted radiant energy from A_i. Then the net radiant flow from A_i to A_k is expressed as

$$Q_{ik} = A_i F_{ik} (J_i - J_k) \qquad i, k = 1, 2, 3, \ldots, n \tag{16.5-4}$$

that is, by Eq. 16.4-9 with substitution of radiosities J_i in place of the black-body emissive powers σT_i^4.

The definition of J_i gives, for an opaque surface,

$$J_i = (1 - e_i) I_i + e_i \sigma T_i^4 \tag{16.5-5}$$

in which I_i is the incident radiant flux on A_i. Elimination of I_i in favor of the net radiant flux Q_{ie}/A_i from A_i into the enclosure gives

$$\frac{Q_{ie}}{A_i} = J_i - I_i = J_i - \frac{J_i - e_i \sigma T_i^4}{1 - e_i} \tag{16.5-6}$$

whence

$$\frac{Q_{ie}}{A_i} = \frac{e_i}{1 - e_i} A_i (\sigma T_i^4 - J_i) \tag{16.5-7}$$

Finally, a steady-state energy balance on each surface gives

$$Q_i = Q_{ie} = \sum_{k=1}^{n} Q_{ik} \tag{16.5-8}$$

Here Q_i is the rate of heat addition to surface A_i by nonradiative means.

---

[1] A. K. Oppenheim, *Trans. ASME*, **78**, 725–735 (1956); for earlier work, see G. Poljak, *Tech. Phys. USSR*, **1**, 555–590 (1935).

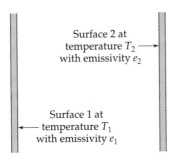

Fig. 16.5-1. Radiation between two infinite, parallel gray surfaces.

Surface 2 at
temperature T_2
with emissivity e_2

Surface 1 at
temperature T_1
with emissivity e_1

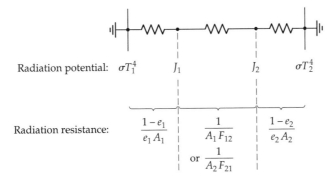

Fig. 16.5-2. Equivalent circuit for system shown in Fig. 16.5-1.

Radiation potential: σT_1^4 J_1 J_2 σT_2^4

Radiation resistance: $\dfrac{1-e_1}{e_1 A_1}$ $\dfrac{1}{A_1 F_{12}}$ or $\dfrac{1}{A_2 F_{21}}$ $\dfrac{1-e_2}{e_2 A_2}$

Equations analogous to Eqs. 16.5-4, 7, and 8 arise in the analysis of direct-current circuits, from Ohm's law of conduction and Kirchhoff's law of charge conservation. Hence we have the following analogies:

Electrical	*Radiative*
Current	Q
Voltage	J or σT^4
Resistance	$(1 - e_i)/e_i A_i$ or $1/A_i F_{ij}$

This analogy allows easy diagramming of equivalent circuits for visualization of simple enclosure radiation problems. For example, the system in Fig. 16.5-1 gives the equivalent circuit shown in Fig. 16.5-2 so that the net radiant heat transfer rate is

$$Q_{12} = \frac{\sigma(T_1^4 - T_2^4)}{\dfrac{1 - e_1}{e_1 A_1} + \dfrac{1}{A_1 F_{12}} + \dfrac{1 - e_2}{e_2 A_2}} \tag{16.5-9}$$

The shortcut solution summarized in Eq. 16.4-15 has been similarly generalized to non-black-walled enclosures giving

$$Q_{12} = A_1 \overline{F}_{12}(J_1 - J_2) \tag{16.5-10}$$

in place of Eq. 16.5-8, for an enclosure with $Q_i = 0$ for $i = 2, 3, \ldots, n$. The result is like that in Eq. 16.5-9, except that $\overline{F}_{12}$ must be used instead of F_{12} to include indirect paths from A_1 to A_2, thus giving a larger heat transfer rate.

EXAMPLE 16.5-1

Radiation Shields

Develop an expression for the reduction in radiant heat transfer between two infinite parallel gray planes having the same area, A, when a thin parallel gray sheet of very high thermal conductivity is placed between them as shown in Fig. 16.5-3.

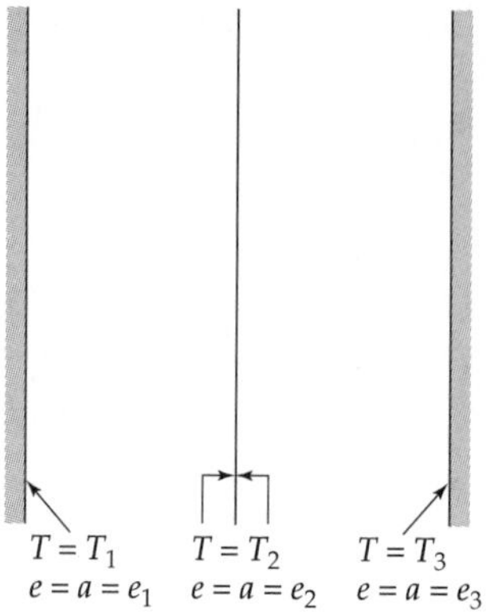

Fig. 16.5-3. Radiation shield.

$T = T_1$ $T = T_2$ $T = T_3$
$e = a = e_1$ $e = a = e_2$ $e = a = e_3$

SOLUTION

The radiation rate between planes 1 and 2 is given by

$$Q_{12} = \frac{A\sigma(T_1^4 - T_2^4)}{\dfrac{1 - e_1}{e_1} + 1 + \dfrac{1 - e_2}{e_2}} = \frac{A\sigma(T_1^4 - T_2^4)}{\dfrac{1}{e_1} + \dfrac{1}{e_2} - 1} \tag{16.5-11}$$

since both planes have the same area A and the view factor is unity. Similarly the heat transfer between planes 2 and 3 is

$$Q_{23} = \frac{A\sigma(T_2^4 - T_3^4)}{\dfrac{1 - e_2}{e_2} + 1 + \dfrac{1 - e_3}{e_3}} = \frac{A\sigma(T_2^4 - T_3^4)}{\dfrac{1}{e_2} + \dfrac{1}{e_3} - 1} \tag{16.5-12}$$

These last two equations may be combined to eliminate the temperature of the radiation shield, T_2, giving

$$Q_{12}\left(\frac{1}{e_1} + \frac{1}{e_2} - 1\right) + Q_{23}\left(\frac{1}{e_2} + \frac{1}{e_3} - 1\right) = A\sigma(T_1^4 - T_3^4) \tag{16.5-13}$$

Then, since $Q_{12} = Q_{23} = Q_{13}$, we get

$$Q_{13} = \frac{A\sigma(T_1^4 - T_3^4)}{\left(\dfrac{1}{e_1} + \dfrac{1}{e_2} - 1\right) + \left(\dfrac{1}{e_2} + \dfrac{1}{e_3} - 1\right)} \tag{16.5-14}$$

Finally, the ratio of radiant energy transfer with a shield to that without one is

$$\frac{(Q_{13})_{\text{with}}}{(Q_{13})_{\text{without}}} = \frac{\left(\dfrac{1}{e_1} + \dfrac{1}{e_3} - 1\right)}{\left(\dfrac{1}{e_1} + \dfrac{1}{e_2} - 1\right) + \left(\dfrac{1}{e_2} + \dfrac{1}{e_3} - 1\right)} \tag{16.5-15}$$

EXAMPLE 16.5-2

Radiation and Free-Convection Heat Losses from a Horizontal Pipe

Predict the total rate of heat loss, by radiation and free convection, from a unit length of horizontal pipe covered with asbestos. The outside diameter of the insulation is 6 in. The outer surface of the insulation is at 560°R, and the walls and air in the room are at 540°R.

SOLUTION

Let the outer surface of the insulation be surface 1 and the walls of the room be surface 2. Then Eq. 16.15-3 gives

$$Q_{12} = \sigma A_1 F_{12}(e_1 T_1^4 - a_1 T_2^4) \tag{16.5-16}$$

Since the pipe surface is convex and completely enclosed by surface 2, F_{12} is unity. From Table 16.2-1, we find $e_1 = 0.93$ at 560°R and $a_1 = 0.93$ at 540 R. Substitution of numerical values into Eq. 16.5-12 then gives for 1 ft of pipe:

$$Q_{12} = (0.1712 \times 10^{-8})(\pi/2)(1.00)[0.93(560)^4 - 0.93(540)^4] \qquad (16.5\text{-}17)$$
$$= 32 \text{ Btu/hr}$$

By adding the convection heat loss from Example 14.5-1, we obtain the total heat loss:

$$Q = Q^{(\text{conv})} + Q^{(\text{rad})} = 21 + 32 = 53 \text{ Btu/hr} \qquad (16.5\text{-}18)$$

Note that in this situation radiation accounts for more than half of the heat loss. If the fluid were *not* transparent, the convection and radiation processes would not be independent, and the convective and radiative contributions could not be added directly.

EXAMPLE 16.5-3

Combined Radiation and Convection

A body directly exposed to a clear night sky will be cooled below ambient temperature by radiation to outer space. This effect can be used to freeze water in shallow trays well insulated from the ground. Estimate the maximum air temperature for which freezing is possible, neglecting evaporation.

SOLUTION

As a first approximation, the following assumptions may be made:

a. All heat received by the water is by free convection from the surrounding air, which is assumed to be quiescent.

b. The heat effect of evaporation or condensation of water is not significant.

c. Steady state has been achieved.

d. The pan of water is square in cross section.

e. Back radiation from the atmosphere is neglected.

The maximum permitted air temperature at the water surface is $T_1 = 492$°R. The rate of *heat loss by radiation* is

$$Q^{(\text{rad})} = \sigma A_1 e_1 T_1^4 = (0.1712 \times 10^{-8})(L^2)(0.95)(402)^4$$
$$= 95 L^2 \text{ Btu/hr} \cdot \text{ft}^2 \qquad (16.5\text{-}19)$$

in which L is the length of one edge of the pan.

To get the *heat gain by convection*, we use the relation

$$Q^{(\text{conv})} = h L^2 (T_{\text{air}} - T_{\text{water}}) \qquad (16.5\text{-}20)$$

in which h is the heat transfer coefficient for free convection. For cooling atmospheric air by a horizontal square facing upward, the heat transfer coefficient is given by[2]

$$h = 0.2(T_{\text{air}} - T_{\text{water}})^{1/4} \qquad (16.5\text{-}21)$$

in which h is expressed in Btu/hr $\cdot$ ft^2 $\cdot$ F and the temperature is given in degrees Rankine.

When the foregoing expressions for heat loss by radiation and heat gain by free convection are equated, we get

$$95 L^2 = 0.2 L^2 (T_{\text{air}} - 492)^{5/4} \qquad (16.5\text{-}22)$$

From this we find that the maximum ambient air temperature is 630°R or 170°F. Except under desert conditions, back radiation and moisture condensation from the surrounding air greatly lower the required air temperature.

---

[2] W. H. McAdams, in *Chemical Engineers' Handbook* (J. H. Perry, Ed.), McGraw-Hill, New York (1950), 3rd edition, p. 474.

§16.6 RADIANT ENERGY TRANSPORT IN ABSORBING MEDIA[1]

The methods given in the preceding sections are applicable only to materials that are completely transparent or completely opaque. To describe energy transport in nontransparent media, we write differential equations for the local rate of change of energy as viewed from both the material and radiation standpoint. That is, we regard a material medium traversed by electromagnetic radiation as two coexisting "phases": a "material phase," consisting of all the mass in the system, and a "photon phase," consisting of the electromagnetic radiation.

In Chapter 11 we have already given an energy balance equation for a system containing no radiation. Here we extend Eq. 11.2-1 for the material phase to take into account the energy that is being interchanged with the photon phase by emission and absorption processes:

$$\frac{\partial}{\partial t} \rho \hat{U} = -(\nabla \cdot \rho \hat{U}\mathbf{v}) - (\nabla \cdot \mathbf{q}) - (\nabla \cdot p\mathbf{v}) - (\boldsymbol{\tau}:\nabla\mathbf{v}) - (\mathscr{E} - \mathscr{A}) \qquad (16.6\text{-}1)$$

Here we have introduced $\mathscr{E}$ and $\mathscr{A}$, which are the local rates of photon emission and absorption per unit volume, respectively. That is, $\mathscr{E}$ represents the energy lost by the material phase resulting from the emission of photons by molecules, and $\mathscr{A}$ represents the local gain of energy by the material phase resulting from photon absorption by the molecules (see Fig. 16.6-1). The $\mathbf{q}$ in Eq. 16.6-1 is the conduction heat flux given by Fourier's law.

For the "photon phase," we may write an equation describing the local rate of change of radiant energy density $u^{(r)}$:

$$\frac{\partial}{\partial t} u^{(r)} = -(\nabla \cdot \mathbf{q}^{(r)}) + (\mathscr{E} - \mathscr{A}) \qquad (16.6\text{-}2)$$

in which $\mathbf{q}^{(r)}$ is the radiant energy flux. This equation may be obtained by writing a radiant energy balance on an element of volume fixed in space. Note that there is no convec-

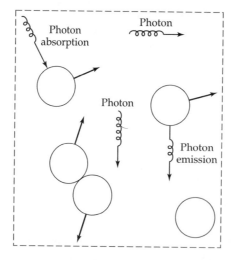

Fig. 16.6-1. Volume element over which energy balances are made; circles represent molecules.

[1] G. C. Pomraning, *Radiation Hydrodynamics*, Pergamon Press, New York (1973); R. Siegel and J. R. Howell, *Thermal Radiation Heat Transfer*, 3rd edition, Hemisphere Publishing Co., New York (1992).

tive term in Eq. 16.6-2, since the photons move independently of the local material velocity. Note further that the term $(\mathscr{E} - \mathscr{A})$ appears with opposite signs in Eqs. 16.6-1 and 2, indicating that a net gain of radiant energy occurs at the expense of molecular energy. Equation 16.6-2 can also be written for the radiant energy within a frequency range ν to $\nu + d\nu$:

$$\frac{\partial}{\partial t} u_\nu^{(r)} = -(\nabla \cdot \mathbf{q}_\nu^{(r)}) + (\mathscr{E}_\nu - \mathscr{A}_\nu) \tag{16.6-3}$$

This expression is obtained by differentiating Eq. 16.6-2 with respect to ν.

For the purpose of simplifying the discussion, we consider a steady-state nonflow system in which the radiation travels only in the positive z direction. Such a system can be closely approximated by passing a collimated light beam through a solution at temperatures sufficiently low that the emission by the solution is unimportant. (If emissions were important, it would be necessary to consider radiation in all directions.) These are the conditions commonly encountered in spectrophotometry. For such a system, Eqs. 16.6-1 and 2 become

$$0 = -\frac{d}{dz} q_z + \mathscr{A} \tag{16.6-4}$$

$$0 = -\frac{d}{dz} q_z^{(r)} - \mathscr{A} \tag{16.6-5}$$

In order to use these equations, we need information about the volumetric absorption rate $\mathscr{A}$. For a unidirectional beam a conventional expression is

$$\mathscr{A} = m_a q_z^{(r)} \tag{16.6-6}$$

in which m_a is known as the *extinction coefficient*. Basically, this states that the rate of photon absorption is proportional to the concentration of photons.

EXAMPLE 16.6-1

Absorption of a Monochromatic Radiant Beam

A monochromatic radiant beam of frequency ν, focused parallel to the z-axis, passes through an absorbing fluid. The local rate of energy absorption is given by $m_{a\nu} q_\nu^{(r)}$, in which $m_{a\nu}$ is the extinction coefficient for radiation of frequency ν. Determine the distribution of the radiant flux $q_\nu^{(r)}(z)$ in the system.

SOLUTION

We neglect refraction and scattering of the incident beam. Also, we assume that the liquid is cooled so that re-radiation can be neglected. Then Eq. 16.6-5 becomes for steady state

$$0 = -\frac{d}{dz} q_\nu^{(r)} - m_{a\nu} q_\nu^{(r)} \tag{16.6-7}$$

Integration with respect to z gives

$$q_\nu^{(r)}(z) = q_\nu^{(r)}(0) \exp(-m_{\alpha\nu} z) \tag{16.6-8}$$

This is *Lambert's law of absorption*,[2] widely used in spectrophotometry. For any given pure material, $m_{a\nu}$ depends in a characteristic way on ν. The shape of the absorption spectrum is therefore a useful tool for qualitative analysis.

---

[2] J. H. Lambert, *Photometria*, Augsburg (1760).

QUESTIONS FOR DISCUSSION

1. The "named laws" in this chapter are important. What is the physical content of the laws associated with the following scientists' names: Stefan and Boltzmann, Planck, Kirchhoff, Lambert, Wien?

2. How are the Stefan–Boltzmann law and the Wien displacement law related to the Planck black-body distribution law?

3. Do black bodies exist? Why is the concept of a black body useful?

4. In specular (mirrorlike) reflection, the angle of incidence equals the angle of reflection. How are these angles related for diffuse reflection?

5. What is the physical significance of the view factor, and how can it be calculated?

6. What are the units of $q^{(e)}$, $q_\nu^{(e)}$, and $q_\lambda^{(e)}$?

7. Under what conditions is the effect of geometry on radiant heat interchange completely expressible in terms of view factors?

8. Which of the equations in this chapter show that the apparent brightness of a black body with a uniform surface temperature is independent of the position (distance and direction) from which it is viewed through a transparent medium?

9. What relation is analogous to Eq. 16.3-2 for an ideal monatomic gas?

10. Check the dimensional consistency of Eq. 16.3-9.

PROBLEMS

16A.1. Approximation of a black body by a hole in a sphere. A thin sphere of copper, with its internal surface highly oxidized, has a diameter of 6 in. How small a hole must be made in the sphere to make an opening that will have an absorptivity of 0.99?

Answer: Radius = 0.70 in.

16A.2. Efficiency of a solar engine. A device for utilizing solar energy, developed by Abbot,[1] consists of a parabolic mirror that focuses the impinging sunlight onto a Pyrex tube containing a high-boiling, nearly black liquid. This liquid is circulated to a heat exchanger in which the heat energy is transferred to superheated water at 25 atm pressure. Steam may be withdrawn and used to run an engine. The most efficient design requires a mirror 10 ft in diameter to generate 2 hp, when the axis of the mirror is pointed directly toward the sun. What is the overall efficiency of the device?

Answer: 15%

16A.3. Radiant heating requirement. A shed is rectangular in shape, with the floor 15 ft by 30 ft and the roof 7.5 ft above the floor. The floor is heated by hot water running through coils. On cold winter days the exterior walls and roof are about −10°F. At what rate must heat be supplied through the floor in order to maintain the floor temperature at 75°F? (Assume that all surfaces of the system are black.)

16A.4. Steady-state temperature of a roof. Estimate the maximum temperature attained by a level roof at 45° north latitude on June 21 in clear weather. Radiation from sources other than the sun may be neglected, and a convection heat transfer coefficient of 2.0 Btu/hr · ft² · F may be assumed. A maximum temperature of 100°F may be assumed for the surrounding air. The solar constant of Example 16.4-1 may be used, and the absorption and scattering of the sun's rays by the atmosphere may be neglected.

(a) Solve for a perfectly black roof.

(b) Solve for an aluminum-coated roof, with an absorptivity of 0.3 for solar radiation and an emissivity of 0.07 at the temperature of the roof.

16A.5. Radiation errors in temperature measurements. The temperature of an air stream in a duct is being measured by means of a thermocouple. The thermocouple wires and junction are cylindrical, 0.05 in. in diameter, and extend across the duct perpendicular to the flow with the junction in the center of the duct. Assuming a junction emissivity $e = 0.8$, estimate the temperature of the gas stream from the following data obtained under steady conditions:

Thermocouple junction temperature	= 500°F
Duct wall temperature	= 300°F
Convection heat transfer coefficient from wire to air	= 50 Btu/hr · ft² · F

The wall temperature is constant at the value given for 20 duct diameters upstream and downstream of the thermocouple installation. The thermocouple leads are positioned so that the effect of heat conduction along them on the junction temperature may be neglected.

16A.6. Surface temperatures on the Earth's moon.

(a) Estimate the surface temperature of our moon at the point nearest the sun by a quasi-steady-state radiant energy balance, regarding the lunar surface as a gray sphere.

---

[1] C. G. Abbot, in *Solar Energy Research* (F. Daniels and J. A. Duffie, eds.), University of Wisconsin Press, Madison (1955), pp. 91–95; see also U.S. Patent No. 2,460,482 (Feb. 1, 1945).

Neglect radiation and reflection from the planets. The solar constant at Earth is given in Example 16.4-1.

(b) Extend part (a) to give the lunar surface temperature as a function of angular displacement from the hottest point.

16B.1. Reference temperature for effective emissivity. Show that, if the emissivity increases linearly with the temperature, Eq. 16.5-3 may be written as

$$Q_{12} = e_1^\circ \sigma A_1 (T_1^4 - T_2^4) \qquad (16B.1\text{-}1)$$

in which e_1° is the emissivity of surface 1 evaluated at a reference temperature T° given by

$$T^\circ = \frac{T_1^5 - T_2^5}{T_1^4 - T_2^4} \qquad (16B.1\text{-}2)$$

16B.2. Radiation across an annular gap. Develop an expression for the radiant heat transfer between two long, gray coaxial cylinders 1 and 2. Show that

$$Q_{12} = \frac{\sigma(T_1^4 - T_2^4)}{\dfrac{1}{A_1 e_1} + \dfrac{1}{A_2}\left(\dfrac{1}{e_2} - 1\right)} \qquad (16B.2\text{-}1)$$

where A_1 is the surface area of the inner cylinder.

16B.3. Multiple radiation shields.

(a) Develop an equation for the rate of radiant heat transfer through a series of n very thin, flat, parallel metal sheets, each having a different emissivity e, when the first sheet is at temperature T_1 and the nth sheet is at temperature T_n. Give your result in terms of the radiation resistances

$$R_{i,i+1} = \frac{\sigma(T_i^4 - T_{i+1}^4)}{Q_{i,i+1}} \qquad (16B.3\text{-}1)$$

for the successive pairs of planes. Edge effects and conduction across the air gaps between the sheets are to be neglected.

(b) Determine the ratio of the radiant heat transfer rate for n identical sheets to that for two identical sheets.

(c) Compare your results for three sheets with that obtained in Example 16.5-1.

The marked reduction in heat transfer rates produced by a number of radiation shields in series has led to the use of multiple layers of metal foils for high-temperature insulation.

16B.4. Radiation and conduction through absorbing media. A glass slab, bounded by planes $z = 0$ and $z = \delta$, is of infinite extent in the x and y directions. The temperatures of the surfaces at $z = 0$ and $z = \delta$ are maintained at T_0 and T_δ, respectively. A uniform monochromatic radiant beam of intensity $q_0^{(r)}$ in the z direction impinges on the face at $z = 0$. Emission within the slab, reflection, and incident radiation in the negative z direction can be neglected.

(a) Determine the temperature distribution in the slab, assuming m_a and k to be constants.

(b) How does the distribution of the conductive heat flux q_z depend on m_a?

16B.5. Cooling of a black body in vacuo. A thin black body of very high thermal conductivity has a volume V, surface area A, density ρ, and heat capacity $\hat{C}_p$. At time $t = 0$, this body at temperature T_1 is placed in a black enclosure, the walls of which are maintained permanently at temperature T_2 (with $T_2 < T_1$). Derive an expression for the temperature T of the black body as a function of time.

16B.6. Heat loss from an insulated pipe. A Schedule 40 two-inch horizontal steel pipe (inside diameter 2.067 in., wall thickness 0.154 in.; $k = 26$ Btu/hr · ft · F) carrying steam is insulated with 2 in. of 85% magnesia ($k = 0.35$ Btu/hr · ft · F) and tightly wrapped with a layer of clean aluminum foil ($e = 0.05$). The pipe is surrounded by air at 1 atm and 80°F and its inner surface is at 250°F.

(a) Compute the conductive heat flow per unit length, $Q^{(\text{cond})}/L$, through the pipe wall and insulation for assumed temperatures, T_0, of 100°F and 250°F at the outer surface of the aluminum foil.

(b) Compute the radiative and free-convective heat losses, $Q^{(\text{rad})}/L$ and $Q^{(\text{conv})}/L$, for the same assumed outer surface temperatures T_0.

(c) Plot or interpolate the foregoing results to obtain the steady-state values of T_0 and $Q^{(\text{cond})}/L = Q^{(\text{rad})}/L + Q^{(\text{conv})}/L$.

16C.1. Integration of the view-factor integral for a pair of disks (Fig. 16C.1). Two parallel, perfectly black disks of radius R are placed a distance H apart. Evaluate the view-factor integrals for this case and show that

$$F_{12} = F_{21} = \frac{1 + 2B^2 - \sqrt{1 + 4B^2}}{2B^2} \qquad (16.1\text{-}1)^2$$

in which $B = R/H$.

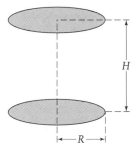

Fig. 16.C-1. Two perfectly black disks.

16D.1. Heat loss from a wire carrying an electric current.[3] An electrically heated wire of length L loses heat to the surroundings by radiative heat transfer. If the ends of the wire are maintained at a constant temperature T_0, obtain an expression for the axial variation in wire temperature. The wire can be considered to be radiating into a black enclosure at temperature T_0.

[2] C. Christiansen, *Wiedemann's Ann. d. Physik*, **19**, 267–283 (1883); see also M. Jakob, *Heat Transfer*, Vol. II, Wiley, New York (1957), p. 14.

[3] H. S. Carslaw and J. C. Jaeger, *Conduction of Heat in Solids*, 2nd edition, Oxford University Press (1959), pp. 154–156.

Part Three

Mass Transport

Chapter 17

Diffusivity and the Mechanisms of Mass Transport

In Chapter 1 we began by stating Newton's law of viscosity, and in Chapter 9 we began with Fourier's law of heat conduction. In this chapter we start by giving Fick's law of diffusion, which describes the movement of one chemical species A through a binary mixture of A and B because of a concentration gradient of A.

The movement of a chemical species from a region of high concentration to a region of low concentration can be observed by dropping a small crystal of potassium permanganate into a beaker of water. The $KMnO_4$ begins to dissolve in the water, and very near the crystal there is a dark purple, concentrated solution of $KMnO_4$. Because of the concentration gradient that is established, the $KMnO_4$ diffuses away from the crystal. The progress of the diffusion can then be followed by observing the growth of the dark purple region.

In §17.1 we give Fick's law for binary diffusion and define the diffusivity $\mathscr{D}_{AB}$ for the pair A–B. Then we discuss briefly the temperature and pressure dependence of the diffusivity. After that we give a summary of the theories available to predict the diffusivity for gases, liquids, colloids, and polymers. At the end of the chapter we discuss the transport of mass of a chemical species by convection, thus paralleling the treatments in Chapters 1 and 9 for momentum and heat transfer. We also introduce molar units and the notation needed for describing diffusion in these units. Finally, we give the Maxwell–Stefan equations for multicomponent gases at low densities.

Before starting the discussion we establish the following conventions. For *multicomponent diffusion*, we designate the species with lower-case Greek letters α, β, γ, ... and their concentrations with the corresponding subscripts. For *binary diffusion* we use the capital italic letters A and B. For *self-diffusion* (diffusion of chemically identical species) we label the species A and A^*. The "tagged" species A^* may differ physically from A by virtue of radioactivity or other nuclear properties such as the mass, magnetic moment, or spin.[1] The use of this system of notation enables one to see at a glance the type of system to which a given formula applies.

§17.1 FICK'S LAW OF BINARY DIFFUSION (MOLECULAR MASS TRANSPORT)

Consider a thin, horizontal, fused-silica plate of area A and thickness Y. Suppose that initially (for time $t < 0$) both horizontal surfaces of the plate are in contact with air, which we regard as completely insoluble in silica. At time $t = 0$, the air below the plate is suddenly replaced by pure helium, which is appreciably soluble in silica. The helium slowly penetrates into the plate by virtue of its molecular motion and ultimately appears in the gas above. This molecular transport of one substance relative to another is known as *diffusion* (also known as *mass diffusion*, *concentration diffusion*, or *ordinary diffusion*). The air above the plate is being replaced rapidly, so that there is no appreciable buildup of helium there. We thus have the situation represented in Fig. 17.1-1; this process is analogous to those described in Fig. 1.1-1 and Fig. 9.1-1 where viscosity and thermal conductivity were defined.

In this system, we will call helium "species A" and silica "species B." The concentrations will be given by the "mass fractions" ω_A and ω_B. The mass fraction ω_A is the mass of helium divided by the mass of helium plus silica in a given microscopic volume element. The mass fraction ω_B is defined analogously.

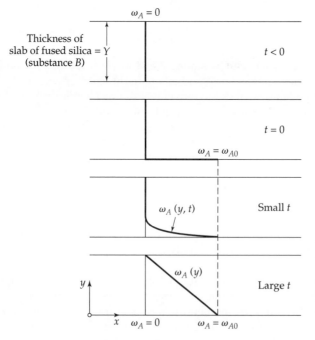

Fig. 17.1-1. Build-up to the steady-state concentration profile for the diffusion of helium (substance A) through fused silica (substance B). The symbol ω_A stands for the mass fraction of helium, and ω_{A0} is the solubility of helium in fused silica, expressed as the mass fraction. See Figs. 1.1-1 and 9.1-1 for the analogous momentum and heat transport situations.

---

[1] E. O. Stejskal and J. E. Tanner, *J. Chem. Phys.*, **42**, 288–292 (1965); P. Stilbs, *Prog. NMR Spectros*, **19**, 1–45 (1987); P. T. Callaghan and J. Stepišnik, *Adv. Magn. Opt. Reson.* **19**, 325–388 (1996).

For time t less than zero, the mass fraction of helium, ω_A, is everywhere equal to zero. For time t greater than zero, at the lower surface, $y = 0$, the mass fraction of helium is equal to ω_{A0}. This latter quantity is the solubility of helium in silica, expressed as mass fraction, just inside the solid. As time proceeds the mass fraction profile develops, with $\omega_A = \omega_{A0}$ at the bottom surface of the plate and $\omega_A = 0$ at the top surface of the plate. As indicated in Fig. 17.1-1, the profile tends toward a straight line with increasing t.

At steady state, it is found that the mass flow w_{Ay} of helium in the positive y direction can be described to a very good approximation by

$$\frac{w_{Ay}}{A} = \rho \mathscr{D}_{AB} \frac{\omega_{A0} - 0}{Y} \tag{17.1-1}$$

That is, the mass flow rate of helium per unit area (or *mass flux*) is proportional to the mass fraction difference divided by the plate thickness. Here ρ is the density of the silica–helium system, and the proportionality factor $\mathscr{D}_{AB}$ is the *diffusivity* of the silica–helium system. We now rewrite Eq. 17.1-1 for a differential element within the slab:

$$j_{Ay} = -\rho \mathscr{D}_{AB} \frac{d\omega_A}{dy} \tag{17.1-2}$$

Here w_{Ay}/A has been replaced by j_{Ay}, the *molecular mass flux* of helium in the positive y direction. Note that the first index, A, designates the chemical species (in this case, helium), and the second index indicates the direction in which diffusive transport is taking place (in this case, the y direction).

Equation 17.1-2 is the one-dimensional form of *Fick's first law of diffusion*.[1] It is valid for any binary solid, liquid, or gas solution, provided that j_{Ay} is defined as the mass flux relative to the mixture velocity v_y. For the system examined in Fig. 17.1-1, the helium is moving rather slowly and its concentration is very small, so that v_y is negligibly different from zero during the diffusion process.

In general, for a binary mixture

$$v_y = \omega_A v_{Ay} + \omega_B v_{By} \tag{17.1-3}$$

Thus **v** is an average in which the species velocities, $\mathbf{v}_A$ and $\mathbf{v}_B$, are weighted according to the mass fractions. This kind of velocity is referred to as the *mass average velocity*. The species velocity $\mathbf{v}_A$ is not the instantaneous molecular velocity of a molecule of A, but rather the arithmetic average of the velocities of all the molecules of A within a tiny volume element.

The mass flux j_{Ay} is then defined, in general, as

$$j_{Ay} = \rho \omega_A (v_{Ay} - v_y) \tag{17.1-4}$$

The mass flux of B is defined analogously. As the two chemical species interdiffuse there is, locally, a shifting of the center of mass in the y direction if the molecular weights of A and B differ. The mass fluxes j_{Ay} and j_{By} are so defined that $j_{Ay} + j_{By} = 0$. In other words, the fluxes j_{Ay} and j_{By} are measured with respect to the motion of the center of mass. This point will be discussed in detail in §§17.7 and 8.

If we write equations similar to Eq. 17.1-2 for the x and z directions and then combine all three equations, we get the vector form of Fick's law:

$$\mathbf{j}_A = -\rho \mathscr{D}_{AB} \nabla \omega_A \tag{17.1-5}$$

---

[1] A. Fick, *Ann. der Physik*, **94**, 59–86 (1855). Fick's second law, the diffusional analog of the heat conduction equation in Eq. 11.2-10, is given in Eq. 19.1-18. **Adolf Eugen Fick** (1829–1901) was a medical doctor who taught in Zürich and Marburg, and later became the Rector of the University of Würzburg. He postulated the laws of diffusion by analogy with heat conduction, not by experiment.

A similar relation can be written for species B:

$$\mathbf{j}_B = -\rho \mathcal{D}_{BA} \nabla \omega_B \tag{17.1-6}$$

It is shown in Example 17.1-2 that $\mathcal{D}_{BA} = \mathcal{D}_{AB}$. Thus for the pair $A–B$, there is just one diffusivity; in general it will be a function of pressure, temperature, and composition.

The mass diffusivity $\mathcal{D}_{AB}$, the thermal diffusivity $\alpha = k/\rho\hat{C}_p$, and the momentum diffusivity (kinematic viscosity) $\nu = \mu/\rho$ all have dimensions of (length)^2/time. The ratios of these three quantities are therefore dimensionless groups:

The Prandtl number:
$$\mathrm{Pr} = \frac{\nu}{\alpha} = \frac{\hat{C}_p \mu}{k} \tag{17.1-7}$$

The Schmidt number:[2]
$$\mathrm{Sc} = \frac{\nu}{\mathcal{D}_{AB}} = \frac{\mu}{\rho \mathcal{D}_{AB}} \tag{17.1-8}$$

The Lewis number:[2]
$$\mathrm{Le} = \frac{\alpha}{\mathcal{D}_{AB}} = \frac{k}{\rho \hat{C}_p \mathcal{D}_{AB}} \tag{17.1-9}$$

These dimensionless groups of fluid properties play a prominent role in dimensionless equations for systems in which competing transport processes occur. (*Note:* Sometimes the Lewis number is defined as the inverse of the expression above.)

In Tables 17.1-1, 2, 3, and 4 some values of $\mathcal{D}_{AB}$ in cm^2/s are given for a few gas, liquid, solid, and polymeric systems. These values can be converted easily to m^2/s by multiplication by 10^{-4}. Diffusivities of gases at low density are almost independent of ω_A, increase with temperature, and vary inversely with pressure. Liquid and solid diffusivities are strongly concentration-dependent and generally increase with temperature. There are numerous experimental methods for measuring diffusivities, and some of these are described in subsequent chapters.[3]

For *gas mixtures*, the Schmidt number can range from about 0.2 to 3, as can be seen in Table 17.1-1. For *liquid mixtures*, values up to 40,000 have been observed.[4]

Up to this point we have been discussing isotropic fluids, in which the speed of diffusion does not depend on the orientation of the fluid mixture. For some solids and structured fluids, the diffusivity will have to be a tensor rather than a scalar, so that Fick's first law has to be modified thus:

$$\mathbf{j}_A = -[\rho \boldsymbol{\Delta}_{AB} \cdot \nabla \omega_A] \tag{17.1-10}$$

in which $\boldsymbol{\Delta}_{AB}$ is the (symmetric) *diffusivity tensor*.[5,6] According to this equation, the mass flux is not necessarily collinear with the mass fraction gradient. We do not pursue this subject further here.

---

[2] These groups were named for: **Ernst Heinrich Wilhelm Schmidt** (1892–1975), who taught at the universities in Gdansk, Braunschweig, and Munich (where he was the successor to Nusselt); **Warren Kendall Lewis** (1882–1975), who taught at MIT and was a coauthor of a pioneering textbook, W. H. Walker, W. K. Lewis, and W. H. McAdams, *Principles of Chemical Engineering*, McGraw-Hill, New York (1923).

[3] For an extensive discussion, see W. E. Wakeham, A. Nagashima, and J. V. Sengers, *Measurement of the Transport Properties of Fluids: Experimental Thermodynamics, Vol. III*, CRC Press, Boca Raton, Fla. (1991).

[4] D. A. Shaw and T. J. Hanratty, *AIChE Journal*, **23**, 28–37, 160–169 (1977); P. Harriott and R. M. Hamilton, *Chem. Eng. Sci.*, **20**, 1073–1078 (1965).

[5] For flowing polymers, theoretical expressions for the diffusion tensor have been derived using kinetic theory; see H. C. Öttinger, *AIChE Journal*, **35**, 279–286 (1989), and C. F. Curtiss and R. B. Bird, *Adv. Polym. Sci.*, 1–101 (1996), §§6 and 15.

[6] M. E. Glicksman, *Diffusion in Solids: Field Theory, Solid State Principles, and Applications*, Wiley, New York (2000).

Table 17.1-1   Experimental Diffusivities[a] and Limiting Schmidt Numbers[b] of Gas Pairs at 1 Atmosphere Pressure

Gas pair A–B	Temperature (K)	$\mathcal{D}_{AB}$ (cm^2/s)	Sc $x_A \to 1$	Sc $x_B \to 1$
CO_2–N_2O	273.2	0.096	0.73	0.72
CO_2–CO	273.2	0.139	0.50	0.96
CO_2–N_2	273.2	0.144	0.48	0.91
	288.2	0.158	0.49	0.92
	298.2	0.165	0.50	0.93
N_2–C_2H_6	298.2	0.148	1.04	0.51
N_2–nC_4H_{10}	298.2	0.0960	1.60	0.33
N_2–O_2	273.2	0.181	0.72	0.74
H_2–SF_6	298.2	0.420	3.37	0.055
H_2–CH_4	298.2	0.726	1.95	0.23
H_2–N_2	273.2	0.674	1.40	0.19
NH_3–H_2[c]	263	0.58	0.19[e]	1.53
NH_3–N_2[c]	298	0.233	0.62[e]	0.65
H_2O–N_2[c]	308	0.259	0.58[e]	0.62
H_2O–O_2[c]	352	0.357	0.56[e]	0.59
C_3H_8–nC_4H_{10}[d]	378.2	0.0768	0.95	0.66
	437.7	0.107	0.91	0.63
C_3H_8–iC_4H_{10}[d]	298.0	0.0439	1.04	0.73
	378.2	0.0823	0.89	0.63
	437.8	0.112	0.87	0.61
C_3H_8–neo-C_5H_{12}[d]	298.1	0.0431	1.06	0.56
	378.2	0.0703	1.04	0.55
	437.7	0.0945	1.03	0.55
nC_4H_{10}–neo-C_5H_{12}[d]	298.0	0.0413	0.76	0.59
	378.2	0.0644	0.78	0.61
	437.8	0.0839	0.80	0.62
iC_4H_{10}–neo-C_5H_{12}[d]	298.1	0.0362	0.89	0.67
	378.2	0.0580	0.89	0.67
	437.7	0.0786	0.87	0.66

[a] Unless otherwise indicated, the values are taken from J. O. Hirschfelder, C. F. Curtiss, and R. B. Bird, *Molecular Theory of Gases and Liquids*, 2nd corrected printing, Wiley, New York (1964), p. 579. All values are given for 1 atmosphere pressure.

[b] Calculated using the Lennard-Jones parameters of Table E.1. The parameters for sulfur hexafluoride were obtained from second virial coefficient data.

[c] Values of $\mathcal{D}_{AB}$ for the water and ammonia mixtures are taken from the tabulation of R. C. Reid, J. M. Prausnitz, and B. E. Poling, *The Properties of Gases and Liquids*, 4th edition, McGraw-Hill, New York (1987).

[d] Values of $\mathcal{D}_{AB}$ for the hydrocarbon–hydrocarbon pairs are taken from S. Gotoh, M. Manner, J. P. Sørensen, and W. E. Stewart, *J. Chem. Eng. Data*, **19**, 169–171 (1974).

[e] Values of μ for water and ammonia were calculated from functions provided by T. E. Daubert, R. P. Danner, H. M. Sibul, C. C. Stebbins, J. L. Oscarson, R. L. Rowley, W. V. Wilding, M. E. Adams, T. L. Marshall, and N. A. Zundel, *DIPPR®*, *Data Compilation of Pure Compound Properties*, Design Institute for Physical Property Data®, AIChE, New York, N.Y. (2000).

Table 17.1-2 Experimental Diffusivities in the Liquid State[a,b]

A	B	T(°C)	x_A	$\mathscr{D}_{AB} \times 10^5$ (cm^2/s)
Chlorobenzene	Bromobenzene	10.10	0.0332	1.007
			0.2642	1.069
			0.5122	1.146
			0.7617	1.226
			0.9652	1.291
		39.92	0.0332	1.584
			0.2642	1.691
			0.5122	1.806
			0.7617	1.902
			0.9652	1.996
Water	n-Butanol	30	0.131	1.24
			0.222	0.920
			0.358	0.560
			0.454	0.437
			0.524	0.267
Ethanol	Water	25	0.026	1.076
			0.266	0.368
			0.408	0.405
			0.680	0.743
			0.880	1.047
			0.944	1.181

[a] The data for the first two pairs are taken from a review article by P. A. Johnson and A. L. Babb, *Chem. Revs.*, **56**, 387–453 (1956). Other summaries of experimental results may be found in: P. W. M. Rutten, *Diffusion in Liquids*, Delft University Press, Delft, The Netherlands (1992); L. J. Gosting, *Adv. in Protein Chem.*, Vol. XI, Academic Press, New York (1956); A. Vignes, *I. E. C. Fundamentals*, **5**, 189–199 (1966).
[b] The ethanol–water data were taken from M. T. Tyn and W. F. Calus, *J. Chem. Eng. Data*, **20**, 310–316 (1975).

Table 17.1-3 Experimental Diffusivities in the Solid State[a]

A	B	T(°C)	$\mathscr{D}_{AB}$ (cm^2/s)
He	SiO$_2$	20	2.4–5.5 × 10^{-10}
He	Pyrex	20	4.5 × 10^{-11}
		500	2 × 10^{-8}
H$_2$	SiO$_2$	500	0.6–2.1 × 10^{-8}
H$_2$	Ni	85	1.16 × 10^{-8}
		165	10.5 × 10^{-8}
Bi	Pb	20	1.1 × 10^{-16}
Hg	Pb	20	2.5 × 10^{-15}
Sb	Ag	20	3.5 × 10^{-21}
Al	Cu	20	1.3 × 10^{-30}
Cd	Cu	20	2.7 × 10^{-15}

[a] It is presumed that in each of the above pairs, component A is present only in very small amounts. The data are taken from R. M. Barrer, *Diffusion in and through Solids*, Macmillan, New York (1941), pp. 141, 222, and 275.

Table 17.1-4   Experimental Diffusivities of Gases in Polymers.[a] Diffusivities, $\mathcal{D}_{AB}$, are given in units of 10^{-6} (cm^2/s). The values for N_2 and O_2 are for 298K, and those for CO_2 and H_2 are for 198K.

	N_2	O_2	CO_2	H_2
Polybutadiene	1.1	1.5	1.05	9.6
Silicone rubber	15	25	15	75
Trans-1,4-polyisoprene	0.50	0.70	0.47	5.0
Polystyrene	0.06	0.11	0.06	4.4

[a] Excerpted from D. W. van Krevelen, *Properties of Polymers*, 3rd edition, Elsevier, Amsterdam (1990), pp. 544–545. Another relevant reference is S. Pauly, in *Polymer Handbook*, 4th edition (J. Brandrup and E. H. Immergut, eds.), Wiley-Interscience, New York (1999), Chapter VI.

In this section we have discussed the diffusion that occurs as a result of a concentration gradient in the system. We refer to this kind of diffusion as *concentration diffusion* or *ordinary diffusion*. There are, however, other kinds of diffusion: *thermal diffusion*, which results from a temperature gradient; *pressure diffusion*, resulting from a pressure gradient; and *forced diffusion*, which is caused by unequal external forces acting on the chemical species. For the time being, we consider only concentration diffusion, and we postpone discussion of the other mechanisms to Chapter 24. Also, in that chapter we discuss the use of activity, rather than concentration, as the driving force for ordinary diffusion.

EXAMPLE 17.1-1

Diffusion of Helium through Pyrex Glass

Calculate the steady-state mass flux j_{Ay} of helium for the system of Fig. 17.1-1 at 500K. The partial pressure of helium is 1 atm at $y = 0$ and zero at the upper surface of the plate. The thickness Y of the pyrex plate is 10^{-2} mm, and its density $\rho^{(B)}$ is 2.6 g/cm^3. The solubility and diffusivity of helium in pyrex are reported[7] as 0.0084 volumes of gaseous helium per volume of glass, and $\mathcal{D}_{AB} = 0.2 \times 10^{-7}$ cm^2/s, respectively. Show that the neglect of the mass average velocity implicit in Eq. 17.1-1 is reasonable.

SOLUTION

The mass concentration of helium in the glass at the lower surface is obtained from the solubility data and the ideal gas law:

$$\rho_{A0} = (0.0084) \frac{p_{A0} M_A}{RT}$$

$$= (0.0084) \frac{(1.0 \text{ atm})(4.00 \text{ g/mole})}{(82.05 \text{ cm}^3 \text{ atm/mole K})(773 \text{K})}$$

$$= 5.3 \times 10^{-7} \text{ g/cm}^3 \tag{17.1-11}$$

The mass fraction of helium in the solid phase at the lower surface is then

$$\omega_{A0} = \frac{\rho_{A0}}{\rho_{A0} + \rho_{B0}} = \frac{5.3 \times 10^{-7}}{5.3 \times 10^{-7} + 2.6} = 2.04 \times 10^{-7} \tag{17.1-12}$$

---

[7] C. C. Van Voorhis, *Phys. Rev.* **23**, 557 (1924), as reported by R. M. Barrer, *Diffusion in and through Solids*, corrected printing, Cambridge University Press (1951).

We may now calculate the flux of helium from Eq. 17.1-1 as

$$j_{Ay} = (2.6 \text{ g/cm}^3)(2.0 \times 10^{-8} \text{ cm}^2/\text{s}) \frac{2.04 \times 10^{-7}}{10^{-3} \text{ cm}}$$
$$= 1.05 \times 10^{-11} \text{ g/cm}^2\text{s} \tag{17.1-13}$$

Next, the velocity of the helium can be obtained from Eq. 17.1-4:

$$v_{Ay} = \frac{j_{Ay}}{\rho_A} + v_y \tag{17.1-14}$$

At the lower surface of the plate ($y = 0$) this velocity has the value

$$v_{Ay}\big|_{y=0} = \frac{1.05 \times 10^{-11} \text{ g/cm}^2 \text{ s}}{5.3 \times 10^{-7} \text{ g/cm}^3} + v_{y0} = 1.98 \times 10^{-5} \text{ cm/s} + v_{y0} \tag{17.1-15}$$

The corresponding value v_{y0} of the mass average velocity of the glass–helium system at $y = 0$ is then obtained from Eq. 17.1-3

$$v_{y0} = (2.04 \times 10^{-7})(1.98 \times 10^{-5} \text{ cm/s} + v_{y0}) + (1 - 2.04 \times 10^{-7})(0) \tag{17.1-16}$$
$$v_{y0} = \frac{(2.04 \times 10^{-7})(1.98 \times 10^{-5} \text{ cm/s})}{1 - (2.04 \times 10^{-7})}$$
$$= 4.04 \times 10^{-12} \text{ cm/s}$$

Thus it is safe to neglect v_y in Eq. 17.1-14, and the analysis of the experiment in Fig. 17.1-1 at steady state is accurate.

EXAMPLE 17.1-2

The Equivalence of $\mathcal{D}_{AB}$ and $\mathcal{D}_{BA}$

Show that only one diffusivity is needed to describe the diffusional behavior of a binary mixture.

SOLUTION

We begin by writing Eq. 17.1-6 as follows:

$$\mathbf{j}_B = -\rho \mathcal{D}_{BA} \nabla \omega_B = +\rho \mathcal{D}_{BA} \nabla \omega_A \tag{17.1-17}$$

The second form of this equation follows from the fact that $\omega_A + \omega_B = 1$. We next use the vector equivalents of Eqs. 17.1-3 and 4 to write

$$\mathbf{j}_A = \rho \omega_A (\mathbf{v}_A - \omega_A \mathbf{v}_A - \omega_B \mathbf{v}_B)$$
$$= \rho \omega_A ((1 - \omega_A)\mathbf{v}_A - \omega_B \mathbf{v}_B)$$
$$= \rho \omega_A \omega_B (\mathbf{v}_A - \mathbf{v}_B) \tag{17.1-18}$$

Interchanging A and B in this expression shows that $\mathbf{j}_A = -\mathbf{j}_B$. Combining this with the second form of Eq. 17.1-17 then gives

$$\mathbf{j}_A = -\rho \mathcal{D}_{BA} \nabla \omega_A \tag{17.1-19}$$

Comparing this with Eq. 17.1-5 gives $\mathcal{D}_{BA} = \mathcal{D}_{AB}$. We find that the order of subscripts is unimportant for a binary system and that only one diffusivity is required to describe the diffusional behavior.

However, it may well be that the diffusivity for a dilute solution of A in B and that for a dilute solution of B in A are *numerically* different. The reason for this is that the diffusivity is concentration-dependent, so that the two limiting values mentioned above are the values of the diffusivity $\mathcal{D}_{BA} = \mathcal{D}_{AB}$ at two different concentrations.

§17.2   TEMPERATURE AND PRESSURE DEPENDENCE OF DIFFUSIVITIES

In this section we discuss the prediction of the diffusivity $\mathcal{D}_{AB}$ for binary systems by corresponding-states methods. These methods are also useful for extrapolating existing data. Comparisons of many alternative methods are available in the literature.[1,2]

For binary gas mixtures at low pressure, $\mathcal{D}_{AB}$ is inversely proportional to the pressure, increases with increasing temperature, and is almost independent of the composition for a given gas pair. The following equation for estimating $\mathcal{D}_{AB}$ at low pressures has been developed[3] from a combination of kinetic theory and corresponding-states arguments:

$$\frac{p\mathcal{D}_{AB}}{(p_{cA}p_{cB})^{1/3}(T_{cA}T_{cB})^{5/12}(1/M_A + 1/M_B)^{1/2}} = a\left(\frac{T}{\sqrt{T_{cA}T_{cB}}}\right)^b \qquad (17.2\text{-}1)$$

Here $\mathcal{D}_{AB}$ [=] cm^2/s, p [=] atm, and T [=] K. Analysis of experimental data gives the dimensionless constants $a = 2.745 \times 10^{-4}$ and $b = 1.823$ for nonpolar gas pairs, excluding helium and hydrogen, and $a = 3.640 \times 10^{-4}$ and $b = 2.334$ for pairs consisting of H_2O and a nonpolar gas. Equation 17.2-1 fits the experimental data at atmospheric pressure within an average deviation of 6 to 8%. If the gases A and B are nonpolar and their Lennard-Jones parameters are known, the kinetic-theory method described in the next section usually gives somewhat better accuracy.

At high pressures, and in the liquid state, the behavior of $\mathcal{D}_{AB}$ is more complicated. The simplest and best understood situation is that of self-diffusion (interdiffusion of labeled molecules of the same chemical species). We discuss this case first and then extend the results approximately to binary mixtures.

A corresponding-states plot of the self-diffusivity $\mathcal{D}_{AA^*}$ for nonpolar substances is given in Fig. 17.2-1.[4] This plot is based on self-diffusion measurements, supplemented by molecular dynamics simulations and by kinetic theory for the low-pressure limit. The ordinate is $c\mathcal{D}_{AA^*}$ at pressure p and temperature T, divided by $c\mathcal{D}_{AA^*}$ at the critical point. This quantity is plotted as a function of the reduced pressure $p_r = p/p_c$ and the reduced temperature $T_r = T/T_c$. Because of the similarity of species A and the labeled species A^*, the critical properties are all taken as those of species A.

From Fig. 17.2-1 we see that $c\mathcal{D}_{AA^*}$ increases strongly with temperature, especially for liquids. At each temperature $c\mathcal{D}_{AA^*}$ decreases toward zero with increasing pressure. With decreasing pressure, $c\mathcal{D}_{AA^*}$ increases toward a low-pressure limit, as predicted by kinetic theory (see §17.3). The reader is warned that this chart is tentative, and that the lines, except for the low-density limit, are based on data for a very few substances: Ar, Kr, Xe, and CH_4.

The quantity $(c\mathcal{D}_{AA^*})_c$ may be estimated by one of the following three methods:

(i)   Given $c\mathcal{D}_{AA^*}$ at a known temperature and pressure, one can read $(c\mathcal{D}_{AA^*})_r$ from the chart and get $(c\mathcal{D}_{AA^*})_c = c\mathcal{D}_{AA^*}/(c\mathcal{D}_{AA^*})_r$.

[1] R. C. Reid, J. M. Prausnitz, and B. E. Poling, *The Properties of Gases and Liquids*, 4th edition, McGraw-Hill, New York (1987), Chapter 11.

[2] E. N. Fuller, P. D. Shettler, and J. C. Giddings, *Ind. Eng. Chem.*, **58**, No. 5, 19–27 (1966); Erratum: *ibid*. **58**, No. 8, 81 (1966). This paper gives a useful method for predicting binary gas diffusivities from the molecular formulas of the two species.

[3] J. C. Slattery and R. B. Bird, *AIChE Journal*, **4**, 137–142 (1958).

[4] Other correlations for self-diffusivity at elevated pressures have appeared in Ref. 3 and in L. S. Tee, G. F. Kuether, R. C. Robinson, and W. E. Stewart, *API Proceedings, Division of Refining*, 235–243 (1966); R. C. Robinson and W. E. Stewart, *IEC Fundamentals*, **7**, 90–95 (1968); J. L. Bueno, J. Dizy, R. Alvarez, and J. Coca, *Trans. Inst. Chem. Eng.*, **68**, Part A, 392–397 (1990).

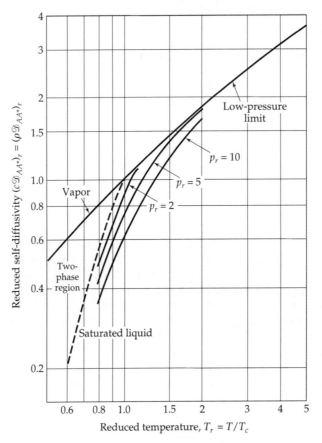

Fig. 17.2-1. A corresponding-states plot for the reduced self-diffusivity. Here $(c\mathscr{D}_{AA^*})_r =$ $(\rho \mathscr{D}_{AA^*})_r$ for Ar, Kr, Xe, and CH$_4$ is plotted as a function of reduced temperature for several values of the reduced pressure. This chart is based on diffusivity data of J. J. van Loef and E. G. D. Cohen, *Physica A*, **156**, 522–533 (1989), the compressibility function of B. I. Lee and M. G. Kesler, *AIChE Journal*, **21**, 510–527 (1975), and Eq. 17.3-11 for the low-pressure limit.

(ii) One can predict a value of $c\mathscr{D}_{AA^*}$ in the low-density region by the methods given in §17.3 and then proceed as in (i).

(iii) One can use the empirical formula (see Problem 17A.9):

$$(c\mathscr{D}_{AA^*})_c = 2.96 \times 10^{-6}\left(\frac{1}{M_A} + \frac{1}{M_{A^*}}\right)^{1/2} \frac{p_{cA}^{2/3}}{T_{cA}^{1/6}} \tag{17.2-2}$$

This equation, like Eq. 17.2-1, should not be used for helium or hydrogen isotopes. Here $c [=]$ g-mole/cm^3, $\mathscr{D}_{AA^*} [=]$ cm^2/s, $T_c [=]$ K, and $p_c [=]$ atm.

Thus far the discussion of high-density behavior has been concerned with self-diffusion. We turn now to the binary diffusion of chemically dissimilar species. In the absence of other information it is suggested that Fig. 17.2-1 may be used for crude estimation of $c\mathscr{D}_{AB}$, with p_{cA} and T_{cA} replaced everywhere by $\sqrt{p_{cA}p_{cB}}$ and $\sqrt{T_{cA}T_{cB}}$ respectively (see Problem 17A.9 for the basis for this empiricism). The ordinate of the plot is then interpreted as $(c\mathscr{D}_{AB})_r = c\mathscr{D}_{AB}/(c\mathscr{D}_{AB})_c$ and Eq. 17.2-2 is replaced by

$$(c\mathscr{D}_{AB})_c = 2.96 \times 10^{-6}\left(\frac{1}{M_A} + \frac{1}{M_B}\right)^{1/2} \frac{(p_{cA}p_{cB})^{1/3}}{(T_{cA}T_{cB})^{1/12}} \tag{17.2-3}$$

With these substitutions, accurate results are obtained in the low-pressure limit. At higher pressures, very few data are available for comparison, and the method must be regarded as provisional.

The results in Fig. 17.2-1, and their extensions to binary systems, are expressed in terms of $c\mathscr{D}_{AA^*}$ and $c\mathscr{D}_{AB}$ rather than $\mathscr{D}_{AA^*}$ and $\mathscr{D}_{AB}$. This is done because the *c*-multiplied diffusion coefficients are more frequently required in mass transfer calculations, and their dependence on pressure and temperature is simpler.

EXAMPLE 17.2-1

Estimation of Diffusivity at Low Density

Estimate $\mathcal{D}_{AB}$ for the system CO–CO$_2$ at 296.1K and 1 atm total pressure.

SOLUTION

The properties needed for Eq. 17.2-1 are (see Table E.1):

Label	Species	M	T_c (K)	p_c (atm)
A	CO	28.01	133	34.5
B	CO$_2$	44.01	304.2	72.9

Therefore,

$$(p_{cA}p_{cB})^{1/3} = (34.5 \times 72.9)^{1/3} = 13.60$$

$$(T_{cA}T_{cB})^{5/12} = (133 \times 304.2)^{5/12} = 83.1$$

$$\left(\frac{1}{M_A} + \frac{1}{M_B}\right)^{1/2} = \left(\frac{1}{28.01} + \frac{1}{44.01}\right)^{1/2} = 0.2417$$

$$a\left(\frac{T}{\sqrt{T_{cA}T_{cB}}}\right)^b = 2.745 \times 10^{-4}\left(\frac{296.1}{\sqrt{133 \times 304.2}}\right)^{1.823} = 5.56 \times 10^{-4}$$

Substitution of these values into Eq. 17.2-1 gives

$$(1.0)\mathcal{D}_{AB} = (5.56 \times 10^{-4})(13.60)(83.1)(24.17) \tag{17.2-4}$$

This gives $\mathcal{D}_{AB} = 0.152$ cm^2/s, in agreement with the experimental value.[5] This is unusually good agreement.

  This problem can also be solved by means of Fig. 17.2-1 and Eq. 17.2-3, together with the ideal gas law $p = cRT$. The result is $\mathcal{D}_{AB} = 0.140$ cm^2/s, in fair agreement with the data.

EXAMPLE 17.2-2

Estimation of Self-Diffusivity at High Density

Estimate $c\mathcal{D}_{AA^*}$ for C^{14}O$_2$ in ordinary CO$_2$ at 171.7 atm and 373K. It is known[6] that $\mathcal{D}_{AA^*} = 0.113$ cm^2/s at 1.00 atm and 298K, at which condition $c = p/RT = 4.12 \times 10^{-5}$ g-mole/cm^3.

SOLUTION

Since a measured value of $\mathcal{D}_{AA^*}$ is given, we use method (i). The reduced conditions of the measurement are $T_r = 298/304.2 = 0.980$ and $p_r = 1.00/72.9 = 0.014$. Then from Fig. 17.2-1 we get the value $(c\mathcal{D}_{AA^*})_r = 0.98$. Hence

$$(c\mathcal{D}_{AA^*})_c = \frac{c\mathcal{D}_{AA^*}}{(c\mathcal{D}_{AA^*})_r} = \frac{(4.12 \times 10^{-5})(0.113)}{0.98}$$

$$= 4.75 \times 10^{-6} \text{ g-mol/cm} \cdot \text{s} \tag{17.2-5}$$

At the conditions of prediction ($T_r = 373/304.2 = 1.23$ and $p_r = 171.7/72.9 = 2.36$), we read $(c\mathcal{D}_{AA^*})_r = 1.21$. The predicted value is then

$$c\mathcal{D}_{AA^*} = (c\mathcal{D}_{AA^*})_r(c\mathcal{D}_{AA^*})_c = (1.21)(4.75 \times 10^{-6})$$

$$= 5.75 \times 10^{-6} \text{ g-mole/cm} \cdot \text{s} \tag{17.2-6}$$

The data of O'Hern and Martin[7] give $c\mathcal{D}_{AA^*} = 5.89 \times 10^{-6}$ g-mole/cm · s at these conditions. This good agreement is not unexpected, inasmuch as their low-pressure data were used in the estimation of $(c\mathcal{D}_{AA^*})_c$.

---

[5] B. A. Ivakin, P. E. Suetin, *Sov. Phys. Tech. Phys.* (English translation), **8**, 748–751 (1964).
[6] E. B. Wynn, *Phys. Rev.*, **80**, 1024–1027 (1950).
[7] H. A. O'Hern and J. J. Martin, *Ind. Eng. Chem.*, **47**, 2081–2086 (1955).

This problem can also be solved by method (iii) without an experimental value of $c\mathcal{D}_{AA^*}$. Equation 17.4-2 gives directly

$$(c\mathcal{D}_{AA^*})_c = 2.96 \times 10^{-6} \left(\frac{1}{44.01} + \frac{1}{46} \right)^{1/2} \frac{(72.9)^{2/3}}{(304.2)^{1/6}}$$

$$= 4.20 \times 10^{-6} \text{ g-mole/cm} \cdot \text{s} \tag{17.2-7}$$

The resulting predicted value of $c\mathcal{D}_{AA^*}$ is 5.1×10^{-6} g-mole/cm · s.

EXAMPLE 17.2-3

Estimation of Binary Diffusivity at High Density

Estimate $c\mathcal{D}_{AB}$ for a mixture of 80 mole% CH_4 and 20 mole% C_2H_6 at 136 atm and 313K. It is known that, at 1 atm and 293K, the molar density is $c = 4.17 \times 10^{-5}$ g-mole/cm³ and $\mathcal{D}_{AB} = 0.163$ cm²/s.

SOLUTION

Figure 17.2-1 is used, with method (i). The reduced conditions for the known data are

$$T_r = \frac{T}{\sqrt{T_{cA}T_{cB}}} = \frac{293}{\sqrt{(190.7)(305.4)}} = 1.22 \tag{17.2-8}$$

$$p_r = \frac{p}{\sqrt{p_{cA}p_{cB}}} = \frac{1.0}{\sqrt{(45.8)(48.2)}} = 0.021 \tag{17.2-9}$$

From Fig. 17.2-1 at these conditions we obtain $(c\mathcal{D}_{AB})_r = 1.21$. The critical value $(c\mathcal{D}_{AB})_c$ is therefore

$$(c\mathcal{D}_{AB})_c = \frac{c\mathcal{D}_{AB}}{(c\mathcal{D}_{AB})_r} = \frac{(4.17 \times 10^{-5})(0.163)}{1.21}$$

$$= 5.62 \times 10^{-6} \text{ g-mol/cm} \cdot \text{s} \tag{17.2-10}$$

Next we calculate the reduced conditions for the prediction ($T_r = 1.30$, $p_r = 2.90$) and read the value $(c\mathcal{D}_{AB})_r = 1.31$ from Fig. 17.2-1. The predicted value of $c\mathcal{D}_{AB}$ is therefore

$$c\mathcal{D}_{AB} = (c\mathcal{D}_{AB})_r(c\mathcal{D}_{AB})_c = (1.31)(5.62 \times 10^{-6})$$

$$= 7.4 \times 10^{-6} \text{ g-mole/cm} \cdot \text{s} \tag{17.2-11}$$

Experimental measurements[8] give $c\mathcal{D}_{AB} = 6.0 \times 10^{-6}$, so that the predicted value is 23% high. Deviations of this magnitude are not unusual in the estimation of $c\mathcal{D}_{AB}$ at high densities.

An alternative solution may be obtained by method (iii). Substitution into Eq. 17.4-3 gives

$$(c\mathcal{D}_{AB})_c = 2.96 \times 10^{-6} \left(\frac{1}{16.04} + \frac{1}{30.07} \right)^{1/2} \frac{(45.8 \times 48.2)^{1/3}}{(190.7 \times 305.4)^{1/12}}$$

$$= 4.78 \times 10^{-6} \text{ g-mole/cm} \cdot \text{s} \tag{17.2-12}$$

Multiplication by $(c\mathcal{D}_{AB})_r$ at the desired condition gives

$$c\mathcal{D}_{AB} = (4.78 \times 10^{-6})(1.31)$$

$$= 6.26 \times 10^{-6} \text{ g-mole/cm} \cdot \text{s} \tag{17.2-13}$$

This is in closer agreement with the measured value.[8]

---

[8] V. J. Berry, Jr., and R. C. Koeller, *AIChE Journal*, **6**, 274–280 (1960).

§17.3   THEORY OF DIFFUSION IN GASES AT LOW DENSITY

The mass diffusivity $\mathcal{D}_{AB}$ for binary mixtures of nonpolar gases is predictable within about 5% by kinetic theory. As in the earlier kinetic theory discussions in §§1.4 and 9.3, we start with a simplified derivation to illustrate the mechanisms involved and then present the more accurate results of the Chapman–Enskog theory.

Consider a large body of gas containing molecular species A and A^*, which are identical except for labeling. We wish to determine the self-diffusivity $\mathcal{D}_{AA^*}$ in terms of the molecular properties on the assumption that the molecules are rigid spheres of equal mass m_A and diameter d_A.

Since the properties of A and A^* are nearly the same, we can use the following results of the kinetic theory for a pure rigid-sphere gas at low density in which the gradients of temperature, pressure, and velocity are small:

$$\bar{u} = \sqrt{\frac{8\kappa T}{\pi m}} \quad = \text{mean molecular speed relative to } \mathbf{v} \tag{17.3-1}$$

$$Z = \tfrac{1}{4}n\bar{u} \quad = \text{wall collision frequency per unit area in a stationary gas} \tag{17.3-2}$$

$$\lambda = \frac{1}{\sqrt{2}\pi d^2 n} = \text{mean free path} \tag{17.3-3}$$

The molecules reaching any plane in the gas have, on the average, had their last collision at a distance a from the plane, where

$$a = \tfrac{2}{3}\lambda \tag{17.3-4}$$

In these equations n is the number density (total number of molecules per unit volume).

To predict the self-diffusivity $\mathcal{D}_{AA^*}$, we consider the motion of species A in the y direction under a mass fraction gradient $d\omega_A/dy$ (see Fig. 17.3-1), where the fluid mixture moves in the y direction at a finite velocity mass average velocity v_y throughout. The temperature T and the total molar mass concentration ρ are considered constant. We assume that Eqs. 17.3-1 to 4 remain valid in this nonequilibrium situation. The net mass flux of species A crossing a unit area of any plane of constant y is found by writing an expression for the mass of A crossing the plane in the positive y direction and subtracting the mass of A crossing in the negative y direction:

$$(\rho\omega_A v_y)\big|_y + [(\tfrac{1}{4}\rho\omega_A\bar{u})\big|_{y-a} - (\tfrac{1}{4}\rho\omega_A\bar{u})\big|_{y+a}] \tag{17.3-5}$$

Here the first term is the mass transport in the y direction because of the mass motion of the fluid—that is, the convective transport—and the last two terms give the molecular transport relative to v_y.

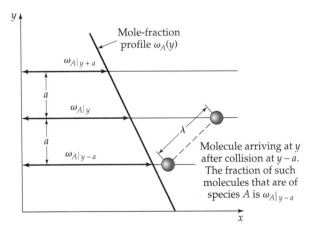

Fig. 17.3-1. Molecular transport of species A from the plane at $(y - a)$ to the plane at y.

It is assumed that the concentration profile $\omega_A(y)$ is very nearly linear over distances of several mean free paths. Then we may write

$$\omega_A\big|_{y\pm a} = \omega_A\big|_y \pm \tfrac{2}{3}\lambda \frac{d\omega_A}{dy} \tag{17.3-6}$$

Combination of the last two equations then gives for the *combined mass flux* at plane y:

$$n_{Ay} = \rho\omega_A v_y - \tfrac{1}{3}\bar{u}\rho\lambda \frac{d\omega_A}{dy}$$

$$\equiv \rho\omega_A v_y - \rho\mathscr{D}_{AA^*}\frac{d\omega_A}{dy} \tag{17.3-7}$$

This is the *convective mass flux* plus the *molecular mass flux*, the latter being given by Eq. 17.1-1. Therefore we get the following expression for the self-diffusivity:

$$\mathscr{D}_{AA^*} = \tfrac{1}{3}\,\bar{u}\lambda \tag{17.3-8}$$

Finally, making use of Eqs. 17.3-1 and 3, we get

$$\mathscr{D}_{AA^*} = \frac{2}{3}\frac{\sqrt{\kappa T/\pi m_A}}{\pi d_A^2}\frac{1}{n} = \frac{2}{3\pi}\frac{\sqrt{\pi m_A \kappa T}}{\pi d_A^2}\frac{1}{\rho} \tag{17.3-9}$$

which can be compared with Eq. 1.4-9 for the viscosity and Eq. 9.3-12 for the thermal conductivity.

The development of a formula for $\mathscr{D}_{AB}$ for rigid spheres of unequal masses and diameters is considerably more difficult. We simply quote the result[1] here:

$$\mathscr{D}_{AB} = \frac{2}{3}\sqrt{\frac{\kappa T}{\pi}}\sqrt{\frac{1}{2}\left(\frac{1}{m_A}+\frac{1}{m_B}\right)}\frac{1}{\pi(\frac{1}{2}(d_A+d_B))^2}\frac{1}{n} \tag{17.3-10}$$

That is, $1/m_A$ is replaced by the arithmetic average of $1/m_A$ and $1/m_B$, and d_A by the arithmetic average of d_A and d_B.

The preceding discussion shows how the diffusivity can be obtained by mean free path arguments. For accurate results the Chapman–Enskog kinetic theory should be used. The Chapman–Enskog results for viscosity and thermal conductivity were given in §§1.4 and 9.3, respectively. The corresponding formula for $c\mathscr{D}_{AB}$ is:[2,3]

$$c\mathscr{D}_{AB} = \frac{3}{16}\sqrt{\frac{2RT}{\pi}\left(\frac{1}{M_A}+\frac{1}{M_B}\right)}\frac{1}{\tilde{N}\sigma_{AB}^2\Omega_{\mathscr{D},AB}}$$

$$= 2.2646 \times 10^{-5}\sqrt{T\left(\frac{1}{M_A}+\frac{1}{M_B}\right)}\frac{1}{\sigma_{AB}^2\Omega_{\mathscr{D},AB}} \tag{17.3-11}$$

Or, if we approximate c by the ideal gas law $p = cRT$, we get for $\mathscr{D}_{AB}$

$$\mathscr{D}_{AB} = \frac{3}{16}\sqrt{\frac{2(RT)^3}{\pi}\left(\frac{1}{M_A}+\frac{1}{M_B}\right)}\frac{1}{\tilde{N}p\sigma_{AB}^2\Omega_{\mathscr{D},AB}}$$

$$= 0.0018583\sqrt{T^3\left(\frac{1}{M_A}+\frac{1}{M_B}\right)}\frac{1}{p\sigma_{AB}^2\Omega_{\mathscr{D},AB}} \tag{17.3-12}$$

In the second line of Eqs. 17.3-11 and 12, $\mathscr{D}_{AB}\,[=]\,cm^2/s$, $\sigma_{AB}\,[=]\,\text{Å}$, $T\,[=]\,K$, and $p\,[=]\,atm$.

---

[1] A similar result is given by R. D. Present, *Kinetic Theory of Gases*, McGraw-Hill, New York (1958), p. 55.

[2] S. Chapman and T. G. Cowling, *The Mathematical Theory of Non-Uniform Gases*, 3rd edition, Cambridge University Press (1970), Chapters 10 and 14.

[3] J. O. Hirschfelder, C. F. Curtiss, and R. B. Bird, *Molecular Theory of Gases and Liquids*, 2nd corrected printing, Wiley, New York (1964), p. 539.

The dimensionless quantity $\Omega_{\mathscr{D},AB}$—the "collisional integral" for diffusion—is a function of the dimensionless temperature $\kappa T/\varepsilon_{AB}$. The parameters σ_{AB} and ε_{AB} are those appearing in the Lennard-Jones potential between one molecule of A and one of B (cf. Eq. 1.4-10):

$$\varphi_{AB}(r) = 4\varepsilon_{AB}\left[\left(\frac{\sigma_{AB}}{r}\right)^{12} - \left(\frac{\sigma_{AB}}{r}\right)^{6}\right] \tag{17.3-13}$$

This function $\Omega_{\mathscr{D},AB}$ is given in Table E.2 and Eq. E.2-2. From these results one can compute that $\mathscr{D}_{AB}$ increases roughly as the 2.0 power of T at low temperatures and as the 1.65 power of T at very high temperatures; see the $p_r \to 0$ curve in Fig. 17.2-1. For rigid spheres, $\Omega_{\mathscr{D},AB}$ would be unity at all temperatures and a result analogous to Eq. 17.3-10 would be obtained.

The parameters σ_{AB} and ε_{AB} could, in principle, be determined directly from accurate measurements of $\mathscr{D}_{AB}$ over a wide range of temperatures. Suitable data are not yet available for many gas pairs, and one may have to resort to using some other measurable property, such as the viscosity[4] of a binary mixture of A and B. In the event that there are no such data, then we can estimate σ_{AB} and ϵ_{AB} from the following combining rules:[5]

$$\sigma_{AB} = \tfrac{1}{2}(\sigma_A + \sigma_B); \qquad \varepsilon_{AB} = \sqrt{\varepsilon_A \varepsilon_B} \tag{17.3-14, 15}$$

for nonpolar gas pairs. Use of these combining rules enables us to predict values of $\mathscr{D}_{AB}$ within about 6% by use of viscosity data on the pure species A and B, or within about 10% if the Lennard-Jones parameters for A and B are estimated from boiling point data by use of Eq. 1.4-12.[6]

For isotopic pairs, $\sigma_{AA^*} = \sigma_A = \sigma_{A^*}$ and $\varepsilon_{AA^*} = \varepsilon_A = \varepsilon_{A^*}$; that is, the intermolecular force fields for the various pairs A–A^*, A^*–A^*, and A–A are virtually identical, and the parameters σ_A and ε_A may be obtained from viscosity data on pure A. If, in addition, M_A is large, Eq. 17.3-11 simplifies to

$$c\mathscr{D}_{AA^*} = 3.2027 \times 10^{-5} \sqrt{\frac{T}{M_A}} \frac{1}{\sigma_A^2 \Omega_{\mathscr{D},AA^*}} \tag{17.3-16}$$

The corresponding equation for the rigid-sphere model is given in Eq. 17.3-9.

Comparison of Eq. 17.3-16 with Eq. 1.4-14 shows that the self-diffusivity $\mathscr{D}_{AA^*}$ and the viscosity μ (or kinematic viscosity ν) are related as follows for heavy isotopic gas pairs at low density:

$$\frac{\mu}{\rho\mathscr{D}_{AA^*}} = \frac{\nu}{\mathscr{D}_{AA^*}} = \frac{5}{6}\frac{\Omega_{\mathscr{D},AA^*}}{\Omega_\mu} \tag{17.3-17}$$

in which $\Omega_\mu \approx 1.1\Omega_{\mathscr{D},AA^*}$ over a wide range of $\kappa T/\varepsilon_A$, as may be seen in Table E.2. Thus $\mathscr{D}_{AA^*} \approx 1.32\nu$ for the *self-diffusivity*. The relation between ν and the *binary diffusivity* $\mathscr{D}_{AB}$ is not so simple, because ν may vary considerably with the composition. The Schmidt number $\mathrm{Sc} = \mu/\rho\mathscr{D}_{AB}$ is in the range from 0.2 to 5.0 for most gas pairs.

Equations 17.3-11, 12, 16, and 17 were derived for monatomic nonpolar gases but have been found useful for polyatomic nonpolar gases as well. In addition, these equations may be used to predict $\mathscr{D}_{AB}$ for interdiffusion of a polar gas and a nonpolar gas by using combining laws different[7] from those given in Eq. 17.3-14 and 15.

---

[4] S. Weissman and E. A. Mason, *J. Chem. Phys.*, **37**, 1289–1300 (1962); S. Weissman, *J. Chem. Phys.*, **40**, 3397–3406 (1964).

[5] J. O. Hirschfelder, R. B. Bird, and E. L. Spotz, *Chem. Revs.*, **44**, 205–231 (1949); S. Gotoh, M. Manner, J. P. Sørensen, and W. E. Stewart, *J. Chem. Eng. Data*, **19**, 169–171 (1974).

[6] R. C. Reid, J. M. Prausnitz, and B. E. Poling, *The Properties of Gases and Liquids*, 4th edition, McGraw-Hill, New York (1987).

[7] J. O. Hirschfelder, C. F. Curtiss, and R. B. Bird, *Molecular Theory of Gases and Liquids*, 2nd corrected printing, Wiley, New York (1964), §8.6b and p. 1201. Polar gases and gas mixtures are discussed by E. A. Mason and L. Monchick, *J. Chem. Phys.* **36**, 2746–2757 (1962).

EXAMPLE 17.3-1

Computation of Mass Diffusivity for Low-Density Gases

Predict the value of $\mathcal{D}_{AB}$ for the system CO–CO_2 at 296.1K and 1.0 atm total pressure.

SOLUTION

From Table E.1 we obtain the following parameters:

CO: $M_A = 28.01$ $\sigma_A = 3.590$ Å $\varepsilon_A/\kappa = 100$K

CO_2: $M_B = 44.01$ $\sigma_B = 3.996$ Å $\varepsilon_B/\kappa = 190$K

The mixture parameters are then estimated from Eqs. 17.3-14 and 15:

$$\sigma_{AB} = \tfrac{1}{2}(3.590 + 3.996) = 3.793 \text{ Å} \tag{17.3-18}$$

$$\varepsilon_{AB}/\kappa = \sqrt{(110)(190)} = 144.6\text{K} \tag{17.3-19}$$

The dimensionless temperature is then $\kappa T/\varepsilon_{AB} = (296.1)/(144.6) = 2.048$. From Table E.2 we can find the collision integral for diffusion, $\Omega_{\mathcal{D},AB} = 1.067$. Substitution of the preceding values in Eq. 17.3-12 gives

$$\mathcal{D}_{AB} = 0.0018583 \sqrt{(296.1)^3 \left(\frac{1}{28.01} + \frac{1}{44.01}\right)} \frac{1}{(1.0)(3.793)^2(1.067)}$$

$$= 0.149 \text{ cm}^2/\text{s} \tag{17.3-20}$$

§17.4 THEORY OF DIFFUSION IN BINARY LIQUIDS

The kinetic theory for diffusion in simple liquids is not as well developed as that for dilute gases, and it cannot presently give accurate analytical predictions of diffusivities.[1–3] As a result our understanding of liquid diffusion depends primarily on the rather crude hydrodynamic and activated-state models. These in turn have spawned a number of empirical correlations, which provide the best available means for prediction. These correlations permit estimation of diffusivities in terms of more easily measured properties such as viscosity and molar volume.

The *hydrodynamic theory* takes as its starting point the Nernst–Einstein equation,[4] which states that the diffusivity of a single particle or solute molecule A through a stationary medium B is given by

$$\mathcal{D}_{AB} = \kappa T(u_A/F_A) \tag{17.4-1}$$

in which u_A/F_A is the "mobility" of a particle A (that is, the steady-state velocity attained by the particle under the action of a unit force). The origin of Eq. 17.4-1 is discussed in §17.5 in connection with the Brownian motion of colloidal suspensions. If the shape and size of A are known, the mobility can be calculated by the solution of the creeping-flow equation of motion[5] (Eq. 3.5-8). Thus, if A is spherical and if one takes into account the possibility of "slip" at the fluid–solid interface, one obtains[6]

$$\frac{u_A}{F_A} = \left(\frac{3\mu_B + R_A\beta_{AB}}{2\mu_B + R_A\beta_{AB}}\right) \frac{1}{6\pi\mu_B R_A} \tag{17.4-2}$$

[1] R. J. Bearman and J. G. Kirkwood, *J. Chem. Phys.*, **28**, 136–145 (1958).

[2] R. J. Bearman, *J. Phys. Chem.*, **65**, 1961–1968 (1961).

[3] C. F. Curtiss and R. B. Bird, *J. Chem. Phys.*, **111**, 10362–10370 (1999).

[4] See §17.7 and E. A. Moelwyn-Hughes, *Physical Chemistry*, 2nd edition, corrected printing, Macmillan, New York (1964), pp. 62–74. See also R. J. Silbey and R. A. Alberty, *Physical Chemistry*, 3rd edtion, Wiley, New York (2001), §20.2. Apparently the Nernst–Einstein equation cannot be generalized to polymeric fluids with appreciable velocity gradients, as has been noted by H. C. Öttinger, *AIChE Journal*, **35**, 279–286 (1989).

[5] S. Kim and S. J. Karrila, *Microhydrodynamics: Principles and Selected Applications*, Butterworth-Heinemann, Boston (1991).

[6] H. Lamb, *Hydrodynamics*, 6th edition, Cambridge University Press (1932), reprinted (1997), §337.

in which μ_B is the viscosity of the pure solvent, R_A is the radius of the solute particle, and β_{AB} is the "coefficient of sliding friction" (formally the same as the μ/ζ of problem 2B.9). The limiting cases of $\beta_{AB} = \infty$ and $\beta_{AB} = 0$ are of particular interest:

a. $\boldsymbol{\beta_{AB} = \infty}$ (no-slip condition)

In this case Eq. 17.4-2 becomes Stokes' law (Eq. 2.6-15) and Eq. 17.4-1 becomes

$$\frac{\mathscr{D}_{AB}\mu_B}{\kappa T} = \frac{1}{6\pi R_A} \tag{17.4-3}$$

which is usually called the *Stokes–Einstein equation*. This equation applies well to the diffusion of very large spherical molecules in solvents of low molecular weight[7] and to suspended particles. Analogous expressions developed for nonspherical particles have been used for estimating the shapes of protein molecules.[8,9]

b. $\boldsymbol{\beta_{AB} = 0}$ (complete slip condition)

In this case Eq. 17.4-1 leads to (see Eq. 4B.3-4)

$$\frac{\mathscr{D}_{AB}\mu_B}{\kappa T} = \frac{1}{4\pi R_A} \tag{17.4-4}$$

If the molecules A and B are identical (that is, for *self-diffusion*) and if they can be assumed to form a cubic lattice with the adjacent molecules just touching, then $2R_A = (\tilde{V}_A/\tilde{N}_A)^{1/3}$ and

$$\frac{\mathscr{D}_{AA}\mu_A}{\kappa T} = \frac{1}{2\pi}\left(\frac{\tilde{N}_A}{\tilde{V}_A}\right)^{1/3} \tag{17.4-5}$$

Equation 17.4-5 has been found[10] to agree with self-diffusion data for a number of liquids, including polar and associated substances, liquid metals, and molten sulfur, to within about 12%. The hydrodynamic model has proven less useful for binary diffusion (that is, for A not identical to B) although the predicted temperature and viscosity dependences are approximately correct.

Keep in mind that the above formulas apply only to dilute solutions of A in B. Some attempts have been made, however, to extend the hydrodynamic model to solutions of finite concentrations.[11]

The *Eyring activated-state theory* attempts to explain transport behavior via a quasi-crystalline model of the liquid state.[12] It is assumed in this theory that there is some unimolecular rate process in terms of which diffusion can be described, and it is further assumed that in this process there is some configuration that can be identified as the "activated state." The Eyring theory of reaction rates is applied to this elementary process in a manner analogous to that described in §1.5 for estimation of liquid viscosity. A modifi-

---

[7] A. Polson, *J. Phys. Colloid Chem.*, **54**, 649–652 (1950).

[8] H. J. V. Tyrrell, *Diffusion and Heat Flow in Liquids*, Butterworths, London (1961), Chapter 6.

[9] Creeping motion around finite bodies in a fluid of infinite extent has been reviewed by J. Happel and H. Brenner, *Low Reynolds Number Hydrodynamics*, Prentice-Hall, Englewood Cliffs, N.J. (1965); see also S. Kim and S. J. Karrila, *Microhydrodynamics: Principles and Selected Applications*, Butterworth-Heinemann, Boston (1991). G. K. Youngren and A. Acrivos, *J. Chem. Phys.* **63**, 3846–3848 (1975) have calculated the rotational friction coefficient for benzene, supporting the validity of the no-slip condition at molecular dimensions.

[10] J. C. M. Li and P. Chang, *J. Chem. Phys.*, **23**, 518–520 (1955).

[11] C. W. Pyun and M. Fixman, *J. Chem. Phys.*, **41**, 937–944 (1964).

[12] S. Glasstone, K. J. Laidler, and H. Eyring, *Theory of Rate Processes*, McGraw-Hill, New York (1941), Chapter IX.

cation of the original Eyring model by Ree, Eyring, and coworkers[13] yields an expression similar to Eq. 17.4-5 for traces of A in solvent B:

$$\frac{\mathscr{D}_{AB}\mu_B}{\kappa T} = \frac{1}{\xi}\left(\frac{\tilde{N}_A}{\tilde{V}_B}\right)^{1/3} \tag{17.4-6}$$

Here ξ is a "packing parameter," which in the theory represents the number of nearest neighbors of a given solvent molecule. For the special case of self-diffusion, ξ is found to be very close to 2π, so that Eqs. 17.4-5 and 6 are in good agreement despite the difference between the models from which they were developed.

The Eyring theory is based on an oversimplified model of the liquid state, and consequently the conditions required for its validity are not clear. However, Bearman has shown[2] that the Eyring model gives results consistent with statistical mechanics for "regular solutions," that is, for mixtures of molecules that have similar size, shape, and intermolecular forces. For this limiting situation, Bearman also obtains an expression for the concentration dependence of the diffusivity,

$$\frac{\mathscr{D}_{AB}\mu_B}{(\mathscr{D}_{AB}\mu_B)_{x_A \to 0}} = \left[1 + x_A\left(\frac{\overline{V}_A}{\overline{V}_B} - 1\right)\right]\left(\frac{\partial \ln a_A}{\partial \ln x_A}\right)_{T,p} \tag{17.4-7}$$

in which $\mathscr{D}_{AB}$ and μ_B are the diffusivity and viscosity of the mixture at the composition x_A, and a_A is the thermodynamic activity of species A. For regular solutions, the partial molar volumes, $\overline{V}_A$ and $\overline{V}_B$, are equal to the molar volumes of the pure components. Bearman suggests on the basis of his analysis that Eq. 17.4-7 should be limited to regular solutions, and it has in fact been found to apply well only to nearly ideal solutions.

Because of the unsatisfactory nature of the theory for diffusion in liquids, it is necessary to rely on *empirical* expressions. For example, the Wilke–Chang equation[14] gives the diffusivity for small concentrations of A in B as

$$\mathscr{D}_{AB} = 7.4 \times 10^{-8} \frac{\sqrt{\psi_B M_B} T}{\mu \tilde{V}_A^{0.6}} \tag{17.4-8}$$

Here $\tilde{V}_A$ is the molar volume of the solute A in cm³/g-mole as liquid at its normal boiling point, μ is the viscosity of the solution in centipoises, ψ_B is an "association parameter" for the solvent, and T is the absolute temperature in K. Recommended values of ψ_B are: 2.6 for water; 1.9 for methanol; 1.0 for benzene, ether, heptane, and other unassociated solvents. Equation 17.4-8 is good only for dilute solutions of nondissociating solutes. For such solutions, it is usually good within ±10%.

Other empiricisms, along with their relative merits, have been summarized by Reid, Prausnitz, and Poling.[15]

EXAMPLE 17.4-1

Estimation of Liquid Diffusivity

Estimate $\mathscr{D}_{AB}$ for a dilute solution of TNT (2,4,6-trinitrotoluene) in benzene at 15°C.

SOLUTION

Use the equation of Wilke and Chang, taking TNT as component A and benzene as component B. The required data are

$$\mu = 0.705\text{cp (the viscosity for pure benzene)}$$
$$\tilde{V}_A = 140 \text{ cm}^3/\text{g-mole (for TNT)}$$

---

[13] H. Eyring, D. Henderson, B. J. Stover, and E. M. Eyring, *Statistical Mechanics and Dynamics*, Wiley, New York (1964), §16.8.

[14] C. R. Wilke, *Chem. Eng. Prog.*, **45**, 218–224 (1949); C. R. Wilke and P. Chang, *AIChE Journal*, **1**, 264–270 (1955).

[15] R. C. Reid, J. M. Prausnitz, and B. E. Poling, *The Properties of Gases and Liquids*, 4th edition, McGraw-Hill, New York (1987), Chapter 11.

$$\psi_B = 1.0 \text{ (for benzene)}$$

$$M_B = 78.11 \text{ (for benzene)}$$

Substitution into Eq. 17.4-8 gives

$$\mathscr{D}_{AB} = 7.4 \times 10^{-8} \frac{\sqrt{(1.0)(78.11)}(273 + 15)}{(0.705)(140)^{0.6}}$$

$$= 1.38 \times 10^{-5} \text{ cm}^2/\text{s} \tag{17.4-9}$$

This result compares well with the measured value of 1.39×10^{-5} cm^2/s.

---

§17.5 THEORY OF DIFFUSION IN COLLOIDAL SUSPENSIONS[1,2,3]

Next we turn to the movement of small colloidal particles in a liquid. Specifically we consider a finely divided, dilute suspension of spherical particles of material A in a stationary liquid B. When the spheres of A are sufficiently small (but still large with respect to the molecules of the suspending medium), the collisions between the spheres and the molecules of B will result in an erratic motion of the spheres. This random motion is referred to as *Brownian motion*.[4]

The movement of each sphere can be described by an equation of motion, called the *Langevin equation*:

$$m \frac{d\mathbf{u}_A}{dt} = -\zeta \mathbf{u}_A + \mathbf{F}(t) \tag{17.5-1}$$

in which $\mathbf{u}_A$ is the instantaneous velocity of the sphere of mass m. The term $-\zeta \mathbf{u}_A$ gives the Stokes' law drag force,[5] $\zeta = 6\pi\mu_B R_A$ being the "friction coefficient." Finally $\mathbf{F}(t)$ is the rapidly oscillating, irregular Brownian motion force. Equation 17.5-1 cannot be "solved" in the usual sense, since it contains the randomly fluctuating force $\mathbf{F}(t)$. Equations such as Eq. 17.5-1 are called "stochastic differential equations."

If it is assumed that (i) $\mathbf{F}(t)$ is independent of $\mathbf{u}_A$ and that (ii) the variations in $\mathbf{F}(t)$ are much more rapid than those of $\mathbf{u}_A$, then it is possible to extract from Eq. 17.5-1 the probability $W(\mathbf{u}_A, t; \mathbf{u}_{A0}) d\mathbf{u}_A$ that at time t the particle will have a velocity in the range of $\mathbf{u}_A$ to $\mathbf{u}_A + d\mathbf{u}_A$. Physical reasoning requires that the probability density $W(\mathbf{u}_A, t; \mathbf{u}_{A0})$ approach a Maxwellian (equilibrium) distribution as $t \to \infty$:

$$W(\mathbf{u}_A, t; \mathbf{u}_{A0}) \to \left(\frac{m}{2\pi\kappa T} \right)^{3/2} \exp(-mu_A^2/2\kappa T) \tag{17.5-2}$$

Here, T is the temperature of the fluid in which the particles are suspended.

---

[1] A. Einstein, *Ann. d. Phys*, **17**, 549–560 (1905), **19**, 371–381 (1906); *Investigations on the Theory of the Brownian Movement*, Dover, New York (1956).

[2] S. Chandrasekhar, *Rev. Mod. Phys.*, **15**, 1–89 (1943).

[3] W. B. Russel, D. A. Saville, and W. R. Schowalter, *Colloidal Dispersions*, Cambridge University Press (1989); H. C. Öttinger, *Stochastic Processes in Polymeric Fluids*, Springer, Berlin (1996).

[4] Named after the botanist R. Brown, *Phil. Mag.* (4), p. 161 (1828); *Ann. d. Phys. u. Chem.*, **14**, 294–313 (1828). Actually the phenomenon had been discovered and reported earlier in 1789 by Jan Ingenhousz (1730–1799) in the Netherlands.

[5] As can be seen from Example 4.2-1, Stokes' law is valid only for the steady, unidirectional motion of a sphere through a fluid. For a sphere moving in an arbitrary manner, there are, in addition to the Stokes' contribution, an inertial term and a memory-integral term (the Basset force). See A. B. Basset, *Phil. Trans.*, **179**, 43–63 (1887); H. Lamb, *Hydrodynamics*, 6th edition, Cambridge University Press (1932), reprinted (1997), p. 644; H. Villat and J. Kravtchenko, *Leçons sur les Fluides Visqueux*, Gauthier-Villars, Paris (1943), p. 213, Eq. (62); L. Landau and E. M. Lifshitz, *Fluid Mechanics*, 2nd edition, Pergamon, New York (1987), p. 94. In applying the Langevin equation to polymer kinetic theory, the role of the Basset force has been investigated by J. D. Schieber, *J. Chem. Phys.*, **94**, 7526–7533 (1991).

Another quantity of interest that can be obtained from the Langevin equation is the probability, $W(\mathbf{r},t;\mathbf{r}_0,\mathbf{u}_{A0})d\mathbf{r}$, that at time t the particle will have a position in the range $\mathbf{r}$ to $\mathbf{r} + d\mathbf{r}$ if its initial position and velocity were $\mathbf{r}_0$ and $\mathbf{u}_{A0}$. For long times, specifically $t >> m/\zeta$, this probability is given by

$$W(\mathbf{r},t;\mathbf{r}_0,\mathbf{u}_{A0})d\mathbf{r} = \left(\frac{\zeta}{4\pi\kappa Tt}\right)^{3/2} \exp(-\zeta(r - r_0)^2/4\kappa Tt)d\mathbf{r} \qquad (17.5\text{-}3)$$

However, this expression turns out to have just the same form as the solution of Fick's second law of diffusion (see Eq. 19.1-18 and Problem 20B.5) for the diffusion from a point source. One simply has to identify W with the concentration c_A, and $\kappa T/\zeta$ with $\mathcal{D}_{AB}$. In this way Einstein (see Ref. 1 on page 531) arrived at the following expression for the diffusivity of a dilute suspension of spherical colloid particles:

$$\mathcal{D}_{AB} = \frac{\kappa T}{\zeta} = \frac{\kappa T}{6\pi\mu_B R_A} \qquad (17.5\text{-}4)$$

Thus, $\mathcal{D}_{AB}$ is related to the temperature and the friction coefficient ζ (the reciprocal of the friction coefficient is called the "mobility"). Equation 17.5-4 was already given in Eq. 17.4-3 for the interdiffusion of liquids.

§17.6   THEORY OF DIFFUSION OF POLYMERS

For a *dilute solution* of a polymer A in a low-molecular-weight solvent B, there is a detailed theory,[1] in which the polymer molecules are modeled as bead–spring chains (see Fig. 8.6-2). Each chain is a linear arrangement of N beads and $N - 1$ Hookean springs. The beads are characterized by a friction coefficient ζ, which describes the Stokes' law resistance to the bead motion through the solvent. The model further takes into account the fact that, as a bead moves around, it disturbs the solvent in the neighborhood of all the other beads; this is referred to as *hydrodynamic interaction*. The theory ultimately predicts that the diffusivity should be proportional to $N^{-1/2}$ for large N. Since the number of beads is proportional to the polymer molecular weight M, the following result is obtained:

$$\mathcal{D}_{AB} \sim \frac{1}{\sqrt{M}} \qquad (17.6\text{-}1)$$

The inverse square-root dependence is rather well borne out by experiment.[2] If hydrodynamic interaction among beads were not included, then one would predict $\mathcal{D}_{AB} \sim 1/M$.

The theory of *self-diffusion* in an undiluted polymer has been studied from several points of view.[3,4] These theories, which are rather crude, lead to the result that

$$\mathcal{D}_{AA^*} \sim \frac{1}{M^2} \qquad (17.6\text{-}2)$$

[1] J. G. Kirkwood, *Macromolecules*, Gordon and Breach, New York (1967), pp. 13, 41, 76–77, 95, 101–102. The original Kirkwood theory has been reexamined and slightly improved by H. C. Öttinger, *J. Chem. Phys.*, **87**, 3156–3165 (1987).

[2] R. B. Bird, C. F. Curtiss, R. C. Armstrong, and O. Hassager, *Dynamics of Polymeric Liquids*, Vol. 2, *Kinetic Theory*, 2nd edition, Wiley, New York (1987), pp. 174–175.

[3] P.-G. de Gennes and L. Léger, *Ann. Rev. Phys. Chem.*, 49–61 (1982); P.-G. de Gennes, *Physics Today*, **36**, 33–39 (1983). De Gennes introduced the notion of *reptation*, according to which the polymer molecules move back and forth along their backbones in a snake-like Brownian motion.

[4] R. B. Bird, C. F. Curtiss, R. C. Armstrong, and O. Hassager, *Dynamics of Polymeric Liquids*, Vol. 2, *Kinetic Theory*, 2nd edition, Wiley, New York (1987), pp. 326–327; C. F. Curtiss and R. B. Bird, *Proc. Nat. Acad. Sci.*, **93**, 7440–7445 (1996).

Experimental data agree more or less with this result,[5] but the exponent on the molecular weight may be as great as 3 for some polymers.

Although a very general theory for diffusion of polymers has been developed,[6] not very much has been done with it. So far it has been used to show that, in flowing dilute solutions of flowing polymers, the diffusivity tensor (see Eq. 17.1-10) becomes anisotropic and dependent on the velocity gradients. It has also been shown how to generalize the Maxwell–Stefan equations (see §17.9 and §24.1) for multicomponent polymeric liquids. Further advances in this subject can be expected through use of molecular simulations.[7]

§17.7 MASS AND MOLAR TRANSPORT BY CONVECTION

In §17.1, the discussion of Fick's (first) law of diffusion was given in terms of *mass units*: mass concentration, mass flux, and the mass average velocity. In this section we extend the previous discussion to include *molar units*. Thus most of this section deals with questions of notation and definitions. One might reasonably wonder whether or not this dual set of notation is really necessary. Unfortunately, it really is. When chemical reactions are involved, molar units are usually preferred. When the diffusion equations are solved together with the equation of motion, mass units are usually preferable. Therefore it is necessary to acquire familiarity with both. In this section we also introduce the concept of the *convective flux* of mass or moles.

Mass and Molar Concentrations

Earlier we defined the *mass concentration* ρ_α as the mass of species α per unit volume of solution. Now we define the *molar concentration* $c_\alpha = \rho_\alpha/M_\alpha$ as the number of moles of α per unit volume of solution.

Similarly, in addition to the *mass fraction* $\omega_\alpha = \rho_\alpha/\rho$, we will use the *mole fraction* $x_\alpha = c_\alpha/c$. Here $\rho = \Sigma_\alpha \rho_\alpha$ is the total mass of all species per unit volume of solution, and $c = \Sigma_\alpha c_\alpha$ is the total number of moles of all species per unit volume of solution. By the word "solution" we mean a one-phase gaseous, liquid, or solid mixture. In Table 17.7-1 we summarize these concentration units and their interrelation for multicomponent systems.

It is necessary to emphasize that ρ_α is the mass concentration of species α in a mixture. We use the notation $\rho^{(\alpha)}$ for the density of pure species α when the need arises.

Mass Average and Molar Average Velocity

In a diffusing mixture, the various chemical species are moving at different velocities. By $\mathbf{v}_\alpha$, the "velocity of species α," we do *not* mean the velocity of an individual molecule of species α. Rather, we mean the average of all the velocities of molecules of species α within a small volume. Then, for a mixture of N species, the local *mass average velocity* $\mathbf{v}$ is defined as

$$\mathbf{v} = \frac{\displaystyle\sum_{\alpha=1}^{N} \rho_\alpha \mathbf{v}_\alpha}{\displaystyle\sum_{\alpha=1}^{N} \rho_\alpha} = \frac{\displaystyle\sum_{\alpha=1}^{N} \rho_\alpha \mathbf{v}_\alpha}{\rho} = \sum_{\alpha=1}^{N} \omega_\alpha \mathbf{v}_\alpha \tag{17.7-1}$$

[5] P. F. Green, in *Diffusion in Polymers* (P. Neogi, ed.), Dekker, New York (1996), Chapter 6. According to T. P. Lodge, *Phys. Rev. Letters*, **86**, 3218–3221 (1999), measurements on undiluted polymers show that the exponent on the molecular weight should be about 2.3.

[6] C. F. Curtiss and R. B. Bird, *Adv. Polym. Sci.*, **125**, 1–101 (1996) and *J. Chem. Phys.*, **111**, 10362–10370 (1999).

[7] D. N. Theodorou, in *Diffusion in Polymers* (P. Neogi, ed.), Dekker, New York (1996), Chapter 2.

Table 17.7-1   Notation for Concentrations

Basic definitions:

ρ_α	= mass concentration of species α	(A)
$\rho = \sum\limits_{\alpha=1}^{N} \rho_\alpha$ = mass density of solution		(B)
$\omega_\alpha = \rho_\alpha/\rho$ = mass fraction of species α		(C)

c_α	= molar concentration of species α	(D)
$c = \sum\limits_{\alpha=1}^{N} c_\alpha$ = molar density of solution		(E)
$x_\alpha = c_\alpha/c$ = mole fraction of species α		(F)

$M = \rho/c$ = molar mean molecular weight of solution   (G)

Algebraic relations:

$c_\alpha = \rho_\alpha/M_\alpha$	(H)	$\rho_\alpha = c_\alpha M_\alpha$	(I)
$\sum\limits_{\alpha=1}^{N} x_\alpha = 1$	(J)	$\sum\limits_{\alpha=1}^{N} \omega_\alpha = 1$	(K)
$\sum\limits_{\alpha=1}^{N} x_\alpha M_\alpha = M$	(L)	$\sum\limits_{\alpha=1}^{N} \omega_\alpha/M_\alpha = 1/M$	(M)
$x_\alpha = \dfrac{\omega_\alpha/M_\alpha}{\sum\limits_{\beta=1}^{N} (\omega_\beta/M_\beta)}$	(N)	$\omega_\alpha = \dfrac{x_\alpha M_\alpha}{\sum\limits_{\beta=1}^{N} (x_\beta M_\beta)}$	(O)

Differential relations:

$$\nabla x_a = -\frac{M^2}{M_\alpha} \sum_{\substack{\gamma=1 \\ \gamma \neq \alpha}}^{N} \left[\frac{1}{M} + \omega_\alpha\left(\frac{1}{M_\gamma} - \frac{1}{M_\alpha}\right)\right]\nabla\omega_\gamma \qquad (P)^a$$

$$\nabla\omega_a = -\frac{M_\alpha}{M^2} \sum_{\substack{\gamma=1 \\ \gamma \neq \alpha}}^{N} [M + x_\alpha(M_\gamma - M_\alpha)]\nabla x_\gamma \qquad (Q)^a$$

^{a}Equations (P) and (Q), simplified for binary ystems, are

$$\nabla x_A = \frac{\dfrac{1}{M_A M_B}\nabla\omega_A}{\left(\dfrac{\omega_A}{M_A} + \dfrac{\omega_B}{M_B}\right)^2} \qquad (P') \qquad\qquad \nabla\omega_A = \frac{M_A M_B \nabla x_A}{(x_A M_A + x_B M_B)^2} \qquad (Q')$$

Note that $\rho\mathbf{v}$ is the local rate at which mass passes through a unit cross section placed perpendicular to the velocity $\mathbf{v}$. This is the local velocity one could measure by means of a Pitot tube or by laser-Doppler velocimetry, and corresponds to the $\mathbf{v}$ used in the equation of motion and in the energy equation in the preceding chapters for pure fluids.

Similarly, one may define a local *molar average velocity* $\mathbf{v}^*$ by

$$\mathbf{v}^* = \frac{\sum\limits_{\alpha=1}^{N} c_\alpha \mathbf{v}_\alpha}{\sum\limits_{\alpha=1}^{N} c_\alpha} = \frac{\sum\limits_{\alpha=1}^{N} c_\alpha \mathbf{v}_\alpha}{c} = \sum\limits_{\alpha=1}^{N} x_\alpha \mathbf{v}_\alpha \qquad (17.7\text{-}2)$$

Note that $c\mathbf{v}^*$ is the local rate at which moles pass through a unit cross section placed perpendicular to the molar velocity $\mathbf{v}^*$. Both the mass average velocity and the molar average velocity will be used extensively throughout the remainder of this book. Still

Table 17.7-2   Notation for Velocities in Multicomponent Systems

Basic definitions:

$\mathbf{v}_\alpha$	velocity of species α with respect to fixed coordinates	(A)
$\mathbf{v} = \displaystyle\sum_{\alpha=1}^{N} \omega_\alpha \mathbf{v}_\alpha$	mass average velocity	(B)
$\mathbf{v}^* = \displaystyle\sum_{\alpha=1}^{N} x_\alpha \mathbf{v}_\alpha$	molar average velocity	(C)

$\mathbf{v}_\alpha - \mathbf{v}$	diffusion velocity of species α with respect to the mass average velocity $\mathbf{v}$	(D)
$\mathbf{v}_\alpha - \mathbf{v}^*$	diffusion velocity of species α with respect to the molar average velocity $\mathbf{v}^*$	(E)

Additional relations:

$$\mathbf{v} - \mathbf{v}^* = \sum_{\alpha=1}^{N} \omega_\alpha(\mathbf{v}_\alpha - \mathbf{v}^*) \quad \text{(F)} \qquad\qquad \mathbf{v}^* - \mathbf{v} = \sum_{\alpha=1}^{N} x_\alpha(\mathbf{v}_\alpha - \mathbf{v}) \quad \text{(G)}$$

other average velocities are sometimes used, such as the *volume average velocity* (see Problem 17C.1). In Table 17.7-2 we give a summary of the various relations among these velocities.

Molecular Mass and Molar Fluxes

In §17.1 we defined the molecular mass flux of α as the flow of mass of α through a unit area per unit time: $\mathbf{j}_\alpha = \rho_\alpha(\mathbf{v}_\alpha - \mathbf{v})$. That is, we include only the velocity of species α relative to the mass average velocity $\mathbf{v}$. Similarly, we define the molecular molar flux of species α as the number of moles of α flowing through a unit area per unit time: $\mathbf{J}_A^* = c_A(\mathbf{v}_A - \mathbf{v}^*)$. Here we include only the velocity of species α relative to the molar average velocity $\mathbf{v}^*$.

Then in §17.1 we presented Fick's (first) law of diffusion, which describes how the mass of species A in a binary mixture is transported by means of molecular motions. This law can also be expressed in molar units. Hence we have the pair of relations for *binary* systems:

Mass units:	$\mathbf{j}_A = \rho_A(\mathbf{v}_A - \mathbf{v}) = -\rho\mathcal{D}_{AB}\nabla\omega_A$	(17.7-3)
Molar units:	$\mathbf{J}_A^* = c_A(\mathbf{v}_A - \mathbf{v}^*) = -c\mathcal{D}_{AB}\nabla x_A$	(17.7-4)

The differences $\mathbf{v}_A - \mathbf{v}$ and $\mathbf{v}_A - \mathbf{v}^*$ are sometimes referred to as *diffusion velocities*. Equation 17.7-4 can be derived from Eq. 17.7-3 by using some of the relations in Tables 17.7-1 and 2.

Convective Mass and Molar Fluxes

In addition to transport by molecular motion, mass may also be transported by the bulk motion of the fluid. In Fig. 9.7-1 we show three mutually perpendicular planes of area dS at a point P where the fluid *mass average velocity* is $\mathbf{v}$. The volume rate of flow across the plane perpendicular to the surface element dS perpendicular to the x-axis is $v_x dS$. The rate at which mass of species α is being swept across the same surface element is then $\rho_\alpha v_x dS$. We can write similar expressions for the mass flows of species α across the surface elements perpendicular to the y- and z-axes as $\rho_\alpha v_y dS$ and $\rho_\alpha v_z dS$, respectively. If we now multiply each of these expressions by the corresponding unit vector, add them, and divide by dS, we get

$$\rho_\alpha \boldsymbol{\delta}_x v_x + \rho_\alpha \boldsymbol{\delta}_y v_y + \rho_\alpha \boldsymbol{\delta}_z v_z = \rho_\alpha \mathbf{v} \qquad (17.7\text{-}5)$$

as the *convective mass flux vector*, which has units of kg/m^2 · s.

If one goes back and repeats the story of the preceding paragraph, but using everywhere molar units and the *molar average velocity* $\mathbf{v}^*$, then we get

$$c_\alpha \boldsymbol{\delta}_x v_x^* + c_\alpha \boldsymbol{\delta}_y v_y^* + c_\alpha \boldsymbol{\delta}_z v_z^* = c_\alpha \mathbf{v}^* \qquad (17.7\text{-}6)$$

as the *convective molar flux vector*, which has units of kg-mole/m^2 · s.

To get the convective mass and molar fluxes across a unit surface whose normal unit vector is $\mathbf{n}$, we form the dot products $(\mathbf{n} \cdot \rho_\alpha \mathbf{v})$ and $(\mathbf{n} \cdot c_\alpha \mathbf{v}^*)$, respectively.

§17.8 SUMMARY OF MASS AND MOLAR FLUXES

In Chapters 1 and 9 we introduced the *combined momentum flux* tensor $\boldsymbol{\phi}$ and the *combined energy flux* vector $\mathbf{e}$, which we found useful in setting up the shell balances and equations of change. We give the corresponding definitions here for the mass and molar flux vectors. We add together the molecular mass flux vector and the convective mass flux vector to get the *combined mass flux* vector, and similarly for the *combined molar flux* vector:

Combined mass flux:	$\mathbf{n}_\alpha = \mathbf{j}_\alpha + \rho_\alpha \mathbf{v}$	(17.8-1)
Combined molar flux:	$\mathbf{N}_\alpha = \mathbf{J}_\alpha^* + c_\alpha \mathbf{v}^*$	(17.8-2)

In the first three lines of Table 17.8-1 we summarize the definitions of the mass and molar fluxes discussed so far. In the shaded squares we also give the definitions of the fluxes $\mathbf{j}_\alpha^*$ (*mass* flux with respect to the *molar* average velocity) and $\mathbf{J}_\alpha$ (*molar* flux with respect to *mass* average velocity). These "hybrid" fluxes should normally not be used.

In the remainder of Table 17.8-1 we give a summary of other useful relations, such as the sums of the fluxes and the interrelations among the fluxes. By using Eqs. (J) and (M) we can rewrite Eqs. 17.8-1 and 2 as

$$\mathbf{j}_\alpha = \mathbf{n}_\alpha - \omega_\alpha \sum_{\beta=1}^{N} \mathbf{n}_\beta \qquad (17.8\text{-}3)$$

$$\mathbf{J}_\alpha^* = \mathbf{N}_\alpha - x_\alpha \sum_{\beta=1}^{N} \mathbf{N}_\beta \qquad (17.8\text{-}4)$$

When simplified for binary systems, these relations can be combined with Eqs. 17.7-3 and 17.7-4, to get Eqs. (C) and (D) of Table 17.8-2, which are equivalent forms of Fick's (first) law. The forms given in Eqs. (E) and (F) of Table 17.8-2, in terms of the relative velocities of the species, are interesting because they involve neither $\mathbf{v}$ nor $\mathbf{v}^*$.

In Chapter 18 we will write Fick's law exclusively in the form of Eq. (D) of Table 17.8-2. It is this form that has generally been used in chemical engineering. In many problems something is known about the relation between $\mathbf{N}_A$ and $\mathbf{N}_B$ from the stoichiometry or from boundary conditions. Therefore $\mathbf{N}_B$ can be eliminated from Eq. (D), giving a direct relation between $\mathbf{N}_A$ and ∇x_A for the particular problem.

In §1.7 we pointed out that the total molecular momentum flux through a surface of orientation $\mathbf{n}$ is the vector $[\mathbf{n} \cdot \boldsymbol{\pi}]$. In §9.7 we mentioned the analogous quantity for the molecular heat flux—namely, the scalar $(\mathbf{n} \cdot \mathbf{q})$. The analogous mass transport quantities are the scalars $(\mathbf{n} \cdot \mathbf{j}_\alpha)$ and $(\mathbf{n} \cdot \mathbf{J}_\alpha^*)$, which give the total mass and molar fluxes through a surface of orientation $\mathbf{n}$. Similarly, for the combined fluxes through a surface of orientation $\mathbf{n}$, we have for momentum $[\mathbf{n} \cdot \boldsymbol{\phi}]$, for energy $(\mathbf{n} \cdot \mathbf{e})$, and for species $(\mathbf{n} \cdot \mathbf{n}_\alpha)$ and $(\mathbf{n} \cdot \mathbf{N}_\alpha)$.

Table 17.8-1 Notation for Mass and Molar Fluxes*

Quantity	With respect to stationary axes		With respect to mass average velocity $\mathbf{v}$		With respect to molar average velocity $\mathbf{v}^*$	
Velocity of species α (cm/s)	$\mathbf{v}_\alpha$	(A)	$\mathbf{v}_\alpha - \mathbf{v}$	(B)	$\mathbf{v}_\alpha - \mathbf{v}^*$	(C)
Mass flux of species α (g/cm² s)	$\mathbf{n}_\alpha = \rho_\alpha \mathbf{v}_\alpha$	(D)	$\mathbf{j}_\alpha = \rho_\alpha(\mathbf{v}_\alpha - \mathbf{v})$	(E)	$\mathbf{j}_\alpha^* = \rho_\alpha(\mathbf{v}_\alpha - \mathbf{v}^*)$	(F)
Molar flux of species α (g-moles/cm² s)	$\mathbf{N}_\alpha = c_\alpha \mathbf{v}_\alpha$	(G)	$\mathbf{J}_\alpha = c_\alpha(\mathbf{v}_\alpha - \mathbf{v})$	(H)	$\mathbf{J}_\alpha^* = c_\alpha(\mathbf{v}_\alpha - \mathbf{v}^*)$	(I)
Sums of mass fluxes	$\sum_{\alpha=1}^{N} \mathbf{n}_\alpha = \rho\mathbf{v}$	(J)	$\sum_{\alpha=1}^{N} \mathbf{j}_\alpha = 0$	(K)	$\sum_{\alpha=1}^{N} \mathbf{j}_\alpha^* = \rho(\mathbf{v} - \mathbf{v}^*)$	(L)
Sums of molar fluxes	$\sum_{\alpha=1}^{N} \mathbf{N}_\alpha = c\mathbf{v}^*$	(M)	$\sum_{\alpha=1}^{N} \mathbf{J}_\alpha = c(\mathbf{v}^* - \mathbf{v})$	(N)	$\sum_{\alpha=1}^{N} \mathbf{J}_\alpha^* = 0$	(O)
Relations between mass and molar fluxes	$\mathbf{n}_\alpha = M_\alpha \mathbf{N}_\alpha$	(P)	$\mathbf{j}_\alpha = M_\alpha \mathbf{J}_\alpha$	(Q)	$\mathbf{j}_\alpha^* = M_\alpha \mathbf{J}_\alpha^*$	(R)
Interrelations among mass fluxes	$\mathbf{n}_\alpha = \mathbf{j}_\alpha + \rho_\alpha \mathbf{v}$	(S)	$\mathbf{j}_\alpha = \mathbf{n}_\alpha - \omega_\alpha \sum_{\beta=1}^{N} \mathbf{n}_\beta$	(T)	$\mathbf{j}_\alpha^* = \mathbf{n}_\alpha - x_\alpha \sum_{\beta=1}^{N} \frac{M_\alpha}{M_\beta} \mathbf{n}_\beta$	(U)
Interrelations among molar fluxes	$\mathbf{N}_\alpha = \mathbf{J}_\alpha^* + c_\alpha \mathbf{v}^*$	(V)	$\mathbf{J}_\alpha = \mathbf{N}_\alpha - \omega_\alpha \sum_{\beta=1}^{N} \frac{M_\beta}{M_\alpha} \mathbf{N}_\beta$	(W)	$\mathbf{J}_\alpha^* = \mathbf{N}_\alpha - x_\alpha \sum_{\beta=1}^{N} \mathbf{N}_\beta$	(X)

* Entries in the shaded boxes, involving the "hybrid fluxes" $\mathbf{j}_\alpha^*$ and $\mathbf{J}_\alpha$, are seldom needed; they are included only for the sake of completeness.

Table 17.8-2 Equivalent Forms of Fick's (First) Law of Binary Diffusion

Flux	Gradient	Form of Fick's Law	
$\mathbf{j}_A$	$\nabla\omega_A$	$\mathbf{j}_A = -\rho\mathcal{D}_{AB}\nabla\omega_A$	(A)
$\mathbf{J}_A^*$	∇x_A	$\mathbf{J}_A^* = -c\mathcal{D}_{AB}\nabla x_A$	(B)
$\mathbf{n}_A$	$\nabla\omega_A$	$\mathbf{n}_A = \omega_A(\mathbf{n}_A + \mathbf{n}_B) - \rho\mathcal{D}_{AB}\nabla\omega_A = \rho_A\mathbf{v} - \rho\mathcal{D}_{AB}\nabla\omega_A$	(C)
$\mathbf{N}_A$	∇x_A	$\mathbf{N}_A = x_A(\mathbf{N}_A + \mathbf{N}_B) - c\mathcal{D}_{AB}\nabla x_A = c_A\mathbf{v}^* - c\mathcal{D}_{AB}\nabla x_A$	(D)
$\rho(\mathbf{v}_A - \mathbf{v}_B)$	$\nabla\omega_A$	$\rho(\mathbf{v}_A - \mathbf{v}_B) = -\dfrac{\rho\mathcal{D}_{AB}}{\omega_A\omega_B}\nabla\omega_A$	(E)
$c(\mathbf{v}_A - \mathbf{v}_B)$	∇x_A	$c(\mathbf{v}_A - \mathbf{v}_B) = -\dfrac{c\mathcal{D}_{AB}}{x_A x_B}\nabla x_A$	(F)

§17.9   THE MAXWELL–STEFAN EQUATIONS FOR MULTICOMPONENT DIFFUSION IN GASES AT LOW DENSITY

For *multicomponent diffusion* in *gases at low density* it has been shown[1,2] that to a very good approximation

$$\nabla x_\alpha = -\sum_{\beta=1}^{N} \frac{x_\alpha x_\beta}{\mathcal{D}_{\alpha\beta}} (\mathbf{v}_\alpha - \mathbf{v}_\beta) = -\sum_{\beta=1}^{N} \frac{1}{c\mathcal{D}_{\alpha\beta}} (x_\beta \mathbf{N}_\alpha - x_\alpha \mathbf{N}_\beta) \qquad \alpha = 1, 2, 3, \ldots, N \qquad (17.9\text{-}1)$$

The $\mathcal{D}_{\alpha\beta}$ here are the *binary* diffusivities calculated from Eq. 17.3-11 or Eq. 17.3-12. Therefore, for an *N*-component system, $\frac{1}{2}N(N-1)$ binary diffusivities are required.

Equations 17.9-1 are referred to as the *Maxwell–Stefan equations*, since Maxwell[3] suggested them for binary mixtures on the basis of kinetic theory, and Stefan[4] generalized them to describe the diffusion in a gas mixture with *N* species. Later Curtiss and Hirschfelder obtained Eqs. 17.9-1 from the multicomponent extension of the Chapman–Enskog theory.

For dense gases, liquids, and polymers it has been shown that the Maxwell–Stefan equations are still valid, but that the strongly concentration-dependent diffusivities appearing therein are *not* the binary diffusivities.[5]

There is an important difference between binary diffusion and multicomponent diffusion.[6] In binary diffusion the movement of species *A* is always proportional to the negative of the concentration gradient of species *A*. In multicomponent diffusion, however, other interesting situations can arise: (i) *reverse diffusion*, in which a species moves against its own concentration gradient; (ii) *osmotic diffusion*, in which a species diffuses even though its concentration gradient is zero; (iii) *diffusion barrier*, when a species does not diffuse even though its concentration gradient is nonzero. In addition, the flux of a species is not necessarily collinear with the concentration gradient of that species.

QUESTIONS FOR DISCUSSION

1. How is the binary diffusivity defined? How is self-diffusion defined? Give typical orders of magnitude of diffusivities for gases, liquids, and solids.
2. Summarize the notation for the molecular, convective, and total fluxes for the three transport processes. How does one calculate the flux of mass, momentum, and energy across a surface with orientation **n**?
3. Define the Prandtl, Schmidt, and Lewis numbers. What ranges of Pr and Sc can one expect to encounter for gases and liquids?
4. How can you estimate the Lennard-Jones potential for a binary mixture, if you know the parameters for the two components of the mixture?
5. Of what value are the hydrodynamic theories of diffusion?
6. What is the Langevin equation? Why is it called a "stochastic differential equation"? What information can be obtained from it?

[1] C. F. Curtiss and J. O. Hirschfelder, *J. Chem. Phys.*, **17**, 550–555 (1949).

[2] For applications to engineering, see E. L. Cussler, *Diffusion: Mass Transfer in Fluid Systems*, 2nd edition, Cambridge University Press (1997); R. Taylor and R. Krishna, *Multicomponent Mass Transfer*, Wiley, New York (1993).

[3] J. C. Maxwell, *Phil. Mag.*, **XIX**, 19–32 (1860); **XX**, 21–32, 33–36 (1868).

[4] J. Stefan, *Sitzungsber. Kais. Akad. Wiss. Wien*, **LXIII**(2), 63–124 (1871); **LXV**(2), 323–363 (1872).

[5] C. F. Curtiss and R. B. Bird, *Ind. Eng. Chem. Res.*, **38**, 2515–2522 (1999); **40**, 1791 (2001); *J. Chem. Phys.*, **111**, 10362–10370 (1999).

[6] H. L. Toor, *AIChE Journal*, **3**, 198–207 (1959).

7. Compare and contrast the relation between binary diffusivity and viscosity for gases and for liquids.

8. How are the Maxwell–Stefan equations for multicomponent diffusion in gases related to the Fick equations for binary systems?

9. In a multicomponent mixture, does the vanishing of N_α imply the vanishing of ∇x_α?

PROBLEMS

17A.1. Prediction of a low-density binary diffusivity. Estimate $\mathscr{D}_{AB}$ for the system methane–ethane at 293K and 1 atm by the following methods:

(a) Equation 17.2-1.

(b) The corresponding-states chart in Fig. 17.2-1 along with Eq. 17.2-3.

(c) The Chapman–Enskog relation (Eq. 17.3-12) with Lennard-Jones parameters from Appendix E.

(d) The Chapman–Enskog relation (Eq. 17.3-12) with the Lennard-Jones parameters estimated from critical properties.

Answers (all in cm^2/s): **(a)** 0.152; **(b)** 0.138; **(c)** 0.146; **(d)** 0.138.

17A.2. Extrapolation of binary diffusivity to a very high temperature. A value of $\mathscr{D}_{AB} = 0.151$ cm^2/s has been reported[1] for the system CO_2–air at 293K and 1 atm. Extrapolate $\mathscr{D}_{AB}$ to 1500K by the following methods:

(a) Equation 17.2-1.

(b) Equation 17.3-10.

(c) Equations 17.3-12 and 15, with Table E.2.

What do you conclude from comparing these results with the experimental value[1] of 2.45 cm^2/s?

Answers (all in cm^2/s): **(a)** 2.96; **(b)** 1.75; **(c)** 2.51

17A.3. Self-diffusion in liquid mercury. The diffusivity of Hg^{203} in normal liquid Hg has been measured,[2] along with viscosity and volume per unit mass. Compare the experimentally measured $\mathscr{D}_{AA^*}$ with the values calculated with Eq. 17.4-5.

T (K)	$\mathscr{D}_{AA^*}$ (cm^2/s)	μ (cp)	$\hat{V}$ (cm^3/g)
275.7	1.52×10^{-5}	1.68	0.0736
289.6	1.68×10^{-5}	1.56	0.0737
364.2	2.57×10^{-5}	1.27	0.0748

17A.4. Schmidt numbers for binary gas mixtures at low density. Use Eq. 17.3-11 and the data given in Problem 1A.4 to compute Sc $= \mu/\rho\mathscr{D}_{AB}$ for binary mixtures of hydrogen and Freon-12 at $x_A = 0.00, 0.25, 0.50, 0.75$, and 1.00, at 25°C and 1 atm.

Sample answers: At $x_A = 0.00$, Sc $= 3.43$; at $x_A = 1.00$, Sc $= 0.407$

17A.5. Estimation of diffusivity for a binary mixture at high density. Predict $c\mathscr{D}_{AB}$ for an equimolar mixture of N_2 and C_2H_6 at 288.2K and 40 atm.

(a) Use the value of $\mathscr{D}_{AB}$ at 1 atm from Table 17.1-1, along with Fig. 17.2-1.

(b) Use Eq. 17.2-3 and Fig. 17.2-1.

Answers: **(a)** 5.8×10^{-6} g-mole/cm · s; **(b)** 5.3×10^{-6} g-mole/cm · s

---

[1] Ts. M. Klibanova, V. V. Pomerantsev, and D. A. Frank-Kamenetskii, *J. Tech. Phys.* (*USSR*), **12**, 14–30 (1942), as quoted by C. R. Wilke and C. Y. Lee, *Ind. Eng. Chem.*, **47**, 1253 (1955).

[2] R. E. Hoffman, *J. Chem. Phys.*, **20**, 1567–1570 (1952).

17A.6. Diffusivity and Schmidt number for chlorine–air mixtures.

(a) Predict $\mathcal{D}_{AB}$ for chlorine–air mixtures at 75°F and 1 atm. Treat air as a single substance with Lennard-Jones parameters as given in Appendix E. Use the Chapman–Enskog theory results in §17.3.

(b) Repeat (a) using Eq. 17.2-1.

(c) Use the results of (a) and of Problem 1A.5 to estimate Schmidt numbers for chlorine–air mixtures at 297K and 1 atm for the following mole fractions of chlorine: 0, 0.25, 0.50, 0.75, and 1.00.

Answers: **(a)** 0.121 cm²/s; **(b)** 0.124 cm²/s; **(c)** Sc = 1.27, 0.832, 0.602, 0.463, 0.372

17A.7. The Schmidt number for self-diffusion.

(a) Use Eqs. 1.3-1b and 17.2-2 to predict the self-diffusion Schmidt number Sc $= \mu/\rho\mathcal{D}_{AA^*}$ at the critical point for a system with $M_A \approx M_{A^*}$.

(b) Use the above result, along with Fig. 1.3-1 and Fig. 17.2-1, to predict Sc $= \mu/\rho\mathcal{D}_{AA^*}$ at the following states:

Phase	Gas	Gas	Gas	Liquid	Gas	Gas
T_r	0.7	1.0	5.0	0.7	1.0	2.0
p_r	0.0	0.0	0.0	saturation	1.0	1.0

17A.8. Correction of high-density diffusivity for temperature. The measured value[3] of $c\mathcal{D}_{AB}$ for a mixture of 80 mole% CH_4 and 20 mole% C_2H_6 at 313K and 136 atm is 6.0×10^{-6} g-mol/cm · s (see Example 17.2-3). Predict $c\mathcal{D}_{AB}$ for the same mixture at 136 atm and 351K, using Fig. 17.2-1.

Answer: 6.3×10^{-6} g-mole/cm · s
Observed:[3] 6.33×10^{-6} g-mol/cm · s

17A.9. Prediction of critical $c\mathcal{D}_{AB}$ values. Figure 17.2-1 gives the low-pressure limit $(c\mathcal{D}_{AA^*})_r = 1.01$ at $T_r = 1$ and $\rho_r \to 0$. At this limit, Eq. 17.2-13 gives

$$1.01(c\mathcal{D}_{AA^*})_c = 2.2646 \times 10^{-5} \sqrt{T_{cA}\left(\frac{1}{M_A} + \frac{1}{M_{A^*}}\right)} \frac{1}{\sigma_{AA^*}^2 \, \Omega_{\mathcal{D},AA^*}} \qquad (17A.9\text{-}1)$$

Here the argument $\kappa T_{cA}/\varepsilon_{AA^*}$ of $\Omega_{\mathcal{D},AA^*}$ is reported[4] as = 1.225 for Ar, Kr, and Xe. We use the value 1/0.77 from Eq. 1.4-11a as a representative average over many fluids.

(a) Combine Eq. 17A.9-1 with the relations

$$\sigma_{AA^*} = 2.44(T_{cA}/p_{cA})^{1/3} \qquad \varepsilon_{AA^*}/\kappa = 0.77T_{cA} \qquad (17A.9\text{-}2, 3)$$

and Table E.2 to obtain Eq. 17.2-2 for $(c\mathcal{D}_{AA^*})_c$.

(b) Show that the approximations

$$\sigma_{AB} = \sqrt{\sigma_A\sigma_B} \qquad \varepsilon_{AB} = \sqrt{\varepsilon_A\varepsilon_B} \qquad (17A.9\text{-}4, 5)$$

for Lennard-Jones parameters for the A–B interaction give

$$\sigma_{AB} = 2.44\left(\frac{T_{cA}T_{cB}}{p_{cA}p_{cB}}\right)^{1/6} \qquad \frac{\varepsilon_{AB}}{\kappa} = 0.77\sqrt{T_{cA}T_{cB}} \qquad (17A.9\text{-}6, 7)$$

when the molecular parameters of each species are predicted according to Eqs. 1.4-11a, c. Combine these expressions with Eq. 17A.9-1 (with A^* replaced by B and T_{cA} by $\sqrt{T_{cA}T_{cB}}$) to obtain Eq. 17.2-3 for $(c\mathcal{D}_{AB})_c$. The corresponding replacement of p_c and T_c in Fig. 17.2-1 by $\sqrt{p_{cA}p_{cB}}$ and $\sqrt{T_{cA}T_{cB}}$ amounts to regarding the A–B collisions as dominant over collisions of like molecules in determining the value of $c\mathcal{D}_{AB}$.

[3] V. J. Berry and R. C. Koeller, *AIChE Journal*, **6**, 274–280 (1960).
[4] J. J. van Loef and E. G. D. Cohen, *Physica A*, **156**, 522–533 (1989).

17A.10. Estimation of liquid diffusivities.

(a) Estimate the diffusivity for a dilute aqueous solution of acetic acid at 12.5°C, using the Wilke–Chang equation. The density of pure acetic acid is 0.937 g/cm^3 at its boiling point.

(b) The diffusivity of a dilute aqueous solution of methanol at 15°C is about 1.28×10^{-5} cm/s. Estimate the diffusivity for the same solution at 100°C.

Answer: (b) 6.7×10^{-5} cm/s

17B.1. Interrelation of composition variables in mixtures.

(a) Using the basic definitions in Eqs. (A) to (G) of Table 17.7-1, verify the algebraic relations in Eqs. (H) to (O).

(b) Verify that, in Table 17.7-1, Eqs. (P) and (Q) simplify to Eqs. (P') and (Q') for binary mixtures.

(c) Derive Eqs. (P') and (Q') from Eqs. (N) and (O).

17B.2. Relations among fluxes in multicomponent systems. Verify Eqs. (K), (O), (T), and (X) of Table 17.8-1 using only the definitions of concentrations, velocities, and fluxes.

17B.3. Relations between fluxes in binary systems. The following equation is useful for interrelating expressions in mass units and those in molar units in two-component systems:

$$\frac{j_A}{\rho \omega_A \omega_B} = \frac{J_A^*}{c x_A x_B} \tag{17B.3-1}$$

Verify the correctness of this relation.

17B.4. Equivalence of various forms of Fick's law for binary mixtures.

(a) Starting with Eq. (A) of Table 17.8-2, derive Eqs. (B), (D), and (F).

(b) Starting with Eq. (A) of Table 17.8-2, derive the folowing flux expressions:

$$j_A = -\rho(M_A M_B / M^2) \mathcal{D}_{AB} \nabla x_A \tag{17B.4-1}$$

$$\nabla x_A = \frac{1}{c \mathcal{D}_{AB}} (x_A N_B - x_B N_A) \tag{17B.4-2}$$

What conclusions can be drawn from these two equations?

(c) Show that Eq. (F) of Table 17.8-2 can be written as

$$\mathbf{v}_A - \mathbf{v}_B = -\mathcal{D}_{AB} \nabla \ln \frac{x_A}{x_B} \tag{17B.4-3}$$

17C.1. Mass flux with respect to volume average velocity. Let the *volume average velocity* in an N-component mixture be defined by

$$\mathbf{v}^\blacksquare = \sum_{\alpha=1}^{N} \rho_\alpha (\overline{V}_\alpha / M_\alpha) \mathbf{v}_\alpha = \sum_{\alpha=1}^{N} c_\alpha \overline{V}_\alpha \mathbf{v}_\alpha \tag{17C.1-1}$$

in which $\overline{V}_\alpha$ is the partial molar volume of species α. Then define

$$j_\alpha^\blacksquare = \rho_\alpha (\mathbf{v}_\alpha - \mathbf{v}^\blacksquare) \tag{17C.1-2}$$

as the mass flux with respect to the volume average velocity.

(a) Show that for a binary system of A and B,

$$j_A^\blacksquare = \rho(\overline{V}_B / M_B) j_A \tag{17C.1-3}$$

To do this you will need to use the identity $c_A \overline{V}_A + c_B \overline{V}_B = 1$. Where does this come from?

(b) Show that Fick's first law then assumes the form

$$j_A^\blacksquare = -\mathcal{D}_{AB} \nabla \rho_A \tag{17C.1-4}$$

To verify this you will need the relation $\overline{V}_A \nabla c_A + \overline{V}_B \nabla c_B = 0$. What is the origin of this?

17C.2. Mass flux with respect to the solvent velocity.

(a) In a system with N chemical species, choose component N to be the solvent. Then define

$$\mathbf{j}_\alpha^N = \rho_\alpha(\mathbf{v}_\alpha - \mathbf{v}_N) \tag{17C.2-1}$$

to be the mass flux with respect to the solvent velocity. Verify that

$$\mathbf{j}_\alpha^N = \mathbf{j}_\alpha - (\rho_\alpha/\rho_N)\mathbf{j}_N \tag{17C.2-2}$$

(b) For a binary system (labeling B as the solvent), show that

$$\mathbf{j}_A^B = (\rho/\rho_B)\mathbf{j}_A = -(\rho^2/\rho_B)\mathcal{D}_{AB}\nabla\omega_A \tag{17C.2-3}$$

How does this result simplify for a very dilute solution of A in solvent B?

17C.3. Determination of Lennard-Jones potential parameters from diffusivity data of a binary gas mixture.

(a) Use the following data[5] for the system H_2O–O_2 at 1 atm pressure to determine σ_{AB} and ε_{AB}/κ:

T (K)	400	500	600	700	800	900	1000	1100
$\mathcal{D}_{AB}$ (cm^2/s)	0.47	0.69	0.94	1.22	1.52	1.85	2.20	2.58

One way to do this is as follows: (i) Plot the data as $\log(T^{3/2}/\mathcal{D}_{AB})$ versus $\log T$ on a thin sheet of graph paper. (ii) Plot $\Omega_{\mathcal{D},AB}$ versus $\kappa T/\varepsilon_{AB}$ on a separate sheet of graph paper to the same scale. (iii) Superpose the first plot on the second, and from the scales of the two overlapping plots, determine the numerical ratios $(T/(\kappa T/\varepsilon_{AB}))$ and $((T^{3/2}/\mathcal{D}_{AB})/\Omega_{\mathcal{D},AB})$. (iv) Use these two ratios and Eq. 17.3-11 to solve for the two parameters σ_{AB} and ε_{AB}/κ.

---

[5] R. E. Walker and A. A. Westenberg, *J. Chem. Phys.*, **32**, 436–442 (1960); R. M. Fristrom and A. A. Westenberg, *Flame Structure*, McGraw-Hill, New York (1965), p. 265.

Chapter 18

Concentration Distributions in Solids and in Laminar Flow

In Chapter 2 we saw how a number of steady-state viscous flow problems can be set up and solved by making a *shell momentum balance*. In Chapter 9 we saw further how steady-state heat-conduction problems can be handled by means of a *shell energy balance*. In this chapter we show how steady-state diffusion problems may be formulated by *shell mass balances*. The procedure used here is virtually the same as that used previously:

a. A mass balance is made over a thin shell perpendicular to the direction of mass transport, and this shell balance leads to a first-order differential equation, which may be solved to get the mass flux distribution.

b. Into this expression we insert the relation between mass flux and concentration gradient, which results in a second-order differential equation for the concentration profile. The integration constants that appear in the resulting expression are determined by the boundary conditions on the concentration and/or mass flux at the bounding surfaces.

In Chapter 17 we pointed out that several kinds of mass fluxes are in common use. For simplicity, we shall in this chapter use the combined flux $\mathbf{N}_A$—that is, the number of moles of A that go through a unit area in unit time, the unit area being fixed in space. We shall relate the molar flux to the concentration gradient by Eq. (D) of Table 17.8-2, which for the z-component is

$$N_{Az} = \underbrace{-c\mathscr{D}_{AB}\frac{\partial x_A}{\partial z}}_{\substack{\text{molecular} \\ \text{flux}}} + \underbrace{x_A(N_{Az} + N_{Bz})}_{\substack{\text{convective} \\ \text{flux}}} \tag{18.0-1}$$

$$\underbrace{\phantom{N_{Az}}}_{\substack{\text{combined} \\ \text{flux}}}$$

Before Eq. 18.0-1 is used, we usually have to eliminate N_{Bz}. This can be done only if something is known beforehand about the ratio N_{Bz}/N_{Az}. In each of the binary diffusion problems discussed in this chapter, we begin by specifying this ratio by physical or chemical reasoning.

In this chapter we study diffusion in both *nonreacting* and *reacting* systems. When chemical reactions occur, we distinguish between two reaction types: *homogeneous*, in which the chemical change occurs in the entire volume of the fluid, and *heterogeneous*, in which the chemical change takes place only in a restricted region, such as the surface of a catalyst. Not only is the physical picture different for homogeneous and heterogeneous reactions, but there is also a difference in the way the two types of reactions are described mathematically. The rate of production of a chemical species by *homogeneous* reaction appears as a source term in the differential equation obtained from the shell balance, just as the thermal source term appears in the shell energy balance. The rate of production by a *heterogeneous* reaction, on the other hand, appears not in the differential equation, but rather in the boundary condition at the surface on which the reaction occurs.

In order to set up problems involving chemical reactions, some information has to be available about the rate at which the various chemical species appear or disappear by reaction. This brings us to the vast subject of *chemical kinetics*, that branch of physical chemistry that deals with the mechanisms of chemical reactions and the rates at which they occur.[1] In this chapter we assume that the reaction rates are described by means of simple functions of the concentrations of the reacting species.

At this point we need to mention the notation to be used for the chemical rate constants. For homogeneous reactions, the molar rate of production of species A may be given by an expression of the form

Homogeneous reaction: $$R_A = k_n''' c_A^n \tag{18.0-2}$$

in which $R_A \, [=] \, \text{moles/cm}^3 \cdot \text{s}$ and $c_A \, [=] \, \text{moles/cm}^3$. The index n indicates the "order" of the reaction;[2] for a first-order reaction, $k_1''' \, [=] \, 1/\text{s}$. For heterogeneous reactions, the molar rate of production at the reaction surface may often be specified by a relation of the form

Heterogeneous reaction: $$N_{Az}|_{\text{surface}} = k_n'' c_A^n|_{\text{surface}} \tag{18.0-3}$$

in which $N_{Az} \, [=] \, \text{moles/cm}^2 \cdot \text{s}$ and $c_A \, [=] \, \text{moles/cm}^3$. Here $k_1'' \, [=] \, \text{cm/s}$. Note that the triple prime on the rate constant indicates a volume source and the double prime a surface source.

We begin in §18.1 with a statement of the shell balance and the kinds of boundary conditions that may arise in solving diffusion problems. In §18.2 a discussion of diffusion through a stagnant film is given, this topic being necessary to the understanding of the film models of diffusional operations in chemical engineering. Then, in §§18.3 and 18.4 we given some elementary examples of diffusion with chemical reaction—both heterogeneous and homogeneous. These examples illustrate the role that diffusion plays in chemical kinetics and the important fact that diffusion can significantly affect the rate of a chemical reaction. In §§18.5 and 6 we turn our attention to forced-convection mass transfer—that is, diffusion superimposed on a flow field. Although we have not in-

---

[1] R. J. Silbey and R. A. Alberty, *Physical Chemistry*, 3rd edition, Wiley, New York (2001), Chapter 18.

[2] Not all rate expressions are of the simple form of Eq. 18.0-2. The reaction rate may depend in a complicated way on the concentration of all species present. Similar remarks hold for Eq. 18.0-3. For detailed information on reaction rates see *Table of Chemical Kinetics, Homogeneous Reactions*, National Bureau of Standards, Circular 510 (1951), Supplement No. 1 to Circular 510 (1956). This reference is now being supplemented by a data base maintained by NIST at "http://kinetics.nist.gov/." For heterogeneous reactions, see R. Mezaki and H. Inoue, *Rate Equations of Solid-Catalyzed Reactions*, U. of Tokyo Press, Tokyo (1991). See also C. G. Hill, *Chemical Engineering Kinetics and Reactor Design: An Introduction*, Wiley, New York (1977).

cluded an example of free-convection mass transfer, it would have been possible to parallel the discussion of free-convection heat transfer given in §10.9. Next, in §18.7 we discuss diffusion in porous catalysts. Finally, in the last section we extend the evaporation problem of §18.2 to a three-component system.

§18.1   SHELL MASS BALANCES; BOUNDARY CONDITIONS

The diffusion problems in this chapter are solved by making mass balances for one or more chemical species over a thin shell of solid or fluid. Having selected an appropriate system, the law of conservation of mass of species A in a binary system is written over the volume of the shell in the form

$$\begin{Bmatrix} \text{rate of} \\ \text{mass of} \\ A \text{ in} \end{Bmatrix} - \begin{Bmatrix} \text{rate of} \\ \text{mass of} \\ A \text{ out} \end{Bmatrix} + \begin{Bmatrix} \text{rate of production of} \\ \text{mass of } A \text{ by} \\ \text{homogeneous reaction} \end{Bmatrix} = 0 \qquad (18.1\text{-}1)$$

The conservation statement may, of course, be expressed in terms of moles. The chemical species A may enter or leave the system by diffusion (i.e., by molecular motion) and also by virtue of the overall motion of the fluid (i.e., by convection), both of these being included in $\mathbf{N}_A$. In addition, species A may be produced or consumed by homogeneous chemical reactions.

After a balance is made on a shell of finite thickness by means of Eq. 18.1-1, we then let the thickness become infinitesimally small. As a result of this process a differential equation for the mass (or molar) flux is generated. If, into this equation, we substitute the expression for the mass (or molar) flux in terms of the concentration gradient, we get a differential equation for the concentration.

When this differential equation has been integrated, constants of integration appear, and these have to be determined by the use of boundary conditions. The boundary conditions are very similar to those used in heat conduction (see §10.1):

a. The concentration at a surface can be specified; for example, $x_A = x_{A0}$.

b. The mass flux at a surface can be specified; for example, $N_{Az} = N_{A0}$. If the ratio N_{Bz}/N_{Az} is known, this is equivalent to giving the concentration gradient.

c. If diffusion is occurring in a solid, it may happen that at the solid surface substance A is lost to a surrounding stream according to the relation

$$N_{A0} = k_c(c_{A0} - c_{Ab}) \qquad (18.1\text{-}2)$$

in which N_{A0} is the molar flux at the surface, c_{A0} is the surface concentration, c_{Ab} is the concentration in the bulk fluid stream, and the proportionality constant k_c is a "mass transfer coefficient." Methods of correlating mass transfer coefficients are discussed in Chapter 22. Equation 18.1-2 is analogous to "Newton's law of cooling" given in Eq. 10.1-2.

d. The rate of chemical reaction at the surface can be specified. For example, if substance A disappears at a surface by a first-order chemical reaction, then $N_{A0} = k_1'' c_{A0}$. That is, the rate of disappearance at a surface is proportional to the surface concentration, the proportionality constant k_1'' being a first-order chemical rate coefficient.

§18.2   DIFFUSION THROUGH A STAGNANT GAS FILM

Let us now analyze the diffusion system shown in Fig. 18.2-1 in which liquid A is evaporating into gas B. We imagine there is some device that maintains the liquid level at $z = z_1$. Right at the liquid–gas interface, the gas-phase concentration of A, expressed as mole

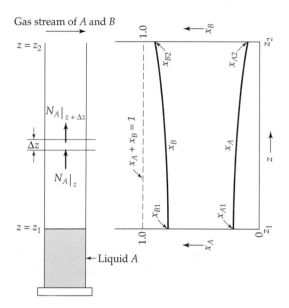

Fig. 18.2-1. Steady-state diffusion of A through stagnant B with the liquid-vapor interface maintained at a fixed position. The graph shows how the concentration profiles deviate from straight lines because of the convective contribution to the mass flux.

fraction, is x_{A1}. This is taken to be the gas-phase concentration of A corresponding to equilibrium[1] with the liquid at the interface. That is, x_{A1} is the vapor pressure of A divided by the total pressure, p_A^{vap}/p, provided that A and B form an ideal gas mixture and that the solubility of gas B in liquid A is negligible.

A stream of gas mixture A–B of concentration x_{A2} flows slowly past the top of the tube, to maintain the mole fraction of A at x_{A2} for $z = z_2$. The entire system is kept at constant temperature and pressure. Gases A and B are assumed to be ideal.

We know that there will be a net flow of gas upward from the gas–liquid interface, and that the gas velocity at the cylinder wall will be smaller than that in the center of the tube. To simplify the problem, we neglect this effect and assume that there is no dependence of the z-component of the velocity on the radial coordinate.

When this evaporating system attains a steady state, there is a net motion of A away from the interface and the species B is stationary. Hence the molar flux of A is given by Eq. 17.0-1 with $N_{Bz} = 0$. Solving for N_{Az}, we get

$$N_{Az} = -\frac{c\mathcal{D}_{AB}}{1 - x_A}\frac{dx_A}{dz} \qquad (18.2\text{-}1)$$

A steady-state mass balance (in molar units) over an increment Δz of the column states that the amount of A entering at plane z equals the amount of A leaving at plane $z + \Delta z$:

$$SN_{Az}|_z - SN_{Az}|_{z+\Delta z} = 0 \qquad (18.2\text{-}2)$$

Here S is the cross-sectional area of the column. Division by $S\Delta z$ and taking the limit as $\Delta z \to 0$ gives

$$-\frac{dN_{Az}}{dz} = 0 \qquad (18.2\text{-}3)$$

---

[1] L. J. Delaney and L. C. Eagleton [*AIChE Journal*, **8**, 418–420 (1962)] conclude that, for evaporating systems, the interfacial equilibrium assumption is reasonable, with errors in the range of 1.3 to 7.0% possible.

Substitution of Eq. 18.2-1 into Eq. 18.2-3 gives

$$\frac{d}{dz}\left(\frac{c\mathcal{D}_{AB}}{1 - x_A}\frac{dx_A}{dz}\right) = 0 \tag{18.2-4}$$

For an ideal gas mixture the equation of state is $p = cRT$, so that at constant temperature and pressure c must be a constant. Furthermore, for gases $\mathcal{D}_{AB}$ is very nearly independent of the composition. Therefore, $c\mathcal{D}_{AB}$ can be moved to the left of the derivative operator to get

$$\frac{d}{dz}\left(\frac{1}{1 - x_A}\frac{dx_A}{dz}\right) = 0 \tag{18.2-5}$$

This is a second-order differential equation for the concentration profile expressed as mole fraction of A. Integration with respect to z gives

$$\frac{1}{1 - x_A}\frac{dx_A}{dz} = C_1 \tag{18.2-6}$$

A second integration then gives

$$-\ln(1 - x_A) = C_1 z + C_2 \tag{18.2-7}$$

If we replace C_1 by $-\ln K_1$ and C_2 by $-\ln K_2$, Eq. 18.2-7 becomes

$$1 - x_A = K_1^z K_2 \tag{18.2-8}$$

The two constants of integration, K_1 and K_2, may then be determined from the boundary conditions

B.C. 1:   at $z = z_1$, $x_A = x_{A1}$ (18.2-9)

B.C. 2:   at $z = z_2$, $x_A = x_{A2}$ (18.2-10)

When the constants have been obtained, we get finally

$$\boxed{\left(\frac{1 - x_A}{1 - x_{A1}}\right) = \left(\frac{1 - x_{A2}}{1 - x_{A1}}\right)^{\frac{z - z_1}{z_2 - z_1}}} \tag{18.2-11}$$

The profiles for gas B are obtained by using $x_B = 1 - x_A$. The concentration profiles are shown in Fig. 18.2-1. It can be seen there that the slope dx_A/dz is not constant although N_{Az} is; this could have been anticipated from Eq. 18.2-1.

    Once the concentration profiles are known, we can get average values and mass fluxes at surfaces. For example, the average concentration of B in the region between z_1 and z_2 is obtained as follows:

$$\frac{x_{B,\text{avg}}}{x_{B1}} = \frac{\int_{z_1}^{z_2}(x_B/x_{B1})dz}{\int_{z_1}^{z_2}dz} = \frac{\int_0^1(x_{B2}/x_{B1})^\zeta\,d\zeta}{\int_0^1 d\zeta} = \frac{(x_{B2}/x_{B1})^\zeta}{\ln(x_{B2}/x_{B1})}\bigg|_0^1 \tag{18.2-12}$$

in which $\zeta = (z - z_1)/(z_2 - z_1)$ is a dimensionless length variable. This average may be rewritten as

$$x_{B,\text{avg}} = \frac{x_{B2} - x_{B1}}{\ln(x_{B2}/x_{B1})} \tag{18.2-13}$$

That is, the average value of x_B is the logarithmic mean, $(x_B)_{\ln}$, of the terminal concentrations.

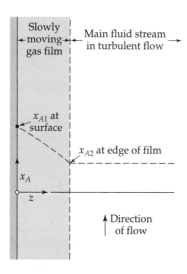

Fig. 18.2-2. Film model for mass transfer; component A is diffusing from the surface into the gas stream through a hypothetical stagnant gas film.

Labels in figure:
- Slowly moving gas film
- Main fluid stream in turbulent flow
- x_{A1} at surface
- x_{A2} at edge of film
- x_A
- z
- Direction of flow

The rate of mass transfer at the liquid–gas interface—that is, the rate of evaporation—may be obtained from Eq. 18.2-1 as follows:

$$N_{Az}\big|_{z=z_1} = -\frac{c\mathcal{D}_{AB}}{1-x_{A1}}\frac{dx_A}{dz}\bigg|_{z=z_1} = +\frac{c\mathcal{D}_{AB}}{x_{B1}}\frac{dx_B}{dz}\bigg|_{z=z_1} = \frac{c\mathcal{D}_{AB}}{z_2-z_1}\ln\!\left(\frac{x_{B2}}{x_{B1}}\right) \tag{18.2-14}$$

By combining Eqs. 18.2-13 and 14 we get finally

$$N_{Az}\big|_{z=z_1} = \frac{c\mathcal{D}_{AB}}{(z_2-z_1)(x_B)_{\ln}}(x_{A1}-x_{A2}) \tag{18.2-15}$$

This expression gives the evaporation rate in terms of the characteristic driving force $x_{A1}-x_{A2}$.

By expanding the solution in Eq. 18.2-15 in a Taylor series, we can get (see §C.2 and Problem 18B.18)

$$N_{Az}\big|_{z=z_1} = \frac{c\mathcal{D}_{AB}(x_{A1}-x_{A2})}{(z_2-z_1)}\left[1 + \tfrac{1}{2}(x_{A1}+x_{A2}) + \tfrac{1}{3}(x_{A1}^2 + x_{A1}x_{A2} + x_{A2}^2) + \cdots\right] \tag{18.2-16}$$

The expression in front of the bracketed expansion is the result that one would get if the convection term were entirely omitted in Eq. 18.0-1. The bracketed expansion then gives the correction resulting from including the convection term. Another way of interpreting this expression is that the simple result corresponds to joining the end points of the x_A curve in Fig. 18.2-1 by a straight line, and the complete result corresponds to using the curve of x_A versus z. If the terminal mole fractions are small, the correction term in brackets in Eq. 18.2-16 is only slightly greater than unity.

The results of this section have been used for experimental determinations of gas diffusivities.[2] Furthermore, these results find use in the "film models" of mass transfer. In Fig. 18.2-2 a solid or liquid surface is shown along which a gas is flowing. Near the surface is a slowly moving film through which A diffuses. This film is bounded by the surfaces $z = z_1$ and $z = z_2$. In this "model" it is assumed that there is a sharp transition from a stagnant film to a well-mixed fluid in which the concentration gradients are negligible. Although this model is physically unrealistic, it has nevertheless proven useful as a simplified picture for correlating mass transfer coefficients.

---

[2] C. Y. Lee and C. R. Wilke, *Ind. Eng. Chem.*, **46**, 2381–2387 (1954).

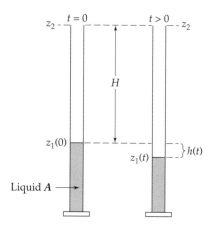

Fig. 18.2-3. Evaporation with quasi-steady-state diffusion. The liquid level goes down very slowly as the liquid evaporates. A gas mixture of composition x_{A2} flows across the top of the tube.

EXAMPLE 18.2-1	We want now to examine a problem that is slightly different from the one just discussed. Instead of maintaining the liquid–gas interface at a constant height, we allow the liquid level to subside as the evaporation proceeds, as shown in Fig. 18.2-3. Since the liquid retreats very slowly, we can use a quasi-steady state method with confidence.

Diffusion with a Moving Interface

SOLUTION

First we equate the molar rate of evaporation of A from the liquid phase with the rate at which moles of A enter the gas phase:

$$-S\frac{\rho^{(A)}}{M_A}\frac{dz_1}{dt} = \frac{c\mathcal{D}_{AB}}{(z_2 - z_1(t))(x_B)_{\ln}}(x_{A1} - x_{A2})S \tag{18.2-17}$$

Here $\rho^{(A)}$ is the density of pure liquid A and M_A is the molecular weight. On the right side of Eq. 18.2-17 we have used the steady-state evaporation rate evaluated at the current liquid column height (this is the quasi-steady-state approximation). This equation can be integrated to give

$$\int_0^h (H + h)\,dh = \frac{c\mathcal{D}_{AB}(x_{A1} - x_{A2})}{(\rho^{(A)}/M_A)(x_B)_{\ln}}\int_0^t dt \tag{18.2-18}$$

in which $h(t) = z_1(0) - z_1(t)$ is the distance that the interface has descended in time t, and $H = z_2 - z_1(0)$ is the initial height of the gas column. When we abbreviate the right side of Eq. 18.2-18 by $\frac{1}{2}Ct$, the equation can be integrated and then solved for h to give

$$h(t) = H(\sqrt{1 + (Ct/H^2)} - 1) \tag{18.2-19}$$

One can use this experiment to get the diffusivity from measurements of the liquid level as a function of time.

EXAMPLE 18.2-2	The diffusivity of the gas pair O_2–CCl_4 is being determined by observing the steady-state evaporation of carbon tetrachloride into a tube containing oxygen, as shown in Fig. 18.2-1. The distance between the CCl_4 liquid level and the top of the tube is $z_2 - z_1 = 17.1$ cm. The total pressure on the system is 755 mm Hg, and the temperature is 0°C. The vapor pressure of CCl_4 at that temperature is 33.0 mm Hg. The cross-sectional area of the diffusion tube is 0.82 cm^2. It is found that 0.0208 cm^3 of CCl_4 evaporate in a 10-hour period after steady state has been attained. What is the diffusivity of the gas pair O_2–CCl_4?

Determination of Diffusivity

SOLUTION

Let A stand for CCl_4 and B for O_2. The molar flux of A is then

$$N_A = \frac{(0.0208\ \text{cm}^3)(1.59\ \text{g/cm}^3)}{(154\ \text{g/g-mole})(0.82\ \text{cm}^2)(3.6 \times 10^4\ \text{s})}$$

$$= 7.26 \times 10^{-9}\ \text{g-mole/cm}^2 \cdot \text{s} \tag{18.2-20}$$

Then from Eq. 18.2-14 we get

$$
\begin{aligned}
\mathcal{D}_{AB} &= \frac{(N_A|_{z=z_1})(z_2 - z_1)}{c \ln(x_{B2}/x_{B1})} \\
&= \frac{(N_A|_{z=z_1})(z_2 - z_1)RT}{p \ln(p_{B2}/p_{B1})} \\
&= \frac{(7.26 \times 10^{-9})(17.1)(82.06)(273)}{(755/760)(2.303 \log_{10}(755/722))} \\
&= 0.0636 \text{ cm}^2/\text{s}
\end{aligned}
\tag{18.2-21}
$$

This method of determining gas-phase diffusivities suffers from several defects: the cooling of the liquid by evaporation, the concentration of nonvolatile impurities at the interface, the climbing of the liquid up the walls of the tube, and the curvature of the meniscus.

EXAMPLE 18.2-3

Diffusion through a Nonisothermal Spherical Film

(a) Derive expressions for diffusion through a spherical shell that are analogous to Eq. 18.2-11 (concentration profile) and Eq. 18.2-14 (molar flux). The system under consideration is shown in Fig. 18.2-4.

(b) Extend these results to describe the diffusion in a nonisothermal film in which the temperature varies radially according to

$$
\frac{T}{T_1} = \left(\frac{r}{r_1}\right)^n
\tag{18.2-22}
$$

where T_1 is the temperature at $r = r_1$. Assume as a *rough* approximation that $\mathcal{D}_{AB}$ varies as the $\frac{3}{2}$-power of the temperature:

$$
\frac{\mathcal{D}_{AB}}{\mathcal{D}_{AB,1}} = \left(\frac{T}{T_1}\right)^{3/2}
\tag{18.2-23}
$$

in which $\mathcal{D}_{AB,1}$ is the diffusivity at $T = T_1$. Problems of this kind arise in connection with drying of droplets and diffusion through gas films near spherical catalyst pellets.

The temperature distribution in Eq. 18.2-22 has been chosen solely for mathematical simplicity. This example is included to emphasize that, in nonisothermal systems, Eq. 18.0-1 is the correct starting point rather than $N_{Az} = -\mathcal{D}_{AB}(dc_A/dz) + x_A(N_{Az} + N_{Bz})$, as has been given in some textbooks.

SOLUTION

(a) A steady-state mass balance on a spherical shell leads to

$$
\frac{d}{dr}(r^2 N_{Ar}) = 0
\tag{18.2-24}
$$

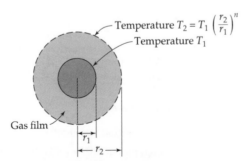

Temperature $T_2 = T_1 \left(\frac{r_2}{r_1}\right)^n$

Temperature T_1

Gas film

r_1

r_2

Fig. 18.2-4. Diffusion through a hypothetical spherical stagnant gas film surrounding a droplet of liquid A.

We now substitute into this equation the expression for the molar flux N_{Ar}, with N_{Br} set equal to zero, since B is insoluble in liquid A. This gives

$$\frac{d}{dr}\left(r^2 \frac{c\mathcal{D}_{AB}}{1-x_A}\frac{dx_A}{dr}\right) = 0 \tag{18.2-25}$$

For *constant temperature* the product $c\mathcal{D}_{AB}$ is constant, and Eq. 18.2-25 may be integrated to give the concentration distribution

$$\left(\frac{1-x_A}{1-x_{A1}}\right) = \left(\frac{1-x_{A2}}{1-x_{A1}}\right)^{\frac{(1/r_1)-(1/r)}{(1/r_1)-(1/r_2)}} \tag{18.2-26}$$

From Eq. 18.2-26 we can then get

$$W_A = 4\pi r_1^2 N_{Ar}|_{r=r_1} = \frac{4\pi c\mathcal{D}_{AB}}{(1/r_1)-(1/r_2)}\ln\left(\frac{1-x_{A2}}{1-x_{A1}}\right) \tag{18.2-27}$$

which is the molar flow of A across any spherical surface of radius r between r_1 and r_2.

(b) For the nonisothermal problem, combination of Eqs. 18.2-22 and 23 gives the variation of diffusivity with position:

$$\frac{\mathcal{D}_{AB}}{\mathcal{D}_{AB,1}} = \left(\frac{r}{r_1}\right)^{3n/2} \tag{18.2-28}$$

When this expression is inserted into Eq. 18.2-25 and c is set equal to p/RT, we get

$$\frac{d}{dr}\left(r^2 \frac{p\mathcal{D}_{AB,1}/RT_1}{1-x_A}\left(\frac{r}{r_1}\right)^{n/2}\frac{dx_A}{dr}\right) = 0 \tag{18.2-29}$$

After integrating between r_1 and r_2, we obtain (for $n \neq -2$)

$$W_A = 4\pi r_1^2 N_{Ar}|_{r=r_1} = \frac{4\pi(p\mathcal{D}_{AB,1}/RT_1)[1+(n/2)]}{[(1/r_1)^{1+(n/2)}-(1/r_2)^{1+(n/2)}]r_1^{n/2}}\ln\left(\frac{1-x_{A2}}{1-x_{A1}}\right) \tag{18.2-30}$$

For $n = 0$, this result simplifies to that in Eq. 18.2-27.

---

§18.3   DIFFUSION WITH A HETEROGENEOUS CHEMICAL REACTION

Let us now consider a simple model for a catalytic reactor, such as that shown in Fig. 18.3-1a, in which a reaction $2A \to B$ is being carried out. An example of a reaction of this type would be the solid-catalyzed dimerization of $CH_3CH = CH_2$.

We imagine that each catalyst particle is surrounded by a stagnant gas film through which A has to diffuse to reach the catalyst surface, as shown in Fig. 18.3-1b. At the catalyst surface we assume that the reaction $2A \to B$ occurs instantaneously, and that the product B then diffuses back out through the gas film to the main turbulent stream composed of A and B. We want to get an expression for the local rate of conversion from A to B when the effective gas-film thickness and the main stream concentrations x_{A0} and x_{B0} are known. We assume that the gas film is isothermal, although in many catalytic reactions the heat generated by the reaction cannot be neglected.

For the situation depicted in Fig. 18.3-1b, there is *one* mole of B moving in the *minus* z direction for every *two* moles of A moving in the *plus* z direction. We know this from the stoichiometry of the reaction. Therefore we know that at steady state

$$N_{Bz} = -\tfrac{1}{2}N_{Az} \tag{18.3-1}$$

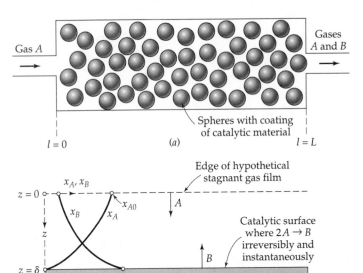

Fig. 18.3-1. (*a*) Schematic diagram of a catalytic reactor in which *A* is being converted to *B*. (*b*) Idealized picture (or "model") of the diffusion problem near the surface of a catalyst particle.

at any value of *z*. This relation may be substituted into Eq. 18.0-1, which may then be solved for N_{Az} to give

$$N_{Az} = -\frac{c\mathcal{D}_{AB}}{1 - \frac{1}{2}x_A}\frac{dx_A}{dz}\tag{18.3-2}$$

Hence, Eq. 18.0-1 plus the stoichiometry of the reaction have led to an expression for N_{Az} in terms of the concentration gradient.

We now make a mass balance on species *A* over a thin slab of thickness Δz in the gas film. This procedure is exactly the same as that used in connection with Eqs. 18.2-2 and 3 and leads once again to the equation

$$\frac{dN_{Az}}{dz} = 0\tag{18.3-3}$$

Insertion of the expression for N_{Az}, developed above, into this equation gives (for constant $\mathcal{D}_{AB}$)

$$\frac{d}{dz}\left(\frac{1}{1 - \frac{1}{2}x_A}\frac{dx_A}{dz}\right) = 0\tag{18.3-4}$$

Integration twice with respect to *z* gives

$$-2\ln(1 - \tfrac{1}{2}x_A) = C_1 z + C_2 = -(2\ln K_1)z - (2\ln K_2)\tag{18.3-5}$$

It is somewhat easier to find the integration constants K_1 and K_2 than C_1 and C_2. The boundary conditions are

B.C. 1:   at $z = 0$, $x_A = x_{A0}$ (18.3-6)

B.C. 2:   at $z = \delta$, $x_A = 0$ (18.3-7)

The final result is then

$$(1 - \tfrac{1}{2}x_A) = (1 - \tfrac{1}{2}x_{A0})^{1-(z/\delta)}\tag{18.3-8}$$

for the concentration profile in the gas film. Equation 18.3-2 may now be used to get the molar flux of reactant through the film:

$$N_{Az} = \frac{2c\mathcal{D}_{AB}}{\delta} \ln\left(\frac{1}{1 - \frac{1}{2}x_{A0}}\right) \qquad (18.3\text{-}9)$$

The quantity N_{Az} may also be interpreted as the local rate of reaction per unit area of catalytic surface. This information can be combined with other information about the catalytic reactor sketched in Fig. 18.3-1(a) to get the overall conversion rate in the entire reactor.

One point deserves to be emphasized. Although the chemical reaction occurs instantaneously at the catalytic surface, the conversion of A to B proceeds at a finite rate because of the diffusion process, which is "in series" with the reaction process. Hence we speak of the conversion of A to B as being *diffusion controlled*.

In the example above we have assumed that the reaction occurs instantaneously at the catalytic surface. In the next example we show how to account for finite reaction kinetics at the catalytic surface.

<table>
<tr><td>

EXAMPLE 18.3-1

Diffusion with a Slow Heterogeneous Reaction

</td><td>

Rework the problem just considered when the reaction $2A \rightarrow B$ is not instantaneous at the catalytic surface at $z = \delta$. Instead, assume that the rate at which A disappears at the catalyst-coated surface is proportional to the concentration of A in the fluid at the interface,

$$N_{Az} = k_1''c_A = k_1''cx_A \qquad (18.3\text{-}10)$$

in which k_1'' is a rate constant for the pseudo-first-order surface reaction.

</td></tr>
<tr><td>

SOLUTION

</td><td>

We proceed exactly as before, except that B.C. 2 in Eq. 18.3-7 must be replaced by

B.C. 2′:   at $z = \delta$, $x_A = \dfrac{N_{Az}}{k_1''c}$ (18.3-11)

N_{Az} being, of course, a constant at steady state. The determination of the integration constants from B.C. 1 and B.C. 2′ leads to

$$(1 - \tfrac{1}{2}x_A) = \left(1 - \frac{1}{2}\frac{N_{Az}}{k_1''c}\right)^{z/\delta} (1 - \tfrac{1}{2}x_{A0})^{1-(z/\delta)} \qquad (18.3\text{-}12)$$

From this we evaluate $(dx_A/dz)|_{z=0}$ and substitute it into Eq. 18.3-2, to get

$$N_{Az} = \frac{2c\mathcal{D}_{AB}}{\delta} \ln\left(\frac{1 - \frac{1}{2}(N_{Az}/k_1''c)}{1 - \frac{1}{2}x_{A0}}\right) \qquad (18.3\text{-}13)$$

This is a transcendental equation for N_{Az} as a function of x_{A0}, k_1'', $c\mathcal{D}_{AB}$, and δ. When k_1'' is large, the logarithm of $1 - \frac{1}{2}(N_{Az}/k_1''c)$ may be expanded in a Taylor series and all terms discarded but the first. We then get

$$N_{Az} = \frac{2c\mathcal{D}_{AB}/\delta}{1 + \mathcal{D}_{AB}/k_1''\delta} \ln\left(\frac{1}{1 - \frac{1}{2}x_{A0}}\right) \quad (k_1 \text{ large}) \qquad (18.3\text{-}14)$$

Note once again that we have obtained the rate of the *combined* reaction and diffusion process. Note also that the dimensionless group $\mathcal{D}_{AB}/k_1''\delta$ describes the effect of the surface reaction kinetics on the overall diffusion-reaction process. The reciprocal of this group is known as the *second Damköhler number*[1] $\mathrm{Da}^{\mathrm{II}} = k_1''\delta/\mathcal{D}_{AB}$. Evidently we get the result in Eq. 18.3-9 in the limit as $\mathrm{Da}^{\mathrm{II}} \rightarrow \infty$.

</td></tr>
</table>

---

[1] G. Damhöhler, Z. *Elektrochem.*, **42**, 846–862 (1936).

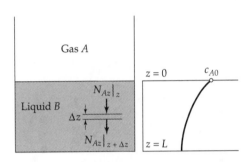

Fig. 18.4-1. Absorption of A by B with a homogeneous reaction in the liquid phase.

§18.4   DIFFUSION WITH A HOMOGENEOUS CHEMICAL REACTION

As the next illustration of setting up a mass balance, we consider the system shown in Fig. 18.4-1. Here gas A dissolves in liquid B in a beaker and diffuses isothermally into the liquid phase. As it diffuses, A also undergoes an irreversible first-order homogeneous reaction: $A + B \rightarrow AB$. An example of such a system is the absorption of CO_2 by a concentrated aqueous solution of NaOH.

We treat this as a binary solution of A and B, ignoring the small amount of AB that is present (the *pseudobinary assumption*). Then the mass balance on species A over a thickness Δz of the liquid phase becomes

$$N_{Az}|_z S - N_{Az}|_{z+\Delta z} S - k_1'''c_A S\Delta z = 0 \tag{18.4-1}$$

in which k_1''' is a first-order rate constant for the chemical decomposition of A, and S is the cross-sectional area of the liquid. The product $k_1'''c_A$ represents the moles of A consumed by the reaction per unit volume per unit time. Division of Eq. 18.4-1 by $S\Delta z$ and taking the limit as $\Delta z \rightarrow 0$ gives

$$\frac{dN_{Az}}{dz} + k_1'''c_A = 0 \tag{18.4-2}$$

If the concentration of A is small, then we may to a good approximation write Eq. 18.0-1 as

$$N_{Az} = -\mathscr{D}_{AB}\frac{dc_A}{dz} \tag{18.4-3}$$

since the total molar concentration c is virtually uniform throughout the liquid. Combining the last two equations gives

$$\mathscr{D}_{AB}\frac{d^2c_A}{dz^2} - k_1'''c_A = 0 \tag{18.4-4}$$

This is to be solved with the following boundary conditions:

B.C. 1: at $z = 0$, $c_A = c_{A0}$ (18.4-5)

B.C. 2: at $z = L$, $N_{Az} = 0$ (or $dc_A/dz = 0$) (18.4-6)

The first boundary condition asserts that the concentration of A at the surface in the liquid remains at a fixed value c_{A0}. The second states that no A diffuses through the bottom of the container at $z = L$.

If Eq. 18.4-4 is multiplied by $L^2/c_{A0}\mathscr{D}_{AB}$, then it can be written in dimensionless variables in the form of Eq. C.1-4

$$\frac{d^2\Gamma}{d\zeta^2} - \phi^2\Gamma = 0 \tag{18.4-7}$$

where $\Gamma = c_A/c_{A0}$ is a dimensionless concentration, $\zeta = z/L$ is a dimensionless length, and $\phi = \sqrt{k_1'''L^2/\mathscr{D}_{AB}}$ is a dimensionless group, known as the *Thiele modulus*.[1] This group represents the relative influence of the chemical reaction $k_1'''c_{A0}$ and diffusion $c_{A0}\mathscr{D}_{AB}/L^2$. Equation 18.4-7 is to be solved with the dimensionless boundary conditions that at $\zeta = 0$, $\Gamma = 1$, and at $\zeta = 1$, $d\Gamma/d\zeta = 0$. The general solution is

$$\Gamma = C_1 \cosh \phi\zeta + C_2 \sinh \phi\zeta \tag{18.4-8}$$

When the constants of integration are evaluated, we get

$$\Gamma = \frac{\cosh \phi \cosh \phi\zeta - \sinh \phi \sinh \phi\zeta}{\cosh \phi} = \frac{\cosh[\phi(1 - \zeta)]}{\cosh \phi} \tag{18.4-9}$$

Then reverting to the original notation

$$\boxed{\frac{c_A}{c_{A0}} = \frac{\cosh[\sqrt{k_1'''L^2/\mathscr{D}_{AB}}(1 - (z/L))]}{\cosh\sqrt{k_1'''L^2/\mathscr{D}_{AB}}}} \tag{18.4-10}$$

The concentration profile thus obtained is plotted in Fig. 18.4-1.

Once we have the complete concentration profile, we may evaluate other quantities, such as the average concentration in the liquid phase

$$\frac{c_{A,\text{avg}}}{c_{A0}} = \frac{\int_0^L (c_A/c_{A0})dz}{\int_0^L dz} = \frac{\tanh \phi}{\phi} \tag{18.4-11}$$

Also, the molar flux at the plane $z = 0$ can be found to be

$$N_{Az}\big|_{z=0} = -\mathscr{D}_{AB}\frac{dc_A}{dz}\bigg|_{z=0} = \left(\frac{c_{A0}\mathscr{D}_{AB}}{L}\right)\phi \tanh \phi \tag{18.4-12}$$

This result shows how the chemical reaction influences the rate of absorption of gas A by liquid B.

The reader may wonder how the solubility c_{A0} and the diffusivity $\mathscr{D}_{AB}$ can be determined experimentally if there is a chemical reaction taking place. First, k_1''' can be measured in a separate experiment in a well-stirred vessel. Then, in principle, c_{A0} and $\mathscr{D}_{AB}$ can be obtained from the measured absorption rates for various liquid depths L.

EXAMPLE 18.4-1

Gas Absorption with Chemical Reaction in an Agitated Tank[2]

Estimate the effect of chemical reaction rate on the rate of gas absorption in an agitated tank (see Fig. 18.4-2). Consider a system in which the dissolved gas A undergoes an irreversible first order reaction with the liquid B; that is, A disappears within the liquid phase at a rate proportional to the local concentration of A. An example of such a system is the absorption of SO_2 or H_2S in aqueous NaOH solutions.

---

[1] E. W. Thiele, *Ind. Eng. Chem.*, **31**, 916–920 (1939). **Ernest William Thiele** (pronounced "tee-lee") (1895–1993) is noted for his work on catalyst effectiveness factors and his part in the development of the "McCabe-Thiele" diagram. After 35 years with Standard Oil of Indiana, he taught for a decade at Notre Dame University.

[2] E. N. Lightfoot, *AIChE Journal*, **4**, 499–500 (1958), **8**, 710–712 (1962).

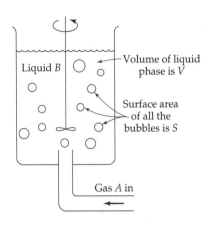

Fig. 18.4-2. Gas-absorption apparatus.

SOLUTION

An exact analysis of this situation is not possible because of the complexity of the gas-absorption process. However, a useful semiquantitative understanding can be obtained by the analysis of a relatively simple model. The model we use involves the following assumptions:

a. Each gas bubble is surrounded by a stagnant liquid film of thickness δ, which is small relative to the bubble diameter.

b. A quasi-steady concentration profile is quickly established in the liquid film after the bubble is formed.

c. The gas A is only sparingly soluble in the liquid, so that we can neglect the convection term in Eq. 18.0-1.

d. The liquid outside the stagnant film is at a concentration $c_{A\delta}$, which changes so slowly with respect to time that it can be considered constant.

The differential equation describing the diffusion with chemical reaction is the same as that in Eq. 18.4-4, but the boundary conditions are now

B.C. 1:	at $z = 0$,	$c_A = c_{A0}$	(18.4-13)
B.C. 2:	at $z = \delta$,	$c_A = c_{A\delta}$	(18.4-14)

The concentration c_{A0} is the interfacial concentration of A in the liquid phase, which is assumed to be at equilibrium with the gas phase at the interface, and $c_{A\delta}$ is the concentration of A in the main body of the liquid. The solution of Eq. 18.4-4 with these boundary conditions is

$$\frac{c_A}{c_{A0}} = \frac{\sinh \phi \cosh \phi \zeta + (B - \cosh \phi \sinh \phi \zeta)}{\sinh \phi} \qquad (18.4\text{-}15)$$

in which $\zeta = z/\delta$, $B = c_{A\delta}/c_{A0}$, and $\phi = \sqrt{k_1''' \delta^2 / \mathcal{D}_{AB}}$. This result is plotted in Fig. 18.4-3.

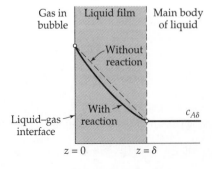

Fig. 18.4-3. Predicted concentration profile in the liquid film near a bubble.

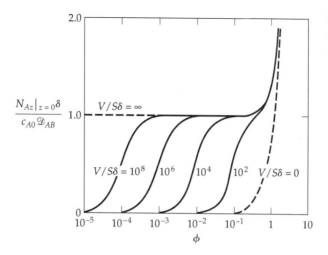

Fig. 18.4-4. Gas absorption accompanied by an irreversible first-order reaction.

Next we use assumption (d) above and equate the amount of A entering the main body of liquid at $z = \delta$ over the total bubble surface S in the tank to the amount of A consumed in the bulk of the liquid by chemical reaction:

$$-S\mathscr{D}_{AB}\frac{dc_A}{dz}\bigg|_{z=\delta} = Vk_1'''c_{A\delta} \tag{18.4-16}$$

Substitution of c_A from Eq. 18.4-15 into Eq. 18.4-16 gives an expression for B:

$$B = \frac{1}{\cosh\phi + (V/S\delta)\phi\sinh\phi} \tag{18.4-17}$$

When this result is substituted into Eq. 18.4-15, we obtain an expression for c_A/c_{A0} in terms of ϕ and $V/S\delta$.

From this expression for the concentration profile we can then get the total rate of absorption with chemical reaction from $N_{Az} = -\mathscr{D}_{AB}(dc_A/dz)$ evaluated at $z = 0$, thus:

$$\check{N} = \frac{N_{Az}\big|_{z=0}\delta}{c_{A0}\mathscr{D}_{AB}} = \frac{\phi}{\sinh\phi}\left(\cosh\phi - \frac{1}{\cosh\phi + (V/S\delta)\phi\sinh\phi}\right) \tag{18.4-18}$$

The result is plotted in Fig. 18.4-4.

It is seen here that the dimensionless absorption rate per unit area of interface, $\check{N}$, increases with ϕ for all finite values of $V/S\delta$. At very low values of ϕ—that is, for very slow reactions—$\check{N}$ approaches zero. For this limiting situation the liquid is nearly saturated with dissolved gas, and the "driving force" for absorption is very small. At large values of ϕ the dimensionless surface mass flux $\check{N}$ increases rapidly with ϕ and becomes very nearly independent of $V/S\delta$. Under the latter circumstances, the reaction is so rapid that almost all of the dissolving gas is consumed within the film. Then B is very nearly zero, and the bulk of the liquid plays no significant role. In the limit as ϕ becomes very large, $\check{N}$ approaches ϕ.

Somewhat more interesting behavior is observed for intermediate values of ϕ. It may be noted that, for moderately large $V/S\delta$, there is a considerable range of ϕ for which $\check{N}$ is very nearly unity. In this region the chemical reaction is fast enough to keep the bulk of the solution almost solute free, but slow enough to have little effect on solute transport in the film. Such a situation will arise when the ratio $V/S\delta$ of bulk to film volume is sufficient to offset the higher volumetric reaction rate in the film. The absorption rate is then equal to the physical absorption rate (that is, the rate for $k_1''' = 0$) for a solute-free tank. This behavior is frequently observed in practice, and operation under such conditions has proven a useful means of characterizing the mass transfer behavior of a variety of gas absorbers.[2]

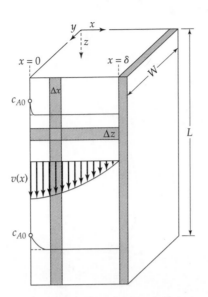

Fig. 18.5-1. Absorption of A into a falling film of liquid B.

§18.5 DIFFUSION INTO A FALLING LIQUID FILM (GAS ABSORPTION)[1]

In this section we present an illustration of *forced-convection* mass transfer, in which viscous flow and diffusion occur under such conditions that the velocity field can be considered virtually unaffected by the diffusion. Specifically, we consider the absorption of gas A by a laminar falling film of liquid B. The material A is only slightly soluble in B, so that the viscosity of the liquid is unaffected. We shall make the further restriction that the diffusion takes place so slowly in the liquid film that A will not "penetrate" very far into the film—that is, that the penetration distance will be small in comparison with the film thickness. The system is sketched in Fig. 18.5-1. An example of this kind of system occurs in the absorption of O_2 in H_2O.

Let us now set up the differential equations describing the diffusion process. First, we have to solve the momentum transfer problem to obtain the velocity profile $v_z(x)$ for the film; this has already been worked out in §2.2 in the absence of mass transfer at the fluid surface, and we know that the result is

$$v_z(x) = v_{\max}\left[1 - \left(\frac{x}{\delta}\right)^2\right] \tag{18.5-1}$$

provided that "end effects" are ignored.

Next we have to establish a mass balance on component A. We note that c_A will be changing with both x and z. Hence, as the element of volume for the mass balance, we select the volume formed by the intersection of a slab of thickness Δz with a slab of thickness Δx. Then the mass balance on A over this segment of a film of width W becomes

$$N_{Az}|_z\,W\Delta x - N_{Az}|_{z+\Delta z}\,W\Delta x + N_{Ax}|_x\,W\Delta z - N_{Ax}|_{x+\Delta x}\,W\Delta z = 0 \tag{18.5-2}$$

Dividing by $W\,\Delta x\,\Delta z$ and performing the usual limiting process as the volume element becomes infinitesimally small, we get

$$\frac{\partial N_{Az}}{\partial z} + \frac{\partial N_{Ax}}{\partial x} = 0 \tag{18.5-3}$$

---

[1] S. Lynn, J. R. Straatemeier, and H. Kramers, *Chem. Engr. Sci.*, **4**, 49–67 (1955).

Into this equation we now insert the expression for N_{Az} and N_{Ax}, making appropriate simplifications of Eq. 18.0-1. For the molar flux in the z direction, we write, assuming constant c,

$$N_{Az} = -\mathscr{D}_{AB}\frac{\partial c_A}{\partial z} + x_A(N_{Az} + N_{Bz}) \approx c_A v_z(x) \tag{18.5-4}$$

We discard the dashed-underlined term, since the transport of A in the z direction will be primarily by convection. We have made use of Eq. (M) in Table 17.8-1 and the fact that $\mathbf{v}$ is almost the same as $\mathbf{v}^*$ in dilute solutions. The molar flux in the x direction is

$$N_{Ax} = -\mathscr{D}_{AB}\frac{\partial c_A}{\partial x} + x_A(N_{Ax} + N_{Bx}) \approx -\mathscr{D}_{AB}\frac{\partial c_A}{\partial x} \tag{18.5-5}$$

Here we neglect the dashed-underlined term because in the x direction A moves predominantly by diffusion, there being almost no convective transport normal to the wall on account of the very slight solubility of A in B. Combining the last three equations, we then get for constant $\mathscr{D}_{AB}$

$$v_z\frac{\partial c_A}{\partial z} = \mathscr{D}_{AB}\frac{\partial^2 c_A}{\partial x^2} \tag{18.5-6}$$

Finally, insertion of Eq. 18.5-1 for the velocity distribution gives

$$v_{\max}\left[1 - \left(\frac{x}{\delta}\right)^2\right]\frac{\partial c_A}{\partial z} = \mathscr{D}_{AB}\frac{\partial^2 c_A}{\partial x^2} \tag{18.5-7}$$

as the differential equation for $c_A(x, z)$.

Equation 18.5-7 is to be solved with the following boundary conditions:

B.C. 1: at $z = 0$, $c_A = 0$ (18.5-8)

B.C. 2: at $x = 0$, $c_A = c_{A0}$ (18.5-9)

B.C. 3: at $x = \delta$, $\dfrac{\partial c_A}{\partial x} = 0$ (18.5-10)

The first boundary condition corresponds to the fact that the film consists of pure B at the top ($z = 0$), and the second indicates that at the liquid–gas interface the concentration of A is determined by the solubility of A in B (that is, c_{A0}). The third boundary condition states that A cannot diffuse through the solid wall. This problem has been solved analytically in the form of an infinite series,[2] but we do not give that solution here. Instead, we seek only a limiting expression valid for "short contact times," that is, for small values of $L/v_{\max}$.

If, as indicated in Fig. 18.5-1, the substance A has penetrated only a short distance into the film, then the species A "has the impression" that the film is moving throughout with a velocity equal to $v_{\max}$. Furthermore if A does not penetrate very far, it does not "sense" the presence of the solid wall at $x = \delta$. Hence, if the film were of infinite thickness moving with the velocity $v_{\max}$, the diffusing material "would not know the difference." This physical argument suggests (correctly) that we will get a very good result if we replace Eq. 18.5-7 and its boundary conditions by

$$v_{\max}\frac{\partial c_A}{\partial z} = \mathscr{D}_{AB}\frac{\partial^2 c_A}{\partial x^2} \tag{18.5-11}$$

B.C. 1: at $z = 0$, $c_A = 0$ (18.5-12)

B.C. 2: at $x = 0$, $c_A = c_{A0}$ (18.5-13)

B.C. 3: at $x = \infty$, $c_A = 0$ (18.5-14)

---

[2] R. L. Pigford, PhD thesis, University of Illinois (1941).

An exactly analogous problem occurred in Example 4.1-1, which was solved by the method of combination of variables. It is therefore possible to take over the solution to that problem just by changing the notation. The solution is[3]

$$\frac{c_A}{c_{A0}} = 1 - \frac{2}{\sqrt{\pi}} \int_0^{x/\sqrt{4\mathscr{D}_{AB}z/v_{\max}}} \exp(-\xi^2)\, d\xi \tag{18.5-15}$$

or

$$\boxed{\frac{c_A}{c_{A0}} = 1 - \operatorname{erf} \frac{x}{\sqrt{4\mathscr{D}_{AB}z/v_{\max}}} = \operatorname{erfc} \frac{x}{\sqrt{4\mathscr{D}_{AB}z/v_{\max}}}} \tag{18.5-16}$$

In these expressions "erf x" and "erfc x" are the "error function" and the "complementary error function" of x, respectively. They are discussed in §C.6 and tabulated in standard reference works.[4]

Once the concentration profiles are known, the local mass flux at the gas–liquid interface may be found as follows:

$$N_{Ax}\big|_{x=0} = -\mathscr{D}_{AB}\frac{\partial c_A}{\partial x}\bigg|_{x=0} = c_{A0}\sqrt{\frac{\mathscr{D}_{AB}v_{\max}}{\pi z}} \tag{18.5-17}$$

Then the total molar flow of A across the surface at $x = 0$ (i.e., being absorbed by a liquid film of length L and width W) is

$$\begin{aligned}
W_A &= \int_0^W \int_0^L N_{Ax}\big|_{x=0}\, dz\, dy \\
&= W c_{A0}\sqrt{\frac{\mathscr{D}_{AB}v_{\max}}{\pi}} \int_0^L \frac{1}{\sqrt{z}}\, dz \\
&= WLc_{A0}\sqrt{\frac{4\mathscr{D}_{AB}v_{\max}}{\pi L}}
\end{aligned} \tag{18.5-18}$$

The same result is obtained by integrating the product $v_{\max}c_A$ over the flow cross section at $z = L$ (see Problem 18C.3).

Equation 18.5-18 shows that the mass transfer rate is directly proportional to the square root of the diffusivity and inversely proportional to the square root of the "exposure time," $t_{\exp} = L/v_{\max}$. This approach for studying gas absorption was apparently first proposed by Higbie.[5]

The problem discussed in this section illustrates the "penetration model" of mass transfer. This model is discussed further in Chapters 20 and 22.

EXAMPLE 18.5-1

Gas Absorption from Rising Bubbles

Estimate the rate at which gas bubbles of A are absorbed by liquid B as the gas bubbles rise at their terminal velocity v_t through a clean quiescent liquid.

---

[3] The solution is worked out in detail by the method of combination of variables in Example 4.1-1.
[4] M. Abramowitz and I. A. Stegun, *Handbook of Mathematical Functions*, Dover, New York, 9th printing (1973), pp. 310 et seq.
[5] R. Higbie, *Trans. AIChE*, **31**, 365–389 (1935). **Ralph Wilmarth Higbie** (1908–1941), a graduate of the University of Michigan, provided the basis for the "penetration model" of mass transfer. He worked at E. I. du Pont de Nemours & Co., Inc., and also at Eagle-Picher Lead Co.; then he taught at the University of Arkansas and the University of North Dakota.

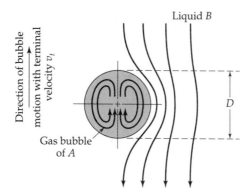

Liquid B

Fig. 18.5-2. Absorption of gas A into liquid B.

Direction of bubble motion with terminal velocity v_t

D

Gas bubble of A

SOLUTION

Gas bubbles of moderate size, rising in liquids free of surface-active agents, undergo a toroidal circulation (Rybczynski–Hadamard circulation) as shown in Fig. 18.5-2. The liquid moves downward relative to each rising bubble, enriched in species A near the interface in the manner of the falling film in Fig. 18.5-1. The depth of penetration of the dissolved gas into the liquid is slight over the major part of the bubble, because of the motion of the liquid relative to the bubble and because of the smallness of the liquid-phase diffusivity $\mathcal{D}_{AB}$. Thus, as a rough approximation, we can use Eq. 18.5-18 to estimate the rate of gas absorption, replacing the exposure time $t_{exp} = L/v_{max}$ for the falling film by D/v_t for the bubble, where D is the instantaneous bubble diameter. This gives an estimate[5] of the molar absorption rate, averaged over the bubble surface, as

$$(N_A)_{avg} = \sqrt{\frac{4\mathcal{D}_{AB}v_t}{\pi D}}\, c_{A0} \qquad (18.5\text{-}19)$$

Here c_{A0} is the solubility of gas A in liquid B at the interfacial temperature and partial pressure of gas A. Interestingly, the result in Eq. 18.5-19 turns out to be correct for potential flow of the liquid around the bubble (see Problem 4B.5). This equation has been approximately confirmed[6] for gas bubbles 0.3 to 0.5 cm in diameter rising through carefully purified water.

This system has also been analyzed for creeping flow[7] and the result is (see Example 20.3-1)

$$(N_A)_{avg} = \sqrt{\frac{4\mathcal{D}_{AB}v_t}{3\pi D}}\, c_{A0} \qquad (18.5\text{-}20)$$

instead of Eq. 18.5-19.

Trace amounts of surface-active agents cause a marked decrease in absorption rates from small bubbles, by forming a "skin" around each bubble and thus effectively preventing internal circulation. The molar absorption rate in the small-diffusivity limit then becomes proportional to the $\frac{1}{3}$ power of the diffusivity, as for a solid sphere (see §§22.2 and 3).

A similar approach has been used successfully for predicting mass transfer rates during drop formation at a capillary tip.[8]

---

[6] D. Hammerton and F. H. Garner, *Trans. Inst. Chem. Engrs.* (*London*), **32**, S18-S24 (1954).

[7] V. G. Levich, *Physicochemical Hydrodynamics*, Prentice-Hall, Englewood Cliffs, N.J. (1962), p. 408, Eq. 72.9. This reference gives many additional results, including liquid–liquid mass transfer and surfactant effects.

[8] H. Groothuis and H. Kramers, *Chem. Eng. Sci.*, **4**, 17–25 (1955).

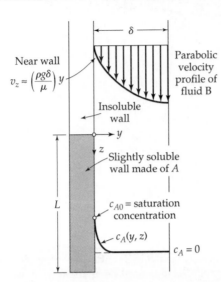

Fig. 18.6-1. Solid A dissolving into a falling film of liquid B, moving with a fully developed parabolic velocity profile.

§18.6   DIFFUSION INTO A FALLING LIQUID FILM (SOLID DISSOLUTION)[1]

We now turn to a falling film problem that is different from the one discussed in the previous section. Liquid B is flowing in laminar motion down a vertical wall as shown in Fig. 18.6-1. The film begins far enough up the wall so that v_z depends only on y for $z \geq 0$. For $0 < z < L$ the wall is made of a species A that is slightly soluble in B.

For short distances downstream, species A will not diffuse very far into the falling film. That is, A will be present only in a very thin boundary layer near the solid surface. Therefore the diffusing A molecules will experience a velocity distribution that is characteristic of the falling film right next to the wall, $y = 0$. The velocity distribution is given in Eq. 2.2-18. In the present situation $\cos \theta = 1$, and $x = \delta - y$, and

$$v_z = \frac{\rho g \delta^2}{2\mu}\left[1 - \left(1 - \frac{y}{\delta}\right)^2\right] = \frac{\rho g \delta^2}{2\mu}\left[2\left(\frac{y}{\delta}\right) - \left(\frac{y}{\delta}\right)^2\right] \tag{18.6-1}$$

At and adjacent to the wall $(y/\delta)^2 \ll (y/\delta)$, so that for this problem the velocity is, to a very good approximation, $v_z = (\rho g \delta / \mu)y \equiv ay$. This means that Eq. 18.5-6, which is applicable here, becomes for short distances downstream

$$ay\frac{\partial c_A}{\partial z} = \mathscr{D}_{AB}\frac{\partial^2 c_A}{\partial y^2} \tag{18.6-2}$$

where $a = \rho g \delta / \mu$. This equation is to be solved with the boundary conditions

B.C. 1:   at $z = 0$, $c_A = 0$ (18.6-3)

B.C. 2:   at $y = 0$, $c_A = c_{A0}$ (18.6-4)

B.C. 3:   at $y = \infty$, $c_A = 0$ (18.6-5)

In the second boundary condition, c_{A0} is the solubility of A in B. The third boundary condition is used instead of the correct one $(\partial c_A / \partial y = 0$ at $y = \delta)$, since for short contact times we feel intuitively that it will not make any difference. After all, since the mole-

---

[1] H. Kramers and P. J. Kreyger, *Chem. Eng. Sci.*, **6**, 42–48 (1956); see also R. L. Pigford, *Chem. Eng. Prog. Symposium Series* No. 17, Vol. 51, pp. 79–92 (1955) for the analogous heat-conduction problem.

cules of A penetrate only slightly into the film, they cannot get far enough to "see" the outer boundary of the film, and hence they cannot distinguish between the true boundary condition and the approximate boundary condition that we use. The same kind of reasoning was encountered in Example 12.2-2 and Problem 12B.4.

The form of the boundary conditions in Eqs. 18.6-3 to 5 suggests the method of combination of variables. Therefore we try $c_A/c_{A0} = f(\eta)$, where $\eta = y(a/9\mathcal{D}_{AB}z)^{1/3}$. This combination of the independent variables can be shown to be dimensionless, and the factor of "9" is included to make the solution look neater.

When this change of variable is made, the partial differential equation in Eq. 18.6-2 reduces to an ordinary differential equation

$$\frac{d^2f}{d\eta^2} + 3\eta^2 \frac{df}{d\eta} = 0 \tag{18.6-6}$$

with the boundary conditions $f(0) = 1$ and $f(\infty) = 0$.

This second-order equation, which is of the form of Eq. C.1-9, has the solution

$$f = C_1 \int_0^\eta \exp(-\bar{\eta}^3)\, d\bar{\eta} + C_2 \tag{18.6-7}$$

The constants of integration can then be evaluated using the boundary conditions, and one obtains finally

$$\frac{c_A}{c_{A0}} = \frac{\displaystyle\int_\eta^\infty \exp(-\bar{\eta}^3)\, d\bar{\eta}}{\displaystyle\int_0^\infty \exp(-\bar{\eta}^3)\, d\bar{\eta}} = \frac{\displaystyle\int_\eta^\infty \exp(-\bar{\eta}^3)\, d\bar{\eta}}{\Gamma(\tfrac{4}{3})} \tag{18.6-8}$$

for the concentration profiles, in which $\Gamma(\tfrac{4}{3}) = 0.8930\ldots$ is the gamma function of $\tfrac{4}{3}$. Next the local mass flux at the wall can be obtained as follows

$$N_{Ay}\big|_{y=0} = -\mathcal{D}_{AB} \frac{\partial c_A}{\partial y}\bigg|_{y=0} = -\mathcal{D}_{AB} c_{A0} \left[\frac{d}{d\eta}\left(\frac{c_A}{c_{A0}}\right)\frac{\partial \eta}{\partial y}\right]\bigg|_{y=0}$$

$$= -\mathcal{D}_{AB} c_{A0} \left[-\frac{\exp(-\eta^3)}{\Gamma(\tfrac{4}{3})}\left(\frac{a}{9\mathcal{D}_{AB}z}\right)^{1/3}\right]\bigg|_{y=0} = +\frac{\mathcal{D}_{AB} c_{A0}}{\Gamma(\tfrac{4}{3})}\left(\frac{a}{9\mathcal{D}_{AB}z}\right)^{1/3} \tag{18.6-9}$$

Then the molar flow of A across the entire mass transfer surface at $y = 0$ is

$$W_A = \int_0^W \int_0^L N_{Ay}\big|_{y=0}\, dz\, dx = \frac{2\mathcal{D}_{AB} c_{A0} WL}{\Gamma(\tfrac{7}{3})}\left(\frac{a}{9\mathcal{D}_{AB}L}\right)^{1/3} \tag{18.6-10}$$

where $\Gamma(\tfrac{7}{3}) = \tfrac{4}{3}\Gamma(\tfrac{4}{3}) = 1.1907\ldots$.

The problem discussed in §18.5 and the one discussed here are examples of two types of asymptotic solutions that are discussed further in §20.2 and §20.3 and again in Chapter 22. It is therefore important that these two problems be thoroughly understood. Note that in §18.5, $W_A \propto (\mathcal{D}_{AB}L)^{1/2}$, whereas in this section $W_A \propto (\mathcal{D}_{AB}L)^{2/3}$. The differences in the exponents reflect the nature of the velocity gradient at the mass transfer interface: in §18.5, the velocity gradient was zero, whereas in this section, the velocity gradient is nonzero.

§18.7   DIFFUSION AND CHEMICAL REACTION INSIDE A POROUS CATALYST

Up to this point we have discussed diffusion in gases and liquids in systems of simple geometry. We now wish to apply the shell mass balance method and Fick's first law to describe diffusion within a porous catalyst pellet. We make no attempt to describe the

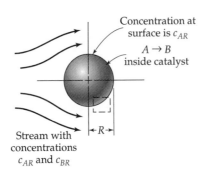

Fig. 18.7-1. A spherical catalyst that is porous. For a magnified version of the inset, see Fig. 18.7-2.

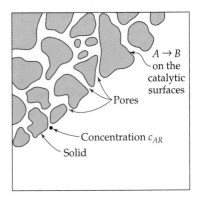

Fig. 18.7-2. Pores in the catalyst, in which diffusion and chemical reaction occur.

diffusion inside the tortuous void passages in the pellet. Instead, we describe the "averaged" diffusion of the reactant in terms of an "effective diffusivity."[1,2,3]

Specifically, we consider a spherical porous catalyst particle of radius R, as shown in Fig. 18.7-1. This particle is in a catalytic reactor, where it is submerged in a gas stream containing the reactant A and the product B. In the neighborhood of the surface of the particular catalyst particle under consideration, we presume that the concentration is c_{AR} moles of A per unit volume. Species A diffuses through the tortuous passages in the catalyst and is converted to B on the catalytic surfaces, as sketched in Fig. 18.7-2.

We start by making a mass balance for species A on a spherical shell of thickness Δr within a single catalyst particle:

$$N_{Ar}|_r \cdot 4\pi r^2 - N_{Ar}|_{r+\Delta r} \cdot 4\pi (r + \Delta r)^2 + R_A \cdot 4\pi r^2 \Delta r = 0 \tag{18.7-1}$$

Here $N_{Ar}|_r$ is the number of moles of A passing in the r direction through an imaginary spherical surface at a distance r from the center of the sphere. The source term $R_A \cdot 4\pi r^2 \Delta r$ is the molar rate of production of A by chemical reaction in the shell of thickness Δr. Dividing by $4\pi \, \Delta r$ and letting $\Delta r \to 0$ gives

$$\lim_{\Delta r \to 0} \frac{(r^2 N_{Ar})|_{r+\Delta r} - (r^2 N_{Ar})|_r}{\Delta r} = r^2 R_A \tag{18.7-2}$$

or, using the definition of the first derivative,

$$\frac{d}{dr}(r^2 N_{Ar}) = r^2 R_A \tag{18.7-3}$$

This limiting process is clearly in conflict with the fact that the porous medium is granular rather than continuous. Consequently, in Eq. 18.7-3 the symbols N_{Ar} and R_A cannot be interpreted as quantities having a meaningful value at a point. Rather we have to interpret them as quantities averaged over a small neighborhood of the point in question—a neighborhood small with respect to the dimension R, but large with respect to the dimensions of the passages within the porous particle.

---

[1] E. W. Thiele, *Ind. Eng. Chem.*, **31**, 916–920 (1939).
[2] R. Aris, *Chem. Eng. Sci.*, **6**, 265–268 (1957).
[3] A. Wheeler, *Advances in Catalysis*, Academic Press, New York (1950), Vol. 3, pp. 250–326.

We now *define* an "effective diffusivity" for species A in the porous medium by

$$N_{Ar} = -\mathscr{D}_A \frac{dc_A}{dr} \tag{18.7-4}$$

in which c_A is the concentration of the gas A contained within the pores. The effective diffusivity $\mathscr{D}_A$ must be measured experimentally. It depends generally on pressure and temperature and also on the catalyst pore structure. The actual mechanism for diffusion in pores is complex, since the pore dimensions may be smaller than the mean free path of the diffusing molecules. We do not belabor the question of mechanism here but assume only that Eq. 18.7-4 can adequately represent the diffusion process (see §24.6).

When the preceding expression is inserted into Eq. 18.7-3, we get, for constant diffusivity

$$\mathscr{D}_A \frac{1}{r^2} \frac{d}{dr} \left(r^2 \frac{dc_A}{dr} \right) = -R_A \tag{18.7-5}$$

We now consider the situation where species A disappears according to a first-order chemical reaction on the catalytic surfaces that form all or part of the "walls" of the winding passages. Let a be the available catalytic surface per unit volume (of solids + voids). Then $R_A = -k_1'' a c_A$, and Eq. 18.7-5 becomes (see Eq. C.1-6)

$$\mathscr{D}_A \frac{1}{r^2} \frac{d}{dr} \left(r^2 \frac{dc_A}{dr} \right) = k_1'' a c_A \tag{18.7-6}$$

This equation is to be solved with the boundary conditions that $c_A = c_{AR}$ at $r = R$, and that c_A is finite at $r = 0$.

Equations containing the operator $(1/r^2)(d/dr)[r^2(d/dr)]$ can frequently be solved by using a "standard trick"—namely, a change of variable $c_A/c_{AR} = (1/r)f(r)$. The equation for $f(r)$ is then

$$\frac{d^2 f}{dr^2} = \left(\frac{k_1'' a}{\mathscr{D}_A} \right) f \tag{18.7-7}$$

This is a standard second-order differential equation, which can be solved in terms of exponentials or hyperbolic functions. When it is solved and the result divided by r we get the following solution of Eq. 18.7-6 in terms of hyperbolic functions (see §C.5):

$$\frac{c_A}{c_{AR}} = \frac{C_1}{r} \cosh \sqrt{\frac{k_1'' a}{\mathscr{D}_A}} r + \frac{C_2}{r} \sinh \sqrt{\frac{k_1'' a}{\mathscr{D}_A}} r \tag{18.7-8}$$

Application of the boundary conditions gives finally

$$\boxed{\frac{c_A}{c_{AR}} = \left(\frac{R}{r} \right) \frac{\sinh \sqrt{k_1'' a / \mathscr{D}_A}\, r}{\sinh \sqrt{k_1'' a / \mathscr{D}_A}\, R}} \tag{18.7-9}$$

In studies on chemical kinetics and catalysis one is frequently interested in the molar flux N_{AR} or the molar flow W_{AR} at the surface $r = R$:

$$W_{AR} = 4\pi R^2 N_{AR} = -4\pi R^2 \mathscr{D}_A \frac{dc_A}{dr} \bigg|_{r=R} \tag{18.7-10}$$

When Eq. 18.7-9 is used in this expression, we get

$$W_{AR} = 4\pi R \mathscr{D}_A c_{AR} \left(1 - \sqrt{\frac{k_1'' a}{\mathscr{D}_A}}\, R \coth \sqrt{\frac{k_1'' a}{\mathscr{D}_A}}\, R \right) \tag{18.7-11}$$

This result gives the rate of conversion (in moles/sec) of A to B in a single catalyst particle of radius R in terms of the parameters describing the diffusion and reaction processes.

If the catalytically active surface were all exposed to the stream of concentration c_{AR}, then the species A would not have to diffuse through the pores to a reaction site. The molar rate of conversion would then be given by the product of the available surface and the surface reaction rate:

$$W_{AR,0} = (\tfrac{4}{3}\pi R^3)(a)(-k_1'' c_{AR}) \tag{18.7-12}$$

Taking the ratio of the last two equations, we get

$$\eta_A = \frac{W_{AR}}{W_{AR,0}} = \frac{3}{\phi^2}\,(\phi \coth \phi - 1) \tag{18.7-13}$$

in which $\phi = \sqrt{k_1'' a/\mathscr{D}_A}\,R$ is the *Thiele modulus*,[1] encountered in §18.4. The quantity η_A is called the *effectiveness factor*.[1-4] It is the quantity by which $W_{AR,0}$ has to be multiplied to account for the intraparticle diffusional resistance to the overall conversion process.

For nonspherical catalyst particles, the foregoing results may be applied approximately by reinterpreting R. We note that for a sphere of radius R the ratio of volume to external surface is $R/3$. For nonspherical particles, we redefine R in Eq. 18.7-13 as

$$R_{\text{nonsph}} = 3\left(\frac{V_P}{S_P}\right) \tag{18.7-14}$$

where V_P and S_P are the volume and external surface of a single catalyst particle. The absolute value of the conversion rate is then given approximately by

$$|W_{AR}| \approx V_P a k_1'' c_{AR} \eta_A \tag{18.7-15}$$

where

$$\eta_A = \frac{1}{3\Lambda^2}\,(3\Lambda \coth 3\Lambda - 1) \tag{18.7-16}$$

in which the quantity $\Lambda = \sqrt{k_1'' a/\mathscr{D}_A}\,(V_P/S_P)$ is a generalized modulus.[2,3]

The particular utility of the quantity Λ may be seen in Fig. 18.7-3. It is clear that when the exact theoretical expressions for η_A are plotted as functions of Λ, the curves

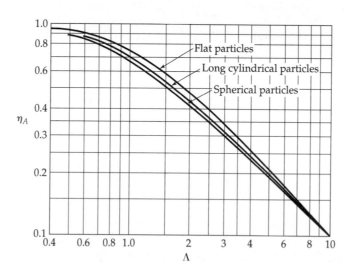

Fig. 18.7-3. Effectiveness factors for porous solid catalysts of various shapes [R. Aris, *Chem. Eng. Sci.*, **6**, 262–268 (1957)].

[4] O. A. Hougen and K. M. Watson, *Chemical Process Principles*, Wiley, New York (1947), Part III, Chapter XIX. See also *CPP Charts*, by O. A. Hougen, K. M. Watson, and R. A. Ragatz, Wiley, New York (1960), Fig. E.

have common asymptotes for large and small Λ and do not differ from one another very much for intermediate values of Λ. Thus Fig. 18.7-3 provides a justification for the use of Eq. 18.7-16 to estimate η_A for nonspherical particles.

§18.8 DIFFUSION IN A THREE-COMPONENT GAS SYSTEM

Up to this point the systems we have discussed have been binary systems, or ones that could be approximated as two-component systems. To illustrate the setting up of multi-component diffusion problems for gases, we rework the initial evaporation problem of §18.2 when liquid water (species 1) is evaporating into air, regarded as a binary mixture of nitrogen (2) and oxygen (3) at 1 atm and 352K. We take the air–water interface to be at $z = 0$ and the top end of the diffusion tube to be at $z = L$. We consider the vapor pressure of water to be known, so that x_1 is known at $z = 0$ (that is, $x_{10} = 341/760 = 0.449$), and the mole fractions of all three gases are known at $z = L$: $x_{1L} = 0.10$, $x_{2L} = 0.75$, $x_{3L} = 0.15$. The diffusion tube has a length $L = 11.2$ cm.

The conservation of mass leads, as in §18.2, to the following expressions:

$$\frac{dN_{\alpha z}}{dz} = 0 \qquad \alpha = 1, 2, 3 \tag{18.8-1}$$

From this it may be concluded that the molar fluxes of the three species are all constants at steady state. Since species 2 and 3 are not moving, we conclude that N_{2z} and N_{3z} are both zero.

Next we need the expressions for the molar fluxes from Eq. 17.9-1. Since $x_1 + x_2 + x_3 = 1$, we need only two of the three available equations, and we select the equations for species 2 and 3. Since $N_{2z} = 0$ and $N_{3z} = 0$, these equations simplify considerably:

$$\frac{dx_2}{dz} = \frac{N_{1z}}{c\mathscr{D}_{12}} x_2; \qquad \frac{dx_3}{dz} = \frac{N_{1z}}{c\mathscr{D}_{13}} x_3 \tag{18.8-2, 3}$$

Note that the diffusivity $\mathscr{D}_{23}$ does not appear here, because there is no relative motion of species 2 and 3. These equations can be integrated from an arbitrary height z to the top of the tube at L, to give for constant $c\mathscr{D}_{\alpha\beta}$

$$\int_{x_2}^{x_{2L}} \frac{dx_2}{x_2} = \frac{N_{1z}}{c\mathscr{D}_{12}} \int_z^L dz; \qquad \int_{x_3}^{x_{3L}} \frac{dx_3}{x_3} = \frac{N_{1z}}{c\mathscr{D}_{13}} \int_z^L dz \tag{18.8-4, 5}$$

Integration then gives

$$\frac{x_2}{x_{2L}} = \exp\left(-\frac{N_{1z}(L - z)}{c\mathscr{D}_{12}}\right); \qquad \frac{x_3}{x_{3L}} = \exp\left(-\frac{N_{1z}(L - z)}{c\mathscr{D}_{13}}\right) \tag{18.8-6, 7}$$

and the mole fraction profile of water vapor in the diffusion column will be

$$\boxed{x_1 = 1 - x_{2L} \exp\left(-\frac{N_{1z}(L - z)}{c\mathscr{D}_{12}}\right) - x_{3L} \exp\left(-\frac{N_{1z}(L - z)}{c\mathscr{D}_{13}}\right)} \tag{18.8-8}$$

When we apply the boundary condition at $z = 0$, we get

$$x_{10} = 1 - x_{2L} \exp\left(-\frac{N_{1z}L}{c\mathscr{D}_{12}}\right) - x_{3L} \exp\left(-\frac{N_{1z}L}{c\mathscr{D}_{13}}\right) \tag{18.8-9}$$

which is a transcendental equation for N_{1z}.

According to Reid, Prausnitz, and Poling,[1] $\mathcal{D}_{12} = 0.364$ cm^2/s and $\mathcal{D}_{13} = 0.357$ cm^2/s at 352K and 1 atm. At these conditions $c = 3.46 \times 10^{-5}$ g-moles/cm^3. To get a quick solution to Eq. 18.8-9, we take both diffusivities to be equal[2] to 0.36 cm^2/s. Then we get

$$0.449 = 1 - 0.90 \exp\left(-\frac{N_{1z}(11.2)}{(3.462 \times 10^{-5})(0.36)}\right) \tag{18.8-10}$$

from which we find that $N_{1z} = 5.523 \times 10^{-7}$ g-moles/cm$^2 \cdot$ s. This can be used as a first guess in solving Eq. 18.8-9 more exactly, if desired. Then the entire profiles can be calculated from Eqs. 18.8-6 to 8.

QUESTIONS FOR DISCUSSION

1. What arguments are used in this chapter for eliminating N_B from Eq. 18.0-1?
2. Suggest ways in which the diffusivity $\mathcal{D}_{AB}$ could be measured by means of the examples in this chapter. Summarize possible sources of error.
3. In what limit do the concentration curves in Fig. 18.2-1 become straight lines?
4. Distinguish between homogeneous and heterogeneous reactions. Which ones are described by boundary conditions and which ones manifest themselves in the differential equations?
5. Discuss the term "diffusion-controlled reaction."
6. What kind of "device" would you suggest in the first sentence of §18.2 for maintaining the level of the interface constant?
7. Why is the left-hand term in Eq. 18.2-15 called the "evaporation rate"?
8. Explain carefully how Eq. 18.2-19 is set up.
9. Criticize Example 18.2-3. To what extent is it "just a schoolbook problem"? What do you learn from the problem?
10. In what sense can the quantity N_{Az} in Eq. 18.3-9 be interpreted as a local rate of chemical reaction?
11. How does the size of a bubble change as it moves upward in a liquid?
12. In what connection have you encountered Eq. 18.5-11 before?
13. What happens if you try to solve Eq. 18.7-8 by using exponentials instead of hyperbolic functions? How can we make the simpler choice ahead of time?
14. Compare and contrast the systems discussed in §§18.5 and 6 as regards the physical problems, the mathematical methods used to solve them, and the final expressions for the molar fluxes.

PROBLEMS   **18A.1 Evaporation rate.** For the system shown in Fig. 18.2-1, what is the evaporation rate in g/hr of CCl_3NO_2 (chloropicrin) into air at 25°C? Make the customary assumption that air is a "pure substance."

Total pressure	770 mm Hg
Diffusivity (CCl_3NO_2–air)	0.088 cm^2/s
Vapor pressure of CCl_3NO_2	23.81 mm Hg
Distance from liquid level to top of tube	11.14 cm
Density of CCl_3NO_2	1.65 g/cm^3
Surface area of liquid exposed for evaporation	2.29 cm^2

Answer: 0.0139 g/hr

[1] R. C. Reid, J. M. Prausnitz, and B. E. Poling, *The Properties of Gases and Liquids*, 4th edition, McGraw-Hill, New York (1987), p. 591.

[2] The solution to ternary diffusion problems in which two of the binary diffusivities are equal was discussed by H. L. Toor, *AIChE Journal*, **3**, 198–207 (1957).

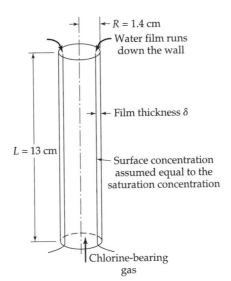

Fig. 18A.4. Schematic drawing of a wetted-wall column.

R = 1.4 cm

Water film runs down the wall

Film thickness δ

L = 13 cm

Surface concentration assumed equal to the saturation concentration

Chlorine-bearing gas

18A.2. Sublimation of small iodine spheres in still air. A sphere of iodine, 1 cm in diameter, is placed in still air at 40°C and 747 mm Hg pressure. At this temperature the vapor pressure of iodine is about 1.03 mm Hg. We want to determine the diffusivity of the iodine–air system by measuring the sublimation rate. To help determine reasonable experimental conditions,

(a) Estimate the diffusivity for the iodine–air system at the temperature and pressure given above, using the intermolecular force parameters in Table E.1.

(b) Estimate the rate of sublimation, basing your calculations on Eq. 18.2-27. (*Hint:* Assume r_2 to be very large.)

This method has been used for measuring the diffusivity, but it is open to question because of the possible importance of free convection.

Answer: **(a)** $\mathcal{D}_{I_2\text{-air}} = 0.0888 \text{ cm}^2/\text{s}$; **(b)** $W_{I_2} = 1.06 \times 10^{-4}$ g-mole/hr

18A.3. Estimating the error in calculating the absorption rate. What is the maximum possible error in computing the absorption rate from Eq. 18.5-18, if the solubility of A in B is known within ±5% and the diffusivity of A in B is known within ±15%? Assume that the geometric quantities and the velocity are known very accurately.

18A.4. Chlorine absorption in a falling film (Fig. 18A.4). Chlorine is being absorbed from a gas in a small experimental wetted-wall tower as shown in the figure. The absorbing fluid is water, which is moving with an average velocity of 17.7 cm/s. What is the absorption rate in g-moles/hr, if the liquid-phase diffusivity of the chlorine–water system is 1.26×10^{-5} cm^2/s, and if the saturation concentration of chlorine in water is 0.823 g chlorine per 100 g water (these are the experimental values at 16°C). The dimensions of the column are given in the figure. (*Hint:* Ignore the chemical reaction between chlorine and water.)

Answer: 0.273 g-moles/hr

18A.5. Measurement of diffusivity by the point-source method (Fig. 18C.1).[1] We wish to design a flow system to utilize the results of Problem 18C.1 for the measure of $\mathcal{D}_{AB}$. The approaching

---

[1] This is the most precise method yet developed for measurements of diffusivity at high temperatures. For a detailed description of the method, see R. E. Walker and A. A. Westenberg, *J. Chem. Phys.*, **29**, 1139–1146, 1147–1153 (1958). For a summary of measured values and comparisons with the Chapman–Enskog theory, see R. M. Fristrom and A. A. Westenberg, *Flame Structure*, McGraw-Hill, New York (1965), Chapter XIII.

stream of pure B will be directed vertically upward, and the gas composition will be measured at several points along the z-axis.

(a) Calculate the gas-injection rate W_A in g-moles/s required to produce a mole fraction $x_A \approx 0.01$ at a point 1 cm downstream of the source, in an ideal gaseous system at 1 atm and 800°C, if $v_0 = 50$ cm/s and $\mathscr{D}_{AB} \approx 5$ cm^2/s.

(b) What is the maximum permissible error in the radial position of the gas-sampling probe, if the measured composition x_A is to be within 1% of the centerline value?

18A.6. Determination of diffusivity for ether–air system. The following data on the evaporation of ethyl ether, with liquid density of 0.712 g/cm^3, have been tabulated by Jost.[2] The data are for a tube of 6.16 mm diameter, a total pressure of 747 mm Hg, and a temperature of 22°C.

Decrease of the ether level (measured from the open end of the tube), in mm	Time, in seconds, required for the indicated decrease of level
from 9 to 11	590
from 14 to 16	895
from 19 to 21	1185
from 24 to 26	1480
from 34 to 36	2055
from 44 to 46	2655

The molecular weight of ethyl ether is 74.12, and its vapor pressure at 22°C is 480 mm Hg. It may be assumed that the ether concentration at the open end of the tube is zero. Jost has given a value of $\mathscr{D}_{AB}$ for the ether–air system of 0.0786 cm^2/s at 0°C and 760 mm Hg.

(a) Use the evaporation data to find $\mathscr{D}_{AB}$ at 747 mm Hg and 22°C, assuming that the arithmetic average gas-column lengths may be used for $z_2 - z_1$ in Fig. 18.2-1. Assume further that the ether–air mixture is ideal and that the diffusion can be regarded as binary.

(b) Convert the result to $\mathscr{D}_{AB}$ at 760 mm Hg and 0°C using Eq. 17-2-1.

18A.7. Mass flux from a circulating bubble.

(a) Use Eq. 18.5-20 to estimate the rate of absorption of CO_2 (component A) from a carbon dioxide bubble 0.5 cm in diameter rising through pure water (component B) at 18°C and at a pressure of 1 atm. The following data[3] may be used: $\mathscr{D}_{AB} = 1.46 \times 10^{-5}$ cm^2/s, $c_{A0} = 0.041$ g-mole/liter, $v_t = 22$ cm/s.

(b) Recalculate the rate of absorption, using the experimental results of Hammerton and Garner,[4] who obtained a surface-averaged k_c of 117 cm/hr (see Eq. 18.1-2).

Answers: **(a)** 1.17×10^{-6} g-mol/cm^2 s; **(b)** 1.33×10^{-6} g-mol/cm^2 s.

18B.1. Diffusion through a stagnant film—alternate derivation. In §18.2 an expression for the evaporation rate was obtained in Eq. 18.2-14 by differentiating the concentration profile found a few lines before. Show that the same results may be derived without finding the concentration profile. Note that at steady state, N_{Az} is a constant according to Eq. 18.2-3. Then Eq. 18.2-1 can be integrated directly to get Eq. 18.2-14.

[2] W. Jost, *Diffusion*, Academic Press, New York (1952), pp. 411–413.

[3] G. Tammann and V. Jessen, *Z. anorg. allgem. Chem.*, **179**, 125–144 (1929); F. H. Garner and D. Hammerton, *Chem. Eng. Sci.*, **3**, 1–11 (1954).

[4] D. Hammerton and F. H. Garner, *Trans. Inst. Chem. Engrs.* (London), **32**, S18-S24 (1954).

18B.2. Error in neglecting the convection term in evaporation.

(a) Rework the problem in the text in §18.2 by neglecting the term $x_A(N_A + N_B)$ in Eq. 18.0-1. Show that this leads to

$$N_{Az} = \frac{c\mathcal{D}_{AB}}{z_2 - z_1}(x_{A1} - x_{A2}) \tag{18B.2-1}$$

This is a useful approximation if A is present only in very low concentrations.

(b) Obtain the result in (a) from Eq. 18.2-14 by making the appropriate approximation.

(c) What error is made in the determination of $\mathcal{D}_{AB}$ in Example 18.2-2 if the result in (a) is used?

Answer: 0.78%

18B.3. Effect of mass transfer rate on the concentration profiles.

(a) Combine the result in Eq. 18.2-11 with that in Eq. 18.2-14 to get

$$\frac{1 - x_A}{1 - x_{A1}} = \exp\left(\frac{N_{Az}(z - z_1)}{c\mathcal{D}_{AB}}\right) \tag{18B.3-1}$$

(b) Obtain the same result by integrating Eq. 18.2-1 directly, using the fact that N_{Az} is constant.

(c) Note what happens when the mass transfer rate becomes small. Expand Eq. 18B.3-1 in a Taylor series and keep two terms only, as is appropriate for small N_{Az}. What happens to the slightly curved lines in Fig. 18.2-1 when N_{Az} is very small?

18B.4. Absorption with chemical reaction.

(a) Rework the problem discussed in the text in §18.4, but take $z = 0$ to be the bottom of the beaker and $z = L$ at the gas–liquid interface.

(b) In solving Eq. 18.4-7, we took the solution to be of the sum of two hyperbolic functions. Try solving the problem by using the equally valid solution $\Gamma = C_1 \exp(\phi\zeta) + C_2 \exp(-\phi\zeta)$.

(c) In what way do the results in Eqs. 18.4-10 and 12 simplify for very large L? For very small L? Interpret the results physically.

18B.5. Absorption of chlorine by cyclohexene. Chlorine can be absorbed from Cl_2–air mixtures by olefins dissolved in CCl_4. It was found[5] that the reaction of Cl_2 with cyclohexene (C_6H_{10}) is second order with respect to Cl_2 and zero order with respect to C_6H_{10}. Hence the rate of disappearance of Cl_2 per unit volume is $k_2''' c_A^2$ (where A designates Cl_2).

Rework the problem of §18.4 where B is a C_6H_{10}–CCl_4 mixture, assuming that the diffusion can be treated as pseudobinary. Assume that the air is essentially insoluble in the C_6H_{10}–CCl_4 mixture. Let the liquid phase be sufficiently deep that L can be taken to be infinite.

(a) Show that the concentration profile is given by

$$\frac{c_{A0}}{c_A} = \left[1 + \sqrt{\frac{k_2''' c_{A0}}{6\mathcal{D}_{AB}}}\, z\right]^2 \tag{18B.5-1}$$

(b) Obtain an expression for the rate of absorption of Cl_2 by the liquid.

(c) Suppose that a substance A dissolves in and reacts with substance B so that the rate of disappearance of A per unit volume is some arbitrary function of the concentration, $f(c_A)$. Show that the rate of absorption of A is given by

$$N_{Az}|_{z=0} = \sqrt{2\mathcal{D}_{AB} \int_0^{c_{A0}} f(c_A)dc_A} \tag{18B.5-2}$$

Use this result to check the result of (b).

---

[5] G. H. Roper, *Chem. Eng. Sci.*, **2**, 18–31, 247–253 (1953).

Fig. 18B.6. Sketch of a two-bulb apparatus for measuring gas diffusivities. The stirrers in the two bulbs maintain uniform concentration in the bulbs.

18B.6. **Two-bulb experiment for measuring gas diffusivity—quasi-steady-state analysis**[6] (Fig. 18B.6). One way of measuring gas diffusivities is by means of a two-bulb experiment. The left bulb and the tube from $z = -L$ to $z = 0$ are filled with gas A. The right bulb and the tube from $z = 0$ to $z = +L$ are filled with gas B. At time $t = 0$ the stopcock is opened, and diffusion begins; then the concentrations of A in the two well-stirred bulbs change. One measures x_A^+ as a function of time, and from this deduces $\mathcal{D}_{AB}$. We wish to derive the equations describing the diffusion.

Since the bulbs are large compared with the tube, x_A^+ and x_A^- change *very slowly* with time. Hence the diffusion in the tube can be treated as a quasi-steady-state problem, with the boundary conditions that $x_A = x_A^-$ and $z = -L$, and that $x_A = x_A^+$ at $z = +L$.

(a) Write a molar balance on A over a segment Δz of the tube (of cross-sectional area S), and show that $N_{Az} = C_1$, a constant.

(b) Show that Eq. 18.0-1 simplifies, for this problem, to

$$N_{Az} = -c\mathcal{D}_{AB}\frac{dx_A}{dz} \tag{18B.6-1}$$

(c) Integrate this equation, using (a). Call the constant of integration C_2.

(d) Evaluate the constant by requiring that $x_A = x_A^+$ at $z = +L$.

(e) Next set $x_A = x_A^-$ (or $1 - x_A^+$) at $z = -L$, and solve for N_{Az} to get finally

$$N_{Az} = (\tfrac{1}{2} - x_A^+)\frac{c\mathcal{D}_{AB}}{L} \tag{18B.6-2}$$

(f) Make a mass balance on substance A over the right bulb to obtain

$$S(\tfrac{1}{2} - x_A^+)\frac{c\mathcal{D}_{AB}}{L} = Vc\frac{dx_A^+}{dt} \tag{18B.6-3}$$

(g) Integrate the equation in (f) to get an expression for x_A^+ which contains $\mathcal{D}_{AB}$:

$$\ln\left(\frac{\tfrac{1}{2} - x_A^+}{\tfrac{1}{2}}\right) = -\frac{S\mathcal{D}_{AB}t}{LV} \tag{18B.6-4}$$

(h) Suggest a method of plotting the experimental data to evaluate $\mathcal{D}_{AB}$.

18B.7. **Diffusion from a suspended droplet** (Fig. 18.2-4). A droplet of liquid A, of radius r_1, is suspended in a stream of gas B. We postulate that there is a spherical stagnant gas film of radius r_2 surrounding the droplet. The concentration of A in the gas phase is x_{A1} at $r = r_1$ and x_{A2} at the outer edge of the film, $r = r_2$.

(a) By a shell balance, show that for steady-state diffusion $r^2 N_{Ar}$ is a constant within the gas film, and set the constant equal to $r_1^2 N_{Ar1}$, the value at the droplet surface.

(b) Show that Eq. 18.0-1 and the result in (a) lead to the following equation for x_A:

$$r_1^2 N_{Ar1} = -\frac{c\mathcal{D}_{AB}}{1 - x_A}r^2\frac{dx_A}{dr} \tag{18B.7-1}$$

---

[6] S. P. S. Andrew, *Chem. Eng. Sci.*, **4**, 269–272 (1955).

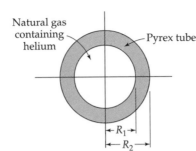

Fig. 18B.8. Diffusion of helium through pyrex tubing. The length of the tubing is L.

(c) Integrate this equation between the limits r_1 and r_2 to get

$$N_{Ar1} = \frac{c\mathcal{D}_{AB}}{r_2 - r_1}\left(\frac{r_2}{r_1}\right)\ln\frac{x_{B2}}{x_{B1}} \tag{18B.7-2}$$

What is the limit of this expression when $r_2 \to \infty$?

18B.8. Method for separating helium from natural gas (Fig. 18B.8). Pyrex glass is almost impermeable to all gases but helium. For example, the diffusivity of He through pyrex is about 25 times the diffusivity of H_2 through pyrex, hydrogen being the closest "competitor" in the diffusion process. This fact suggests that a method for separating helium from natural gas could be based on the relative diffusion rates through pyrex.[7]

Suppose a natural gas mixture is contained in a pyrex tube with dimensions shown in the figure. Obtain an expression for the rate at which helium will "leak" out of the tube, in terms the diffusivity of helium through pyrex, the interfacial concentrations of the helium in the pyrex, and the dimensions of the tube.

Answer: $W_{He} = 2\pi L \dfrac{\mathcal{D}_{He\text{-}Pyrex}(c_{He,1} - c_{He,2})}{\ln(R_2/R_1)}$

18B.9. Rate of leachng (Fig. 18B.9). In studying the rate of leaching of a substance A from solid particles by a solvent B, we may postulate that the rate-controlling step is the diffusion of A from the particle surface through a stagnant liquid film thickness δ out into the main stream. The molar solubility of A in B is c_{A0}, and the concentration in the main stream is $c_{A\delta}$.

(a) Obtain a differential equation for c_A as a function of z by making a mass balance on A over a thin slab of thickness Δz. Assume that $\mathcal{D}_{AB}$ is constant and that A is only slightly soluble in B. Neglect the curvature of the particle.

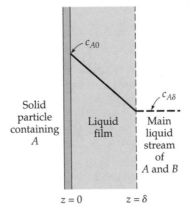

Fig. 18B.9. Leaching of A by diffusion into a stagnant liquid film of B.

[7] *Scientific American*, **199**, 52 (1958) describes briefly the method developed by K. B. McAfee of Bell Telephone Laboratories.

(b) Show that, in the absence of chemical reaction in the liquid phase, the concentration profile is linear.

(c) Show that the rate of leaching is given by

$$N_{Az} = \mathcal{D}_{AB}(c_{A0} - c_{A\delta})/\delta \qquad (18\text{B}.9\text{-}1)$$

18B.10 **Constant-evaporating mixtures.** Toluene (1) and ethanol (2) are evaporating at $z = 0$ in a vertical tube, from a binary liquid mixture of uniform composition x_1 through stagnant nitrogen (3), with pure nitrogen at the top. The unequal diffusivities of toluene and ethanol through nitrogen shift the relative evaporation rates in favor of ethanol. Analyze this effect for an isothermal system at 60 F and 760 mm Hg total pressure, if the predicted[8] diffusivities at 60° F are $c\mathcal{D}_{12} = 1.53 \times 10^{-6}$, $c\mathcal{D}_{13} = 2.98 \times 10^{-6}$, $c\mathcal{D}_{23} = 4.68 \times 10^{-6}$ g-moles/cm · s.

(a) Use the Maxwell-Stefan equations to obtain the steady-state vapor-phase mole fraction profiles $y_\alpha(z)$ in terms of the molar fluxes $N_{\alpha z}$ in this ternary system. The molar fluxes are known to be constants from the equations of continuity for the three species. Since nitrogen has a negligible solubility in the liquid at the conditions given, $N_{3z} = 0$. As boundary conditions, set $y_1 = y_2 = 0$ at $z = L$, and let $y_1 = y_{10}$ and $y_2 = y_{20}$ at $z = 0$; the latter values remain to be determined. Show that

$$y_3(z) = e^{-A(L-z)}; \qquad y_1(z) = \frac{D}{A-B}e^{-A(L-z)} - \left(\frac{C}{B} + \frac{D}{A-B}\right)e^{-B(L-z)} + \frac{C}{B} \qquad (18\text{B}.10\text{-}1)$$

$$A = \frac{N_{1z}}{c\mathcal{D}_{13}} + \frac{N_{2z}}{c\mathcal{D}_{23}}; \qquad B = \frac{N_{1z} + N_{2z}}{c\mathcal{D}_{12}}; \qquad C = \frac{N_{1z}}{c\mathcal{D}_{12}}; \qquad D = \frac{N_{1z}}{c\mathcal{D}_{12}} - \frac{N_{1z}}{c\mathcal{D}_{13}} \qquad (18\text{B}.10\text{-}2)$$

(b) A constant evaporating liquid mixture is one whose composition is the same as that of the evaporated material, that is, for which $N_{1z}/(N_{1z} + N_{2z}) = x_1$. Use the results of part (a) along with the equilibrium data in the table below to calculate the constant-evaporating liquid composition at a total pressure of 760 mm Hg. In the table, row I gives liquid-phase compositions. Row II gives vapor-phase compositions in two-component experiments; these are expressed as nitrogen-free values $y_1/(y_1 + y_2)$ for the ternary system. Row III gives the sum of the partial pressures of toluene and ethanol.

I:	x_1	0.096	0.155	0.233	0.274	0.375
II:	$y_1/(y_1 + y_2)$	0.147	0.198	0.242	0.256	0.277
III:	$p_1 + p_2$ (mm Hg)	388	397	397	395	390

A suggested strategy for the calculation is as follows: (i) guess a liquid composition x_1; (ii) calculate y_{10}, y_{20}, and y_{30} using lines 2 and 3 of the table; (iii) calculate A from Eq. 18B.10-1, with $z = 0$; (iv) use the result of iii to calculate LN_{2z}, LB, LC, and LD, and finally $y_1(0)$ for assumed values of LN_{1z}; (v) interpolate the results of iv to $y_1(0) = y_{10}$ to obtain the correct LN_{1z} and LN_{2z} for the guessed x_1. Repeat steps i–v with improved guesses for x_1 until $N_{1z}/(N_{1z} + N_{2z})$ converges to x_1. The final x_1 is the constant evaporating composition.

18B.11. **Diffusion with fast second-order reaction** (Figs. 18.2-2 and 18B.11). A solid A is dissolving in a flowing liquid stream S in a steady-state, isothermal flow system. Assume in accordance with the film model that the surface of A is covered with a stagnant liquid film of thickness δ and that the liquid outside the film is well mixed (see Fig. 18.2-2).

(a) Develop an expression for the rate of dissolution of A into the liquid if the concentration of A in the main liquid stream is negligible.

(b) Develop a corresponding expression for the dissolution rate if the liquid contains a substance B, which, at the plane $z = \kappa\delta$, reacts instantaneously and irreversibly with A: $A + B \rightarrow P$. (An example of such a system is the dissolution of benzoic acid in an aqueous NaOH solution.) The main liquid stream consists primarily of B and S, with B at a mole fraction of $x_{B\infty}$.

---

[8] L. Monchick and E. A. Mason, *J. Chem. Phys.*, **35**, 1676–1697 (1961), with δ read as δ_{max} in Table IV; E. A. Mason and L. Monchick, *J. Chem. Phys.*, **36**, 2746–2757 (1962); L. S. Tee, S. Gotoh, and W. E. Stewart, *Ind. Eng. Chem. Fundam.*, **5**, 356–362 (1966).

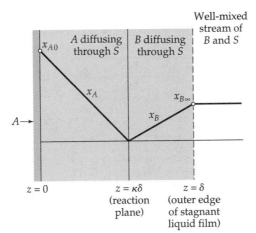

Fig. 18B.11. Concentration profiles for diffusion with rapid second-order reaction. The concentration of product P neglected.

(*Hint:* It is necessary to recognize that species A and B both diffuse toward a thin reaction zone as shown in Fig. 18B.11.)

Answers: **(a)** $N_{Az}|_{z=0} = \left(\dfrac{c\mathcal{D}_{AS}x_{A0}}{\delta}\right)$; **(b)** $N_{Az}|_{z=0} = \left(\dfrac{c\mathcal{D}_{AS}x_{A0}}{\delta}\right)\left(1 + \dfrac{x_{B\infty}\mathcal{D}_{BS}}{x_{A0}\mathcal{D}_{AS}}\right)$

18B.12. **A sectioned-cell experiment[9] for measuring gas-phase diffusivity** (Fig. 18B.12). Liquid A is allowed to evaporate through a stagnant gas B at 741 mm Hg total pressure and 25°C. At that temperature, the vapor pressure of A is known to be 600 mm Hg. After steady state has been

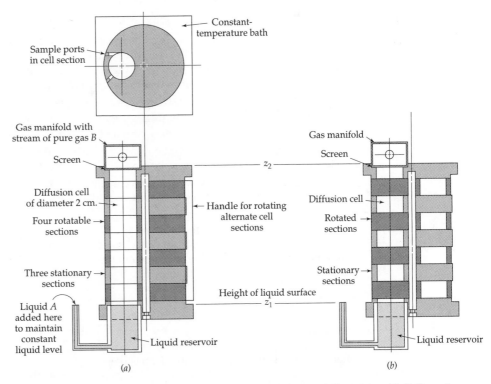

Fig. 18B.12. A sectioned-cell experiment for measuring gas diffusivities. (*a*) Cell configuration during the approach to steady-state. (*b*) Cell configuration for gas sampling at the end of the experiment.

[9] E. J. Crosby, *Experiments in Transport Phenomena*, Wiley, New York (1961), Experiment 10.a.

attained, the cylindrical column of gas is divided into sections as shown. For a 4-section apparatus with total height 4.22 cm, the analysis of the gas samples thus obtained gives the following results:

	$(z - z_1)$ in cm		
Section	Bottom of section	Top of section	Mole fraction of A
I	0.10	1.10	0.757
II	1.10	2.10	0.641
III	2.10	3.10	0.469
IV	3.10	4.10	0.215

The measured evaporation rate of A at steady state is 0.0274 g-moles/hr. Ideal gas behavior may be assumed.

(a) Verify the following expression for the concentration profile at steady state:

$$\ln \frac{x_{B2}}{x_B} = \frac{N_{Az}(z_2 - z)}{c \mathcal{D}_{AB}} \tag{18B.12-1}$$

(b) Plot the mole fraction x_B in each cell versus the value of z at the midplane of the cell on semilogarithmic graph paper. Is a straight line obtained? What are the intercepts at z_1 and z_2? Interpret these results.

(c) Use the concentration profile of Eq. 18B.12-1 to find analytical expressions for the average concentrations in each section of the tube.

(d) Find the best value of $\mathcal{D}_{AB}$ from this experiment.

Answer: **(d)** 0.155 cm^2/s

18B.13. **Tarnishing of metal surfaces.** In the oxidation of most metals (excluding the alkali and alkaline-earth metals) the volume of oxide produced is greater than that of the metal consumed. This oxide thus tends to form a compact film, effectively insulating the oxygen and metal from each other. For the derivations that follow, it may be assumed that

(a) For oxidation to proceed, oxygen must diffuse through the oxide film and that this diffusion follows Fick's law.

(b) The free surface of the oxide film is saturated with oxygen from the surrounding air.

(c) Once the film of oxide has become reasonably thick, the oxidation becomes diffusion controlled; that is, the dissolved oxygen concentration is essentially zero at the oxide-metal surface.

(d) The rate of change of dissolved oxygen content of the film is small compared to the rate of reaction. That is, quasi-steady-state conditions may be assumed.

(e) The reaction involved is $\frac{1}{2}xO_2 + M \rightarrow MO_x$.

We wish to develop an expression for rate of tarnishing in terms of oxygen diffusivity through the oxide film, the densities of the metal and its oxide, and the stoichiometry of the reaction. Let c_O be the solubility of oxygen in the film, c_f the molar density of the film, and z_f the thickness of the film. Show that the film thickness is

$$z_f = \sqrt{\frac{2\mathcal{D}_{O_2-MO_x} t}{x} \frac{c_O}{c_f}} \tag{18B.13-1}$$

This result, the so-called "quadratic law," gives a satisfactory empirical correlation for a number of oxidation and other tarnishing reactions.[10] Most such reactions are, however, much more complex than the mechanism given above.[11]

---

[10] G. Tammann, Z. *anorg. allgem. Chemie*, **124**, 25–35 (1922).

[11] W. Jost, Diffusion, *Academic Press*, New York (1952), Chapter IX. For a discussion of the oxidation of silicon, see R. Ghez, *A Primer of Diffusion Problems*, Wiley, New York (1988), §2.3.

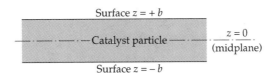

Fig. 18B.14. Side view of a disk-shaped catalyst particle.

18B.14. Effectiveness factors for thin disks (Fig. 18B.14). Consider porous catalyst particles in the shape of thin disks, such that the surface area of the edge of the disk is small in comparison with that of the two circular faces. Apply the method of §18.7 to show that the steady-state concentration profile is

$$\frac{c_A}{c_{As}} = \frac{\cosh\sqrt{k_1''a/\mathcal{D}_A}\,z}{\cosh\sqrt{k_1''a/\mathcal{D}_A}\,b} \tag{18B.14-1}$$

where z and b are described in the figure. Show that the total mass transfer rate at the surfaces $z = \pm b$ is

$$|W_A| = 2\pi R^2 c_{As}\mathcal{D}_A\lambda \tanh \lambda b \tag{18B.14-2}$$

in which $\lambda = \sqrt{k_1''a/\mathcal{D}_A}$. Show that, if the disk is sliced parallel to the xy-plane into n slices, the total mass transfer rate becomes

$$|W_A^{(n)}| = 2\pi R^2 c_{As}\mathcal{D}_A\lambda n \tanh(\lambda b/n) \tag{18B.14-3}$$

Obtain the expression for the effectiveness factor by taking the limit

$$\eta_A = \lim_{n\to\infty}\frac{|W_A|}{|W_A^{(n)}|} = \frac{\tanh \lambda b}{\lambda b} \tag{18B.14-4}$$

Express this result in terms of the parameter Λ defined in §18.6.

18B.15. Diffusion and heterogeneous reaction in a slender cylindrical tube with a closed end (Fig. 18B.15). A slender cylindrical pore of length L, cross-sectional area S, and perimeter P, is in contact at its open end with a large body of well-mixed fluid, consisting of species A and B. Species A, a minor constituent of this fluid, disappears into the pore, diffuses in the z direction and reacts on its walls. The rate of this reaction may be expressed as $(\mathbf{n}\cdot\mathbf{n}_A)|_{\text{surface}} = f(\omega_{A0})$; that is, at the wall the mass flux normal to the surface is some function of the mass fraction, ω_{A0}, of A in the fluid adjacent to the solid surface. The mass fraction ω_{A0} depends on z, the distance from the inlet. Because A is present in low concentration, the fluid temperature and density may be considered constant, and the diffusion flux is adequately described by $\mathbf{j}_A = -\rho\mathcal{D}_{AB}\nabla\omega_A$,

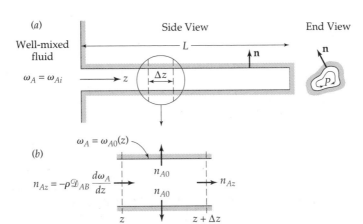

Fig. 18B.15. (a) Diffusion and heterogeneous reaction in a long, noncircular cylinder. (b) Region of thickness Δz over which the mass balance is made.

where the diffusivity may be regarded as a constant. Because the pore is long compared to its lateral dimension, concentration gradients in the lateral directions may be neglected. Note the similarity with the problem discussed in §10.7.

(a) Show by means of a shell balance that, at steady state,

$$-\frac{dn_{Az}}{dz} = \frac{P}{S} f(\omega_{A0}) \tag{18B.15-1}$$

(b) Show that the steady-state mass average velocity v_z is zero for this system.

(c) Substitute the appropriate form of Fick's law into Eq. 18.15-1, and integrate the resulting differential equation for the special case that $f(\omega_{A0}) = k_1'' \omega_{A0}$. To obtain a boundary condition at $z = L$, neglect the rate of reaction on the closed end of the cylinder; why is this a reasonable approximation?

(d) Develop an expression for the total rate w_A of disappearance of A in the cylinder.

(e) Compare the results of parts (c) and (d) with those of §10.7 both from the standpoint of the mathematical development and the nature of the assumptions made.

Answers: (c) $\dfrac{\omega_A}{\omega_{Ai}} = \dfrac{\cosh N[1 - (z/L)]}{\cosh N}$, where $N = \sqrt{\dfrac{PL^2 k_1''}{S\rho \mathcal{D}_{AB}}}$; (d) $w_A = (S\rho \mathcal{D}_{AB}\omega_{Ai}/L)N \tanh N$

18B.16. Effect of temperature and pressure on evaporation rate.

(a) In §18.2 what is the effect of a change of temperature and pressure on the quantity x_{A1}?

(b) If the pressure is doubled, how is the evaporation rate in Eq. 18.2-14 affected?

(c) How does the evaporation rate change when the system temperature is raised from T to T'?

18B.17. Reaction rates in large and small particles.

(a) Obtain the following limits for Eq. 18.7-11:

$R \to 0$: $\qquad\qquad\qquad W_{AR} = -(\tfrac{4}{3}\pi R^3)(k_1''a)c_{AR}$ $\qquad\qquad$ (18B.17-1)

$R \to \infty$: $\qquad\qquad\qquad W_{AR} = -(4\pi R^2)(k_1''a\mathcal{D}_A)^{1/2} c_{AR}$ $\qquad$ (18B.17-2)

Interpret these results physically.

(b) Obtain the corresponding asymptotes for the system discussed in Problem 18B.14. Compare them with the results in (a).

18B.18. Evaporation rate for small mole fraction of the volatile liquid. In Eq. 18.2-15, expand

$$\frac{1}{(x_B)_{\ln}} = \left(\frac{1}{x_{A1} - x_{A2}}\right)\left(\ln \frac{1 - x_{A2}}{1 - x_{A1}}\right) \tag{18B.18-1}$$

in a Taylor series appropriate for small mole fractions of A. First rewrite the logarithm of the quotient as the difference of the logarithms. Then expand $\ln(1 - x_{A1})$ and $\ln(1 - x_{A2})$ in Taylor series about $x_{A1} = 1$ and $x_{A2} = 1$, respectively. Verify that Eq. 18.2-16 is correct.

18B.19. Oxygen uptake by a bacterial aggregate. Under suitable circumstances the rate of oxygen metabolism by bacterial cells is very nearly zero order with respect to oxygen concentration. We examine such a case here and focus our attention on a spherical aggregate of cells, which has a radius R. We wish to determine the total rate of oxygen uptake by the aggregate as a function of aggregate size, oxygen mass concentration ρ_0 at the aggregate surface, the metabolic activity of the cells, and the diffusional behavior of the oxygen. For simplicity we consider the aggregate to be homogeneous. We then approximate the metabolic rate by an effective volumetric reaction rate $r_{O_2} = -k_0'''$ and the diffusional behavior by Fick's law, with an effective pseudobinary diffusivity $\mathcal{D}_{O_2m}$. Because the solubility of oxygen is very low in this system, both convective oxygen transport and transient effects may be neglected.[12]

---

[12] J. A. Mueller, W. C. Boyle, and E. N. Lightfoot, *Biotechnol. and Bioengr.*, **10**, 331–358 (1968).

(a) Show by means of a shell mass balance that the quasi-steady-state oxygen concentration profile is described by the differential equation

$$\frac{1}{\xi^2}\frac{d}{d\xi}\left(\xi^2\frac{d\chi}{d\xi}\right) = N \tag{18B.19-1}$$

where $\chi = \rho_{O_2}/\rho_0$, $\xi = r/R$, and $N = k_0''' R^2/\rho_0 \mathcal{D}_{O_2 m}$.

(b) There may be an oxygen-free core in the aggregate, if N is sufficiently large, such that $\chi = 0$ for $\xi < \xi_0$. Write sufficient boundary conditions to integrate Eq. 18B.19-1 for this situation. To do this, it must be recognized that both χ and $d\chi/d\xi$ are zero at $\xi = \xi_0$. What is the physical significance of this last statement?

(c) Perform the integration of Eq. 18B.19-1 and show how ξ_0 may be determined.

(d) Sketch the total oxygen uptake rate and ξ_0 as functions of N, and discuss the possibility that no oxygen-free core exists.

Answer: **(c)** $\chi = 1 - \dfrac{N}{6}(1 - \xi^2) + \dfrac{N}{3}\xi_0^3\left(\dfrac{1}{\xi} - 1\right)$ for $\xi \geq \xi_0 \geq 0$, where ξ_0 is determined as a function of N from

$$\xi_0^3 - \frac{3}{2}\xi_0^2 + \left(\frac{1}{2} - \frac{3}{N}\right) = 0$$

18C.1. **Diffusion from a point source in a moving stream** (Fig. 18C.1). A stream of fluid B in laminar motion has a uniform velocity v_0. At some point in the stream (taken to be the origin of coordinates) species A is injected at a small rate W_A g-moles/s. This rate is assumed to be sufficiently small that the mass average velocity will not deviate appreciably from v_0. Species A is swept downstream (in the z direction), and at the same time it diffuses both axially and radially.

(a) Show that a steady-state mass balance on species A over the indicated ring-shaped element leads to the following partial differential equation if $\mathcal{D}_{AB}$ is assumed to be constant:

$$v_0\frac{\partial c_A}{\partial z} = \mathcal{D}_{AB}\left[\frac{1}{r}\frac{\partial}{\partial r}\left(r\frac{\partial c_A}{\partial r}\right) + \frac{\partial^2 c_A}{\partial z^2}\right] \tag{18C.1-1}$$

(b) Show that Eq. 18C.1-1 can also be written as

$$v_0\left(\frac{z}{s}\frac{\partial c_A}{\partial s} + \frac{\partial c_A}{\partial z}\right) = \mathcal{D}_{AB}\left[\frac{1}{s^2}\frac{\partial}{\partial s}\left(s^2\frac{\partial c_A}{\partial s}\right) + \frac{\partial^2 c_A}{\partial z^2} + 2\frac{z}{s}\frac{\partial^2 c_A}{\partial s\partial z}\right] \tag{18C.1-2}$$

in which $s^2 = r^2 + z^2$.

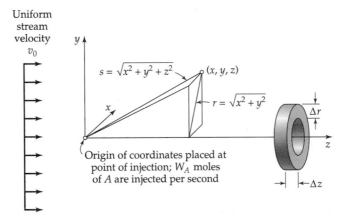

Uniform stream velocity v_0

$s = \sqrt{x^2 + y^2 + z^2}$ (x, y, z)

$r = \sqrt{x^2 + y^2}$

Δr

Origin of coordinates placed at point of injection; W_A moles of A are injected per second

Δz

Fig. 18C.1. Diffusion of A from a point source into a stream of B that moves with a uniform velocity.

(c) Verify (lengthy!) that the solution

$$c_A = \frac{W_A}{4\pi \mathscr{D}_{AB}s} \exp[-(v_0/2\mathscr{D}_{AB})(s-z)] \tag{18C.1-3}^{[13]}$$

satisfies the differential equation above.

(d) Show further that the following boundary conditions are also satisfied by Eq. 18C.1-3:

B.C. 1:   at $s = \infty$, $c_A = 0$ (18C.1-4)

B.C. 2:   as $s \to 0$, $-4\pi s^2 \mathscr{D}_{AB} \dfrac{\partial c_A}{\partial s} \to W_A$ (18C.1-5)

B.C. 3:   at $r = 0$, $\dfrac{\partial c_A}{\partial r} = 0$ (18C.1-6)

Explain the physical meaning of each of these boundary conditions.

(e) Show how data on $c_A(r, z)$ for given v_0 and $\mathscr{D}_{AB}$ may be plotted, when the preceding solution applies, to give a straight line with slope $v_0/2\mathscr{D}_{AB}$ and intercept $\ln \mathscr{D}_{AB}$.

18C.2. Diffusion and reaction in a partially impregnated catalyst. Consider a catalytic sphere like that in §18.7, except that the active ingredient of the catalyst is present only in the annular region between $r = \kappa R$ and $r = R$:

In region I ($0 < r < \kappa R$), $k_1'' a = 0$ (18C.2-1)
In region II ($\kappa R < r < R$), $k_1'' a = \text{constant} > 0$ (18C.2-2)

Such a situation may arise when the active ingredient is put on the particles after pelleting, as is done for many commercial catalysts.

(a) Integrate Eq. 18.7-6 separately for the active and inactive regions. Then apply the appropriate boundary conditions to evaluate the integration constants, and solve for the concentration profile in each region. Give qualitative sketches to illustrate the forms of the profiles.

(b) Evaluate W_{AR}, the total molar rate of conversion of A in a single particle.

18C.3. Absorption rate in a falling film. The result in Eq. 18.5-18 may be obtained by an alternative procedure.

(a) According to an overall mass balance on the film, the total moles of A transferred per unit time across the gas–liquid interface must be the same as the total molar rate of flow of A across the plane $z = L$. The latter rate is calculated as follows:

$$W_A = \lim_{\delta \to \infty}(W\delta v_{\max})\left(\frac{1}{\delta}\int_0^\delta c_A|_{z=L}\, dx\right) = Wv_{\max}\int_0^\infty c_A|_{z=L}\, dx \tag{18C.3-1}$$

Explain this procedure carefully.

(b) Insert the solution for c_A in Eq. 18.5-15 into the result of (a) to obtain:

$$W_A = Wv_{\max}c_{A0}\frac{2}{\sqrt{\pi}}\int_0^\infty\left(\int_{x/\sqrt{4\mathscr{D}_{AB}L/v_{\max}}}^\infty \exp(-\xi^2)\, d\xi\right)dx$$

$$= Wv_{\max}c_{A0}\frac{2}{\sqrt{\pi}}\sqrt{\frac{4\mathscr{D}_{AB}L}{v_{\max}}}\int_0^\infty\left(\int_u^\infty \exp(-\xi^2)\, d\xi\right)du \tag{18C.3-2}$$

In the second line, the new variable $u = x/\sqrt{4\mathscr{D}_{AB}L/v_{\max}}$ has been introduced.

(c) Change the order of integration in the double integral, to get

$$W_A = WLc_{A0}\sqrt{\frac{4\mathscr{D}_{AB}v_{\max}}{\pi L}} \cdot 2\int_0^\infty \exp(-\xi^2)\left(\int_0^\xi du\right)d\xi \tag{18C.3-3}$$

Explain by means of a carefully drawn sketch how the limits are chosen for the integrals The integrals may now be done analytically to get Eq. 18.5-18.

18C.4. **Estimation of the required length of an isothermal reactor** (Fig. 18.3-1). Let a be the area of catalyst surface per unit volume of a packed-bed catalytic reactor and S be the cross-sectional area of the reactor. Suppose that the rate of mass flow through the reactor is w (in lb_m/hr, for example).

(a) Show that a steady-state mass balance on substance A over a length dl of the reactor leads to

$$\frac{d\omega_{A0}}{dl} = -\frac{SaN_AM_A}{w} \tag{18C.4-1}$$

(b) Use the result of (a) and Eq. 18.3-9, with the assumptions of constant δ and $\mathscr{D}_{AB}$, to obtain an expression for the reactor length L needed to convert an inlet stream of composition $x_A(0)$ to an outlet stream of composition $x_A(L)$.

(*Hint:* Equation (P) of Table 17.8-1 may be useful.)

Answer: (b) $L = \left(\dfrac{w\delta M_B}{2Sac\,\mathscr{D}_{AB}}\right) \displaystyle\int_{x_A(0)}^{x_A(L)} \dfrac{dx_{A0}}{[M_Ax_{A0} + M_B(1 - x_{A0})]^2 \ln(1 - \frac{1}{2}x_{A0})}$

18C.5. **Steady-state evaporation.** In a study of the evaporation of a mixture of methanol (1) and acetone (2) through air (3), the concentration profiles of the three species in the tube were measured[14] after attainment of steady state. In this situation, species 3 is not moving, and species 1 and 2 are diffusing upward, with the molar fluxes N_{z1} and N_{z2}, measured in the experiments. The interfacial concentrations of these two species, x_{10} and x_{20}, were also measured. In addition, the three binary diffusion coefficients were known. The interface was located at $z = 0$ and the upper end of the diffusion tube was at $z = L$.

(a) Show that the Maxwell–Stefan equation for species 3 can be solved to get

$$x_3 = x_{30}e^{A\zeta} \tag{18C.5-1}$$

in which $A = \nu_{113} + \nu_{223}$, with $\nu_{\alpha\beta\gamma} = N_\alpha L/c\mathscr{D}_{\beta\gamma}$ and $\zeta = z/L$.

(b) Next verify that the equation for species 2 can be solved to get

$$x_2 = x_{20}e^{B\zeta} + \frac{\nu_{212}}{B}(1 - e^{B\zeta}) + \frac{Cx_{30}}{A - B}(e^{A\zeta} - e^{B\zeta}) \tag{18C.5-2}$$

where $B = \nu_{112} + \nu_{212}$ and $C = \nu_{212} - \nu_{223}$.

(c) Compare the above equations with the published results.

(d) How well do Eqs. 18C.5-1 and 2 fit the experimental data?

18D.1. **Effectiveness factors for long cylinders.** Derive the expression for η_A for long cylinders analogous to Eq. 18.7-16. Neglect the diffusion through the ends of the cylinders.

Answer: $\eta_A = \dfrac{I_1(2\Lambda)}{\Lambda I_0(2\Lambda)}$, where I_0 and I_1 are "modified Bessel functions"

18D.2. **Gas absorption in a falling film with chemical reaction.** Rework the problem discussed in §18.5 and described in Fig. 18.5-1, when gas A reacts with liquid B by a first-order irreversible chemical reaction in the liquid phase, with rate constant k_1'''. Specifically, find the expression for the total absorption rate analogous to that given in Eq. 18.5-18. Show that the result for absorption with reaction properly simplifies to that for absorption without reaction.

Answer: $W_A = Wc_{A0}v_{max}\sqrt{\dfrac{\mathscr{D}_{AB}}{k_1'''}}\left[(\frac{1}{2} + u)\,\mathrm{erf}\sqrt{u} + \sqrt{\dfrac{u}{\pi}}\,e^{-u}\right]$ in which $u = k_1'''L/v_{max}$

---

[13] H. A. Wilson, *Proc. Camb. Phil. Soc.*, **12**, 406–423 (1904).

[14] R. Carty and T. Schrodt, *Ind. Eng. Chem.*, **14**, 276–278 (1975).

Chapter 19

Equations of Change for Multicomponent Systems

§19.1 The equations of continuity for a multicomponent mixture

§19.2 Summary of the multicomponent equations of change

§19.3 Summary of the multicomponent fluxes

§19.4 Use of the equations of change for mixtures

§19.5 Dimensional analysis of the equations of change for binary mixtures

In Chapter 18, problems in diffusion were formulated by making shell mass balances on one or more of the diffusing species. In this chapter we start by making a mass balance over an arbitrary differential fluid element to establish the equation of continuity for the various species in a multicomponent mixture. Then insertion of mass flux expressions gives the diffusion equations in a variety of forms. These diffusion equations can be used to set up any of the problems in Chapter 18 and more complicated ones as well.

Then we summarize all of the equations of change for mixtures: the equations of continuity, the equation of motion, and the equation of energy. These include the equations of change that were given in Chapters 3 and 11. Next we summarize the flux expressions for mixtures. All these equations are given in general form, although for problem solving we generally use simplified versions of them.

The remainder of the chapter is devoted to analytical solutions and dimensional analyses of mass transfer systems.

§19.1 THE EQUATIONS OF CONTINUITY FOR A MULTICOMPONENT MIXTURE

In this section we apply the law of conservation of mass to each species α in a mixture, where $\alpha = 1, 2, 3, \ldots, N$. The system we consider is a volume element $\Delta x\, \Delta y\, \Delta z$ fixed in space, through which the fluid mixture is flowing (see Fig. 3.1-1). Within this mixture, reactions among the various chemical species may be occurring, and we use the symbol r_α to indicate the rate at which species α is being produced, with dimensions of mass/volume · time.

The various contributions to the mass balance are

rate of increase of mass of α in the volume element	$(\partial \rho_\alpha / \partial t)\Delta x\, \Delta y\, \Delta z$	(19.1-1)
rate of addition of mass of α across face at x	$n_{\alpha x}\vert_x \Delta y\, \Delta z$	(19.1-2)

rate of removal of mass of
α across face at $x + \Delta x$
$$n_{\alpha x}|_{x+\Delta x} \Delta y \, \Delta z \qquad (19.1\text{-}3)$$

rate of production of mass of
α by chemical reactions
$$r_\alpha \Delta x \, \Delta y \, \Delta z \qquad (19.1\text{-}4)$$

The combined mass flux $n_{\alpha x}$ includes both the molecular flux and the convective flux. There are also addition and removal terms in the y and z directions. When the entire mass balance is written down and divided by $\Delta x \, \Delta y \, \Delta z$, one obtains, after letting the size of the volume element decrease to zero,

$$\frac{\partial \rho_\alpha}{\partial t} = -\left(\frac{\partial n_{\alpha x}}{\partial x} + \frac{\partial n_{\alpha y}}{\partial y} + \frac{\partial n_{\alpha z}}{\partial z} \right) + r_\alpha \qquad \alpha = 1, 2, 3, \ldots, N \qquad (19.1\text{-}5)$$

This is the *equation of continuity for species* α in a multicomponent reacting mixture. It describes the change in mass concentration of species α with time at a fixed point in space by the diffusion and convection of α, as well as by chemical reactions that produce or consume α. The quantities $n_{\alpha x}, n_{\alpha y}, n_{\alpha z}$ are the Cartesian components of the mass flux vector $\mathbf{n}_\alpha = \rho_\alpha \mathbf{v}_\alpha$ given in Eq. (D) of Table 17.8-1.

Equation 19.1-5 may be rewritten in vector notation as

$$\frac{\partial \rho_\alpha}{\partial t} = -(\nabla \cdot \mathbf{n}_\alpha) + r_\alpha \qquad \alpha = 1, 2, 3, \ldots, N \qquad (19.1\text{-}6)$$

Alternatively we can use Eq. (S) of Table 17.8-1 to write

$$\frac{\partial \rho_\alpha}{\partial t} = -(\nabla \cdot \rho_\alpha \mathbf{v}) - (\nabla \cdot \mathbf{j}_\alpha) + r_\alpha \qquad \alpha = 1, 2, 3, \ldots, N \qquad (19.1\text{-}7)[1]$$

rate of increase of mass of A per unit volume	net rate of addition of mass of A per unit volume by convection	net rate of addition of mass of A per unit volume by diffusion	rate of production of mass of A per unit volume by reaction

Addition of all N equations in either Eq. 19.1-6 or 7 gives

$$\frac{\partial \rho}{\partial t} = -(\nabla \cdot \rho \mathbf{v}) \qquad (19.1\text{-}8)$$

which is the *equation of continuity for the mixture*. This equation is identical to the equation of continuity for a pure fluid given in Eq. 3.1-4. In obtaining Eq. 19.1-8 we had to use Eq. (J) of Table 17.8-1 and also the fact that the law of conservation of total mass gives $\Sigma_\alpha r_\alpha = 0$. Finally we note that Eq. 19.1-8 becomes

$$(\nabla \cdot \mathbf{v}) = 0 \qquad (19.1\text{-}9)$$

for a fluid mixture of *constant mass density* ρ.

In the preceding discussion we used mass units. However, a corresponding derivation is also possible in molar units. The equation of continuity for species α in molar quantities is

$$\frac{\partial c_\alpha}{\partial t} = -(\nabla \cdot \mathbf{N}_\alpha) + R_\alpha \qquad \alpha = 1, 2, 3, \ldots, N \qquad (19.1\text{-}10)$$

---

[1] J. Crank, *The Mathematics of Diffusion*, 2nd edition, Oxford University Press (1975).

where R_α is the molar rate of production of α per unit volume. This equation can be rewritten by use of Eq. (V) of Table 17.8-1 to give

$$\frac{\partial c_\alpha}{\partial t} = -(\nabla \cdot c_\alpha \mathbf{v}^*) - (\nabla \cdot \mathbf{J}_\alpha^*) + R_\alpha \qquad \alpha = 1, 2, 3, \dots, N \qquad (19.1\text{-}11)$$

rate of	net rate of	rate of	rate of
increase	addition	addition	production
in moles	in moles of	of moles of	of moles of
of A per	A per unit	A per unit	A per unit
unit	volume by	volume by	volume by
volume	convection	diffusion	reaction

When all N equations in Eq. 19.1-10 or 11 are added we get

$$\frac{\partial c}{\partial t} = -(\nabla \cdot c\mathbf{v}^*) + \sum_{\alpha=1}^{N} R_\alpha \qquad (19.1\text{-}12)$$

for the equation of continuity for the mixture. To get this we used Eq. (M) of Table 17.8-1. We also note that the chemical reaction term does not drop out because the number of moles is not necessarily conserved in a chemical reaction. Finally we note that

$$(\nabla \cdot \mathbf{v}^*) = \frac{1}{c} \sum_{\alpha=1}^{N} R_\alpha \qquad (19.1\text{-}13)$$

for a fluid mixture of constant *molar density c*.

    We have thus seen that the equation of continuity for species α may be written in two forms, Eq. 19.1-7 and Eq. 19.1-11. Using the continuity relations in Eqs. 19.1-8 and 19.1-12 the reader may verify that the equation of continuity for species α can be put into two additional, equivalent forms:

$$\rho\left(\frac{\partial \omega_\alpha}{\partial t} + (\mathbf{v} \cdot \nabla \omega_\alpha)\right) = -(\nabla \cdot \mathbf{j}_\alpha) + r_\alpha \qquad \alpha = 1, 2, 3, \dots, N \qquad (19.1\text{-}14)$$

$$c\left(\frac{\partial x_\alpha}{\partial t} + (\mathbf{v}^* \cdot \nabla x_\alpha)\right) = -(\nabla \cdot \mathbf{J}_\alpha^*) + R_\alpha - x_\alpha \sum_{\beta=1}^{N} R_\beta \qquad \alpha = 1, 2, 3, \dots, N \qquad (19.1\text{-}15)$$

These two equations express exactly the same physical content, but they are written in two different sets of notation—the first in mass quantities and the second in molar quantities. To use these equations we have to insert the appropriate expressions for the fluxes and the chemical reaction terms. In this chapter we give only the results for *binary systems* with constant $\rho \mathcal{D}_{AB}$, with constant $c\mathcal{D}_{AB}$, or with zero velocity.

Binary Systems with Constant $\rho\mathcal{D}_{AB}$

For this assumption, Eq. 19.1-14 becomes, after inserting Fick's law from Eq. (A) of Table 17.8-2,

$$\rho\left(\frac{\partial \omega_A}{\partial t} + (\mathbf{v} \cdot \nabla \omega_A)\right) = \rho \mathcal{D}_{AB} \nabla^2 \omega_A + r_A \qquad (19.1\text{-}16)$$

with a corresponding equation for species B. This equation is appropriate for describing the diffusion in *dilute liquid solutions* at constant temperature and pressure. The left side can be written as $\rho D\omega_A/Dt$. Equation 9.1-16 without the r_A term is of the same form as Eq. 11.2-8 or 9. This similarity is quite important, since it is the basis for the analogies that are frequently drawn between heat and mass transport in flowing fluids with constant physical properties.

Binary Systems with Constant $c\mathcal{D}_{AB}$

For this assumption, Eq. 19.1-15 becomes, after inserting Fick's law from Eq. (B) of Table 17.8-2,

$$c\left(\frac{\partial x_A}{\partial t} + (\mathbf{v}^* \cdot \nabla x_A)\right) = c\mathcal{D}_{AB}\nabla^2 x_A + (x_B R_A - x_A R_B) \tag{19.1-17}$$

with a corresponding equation for species B. This equation is useful for *low-density gases* at constant temperature and pressure. The left side can *not* be written as cDx_α/Dt because of the appearance of $\mathbf{v}^*$ rather than $\mathbf{v}$.

Binary Systems with Zero Velocity

If there are no chemical reactions occurring, then the chemical production terms are all zero. If, in addition $\mathbf{v}$ is zero and ρ constant in Eq. 19.1-16, or $\mathbf{v}^*$ is zero and c constant in Eq. 19.1-17, then we get

$$\frac{\partial c_A}{\partial t} = \mathcal{D}_{AB}\nabla^2 c_A \tag{19.1-18}$$

which is called *Fick's second law of diffusion*, or sometimes simply *the diffusion equation*. This equation is usually used for diffusion in *solids* or *stationary liquids* (that is, $\mathbf{v} = 0$ in Eq. 19.1-16) and for *equimolar counter-diffusion* in gases (that is, $\mathbf{v}^* = 0$ in Eq. 19.1-17). By equimolar counter-diffusion we mean that the net molar flux with respect to stationary coordinates is zero; in other words, that for every mole of A that moves, say, in the positive z direction, there is a mole of B that moves in the negative z direction.

Note that Eq. 19.1-18 has the same form as the *heat conduction equation* in Eq. 11.2-10. This similarity is the basis for analogies between many heat conduction and diffusion problems in solids. Keep in mind that many hundreds of problems described by Fick's second law have been solved. Solutions are tabulated in the monographs of Crank[1] and of Carslaw and Jaeger.[2]

In Tables B-10 and 11 we give Eq. 19.1-14 (multicomponent equation of continuity in terms of $\mathbf{j}_\alpha$) and Eq. 19.1-16 (binary diffusion equation for constant ρ and $\mathcal{D}_{AB}$) in the three standard coordinate systems. Other forms of the equation of continuity can be patterned after these.

EXAMPLE 19.1-1 *Diffusion, Convection, and Chemical Reaction*[3]	In Fig. 19.1-1 we show a system in which a liquid, B, moves slowly upward through a slightly soluble porous plug of A. Then A slowly disappears by a first-order reaction after it has dissolved. Find the steady-state concentration profile $c_A(z)$, where z is the coordinate upward from the plug. Assume that the velocity profile is approximately flat across the tube. Assume further that c_{A0} is the solubility of unreacted A in B. Neglect temperature effects associated with the heat of reaction.
SOLUTION	Equation 19.1-16 is appropriate for dilute liquid solutions. Dividing this equation by the molecular weight M_A and specializing for the one-dimensional steady-state problem at hand, we get for constant ρ:

$$v_0 \frac{dc_A}{dz} = \mathcal{D}_{AB}\frac{d^2 c_A}{dz^2} - k_1''' c_A \tag{19.1-19}$$

[2] H. S. Carslaw and J. C. Jaeger, *Conduction of Heat in Solids*, 2nd edition, Oxford University Press (1959).
[3] W. Jost, *Diffusion*, Academic Press, New York (1952), pp. 58–59.

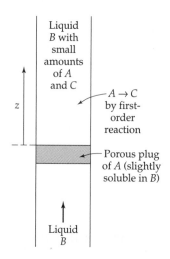

Fig. 19.1-1. Simultaneous diffusion, convection, and chemical reaction.

This is to be solved with the boundary conditions that $c_A = c_{A0}$ at $z = 0$ and $c_A = 0$ at $z = \infty$. Equation 19.1-19 is a standard second-order linear differential equation (Eq. C.7) for which there is a well-known method of solution.

A trial function $c_A = e^{az}$ leads to two values of a, one of which violates the boundary condition at $z = \infty$. The final solution is then

$$\frac{c_A}{c_{A0}} = \exp[-(\sqrt{1 + (4k_1''' \mathcal{D}_{AB}/v_0^2)} - 1)(v_0 z/2\mathcal{D}_{AB})] \tag{19.1-20}$$

This example illustrates the use of the equation of continuity of A for setting up a diffusion problem with convection and chemical reaction.

§19.2   SUMMARY OF THE MULTICOMPONENT EQUATIONS OF CHANGE

In the three main parts of this book we have by stages introduced the conservation laws known as the equations of change. In Chapter 3 conservation of mass and conservation of momentum in pure fluids were presented. In Chapter 11 we added the conservation of energy in pure fluids. In §19.1 we added mass conservation equations for the various species present. We now want to summarize the conservation equations for multicomponent systems.

We start, in Table 19.2-1, by giving the equations of change for a mixture of N chemical species in terms of the combined fluxes with respect to stationary axes. The equation numbers indicate where each equation first appeared. By tabulating the equations of change in this way, we can gain an appreciation for the unity of the subject. The only assumption made here is that all the species are acted on by the same external force per unit mass, $\mathbf{g}$; note (b) of Table 19.2-1 explains the modifications needed when this is not the case.

The important feature of these equations is that they are all of the form

$$\begin{Bmatrix} \text{rate of} \\ \text{increase of} \\ \text{entity} \end{Bmatrix} = \begin{Bmatrix} \text{net rate} \\ \text{of addition} \\ \text{of entity} \end{Bmatrix} + \begin{Bmatrix} \text{rate of} \\ \text{production} \\ \text{of entity} \end{Bmatrix} \tag{19.2-1}$$

in which "entity" stands for mass, momentum, or energy, respectively. In each equation the net rate of addition of the entity per unit volume is the negative of a divergence term. The "rates of production" arise from chemical reactions in the first equation and from the external force field in the other two. Each equation is a statement of a *conservation law*. Usually we think of the conservation statements as laws that have gradually

Table 19.2-1   Equations of Change for Multicomponent Mixtures in Terms of the Combined Fluxes

Mass of α: $(\alpha = 1, 2, \ldots, N)$	$\dfrac{\partial}{\partial t} \rho \omega_\alpha = -(\nabla \cdot \mathbf{n}_\alpha) + r_\alpha$	(A)[a] (Eq. 19.1-6)
Momentum:	$\dfrac{\partial}{\partial t} \rho \mathbf{v} = -[\nabla \cdot \boldsymbol{\phi}] + \rho \mathbf{g}$	(B)[b] (Eq. 3.2-8)
Energy:	$\dfrac{\partial}{\partial t} \rho(\hat{U} + \tfrac{1}{2}v^2) = -(\nabla \cdot \mathbf{e}) + (\rho \mathbf{v} \cdot \mathbf{g})$	(C)[b] (Eq. 11.1-6)

[a] When all N equations of continuity are added, the equation of continuity for the fluid mixture

$$\frac{\partial}{\partial t} \rho = -(\nabla \cdot \rho \mathbf{v}) \qquad \begin{array}{l}\text{(D)}\\ \text{(Eq. 3.1-4)}\end{array}$$

is obtained. Here $\mathbf{v}$ is the mass average velocity defined in Eq. 17.7-1.

[b] If species α is acted on by a force per unit volume given by $\mathbf{g}_\alpha$, then $\rho \mathbf{g}$ has to be replaced by $\Sigma_\alpha \rho_\alpha \mathbf{g}_\alpha$ in Eq. (B), and $(\rho \mathbf{v} \cdot \mathbf{g})$ has to be replaced by $\Sigma_\alpha (\mathbf{n}_\alpha \cdot \mathbf{g}_\alpha)$ in Eq. (C). These replacements are required, for example, if some of the species are ions with different charges on them, acted on by an electric field. Problems of this sort are discussed in Chapter 24.

evolved by experience and experiment and therefore are generally accepted by the scientific community.[1]

The three "combined fluxes," which appear in Eqs. (A) to (C) of Table 19.2-1, can be written as the *convective fluxes* plus the *molecular* (or *diffusive*) *fluxes*. These various fluxes are displayed in Table 19.2-2, where the equation numbers corresponding to their first appearance are given.

When the flux expressions of Table 19.2-2 are substituted into the conservation equations of Table 19.2-1 and then converted to the D/Dt form by means of Eqs. 3.5-4 and 5, we get the multicomponent equations of change in their usual forms. These are tabulated in Table 19.2-3.

In addition to these conservation equations, one needs also to have the expressions for the fluxes in terms of the gradients and the transport properties (the latter being functions of temperature, density, and composition). Finally one needs also the thermal equation of state, $p = p(\rho, T, x_\alpha)$, and the caloric equation of state, $\hat{U} = \hat{U}(\rho, T, x_\alpha)$, and information about the rates of any homogeneous chemical reactions occurring.[2]

---

[1] Actually the conservation laws for energy, momentum, and angular momentum follow from Lagrange's equation of motion, together with the homogeneity of time, the homogeneity of space, and the isotropy of space, respectively (*Noether's theorem*). Thus there is something very fundamental about these conservation laws, more than is apparent at first sight. For more on this, see L. Landau and E. M. Lifshitz, *Mechanics*, Addison-Wesley, Reading, Mass. (1960), Chapter 2, and Emmy Noether, *Nachr. Kgl. Ges. Wiss. Göttingen (Math.-phys. Kl.)* (1918), pp. 235–257. **Amalie Emmy Noether** (1882–1935), after doing the doctorate at the University of Erlangen, was a protégée of Hilbert in Göttingen until Hitler's purge of 1933 forced her to move to the United States, where she became a professor of mathematics at Bryn Mawr College; a crater on the moon is named after her.

[2] One might wonder whether or not we need separate equations of motion and energy for species α. Such equations can be derived by continuum arguments, but the species momentum and energy fluxes are not measurable quantities and molecular theory is required in order to clarify their meanings. These separate species equations are not needed for solving transport problems. However, the species equations of motion have been helpful for deriving kinetic expressions for the mass fluxes in multicomponent systems [see C. F. Curtiss and R. B. Bird, *Proc. Nat. Acad. Sci. USA*, **93**, 7440–7445 (1996) and *J. Chem. Phys.*, **111**, 10362–10370 (1999)].

Table 19.2-2   The Combined, Molecular, and Convective Fluxes for Multicomponent Mixtures (all with the same sign convention)

Entity	Combined flux	=	Molecular flux	+	Convective flux	
Mass ($\alpha = 1, 2, \ldots, N$)	$\mathbf{n}_\alpha$	=	$\mathbf{j}_\alpha$	+	$\rho \mathbf{v} \omega_\alpha$	(A)[a] (Eq. 17.8-1)
Momentum	$\boldsymbol{\phi}$	=	$\boldsymbol{\pi}$	+	$\rho \mathbf{v v}$	(B)[b] (Eq. 1.7-1)
Energy	$\mathbf{e}$	=	$\mathbf{q} + [\boldsymbol{\pi} \cdot \mathbf{v}]$	+	$\rho \mathbf{v}(\hat{U} + \frac{1}{2}v^2)$	(C)[c] (Eq. 9.8-5)

[a] The velocity $\mathbf{v}$ appearing in all these expressions is the mass average velocity, defined in Eq. 17.7-1.

[b] The molecular momentum flux consists of two parts: $\boldsymbol{\pi} = p\boldsymbol{\delta} + \boldsymbol{\tau}$.

[c] The molecular energy flux is made up of the heat flux vector $\mathbf{q}$ and the work flux vector $[\boldsymbol{\pi} \cdot \mathbf{v}] = p\mathbf{v} + [\boldsymbol{\tau} \cdot \mathbf{v}]$, the latter occurring only in flow systems.

Table 19.2-3   Equations of Change for Multicomponent Mixtures in Terms of the Molecular Fluxes

Total mass:
$$\frac{D\rho}{Dt} = -\rho(\nabla \cdot \mathbf{v})$$
$$\text{(A)}$$
$$\text{(Eq. (A) of Table 3.5-1)}$$

Species mass: ($\alpha = 1, 2, \cdots, N$)
$$\rho \frac{D\omega_\alpha}{Dt} = -(\nabla \cdot \mathbf{j}_\alpha) + r_\alpha$$
$$\text{(B)}^a$$
$$\text{(Eq. 19.1-7a)}$$

Momentum:
$$\rho \frac{D\mathbf{v}}{Dt} = -\nabla p - [\nabla \cdot \boldsymbol{\tau}] + \rho \mathbf{g}$$
$$\text{(C)}^b$$
$$\text{(Eq. (B) of Table 3.5-1)}$$

Energy:
$$\rho \frac{D}{Dt}(\hat{U} + \tfrac{1}{2}v^2) = -(\nabla \cdot \mathbf{q}) - (\nabla \cdot p\mathbf{v}) - (\nabla \cdot [\boldsymbol{\tau} \cdot \mathbf{v}]) + (\rho \mathbf{v} \cdot \mathbf{g})$$
$$\text{(D)}^b$$
$$\text{(Eq. (E) of Table 11.4-1)}$$

[a] Only $N - 1$ of these equations are independent, since the sum of the N equations gives $0 = 0$.

[b] See note (b) of Table 19.2-1 for the modifications needed when the various species are acted on by different forces.

We conclude this discussion with a few remarks about special forms of the equations of motion and energy. In §11.3 it was pointed out that the *equation of motion* as presented in Chapter 3 is in suitable form for setting up forced-convection problems, but that an alternate form (Eq. 11.3-2) is desirable for displaying explicitly the buoyant forces resulting from temperature inequalities in the system. In binary systems with concentration inequalities as well as temperature inequalities, we write the equation of motion as in Eq. (B) of Table 3.5-1 and use an approximate equation of state formed by making a double Taylor expansion of $\rho(T, \omega_A)$ about the state $\bar{T}, \bar{\omega}_A$:

$$\rho(T, \omega_A) = \bar{\rho} + \frac{\partial \rho}{\partial T}\bigg|_{\bar{T}, \bar{\omega}_A} (T - \bar{T}) + \frac{\partial \rho}{\partial \omega_A}\bigg|_{\bar{T}, \bar{\omega}_A} (\omega_A - \bar{\omega}_A) + \cdots$$
$$\approx \bar{\rho} - \bar{\rho}\bar{\beta}(T - \bar{T}) - \bar{\rho}\bar{\zeta}(\omega_A - \bar{\omega}_A) \tag{19.2-2}$$

Table 19.2-4 The Equations of Energy for Multicomponent Systems, with Gravity as the Only External Force[a,b]

$$\rho \frac{D}{Dt}(\hat{U} + \hat{\Phi} + \tfrac{1}{2}v^2) = -(\nabla \cdot \mathbf{q}) - (\nabla \cdot [\boldsymbol{\pi} \cdot \mathbf{v}]) \tag{A}^c$$

$$\rho \frac{D}{Dt}(\hat{U} + \tfrac{1}{2}v^2) = -(\nabla \cdot \mathbf{q}) - (\nabla \cdot [\boldsymbol{\pi} \cdot \mathbf{v}]) + (\mathbf{v} \cdot \rho \mathbf{g}) \tag{B}$$

$$\rho \frac{D}{Dt}(\tfrac{1}{2}v^2) = -(\mathbf{v} \cdot [\nabla \cdot \boldsymbol{\pi}]) + (\mathbf{v} \cdot \rho \mathbf{g}) \tag{C}$$

$$\rho \frac{D\hat{U}}{Dt} = -(\nabla \cdot \mathbf{q}) - (\boldsymbol{\pi} : \nabla \mathbf{v}) \tag{D}$$

$$\rho \frac{D\hat{H}}{Dt} = -(\nabla \cdot \mathbf{q}) - (\boldsymbol{\tau} : \nabla \mathbf{v}) + \frac{Dp}{Dt} \tag{E}$$

$$\rho \hat{C}_p \frac{DT}{Dt} = -(\nabla \cdot \mathbf{q}) - (\boldsymbol{\tau} : \nabla \mathbf{v}) + \left(\frac{\partial \ln \hat{V}}{\partial \ln T}\right)_{p,x_\alpha} \frac{Dp}{Dt} + \sum_{\alpha=1}^{N} \overline{H}_\alpha [(\nabla \cdot \mathbf{J}_\alpha) - R_\alpha] \tag{F}^d$$

$$\rho \hat{C}_V \frac{DT}{Dt} = -(\nabla \cdot \mathbf{q}) - (\boldsymbol{\pi} : \nabla \mathbf{v}) + \left(1 - \left(\frac{\partial \ln p}{\partial \ln T}\right)_{\rho,x_\alpha}\right) p(\nabla \cdot \mathbf{v})$$
$$+ \sum_{\alpha=1}^{N} \left(\overline{U}_\alpha + \left(1 - \left(\frac{\partial \ln p}{\partial \ln T}\right)_{\rho,x_\alpha}\right) p\overline{V}_\alpha\right)[(\nabla \cdot \mathbf{J}_\alpha) - R_\alpha] \tag{G}$$

$$\frac{\partial}{\partial t}\sum_{\alpha=1}^{N} c_\alpha \overline{H}_\alpha + \left(\nabla \cdot \sum_{\alpha=1}^{N} \mathbf{N}_\alpha \overline{H}_\alpha\right) = (\nabla \cdot k\nabla T) - (\boldsymbol{\tau} : \nabla \mathbf{v}) + \frac{Dp}{Dt} \tag{H}^e$$

[a] For multicomponent mixtures $\mathbf{q} = -k\nabla T + \sum_{\alpha=1}^{N} \frac{\overline{H}_\alpha}{M_\alpha}\mathbf{j}_\alpha + \mathbf{q}^{(x)}$, where $\mathbf{q}^{(x)}$ is a usually negligible term associated with the diffusion-thermo effect (see Eq. 24.2-6).

[b] The equations in this table are valid only if the same external force is acting on all species. If this is not the case, then $\Sigma_\alpha(\mathbf{j}_\alpha \cdot \mathbf{g}_\alpha)$ must be added to Eq. (A) and Eqs. (D–H), the last term in Eq. (B) has to be replaced by $\Sigma_\alpha(\mathbf{n}_\alpha \cdot \mathbf{g}_\alpha)$, and the last term in Eq. (C) has to be replaced by $\Sigma_\alpha(\mathbf{v} \cdot \rho_\alpha \mathbf{g}_\alpha)$.

[c] Exact only if $\partial \hat{\Phi}/\partial t = 0$.

[d] L. B. Rothfeld, PhD thesis, University of Wisconsin (1961); see also Problem 19D.2.

[e] The contribution of $\mathbf{q}^{(x)}$ to the heat flux vector has been omitted in this equation.

Here the coefficient $\overline{\zeta} = -(1/\rho)(\partial\rho/\partial\omega_A)$ evaluated at $\overline{T}$ and $\overline{\omega}_A$ relates the density to the composition. This coefficient is the mass transfer analog of the coefficient $\overline{\beta}$ introduced in Eq. 11.3-1. When this approximate equation of state is substituted into the $\rho\mathbf{g}$ term (but not into the $\rho D\mathbf{v}/Dt$ term) of the equation of motion, we get the *Boussinesq equation of motion* for a binary mixture, with gravity as the only external force:

$$\rho \frac{D\mathbf{v}}{Dt} = (-\nabla p + \overline{\rho}\mathbf{g}) - [\nabla \cdot \boldsymbol{\tau}] - \overline{\rho}\mathbf{g}\overline{\beta}(T - \overline{T}) - \overline{\rho}\mathbf{g}\overline{\zeta}(\omega_A - \overline{\omega}_A) \tag{19.2-3}$$

The last two terms in this equation describe the buoyant force resulting from the temperature and composition variations within the fluid.

Next we turn to the *equation of energy*. Recall that in Table 11.4-1 the energy equation for pure fluids was given in a variety of forms. The same can be done for mixtures, and a representative selection of the many possible forms of this equation is given in Table 19.2-4. Note that it is not necessary to add a term S_c (as we did in Chapter 10) to describe the thermal energy released by homogeneous chemical reactions. This information is included implicitly in the functions $\hat{H}$ and $\hat{U}$, and appears explicitly as $-\Sigma_\alpha\overline{H}_\alpha R_\alpha$ and $-\Sigma_\alpha\overline{U}_\alpha R_\alpha$ in Eqs. (F) and (G). Remember that in calculating $\hat{H}$ and $\hat{U}$, the energies of formation and mixing of the various species must be included (see Example 23.5-1).

§19.3   SUMMARY OF THE MULTICOMPONENT FLUXES

The equations of change have been given in terms of the fluxes of mass, momentum, and energy. To solve these equations, we have to replace the fluxes by expressions involving the transport properties and the gradients of concentration, velocity, and temperature. Here we summarize the flux expressions for mixtures:

Mass: $\qquad\qquad\qquad \mathbf{j}_A = -\rho \mathcal{D}_{AB} \nabla \omega_A \qquad$ binary only $\qquad\qquad$ (19.3-1)

Momentum: $\qquad\quad \boldsymbol{\tau} = -\mu[\nabla \mathbf{v} + (\nabla \mathbf{v})^\dagger] + (\tfrac{2}{3}\mu - \kappa)(\nabla \cdot \mathbf{v})\boldsymbol{\delta} \qquad$ (19.3-2)

Energy: $\qquad\qquad \mathbf{q} = -k\nabla T + \sum_{\alpha=1}^{N} \dfrac{\overline{H}_\alpha}{M_\alpha} \mathbf{j}_\alpha \qquad\qquad\qquad$ (19.3-3)

Now we append a few words of explanation:

a. The *mass flux* expression given here is for binary mixtures only. For multicomponent gas mixtures at moderate pressures, we can use the Maxwell–Stefan equations of Eq. 17.9-1. There are additional contributions to the mass flux corresponding to driving forces other than the concentration gradients: *forced diffusion*, which occurs when the various species are subjected to different external forces; *pressure diffusion*, proportional to ∇p; and *thermal diffusion*, proportional to ∇T. These other diffusion mechanisms, the first two of which can be quite important, are covered in Chapter 24.

b. The *momentum flux* expression is the same for multicomponent mixtures as for pure fluids. Once again we point out that the contribution containing the dilatational viscosity κ is seldom important. Of course, for polymers and other viscoelastic fluids, Eq. 19.3-2 has to be replaced by more complex models, as explained in Chapter 8.

c. The *energy-flux* expression given here for multicomponent fluids consists of two terms: the first term is the heat transport by conduction which was given for pure materials in Eq. 9.1-4, and the second term describes the heat transport by each of the diffusing species. The quantity $\overline{H}_\alpha$ is the partial molar enthalpy of species α. There is actually one further contribution to the energy flux, related to a concentration driving force—usually quite small—and this *diffusion-thermo effect* will be discussed in Chapter 24. The thermal conductivity of a mixture—the k in Eq. 19.3-3—is defined as the proportionality constant between the heat flux and the temperature gradient in the absence of any mass fluxes.

We conclude this discussion with a few comments about the combined energy flux $\mathbf{e}$. By substituting Eq. 19.3-3 into Eq. (C) of Table 19.2-2, we get after some minor rearranging:

$$\mathbf{e} = \rho(\hat{U} + \tfrac{1}{2}v^2)\mathbf{v} + \mathbf{q} + p\mathbf{v} + [\boldsymbol{\tau} \cdot \mathbf{v}]$$

$$= \rho(\hat{U} + \tfrac{1}{2}v^2)\mathbf{v} - k\nabla T + \sum_{\alpha=1}^{N} \frac{\overline{H}_\alpha}{M_\alpha} \mathbf{j}_\alpha + p\mathbf{v} + [\boldsymbol{\tau} \cdot \mathbf{v}]$$

$$= -k\nabla T + \sum_{\alpha=1}^{N} \overline{H}_\alpha \mathbf{J}_\alpha + \rho(\hat{U} + p\hat{V})\mathbf{v} + \tfrac{1}{2}\rho v^2 \mathbf{v} + [\boldsymbol{\tau} \cdot \mathbf{v}] \qquad (19.3\text{-}4)$$

In some situations, notably in films and low-velocity boundary layers, the contributions $\tfrac{1}{2}\rho v^2 \mathbf{v}$ and $[\boldsymbol{\tau} \cdot \mathbf{v}]$ are negligible. Then the dashed-underlined terms may be discarded. This leads to

$$\mathbf{e} = -k\nabla T + \sum_{\alpha=1}^{N} \overline{H}_\alpha \mathbf{J}_\alpha + \rho \hat{H}_\alpha \mathbf{v}$$

$$= -k\nabla T + \sum_{\alpha=1}^{N} \overline{H}_\alpha \mathbf{J}_\alpha + \sum_{\alpha=1}^{N} c_\alpha \overline{H}_\alpha \mathbf{v} \qquad (19.3\text{-}5)$$

Then use of Eqs. (G) and (H) of Table 17.8-1 leads finally to

$$\mathbf{e} = -k\nabla T + \sum_{\alpha=1}^{N} \overline{H}_\alpha \mathbf{N}_\alpha \tag{19.3-6}$$

Finally, for ideal gas mixtures, this expression can be further simplified by replacing the partial molar enthalpies $\overline{H}_\alpha$ by the molar enthalpies $\tilde{H}_\alpha$. Equation 19.3-6 provides a standard starting point for solving one-dimensional problems in simultaneous heat and mass transfer.[1]

EXAMPLE 19.3-1

The Partial Molar Enthalpy

The partial molar enthalpy $\overline{H}_\alpha$, which appears in Eqs. 19.3-3 and 19.3-6, is defined for a multicomponent mixture as

$$\overline{H}_\alpha = \left(\frac{\partial H}{\partial n_\alpha} \right)_{T,p,n_\beta} \tag{19.3-7}$$

in which n_α is the number of moles of species α in the mixture, and the subscript n_β indicates that the derivative is to be taken holding the number or moles of each species other than α constant. The enthalpy $H(n_1, n_2, n_3, \ldots)$ is an "extensive property," since, if the number of moles of each component is multiplied by k, the enthalpy itself will be multiplied by k:

$$H(kn_1, kn_2, kn_3, \cdots) = kH(n_1, n_2, n_3, \cdots) \tag{19.3-8}$$

Mathematicians refer to this kind of function as being "homogeneous of degree 1." For such functions Euler's theorem[2] can be used to conclude that

$$H = \sum_\alpha n_\alpha \overline{H}_\alpha \tag{19.3-9}$$

(a) Prove that, for a binary mixture, the partial molar enthalpies at a given mole fraction can be determined by plotting the enthalpy per mole as a function of mole fraction, and then determining the intercepts of the tangent drawn at the mole fraction in question (see Fig. 19.3-1). This shows one way to get the partial molar enthalpy from data on the enthalpy of the mixture.

(b) How else could one get the partial molar enthalpy?

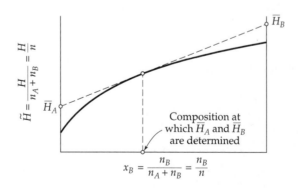

Fig. 19.3-1. The "method of intercepts" for determining partial molar quantities in a binary mixture.

[1] T. K. Sherwood, R. L. Pigford, and C. R. Wilke, *Mass Transfer*, McGraw-Hill, New York (1975), Chapter 7. **Thomas Kilgore Sherwood** (1903–1976) was a professor at MIT for nearly 40 years, and then taught at the University of California in Berkeley. Because of his many contributions to the field of mass transfer, the Sherwood number (Sh) was named after him.

[2] M. D. Greenberg, *Foundations of Applied Mathematics*, Prentice-Hall, Englewood Cliffs, N.J. (1978), p. 128; R. J. Silbey and R. A. Alberty, *Physical Chemistry*, 3rd edition, Wiley, New York (2001), §§1.10, 4.9, and 6.10.

SOLUTION

(a) Throughout this example, for brevity we omit the subscripts p, T indicating that these quantities are held constant. First we write expressions for the intercepts as follows:

$$\overline{H}_A = \tilde{H} - x_B\left(\frac{\partial \tilde{H}}{\partial x_B}\right)_n; \qquad \overline{H}_B = \tilde{H} + x_A\left(\frac{\partial \tilde{H}}{\partial x_B}\right)_n \tag{19.3-10, 11}$$

in which $\tilde{H} = H/(n_A + n_B) = H/n$. To verify the correctness of Eq. 19.3-10, we rewrite the expression in terms of H:

$$\overline{H}_A = \frac{H}{n} - \frac{x_B}{n}\left(\frac{\partial H}{\partial x_B}\right)_n \tag{19.3-12}$$

Now the expression $\overline{H}_A = (\partial H/\partial n_A)_{n_B}$ implies that H is a function of n_A and n_B, whereas $(\partial H/\partial x_A)_n$ implies that H is a function of x_A and n. The relation between these kinds of derivatives is given by the chain rule of partial differentiation. To apply this rule we need the relation between the independent variables, which, in this problem, are

$$n_A = (1 - x_B)n; \qquad n_B = x_B n \tag{19.3-13, 14}$$

Therefore we may write

$$\left(\frac{\partial H}{\partial x_B}\right)_n = \left(\frac{\partial H}{\partial n_A}\right)_{n_B}\left(\frac{\partial n_A}{\partial x_B}\right)_n + \left(\frac{\partial H}{\partial n_B}\right)_{n_A}\left(\frac{\partial n_B}{\partial x_B}\right)_n$$
$$= \overline{H}_A(-n) + \overline{H}_B(+n) \tag{19.3-15}$$

Substitution of this into Eq. 19.3-12 and use of Euler's theorem ($H = n_A\overline{H}_A + n_B\overline{H}_B$) then gives an identity. This proves the validity of Eq. 19.3-10, and the correctness of Eq. 19.3-11 can be proved similarly.

(b) One can also get $\overline{H}_A$ by using the definition in Eq. 19.3-7 and measuring the slope of the curve of H versus n_A, holding n_B constant. One can also get $\overline{H}_A$ by measuring the enthalpy of mixing and using

$$H = n_A\overline{H}_A + n_B\overline{H}_B = n_A\tilde{H}_A + n_B\tilde{H}_B + \Delta H_{\text{mix}} \tag{19.3-16}$$

Often the enthalpy of mixing is neglected and the enthalpies of the pure substances are given as $\tilde{H}_A \approx \tilde{C}_{pA}(T - T^\circ)$ and a similar expression for $\tilde{H}_B$. This is a standard assumption for gas mixtures at low to moderate pressures.

Other methods for evaluating partial molar quantities may be found in current textbooks on thermodynamics.

§19.4 USE OF THE EQUATIONS OF CHANGE FOR MIXTURES

The equations of change in §19.2 can be used to solve all the problems of Chapter 18, and more difficult ones as well. Unless the problems are idealized or simplified, mixture transport phenomena are quite complicated and usually numerical techniques are required. Here we solve a few introductory problems by way of illustration.

EXAMPLE 19.4-1

Simultaneous Heat and Mass Transport[1]

(a) Develop expressions for the mole fraction profile $x_A(y)$ and the temperature profile $T(y)$ for the system pictured in Fig. 19.4-1, given the mole fractions and temperatures at both film boundaries ($y = 0$ and $y = \delta$). Here a hot condensable vapor, A, is diffusing at steady state through a stagnant film of noncondensable gas, B, to a cold surface at $y = 0$, where A condenses. Assume ideal gas behavior and uniform pressure. Furthermore assume the physical

---

[1] A. P. Colburn and T. B. Drew, *Trans. Am. Inst. Chem. Engrs.*, **38**, 197–212 (1937).

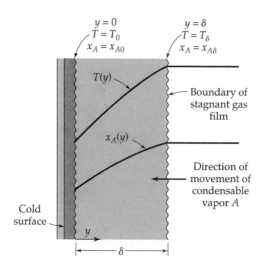

Fig. 19.4-1. Condensation of a hot vapor A on a cold surface in the presence of a non-condensable gas B.

properties to be constant, evaluated at some mean temperature and composition. Neglect radiative heat transfer.

(b) Generalize the result for the situation where both A and B are condensing on the wall, and allow for unequal film thicknesses for heat and mass transport.

SOLUTION

(a) To determine the desired quantities, we must solve the equations of continuity and energy for this system. Simplification of Eq. 19.1-10 and Eq. C of Table 19.2-1 for steady, one-dimensional transport, in the absence of chemical reactions and external forces, gives

Continuity of A:
$$\frac{dN_{Ay}}{dy} = 0 \qquad (19.4\text{-}1)$$

Energy:
$$\frac{de_y}{dy} = 0 \qquad (19.4\text{-}2)$$

Therefore, both N_{Ay} and e_y are constant throughout the film.

To determine the *mole fraction profile*, we need the molar flux for diffusion of A through stagnant B:

$$N_{Ay} = -\frac{c\mathcal{D}_{AB}}{1 - x_A} \frac{dx_A}{dy} \qquad (19.4\text{-}3)$$

Insertion of Eq. 19.4-3 into Eq. 19.4-1 and integration gives the mole fraction profile (see §18.2)

$$\left(\frac{1 - x_A}{1 - x_{A0}}\right) = \left(\frac{1 - x_{A\delta}}{1 - x_{A0}}\right)^{y/\delta} \qquad (19.4\text{-}4)$$

Here we have taken $c\mathcal{D}_{AB}$ to be constant, at the value for the mean film temperature. We can then evaluate the constant flux N_{Ay} from Eqs. 19.4-3 and 4:

$$N_{Ay} = \frac{c\mathcal{D}_{AB}}{\delta} \ln \frac{1 - x_{A\delta}}{1 - x_{A0}} \qquad (19.4\text{-}5)$$

Note that N_{Ay} is negative because species A is condensing. The last two expressions may be combined to put the concentration profiles in an alternative form:

$$\frac{x_A - x_{A0}}{x_{A\delta} - x_{A0}} = \frac{1 - \exp[(N_{Ay}/c\mathcal{D}_{AB})y]}{1 - \exp[(N_{Ay}/c\mathcal{D}_{AB})\delta]} \qquad (19.4\text{-}6)$$

To get the *temperature profile*, we use the energy flux from Eq. 19.3-6 for an ideal gas along with Eq. 9.8-8:

$$e_y = -k\frac{dT}{dy} + (\tilde{H}_A N_{Ay} + \tilde{H}_B N_{By})$$

$$= -k\frac{dT}{dy} + N_{Ay}\tilde{C}_{pA}(T - T_0) \tag{19.4-7}$$

Here we have chosen T_0 as the reference temperature for the enthalpy. Insertion of this expression for e_y into Eq. 19.4-2 and integration between the limits $T = T_0$ at $y = 0$, and $T = T_\delta$ at $y = \delta$ gives

$$\frac{T - T_0}{T_\delta - T_0} = \frac{1 - \exp[(N_{Ay}\tilde{C}_{pA}/k)y]}{1 - \exp[(N_{Ay}\tilde{C}_{pA}/k)\delta]} \tag{19.4-8}$$

It can be seen that the temperature profile is not linear for this system except in the limit as $N_{Ay}\tilde{C}_{pA}/k \to 0$. Note the similarity between Eqs. 19.4-6 and 8.

The conduction energy flux at the wall is greater here than in the absence of mass transfer. Thus, using a superscript zero to indicate the conditions in the absence of mass transfer, we may write

$$\frac{-k(dT/dy)|_{y=0}}{-k(dT/dy)^0|_{y=0}} = \frac{-(N_{Ay}\tilde{C}_{pA}/k)\delta}{1 - \exp[(N_{Ay}\tilde{C}_{pA}/k)\delta]} \tag{19.4-9}$$

We see then that the rate of heat transfer is directly affected by simultaneous mass transfer, whereas the mass flux is not directly affected by simultaneous heat transfer. In applications at temperatures below the normal boiling point of species A, the quantity $N_{Ay}\tilde{C}_{pA}/k$ is small, and the right side of Eq. 19.4-9 is very nearly unity (see Problem 19A.1). The interaction between heat and mass transfer is further discussed in Chapter 22.

(b) If both A and B are condensing at the wall, then Eqs. 19.4-1 and 2, when integrated, lead to $N_{Ay} = N_{A0}$ and $e_y = e_0$, where the subscript "0" quantities are evaluated at $y = 0$. We also integrate the analog of Eq. 19.4-1 for B to get $N_{By} = N_{B0}$ and obtain

$$-c\mathcal{D}_{AB}\frac{dx_A}{dy} + x_A(N_{A0} + N_{B0}) = N_{A0} \tag{19.4-10}$$

$$-k\frac{dT}{dy} + (N_{A0}\overline{H}_A + N_{B0}\overline{H}_B) = e_0 \tag{19.4-11}$$

In the second of these equations, we replace $\overline{H}_A$ by $\tilde{C}_{pA}(T - T_0)$ and $\overline{H}_B$ by $\tilde{C}_{pB}(T - T_0)$, and since the reference temperature is T_0, we may replace e_0 by q_0, the conductive heat flux at the wall. In the first equation, we subtract $x_{A0}(N_{A0} + N_{B0})$ from both sides to make the equation similar in form to the temperature equation just obtained. Thus

$$-c\mathcal{D}_{AB}\frac{dx_A}{dy} + (N_{A0} + N_{B0})(x_A - x_{A0}) = N_{A0} - x_{A0}(N_{A0} + N_{B0}) \tag{19.4-12}$$

$$-k\frac{dT}{dy} + (N_{A0}\tilde{C}_{pA} + N_{B0}\tilde{C}_{pB})(T - T_0) = q_0 \tag{19.4-13}$$

Integration with respect to y and application of the boundary conditions at $y = 0$ gives

$$\frac{(N_{A0} + N_{B0})(x_A - x_{A0})}{N_{A0} - x_{A0}(N_{A0} + N_{B0})} = 1 - \exp\left[(N_{A0} + N_{B0})\frac{y}{c\mathcal{D}_{AB}}\right] \tag{19.4-14}$$

$$\frac{(N_{A0}\tilde{C}_{pA} + N_{B0}\tilde{C}_{pB})(T - T_0)}{q_0} = 1 - \exp\left[(N_{A0}\tilde{C}_{pA} + N_{B0}\tilde{C}_{pB})\frac{y}{k}\right] \tag{19.4-15}$$

These are the concentration and temperature profiles in terms of the mass and heat fluxes. Applications of the boundary conditions at the outer edges of the films—that is, at $y = \delta_x$ and $y = \delta_T$, respectively—give

$$\frac{(N_{A0} + N_{B0})(x_{A\delta} - x_{A0})}{N_{A0} - x_{A0}(N_{A0} + N_{B0})} = 1 - \exp\left[(N_{A0} + N_{B0})\frac{\delta_x}{c\mathcal{D}_{AB}}\right] \tag{19.4-16}$$

$$\frac{(N_{A0}\tilde{C}_{pA} + N_{B0}\tilde{C}_{pB})(T_\delta - T_0)}{q_0} = 1 - \exp\left[(N_{A0}\tilde{C}_{pA} + N_{B0}\tilde{C}_{pB})\frac{\delta_T}{k}\right] \tag{19.4-17}$$

These equations relate the fluxes to the film thicknesses and the transport properties. When Eq. 19.4-14 is divided by Eq. 19.4-16 and Eq. 19.4-15 is divided by Eq. 19.4-17, we get the concentration profiles in terms of the transport coefficients (analogously to Eqs. 19.4-6 and 8). Equations 19.4-16 and 17 will be encountered again in §22.8.

EXAMPLE 19.4-2

Concentration Profile in a Tubular Reactor

A catalytic tubular reactor is shown in Fig. 19.4-2. A dilute solution of solute A in a solvent S is in fully developed, laminar flow in the region $z < 0$. When it encounters the catalytic wall in the region $0 \leq z \leq L$, solute A is instantaneously and irreversibly rearranged to an isomer B. Write the diffusion equation appropriate for this problem, and find the solution for short distances into the reactor. Assume that the flow is isothermal and neglect the presence of B.

SOLUTION

For the conditions stated above, the flowing liquid will always be very nearly pure solvent S. The product $\rho\mathcal{D}_{AS}$ can be considered constant, and the diffusion of A in S can be described by the steady-state version of Eq. 19.1-14 (ignoring the presence of a small amount of the reaction product B). The relevant equations of change for the system are then

Continuity of A:
$$v_z\frac{\partial c_A}{\partial z} = \mathcal{D}_{AS}\left[\frac{1}{r}\frac{\partial}{\partial r}\left(r\frac{\partial c_A}{\partial r}\right) + \frac{\partial^2 c_A}{\partial z^2}\right] \tag{19.4-18}$$

Motion:
$$0 = -\frac{d\mathcal{P}}{dz} + \mu\frac{1}{r}\frac{d}{dr}\left(r\frac{dv_z}{dr}\right) \tag{19.4-19}$$

We make the usual assumption that axial diffusion can be neglected with respect to axial convection, and therefore delete the dashed-underlined term (compare with Eqs. 10.8-11 and 12). Equation 19.4-19 can be solved to give the parabolic velocity profile $v_z(r) = v_{z,\max}[1 - (r/R)^2]$. When this result is substituted into Eq. 19.4-18, we get

$$v_{z,\max}\left[1 - \left(\frac{r}{R}\right)^2\right]\frac{\partial c_A}{\partial z} = \mathcal{D}_{AS}\frac{1}{r}\frac{\partial}{\partial r}\left(r\frac{\partial c_A}{\partial r}\right) \tag{19.4-20}$$

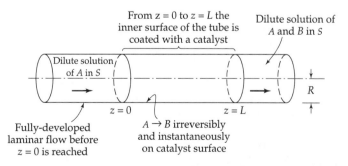

From $z = 0$ to $z = L$ the inner surface of the tube is coated with a catalyst

Dilute solution of A and B in S

Dilute solution of A in S

$z = 0$ $z = L$

R

Fully-developed laminar flow before $z = 0$ is reached

$A \rightarrow B$ irreversibly and instantaneously on catalyst surface

Fig. 19.4-2. Boundary conditions for a tubular reactor.

This is to be solved with the boundary conditions

B.C. 1:   at $z = 0$, $c_A = c_{A0}$ (19.4-21)

B.C. 2:   at $r = R$, $c_A = 0$ (19.4-22)

B.C. 3:   at $r = 0$, $c_A = $ finite   (19.4-23)

For short distances z into the reactor, the concentration c_A differs from c_{A0} only near the wall, where the velocity profile is practically linear. Hence we can introduce the variable $y = R - r$, neglect curvature terms, and replace B.C. 3 by a fictitious boundary condition at $y = \infty$ (see Example 12.2-2 for a detailed discussion of this method of treating the entrance region of the tube).

The reformulated problem statement is then

$$2v_{z,\max} \frac{y}{R} \frac{\partial c_A}{\partial z} = \mathscr{D}_{AS} \frac{\partial^2 c_A}{\partial y^2}$$ (19.4-24)

with the boundary conditions

B.C. 1:   at $z = 0$, $c_A = c_{A0}$ (19.4-25)

B.C. 2:   at $y = 0$, $c_A = 0$ (19.4-26)

B.C. 3:   at $y = \infty$, $c_A = c_{A0}$ (19.4-27)

This problem can be solved by the method of combination of independent variables by seeking a solution of the form $c_A/c_{A0} = f(\eta)$, where $\eta = (y/R)(2v_{z,\max}R^2/9\mathscr{D}_{AS}z)^{1/3}$. One thus obtains the ordinary differential equation $f'' + 3\eta^2 f' = 0$, which can be integrated to give (see Eq. C.1-9)

$$\frac{c_A}{c_{A0}} = \frac{\int_0^\eta \exp(-\bar{\eta}^3)\, d\bar{\eta}}{\int_0^\infty \exp(-\bar{\eta}^3)\, d\bar{\eta}} = \frac{\int_0^\eta \exp(-\bar{\eta}^3)\, d\bar{\eta}}{\Gamma(\tfrac{4}{3})}$$ (19.4-28)

This problem is mathematically analogous to the Graetz problem of Problem 12D.4, Θ of that problem being analogous to $1 - (c_A/c_{A0})$ here.

Experiments of the type described here have proved useful for obtaining mass transfer data at high Schmidt numbers.[2] A particularly attractive reaction is the reduction of ferricyanide ions on metallic surfaces according to the reaction

$$Fe(CN)_6^{-3} + e^{-1} \rightarrow Fe(CN)_6^{-4}$$ (19.4-29)

in which ferricyanide and ferrocyanide take the place of A and B in the above development. This electrochemical reaction is quite rapid under properly chosen conditions. Furthermore, since it involves only electron transfer, the physical properties of the solution are almost entirely unaffected. The forced diffusion effects neglected here may be suppressed by the addition of an indifferent electrolyte in excess.[3,4]

EXAMPLE 19.4-3

Catalytic Oxidation of Carbon Monoxide

Figure 19.4-3 shows schematically how oxygen and carbon monoxide combine at a catalytic surface (palladium) to make carbon dioxide, according to the technologically important reaction[5]

$$O_2 + 2CO \rightarrow 2CO_2$$ (19.4-30)

---

[2] D. W. Hubbard and E. N. Lightfoot, *Ind. Eng. Chem. Fundam.*, **5**, 370–379 (1966).

[3] J. S. Newman, *Electrochemical Systems*, 2nd edition, Prentice-Hall, Englewood Cliffs, N.J. (1991), §1.10.

[4] J. R. Selman and C. W. Tobias, *Advances in Chemical Engineering*, **10**, Academic Press, New York, N.Y. (1978), pp. 212–318.

[5] B. C. Gates, *Catalytic Chemistry*, Wiley, New York (1992), pp. 356–362; C. N. Satterfield, *Heterogeneous Catalysis in Industrial Practice*, McGraw-Hill, New York, 2nd edition (1991), Chapter 8.

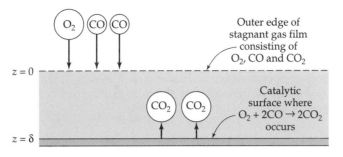

Fig. 19.4-3. Three-component system with a catalytic chemical reaction.

Outer edge of stagnant gas film consisting of O_2, CO and CO_2

$z = 0$

Catalytic surface where $O_2 + 2CO \rightarrow 2CO_2$ occurs

$z = \delta$

For this analysis, the reaction is assumed to occur instantaneously and irreversibly at the catalytic surface. The gas composition at the outer edge of the film (at $z = 0$) is presumed known, and the catalyst surface is at $z = \delta$. The temperature and pressure are assumed to be independent of position throughout the film. We label the chemical species by: $O_2 = 1$, $CO = 2$, $CO_2 = 3$.

SOLUTION

For steady-state, one-dimensional diffusion without homogeneous reactions, Eq. 19.1-10 gives

$$\frac{dN_{1z}}{dz} = 0; \qquad \frac{dN_{2z}}{dz} = 0; \qquad \frac{dN_{3z}}{dz} = 0 \qquad (19.4\text{-}31)$$

which tells us that all of the molar fluxes are constants across the film. From boundary conditions provided by the stoichiometry of the problem we further know that

$$N_{1z} = \tfrac{1}{2}N_{2z} = -\tfrac{1}{2}N_{3z} \qquad (19.4\text{-}32)$$

The Maxwell–Stefan equations of Eq. 17.9-1 then give:

$$\frac{dx_3}{dz} = -\frac{1}{c\mathscr{D}_{13}}(x_1 N_{3z} - x_3 N_{1z}) - \frac{1}{c\mathscr{D}_{23}}(x_2 N_{3z} - x_3 N_{2z})$$

$$= -\frac{N_{3z}}{c\mathscr{D}_{13}}(1 + \tfrac{1}{2}x_3) \qquad (19.4\text{-}33)$$

$$\frac{dx_1}{dz} = -\frac{1}{c\mathscr{D}_{12}}(x_2 N_{1z} - x_1 N_{2z}) - \frac{1}{c\mathscr{D}_{13}}(x_3 N_{1z} - x_1 N_{3z})$$

$$= \frac{N_{3z}}{2c\mathscr{D}_{12}}(1 - 3x_1 - x_3) + \frac{N_{3z}}{2c\mathscr{D}_{13}}(2x_1 + x_3) \qquad (19.4\text{-}34)$$

These equations have been simplified by using Eq. 19.4-32, and by using the fact that $\mathscr{D}_{23} \approx \mathscr{D}_{13}$ over a wide range of temperature. The latter may be seen by using Appendix E to show that $\sigma_{23} = 3.793\text{Å}$ and $\sigma_{13} = 3.714\text{Å}$, and that $\varepsilon_{23}/K = 145K$ and $\varepsilon_{13}/K = 146K$. Since only the mole fraction x_3 appears in Eq. 19.4-33, this equation may be integrated[6] at once to give

$$x_3 = -2 + (x_{30} + 2)\exp\left(-\frac{N_{3z}z}{2c\mathscr{D}_{13}}\right) \qquad (19.4\text{-}35)$$

Combination of the last two equations then gives, after integration

$$x_1 = 1 - \frac{1}{3}(x_{30} + 2)\exp\left(-\frac{N_{3z}z}{2c\mathscr{D}_{13}}\right) - \left(\frac{1}{3} - x_{10} - \frac{1}{3}x_{30}\right)\exp\left[-\left(\frac{3}{2}\frac{\mathscr{D}_{13}}{\mathscr{D}_{12}} - 1\right)\left(\frac{N_{3z}z}{c\mathscr{D}_{13}}\right)\right] \qquad (19.4\text{-}36)$$

---

[6] Three-component problems with two diffusivities equal have been discussed by H. L. Toor, *AIChE Journal*, **3**, 198–207 (1957).

From this equation and a similar one for x_2, we can get x_3 at $z = \delta$. Then from Eq. 19.4-35 we get

$$N_{3z} = -\frac{c\mathcal{D}_{13}}{\delta}\ln\left(\frac{x_{3\delta} + 2}{x_{30} + 2}\right) \tag{19.4-37}$$

which gives the rate of production of carbon dioxide at the catalytic surface. This result can then be substituted into Eqs. 19.4-35 and 36 and the three mole fractions can be calculated as functions of z.

EXAMPLE 19.4-4

Thermal Conductivity of a Polyatomic Gas

In §9.3 we pointed out that the thermal conductivities of polyatomic gases deviate from the formula for monatomic gases, because of the effects of the internal degrees of freedom in the complex molecules. When the Eucken formula for polyatomic gases (Eq. 9.3-15) is divided by the formula for monatomic gases (Eq. 9.3-14) and use is made of the ideal gas law, one can write the ratio of the polyatomic gas thermal conductivity to that of a monatomic gas as

$$\frac{k_{\text{poly}}}{k_{\text{mon}}} = \frac{3}{5} + \frac{4}{15}\frac{\tilde{C}_v}{R} \tag{19.4-38}$$

Derive a result of this form by modeling the polyatomic gas as an interacting gas mixture, in which the various "species" are the polyatomic gas molecules in the various rotational and vibrational states.

SOLUTION

The heat flux for a gas mixture is given in Eq. 19.3-3. All "species" will have the same thermal conductivity because they differ only in their internal quantum states. Therefore we expect each k_α to be k_{mon}. Similarly, the mass flux for each "species" should be given by Fick's law for a pure gas $\mathbf{j}_\alpha = -\rho\mathcal{D}_{\alpha\alpha}\nabla\omega_\alpha$, with all the $\mathcal{D}_{\alpha\alpha}$ having a common value $\mathcal{D}_{\text{mon}}$. Thus we get

$$\mathbf{q}_{\text{poly}} = -k_{\text{mon}}\nabla T - \sum_{\alpha=1}^{N}\frac{\tilde{H}_\alpha}{M_\alpha}\rho\mathcal{D}_{\alpha\alpha}\nabla\omega_\alpha$$

$$= -k_{\text{mon}}\nabla T - c\mathcal{D}_{\text{mon}}\sum_{\alpha=1}^{N}\tilde{H}_\alpha\nabla x_\alpha \tag{19.4-39}$$

since the molecular weights of all the "species" are the same.

If now it is postulated that the distribution over the various quantum states is in equilibrium with the local temperature, then $\nabla x_\alpha = (dx_\alpha/dT)\nabla T$. Then we can define the *effective thermal conductivity* of the mixture by

$$\mathbf{q}_{\text{poly}} = -k_{\text{mon}}\nabla T - c\mathcal{D}_{\text{mon}}\sum_{\alpha=1}^{N}\tilde{H}_\alpha(dx_\alpha/dT)\nabla T \equiv -k_{\text{poly}}\nabla T \tag{19.4-40}$$

and write

$$\frac{k_{\text{poly}}}{k_{\text{mon}}} = 1 + \left(\frac{c\mathcal{D}_{\text{mon}}}{k_{\text{mon}}}\right)\left(\frac{d}{dT}\sum_{\alpha=1}^{N}\tilde{H}_\alpha x_\alpha - \sum_{\alpha=1}^{N}x_\alpha\left(\frac{d\tilde{H}_\alpha}{dT}\right)\right)$$

$$= 1 + \left(\frac{c\mathcal{D}_{\text{mon}}}{k_{\text{mon}}}\right)(\tilde{C}_{p,\text{poly}} - \tilde{C}_{p,\text{mon}})$$

$$= 1 + \tfrac{4}{5}A[(\tilde{C}_{p,\text{poly}}/\tilde{C}_{p,\text{mon}}) - 1] \tag{19.4-41}$$

Here the temperature-dependent quantity

$$A = \tfrac{5}{4}(c\mathcal{D}_{\text{mon}}\tilde{C}_{p,\text{mon}}/k_{\text{mon}}) \tag{19.4-42}$$

can be calculated from the kinetic theory of gases at low density. It varies only very slowly with temperature, and a suitable mean value is 1.106. The quantity $\tilde{C}_{p,\text{poly}} = d\tilde{H}/dT$ is the heat capacity for a gas in which the equilibrium among the various quantum states is maintained during the change of temperature, whereas $\tilde{C}_{p,\text{mon}}$ is the heat capacity for a gas in which transitions between quantum states are not allowed, so that $\tilde{C}_{p,\text{mon}} = \frac{5}{2}R$. When the numerical value $A = 1.106$ is inserted in Eq. 19.4-41, we get finally

$$\frac{k_{\text{poly}}}{k_{\text{mon}}} = 0.115 + 0.354\left(\frac{\tilde{C}_{p,\text{poly}}}{R}\right) = 0.469 + 0.354\left(\frac{\tilde{C}_{V,\text{poly}}}{R}\right) \tag{19.4-43}$$

which is the formula recommended by Hirschfelder.[7] Although the predictions of Eq. 19.4-43 are not much better than those of the older Eucken formula, the above development does at least give some feel for the role of the internal degrees of freedom in heat conduction.[8,9]

§19.5   DIMENSIONAL ANALYSIS OF THE EQUATIONS OF CHANGE FOR NONREACTING BINARY MIXTURES

In this section we dimensionally analyze the equations of change summarized in §19.2, using special cases of the flux expressions of §19.3. The discussion parallels that of §11.5 and serves analogous purposes: to identify the controlling dimensionless parameters of representative mass transfer problems, and to provide an introduction to the mass transfer correlations of Chapter 22.

Once again we restrict the discussion primarily to systems of constant physical properties. The equation of continuity for the mixture then takes the familiar form

Continuity: $$(\nabla \cdot \mathbf{v}) = 0 \tag{19.5-1}$$

The equation of motion may be approximated in the manner of Boussinesq (see §11.3) by putting Eqs. 19.3-2 and 19-5.1 into Eq. 19.2-3, and replacing $-\nabla p + \bar{\rho}\mathbf{g}$ by $-\nabla\mathscr{P}$. For a constant-viscosity Newtonian fluid this gives

Motion: $$\bar{\rho}\frac{D\mathbf{v}}{Dt} = \mu\nabla^2\mathbf{v} - \nabla\mathscr{P} - \bar{\rho}\mathbf{g}\bar{\beta}(T - \overline{T}) - \bar{\rho}\mathbf{g}\bar{\zeta}(\omega_A - \overline{\omega}_A) \tag{19.5-2}$$

The energy equation, in the absence of chemical reactions, viscous dissipation, and external forces other than gravity, is obtained from Eq. (F) of Table 19.2-4, with Eq. 19.3-3. In using the latter we further neglect the diffusional transport of energy relative to the mass average velocity. For constant thermal conductivity this leads to

Energy: $$\frac{DT}{Dt} = \alpha\nabla^2 T \tag{19.5-3}$$

in which $\alpha = k/\rho\tilde{C}_p$ is the thermal diffusivity. For nonreacting binary mixtures with constant ρ and $\mathscr{D}_{AB}$, Eq. 19.1-14 takes the form

Continuity of A: $$\frac{D\omega_A}{Dt} = \mathscr{D}_{AB}\nabla^2\omega_A \tag{19.5-4}$$

For the assumptions that have been made, the analogy between Eqs. 19.5-3 and 4 is clear.

---

[7] J. O. Hirschfelder, *J. Chem. Phys.*, **26**, 274–281 (1957); see also D. Secrest and J. O. Hirschfelder, *Physics of Fluids*, **4**, 61–73 (1961) for further development of the theory, in which equilibrium among the various quantum states is not assumed.

[8] For a comparison of the two formulas with experimental data, see Reid, Prausnitz, and Poling, op. cit., p. 497. The Hirschfelder formula in Eq. 19.4-42 and the Eucken formula of Eq. 9.3-15 tend to bracket the observed conductivity values.

[9] J. H. Ferziger and H. G. Kaper, *Mathematical Theory of Transport Processes in Gases*, North Holland, Amsterdam (1977), §§11.2 and 3.

We now introduce the reference quantities l_0, v_0, and $\mathcal{P}_0$, used in §3.7 and §11.5, the reference temperatures T_0 and T_1 of §11.5, and the analogous reference mass fractions ω_{A0} and ω_{A1}. Then the dimensionless quantities we will use are

$$\check{x} = \frac{x}{l_0} \qquad \check{y} = \frac{y}{l_0} \qquad \check{z} = \frac{z}{l_0} \qquad \check{t} = \frac{v_0 t}{l_0} \tag{19.5-5}$$

$$\check{\mathbf{v}} = \frac{\mathbf{v}}{v_0} \qquad \check{\nabla} = l_0 \nabla \qquad \frac{D}{D\check{t}} = \left(\frac{l_0}{v_0}\right)\frac{D}{Dt} \qquad \check{\mathcal{P}} = \frac{\mathcal{P} - \mathcal{P}_0}{\bar{\rho} v_0^2} \tag{19.5-6}$$

$$\check{T} = \frac{T - T_0}{T_1 - T_0} \qquad \check{\omega}_A = \frac{\omega_A - \omega_{A0}}{\omega_{A1} - \omega_{A0}} \tag{19.5-7}$$

Here it is understood that $\mathbf{v}$ is the mass average velocity of the mixture. It should be recognized that for some problems other choices of dimensionless variables may be preferable.

In terms of the dimensionless variables listed above, the equations of change may be expressed as

Continuity:
$$(\check{\nabla} \cdot \check{\mathbf{v}}) = 0 \tag{19.5-8}$$

Motion:
$$\frac{D\check{\mathbf{v}}}{D\check{t}} = \frac{1}{\mathrm{Re}} \check{\nabla}^2 \check{\mathbf{v}} - \check{\nabla}\check{\mathcal{P}} - \frac{\mathrm{Gr}}{\mathrm{Re}^2}\frac{\mathbf{g}}{g}(\check{T} - \bar{\check{T}}) - \frac{\mathrm{Gr}_\omega}{\mathrm{Re}^2}\frac{\mathbf{g}}{g}(\check{\omega}_A - \bar{\check{\omega}}_A) \tag{19.5-9}$$

Energy:
$$\frac{D\check{T}}{D\check{t}} = \frac{1}{\mathrm{RePr}} \check{\nabla}^2 \check{T} \tag{19.5-10}$$

Continuity of A:
$$\frac{D\check{\omega}_A}{D\check{t}} = \frac{1}{\mathrm{ReSc}} \check{\nabla}^2 \check{\omega}_A \tag{19.5-11}$$

The Reynolds, Prandtl, and thermal Grashof numbers have been given in Table 11.5-1. The other two numbers are new:

$$\mathrm{Sc} = \left[\left[\frac{\mu}{\rho \mathcal{D}_{AB}}\right]\right] = \left[\left[\frac{\nu}{\mathcal{D}_{AB}}\right]\right] = \text{Schmidt number} \tag{19.5-12}$$

$$\mathrm{Gr}_\omega = \left[\left[\frac{g\bar{\zeta}(\omega_{A1} - \omega_{A0})l_0^3}{\nu^2}\right]\right] = \text{diffusional Grashof number} \tag{19.5-13}$$

The Schmidt number is the ratio of momentum diffusivity to mass diffusivity and represents the relative ease of molecular momentum and mass transfer. It is analogous to the Prandtl number, which represents the ratio of the momentum diffusivity to the thermal diffusivity. The diffusional Grashof number arises because of the buoyant force caused by the concentration inhomogeneities. The products RePr and ReSc in Eqs. 19.5-10 and 11 are known as Péclet numbers, Pé and Pé$_{AB}$, respectively.

The dimensional analysis of mass transfer problems parallels that for heat transfer problems. We illustrate the technique by three examples: (i) The strong similarity between Eqs. 19.5-10 and 11 permits the solution of many mass transfer problems by analogy with previously solved heat transfer problems; such an analogy is used in Example 19.5-1. (ii) Frequently the transfer of mass requires or releases energy, so that the heat and mass transfer must be considered simultaneously, as is illustrated in Example 19.5-2. (iii) Sometimes, as in many industrial mixing operations, diffusion plays a subordinate role in mass transfer and need not be given detailed consideration; this situation is illustrated in Example 19.5-3.

We shall see then that, just as for heat transfer, the use of dimensional analysis for the solution of practical mass transfer problems is an art. This technique is normally most useful when the effects of at least some of the many dimensionless ratios can be neglected. Estimation of the relative importance of pertinent dimensionless groups normally requires considerable experience.

EXAMPLE 19.5-1

Concentration Distribution about a Long Cylinder

We wish to predict the concentration distribution about a long isothermal cylinder of a volatile solid A, immersed in a gaseous stream of a species B, which is insoluble in solid A. The system is similar to that pictured in Fig. 11.5-1, except that here we consider the transfer of mass rather than heat. The vapor pressure of the solid is small compared to the total pressure in the gas, so that the mass transfer system is virtually isothermal.

Can the results of Example 11.5-1 be used to make the desired prediction?

SOLUTION

The results of Example 11.5-1 are applicable if it can be shown that suitably defined dimensionless concentration profiles in the mass transfer system are identical to the temperature profiles in the heat transfer system:

$$\breve{\omega}_A(\breve{x}, \breve{y}, \breve{z}) = \breve{T}(\breve{x}, \breve{y}, \breve{z}) \tag{19.5-14}$$

This equality will be realized if the differential equations and boundary conditions for the two systems can be put into identical form.

We therefore begin by choosing the same reference length, velocity, and pressure as in Example 11.5-1, and an analogous composition function: $\breve{\omega}_A = (\omega_A - \omega_{A0})/(\omega_{A\infty} - \omega_{A0})$. Here ω_{A0} is the mass fraction of A in the gas adjacent to the interface, and $\omega_{A\infty}$ is the value far from the cylinder. We also specify that $\overline{\omega}_A = \omega_{A0}$ so that $\breve{\overline{\omega}}_A = 0$. The equations of change needed here are then Eqs. 19.5-8, 9, and 11. Thus the differential equations here and in Problem 11.5-1 are analogous except for the viscous heating term in Eq. 11.5-3.

As for the boundary conditions, we have here:

B.C. 1:   as $\breve{x}^2 + \breve{y}^2 \rightarrow \infty$, $\breve{\mathbf{v}} \rightarrow \boldsymbol{\delta}_x$ $\breve{\omega}_A \rightarrow 1$ (19.5-15)

B.C. 2:   at $\breve{x}^2 + \breve{y}^2 = \frac{1}{4}$, $\breve{\mathbf{v}} = \dfrac{1}{\text{ReSc}} \dfrac{(\omega_{A0} - \omega_{A\infty})}{(1 - \omega_{A0})} \breve{\nabla}\breve{\omega}_A$ $\breve{\omega}_A = 0$ (19.5-16)

B.C. 3:   at $\breve{x}^2 + \breve{y}^2 = \infty$ and $\breve{y} = 0$, $\breve{\mathscr{P}} = 0$ (19.5-17)

The boundary condition on $\breve{\mathbf{v}}$, obtained with the help of Fick's first law, states that there is an interfacial radial velocity resulting from the sublimation of A.

If we compare the above description with that for heat transfer in Example 11.5-1, we see that there is no mass transfer counterpart of the viscous dissipation term in the energy equation and no heat transfer counterpart to the interfacial radial velocity component in the boundary condition of Eq. 19.5-16. The descriptions are otherwise analogous, however, with $\breve{\omega}_A$, Sc, and Gr_ω taking the places of $\breve{T}$, Pr, and Gr.

When the Brinkman number is sufficiently small, viscous dissipation will be unimportant, and that term in the energy equation can be neglected. Neglecting the Brinkman number term is appropriate, except for flows of very viscous fluids with large velocity gradients, or in hypersonic boundary layers (§10.4). Similarly, when $(1/\text{ReSc})[(\omega_{A0} - \omega_{A\infty})/(1 - \omega_{A0})]$ is very small, it may be set equal to zero without introducing appreciable error. If these limiting conditions are met, analogous behavior will be obtained for heat and mass transfer. More precisely, the dimensionless concentration $\breve{\omega}_A$ will have the same dependence on $\breve{x}, \breve{y}, \breve{z}, \breve{t}$, Re, Pr, and Gr_ω as the dimensionless temperature $\breve{T}$ will have on $\breve{x}, \breve{y}, \breve{z}, \breve{t}$, Re, Pr, and Gr. The concentration and temperature profiles will then be identical at a given Re whenever Sc = Pr and Gr_ω = Gr.

The thermal Grashof number can, at least in principle, be varied at will by changing $T_0 - T_\infty$. Hence it is likely that the desired Grashof numbers can be obtained. However, it can be seen from Tables 9.1-1 and 17.1-1 that Schmidt numbers for gases can vary over a considerably wide range than can the Prandtl numbers. Hence it may be difficult to obtain a satisfactory thermal model of the mass transfer process, except in a limited range of the Schmidt number.

Another possibly serious obstacle to achieving similar heat and mass transfer behavior is the possible nonuniformity of the surface temperature. The heat of sublimation must be obtained from the surrounding gas, and this in turn will cause the solid temperature to become lower than that of the gas. Hence it is necessary to consider both heat and mass transfer simultaneously. A very simple analysis of simultaneous heat and mass transfer is discussed in the next example.

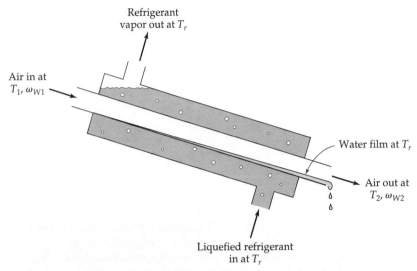

Fig. 19.5-1. Schematic representation of a dehumidifier. Air enters with inlet temperature T_1 and humidity ω_{W1} (the mass fraction of water vapor). It leaves with outlet temperature T_2 and humidity ω_{W2}. Because the heat transfer to the refrigerant is very effective, the temperature at the air–condensate interface is close to the refrigerant temperature T_r.

EXAMPLE 19.5-2

Fog Formation during Dehumidification

Wet air is being simultaneously cooled and dehumidified by passage through a metal tube chilled by the boiling of a liquid refrigerant. The tube surface is below the dew point of the entering air and therefore becomes covered with a water film. Heat transfer from the refrigerant to this condensate layer is sufficiently effective that the free water surface may be considered isothermal and at the boiling point of the refrigerant. This system is shown in Fig. 19.5-1.

We wish to determine the range of refrigerant temperatures that may be used without danger of fog formation. Fog is undesirable, because most of the tiny water droplets constituting the fog will pass through the cooling tube along with the air unless special collectors are provided. Fog can form if the wet air become supersaturated at any point in the system.

SOLUTION

Let species A be air and W be water. It is convenient here to choose the dimensionless variables

$$\check{T} = \frac{T - T_r}{T_1 - T_r}; \qquad \check{\omega}_W = \frac{\omega_W - \omega_{Wr}}{\omega_{W1} - \omega_{Wr}} \tag{19.5-18}$$

The subscripts are further defined in Fig. 19.5-1.

For the air–water system at moderate temperatures, the assumption of constant ρ and $\mathcal{D}_{AW}$ is reasonable, with air regarded as a single species. The heat capacities of water vapor and air are unequal, but the diffusional transport of energy is expected to be small. Hence Eqs. 19.5-9 to 11 provide a reasonably reliable description of the dehumidification process. The boundary conditions needed to integrate these equations include $\check{\omega}_W = \check{T} = 1$ at the tube inlet, $\check{\omega}_W = \check{T} = 0$ at the gas–liquid boundary, and no-slip and inlet conditions on the velocity $\check{\mathbf{v}}$.

We find then that the dimensionless profiles are related by

$$\check{\omega}_W(\check{x}, \check{y}, \check{z}, \text{Re}, \text{Gr}_\omega, \text{Gr}, \text{Sc}, \text{Pr}) = \check{T}(\check{x}, \check{y}, \check{z}, \text{Re}, \text{Gr}, \text{Gr}_\omega, \text{Pr}, \text{Sc}) \tag{19.5-19}$$

Thus $\check{\omega}_W$ is the same function of its arguments as $\check{T}$ is of its arguments *in the exact order given*. Since in general Gr_ω is not equal to Gr and Sc is not equal to Pr, the two profiles are not similar. This general result is too complex to be of much value.

However, for the air–water system, at moderate temperatures and near-atmospheric pressure, Sc is about 0.6 and Pr is about 0.71.

If we assume for the moment that Sc and Pr are equal, the dimensional analysis becomes much simpler. For this special situation, the energy and species continuity equations are iden-

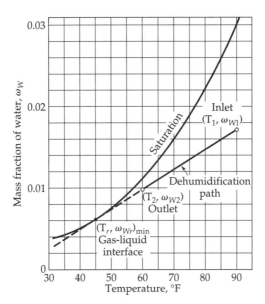

Fig. 19.5-2. A representative dehumidification path. The dehumidification path shown here corresponds to $T_{r,\text{min}}$, the lowest refrigerant temperature ensuring the absence of fog. The dehumidification path for this situation is a tangent to the saturation curve through the point (ω_{W1}, T_1), representing the given inlet-air conditions. Calculated dehumidification paths for lower refrigerant temperatures would cross the saturation curve. Saturation water vapor concentrations would then be exceeded, making fog formation possible.

tration and temperature profiles are then identical. It should be noted that equality of Gr_ω and Gr is not required. This is because the Grashof numbers affect the concentration and temperature profiles only by way of the velocity $\mathbf{v}$, which appears in both the continuity equation and the energy equation in the same way.

Therefore, with the assumption that $\text{Sc} = \text{Pr}$, we have

$$\check{\omega}_W = \check{T} \tag{19.5-20}$$

at each point in the system. This means, in turn, that *every* concentration-temperature pair in the tube lies on a straight line between (ω_{W1}, T_1) and (ω_{Wr}, T_r) on a psychrometric chart. This is shown graphically in Fig. 19.5-2 for a representative set of conditions. Note that (ω_{Wr}, T_r) must lie on the saturation curve, since equilibrium is very closely approximated.

It follows that there can be no fog formation if a straight line drawn between (ω_{W1}, T_1) and (ω_{Wr}, T_r) does not cross the saturation curve. Then the lowest refrigerant temperature that cannot produce fog is represented by the point of tangency of a straight line through (ω_{W1}, T_1) with the saturation curve.

It should be noted that *all* of the conditions along the line from the inlet (ω_{W1}, T_1) to (ω_{Wr}, T_r) will occur in the gas even though the bulk or cup-mixing conditions vary only from (ω_{W1}, T_1) to (ω_{W2}, T_2). Thus some fog can form even if saturation is not reached in the bulk of the flowing gas. For air entering at 90°F and 50% relative humidity, the minimum safe refrigerant temperature is about 45°F. It may also be seen from Fig. 19.5-2 that it is not necessary to bring all of the wet air to its dew point in order to dehumidify it. It is only necessary that the air be saturated at the cooling surface. The exit bulk conditions (ω_{W2}, T_2) can be anywhere along the dehumidification path between (ω_{W1}, T_1) and (ω_{Wr}, T_r), depending on the effectiveness of the apparatus used. Calculations based on the assumed equality of Sc and Pr have proven very useful for the air–water system.

In addition, it can be seen, by considering the physical significance of the Schmidt and Prandtl numbers, that the above-outlined calculation procedure is conservative. Since the Schmidt number is slightly smaller than the Prandtl number, dehumidification will proceed proportionally faster than cooling, and concentration–temperature pairs will lie slightly below the dehumidification path drawn in Fig. 19.5-2. In condensing organic vapors from air, the reverse situation often occurs. Then the Schmidt numbers tend to be higher than the Prandtl numbers, and cooling proceeds faster than condensation. Conditions then lie above the straight line of Fig. 19.5-2, and the danger of fog formation is increased.

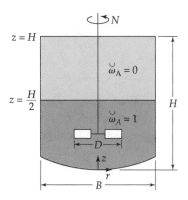

Fig. 19.5-3. Blending of miscible fluids. At zero time, the upper half of this tank is solute free, and the lower half contains a uniform distribution of solute at a dimensionless concentration of unity, and the fluid is motionless. The impeller is caused to turn at a constant rate of rotation N for all time greater than zero. Positions in the tank are given by the coordinates r, θ, z, with r measured radially from the impeller axis, and z upward from the bottom of the tank.

EXAMPLE 19.5-3

Blending of Miscible Fluids

Develop by dimensional analysis the general form of a correlation for the time required to blend two miscible fluids in an agitated tank. Consider a tank of the type described in Fig. 19.5-3, and assume that the two fluids and their mixtures have essentially the same physical properties.

SOLUTION

It will be assumed that the achievement of "equal degrees of blending" in any two mixing operations means obtaining the same dimensionless concentration profile in each. That is, the dimensionless solute concentration $\breve{\omega}_A$ is the same function of suitable dimensionless coordinates ($\breve{r}, \theta, \breve{z}$) of the two systems when the degrees of blending are equal. These concentration profiles will depend on suitably defined dimensionless groups appearing in the pertinent conservation equations and their boundary conditions, and on a dimensionless time.

In this problem we select the following definitions for the dimensionless variables:

$$\breve{r} = \frac{r}{D} \qquad \breve{z} = \frac{z}{D} \qquad \breve{\mathbf{v}} = \frac{\mathbf{v}}{ND} \qquad \breve{t} = Nt \qquad \breve{p} = \frac{p - p_0}{\rho N^2 D^2} \tag{19.5-21}$$

Here D is the impeller diameter, N is the rate of rotation of the impeller in revolutions per unit time, and p_0 is the prevailing atmospheric pressure. The dimensionless pressure $\breve{p}$ is used here rather than the quantity $\breve{\mathscr{P}}$ defined in §3.7; the formulation with $\breve{p}$ is simpler and gives equivalent results. Note that $\breve{t}$ is equal to the total number of turns of the impeller since the start of mixing.

The conservation equations describing this system are Eqs. 19.5-8, 9, and 11 with zero Grashof numbers. The dimensionless groups arising in these equations are Re, Fr, and Sc. The boundary conditions include the vanishing of $\mathbf{v}$ on the tank wall and of p on the free liquid surface. In addition we have to specify the initial conditions

C. 1: $\qquad$ at $\breve{t} \le 0$, $\qquad \breve{\omega}_A = 0 \qquad$ for $\dfrac{1}{2}\dfrac{H}{D} < \breve{z} < \dfrac{H}{D}$ $\qquad$ (19.5-22)

C. 2: $\qquad$ at $\breve{t} \le 0$, $\qquad \breve{\omega}_A = 1 \qquad$ for $0 < \breve{z} < \dfrac{1}{2}\dfrac{H}{D}$ $\qquad$ (19.5-23)

C. 3: $\qquad$ at $\breve{t} \le 0$, $\qquad \breve{\mathbf{v}} = 0 \qquad$ for $0 < \breve{z} < \dfrac{H}{D}$ and $0 < \breve{r} < \dfrac{1}{2}\dfrac{B}{D}$ $\qquad$ (19.5-24)

and the requirement of no slip on the impeller (see Eq. 3.7-34).

We find then that the concentration profiles are functions of Re, Sc, Fr, the dimensionless time $\breve{t}$, the tank geometry (via H/D and B/D), and the relative proportions of the two fluids. That is,

$$\breve{\omega}_A = f(\text{Re, Fr, Sc}, \breve{t}, \text{geometry, initial conditions}) \tag{19.5-25}$$

It is frequently possible to reduce the number of variables to be investigated.

It has been observed that, if the tank is properly baffled,[1] no vortices of importance occur; that is, the free liquid surface is effectively level. Under these circumstances, or in the absence of a free liquid surface, the Froude number does not appear in the system description, as we found in §3.7.

It is further found, in most operations on low-viscosity liquids, that the rate-limiting step is the creation of a finely divided dispersion of one fluid in the other. In such a dispersion, the diffusional processes take place over very small distances. As a result, molecular diffusion is not rate limiting, and the Schmidt number (Sc) has little importance. It is further found that the effect of the Reynolds number (Re) is negligible under most commonly encountered conditions. This is because most of the mixing takes place in the interior of the tank where viscous effects are small, rather than in the boundary layers adjacent to the tank and impeller surfaces, where they are large.[2]

For most impeller–tank combinations in common use, the Reynolds number (Re) is unimportant when its value is above about 10^4. This behavior has been substantiated by a number of investigators.[3]

We thus arrive, *after extensive experimentation*, at a surprisingly simple result. When all of the assumptions above are valid, the concentration profile depends only on $\check{t}$. Hence *the dimensionless time required to produce any desired degree of mixing is a constant* for a given system geometry. In other words, the total number of turns of the impeller during the mixing process determines the degree of blending, independently of Re, Fr, Sc, and tank size—provided, of course, that the tanks and impellers are geometrically similar.

For the same reasons, in a properly baffled tank, the dimensionless velocity distribution and the volumetric pumping efficiency of the impeller are nearly independent of the Froude number (Fr) and of the Reynolds number (Re), when Re > 10^4.

QUESTIONS FOR DISCUSSION

1. How do the various equations of change given in Chapters 3 and 11 have to be modified for reacting mixtures?
2. What modifications in the flux expressions given in Chapters 3 and 11 are needed to describe chemically reacting mixtures?
3. Under what conditions is $(\nabla \cdot \mathbf{v}) = 0$? $(\nabla \cdot \mathbf{v}^*) = 0$?
4. Equations 19.1-14 and 15 are physically equivalent. For what kinds of problems is there a preference for one form over the other?
5. Interpret physically each term in the equations in Table 19.2-3.
6. The thermal conductivity of a mixture is defined as the ratio of the heat flux to the negative of the temperature gradient when all the diffusional mass fluxes are zero. Interpret this statement in terms of Eq. 19.3-3.

---

[1] A common and effective baffling arrangement for vertical cylindrical tanks with axially mounted impellers is a set of four evenly spaced strips along the tank wall, with their flat surfaces in planes through the tank axis, extending from the top to the bottom of the tank and at least two-tenths of the distance to the tank center.

[2] The insensitivity of the required mixing time to the Reynolds number can be seen intuitively from the fact that the term $(1/\mathrm{Re})\check{\nabla}^2\check{\mathbf{v}}$ in Eq. 19.5-9 becomes small compared to the acceleration term $D\check{\mathbf{v}}/D\check{t}$ at large Re. Such intuitive arguments are dangerous, however, and the effect of Re is always important in the immediate neighborhood of solid surfaces. Here the amount of mixing taking place in the immediate neighborhood of solid surfaces is small and can be neglected.

The insensitivity of the required mixing time to the Schmidt number can be seen from the time-averaged equation of continuity in Chapter 21. At large Re, the turbulent mass flux is much greater than that due to molecular diffusion, except in the immediate neighborhood of the solid surfaces.

[3] E. A. Fox and V. E. Gex, *AIChE Journal*, **2**, 539–544 (1956); H. Kramers, G. M. Baars, and W. H. Knoll, *Chem. Eng Sci*, **2**, 35–42 (1955); J. G. van de Vusse, *Chem. Eng. Sci.*, **4**, 178–200, 209–220 (1955).

7. Discuss the similarities and differences between heat transfer and mass transfer.

8. Go through all the steps in converting Eq. 19.3-4 into Eq. 19.3-6. Why is the latter (approximate) result important?

9. Comment on the statement at the end of Example 19.4-1 that the rate of heat transfer is directly affected by simultaneous mass transfer, whereas the reverse is not true.

PROBLEMS   **19A.1.**   **Dehumidification of air** (Fig. 19.4-1). For the system of Example 19.4-1, let the vapor be H_2O and the stagnant gas be air. Assume the following conditions (which are representative in air conditioning): (i) at $y = \delta$, $T = 80°F$ and $x_{H_2O} = 0.018$; (ii) at $y = 0$, $T = 50°F$.

(a) For $p = 1$ atm, calculate the right side of Eq. 19.4-9.

(b) Compare the conductive and diffusive heat flux at $y = 0$. What is the physical significance of your answer?

Answer: (a) 1.004

19B.1.   **Steady-state evaporation** (Fig. 18.2-1). Rework the problem solved in §18.2, dealing with the evaporation of liquid A into gas B, starting from Eq. 19.1-17.

(a) First obtain an expression for $\mathbf{v}^*$, using Eq. (M) of Table 17.8-1, as well as Fick's law in the form of Eq. (D) of Table 17.8-2.

(b) Show that Eq. 19.1-17 then becomes the following nonlinear second-order differential equation:

$$\frac{d^2 x_A}{dz^2} + \frac{1}{1 - x_A}\left(\frac{dx_A}{dz}\right)^2 = 0 \tag{19B.1-1}$$

(c) Solve this equation to get the mole fraction profile given in Eq. 18.2-11.

19B.2.   **Gas absorption with chemical reaction** (Fig. 18.4-1). Rework the problem solved in §18.4, by starting with Eq. 19.1-16. What assumptions do you have to make in order to get Eq. 18.4-4?

19B.3.   **Concentration-dependent diffusivity.** A stationary liquid layer of B is bounded by planes $z = 0$ (a solid wall) and $z = b$ (a gas–liquid interface). At these planes the concentration of A is c_{A0} and c_{Ab} respectively. The diffusivity $\mathscr{D}_{AB}$ is a function of the concentration of A.

(a) Starting from Eq. 19.1-5 derive a differential equation for the steady-state concentration distribution.

(b) Show that the concentration distribution is given by

$$\frac{\displaystyle\int_{c_A}^{c_{A0}} \mathscr{D}_{AB} dc_A}{\displaystyle\int_{c_{Ab}}^{c_{A0}} \mathscr{D}_{AB} dc_A} = \frac{z}{b} \tag{19B.3-1}$$

(c) Show that the molar flux at the solid–liquid surface is

$$N_{Az}\big|_{z=0} = \frac{1}{b}\int_{c_{Ab}}^{c_{A0}} \mathscr{D}_{AB}(c_A) dc_A \tag{19B.3-2}$$

(d) Now assume that the diffusivity can be expressed as a Taylor series in the concentration

$$\mathscr{D}_{AB}(c_A) = \overline{\mathscr{D}}_{AB}[1 + \beta_1(c_A - \bar{c}_A) + \beta_2(c_A - \bar{c}_A)^2 + \cdots] \tag{19B.3-3}$$

in which $\bar{c}_A = \frac{1}{2}(c_{A0} + c_{Ab})$ and $\overline{\mathscr{D}}_{AB} = \mathscr{D}_{AB}(\bar{c}_A)$. Then, show that

$$N_{Az}\big|_{z=0} = \frac{\overline{\mathscr{D}}_{AB}}{b}(c_{A0} - c_{Ab})[1 + \frac{1}{12}\beta_2(c_{A0} - c_{Ab})^2 + \cdots] \tag{19B.3-4}$$

(e) How does this result simplify if the diffusivity is a linear function of the concentration?

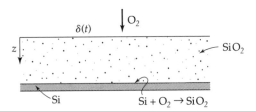

Fig. 19B.4. Oxidation of silicon.

19B.4. **Oxidation of silicon** (Fig. 19B.4).[1] A slab of silicon is exposed to gaseous oxygen (species A) at pressure p, producing a layer of silicon dioxide (species B). The layer extends from the surface $z = 0$, where the oxygen dissolves with concentration $c_{A0} = Kp$, to the surface at $z = \delta(t)$, where the oxygen and silicon undergo a first-order reaction with rate coefficient k_1''. The thickness $\delta(t)$ of the growing oxide layer is to be predicted. A quasi-steady-state method is useful here, inasmuch as the advancement of the reaction front is very slow.

(a) First solve the diffusion equation of Eq. 19.1-18, with the term $\partial c_A / \partial t$ neglected, and apply the boundary conditions to obtain

$$c_A = c_{A0} - (c_{A0} - c_{A\delta}) \frac{z}{\delta} \tag{19B.4-1}$$

in which the concentration $c_{A\delta}$ at the reaction plane is as yet unknown.

(b) Next use an unsteady-state molar O_2 balance on the region $0 < z < \delta(t)$ to obtain, with the aid of the Leibniz formula of §C.3,

$$c_{A\delta} \frac{d\delta}{dt} = -\mathcal{D}_{AB} \frac{dc_A}{dz} - k_1'' c_{A\delta} \tag{19B.4-2}$$

(c) Now write an unsteady-state molar balance on SiO_2 in the same region to obtain

$$+k_1'' c_{A\delta} = \frac{1}{\tilde{V}_B} \frac{d\delta}{dt} \tag{19B.4-3}$$

(d) In Eq. 19B.4-2, evaluate $d\delta/dt$ from Eq. 19B.4-3 and dc_A/dz from Eq. 19B.4-1. This will yield an equation for $c_{A\delta}$:

$$\frac{k_1'' \delta \tilde{V}_B}{\mathcal{D}_{AB}} c_{A\delta}^2 + \left(1 + \frac{k_1'' \delta}{\mathcal{D}_{AB}}\right) c_{A\delta} = c_{A0} \tag{19B.4-4}$$

Inserting numerical values into Eq. 19B.4-4 shows that the quadratic term can safely be neglected.[1]

(e) Combine Eqs. 19B.4-3 and 19B.4-4 (without the quadratic term) to get a differential equation for $\delta(t)$. Show that this leads to

$$\frac{\delta^2}{2\mathcal{D}_{AB}} + \frac{\delta}{k_1''} = \tilde{V}_B c_{A0} t \tag{19B.4-5}$$

which agrees with experimental data.[1] Interpret the result.

19B.5. **The Maxwell–Stefan equations for multicomponent gas mixtures.** In Eq. 17.9-1 the Maxwell-Stefan equations for the mass fluxes in a multicomponent gas system are given. Show that these equations simplify for a binary system to Fick's first law, as given in Eq. 17.1-5.

19B.6. **Diffusion and chemical reaction in a liquid.**

(a) A solid sphere of substance A is suspended in a liquid B in which it is slightly soluble, and with which A undergoes a first-order chemical reaction with rate constant k_1'''. At steady

---

[1] R. Ghez, *A Primer of Diffusion Problems*, Wiley-Interscience, New York (1988), pp. 46–55; this book discusses a number of problems that arise in the microelectronics field.

state the diffusion is exactly balanced by the chemical reaction. Show that the concentration profile is

$$\frac{c_A}{c_{A0}} = \frac{R}{r}\frac{e^{-br/R}}{e^{-b}} \tag{19B.6-1}$$

in which R is the radius of the sphere, c_{A0} is the molar solubility of A in B, and $b^2 = k_1'''R^2/\mathcal{D}_{AB}$.

(b) Show by quasi-steady-state arguments how to calculate the gradual decrease in diameter of the sphere as A dissolves and reacts. Show that the radius of the sphere is given by

$$\sqrt{\frac{k_1'''}{\mathcal{D}_{AB}}}(R - R_0) - \ln\frac{1 + \sqrt{k_1'''/\mathcal{D}_{AB}}R}{1 + \sqrt{k_1'''/\mathcal{D}_{AB}}R_0} = -\frac{k_1'''c_{A0}M_A}{\rho_{sph}}(t - t_0) \tag{19B.6-2}$$

in which R_0 is the sphere radius at time t_0, and ρ_{sph} is the density of the sphere.

19B.7.   Various forms of the species continuity equation.

(a) In this chapter the species equation of continuity is given in three different forms: Eq. 19.1-7, Eq. (A) of Table 19.2-1, and Eq. (B) in Table 19.2-3. Show that these three equations are equivalent.

(b) Show how to get Eq. 19.1-15 from Eq. 19.1-11.

19C.1.   Alternate form of the binary diffusion equation. In the absence of chemical reactions, Eq. 19.1-17 can be written in terms of $\mathbf{v}$ rather than $\mathbf{v}^*$ by using a different measure of concentration—namely, the logarithm of the mean molecular weight:[2]

$$\frac{\partial}{\partial t}\ln M + (\mathbf{v} \cdot \nabla \ln M) = \mathcal{D}_{AB}\nabla^2 \ln M \tag{19C.1-1}$$

in which $M = x_A M_A + x_B M_B$. (*Caution:* Solution is lengthy.)

Equation 19C.1-1 is difficult to solve even for the stagnant gas film of §18.2, because of the variable mass density ρ that appears in the continuity equation (Eq. A of Table 19.2-3).

19D.1.   Derivation of the equation of continuity. In §19.1 the species equation of continuity is derived by making a mass balance on a small rectangular volume $\Delta x\, \Delta y\, \Delta z$ fixed in space.

(a) Repeat the derivation for an arbitrarily shaped volume element V with a sufficiently smooth fixed boundary S. Show that the species mass balance can be written as

$$\frac{d}{dt}\int_V \rho_A dV = -\int_S (\mathbf{n} \cdot \mathbf{n}_A)dS + \int_V r_A dV \tag{19D.1-1}$$

Use the Gauss divergence theorem to convert the surface integral to a volume integral, and then obtain Eq. 19.1-6.

(b) Repeat the derivation using a region of fluid contained within a surface, each point of which is moving with local mass average velocity.

19D.2.   Derivation of the equation of change for temperature for a multicomponent system. Derive Eq. (F) in Table 19.2-4 from Eq. (E). We suggest the following sequence of steps:

(a) Since the enthalpy is an extensive thermodynamic property, we can write

$$H(m_1, m_2, m_3, \ldots, m_N) = m\hat{H}(\omega_1, \omega_2, \omega_3, \ldots, \omega_{N-1}) \tag{19D.2-1}$$

in which the m_α are the masses of the various species, m is the sum of the m_α, and the $\omega_\alpha = m_\alpha/m$ are the corresponding mass fractions. Both H and $\hat{H}$ are understood to be functions of T and p as well as of composition. Use the chain rule of partial differentiation to show that

$$(\alpha \neq N)\qquad \left(\frac{\partial H}{\partial m_\alpha}\right)_{m_\gamma} = \sum_{\beta=1}^{N-1}\left(\frac{\partial \hat{H}}{\partial \omega_\beta}\right)_{\omega_\gamma}\left(\delta_{\alpha\beta} - \frac{m_\beta}{m}\right) + \hat{H} \tag{19D.2-2}$$

---

[2] C. H. Bedingfield, Jr., and T. B. Drew, *Ind. Eng. Chem.*, **42**, 1164–1173 (1950).

$$(\alpha = N) \qquad \left(\frac{\partial H}{\partial m_N}\right)_{m_\gamma} = \sum_{\beta=1}^{N-1}\left(\frac{\partial \hat{H}}{\partial \omega_\beta}\right)_{\omega_\gamma}\left(-\frac{m_\beta}{m}\right) + \hat{H} \tag{19D.2-3}$$

Subtraction then gives for $\alpha \neq N$

$$\left(\frac{\partial H}{\partial m_\alpha}\right)_{m_\gamma} - \left(\frac{\partial H}{\partial m_N}\right)_{m_\gamma} = \left(\frac{\partial \hat{H}}{\partial \omega_\alpha}\right)_{\omega_\gamma} \tag{19D.2-4}$$

The subscript ω_γ means "holding all other mass fractions constant."

(b) The left side of Eq. (E) can be expanded by regarding the enthalpy per unit mass to be a function of p, T, and the first $(N-1)$ mass fractions:

$$\rho\frac{D\hat{H}}{Dt} = \rho\left(\frac{\partial \hat{H}}{\partial p}\right)_{T,\omega_\gamma}\frac{Dp}{Dt} + \rho\left(\frac{\partial \hat{H}}{\partial T}\right)_{p,\omega_\gamma}\frac{DT}{Dt} + \rho\sum_{\alpha=1}^{N-1}\left(\frac{\partial \hat{H}}{\partial \omega_\alpha}\right)_{p,T,\omega_\gamma}\frac{D\omega_\alpha}{Dt} \tag{19D.2-5}$$

Next, verify that the coefficients of the substantial derivatives can be identified as

$$\rho\left(\frac{\partial \hat{H}}{\partial p}\right)_{T,\omega_\gamma} = 1 - \left(\frac{\partial \ln \hat{V}}{\partial \ln T}\right)_{p,\omega_\gamma} \tag{19D.2-6}$$

$$\rho\left(\frac{\partial \hat{H}}{\partial T}\right)_{p,\omega_\gamma} = \rho\hat{C}_p \tag{19D.2-7}$$

The coefficient of $\rho(D\omega_\alpha/Dt)$ has already been given in Eq. 19D.2-4.

(c) Substitute the coefficients into Eq. 19D.2-5, and then use Eq. 19.1-14 to eliminate $\rho(D\omega_\alpha/Dt)$, and verify that $(\partial H/\partial m_\alpha)_{p,T,m_\gamma}$ is the same as $(\overline{H}_\alpha/M_\alpha)$. The summation on α, which goes from 1 to $N-1$, now has to be appropriately rewritten as a summation from 0 to N, by using Eq. (K) of Table 17.8-1 and the fact that $\Sigma_\alpha r_\alpha = 0$.

(d) Then combine the results of (a), (b), and (c) with Eq. (E) to get Eq. (F).

19D.3. Gas separation by atmolysis or "sweep diffusion" (Fig. 19D.3). When two gases A and B are forced to diffuse through a third gas C, there is a tendency of A and B to separate because of the difference in their diffusion rates. This phenomenon was first studied by Hertz,[3] and later by Maier.[4] Benedict and Boas[5] studied the economics of the process particularly with regard to isotope separation. Keyes and Pigford[6] contributed further to both theory and experiment.

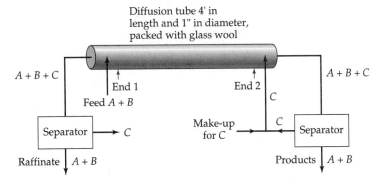

Fig. 19D.3. The Keyes–Pigford experiment for studying atmolysis.

[3] G. Hertz, *Zeits. f. Phys.*, **91**, 810–815 (1934).
[4] G. G. Maier, *Mechanical Concentration of Gases*, U.S. Bureau of Mines Bulletin 431 (1940).
[5] M. Benedict and A. Boas, *Chem. Eng. Prog.*, **47**, 51–62, 111–122 (1951).
[6] J. J. Keys, Jr., and R. L. Pigford, *Chem. Eng. Sci.*, **6**, 215–226 (1957).

In their experimental arrangement, C was a condensable vapor, which could be separated from A and B by lowering the temperature so that C would be liquefied.

We want to study the details of the three-component diffusion taking place in the diffusion tube of length L, when the apparatus is operated at steady state. Obtain an expression relating the concentrations x_{A1} and x_{B1} at the feed end of the tube to the concentrations x_{A2} and x_{B2} at the product end. This expression will contain the molar fluxes of the three species, which are controlled by the rates of addition of materials in the two entering streams.

Use the following notation for dimensionless quantities: $\zeta = z/L$ for the distance down the tube from the feed entrance; $r_A = \mathcal{D}_{AB}/\mathcal{D}_{AC}$ and $r_B = \mathcal{D}_{AB}/\mathcal{D}_{BC}$ for the diffusivity ratios; and $\nu_\alpha = N_{\alpha z}L/c\mathcal{D}_{AB}$ for the molar fluxes (with $\alpha = A, B, C$).

(a) Shows that, in terms of these dimensionless quantities, the Maxwell–Stefan equations for the diffusion are

$$\frac{dx_A}{d\zeta} = Y_{AA}x_A + Y_{AB}x_B + Y_A \tag{19D.3-1}$$

$$\frac{dx_B}{d\zeta} = Y_{BA}x_A + Y_{BB}x_B + Y_B \tag{19D.3-2}$$

where $Y_{AA} = \nu_B + r_A(\nu_A + \nu_C)$, $Y_{AB} = \nu_A(r_A - 1)$, and $Y_A = -r_A\nu_A$, and the remaining quantities are obtained by interchanging A and B.

(b) By using Laplace transforms, solve Eqs. 19D.3-1 and 2 to get the concentration profiles for A and B in the tube.

(c) Show that the terminal concentrations are interrelated thus,

$$x_{A2} = \frac{X_A(x_{A1}, x_{B1}; 0)}{p_+p_-} + \frac{X_A(x_{A1}, x_{B1}; p_+)\exp p_+}{p_+(p_+ - p_-)} + \frac{X_A(x_{A1}, x_{B1}; p_-)\exp p_-}{p_-(p_- - p_+)} \tag{19D.3-3}$$

in which

$$p_\pm = \tfrac{1}{2}[(Y_{AA} + Y_{BB}) \pm \sqrt{(Y_{AA} + Y_{BB})^2 + 4Y_{AB}Y_{BA}}] \tag{19D.3-4}$$

$$X_A(x_{A1}, x_{B1}, p) = p^2x_{A1} + p(Y_A - x_{A1}Y_{BB} + x_{B1}Y_{AB}) + (Y_{AB}Y_B - Y_{BB}Y_A) \tag{19D.3-5}$$

A similar expression may be obtained for x_{B2}. Keyes and Pigford[6] give further results for special cases.

19D.4. **Steady-state diffusion from a rotating disk.**[7] A large disk is rotating with an angular velocity Ω in an infinite expanse of liquid B. The surface is coated with a material A that is slightly soluble in B. Find the rate at which A dissolves in B. (The solution to this problem can be applied to a disk of finite radius R with negligible error.)

The fluid dynamics of this problem was developed by von Kármán[8] and later corrected by Cochran.[9] It was found that the velocity components can be expressed, except near the edge, as

$$v_r = \Omega r F(\zeta); \qquad v_\theta = \Omega r G(\zeta); \qquad v_z = \sqrt{\Omega\nu}H(\zeta) \tag{19D.4-1}$$

in which $\zeta = z\sqrt{\Omega/\nu}$. The functions F, G, and H have the following expansions,[8]

$$F = a\zeta - \tfrac{1}{2}\zeta^2 - \tfrac{1}{3}b\zeta^3 - \tfrac{1}{12}b^2\zeta^4 - \cdots \tag{19D.4-2}$$

$$G = 1 + b\zeta + \tfrac{1}{3}a\zeta^3 + \tfrac{1}{12}(ab - 1)\zeta^4 - \cdots \tag{19D.4-3}$$

$$H = -a\zeta^2 + \tfrac{1}{3}\zeta^3 + \tfrac{1}{6}\zeta^4 + \cdots \tag{19D.4-4}$$

in which $a = 0.510$ and $b = -0.616$. It is further known that, in the limit as $\zeta \to \infty$, $H \to -0.886$, and F, G, F', and G' all approach zero. Also it is known that the boundary layer thickness is proportional to $\sqrt{\nu/\Omega}$, except near the edge of the disk.

---

[7] V. G. Levich, *Physicochemical Hydrodynamics*, Prentice-Hall, Englewood Cliffs, N.J. (1962), §11.

[8] T. von Kármán, *Zeits. f. angew. Math. u. Mech.*, **1**, 244–247 (1921).

[9] W. G. Cochran, *Proc. Camb. Phil. Soc.*, **30**, 365–375 (1934).

The diffusion equation of Eq. 19.1-16 with the known velocity components is to be solved under the boundary conditions that: $\rho_A = \rho_{A0}$ at $z = 0$; $\rho_A = 0$ at $z = \infty$; and $\partial \rho_A / \partial r = 0$ at $r = 0, \infty$. Since there can be but one solution to this linear problem, it may be seen that a solution of the form $\rho_A(z)$ can be found that satisfies the differential equation and all the boundary conditions. Thus, the solution for ρ_A does not depend on the radial coordinate in the region considered.

(a) Show that at steady-state Eq. 19.1-16 gives

$$H(\zeta)\frac{d\rho_A}{d\zeta} = \frac{1}{Sc}\frac{d^2\rho_A}{d\zeta^2} \tag{19D.4-5}$$

(b) Solve Eq. 19D.4-5 to get, for large Schmidt number,

$$\frac{\rho_A}{\rho_{A0}} = 1 - \frac{(\tfrac{1}{3}a Sc)^{1/3}}{\Gamma(\tfrac{4}{3})}\int_0^\zeta \exp(-\tfrac{1}{3}a Sc\,\bar{\zeta}^3)d\bar{\zeta} \tag{19D.4-6}$$

(c) Show that the mass flux at the surface of the disk is[7]

$$j_{Az}|_{z=0} = 0.620\frac{\rho_{A0}\mathscr{D}_{AB}^{2/3}\Omega^{1/2}}{\nu^{1/6}} \tag{19D.4-7}$$

for large Schmidt number. Clearly, if desired, one could use higher terms in the series expansion for H and extend the Schmidt-number range.[10] This system has been used for studying the removal of solid behenic acid from stainless-steel surfaces.[11]

---

[10] D. Schuhmann, *Physicochemical Hydrodynamics* (*V. G. Levich Fextschrift*), Vol. 1 (D. B. Spalding ed.), Advance Publications Ltd., London (1977), pp. 445–459; see also K.-T. Liu and W. E. Stewart, *Intl. Jnl. Heat and Mass Trf.*, **15**, 187–189 (1972).

[11] C. S. Grant, A. T. Perka, W. D. Thomas, and R. Caton, *AIChE Journal*, **42**, 1465–1476 (1996).

Chapter 20

Concentration Distributions with More Than One Independent Variable

Most of the diffusion problems discussed in the preceding two chapters led to ordinary differential equations for the concentration profiles. In this chapter we use the general equations of Chapter 19 to set up and solve some diffusion problems that lead to partial differential equations.

A large number of diffusion problems can be solved by simply looking up the solutions to the analogous problems in heat conduction. When the differential equations and the boundary and initial conditions for the diffusion process are of exactly the same form as those for the heat conduction process, then the heat conduction solution may be taken over with appropriate changes in notation. In Table 20.0-1 the three main heat transport equations used in Chapter 12 are shown along with their mass transport analogs. Many solutions to the nonflow equations may be found in the monographs of Carslaw and Jaeger[1] and of Crank.[2]

Because the diffusion problems described by the equations in Table 20.0-1 are analogous to the problems of Chapter 12, we do not discuss them extensively here. Instead, we focus primarily on problems involving diffusion with chemical reactions, diffusion with a moving interface, and diffusion with rapid mass transfer.

In §20.1 we discuss a variety of time-dependent diffusion problems. In §20.2 we present some steady-state boundary layer problems involving binary mixtures. This is followed by two boundary layer analyses for more complicated systems: the diffusion in steady flow around arbitrary objects in §20.3, and the diffusion in flows with complex interfacial motion in §20.4. Finally, in §20.5 we explore an asymptotic solution to the "Taylor dispersion" problem.

---

[1] H. S. Carslaw and J. C. Jaeger, *Conduction of Heat in Solids*, 2nd edition, Oxford University Press (1959).

[2] J. Crank, *The Mathematics of Diffusion*, 2nd edition, Clarendon Press, Oxford (1975).

Table 20.0-1   Analogies Between Special Forms of the Heat Conduction and Diffusion Equations

Process		Unsteady-state nonflow	Steady-state flow	Steady-state nonflow
	Solution Given in	§12.1—Exact solutions	§12.2—Exact solutions §12.4—Boundary layer solutions	§12.3—Exact solutions in two dimensions by analytic functions
Heat conduction	Equations	$\dfrac{\partial T}{\partial t} = \alpha \nabla^2 T$	$(\mathbf{v} \cdot \nabla T) = \alpha \nabla^2 T$	$\nabla^2 T = 0$
	Applications	Heat conduction in solids	Heat conduction in laminar incompressible flow	Steady heat conduction solids
	Assumptions	1. k = constant 2. $\mathbf{v} = 0$	1. k, ρ = constants 2. No viscous dissipation 3. Steady state	1. k = constant 2. $\mathbf{v} = 0$ 3. Steady state
Diffusion	Equations	$\dfrac{\partial c_A}{\partial t} = \mathscr{D}_{AB} \nabla^2 c_A$	$(\mathbf{v} \cdot \nabla c_A) = \mathscr{D}_{AB} \nabla^2 c_A$	$\nabla^2 c_A = 0$
	Applications	Diffusion of traces of A through B	Diffusion in laminar flow (dilute solutions of A in B)	Steady diffusion in solids
	Assumptions	1. $\mathscr{D}_{AB}, \rho$ = constants 2. $\mathbf{v} = 0$ 3. No chemical reactions OR Equimolar counter-diffusion in low density gases 1. $\mathscr{D}_{AB}, c$ = constants 2. $\mathbf{v}^* = 0$ 3. No chemical reactions	1. $\mathscr{D}_{AB}, \rho$ = constants 2. Steady state 3. No chemical reactions	1. $\mathscr{D}_{AB}, \rho$ = constants 2. Steady state 3. No chemical reactions 4. $\mathbf{v} = 0$

§20.1 TIME-DEPENDENT DIFFUSION

In this section we give four examples of time-dependent diffusion. The first deals with evaporation of a volatile liquid and illustrates the deviations from Fick's second law that arise at high mass-transfer rates. The second and third examples deal with unsteady-state diffusion with chemical reactions. In the last example we examine the role of interfacial-area changes in diffusion. The method of combination of variables is used in Examples 20.1-1, 2, and 4, and Laplace transforms are used in Example 20.1-3.

EXAMPLE 20.1-1 *Unsteady-State Evaporation of a Liquid (the "Arnold Problem")*	We wish to predict the rate at which a volatile liquid A evaporates into pure B in a tube of infinite length. The liquid level is maintained at $z = 0$ at all times. The temperature and pressure are assumed constant, and the vapors of A and B form an ideal gas mixture. Hence the molar density c is constant throughout the gas phase, and $\mathscr{D}_{AB}$ may be considered to be constant. It is further assumed that species B is insoluble in liquid A, and that the molar average velocity in the gas phase does not depend on the radial coordinate.

For this system the equation of continuity for the mixture, given in Eq. 19.1-12, becomes

$$\frac{\partial v_z^*}{\partial z} = 0 \tag{20.1-1}$$

in which v_z^* is the z-component of the molar average velocity. Integration with respect to z gives

$$v_z^* = v_{z0}^*(t) \tag{20.1-2}$$

Here and elsewhere in this problem, the subscript "0" indicates a quantity evaluated at $z = 0$. According to Eq. (M) of Table 17.8-1, this velocity can be written in terms of the molar fluxes of A and B as

$$v_z^* = \frac{N_{Az0} + N_{Bz0}}{c} \tag{20.1-3}$$

However, N_{Bz0} is zero because of the insolubility of species B in liquid A. Then use of Eq. (D) of Table 17.8-2 gives finally

$$v_z^* = -\frac{\mathcal{D}_{AB}}{1 - x_{A0}} \frac{\partial x_A}{\partial z}\bigg|_{z=0} \tag{20.1-4}$$

in which x_{A0} is the interfacial gas-phase concentration, evaluated here on the assumption of interfacial equilibrium. For an ideal gas mixture this is just the vapor pressure of pure A divided by the total pressure.

The equation of continuity of Eq. 19.1-17 then becomes

$$\frac{\partial x_A}{\partial t} - \left(\frac{\mathcal{D}_{AB}}{1 - x_{A0}} \frac{\partial x_A}{\partial z}\bigg|_{z=0}\right) \frac{\partial x_A}{\partial z} = \mathcal{D}_{AB} \frac{\partial^2 x_A}{\partial z^2} \tag{20.1-5}$$

This is to be solved with the initial and boundary conditions:

I.C.:   at $t = 0$, $x_A = 0$ (20.1-6)

B.C. 1:   at $z = 0$, $x_A = x_{A0}$ (20.1-7)

B.C. 2:   at $z = \infty$, $x_A = 0$ (20.1-8)

We can try the same kind of combination of variables used in Example 4.1-1; namely, $X = x_A/x_{A0}$ and $Z = z/\sqrt{4\mathcal{D}_{AB}t}$. However, since Eq. 20.1-5 contains the parameter x_{A0}, we can anticipate that X will depend not only on Z but also parametrically on x_{A0}.

In terms of these dimensionless variables, Eq. 20.1-5 can be written as

$$\frac{d^2X}{dZ^2} + 2(Z - \varphi) \frac{dX}{dZ} = 0 \tag{20.1-9}$$

Here the quantity

$$\varphi(x_{A0}) = -\frac{1}{2} \frac{x_{A0}}{1 - x_{A0}} \frac{dX}{dZ}\bigg|_{Z=0} \tag{20.1-10}$$

is a dimensionless molar average velocity, $\varphi = v_z^* \sqrt{t/\mathcal{D}_{AB}}$, as can be seen by comparing Eqs. 20.1-10 and 20.1-4. The initial and boundary conditions in Eqs. 20.1-6 to 8 now become

B.C. 1:   at $Z = 0$, $X = 1$ (20.1-11)

B.C. 2 and I.C.:   at $Z = \infty$, $X = 0$ (20.1-12)

Equation 20.1-9 can be attacked by first letting $dX/dZ = Y$. This gives a first-order differential equation for Y that can be solved to obtain

$$Y = C_1 \exp[-(Z - \varphi)^2] \equiv \frac{dX}{dZ} \tag{20.1-13}$$

This gives on integration

$$X = C_1 \int_0^Z \exp[-(\overline{Z} - \varphi)^2]d\overline{Z} + C_2 \tag{20.1-14}$$

Combining this result with Eqs. 20.1-11 and 12, we get

$$X(Z) = 1 - \frac{\int_0^Z \exp[-(\overline{Z} - \varphi)^2]d\overline{Z}}{\int_0^\infty \exp[-(\overline{Z} - \varphi)^2]d\overline{Z}} = 1 - \frac{\int_{-\varphi}^{Z-\varphi} \exp(-W^2)dW}{\int_{-\varphi}^\infty \exp(-W^2)dW} \tag{20.1-15}$$

Then we use the definition of the error function and some of the properties of this function, in particular, $-\mathrm{erf}(-\varphi) = \mathrm{erf}\,\varphi$ and $\mathrm{erf}\,\infty = 1$ (see §C.6). This leads to the final expression for the mole fraction distribution:[1]

$$X(Z) = 1 - \frac{\mathrm{erf}(Z - \varphi) + \mathrm{erf}\,\varphi}{\mathrm{erf}\,\infty + \mathrm{erf}\,\varphi} = \frac{1 - \mathrm{erf}(Z - \varphi)}{1 + \mathrm{erf}\,\varphi} \tag{20.1-16}$$

To get the function $\varphi(x_{A0})$, this mole fraction distribution has to be substituted into Eq. 20.1-10. This gives

$$\varphi = \frac{1}{\sqrt{\pi}} \frac{x_{A0}}{1 - x_{A0}} \frac{\exp(-\varphi^2)}{1 + \mathrm{erf}\,\varphi} \tag{20.1-17}$$

Rather than solving this to get φ as a function of x_{A0}, it is easier to evaluate x_{A0} as a function of φ:

$$x_{A0} = \frac{1}{1 + [\sqrt{\pi}(1 + \mathrm{erf}\,\varphi)\varphi \exp \varphi^2]^{-1}} \tag{20.1-18}$$

A small table of $\varphi(x_{A0})$ is given in Table 20.1-1, and the concentration profiles are shown in Fig. 20.1-1.

We can now calculate the rate of production of vapor from a surface of area S. If V_A is the volume of A produced by evaporation up to time t, then

$$\frac{dV_A}{dt} = \frac{N_{Az0}S}{c} = S\varphi \sqrt{\frac{\mathcal{D}_{AB}}{t}} \tag{20.1-19}$$

Table 20.1-1   Table[1] of $\varphi(x_{A0})$ and $\psi(x_{A0})$

x_{A0}	φ	$\psi = \varphi\sqrt{\pi}/x_{A0}$
0.00	0.0000	1.000
0.25	0.1562	1.108
0.50	0.3578	1.268
0.75	0.6618	1.564
1.00	∞	∞

---

[1] J. H. Arnold, *Trans. AIChE*, **40**, 361–378 (1944). **Jerome Howard Arnold** (1907–1974) taught at MIT, the University of Minnesota, the University of North Dakota, and the University of Iowa; he worked for Standard Oil of California (1944–1948) and was the director of the Contra Costa Transit District (1956–1960).

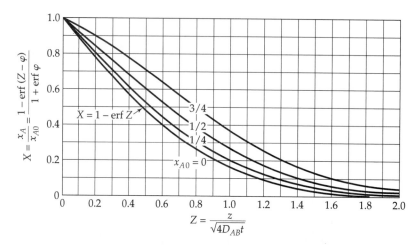

Fig. 20.1-1. Concentration profiles in time-dependent evaporation, showing that the deviation from Fick's law increases with the volatility of the evaporating liquid.

Integration with respect to t then gives

$$V_A = S\varphi\sqrt{4\mathcal{D}_{AB}t} \tag{20.1-20}$$

This relation can be used to calculate the diffusivity from the rate of evaporation (see Problem 20A.1).

We can now assess the importance of including the convective transport of species A in the tube. If Fick's second law (Eq. 19.1-18) had been used to determine X, we would have obtained

$$V_A^{\text{Fick}} = Sx_{A0}\sqrt{\frac{4\mathcal{D}_{AB}t}{\pi}} \tag{20.1-21}$$

Thus we can rewrite Eq. 20.1-20 as

$$V_A = Sx_{A0}\sqrt{\frac{4\mathcal{D}_{AB}t}{\pi}} \cdot \psi \tag{20.1-22}$$

The factor $\psi = \varphi\sqrt{\pi}/x_{A0}$, tabulated in Table 20.1-1, is a correction for the deviation from the Fick's second law results caused by the nonzero molar average velocity. We see that the deviation becomes especially significant when x_{A0} is large—that is, for liquids with large volatility.

In the preceding analysis it is assumed that the system is isothermal. Actually, the interface will be cooled by the evaporation, particularly at large values of x_{A0}. This effect can be minimized by using a small-diameter tube made of a good thermal conductor. For application to other mass transfer systems, however, the analysis given here needs to be extended by including the solution to the energy equation, so that the interfacial temperature and compositions can be calculated (see Problem 20B.2).

This analysis can be extended[2] to include interphase transfer of both species, with any time-independent flux ratio N_{Az0}/N_{Bz0} and any initial gas composition $x_{A\infty}$. A simple example of such a system is the diffusion-controlled reaction $2A \rightarrow B$ on a catalytic solid at $z = 0$, with

---

[2] W. E. Stewart, J. B. Angelo, and E. N. Lightfoot, *AIChE Journal*, **16**, 771–786 (1970), have generalized this example and the following one to forced convection in three-dimensional flows, including turbulent systems.

the heat of reaction removed through the solid. The concentration profile is a generalization of that in Eq. 20.1-16:

$$\Pi \equiv \frac{x_A - x_{A0}}{x_{A\infty} - x_{A0}} = \frac{\text{erf}(Z - \varphi) + \text{erf}\,\varphi}{1 + \text{erf}\,\varphi} \tag{20.1-23}$$

The dimensionless flux φ now depends on x_{A0}, $x_{A\infty}$, and the ratio N_{Bz0}/N_{Az0}:

$$\varphi = \frac{1}{2} \frac{(x_{A0} - x_{A\infty})(N_{Az0} + N_{Bz0})}{N_{Az0} - x_{A0}(N_{Az0} + N_{Bz0})} \frac{d\Pi}{dZ}\bigg|_{Z=0} \tag{20.1-24}$$

The relation between the interfacial fluxes and the terminal compositions is

$$\frac{(x_{A0} - x_{A\infty})(N_{Az0} + N_{Bz0})}{N_{Az0} - x_{A0}(N_{Az0} + N_{Bz0})} = \sqrt{\pi}(1 + \text{erf}\,\varphi)\varphi \exp \varphi^2 \tag{20.1-25}$$

Equations 20.1-16, 10, and 18 are included as special cases of the last three equations. The last one is a key result for mass transfer calculations (see §22.8).

EXAMPLE 20.1-2

Gas Absorption with Rapid Reaction[3,4]

Gas A is absorbed by a stationary liquid solvent S, the latter containing solute B. Species A reacts with B in an instantaneous irreversible reaction according to the equation $aA + bB \rightarrow$ Products. It may be assumed that Fick's second law adequately describes the diffusion processes, since A, B, and the reaction products are present in S in low concentrations. Obtain expressions for the concentration profiles.

SOLUTION

Because of the instantaneous reaction of A and B, there will be a plane parallel to the liquid–vapor interface at a distance z_R from it, which separates the region containing no A from that containing no B. The distance z_R is a function of t, since the boundary between A and B retreats as B is used up in the chemical reaction.

The differential equations for c_A and c_B are then

$$\frac{\partial c_A}{\partial t} = \mathcal{D}_{AS} \frac{\partial^2 c_A}{\partial z^2} \qquad \text{for } 0 \leq z \leq z_R(t) \tag{20.1-26}$$

$$\frac{\partial c_B}{\partial t} = \mathcal{D}_{BS} \frac{\partial^2 c_B}{\partial z^2} \qquad \text{for } z_R(t) \leq z < \infty \tag{20.1-27}$$

These are to be solved with the following initial and boundary conditions:

I.C.:	at $t = 0$,	$c_B = c_{B\infty}$	for $z > 0$	(20.1-28)
B.C. 1:	at $z = 0$,	$c_A = c_{A0}$		(20.1-29)
B.C. 2, 3:	at $z = z_R(t)$,	$c_A = c_B = 0$		(20.1-30)
B.C. 4:	at $z = z_R(t)$,	$-\dfrac{1}{a}\mathcal{D}_{AS}\dfrac{\partial c_A}{\partial z} = +\dfrac{1}{b}\mathcal{D}_{BS}\dfrac{\partial c_B}{\partial z}$		(20.1-31)
B.C. 5:	at $z = \infty$,	$c_B = c_{B\infty}$		(20.1-32)

Here c_{A0} is the interfacial concentration of A, and $c_{B\infty}$ is the original concentration of B. The fourth boundary condition is the stoichiometric requirement that a moles of A consume b moles of B (see Problem 20B.2).

---

[3] T. K. Sherwood, R. L. Pigford, and C. R. Wilke, *Absorption and Extraction*, 3rd edition, McGraw-Hill, New York (1975), Chapter 8. See also G. Astarita, *Mass Transfer with Chemical Reaction*, Elsevier, Amsterdam (1967), Chapter 5.

[4] For related problems with moving boundaries associated with phase changes, see H. S. Carslaw and J. C. Jaeger, *Conduction of Heat in Solids*, 2nd edition, Oxford University Press (1959). See also S. G. Bankoff, *Advances in Chemical Engineering*, Academic Press, New York (1964), Vol. 5, pp. 76–150; J. Crank, *Free and Moving Boundary Problems*, Oxford University Press (1984).

The absence of a characteristic length in this problem, and the fact that $c_B = c_{B\infty}$ both at $t = 0$ and $z = \infty$, suggests trying a combination of variables. Comparison with the previous example (without the v_z^* term) suggests the following trial solutions:

$$\frac{c_A}{c_{A0}} = C_1 + C_2 \operatorname{erf} \frac{z}{\sqrt{4\mathscr{D}_{AS}t}} \qquad \text{for } 0 \leq z \leq z_R(t) \tag{20.1-33}$$

$$\frac{c_B}{c_{B\infty}} = C_3 + C_4 \operatorname{erf} \frac{z}{\sqrt{4\mathscr{D}_{BS}t}} \qquad \text{for } z_R(t) \leq z < \infty \tag{20.1-34}$$

These functions satisfy the differential equations, and if the constants of integration, C_1 to C_4, can be so chosen that the initial and boundary conditions are satisfied, we will have the complete solution to the problem.

Application of the initial condition and the first three boundary conditions permits the evaluation of the integration constants in terms of $z_R(t)$, thereby giving

$$\frac{c_A}{c_{A0}} = 1 - \frac{\operatorname{erf}(z/\sqrt{4\mathscr{D}_{AS}t})}{\operatorname{erf}(z_R/\sqrt{4\mathscr{D}_{AS}t})} \qquad \text{for } 0 \leq z \leq z_R(t) \tag{20.1-35}$$

$$\frac{c_B}{c_{B\infty}} = 1 - \frac{1 - \operatorname{erf}(z/\sqrt{4\mathscr{D}_{BS}t})}{1 - \operatorname{erf}(z_R/\sqrt{4\mathscr{D}_{BS}t})} \qquad \text{for } z_R(t) \leq z < \infty \tag{20.1-36}$$

B.C. 5 is then automatically satisfied. Finally, insertion of these solutions into B.C. 4 gives the following implicit equation from which $z_R(t)$ can be obtained:

$$1 - \operatorname{erf}\sqrt{\frac{\gamma}{\mathscr{D}_{BS}}} = \frac{ac_{B\infty}}{bc_{A0}}\sqrt{\frac{\mathscr{D}_{BS}}{\mathscr{D}_{AS}}} \operatorname{erf}\sqrt{\frac{\gamma}{\mathscr{D}_{AS}}} \exp\!\left(\frac{\gamma}{\mathscr{D}_{AS}} - \frac{\gamma}{\mathscr{D}_{BS}}\right) \tag{20.1-37}$$

Here γ is a constant equal to $z_R^2/4t$. Thus z_R increases as $\sqrt{t}$.

To calculate the concentration profiles, one first solves Eq. 20.1-37 for $\sqrt{\gamma}$, and then inserts this value for $z_R/\sqrt{4t}$ in Eqs. 20.1-35 and 36. Some calculated concentration profiles are shown in Fig. 20.1-2 (for $a = b$), to illustrate the rate of movement of the reaction zone.

From the concentration profiles we can calculate the rate of mass transfer at the interface:

$$N_{Az0} = -\mathscr{D}_{AS}\frac{\partial c_A}{\partial z}\Big|_{z=0} = \frac{c_{A0}}{\operatorname{erf}\sqrt{\gamma/\mathscr{D}_{AS}}}\sqrt{\frac{\mathscr{D}_{AS}}{\pi t}} \tag{20.1-38}$$

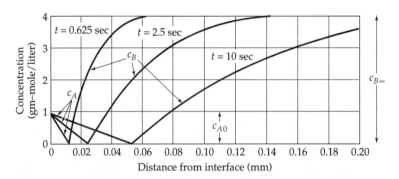

Fig. 20.1-2. Gas absorption with rapid chemical reaction, with concentration profiles given by Eqs. 20.1-35 to 37 (for $a = b$). This calculation was made for $\mathscr{D}_{AS} = 3.9 \times 10^{-5}$ ft^2/hr and $\mathscr{D}_{BS} = 1.95 \times 10^{-5}$ ft^2/hr [T. K. Sherwood and R. L. Pigford, *Absorption and Extraction*, McGraw-Hill, New York (1952), p. 336].

The average rate of absorption up to time t is then

$$N_{Az0,\text{avg}} = \frac{1}{t} \int_0^t N_{Az0} dt = 2 \frac{c_{A0}}{\text{erf}\sqrt{\gamma/\mathscr{D}_{AS}}} \sqrt{\frac{\mathscr{D}_{AS}}{\pi t}} \qquad (20.1\text{-}39)$$

Hence the average rate up to time t is just twice the instantaneous rate.

EXAMPLE 20.1-3

Unsteady Diffusion with First-Order Homogeneous Reaction[5–8]

When species A diffuses in a liquid medium B and reacts with it irreversibly ($A + B \rightarrow C$) according to a pseudo-first-order reaction, then the process of diffusion plus reaction is described by

$$\frac{\partial \omega_A}{\partial t} + (\mathbf{v} \cdot \nabla \omega_A) = \mathscr{D}_{AB} \nabla^2 \omega_A - k_1''' \omega_A \qquad (20.1\text{-}40)$$

provided that the solution of A is dilute and that not much C is produced. Here k_1''' is the rate constant for the homogeneous reaction. Equation 20.1-40 is frequently encountered with the initial and boundary conditions

I.C. at $t = 0$: $\qquad\qquad\qquad \omega_A = \omega_{AI}(x, y, z) \qquad\qquad\qquad (20.1\text{-}41)$

B.C. at bounding surfaces: $\qquad\quad \omega_A = \omega_{A0}(x, y, z) \qquad\qquad\qquad (20.1\text{-}42)$

and with a velocity profile independent of time. For such problems show that the solution is

$$\omega_A = g \exp(-k_1''' t) + \int_0^t \exp(-k_1''' t) \frac{\partial}{\partial t'} f(x, y, z, t') dt' \qquad (20.1\text{-}43)$$

Here f is the solution of Eqs. 20.1-40 to 42 with $k_1''' = 0$ and $\omega_{AI} = 0$, whereas g is the solution with $k_1''' = 0$ and $\omega_{A0} = 0$.

SOLUTION

This problem is linear in ω_A. It may, therefore, be solved by a superposition of two simpler problems:

$$\omega_A = \omega_A^{(1)} + \omega_A^{(2)} \qquad (20.1\text{-}44)$$

with $\omega_A^{(1)}$ described by the equations

P.D.E.: $\qquad\qquad \dfrac{\partial \omega_A^{(1)}}{\partial t} + (\mathbf{v} \cdot \nabla \omega_A^{(1)}) = \mathscr{D}_{AB} \nabla^2 \omega_A^{(1)} - k_1''' \omega_A^{(1)} \qquad (20.1\text{-}45)$

I.C. at $t = 0$: $\qquad\qquad\qquad \omega_A^{(1)} = \omega_{AI}(x, y, z) \qquad\qquad\qquad (20.1\text{-}46)$

B.C. at surfaces: $\qquad\qquad\quad \omega_A^{(1)} = 0 \qquad\qquad\qquad\qquad\qquad (20.1\text{-}47)$

[5] P. V. Danckwerts, *Trans. Faraday Soc.*, **47**, 1014–1023 (1951). **Peter Victor Danckwerts** (1916–1984) was bomb disposal officer for the Port of London during "the Blitz" and was wounded in a mine field in Italy during WWII; while teaching at Imperial College in London and at Cambridge University he directed research on residence-time distribution, diffusion and chemical reaction, and the role of diffusion in gas absorption.

[6] A. Giuliani and F. P. Foraboschi, *Atti. Acad. Sci. Inst. Bologna*, **9**, 1–16 (1962); F. P. Foraboschi, *ibid.*, **11**, 1–14 (1964); F. P. Foraboschi, *AIChE Journal*, **11**, 752–768 (1965).

[7] E. N. Lightfoot, *AIChE Journal*, **10**, 278–284 (1964).

[8] W. E. Stewart, *Chem. Eng. Sci.*, **23**, 483–487 (1968); corrigenda, *ibid.*, **24**, 1189–1190 (1969). There this approach was generalized to time-dependent flows with homogeneous and heterogeneous first-order reactions.

and $\omega_A^{(2)}$ described by

P.D.E.:
$$\frac{\partial \omega_A^{(2)}}{\partial t} + (\mathbf{v} \cdot \nabla \omega_A^{(2)}) = \mathcal{D}_{AB} \nabla^2 \omega_A^{(2)} - k_1''' \omega_A^{(2)} \tag{20.1-48}$$

I.C. at $t = 0$:
$$\omega_A^{(2)} = 0 \tag{20.1-49}$$

B.C. at surfaces:
$$\omega_A^{(2)} = \omega_{A0}(x, y, z) \tag{20.1-50}$$

We now proceed to solve these two auxiliary problems by means of Laplace transform.
Taking the Laplace transform of the equations for $\omega_A^{(1)}$ gives

P.D.E. + I.C.:
$$(p + k_1''')\overline{\omega}_A^{(1)} - \omega_{AI}(x, y, z) + (\mathbf{v} \cdot \nabla \overline{\omega}_A^{(1)}) = \mathcal{D}_{AB} \nabla^2 \overline{\omega}_A^{(1)} \tag{20.1-51}$$

B.C. at surfaces:
$$\overline{\omega}_A^{(1)} = 0 \tag{20.1-52}$$

Now, the function g in Eq. 20.1-43 is the solution for $\omega_A^{(1)}$ with k_1''' replaced by zero. Correspondingly the Laplace transform $\overline{g}$ satisfies Eqs. 20.1-51 and 52 with $p + k_1'''$ replaced by p:

$$\overline{\omega}_A^{(1)}(p, x, y, z) = \overline{g}(p + k_1''', x, y, z) \tag{20.1-53}$$

Hence by taking the inverse Laplace transform we get

$$\omega_A^{(1)} = g \exp(-k_1''' t) \tag{20.1-54}$$

which is the first part of the solution.
Next, taking the Laplace transform of Eqs. 20.1-48 to 50 gives

P.D.E. + I.C.:
$$(p + k_1''')\overline{\omega}_A^{(2)} + (\mathbf{v} \cdot \nabla \overline{\omega}_A^{(2)}) = \mathcal{D}_{AB} \nabla^2 \overline{\omega}_A^{(2)} \tag{20.1-55}$$

B.C. at surfaces:
$$\overline{\omega}_A^{(2)} = \frac{1}{p} \omega_{A0}(x, y, z) \tag{20.1-56}$$

The Laplace transform $\overline{f}$ satisfies the same two equations with k_1''' replaced by zero. That is, if we now use s for the transform variable in lieu of p, we have

P.D.E. + I.C.:
$$s\overline{f} + (\mathbf{v} \cdot \nabla \overline{f}) = \mathcal{D}_{AB} \nabla^2 \overline{f} \tag{20.1-57}$$

B.C. at surfaces:
$$\overline{f} = \frac{1}{s} \omega_{A0}(x, y, z) \tag{20.1-58}$$

We see that the function $s\overline{f}$ satisfies the same boundary condition as $p\overline{\omega}_A^{(2)}$ and that the differential equations for $s\overline{f}$ and $p\overline{\omega}_A^{(2)}$ are identical when $s = p + k'''$. Hence

$$p\overline{\omega}_A^{(2)}\big|_p = (s\overline{f})\big|_{s=p+k_1'''} \tag{20.1-59}$$

or

$$\overline{\omega}_A^{(2)}(p, x, y, z) = \frac{p + k_1'''}{p} \overline{f}(p + k_1''', x, y, z) \tag{20.1-60}$$

Taking the inverse transform then gives

$$\omega_A^{(2)} = \int_0^t \exp(-k_1''' t') \frac{\partial}{\partial t'} f(x, y, z, t') dt' \tag{20.1-61}$$

as the second part of the solution. Addition of the two parts of the solution, $\omega_A^{(1)}$ and $\omega_A^{(2)}$, then gives Eq. 20.1-43 directly.

Equation 20.1-43 provides a means for predicting concentration profiles in reacting systems from calculations or experiments on nonreacting systems at the same flow conditions. Several extensions of this treatment are available, including multicomponent systems,[9] turbulent flow,[8,9] and more general boundary conditions.[7–9]

---

[9] Y.-H. Pao, *AIAA Journal*, **2**, 1550–1559 (1964); *Chem. Eng. Sci.*, **19**, 694–696 (1964); *ibid.*, **20**, 665–669 (1965).

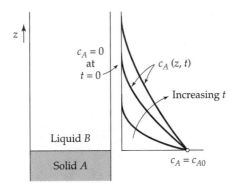

Fig. 20.1-3. Time-dependent diffusion from a soluble wall of A into a semi-infinite column of liquid B.

EXAMPLE 20.1-4	Figure 20.1-3 shows schematically the concentration profiles for the diffusion of A from a slightly soluble wall into a semi-infinite body of liquid above it. If the density and diffusivity are constants, then this problem is the mass transfer analog of the problems discussed in §§4.1 and 12.1. The diffusion is described by the one-dimensional version of Fick's second law, Eq. 19.1-18,

Influence of Changing Interfacial Area on Mass Transfer at an Interface[10,11]

$$\frac{\partial c_A}{\partial t} = \mathcal{D}_{AB} \frac{\partial^2 c_A}{\partial z^2} \tag{20.1-62}$$

along with the initial condition that $c_A = 0$ throughout the liquid, and the boundary conditions that $c_A = c_{A0}$ at the solid–liquid interface and $c_A = 0$ infinitely far from the interface. The solution to this problem is

$$\frac{c_A}{c_{A0}} = 1 - \operatorname{erf} \frac{z}{\sqrt{4\mathcal{D}_{AB}t}} \tag{20.1-63}$$

from which we can get the interfacial flux

$$N_{Az0} = c_{A0} \sqrt{\frac{\mathcal{D}_{AB}}{\pi t}} \tag{20.1-64}$$

Equation 20.1-63 is the mass transfer analog of Eqs. 4.1-15 and 12.1-8.

In Fig. 20.1-4 we depict a similar problem in which the interfacial area is changing with time as the liquid spreads out in the x and y directions, so that the interfacial area is a function of time, $S(t)$. The initial and boundary conditions for the concentration are kept the same. We wish to know the function $c_A(z, t)$ for this system.

SOLUTION

The velocity distribution for this varying interfacial area problem is $v_x = +\frac{1}{2}ax$, $v_y = +\frac{1}{2}ay$, $v_z = -az$, where $a = d \ln S/dt$. Then the diffusion equation for this system is

$$\frac{\partial c_A}{\partial t} - \left(\frac{d}{dt} \ln S\right) z \frac{\partial c_A}{\partial z} = \mathcal{D}_{AB} \frac{\partial^2 c_A}{\partial z^2} \tag{20.1-65}$$

---

[10] D. Ilkovič, *Collec. Czechoslov. Chem. Comm.*, **6**, 498–513 (1934). The final result in this section was obtained by Ilkovič in connection with his work on the dropping-mercury electrode.

[11] V. G. Levich, *Physicochemical Hydrodynamics*, 2nd edition (English translation), Prentice-Hall, Englewood Cliffs N.J. (1962), §108. This book contains a wealth of theoretical and experimental results on diffusion and flow phenomena in liquids and two-phase systems.

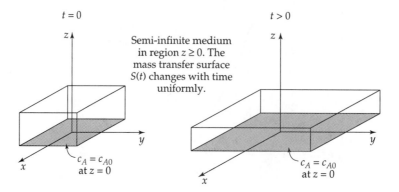

Fig. 20.1-4. Time-dependent diffusion across a mass transfer interface $S(t)$ that is changing with time. The liquid B, in the region above the plane $z = 0$, has a velocity distribution $v_x = +\frac{1}{2}ax$, $v_y = +\frac{1}{2}ay$, and $v_z = -az$, where $a = d \ln S/dt$.

Since Eq. 20.1-62 is solved by the method of combination of variables, the same technique can be tried here. We postulate

$$\frac{c_A}{c_{A0}} = g(\zeta) \qquad \text{with } \zeta = \frac{z}{\delta(t)} \tag{20.1-66}$$

Substitution of this trial solution into Eq. 20.1-65 gives

$$\frac{d^2 g}{d\zeta^2} + 2\left[\frac{\delta^2}{2\mathscr{D}_{AB}}\left(\frac{d}{dt}\ln (S\delta)\right)\right]\zeta\frac{dg}{d\zeta} = 0 \tag{20.1-67}$$

If we set the expression within the brackets equal to unity, then we accomplish two things: (i) we obtain an equation for g that has the same form as Eq. 4.1-9, to which the solution is known; (ii) we get an equation for δ as a function of t:

$$\frac{d}{dt}\ln (S\delta) = \frac{2\mathscr{D}_{AB}}{\delta^2} \tag{20.1-68}$$

This equation may be integrated to give

$$\int_0^{S(t)\delta(t)} d(S\delta)^2 = 4\mathscr{D}_{AB}\int_0^t S^2(\bar{t})d\bar{t} \tag{20.1-69}$$

The lower limit on the left side is chosen so as to ensure that $c_A = 0$ initially throughout the fluid. This choice then leads to

$$\delta(t) = \sqrt{4\mathscr{D}_{AB}\int_0^t [S(\bar{t})/S(t)]^2 d\bar{t}} \tag{20.1-70}$$

and we get finally for the concentration profiles

$$\frac{c_A}{c_{A0}} = 1 - \text{erf}\frac{z}{\sqrt{4\mathscr{D}_{AB}\int_0^t [S(\bar{t})/S(t)]^2 d\bar{t}}} \tag{20.1-71}$$

The interfacial mass flux is then obtained by differentiating Eq. 20.1-71 to get

$$N_{Az0} = c_{A0}\sqrt{\frac{\mathscr{D}_{AB}}{\pi t}}\left(\frac{1}{t}\int_0^t [S(\bar{t})/S(t)]^2 d\bar{t}\right)^{-1/2} \tag{20.1-72}$$

The total number of moles of A that have crossed the interface at time t through the surface $S(t)$ can be obtained from integration of Eq. 20.1-71 as follows:

$$M_A(t) = \iiint_0^\infty c_A dz\, dy\, dx = S(t)c_{A0} \int_0^\infty (1 - \mathrm{erf}(z/\delta))dz$$

$$= S(t)c_{A0} \frac{2}{\sqrt{\pi}} \int_0^\infty \int_{z/\delta}^\infty \exp\left(-\zeta^2\right) d\zeta\, dz$$

$$= S(t)c_{A0} \frac{2}{\sqrt{\pi}} \delta \int_0^\infty \exp\left(-\zeta^2\right) \zeta\, d\zeta$$

$$= S(t)c_{A0} \frac{1}{\sqrt{\pi}} \sqrt{4\mathscr{D}_{AB} \int_0^t [S(\bar{t})/S(t)]^2\, d\bar{t}}$$

$$= c_{A0}\sqrt{\frac{4\mathscr{D}_{AB}}{\pi} \int_0^t [S(\bar{t})]^2\, d\bar{t}} \tag{20.1-73}$$

An equivalent expression can be obtained by integrating Eq. 20.1-72:

$$M_A(t) = \int_0^t S(\bar{t}) N_{Az0}(\bar{t}) d\bar{t}$$

$$= c_{A0}\sqrt{\frac{\mathscr{D}_{AB}}{\pi}} \int_0^t \frac{[S(\bar{t})]^2}{\sqrt{\int_0^{\bar{t}} [S(\bar{\bar{t}})/S(\bar{t})]^2 d\bar{\bar{t}}}}\, d\bar{t} \tag{20.1-74}$$

Both Eq. 21.1-73 and Eq. 21.1-74 can be checked by verifying that $dM_A/dt = N_{Az0}(t)S(t)$.

If $S(t) = at^n$, where a is a constant, the above results simplify to

$$N_{Az0}(t) = c_{A0}\sqrt{\frac{(2n+1)\mathscr{D}_{AB}}{\pi t}} \tag{20.1-75}$$

$$M_A(t) = c_{A0}a\sqrt{\frac{4\mathscr{D}_{AB}t^{2n+1}}{(2n+1)\pi}} \tag{20.1-76}$$

For the diffusion into the surrounding liquid from a gas bubble whose volume is increasing linearly with time, $n = \frac{2}{3}$ and $2n + 1 = \frac{7}{3}$. This is of course an approximate result, in which curvature has been neglected, and is therefore valid only for short contact times. Related results have been obtained for interfaces of arbitrary shapes,[2,12] and experimentally verified for several laminar and turbulent systems.[2,13]

§20.2 STEADY-STATE TRANSPORT IN BINARY BOUNDARY LAYERS

In §12.4 we discussed the application of boundary layer analysis to nonisothermal flow of pure fluids. The equations of continuity, motion, and energy were presented in boundary layer form and were solved for some simple situations. In this section we extend the set of boundary layer equations to binary reacting mixtures, adding the equation of continuity for species A so that the concentration profiles can be evaluated. Then we analyze three examples for the flat-plate geometry: one on forced convection with a homogeneous reaction, one on rapid mass transfer, and one on analogies for small mass-transfer rates.

[12] J. B. Angelo, E. N. Lightfoot, and D. W. Howard, *AIChE Journal*, **12**, 751–760 (1966).

[13] W. E. Stewart, in *Physicochemical Hydrodynamics* (D. B. Spalding, ed.), Advance Publications Ltd., London, Vol. 1 (1977), pp. 22–63.

Consider the steady, two-dimensional flow of a binary fluid around a submerged object, such as that in Fig. 4.4-1. In the vicinity of the solid surface, the equations of change given in §§18.2 and 3 may be simplified as follows, provided that ρ, μ, k, $\hat{C}_p$, and $\mathcal{D}_{AB}$ are essentially constant (except in the ρg term), and that viscous dissipation can be neglected:

Continuity:
$$\frac{\partial v_x}{\partial x} + \frac{\partial v_y}{\partial y} = 0 \tag{20.2-1}$$

Motion:
$$\rho\left(v_x \frac{\partial v_x}{\partial x} + v_y \frac{\partial v_x}{\partial y}\right) = \rho v_e \frac{dv_e}{dx} + \mu \frac{\partial^2 v_x}{\partial y^2}$$
$$+ \bar{\rho} g_x \bar{\beta}(T - T_\infty) + \bar{\rho} g_x \bar{\zeta}(\omega_A - \omega_{A\infty}) \tag{20.2-2}$$

Energy:
$$\rho\hat{C}_p\left(v_x \frac{\partial T}{\partial x} + v_y \frac{\partial T}{\partial y}\right) = k \frac{\partial^2 T}{\partial y^2} - \left(\frac{\bar{H}_A}{M_A} - \frac{\bar{H}_B}{M_B}\right)r_A \tag{20.2-3}$$

Continuity of A:
$$\rho\left(v_x \frac{\partial \omega_A}{\partial x} + v_y \frac{\partial \omega_A}{\partial y}\right) = \rho\mathcal{D}_{AB} \frac{\partial^2 \omega_A}{\partial y^2} + r_A \tag{20.2-4}$$

The equation of continuity is the same as Eq. 12.4-1. The equation of motion, obtained from Eq. 19.2-3, differs from Eq. 12.4-2 by the addition of the binary buoyant force term $\bar{\rho} g_x \bar{\zeta}(\omega_A - \omega_{A\infty})$. The energy equation, obtained from Eq. (F) of Table 19.2-4, differs from Eq. 12.4-3 by the addition of the chemical heat-source term $-[(\bar{H}_A/M_A) - (\bar{H}_B/M_B)]r_A$. Equation 20.2-4 is obtained from Eq. 19.1-16 by setting $\omega_A = \omega_A(x, y)$ and neglecting the diffusion in the x direction. More complete equations, valid for high-velocity, variable-property boundary layers, are available elsewhere.[1]

The usual boundary conditions on v_x are that $v_x = 0$ at the solid surface, and $v_x = v_e(x)$ at the outer edge of the *velocity boundary layer*. The usual boundary conditions on T in Eq. 20.2-3 are that $T = T_0(x)$ at the solid surface, and $T = T_\infty$ at the outer edge of the *thermal boundary layer*. The corresponding boundary conditions on ω_A in Eq. 20.2-4 are that $\omega_A = \omega_{A0}(x)$ at the surface and $\omega_A = \omega_{A\infty}$ at the outer edge of the *diffusional boundary layer*. Thus there are now three boundary layers to consider, each with its own thickness. In fluids with constant physical properties and large Prandtl and Schmidt numbers, the thermal and diffusional boundary layers usually lie within the velocity boundary layer, whereas for Pr < 1 and Sc < 1 they may extend beyond it.

For mass transfer systems the velocity v_y at the surface is usually not zero, but depends on x. Hence we set $v_y = v_0(x)$ at $y = 0$. This boundary condition is appropriate whenever there is a net mass flux between the surface and the stream, as in melting, drying, sublimation, combustion of the wall, or transpiration of the fluid through a porous wall. Clearly, some of these processes are possible with pure fluids, but for simplicity we have deferred their consideration to this chapter (see also §§18.3 and 22.8 for related analyses).

With the help of the equation of continuity, Eqs. 20.2-1 to 4 can be formally integrated, with the boundary conditions just given, to obtain the following set of boundary layer balances:

Continuity + motion:
$$\mu \left.\frac{\partial v_x}{\partial y}\right|_{y=0} = \frac{d}{dx}\int_0^\infty \rho v_x(v_e - v_x)dy + \frac{dv_e}{dx}\int_0^\infty \rho(v_e - v_x)dy$$
$$- \int_0^\infty \rho g_x \beta(T - T_\infty)dy - \int_0^\infty \rho g_x \zeta(\omega_A - \omega_{A\infty})dy + \rho v_0 v_e \tag{20.2-5}$$

Continuity + energy:
$$k \left.\frac{\partial T}{\partial y}\right|_{y=0} = \frac{d}{dx}\int_0^\infty \rho v_x \hat{C}_p(T_\infty - T)\,dy - \int_0^\infty \left(\frac{\bar{H}_A}{M_A} - \frac{\bar{H}_B}{M_B}\right)r_A dy - \rho v_0 \hat{C}_p(T_\infty - T_0) \tag{20.2-6}$$

[1] See, for example, W. H. Dorrance, *Viscous Hypersonic Flow*, McGraw-Hill, New York (1962), and K. Stewartson, *The Theory of Laminar Boundary Layers in Compressible Fluids*, Oxford University Press (1964).

Continuity + continuity of A:

$$\rho \mathcal{D}_{AB} \frac{\partial \omega_A}{\partial y}\bigg|_{y=0} = \frac{d}{dx}\int_0^\infty \rho v_x(\omega_{A\infty} - \omega_A)\,dy + \int_0^\infty r_A dy - \rho v_0(\omega_{A\infty} - \omega_{A0}) \quad (20.2\text{-}7)$$

These equations are extensions of the *von Kármán balances* of §§4.4 and 12.4 and may be similarly applied, as shown in Example 20.2-1.

Boundary layer techniques have been of considerable value in developing the theory of high-speed flight, separations processes, chemical reactors, and biological mass transfer systems. A few of the interesting problems that have been studied are chemical reactions in hypersonic boundary layers,[1] mass transfer from droplets,[2] electrode polarization in forced convection[2] and free convection,[3] reverse-osmosis water desalination,[4] and interphase transfer in packed-bed reactors and distillation columns.[5]

EXAMPLE 20.2-1

Diffusion and Chemical Reaction in Isothermal Laminar Flow Along a Soluble Flat Plate

An appropriate mass transfer analog to the problem discussed in Example 12.4-1 would be the flow along a flat plate that contains a species A slightly soluble in the fluid B. The concentration at the plate surface would be c_{A0}, the solubility of A in B, and the concentration of A far from the plate would be $c_{A\infty}$. In this example we let $c_{A\infty} = 0$ and break the analogy with Example 12.4-1 by letting A react with B by an nth order homogeneous reaction, so that $R_A = -k_n''' c_A^n$. The concentration of dissolved A is assumed to be small, so that the physical properties μ, ρ, and $\mathcal{D}_{AB}$ are virtually constant throughout the fluid. We wish to analyze the system, sketched in Fig. 20.2-1, by the von Kármán method.

SOLUTION

We begin by postulating forms for the velocity and concentration profiles. To minimize the algebra and still illustrate the method, we select simple functions (clearly one can suggest more realistic functions):

$$\begin{cases} \dfrac{v_x}{v_\infty} = \dfrac{y}{\delta} & y \le \delta(x) \\[2mm] \dfrac{v_x}{v_\infty} = 1 & y \ge \delta(x) \end{cases} \qquad (20.2\text{-}8)$$

$$\begin{cases} \dfrac{c_A}{c_{A0}} = 1 - \dfrac{y}{\delta_c} & y \le \delta_c(x) \\[2mm] \dfrac{c_A}{c_{A0}} = 0 & y \ge \delta_c(x) \end{cases} \qquad (20.2\text{-}9)$$

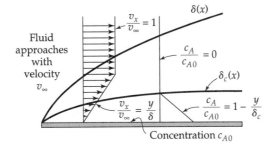

Fig. 20.2-1. Assumed velocity and concentration profiles for the laminar boundary layer with homogeneous chemical reaction.

---

[2] V. G. Levich, *Physicochemical Hydrodynamics*, 2nd edition (English translation), Prentice-Hall, Englewood Cliffs, N.J. (1962).

[3] C. R. Wilke, C. W. Tobias, and M. Eisenberg, *Chem. Eng. Prog.*, **49**, 663–674 (1953).

[4] W. N. Gill, D. Zeh, and C. Tien, *Ind. Eng. Chem. Fund.*, **4**, 433–439 (1965); *ibid.*, **5**, 367–370 (1966). See also P. L. T. Brian, *ibid.*, **4**, 439–445 (1965).

[5] J. P. Sørensen and W. E. Stewart, *Chem. Eng. Sci.*, **29**, 833–837 (1974); W. E. Stewart and D. L. Weidman, *ibid.*, **45**, 2155–2160 (1990); T. C. Young and W. E. Stewart, *AIChE Journal*, **38**, 592–602, 1302 (1992).

Note that we use different thicknesses, δ and δ_c, for the velocity and concentration boundary layers. In order to relate this problem to that of Example 12.4-1, we introduce the quantity $\Delta = \delta_c/\delta$, which in this case is a function of x because of the chemical reaction occurring. We restrict the discussion to $\Delta \leq 1$, for which the concentration boundary layer lies entirely within the velocity boundary layer. We can also neglect the interfacial velocity $v_0 = v_y|_{y=0}$, which is small here because of the small solubility of A. Insertion of these expressions into Eqs. 20.2-5 and 7 then gives the differential equations

$$\frac{\mu v_\infty}{\delta} = \frac{d}{dx}\left(\tfrac{1}{6}\rho v_\infty^2 \delta\right) \tag{20.2-10}$$

$$-\frac{\mathscr{D}_{AB}c_{A0}}{\delta\Delta} = \frac{d}{dx}\left(-\tfrac{1}{6}c_{A0}v_\infty\delta\Delta^2\right) - \frac{k_n''' c_{A0}^n \delta\Delta}{n+1} \tag{20.2-11}$$

for the boundary layer thicknesses δ and $\delta_c = \delta\Delta$.

Equation 20.2-10 is readily integrated to give

$$\delta = \sqrt{12\frac{\nu x}{v_\infty}} \tag{20.2-12}$$

Insertion of this result into Eq. 20.2-11 and multiplication by $-\delta\Delta/\nu c_{A0}$ gives

$$\frac{1}{Sc} = \frac{4}{3}x\frac{d}{dx}\Delta^3 + \Delta^3 + 12\left[\frac{k_n''' c_{A0}^{n-1} x}{(n+1)v_\infty}\right]\Delta^2 \tag{20.2-13}$$

as the differential equation for Δ. Thus Δ depends on the Schmidt number, $Sc = \mu/\rho\mathscr{D}_{AB}$, and on the dimensionless position coordinate shown in the square brackets. The bracketed quantity is $1/(n+1)$ times the *first Damköhler number*[6] based on the distance x.

When *no reaction* is occurring, k_n''' is zero, and Eq. 20.2-13 becomes a linear first-order reaction for Δ^3. When that equation is integrated, we get

$$\Delta^3 = \frac{1}{Sc} + \frac{C}{x^{3/4}} \tag{20.2-14}$$

in which C is a constant of integration. Because Δ does not become infinite as $x \to 0$, we obtain in the absence of chemical reaction (cf. Eq. 12.4-15):

$$\Delta = Sc^{-1/3} \qquad \Delta < 1 \tag{20.2-15}$$

That is, when there is no reaction and $Sc > 1$, the concentration and velocity boundary layer thicknesses bear a constant ratio to one another, dependent only on the value of the Schmidt number.

When a *slow reaction* occurs (or when x is small), a series solution to Eq. 20.2-13 can be obtained:

$$\Delta = Sc^{-1/3}(1 + a_1\xi + a_2\xi^2 + \cdots) \tag{20.2-16}$$

in which

$$\xi = 12\left[\frac{k_n''' c_{A0}^{n-1} x}{(n+1)v_\infty}\right] \tag{20.2-17}$$

Substitution of this expression into Eq. 20.2-13 gives

$$a_1 = -\tfrac{1}{7}Sc^{1/3}, \qquad a_2 = +\tfrac{3}{539}Sc^{2/3}, \qquad \text{etc.} \tag{20.2-18}$$

Because a_1 is negative, the concentration boundary layer thickness is diminished by the chemical reaction.

---

[6] G. Damköhler, *Zeits. f. Electrochemie*, **42**, 846–862 (1936); W. E. Stewart, *Chem. Eng. Prog. Symp. Series*, #58, **61**, 16–27 (1965).

When a *fast reaction* occurs (or when x is very large), a series solution in $1/\xi$ is more appropriate. For large ξ, we assume that the dominant term is of the form $\Delta = \text{const.} \cdot \xi^m$ where $m < 0$. Substitution of this trial function into Eq. 20.2-13 then shows that

$$\Delta = (Sc\xi)^{-1/2} \qquad \text{for large } \xi \tag{20.2-19}$$

Combination of Eqs. 20.2-12 and 19 shows that, at large distances from the leading edge, the concentration boundary layer thickness $\delta_c = \delta\Delta$ becomes a constant independent of v_∞ and ν.

Once $\Delta(\xi, Sc)$ is known, then the concentration profiles and the mass transfer rate at the surface may be found. A more refined treatment of this problem has been given elsewhere.[7]

EXAMPLE 20.2-2	The laminar boundary layer on a flat plate (see Fig. 20.2-2) has been a popular system for heat and mass transfer studies. In this example, we give an analysis of subsonic forced convection in this geometry at high mass-transfer rates, and discuss the analogies that hold in this situation. This example is an extension of Example 4.4-2.
Forced Convection from a Flat Plate at High Mass-Transfer Rates	

SOLUTION

Consider the nonisothermal, steady, two-dimensional flow of a binary fluid in the system of Fig. 20.2-2. The fluid properties ρ, μ, $\hat{C}_p$, k, and $\mathscr{D}_{AB}$ are considered constant, viscous dissipation is neglected, and there are no homogeneous chemical reactions. The Prandtl boundary layer equations for the laminar region are

Continuity:
$$\frac{\partial v_x}{\partial x} + \frac{\partial v_y}{\partial y} = 0 \tag{20.2-20}$$

Motion:
$$v_x \frac{\partial v_x}{\partial x} + v_y \frac{\partial v_x}{\partial y} = \nu \frac{\partial^2 v_x}{\partial y^2} \tag{20.2-21}$$

Energy:
$$v_x \frac{\partial T}{\partial x} + v_y \frac{\partial T}{\partial y} = \alpha \frac{\partial^2 T}{\partial y^2} \tag{20.2-22}$$

Continuity of A:
$$v_x \frac{\partial \omega_A}{\partial x} + v_y \frac{\partial \omega_A}{\partial y} = \mathscr{D}_{AB} \frac{\partial^2 \omega_A}{\partial y^2} \tag{20.2-23}$$

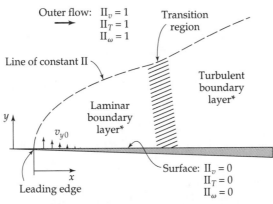

Fig. 20.2-2. Tangential flow along a sharp-edged semi-infinite flat plate with mass transfer into the stream. The laminar-turbulent transition usually occurs at a length Reynolds number $(xv_\infty/\nu)_{crit}$ on the order of 10^5 to 10^6.

---

[7] P. L. Chambré and J. D. Young, *Physics of Fluids*, **1**, 48–54 (1958). Catalytic surface reactions in boundary layers have been studied by P. L. Chambré and A. Acrivos, *J. Appl. Phys.*, **27**, 1322–1328 (1956).

The boundary conditions are taken to be:

$$\text{at } x \leq 0 \text{ or } y = \infty, \quad v_x = v_\infty$$
$$T = T_\infty$$
$$\omega_A = \omega_{A\infty} \tag{20.2-24}$$
$$\text{at } y = 0, \quad v_x = 0$$
$$T = T_0$$
$$\omega_A = \omega_{A0} \tag{20.2-25}$$
$$\text{at } y = 0, \quad v_y = v_0(x) \tag{20.2-26}$$

Here the function $v_0(x)$ stands for $v_y(x, y)$ evaluated at $y = 0$ and describes the distribution of mass transfer rate along the surface. This function will be specified later.

Equation 20.2-20 can be integrated, with the boundary condition of Eq. 20.2-26, to give

$$v_y = v_0(x) - \frac{\partial}{\partial x} \int_0^y v_x dy \tag{20.2-27}$$

This expression is to be inserted for v_y into Eqs. 20.2-21 to 23.

To capitalize on the analogous form of Eqs. 20.2-21 to 23 and the first six boundary conditions, we define the dimensionless profiles

$$\Pi_v = \frac{v_x}{v_\infty} \qquad \Pi_T = \frac{T - T_0}{T_\infty - T_0} \qquad \Pi_\omega = \frac{\omega_A - \omega_{A0}}{\omega_{A\infty} - \omega_{A0}} \tag{20.2-28}$$

and the dimensionless physical property ratios

$$\Lambda_v = \frac{\nu}{\nu} = 1 \qquad \Lambda_T = \frac{\nu}{\alpha} = \text{Pr} \qquad \Lambda_\omega = \frac{\nu}{\mathscr{D}_{AB}} = \text{Sc} \tag{20.2-29}$$

With these definitions, and the above equation for v_y, Eqs. 20.2-21 to 23 all take the form

$$\Pi_v \frac{\partial \Pi}{\partial x} + \left(\frac{v_0(x)}{v_\infty} - \frac{\partial}{\partial x} \int_0^y \Pi_v dy \right) \frac{\partial \Pi}{\partial y} = \frac{\nu}{v_\infty \Lambda} \frac{\partial^2 \Pi}{\partial y^2} \tag{20.2-30}$$

and the boundary conditions on the dependent variables reduce to the following:

$$\text{at } x \leq 0 \text{ or } y = \infty, \quad \Pi = 1 \tag{20.2-31}$$
$$\text{at } y = 0, \quad \Pi = 0 \tag{20.2-32}$$

Thus the dimensionless velocity, temperature, and composition profiles all satisfy the same equation, but with their individual values of Λ.

The form of the boundary conditions on Π suggests that a combination of variables be tried. By analogy with Eq. 4.4-20 we select the combination:

$$\eta = y \sqrt{\frac{1}{2} \frac{v_\infty}{\nu x}} \tag{20.2-33}$$

Then by treating Π and Π_v as functions of η (see Problem 20B.3), we obtain the differential equation

$$\left(\frac{v_0(x)}{v_\infty} \sqrt{2 \frac{v_\infty x}{\nu}} - \int_0^\eta \Pi_v d\eta \right) \frac{d\Pi}{d\eta} = \frac{1}{\Lambda} \frac{d^2 \Pi}{d\eta^2} \tag{20.2-34}$$

with the boundary conditions

$$\text{at } \eta = \infty, \quad \Pi = 1 \tag{20.2-35}$$
$$\text{at } \eta = 0, \quad \Pi = 0 \tag{20.2-36}$$

From the last three equations we conclude that the profiles will be expressible in terms of the single coordinate η, if and only if the interfacial velocity $v_0(x)$ is of the form

$$\frac{v_0(x)}{v_\infty} \sqrt{2 \frac{v_\infty x}{\nu}} = K = \text{const.} \tag{20.2-37}$$

Any other functional form for $v_0(x)$ would cause the left side of Eq. 20.2-34 to depend on both x and η, so that a combination of variables would not be possible. The boundary layer equations would then require integration in two dimensions, and the calculations would become more difficult. Equation 20.2-37 specifies that $v_0(x)$ vary as $1/\sqrt{x}$, and thus, inversely with the boundary layer thickness δ of Eq. 4.4-17.This equation has the same range of validity as Eq. 20.2-34, that is, $1 << (v_\infty x/\nu) < (v_\infty x/\nu)_{\text{crit}}$ (see Fig. 20.2-2).

Fortunately the condition in Eq. 20.3-37 is a useful one. It corresponds to a direct proportionality of ρv_0 to the interfacial fluxes τ_0, q_0, and j_{A0}. Conditions of this type arise naturally in diffusion-controlled surface reactions, and also in certain cases of drying and transpiration cooling. The determination of K for these situations is considered at the end of this example. Until then we treat K as given.

With the specification of $v_0(x)$ according to Eq. 20.2-37, the problem statement is complete, and we are ready to discuss the calculation of the profiles. This is best done by numerical integration, with specified values of the parameters Λ and K.

The first step in the solution is to evaluate the velocity profile Π_v. For this purpose it is convenient to introduce the function

$$f = -K + \int_0^\eta \Pi_v d\eta \tag{20.2-38}$$

which is a generalization of the dimensionless stream function f used in Example 4.4-2. Then setting $\Lambda = 1$ in Eq. 20.2-34 and making the substitutions $f' = df/d\eta = \Pi_v$, $f'' = d^2f/d\eta^2 = d\Pi_v/d\eta$, and so on, gives the equation of motion in the form

$$-ff'' = f''' \tag{20.2-39}$$

and Eqs. 20.2-35, 36, and 38 give the boundary conditions

$$\text{at } \eta = \infty, \quad f' = 1 \tag{20.2-40}$$
$$\text{at } \eta = 0, \quad f' = 0 \tag{20.2-41}$$
$$\text{at } \eta = 0, \quad f = -K \tag{20.2-42}$$

Equation 20.2-39 can be solved numerically with these boundary conditions to obtain f as a function of η for various values of K.

Once the function $f(\eta, K)$ has been evaluated, we can integrate Eq. 20.2-34 with the boundary conditions in Eqs. 20.2-35 and 36 to obtain

$$\Pi(\eta, \Lambda, K) = \frac{\int_0^\eta \exp\left(-\Lambda \int_0^{\bar\eta} f(\bar{\bar\eta}, K) d\bar{\bar\eta}\right) d\bar\eta}{\int_0^\infty \exp\left(-\Lambda \int_0^{\bar\eta} f(\bar{\bar\eta}, K) d\bar{\bar\eta}\right) d\bar\eta} \tag{20.2-43}$$

Some profiles calculated from this equation by numerical integration are given in Fig. 20.2-3. The velocity profiles are given by the curves for $\Lambda = 1$. The temperature and composition profiles for various Prandtl and Schmidt numbers are given by the curves for the corresponding values of Λ. Note that the velocity, temperature, and composition boundary layers get thicker when K is positive (as in evaporation) and thinner when K is negative (as in condensation).

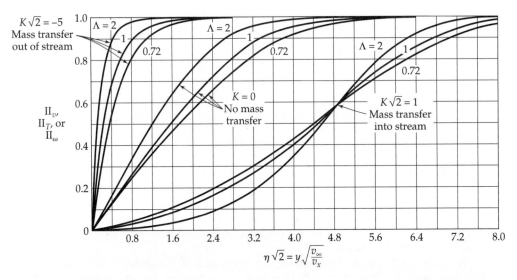

Fig. 20.2-3. Velocity, temperature, and composition profiles in the laminar boundary layer on a flat plate with mass transfer at the wall [H. S. Mickley, R. C. Ross, A. L. Squyers, and W. E. Stewart, *NACA Technical Note 3208* (1954).]

The gradients of the velocity, temperature, and composition at the wall are obtainable from the derivative of Eq. 20.2-43:

$$\Pi'(0, \Lambda, K) = \frac{d\Pi(\eta, \Lambda, K)}{d\eta}\bigg|_{\eta=0} = \frac{1}{\int_0^\infty \exp\left(-\Lambda \int_0^\eta f(\overline{\eta}, K)d\overline{\eta}\right)d\eta} \tag{20.2-44}$$

Some values computed from this formula by numerical integration are then given in Table 20.2-1.

Table 20.2-1   Dimensionless Gradients of Velocity, Temperature, and Composition in Laminar Flow Along a Flat Plate[a]

K	$\Lambda = 0.1$	$\Lambda = 0.2$	$\Lambda = 0.4$	$\Lambda = 0.6$	$\Lambda = 0.7$	$\Lambda = 0.8$	$\Lambda = 1.0$	$\Lambda = 1.4$	$\Lambda = 2.0$	$\Lambda = 5.0$
−3.0	0.4491	0.7681	1.3722	1.9648	2.2600	2.5550	3.1451	4.3273	6.1064	15.0567
−2.0	0.3664	0.5956	1.0114	1.4100	1.6070	1.8032	2.1945	2.9764	4.1524	10.0863
−1.0	0.2846	0.4282	0.6658	0.8799	0.9829	1.0842	1.2836	1.6754	2.2568	5.1747
−0.5	0.2427	0.3452	0.4999	0.6291	0.6890	0.7468	0.8579	1.0688	1.3707	2.8194
−0.2	0.2165	0.2948	0.4024	0.4849	0.5213	0.5555	0.6190	0.7333	0.8861	1.5346
0.0	0.1980	0.2604	0.3380	0.3917	0.4139	0.4340	0.4696	0.5281	0.5972	0.8156
0.2	0.1783	0.2246	0.2736	0.3011	0.3108	0.3187	0.3305	0.3439	0.3496	0.3015
0.5	0.1441	0.1657	0.1751	0.1701	0.1656	0.1603	0.1485	0.1240	0.09096	0.01467
0.75	0.1032	0.1023	0.0840	0.0638	0.0549	0.0471	0.0340	0.0172	0.00571	0.0000152
0.87574[b]	0	0	0	0	0	0	0	0	0	0

[a] Taken from the following sources: E. Elzy and R. M. Sisson, *Engineering Experiment Station Bulletin No. 40*, Oregon State University, Corvallis, Or. (1967); H. L. Evans, *Int. J. Heat and Mass Transfer*, **3**, 321–339 (1961); W. E. Stewart and R. Prober, *Int. J. Heat and Mass Transfer*, **5**, 1149–1163 (1962) and **6**, 872 (1963). More complete results, and reviews of earlier work, are given in these references.

[b] The value K = 0.87574 is the largest positive mass transfer rate attainable in this geometry with steady laminar flow. See H. W. Emmons and D. C. Leigh, *Interim Technical Report No. 9*, Combustion Aerodynamics Laboratory, Harvard University (1953).

The molecular fluxes of momentum, energy, and mass at the wall are then given by the dimensionless expressions

$$\frac{\tau_0}{\rho v_\infty (v_\infty - 0)} = \Pi'(0, 1, K) \sqrt{\frac{\nu}{2v_\infty x}} \tag{20.2-45}$$

$$\frac{q_0}{\rho \hat{C}_p v_\infty (T_0 - T_\infty)} = \frac{\Pi'(0, \Pr, K)}{\Pr} \sqrt{\frac{\nu}{2v_\infty x}} \tag{20.2-46}$$

$$\frac{j_{A0}}{\rho v_\infty (\omega_{A0} - \omega_{A\infty})} = \frac{\Pi'(0, \mathrm{Sc}, K)}{\mathrm{Sc}} \sqrt{\frac{\nu}{2v_\infty x}} \tag{20.2-47}$$

with the tabulated values of $\Pi'(0, \Lambda, K)$. Thus the fluxes can be computed directly when K is known. These expressions are obtained from the flux expressions of Newton, Fourier, and Fick, and the profiles as given in Eq. 20.2-43. The energy flux q_0 here corresponds to the conduction term $-k\nabla T$ of Eq. 19.3-3; the diffusive flux j_{A0} is obtained by using Eq. 20.2-47 above.

The fluid properties ρ, μ, $\hat{C}_p$, k, and $\mathscr{D}_{AB}$ have been treated as constants in this development. However, Eqs. 20.2-45 to 47 have been found to agree closely with the corresponding variable-property calculations,[8-10] provided that K is generalized as follows,

$$K = \frac{\rho_0 v_0(x)}{\rho v_\infty} \sqrt{2 \frac{v_\infty x}{\nu}} \tag{20.2-48}$$

and that ρ, μ, $\hat{C}_p$, k, and $\mathscr{D}_{AB}$ are evaluated at the "reference conditions" $T_f = \frac{1}{2}(T_0 + T_\infty)$ and $\omega_{Af} = \frac{1}{2}(\omega_{A0} + \omega_{A\infty})$.

In many situations, one of the following dimensionless quantities

$$R_v = \frac{(n_{A0} + n_{B0})(v_\infty - 0)}{\tau_0} \tag{20.2-49}$$

$$R_T = \frac{(n_{A0} + n_{B0})\hat{C}_p(T_0 - T_\infty)}{q_0} \tag{20.2-50}$$

$$R_\omega = \frac{(n_{A0} + n_{B0})(\omega_{A0} - \omega_{A\infty})}{j_{A0}} = \frac{(\omega_{A0} - \omega_{A\infty})(n_{A0} + n_{B0})}{n_{A0} - \omega_{A0}(n_{A0} + n_{B0})} \tag{20.2-51}$$

is known or readily computed. These flux ratios, R, are independent of x under the present boundary conditions and are related to Λ and K as follows,

$$R = \frac{K\Lambda}{\Pi'(0, \Lambda, K)} \tag{20.2-52}$$

according to Eqs. 20.2-45 to 51. From Eq. 20.2-52 we see that the dimensionless interfacial mass flux K can be tabulated as a function of R and Λ, by use of the results in Table 20.2-1. Then K can be found by interpolation if the numerical values of R and Λ are given for one of the three profiles (i.e., if we can specify R_v, or R_T and Pr, or R_ω and Sc.) Convenient plots of these relations are given in Figures 22.8-5 to 7.

As a simple illustration, suppose that the flat plate is porous and is saturated with liquid A, which vaporizes into a gaseous stream of A and B. Suppose also that gas B is noncondensable and insoluble in liquid A, and that ω_{A0} and $\omega_{A\infty}$ are given. Then R_ω can be calculated from

---

[8] For calculations of momentum and energy transfer in gas flows with $K = 0$, see E. R. G. Eckert, *Trans. A.S.M.E.*, **78**, 1273–1283 (1956).

[9] For calculations of momentum and mass transfer in binary and multicomponent gas mixtures, see W. E. Stewart and R. Prober, *Ind. Eng. Chem. Fundamentals*, **3**, 224–235 (1964); improved reference conditions are provided by T. C. Young and W. E. Stewart, *ibid.*, **25**, 276–482 (1986), as noted in §20.9.

[10] For other methods of applying Eq. 20.2-47 to variable-property fluids, see O. T. Hanna, *AIChE Journal*, **8**, 278–279 (1962); **11**, 706–712 (1965).

Table 20.2-2   Coefficients for the Approximate Flat-Plate Formulas,[a] Eqs. 20.2-54 and 55

Λ	0	0.1	0.2	0.5	0.7	1.0	2	5	10	100	∞
$a(\Lambda)$	$\left(\dfrac{2}{\pi}\right)^{1/2}\Lambda^{1/6}$	0.4266	0.4452	0.4620	0.4662	0.4696	0.4740	0.4769	0.4780	0.4789	0.4790
$b(\Lambda)$	1.308	0.948	0.874	0.783	0.752	0.723	0.676	0.632	0.610	0.577	0.566

[a] Taken from H. J. Merk, *Appl. Sci. Res.*, **A8**, 237–277 (1959), and R. Prober and W. E. Stewart, *Int. J. Heat and Mass Transfer*, **6**, 221–229, 872 (1963).

Eq. 20.2-51 with $n_{B0} = 0$, and K can be found by interpolating the function $K(R, \Lambda)$ to $R = R_\omega$ and $\Lambda = \mu/\rho\mathcal{D}_{AB}$.

For moderate values of K, the calculations can be simplified by representing $\Pi'(0, \Lambda, K)$ as a truncated Taylor series in the parameter K:

$$\Pi'(0, \Lambda, K) = \Pi'(0, \Lambda, 0) + K\frac{\partial}{\partial K}\Pi'(0, \Lambda, K)\bigg|_{K=0} \tag{20.2-53}$$

This expansion can be written more compactly as

$$\Pi'(0, \Lambda, K) = a\Lambda^{1/3} - bK\Lambda \tag{20.2-54}$$

in which a and b are slowly varying functions of Λ, given in Table 20.2-2. Insertion of Eq. 20.2-54 into Eq. 20.2-52 gives the convenient expression for the dimensionless interfacial mass flux K

$$K = a\Lambda^{-2/3}\frac{R}{1 + bR} \tag{20.2-55}$$

for calculations with unknown parameter K. This result is easy to use and fairly accurate. The predicted function $K(R, \Lambda)$ is within 1.6% of that found from Table 20.2-1 for $|R| < 0.25$ and $\Lambda > 0.1$.

This example illustrates the related effects of the interfacial velocity v_0 on the velocity, temperature, and composition profiles. The effect of v_0 on a given profile, Π, is small if $R <\!< 1$ for that profile (as in most separation processes) and large if $R \geq 1$ (as in many combustion and transpiration cooling processes). Some applications are given in Chapter 22.

EXAMPLE 20.2-3 *Approximate Analogies for the Flat Plate at Low Mass-Transfer Rates*	Pohlhausen[11] solved the energy equation for the system of Example 12.1-2 and curve-fitted his results for the heat transfer rate Q (see third line of Table 12.4-1). Compare his result with Eq. 20.2-46, and derive the corresponding results for the momentum and mass fluxes. **SOLUTION** By inserting the coefficient 0.664 in place of $\sqrt{148/315}$ in Eq. 12.4-17, and setting $2Wq_0(x) = (dQ/dL)\vert_{L=x}$, we get

$$\frac{q_0}{\rho\hat{C}_p v_\infty(T_0 - T_\infty)} \cong 0.332\,\mathrm{Pr}^{2/3}\sqrt{\frac{\nu}{v_\infty x}} \tag{20.2-56}$$

This result is subject to the boundary condition $v_0(x) = 0$, which corresponds to $K = 0$ in the system of Example 20.2-2.

---

[11] E. Pohlhausen, *Zeits. f. angew. Math. Mech.*, **1**, 115–121 (1921).

Equation 20.2-56 is obtainable from Eq. 20.2-46 when $K = 0$ by setting $\Pi'(0, \text{Pr}, 0) \cong 0.4696\text{Pr}^{1/3}$; this agrees with Table 20.2-2 at $\Lambda = 1$. Making comparable substitutions in Eqs. 20.2-45 and 46, we get the convenient analogy

$$\frac{\tau_0}{\rho v_\infty(v_\infty - 0)} \cong \frac{q_0}{\rho \hat{C}_p v_\infty(T_0 - T_\infty)} \text{Pr}^{2/3} \cong \frac{j_{A0}}{\rho v_\infty(\omega_{A0} - \omega_{A\infty})} \text{Sc}^{2/3}$$

$$\cong 0.332 \sqrt{\frac{\nu}{v_\infty x}} \tag{20.2-57}$$

which has been recommended by Chilton and Colburn[12] for this flow situation (cf. §§14.3 and 22.3). The expression for τ_0 agrees with the exact solution at $K = 0$, and the results for q_0 and j_{A0} are accurate within $\pm 2\%$ at $K = 0$ for $\Lambda > 0.5$.

§20.3 STEADY-STATE BOUNDARY LAYER THEORY FOR FLOW AROUND OBJECTS

In §§18.5 and 6 we discussed two related mass transfer problems of boundary layer type. Now we want to enlarge[1-7] on the ideas presented there and consider the flow around objects of other shapes such as the one shown in Fig. 12.4-2. Although we present the material in this section in terms of mass transfer, it is understood that the results can be taken over directly for the analogous heat transfer problem by appropriate changes of notation. The concentration boundary layer is presumed to be very thin, which means that the results are restricted either to small diffusivity or to short exposure times. The results are applicable only to the region between the forward stagnation locus (from which x is measured) and the region of separation or turbulence, if any, as indicated in Figure 12.4-2.

The concentration of the diffusing species is called c_A, and its concentration at the surface of the object is c_{A0}. Outside the concentration boundary layer, the concentration of A is zero.

Proceeding as in Example 12.4-3, we adopt an orthogonal coordinate system for the concentration boundary layer, in which x is measured along the surface everywhere in the direction of the streamlines. The y-coordinate is perpendicular to the surface, and the z-coordinate is measured along the surface perpendicular to the streamlines. These are "general orthogonal coordinates," as described in Eqs. A.7-10 to 18, but with $h_y = 1$, and $h_x = h_x(x, z)$ and $h_z = h_z(x, z)$. Since the flow near the interface does not have a velocity component in the z direction, the equation of continuity there is

$$\frac{\partial}{\partial x}(h_z v_x) + h_x h_z \frac{\partial}{\partial y} v_y = 0 \tag{20.3-1}$$

[12] T. H. Chilton and A. P. Colburn, *Ind. Eng. Chem.*, **26**, 1183–1187 (1934).

[1] A. Acrivos, *Chem. Eng. Sci.*, **17**, 457–465 (1962).

[2] W. E. Stewart, *AIChE Journal*, **9**, 528–535 (1963).

[3] D. W. Howard and E. N. Lightfoot, *AIChE Journal*, **14**, 458–467 (1968).

[4] W. E. Stewart, J. B. Angelo, and E. N. Lightfoot, *AIChE Journal*, **16**, 771–786 (1970).

[5] E. N. Lightfoot, in *Lectures in Transport Phenomena*, American Institute of Chemical Engineers, New York (1969).

[6] E. Ruckenstein, *Chem. Eng. Sci.*, **23**, 363–371 (1968).

[7] W. E. Stewart, in *Physicochemical Hydrodynamics*, Vol. 1 (D. B. Spalding, ed.), Advance Publications, Ltd., London (1977), pp. 22–63.

according to Eq. A.7-16. The diffusion equation for the concentration boundary layer is then

$$v_x \frac{1}{h_x} \frac{\partial c_A}{\partial x} + v_y \frac{\partial c_A}{\partial y} = \mathcal{D}_{AB} \frac{\partial^2 c_A}{\partial y^2} \tag{20.3-2}$$

where Eqs. A.7-15 and 17 have been used. In writing these equations it has been assumed that: (i) the x- and z-components of the diffusion flux are negligible, (ii) the boundary layer thickness is small compared to the local interfacial radii of curvature, and (iii) the density and diffusivity are constant. We now want to get formal expressions for the concentration profiles and mass fluxes for two cases that are generalizations of the problems solved in §18.5 and §18.6. When we get the expressions for the local molar flux at the interface, we will find that the dependences on the diffusivity ($\frac{1}{2}$-power in §18.5 and the $\frac{2}{3}$-power in §18.6) correspond to cases (a) and (b) below. This turns out to be of great importance in the establishment of dimensionless correlations for mass transfer coefficients, as we shall see in Chapter 22.

Zero Velocity Gradient at the Mass Transfer Surface

This situation arises in a surfactant-free liquid flowing around a gas bubble. Here v_x does not depend on y, and v_y can be obtained from the equation of continuity given above. Therefore, for small mass-transfer rates we can write general expressions for the velocity components as

$$v_x = v_s(x, z) \tag{20.3-3}$$

$$v_y = -\frac{y}{h_x h_z} \frac{\partial}{\partial x} (h_z v_s) \equiv -\gamma y \tag{20.3-4}$$

where γ depends on x and z. When this is used in Eq. 20.3-2, we get for the diffusion in the liquid phase

$$v_s \frac{1}{h_x} \frac{\partial c_A}{\partial x} - \gamma y \frac{\partial c_A}{\partial y} = \mathcal{D}_{AB} \frac{\partial^2 c_A}{\partial y^2} \tag{20.3-5}$$

which is to be solved with the boundary conditions

B.C. 1:	at $x = 0$,	$c_A = 0$	(20.3-6)
B.C. 2:	at $y = 0$,	$c_A = c_{A0}$	(20.3-7)
B.C. 3:	as $y \to \infty$,	$c_A \to 0$	(20.3-8)

The nature of the boundary conditions suggests that a combination of variables treatment might be appropriate. However, it is far from obvious how to construct an appropriate dimensionless combination. Hence we try the following: let $c_A/c_{A0} = f(\eta)$, where $\eta = y/\delta_A(x, z)$, and $\delta_A(x, z)$ is the boundary layer thickness for species A, to be determined later.

When the indicated combination of variables is introduced into Eq. 20.3-5, the equation becomes

$$\frac{d^2 f}{d\eta^2} + \frac{1}{\mathcal{D}_{AB}} \left(\frac{v_s}{h_x} \delta_A \frac{\partial \delta_A}{\partial x} + \gamma \delta_A^2 \right) \eta \frac{df}{d\eta} = 0 \tag{20.3-9}$$

with the boundary conditions: $f(0) = 1$ and $f(\infty) = 0$. If, now, the coefficient of the $\eta(df/d\eta)$ term were a constant, then Eq. 20.3-9 would have the same form as Eq. 4.1-9, which we know how to solve. For convenience we specify the constant as

$$\frac{1}{\mathcal{D}_{AB}} \left(\frac{v_s}{h_x} \delta_A \frac{\partial \delta_A}{\partial x} + \gamma \delta_A^2 \right) = 2 \tag{20.3-10}$$

Next we insert the expression for γ from Eq. 20.3-4 and rearrange the equation thus:

$$\frac{\partial}{\partial x} \delta_A^2 + \left(\frac{\partial}{\partial x} \ln (h_z v_s)^2 \right) \delta_A^2 = \frac{4 \mathcal{D}_{AB} h_x}{v_s} \tag{20.3-11}$$

This is a linear, first-order equation for δ_A^2, which has to be solved with the boundary condition $\delta_A = 0$ at $x = 0$. Integration of Eq. 20.3-11 gives

$$\delta_A(x, z) = 2\sqrt{\dfrac{\mathscr{D}_{AB}\displaystyle\int_0^x h_x h_z^2 v_s x}{h_z^2 v_s^2}} \tag{20.3-12}$$

as the thickness function for the diffusional boundary layer. Since Eq. 20-3-9 and the boundary conditions then contain η as the only independent variable, the postulated combination of variables is valid, and the concentration profiles are given by the solution of Eq. 20.3-9:

$$f(\eta) = 1 - \frac{2}{\sqrt{\pi}}\int_0^\eta \exp(-\bar{\eta}^2)\, d\bar{\eta} = 1 - \mathrm{erf}\,\eta \tag{20.3-13}$$

Equations 20.3-12 and 13 are the solution to the problem at hand.

Next, we combine this solution with Fick's first law to evaluate the molar flux of species A at the interface:

$$N_{Ay}|_{y=0} = -\mathscr{D}_{AB}\frac{\partial c_A}{\partial y}\bigg|_{y=0} = -\mathscr{D}_{AB}c_{A0}\left(\frac{df}{d\eta}\frac{\partial\eta}{\partial y}\right)\bigg|_{y=0}$$

$$= +\mathscr{D}_{AB}c_{A0}\frac{2}{\sqrt{\pi}}\left(\exp(-\eta^2)\frac{1}{\delta_A}\right)\bigg|_{y=0} = c_{A0}\sqrt{\frac{\mathscr{D}_{AB}}{\pi}\frac{h_z^2 v_s^2}{\displaystyle\int_0^x h_x h_z^2 v_s\, d\bar{x}}} \tag{20.3-14}$$

This result shows the same dependence of the mass flux on the $\frac{1}{2}$-power of the diffusivity that arose in Eq. 18.5-17, for the much simpler gas absorption problem solved there. In fact, if we set the scale factors h_x and h_z equal to unity and replace v_s by $v_{\max}$, we recover Eq. 18.5-17 exactly.

Linear Velocity Profile Near the Mass-Transfer Surface

This velocity function is appropriate for mass transfer at a solid surface (see Example 12.4-3) when the concentration boundary layer is very thin. Here v_x depends linearly on y within the concentration boundary layer, and v_y can be obtained from the equation of continuity. Consequently, when the net mass flux through the interface is small, the velocity components in the concentration boundary layer are

$$v_x = \beta(x, z)y \tag{20.3-15}$$

$$v_y = -\frac{y^2}{2h_x h_z}\frac{\partial}{\partial x}(h_z\beta) \equiv -\gamma y^2 \tag{20.3-16}$$

in which γ depends on x and z. Substituting these expressions into Eq. 20.3-2 gives the diffusion equation for the liquid phase

$$\frac{\beta y}{h_x}\frac{\partial c_A}{\partial x} - \gamma y^2\frac{\partial c_A}{\partial y} = \mathscr{D}_{AB}\frac{\partial^2 c_A}{\partial y^2} \tag{20.3-17}$$

which is to be solved with the boundary conditions

B.C. 1:	at $x = 0$,	$c_A = 0$	(20.3-18)
B.C. 2:	at $y = 0$,	$c_A = c_{A0}$	(20.3-19)
B.C. 3:	as $y \to \infty$,	$c_A \to 0$	(20.3-20)

Once again, we use the method of combination of variables, by setting $c_A/c_{A0} = f(\eta)$, where $\eta = y/\delta_A(x, z)$.

When the change of variables is made, the diffusion equation becomes

$$\frac{d^2 f}{d\eta^2} + \frac{1}{\mathscr{D}_{AB}}\left(\frac{\beta}{h_x}\delta_A^2\frac{\partial\delta_A}{\partial x} + \gamma\delta_A^3\right)\eta^2\frac{df}{d\eta} = 0 \tag{20.3-21}$$

with the boundary conditions: $f(0) = 1$ and $f(\infty) = 0$. A solution of the form $f(\eta)$ is possible only if the factor in parentheses is a constant. Setting the constant equal to 3 reduces Eq. 20.3-21 to Eq. 18.6-6, for which the solution is known. Therefore we now get the boundary layer thickness by requiring that

$$\frac{1}{\mathscr{D}_{AB}} \left(\frac{\beta}{h_x} \delta_A^2 \frac{\partial \delta_A}{\partial x} + \gamma \delta_A^3 \right) = 3 \tag{20.3-22}$$

or

$$\frac{\partial}{\partial x} \delta_A^3 + \left(\frac{\partial}{\partial x} \ln (h_z \beta)^{3/2} \right) \delta_A^3 = \frac{9 \mathscr{D}_{AB} h_x}{\beta} \tag{20.3-23}$$

The solution of this first-order, linear equation for δ_A^3 is

$$\delta_A = \frac{1}{\sqrt{h_z \beta}} \sqrt[3]{9 \mathscr{D}_{AB} \int_0^x \sqrt{h_z \beta} \, h_x h_z \, dx} \tag{20.3-24}$$

Hence the solution to the problem in this subsection is

$$\frac{c_A}{c_{A0}} = f(\eta) = \frac{\int_\eta^\infty \exp (-\overline{\eta}^3) \, d\overline{\eta}}{\Gamma(\tfrac{4}{3})} \tag{20.3-25}$$

which reduces to Eq. 18.6-10 for the system considered there.

Finally, we get the expression for the molar flux at the interface, which is

$$N_{Ay}\big|_{y=0} = -\mathscr{D}_{AB} \frac{\partial c_A}{\partial y}\bigg|_{y=0} = -\mathscr{D}_{AB} c_{A0} \left(\frac{df}{d\eta} \frac{\partial \eta}{\partial y} \right)\bigg|_{y=0}$$

$$= \frac{\mathscr{D}_{AB} c_{A0}}{\Gamma(\tfrac{4}{3})} \frac{\sqrt{h_z \beta}}{\sqrt[3]{9 \mathscr{D}_{AB} \int_0^x \sqrt{h_z \beta} \, h_x h_z \, dx}} \tag{20.3-26}$$

For a plane surface, with $h_x = h_z = 1$ and $\beta = $ constant, Eq. 20.3-26 reduces to Eq. 18.6-11.

EXAMPLE 20.3-1

Mass Transfer for Creeping Flow Around a Gas Bubble

A liquid B is flowing very slowly around a spherical bubble of gas A of radius R. Find the rate of mass transfer of A into the surrounding fluid, if the solubility of gas A in liquid B is c_{A0}.
(a) Show how to use Eq. 20.3-14 to get the mass flux at the gas–liquid interface for this system.
(b) Then get the average mass flux over the entire spherical surface.

SOLUTION

(a) Select as the origin of coordinates the upstream stagnation point, and define the coordinates x and z as follows: $x = R\theta$ and $z = R(\sin \theta)\phi$, in which θ and ϕ are the usual spherical coordinates. The y direction is then the same as the r direction of spherical coordinates. The interfacial velocity is obtained from Eq. 4B.3-3 as $v_s = \tfrac{1}{2} v_\infty \sin \theta$, where v_∞ is the approach velocity.

When these quantities are inserted into Eq. 20.3-14 we get

$$N_{Ay}\big|_{y=0} = c_{A0} \sqrt{\frac{\mathscr{D}_{AB}}{\pi} \frac{(R \sin \theta)^2 (\tfrac{1}{2} v_\infty \sin \theta)^2}{\int_0^{R\theta} (R)(R \sin \theta)^2 (\tfrac{1}{2} v_\infty \sin \theta) d(R\theta)}}$$

$$= c_{A0} \sin^2 \theta \sqrt{\frac{\mathscr{D}_{AB} v_\infty}{2\pi R} \frac{1}{\int_0^\theta \sin^3 \theta \, d\theta}}$$

$$= c_{A0} \sin^2 \theta \sqrt{\frac{3 \mathscr{D}_{AB} v_\infty}{2\pi R}} \frac{1}{\sqrt{\cos^3 \theta - 3 \cos \theta + 2}} \tag{20.3-27}$$

(b) To get the surface-averaged value of the mass flux, we integrate the above expression over all θ and ϕ and divide by the sphere surface:

$$
\begin{aligned}
N_{A0,\text{avg}} &= \frac{1}{4\pi R^2} \int_0^{2\pi} \int_0^{\pi} N_{Ay}\bigg|_{y=0} R \sin\theta \, d\theta \, d\phi \\
&= \frac{2\pi R^2 c_{A0}}{4\pi R^2} \sqrt{\frac{3\mathscr{D}_{AB} v_\infty}{2\pi R}} \int_0^{\pi} \frac{\sin^3\theta \, d\theta}{\sqrt{\cos^3\theta - 3\cos\theta + 2}} \\
&= \frac{c_{A0}}{2} \sqrt{\frac{3\mathscr{D}_{AB} v_\infty}{2\pi R}} \int_{-1}^{+1} \frac{(1 - u^2)\,du}{\sqrt{u^3 - 3u + 2}} \\
&= \frac{c_{A0}}{2} \sqrt{\frac{3\mathscr{D}_{AB} v_\infty}{2\pi R}} \int_{-1}^{+1} \frac{(1 + u)\,du}{\sqrt{2 + u}} \\
&= \frac{c_{A0}}{2} \sqrt{\frac{3\mathscr{D}_{AB} v_\infty}{2\pi R}} \left(\frac{4}{3}\right) = \sqrt{\frac{4\mathscr{D}_{AB} v_\infty}{3\pi D}}\, c_{A0}
\end{aligned}
\tag{20.3-28}
$$

In going from the second to the third line, we made the change of variable $\cos\theta = u$, and to get the fourth line, we factored out $(1 - u)$ from the numerator and denominator. Equation 20.3-28 was cited in Eq. 18.5-20 in connection with absorption from gas bubbles.[8] This equation is referred to again in Chapter 22 in connection with the subject of mass transfer coefficients.

§20.4 BOUNDARY LAYER MASS TRANSPORT WITH COMPLEX INTERFACIAL MOTION[1-3]

Time-dependent interfacial motions and turbulence are common in fluid–fluid transfer operations. Boundary layer theory gives useful insight and asymptotic relations for these systems, utilizing the thinness of the concentration boundary layers for small $\mathscr{D}_{AB}$ (as in liquids) or for flows with frequent boundary layer separation (as at rippling or oscillating interfaces). Mass transfer with simple interfacial motions has been discussed in §18.5 for a laminar falling film and a circulating bubble, and in Example 20.1-4 for a uniformly expanding interface. Here we consider mass transfer with more general interfacial motions.

Consider the time-dependent transport of species A between two fluid phases, with initially uniform but different compositions. We start with the binary continuity equation for constant ρ and $\mathscr{D}_{AB}$ (Eq. 19.1-16, divided by ρ):

$$
\frac{D\omega_A}{Dt} = \mathscr{D}_{AB}\nabla^2\omega_A + \frac{1}{\rho} r_A
\tag{20.4-1}
$$

We now want to reduce this to boundary layer form for small $\mathscr{D}_{AB}$, and then present solutions for various forced-convection problems with controlling resistance in one phase.

We use the following boundary layer approximations:

(i) that the diffusive mass flux is collinear with the unit vector $\mathbf{n}$ normal to the nearest interfacial element. (This approximation is used throughout the boundary layer sections of this book. Higher-order approximations,[4] not treated here, are appropriate for describing boundary layer diffusion near edges, wakes, and separation loci.)

[8] V. G. Levich, *Physicochemical Hydrodynamics*, Prentice-Hall, Englewood Cliffs, N.J. (1962), p. 408, Eq. 72.9.

[1] J. B. Angelo, E. N. Lightfoot, and D. W. Howard, *AIChE Journal*, **12**, 751–760 (1966).

[2] W. E. Stewart, J. B. Angelo, and E. N. Lightfoot, *AIChE Journal*, **16**, 771–786 (1970).

[3] W. E. Stewart, *AIChE Journal*, **33**, 2008–2016 (1987); **34**, 1030 (1988).

[4] J. Newman, *Electroanal. Chem. and Interfacial Electrochem.*, **6**, 187–352 (1973).

(ii) that the tangential fluid velocity relative to the interface is negligible within the concentration boundary layer. (This approximation is satisfactory for fluid–fluid systems free of surfactants, when the interfacial drag is not too large.)

(iii) that the concentration boundary layer along each interface is thin relative to the local radii of interfacial curvature.

(iv) that the concentration boundary layers on nonadjacent interfacial elements do not overlap.

Each of these approximations is asymptotically valid for small $\mathcal{D}_{AB}$ in nonrecirculating flows with nonrigid interfaces and nonzero $D\omega_A/Dt$—that is, with time-dependent concentration as viewed by an observer moving with the fluid. The systems considered in part (a) of §20.3 are thus included, because they are time-dependent for such an observer (though steady for a stationary one).

Interfacially embedded coordinates are used in this discussion, with a piecewise smooth interfacial grid as in Fig. 20.4-1. Each interfacial element in the system is permanently labeled with surface coordinates (u, w), and its position vector is $\mathbf{r}_s(u, w, t)$. Each point in a boundary layer is identified by its distance y from the nearest interfacial point, together with the surface coordinates (u, w) of that point. The instantaneous position vector of each point (u, w, y) at time t is then

$$\mathbf{r}(u, w, y, t) = \mathbf{r}_s(u, w, t) + y\mathbf{n}(u, w, t) \tag{20.4-2}$$

relative to a stationary origin, as illustrated in Fig. 20.4-2. The function $\mathbf{r}_s(u, w, t)$ gives the trajectory of each interfacial point $(u, w, 0)$, and the associated function $\mathbf{n}(u, w, t) = (\partial/\partial y)\mathbf{r}$ gives the instantaneous normal vector from each surface element toward its positive side. These functions are computable from fluid mechanics for simple flows, and provide a framework for analyzing experiments in complex flows.

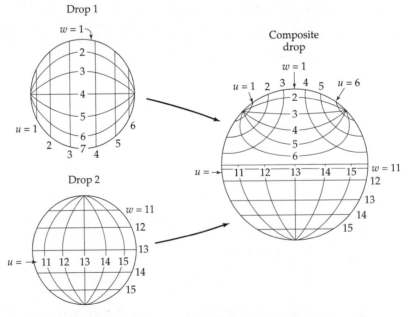

Fig. 20.4-1. Schematic illustration of embedded coordinates in a simple coalescence process. W.E. Stewart, J.B. Angelo, and E.N. Lightfoot, AIChE Journal, **14**, 458–467 (1968).

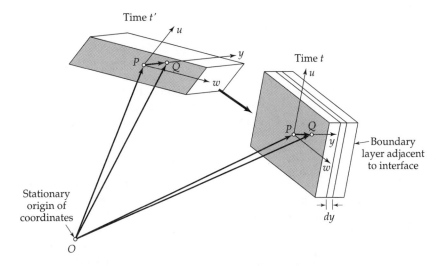

Fig. 20.4-2. Element dS (shaded) of a deforming interfacial area shown at two different times, t' and t, with the adjacent boundary layer. The vectors are (at time t):

$$\vec{OP} = \mathbf{r}_s(u, w, t) \quad = \text{position vector of a point on the interface}$$

$$\vec{PQ} = y\mathbf{n}(u, w, t) \quad = \text{vector of length } y \text{ normal to the interface locating a point in the boundary layer}$$

$$\vec{OQ} = \mathbf{r}(u, w, y, t) = \text{position vector for a point in the boundary layer}$$

The element of interfacial area consists of the same material particles as it moves through space. The magnitude of the area changes with time and is given by $dS = \left| \dfrac{\partial \mathbf{r}_s}{\partial u}\, du \times \dfrac{\partial \mathbf{r}_s}{\partial w}\, dw \right|$. Similarly, the magnitude of the volume of that part of the boundary layer between y and $y + dy$ is $dV = \left| \left[\dfrac{\partial \mathbf{r}_s}{\partial u}\, du \times \dfrac{\partial \mathbf{r}_s}{\partial w}\, dw \right] \cdot \mathbf{n}\, dy \right|$.

The instantaneous volume of a spatial element $du\, dw\, dy$ in the boundary layer (see Fig. 20.4-2) is

$$dV = \sqrt{g(u, w, y, t)}\; du\, dw\, dy \tag{20.4-3}$$

in which $\sqrt{g(u, w, y, t)}$ is the following product of the local interfacial base vectors, $(\partial/\partial u)\mathbf{r}_s$ and $(\partial/\partial v)\mathbf{r}_s$, and the normal unit vector $(\partial/\partial y)\mathbf{r}_s = \mathbf{n}$,

$$\sqrt{g} = \left| \left[\frac{\partial \mathbf{r}_s}{\partial u} \times \frac{\partial \mathbf{r}_s}{\partial w} \right] \cdot \mathbf{n} \right| = \left| \left[\frac{\partial \mathbf{r}_s}{\partial u} \times \frac{\partial \mathbf{r}_s}{\partial w} \right] \right| \tag{20.4-4}$$

and is considered nonnegative in this discussion. The second equality follows because $\mathbf{n}$ is collinear with the vector product of the local interfacial base vectors, which lie in the plane of the interface. Correspondingly, the instantaneous area of the interfacial element $du\, dw$ in Fig. 20.4-2 is

$$dS = s(u, w, t)du\, dw \tag{20.4-5}$$

in which $s(u, w, t)$ is the following product of the interfacial basis vectors:

$$s(u, w, t) = \sqrt{g(u, w, 0, t)} = \left| \left[\frac{\partial \mathbf{r}_s}{\partial u} \times \frac{\partial \mathbf{r}_s}{\partial w} \right] \right| \tag{20.4-6}$$

In these interfacially embedded coordinates, the mass average velocity $\mathbf{V}$ relative to stationary coordinate axes takes the form

$$\mathbf{V}(u, w, y, t) = \mathbf{v}(u, w, y, t) + \frac{\partial}{\partial t} \mathbf{r}(u, w, y, t) \qquad (20.4\text{-}7)$$

In this section, $\mathbf{v}$ is the mass average fluid velocity relative to an observer at (u, w, y), and $(\partial/\partial t)\mathbf{r}(u, w, y, t)$ is the velocity of that observer relative to the stationary origin. Taking the divergence of this equation gives the corollary[2] (see Problem 20D.5)

$$(\nabla \cdot \mathbf{V}(\mathbf{r}, t)) = (\nabla \cdot \mathbf{v}(u, w, y, t)) + \frac{\partial \ln \sqrt{g(u, w, y, t)}}{\partial t} \qquad (20.4\text{-}8)$$

This equation states that the divergence of $\mathbf{V}$ differs from that of $\mathbf{v}$ by the local rate of expansion or contraction of the embedded coordinate frame.

The last term in Eq. 20.4-8 arises when interfacial deformation occurs. Its omission in such problems gives inaccurate predictions, which Higbie[5] and Danckwerts[6,7] then adjusted by introducing hypothetical surface residence times[5,6] or surface rejuvenation.[7] Such hypotheses are not needed in the present analysis.

Application of Eq. 20.4-8 at $y = 0$ and use of the constant-density condition

$$(\nabla \cdot \mathbf{V}) = 0 \qquad (20.4\text{-}9)$$

along with the no-slip condition on the tangential part of $\mathbf{v}$, gives the derivative

$$\left. \frac{\partial v_y}{\partial y} \right|_{y=0} = -\frac{\partial \ln s(u, w, t)}{\partial t} \qquad (20.4\text{-}10)$$

Hence, the truncated Taylor expansion

$$v_y = v_{y0} - y \frac{\partial \ln s(u, w, t)}{\partial t} + O(y^2) \qquad (20.4\text{-}11)$$

describes the normal component of $\mathbf{v}$ in an incompressible fluid near a deforming interface.

The corresponding expansion for the tangential part of $\mathbf{v}$ gives

$$\mathbf{v}_{\parallel} = y\mathbf{B}_{\parallel}(u, w, t) + O(y^2) \qquad (20.4\text{-}12)$$

in which $\mathbf{B}_{\parallel}(u, w, t)$ is the interfacial y-derivative of $\mathbf{v}_{\parallel}$. With these results (neglecting the $O(y^2)$ terms) and approximation (i), we can write Eq. 20.4-1 for $\omega_A(u, w, y, t)$ as

$$\frac{\partial \omega_A}{\partial t} + (y\mathbf{B}_{\parallel} \cdot \nabla \omega_A) + \left(v_{y0} - y \frac{\partial \ln s}{\partial t} \right) \frac{\partial \omega_A}{\partial y} = \mathscr{D}_{AB}(\nabla \cdot \mathbf{n}\nabla \omega_A) + \frac{1}{\rho} r_A$$

$$= \mathscr{D}_{AB} \left[\frac{\partial^2 \omega_A}{\partial y^2} + (\nabla_0 \cdot \mathbf{n}) \frac{\partial \omega_A}{\partial y} + \cdots \right] + \frac{1}{\rho} r_A \quad (20.4\text{-}13)$$

Here $(\nabla_0 \cdot \mathbf{n})$ is the surface divergence of $\mathbf{n}$ at the nearest interfacial point and is the sum of the principal curvatures of the surface there. The $+ \cdots$ stands for terms of higher order, which are here neglected.

To select the dominant terms in Eq. 20.4-13, we introduce a dimensionless coordinate

$$Y = y/\kappa(\mathscr{D}_{AB}) \qquad (20.4\text{-}14)$$

[5] R. Higbie, *Trans. AIChE*, **31**, 365–389 (1935).
[6] P. V. Danckwerts, *Ind. Eng. Chem.*, **43**, 1460–1467 (1951).
[7] P. V. Danckwerts, *AIChE Journal*, **1**, 456–463 (1955).

in which κ is an average thickness of the concentration boundary layer. When Eq. 20.4-14 is written in terms of this new variable, we get

$$
\frac{\partial \omega_A}{\partial t} + \kappa(Y\mathbf{B}_\parallel \cdot \nabla \omega_A) + \left(\frac{v_{y0}}{\kappa} - Y \frac{\partial \ln s}{\partial t} \right) \frac{\partial \omega_A}{\partial Y}
$$

$$
= \frac{\mathscr{D}_{AB}}{\kappa^2} \left[\frac{\partial^2 \omega_A}{\partial Y^2} + \kappa(\nabla_0 \cdot \mathbf{n}) \frac{\partial \omega_A}{\partial Y} + \cdots \right] + \frac{1}{\rho} r_A \tag{20.4-15}
$$

for ω_A in terms of u, w, Y, and t. Since, on physical grounds, κ will decrease with decreasing $\mathscr{D}_{AB}$, the dominant terms for small $\mathscr{D}_{AB}$ are those of lowest order in κ—namely, all but the $\mathbf{B}_\parallel$ and $(\nabla_0 \cdot \mathbf{n})$ contributions. The subdominance of the latter terms confirms the asymptotic validity of approximations (ii) and (iii) in non-recirculating flows.

Now, the coefficients of all the dominant terms must be proportional over the range of $\mathscr{D}_{AB}$, in order that these terms remain of comparable size in the small-$\mathscr{D}_{AB}$ limit. Such a "dominant balance principle" was applied previously in §13.6. Here it gives the orders of magnitude

$$
\mathscr{D}_{AB}/\kappa^2 = \mathrm{O}(1) \qquad \text{and} \qquad v_{y0}/\kappa = \mathrm{O}(1) \tag{20.4-16, 17}
$$

for the terms of the lowest order with respect to κ. Equation 20.4-16 is consistent with the previous examples of $\frac{1}{2}$-power dependence of the diffusional boundary layer thickness on $\mathscr{D}_{AB}$ in free-surface flows. It also confirms the asymptotic correctness of assumption (iv) for small values of $\mathscr{D}_{AB}$. Equation 20.4-17 is consistent with the proportionality of v_z^* to $\sqrt{\mathscr{D}_{AB}}$ shown under Eq. 20.1-10 for the Arnold problem. Thus, the boundary layer equation for ω_A in either phase near a deforming interface is

$$
\frac{\partial \omega_A}{\partial t} + \left(v_{y0} - y \frac{\partial \ln s}{\partial t} \right) \frac{\partial \omega_A}{\partial y} = \mathscr{D}_{AB} \frac{\partial^2 \omega_A}{\partial y^2} + \frac{1}{\rho} r_A \tag{20.4-18}
$$

to lowest order in κ. At the next order of approximation, terms proportional to κ would appear, and these involve the tangential velocity $y\mathbf{B}_\parallel$ and the interfacial curvature $(\nabla_0 \cdot \mathbf{n})$. The latter term appears in Problems 20C.1 and 20C.2.

Multiplication of Eq. 20.4-18 by ρ/M_A (a constant for the assumptions made here), and use of z as the coordinate normal to the interface as in Example 20.1-1, give the corresponding equation for the molar concentration $c_A(u, w, z, t)$

$$
\frac{\partial c_A}{\partial t} + \left(v_{z0} - z \frac{\partial \ln s}{\partial t} \right) \frac{\partial c_A}{\partial z} = \mathscr{D}_{AB} \frac{\partial^2 c_A}{\partial z^2} + R_A \tag{20.4-19}
$$

which allows convenient extension of several earlier examples. Another useful corollary is the binary boundary layer equation in terms of x_A and $\mathbf{v}^*$

$$
\frac{\partial x_A}{\partial t} + \left(v_{z0}^* - z \frac{\partial \ln s}{\partial t} \right) \frac{\partial x_A}{\partial z} = \mathscr{D}_{AB} \frac{\partial^2 x_A}{\partial z^2} + \frac{1}{c} [R_A - x_A(R_A + R_B)] \tag{20.4-20}
$$

in which c and $\mathscr{D}_{AB}$ have been treated as constants, as in Example 20.1-1.

EXAMPLE 20.4-1

Mass Transfer with Nonuniform Interfacial Deformation

Equation 20.4-19 readily gives a generalized form of Eq. 20.1-65, by omitting the reaction source term R_A and neglecting the normal velocity term v_{z0} (thus assuming the interfacial net mass flux to be small). The equation thus obtained has the form of Eq. 20.1-65, except that the total surface growth rate $d \ln S/dt$ is replaced by the local growth rate, given by $\partial \ln s(u, w, t)/\partial t$. The resulting partial differential equation has two additional space variables (u and w), but is solvable in the same manner, since no derivatives with respect to the added variables appear.

SOLUTION

Rewriting Eq. 20.1-66 with a boundary layer thickness function $\delta(u, w, t)$ leads by analogous steps to the relation

$$\delta(u, w, t) = \sqrt{4\mathscr{D}_{AB} \int_0^t [s(u, w, \bar{t})/s(u, w, t)]^2 \, d\bar{t}} \tag{20.4-21}$$

and the corresponding generalizations of Eqs. 20.1-71 and 72:

$$\frac{c_A}{c_{A0}} = 1 - \operatorname{erf} \frac{z}{\sqrt{4\mathscr{D}_{AB} \int_0^t [s(u, w, \bar{t})/s(u, w, t)]^2 \, d\bar{t}}} \tag{20.4-22}$$

$$N_{Az0} = c_{A0} \sqrt{\frac{\mathscr{D}_{AB}}{\pi t}} \left(\frac{1}{t} \int_0^t [s(u, w, \bar{t})/s(u, w, t)]^2 \, d\bar{t} \right)^{-1/2} \tag{20.4-23}$$

These solutions, unlike Eq. 20.1-71 and Eq. 20.1-72, include the spatial variations of the boundary layer thickness and interfacial molar flux N_{Az0} that occur in nonuniform flows. Local stretching of the interface (as at stagnation loci) thins the boundary layer and enhances N_{Az0}. Local interfacial shrinkage (at separation loci) diminishes N_{Az0}, but also ejects stale fluid from the boundary layer, allowing its mixing into the interior of the same phase. Observations of mass transfer enhancement by such mixing have been interpreted by some workers as "surface renewal," even though creation of new surface elements in an existing surface is not permitted in continuum fluid mechanics.

These results, and others for negligible v_{z0}, are obtainable conveniently by introducing the following new variables into Eq. 20.4-19:

$$Z = zs(u, w, t) \quad \text{and} \quad \tau = \int_0^t s^2(u, w, \bar{t}) d\bar{t} \tag{20.4-24, 25}$$

In the absence of chemical reactions, the resulting differential equation for the concentration function $c_A(u, w, Z, \tau)$ becomes

$$\frac{\partial c_A}{\partial \tau} = \mathscr{D}_{AB} \frac{\partial^2 c_A}{\partial Z^2} \tag{20.4-26}$$

This is a generalization of Fick's second law to an asymptotic relation for forced convection in free-surface flows.

EXAMPLE 20.4-2

Gas Absorption with Rapid Reaction and Interfacial Deformation

Show how to generalize Example 20.1-2 to flow systems, by using Eq. 20.4-26 for the two reaction-free zones.

SOLUTION

Using Eq. 20.4-26, we get the following replacements for Eqs. 20.1-26 and 27:

$$\frac{\partial c_A}{\partial \tau} = \mathscr{D}_{AS} \frac{\partial^2 c_A}{\partial Z^2} \qquad \text{for } 0 \leq Z \leq Z_R \tag{20.4-27}$$

$$\frac{\partial c_B}{\partial \tau} = \mathscr{D}_{BS} \frac{\partial^2 c_B}{\partial Z^2} \qquad \text{for } Z_R \leq Z < \infty \tag{20.4-28}$$

Now the reaction *plane* $z = z_R$ of the original example is a *time-dependent surface*, $Z = Z_R$, or $z_R(u, w, t) = Z_R/s(u, w, t)$. The initial and boundary conditions remain as before, subject to this generalization of the reaction-front location.

The solutions for c_A and c_B then take the forms in Eqs. 20.1-35 and 36, with $z/\sqrt{t}$ replaced by $Z/\sqrt{\tau}$, and $z_R/\sqrt{t}$ by $\sqrt{\gamma}$. The latter constant is again given by Eq. 20.1-37. The enhancement of the absorption rate by the chemical reaction accordingly parallels the expressions that will be given in Eq. 22.5-10, and simplified in Eqs. 22.5-11 through 13.

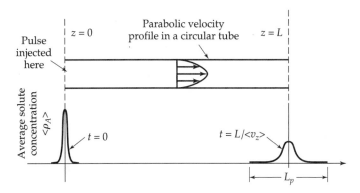

Fig. 20.5-1. Sketch showing the axial spreading of a concentration pulse in Taylor dispersion in a circular tube.

§20.5 "TAYLOR DISPERSION" IN LAMINAR TUBE FLOW

Here we discuss the transport and spreading of a solute "pulse" of material A introduced into fluid B in steady laminar flow through a long, straight tube of radius R, as shown in Fig. 20.5-1. A pulse of mass m_A is introduced at the inlet $z = 0$ over a very short period near time $t = 0$, and its progress through the tube is to be analyzed in the long-time limit. Problems of this type arise frequently in process control (see Problem 20C.4), medical diagnostic procedures,[1] and in a variety of environmental applications.[2]

A short distance downstream from the inlet, the θ-dependence of the mass fraction distribution will die out. Then the diffusion equation for $\omega_A(r, z, t)$ in Poiseuille flow with constant μ, ρ, and $\mathscr{D}_{AB}$ takes the form

$$\frac{\partial \omega_A}{\partial t} + v_{z,\max}\left[1 - \left(\frac{r}{R}\right)^2\right]\frac{\partial \omega_A}{\partial z} = \mathscr{D}_{AB}\left(\frac{1}{r}\frac{\partial}{\partial r}\left(r\frac{\partial \omega_A}{\partial r}\right) + \frac{\partial^2 \omega_A}{\partial z^2}\right) \tag{20.5-1}$$

This equation is to be solved with the boundary conditions

B.C. 1 and 2: at $r = 0$ and at $r = R$, $\dfrac{\partial \omega_A}{\partial r} = 0$ (20.5-2)

which express the radial symmetry of the mass fraction profile and the impermeability of the tube wall to diffusion. For this long-time analysis it is not necessary to specify the exact shape of the pulse injected at $t = 0$. No exact analytical solution is available for the mass fraction profile $\omega_A(r, z, t)$—even if an initial condition were clearly formulated—but Taylor[3,4] gave a useful approximate analysis that we summarize here. This involves getting from Eq. 20.5-1 a partial differential equation for the cross-sectional average mass fraction

$$\langle \omega_A \rangle = \frac{\int_0^{2\pi}\int_0^R \omega_A r\, dr\, d\theta}{\int_0^{2\pi}\int_0^R r\, dr\, d\theta} = \frac{2}{R^2}\int_0^R \omega_A r\, dr \tag{20.5-3}$$

which can then be solved to describe the behavior at long times.

[1] J. B. Bassingthwaighte and C. A. Goresky, in Section 2, Volume 3 of *Handbook of Physiology*, 2nd edition, American Physiological Society, Bethesda, Md. (1984).

[2] P. C. Chatwin and C. M. Allen, *Ann. Rev. Fluid Mech.*, **17**, 119–150 (1985); B. E. Logan, *Environmental Transport Processes*, Wiley-Interscience, New York (1999), Chapters 10 and 11; J. H. Seinfeld, *Advances in Chemical Engineering*, Academic Press, New York (1983), pp. 209–299.

[3] G. I. Taylor, *Proc. Roy. Soc.* **A219**, 186–203 (1953).

[4] G. I. Taylor, *Proc. Roy. Soc.*, **A225**, 473–477 (1954).

Taylor began by neglecting the axial molecular diffusion term (dashed underlined term in Eq. 20.5-1), and subsequently showed[4] that this is permissible if the Péclet number $\text{Pé}_{AB} = R\langle v_z \rangle / \mathscr{D}_{AB}$ is of the order of 70 or greater, and if the length $L_p(t)$ of the region occupied by the pulse, measured visually in Taylor's experiments,[3] is of the order of $170R$ or greater. Here $\langle v_z \rangle = \frac{1}{2} v_{z,\max}$ is the mean speed of the flow.

Taylor sought a solution valid for long times. He estimated the condition for the validity of his result to be

$$\frac{L_p}{v_{z,\max}} >> \frac{R^2}{(3.8)^2 \mathscr{D}_{AB}} \tag{20.5-4}$$

When the pulse length L_p attains this range, enough time has elapsed that the initial shape of the pulse no longer matters.

In order to follow the development of the concentration profile as the fluid moves downstream, it is useful to introduce the shifted axial coordinate

$$\bar{z} = z - \langle v_z \rangle t \tag{20.5-5}$$

When this is used in Eq. 20.5-1 (without the dashed-underlined term), we get the following diffusion equation for $\omega_A(r, \bar{z}, t)$,

$$\frac{\partial \omega_A}{\partial t} + v_{z,\max}(\tfrac{1}{2} - \xi^2) \frac{\partial \omega_A}{\partial \bar{z}} = \frac{\mathscr{D}_{AB}}{R^2} \frac{1}{\xi} \frac{\partial}{\partial \xi}\left(\xi \frac{\partial \omega_A}{\partial \xi} \right) \tag{20.5-6}$$

in which $\xi = r/R$ is the dimensionless radial coordinate. The time derivative here is understood to be taken at constant $\bar{z}$, and, under the condition of Eq. 20.5-4, it may be neglected relative to the radial diffusion term. As a result we have a quasi-steady-state equation

$$\frac{1}{\xi} \frac{\partial}{\partial \xi}\left(\xi \frac{\partial \omega_A}{\partial \xi} \right) = \frac{R^2 v_{z,\max}}{\mathscr{D}_{AB}} (\tfrac{1}{2} - \xi^2) \frac{\partial \omega_A}{\partial \bar{z}} \tag{20.5-7}$$

For the condition of Eq. 20.5-4, the mass fraction can be expressed as

$$\omega_A(\xi, \bar{z}, t) = \langle \omega_A \rangle + \omega_A'(\xi, \bar{z}, t) \qquad \text{with } |\omega_A'| << \langle \omega_A \rangle \tag{20.5-8}$$

where $\langle \omega_A \rangle$ is a function of $\bar{z}$ and t. Substituting this expression into the right side of Eq. 20.5-7, and accordingly neglecting ω_A', we then get

$$\frac{1}{\xi} \frac{\partial}{\partial \xi}\left(\xi \frac{\partial \omega_A}{\partial \xi} \right) = \frac{R^2 v_{z,\max}}{\mathscr{D}_{AB}} (\tfrac{1}{2} - \xi^2) \frac{\partial \langle \omega_A \rangle}{\partial \bar{z}} \tag{20.5-9}$$

from which the radial dependence of the mass fraction can be obtained under the condition of Eq. 20.5-4.

Integration of Eq. 20.5-9 with the boundary conditions of Eq. 20.5-2 then yields

$$\omega_A(\xi, \bar{z}) = \frac{R^2 v_{z,\max}}{8 \mathscr{D}_{AB}} \frac{\partial \langle \omega_A \rangle}{\partial \bar{z}} (\xi^2 - \tfrac{1}{2}\xi^4) + \omega_A(0, \bar{z}) \tag{20.5-10}$$

The average of this profile over the cross section is

$$\langle \omega_A \rangle = \frac{\displaystyle\int_0^1 \omega_A \xi d\xi}{\displaystyle\int_0^1 \xi d\xi} = \frac{R^2 v_{z,\max}}{24 \mathscr{D}_{AB}} \frac{\partial \langle \omega_A \rangle}{\partial \bar{z}} + \omega_A(0, \bar{z}) \tag{20.5-11}$$

Subtracting this equation from the previous one, and replacing $v_{z,\max}$ by $2\langle v_z \rangle$, gives finally

$$\omega_A - \langle \omega_A \rangle = \frac{R^2 \langle v_z \rangle}{2 \mathscr{D}_{AB}} \frac{\partial \langle \omega_A \rangle}{\partial \bar{z}} (-\tfrac{1}{3} + \xi^2 - \tfrac{1}{2}\xi^4) \tag{20.5-12}$$

as Taylor's approximate solution of Eq. 20.5-6.

The total mass flow of A through a plane of constant $\bar{z}$ (that is, the flow relative to the average velocity $\langle v_z \rangle$) is

$$\pi R^2 \rho \langle \omega_A(v_z - \langle v_z \rangle) \rangle = \frac{\pi R^4 \rho \langle v_z \rangle^2}{\mathscr{D}_{AB}} \frac{\partial \langle \omega_A \rangle}{\partial \bar{z}} \int_0^1 (-\tfrac{1}{3} + \xi^2 - \tfrac{1}{2}\xi^4)(\tfrac{1}{2} - \xi^2)\xi \, d\xi$$

$$= -\frac{\pi R^4 \rho \langle v_z \rangle^2}{48 \mathscr{D}_{AB}} \frac{\partial \langle \omega_A \rangle}{\partial \bar{z}} \tag{20.5-13}$$

Next we note that, with the assumption of $\rho = $ constant, $\rho \langle \omega_A \langle v_z \rangle \rangle = \langle \rho_A \rangle \langle v_z \rangle$ and $\rho \langle \omega_A v_z \rangle \approx \langle \rho_A v_{Az} \rangle = \langle n_{Az} \rangle$. (Replacing v_z by v_{Az} is allowed here because, with axial molecular diffusion neglected, species A and B are moving with the same axial speed). Therefore when Eq. 20.5-13 is divided by πR^2, we obtain the averaged mass flux expression

$$\langle n_{Az} \rangle = \langle \rho_A \rangle \langle v_z \rangle - K \frac{\partial \langle \rho_A \rangle}{\partial \bar{z}} = \langle \rho_A \rangle \langle v_z \rangle - K \frac{\partial \langle \rho_A \rangle}{\partial z} \tag{20.5-14}$$

relative to stationary coordinates. Here K is an *axial dispersion coefficient*, given by Taylor's analysis as

$$K = \frac{R^2 \langle v_z \rangle^2}{48 \mathscr{D}_{AB}} = \frac{1}{48} \mathscr{D}_{AB} \text{Pé}_{AB}^2 \tag{20.5-15}$$

This formula indicates that axial dispersion (in the range Pé $>> 1$ considered so far) is enhanced by the radial variation of v_z and reduced by radial molecular diffusion.

Although Eq. 20.5-14 has the form of Fick's law in Eq. (C) of Table 17.8-2, the present equation does not include any axial molecular diffusion. Also it should be emphasized that K is not a property of the fluid mixture, but depends on R and $\langle v_z \rangle$ as well as on $\mathscr{D}_{AB}$.

Next we write the equation of continuity of Eq. 19.1-6, averaged over the tube cross section, as

$$\frac{\partial}{\partial t} \langle \rho_A \rangle = -\frac{\partial}{\partial z} \langle n_{Az} \rangle \tag{20.5-16}$$

When the expression for the mass flux of A from Eq. 20.5-14 is inserted, we get the following *axial dispersion equation*:

$$\frac{\partial}{\partial t} \langle \rho_A \rangle + \langle v_z \rangle \frac{\partial}{\partial z} \langle \rho_A \rangle = K \frac{\partial^2}{\partial z^2} \langle \rho_A \rangle \tag{20.5-17}$$

This equation can be solved to get the shape of the traveling pulse resulting from a δ-function input of a mass m_A of solute A into a stream of otherwise pure B:

$$\langle \rho_A \rangle = \frac{m_A}{2\pi R^2 \sqrt{\pi K t}} \exp\left(-\frac{(z - \langle v_z \rangle t)^2}{4Kt}\right) \tag{20.5-18}$$

This can be used along with Eq. 20.5-15 to extract $\mathscr{D}_{AB}$ from data on the concentrations in the traveling pulse. In fact, this is probably the best method for reasonably quick measurements of liquid diffusivities.

Taylor's development laid the foundation for an extensive literature on convective dispersion. However, it remained to study the approximations made and to determine their range of validity. Aris[5] gave a detailed treatment of dispersion in tubes and ducts, covering the full range of t and including diffusion in the z and θ directions. His long-time asymptote

$$K = \mathscr{D}_{AB} + \frac{R^2 \langle v_z \rangle^2}{48 \mathscr{D}_{AB}} = \mathscr{D}_{AB}\left(1 + \frac{1}{48} \text{Pé}_{AB}^2\right) \tag{20.5-19}$$

---

[5] R. Aris, *Proc. Roy. Soc.*, **A235**, 67–77 (1956).

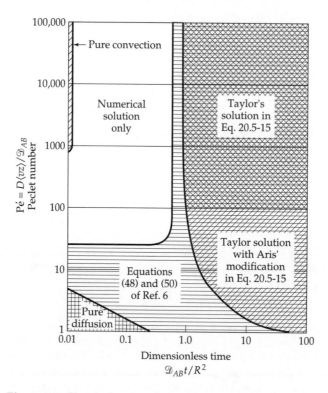

Fig. 20.5-2. Sketch showing the limits of the Taylor (Eq. 20.3-15) and Aris (Eq. 20.5-19) expressions for the axial dispersion coefficient. This figure is patterned after one in Ref. 6.

is an important extension of Eq. 20.5-15. From this result, we see that molecular diffusion enhances the axial dispersion when the Péclet number $\text{Pé} = R\langle v_z \rangle / \mathscr{D}_{AB}$ is less than $\sqrt{48}$ and inhibits axial dispersion at larger Péclet numbers, where Taylor's mode of transport predominates.

The ranges of validity of the Taylor and Aris dispersion formulas have been studied thoroughly by finite difference calculations[6] and by orthogonal collocation.[7] Figure 20.5-2 shows the useful ranges of Eq. 20.5-15 and 19. The latter formula has been widely used for measurements of binary diffusivities, and an extension of it[8] has been used to measure ternary diffusivities in liquids.

Several further investigations on convective dispersion will be mentioned here. *Coiled tubes* give reduced longitudal dispersion, as shown by the experiments of Koutsky and Adler[9] and analyzed for laminar flow by Nunge, Lin, and Gill.[10] This effect is important in chemical reactor design and in diffusivity measurements, where coiling is often necessary to get enough tube length into a compact apparatus.

Extra-column dispersion, caused by the pump and connecting tubing of chromatographic systems, was investigated by Shankar and Lenhoff[11] with detailed predic-

[6] V. Ananthakrishnan, W. N. Gill, and A. J. Barduhn, *AIChE Journal*, **11**, 1063–1072 (1965).

[7] J. C. Wang and W. E. Stewart, *AIChE Journal*, **29**, 493–497 (1983).

[8] Ph. W. M. Rutten, *Diffusion in Liquids*, Delft University Press, Delft, The Netherlands (1992).

[9] J. A. Koutsky and R. J. Adler, *Can. J. Chem. Eng.*, **42**, 239–246 (1964).

[10] R. J. Nunge, T. S. Lin, and W. N. Gill, *J. Fluid Mech.*, **51**, 363–382 (1972).

[11] A. Shankar and A. M. Lenhoff, *J. Chromatography*, **556**, 235–248 (1991).

tions and precise experiments. Their experiments showed that the form of radial averaging is important at times shorter than the recommended range shown in Fig. 20.5-2 for the Taylor–Aris formula. Depending on the type of analyzer used, the data may be better described either by a cup-mixing average ρ_{Ab} or by the area average $\langle \rho_A \rangle$ used above.

Hoagland and Prud'homme[12] have analyzed laminar longitudinal dispersion in *tubes of sinusoidally varying radius*, $R(z) = R_0(1 + \varepsilon \sin(2\pi z/\lambda))$, to model dispersion in packed-bed processes. Their results parallel Eq. 20.5-19, when the variations have small relative amplitude ε and long relative wavelength λ/R_0. One might think that the axial dispersion in a packed column would be similar to that in tubes of sinusoidally varying radius, but that is not the case. Instead of Eq. 20.3-19, one finds $K \approx 2.5 \mathscr{D}_{AB} \text{Pé}_{AB}$, with the first power of the Péclet number appearing, instead of the second power and with K independent of $\mathscr{D}_{AB}$.[13] Brenner and Edwards[14] have given analyses of convective dispersion and reaction in various geometries, including tubes and spatially periodic packed beds.

Dispersion has also been investigated in more complex flows. For *turbulent flows* in straight tubes, Taylor[15] derived and experimentally verified the axial dispersion formula $K/Rv^* = 10.1$, where v^* is the friction velocity used in Eq. 5.3-2. Bassingthwaighte and Goresky[1] investigated models of solute and water exchange in the cardiovascular system, and Chatwin and Allen[2] give mathematical models of turbulent dispersion in rivers and estuaries.

Equations 20.5-1 and 19 are limited to the conditions of Eqs. 20.5-2 and 4. Therefore, they are *not* appropriate for describing entrance regions of steady-state reactor operations or systems with heterogeneous reactions. Equation 20.5-1 is a better starting point for laminar flows.

QUESTIONS FOR DISCUSSION

1. What experimental difficulties might be encountered in using the system in Example 20.1-1 to measure gas-phase diffusivities?
2. What problems do you foresee in using the Taylor dispersion technique of §20.5 for measuring liquid-phase diffusivities?
3. Show that Eq. 20.1-16 satisfies the partial differential equation as well as the initial and boundary conditions.
4. What do you conclude from Table 20.1-1?
5. Why are Laplace transforms useful in solving the problem in Example 20.1-3? Could Laplace transforms be used to solve the problem in Example 20.1-1?
6. How is the velocity distribution in Example 20.1-4 obtained?
7. Describe the method of solving the variable surface area problem in Example 20.1-4.
8. Perform the check suggested after Eq. 20.1-74.
9. What effects do chemical reactions have on the boundary layer?
10. Discuss the Chilton–Colburn expressions in Eq. 20.2-57. Would you expect these same relations to be valid for flows around cylinders and spheres?

[12] D. A. Hoagland and R. K. Prud'homme, *AIChE Journal*, **31**, 236–244 (1985).
[13] A. M. Athalye, J. Gibbs, and E. N. Lightfoot, *J. Chromatog.* **589**, 71–85 (1992).
[14] H. Brenner and D. A. Edwards, *Macrotransport Processes*, Butterworth-Heinemann, Boston (1993).
[15] G. I. Taylor, *Proc. Roy. Soc.*, **A223**, 446–467 (1954).

PROBLEMS   **20A.1. Measurement of diffusivity by unsteady-state evaporation.** Use the following data to determine the diffusivity of ethyl propionate (species A) into a mixture of 20 mole% air and 80 mole% hydrogen (this mixture being treated as a pure gas B).[1]

Increase in vapor volume (cm^3)	$\sqrt{t}$ (s$^{1/2}$)
0.01	15.5
0.11	19.4
0.22	23.4
0.31	26.9
0.41	30.5
0.50	34.0
0.60	37.5
0.70	41.5

These data were obtained[1] by using a glass tube 200 cm long, with an inside diameter 1.043 cm; the temperature was 27.9°C and the pressure 761.2 mm Hg. The vapor pressure of ethyl propionate at this temperature is 41.5 mm Hg. Note that t is the actual time from the start of the evaporation, whereas the volume increase is measured from $t \approx 240$ s.

20A.2. Absorption of oxygen from a growing bubble (Fig. 20A.2). Oxygen is being injected into pure water from a capillary tube. The system is virtually isothermal and isobaric at 25°C and 1 atm. The solubility of oxygen in the liquid phase is $\omega_{A0} = 7.78 \times 10^{-4}$, and the liquid-phase diffusivity for the oxygen–water pair is $\mathcal{D}_{AB} = 2.60 \times 10^{-5}$ cm^2/s. Calculate the instantaneous total absorption rate in g/s, for a bubble of 1 mm diameter and age $t = 2$ s, assuming

(a) Constant volumetric growth rate

(b) Constant radial growth rate dr_s/dt

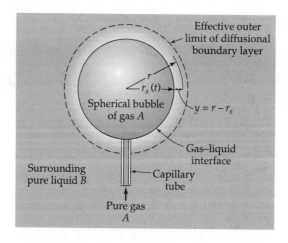

Fig. 20A.2.   Gas absorption from a growing bubble, idealized as a sphere.

Answers: **(a)** 7.6×10^{-8} g/s; **(b)** 1.11×10^{-7} g/s

20A.3. Rate of evaporation of *n*-octane. At 20°C, how many grams of liquid *n*-octane will evaporate into N_2 in 24.5 hr in a system such as that studied in Example 20.1-1 at system pressures of **(a)** 1 atm, and **(b)** 2 atm? The area of the liquid surface is 1.29 cm^2, and the vapor pressure of *n*-octane at 20°C is 10.45 mm Hg.

Answer: **(a)** 6.71 mg

---

[1] D. F. Fairbanks and C. R. Wilke, *Ind. Eng. Chem.* **42**, 471–475 (1950).

20A.4. **Effect of bubble size on interfacial composition** (Fig. 20A.2). Here we examine the assumption of time-independent interfacial composition, ω_{A0}, for the system in Fig. 20A.2. We note that, because of the interfacial tension, the gas pressure p_A depends on the instantaneous bubble radius r_s. The equilibrium expression

$$p_A = p_\infty + \frac{2\sigma}{r_s} \tag{20A.4-1}$$

is adequate unless dr_s/dt is very large. Here p_∞ is the ambient liquid pressure at the mean elevation of the bubble, and σ is the interfacial tension.

For a sparingly soluble solute, the interfacial liquid composition ω_{A0} depends on p_A according to Henry's law

$$\omega_{A0} = H p_A \tag{20A.4-2}$$

in which the Henry's law constant, H, depends on the two species and on the liquid temperature and pressure. This expression may be combined with Eq. 20A.4-1 to obtain the dependence of ω_{A0} on r_s.

For a gas bubble dissolving in liquid water to $T = 25°C$ and $p_\infty = 1$ atm, how small must the bubble be in order to obtain a 10% increase in ω_{A0} above the value for a very large bubble? Assume $\sigma = 72$ dyn/cm over the relevant composition range.

Answer: 1.4 microns

20A.5. **Absorption with rapid second-order reaction** (Fig. 20.1-2). Make the following calculations for the reacting system depicted in the figure:

(a) Verify the location of the reaction zone, using Eq. 20.1-37.

(b) Calculate N_{A0} at $t = 2.5$ s.

20A.6. **Rapid forced-convection mass transfer into a laminar boundary layer.** Calculate the evaporation rate $n_{A0}(x)$ for the system described under Eq. 20.2-52, given that $\omega_{A0} = 0.9$, $\omega_{A\infty} = 0.1$, $n_{B0}(x) = 0$ and Sc $= 2.0$. Use Fig. 22.8-5 with R calculated as R_ω from Eq. 20.2-51, to find the dimensionless mass flux ϕ (denoted by ϕ_ω for diffusional calculations with mass fractions). Then use Eq. 22.8-21 and Table 20.2-1 to calculate K, and Eq. 20.2-48 to calculate $n_{A0}(x)$.

Answer: $n_{A0}(x) = 0.33\sqrt{\rho v_\infty \mu / x}$

20A.7. **Slow forced-convection mass transfer into a laminar boundary layer.** This problem illustrates the use of Eqs. 20.2-55 and 57 and tests their accuracy against that of Eq. 20.2-47.

(a) Estimate the local evaporation rate, n_{A0}, as a function of x for the drying of a porous water-saturated slab, shaped as in Fig. 20.2-2. The slab is being dried in a rapid current of air, under conditions such that $\omega_{A0} = 0.05$, $\omega_{A\infty} = 0.01$, and Sc $= 0.6$. Use Eq. 20.2-55 for the calculation.

(b) Make an alternate calculation of n_{A0} using Eq. 20.2-57.

(c) For comparison with the preceding approximate results, calculate n_{A0} from Eq. 20.2-47 and Table 20.2-1. The K values found in (a) will be sufficiently accurate for looking up $\Pi'(0, \text{Sc}, K)$.

Answers: **(a)** $n_{A0}(x) = 0.0188\sqrt{\rho v_\infty \mu / x}$; **(b)** $n_{A0}(x) = 0.0196\sqrt{\rho v_\infty \mu / x}$; **(c)** $n_{A0}(x) = 0.0188\sqrt{\rho v_\infty \mu / x}$

20B.1. **Extension of the Arnold problem to account for interphase transfer of both species.** Show how to obtain Eqs. 20.1-23, 24, and 25 starting with the equations of continuity for species A and B (in molar units) and the appropriate initial and boundary conditions.

20B.2. **Extension of the Arnold problem to nonisothermal diffusion.** In the situation described in Problem 20B.1, find the analogous result for the temperature distribution $T(z, t)$.

(a) Show that the energy equation [Eq. (H) of Table 19.2-4] reduces to

$$\frac{\partial T}{\partial t} + v_z^* \frac{\partial T}{\partial z} = \alpha \frac{\partial^2 T}{\partial z^2} \tag{20B.2-1}$$

provided that k, p, and c (or ρ) are essentially constant, and that $\overline{H}_\alpha = \tilde{H}_\alpha(p, T)$ and $\tilde{C}_{pA} = \tilde{C}_{pB} =$ constant; consequently α is then a constant. Here the dissipation term $(\tau:\nabla v)$ and the work term $\Sigma_\alpha(j_\alpha \cdot g_\alpha)$ are appropriately neglected. (*Hint:* Use the species equation of continuity of Eq. 19.1-10.)

(b) Show that the solution of Eq. 20B.2-1, with the initial condition that $T = T_\infty$ at $t = 0$, and the boundary conditions that $T = T_0$ at $z = 0$ and $T = T_\infty$ at $z = \infty$, is

$$\frac{T - T_0}{T_\infty - T_0} = \Pi_T(Z_T) = \frac{\text{erf}(Z_T - \varphi_T) + \text{erf}\varphi_T}{1 + \text{erf}\varphi_T} \tag{20B.2-2}$$

with

$$Z_T = \frac{z}{\sqrt{4\alpha t}} \quad \text{and} \quad \varphi_T = v_z^* \sqrt{\frac{t}{\alpha}} \tag{20B.2-3}$$

(c) Show that the interfacial mass and energy fluxes are related to T_0 and T_∞ by

$$\frac{N_{A0} + N_{B0}}{[q_0/\tilde{C}_p(T_0 - T_\infty)]} = \sqrt{\pi}(1 + \text{erf } \varphi_T)\varphi_T \exp \varphi_T^2 \tag{20B.2-4}$$

so that N_{A0}/q_0 and N_{B0}/q_0 are constant for $t > 0$. This nifty result arises because there is no characteristic length or time in the mathematical model of the system.

20B.3. **Stoichiometric boundary condition for rapid irreversible reaction.** The reactant fluxes in Example 20.1-2 must satisfy the stoichiometric relation

$$\text{at } z = z_R(t), \quad \frac{1}{a}c_A(v_{Az} - v_R) = -\frac{1}{b}c_B(v_{Bz} - v_R) \tag{20B.3-1}$$

in which $v_R = dz_R/dt$. Show that this relation leads to Eq. 20.1-31 when use is made of Fick's first law, with the assumptions of constant c and instantaneous irreversible reaction.

20B.4. **Taylor dispersion in slit flow** (Fig. 2B.3). Show that, for laminar flow in a plane slit of width $2B$ and length L, the Taylor dispersion coefficient is

$$K = \frac{2B^2\langle v_z\rangle^2}{105\mathscr{D}_{AB}} \tag{20B.4-1}$$

20B.5. **Diffusion from an instantaneous point source.** At time $t = 0$, a mass m_A of species A is injected into a large body of fluid B. Take the point of injection to be the origin of coordinates. The material A diffuses radially in all directions. The solution may be found in Carslaw and Jaeger:[2]

$$\rho_A = \frac{m_A}{(4\pi\mathscr{D}_{AB}t)^{3/2}} \exp\left(-r^2/4\mathscr{D}_{AB}t\right) \tag{20B.5-1}$$

(a) Verify that Eq. 20B.5-1 satisfies Fick's second law.

(b) Verify that Eq. 20B.5-1 satisfies the boundary conditions at $r = \infty$.

(c) Show that Eq. 20B.5-1, when integrated over all space, gives m_A, as required.

(d) What happens to Eq. 20B.5-1 when $t \to 0$?

20B.6. **Unsteady diffusion with first-order chemical reaction.** Use Eq. 20.1-43 to obtain the concentration profile for the following situations:

---

[2] H. S. Carslaw and J. C. Jaeger, *Conduction of Heat in Solids*, 2nd edition, Oxford University Press (1959), p. 257.

(a) The catalyst particle of Problem 18B.14, in time-dependent operation with the boundary conditions as given before, but with the initial condition that $c_A = 0$ at $t = 0$. The differential equation for c_A is

$$\varepsilon \frac{\partial c_A}{\partial t} = \mathcal{D}_A \frac{\partial^2 c_A}{\partial z^2} - k_1'' a c_A \tag{20B.6-1}$$

where ε is the interior void fraction for the particle. The necessary solution with $k_1'' a = 0$ may be found from the result of Example 12.1-2.

(b) Diffusion and reaction of a solute, A, injected at $t = 0$ at the point $r = 0$ (in spherical coordinates) in an infinite stationary medium. Here the function g of Eq. 20.1-43 is given as

$$g = \frac{1}{(4\pi \mathcal{D}_{AB} t)^{3/2}} \exp(-r^2/4\mathcal{D}_{AB} t) \tag{20B.6-2}$$

and the function f vanishes.

20B.7. Simultaneous momentum, heat, and mass transfer: alternate boundary conditions (Fig. 20B.7). The dimensionless profiles $\Pi(\eta, \Lambda, K)$ in Eq. 20.2-43 are applicable to a variety of situations. Use Eqs. 20.2-49 to 52 to obtain implicit equations for the evaluation of the dimensionless net mass flux K for the following steady-state operations:

(a) Evaporation of pure liquid A from a saturated porous plate into a gaseous stream of A and B. Substance B is insoluble in liquid A.

(b) Instantaneous irreversible reaction of gas A with a solid plate of C to give gaseous B, according to the reaction $A + C \rightarrow 2B$. The molecular weights of A and B are equal.

(c) Transpiration cooling of a porous-walled hollow plate, as shown in the figure. The fluid is pure A throughout, and the injected fluid is distributed so as to maintain the whole outer surface of the plate at a uniform temperature T_0.

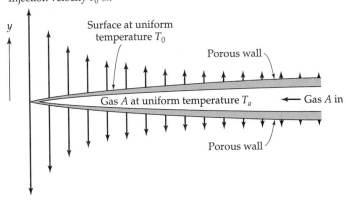

Fig. 20B.7.
A transpiration-cooled porous plate.

Answers: **(a)** $K = \dfrac{1}{Sc}\left(\dfrac{\omega_{A0} - \omega_{A\infty}}{1 - \omega_{A0}}\right)\Pi'(0, Sc, K);$ **(b)** $K = \dfrac{1}{Sc}\,\omega_{A\infty}\,\Pi'(0, Sc, K)$

(c) $K = \dfrac{1}{Pr}\left(\dfrac{T_0 - T_\infty}{T_a - T_0}\right)\Pi'(0, Pr, K)$

20B.8. Absorption from a pulsating bubble. Use the results of Example 20.1-4 to calculate $\delta(t)$ and $N_{A0}(t)$ for a bubble whose radius undergoes a square-wave pulsation:

$$r_s = R_1 \qquad \text{for } 2n < \omega t < 2n + 1$$
$$r_s = R_2 \qquad \text{for } 2n + 1 < \omega t < 2n + 2 \qquad \text{(20B.10-1)}$$

Here ω is a characteristic frequency, and $n = 0, 1, 2, \ldots$.

20B.9. Verification of the solution of the Taylor-dispersion equation. Show that the solution to Eq. 20.5-17, given in Eq. 20.5-18, satisfies the differential equation, the initial condition, and the boundary conditions.[3] The latter are that at $z = \pm\infty$,

$$\langle \rho_A \rangle = 0 \quad \text{and} \quad \frac{\partial}{\partial z} \langle \rho_A \rangle = 0 \qquad \text{(20B.9-1)}$$

The initial condition is that, at $t = 0$, the solute pulse, of mass m_A, is concentrated at $z = 0$, with no solute anywhere else in the tube, so that for all times,

$$\pi R^2 \int_{-\infty}^{+\infty} \langle \rho_A \rangle dz = m_A \qquad \text{(20B.9-2)}$$

(a) Show that Eq. 20.5-17 can be reduced to the one-dimensional form of Fick's second law by the coordinate transformation

$$\bar{z} = z - \langle v_z \rangle t \qquad \text{(20B.9-3)}$$

(b) Show that Eq. 20.5-18 satisfies the equation derived in (a).

(c) Show that Eqs. 20B.9-1 and 2 are also satisfied.

20C.1. Order-of-magnitude analysis of gas absorption from a growing bubble (Fig. 20A.2).

(a) For the growth of the spherical bubble of Problem 20A.2(a) in a liquid of constant density, show that in the liquid phase the radial velocity is $v_r = C_0/r^2$ according to the equation of continuity. Then use the boundary condition that $v_r = dr_s/dt$ at $r = r_s(t)$ to obtain

$$v_r = \frac{r_s^2}{r^2} \frac{dr_s}{dt} \qquad \text{(20C.1-1)}$$

(b) Next, using the species equation of continuity in spherical coordinates with diffusion in the radial direction only, show that

$$\frac{\partial \omega_A}{\partial t} + \left(\frac{r_s^2}{r^2} \frac{dr_s}{dt} \right) \frac{\partial \omega_A}{\partial r} = \mathcal{D}_{AB} \frac{1}{r^2} \frac{\partial}{\partial r} \left(r^2 \frac{\partial \omega_A}{\partial r} \right) \qquad (r > r_s(t)) \qquad \text{(20C.1-2)}$$

and indicate suitable initial and boundary conditions.

(c) For short contact times, the effective diffusion zone is a relatively thin layer, so that it is convenient to introduce a variable $y = r - r_s(t)$. Show that this leads to

$$\underset{(1)}{\frac{\partial \omega_A}{\partial t}} + \underset{(2)\quad(3)}{\left(-\frac{2y}{r_s} + \frac{3y^2}{r_s^2} + \cdots \right)} \frac{dr_s}{dt} \frac{\partial \omega_A}{\partial y} = \mathcal{D}_{AB} \left[\underset{(4)}{\frac{\partial^2 \omega_A}{\partial y^2}} + \underset{(5)\quad(6)\quad(7)}{\frac{2}{r_s} \left(1 - \frac{y}{r_s} + \frac{y^2}{r_s^2} - \cdots \right)} \frac{\partial \omega_A}{\partial y} \right] \qquad \text{(20C.1-3)}$$

(d) From Example 20.1-4 we can see that the contributions of terms (1), (2), and (4) are all of the same order of magnitude in the concentration boundary layer, that is, at $y = O(\delta_\omega) = O(\sqrt{\mathcal{D}_{AB}t})$. Taking these terms to be of order $O(1)$, estimate the orders of magnitude of the remaining terms shown in Eq. 20C.1-3.

---

[3] See, for example, H. S. Carslaw and J. C. Jaeger, *Heat Conduction in Solids*, 2nd edition, Oxford University Press (1959), §10.3. For the effects of finite tube length, see H. Brenner, *Chem. Eng. Sci.*, **17**, 229–243 (1961).

(e) Show that the terms of the two leading orders in Eq. 20C.1-3 give

$$\frac{\partial \omega_A}{\partial t} + \left[-\frac{2y}{r_s} + \underline{\frac{3y^2}{r_s^2}} \right] \frac{dr_s}{dt} \frac{\partial \omega_A}{\partial r} = \mathscr{D}_{AB} \left[\frac{\partial^2 \omega_A}{\partial y^2} + \underline{\frac{2}{r_s} \frac{\partial \omega_A}{\partial y}} \right] \tag{20C.1-4}$$

the second-order terms being designated by dashed underlines.

(e) This equation has been analyzed thoroughly in the electrochemical literature.[4] The results for n_{A0} are further considered in Problem 20C.2.

20C.2. **Effect of surface curvature on absorption from a growing bubble** (Fig. 20A.2). Pure gas A is flowing from a small capillary into a large reservoir of initially pure liquid B at a constant molar flow rate W_A. The interfacial molar flux of A into the liquid is predictable from the *Levich-Koutecký-Newman equation*

$$N_{A0} = c_{A0} \sqrt{\frac{7\mathscr{D}_{AB}}{3\pi t}} \left(1 + \frac{16}{11} \sqrt{\frac{3}{7} \frac{\Gamma(\frac{15}{14})}{\Gamma(\frac{11}{7})} \frac{\mathscr{D}_{AB}^{1/2} t^{1/6}}{\gamma}} \right) \tag{20C.2-1}$$

in which

$$\gamma = \frac{r(t)}{t^{1/3}} = \left(\frac{3W_A}{4\pi c} \right)^{1/3} \tag{20C.2-2}$$

for purely radial motion and a spherical bubble. Equation 20C.2-1 is a consequence of Eq. 20C.1-4.

(a) Give an expression for the number of moles of A absorbed over a bubble lifetime t_0.

(b) Use Eq. 20C.2-1 to obtain more accurate results for the absorption rates in Problem 20A.2.

20C.3. **Absorption with chemical reaction in a semi-infinite medium.** A semi-infinite medium of material B extends from the plane boundary $x = 0$ to $x = \infty$. At time $t = 0$ substance A is brought into contact with this medium at the plane $x = 0$, the surface concentration being c_{A0} (for absorption of gas A by liquid B, for example, c_{A0} would be the saturation concentration). Substances A and B react to produce C according to the irreversible first-order reaction $A + B \rightarrow C$. It is assumed that A is present in such a small concentration that the equation describing the diffusion plus chemical reaction process is

$$\frac{\partial c_A}{\partial t} = \mathscr{D}_{AB} \frac{\partial^2 c_A}{\partial x^2} - k_1''' c_A \tag{20C.3-1}$$

in which k_1''' is the first-order rate constant. This equation has been solved for the initial condition that $c_A = 0$ at $t = 0$, and the boundary conditions that $c_A = c_{A0}$ at $x = 0$, and $c_A = 0$ at $x = \infty$. The solution is[5]

$$\frac{c_A}{c_{A0}} = \frac{1}{2} \exp\left(-\sqrt{\frac{k_1''' x^2}{\mathscr{D}_{AB}}} \right) \mathrm{erfc}\left(\frac{x}{\sqrt{4\mathscr{D}_{AB}t}} - \sqrt{k_1''' t} \right)$$
$$+ \frac{1}{2} \exp\left(\sqrt{\frac{k_1''' x^2}{\mathscr{D}_{AB}}} \right) \mathrm{erfc}\left(\frac{x}{\sqrt{4\mathscr{D}_{AB}t}} + \sqrt{k_1''' t} \right) \tag{20C.3-2}$$

(a) Verify that Eq. 20C.3-2 satisfies the differential equation and the boundary conditions.

(b) Show that the molar flux at the interface $x = 0$ is

$$N_{Ax}|_{x=0} = c_{A0} \sqrt{\mathscr{D}_{AB} k_1'''} \left(\mathrm{erf}\sqrt{k_1''' t} + \frac{\exp(-k_1''' t)}{\sqrt{\pi k_1''' t}} \right) \tag{20C.3-3}$$

[4] J. Koutecký, *Czech. J. Phys.*, **2**, 50–55 (1953). See also V. Levich, *Physicochemical Hydrodynamics*, 2nd edition, Prentice-Hall, Englewood Cliffs, N.J. (1962). The right sides of Levich's Eqs. 108.17 and 108.18 should be multiplied by $t^{2/3}$. See also J. S. Newman, *Electrochemical Systems*, 2nd edition, Prentice-Hall, Englewood Cliffs, N.J. (1991).

[5] P. V. Danckwerts, *Trans. Faraday Soc.*, **46**, 300–304 (1950).

(c) Show further that the total moles absorbed across area A up to time t is

$$M_A = Ac_{A0}\sqrt{\mathscr{D}_{AB}t}\left[\left(\sqrt{k_1'''t} + \frac{1}{2\sqrt{k_1'''t}}\right)\text{erf}\sqrt{k_1'''t} + \frac{\exp(-k_1'''t)}{\sqrt{\pi}}\right] \tag{20C.3-4}$$

(d) Show that, for large values of $k_1'''t$, the expression in (c) reduces asymptotically to

$$M_A = Ac_{A0}\sqrt{\mathscr{D}_{AB}k_1'''}\left(t + \frac{1}{2k_1'''}\right) \tag{20C.3-5}$$

This result[6] is good within 2% for values of $k_1'''t$ greater than 4.

20C.4. **Design of fluid control circuits.** It is desired to control a reactor via continuous analysis of a side stream. Calculate the maximum frequency of concentration changes that can be detected as a function of the volumetric withdrawal rate, if the stream is drawn through a 10 cm length of tubing with an internal diameter of 0.5 mm. *Suggestion:* Use as a criterion that the standard deviation of a pulse duration be no more than 5% of the cycle time $t_0 = 2\pi/\omega$, where ω is the frequency it is desired to detect.

20C.5. **Dissociation of a gas caused by a temperature gradient.** A dissociating gas (for example, $Na_2 \rightleftharpoons 2Na$) is enclosed in a tube, sealed at both ends, and the two ends are maintained at different temperatures. Because of the temperature gradient established, there will be a continuous flow of Na_2 molecules from the cold end to the hot end, where they dissociate into Na atoms, which in turn flow from the hot end to the cold end. Set up the equations to find the concentration profiles. Check your results against those of Dirac.[7]

20D.1. **Two-bulb experiment for measuring gas diffusivities—analytical solution** (Fig. 18B.6). This experiment, described in Problem 18B.6, is analyzed there by a quasi-steady-state method. The method of separation of variables gives the exact solution[8] for the compositions in the two bulbs as

$$x_A^{\pm} = \frac{1}{2}\left[1 \pm \sum_{n=1}^{\infty}(-1)^{n+1}\left(\frac{2N}{\gamma_n}\right)\frac{\sqrt{\gamma_n^2 + N^2}}{\gamma_n^2 + N^2 + N}\exp\left(-\frac{\gamma_n^2\mathscr{D}_{AB}t}{L^2}\right)\right] \tag{20D.1-1}$$

in which γ_n is the nth root of $\gamma \tan \gamma = N$, and $N = SL/V$. Here the $\pm$ sign corresponds to the reservoirs attached at $\pm L$. Make a numerical comparison between Eq. 20D.1-1 and the experimental measurements of Andrew.[9] Also compare Eq. 20D.1-1 with the simpler result in Eq. 18B.6-4.

20D.2. **Unsteady-state interphase diffusion.** Two immiscible solvents I and II are in contact at the plane $z = 0$. At time $t = 0$ the concentration of A is $c_I = c_I^0$ in phase I and $c_{II} = c_{II}^0$ in phase II. For $t > 0$ diffusion takes place across the liquid–liquid interface. It is to be assumed that the solute is present only in small concentration in both phases, so that Fick's second law of diffusion is applicable. We therefore have to solve the equations

$$\frac{\partial c_I}{\partial t} = \mathscr{D}_I\frac{\partial^2 c_I}{\partial z^2} \qquad -\infty < z < 0 \tag{20D.2-1}$$

$$\frac{\partial c_{II}}{\partial t} = \mathscr{D}_{II}\frac{\partial^2 c_{II}}{\partial z^2} \qquad 0 < z < +\infty \tag{20D.2-2}$$

---

[6] R. A. T. O. Nijsing, *Absorptie van gassen in vloeistoffen, zonder en met chemische reactie*, Academisch Proefschrift, Technische Universiteit Delft (1957).

[7] P. A. M. Dirac, *Proc. Camb. Phil. Soc.*, **22**, Part II, 132–137 (1924). This was Dirac's first publication, written while he was a graduate student.

[8] R. B. Bird, *Advances in Chemical Engineering*, Vol. 1, Academic Press, New York (1956), pp. 156–239; errata, Vol. 2 (1958), p. 325. The result at the bottom of p. 207 is in error, since the factor of $(-1)^{n+1}$ is missing. See also H. S. Carslaw and J. C. Jaeger, *Conduction of Heat in Solids*, 2nd edition, Oxford University Press (1959), p. 129.

[9] S. P. S. Andrew, *Chem. Eng. Sci.*, **4**, 269–272 (1955).

in which c_I and c_{II} are the concentrations of A in phases I and II, and $\mathcal{D}_I$ and $\mathcal{D}_{II}$ are the corresponding diffusivities. The initial and boundary conditions are:

I. C. 1:	at $t = 0$,	$c_I = c_I^o$	(20D.2-3)
I. C. 2:	at $t = 0$,	$c_{II} = c_{II}^o$	(20D.2-4)
B. C. 1:	at $z = 0$,	$c_{II} = mc_I$	(20D.2-5)
B. C. 2:	at $z = 0$,	$-\mathcal{D}_I \dfrac{\partial c_I}{\partial z} = -\mathcal{D}_{II} \dfrac{\partial c_{II}}{\partial z}$	(20D.2-6)
B. C. 3:	at $z = -\infty$,	$c_I = c_I^o$	(20D.2-7)
B. C. 4:	at $z = +\infty$,	$c_{II} = c_{II}^o$	(20D.2-8)

The first boundary condition at $z = 0$ is the statement of equilibrium at the interface, m being the "distribution coefficient" or "Henry's law constant." The second boundary condition is a statement that the molar flux calculated at $z = 0^-$ is the same as that at $z = 0^+$; that is, there is no loss of A at the liquid–liquid interface.

(a) Solve the equations simultaneously by Laplace transform or other appropriate means to obtain:

$$\frac{c_I - c_I^o}{c_{II}^o - mc_I^o} = \frac{1 + \mathrm{erf}(z/\sqrt{4\mathcal{D}_I t})}{m + \sqrt{\mathcal{D}_I/\mathcal{D}_{II}}} \tag{20D.2-9}$$

$$\frac{c_{II} - c_{II}^o}{c_I^o - (1/m)c_{II}^o} = \frac{1 - \mathrm{erf}(z/\sqrt{4\mathcal{D}_{II} t})}{(1/m) + \sqrt{\mathcal{D}_{II}/\mathcal{D}_I}} \tag{20D.2-10}$$

(b) Obtain the expression for the mass transfer rate at the interface.

20D.3. **Critical size of an autocatalytic system.** It is desired to use the result of Example 20.1-3 to discuss the critical size of a system in which an "autocatalytic reaction" is occurring. In such a system the reaction products increase the rate of reaction. If the ratio of the system surface to the system volume is large, then the reaction products tend to escape from the boundaries of the system. If the surface to volume ratio is small, however, the rate of escape may be less than the rate of creation, and the reaction rate will increase rapidly. For a system of a given shape, there will be a critical size for which the rate of production just equals the rate of removal.

One example is that of nuclear fission. In a nuclear pile the rate of fission depends on the local neutron concentration. If neutrons are produced at a rate that exceeds the rate of escape by diffusion, the reaction is self-sustaining and a nuclear explosion occurs.

Similar behavior is also encountered in many chemical systems, although the behavior here is generally more complicated. An example is the thermal decomposition of acetylene gas, which is thermodynamically unstable according to the overall reaction.

$$H - C \equiv C - H \rightarrow H_2 + 2C \tag{20D.3-1}$$

This reaction appears to proceed by a branched-chain, free-radical mechanism, in which the free radicals behave qualitatively as the neutrons in the preceding paragraph, so that the decomposition is autocatalytic.

However, the free radicals are effectively neutralized by contact with an iron surface, so that the free-radical concentration is maintained near zero at such a surface. Acetylene gas can then be stored safely in an iron pipe below a "critical" diameter, which is smaller the higher the pressure or temperature of the gas. If the pipe is too large, the formation of even one free radical is likely to cause a rapidly increasing rate of decomposition, which may result in a serious explosion.

(a) Consider a system enclosed in a long cylinder in which the diffusion and reaction process is described by

$$\frac{\partial c_A}{\partial t} = \mathcal{D}_{AB} \frac{1}{r} \frac{\partial}{\partial r}\left(r \frac{\partial c_A}{\partial r}\right) + k_1''' c_A \tag{20D.3-2}$$

with $c_A = 0$ at $r = R$, and $c_A = f(r)$ at $t = 0$, in which $f(r)$ is some function of r. Use the result of Example 20.1-3 to get a solution for $c_A(r, t)$.

(b) Show that the critical radius for the system is

$$R_{\text{crit}} = \alpha_1 \sqrt{\frac{\mathscr{D}_{AB}}{k_1'''}} \tag{20D.3-3}$$

in which α_1 is the first zero of the zero-order Bessel function J_0.

(c) For a bare cylindrical nuclear reactor core,[10] the effective value of $k_1'''/\mathscr{D}_{AB}$ is 9×10^{-3} cm^{-2}. What is the critical radius?

Answer: **(c)** $R_{\text{crit}} = 25.3$ cm

20D.4. Dispersion of a broad pulse in steady, laminar axial flow in a tube. In the Taylor dispersion problem, consider a distributed solute pulse of substance A introduced into a tube of length L containing a fluid in steady, laminar flow. Now the inlet boundary condition is that

$$\text{at } t = 0, \qquad \frac{d}{dt} m_A = f(t) \tag{20D.4-1}$$

with the same constraints of negligible diffusion across the tube inlet and outlet as in Problem 20B.9. Note now that each element of solute acts independently of all the others.

(a) Using the result of Problem 20B.9, show that the exit concentration is given by

$$\langle \rho_A \rangle |_{z=L} = \frac{1}{\sqrt{4\pi R^4 \mathscr{D}_{AB}}} \int_{-\infty}^{t} f(t') \frac{\exp[-(L - \langle v_z \rangle(t - t'))/\sqrt{4\mathscr{D}_{AB}(t - t')}]}{\sqrt{t'}} dt' \tag{20D.4-2}$$

(b) Specialize this result for a square pulse:

$$f = f_0 \quad \text{for } 0 < t < t_0; \qquad f = 0 \quad \text{for } t > t_0 \tag{20D.4-3}$$

Sketch the result for several values of $\langle v_z \rangle t_0/L$.

20D.5. Velocity divergence in interfacially embedded coordinates. Consider a closed domain $D(u, w, y)$ in the interfacially embedded coordinates of Fig. 20.4-2.

(a) Integrate Eq. 20.4-7 over the boundary surface of D to obtain

$$\int_{S_D} (\mathbf{V} \cdot d\mathbf{S}_D) = \int_{S_D} (\mathbf{v} \cdot d\mathbf{S}_D) + \int_{S_D} \left(\frac{\partial \mathbf{r}(u, w, y, t)}{\partial t} \cdot d\mathbf{S}_D\right) \tag{20D.5-1}$$

in which $d\mathbf{S}_D$ is a vector element of area, having magnitude dS_D and the direction of the outward normal to the boundary of the domain D.

(b) The integrand of the last term is the velocity of the boundary element $d\mathbf{S}_D$. Hence, the last integral is the rate of change of the volume of D. Rewrite this integral accordingly with the aid of Eq. 20.4-3, giving

$$\int_{S_D} \left(\frac{\partial \mathbf{r}(u, w, y, t)}{\partial t} \cdot d\mathbf{S}_D\right) = \frac{d}{dt} \int_D \sqrt{g(u, w, y, t)}\, du\, dw\, dy$$

$$= \int_D \frac{\partial \sqrt{g(u, w, y, t)}}{\partial t}\, du\, dw\, dy \tag{20D.5-2}$$

The second equality is obtained by the Leibniz rule, noting that u, w, and y are independent of t on each surface element $d\mathbf{S}_D$.

(c) Use the result of (b) and the Gauss–Ostrogradskii divergence theorem of §A.5 to express Eq. 20D.5-1 as the vanishing of a sum of three volume integrals over $D(u, w, y)$. Show that this result, and the arbitrariness of the choice of D, yield Eq. 20.4-8.

---

[10] R. L. Murray, *Nuclear Reactor Physics*, Prentice-Hall, Englewood Cliffs, N.J. (1957), pp. 23, 30, 53.

Chapter 21

Concentration Distributions in Turbulent Flow

In preceding chapters we have derived the equations for diffusion in a fluid or solid, and we have shown how one can obtain expressions for the concentration distribution, provided no fluid turbulence is involved. Next we turn our attention to mass transport in turbulent flow.

The discussion here is quite similar to that in Chapter 13, and much of that material can be taken over by analogy. Specifically, §§13.4, 13.5, and 13.6 can be taken over directly by replacing heat transfer quantities by mass transfer quantities. In fact, the problems discussed in those sections have been tested more meaningfully in mass transfer, since the range of experimentally accessible Schmidt numbers is considerably greater than that for Prandtl numbers.

We restrict ourselves here to isothermal binary systems, and make the assumption of constant mass density and diffusivity. Therefore the partial differential equation describing diffusion in a flowing fluid (Eq. 19.1-16) is of the same form as that for heat conduction in a flowing fluid (Eq. 11.2-9), except for the inclusion of the chemical reaction term in the former.

§21.1   CONCENTRATION FLUCTUATIONS AND THE TIME-SMOOTHED CONCENTRATION

The discussion in §13.1 about temperature fluctuations and time-smoothing can be taken over by analogy for the molar concentration c_A. In a turbulent stream, c_A will be a rapidly oscillating function that can be written as the sum of a time-smoothed value $\bar{c}_A$ and a turbulent concentration fluctuation c'_A

$$c_A = \bar{c}_A + c'_A \tag{21.1-1}$$

which is analogous to Eq. 13.1-1 for the temperature. By virtue of the definition of c'_A we see that $\overline{c'_A} = 0$. However, quantities such as $\overline{v'_x c'_A}$, $\overline{v'_y c'_A}$, and $\overline{v'_z c'_A}$ are not zero, because the local fluctuations in concentration and velocity are not independent of one another.

The time-smoothed concentration profiles $\bar{c}_A(x, y, z, t)$ are those measured, for example, by the withdrawal of samples from the fluid stream at various points and various times. In tube flow with mass transfer at the wall, one expects that the time-smoothed concentration $\bar{c}_A$ will vary only slightly with position in the turbulent core, where the transport by turbulent eddies predominates. In the slowly moving region near the boundary surface, on the other hand, the concentration $\bar{c}_A$ will be expected to change within a small distance from its turbulent-core value to the wall value. The steep concentration gradient is then associated with the slow molecular diffusion process in the viscous sublayer in contrast to the rapid eddy transport in the turbulent core.

§21.2 TIME-SMOOTHING OF THE EQUATION OF CONTINUITY OF A

We begin with the equation of continuity for species A, which we presume is disappearing by an nth-order chemical reaction.[1] Equation 19.1-16 then gives, in rectangular coordinates,

$$\frac{\partial c_A}{\partial t} = -\left(\frac{\partial}{\partial x}v_x c_A + \frac{\partial}{\partial y}v_y c_A + \frac{\partial}{\partial z}v_z c_A\right) + \mathcal{D}_{AB}\left(\frac{\partial^2 c_A}{\partial x^2} + \frac{\partial^2 c_A}{\partial y^2} + \frac{\partial^2 c_A}{\partial z^2}\right) - k_n''' c_A^n \quad (21.2\text{-}1)$$

Here k_n''' is the reaction rate coefficient for the nth-order chemical reaction, and is presumed to be independent of position. In subsequent equations we shall consider $n = 1$ and $n = 2$ to emphasize the difference between reactions of first and higher order.

When c_A is replaced by $\bar{c}_A + c_A'$, and v_i by $\bar{v}_i + v_i'$, we obtain after time-averaging

$$\frac{\partial \bar{c}_A}{\partial t} = -\left(\frac{\partial}{\partial x}\bar{v}_x\bar{c}_A + \frac{\partial}{\partial y}\bar{v}_y\bar{c}_A + \frac{\partial}{\partial z}\bar{v}_z\bar{c}_A\right) - \left(\frac{\partial}{\partial x}\overline{v_x'c_A'} + \frac{\partial}{\partial y}\overline{v_y'c_A'} + \frac{\partial}{\partial z}\overline{v_z'c_A'}\right)$$

$$+ \mathcal{D}_{AB}\left(\frac{\partial^2 \bar{c}_A}{\partial x^2} + \frac{\partial^2 \bar{c}_A}{\partial y^2} + \frac{\partial^2 \bar{c}_A}{\partial z^2}\right) - \begin{cases} k_1'''\bar{c}_A & \text{or} \\ k_2'''(\bar{c}_A^2 + \overline{c_A'^2}) \end{cases} \quad (21.2\text{-}2)$$

Comparison of this equation with Eq. 21.2-1 indicates that the time-smoothed equation differs in the appearance of some extra terms, marked here with dashed underlines. The terms containing $\overline{v_i'c_A'}$ describe the turbulent mass transport and we designate them by $\bar{J}_{Ai}^{(t)}$, the ith component of the turbulent molar flux vector. We have now met the third of the turbulent fluxes, and we may summarize their components thus:

turbulent molar flux (vector)	$\bar{J}_{Ai}^{(t)} = \overline{v_i'c_A'}$	(21.2-3)
turbulent momentum flux (tensor)	$\bar{\tau}_{ij}^{(t)} = \rho\overline{v_i'v_j'}$	(21.2-4)
turbulent heat flux (vector)	$\bar{q}_i^{(t)} = \rho C_p\overline{v_i'T'}$	(21.2-5)

All of these are defined as fluxes with respect to the mass average velocity.

It is interesting to note that there is an essential difference between the behaviors of chemical reactions of different orders. The first-order reaction has the same form in the time-smoothed equation as in the original equation. The second-order reaction, on the other hand, contributes on time-smoothing an extra term $-k_2'''\overline{c_A'^2}$, this being the manifestation of the interaction between the chemical kinetics and the turbulent fluctuations.

We now summarize all three of the time-smoothed equations of change for turbulent flow of an isothermal, binary fluid mixture with constant ρ, $\mathcal{D}_{AB}$, and μ:

continuity	$(\nabla \cdot \bar{\mathbf{v}}) = 0$	(21.2-6)
motion	$\rho\dfrac{D\bar{\mathbf{v}}}{Dt} = -\nabla\bar{p} - [\nabla \cdot (\bar{\boldsymbol{\tau}}^{(v)} + \bar{\boldsymbol{\tau}}^{(t)})] + \rho\mathbf{g}$	(21.2-7)
continuity of A	$\dfrac{D\bar{c}_A}{Dt} = -(\nabla \cdot (\bar{\mathbf{J}}_A^{(v)} + \bar{\mathbf{J}}_A^{(t)})) - \begin{cases} k_1'''\bar{c}_A & \text{or} \\ k_2'''(\bar{c}_A^2 + \overline{c_A'^2}) \end{cases}$	(21.2-8)

Here $\bar{\mathbf{J}}_A^{(v)} = -\mathcal{D}_{AB}\bar{c}_A$, and it is understood that the operator D/Dt is to be written with the time-smoothed velocity $\bar{\mathbf{v}}$ in it.

---

[1] S. Corrsin, *Physics of Fluids*, **1**, 42–47 (1958).

§21.3   SEMI-EMPIRICAL EXPRESSIONS FOR THE TURBULENT MASS FLUX

In the preceding section we showed that the time-smoothing of the equation of continuity of A gives rise to a turbulent mass flux, with components $\bar{J}_{Ai}^{(t)} = \overline{v_i' c_A'}$. To solve mass transport problems in turbulent flow, it may be useful to postulate a relation between $\bar{J}_{Ai}^{(t)}$ and the time-smoothed concentration gradient. A number of empirical expressions can be found in the literature, but we present here only the two most popular ones.

Eddy Diffusivity

By analogy with Fick's first law of diffusion, we may write

$$\bar{J}_{Ay}^{(t)} = -\mathscr{D}_{AB}^{(t)} \frac{d\bar{c}_A}{dy} \tag{21.3-1}$$

as the defining equation for the *turbulent diffusivity* $\mathscr{D}_{AB}^{(t)}$, also called the *eddy diffusivity*. As is the case with the eddy viscosity and the eddy thermal conductivity, the eddy diffusivity is not a physical property characteristic of the fluid, but depends on position, direction, and the nature of the flow field.

The eddy diffusivity $\mathscr{D}_{AB}^{(t)}$ and the eddy kinematic viscosity $\nu^{(t)} = \mu^{(t)}/\rho$ have the same dimensions—namely, length squared divided by time. Their ratio

$$\mathrm{Sc}^{(t)} = \frac{\nu^{(t)}}{\mathscr{D}_{AB}^{(t)}} \tag{21.3-2}$$

is a dimensionless quantity, known as the *turbulent Schmidt number*. As is the case with the turbulent Prandtl number, the turbulent Schmidt number is of the order of unity (see the discussion in §13.3). Thus the eddy diffusivity may be estimated by replacing it by the turbulent kinematic viscosity, about which a fair amount is known. This is done in §21.4, which follows.

The Mixing-Length Expression of Prandtl and Taylor

According to the mixing-length theory of Prandtl, momentum, energy, and mass are all transported by the same mechanism. Hence by analogy with Eqs. 5.4-4 and 13.3-3 we may write

$$\bar{J}_{Ay}^{(t)} = -l^2 \left| \frac{d\bar{v}_x}{dy} \right| \frac{d\bar{c}_A}{dy} \tag{21.3-3}$$

where l is the Prandtl mixing length introduced in Chapter 5. The quantity $l^2 |d\bar{v}_x/dy|$ appearing here corresponds to $\mathscr{D}_{AB}^{(t)}$ of Eq. 21.3-1, and to the expressions for $\nu^{(t)}$ and $\alpha^{(t)}$ implied by Eqs. 5.4-4 and 13.3-3. Thus, the mixing-length theory satisfies the *Reynolds analogy* $\nu^{(t)} = \alpha^{(t)} = \mathscr{D}_{AB}^{(t)}$, or $\mathrm{Pr}^{(t)} = \mathrm{Sc}^{(t)} = 1$.

§21.4   ENHANCEMENT OF MASS TRANSFER BY A FIRST-ORDER REACTION IN TURBULENT FLOW[1]

We now examine the effect of the chemical reaction term in the turbulent diffusion equation. Specifically we study the effect of the reaction on the rate of mass transfer at the wall for steadily driven turbulent flow in a tube, where the wall (of material A) is slightly

---

[1] O. T. Hanna, O. C. Sandall, and C. L. Wilson, *Ind. Eng. Chem. Research*, **26**, 2286–2290 (1987). An analogous problem dealing with falling films is given by O. C. Sandall, O. T. Hanna, and F. J. Valeri, *Chem. Eng. Communications*, **16**, 135–147 (1982).

soluble in the fluid (a liquid B) flowing through the tube. Material A dissolves in liquid B and then disappears by a first-order reaction. We shall be particularly interested in the behavior with high Schmidt numbers and rapid reaction rates.

For tube flow with axial symmetry and with $\bar{c}_A$ independent of time, Eq. 21.2-8 becomes

$$\bar{v}_z \frac{\partial \bar{c}_A}{\partial z} = \frac{1}{r}\frac{\partial}{\partial r}\left(r(\mathcal{D}_{AB} + \mathcal{D}_{AB}^{(t)})\frac{\partial \bar{c}_A}{\partial r}\right) - k_1''' \bar{c}_A \tag{21.4-1}$$

Here we have made the customary assumption that the axial transport by both molecular and turbulent diffusion can be neglected. We want to find the mass transfer rate at the wall

$$+\mathcal{D}_{AB}\frac{\partial \bar{c}_A}{\partial r}\bigg|_{r=R} = k_c(c_{A0} - \bar{c}_{A,\text{axis}}) \tag{21.4-2}$$

where c_{A0} and $\bar{c}_{A,\text{axis}}$ are the concentrations of A at the wall and at the tube axis. As pointed out in the preceding section, the turbulent diffusivity is zero at the wall, and consequently does not appear in Eq. 21.4-2. The quantity k_c is a *mass transfer coefficient*, analogous to the heat transfer coefficient h. The coefficient h was discussed in Chapter 14 and mentioned in Chapter 9 in connection with "Newton's law of cooling." As a first approximation[1] we take $\bar{c}_{A,\text{axis}}$ to be zero, assuming that the reaction is sufficiently rapid that the diffusing species never reaches the tube axis; then $\partial \bar{c}_A/\partial r$ must also be zero at the tube axis. After analyzing the system under this assumption, we will relax the assumption and give computations for a wider range of reaction rates.

We now define the dimensionless reactant concentration $C = \bar{c}_A/c_{A0}$. Then under the further assumption[1] that, for large z, the concentration will be independent of z, Eq. 21.4-1 becomes

$$\frac{1}{r}\frac{\partial}{\partial r}\left(r(\mathcal{D}_{AB} + \mathcal{D}_{AB}^{(t)})\frac{\partial C}{\partial r}\right) = k_1''' C \tag{21.4-3}$$

This equation may now be multiplied by r and integrated from an arbitrary position to the tube wall to give

$$k_c R - r(\mathcal{D}_{AB} + \mathcal{D}_{AB}^{(t)})\frac{\partial C}{\partial r} = k_1''' \int_r^R \bar{r} C(\bar{r}) d\bar{r} \tag{21.4-4}$$

Here the boundary conditions at $r = 0$ have been used, as well as the definition of the mass transfer coefficient. Then a second integration from $r = 0$ to $r = R$ gives

$$k_c R \int_0^R \frac{1}{r(\mathcal{D}_{AB} + \mathcal{D}_{AB}^{(t)})} dr - 1 = k_1''' \int_0^R \frac{1}{r(\mathcal{D}_{AB} + \mathcal{D}_{AB}^{(t)})}\left[\int_r^R \bar{r} C(\bar{r}) d\bar{r}\right] dr \tag{21.4-5}$$

Here we have used the boundary conditions $C = 0$ at $r = 0$ and $C = 1$ at $r = R$.

Next we introduce the variable $y = R - r$, since the region of interest is right next to the wall. Then we get

$$k_c R \int_0^R \frac{1}{(R - y)(\mathcal{D}_{AB} + \mathcal{D}_{AB}^{(t)})} dy - 1 = k_1''' \int_0^R \frac{1}{(R - y)(\mathcal{D}_{AB} + \mathcal{D}_{AB}^{(t)})}\left[\int_0^y (R - \bar{y})C(\bar{y})d\bar{y}\right] dy \tag{21.4-6}$$

in which $C(\bar{y})$ is not the same function of $\bar{y}$ as $C(\bar{r})$ is of $\bar{r}$. For large Sc the integrands are important only in the region where $y \ll R$, so that $R - y$ may be safely approximated by R. Furthermore, we can use the fact that the turbulent diffusivity in the neighborhood of the wall is proportional to the third power of the distance from the wall. When the integrals are rewritten in terms of $\sigma = y/R$, we get the dimensionless equation

$$\frac{1}{2}\left(\frac{k_c D}{\mathscr{D}_{AB}}\right)\left(\frac{\mathscr{D}_{AB}}{\nu}\right)\int_0^1 \frac{1}{(\mathscr{D}_{AB}/\nu) + K\sigma^3}\,d\sigma - 1 = \left(\frac{k_1''' R^2}{\nu}\right)\int_0^1 \frac{1}{(\mathscr{D}_{AB}/\nu) + K\sigma^3}\left[\int_0^\sigma C(\bar\sigma)d\bar\sigma\right]d\sigma$$

(21.4-7)

This equation contains several dimensionless groupings: the Schmidt number $\mathrm{Sc} = \nu/\mathscr{D}_{AB}$, a dimensionless reaction-rate parameter $\mathrm{Rx} = k_1''' R^2/\nu$, and a dimensionless mass transfer coefficient $\mathrm{Sh} = k_c D/\mathscr{D}_{AB}$ known as the Sherwood number (D being the tube diameter).

In the limit that $\mathrm{Rx} \to \infty$, the solution to Eq. 21.4-3 under the given boundary conditions is $C = \exp(-\mathrm{Sh}\sigma/2)$. Substitution of this solution into Eq. 21.4-7 then gives after straightforward integration

$$\frac{1}{2}\frac{\mathrm{Sh}}{\mathrm{Sc}}I_0 - 1 = 2\frac{\mathrm{Rx}}{\mathrm{Sh}}I_0 - 2\frac{\mathrm{Rx}}{\mathrm{Sh}}I_1$$

(21.4-8)

in which

$$I_0 = \int_0^1 \frac{1}{\mathrm{Sc}^{-1} + K\sigma^3}\,d\sigma$$

(21.4-9)

$$I_1 = \int_0^1 \frac{\exp(-\mathrm{Sh}\sigma/2)}{\mathrm{Sc}^{-1} + K\sigma^3}\,d\sigma$$

(21.4-10)

This can be solved[1] to give Sh as a function of Sc, Rx, and K.

The foregoing solution of Eq. 21.4-3 is reasonable when Sc, Rx, and z are sufficiently large, and is an improvement over the result given by Vieth, Porter and Sherwood.[2] However, in the absence of chemical reaction, Eq. 21.4-3 fails to describe the downstream increase of C caused by the transfer of species A into the fluid. Thus, the mass-transfer enhancement by the chemical reaction cannot be assessed realistically from the results of either Ref. 1 or Ref. 2.

For a better analysis of the enhancement problem, we use Eq. 21.4-1 to get a more complete differential equation for C:

$$\bar{v}_z \frac{\partial C}{\partial z} = \frac{1}{r}\frac{\partial}{\partial r}\left(r(\mathscr{D}_{AB} + \mathscr{D}_{AB}^{(t)})\frac{\partial C}{\partial r}\right) - k_1''' C$$

(21.4-11)

The assumption that $C = 0$ at $r = 0$ is then replaced by the zero-flux condition $\partial C/\partial r = 0$ there. We represent $\mathscr{D}_{AB}^{(t)}$ in this geometry as $l^2|d\bar{v}_z/dr|$ for fully developed flow, by use of a position-dependent mixing length l as in Eq. 21.3-3. Introducing dimensionless notations $v^+ = \bar{v}_z/v_*$, $z^+ = zv_*/\nu$, $r^+ = rv_*/\nu$, and $l^+ = lv_*/\nu$ based on the friction velocity $v_* = \sqrt{\tau_0/\rho}$ of §5.3, we can then express Eq. 21.4-11 in the dimensionless form

$$v^+ \frac{\partial C}{\partial z^+} = \frac{1}{r^+}\frac{\partial}{\partial r^+}\left(r^+\left(\frac{\mathscr{D}_{AB} + \mathscr{D}_{AB}^{(t)}}{\nu}\right)\frac{\partial C}{\partial r^+}\right) - \left[\frac{k_1'''\nu}{v_*^2}\right]C$$

(21.4-12)

$$= \frac{1}{r^+}\frac{\partial}{\partial r^+}\left(r^+\left(\frac{1}{\mathrm{Sc}} + (l^+)^2\left|\frac{dv^+}{dr^+}\right|\right)\frac{\partial C}{\partial r^+}\right) - \mathrm{Da}\,C$$

in which a Damkohler number $\mathrm{Da} = k_1'''\nu/v_*^2$ has been introduced.

An excellent model for the mixing length l is available in Eq. 5.4-7, developed by Hanna, Sandall, and Mazet[3] by modifying the model given by van Driest.[4] This model

[2] W. R. Vieth, J. H. Porter, and T. K. Sherwood, *Ind. Eng. Chem. Fundam.*, **2**, 1–3 (1963).

[3] O. T. Hanna, O. C. Sandall, and P. R. Mazet, *AIChE Journal*, **27**, 693–697 (1981).

[4] E. R. van Driest, *J. Aero. Sci.*, **23**, 1007–1011, 1036 (1956).

will give smooth concentration profiles, provided that we use a velocity function with continuous radial derivative, rather than the piecewise continuous expressions given in Fig. 5.5-3. Such a function is obtainable by integrating the differential equation

$$(l^+)^2 \left(\frac{dv^+}{dy^+} \right)^2 + \frac{dv^+}{dy^+} = 1 - \frac{y^+}{R^+} \quad \text{for } 0 \leqslant y^+ \leqslant R^+ \tag{21.4-13}$$

in the dimensionless variables $v^+ = \bar{v}_z/v_*$ and $y^+ = yv_*/\nu$ of Fig. 5.5-3, with the boundary conditions $v^+ = 0$ at $y^+ = 0$ (the wall) and $dv^+/dy^+ = 0$ at $y^+ = R^+$ (the centerline). Equation 21.4-13 is obtained (see Problem 21B.5) by combining the cylindrical-coordinate versions of Eqs. 5.5-3 and 5.4-4 with the dimensionless form

$$l^+ = \frac{lv_*}{\nu} = 0.4y^+ \frac{1 - \exp(-y^+/26)}{\sqrt{1 - \exp(-0.26y^+)}} \quad \text{for } 0 \leqslant y^+ \leqslant R^+ \tag{21.4-14}$$

of the mixing-length model shown in Eq. 5.4-7. Equation 21.4-13 is solvable via the quadratic formula to give

$$\frac{dv^+}{dy^+} = \begin{cases} \dfrac{-1 + \sqrt{1 + 4(l^+)^2[1 - y^+/R^+]}}{2(l^+)^2} & \text{if } y^+ > 0; \\ 1 & \text{if } y^+ = 0 \end{cases} \tag{21.4-15}$$

and v^+ is then computable by quadrature using, for example, the subroutines trapzd and qtrap of Press et al.[5] The resulting v^+ function closely resembles the plotted line in Fig. 5.5-3, with small changes near $y^+ = 30$ where the plotted line has a slope discontinuity, and near the centerline where the calculated v^+ function attains a maximum value dependent on the dimensionless wall radius R^+ whereas the line in Fig. 5.5-3 improperly does not.

Equations 21.4-12 through 15 were solved numerically[6] for fully developed flow of a fluid of kinematic viscosity $\nu = 0.6581$ cm^2/s in a smooth tube of 3 cm inner diameter, at Re = 10,000, Sc = 200 and various Damkohler numbers Da. These calculations were done with the software package Athena Visual Workbench.[7] The resulting Sherwood numbers Sh = $k_c D/\mathscr{D}_{AB}$, based on k_c as defined in Eq. 21.4-2, are plotted in Fig. 21.4-1 as

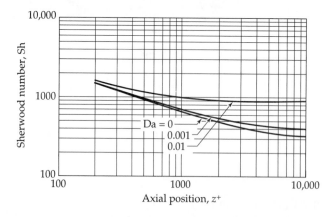

Fig. 21.4-1. Calculated Sherwood numbers, Sh = $k_c D/\mathscr{D}_{AB}$, for turbulent mass transfer from the wall of a tube, with and without homogeneous first-order chemical reaction. Results calculated at Re = 10,000 and Sc = 200, as functions of axial position $z^+ = zv_*/D$ and Damkohler number Da = $k_1''' \nu/v_*^2$.

---

[5] W. H. Press, S. A. Teukolsky, W. T. Vettering, and B. P. Flannery, *Numerical Recipes in FORTRAN*, Cambridge University Press, 2nd edition (1992).

[6] M. Caracotsios, personal communication.

[7] Information on this package is available at www.athenavisual.com and from stewart_associates.msn.com.

functions of z^+ for various values of the Damkohler number Da. These results lead to the following conclusions:

1. In the absence of reaction (that is, when Da = 0), the Sherwood number falls off rapidly with increasing distance into the mass-transfer region. This behavior is consistent with the results of Sleicher and Tribus[8] for a corresponding heat transfer problem, and confirms that the convection term of Eq. 21.4-11 is essential for this system. This term was neglected in References 2 and 3 by regarding the concentration profiles as "fully developed."

2. In the presence of a pseudo-first-order homogeneous reaction of the solute (that is, when Da > 0), the Sherwood number falls off downstream less rapidly, and ultimately attains a constant asymptote that depends on the Damkohler number. Thus, the enhancement factor, defined as Sh (with reaction)/Sh (without reaction), can increase considerably with increasing distance into the mass-transfer region.

§21.5 TURBULENT MIXING AND TURBULENT FLOW WITH SECOND-ORDER REACTION

We now consider processes occurring within turbulent fluid systems, with particular reference to the two mixers shown in Fig. 12.5-1. In Fig. 12.5-1(a) is shown a *steady state system*, in which two input streams enter a system of fixed geometry at constant rates, and in Fig. 12.5-1(b) an *unsteady state system*, in which two initially stationary, segregated, miscible fluids are mixed by turning an impeller at a constant angular velocity, starting at time $t = 0$. One stream [in (a)] or one initial region [in (b)] contains solute A in solvent S, and the other contains solute B in solvent S. All solutions are sufficiently dilute that the solutes do not appreciably affect the viscosity, density, or species diffusivities. Then the behavior of the solute (A or B) in either system [(a) or (b)] is described by the non-time-smoothed diffusion equations

$$\frac{Dc_A}{Dt} = \mathscr{D}_{AS}\nabla^2 c_A + R_A \qquad \frac{Dc_B}{Dt} = \mathscr{D}_{BS}\nabla^2 c_B + R_B \qquad (21.5\text{-}1, 2)$$

with suitable initial and boundary conditions.

For these systems, we may write that at $z = 0$ [in (a)] or $t = 0$ [in (b)]

$$c_A = c_{A0} \quad \text{and} \quad c_B = 0 \qquad (21.5\text{-}3, 4)$$

over the A inlet port [in (a)] or the initial region [in (b)], and

$$c_B = c_{B0} \quad \text{and} \quad c_A = 0 \qquad (21.5\text{-}5, 6)$$

over the B inlet port [in (a)] or the initial region [in (b)]. In addition, we consider all confining surfaces to be inert and impenetrable.[1]

No Reaction Occurring

For this situation, the terms R_A and R_B are identically zero. We now define a single new independent variable

$$\Gamma = \frac{c_{A0} - c_A}{c_{A0}} = \frac{c_B}{c_{B0}} \qquad (21.5\text{-}7)$$

---

[8] C. A. Sleicher and M. Tribus, *Trans. ASME*, **79**, 789–797 (1957).
[1] In system (a), these boundary conditions are only approximations. The indicated values of c_A and c_B are regarded as asymptotic values for $z \ll 0$.

Then both Eqs. 21.5-1 and 2 take the following form over the whole system:

$$\frac{D\Gamma}{Dt} = \mathcal{D}_{iS}\nabla^2\Gamma \tag{21.5-8}$$

Here the subscript i can represent either solute A or solute B, and

$\Gamma = 0$ for (a) the entering A-rich stream, or
(b) initially A-rich region   (21.5-9)

$\Gamma = 1$ for (a) the entering B-rich stream, or
(b) initially B-rich region   (21.5-10)

It follows that, for equal diffusivities, the time-smoothed concentration profiles, $\bar{\Gamma}(x, y, z, t)$ are identical for both solutes, where

$$\bar{\Gamma} = \frac{c_{A0} - \bar{c}_A}{c_{A0}} = \frac{\bar{c}_B}{c_{B0}} \tag{21.5-11}$$

However, the fluctuating quantities Γ' are also of interest, as they are measures of "unmixedness." These can be equal only in a statistical sense. To show this, we subtract Eq. 21.5-11 from Eq. 21.5-7, and then square the result and time-smooth it to give

$$\overline{\left(\frac{c'_A}{c_{A0}}\right)^2} = \overline{\left(\frac{c'_B}{c_{B0}}\right)^2} = d^2 \tag{21.5-12}$$

Here $d(x, y, z, t)$ is a dimensionless *decay function*, which decreases toward zero at large z [for the motionless mixer in Fig. 21.5-1(a)], or at large t [for the mixing tank of Fig. 21.5-1(b)]. Cross-sectional averages of this quantity can be measured, and are shown in Fig. 21.5-2.

It remains to determine the functional dependence of the decay function, and to do this we introduce the dimensionless variables:

$$\check{\mathbf{v}} = \frac{\mathbf{v}}{v_0}; \qquad \check{t} = \frac{v_0 t}{l_0}; \qquad \check{\nabla} = l_0 \nabla \tag{21.5-13}$$

Then Eq. 21.5-8 becomes

$$\frac{D\Gamma}{D\check{t}} = \frac{1}{\text{ReSc}} \check{\nabla}^2 \Gamma \tag{21.5-14}$$

in which $\text{Re} = l_0 v_0 \rho / \mu$.

In order to be able to draw specific conclusions, we now focus our attention on mixing tanks [see Fig. 21.5(b)], and further assume low-viscosity liquids and low-molecular-weight solutes. For these systems l_0 is normally chosen to be the diameter of the impeller, and v_0 to be $l_0 N$, where N is the rate of impeller rotation in revolutions per unit time.

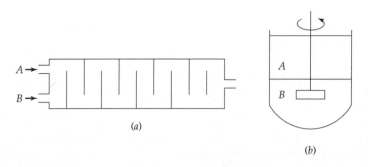

(a)

(b)

Fig. 21.5-1. Two types of mixers: (a) a baffled mixer with no moving parts; (b) a batch mixer with a stirrer.

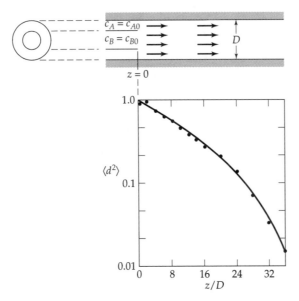

Fig. 21.5-2. The decay function for a specific device for the mixing of two streams emerging from a tube and from an annular region. This figure is patterned after one by E. L. Cussler, *Diffusion: Mass Transfer in Fluid Systems*, Cambridge University Press (1997), p. 422, based on data of R. S. Brodkey, *Turbulence in Mixing Operations*, Academic Press, New York (1975), p. 65, Fig. 6, upper curve. The radius of the outer tube is $\sqrt{2}$ times that of the inner one.

It is now useful to consider experience gained in the study of such systems and to classify the overall mixing process as follows:[2]

(i) *macromixing*, in which large-scale motions spread the A-rich and B-rich fluids over the entire tank region, into subregions that are large compared to the distances solute molecules have moved by diffusion.

(ii) *micromixing*, in which diffusion provides the final blending over scales of molecular dimensions.

It has been found[1] that macromixing is normally much the slower process, and this observation can be explained in terms of dimensional analysis. This finding is consistent with experience in large-scale mixing.

For industrial systems, Reynolds numbers are normally well over 10^4 and Schmidt numbers on the order of 10^5. The diffusion term in Eq. 21.5-14 thus tends to be small almost everywhere in the system. This term is negligible during the period of macromixing, where diffusion, and hence the Schmidt number, have no significant effect. Then for many practical purposes one may write

$$\frac{D\Gamma}{D\tilde{t}} \approx 0 \qquad (\text{ReSc} \gg 1) \qquad (21.5\text{-}15)$$

We may then relax the requirement of equal diffusivities in extrapolating experience to a new system. It follows that Reynolds numbers as well as Schmidt numbers should have no significant effect on the macromixing process, and that the effective degree of unmixedness, d^2, depends mainly on the dimensionless time.

For large-scale mixing tanks, this prediction is amply confirmed.[3] These normally operate at large Reynolds numbers (typically greater than 10^4), where the large-scale motions, expressed in terms of $\check{\mathbf{v}}(\check{x}, \check{y}, \check{z}, \check{t})$, are observed to be independent of both Reynolds number and system size. Thus a very large number of investigators have observed using

[2] M. L. Hanks and H. L. Toor, *Ind. Eng. Chem. Res.*, **34**, 3252–3256 (1995).

[3] J. Y. Oldshue, *Fluid Mixing Technology*, McGraw-Hill, New York (1983); H. Benkreira, *Fluid Mixing*, Institution of Chemical Engineers, Rugby, UK, Vol. 4 (1990), Vol. 6 (1999); I. Bouwmans and H. E. A. van den Akker, in Vol. 4 of *Fluid Mixing*, Institution of Chemical Engineers, Rugby, UK (1990), pp. 1–12.

many different mixer geometries, that the product of the required mixing time t_{mix} and rotation rate N is a constant independent of mixer size and Reynolds number:

$$Nt_{mix} = K(\text{geometry}) \quad \text{or} \quad t_{mix} = K/N \tag{21.5-16}$$

That is, the required mixing time t_{mix} corresponds, for a given tank geometry, essentially to the required number of turns of the impeller. This expectation is confirmed by experience.

This finding is consistent with observations[2] that both the dimensionless volume flow rate through the impeller, Q/ND^3, and the tank friction factor, $P/\rho N^3 D^5$, are constants, depending only on the tank and impeller geometries (see Problem 6C.3). Here Q is the volumetric flow in the jet produced by the impeller, and P is the power required to turn it.

Similar remarks usually apply to motionless mixers, where increasing the flow velocities typically has little effect on the degree of mixing. However, approximations like this must be tested, and such tests should be considered as first steps in an experimental program. As a practical matter, these approximations are almost always reliable on scale-up, since Reynolds numbers normally increase with equipment size.

Reaction Occurring

We next consider the effects of a homogeneous, irreversible chemical reaction, and for simplicity we write this as $A + B \rightarrow$ products. Again we assume dilute solutions, so that the heat of reaction and the presence of reaction products have no significant effect. In addition, we assume equal diffusivities for the two solutes.

We next define

$$\Gamma_{\text{reaction}} = \frac{c_{A0} - (c_A - c_B)}{c_{A0} + c_{B0}} \tag{21.5-17}$$

Then when we subtract Eq. 21.5-2 from Eq. 21.5-1, we find that the description of Γ_{reaction} is identical to that for its nonreactive counterpart. Hence

$$\left(\frac{c_{A0} - (c_A - c_B)}{c_{A0} + c_{B0}}\right)_{\text{reactive}} = \left(\frac{c_{A0} - c_A}{c_{A0}}\right)_{\text{nonreactive}} \left(\frac{c_B}{c_{B0}}\right)_{\text{nonreactive}} \tag{21.5-18}$$

By subtracting from this its time-smoothed counterpart, we find that an equation like Eq. 21.5-18 must hold for the fluctuations:

$$\left(\frac{c_A' - c_B'}{c_{A0} + c_{B0}}\right)_{\text{reactive}} = \left(\frac{c_A'}{c_{A0}}\right)_{\text{nonreactive}} \tag{21.5-19}$$

The time-smoothed mean square of the quantity on the right side is equal to d^2, which is measurable as illustrated in Fig. 21.5-2, and therefore we have a way of predicting the corresponding quantity for reacting systems.

Equation 21.5-19 suggests that the fluctuations in c_A and c_B in reactive problems occur on the same time and distance scales as for nonreactive problems. Note that this is true for arbitrary geometry, flow conditions, and reaction kinetics. We are now ready to consider special cases.

We begin with a *fast reaction*, for which the two solutes cannot coexist, and the rate of the reaction is controlled by the diffusion of the species toward each other. Then, for the first (macromixing) stage of the blending process, where diffusion is very slow compared to the larger-scale convective processes, there is no significant reaction. In this, typically dominant, stage of the blending process

$$\left(\frac{c_A'}{c_{A0}}\right)_{\text{reactive}} = \left(\frac{c_A'}{c_{A0}}\right)_{\text{nonreactive}} \tag{21.5-20}$$

It has been suggested[4] that Eq. 21.5-20 is also true for the micromixing stage. Where this can be assumed (e.g., in the common situation where macromixing is rate controlling), it follows that reactive and nonreactive processes lead to identical descriptions of solute fluctuations.

In practice, fast reactions (e.g., neutralization of acids with bases) are often used to determine the effectiveness of mixers, as these are much easier to follow experimentally than nonreactive mixing. Frequently one can use simple macroscopic measures such as temperature rise or an indicator color change. However, the measurement of concentration fluctuations can provide more insight into the nature and the course of the mixing process.

Slow reactions are also important, and we consider the special case of irreversible second-order kinetics, defined by

$$R_A = -k_2''' c_A c_B \tag{21.5-21}$$

When this is time-smoothed, we get

$$\overline{R}_A = -k'''(\overline{c}_A \overline{c}_B + \overline{c'_A c'_B}) \tag{21.5-22}$$

We find, therefore, that the fluctuations in solute concentration increase the time-smoothed reaction rate relative to that when a simple product of time-smoothed concentrations is used. It is, however, difficult to assess the practical importance of this effect.

We illustrate this point by a simple order-of-magnitude analysis, beginning with the definition of a reaction time constant t_A for one of the reactants, here solute A:

$$t_A = c_{A0}/R_A \tag{21.5-23}$$

To a first approximation, we may write

$$t_A \approx 1/k_2''' c_{B0} \tag{21.5-24}$$

Fast and slow reactions may then be defined as those for which

$$t_{\text{mix}} >> t_A \quad \text{fast reaction} \tag{21.5-25}$$

$$t_{\text{mix}} << t_A \quad \text{slow reaction} \tag{21.5-26}$$

We have already discussed the case of fast reaction. For slow reactions, turbulence has no significant effect, because fluctuations become negligible before any appreciable reaction has taken place.

If the mixing and reaction time constants are of the same order of magnitude, a deeper analysis than the above is needed. Such an analysis must include a model for the turbulent motion, and does not appear to be presently available.

QUESTIONS FOR DISCUSSION

1. Discuss the similarities and differences between turbulent heat and mass transport.
2. Discuss the behavior of first- and higher-order reactions in the time-smoothing of the equation of continuity for a given species. What are the consequences of this?
3. To what extent are the turbulent momentum flux, heat flux, and mass flux similar in form?
4. What empiricisms are available for describing the turbulent mass flux?
5. How can eddy diffusivities be measured, and on what do they depend?
6. Would you expect to get trustworthy results for mass transfer in turbulent tube flow without chemical reaction just by setting Rx = 0 in Eq. 21.4-8?

---

[4] K.-T. Li and H. L. Toor, *Ind. Eng. Chem. Fundam.*, **25**, 719–723 (1986).

PROBLEMS   **21A.1. Determination of eddy diffusivity** (Figs. 18C.1 and 21A.1). In Problem 18C.1 we gave the formula for the concentration profiles in diffusion from a point source in a moving stream. In isotropic highly turbulent flow, Eq. 18C.1-2 may be modified by replacing $\mathscr{D}_{AB}$ by the eddy diffusivity $\mathscr{D}_{AB}^{(t)}$. This equation has been found to be useful for determining the eddy diffusivity. The molar flow rate of carbon dioxide is $1/1000$ that of air.

(a) Show that if one plots $\ln s c_A$ versus $s - z$ the slope is $-v_0/2\mathscr{D}_{AB}^{(t)}$.

(b) Use the data on the diffusion of CO_2 from a point source in a turbulent air stream shown in Fig. 21A.1 to get $\mathscr{D}_{AB}^{(t)}$ for these conditions: pipe diameter, 15.24 cm; $v_0 = 1512$ cm/s.

(c) Compare the value of $\mathscr{D}_{AB}^{(t)}$ with the molecular diffusivity $\mathscr{D}_{AB}$ for the system CO_2–air.

(d) List all assumptions made in the calculations.

Answer: **(b)** $\mathscr{D}_{AB}^{(t)} = 19$ cm^2/s

21A.2. Heat and mass transfer analogy. Write the mass transfer analog of Eq. 13.4-19. What are the limitations of the resulting equation?

21B.1. Wall mass flux for turbulent flow with no chemical reactions. Use the diffusional analog of Eq. 13.4-20 for turbulent flow in circular tubes, and the Blasius formula for the friction factor, to obtain the following expression for the Sherwood number,

$$\text{Sh} = 0.0160\,\text{Re}^{7/8}\,\text{Sc}^{1/3} \tag{21B.1-1}$$

valid for large Schmidt numbers.[1]

21B.2. Alternate expressions for the turbulent mass flux. Seek an asymptotic expression for the turbulent mass flux for long circular tubes with a boundary condition of constant wall mass flux. Assume that the net mass transfer across the wall is small.

(a) Parallel the approach to laminar flow heat transfer in §10.8 to write

$$\Pi(\xi, \zeta) = -\frac{\omega_A - \omega_{A1}}{j_{A0}D/\rho\mathscr{D}_{AB}} = C_1\zeta + \Pi_\infty(\xi) + C_2 \tag{21B.2-1}$$

in which $\xi = r/D$, $\zeta = (z/D)/\text{ReSc}$, ω_{A1} is the inlet mass fraction of A, and j_{A0} is the interfacial mass flux of A into the fluid.

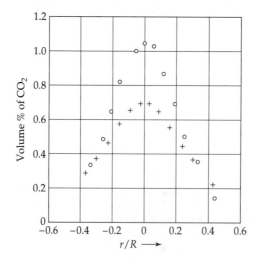

Fig. 21A.1. Concentration traverse data for CO_2 injected into a turbulent air stream with Re = 119,000 in a tube of diameter 15.24 cm. The circles are concentrations at a distance $z = 112.5$ cm downstream from the injection point, and the crosses are concentrations at $z = 152.7$ cm. [Experimental data are taken from W. L. Towle and T. K. Sherwood, *Ind. Eng. Chem.*, **31**, 457–462 (1939).]

[1] O. T. Hanna, O. C. Sandall, and C. L. Wilson, *Ind. Eng. Chem. Res.*, **28**, 2286–2290 (1987).

(b) Next use the equation of continuity for species A to obtain

$$-4\frac{v_z}{\langle v_z\rangle} = \frac{1}{\xi}\frac{d}{d\xi}\left[\left(1 + \frac{Sc}{Sc^{(t)}}\frac{\mu^{(t)}}{\mu}\right)\xi\frac{d\Pi_\infty}{d\xi}\right] \tag{21B.2-2}$$

in which $Sc^{(t)} = \mu^{(t)}/\rho\mathcal{D}_{AB}^{(t)}$. This equation is to be integrated with the boundary conditions that Π_∞ is finite at $\xi = 0$ and $d\Pi_\infty/d\xi = -1$ at $\xi = \frac{1}{2}$.

(c) Integrate once with respect to ξ to obtain

$$-\frac{d\Pi_\infty}{d\xi} = \frac{\frac{1}{2} - 4\int_\xi^{1/2}(v_z/\langle v_z\rangle)\xi d\xi}{\xi[1 + (Sc/Sc^{(t)})(\mu^{(t)}/\mu)]} \tag{21B.2-3}$$

21B.3. An asymptotic expression for the turbulent mass flux.[1] Start with the final result of Problem 21B.2, and note that for sufficiently high Sc all curvature of the concentration profile will take place very near the wall, where $v_z/\langle v_z\rangle \approx 0$ and $\xi \approx \frac{1}{2}$. Assume that $Sc^{(t)} = 1$ and use Eq. 5.4-2 to obtain

$$-\frac{d\Pi_\infty}{d\xi} = \frac{1}{[1 + Sc(\mu^{(t)}/\mu)]} = \frac{1}{1 + Sc(yv_*/14.5\nu)^3} \tag{21B.3-1}$$

Introduce the new coordinate $\eta = Sc^{1/3}(yv_*/14.5\nu)$ into Eq. 21B.3-1 to get an equation for $d\Pi/d\eta$ valid within the laminar sublayer. Then integrate from $\eta = 0$ (where $\omega_A = \omega_{A0}$) to $\eta = \infty$ (where $\omega_A \approx \omega_{Ab}$) to obtain an explicit relation for the wall mass flux j_{A0}. Compare with the analog of Eq. 13.4-20 obtained in Problem 21A.2.

21B.4. Deposition of silver from a turbulent stream (Fig. 21B.3). An approximately 0.1 N solution of KNO_3 containing 1.00×10^{-6} g-equiv. $AgNO_3$ per liter is flowing between parallel Ag plates,

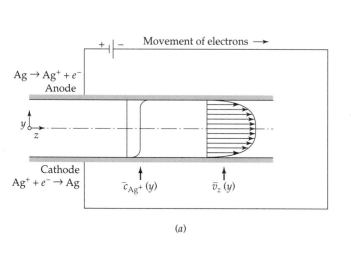

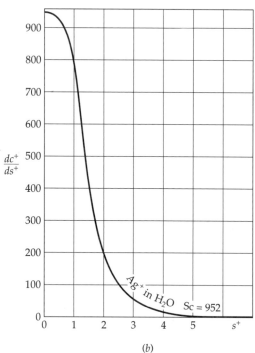

Fig. 21B.3. (a) Electrodeposition of Ag^+ from a turbulent stream flowing in the positive z direction between two parallel plates. (b) Concentration gradients in electrodeposition of Ag at an electrode.

[1] C. S. Lin, R. W. Moulton, and G. L. Putnam, *Ind. Eng. Chem.*, **45**, 636 (1953).

as shown in Fig. 21B.3(a). A small voltage is applied across the plates to produce a deposition of Ag on the cathode (lower plate) and to polarize the circuit completely (that is, to maintain the Ag^+ concentration at the cathode very nearly zero). Forced diffusion may be ignored, and the Ag^+ may be considered to be moving to the cathode by ordinary (that is, Fickian) diffusion and eddy diffusion only. Furthermore, this solution is sufficiently dilute that the effects of the other ionic species on the diffusion of Ag^+ are negligible.

(a) Calculate the Ag^+ concentration profile, assuming that (i) the effective binary diffusivity of Ag^+ through water is 1.06×10^{-5} cm²/s; (ii) the truncated Lin, Moulton, and Putnam expression of Eq. 5.4-2 for the turbulent velocity distribution in round tubes is valid for "slit flow" as well, if four times the hydraulic radius is substituted for the tube diameter; (iii) the plates are 1.27 cm apart, and $\sqrt{\tau_0/\rho}$ is 11.4 cm/s.

(b) Estimate the rate of deposition of Ag on the cathode, neglecting all other electrode reactions.

(c) Does the method of calculation in part (a) predict a discontinuous slope for the concentration profile at the center plane of the system? Explain.

Answers: **(a)** See Fig. 21B.3(b); **(b)** 6.7×10^{-12} equiv/cm² · s

21B.5. **Mixing-length expression for the velocity profile.**

(a) Start with Eq. 5.5-3, and show that for steadily driven, fully developed turbulent flow in a tube

$$\frac{\bar{\tau}_{rz}}{\tau_0} = \frac{r}{R} = 1 - \frac{y}{R} \tag{21B.5-1}$$

(b) Next set $\bar{\tau}_{rz} = \bar{\tau}_{rz}^{(v)} + \bar{\tau}_{rz}^{(t)}$, where $\bar{\tau}_{rz}^{(v)}$ is given by the cylindrical coordinate analog of Eq. 5.2-9, and $\bar{\tau}_{rz}^{(t)}$ by Eq. 5.5-5. Show that Eq. 21B.5-1 then becomes

$$\rho l^2 \left(\frac{d\bar{v}_z}{dy}\right)^2 + \mu\left(\frac{d\bar{v}_z}{dy}\right) = \tau_0 \left(1 - \frac{y}{R}\right) \text{ for } 0 \le y \le R \tag{21B.5-2}$$

(c) Obtain Eq. 21.4-13 from Eq. 21B.5-2 by introducing the dimensionless symbols used in the former equation.

Chapter 22

Interphase Transport in Nonisothermal Mixtures

§22.1 Definition of transfer coefficients in one phase

§22.2 Analytical expressions for mass transfer coefficients

§22.3 Correlation of binary transfer coefficients in one phase

§22.4 Definition of transfer coefficients in two phases

§22.5$^\text{O}$ Mass transfer and chemical reactions

§22.6$^\text{O}$ Combined heat and mass transfer by free convection

§22.7$^\text{O}$ Effects of interfacial forces on heat and mass transfer

§22.8$^\text{O}$ Transfer coefficients at high net mass transfer rates

§22.9$^\bullet$ Matrix approximations for multicomponent mass transport

Here we build on earlier discussions of binary diffusion to provide means for predicting the behavior of mass transfer operations such as distillation, absorption, adsorption, extraction, drying, membrane filtrations, and heterogeneous chemical reactions. This chapter has many features in common with Chapters 6 and 14. It is particularly closely related to Chapter 14, because there are many situations where the analogies between heat and mass transfer can be regarded as exact.

There are, however, important differences between heat and mass transfer, and we will devote much of this chapter to exploring these differences. Since many mass transfer operations involve fluid–fluid interfaces, we have to deal with distortions of the interfacial shape by viscous drag and by surface tension gradients resulting from inhomogeneities in temperature and composition. In addition, there may be interactions between heat and mass transfer, and there may be chemical reactions occurring. Furthermore, at high mass transfer rates, the temperature and concentration profiles may be distorted. These effects complicate and sometimes invalidate the neat analogy between heat and mass transfer that one might otherwise expect.

In Chapter 14 the interphase heat transfer involved the movement of heat to or from a solid surface, or the heat transfer between two fluids separated by a solid surface. Here we will encounter heat and mass transfer between two contiguous phases: fluid–fluid or fluid–solid. This raises the question as to how to account for the resistance to diffusion provided by the fluids on both sides of the interface.

We begin the chapter by defining, in §22.1, the mass and heat transfer coefficients for binary mixtures in one phase (liquid or gas). Then in §22.2 we show how analytical solutions to diffusion problems lead to explicit expressions for mass transfer coefficients. In that section we give some analytic expressions for mass transfer coefficients at high

Schmidt numbers for a number of relatively simple systems. We emphasize the different behavior of systems with fluid–fluid and solid–fluid interfaces.

In §22.3 we show how dimensional analysis leads to predictions involving the Sherwood number (Sh) and the Schmidt number (Sc), which are the analogs of the Nusselt number (Nu) and the Prandtl number (Pr) defined in Chapter 14. Here the emphasis is on the analogies between heat transfer in pure fluids and mass transfer in binary mixtures. Then in §22.4 we proceed to the definition of mass transfer coefficients for systems with diffusion in two adjoining phases. We show there how to apply the information about mass transfer in single phases to the understanding of mass transfer between two phases.

Finally, in the last five sections of the chapter, we take up some effects that are peculiar to mass transfer systems: mass transfer with chemical reactions (§22.5), the interaction of heat and mass transfer processes in free convection (§22.6), the complicating factors of interfacial tension forces and Marangoni effects (§22.7), the distortions of temperature and concentration profiles that arise in systems with large net mass transfer rates across the interface (§22.8); and finally the matrix analysis of mass transport in multicomponent systems. In this chapter the emphasis is on the non-analogous behavior of heat and mass transfer systems.

In this chapter we have limited the discussion to a few key topics on mass transfer and transfer coefficient correlations. Further information is available in specialized textbooks on these and related topics.[1-4]

§22.1   DEFINITION OF TRANSFER COEFFICIENTS IN ONE PHASE

In this chapter we relate the rates of mass transfer across phase boundaries to the relevant concentration differences, mainly for binary systems. These relations are analogous to the heat transfer correlations of Chapter 14 and contain *mass transfer coefficients* in place of the heat transfer coefficients of that chapter. The system may have a true phase boundary, as in Fig. 22.1-1, 2, or 4, or an abrupt change in hydrodynamic properties, as in the system of Fig. 22.1-3, containing a porous solid. Figure 22.1-1 shows the evaporation of a volatile liquid, often used in experiments to develop mass transfer correlations. Figure 22.1-2 shows a permselective membrane, in which a selectively permeable surface permits more effective transport of solvent than of a solute that is to be retained, as in ultrafiltration of protein solutions and the desalting of sea water. Figure 22.1-3 shows a macroscopically porous solid, which can serve as a mass transfer surface or can provide sites for adsorption or reaction. Figure 22.1-4 shows an idealized liquid–vapor contactor where the mass transfer interface may be distorted by viscous or surface-tension forces.

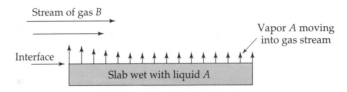

Stream of gas *B*

Interface

Vapor *A* moving into gas stream

Slab wet with liquid *A*

Fig. 22.1-1. Example of mass transfer across a plane boundary: drying of a saturated slab.

[1] T. K. Sherwood, R. L. Pigford, and C. R. Wilke, *Mass Transfer*, McGraw-Hill, New York (1975).

[2] R. E. Treybal, *Mass Transfer Operations*, 3rd edition, McGraw-Hill, New York (1980).

[3] E. L. Cussler, *Diffusion: Mass Transfer in Fluid Systems*, 2nd edition, Cambridge University Press (1997).

[4] D. E. Rosner, *Transport Processes in Chemically Reacting Flow Systems* (Unabridged), Dover, New York (2000).

Representative Membrane Processes

Pe' << 1:
Dialysis
Blood oxygenation

Pe' >> 1:
Microfiltration
Ultrafiltration
Nanofiltration
Reverse osmosis

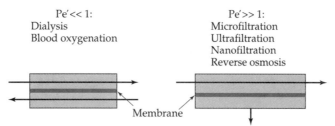

Fig. 22.1-2. Two rather typical kinds of membrane separators, classified here according to a Péclet number, Pé $= \delta v/\mathscr{D}_{eff}$, for the flow through the membrane. Here δ is the membrane thickness, v is the velocity at which solvent passes through the membrane, and $\mathscr{D}_{eff}$ is the effective solute diffusivity through the membrane. The heavy line represents the membrane, and the arrows represent the flow along or through the membrane.

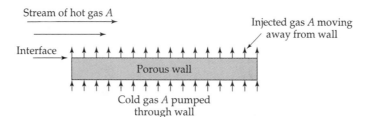

Stream of hot gas A

Injected gas A moving away from wall

Interface

Porous wall

Cold gas A pumped through wall

Fig. 22.1-3. Example of mass transfer through a porous wall: transpiration cooling.

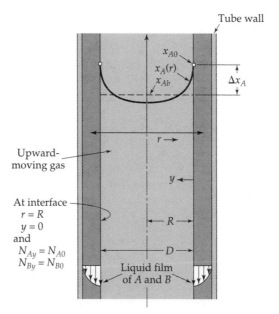

Tube wall

x_{A0}
$x_A(r)$
x_{Ab}

Δx_A

$r \rightarrow$

Upward-moving gas

$y \leftarrow$

At interface
$r = R$
$y = 0$
and
$N_{Ay} = N_{A0}$
$N_{By} = N_{B0}$

$\leftarrow R \rightarrow$

$\leftarrow D \rightarrow$

Liquid film of A and B

Fig. 22.1-4. Example of a gas–liquid contacting device: the wetted-wall column. Two chemical species A and B are moving from the downward-flowing liquid stream into the upward-flowing gas stream in a cylindrical tube.

In each of these systems, there will be both heat and mass transfer at the interface, and each of these fluxes will have a molecular (diffusive) and a convective term (here we have moved the convective term to the left side of the equation):

$$N_{A0} - x_{A0}(N_{A0} + N_{B0}) = -\left(c\mathcal{D}_{AB}\frac{\partial x_A}{\partial y}\right)\Bigg|_{y=0} \tag{22.1-1}$$

$$e_0 - (N_{A0}\overline{H}_{A0} + N_{B0}\overline{H}_{B0}) = -\left(k\frac{\partial T}{\partial y}\right)\Bigg|_{y=0} \tag{22.1-2}$$

These equations are just Eq. 18.0-1 and Eq. 19.3-6 written at the mass transfer interface ($y = 0$). They describe the interphase molar flux of species A and the interphase flux of energy (excluding the kinetic energy and the contribution from $[\tau \cdot \mathbf{v}]$). Both N_{A0} and e_0 are defined as positive for transfer *into* the local phase except in §22.4 where the fluxes in each phase are defined as positive for transfer toward the liquid.

In Chapter 14 we defined the heat transfer coefficient in the absence of mass transfer by Eq. 14.1-1 ($Q = hA\,\Delta T$). For surfaces with mass and heat transfer, Eqs. 22.1-1 and 2 suggest that the following definitions are appropriate:

$$W_{A0} - x_{A0}(W_{A0} + W_{B0}) = k_{xA}A\,\Delta x_A \tag{22.1-3}$$

$$E_0 - (W_{A0}\overline{H}_{A0} + W_{B0}\overline{H}_{B0}) = hA\,\Delta T \tag{22.1-4}$$

Here W_{A0} is the number of moles of species A per unit time going through the transfer surface at $y = 0$, and E_0 is the total amount of energy going through the surface. The transfer coefficients k_{xA} and h are not defined until the area A and the driving forces Δx_A and ΔT have been specified. All the comments in Chapter 14 regarding these definitions may be taken over in this chapter, with the result that a subscript 1, ln, a, m, or loc can be added to make clear the type of driving force that is used. In this chapter, however, we shall mainly use the local transfer coefficients and occasionally the mean transfer coefficients. Also, in this chapter, molar fluxes of the species will be used, since in chemical engineering this is traditional. The relations between the mass-transfer expressions in molar and mass units are summarized in Table 22.2-1.

Local transfer coefficients are defined by writing Eqs. 22.1-3 and 4 for a differential area. Since $dW_{A0}/dA = N_{A0}$ and $dE_0/dA = e_0$, we get the definitions

$$N_{A0} - x_{A0}(N_{A0} + N_{B0}) = k_{xA,\text{loc}}\Delta x_A \tag{22.1-5}$$

$$e_0 - (N_{A0}\overline{H}_{A0} + N_{B0}\overline{H}_{B0}) = h_{\text{loc}}\Delta T \tag{22.1-6}$$

Next, we note that the left side of Eq. 22.1-5 is J_{A0}^*, and that the left side of a similar equation written for species B is J_{B0}^*. However, since $J_{A0}^* = -J_{B0}^*$ and $\Delta x_A = -\Delta x_B$, we find that $k_{xA,\text{loc}} = k_{xB,\text{loc}}$, and therefore we can write both mass transfer coefficients as $k_{x,\text{loc}}$, which has units of (moles)/(area)(time). Furthermore, if the heat of mixing is zero (as in ideal gas mixtures), we can replace $\overline{H}_{A0}$ by $\tilde{C}_{pA,0}(T_0 - T^\circ)$, where T° is an arbitrarily chosen reference temperature, as explained in Example 19.3-1. A similar replacement may be made for $\overline{H}_{B0}$. With these changes we get

$$N_{A0} - x_{A0}(N_{A0} + N_{B0}) = k_{x,\text{loc}}\Delta x_A \tag{22.1-7}$$

$$e_0 - (N_{A0}\tilde{C}_{pA,0} + N_{B0}\tilde{C}_{pB,0})(T_0 - T^\circ) = h_{\text{loc}}\Delta T \tag{22.1-8}$$

We remind the reader that rapid mass transfer across phase boundaries can distort the velocity, temperature, and concentration profiles, as we have already seen in §18.2 and in Example 19.4-1. The correlations provided in §22.2, as well as their analogs in Chapters 6 and 14, are all for small net mass-transfer rates, that is, for situations in which the convective terms in Eqs. 22.1-7 and 8 are negligible relative to the first term. Such situations are common, and most correlations in the literature suffer from the same limitation. In §22.8 we consider the deviations associated with high net mass-transfer rates and decorate the transfer coefficients at these conditions with a superscript "•" (see §22.8).

In much of the chemical engineering literature, the mass transfer coefficients are defined by

$$N_{A0} = k^0_{x,\text{loc}} \Delta x_A \tag{22.1-9}$$

The relation of this "apparent" mass transfer coefficient to that defined by Eq. 22.1-7 is

$$k^0_{x,\text{loc}} = \frac{k_{x,\text{loc}}}{[1 - x_{A0}(1 + r)]} \tag{22.1-10}$$

in which $r = N_{B0}/N_{A0}$. Other widely used mass transfer coefficients are defined by

$$N_{A0} = k^0_{c,\text{loc}} \Delta c_A \quad \text{and} \quad n_{A0} = k^0_{\rho,\text{loc}} \Delta \rho_A \tag{22.1-11}$$

for liquids and

$$N_{A0} = k^0_{p,\text{loc}} \Delta p_A \tag{22.1-12}$$

for gases. In the limit of low solute concentrations and low net mass transfer rates, for which most correlations have been obtained,

$$\lim_{x_{A0}(1+r) \to 0} \begin{Bmatrix} k^0_{x,\text{loc}} \\ c k^0_{c,\text{loc}} \\ p k^0_{p,\text{loc}} \\ \rho k^0_{\rho,\text{loc}} \end{Bmatrix} = k_{x,\text{loc}} \tag{22.1-13}$$

The superscript 0 indicates that these quantities are applicable only for *small mass-transfer rates* and *small mole fractions of species A*.

In many industrial contactors, the true interfacial area is not known. An example of such a system would be a column containing a random packing of irregular solid particles. In such a situation, one can define a volumetric mass transfer coefficient, $k_x a$, incorporating the interfacial area for a differential region of the column. The rate at which moles of species A are transferred to the interstitial fluid in a volume $S\,dz$ of the column is then given by

$$\begin{aligned} dW_{A0} &= (k_x a)(x_{A0} - x_{Ab})S\,dz + x_{A0}(dW_{A0} + dW_{B0}) \\ &\approx (k^0_x a)(x_{A0} - x_{Ab})S\,dz \end{aligned} \tag{22.1-14}$$

Here the interfacial area, a, per unit volume is combined with the mass transfer coefficient, S is the total column cross section, and z is measured in the primary flow direction. Correlations for predicting values of these coefficients are available, but they should be used with caution. Rarely do they include all the important parameters, and as a result they cannot be safely extrapolated to new systems. Furthermore, although they are usually described as "local," they actually represent a poorly defined average over a wide range of interfacial conditions.[1-5]

We conclude this section by defining a dimensionless group widely used in the mass-transfer literature and in the remainder of this book:

$$\text{Sh} = \frac{k_x l_0}{c \mathscr{D}_{AB}} \tag{22.1-15}$$

which is called the Sherwood number based on the characteristic length l_0. This quantity can be "decorated" with subscripts 1, a, m, ln, and loc in the same manner as h.

[1] J. Stichlmair and J. F. Fair, *Distillation Principles and Practice*, Wiley, New York (1998).

[2] H. Z. Kister, *Distillation Design*, McGraw-Hill, New York (1992).

[3] J. C. Godfrey and M. M. Slater, *Liquid–Liquid Extraction Equipment*, Wiley, New York (1994).

[4] R. H. Perry and D. W. Green, *Chemical Engineers' Handbook*, 8th edition, McGraw-Hill, New York (1997).

[5] J. E. Vivian and C. J. King, in *Modern Chemical Engineering* (A. Acrivos, ed.), Reinhold, New York (1963).

§22.2   ANALYTICAL EXPRESSIONS FOR MASS TRANSFER COEFFICIENTS

In the preceding chapters we obtained a number of analytical solutions for concentration profiles and for the associated molar fluxes. From these solutions we can now derive the corresponding mass transfer coefficients. These are usually presented in dimensionless form in terms of Sherwood numbers. We summarize these analytical expressions here for use in later sections of this chapter. All of the results given in this section are for systems with a slightly soluble component A, small diffusivities $\mathcal{D}_{AB}$, and small net mass-transfer rates, as defined in §§22.1 and 8. It may be helpful at this point to refer to Table 22.2-1, where the dimensionless groups for heat and mass transfer have been summarized.

Mass Transfer in Falling Films on Plane Surfaces

For the absorption of a slightly soluble gas A into a falling film of pure liquid B, we can put the result of Eq. 18.5-18 into the form of Eq. 22.1-3 (appropriately modified for molar concentration units in the manner of Eq. 22.1-11), thus

$$W_{A0} = \left(\sqrt{\frac{4\mathcal{D}_{AB}v_{max}}{\pi L}} \right)(WL)(c_{A0} - 0) \equiv k_{c,m}^0 A \Delta c_A \qquad (22.2\text{-}1)$$

Table 22.2-1   Analogies Among Heat and Mass Transfer at Low Mass-Transfer Rates

	Heat transfer quantities (pure fluids)	Binary mass transfer quantities (isothermal fluids, molar units)	Binary mass transfer quantities (isothermal fluids, mass units)
Profiles	T	x_A	ω_A
Diffusivity	$\alpha = k/\rho \hat{C}_p$	$\mathcal{D}_{AB}$	$\mathcal{D}_{AB}$
Effect of profiles on density	$\beta = -\dfrac{1}{\rho}\left(\dfrac{\partial \rho}{\partial T}\right)_p$	$\xi = -\dfrac{1}{\rho}\left(\dfrac{\partial \rho}{\partial x_A}\right)_{p,T}$	$\zeta = -\dfrac{1}{\rho}\left(\dfrac{\partial \rho}{\partial \omega_A}\right)_{p,T}$
Flux	$\mathbf{q}$	$\mathbf{J}_A^* = \mathbf{N}_A + x_A(\mathbf{N}_A + \mathbf{N}_B)$	$\mathbf{j}_A = \mathbf{n}_A + \omega_A(\mathbf{n}_A + \mathbf{n}_B)$
Transfer rate	Q	$W_{A0} - x_{A0}(W_{A0} + W_{B0})$	$w_{A0} - \omega_{A0}(w_{A0} + w_{B0})$
Transfer coefficient	$h = \dfrac{Q}{A\,\Delta T}$	$k_x = \dfrac{W_{A0} - x_{A0}(W_{A0} + W_{B0})}{A\,\Delta x_A}$	$k_\omega = \dfrac{w_{A0} - \omega_{A0}(w_{A0} + w_{B0})}{A\,\Delta \omega_A}$
Dimensionless groups common to all three correlations	$Re = l_0 v_0 \rho/\mu$ $Fr = v_0^2/g l_0$	$Re = l_0 v_0 \rho/\mu$ $Fr = v_0^2/g l_0$	$Re = l_0 v_0 \rho/\mu$ $Fr = v_0^2/g l_0$
Dimensionless groups that are different	$Nu = h l_0/k$ $Pr = \hat{C}_p \mu/k$ $Gr = l_0^3 \rho^2 g \beta \Delta T/\mu^2$ $Pé = RePr = l_0 v_0 \hat{C}_p/k$	$Sh = k_x l_0/c\mathcal{D}_{AB}$ $Sc = \mu/\rho \mathcal{D}_{AB}$ $Gr_x = l_0^3 \rho^2 g \xi \Delta x_A/\mu^2$ $Pé = ReSc = l_0 v_0/\mathcal{D}_{AB}$	$Sh = k_\omega l_0/\rho \mathcal{D}_{AB}$ $Sc = \mu/\rho \mathcal{D}_{AB}$ $Gr_\omega = l_0^3 \rho^2 g \zeta \Delta \omega_A/\mu^2$ $Pé = ReSc = l_0 v_0/\mathcal{D}_{AB}$
Chilton–Colburn j-factors	$j_H = NuRe^{-1}Pr^{-1/3}$ $= \dfrac{h}{\rho \hat{C}_p v_0}\left(\dfrac{\hat{C}_p \mu}{k}\right)^{2/3}$	$j_D = ShRe^{-1}Sc^{-1/3}$ $= \dfrac{k_x}{c v_0}\left(\dfrac{\mu}{\rho \mathcal{D}_{AB}}\right)^{2/3}$	$j_D = ShRe^{-1}Sc^{-1/3}$ $= \dfrac{k_\omega}{\rho v_0}\left(\dfrac{\mu}{\rho \mathcal{D}_{AB}}\right)^{2/3}$

Notes: (a) The subscript 0 on l_0 and v_0 indicates the characteristic length and velocity respectively, whereas the subscript 0 on the mole (or mass) fraction and molar (or mass) flux means "evaluated at the interface." (b) All three of these Grashof numbers can be written as $Gr = l_0^3 \rho g\,\Delta\rho/\mu^2$, provided that the density change is caused only by a difference of temperature or composition.

Then, when the characteristic area is chosen to be the area of the interface WL, we see that

$$Sh_m = \frac{k^0_{c,m}L}{\mathcal{D}_{AB}} = \sqrt{\frac{4Lv_{max}}{\pi\mathcal{D}_{AB}}} = \sqrt{\frac{4}{\pi}\left(\frac{Lv_{max}\rho}{\mu}\right)\left(\frac{\mu}{\rho\mathcal{D}_{AB}}\right)}$$
$$= 1.128(ReSc)^{1/2} \tag{22.2-2}$$

This equation expresses the Sherwood number (the dimensionless mass transfer coefficient) in terms of the Reynolds number and the Schmidt number, with Re defined in terms of the maximum velocity v_{max} in the film and the film length L. The Reynolds number could also be defined in terms of the average film velocity with a different numerical coefficient.

Similarly, for the dissolution of a slightly soluble material A from the wall into a falling liquid film of pure B, we can put Eq. 18.6-10 into the form of Eq. 22.1-3 as follows:

$$W_{A0} = \left(\frac{2\mathcal{D}_{AB}}{\Gamma(\frac{4}{3})}\sqrt[3]{\frac{a}{9\mathcal{D}_{AB}L}}\right)(WL)(c_{A0} - 0) \equiv k^0_{c,m}A\,\Delta c_A \tag{22.2-3}$$

Then, using the definition $a = \rho g\delta/\mu$ given just after Eq. 18.6-1 and the expression for the maximum velocity in the film in Eq. 2.2-19, we find the Sherwood number as follows:

$$Sh_m = \frac{k^0_{c,m}L}{\mathcal{D}_{AB}} = \frac{2}{\Gamma(\frac{4}{3})}\sqrt[3]{\frac{(2v_{max}/\delta)L^2}{9\mathcal{D}_{AB}}} = \frac{1}{\Gamma(\frac{4}{3})}\sqrt[3]{\frac{16}{9}\left(\frac{L}{\delta}\right)\left(\frac{Lv_{max}\rho}{\mu}\right)\left(\frac{\mu}{\rho\mathcal{D}_{AB}}\right)}$$
$$= 1.017\sqrt[3]{\left(\frac{L}{\delta}\right)}(ReSc)^{1/3} \tag{22.2-4}$$

In this instance we have not only the Reynolds number and Schmidt number appearing, but also the ratio of the film length to the film thickness.

These two problems—gas absorption by a falling film and the dissolution of a solid wall into a falling film—illustrate two important situations. In the first problem, there is no velocity gradient at the gas–liquid interface, and the quantity ReSc appears to the $\frac{1}{2}$-power in the expression for the Sherwood number. In the second problem, there is a velocity gradient at the solid–liquid interface, and the quantity ReSc appears to the $\frac{1}{3}$-power in the Sherwood number expression.

Mass Transfer for Flow Around Spheres

Next we consider the diffusion that occurs in the creeping flow around a spherical gas bubble and around a solid sphere of diameter D. This pair of systems parallels the two systems discussed in the previous subsection.

For the gas absorption from a gas bubble surrounded by a liquid in creeping flow, we can put Eq. 20.3-28 in the form of Eq. 22.1-5 thus:

$$N_{A0,avg} = \sqrt{\frac{4}{3\pi}\frac{\mathcal{D}_{AB}v_\infty}{D}}(c_A - 0) \equiv k^0_{c,m}\Delta c_A \tag{22.2-5}$$

The Sherwood number is then

$$Sh_m = \frac{k^0_{c,m}D}{\mathcal{D}_{AB}} = \sqrt{\frac{4}{3\pi}\frac{Dv_\infty}{\mathcal{D}_{AB}}} = \sqrt{\frac{4}{3\pi}\left(\frac{Dv_\infty\rho}{\mu}\right)\left(\frac{\mu}{\rho\mathcal{D}_{AB}}\right)}$$
$$= 0.6415(ReSc)^{1/2} \tag{22.2-6}$$

Here the Reynolds number is defined using the approach velocity v_∞ of the fluid (or, alternatively, the terminal velocity of the rising bubble).

For the creeping flow around a solid sphere with a slightly soluble coating that dissolves into the approaching fluid, we may modify the result in Eq. 12.4-34 to get

$$N_{A0,avg} = \frac{(3\pi)^{2/3}}{2^{7/3}\Gamma(\frac{4}{3})} \sqrt[3]{\frac{\mathcal{D}_{AB}^2 v_\infty}{D^2}} (c_A - 0) \equiv k_{c,m}^0 \Delta c_A \tag{22.2-7}$$

This result may be rewritten in terms of the Sherwood number as

$$\mathrm{Sh}_m = \frac{k_{c,m}^0 D}{\mathcal{D}_{AB}} = \frac{(3\pi)^{2/3}}{2^{7/3}\Gamma(\frac{4}{3})} \sqrt[3]{\frac{D v_\infty}{\mathcal{D}_{AB}}} = \frac{(3\pi)^{2/3}}{2^{7/3}\Gamma(\frac{4}{3})} \sqrt[3]{\left(\frac{D v_\infty \rho}{\mu}\right)\left(\frac{\mu}{\rho \mathcal{D}_{AB}}\right)}$$

$$= 0.991(\mathrm{ReSc})^{1/3} \tag{22.2-8}$$

As in the preceding subsection we have ReSc to the $\frac{1}{2}$-power for the gas–liquid system and ReSc to the $\frac{1}{3}$-power for the liquid–solid system.

Both Eq. 22.2-6 and Eq. 22.2-8 are valid only for creeping flow. However, they are not valid in the limit that Re goes to zero. As we know from Problem 10B.1 and Eq. 14.4-5, if there is no flow past the solid sphere or the spherical bubble, $\mathrm{Sh}_m = 2$. It has been found that a satisfactory description of the mass transfer all the way down to Re = 0 can be obtained by using the simple superpositions: $\mathrm{Sh}_m = 2 + 0.6415(\mathrm{ReSc})^{1/2}$ and $\mathrm{Sh}_m = 2 + 0.991(\mathrm{ReSc})^{1/3}$ in lieu of Eqs. 22.2-6 and 8.

Mass Transfer in Steady, Nonseparated Boundary Layers on Arbitrarily Shaped Objects

For systems with a fluid-fluid interface and no velocity gradient at the interface, we found the mass flux at the surface to be given by Eq. 20.3-14:

$$N_{A0} = \sqrt{\frac{\mathcal{D}_{AB}}{\pi} \frac{h_z^2 v_s^2}{\int_0^x h_x h_z^2 v_s d\bar{x}}} (c_{A0} - 0) \equiv k_{c,loc}^0 \Delta c_A \tag{22.2-9}$$

The local Sherwood number is

$$\mathrm{Sh}_{loc} = \frac{k_{c,loc}^0 l_0}{\mathcal{D}_{AB}} = \frac{1}{\sqrt{\pi}} (\mathrm{ReSc})^{1/2} \sqrt{\frac{h_z^2 v_s^2}{\int_0^x h_x h_z^2 v_s d\bar{x}} \frac{l_0}{v_0}} \tag{22.2-10}$$

in which the constant, $1/\sqrt{\pi}$, is equal to 0.5642 and Re $= l_0 v_0 \rho / \mu$.

Similarly for systems with fluid–solid interfaces and a velocity gradient at the interface, the mass flux expression is given in Eq. 20.3-26 as

$$N_{A0} = \frac{\mathcal{D}_{AB}}{\Gamma(\frac{4}{3})} \sqrt[3]{\frac{(h_z \beta)^{3/2}}{9\mathcal{D}_{AB} \int_0^x \sqrt{h_z \beta}\, h_x h_z\, dx}} (c_{A0} - 0) \equiv k_{c,loc}^0 \Delta c_A \tag{22.2-11}$$

The analogous Sherwood number expression is

$$\mathrm{Sh}_{loc} = \frac{k_{c,loc}^0 l_0}{\mathcal{D}_{AB}} = \frac{1}{9^{1/3}\Gamma(\frac{4}{3})} (\mathrm{ReSc})^{1/3} \sqrt[3]{\frac{(h_z \beta)^{3/2}}{\int_0^x \sqrt{h_z \beta}\, h_x h_z\, dx} \frac{l_0^2}{v_0}} \tag{22.2-12}$$

where the numerical coefficient has the value 0.5384. In these equations l_0 and v_0 are a characteristic length and a characteristic velocity that can be chosen after the shape of the body has been defined. Here again we see that the $\frac{1}{2}$-power on ReSc appears in the fluid–fluid system, and the $\frac{1}{3}$-power on ReSc appears in the fluid–solid system—regardless of the shape. The radicands of the Sherwood number expressions are dimensionless.

Mass Transfer in the Neighborhood of a Rotating Disk

For a disk of diameter D coated with a slightly soluble material A rotating with angular velocity Ω in a large region of liquid B, the mass flux at the surface of the disk is independent of position. According to Eq. 19D.4-7 we have

$$N_{A0} = 0.620 \left(\frac{\mathscr{D}_{AB}^{2/3} \Omega^{1/2}}{\nu^{1/6}} \right) (c_{A0} - 0) \equiv k_{c,m}^0 \Delta c_A \tag{22.2-13}$$

This may be expressed in terms of the Sherwood number as

$$\mathrm{Sh}_m = \frac{k_{c,m}^0 D}{\mathscr{D}_{AB}} = 0.620 \left(\frac{D\Omega^{1/2} \rho^{1/6}}{\mathscr{D}_{AB}^{1/3} \mu^{1/6}} \right) = 0.620 \sqrt{\frac{D(D\Omega)\rho}{\mu}} \sqrt[3]{\frac{\mu}{\rho \mathscr{D}_{AB}}}$$

$$= 0.620 \, \mathrm{Re}^{1/2} \mathrm{Sc}^{1/3} \tag{22.2-14}$$

Here the characteristic velocity in the Reynolds number is chosen to be $D\Omega$.

§22.3   CORRELATION OF BINARY TRANSFER COEFFICIENTS IN ONE PHASE

In this section we show that correlations for binary mass transfer coefficients at low mass-transfer rates can be obtained directly from their heat transfer analogs simply by a change of notation. These correspondences are quite useful, and many heat transfer correlations have, in fact, been obtained from their mass transfer analogs.

To illustrate the background of these useful analogies and the conditions under which they apply, we begin by presenting the diffusional analog of the dimensional analysis given in §14.3. Consider the steadily driven, laminar or turbulent isothermal flow of a liquid solution of A in B, in the tube shown in Fig. 22.3-1. The fluid enters the tube at $z = 0$ with velocity uniform out to very near the wall and with a uniform inlet composition x_{A1}. From $z = 0$ to $z = L$, the tube wall is coated with a solid solution of A and B, which dissolves slowly and maintains the interfacial liquid composition constant at x_{A0}. For the moment we assume that the physical properties ρ, μ, c, and $\mathscr{D}_{AB}$ are constant.

The mass transfer situation just described is mathematically analogous to the heat transfer situation described at the beginning of §14.3. To emphasize the analogy, we present the equations for the two systems together. Thus the rate of heat addition by conduction between 1 and 2 in Fig. 14.3-1 and the molar rate of addition of species A by

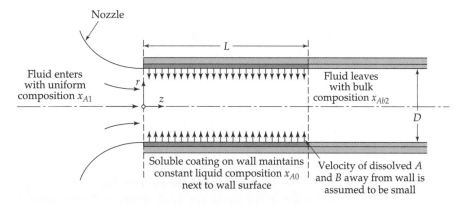

Fig. 22.3-1. Mass transfer in a pipe with a soluble wall.

diffusion between 1 and 2 in Fig. 22.3-1 are given by the following expressions, valid for either laminar or turbulent flow:

heat transfer:
$$Q(t) = \int_0^L \int_0^{2\pi} \left(+k \frac{\partial T}{\partial r} \bigg|_{r=R} \right) R \, d\theta \, dz \tag{22.3-1}$$

mass transfer:
$$W_{A0}(t) - x_{A0}(W_{A0}(t) + W_{B0}(t)) = \int_0^L \int_0^{2\pi} \left(+c\mathcal{D}_{AB} \frac{\partial x_A}{\partial r} \bigg|_{r=R} \right) R \, d\theta \, dz \tag{22.3-2}$$

Equating the left sides of these equations to $h_1(\pi DL)(T_0 - T_1)$ and $k_{x1}(\pi DL)(x_{A0} - x_{A1})$ respectively, we get for the transfer coefficients

heat transfer:
$$h_1(t) = \frac{1}{\pi DL(T_0 - T_1)} \int_0^L \int_0^{2\pi} \left(+k \frac{\partial T}{\partial r} \bigg|_{r=R} \right) R \, d\theta \, dz \tag{22.3-3}$$

mass transfer:
$$k_{x1}(t) = \frac{1}{\pi DL(x_{A0} - x_{A1})} \int_0^L \int_0^{2\pi} \left(+c\mathcal{D}_{AB} \frac{\partial x_A}{\partial r} \bigg|_{r=R} \right) R \, d\theta \, dz \tag{22.3-4}$$

We now introduce the dimensionless variables $\check{r} = r/D$, $\check{z} = z/D$, $\check{T} = (T - T_0)/(T_1 - T_0)$, and $\check{x}_A = (x_A - x_{A0})/(x_{A1} - x_{A0})$ and rearrange to obtain

heat transfer:
$$\mathrm{Nu}_1(t) = \frac{h_1 D}{k} = \frac{1}{2\pi L/D} \int_0^{L/D} \int_0^{2\pi} \left(-\frac{\partial \check{T}}{\partial \check{r}} \bigg|_{\check{r}=\frac{1}{2}} \right) d\theta \, d\check{z} \tag{22.3-5}$$

mass transfer:
$$\mathrm{Sh}_1(t) = \frac{k_{x1} D}{c\mathcal{D}_{AB}} = \frac{1}{2\pi L/D} \int_0^{L/D} \int_0^{2\pi} \left(-\frac{\partial \check{x}_A}{\partial \check{r}} \bigg|_{\check{r}=\frac{1}{2}} \right) d\theta \, d\check{z} \tag{22.3-6}$$

Here Nu is the Nusselt number for heat transfer without mass transfer, and Sh is the Sherwood number for isothermal mass transfer at small mass-transfer rates. The Nusselt number is a dimensionless temperature gradient integrated over the surface, and the Sherwood number is a dimensionless concentration gradient integrated over the surface.

These gradients can, in principle, be evaluated from Eqs. 11.5-7, 8, and 9 (for heat transfer) and Eqs. 19.5-8, 9, and 11 (for mass transfer), under the following boundary conditions (with $\check{\mathbf{v}}$ and $\check{\mathscr{P}}$ defined as in §14.3 and with time averaging of the solutions if the flow is turbulent):

velocity and pressure:

$$\text{at } \check{z} = 0, \check{\mathbf{v}} = \boldsymbol{\delta}_z \qquad \text{for } 0 \leq \check{r} < \tfrac{1}{2} \tag{22.3-7}$$
$$\text{at } \check{r} = \tfrac{1}{2}, \check{\mathbf{v}} = 0 \qquad \text{for } \check{z} \geq 0 \tag{22.3-8}$$
$$\text{at } \check{r} = 0 \text{ and } \check{z} = 0, \check{\mathscr{P}} = 0 \tag{22.3-9}$$

temperature:

$$\text{at } \check{z} = 0, \check{T} = 1 \qquad \text{for } 0 \leq r < \tfrac{1}{2} \tag{22.3-10}$$
$$\text{at } \check{r} = \tfrac{1}{2}, \check{T} = 0 \qquad \text{for } 0 \leq z \leq L/D \tag{22.3-11}$$

concentration:

$$\text{at } \check{z} = 0, \check{x}_A = 1 \qquad \text{for } 0 \leq r < \tfrac{1}{2} \tag{22.3-12}$$
$$\text{at } \check{r} = \tfrac{1}{2}, \check{x}_A = 0 \qquad \text{for } 0 \leq z \leq L/D \tag{22.3-13}$$

The boundary condition in Eq. 22.3-8, on the velocity at the wall, is accurate for the heat-transfer system and also for the mass-transfer system provided that $x_{A0}(W_{A0} + W_{B0})$ is small; the latter criterion is discussed in §§22.1 and 8. No boundary conditions are needed at the outlet plane, $z = L/D$, when we neglect the $\partial^2/\partial z^2$ terms of the conservation equations in the manner of §4.4 and §14.3.

If we can neglect the heat production by viscous dissipation in Eq. 11.5-9 and if there is no production of A by chemical reaction as in Eq. 19.5-11, then the differential equations for heat and mass transport are analogous along with the boundary conditions. It follows then that the dimensionless profiles of temperature and concentration (time smoothed, when necessary) are similar,

$$\check{T} = F(\check{r}, \theta, \check{z}, Re, Pr); \qquad \check{x}_A = F(\check{r}, \theta, \check{z}, Re, Sc) \qquad (22.3\text{-}14, 15)$$

with the same form of F in both systems. Thus, to get the concentration profiles from the temperature profiles, one replaces $\check{T}$ by $\check{x}_A$ and Pr by Sc.

Finally, inserting the profiles into Eqs. 22.3-5 and 6 and performing the integrations and then time-averaging give for *forced convection*

$$Nu_1 = G(Re, Pr, L/D); \qquad Sh_1 = G(Re, Sc, L/D) \qquad (22.3\text{-}16, 17)$$

Here G is the same function in both equations. The same formal expression is obtained for Nu_a, Nu_{ln}, Nu_{loc} as well as for the corresponding Sherwood numbers. This important analogy permits one to write down a mass transfer correlation from the corresponding heat transfer correlation merely by replacing Nu by Sh, and Pr by Sc. The same can be done for any geometry and for both laminar and turbulent flow. Note, however, that to get this analogy one has to assume (i) constant physical properties, (ii) small net mass-transfer rates, (iii) no chemical reactions, (iv) no viscous dissipation heating, (v) no absorption or emission of radiant energy, and (vi) no pressure diffusion, thermal diffusion, or forced diffusion. Some of these effects will be discussed in subsequent sections of this chapter; others will be treated in Chapter 24.

For *free convection* around objects of any given shape, a similar analysis shows that

$$Nu_m = H(Gr, Pr); \qquad Sh_m = H(Gr_x, Sc) \qquad (22.3\text{-}18, 19)$$

Here H is the same function in both cases, and the Grashof numbers for both processes are defined analogously (see Table 22.2-1 for a summary of the analogous quantities for heat and mass transfer).

To allow for the variation of physical properties in mass transfer systems, we extend the procedures introduced in Chapter 14 for heat transfer systems. That is, we generally evaluate the physical properties at some kind of mean film composition and temperature, except for the viscosity ratio μ_b/μ_0.

We now give three illustrations of how to "translate" from heat transfer to mass transfer correlations:

Forced Convection Around Spheres

For forced convection around a solid sphere, Eq. 14.4-5 and its mass-transfer analog are:

$$Nu_m = 2 + 0.60\, Re^{1/2}\, Pr^{1/3}; \qquad Sh_m = 2 + 0.60\, Re^{1/2}\, Sc^{1/3} \qquad (22.3\text{-}20, 21)$$

Equations 22.3-20 and 21 are valid for constant surface temperature and composition, respectively, and for small net mass-transfer rates. They may be applied to simultaneous heat and mass transfer under restrictions (i)–(vi) given after Eq. 22.3-17.

Forced Convection along a Flat Plate

As another illustration of the use of analogies, we can cite the extension of Eq. 14.4-4 for the laminar boundary layer along a flat plate, to include mass transfer:

$$j_{H,\text{loc}} = j_{D,\text{loc}} = \tfrac{1}{2} f_{\text{loc}} = 0.332\, Re_x^{-1/2} \qquad (22.3\text{-}22)$$

The Chilton–Colburn j-factors, one for heat transfer and one for diffusion, are defined as[1]

$$j_{H,\text{loc}} = \frac{\text{Nu}_{\text{loc}}}{\text{RePr}^{1/3}} = \frac{h_{\text{loc}}}{\rho \hat{C}_p v_\infty} \left(\frac{\hat{C}_p \mu}{k} \right)^{2/3} \tag{22.3-23}$$

$$j_{D,\text{loc}} = \frac{\text{Sh}_{\text{loc}}}{\text{ReSc}^{1/3}} = \frac{k_{x,\text{loc}}}{c v_\infty} \left(\frac{\mu}{\rho \mathcal{D}_{AB}} \right)^{2/3} \tag{22.3-24}$$

The three-way analogy in Eq. 22.3-22 is accurate for Pr and Sc near unity (see Table 12.4-1) within the limitations mentioned after Eq. 22.3-17. For flow around other objects, the friction factor part of the analogy is not valid because of the form drag, and even for flow in circular tubes the analogy with $\frac{1}{2} f_{\text{loc}}$ is only approximate (see §14.4).

The Chilton–Colburn Analogy

The more widely applicable empirical analogy

$$j_H = j_D = \text{a function of Re, geometry, and boundary conditions} \tag{22.3-25}$$

has proven to be useful for transverse flow around cylinders, flow through packed beds, and flow in tubes at high Reynolds numbers. For flow in ducts and packed beds, the "approach velocity" v_∞ has to be replaced by the interstitial velocity or the superficial velocity. Equation 22.3-25 is the usual form of the *Chilton–Colburn analogy*. It is evident from Eqs. 22.3-20 and 21, however, that the analogy is valid for flow around spheres only when Nu and Sh are replaced by (Nu − 2) and (Sh − 2).

It would be very misleading to leave the impression that all mass transfer coefficients can be obtained from the analogous heat transfer coefficient correlations. For mass transfer we encounter a much wider variety of boundary conditions and other ranges of the relevant variables. Non-analogous behavior is addressed in §§22.5-8.

EXAMPLE 22.3-1

Evaporation from a Freely Falling Drop

A spherical drop of water, 0.05 cm in diameter, is falling at a velocity of 215 cm/s through dry, still air at 1 atm pressure with no internal circulation. Estimate the instantaneous rate of evaporation from the drop, when the drop surface is at $T_0 = 70°F$ and the air (far from the drop) is at $T_\infty = 140°F$. The vapor pressure of water at 70°F is 0.0247 atm. Assume quasi-steady state conditions.

SOLUTION

Designate water as species A and air as species B. The solubility of air in water may be neglected, so that $W_{B0} = 0$. Then assuming that the evaporation rate is small, we may write Eq. 22.1-3 for the entire spherical surface as

$$W_{A0} = k_{xm}(\pi D^2) \frac{x_{A0} - x_{A\infty}}{1 - x_{A0}} \tag{22.3-26}$$

The mean mass transfer coefficient, k_{xm}, may be predicted from Eq. 22.3-21 in the assumed absence of internal circulation.

The film conditions needed for estimating the physical properties are obtained as follows:

$$T_f = \tfrac{1}{2}(T_0 + T_\infty) \quad = \tfrac{1}{2}(70 + 140) \quad = 105°F \tag{22.3-27}$$

$$x_{Af} = \tfrac{1}{2}(x_{A0} + x_{A\infty}) = \tfrac{1}{2}(0.0247 + 0) = 0.0124 \tag{22.3-28}$$

---

[1] T. H. Chilton and A. P. Colburn, *Ind. Eng. Chem.*, **26**, 1183–1187 (1934).

In computing x_{Af}, we have assumed ideal gas behavior, equilibrium at the interface, and complete insolubility of air in water. The mean mole fraction, x_{Af}, of the water vapor is sufficiently small that it can be neglected in evaluating the physical properties at the film conditions:

$$c = 3.88 \times 10^{-5}\,\text{g-moles/cm}^3$$

$$\rho = 1.12 \times 10^{-3}\,\text{g/cm}^3$$

$$\mu = 1.91 \times 10^{-4}\,\text{g/cm} \cdot \text{s (from Table 1.1-1)}$$

$$\mathscr{D}_{AB} = 0.292\,\text{cm}^2/\text{s (from Eq. 17.2-1)}$$

$$\text{Sc} = \left(\frac{\mu}{\rho \mathscr{D}_{AB}}\right) = \frac{1.91 \times 10^{-4}}{(1.12 \times 10^{-3})(0.292)} = 0.58$$

$$\text{Re} = \left(\frac{D v_\infty \rho}{\mu}\right) = \frac{(0.05)(215)(1.12 \times 10^{-3})}{1.91 \times 10^{-4}} = 63$$

When these values are used in Eq. 22.3-21 we get

$$\text{Sh}_m = 2 + 0.60(63)^{1/2}(0.58)^{1/3} = 5.96 \tag{22.3-29}$$

and the mean mass transfer coefficient is then

$$k_{xm} = \frac{c\mathscr{D}_{AB}}{D}\,\text{Sh}_m = \frac{(3.88 \times 10^{-5})(0.292)}{0.05}\,(5.96)$$

$$= 1.35 \times 10^{-3}\,\text{g-mol/s} \cdot \text{cm}^2 \tag{22.3-30}$$

Then from Eq. 22.3-26 the evaporation rate is found to be

$$W_{A0} = (1.35 \times 10^{-3})(\pi)(0.05)^2\,\frac{0.0247 - 0}{1 - 0.0247}$$

$$= 2.70 \times 10^{-7}\,\text{g-mole/s} \tag{22.3-31}$$

This result corresponds to a decrease of 1.23×10^{-3} cm/s in the drop diameter and indicates that a drop of this size will fall a considerable distance before it evaporates completely.

In this example, for simplicity, the velocity and surface temperature of the drop were given. In general, these conditions must be calculated from momentum and energy balances, as discussed in Problem 22B.1.

EXAMPLE 22.3-2

The Wet and Dry Bulb Psychrometer

We next turn to a problem for which the analogy between heat and mass transfer leads to a surprisingly simple and useful, if approximate, result. The system, shown in Fig. 22.3-2, is a pair of thermometers, one of which is covered with a cylindrical wick kept saturated with water. The wick will cool by evaporation into the moving air stream and for steady operation will approach an asymptotic value known as the *wet bulb temperature*. The bare thermometer, on the other hand, will tend to approach the actual temperature of the approaching air, and this value is called the *dry bulb temperature*. Develop an expression for determining the humidity of the air from the wet and dry bulb temperature readings neglecting radiation and assuming that the replacement of the evaporating water has no significant effect on the wet bulb temperature measurement. In Problem 22B.2 we will see how radiation can be taken into account.

SOLUTION

For simplicity, we assume that the fluid velocity is high enough that the thermometer readings are unaffected by radiation and by heat conduction along the thermometer stems, but not so high that viscous dissipation heating effects become significant. These assumptions are usually satisfactory for glass thermometers and for gas velocities of 30 to 100 ft/s. The dry bulb temperature is then the same as the temperature T_∞ of the approaching gas, and the wet bulb temperature is the same as the temperature T_0 of the outside of the wick.

Let species A be water and species B be air. An energy balance is made on a system that contains a length L of the wick (the distance between planes 1 and 2 in the figure). The rate of

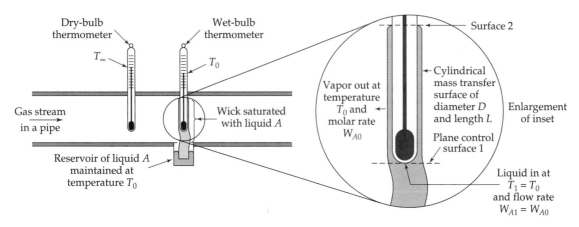

Fig. 22.3-2. Sketch of a wet-bulb and dry-bulb psychrometer installation. It is assumed that no heat or mass moves across plane 2.

heat addition to the system by the gas stream is $h_m(\pi DL)(T_\infty - T_0)$. Enthalpy also enters via plane 1 at a rate $W_{A1}\overline{H}_{A1}$ in the liquid phase and leaves at the mass transfer surface at a rate $W_{A0}\overline{H}_{A0}$, both of these occurring at a temperature T_0. Hence the energy balance gives

$$h_m(\pi DL)(T_\infty - T_0) = W_{A0}(\overline{H}_{A1} - \overline{H}_{A0}) \tag{22.3-32}$$

since the water enters the system at plane 1 at the same rate that it leaves as water vapor at the mass transfer interface 0. To a very good approximation, $\overline{H}_{A1} - \overline{H}_{A0}$ may be replaced by $\Delta\tilde{H}_{vap}$, the molar heat of vaporization of water.

From the definition of the mass transfer coefficient

$$W_{A0} - x_{A0}(W_{A0} + W_{B0}) = k_{xm}(\pi DL)(x_{A0} - x_{A\infty}) \tag{22.3-33}$$

in which $W_{B0} = 0$ as in the preceding example. Combination of Eqs. 22.2-32 and 33 gives then

$$\frac{(x_{A0} - x_{A\infty})}{(T_\infty - T_0)(1 - x_{A0})} = \frac{h_m}{k_{xm}\Delta\tilde{H}_{vap}} \tag{22.3-34}$$

Then using the definitions of Nu_m and Sh_m, and noting that $\rho\hat{C}_p = c\tilde{C}_p$, we may rewrite Eq. 22.3-34 as

$$\frac{(x_{A0} - x_{A\infty})}{(T_\infty - T_0)(1 - x_{A0})} = \frac{\mathrm{Nu}_m}{\mathrm{Sh}_m}\left(\frac{\mathrm{Sc}}{\mathrm{Pr}}\right)\frac{\tilde{C}_p}{\Delta\tilde{H}_{vap}} \tag{22.3-35}$$

Because of the analogy between heat and mass transfer, we can expect that the mean Nusselt and Sherwood numbers will be of the same form:

$$\mathrm{Nu}_m = F(\mathrm{Re})\mathrm{Pr}^n; \qquad \mathrm{Sh}_m = F(\mathrm{Re})\mathrm{Sc}^n \tag{22.3-36, 37}$$

where F is the same function of Re in both expressions. Therefore, knowing the dry and wet bulb temperatures and the mole fraction of the water vapor adjacent to the wick (x_{A0}), we can calculate the upstream composition $x_{A\infty}$ of the air stream from

$$\frac{(x_{A0} - x_{A\infty})}{(T_\infty - T_0)(1 - x_{A0})} = \left(\frac{\mathrm{Sc}}{\mathrm{Pr}}\right)^{1-n}\frac{\tilde{C}_p}{\Delta\tilde{H}_{vap}} \tag{22.3-38}$$

The exponent n depends to a slight extent on the geometry, but is not far from $\frac{1}{3}$, and the quantity $(\mathrm{Sc}/\mathrm{Pr})^{1-n}$ is not far from unity.[2] Furthermore, the wet bulb temperature is seen to be

---

[2] A somewhat different equation, with $1 - n = 0.56$, was recommended for measurements in air by C. H. Bedingfield and T. B. Drew, *Ind. Eng. Chem.*, **42**, 1164–1173 (1950).

independent of the Reynolds number under the assumption introduced in Eqs. 22.3-36 and 37. This result would also have been obtained by using the Chilton–Colburn relations, which would give $n = \frac{1}{3}$ directly.

The interfacial gas composition x_{A0} can be accurately predicted, at low mass-transfer rates, by neglecting the heat and mass transfer resistance of the interface itself (see §22.4 for further discussion of this point). One can then represent x_{A0} by the vapor–liquid equilibrium relationship:

$$x_{A0} = x_{A0}(T_0, p) \tag{22.3-39}$$

A relation of this kind will hold for given species A and B if the liquid is pure A as assumed above A commonly used approximation of this relationship is

$$x_{A0} = \frac{p_{A,\mathrm{vap}}}{p} \tag{22.3-40}$$

in which $p_{A,\mathrm{vap}}$ is the vapor pressure of pure A at temperature T_0. This relation assumes tacitly that the presence of B does not alter the partial pressure of A at the interface, and that A and B form an ideal gas mixture.

If an air–water mixture at 1 atm pressure gives a wet bulb temperature of 70°F and a dry bulb temperature of 140°F, then

$$p_{A,\mathrm{vap}} = 0.0247 \text{ atm}$$
$$x_{A0} = 0.0247, \text{ from Eq. 22.3-40}$$
$$\tilde{C}_p = 6.98 \text{ Btu/lb-mole} \cdot \text{F at 105°F, the film temperature}$$
$$\Delta \tilde{H}_\mathrm{vap} = 18{,}900 \text{ Btu/lb-mole at 70°F}$$
$$\mathrm{Sc} = 0.58 \text{ (see Example 22.2-1)}$$
$$\mathrm{Pr} = 0.74, \text{ from Eq. 9.3-16}$$

Substitution into Eq. 22.3-37, with $n = \frac{1}{3}$, then gives

$$\frac{(0.0247 - x_{A\infty})}{(140 - 70)(1 - 0.0247)} = \left(\frac{0.58}{0.74}\right)^{2/3} \frac{6.98}{18{,}900} \tag{22.3-41}$$

From this the mole fraction of water in the approaching air is

$$x_{A\infty} = 0.0033 \tag{22.3-42}$$

Since we assumed that the film concentration was $x_A = 0$ as a first approximation, we could go back and make a second approximation by using an average film concentration of $\frac{1}{2}(0.0247 + 0.0033) = 0.0140$ in the physical property calculations. The physical properties are not known accurately enough here to justify recalculation.

The calculated result in Eq. 22.3-43 is in only fair agreement with published humidity charts, because these are typically based on the adiabatic saturation temperature rather than the wet bulb temperature.[3]

EXAMPLE 22.3-3

Mass Transfer in Creeping Flow Through Packed Beds

Many important adsorptive operations, from purification of proteins in modern biotechnology to the recovery of solvent vapor by dry-cleaning establishments, occur in dense particulate beds and are typically carried out in steady creeping flow—that is, at $\mathrm{Re} = D_p v_0 \rho / \mu < 20$. Here D_p is the effective particle diameter and v_0 is the superficial velocity, defined as volumetric flow rate divided by the total cross section of the bed (see §6.4). It follows that the dimensionless velocity $\mathbf{v}/v_0$ will have a spatial distribution independent of the Reynolds number. Detailed information is available only for spherical packing particles.

---

[3] O. A. Hougen, K. M. Watson, and R. A. Ragatz, *Chemical Process Principles*, Part I, 2nd edition, Wiley, New York, (1954), p. 120.

Using the dimensional analysis discussion at the beginning of this section, predict the form of the steady-state mass transfer coefficient correlation for creeping flow.

SOLUTION

The dimensional analysis procedure in §19.5 may be used, with D_p as the characteristic length and v_0 the characteristic velocity. Then, from Eq. 19.5-11, we see that the dimensionless concentration depends only on the product ReSc, in addition to the dimensionless position coordinates and the geometry of the bed.

The most extensive data are for creeping flow at large Péclet numbers. Experimental data on the dissolution of benzoic acid spheres in water[4] have yielded the result

$$Sh_m = \frac{1.09}{\varepsilon}(ReSc)^{1/3} \qquad ReSc \gg 1 \tag{22.3-43}$$

where ε is the volume fraction of the bed occupied by the flowing fluid. Equation 22.3-43 is reasonably consistent with the relation

$$Sh_m = 2 + 0.991(ReSc)^{1/3} \tag{22.3-44}$$

which incorporates the creeping flow solution for flow around an isolated sphere[5] ($\varepsilon = 1$) (see §§22.2b). This suggests that the flow pattern around an isolated sphere is not much different from that around a sphere surrounded by other spheres, particularly near the sphere surface where most of the mass transport takes place.

No reliable data are available for the limiting behavior at very low values of ReSc, but numerical calculations for a regular packing[6] predict that the Sherwood number asymptotically approaches a constant near 4.0 if based on a local difference between interfacial and bulk compositions.

Behavior within the solid phase is far more complex, and no simple approximation is wholly trustworthy. However, experiments to date[7] show that where intraparticle mass transport is described by Fick's second law, one can use the approximation

$$Sh_m = \frac{k_{c,s}D_p}{\mathcal{D}_{As}} \approx 10 \tag{22.3-45}$$

where $k_{c,s}$ is the effective mass transfer coefficient within the solid phase and $\mathcal{D}_{As}$ is the diffusivity of A in the solid phase. The equation is for "slow" changes in the solute concentration bathing the particle. This is an asymptotic solution for a linear change of surface concentration with time,[8] and has been justified[9] by calculations. For a Gaussian (bell-shaped) concentration wave, "slow" means that the passage time (temporal standard deviation) of the wave is long relative to the particle diffusional response time, which is of the order of $D_p^2/6\mathcal{D}_{As}$. Fick's second law must be solved with the detailed history of surface concentration when this inequality is not satisfied.

In packed beds, as with tube flow, one must keep in mind the fact that there will be nonuniformities in the concentration as a function of the radial coordinate. This was discussed in §14.5 and §20.3.

[4] E. J. Wilson and C. J. Geankopolis, *Ind. Eng. Chem. Fundamentals*, **5**, 9–14 (1966). See also J. R. Selman and C. W. Tobias, *Advances in Chemical Engineering*, **10**, 212–318 (1978), for an extensive summary of mass transfer coefficient correlations obtained by electrochemical measurements.

[5] V. G. Levich, *Physicochemical Hydrodynamics*, Prentice-Hall, Englewood Cliffs, N.J. (1962), §14.

[6] J. P. Sørensen and W. E. Stewart, *Chem. Eng. Sci.*, **29**, 811–837 (1974).

[7] A. M. Athalye, J. Gibbs, and E. N. Lightfoot, *J. Chromatography*, **589**, 71–85 (1992).

[8] H. S. Carslaw and J. C. Jaeger, *Conduction of Heat in Solids*, 2nd edition, Oxford University Press (1959), §9.3, Eqs. 10 and 11.

[9] J. F. Reis, E. N. Lightfoot, P. T. Noble, and A. S. Chiang, *Sep. Sci. Tech.*, **14**, 367–394 (1979).

EXAMPLE 22.3-4

Mass Transfer to Drops and Bubbles

In both gas–liquid[10] and liquid–liquid[11] contactors, sprays of liquid drops or clouds of bubbles are frequently encountered. Contrast their mass transfer behavior with that of solid spheres.

SOLUTION

Many different types of behavior are encountered, and surface forces can play a very important role. We discuss surface forces in some detail in §22.7. Here we consider only some limiting cases and refer readers to the above-cited references.

Very small drops and bubbles behave like solid spheres and can be treated by the correlations in Example 22.3-3 and in Chapter 14. However, if both adjacent phases are free of surfactants and small particulate contaminants, the interior phase circulates and carries the adjacent regions of the exterior phase along. This stress-driven "Hadamard–Rybczinski circulation"[12] increases the mass transfer rates markedly, often by almost an order of magnitude, and the rates can be estimated from extensions[13–16] of the "penetration model" discussed in §18.5. Thus, for a spherical bubble of gas A with diameter D rising through a clean liquid B, the Sherwood number on the liquid side lies in the range[16]

$$\sqrt{\frac{4}{3\pi}\frac{Dv_t}{\mathscr{D}_{AB}}} < \text{Sh}_m < \sqrt{\frac{4}{\pi}\frac{Dv_t}{\mathscr{D}_{AB}}} \tag{22.3-46}$$

where v_t is the terminal velocity (see Eqs. 18.5-19 and 20).

The size at which the transition from the solid-like behavior to circulation occurs depends on degree of surface contamination and is not easily predicted.

Very large drops or bubbles oscillate,[13] and both phases follow a modified penetration model,

$$\text{Sh}_m \approx \sqrt{\frac{4.8D^2\omega}{\pi\mathscr{D}_{AB}}} \tag{22.3-47}$$

with angular frequency of oscillation[17]

$$\omega = \sqrt{\frac{192\sigma}{D^3(3\rho_D + 2\rho_C)}} \tag{22.3-48}$$

where σ is the interfacial tension, and ρ_D and ρ_C are the densities of the drops and the continuous medium.

The success of this model implies that the boundary layer is refreshed once every oscillation, but there is also a small effect of periodic stretching of the surface.

§22.4   DEFINITION OF TRANSFER COEFFICIENTS IN TWO PHASES

Recall that in §10.6 we introduced the concept of an overall heat transfer coefficient, U, to describe the heat transfer between two streams separated from each other by a wall. This overall coefficient accounted for the thermal resistance of the wall itself, as well as the thermal resistance in the fluids on either side of the wall.

---

[10] J. Stichlmair and J. F. Fair, *Distillation Principles and Practice*, Wiley, New York (1998).

[11] J. C. Godfrey and M. M. Slater, *Liquid–Liquid Extraction Equipment*, Wiley, New York (1994).

[12] J. Happel and H. Brenner, *Low Reynolds Number Hydrodynamics*, Martinus Nijhoff, The Hague (1983).

[13] J. B. Angelo, E. N. Lightfoot, and D. W. Howard, *AIChE Journal*, **12**, 751–760 (1966).

[14] J. B. Angelo and E. N. Lightfoot, *AIChE Journal*, **14**, 531–540 (1968).

[15] W. E. Stewart, J. B. Angelo, and E. N. Lightfoot, *AIChE Journal*, **16**, 771–786 (1970).

[16] R. Higbie, *Trans. AIChE*, **31**, 365–389 (1935).

[17] R. R. Schroeder and R. C. Kintner, *AIChE Journal*, **11**, 5–8 (1965).

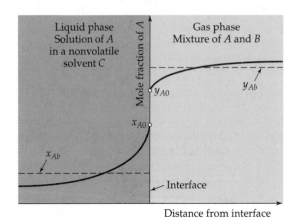

Fig. 22.4-1. Concentration profiles in the neighborhood of a gas–liquid interface

We now treat the analogous situation for mass transfer, except that here we are concerned with two fluids in intimate contact with one another, so that there is no wall resistance or interfacial resistance. This is the situation most commonly met in practice. Since the interface itself contains no significant mass, we may begin by assuming continuity of the total mass flux at the interface for any species being transferred. Then for the system shown in Fig. 22.4-1 we write

$$N_{A0}|_{\text{gas}} = N_{A0}|_{\text{liquid}} = N_{A0} \tag{22.4-1}$$

for the interfacial flux of A toward the liquid phase. Then using the definition given in Eq. 22.1-9, we get

$$N_{A0} = k^0_{y,\text{loc}}(y_{Ab} - y_{A0}) = k^0_{x,\text{loc}}(x_{A0} - x_{Ab}) \tag{22.4-2}$$

in which we are now following the tradition of using x for mole fractions in the liquid phase and y for mole fractions in the gas phase. We now have to interrelate the interfacial compositions in the two phases.

In nearly all situations this can be done by assuming equilibrium across the interface, so that adjacent gas and liquid compositions lie on the equilibrium curve (see Fig. 22.4-2), which is regarded as known from solubility data:

$$y_{A0} = f(x_{A0}) \tag{22.4-3}$$

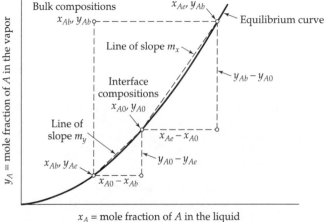

Fig. 22.4-2. Relations among gas- and liquid-phase compositions, and the graphical interpretations of m_x and m_y.

Exceptions to this are: (i) extremely high mass-transfer rates, observed for gas phases at high vacuum, where N_{A0} approaches $p_{A0}/\sqrt{2\pi M_A RT}$, the equilibrium rate at which gas molecules impinge on the interface; and (ii) interfaces contaminated with high concentrations of adsorbed particles or surfactant molecules. Situation (i) is quite rare, and situation (ii) normally acts indirectly by changing the flow behavior rather than causing deviations from equilibrium. In extreme cases surface contamination can provide additional transport resistances.

To describe rates of interphase transport, one can either use Eqs. 22.4-2 and 3 to calculate interface concentrations and then proceed to use the single-phase coefficients, or else work with overall mass transfer coefficients

$$N_{A0} = K^0_{y,\text{loc}}(y_{Ab} - y_{Ae}) = K^0_{x,\text{loc}}(x_{Ae} - x_{Ab}) \tag{22.4-4}$$

Here y_{Ae} is the gas phase composition in equilibrium with a liquid at composition x_{Ab}, and x_{Ae} is the liquid phase composition in equilibrium with a gas at composition y_{Ab}. The quantity $K^0_{y,\text{loc}}$ is the overall mass transfer coefficient "based on the gas phase," and $K^0_{x,\text{loc}}$ is the overall mass transfer coefficient "based on the liquid phase." Here again the molar flux N_{A0} is taken to be positive for transfer to the liquid phase.

Equating the quantities in Eqs. 22.4-2 and 4 gives two relations

$$K^0_{x,\text{loc}}(x_{Ae} - x_{Ab}) = k^0_{x,\text{loc}}(x_{A0} - x_{Ab}) \tag{22.4-5}$$
$$K^0_{y,\text{loc}}(y_{Ab} - y_{Ae}) = k^0_{y,\text{loc}}(y_{Ab} - y_{A0}) \tag{22.4-6}$$

connecting the two-phase coefficients with the single-phase coefficients.

The quantities x_{Ae} and y_{Ae} introduced in the above three relations may be used to define quantities m_x and m_y as follows:

$$m_x = \frac{y_{Ab} - y_{A0}}{x_{Ae} - x_{A0}}; \qquad m_y = \frac{y_{A0} - y_{Ae}}{x_{A0} - x_{Ab}} \tag{22.4-7, 8}$$

As we can see from Fig. 22.4-2, m_x is the slope of the line connecting points (x_{A0}, y_{A0}) and (x_{Ae}, y_{Ab}) on the equilibrium curve, and m_y is the slope of the line from (x_{Ab}, y_{Ae}) to (x_{A0}, y_{A0}).

From the above relations we can then eliminate the concentrations and get relations among the single-phase and two-phase mass transfer coefficients:

$$\frac{k^0_{x,\text{loc}}}{K^0_{x,\text{loc}}} = 1 + \frac{k^0_{x,\text{loc}}}{m_x k^0_{y,\text{loc}}}; \qquad \frac{k^0_{y,\text{loc}}}{K^0_{y,\text{loc}}} = 1 + \frac{m_y k^0_{y,\text{loc}}}{k^0_{x,\text{loc}}} \tag{22.4-9, 10}$$

The first of these was obtained from Eqs. 22.4-5, 2, and 7, and the second from Eqs. 22.4-6, 2, and 8. If the equilibrium curve is nearly linear over the range of interest, then $m_x = m_y = m$, which is the local slope of the curve at the interfacial conditions. We see, then, that the expressions in Eqs. 22.4-9, 10 both contain a ratio of single-phase coefficients weighted with a quantity m. This quantity is of considerable importance:

(i)   If $k^0_{x,\text{loc}}/m k^0_{y,\text{loc}} \ll 1$, the mass-transport resistance of the gas phase has little effect, and it is said that the mass transfer is *liquid-phase controlled*. In practice, this means that the system design should favor liquid-phase mass transfer.

(ii)   If $k^0_{x,\text{loc}}/m k^0_{y,\text{loc}} \gg 1$, then the mass transfer is *gas-phase controlled*. In a practical situation, this means that the system design should favor gas-phase mass transfer.

(iii)   If $0.1 < k^0_{x,\text{loc}}/m k^0_{y,\text{loc}} < 10$, roughly, one must be careful to consider the interactions of the two phases in calculating the two-phase transfer coefficients. Outside this range the interactions are usually unimportant. We return to this point in the example below.

The mean two phase mass transfer coefficients must be defined carefully, and we consider here only the special case where bulk concentrations in the two adjacent phases do not change significantly over the total mass-transfer surface S. We may then define K_{xm}^0 by

$$(N_{A0})_m = \frac{1}{S}\int_S K_{x,\text{loc}}^0 (x_{Ae} - x_{Ab})dS = K_{xm}^0 (x_{Ae} - x_{Ab}) \tag{22.4-11}$$

so that, when Eq. 22.3-9 is used,

$$K_{xm}^0 = \frac{1}{S}\int_S \frac{1}{(1/k_{x,\text{loc}}^0) + (1/m_x k_{y,\text{loc}}^0)}\,dS \tag{22.4-12}$$

Frequently area mean overall mass transfer coefficients are calculated from area mean coefficients for the two adjoining phases:

$$K_{x,\text{approx}}^0 = \frac{1}{(1/k_{xm}^0) + (1/m_x k_{ym}^0)} \tag{22.4-13}$$

The two mean values in Eqs. 22.4-12 and 13 can be significantly different (see Example 22.4-3).

EXAMPLE 22.4-1

Determination of the Controlling Resistance

Oxygen is to be removed from water using nitrogen gas at atmospheric pressure and 20°C in the form of bubbles exhibiting internal circulation, as shown in Fig. 22.4-3. Estimate the relative importance of the two mass transfer coefficients $k_{x,\text{loc}}^0$ and $k_{y,\text{loc}}^0$. Let A stand for O_2, B for H_2O, and C for N_2.

SOLUTION

We can do this by assuming that the penetration model (see §18.5) holds in each phase, so that

$$k_{x,\text{loc}}^0 \approx k_{x,\text{loc}} = c_l\sqrt{\frac{\mathscr{D}_{AB}}{\pi t_{\text{exp}}}}; \qquad k_{y,\text{loc}}^0 \approx k_{y,\text{loc}} = c_g\sqrt{\frac{\mathscr{D}_{AC}}{\pi t_{\text{exp}}}} \tag{22.4-14}$$

where c_l and c_g are the total molar concentrations in the liquid and gas phases, respectively. The effective exposure time, t_{exp}, is the same for each of the phases.

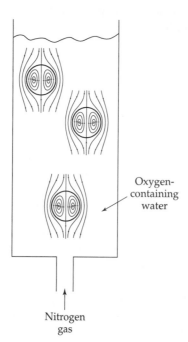

Oxygen-containing water

Nitrogen gas

Fig. 22.4-3. Schematic diagram of an oxygen stripper, in which oxygen from the water diffuses into the nitrogen gas bubbles.

The solubility of O_2 in water at 20°C is 1.38×10^{-3} moles per liter at an oxygen partial pressure of 760 mm Hg, the vapor pressure of water is 17.535 mm Hg, and the total pressure in the solubility measurements is 777.5 mm Hg. At 20°C, the diffusivity of O_2 in water is $\mathscr{D}_{AB} = 2.1 \times 10^{-5}$ cm^2/s, and in the gas phase the diffusivity for $O_2 - N_2$ is $\mathscr{D}_{AC} = 0.2$ cm^2/s. We can then write

$$\frac{k_{x,\text{loc}}^0}{m k_{y,\text{loc}}^0} = \frac{c_l}{c_g} \sqrt{\frac{\mathscr{D}_{AB}}{\mathscr{D}_{AC}}} \cdot \frac{1}{m} \tag{22.4-15}$$

Into this we must substitute

$$\frac{c_l}{c_g} = \frac{c_l}{(p/RT)} = \frac{1000/18}{(777.5/760)/(0.08206)(293.5)} = 1308 \tag{22.4-16}$$

$$\sqrt{\frac{\mathscr{D}_{AB}}{\mathscr{D}_{AC}}} = \sqrt{\frac{2.1 \times 10^{-5}}{0.2}} = 0.01 \tag{22.4-17}$$

$$\frac{1}{m} = \frac{1.38 \times 10^{-3}/55.5}{760/777.5} = 2.54 \times 10^{-5} \tag{22.4-18}$$

It follows that

$$\frac{k_{x,\text{loc}}^0}{m k_{y,\text{loc}}^0} = (1308)(0.01)(2.54 \times 10^{-5}) = 3.32 \times 10^{-4} \tag{22.4-19}$$

Therefore, only the liquid-phase resistance is significant, and the assumption of penetration behavior in the gas phase is not critical to the determination of liquid-phase control. It may also be seen that the dominant factor is the low solubility of oxygen in water. One may generalize and state that absorption or desorption of sparingly soluble gases is almost always liquid-phase controlled. Correction of the gas-phase coefficient for net mass transfer is clearly not significant, and the correction for the liquid phase is negligible.

EXAMPLE 22.4-2

Interaction of Phase Resistances

There are many situations for which the one-phase transfer coefficients are not available for the boundary conditions of the two-phase mass transfer problem, and it is common practice to use one-phase models in which interfacial boundary conditions are assumed, without regard to the interaction of the diffusion processes in the two phases. Such a simplification can introduce significant errors. Test this approximate procedure for the leaching of a solute A from a solid sphere of B of radius R in an incompletely stirred fluid C, so large in volume that the bulk fluid concentration of A can be neglected.

SOLUTION

The exact description of the leaching process is given by the solution of Fick's second law written for the concentration of A in the solid in the region $0 < r < R$:

$$\frac{\partial c_{As}}{\partial t} = \mathscr{D}_{AB} \frac{1}{r^2} \frac{\partial}{\partial r}\left(r^2 \frac{\partial c_{As}}{\partial r}\right) \tag{22.4-20}$$

The boundary and initial conditions are:

B.C. 1:	at $r = 0$,	c_{As} is finite	(22.4-21)
B.C. 2:	at $r = R$,	$c_{As} = m c_{Al} + b$	(22.4-22)
I.C.:	at $t = 0$,	$c_{As} = c_0$	(22.4-23)

The diffusional process on the liquid side of the solid–liquid interface is described in terms of a mass transfer coefficient, defined by

$$-\mathscr{D}_{AB} \left.\frac{\partial c_{As}}{\partial r}\right|_{r=R} = k_c(c_{Al} - 0) \tag{22.4-24}$$

in which $c_{Al}(t)$ is the concentration in the liquid phase adjacent to the interface. The behavior of the diffusion in the two phases is coupled through Eq. 22.4-22, which describes the equilibrium at the interface. Because of the coupling, it is convenient to use the method of Laplace transform. First, however, we restate the problem in dimensionless form, using $\xi = r/R$, $\tau = \mathcal{D}_{AB}t/R^2$, $C_s = c_{As}/c_0$, $C_l = (mc_{Al} + b)/c_0$, and $N = k_cR/m\mathcal{D}_{AB}$. Eqs. 22.3-20 and 24 become

$$\frac{\partial C_s}{\partial \tau} = \frac{1}{\xi^2}\frac{\partial}{\partial \xi}\left(\xi^2 \frac{\partial C_s}{\partial \xi}\right); \qquad -\frac{\partial C_s}{\partial \xi}\bigg|_{\xi=1} = NC_l \qquad (22.4\text{-}25, 26)$$

with C_s finite at the sphere center, $C_s = C_l$ at the sphere surface, and $C_s = 1$ throughout the sphere initially.

When we take the Laplace transform of this problem, we get

$$p\overline{C}_s - 1 = \frac{1}{\xi^2}\frac{d}{d\xi}\left(\xi^2 \frac{d\overline{C}_s}{d\xi}\right); \qquad -\frac{\partial \overline{C}_s}{\partial \xi}\bigg|_{\xi=1} = N\overline{C}_l \qquad (22.4\text{-}27, 28)$$

with $\overline{C}_s$ finite at the sphere center, and $\overline{C}_s = \overline{C}_l$ at the sphere surface. The solution of Eqs. 22.4-27 (which is a nonhomogeneous analog of Eq. C.1-6a) and 28 is

$$\overline{C}_s = -\frac{N}{p[\sqrt{p}\cosh\sqrt{p} - (1-N)\sinh\sqrt{p}]}\frac{\sinh\sqrt{p}\xi}{\xi} + \frac{1}{p} \qquad (22.4\text{-}29)$$

The Laplace transform of M_A, the total amount of A within the sphere at any time t, is

$$\frac{\overline{M}_A}{4\pi R^3 c_0} = \int_0^1 \overline{C}_s \xi d\xi = \frac{N^2}{p^2(\sqrt{p}\coth\sqrt{p} - (1-N))} - \frac{N}{p^2} + \frac{1}{3p} \qquad (22.4\text{-}30)$$

Inversion by using the Heaviside partial fractions expansion theorem for repeated roots[1] gives

$$\frac{M_A(t)}{\frac{4}{3}\pi R^3 c_{A0}} = 6\sum_{n=1}^{\infty} B_n \exp(-\lambda_n^2 \mathcal{D}_{AB}t/R^2) \qquad (22.4\text{-}31)$$

The constants λ_n and B_n are found to be, for finite k_c (or N),

$$\lambda_n \cot \lambda_n - (1-N) = 0; \qquad B_n = \frac{N^2}{\lambda_n^3}\frac{\sin^2 \lambda_n}{(\lambda_n - \sin \lambda_n \cos \lambda_n)} \qquad (22.4\text{-}32, 33)$$

and for infinite k_c (or N),

$$\lambda_n = n\pi; \qquad B_n = \left(\frac{1}{n\pi}\right)^2 \qquad (22.4\text{-}34, 35)$$

Note that we have succeeded in getting the total amount of A transferred across the interface, $M_A(t)$, without finding the expression for the concentration profile in the system. This is an advantage in using the Laplace transform.

We may now define two overall mass transfer coefficients: (i) the correct overall coefficient for this system based on the solid phase

$$K_s = \frac{N_{Ar}|_{r=R}}{c_{Ab}} = -\frac{R}{3}\frac{1}{M_A}\frac{dM_A}{dt} \qquad (22.4\text{-}36)$$

where c_{Ab} is the volume-average concentration of A in the solid phase, and (ii) an approximate overall coefficient, based on the separately calculated behavior of the two phases, calculated by Eq. 22.4-13,

$$\frac{1}{K_{s,approx}} = -\frac{3}{R}\frac{M_A^0}{dM_A^0/dt} + \frac{m}{k_c} \qquad (22.4\text{-}37)$$

where the superscript 0 indicates "zero external resistance" and k_c is the liquid-phase transfer coefficient.

---

[1] A. Erdélyi, W. Magnus, F. Oberhettinger, and F. G. Tricomi, *Tables of Integral Transforms*, McGraw-Hill (1954), p. 232, Formula 21.

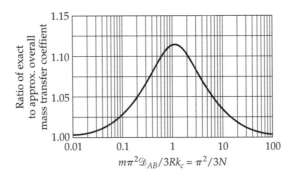

Fig. 22.4-4. Ratio of exact to approximate overall mass transfer coefficient in the leaching of a solute from a solute from a sphere, for large $\mathcal{D}_{AB}t/R^2$, plotted versus the dimensionless ratio $m\pi^2\mathcal{D}_{AB}/3Rk_c$.

We can now make a comparison between K_s and $K_{s,\text{approx}}$. We do this only for large values of $\mathcal{D}_{AB}t/R^2$, for which the leading term of the sum in Eq. 22.4-31 suffices. For this situation, we obtain

$$\frac{1}{K_{s,\text{approx}}} = \frac{3R}{\pi^2\mathcal{D}_{AB}} + \frac{m}{k_c} \tag{22.4-38}$$

and

$$\frac{K_s}{K_{s,\text{approx}}} = \frac{\lambda_1^2}{\pi^2}\left(1 + \frac{\pi^2 m\mathcal{D}_{AB}}{3Rk_c}\right) \tag{22.4-39}$$

where λ_1 is to be calculated for the actual value of k_c; keep in mind that λ_1 is obtained from Eq. 22.4-32, in which $N = k_cR/m\mathcal{D}_{AB}$. A plot of Eq. 22.4-39 is shown in Fig. 22.4-4. There we see that the maximum error in the two-film model occurs near $\pi^2/3N = 1$, and that departures from the two-film theory are appreciable but not very large.

EXAMPLE 22.4-3

Area Averaging[2]

Consider a characteristic section of a packed tower for which the separately measured single-phase mass transfer coefficients yield a calculated ratio

$$\frac{k_{xm}^0}{mk_{ym}^0} = 10 \tag{22.4-40}$$

but in which the liquid phase wets only half of the packing surface. Here the subscript m refers to the mean value over a typical area S of the packing surface. The gas-phase transfer coefficient, on the other hand, is uniform over the entire surface. This hypothetical example is a special case of nonuniform wetting. Calculate the true and approximate values of k_{xm}^0/K_{xm}^0 according to Eqs. 12.4-12 and 13.

SOLUTION

We begin with Eq. 22.4-12 and note that for half of the area $k_{x,\text{loc}}^0 = 0$, and that over the other half

$$k_{x,\text{loc}}^0 = 2k_{xm}^0 \tag{22.4-41}$$

whereas, for the gas phase

$$k_{y,\text{loc}}^0 = k_{ym}^0 \tag{22.4-42}$$

Eq. 22.4-12 thus yields

$$K_{xm}^0 = \frac{1}{S}\left[\frac{\tfrac{1}{2}S}{(1/2k_{xm}^0) + (1/mk_{ym}^0)}\right] \tag{22.4-43}$$

---

[2] C. J. King, *AIChE Journal*, **10**, 671–677 (1964).

From this and Eqs. 22.4-40 and 22.4-9, we find that the correct value for k_{xm}^0/K_{xm}^0 is

$$\frac{k_{xm}^0}{K_{xm}^0} = 1 + 2\,\frac{k_{xm}^0}{mk_{ym}^0} = 21 \qquad (22.4\text{-}44)$$

whereas the approximate value from Eq. 22.4-13 is

$$\frac{k_{xm}^0}{K_{xm}^0} = 1 + \frac{k_{xm}^0}{mk_{ym}^0} = 11 \qquad (22.4\text{-}45)$$

Thus the maldistribution of the liquid-phase mass transfer coefficient halves the rate of mass transfer, even though the liquid phase resistance "on the average" is very low. The general unavailability of such detailed information is one more reason for the uncertainty in predicting the behavior of complex contactors.

§22.5 MASS TRANSFER AND CHEMICAL REACTIONS

Many mass transfer operations are accompanied by chemical reactions, and the reaction kinetics can have a profound effect on transport rates. Important examples include absorption of reactive gases and reactive distillation. There are two situations of particular interest:

(i)   Absorption of a sparingly soluble substance A into a phase containing a second reactant B in large concentration. Absorption of carbon dioxide into NaOH or amine solutions is an industrially important example, and here the reaction may be considered pseudo-first-order because reactant B is present in great excess:

$$R_A = -(k_2''' c_B)c_A = -k_1''' c_A \qquad (22.5\text{-}1)$$

An example of this type of problem was given in §18.4.

(ii) Absorption of a rapidly reacting solute A into a solution of B. Here to a first approximation it may be assumed that the two species react so rapidly that they cannot coexist. An illustration of this was given in Example 20.1-2.

We shall be particularly interested in liquid boundary layers, and heat-of-reaction effects tend to be modest because the ratio of Sc to Pr is usually very large. Macroscopic heating effects do occur, and these are discussed in Chapter 23. Here we limit ourselves to a few illustrative examples showing how one can use models of absorption with chemical reaction to predict the performance of operating equipment.[1]

EXAMPLE 22.5-1

Estimation of the Interfacial Area in a Packed Column

Mass transfer measurements with irreversible first-order reaction have often been used to estimate interfacial area in complex mass transfer equipment. Show here how this method can be justified.

SOLUTION

The system we consider here is the absorption of carbon dioxide into a caustic solution, which is limited by hydration of dissolved CO_2 according to the reaction

$$CO_2(aq) + H_2O \rightleftharpoons H_2CO_3 \qquad (22.5\text{-}2)$$

---

[1] T. K. Sherwood, R. L. Pigford, and C. R. Wilke, *Mass Transfer*, McGraw-Hill, New York (1975), Chapter 8.

The carbonic acid then reacts with NaOH at a rate proportional to carbon dioxide concentration. The kinetics of this reaction are well characterized.[1]

The solution of this diffusion problem has been given in Problem 20C.3. From Eq. 20.3-3, we find that for long times[2,3]

$$W_{A0} = Ac_{A0}\sqrt{\mathscr{D}_{AB}k_1'''}$$ (22.5-3)

which can be solved for the total surface area. It follows that the total surface area A under consideration is given by

$$A = \frac{1}{c_{A0}\sqrt{\mathscr{D}_{AB}k_1'''}}\frac{dM_{A,\text{tot}}}{dt}$$ (22.5-4)

here $M_{A,\text{tot}}$ is the number of moles of carbon dioxide absorbed by time t.

This development is readily extended to a falling film of length L and surface velocity v_s, provided that $k_1 L/v_s \gg 1$. First-order reaction in mass transfer boundary layers is discussed in Example 18.4-1 for a simple film model and in Example 20.1-3. The development can be further extended to estimate the interfacial area in packed columns, in which the liquid phase is supported as a falling film on solid surfaces, a common design.

EXAMPLE 22.5-2 *Estimation of Volumetric Mass Transfer Coefficients*	We next consider gas absorption with first-order reaction in an agitated tank and take as a starting point the reaction $$O_2 + 2Na_2SO_3 \rightarrow 2Na_2SO_4 \qquad (22.5\text{-}5)$$ already discussed in Example 18.4-1, using a thin stagnant film of liquid as a mass transfer model.
SOLUTION	This is not a realistic model, but the development in Example 18.4-1 can be rephrased in a *model-insensitive form* by writing $$k_c = \frac{\mathscr{D}_{AB}}{\delta} \qquad (22.5\text{-}6)$$ so that

$$\phi = \sqrt{\frac{k_1'''\delta^2}{\mathscr{D}_{AB}}} \rightarrow \sqrt{\frac{k_1'''\mathscr{D}_{AB}}{k_c^2}} \quad \text{and} \quad \frac{V}{A\delta} \rightarrow \frac{Vk_c}{A\mathscr{D}_{AB}}$$ (22.5-7)

The subscript AB should be changed to O_2S, where S represents the sulfite solution.

One can now test the *model sensitivity* of the system by comparing the film model with the penetration model. This is done in Fig. 22.5-1, where it can be seen that there is no significant difference between the two.[4] Moreover, there is a substantial region of parameter space where the predicted rate of oxygen absorption is identical to that for physical absorption in an oxygen-free tank. This chemical system has therefore proven a popular means for estimating volumetric mass transfer coefficients. It has long been used to characterize the oxygenation effectiveness of aerobic bioreactors.[5]

[2] P. V. Danckwerts, *Trans. Faraday Soc.*, **46**, 300–304 (1950).

[3] R. A. T. O. Nijsing, *Absorptie van gassen in vloeistoffen, zonder en met chemische reactie*, Academisch Proefschrift, Delft (1957).

[4] E. N. Lightfoot, *AIChE Journal*, **8**, 710–712 (1962).

[5] A. M. Friedman and E. N. Lightfoot, *Ind. Eng. Chem.*, **49**, 1227–1230 (1957); J. E. Bailey and D. F. Ollis, *Biochemical Engineering Fundamentals*, McGraw-Hill, New York (1986); V. Linek, P. Benes, and J. Sinkule, *Biotechnol.-Bioeng.*, **35**, 766–770 (1990).

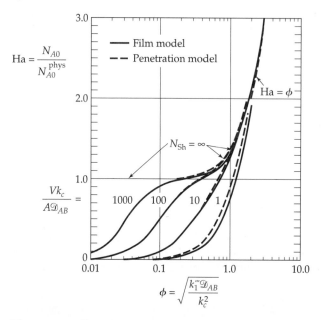

$$\text{Ha} = \frac{N_{A0}}{N_{A0}^{\text{phys}}}$$

$$\frac{Vk_c}{A\mathscr{D}_{AB}} =$$

$$\phi = \sqrt{\frac{k_1''' \mathscr{D}_{AB}}{k_c^2}}$$

Fig. 22.5-1. Effect of irreversible first-order reaction on pseudosteady-state absorption of a sparingly soluble gas in an agitated tank. Comparison of the penetration and stagnant-film models.

EXAMPLE 22.5-3	Next consider the absorption with rapid irreversible reaction, and seek to simplify and generalize the discussion of Example 20.1-2. Do this in terms of the *Hatta number*[6] defined as

Model-Insensitive Correlations for Absorption with Rapid Reaction

$$\text{Ha} = \frac{N_{A0}}{N_{A0}^{\text{phys}}} \tag{22.5-8}$$

here the superscript phys denotes absorption of solute A in the same system but without reaction. This dimensionless group provides a convenient measure of the promoting effect of chemical reaction on the rate of absorption.

SOLUTION

In the absence of solute B, species A would undergo physical absorption (that is, absorption without reaction) at a rate

$$N_{A0}^{\text{phys}} = c_{A0}\sqrt{\frac{\mathscr{D}_{AS}}{\pi t}} \tag{22.5-9}$$

since $\text{erf}\sqrt{\gamma/\mathscr{D}_{AS}}$ goes to unity with decreasing $c_{B\infty}/c_{A0}$. We now divide the result in Eq. 20.1-39 by Eq. 22.5-9 to get

$$\text{Ha} = \frac{1}{\text{erf}\sqrt{\gamma/\mathscr{D}_{AS}}} \tag{22.5-10}$$

which can be further simplified in the following cases:[7,8]

[6] S. Hatta, *Technological Reports of Tôhoku University*, **10**, 613–662 (1932). **Shirôji Hatta** (1895–1973) taught at Tôhoku University from 1925 to 1958 and in 1954 he was appointed Dean of Engineering; after "retiring" he accepted a position at Chiyoda Chemical Engineering and Construction Co. He served as editor-in-chief of *Kagaku Kôgaku* and as president of Kagaku Kôgakkai.

[7] E. N. Lightfoot, *Chem. Eng. Sci.*, **17**, 1007–1011 (1962).

[8] D. H. Cho and W. E. Ranz, *Chem. Eng. Prog. Symposium Series* # 72, **63**, 37–45 and 46–58 (1967).

(i)   *For small values of $c_{B\infty}/c_{A0}$, or for equal diffusivities,*

$$\text{Ha} = 1 + \frac{ac_{B\infty}}{bc_{A0}} \tag{22.5-11}$$

(ii)   *For large values of $c_{B\infty}/c_{A0}$,*

$$\text{Ha} = \left(1 + \frac{ac_{B\infty}\mathscr{D}_{BS}}{bc_{A0}\mathscr{D}_{AS}}\right)\sqrt{\frac{\mathscr{D}_{AS}}{\mathscr{D}_{BS}}} \tag{22.5-12}$$

(iii)   *For all values of $c_{B\infty}/c_{A0}$ (approximate),*

$$\text{Ha} \approx 1 + \frac{ac_{B\infty}}{bc_{A0}}\sqrt{\frac{\mathscr{D}_{BS}}{\mathscr{D}_{AS}}} \tag{22.5-13}$$

Equation 22.5-11 is particularly useful, since it is accurate for the common situation of nearly equal diffusivities as well as for small $c_{B\infty}/c_{A0}$. Equation 22.5-13 is useful because it is valid for both large and very small values of $ac_{B\infty}\mathscr{D}_{BS}/bc_{A0}\mathscr{D}_{AS}$. In addition, the exact solution always lies in the space between the curve of Eq. 22.5-13 and those portions of the curves of Eqs. 22.5-11 and 12 that are closest to it. This is shown in Fig. 22.5-2, where these bounding approximations are compared to the exact solution.

We next note that we can replace the diffusivity ratio by the corresponding ratio of nonreactive Sherwood numbers,

$$\sqrt{\frac{\mathscr{D}_{AS}}{\mathscr{D}_{BS}}} = \frac{\text{Sh}_B^0}{\text{Sh}_A^0} \tag{22.5-14}$$

where the superscript 0 denotes the observed Sherwood number in the absence of chemical reaction. We may thus obtain a set of model-insensitive bounding solutions

$$
\begin{aligned}
\text{Ha} &\approx 1 + \frac{ac_{B\infty}}{bc_{A0}} \\
&\approx \left[1 + \left(\frac{ac_{B\infty}}{bc_{A0}}\right)\left(\frac{\mathscr{D}_{BS}}{\mathscr{D}_{AS}}\right)\right]\frac{\text{Sh}_B^0}{\text{Sh}_A^0} \\
&\approx 1 + \left(\frac{ac_{B\infty}}{bc_{A0}}\right)\frac{\text{Sh}_A^0}{\text{Sh}_B^0}
\end{aligned}
\tag{22.5-15}
$$

These equations have been shown[7] to provide convenient bounds for laminar and turbulent boundary layers as well as the penetration model of Example 20.1-2. They thus form a highly model-insensitive correlation and are widely useful.

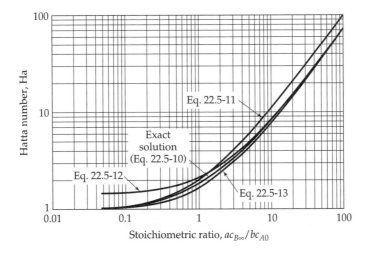

Fig. 22.5-2. Model-insensitive correlations for absorption with rapid chemical reaction, derived from the penetration model, for the case that $\mathscr{D}_{AS} = 2\mathscr{D}_{BS}$.

§22.6   COMBINED HEAT AND MASS TRANSFER BY FREE CONVECTION

In this section we consider briefly some important interactions among the transfer processes, with emphasis on free convection. This is an extension of our earlier discussion of free-convection heat transfer in §14.6 and is reasonably well understood.

Combined heat and mass transfer by free convection is among the simple examples of interaction between all three transport phenomena. The dimensionless equations describing them have been given in Eqs. 19.5-8 to 11. Numerical integration of these equations is possible,[1] but we can obtain simple, useful results via boundary layer theory. We consider two particularly simple problems in the examples that follow.

EXAMPLE 22.6-1

Additivity of Grashof Numbers

Develop an expression for the combined free-convection heat and mass transfer for the special case of equal Prandtl and Schmidt numbers. Assume that transfer is between a surface of constant temperature and composition, and a large uniform surrounding fluid.

SOLUTION

This is a direct extension of the boundary conditions of Example 11.4-5. Then if the dimensionless temperature and composition are defined analogously, it follows that $\check{T} = \check{x}_A$ everywhere within the system under investigation.

It then follows that the solution of this mixed convection problem is identical to that for heat or mass transfer alone, but with Gr or Gr_ω replaced by the sum $(Gr + Gr_\omega)$. This simplification is widely used for the air–water system, where the small difference between the Sc and Pr numbers does not have a significant effect.

Thus for evaporation from a water-wetted vertical plate (with $Sc = 0.61$ and $Pr = 0.73$), one may use Eq. 11.4-11 with $C = 0.518$ to obtain

$$Nu_m \approx 0.518[0.73(Gr + Gr_\omega)]^{1/4} \tag{22.6-1}$$

$$Sh_m \approx 0.518[0.61(Gr + Gr_\omega)]^{1/4} \tag{22.6-2}$$

Note that the $\frac{1}{4}$-th powers of Pr and Sc are 0.92 and 0.88, respectively. This difference is hardly significant in view of the uncertainties of any actual situation and the boundary layer model on which these results are based. Note also that the thermal Grashof number is normally by far the larger, so that neglect of this interaction would greatly underestimate the evaporation rates.

EXAMPLE 22.6-2

Free-Convection Heat Transfer as a Source of Forced-Convection Mass Transfer

There are many situations—for example, the evaporation of solvents with low volatility—where thermal Grashof numbers are much larger than their mass transfer counterparts $(Gr > Gr_\omega)$ and the Schmidt numbers exceed the Prandtl numbers $(Sc > Pr)$. Under these conditions, the thermal buoyant forces provide a momentum source, which in turn provides a convective flow to drive mass transfer. It has been shown[2] that the thermally induced gradient of upward velocity at the surface of a vertical flat plate of length L is given by

$$\left.\frac{\partial v_z}{\partial y}\right|_{y=0} \approx 1.08\left(\frac{4}{3}\frac{z}{L}\right)^{1/4}\left(\frac{Gr}{Pr}\right)^{1/4}\sqrt{\frac{g\beta\Delta T}{L}} \tag{22.6-3}$$

Here z is the distance measured upward along the plate, y is measured outward into the fluid, and ΔT is the difference between the plate temperature and the temperature of the surroundings. This is an asymptotic expression for large Prandtl number, but is also useful for gases. Develop expressions for the local and mean Sherwood numbers.

---

[1] W. R. Wilcox, *Chem. Eng. Sci.*, **13**, 113–119 (1961).
[2] A. Acrivos, *Phys. Fluids*, **3**, 657–658 (1960).

SOLUTION

The thermal free convection provides a velocity field within which the mass transfer boundary layer develops. Given this velocity field, we may use the mass transfer analog of Eqs. 12.4-30 and 29 along with the definition $\text{Nu}_{loc} = D/\Gamma(\tfrac{4}{3})\delta_T$ to obtain a description of the mass transfer rate in two-dimensional flow:

$$\text{Sh}_{loc} = \frac{(\text{ReSc})^{1/3}}{9^{1/3}\Gamma(\tfrac{4}{3})} \frac{\sqrt{\Gamma_0}}{\left(\displaystyle\int_0^{z/L} \sqrt{\Gamma_0(u)}\,du\right)^{1/3}} \tag{22.6-4}$$

Here

$$\Gamma_0 = \frac{l_0}{v_0}\frac{\partial v_z}{\partial y}\bigg|_{y=0} \tag{22.6-5}$$

is the dimensionless velocity gradient at the wall (l_0 and v_0 are arbitrary reference quantities used in the definition of the Reynolds number). For free convection it is convenient to use the plate height L for l_0 and ν/L for v_0. Then the Reynolds number is unity, and the quantity Γ_0 is $\Gamma_0 = (L^2/\nu)(\partial v_z/\partial y)|_{y=0}$. Then Eq. 22.6-4 becomes

$$\text{Sh}_{loc} = \left(\frac{1.08}{8}\right)^{1/3} \frac{(\tfrac{4}{3})^{1/12}}{\Gamma(\tfrac{4}{3})} (\text{GrSc})^{1/4}\left(\frac{\text{Sc}}{\text{Pr}}\right)^{1/12}\left(\frac{L}{z}\right)^{1/4}$$

$$\approx 0.59(\text{GrSc})^{1/4}\left(\frac{\text{Sc}}{\text{Pr}}\right)^{1/12}\left(\frac{L}{z}\right)^{1/4} \tag{22.6-6}$$

The mean Sherwood number, obtained by averaging over the plate surface, is

$$\text{Sh}_m \approx 0.79(\text{GrSc})^{1/4}\left(\frac{\text{Sc}}{\text{Pr}}\right)^{1/12} \tag{22.6-7}$$

Note that these last two equations show features of both free and forced convection in laminar boundary layers: the $\tfrac{1}{4}$-power of the Grashof number for free convection and the $\tfrac{1}{3}$-power of the Schmidt number for forced convection.

Moreover, we can now test the effect of Sc/Pr, because we know from the preceding example and Table 14.6-1 that, for Pr = Sc,

$$\text{Sh}_m = 0.67(\text{GrSc})^{1/4} \tag{22.6-8}$$

in which the coefficient is lower than that in Eq. 22.6-7 by the ratio 0.85. The Sherwood number Sh_m will lie between the predictions of Eqs. 22.6-7 and 8 for $\text{Sc} \geq \text{Pr}$ and $\text{Pr} \gg 1$.

Arguments similar to those used in Eq. 14.6-6 now suggest the following extension of Eqs. 22.67 and 68,

$$\text{Sh}_m \approx 0.73(1 \pm 0.1)\frac{(\text{GrSc})^{1/4}(\text{Sc}/\text{Pr})^{1/12}}{[1 + (0.492/\text{Pr})^{9/16}]^{4/9}} \tag{22.6-9}$$

for $\text{Sc} \geq \text{Pr}$ and $\text{Pr} \geq 0.73$. This result is correct for the limits $\text{Pr} = 0.73$ and $\text{Pr} = \infty$ and hence can include the evaporation of solvents in air. This analysis can also be extended to other shapes.

§22.7   EFFECTS OF INTERFACIAL FORCES ON HEAT AND MASS TRANSFER

In this section we consider briefly some important interactions among the three transfer processes, with emphasis on the effects of variable interfacial tension (*Marangoni effects*). The importance of this subject stems from the prevalence of direct fluid–fluid contact in mass transfer systems, but it can also be important in similar heat transfer operations. Still poorly understood diffusional processes permit violation of the no-slip condition on fluid flow over solid surfaces in the neighborhood of advancing menisci.[1] As for the distorting effects of surface tension gradients on mass and heat transfer in gas–liquid contacting, these will enter through a description of the boundary conditions.

According to Eq. 11C.6-4, if the stresses in the gas (phase II) are ignored, the interfacial tangential stresses acting on an interface with normal unit vector **n** are given by[2]

$$[(\boldsymbol{\delta} - \mathbf{nn}) \cdot [\mathbf{n} \cdot \boldsymbol{\tau}]] = -\nabla^s \sigma \tag{22.7-1}$$

where σ is the surface tension ∇^s is the two-dimensional gradient operator in the interface, and $(\boldsymbol{\delta} - \mathbf{nn})$ is a "projection operator" that selects those components of $[\mathbf{n} \cdot \boldsymbol{\tau}]$ that lie in the interfacial tangent plane. For example, if **n** is taken to be the unit vector in the z direction, Eq. 22.7-1 gives

$$\tau_{zx} = -\frac{\partial \sigma}{\partial x} \qquad \tau_{zy} = -\frac{\partial \sigma}{\partial y} \tag{22.7-2, 3}$$

which are the interfacial tension forces in the x and y directions acting in the xy-plane.

The surface-tension-induced stresses are typically of the same order as their hydrodynamic counterparts, and the flow phenomena that may result from them are known collectively as *Marangoni effects*.[3] It has been shown[4] that mass transfer rates can be increased up to threefold by Marangoni effects, but can also be reduced in other circumstances.

The nature and extent of Marangoni effects depend strongly on the system geometry and the transport properties, and it will be convenient to consider here four specific examples:

(i)   drops and bubbles surrounded by a liquid continuum

(ii)   sprays of drops in a gaseous continuum

(iii)   supported liquid films in a gaseous or liquid continuum

(iv)   foams of gas bubbles in a liquid continuum

These systems, each important in practice, show very different behavior from one another.

For drops and bubbles moving through a liquid continuum, the primary problems are surfactants or microscopic particles that can reduce or eliminate the "Hadamard–Rybczinski circulation" and also hinder the periodic mixing accompanying oscillation in

---

[1] V. Ludviksson and E. N. Lightfoot, *AIChE Journal*, **14**, 674–677 (1968); P. A. Thompson and S. M. Troian, *Phys. Rev. Letters*, **63**, 766–769 (1997); A. Marmur, in *Modern Approach to Wettability: Theory and Applications* (M. E. Schrader and G. Loeb, eds.), Plenum Press (1992); D. Schaeffer and P.-Z. Wong, *Phys. Rev. Letters*, **80**, 3069–3072 (1998).

[2] In Eq. 3.2-6 of D. A. Edwards, H. Brenner, and D. T. Wasan, *Interfacial Transport Processes and Rheology*, Butterworth-Heinemann, Boston (1991), the operator $(\boldsymbol{\delta} - \mathbf{nn})$ is called the "dyadic surface idemfactor"; the same quantity is called the "projection tensor" by J. C. Slattery, *Interfacial Transport Phenomena*, Springer Verlag, New York (1990), p. 1086. Both books contain a wealth of information on surface tension, surface viscosity, surface viscoelasticity, and other properties of interfaces and their methods of measurement.

[3] C. G. M. Marangoni, *Tipographia dei fratelli Fusi*, Pavia (1865); *Ann. Phys.* (Poggendorf), **143**, 337–354 (1871). Historical articles on the Marangoni effects are L. E. Scriven and C. V. Sternling, *Nature*, **187**, 186–188 (1960), and S. Ross and P. Becher, *J. Coll. Interfac. Sci.*, **149**, 575–579 (1992).

[4] A good overview of Marangoni effects and related phenomena, with emphasis on liquid–liquid systems, is provided in J. C. Godfrey and M. J. Slater, *Liquid–Liquid Extraction Equipment*, Wiley, New York (1994), pp. 68–75. A theory offered by C. V. Sternling and L. E. Scriven, *AIChE Journal*, **5**, 514–523 (1959), provides useful insight but is considered too simple to give reliable predictions of the onset of instabilities.

larger drops or bubbles.[5] These are discussed briefly in Example 22.3-4. These situations are important in gas absorbers and liquid extractors. For sprays of drops in a gas, important in large distillation columns, Marangoni forces play no significant role.[6]

Foam beds, important in smaller distillation columns, and supported films, important in a wide variety of packed columns, are particularly interesting. Both are strongly affected by surface-tension gradients resulting from the changes of surface tension with composition of the adjoining streams.

Foam beds are stabilized when the bulk liquid has a lower surface tension than that in equilibrium with the bulk gas, called a "positive system." In such a situation, interfacial tension tends to be higher where bubbles are close together than where they are far apart, and the shrinking of high-surface-tension regions tends to drive the bubbles apart, thus stabilizing the foam. Where there are only small differences in surface tension, or where the direction is reversed, a "negative system," there is no stabilizing effect and the foaming is poor. Concentration of ethanol from water is interesting, because it has strong positive surface tension gradients where the relative volatility is high, but becomes very nearly neutral as the azeotrope is approached. Thus, for a bubble-cap column, stage efficiencies are high where least needed and low as the azeotropic composition is approached.

In packed columns, where the descending liquid is supported on solid surfaces as thin films, the situation is quite different. Here the surface tension of the descending liquid decreases downward for a positive system and is subject to hydrodynamic instability to form narrow rivulets. These markedly decrease interfacial area and mass transfer effectiveness. In negative systems, on the other hand, films are stabilized, and mass transfer is more effective than for neutral systems. No quantitative analysis of this situation appears to be available, but it has been shown that instabilities found by Zuiderweg and Harmens for wetted-wall columns can be predicted by linearized stability analysis.[7] Stability analysis also suggests that the presence of a positive surface-tension gradient should improve the efficiency of condensers. Another study of stability for very small films opens up new possibilities for microfluidic processors.[8]

EXAMPLE 22.7-1	The presence of surfactants can stop Hadamard–Rybczinski circulation in a rising gas bubble. Explain this phenomenon (see Fig. 22.7-1).
Elimination of Circulation in a Rising Gas Bubble	

SOLUTION

Circulation results in stretching of the surface at the top of a rising bubble and shrinking of the surface at the bottom. As a result, surfactant accumulates at the bottom, producing a higher than average concentration there, whereas a lower-than-average concentration exists at the top. Since surfactants reduce surface tension, this results in a surface-tension-induced stress (in spherical coordinates)

$$\tau_{r\theta,s}\big|_{r=R} = \frac{1}{R}\frac{\partial \sigma}{\partial \theta} \tag{22.7-4}$$

tending to oppose the interfacial deformation (see Eq. 22.7-1). If the magnitude of this stress reaches the value that would occur on a rising solid sphere (see Eq. 2.6-6)

$$\tau_{r\theta}\big|_{r=R} = \frac{3}{2}\frac{\mu v_\infty}{R}\sin\theta \tag{22.7-5}$$

circulation will stop.

---

[5] J. B. Angelo and E. N. Lightfoot, *AIChE Journal*, **12**, 751–760 (1966).
[6] F. J. Zuiderweg and A. Harmens, *Chem. Eng. Sci.*, **9**, 89–103 (1958).
[7] K. H. Wang, V. Ludviksson, and E. N. Lightfoot, *AIChE Journal*, **17**, 1402–1408 (1971).
[8] D. E. Kataoka and M. S. Troian, *Nature*, **402**, 794–797 (16 December 1999).

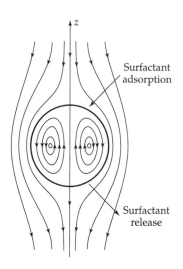

Fig. 22.7-.1 Surfactant transport during Hadamard–Rybczynski circulation

As a practical matter, even small amounts of surfactant prevent circulation. Small concentration of microscopic suspended particulates have a similar effect, being swept to the rear of bubbles and forming a rigid surface.

EXAMPLE 22.7-2

Marangoni Instability in a Falling Film

Among the simplest mass-transfer-induced Marangoni effects is instability in a falling film resulting from counterflow adsorption of vapors with a high heat of solution. An important representative example is the counterflow absorption of HCl vapor into water, which is so inefficient that cocurrent flow is preferable. Explain this effect.

SOLUTION

This situation can be simulated by allowing a film of water to flow down a plate that is colder at the top than at the bottom. If sufficient care is taken, one can obtain a sinusoidally varying film thickness, as shown[9] by interferometry in Fig. 22.7-2(*a*). Here each new dark line represents a line of constant thickness, differing from its neighbors by one-half wavelength of light in the water.

This situation corresponds to a series of parallel *roll cells* of the type pictured in Fig. 22.7-2(*b*), driven by lateral surface tension gradients. These gradients, in turn, result from small variations in film thickness caused by inevitable small spatial variations of surface velocity: the thicker regions move faster and thus tend to be colder than the thin regions. A simple perturbation analysis[1] shows that perturbations of some widths grow faster than others, and the fastest growing ones tend to dominate. The periods of the sinusoidal lines in Fig. 22.7-2(*a*) correspond to these fastest growing disturbances.

Such regularity is, however, seldom observed in practice. More commonly one sees occasional thick rivulets surrounded by large thin regions. These thin regions, taking up most of the available surface, are both slowly moving and quickly saturated and are thus ineffective for mass transfer. Only the rivulets are effective, and their total surface area is very small. Similar behavior is observed for surface-tension gradients caused by vertical variations in composition. However, in that case the behavior is more complicated and requires an analysis of interphase mass transfer.[7]

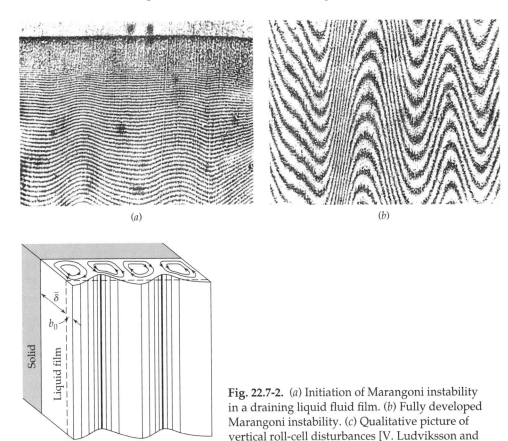

Fig. 22.7-2. (*a*) Initiation of Marangoni instability in a draining liquid fluid film. (*b*) Fully developed Marangoni instability. (*c*) Qualitative picture of vertical roll-cell disturbances [V. Ludviksson and E. N. Lightfoot, *AIChE Journal*, **14**, 620–626 (1968)].

§22.8   TRANSFER COEFFICIENTS AT HIGH NET MASS TRANSFER RATES

High net mass transfer rates across phase boundaries distort the boundary-layer profiles of velocity and temperature as well as species concentration, and they also alter the boundary layer thicknesses. Both of these effects tend to increase friction factors and the heat and mass transfer coefficients, if the mass transfer is toward the boundary, and to reduce them in the reverse situation. These usual trends are reversed, however, in free convection and in flows driven by a rotating surface. The magnitudes of such changes are dependent on the system geometry, boundary conditions, and the magnitudes of the governing parameters such as the Reynolds, Prandtl, and Schmidt numbers, and they are accompanied by the effects of changes in physical properties. They can also either increase or decrease the hydrodynamic stability. Accurate allowance for the effects of net mass transfer thus requires extensive calculation and/or experimentation, but some of the more salient features can be illustrated by using idealized physical models, and this is the approach we follow here.

We begin with the classic *stagnant-film model*, which provides simple estimates of the profile distortion, but is incapable of predicting changes in the effective film thickness. We then discuss the *penetration model* and the flat-plate *laminar boundary layer model*. We conclude with several illustrative examples, the last of which is a complete numerical example of boundary layers on a spinning disk. This example will provide a useful appraisal of model sensitivity.

As pointed out in §22.1, when high net mass transfer rates are being considered, we introduce a modified notation for the transfer coefficients:

$$N_{A0} - x_{A0}(N_{A0} + N_{B0}) = k^{\bullet}_{x,\text{loc}}\Delta x_A \tag{22.8-1a}$$

$$e_0 - (N_{A0}\tilde{C}_{pA,0} + N_{B0}\tilde{C}_{pB,0})(T_0 - T^\circ) = h^{\bullet}_{\text{loc}}\Delta T \tag{22.8-1b}$$

The black dots in $k^{\bullet}_{x,\text{loc}}$ and $h^{\bullet}_{\text{loc}}$ imply that the distortions of the concentration and temperature profiles resulting from high net mass transfer rates are being included.

The relations between these transfer coefficients and those defined in Eqs. 22.1-7 and 8 are

$$k_{x,\text{loc}} = \lim_{N_{A0}+N_{B0}\to 0} k^{\bullet}_{x,\text{loc}} \tag{22.8-2a}$$

$$h_{\text{loc}} = \lim_{N_{A0}\tilde{C}_{pA,0}+N_{B0}\tilde{C}_{pB,0}\to 0} h^{\bullet}_{\text{loc}} \tag{22.8-2b}$$

This shows explicitly the limiting process that relates the two types of transfer coefficients.

The Stagnant-Film Model[1-4]

We have already discussed this model briefly in §18.2 and more fully in Example 19.4-1. By combining the expressions in Eqs. 19.4-16 and 17 with the definitions in Eqs. 22.8-1a and 1b, we get for the system in Fig. 22.8-1

$$1 + \frac{(N_{A0} + N_{B0})}{k^{\bullet}_{x,\text{loc}}} = \exp\left[(N_{A0} + N_{B0})\frac{\delta_x}{c\mathcal{D}_{AB}}\right] \tag{22.8-3}$$

$$1 + \frac{(N_{A0}\tilde{C}_{pA} + N_{B0}\tilde{C}_{pB})}{h^{\bullet}_{\text{loc}}} = \exp\left[(N_{A0}\tilde{C}_{pA} + N_{B0}\tilde{C}_{pB})\frac{\delta_T}{k}\right] \tag{22.8-4}$$

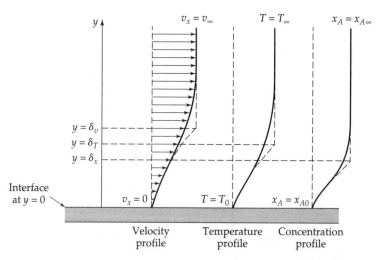

Fig. 22.8-1. Steady flow along a flat surface with rapid mass transfer into the stream. The unbroken curves represent the true profiles, and the broken curves are predicted by the film model.

[1] W. K. Lewis and K. C. Chang, *Trans. AIChE*, **21**, 127–136 (1928).
[2] G. Ackerman, *Forschungsheft*, **382**, 1–16 (1937).
[3] A. P. Colburn and T. B. Drew, *Trans. AIChE*, **33**, 197–212 (1937).
[4] H. S. Mickley, R. C. Ross, A. L. Squyers, and W. E. Stewart, *NACA Tech. Note 3208* (1954).

By following the limiting processes indicated in Eqs. 22.8-2a and 2b, we then get expressions for the transfer coefficients in the low net mass transfer limit:

$$\frac{1}{k_{x,\mathrm{loc}}} = \frac{\delta_x}{c\mathscr{D}_{AB}} \tag{22.8-5}$$

$$\frac{1}{h_{\mathrm{loc}}} = \frac{\delta_T}{k} \tag{22.8-6}$$

These limiting values are found by expanding the right sides of Eqs. 22.8-3 and 4 in Taylor series and retaining two terms. Substitution of Eqs. 22.7-5 and 6 into Eqs. 19.4-16 and 17 enables us to eliminate the film thicknesses (which are ill-defined) in favor of the transfer coefficients at low mass-transfer rates (which are measurable):

$$1 + \frac{(N_{A0} + N_{B0})(x_{A0} - x_{A\infty})}{N_{A0} - x_{A0}(N_{A0} + N_{B0})} = \exp\left(\frac{N_{A0} + N_{B0}}{k_{x,\mathrm{loc}}}\right) \tag{22.8-7}$$

$$1 + \frac{(N_{A0}\tilde{C}_{pA} + N_{B0}\tilde{C}_{pB})(T_0 - T_\infty)}{q_0} = \exp\left(\frac{N_{A0}\tilde{C}_{pA} + N_{B0}\tilde{C}_{pB}}{h_{\mathrm{loc}}}\right) \tag{22.8-8}$$

These equations are the principal results of the film model. They show how the conductive energy flux and the diffusion flux at the wall depend on N_{A0} and N_{B0}. In this model, the effects of net mass transfer on the conductive and diffusive interfacial fluxes are clearly analogous. Although these relations were derived for laminar flow and constant physical properties, they are also useful for turbulent flow and for variable physical properties (see Problem 22B.3).

The results for heat and mass transfer can be summarized in two equations:

$$\theta = \frac{\phi}{R} = \frac{\phi}{e^\phi - 1} = \frac{\ln(1 + R)}{R} \tag{22.8-9}$$

$$\Pi = \frac{e^{\phi\eta} - 1}{e^\phi - 1} = \frac{(1 + R)^\eta}{R} \qquad (\eta \le 1) \tag{22.8-10}$$

Equation 22.8-9 gives the correction factors θ_x and θ_T by which the coefficients $k_{x,\mathrm{loc}}$ and h_{loc} must be multiplied to obtain the coefficients at high net mass transfer rates. Equation 22.8-10 gives the concentration and temperature profiles. The meanings of the symbols are summarized in Table 22.8-1.

Table 22.8-1 Summary of Dimensionless Quantities to be Used for All Models Discussed in §22.8. Mass-based versions appear in §20.2 and §22.9.

θ = correction factors	R = flux ratios
ϕ = rate factors	Π = profiles
η = dimensionless distance from wall	

	Mass transfer	Heat transfer
θ	$\theta_x = \dfrac{k^\bullet_{x,\mathrm{loc}}}{k_{x,\mathrm{loc}}}$	$\theta_T = \dfrac{h^\bullet_{\mathrm{loc}}}{h_{\mathrm{loc}}}$
ϕ	$\phi_x = \dfrac{N_{A0} + N_{B0}}{k_{x,\mathrm{loc}}}$	$\phi_T = \dfrac{N_{A0}\tilde{C}_{pA} + N_{B0}\tilde{C}_{pB}}{h_{\mathrm{loc}}}$
R	$R_x = \dfrac{(N_{A0} + N_{B0})(x_{A0} - x_{A\infty})}{N_{A0} - x_{A0}(N_{A0} + N_{B0})}$	$R_T = \dfrac{(N_{A0}\tilde{C}_{pA} + N_{B0}\tilde{C}_{pB})(T_0 - T_\infty)}{q_0}$
Π	$\Pi_x = \dfrac{x_A - x_{A0}}{x_{A\infty} - x_{A0}}$	$\Pi_T = \dfrac{T - T_0}{T_\infty - T_0}$
η	$\eta_x = \dfrac{y}{\delta_x} = \dfrac{y}{c\mathscr{D}_{AB}/k_{x,\mathrm{loc}}}$	$\eta_T = \dfrac{y}{\delta_T} = \dfrac{y}{k/h_{\mathrm{loc}}}$

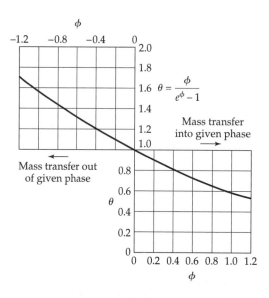

Fig. 22.8-2. The variation of the transfer coefficients with mass transfer rate, as given by the film model (see Eq. 22.8-9).

$$\theta = \frac{\phi}{e^\phi - 1}$$

Equation 22.8-9 is given graphically in Fig. 22.8-2. This shows that for net transfer of A and B into the stream (positive ϕ), the transfer coefficients decrease, whereas net transfer of A and B out of the stream (negative ϕ) causes the transfer coefficients to increase.

Some sample profiles from Eq. 22.8-10 are shown in Fig. 22.8-3. In the limit of small mass-transfer rates (i.e., $\phi \to 0$ or $R \to 0$), Eq. 22.8-10 becomes simply $\Pi = \eta$. The film model regards the region outside the film as perfectly mixed, thus giving a profile that is flat beyond $\eta = 1$.

The Penetration Model

We next turn to the transfer coefficient at large net mass transfer rates for systems in which there is no significant drag at the interface. We have already studied several systems of this type: gas absorption into a falling liquid film and from a rising bubble (§18.5), and unsteady-state evaporation (§20.1). These systems are generally lumped together under the heading of *penetration theory*.

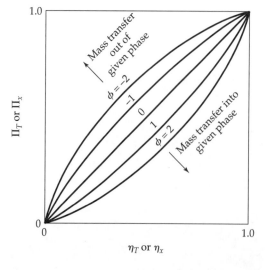

Fig. 22.8-3. Temperature and concentration profiles in a laminar film, as calculated by the film model (see Eq. 22.8-10).

A falling film system is shown in Fig. 22.8-4. The time of travel from the liquid inlet to the liquid outlet (the "exposure time") is sufficiently short that the diffusing species does not penetrate very far into the liquid. In such a situation, we can (from a mathematical point of view) regard the falling film as infinitely thick. We may then take over the results from Example 20.1-1.

Equation 20.1-23 gives the concentration profiles for a corresponding unsteady-state system with large net mass transfer rate, and an analogous equation can be written down for the temperature profiles:

$$\Pi_x \equiv \frac{x_A - x_{A0}}{x_{A\infty} - x_{A0}} = \frac{\operatorname{erf}(\eta_x - \varphi_x) + \operatorname{erf}\varphi_x}{1 + \operatorname{erf}\varphi_x} \tag{22.8-11}$$

$$\Pi_T \equiv \frac{T - T_0}{T_\infty - T_0} = \frac{\operatorname{erf}(\eta_T - \varphi_T) + \operatorname{erf}\varphi_T}{1 + \operatorname{erf}\varphi_T} \tag{22.8-12}$$

Here $\eta_x = y/\sqrt{4\mathscr{D}_{AB}t}$ and $\eta_T = y/\sqrt{4\alpha t}$ are dimensionless distances from the interface, and φ in each formula is a dimensionless molar average velocity at the interface:

$$\varphi_x = \frac{N_{A0} + N_{B0}}{c}\sqrt{\frac{t}{\mathscr{D}_{AB}}} \qquad \varphi_T = \frac{N_{A0} + N_{B0}}{c}\sqrt{\frac{t}{\alpha}} \tag{22.8-13a, b}$$

From these results and the definitions for the transfer coefficients in Eqs. 22.8-1 and 2, we may now get the rate factors ϕ, the flux ratios R, and the correction factors θ, defined in the preceding subsection:

$$\phi = \sqrt{\pi}\varphi \tag{22.8-14}$$

$$R = \left(1 + \operatorname{erf}\frac{\phi}{\sqrt{\pi}}\right)\phi \exp\left(\frac{\phi^2}{\pi}\right) \tag{22.8-15}$$

$$\theta = \frac{\exp(-\phi^2/\pi)}{1 + \operatorname{erf}(\phi/\sqrt{\pi})} \tag{22.8-16}$$

From the definitions in Eqs. 22.8-1 and 2 and the profiles in Eqs. 22.8-11 and 12, we can also get the expressions for the transfer coefficients at low net mass transfer rates:

$$k_{x,\text{loc}} = c\sqrt{\frac{\mathscr{D}_{AB}}{\pi t}}; \qquad h_{\text{loc}} = \rho\hat{C}_p\sqrt{\frac{\alpha}{\pi t}} \tag{22.8-17, 18}$$

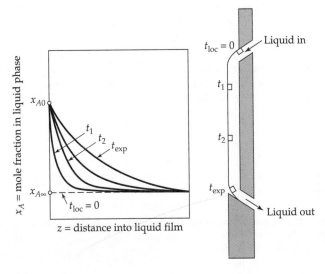

Fig. 22.8-4. Diffusion into a falling liquid film. Here t_{exp} is the total time of exposure of a typical element of volume near the surface.

The corresponding coefficients at high net mass transfer rates can be obtained by multiplying by the correction factor in Eq. 22.8-16.

From the last two equations we get the relation

$$h_{\text{loc}} = k_{x,\text{loc}} \tilde{C}_p \left(\frac{\alpha}{\mathscr{D}_{AB}} \right)^{1/2} \tag{22.8-19}$$

A similar relation, with an exponent of $\frac{2}{3}$ (instead of $\frac{1}{2}$) is obtained from the Chilton–Colburn relations given in Eqs. 22.3-23 to 25. The latter are valid for flows adjacent to rigid boundaries, whereas Eq. 22.8-19 pertains to fluid-fluid systems with no velocity gradient at the interface.

The proportionality of $k_{x,\text{loc}}$ to the square root of the diffusivity, given in Eq. 22.8-17, has been confirmed experimentally for the liquid phase in several gas–liquid mass transfer systems, including short wetted-wall columns, packed columns, and liquids around gas bubbles in certain instances. The penetration model has also been applied to absorption with chemical reactions (see Example 20.1-2).

The Flat-Plate Boundary Layer Model

The steady-state transport in the boundary layer along a flat plate for a fluid with constant physical properties was discussed in §20.2. The general expression for the profiles, $\Pi(\eta, \Lambda, K)$, was given in Eq. 20.2-43. There $\eta = y\sqrt{v_\infty/2\nu x}$ is a dimensionless position coordinate measured from the plate, Λ is the physical property group (i.e., 1, Pr, or Sc), and $K = (v_{y0}/v_\infty)\sqrt{2v_\infty x/\nu}$ is a dimensionless net mass flux from the plate.

Once again we introduce the notations defined in Table 22.8-1. Then for the boundary layer calculation we have

$$R = \frac{K\Lambda}{\Pi'(0, \Lambda, K)} \qquad \phi = \frac{K\Lambda}{\Pi'(0, \Lambda, 0)} \qquad \theta = \frac{\Pi'(0, \Lambda, K)}{\Pi'(0, \Lambda, 0)} \tag{22.8-20, 21, 22}$$

In the boundary layer calculation it was assumed that the heat capacities of both species are identical.

The momentum, heat, and mass fluxes for the flat plate are given in Fig. 22.8-5. Then in the following two figures, Figs. 22.8-6 and 22.8-7, two plots are given, comparing the correction factors, θ, for the film model, the penetration model, and the boundary layer model. The boundary layer model gives a dependence on Λ that is not found in the other models, because this model includes the effect of the tangential velocity profiles on the temperature and concentration profiles. The film model predicts the smallest dependence of the transfer coefficients on net mass-transfer rate.

Correction factors considerably different from 1 arise when either ϕ or R is of magnitude 1 or greater for T or x_A; see Figures 22.8-6, 7 and 8, and the relation $\theta = \phi/R$. Large net interfacial mass fluxes, by these measures, are common when the mass transfer is mechanically driven as in ultrafiltration (Example 22.8-5) and transpiration cooling (Problem 20B.7(c)). Large net mass fluxes can also occur in vaporization, condensation, melting and other changes of state, and in heterogeneous chemical reactions, when accompanied by correspondingly large temperature differences or radiation intensities to transfer the requisite latent heat or energy of reaction. More moderate net fluxes, and correction factors near unity, are common in multistage and packed-column separation processes, where the differences of temperature and composition within a separation stage or flow cross-section are normally rather small. The energy flux ratio R_T is an important criterion for assessing the net-flux corrections, as illustrated in Examples 22.8-2 and 3.

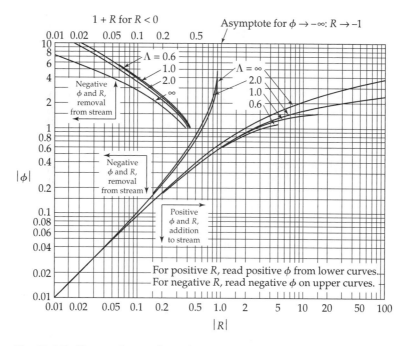

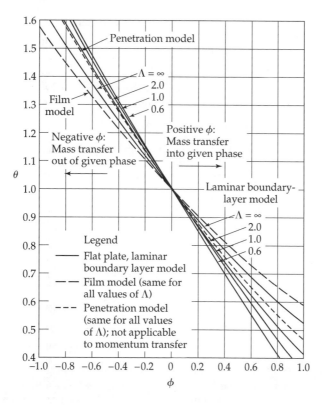

Fig. 22.8-5. Heat and mass fluxes between a flat plate and a laminar boundary layer [W. E. Stewart, ScD thesis, Massachusetts Institute of Technology (1951)].

Fig. 22.8-6. The variation of the transfer coefficients with mass transfer rate as predicted by various models. The line for $\Lambda \to \infty$ holds for the nonseparated, steady state boundary-layer regions on rigid surfaces, whatever their geometry.

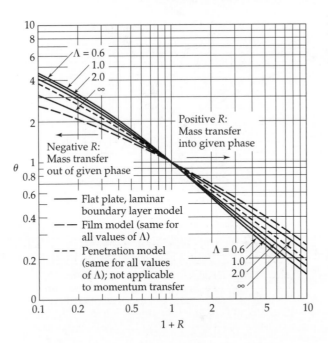

Fig. 22.8-7. The variation of the transfer coefficients with the flux ratio R as predicted by various models. The line for $\Lambda \to \infty$ holds for the nonseparated, steady state boundary-layer regions on smooth rigid surfaces, whatever their geometry.

EXAMPLE 22.8-1

Rapid Evaporation of a Liquid from a Plane Surface

Solvent A is evaporating out of a coat of lacquer on a plane surface exposed to a tangential stream of noncondensable gas B. At a given point on the surface, the gas-phase mass transfer coefficient $k_{x,\mathrm{loc}}$ at the prevailing average fluid properties is given as 0.1 lb-mole/hr · ft²; the Schmidt number is Sc = 2.0. The interfacial gas composition is $x_{A0} = 0.80$. Estimate the local rate of evaporation, using **(a)** the stagnant film model, **(b)** the flat-plate boundary layer model, and **(c)** the uncorrected mass transfer coefficient $k_{x,\mathrm{loc}}$.

SOLUTION

(a) Since B is noncondensable, $N_{B0} = 0$. Application of Eq. 22.8-7 (which is the same as $1 + R_x = \exp \phi_x$) to the gas phase then gives

$$1 + \frac{(N_{A0} + 0)(0.80 - 0)}{N_{A0} - 0.80(N_{A0} + 0)} = \exp\left(\frac{N_{A0} + 0}{0.1}\right) \tag{22.8-23}$$

From this we get, after taking the logarithm,

$$N_{A0} = 0.1 \ln(1 + 4.0) = 0.161 \text{ lb-mole/hr} \cdot \text{ft}^2 \tag{22.8-24}$$

as the result of the stagnant-film model. This corresponds to a correction factor $\theta_x = \phi_x/R_x = 0.40$.

(b) As in part (a), $R_x = 4.0$. Then from Fig. 22.8-5, at $R_x = 4.0$ and $\Lambda_x = 2.0$, we find that $\phi_x = 1.3$. By setting $N_{B0} = 0$ in the formula for ϕ_x in Table 22.8-1, we get

$$N_{A0} = k_{x,\mathrm{loc}}\phi_x = (0.1)(1.3) = 0.13 \text{ lb-mole/hr} \cdot \text{ft}^2 \tag{22.8-25}$$

as the result of the flat-plate boundary layer model. The corresponding correction factor θ_x is 0.33.

(c) If the mass transfer coefficient $k_{x,\mathrm{loc}}$ is used without correction for the net interfacial flux, we get from Eq. 22.1-5, with $N_{B0} = 0$

$$N_{A0} - 0.80(N_{A0} + 0) = (0.1)(0.80 - 0) \tag{22.8-26}$$

whence $N_{A0} = 0.400$. This result is much too high and shows that the corrections for net molar flux are important at these conditions. The boundary layer solution in part (b) should be accurate if the flow is laminar and the variation in the physical properties is not too great.

EXAMPLE 22.8-2

Correction Factors in Droplet Evaporation

Adjust the results of Example 22.3-1 for the net molar flux by applying the correction factors θ_x and θ_T from the film model and from the flat-plate boundary layer model.

SOLUTION

In Example 22.3-3 the molar flux ratio R_x at any point on the surface of the drop is

$$
\begin{aligned}
R_x &= \frac{(N_{A0} + N_{B0})(x_{A0} - x_{A\infty})}{N_{A0} - x_{A0}(N_{A0} + N_{B0})} \\
&= \frac{(N_{A0} + 0)(0.0247 + 0)}{N_{A0} - 0.0247(N_{A0} + 0)} = \frac{0.0247}{1 - 0.0247} = 0.0253
\end{aligned}
\tag{22.8-27}
$$

From Eq. 22.8-9 (film model) or Fig. 22.8-7 (flat-plate boundary layer model), the predicted correction factor θ_x is about 0.99 at all points on the drop. Hence the corrected mass transfer rate is (by adjustment of Eq. 22.3-31)

$$
\begin{aligned}
W_{A0} &= \theta_x k_{xm} \pi D^2 \frac{x_{A0} - x_{A\infty}}{1 - x_{A0}} = (0.987)(2.70 \times 10^{-7}) \\
&= 2.66 \times 10^{-7} \text{ g-mole/s} \cdot \text{cm}^2
\end{aligned}
\tag{22.8-28}
$$

This result differs only slightly from that obtained in Example 22.3-1. Thus the assumption of a small mass-transfer rate was satisfactory under the given conditions.

EXAMPLE 22.8-3

Wet-Bulb Performance Corrected for Mass-Transfer Rate

Extend the analysis of Example 22.3-2 to include the corrections for net mass-transfer rate, using the stagnant film model.

SOLUTION

By rewriting the energy balance, Eq. 22.3-32, for any point on the wick, we obtain for finite mass-transfer rate

$$
N_{A0} \Delta \tilde{H}_{A,\text{vap}} = h^{\bullet}_{\text{loc}}(T_\infty - T_0)
\tag{22.8-29}
$$

Multiplication of both sides by $\tilde{C}_{pA}/(\Delta \tilde{H}_{A,\text{vap}} h^{\bullet}_{\text{loc}})$ gives, since $N_{B0} = 0$,

$$
R_T = \frac{N_{A0}\tilde{C}_{pA}}{h^{\bullet}_{\text{loc}}} = \frac{\tilde{C}_{pA}(T_\infty - T_0)}{\Delta \tilde{H}_{A,\text{vap}}}
\tag{22.8-30}
$$

The right-hand member of this equation is easily calculated if T_0, T_∞, and p are given.

Next we write the expression $\phi = \ln(1 + R)$ for both heat and mass transfer, taking into account the fact that $N_{B0} = 0$:

$$
\frac{N_{A0}\tilde{C}_{pA}}{h_{\text{loc}}} = \ln(1 + R_T); \qquad \frac{N_{A0}}{k_{x,\text{loc}}} = \ln(1 + R_x)
\tag{22.8-31, 32}
$$

Solving both equations for N_{A0} and equating the resulting expressions gives

$$
\ln(1 + R_x) = \frac{h_{\text{loc}}}{k_{x,\text{loc}}\tilde{C}_{pA}} \ln(1 + R_T)
\tag{22.8-33}
$$

Then substituting the expressions for R_x and R_T from Table 22.8-1 yields

$$
\ln\left(1 + \frac{x_{A0} - x_{A\infty}}{1 - x_{A0}}\right) = \frac{h_{\text{loc}}}{k_{x,\text{loc}}\tilde{C}_{pA}} \ln\left(1 + \frac{\tilde{C}_{pA}(T_\infty - T_0)}{\Delta \tilde{H}_{A,\text{vap}}}\right)
\tag{22.8-34}
$$

This equation shows that x_{A0} and T_0 will be constant over the surface of the wick if $h_{\text{loc}}/(k_{x,\text{loc}}\tilde{C}_{pA})$ is constant and thus equal to $h_m/(k_{xm}\tilde{C}_{pA})$. This constancy is assumed here for simplicity. Such an assumption is particularly satisfactory for the water–air system, for which Pr and Sc are nearly equal. With this substitution, Eq. 22.8-34 becomes

$$\ln\left(1 + \frac{x_{A0} - x_{A\infty}}{1 - x_{A0}}\right) = \frac{h_m}{k_{xm}\tilde{C}_{pA}}\ln\left(1 + \frac{\tilde{C}_{pA}(T_\infty - T_0)}{\Delta\tilde{H}_{A,\text{vap}}}\right) \tag{22.8-35}$$

This solution simplifies exactly to Eq. 22.3-35 at low mass-transfer rates.

For the numerical problem in Example 22.3-2, the following values apply:

$$x_{A0} = 0.0247$$
$$\tilde{C}_{pA} = 8.03 \text{ Btu/lb-mole} \cdot \text{F for water vapor at } 105°\text{F}$$
$$h_m/k_{xm} = 5.93 \text{ Btu/lb-mole} \cdot \text{F from the Chilton–Colburn analogy (Eq. 22.3-25)}$$
$$\frac{\tilde{C}_{pA}(T_\infty - T_0)}{\Delta\tilde{H}_{A,\text{vap}}} = \frac{(8.03)(140 - 70)}{18,900} = 0.0297$$

Insertion of these values into Eq. 22.8-35 gives

$$\ln\left(1 + \frac{0.0247 - x_{A\infty}}{1 - 0.0247}\right) = \frac{5.93}{8.03}\ln(1.0297) = 0.0216 \tag{22.8-36}$$

Solving this equation, we get

$$x_{A\infty} = 0.0034 \tag{22.8-37}$$

This result differs only slightly from the value 0.0033 obtained in Example 22.3-2 and justifies the previous omission of the correction factors under the given conditions.

Numerical studies indicate that the simple Eq. 22.3-34 gives a close approximation to Eq. 22.8-35 for the air–water system under all likely wet-bulb conditions. Eqs. 22.3-32 and 33 overestimate the mass transfer rate almost equally, and when these equations are combined, the errors largely compensate.

EXAMPLE 22.8-4

Comparison of Film and Penetration Models for Unsteady Evaporation in a Long Tube

Compare the effects of net mass transfer for the unsteady evaporation system described in Example 20.1-1 with the predictions of **(a)** the generalized penetration model, and **(b)** the stagnant-film model introduced above. The latter calculation amounts to a quasi-steady-state treatment of this time-dependent system.

SOLUTION

We begin by noting that for this system $x_{A\infty} = 0$ and $N_{B0} = 0$. It follows from Eq. 22.8-1a and Table 22.8-1 that

$$N_{A0} = \theta_x k_x \frac{x_{A0} - 0}{1 - x_{A0}} \tag{22.8-38}$$

The correction factor θ_x is thus the ratio of the flux corrected for net mass transfer to the uncorrected flux.

(a) *The penetration model.* We note that the concentration gradient at the liquid surface can be obtained by differentiating Eq. 20.1-16 and rewriting the result in terms of x_A and z. The result is

$$\left.\frac{\partial x_A}{\partial z}\right|_{z=0} = -\frac{2}{\sqrt{\pi}}\frac{\exp(-\varphi^2)}{1 + \text{erf }\varphi}\frac{x_{A0}}{\sqrt{4\mathcal{D}_{AB}t}} \tag{22.8-39}$$

For negligible net mass transfer, $\varphi = 0$. Thus, the ratio of the mass flux in the presence of net mass transfer to the flux in the absence of net mass transfer is

$$\theta_x = \frac{\exp(-\varphi^2)}{1 + \text{erf }\varphi} \tag{22.8-40}$$

Table 22.8-2 Comparison of Film and Penetration Models.

x_{A0}	θ_x from penetration model (Eq. 22.8-41)	θ_x from film model (Eq. 22.8-42)
0.00	1.000	1.000
0.25	0.831	0.863
0.50	0.634	0.693
0.75	0.391	0.462
1.00	0.000	0.000

in agreement with Eqs. 22.8-14 and 16. To get θ_x as a function of x_{A0}, we may use Fig. 22.8-7, or use Eq. 20.1-17 to write

$$\theta_x = (1 - x_{A0})\psi(x_{A0}) \qquad \text{(penetration model)} \qquad (22.8\text{-}41)$$

where $\psi(x_{A0})$ is the quantity defined just after Eq. 20.1-22 and given in Table 20.1-1.

(b) *The stagnant-film model.* The film model result may be obtained from Eq. 22.8-9 in the form $\theta = (1/R)\ln(1 + R)$ to obtain

$$\theta_x = \frac{1 - x_{A0}}{x_{A0}} \ln\left(\frac{1}{1 - x_{A0}}\right) \qquad \text{(film model)} \qquad (22.8\text{-}42)$$

Numerical values for both models are provided in Table 22.8-2 and also in Fig. 22.8-7.

It is seen that the penetration model predicts a stronger correction θ_x for net mass transfer than does the film model. This is in part because the net flow thickens the boundary layer, an effect that the film model does not consider. It may also be noted that this example is a realistic use of the penetration model, as there is little effect of solute concentration on the physical properties in this simple isothermal system. A much different situation is seen in the next example.

EXAMPLE 22.8-5

Concentration Polarization in Ultrafiltration

Ultrafiltration of proteins is a concentration process, in which water from an aqueous protein solution is forced through a membrane impermeable to the protein but permeable to water and small solutes such as inorganic salts. Protein then accumulates in a *polarization layer*, or region of high protein concentration adjacent to the membrane surface, as indicated in Fig. 22.8-8. Determine the relation between water permeation velocity and the transmembrane

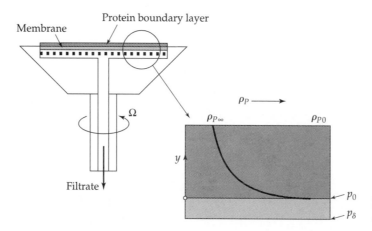

Fig. 22.8-8. A spinning-disk ultrafilter.

pressure difference. Describe the effect of net mass transfer on the mass transfer coefficient for protein transport. Assume that the membrane is completely impermeable to protein so that the net transport of protein across the membrane surface is zero.

For simplicity we choose a spinning-disk geometry as shown in Fig. 22.8-8, for which the protein concentration will be a function only of the distance y from the disk surface and not of radial position[5] (see Problem 19D.4). However, we will have to consider the dependence of density, viscosity, and protein–water diffusivity on the protein concentration, and we will need the concept of osmotic pressure.[6]

The basis for our solution is the concept of *hydraulic permeability* of the filtration membrane:

$$v_\delta = K_H(p_0 - p_\delta - \pi) \tag{22.8-43}$$

Here v_δ is the velocity, or volumetric flux, of the solvent leaving the downstream surface of the membrane. Equation 22.8-43 defines K_H, the hydraulic permeability of the membrane. The quantities p_0 and p_δ are the hydrodynamic pressures against the membrane as indicated in Fig. 22.8-8, and π is the *osmotic pressure* at the upstream surface of the membrane. The inclusion of π recognizes that it is really the total thermodynamic potential that drives the transmembrane transport (this point will be discussed further in Chapter 24.)

For this situation, the interfacial protein velocity is zero, so that a solvent mass balance across the protein boundary layer gives

$$\rho^{(S)}v_\delta = -(\rho_S v_{Sy})|_{y=0} \tag{22.8-44}$$

in which y is the distance from the upstream membrane surface into the protein boundary layer. The quantity $\rho^{(S)}$ is the density of the pure solvent, and $\rho_{S0} = \rho_S|_{y=0}$ and $v_{S0} = v_{Sy}|_{y=0}$ are the mass concentration and velocity of solvent at the upstream membrane surface.

The osmotic pressure π is a function of the protein concentration ρ_P, and we will provide an example of this in Problem 22C.1. We find then that the water flux across the membrane depends on the protein concentration at the membrane surface as well as the hydrodynamic pressure drop across the membrane. This concentration, in turn, can be related to v_δ through the membrane impermeability condition for the protein and the definition of the mass transfer coefficient. Then at $y = 0$, we describe the impermeability of the membrane to protein by

$$n_{Py} = 0 = k_p^\bullet(\rho_{P0} - \rho_{P\infty}) + \omega_{P0}(0 + \rho_{S0}v_{S0}) \tag{22.8-45}$$

where $k_p^\bullet$ has been defined analogously to $k_c^\bullet$. Combination with Eq. 22.8-44 then gives

$$-\rho^{(S)}v_\delta = k_p^\bullet(\rho_{P0} - \rho_{P\infty})/\omega_{P0} \tag{22.8-46}$$

This equation may now be solved for the filtrate velocity:

$$v_\delta = k_p\theta\left(\frac{\rho_0}{\rho^{(S)}}\right)\left(1 - \frac{\rho_{P\infty}}{\rho_{P0}}\right) \tag{22.8-47}$$

Here $\rho_0 = \rho_{P0} + \rho_{S0}$ and $\theta = k_p^\bullet/k_p$ is a mass transfer correction factor, analogous to θ_x, which now must include the effects of property changes as well as the net velocity correction introduced in Table 22.8-1. We return to a discussion of this quantity below (see Eq. 22.8-48). The term ρ_0 is the solution density at the upstream membrane surface.

We can now calculate the desired quantities, v_δ and the transmembrane pressure difference, if we have sufficient information about the transport and equilibrium properties. Here we consider the approaching protein concentration $\rho_{P\infty}$ to be given, and for convenience we

[5] D. R. Olander, *J. Heat Transfer*, **84**, 185 (1972).
[6] R. J. Silbey and R. A. Alberty, Physical Chemistry, 3rd edition, Wiley, New York (2001), p. 206.

begin by selecting values of the protein concentration ρ_{P0} at the membrane surface over the range between $\rho_{p\infty}$ and the solubility limit of the protein:

(i)   For any chosen value of ρ_{P0}, we can calculate the corresponding value of v_δ from Eq. 22.8-47 with appropriate values for k_ρ and θ. These values also permit calculation of osmotic pressure π from the appropriate equilibrium relationship.

(ii)   We may then calculate the transmembrane pressure difference required for this flow from Eq. 22.8-43 and an appropriate value of K_H.

The strong effects of protein concentration on system properties mean that the solution must be obtained numerically.

We content ourselves here to summarize the results of Kozinski and Lightfoot[7] for bovine serum albumin; they were the first to make such calculations and seem still to have provided the best documentation. In their publications it is shown that the effective mass transfer coefficient can be expressed as the product of two factors, one accounting for the concentration effects and another taking account of the additional effect of property variations:

$$\theta = \theta_c\theta_p \tag{22.8-48}$$

where, over the parameter space investigated,

$$\theta_c = 1.6\left(\frac{\rho_{P0}}{\rho_{P\infty}}\right)^{1/3}; \qquad \theta_p = \mathscr{D}_{\text{rel}}^{2/3}\left(\frac{1}{\nu}\right)_{\text{rel}}^{1/3} \tag{22.8-49, 50}$$

and

$$\mathscr{D}_{\text{rel}} = \frac{1}{2}\left(1 + \frac{\mathscr{D}_{PS}(0)}{\mathscr{D}_{PS}(\infty)}\right); \qquad \left(\frac{1}{\nu}\right)_{\text{rel}} = \frac{1}{2}\left(1 + \frac{\nu(\infty)}{\nu(0)}\right) \tag{22.8-51, 52}$$

Equations 22.8-49 to 52 must be considered empirical. Equation 22.8-47 overpredicts v_δ for small polarization levels, but for that situation the effect of osmotic pressure on flow is small. The subscript rel means "relative to the free-stream value."

The mass transfer coefficient in the limit of slow mass transfer and small property variations is given[7] as

$$\text{Sh}_m = \text{Sh}_{\text{loc}} = \frac{k_\rho L}{\mathscr{D}_{PS}(\infty)} = 0.6205\left(\frac{L^2\Omega}{\nu(\infty)}\right)^{1/2}\left(\frac{\nu(\infty)}{\mathscr{D}_{PS}(\infty)}\right)^{1/3} \tag{22.8-53}$$

in which L is the disk diameter and Ω is the rate of rotation in radians per unit time. The independence of mass transfer rate on disk size is the reason that this geometry is so popular for careful mass transfer studies. Other geometries are considered briefly by Kozinski and Lightfoot.[7]

A comparison of a priori predictions from the above model with experimental data is shown in Fig. 22.8-9, where we see that the two agree well. This good agreement may result in part because the albumin molecules behave much like incompressible particles at the high solvent ionic strength at which the data were taken. It may also be seen that osmotic effects are negligible below pressure drops of about 5 psi; here the predicted behavior is indistinguishable from that of the protein-free solvent, essentially water. It is only in this unimportant region that Eq. 22.8-48 is unreliable. Details of the calculations are provided in Problem 22C.1.

The effect of increasing pressure difference across the protein boundary layer is quite different from that for a nonselective membrane. At first, the concentration boundary layer gets thinner, as would be expected, and the mass transfer coefficient $k_\rho^{\cdot}$ increases. However, with

---

[7] A. A. Kozinski, PhD thesis, University of Wisconsin (1971); A. A. Kozinski and E. N. Lightfoot, *AIChE Journal*, **18**, 1030–1040 (1972).

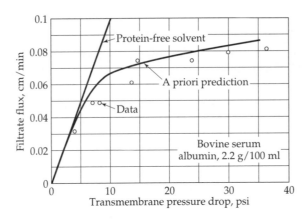

Fig. 22.8-9. Protein ultrafiltration with a spinning disk at 273 rpm.

further increase in the pressure difference the boundary layer thickness, $k_\rho^\bullet$, and θ_c all approach asymptotic limits. In practice, these asymptotes are closely approached before the effect of polarization becomes appreciable, relative to the membrane flow resistance, and these asymptotes suffice to predict the relation between the transmembrane pressure difference and transmembrane flow.

The behavior can be seen more clearly inserting Eq. 22.8-48 and 49 and the approximate formula

$$1.6\left(\frac{\rho_{P0}}{\rho_{P\infty}}\right)^{1/3}\left(1 - \frac{\rho_{P\infty}}{\rho_{P0}}\right) \approx 1.39 \ln \frac{\rho_{P0}}{\rho_{P\infty}} \tag{22.8-54}$$

into Eq. 22.8-47. Then, to a surprisingly good approximation, Eq. 22.8-47 takes the form

$$v_\delta = \left(1.39 k_\rho \ln \frac{\rho_{P0}}{\rho_{P\infty}}\right)\left(\frac{\rho_0}{\rho^{(S)}}\right)\theta_p \tag{22.8-55}$$

The quantity in the first set of parentheses has the form of the simple film model, but with k_ρ multiplied by 1.39. It is probably Eq. 22.8-55 that has made the simple film model attractive to many for correlating ultrafiltration and reverse osmosis data. However, neglect of the multiplier 1.39 has caused corresponding underestimation of v_δ, even before addressing the effects of property variations.

§22.9   MATRIX APPROXIMATIONS FOR MULTICOMPONENT MASS TRANSPORT

Multicomponent mass transport occurs widely in chemical, physiological, biological, and environmental processes and is analyzed by various mathematical methods. Here we review some matrix approximation methods for mass transport by convection and ordinary diffusion in multicomponent gases. A fuller treatment, including mass transport in liquids, is given in the text by Taylor and Krishna.[1]

Multicomponent mass transport problems are commonly approximated by linearization—that is, by replacing the variable properties in the governing equations with constant reference values. This approach is a useful complement to purely numerical methods, especially for complex flows, and can give good predictions when the property

---

[1] R. Taylor and R. Krishna, *Multicomponent Mass Transfer*, Wiley, New York (1993).

variations are not too large. Multicomponent analyses of this sort have been presented by many investigators, for quiescent media[2] and for forced-convection systems.[3-6]

We begin with the species continuity equations as given in Eq. 19.1-15, and apply them to an N-component gas system with $N - 1$ independent mole fractions x_α and an equal number of independent diffusion fluxes J_α^*. Let $[x]$ and $[J^*]$ denote, respectively, the column arrays of independent mole fractions $x_1, \ldots, x_{N-1}$ and independent diffusion fluxes $J_1^*, \ldots, J_{N-1}^*$; then approximating the molar density c in Eq. 19.1-15 by a reference value c_{ref} gives the linearized equation system

$$c_{ref}\left(\frac{\partial}{\partial t}[x] + (\mathbf{v}^* \cdot \nabla[x])\right) = -(\nabla \cdot [J^*]) \tag{22.9-1}$$

for laminar or turbulent flows free of homogeneous chemical reactions.

For multicomponent ordinary diffusion, the flux expression may be written either as a matrix generalization[2,4] of Fick's first law (Eq. B of Table 17.8-2),

$$[J^*] = -c[\mathbf{D}]\nabla[x] \tag{22.9-2}$$

or as a matrix statement[3,5] of the Maxwell–Stefan equation (Eq. 17.9-1):

$$c\nabla[x] = -[\mathbf{A}][J^*] \tag{22.9-3}$$

The matrices $[\mathbf{D}]$ and $[\mathbf{A}]$ must be $(N - 1) \times (N - 1)$ and nonsingular to give the stated number of independent fluxes (in Eq. 22.9-2), and of independent mole fractions (in Eq. 22.9-3). Consistency of these two equations then requires that $[\mathbf{D}] = [\mathbf{A}]^{-1}$ at any given state.

In the moderate-density gas region, the elements of the matrix $[\mathbf{A}]$ are predictable accurately from Eq. 17.9-1, giving

$$\left.\begin{aligned} A_{\alpha\beta} &= \frac{x_\alpha}{\mathcal{D}_{\alpha N}} - \frac{x_\alpha}{\mathcal{D}_{\alpha\beta}} \quad \text{for } \beta \neq \alpha \\ A_{\alpha\alpha} &= \frac{x_\alpha}{\mathcal{D}_{\alpha N}} + \sum_{\substack{\beta=1 \\ \beta \neq \alpha}}^{N} \frac{x_\alpha}{\mathcal{D}_{\alpha\beta}} \end{aligned}\right\} \tag{22.9-4}$$

in which the divisors $\mathcal{D}_{\alpha\beta}$ are the *binary* diffusivities of the corresponding pairs of species. In the first approximation of the Chapman–Enskog kinetic theory of gases, the coefficient for a given pair α, β depends only on c and T, as in Eq. 17.3-11. These simple expressions lead us to prefer Eq. 22.9-3 over Eq. 22.9-2, unless measurements of $[\mathbf{D}]$ are available at the desired conditions. Formally similar equations may be written in mass- or volume-based compositions and fluxes, after appropriate transformation of the coefficient matrix $[\mathbf{A}]$ or $[\mathbf{D}]$. Mass units are preferred if the equation of motion is included in the problem formulation, since the mass average velocity is then essential as indicated in §19.2.

---

[2] L. Onsager, *Ann. N.Y. Acad. Sci.*, **46**, 241–265 (1948); P. J. Dunlop and L. J. Gosting, *J. Phys. Chem.*, **63**, 86–93 (1959); J. S. Kirkaldy, *Can. J. Phys.*, **37**, 30–34 (1959); S. R. de Groot and P. Mazur, *Non-Equilibrium Thermodynamics*, North-Holland, Amsterdam (1961); J. S. Kirkaldy, D. Weichert, and Zia-Ul-Haq, *Can. J. Phys.*, **41**, 2166–2173 (1963); E. L. Cussler, Jr., and E. N. Lightfoot, *AIChE Journal*, **10**, 702–703, 783–785 (1963); H. T. Cullinan, *Ind. Eng. Chem. Fund.*, **4**, 133–139 (1965).
[3] R. Prober, PhD thesis, Univ. of Wisconsin (1961).
[4] H. L. Toor, *AIChE Journal*, **10**, 460–465 (1964).
[5] W. E. Stewart and R. Prober, *Ind. Eng. Chem. Fund.*, **3**, 224–235 (1964).
[6] V. Tambour and B. Gal-Or, *Physics of Fluids*, **19**, 219–225 (1976).

For multicomponent systems ($N \geq 3$), each of these flux expressions normally has a nondiagonal coefficient matrix, giving a coupled system of diffusion equations. Equation 22.9-3 can be decoupled by use of the transformation

$$[\mathbf{P}]^{-1}[\mathbf{A}][\mathbf{P}] = \begin{bmatrix} \check{A}_1 & & \\ & \ddots & \\ & & \check{A}_{N-1} \end{bmatrix} \tag{22.9-5}$$

in which $[\mathbf{P}]$ is the matrix of column eigenvectors of $[\mathbf{A}]$, and $\check{A}_1, \ldots, \check{A}_{N-1}$ are the corresponding eigenvalues. These eigenvalues, the roots of the equation $\det[\mathbf{A} - \lambda\mathbf{I}] = 0$, are positive at any locally stable state of the mixture; they are also invariant to similarity transformations of $[\mathbf{A}]$ to other composition units. Here $\mathbf{I}$ is the unit matrix of order $N - 1$. The matrix $[\mathbf{D}]$, when used, is reducible in like manner with the same matrix $[\mathbf{P}]$, and its eigenvalues $\check{D}_1, \ldots, \check{D}_{N-1}$ are the reciprocals of $\check{A}_1, \ldots, \check{A}_{N-1}$. For economy of effort, $[\mathbf{A}]$ (or $[\mathbf{D}]$) and the arrays derived therefrom will always be evaluated at reference property values, so will not need the subscript $_{\text{ref}}$; however, a subscript ω will be added in $[\mathbf{A}]$, $[\mathbf{D}]$, $[\mathbf{P}]$, and $[\mathbf{P}]^{-1}$ when these arrays are based on quantities in mass units.

Equation 22.9-5 suggests that the following transformed compositions and transformed diffusion fluxes should be useful:

$$[\check{x}] = [\mathbf{P}]^{-1}[x] = \begin{bmatrix} \check{x}_1 \\ \vdots \\ \check{x}_{N-1} \end{bmatrix}; \qquad \begin{bmatrix} x_1 \\ \vdots \\ x_{N-1} \end{bmatrix} = [\mathbf{P}][\check{x}] \tag{22.9-6, 7}$$

$$[\check{\mathbf{J}}^*] = [\mathbf{P}]^{-1}[\mathbf{J}^*] = \begin{bmatrix} \check{J}_1^* \\ \vdots \\ \check{J}_{N-1}^* \end{bmatrix}; \qquad \begin{bmatrix} J_1^* \\ \vdots \\ J_{N-1}^* \end{bmatrix} = [\mathbf{P}][\check{\mathbf{J}}^*] \tag{22.9-8, 9}$$

Hereafter, an accent ($\check{\ }$) will be placed on such transformed variables and on the corresponding diagonal matrix elements, including the eigenvalues $\check{A}_\alpha$ and $\check{D}_\alpha$. Premultiplication of Eq. 22.9-3 by $[\mathbf{P}]^{-1}$ and use of Eqs. 22.9-5 through 9 then gives uncoupled flux equations

$$c_{\text{ref}}\nabla\check{x}_\alpha = -\check{A}_\alpha\check{J}_\alpha^* \qquad (\alpha = 1, \ldots, N-1) \tag{22.9-10}$$

formally equivalent to Fick's first law for $N - 1$ binary systems. The multicomponent continuity equation 22.9-1 correspondingly transforms to

$$c_{\text{ref}}\frac{\partial\check{x}_\alpha}{\partial t} + c_{\text{ref}}(\check{\mathbf{v}}^* \cdot \nabla x_\alpha) = -(\nabla \cdot \check{\mathbf{J}}_\alpha^*) \qquad (\alpha = 1, \ldots, N-1) \tag{22.9-11}$$

Thus, the transformed compositions $\check{x}_\alpha$ and fluxes $\check{J}_\alpha^*$ for each α satisfy the continuity and flux equations of a binary problem with the same $\mathbf{v}^*$ function (laminar or turbulent) as the multicomponent system, and with a diffusivity $\mathscr{D}_{AB}$ equal to the eigenvalue $\check{D}_\alpha = 1/\check{A}_\alpha$.

The initial and boundary conditions on $[\check{x}]$ and $[\check{\mathbf{J}}^*]$ are obtained from those on $[x]$ and $[\mathbf{J}^*]$ by application of Eqs. 22.9-6 and 8. The resulting quasi-binary problems may then be solved, using theory or correlations of experiments, and the results combined[5] via Eqs. 22.9-7 and 9 to get the multicomponent solution in terms of $[x]$ and $[\mathbf{J}^*]$.

Local mass transfer rates in binary systems are expressible in the form

$$N_{A0} - x_{A0}(N_{A0} + N_{B0}) = k_{x,\text{loc}}^\bullet(\mathscr{D}_{AB}, \ldots)(x_{A0} - x_{Ab}) \tag{22.9-12}$$

as indicated in Eq. 22.1-7 and §22.8. The notation $\ldots$ after $\mathscr{D}_{AB}$ stands for any additional variables (such as ϕ_x of §22.8) on which the binary mass transfer coefficient $k_x^\bullet$ may depend. The corresponding set of equations in the notation of Eqs. 22.9-10 and 11 is

$$\check{J}_{\alpha 0}^* = \check{k}_{x,\text{loc}}^\bullet(\check{D}_\alpha, \ldots)(\check{x}_{\alpha 0} - \check{x}_{\alpha b}) \qquad (\alpha = 1, \ldots, N-1) \tag{22.9-13}$$

or in matrix form,

$$
\begin{bmatrix} \check{J}_{1,0}^* \\ \vdots \\ \check{J}_{N-1,0}^* \end{bmatrix} = \begin{bmatrix} \check{k}_x(\check{D}_{1'} \dots & & \\ & \ddots & \\ & & \check{k}_x(\check{D}_{N-1'} \dots) \end{bmatrix} \begin{bmatrix} \check{x}_{1,0} - \check{x}_{1b} \\ \vdots \\ \check{x}_{n-1,0} - \check{x}_{N-1,b} \end{bmatrix} \tag{22.9-14}
$$

Transformation of this result into the original variables gives the interfacial diffusion fluxes $J_{1,0}^*, \dots, J_{N-1,0}^*$ into the gas phase as

$$
[J_0^*] = [\mathbf{P}][\check{\mathbf{k}}_x][\mathbf{P}]^{-1}[x_0 - x_b] \tag{22.9-15}
$$

or the composition differences for given fluxes $J_{\alpha0}$ as

$$
[x_0 - x_b] = [\mathbf{P}][\check{\mathbf{k}}_x]^{-1}[\mathbf{P}]^{-1}[J_0^*] \tag{22.9-16}
$$

Here $[\check{\mathbf{k}}_x]$ is the diagonal matrix shown in Eq. 22.9-14, and $[\check{\mathbf{k}}_x]^{-1}$ is formed from the reciprocals of the same diagonal elements.

As for binary systems, further information is needed to calculate the species fluxes $N_{\alpha0}$ relative to the interface, which give the local transfer rates. A flux ratio $r = N_{A0}/N_{B0}$ was specified in Eq. 21.1-9 to solve for N_{A0}; analogous specifications are required for multicomponent systems. The calculation of the fluxes $N_{\alpha0}$ from diffusion fluxes $J_{\alpha0}^*$ and relative transfer rates is called the "bootstrap problem,"[1,7] and is treated well in Ref. 1. This problem becomes simpler if Eq. 22.9-14 is rewritten as follows, using the array $[N_0]$ of interfacial molar fluxes $N_{1,0}, \dots, N_{N-1,0}$ relative to the interface,

$$
\left[N_0 - x_0 \sum_{\alpha=1}^{N} N_{\alpha0} \right] = [\mathbf{P}][\check{\mathbf{k}}_x][\mathbf{P}]^{-1}[x_0 - x_b] \tag{22.9-17}
$$

to allow direct insertion of relations among the species transfer rates. The corresponding result for the array $[n_0]$ of interfacial mass fluxes $n_{1,0}, \dots, n_{N-1,0}$ relative to the interface is:

$$
\left[n_0 - \omega_0 \sum_{\alpha=1}^{N} n_{\alpha,0} \right] = [\mathbf{P}_\omega][\check{\mathbf{k}}_\omega][\mathbf{P}_\omega]^{-1}[\omega_0 - \omega_\infty] \tag{22.9-18}
$$

Several special forms of these results will now be given.

For systems with *no net molar interfacial flux*, the N-term summation in Eq. 22.9-17 vanishes, and this equation takes the convenient form

$$
[N_0] = [\mathbf{P}][\check{\mathbf{k}}_x][\mathbf{P}]^{-1}[x_0 - x_b] \tag{22.9-19}
$$

in which the diagonal array $[\check{\mathbf{k}}_x]$ needs *no net-flux correction*. This result can be extended to *moderate net molar interfacial flux* by approximating each transfer coefficient $k_x(\check{D}_\alpha, \check{\phi}_{x\alpha})$ in Eq. 22.9-14 as a linear function of the net molar interfacial flux, using the tangent line at $\phi = 0$ of the θ-curve in Fig. 22.8-2 for the chosen mass transfer model. This gives the linear equation system[8]

$$
\left[N_0 - 0.5(x_0 + x_b) \sum_{\alpha=1}^{N} N_{\alpha0} \right] = [\mathbf{P}][\check{\mathbf{k}}_x][\mathbf{P}]^{-1}[x_0 - x_b] \tag{22.9-20}
$$

for the *stagnant-film model* given in §22.8. In the same manner, one obtains

$$
\left[N_0 - (0.363x_0 + 0.637x_b) \sum_{\alpha=1}^{N} N_{\alpha0} \right] = [\mathbf{P}][\check{\mathbf{k}}_x][\mathbf{P}]^{-1}[x_0 - x_b] \tag{22.9-21}
$$

---

[7] R. Krishna and G. L. Standart, *Chem. Eng. Commun.*, **3**, 201–275 (1979).
[8] W. E. Stewart, *AIChE Journal*, **19**, 398–400 (1973); Erratum, **25**, 208 (1979).

for the *penetration model* given in §20.4 and §22.8, and

$$\left[N_0 - (0.434x_0 + 0.566x_b) \sum_{\alpha=1}^{N} N_{\alpha 0}\right] = [\mathbf{P}][\check{\mathbf{k}}_x][\mathbf{P}]^{-1}[x_0 - x_b] \tag{22.9-22}$$

for the limit $\Lambda \to \infty$ in *laminar boundary layers*, shown in Figs. 22.8-5, 6 and valid for non-separated boundary layers in three-dimensional steady flows.[9]

In systems with *no net mass interfacial flux*, as in steady-state solid-catalyzed reactions, Eq. 22.9-18 reduces to

$$[n_0] = [\mathbf{P}_\omega][\check{\mathbf{k}}_\omega][\mathbf{P}_\omega]^{-1}[\omega_0 - \omega_b] \tag{22.9-23}$$

The elements of the matrix $\check{k}_\omega$ can be predicted from expressions for the binary Sherwood number or j_D factor as defined for mass-based units in Table 22.2-1, with eigenvalues $\check{D}_\alpha$ inserted in place of binary diffusivities $\mathscr{D}_{AB}$.

For a given flow field, the product $[\mathbf{P}][\check{\mathbf{k}}_x][\mathbf{P}]^{-1}$ in Eqs. 22.9-19 through 22 is a function of the matrix $[\mathbf{A}]$. This matrix triple product, here called $[\mathbf{k}_x]$, is non-diagonal for $N \geq 3$ whereas $[\check{\mathbf{k}}_x]$ is diagonal as noted above. A simple, efficient method for approximating such functions has been developed by Alopaeus and Nordén.[10] Let f be a scalar real-valued function defined on the eigenvalues of a matrix $[\mathbf{A}]$, in which the diagonal elements are dominant as in Eq. 22.9-4. The proposed approximations to the elements of the matrix $[\mathbf{B}] = f[\mathbf{A}]$ are then as follows:

$$\text{for diagonal elements, } B_{ii} = f(A_{ii}) \tag{22.9-24}$$

$$\text{for off-diagonal elements, } B_{ij} = \begin{cases} A_{ii} \dfrac{df(A_{ii})}{dA_{ii}} & \text{if } A_{ii} \approx A_{jj} \\ A_{ij} \dfrac{f(A_{ii}) - f(A_{jj})}{A_{ii} - A_{jj}} & \text{otherwise.} \end{cases} \tag{22.9-25}$$

Alopaeus and Nordén[10] tested these approximations to mass-transfer coefficient matrices $[\mathbf{k}_x]$ of the form $b[\mathbf{D}]^{1-p}$ or the form $b[\mathbf{A}]^{p-1}$, and to the corresponding fluxes $N_{\alpha 0}$, in systems of 3 to 25 gaseous species. Exponents p from 0.25 to 0.66 were used; values from 0 to 0.5 appear in the mass transfer expressions of this chapter. Comparisons were made against exact calculations of elements $k_{x\alpha\beta}$ and $N_{\alpha 0}$ via Eq. 22.9-19, and against a film model given by Krishna and Standart[11] in which each element $k_{x\alpha\beta}$ is calculated independently with the corresponding binary diffusivity $\mathscr{D}_{\alpha\beta}$. The calculations from Eqs. 22.9-24 and 25 were 3 to 5 times quicker than those with Eq. 22.9-19 and proved quite accurate (relative errors typically less than 1% and seldom as large as 10%), especially when done directly from the diagonally dominant Stefan-Maxwell matrix $[\mathbf{A}]$ rather than from its inverse, $[\mathbf{D}]$. Calculations with the Krishna-Standart film model were slower than those with Eqs. 22.9-24 and 25, and the typical errors were several times as large. Therefore, Eqs. 22.9-24 and 25 are recommended as practical approximations to the elements of the product matrix $[\mathbf{B}] = [\mathbf{P}][\check{\mathbf{k}}_x][\mathbf{P}]^{-1}$ in Eqs. 22.9-19 through 22 whenever Eq. 22.9-4 is used. This approximation may be used in Eq. 22.9-23 also, with $[\mathbf{B}]$ transformed at the end into mass-based units; however, Eq. 22.9-20 or 22 will be more convenient and comparably accurate at the moderate net molar fluxes normally encountered in heterogeneous catalysis.

The accuracy of the linearized solutions depends on the choice of the reference property values, especially when the property variations are large. In the following discussion all properties are evaluated at a common reference state, with composition given as a mole fraction

$$[x_{\text{ref}}] = a_x[x_b] + (1 - a_x)[x_0] \tag{22.9-26}$$

[9] W. E. Stewart, *AIChE Journal*, **9**, 528–535 (1963).

[10] V. Alopaeus and H. V. Nordén, *Computers & Chemical Engineering*, **23**, 1177–1182 (1999).

[11] R. Krishna and G. L. Standart, *AIChE Journal*, **22**, 383–389 (1976).

or a mass fraction

$$[\omega_{\text{ref}}] = a_\omega[\omega_b] + (1 - a_\omega)[\omega_0]$$
(22.9-27)

Note that $[x_{\text{ref}}]$ remains open to choice even for Eq. 22.9-20, 21, or 22, since the average compositions shown there provide net-flux corrections and not physical property values.

Equations 22.9-17, 18 and several other approximations for multicomponent mass transfer have been tested[12] against detailed variable-property integrations for isothermal systems. The conclusions from this study were as follows:

1. For twenty problems of unsteady-state gaseous diffusion, covering a wide range of net mass transfer rates, linearization in molar units approximated the exact solutions best. Rates of isobutane evaporation and condensation, for the system i-C_4H_{10}-N_2-H_2 in the geometry of Example 20.1-1, were approximated with a standard deviation of 1.6% by Eq. 22.9-17, using reference mole fractions calculated from Eq. 22.9-26 with $a_x = 0.5$. Linearization in mass-based units, via Eq. 22.9-18, proved inferior because of the large variations in ρ and $[\mathbf{A}_\omega]$. This method, with its preferred a_ω value of 0.8, gave a standard deviation of 3.8% for the interfacial fluxes $N_{\alpha 0}$ of the single transferable species (isobutane). Quasi-steady-state film approximations proved less accurate; use of correction factors $\theta_{x\alpha} = \phi_{x\alpha}/(\exp\phi_{x\alpha} - 1)$ (as given by Stewart and Prober[5] for the film model of §22.8) gave a standard deviation of 7.88% with a_x optimized to 1.0. The film model of Krishna and Standart,[11] which does not use linearization, gave a standard deviation of 14.3% independent of a_x and a_ω. These results favor the use of Eq. 22.9-17 (or, for moderate transfer rates, Eq. 22.9-21) with $a_x = 0.5$ for the gas phase in transfer operations described by a penetration model.

2. For twenty problems of momentum and mass transfer in laminar gaseous boundary layers of H_2, N_2 and CO_2 on a porous flat plate, solved accurately by Prober,[3] linearization in mass-based units approximated the exact solutions best. The detailed variable-property solutions for $n_{\alpha 0}$ were approximated[12] for all three species with a standard deviation of 0.55% by Eq. 22.9-18, using mass transfer coefficients $\check{k}_\alpha$ predicted via Eq. 20.2-47 and 22.9-27 with a_ω optimized to 0.4. The film models of Stewart and Prober[5] and of Standart and Krishna[12] gave standard deviations of 4.78% (with $a_x = 1.0$) and 8.25%, respectively, for the species transfer rates.

The methods presented here are coming into widespread use in the engineering of multicomponent separation processes. Advances in computing technology have facilitated the use of these methods and stimulated investigations toward better ones, to deal with nonlinear phenomena including complex chemical reactions.

QUESTIONS FOR DISCUSSION

1. Under what conditions can the analogies in Table 22.2-1 be applied? Can they be applied in systems with chemical reaction?
2. Why is the heat transfer coefficient in Eq. 22.1-6 defined differently from that in Eq. 14.1-1—or is it?
3. Some of the mass transfer coefficients in this chapter have a superscript 0 and others have a superscript •. Explain carefully what these superscripts denote.
4. What conclusions can you draw from the analytical calculations of mass transfer coefficients in §22.2?

---

[12] T. C. Young and W. E. Stewart, *Ind. Eng. Chem. Res.*, **25**, 476–482 (1986).

5. What is the significance of the 2 in Eqs. 22.3-20 and 21?
6. What is the meaning of the subscripts 0, e, and b in §22.4?
7. What is meant by the term "model insensitive"?
8. In what way does surface tension have an influence on interphase mass transfer? How is surface tension defined? How does surface tension depend on temperature?
9. Discuss the physical basis for the film model, the penetration model, and the boundary layer model for heat and mass transfer.
10. How are the heat and mass transfer coefficients affected by high mass-transfer rates across the interface?

PROBLEMS

22A.1. Prediction of mass transfer coefficients in closed channels. Estimate the gas-phase mass transfer coefficients for water vapor evaporating into air at 2 atm and 25°C, and a mass flow rate of 1570 lb_m/hr, in the systems that follow. Take $\mathscr{D}_{AB} = 0.130$ cm^2/s.

(a) A 6-in. i.d. vertical pipe with a falling film of water on the wall. Use the following correlation[1] for gases in a wetted-wall column:

$$Sh_{loc} = 0.023\ Re^{0.83}Sc^{0.44} \quad (Re > 2000) \tag{22A.1-1}$$

(b) a 6-in.-diameter packed bed of water-saturated spheres, with $a = 100$ ft^{-1}.

22A.2. Calculation of gas composition from psychrometric data. A stream of moist air has a wet-bulb temperature of 80°F and a dry-bulb temperature of 130°F, measured at 800 mm Hg total pressure and high air velocity. Compute the mole fraction of water vapor in the air stream. For simplicity, consider water as a trace component in estimating the film properties.

Answer: $x_{A\infty} = 0.0158$ (using n = 0.44 in Eq. 22.3-38)

22A.3. Calculating the inlet air temperature for drying in a fixed bed. A shallow bed of water-saturated granular solids is to be dried by blowing dry air through it at 1.1 atm pressure and a superficial velocity of 15 ft/s. What air temperature is required initially to keep the solids at a surface temperature of 60°F? Neglect radiation. See §14.5 for forced-convection heat transfer coefficients in fixed beds.

22A.4. Rate of drying of granular solids in a fixed bed. Calculate the initial rate of water removal in the drying operation described in Problem 22A.3, if the solids are cylinders with $a = 180$ ft^{-1}.

22B.1. Evaporation of a freely falling drop. A drop of water, 1.00 mm in diameter, is falling freely through dry, still air at pressure of 1 atm and a temperature of 100°F with no internal circulation. Assume quasi-steady-state behavior and a small mass-transfer rate to compute (a) the velocity of the falling drop, (b) the surface temperature of the drop, and (c) the rate of change of the drop diameter in cm/s. Assume that the film properties are those of dry air at 80°F.

Answers: (a) 390 cm/s; (b) 54°F; (c) 5.6×10^{-4} cm/s

22B.2. Effect of radiation on psychrometric measurements. Suppose that a wet-bulb and dry-bulb thermometer are installed in a long duct with constant inside surface temperature T_s and that the gas velocity is small. Then the dry-bulb temperature T_{db} and the wet-bulb temperature T_{wb} should be corrected for radiation effects. We assume, as in Example 22.3-2, that the thermometers are so installed that the heat conduction along the glass stems can be neglected.

(a) Make an energy balance on a unit area of the dry bulb to obtain an equation for the gas temperature T_∞ in terms of T_{db}, T_s, h_{db}, e_{db}, and a_{db} (these last two are the emissivity and absorptivity of the dry bulb).

---

[1] E. R. Gilliland and T. K. Sherwood, *Ind. Eng. Chem.*, **26**, 516–523 (1934).

(b) Make an energy balance on a unit area of the wet bulb and obtain an expression for the evaporation rate.

(c) Compute $x_{A\infty}$ for the pressure and thermometer readings of Example 22.3-2, with the additional information that $v_\infty = 15$ ft/s, $T_s = 130°F$, $e_{db} = a_{db} = e_{wb} = a_{wb} = 0.93$, dry-bulb diameter = 0.1 in., and wet-bulb diameter = 0.15 in. including the wick.

Answer: **(c)** $x_{A\infty} = 0.0021$

22B.3. **Film theory with variable transport properties.**

(a) Show that for systems in which the transport properties are functions of y, Eqs. 19.4-12 and 13 may be integrated to give for $y \leq \delta_x$ or $y \leq \delta_T$, respectively,

$$1 - \frac{(N_{A0} + N_{B0})(x_A - x_A)}{N_{A0} - x_{A0}(N_{A0} + N_B)} = \exp\left[(N_{A0} + N_{B0})\int_0^y \frac{dy}{c\mathcal{D}_{AB}}\right] \tag{22B.3-1}$$

$$1 - \frac{(N_{A0}\tilde{C}_{pA} + N_{B0}\tilde{C}_{pB})(T - T_0)}{q_0} = \exp\left[(N_{A0}\tilde{C}_{pA} + N_{B0}\tilde{C}_{pB})\int_0^y \frac{dy}{k}\right] \tag{22B.3-2}$$

(b) Make the corresponding changes in Eqs., 19.4-16 and 17 as well as in Eqs. 22.8-5 and 6. Then verify that Eqs. 22.8-7 and 8 remain valid. Thus it is not necessary to work with the integrals in calculating transfer rates if h_{loc} and $k_{x,loc}$ can be predicted.

(c) Show that h_{loc} and $k_{x,loc}$ have to be evaluated in terms of the physical properties and flow regime (laminar or turbulent) that prevail at the conditions for which h_{loc} and $k_{x,loc}$ are desired.

22B.4. **An evaporative ice maker.** Consider a circular shallow dish of water 0.5 m in diameter and filled to the brim, resting on an insulating layer, such as loose straw, and in a windless area. At what air temperature can the water be cooled to freezing if the relative humidity of the air is 30%? Make the following assumptions: (i) neglect radiation, (ii) consider radiation to a night sky of effective temperature 150K, and (iii) assume that the dish has a lip around the edge 2 mm high.

22B.5. **Oxygen stripping.** Calculate the rate at which oxygen transfers from quiescent oxygen-saturated water at 20°C to a bubble of pure nitrogen 1 mm in diameter, if the bubble acts as a rigid sphere. Note that it will first be necessary to determine the bubble velocity of rise through the water.

22B.6. **Controlling diffusional resistance.** Water drops 2 mm in diameter are being oxygenated by falling freely through pure oxygen at 20°C and a pressure of 1 atm. Do you need to know the gas-phase diffusivity to calculate the rate of oxygen transport? Why? The solubility of oxygen under these conditions is 1.39 mmols/liter, and its diffusivity in the liquid phase is about 2.1×10^{-5} cm^2/s.

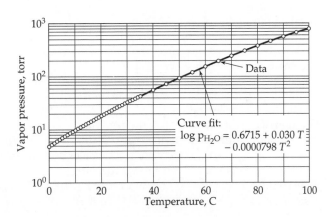

Fig. 22B.7. Water vapor pressure under its own vapor data from *Lange's Handbook of Chemistry* (J. Dean, ed.), 15th edition, McGraw-Hill, New York (1999).

22B.7. **Determination of diffusivity** (Fig. 22B.7). The diffusivity of water vapor in nitrogen is to be determined at a pressure of 1 atm over the temperature range from 0°C to 100°C by means of the "Arnold experiment" of Example 20.1-1. It will, therefore, be necessary to use the correction factor θ_{AB} to the penetration model. Calculate this factor as a function of temperature. The vapor pressure of water in this range may be obtained from Fig. 22B.7 or calculated from

$$\log_{10} p_{H_2O} = 0.6715 + 0.030T - 0.00008T^2 \qquad (22B.7\text{-}1)$$

where p_{H_2O} is the vapor pressure in mm Hg, and T is the temperature in degrees centigrade.

22B.8. **Marangoni effects in condensation of vapors.** In many situations the heat transfer coefficient for condensing vapors is given as $h = k/\delta$, where k is the thermal conductivity of the condensate film, and δ is the film thickness. Correlations available in the literature are normally based on the assumption of zero shear stress at the free surface of the film, but if the surface temperature decreases downward, there will be a shear stress $\tau_s = \partial\sigma/\partial z$, where σ is the surface tension, and z is measured downward, that is, in the direction of flow. How much will this effect change a heat transfer coefficient of 5000 kcal/hr · m² · C for a water film? The kinematic viscosity of water may be assumed to be 0.0029 cm²/s, the density is 0.96 g/cm³, the thermal conductivity 0.713 kcal/hr · m · C, and $d\sigma/dT = -0.2$ dynes/cm · C for the purposes of this problem.

Partial Answer: $\rho\langle v_z\rangle = \left(\dfrac{\rho^2 g\delta^2}{3\mu}\right)\left(1 + \dfrac{3}{2}\dfrac{\tau_s}{\rho g\delta}\right)$

The term in τ_s represents the effect of surface tension gradients, and when this term is small, its denominator will be near the value for no gradient. For the conditions of this problem, $\rho g\delta = 14.3$ dyn/cm². Surface tension effects will thus be small for systems such as the one under consideration, where the surface tension increases downward. In the opposite case, however, even small gradients can cause hydrodynamic instabilities and thus can have major effects.

22B.9. **Film model for spheres.** Derive the results that correspond to Eqs. 22.8-3, 4 for simultaneous heat and mass transfer in a system with spherical symmetry. That is, assume a spherical mass transfer surface and assume that T and x_A depend only on the radial coordinate r. Show that Eqs. 22.8-7 and 8 do not need to be changed. What difficulties would be encountered if one tried to use the film theory to calculate the drag on a sphere?

22B.10. **Film model for cylinders.** Derive the results that correspond to Eqs. 22.8-3, 4 for a system with cylindrical symmetry. That is, assume a cylindrical mass transfer surface and assume that T and x_A depend only on r. Verify that Eqs. 22.8-7, 8 do not need to be changed.

22C.1. **Calculation of ultrafiltration rates.** Check the accuracy of the predictions shown in Fig. 22.8-9 for the following data and physical properties:

> Physical system:
>> Rotation rate of disk filter = 273 rpm
>> Bovine serum albumin at $\rho_P = 2.2$ g/100 ml
>> Diffusivity in phosphate buffer (at pH 6.7) = 7.1×10^{-7} cm²/s
>> Kinematic viscosity of buffer = 0.01 cm²/s
>> Partial specific volumes of protein and buffer are 0.75 and 1.00 ml/g, respectively
>> Hydraulic permeability, $K_H = 0.0098$ cm/min · psi
>
> Effect of protein concentration:
>> Solution density $\rho = 0.997 + 0.224\rho_P$ in g/ml
>> Protein-buffer diffusivity ratio $\mathcal{D}_{PS}(0)/\mathcal{D}_{PS}(\rho_P) = 21.3\phi_P/\tanh(21.3\phi_P)$, where $\phi_P = \omega_P\hat{V}_P/(\omega_P\hat{V}_P + \omega_S\hat{V}_S)$ is the volume fraction of protein, with $\hat{V}_P$ and $\hat{V}_S$ being the partial specific volumes of protein and solvent
>> Protein-buffer viscosity ratio $\mu(0)/\mu(\rho_P) = 1.11 - 0.054\rho_P + 0.00067\rho_P^2$, with ρ_P in g/100 ml
>> Osmotic pressure $\pi = 0.013\rho_P^2$ in psi (100 ml/g)²

The operating data are as follows:

Transmembrane pressure difference, $(p_0 - p_\delta)$, psi	Percolation velocity v_δ, cm/min
4.0	0.032
7.1	0.049
8.4	0.049
13.6	0.061
14.7	0.066
23.9	0.074
29.7	0.078
30.0	0.079
36.4	0.081
47.0	0.082

Chapter 23

Macroscopic Balances for Multicomponent Systems

Applications of the laws of the conservation of mass, momentum, and energy to engineering flow systems have been discussed in Chapter 7 (isothermal systems) and Chapter 15 (nonisothermal systems). In this chapter we continue the discussion by introducing three additional factors not encountered in the earlier chapters: (a) the fluid in the system is composed of more than one chemical species; (b) chemical reactions may be occurring, along with changes of composition and production or consumption of heat; and (c) mass may be entering the system through the bounding surfaces (that is, across surfaces other than planes 1 and 2). Various mechanisms by which mass may enter or leave through the bounding surfaces of the system are shown in Fig. 23.0-1.

Fig. 23.0-1. Ways in which mass may enter or leave through boundary surfaces: (*a*) benzoic acid enters system by dissolution of the wall; (*b*) water vapor enters the system, defined as the gas phase, by evaporation, and ammonia vapor leaves by absorption; (*c*) oxygen enters the system by transpiration through a porous wall.

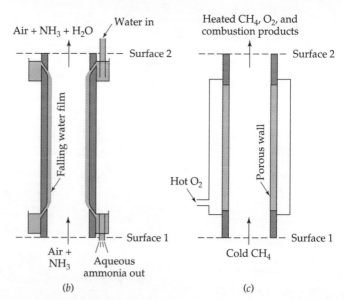

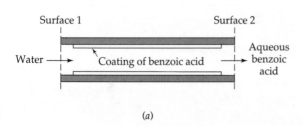

(*a*)

(*b*)

(*c*)

In this chapter we summarize the macroscopic balances for the more general situation described above. Each of these balances will now contain one extra term, to account for mass, momentum, or energy transport across the bounding surfaces. The balances thus obtained are capable of describing industrial mass transfer processes, such as absorption, extraction, ion exchange, and selective adsorption. Inasmuch as entire treatises have been devoted to these topics, all we try to do here is to show how the material discussed in the preceding chapters paves the way for the study of mass transfer operations. The reader interested in pursuing these topics further should consult the available textbooks and treatises.[1-8]

The main emphasis on this chapter is on the mass balances for mixtures. For that reason, §23.1 is accompanied by five examples, which illustrate problems arising in environmental science, isotope separation, economic evaluation, and biomedical science. In §§23.2 to 23.4 the other macroscopic balances are given. In Table 23.5-1 they are summarized for systems with multiple inlets and outlets. The last two sections of the chapter illustrate applications of the macroscopic balances to more complex systems.

§23.1 THE MACROSCOPIC MASS BALANCES

The statement of the law of conservation of mass of chemical species α in a multicomponent macroscopic flow system is

$$\frac{dm_{\alpha,\text{tot}}}{dt} = -\Delta w_\alpha + w_{\alpha,0} + r_{\alpha,\text{tot}} \qquad \alpha = 1, 2, 3, \ldots, N \tag{23.1-1}$$

This is a generalization of Eq. 7.1-2. Here $m_{\alpha,\text{tot}}$ is the instantaneous total mass of α in the system, and $-\Delta w_\alpha = w_{\alpha 1} - w_{\alpha 2} = \rho_{\alpha 1}\langle v_1 \rangle S_1 - \rho_{\alpha 2}\langle v_2 \rangle S_2$ is the difference between the mass rates of flow of species α across planes 1 and 2. The quantity $w_{\alpha,0}$ is the mass rate of addition of species α to the system by mass transfer across the bounding surface. Note that $w_{\alpha,0}$ is positive when mass is *added* to the system, just as Q and W_m are taken to be positive in the total energy balance when heat is added to the system and work is done on the system by moving parts. Finally, the symbol $r_{\alpha,\text{tot}}$ stands for the net rate of production of species α by homogeneous and heterogeneous reactions within the system.[1]

Recall that in Table 15.5-1 the molecular and eddy transport of momentum and energy across surfaces 1 and 2 in the direction of flow were neglected with respect to the convective transport. The same is done everywhere in this chapter—in Eq. 23.1-1 and in the other macroscopic balances presented here.

---

[1] W. L. McCabe, J. C. Smith, and P. Harriot, *Unit Operations of Chemical Engineering*, McGraw-Hill, New York, 6th edition (2000).

[2] T. K. Sherwood, R. L. Pigford, and C. R. Wilke, *Mass Transfer*, McGraw-Hill, New York (1975).

[3] R. E. Treybal, *Mass Transfer Operations*, 3rd edition, McGraw-Hill, New York (1980).

[4] C. J. King, *Separation Processes*, McGraw-Hill, New York (1971).

[5] C. D. Holland, *Multicomponent Distillation*, McGraw-Hill, New York (1963).

[6] T. C. Lo, M. H. I. Baird, and C. Hanson, eds., *Handbook of Solvent Extraction*, Wiley-Interscience, New York (1983).

[7] R. T. Yang, *Gas Separations by Adsorption Processes*, Butterworth, Boston (1987).

[8] J. D. Seader and E. J. Henley, *Separation Process Principles*, Wiley, New York (1998).

[1] The quantities $m_{\alpha,\text{tot}}$, $w_{\alpha,0}$, and $r_{\alpha,\text{tot}}$ may be expressed as integrals:

$$m_{\alpha,\text{tot}} = \int_V \rho_\alpha dV; \qquad w_{\alpha,0} = -\int_{S_0} (\mathbf{n} \cdot \rho_\alpha \mathbf{v}_\alpha) dS; \qquad r_{\alpha,\text{tot}} = \int_V r_\alpha dV + \int_{S_0} r_\alpha^{(s)} dS \tag{23.1-1a, b, c}$$

in which $\mathbf{n}$ is the outwardly directed unit normal vector, and S_0 is that portion of the bounding surface on which mass transfer occurs. The integrands in $r_{\alpha,\text{tot}}$ are the net rates of production of species α by homogeneous and heterogeneous reactions, respectively.

If all N equations in Eq. 23.1-1 are summed, we get

$$\frac{dm_{tot}}{dt} = -\Delta w + w_0 \tag{23.1-2}$$

in which $w_0 = \Sigma_\alpha w_{\alpha,0}$, and use has been made of the law of conservation of mass in the form $\Sigma_\alpha r_{\alpha,tot} = 0$.

It is often convenient to write Eq. 23.1-1 in molar units:

$$\boxed{\frac{dM_{\alpha,tot}}{dt} = -\Delta W_\alpha + W_{\alpha,0} + R_{\alpha,tot} \qquad \alpha = 1, 2, 3, \ldots, N} \tag{23.1-3}$$

Here the capital letters represent the molar counterparts of the lowercase symbols in Eq. 23.1-1. When Eq. 23.1-3 is summed over all species, the result is

$$\frac{dM_{tot}}{dt} = -\Delta W + W_0 + \sum_{\alpha=1}^{N} R_{\alpha,tot} \tag{23.1-4}$$

Note that the last term is not in general zero, because moles are produced or consumed in many reaction systems.

In some applications, such as spatially continuous mass transfer operations, it is customary to rewrite Eq. 23.1-1 or 3 for a differential element of the system (that is, in the "d-form" discussed in §15.4). Then the differentials $dw_{\alpha,0}$ or $dW_{\alpha,0}$ can be expressed in terms of local mass transfer coefficients.

EXAMPLE 23.1-1

Disposal of an Unstable Waste Product

A fluid stream emerges from a chemical plant with a constant mass flow rate w and discharges into a river (Fig. 23.1-1a). It contains a waste material A at mass fraction ω_{A0}, which is unstable and decomposes at a rate proportional to its concentration according to the expression $r_A = -k_1''' \rho_A$—that is, by a first-order reaction.

To reduce pollution it is decided to allow the effluent stream to pass through a holding tank of volume V, before discharging into the river (Fig. 23.1-1b). The tank is equipped with

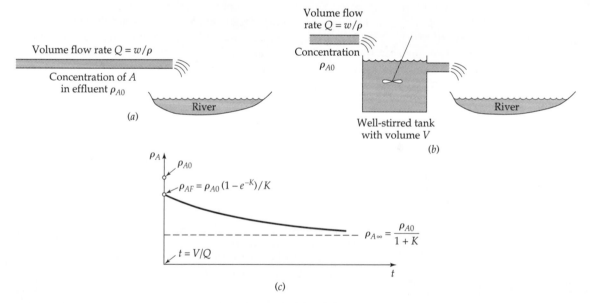

(c)

Fig. 23.1-1. (a) Waste stream with unstable pollutant emptying directly into a river. (b) Waste stream with holding tank that allows the unstable pollutant to decay prior to going into the river. (c) Sketch showing the concentration of pollutant being discharged into the river after the holding tank has been filled (the dimensionless quantity K is $k_1''' V/Q$).

an efficient stirrer that keeps the fluid in the tank at very nearly uniform composition. At time $t = 0$ the fluid begins to flow into the empty tank. No liquid flows out until the tank has been filled up to the volume V.

Develop an expression for the concentration of the fluid in the tank as a function of time, both during the tank-filling process and after the tank has been completely filled.

SOLUTION

(a) We begin by considering the period during which the tank is being filled—that is the period $t \leq \rho V/w$, where ρ is the density of the fluid mixture. We apply the macroscopic mass balance of Eq. 23.1-1 to the holding tank. The quantity $m_{A,tot}$ on the left side is $wt\omega_A$ at time t. The mass rate of flow entering the tank is $w\omega_{A0}$, and there is no outflow during the tank-filling stage. No A is entering or leaving through a mass transfer interface. The mass rate of production of species A is $r_{A,tot} = (wt/\rho)(-k_1'''\rho_A) = -k_1'''m_{A,tot}$. Therefore the macroscopic mass balance for species A during the filling period is

$$\frac{d}{dt} m_{A,tot} = w\omega_{A0} - k_1''' m_{A,tot} \tag{23.1-5}$$

This first-order differential equation can be solved with the initial condition that $m_{A,tot} = 0$ at $t = 0$ to give

$$m_{A,tot} = \frac{w\omega_{A0}}{k_1'''}(1 - \exp(-k_1''' t)) \tag{23.1-6}$$

This may be written in terms of the instantaneous mass fraction of A in the tank by using the relation $m_{A,tot} = wt\omega_A$:

$$\frac{\omega_A}{\omega_{A0}} = \frac{1 - \exp(-k_1''' t)}{k_1''' t} \qquad \left(t \leq \frac{\rho V}{w}\right) \tag{23.1-7}$$

The mass fraction of A at the instant when the tank is full, ω_{AF}, is then given by

$$\frac{\omega_{AF}}{\omega_{A0}} = \frac{1 - e^{-K}}{K} \tag{23.1-8}$$

in which $K = k_1'''\rho V/w = k_1'''V/Q$.

(b) The mass balance on the tank after it has been filled is

$$\frac{d}{dt}(\rho_A V) = w\omega_{A0} - w\omega_A - k_1'''\rho_A V \tag{23.1-9}$$

or, in dimensionless form, with $\tau = (w/\rho V)t$,

$$\frac{d\omega_A}{d\tau} + (1 + K)\omega_A = \omega_{A0} \tag{23.1-10}$$

This first-order differential equation can be solved with the initial condition that $\omega_A = \omega_{AF}$ at $\tau = 1$ to give

$$\frac{\omega_A - [\omega_{A0}/(1 + K)]}{\omega_{AF} - [\omega_{A0}/(1 + K)]} = e^{-(1+K)(\tau-1)} \qquad \left(t \geq \frac{\rho V}{w}\right) \tag{23.1-11}$$

This shows that as time progresses the mass fraction of the pollutant being discharged into the river decreases exponentially, with a limiting value of

$$\omega_{A\infty} = \frac{\omega_{A0}}{1 + K} = \frac{\omega_{A0}}{1 + (k_1'''\rho V/w)} \tag{23.1-12}$$

The curve for the mass concentration as a function of time after the filling of the tank is shown in Fig. 23.1-1(c). This curve can be used to determine conditions such that the effluent concentration will be in the permitted range. Equation 23.1-12 can be used to decide on the size of holding tank that is required.

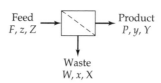

Feed $\longrightarrow$ Product   **Fig. 23.1-2.** Binary splitter, in which a feed stream is split into
F, z, Z P, y, Y a product stream and a waste stream.

Waste
W, x, X

EXAMPLE 23.1-2

Binary Splitters

Describe the operation of a binary splitter, one of the commonest and simplest separation devices (see Fig. 23.1-2). Here a binary mixture of A and B enters the apparatus in a feed stream at a molar rate F, and by some separation mechanism it is split into a product stream with a molar rate P and a waste stream with molar rate W. The mole fraction of A (the desired component) in the feed stream is z, and the mole fractions in the product and waste streams are y and x, respectively.

SOLUTION

We start by writing the steady-state macroscopic mass balances for component A and for the entire fluid as

$$zF = yP + xW \tag{23.1-13}$$
$$F = P + W \tag{23.1-14}$$

It is customary to define the ratio $\theta = P/F$ of the molar rates of the product and feed streams as the *cut*. Equation 23.1-13 then becomes, after eliminating W by use of Eq. 23.1-14,

$$z = \theta y + (1 - \theta)x \tag{23.1-15}$$

Normally the cut θ and the feed composition z are taken to be known.

We now need a relation between the feed and waste compositions, and it is conventional to write an equation relating the compositions of the two outgoing streams:

$$Y = \alpha X \tag{23.1-16}$$

Here α is known as the *separation factor*, also usually taken as known, and which characterizes the separation capability of the splitter. Here Y and X are the mole ratios defined by

$$X = \frac{y}{1 - y} \quad \text{and} \quad X = \frac{x}{1 - x} \tag{23.1-17, 18}$$

In terms of the mole fractions, Eq. 23.1-16 may be written as

$$y = \frac{\alpha x}{1 + (\alpha - 1)x} \quad \text{or} \quad x = \frac{y}{\alpha - (\alpha - 1)y} \tag{23.1-19, 20}$$

Equations 23.1-15 and 19 (or 20) describe completely the splitter operation.

For vapor–liquid splitting—that is, equilibrium distillation—it is typical to define the *ideal* splitter in terms of an operation in which the product and waste streams are in equilibrium. For this situation, α is the *relative volatility*, and for thermodynamically ideal systems, it is just the ratio of the component vapor pressures. Even for nonideal systems, α changes relatively slowly with composition.

For *real* splitters one can then define α in terms of an empirical correction factor—for example, the *efficiency*—defined by

$$\alpha = E\alpha^* \tag{23.1-21}$$

where α^* is the separation factor for the ideal model, and E is a correction factor that accounts for the failure of the actual system to meet the ideal behavior.

We thus find that, for a given feed composition, the *enrichment* $(y - z)/z$ produced by the splitter is a function of the cut θ and the separation factor α. The enrichment can be calculated from the following equation, which is obtained by combining Eqs. 23.1-15 and 20:

$$z = \theta y + (1 - \theta) \frac{y}{\alpha - (\alpha - 1)y} \tag{23.1-22}$$

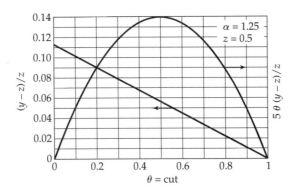

Fig. 23.1-3. Behavior of a binary splitter.

This is a quadratic equation for y that can be solved when z is given, and then the enrichment $(y - z)/z$ is obtained. An example is given in Fig. 23.1-3 where both $(y - z)/z$ and $5\theta(y - z)/z$ are plotted as functions of θ for $z = \frac{1}{2}$ and $\alpha = 1.25$ (a reasonable value for many processes). It may be seen that, whereas the maximum enrichment $(y - z)/z$ is obtained for vanishingly small cuts, the product of enrichment and product rate is greatest at an intermediate θ value. Finding an optimum θ value is a problem that must be addressed on economic grounds.

Simple splitters of the general type pictured in Fig. 23.1-2 are very widely used as building blocks in multistage separation processes. These include evaporators and crystallizers, which typically have a very high separation factor α per stage, and systems for distillation, gas absorption, and liquid extraction, where α can vary widely. All of these applications are well covered in standard texts on unit operations.

Membrane processes are rapidly increasing in importance, and many of the design principles were developed for the isotope fractionation industry.[2] Discussions of modern applications are also available.[3]

EXAMPLE 23.1-3

The Macroscopic Balances and Dirac's "Separative Capacity" and "Value Function"

During the Manhattan Project of World War II, the British physicist Dirac[2,4,5] used the macroscopic mass balances for a binary splitter to develop a criterion for comparing the effectiveness of different separation processes—for example, thermal diffusion and centrifugation. The same criterion has also proven useful in the evaluation of bioseparations.

We imagine the simple separation system shown in Fig. 23.1-2 in which F is the molar rate of flow of the feed stream, which contains a binary mixture of A and B, and P and W are the molar rates of flow of the product and waste streams. The mole fractions of species A in the three streams are z, y, and x, respectively.

In the system there is some mechanism (for example, a membrane) for increasing the concentration of A in the product stream and decreasing it in the waste stream. We then may define a *separation factor* α as in Eqs. 23.1-16 to 18

$$\alpha = \frac{y/(1 - y)}{x/(1 - x)} = 1 + \frac{y - x}{x(1 - y)} \tag{23.1-23}$$

---

[2] E. Von Halle and J. Schacter, *Diffusion Separation Methods*, in Volume 8 of *Kirk-Othmer Encyclopedia of Chemical Technology* (M. Howe-Grant, ed.), 4th edition, Wiley, New York (1993), pp. 149–203.

[3] W. S. W. Ho and K. K. Sirkar, *Membrane Handbook*, Van Nostrand Reinhold, New York (1992), p. 954; R. D. Noble and S. A. Stern, *Membrane Separations Technology*, Elsevier, Amsterdam (1995), p. 718.

[4] P. A. M. Dirac, British Ministry of Supply (1941); this is reprinted in *The Collected Works of P. A. M. Dirac (1924–1948)*, (R. H. Dalitz, ed.) Cambridge University Press (1995). Nobel Laureate **Paul Adrien Maurice Dirac** (1902–1984), one of the leaders in the development of quantum mechanics, developed the relativistic wave equation and predicted the existence of the positron.

[5] K. Cohen, *Theory of Isotope Separation*, McGraw-Hill, New York (1951).

We have written this in a second form, because we will consider only systems in which there is only a slight enrichment of species A, so that $\alpha - 1$ is a very small quantity. When Eq. 23.1-23 is solved for y as a function of x we then get

$$y = \frac{x + (\alpha - 1)x}{1 + (\alpha - 1)x} \approx x + (\alpha - 1)x(1 - x) \tag{23.1-24}$$

Next we define the *Dirac separative capacity* Δ of the system as the net increase in "value" (this could, for example, be the monetary value) of the streams that are participating in the system:

$$\Delta = Pv(y) + Wv(x) - Fv(z) \tag{23.1-25}$$

in which $v(x)$ is the *Dirac value function*. (In the separation science literature, the separative capacity is often given the symbol δU.)

Show how the separative capacity and value function can be obtained by using the definition in Eq. 23.1-25 along with the mass balances for the system.

SOLUTION   The total mass balance and the mass balance for species A are:

$$F = P + W; \qquad Fz = Py + Wx \tag{23.1-26, 27}$$

We now divide Eq. 23.1-27 by F, and then use Eq. 23.1-26 to eliminate W. Then introducing the quantity $\theta = P/F$ (called the "cut"), we get

$$z = \theta y + (1 - \theta)x \quad \text{or} \quad z - x = \theta(y - x) \tag{23.1-28}$$

Next we divide Eq. 23.1-25 by F and introduce θ to get

$$\frac{\Delta}{F} = \theta v(y) + (1 - \theta)v(x) - v(z) \tag{23.1-29}$$

Inasmuch as the differences between the concentrations of the streams are quite small, we can expand $v(y)$ and $v(x)$ about z and get

$$v(x) = v(z) + (x - z)v'(z) + \tfrac{1}{2}(x - z)^2 v''(z) + \cdots \tag{23.1-30}$$
$$v(y) = v(z) + (y - z)v'(z) + \tfrac{1}{2}(y - z)^2 v''(z) + \cdots \tag{23.1-31}$$

where the primes indicate differentiation with respect to z. When these expressions are put into Eq. 23.1-29 and we use Eq. 23.1-28, we get

$$\frac{\Delta}{F} = \tfrac{1}{2}\theta(1 - \theta)(y - z)^2 v''(z) \tag{23.1-32}$$

When we use Eq. 23.1-24, this last equation becomes

$$\frac{\Delta}{F} = \tfrac{1}{2}\theta(1 - \theta)(\alpha - 1)z^2(1 - z)^2 v''(z) \tag{23.1-33}$$

We now assume that the separative capacity of the system is virtually independent of concentration. Therefore we set the concentration-dependent factor in Eq. 23.1-33 equal to unity, so that

$$\frac{\Delta}{F} = \tfrac{1}{2}\theta(1 - \theta)(\alpha - 1) \tag{23.1-34}$$

is the final expression for the separative capacity. According to this expression, the separative capacity has a maximum when the system is operated at $\theta = \tfrac{1}{2}$.

It remains to obtain the Dirac value function, which must satisfy the differential equation

$$\frac{d^2 v}{dz^2} = \frac{1}{z^2(1 - z)^2} \tag{23.1-35}$$

When this equation is integrated, we get

$$v = (2z - 1) \ln\left(\frac{z}{1-z}\right) - 2 + C_1 z + C_2 \tag{23.1-36}$$

The two integration constants may be assigned arbitrarily, and several different choices have been used. However, the most common choice is $v(\frac{1}{2}) = 0$ and $v'(\frac{1}{2}) = 0$. This leads to

$$v = (2z - 1) \ln\left(\frac{z}{1-z}\right) \tag{23.1-37}$$

which is the symmetrical solution, in the sense that $v(1 - z) = v(z)$ and $v'(1 - z) = -v'(z)$.

The value function $v(z)$ and the separative capacity Δ have proven useful in comparing separations made in different kinds of equipment as well as different concentration ranges. From an economic standpoint $v(z)$ as given by Eq. 23.1-37 has been found useful for determining price differences for isotope mixtures of differing purity.

EXAMPLE 23.1-4

Compartmental Analysis

One of the simplest and most useful applications of the species macroscopic mass balance is *compartmental analysis*, in which a complex system is treated as a network of perfect mixers, each of constant volume, connected by ducts of negligible volume, with no dispersion occurring in the connecting ducts. Imagine mixing units, labeled $1, 2, 3, \ldots, n, \ldots, N$, containing various species (labeled with indices $\alpha, \beta, \gamma, \ldots$). Then the mass concentration $\rho_{\alpha n}$ of species α in unit n changes with time according to the equation

$$V_n \frac{d\rho_{\alpha n}}{dt} = \sum_{m=1}^{N} Q_{mn}(\rho_{\alpha m} - \rho_{\alpha n}) + V_n r_{\alpha n} \tag{23.1-38}$$

Here V_n is the volume of unit n, Q_{mn} is the volumetric flow rate of solvent flow from unit m to unit n, and $r_{\alpha n}$ is the rate of formation of species α per unit volume in unit n.

Show how such a model can be specialized to describe the removal of toxic metabolic products (that is, the toxic materials resulting from the human metabolism) from a patient by *hemodialysis*. Hemodialysis is the periodic removal of toxic metabolites achieved by contacting the blood and a dialysis fluid in countercurrent flow, separated by a cellophane membrane that is permeable to the metabolite.

SOLUTION

The simple two-compartment model of Fig. 23.1-4 has been found to be adequate for representing the hemodialysis system. Here the large block, or compartment 1 (labeled "body") represents the combined body fluids, except for those in the blood, which are represented by compartment 2. The blood circulates via a branching system of vessels through compartment 1 at a volumetric rate Q, and in the process extracts solute across the vessel walls. This process is highly efficient, and a single solute is assumed to leave compartment 1 at concentration ρ_1, equal to the concentration throughout that compartment. At the same time, the solute is being formed within the body fluids at a constant rate G, and during dialysis it is being extracted from the blood by the dialyzer at a rate $D\rho_2$. The proportionality constant D is known as the "dialyzer clearance" and is fixed by the dialyzer design and operating conditions.

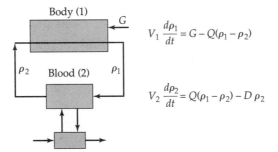

$$V_1 \frac{d\rho_1}{dt} = G - Q(\rho_1 - \rho_2)$$

$$V_2 \frac{d\rho_2}{dt} = Q(\rho_1 - \rho_2) - D\rho_2$$

Fig. 23.1-4. Two-compartment model used to analyze the functioning of a dialyzer.

The very complex process actually taking place is modeled by the two equations

$$V_1 \frac{d\rho_1}{dt} = -Q(\rho_1 - \rho_2) + G \tag{23.1-39}$$

$$V_2 \frac{d\rho_2}{dt} = Q(\rho_1 - \rho_2) - D\rho_2 \tag{23.1-40}$$

with $D = 0$ between the dialysis periods. Because we are considering a single solute, the concentrations have only one subscript to indicate the compartment. We measure the time t from the start of a dialysis procedure, when the blood and body fluids are very nearly in equilibrium with each other, so that we may write the initial conditions as

I. C.:   at $t = 0$, $\qquad\qquad\qquad \rho_1 = \rho_2 = \rho_0$ $\hfill$ (23.1-41)

where ρ_0 is a constant. We now want to get an explicit expression for the toxic metabolite concentration in the blood as a function of time.

We start by adding Eqs. 23.1-39 and 40 and solving for $d\rho_1/dt$. The latter is then substituted into the time derivative of Eq. 23.1-40 to obtain a differential equation for the metabolite concentration in the blood:

$$\frac{d^2\rho_2}{dt^2} + \left(\frac{Q}{V_1} + \frac{Q}{V_2} + \frac{D}{V_2}\right)\frac{d\rho_2}{dt} + \frac{QD}{V_1 V_2}\rho_2 = \frac{QG}{V_1 V_2} \tag{23.1-42}$$

with

I. C.:   at $t = 0$, $\qquad\qquad \rho_2 = \rho_0 \quad\text{and}\quad \dfrac{d\rho_2}{dt} = -\dfrac{D\rho_0}{V_2}$ $\hfill$ (23.1-43)

The second initial condition is obtained by use of Eqs. 23.1-40 and 41.

This equation is now to be solved with the following specific parameter values, which are typical for the removal of creatinine from a 70-kg adult human:

Quantity	V_1 (liters)	V_2 (liters)	Q (liters per min)	D (liters per min)	G (g/min)	ρ_0 (g per liter)
Magnitude	43	4.5	5.4	0.3	0.0024	0.140

The differential equation and initial conditions now take the form:

$$\frac{d^2\rho_2}{dt^2} + (1.3922)\frac{d\rho_2}{dt} + (0.00837)\rho_2 = 6.70 \times 10^{-5} \tag{23.1-44}$$

I. C.:   at time $t = 0$, $\qquad \rho_2 = \rho_0 \quad\text{and}\quad \dfrac{d\rho_2}{dt} = -0.00933$ $\hfill$ (23.1-45)

in which concentration is in grams per liter and time is in minutes. The complementary function that satisfies the associated homogeneous equation is

$$\rho_{2,cf} = C_1 \exp(0.006043t) + C_2 \exp(1.386t) \tag{23.1-46}$$

and the particular integral is

$$\rho_{2,pi} = 0.0080 \tag{23.1-47}$$

The complete solution to the nonhomogeneous equation is given by the sum of the complementary function and the particular integral. When the constants of integration are determined from the initial conditions, we get

$$\rho_2 = 0.1258 \exp(0.006043t) + (0.0062 \exp(1.386t) + 0.0080 \tag{23.1-48}$$

$$\frac{d\rho_2}{dt} = -0.000760 \exp(0.006043t) - 0.0086 \exp(1.386t) \tag{23.1-49}$$

during the dialysis period.

For the recovery period following dialysis, we assume here that the patient has no kidney function, so the clearance D is zero. Equation 23.1-42 takes the simpler form

$$\frac{d^2\rho_2'}{dt^2} + Q\left(\frac{V_1 + V_2}{V_1 V_2}\right)\frac{d\rho_2'}{dt} = \frac{QG}{V_1 V_2} \tag{23.1-50}$$

where ρ' is the concentration during the recovery period. The complementary function and particular integral are

$$\rho_{2,cf}' = C_3 \exp\left[-Q\left(\frac{V_1 + V_2}{V_1 V_2}\right)t'\right] + C_4 \tag{23.1-51}$$

$$\rho_{2,pi}' = \frac{Gt'}{V_1 + V_2} \tag{23.1-52}$$

in which t' is the time measured from the start of the recovery period. Inserting the numerical values, we then get for the concentration during the recovery period and its time derivative

$$\rho_2' = C_3 \exp(-1.325t') + (5.05 \times 10^{-5})t' + C_4 \tag{23.1-53}$$

$$\frac{d\rho_2'}{dt'} = -1.325 C_3 \exp(-1.325t') + (5.05 \times 10^{-5}) \tag{23.1-54}$$

The integration constants are to be determined from the matching conditions at $t' = 0$,

at $t' = 0$, $\qquad\qquad\qquad \rho_2' = \rho_2 \quad \text{and} \quad \rho_1' = \rho_1 \tag{23.1-55, 56}$

We need a second initial condition for determining the integration constants in Eq. 23.1-53. This can be obtained from Eq. 23.1-40 and the corresponding equation for ρ_2' (i.e., with $D = 0$), combined with the two relations in Eqs. 23.1-55 and 56. This relation is

at $t' = 0$, $\qquad\qquad\qquad \dfrac{d\rho_2'}{dt} = \dfrac{d\rho_2}{dt} + \dfrac{D\rho_2}{V_2} \tag{23.1-57}$

For illustrative purposes, we shall end the dialysis at 50 min, for which

$$\rho_2(t = 50) = 0.099239 = \rho_2' \tag{23.1-58}$$

We now have enough information to determine the constants of integration, and therefore we get for the concentration in the blood during the recovery period

$$\rho_2' = 0.0972 - 0.00422 \exp(-1.325t') + (5.05 \times 10^{-5})t' \tag{23.1-59}$$

Equations 23.1-48 and 59 are plotted in Fig. 23.1-5.

Of perhaps more interest is Fig. 23.1-6, which shows the application of Eqs. 23.1-39 and 40 to an actual patient. Here the points represent data and the lines are the model predictions. Here only the dialyzer clearance and the creatinine concentrations are known, and the data of

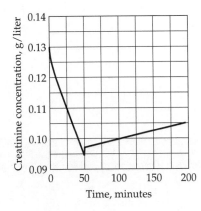

Fig. 23.1-5. Pharmacokinetics of dialysis: model prediction.

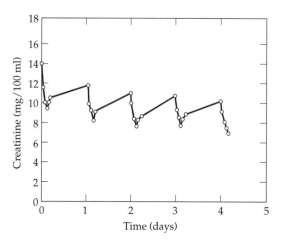

Fig. 23.1-6. Experimental (dots) and simulated creatinine data (solid curve) for a dialysis patient [R. L. Bell, K. Curtiss, and A. L. Babb, *Trans. Amer. Soc. Artificial Internal Organs*, **11**, 183 (1965)].

the first cycle are used to estimate the remaining parameters. The resulting model is then used to predict the next three cycles. We see that this approach does an excellent job of correlating the data and has predictive value. Note that the sudden rise in creatinine concentration at 50 min results from the fact that the dialyzer is no longer removing it from the blood. As a result, the disequilibrium between the blood and the rest of the body then becomes smaller.

Similar compartmental models have wide application in medicine, where they are referred to as *pharmacokinetic models*.[6] A priori pharmacokinetic modeling, where model parameters are determined separately from the process being modeled, was pioneered by Bischoff and Dedrick.[7]

EXAMPLE 23.1-5

Time Constants and Model Insensitivity

In the foregoing example it is clear, even on casual inspection, that neither the body fluids nor the circulating blood have much in common with ideal mixing tanks, and it is therefore of some interest to examine the success of the simple compartmental model critically. To make a start in that direction, compare the response (that is, the output concentration) of two quite different systems in Fig. 23.1-7 to an exponentially decaying solute input: one in which the en-

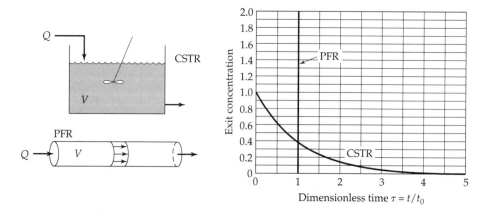

Fig. 23.1-7. Responses of the PFR and the CSTR to a pulse input.

---

[6] P. G. Welling, *Pharmacokinetics*, American Chemical Society (1997).

[7] K. B. Bischoff and R. L. Dedrick, *J. Pharm. Sci.*, **87**, 1347–1357 (1968); *AIChE Symposium Series*, **64**, 32–44 (1968).

tering fluid moves through in plug flow (*a plug flow reactor*, PFR), and another that acts as a perfect mixer (or *continuous stirred tank reactor*, CSTR). As shown in Fig. 23.1-7, the responses to a pulse input are quite different for the PFR and the CSTR. Assume steady flow at a volumetric flow rate Q through each system, and further assume that the tracer being followed is too dilute to affect the flow behavior of the carrier solvent. Assume that no reaction is occurring.

SOLUTION

For both systems we assume that the concentration is initially zero throughout and that the concentration of species α in the inlet stream is

$$\rho_\alpha = \rho_0 \exp(-t/t_0) \tag{23.1-60}$$

where ρ_0 and t_0 are constants, specific to the problem.

For the PFR, the exit stream concentration shows only a time delay and decay, and we may write at once for $X = \rho_\alpha/\rho_0$

$$X = 0 \qquad \text{for } t < t_{\text{res}} \tag{23.1-61}$$

where $t_{\text{res}} = V/Q$ is the *mean solute residence time*, a second time constant imposed on the system. The result for longer time is

$$X = \exp[-(t - t_{\text{res}})/t_0] \qquad \text{for } t > t_{\text{res}} \tag{23.1-62}$$

which is of more interest to us here.

For the CSTR we begin with the basic differential equation

$$V\frac{dX}{dt} = Q(\exp(-t/t_0) - X) \tag{23.1-63}$$

with the initial condition that $X = 0$ at $t = 0$. This first-order linear differential equation has the solution

$$X = \left(\frac{\alpha}{\alpha - 1}\right)(e^{-\tau} - e^{-\alpha\tau}) \qquad (\alpha \neq 1) \tag{23.1-64}$$

$$X = \tau e^{-\tau} \qquad (\alpha = 1) \tag{23.1-65}$$

in which $\alpha = t_0/t_{\text{res}}$ and $\tau = t/t_0$. Exit concentrations are plotted in Fig. 23.1-8 as functions of the dimensionless time $\tau = t/t_0$ for each reactor and for $1/\alpha = t_{\text{res}}/t_0$ of 0.1 and 1.0.

It may be seen that for $1/\alpha = t_{\text{res}}/t_0 = 1.0$ the two reactors produce much different effluent concentrations, as one might expect. However, for $t_{\text{res}}/t_0 \ll 1$ and t significantly greater than t_{res}, the effluent curves for the two reactors are virtually indistinguishable. This is the region of validity for compartmental analysis, and we see that in addition to the time constants imposed by the system itself, there is also another time constant t_{obs}, the time at which the ob-

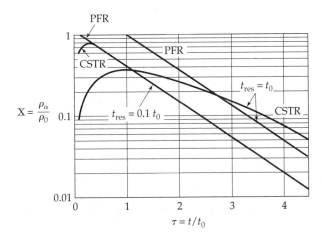

Fig. 23.1-8. Response of the PFR and the CSTR to an exponentially decaying input.

servations of the effluent concentration begin. We may then define the range of validity of compartmental analysis by the inequalities

$$t_0 \gg t_{res} \text{ and } t_{obs} \geq t_{res} \tag{23.1-66}$$

Thus, compartmental analysis is most useful as a long-time approximate description of a system that responds slowly relative to solute residence times of its component units. It may immediately be seen that these conditions are met in Example 23.1-4, where the long-time metabolite concentrations are of primary interest.

Equation 23.1-66 summarizes the requirements for pharmacokinetics, which are met in a very wide variety of biological transport-reaction problems. They are also satisfied in a great many environmental situations.[8]

§23.2 THE MACROSCOPIC MOMENTUM AND ANGULAR MOMENTUM BALANCES

The macroscopic statements of the laws of conservation of momentum and angular momentum for a fluid mixture, with gravity as the only external force, are

$$\frac{d\mathbf{P}_{tot}}{dt} = -\Delta\left(\frac{\langle v^2 \rangle}{\langle v \rangle} w + pS\right)\mathbf{u} + \mathbf{F}_{s \to f} + \mathbf{F}_0 + m_{tot}\mathbf{g} \tag{23.2-1}$$

$$\frac{d\mathbf{L}_{tot}}{dt} = -\Delta\left(\frac{\langle v^2 \rangle}{\langle v \rangle} w + pS\right)[\mathbf{r} \times \mathbf{u}] + \mathbf{T}_{s \to f} + \mathbf{T}_0 + \mathbf{T}_{ext} \tag{23.2-2}$$

These (seldom used) equations are the same as Eqs. 7.2-2 and 7.3-2, except for the addition of the terms $\mathbf{F}_0$ and $\mathbf{T}_0$, which are the net influxes[1] of momentum and angular momentum into the system by mass transfer. For most mass transfer processes these terms are so small that they can be safely neglected.

§23.3 THE MACROSCOPIC ENERGY BALANCE

For a fluid mixture, the macroscopic statement of the law of conservation of energy is

$$\frac{d}{dt}(U_{tot} + K_{tot} + \Phi_{tot}) = -\Delta\left[\left(\hat{U} + p\hat{V} + \frac{1}{2}\frac{\langle v^3 \rangle}{\langle v \rangle} + \hat{\Phi}\right)w\right] + Q_0 + Q + W_m \tag{23.3-1}$$

This equation is the same as Eq. 15.1-2, except that an additional term Q_0 has been added.[1] This term accounts for addition of energy to the system as a result of mass trans-

---

[8] F. H. Shair and K. L. Heitner, *Envir. Sci. and Tech.*, **8**, 444–451 (1974).

[1] These terms may be written as integrals,

$$\mathbf{F}_0 = -\int_{S_0} [\mathbf{n} \cdot \rho\mathbf{v}\mathbf{v}] \, dS; \qquad \mathbf{T}_0 = -\int_{S_0} [\mathbf{n} \cdot \{\mathbf{r} \times \rho\mathbf{v}\mathbf{v}\}] \, dS \tag{23.2-1a, b}$$

in which $\mathbf{n}$ is the outwardly directed unit normal vector.

[1] This term may be written as an integral,

$$Q_0 = -\int_{S_0} \left(\mathbf{n} \cdot \left\{\frac{1}{2}\rho v^2\mathbf{v} + \rho\hat{\Phi}\mathbf{v} + \sum_{\alpha=1}^{N} c_\alpha \overline{H}_\alpha \mathbf{v}_\alpha\right\}\right) dS \tag{23.3-1a}$$

in which $\mathbf{n}$ is the outwardly directed unit normal vector. The origin of this term may be seen by referring to Eq. 19.3-5 and Eq. (H) of Table 17-8.1.

fer. It may be of considerable importance, particularly if material is entering through the bounding surface at a much higher or lower temperature than that of the fluid inside the flow system, or if it reacts chemically in the system.

When chemical reactions are occurring, considerable heat may be released or absorbed. This heat of reaction is automatically taken into account in the calculation of the enthalpies of the entering and leaving streams (see Example 23.5-1).

In some applications, in which the energy transfer rates across the surface are functions of position, it is more convenient to rewrite Eq. 23.3-1 in the d-form—that is, over a differential portion of the flow system as described in §15.4. Then the increment of heat added, dQ, is expressible in terms of a local heat transfer coefficient.

§23.4   THE MACROSCOPIC MECHANICAL ENERGY BALANCE

A careful examination of the derivation of the mechanical energy balance in §7.8 shows that the result obtained there applies to mixtures as well as to pure fluids. If we now include the surface S_0, then we get

$$\frac{d}{dt}(K_{tot} + \Phi_{tot}) = -\Delta\left[\left(\frac{1}{2}\frac{\langle v^3\rangle}{\langle v\rangle} + \hat{\Phi} + \frac{p}{\rho}\right)w\right] + B_0 + W_m - E_c - E_v \tag{23.4-1}$$

This is the same as Eq. 7.4-2, except for the addition of the term B_0, which accounts for the mechanical energy transport across the mass transfer boundary.[1] The use of this equation is illustrated in Example 22.5-3.

§23.5   USE OF THE MACROSCOPIC BALANCES
TO SOLVE STEADY-STATE PROBLEMS

The macroscopic balances are summarized in Table 23.5-1 for systems with more than one entry and exit plane. The terms with subscript 0 describe the addition or removal of mass, momentum, angular momentum, energy, and mechanical energy at mass-transfer surfaces. Usually these balances are not used in their entirety, but it is convenient to have a complete listing of them for problem-solving purposes. For steady-state problems, the left sides of the equations may be omitted. As we saw in Chapters 7 and 15, considerable intuition is required in using the macroscopic balances, and sometimes it is necessary to supplement the equations with experimental observations.

EXAMPLE 23.5-1 *Energy Balances for a Sulfur Dioxide Converter*	Hot gases from a sulfur burner enter a converter, in which the sulfur dioxide present is to be oxidized catalytically to sulfur trioxide, according to the reaction $SO_2 + \frac{1}{2}O_2 \rightleftarrows SO_3$. How much heat must be removed from the converter per hour to permit a 95% conversion of the SO_2 for the conditions shown in Fig. 23.5-1? Assume that the converter is large enough for the components of the exit gas to be in thermodynamic equilibrium with one another. That is, the partial pressures of the exit gases are related by the equilibrium constraint

$$K_p = \frac{p_{SO_3}}{p_{SO_2}p_{O_2}^{1/2}} \tag{23.5-1}$$

---

[1] In terms of a surface integral, this term is given by

$$B_0 = -\int_{S_0} (\mathbf{n} \cdot [\tfrac{1}{2}\rho v^2 + \rho\hat{\Phi} + p]\mathbf{v}) \, dS \tag{23.4-1a}$$

Table 23.5-1   Unsteady-State Macroscopic Balances for Nonisothermal Multicomponent Systems

Mass:

$$\frac{d}{dt} m_{\text{tot}} = \Sigma w_1 - \Sigma w_2 + w_0 = \Sigma \rho_1 \langle v_1 \rangle S_1 - \Sigma \rho_2 \langle v_2 \rangle S_2 + w_0 \tag{A}$$

Mass of species α:

$$\frac{d}{dt} m_{\alpha,\text{tot}} = \Sigma w_{\alpha 1} - \Sigma w_{\alpha 2} + w_{\alpha 0} + r_{\alpha,\text{tot}} \qquad \alpha = 1, 2, 3, \ldots N \tag{B}$$

Momentum:

$$\frac{d}{dt} \mathbf{P}_{\text{tot}} = \Sigma \left(\frac{\langle v_1^2 \rangle}{\langle v_1 \rangle} w_1 + p_1 S_1 \right) \mathbf{u}_1 - \Sigma \left(\frac{\langle v_2^2 \rangle}{\langle v_2 \rangle} w_2 + p_2 S_2 \right) \mathbf{u}_2 + m_{\text{tot}} \mathbf{g} + \mathbf{F}_0 - \mathbf{F}_{f \to s} \tag{C}$$

Angular momentum:

$$\frac{d}{dt} \mathbf{L}_{\text{tot}} = \Sigma \left(\frac{\langle v_1^2 \rangle}{\langle v_1 \rangle} w_1 + p_1 S_1 \right) [\mathbf{r}_1 \times \mathbf{u}_1] - \Sigma \left(\frac{\langle v_2^2 \rangle}{\langle v_2 \rangle} w_2 + p_2 S_2 \right) [\mathbf{r}_2 \times \mathbf{u}_2] + \mathbf{T}_{\text{ext}} + \mathbf{T}_0 - \mathbf{T}_{f \to s} \tag{D}$$

Mechanical energy:

$$\frac{d}{dt} (K_{\text{tot}} + \Phi_{\text{tot}}) = \Sigma \left(\frac{1}{2} \frac{\langle v_1^3 \rangle}{\langle v_1 \rangle} + gh_1 + \frac{p_1}{\rho_1} \right) w_1 - \Sigma \left(\frac{1}{2} \frac{\langle v_2^3 \rangle}{\langle v_2 \rangle} + gh_2 + \frac{p_2}{\rho_2} \right) w_2 + W_m + B_0 - E_c - E_v \tag{E}$$

(Total) energy:

$$\frac{d}{dt} (K_{\text{tot}} + \Phi_{\text{tot}} + U_{\text{tot}}) = \Sigma \left(\frac{1}{2} \frac{\langle v_1^3 \rangle}{\langle v_1 \rangle} + gh_1 + \hat{H}_1 \right) w_1 - \Sigma \left(\frac{1}{2} \frac{\langle v_2^3 \rangle}{\langle v_2 \rangle} + gh_2 + \hat{H}_2 \right) w_2 + W_m + Q_0 + Q \tag{F}$$

Notes:

(a) $\Sigma w_{\alpha 1} = w_{\alpha 1a} + w_{\alpha 1b} + w_{\alpha 1c} + \cdots$, where $w_{\alpha 1a} = \rho_{\alpha 1a} v_{1a} S_{1a}$, and so on; Equations (A) and (B) can be written in molar units by replacing the lowercase symbols by capital letters and adding to Eq. (A) the term $\Sigma_\alpha R_{\alpha,\text{tot}}$ to account for the fact that moles need not be conserved in a chemical reaction.

(b) h_1 and h_2 are elevations above an arbitrary datum plane.

(c) $\hat{H}_1$ and $\hat{H}_2$ are enthalpies per unit mass (for the mixture) relative to some arbitrarily chosen reference state; see Example 19.3-1.

(d) All equations are written for compressible flow; for incompressible flow, $E_c = 0$. The quantities E_c and E_v are defined in Eqs. 7.3-3 and 4.

(e) $\mathbf{u}_1$ and $\mathbf{u}_2$ are unit vectors in the direction of flow.

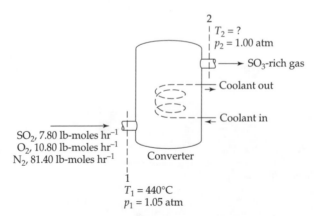

Fig. 23.5-1. Catalytic oxidation of sulfur dioxide.

Approximate values of K_p for this reaction are

T (K)	600	700	800	900
$K_p(\text{atm}^{-1/2})$	9500	880	69.5	9.8

SOLUTION

It is convenient to divide this problem into two parts: **(a)** first we use the mass balance and equilibrium expression to find the desired exit temperature, and then **(b)** we use the energy balance to determine the required heat removal.

(a) *Determination[1] of T_2.* We begin by writing the steady-state macroscopic mass balance, Eq. 23.1-3, for the various constituents in the two streams in the form:

$$W_{\alpha2} = W_{\alpha1} + R_{\alpha,\text{tot}} \tag{23.5-2}$$

In addition, we take advantage of the two stoichiometric relations

$$R_{SO_2,\text{tot}} = -R_{SO_3,\text{tot}} \tag{23.5-3}$$

$$R_{O_2,\text{tot}} = \tfrac{1}{2}R_{SO_2,\text{tot}} \tag{23.5-4}$$

We can now get the desired molar flow rates through surface 2:

$$W_{SO_2,2} = 7.80 - (0.95)(7.80) = 0.38 \text{ lb-mole/hr} \tag{23.5-5}$$

$$W_{SO_3,2} = 0 + (0.95)(7.80) = 7.42 \text{ lb-mole/hr} \tag{23.5-6}$$

$$W_{O_2,2} = 10.80 - \tfrac{1}{2}(0.95)(7.80) = 7.09 \text{ lb-mole/hr} \tag{23.5-7}$$

$$W_{N_2,2} = W_{N_2,1} = 81.40 \text{ lb-mole/hr} \tag{23.5-8}$$

$$W_2 = 0.38 + 7.42 + 7.09 + 81.40 = 96.29 \text{ lb-mole/hr} \tag{23.5-9}$$

Next, substituting numerical values into the equilibrium expression Eq. 23.5-1 gives

$$K_p = \frac{(7.42/96.29)}{(0.38/96.29)(7.09/96.29)^{1/2}} = 72.0 \text{ atm}^{-1/2} \tag{23.5-10}$$

This value of K_p corresponds to an exit temperature T_2 of about 510°C, according to the equilibrium data given above.

(b) *Calculation of the required heat removal.* As indicated by the results of Example 15.3-1, changes in kinetic and potential energy may be neglected here in comparison with changes in enthalpy. In addition, for the conditions of this example, we may assume ideal gas behavior. Then, for each constituent, $\bar{H}_\alpha = \tilde{H}_\alpha(T)$. We may then write the macroscopic energy balance, Eq. 23.3-1, as

$$-Q = \sum_{\alpha=1}^{N} (W_\alpha \tilde{H}_\alpha)_1 - \sum_{\alpha=1}^{N} (W_\alpha \tilde{H}_\alpha)_2 \tag{23.5-11}$$

For each of the individual constituents we may write

$$\tilde{H}_\alpha = \tilde{H}_\alpha^\circ + (\tilde{C}_{p\alpha})_{\text{avg}}(T - T^\circ) \tag{23.5-12}$$

Here $\tilde{H}_\alpha^\circ$ is the standard enthalpy of formation[2] of species α from its constituent elements at the enthalpy reference temperature T°, and $(\tilde{C}_{p\alpha})_{\text{avg}}$ is the enthalpy-mean heat capacity[2] of the species between T and T°. For the conditions of this problem, we may use the

[1] See O. A. Hougen, K. M. Watson, and R. A. Ragatz, *Chemical Process Principles*, Part II, 2nd edition, Wiley, New York (1959), pp. 1017–1018.

[2] See, for example, O. A. Hougen, K. M. Watson, and R. A. Ragatz, *Chemical Process Principles*, Part I, 2nd edition, Wiley, New York (1959), pp. 257, 296.

following[2] numerical values for these physical properties (the last two columns are obtained from Eq. 23.5-12):

Species	$\tilde{H}_\alpha^\circ$ cal/g-mole at 25°C	$(\tilde{C}_{p\alpha})_{avg}$ [cal/g-mole · C] from 25°C to 440°C	$(\tilde{C}_{p\alpha})_{avg}$ [cal/g-mole · C] from 25°C to 510°C	$(W_\alpha\tilde{H}_\alpha)_1$ Btu/hr	$(W_\alpha\tilde{H}_\alpha)_2$ Btu/hr
SO$_2$	$-70{,}960$	11.05	11.24	$-931{,}900$	$-44{,}800$
SO$_3$	$-94{,}450$	—	15.87	0	$-1{,}158{,}700$
O$_2$	0	7.45	7.53	60,100	46,600
N$_2$	0	7.12	7.17	433,000	509,500
			Totals	$-438{,}800$	$-647{,}400$

Substitution of the preceding values into Eq. 23.5-11 gives the required rate of heat removal:

$$-Q = (-438{,}800) - (-647{,}400) = 208{,}600 \text{ Btu/hr} \tag{23.5-13}$$

EXAMPLE 23.5-2

Height of a Packed-Tower Absorber[3]

It is desired to remove a soluble gas A from a mixture of A and an insoluble gas B by contacting the mixture with a nonvolatile liquid solvent L in the apparatus shown in Fig. 23.5-2. The apparatus consists essentially of a vertical pipe filled with a randomly arranged packing of small rings of a chemically inert material. The liquid L is sprayed evenly over the top of the packing and trickles over the surfaces of these small rings. In so doing, it is intimately contacted with the gas mixture that is passing up the tower. This direct contacting between the two streams permits the transfer of A from the gas to the liquid.

The gas and liquid streams enter the apparatus at molar rates of $-W_G$ and W_L, respectively, on an A-free basis. Note that the gas rate is negative, because the gas stream is flowing from plane 2 to plane 1 in this problem. The molar ratio of A to G in the entering gas stream is $Y_{A2} = y_{A2}/(1 - y_{A2})$, and the molar ratio of A to L in the entering liquid stream is $X_{A1} = x_{A1}/(1 - x_{A1})$. Develop an expression for the tower height z required to reduce the molar ratio Y_A in the gas stream from Y_{A2} to Y_{A1}, in terms of the mass transfer coefficients in the two streams and the stream rates and compositions.

Assume that the concentration of A is always small in both streams, so that the operation may be considered isothermal and so that the high mass-transfer rate corrections to the mass transfer coefficients are not needed, and the mass transfer coefficients, k_x^0 and k_y^0, defined in the second line of Eq. 22.2-14 can be used.

SOLUTION

Since the behavior of a packed tower is quite complex, we replace the true system by a hypothetical model. We consider the system to be equivalent to two streams flowing side-by-side with no back-mixing, as shown in Fig. 23.5-3, and in contact with one another across an interfacial area a per unit volume of packed column (see Eq. 22.1-14).

We further assume that the fluid velocity and composition of each stream are uniform over the tower cross section, and neglect both eddy and molecular transport in the flow direction. We also consider the concentration profiles in the direction of flow to be continuous curves, not appreciably affected by the placement of the individual packing particles.

The model resulting from these simplifying assumptions is probably not a very satisfactory description of a packed tower. The neglect of back-mixing and fluid-velocity nonuniformity are probably particularly serious. However, the presently available correlations for mass

---

[3] J. D. Seader and E. J. Henley, *Separation Process Principles*, Wiley, New York (1998).

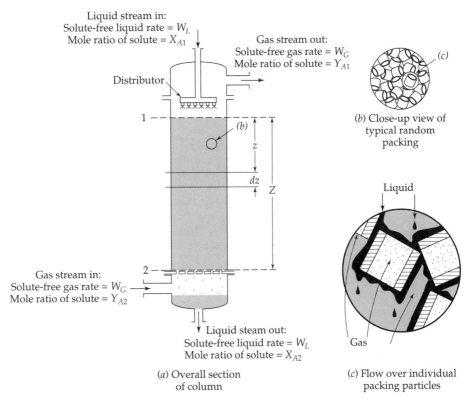

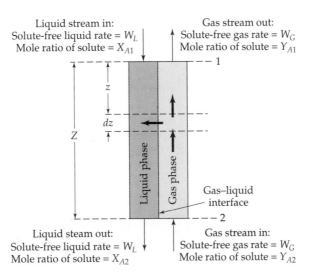

Fig. 23.5-2. A packed-column mass transfer apparatus in which the descending phase is dispersed. Note that in this drawing W_G is negative; that is, the gas is flowing from 2 toward 1.

transfer coefficients have been calculated on the basis of this model, which should therefore be employed when these correlations are used.

We are now in a position to develop an expression for the column height, and we do this in two stages: **(a)** First we use the overall macroscopic mass balance to determine the exit liquid-phase composition and the relation between bulk compositions of the two phases at

Fig. 23.5-3. Schematic representation of a packed-tower absorber, showing a differential element on which a mass balance is made.

each point in the tower. **(b)** We then use these results along with the differential form of the macroscopic mass balance to determine the interfacial conditions and the required tower height.

(a) *Overall macroscopic mass balances.* For the solute A we write the macroscopic mass balance of Eq. 23.1-3 for each stream of the system between planes 1 and 2 as

liquid stream
$$W_{Al2} - W_{Al1} = W_{Al,0} \tag{23.5-14}$$

gas stream
$$W_{Ag2} - W_{Ag1} = W_{Ag,0} \tag{23.5-15}$$

Here the subscripts Al and Ag refer to the solute A in the liquid and gas streams, respectively. Since the number of moles leaving the liquid stream must enter the gas stream across the interface, $W_{Al,0} = -W_{Ag,0}$, and Eqs. 23.5-14 and 15 may be combined to give

$$W_{Al2} - W_{Al1} = -(W_{Ag2} - W_{Ag1}) \tag{23.5-16}$$

This can now be rewritten in terms of the compositions of the entering and leaving streams by setting $W_{Al2} = W_L X_{A2}$, and so on, and then rearrangement gives

$$X_{A2} = X_{A1} - \frac{W_G}{W_L}(Y_{A2} - Y_{A1}) \tag{23.5-17}$$

Thus we have found the concentration of A in the exit liquid stream.

By replacing plane 2 by a plane at a distance z down the column, Eq. 23.5-17 may be used to obtain an expression relating bulk stream compositions at any point in the tower:

$$X_A = X_{A1} - \frac{W_G}{W_L}(Y_A - Y_{A1}) \tag{23.5-18}$$

Equation 23.5-18 (the "operating line") is shown in Fig. 23.5-4 along with the equilibrium distribution for the conditions of Problem 23A.2.

(b) *Application of the macroscopic-balances in the d-form.* We now apply Eq. 23.1-3 to a differential increment dz of the tower, first to estimate the interfacial conditions and then determine the tower height required for a given separation.

(i) *Determination of interfacial conditions.* Because only A is transferred across the interface, we may write, according to the second line of Eq. 22.1-14 (which presumes low concentrations of A and small mass-transfer rates):

$$dW_{Al,0} = (k_x^0 a)(x_{A0} - x_A)S\,dz \tag{23.5-19}$$

$$dW_{Ag,0} = (k_y^0 a)(y_{A0} - y_A)S\,dz \tag{23.5-20}$$

Here a is the interfacial area per unit volume of the packed bed tower, S is the cross-sectional area of the tower, x_{A0} and y_{A0} are the interfacial mole fractions of A in the liquid and gas phases, respectively, and x_A and y_A are the corresponding bulk concentrations (the index b is being omitted here, so that x_A, y_A, X_A, and Y_A are all bulk compositions).

Then, since (for the dilute solutions considered here) $x_A = X_A/(X_A + 1) \approx X_A$ and $y_A = Y_A/(Y_A + 1) \approx Y_A$, Eqs. 23.5-19 and 20 may be combined to give

$$\frac{Y_A - Y_{A0}}{X_A - X_{A0}} = -\frac{(k_x^0 a)}{(k_y^0 a)} \tag{23.5-21}$$

This equation enables us to determine Y_{A0} as a function of Y_A. For any Y_A, one may locate X_A on the operating line (mass balance). One then draws a straight line of slope $-(k_x^0 a)/(k_y^0 a)$ through the point (Y_A, X_A), as shown in Fig. 23.5-4. The intersection of this line with the equilibrium curve then gives the local interfacial compositions (Y_{A0}, X_{A0}).

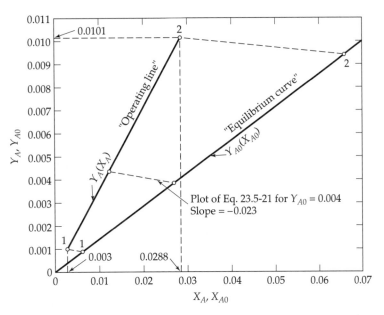

Fig. 23.5-4. Calculation of interfacial conditions in the absorption of cyclohexane from air in a packed column (see Problem 23A.2).

(ii) *Determination of required column height.* Application of Eq. 23.1-1 to the gas stream in a volume $S\,dz$ of the tower gives

$$W_G dY_A = dW_{Ag,0} \tag{23.5-22}$$

This expression may be combined with Eq. 23.5-20 for the dilute solutions being considered to obtain

$$-W_G dY_A = (k_y^0 a)(Y_A - Y_{A0})S\,dz \tag{23.5-23}$$

This equation may now be rearranged and integrated from $z = 0$ to $z = Z$:

$$Z = -\frac{W_G}{S(k_y^0 a)} \int_{Y_{A1}}^{Y_{A2}} \frac{dY_A}{Y_A - Y_{A0}} \tag{23.5-24}$$

Equation 23.5-24 is the desired expression for the column height required to effect the specified separation. In writing Eq. 23.5-24 we have neglected the variation of the mass transfer coefficient k_y^0 with composition. This is usually permissible only for dilute solutions.

In general, Eq. 23.5-24 must be integrated by numerical or graphical procedures. However, for dilute solutions, it may frequently be assumed that the operating and equilibrium lines of Fig. 23.5-4 are straight. If, in addition, the ratio k_x^0/k_y^0 is constant, then $Y_A - Y_{A0}$ varies linearly with Y_A. We may then integrate Eq. 23.5-24 to obtain (see Problem 23B.1)

$$Z = \frac{W_G}{S(k_y^0 a)} \frac{Y_{A2} - Y_{A1}}{(Y_{A0} - Y_A)_{\ln}} \tag{23.5-25}$$

where

$$(Y_{A0} - Y_A)_{\ln} = \frac{(Y_{A0} - Y_A)_2 - (Y_{A0} - Y_A)_1}{\ln[(Y_{A0} - Y_A)_2/(Y_{A0} - Y_A)_1]} \tag{23.5-26}$$

Equation 23.5-25 can be rearranged to give

$$W_{Ag,0} = W_G(Y_{A2} - Y_{A1}) = (k_y^0 a)ZS(Y_{A0} - Y_A)_{\ln} \tag{23.5-27}$$

Comparison of Eq. 23.5-27 and Eq. 15.4-15 shows the close analogy between packed towers and simple heat exchangers. Expressions analogous to Eq. 23.5-24 but containing the overall mass transfer coefficient K_y^0 may also be derived (see Problem 23B.1). Again, we may use the final results, Eqs. 23.5-25 or 27, for either cocurrent or countercurrent flow. Keep in mind, however, that the simplified model used to describe the packed tower is not as reliable as the corresponding one used for heat exchangers.

We saw in Example 23.1-2 that the degree of separation possible in a simple binary splitter can be quite limited, and it is therefore often desirable to combine individual splitters in a countercurrent *cascade* such as that shown in Fig. 23.5-5. Here the feed to any splitter stage is the sum of the waste stream from the splitter immediately above it and the product from the splitter immediately below.

Show how such an arrangement can increase the degree of separation relative to that obtained in a single splitter.

SOLUTION

For the system as a whole we can write a mass balance for the desired product and for the solution as a whole. That is, we treat the entire system as a splitter and write

$$z_F F = y_P P + x_W W \qquad F = P + W \qquad (23.5\text{-}28, 29)$$

It will be assumed here that all of quantities in these equations are given, so that the problem is specified as far as the overall mass balances are concerned. It remains for us to determine the number of stages required to meet these conditions.

We begin by writing a set of mass balances over the top portion of the column, here the top two stages for illustrative purposes (see Fig. 23.5-5):

$$y_3 U_3 - x_2 D_2 = y_P P \qquad U_3 - D_2 = P \qquad (23.5\text{-}30, 31)$$

Here U_n and D_n are the upflowing and downflowing streams from stage n, and y_n and x_n are the corresponding mole fractions of the desired solute. When P is eliminated between Eqs. 23.5-3 and 31, we get

$$\frac{y_2 - y_P}{x_2 - y_P} = \frac{D_2}{U_3} \qquad (23.5\text{-}32)$$

This equation gives the relation between the compositions of the downflowing and upflowing streams passing each other at any column cross section above the feed stage, in terms of the corresponding flow rates. This relation, when shown on an x-y plot (which is called a *McCabe–Thiele diagram*[3,4]) is known as the *operating line* for the system. We concentrate for the moment on compositions and return later to the problem of determining stream rate ratios.

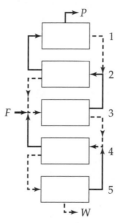

Fig. 23.5-5. A linear cascade. Upward flows are shown by solid lines, and downward flows by dashed lines.

[4] W. L. McCabe and E. W. Thiele, *Ind. Eng. Chem.*, **17**, 605–611 (1925).

The phase compositions in each stage are assumed to satisfy an equilibrium relation such as (see Eq. 23.1-19)

$$y_n = \frac{ax_n}{1 + (\alpha - 1)x_n} \tag{23.5-33}$$

or, more generally, $y_n = f(x_n)$, where $f(x)$ is taken to be a known function.

Equations 23.5-32 and 33 (or its generalization) now permit determination of all compositions in the portion of the column above the feed point, usually known as the *rectifying section*, and similar calculations can be made for the *stripping section*, the portion below the feed point. We may then determine the number of stages required for the separation under consideration and the proper location of the feed stage.

First, however, we need to determine the stream rate ratios required in Eq. 23.5-32, and we consider three special cases here:

(a) *Total reflux.* This special mode of operation, in which P and W are zero, is important, as it provides the smallest possible number of stages that can yield the desired output compositions. Here

$$U_n = D_{n-1} \tag{23.5-34}$$

for all n, and the operating line is given by

$$y_n = x_{n-1} \tag{23.5-35}$$

This simple relation holds for all physical systems. The stage compositions are plotted in Fig. 23.5-6 (for a product mole fraction of 0.9 and a waste mole fraction of 0.1), along with an equilibrium curve of the form of Eq. 23.5-33 with $\alpha = 2.5$.

The steplike lines between the equilibrium and operating lines in this figure suggest a graphical method of determining stage compositions: each "step" between the equilibrium and operating lines represents an incremental one-component splitter or stage. The diagram thus suggests that six stages are required for this rather simple separation. However, for the situation of total reflux and constant relative volatility α, it is simplest to recognize that

$$Y_n = Y_{n-1}/\alpha \tag{23.5-36a}$$

so that

$$Y_N = Y_1/\alpha^N \tag{23.5-36b}$$

For the situation pictured in Fig. 23.5-6, we have then

$$\log\left(\frac{0.9/0.1}{0.1/0.9}\right) = (N - 1)\ \log\ 2.5 \tag{23.5-37}$$

or

$$N = 1 + \frac{\log\ 81}{\log\ 2.5} = 5.796 \tag{23.5-38}$$

which is more accurate but virtually equal to the graphical estimate.

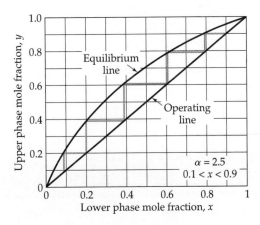

Fig. 23.5-6. McCabe–Thiele diagram for total reflux, with $\alpha = 2.5$ and $0.1 < x < 0.9$.

If products are to be withdrawn, it is necessary to calculate the stream-rate ratios, and the means for doing so vary with the specific operation considered.

(b) *Thermodynamic constraints: adiabatic cascades and minimum reflux.* For most of the common stagewise operations, stream ratios are determined by thermodynamic constraints, and these are thoroughly discussed in a wide variety of unit operations texts. We need not repeat this readily accessible information here, but we briefly consider distillation, the most widely used of all, by way of example. In principle, stream ratios in distillation are determined by assuming adiabatic columns and a set of "enthalpy balances" (see last paragraph of §15.1) corresponding to the mass balances just introduced.

However, it is very often permissible to assume equal molar heats of vaporization for the various species and to neglect "sensible heats" (i.e., the $\tilde{C}_p \Delta T$ contributions to $\Delta \tilde{H}$). With these simplifications the stream rates U_n and D_n are constants. We may then write for any position above the feed plate

$$U = D + P \quad \text{and} \quad y_{n-1}U = x_n D + y_P P \qquad (23.5\text{-}39, 40)$$

and below the feed plate

$$D = U + W \quad \text{and} \quad x_m D = y_{m+1}U + x_W W \qquad (23.5\text{-}41, 42)$$

Here the stage indices n and m refer respectively to the upper or rectifying section (above the feed point) and to the lower or stripping section of the column (below the feed point).

By way of example we consider the system in part (a) for a saturated liquid feed, equimolar in the two species involved, and operated at minimum reflux: the smallest amount of returning liquid from the top plate that can produce the desired separation. This situation will occur when the operating line touches the equilibrium curve, and in the system being considered, this "pinch" will occur first at the feed plate. The vapor composition on the feed plate is then given by

$$Y_F = 2.5X_F = 2.5 \quad \text{or} \quad y_F = \frac{Y_F}{1 + Y_F} = 0.7143 \qquad (23.5\text{-}43, 44)$$

The operating line then has two branches, one above and one below the feed plate, as shown in Fig. 23.5-7.

Any real column must operate between the limits of total and minimum reflux, but normal operation is just a few percent above the minimum. This is because the cost of individual plates tends to be much lower than the costs associated with increasing the reflux (the liquid returned to the column by condensation of vapor from the top plate): increasing the steam load required to return vapor from the liquid leaving the bottom plate, the condenser load to return the overhead vapor, and the capital costs of larger column diameter, larger reboiler, to return vapor at the bottom, and condenser, to return liquid at the top.

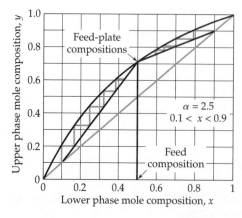

Fig. 23.5-7. McCabe–Thiele diagram for minimum reflux, with $\alpha = 2.5$ and $0.1 < x < 0.9$.

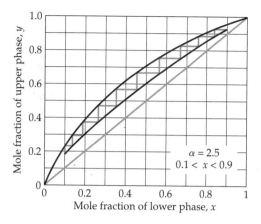

Fig. 23.5-8. McCabe–Thiele diagram for an ideal cascade, with $\alpha = 2.5$ and $0.1 < x < 0.9$.

(c) Transport constraints and ideal cascades. For separation via selectively permeable membranes, the ratio of the product to waste streams is governed by the pressure exerted across the membrane, and the energy required to produce this pressure must be renewed for every stage of the cascade. This gives the designer an extra degree of freedom and has led to a wide variety of cascade configurations. First developed for isotopes,[5] membrane cascades have now been developed for industrial gas separations[6] and appear promising for many other applications.

We consider here by example *ideal* cascades, which are those in which only streams of identical composition are mixed. In the terms of this example, that means

$$Y_{n+1} = X_{n-1} = \frac{Y_{n-1}}{\alpha} \tag{23.5-45}$$

and, by extension,

$$Y_{n+1} = \frac{Y_n}{\sqrt{\alpha}} \tag{23.5-46}$$

It follows that just twice as many stages are needed as at total reflux, and that the operating line lies halfway between the "equilibrium" curve and the 45° line. As shown in Fig. 23.5-8, the operating line has a continuous derivative across the feed stage.

Ideal cascades provide the smallest possible total stage stream flows, but the flows now vary with position: they are highest at the feed stage and decrease toward the ends of the cascade. For this reason these systems are known as *tapered cascades* (see Problem 23B.6).

EXAMPLE 23.5-4

Expansion of a Reactive Gas Mixture Through a Frictionless Adiabatic Nozzle

An equimolar mixture of CO_2 and H_2 is confined at 1000K and 1.50 atm in the large insulated pressure tank shown in Fig. 15.5-9. Under these conditions the reaction

$$CO_2 + H_2 \Leftrightarrow CO + H_2O \tag{23.5-47}$$

may take place. After being stored in the tank long enough for the reaction to proceed to equilibrium, the gas is allowed to escape through the small converging nozzle shown to the ambient pressure of 1 atm.

[5] E. von Halle and J. Schacter, *Diffusion Separation Methods*, in *Kirk-Othmer Encyclopedia of Chemical Technology*, Volume 8, Wiley, New York (1993), pp. 149–203.
[6] R. Agrawal, *Ind. Eng. Chem. Research*, **35**, 3607–3617 (1996); R. Agrawal and J. Xu, *AIChE Journal*, **42**, 2141–2154 (1996).

Estimate the temperature and velocity of the escaping gas through the nozzle throat (*a*) assuming that no appreciable reaction takes place during passage of gas through the nozzle, and (*b*) assuming instant attainment of thermodynamic equilibrium at all points in the nozzle. In each case, assume that the expansion is adiabatic and frictionless.

SOLUTION

We begin by assuming quasi-steady-state operation, flat velocity profiles, and negligible changes in potential energy. We also assume constant heat capacities and ideal gas behavior, and we neglect diffusion in the direction of flow. We may then write the macroscopic energy balance, Eq. 23.3-1, in the form

$$\tfrac{1}{2}v_2^2 = \hat{H}_1 - \hat{H}_2 \tag{23.5-48}$$

Here the subscripts 1 and 2 refer to conditions in the tank and at the nozzle throat, respectively, and, as in Example 15.5-4, the fluid velocity in the tank is assumed to be zero.

To determine the enthalpy change, we equate $d(\tfrac{1}{2}v^2)$ from the *d*-form of the steady-state energy balance (Eq. 23.3-1) to $d(\tfrac{1}{2}v^2)$ of the *d*-form of the steady-state mechanical energy balance (Eq. 23.4-1) to get

$$\frac{1}{\rho}\,dp = d\hat{H} \tag{23.5-49}$$

This result also follows from Eq. E of Table 19.2-4. In addition to Eq. 23.5-49, we use the ideal gas law and an expression for $\hat{H}(T)$, obtained with the help of Table 17.1-1, Eq. 19.3-16, and the relation $\rho\hat{H} = c\tilde{H}$, to get

$$p = cRT = \frac{\rho RT}{M} = \frac{\rho RT}{\displaystyle\sum_{\alpha=1}^{N} x_\alpha M_\alpha} \tag{23.5-50}$$

$$\hat{H} = \frac{\tilde{H}}{\rho/c} = \frac{\displaystyle\sum_{\alpha=1}^{N} x_\alpha \overline{H}_\alpha}{M} = \frac{\displaystyle\sum_{\alpha=1}^{N} x_\alpha[\tilde{H}_\alpha^o + \tilde{C}_{p\alpha}(T - T^o)]}{\displaystyle\sum_{\alpha=1}^{N} x_\alpha M_\alpha} \tag{23.5-51}$$

Here x_α is the mole fraction of the species α at temperature T, and $\tilde{H}_\alpha^o$ is the molar enthalpy of species α at the reference temperature T^o. The evaluation of $\hat{H}$ is discussed separately for the two approximations.

Approximation (a): Assumption of very slow chemical reaction. Here the x_α are constant at the equilibrium values for 1000K, and we may write Eq. 23.5-51 as

$$d\hat{H} = \left(\frac{\Sigma_\alpha x_\alpha \tilde{C}_{p\alpha}}{\Sigma_\alpha x_\alpha M_\alpha}\right)dT \tag{23.5-52}$$

Hence we may write Eq. 23.5-49 in the form

$$d\ln\,p = \left(\frac{\Sigma_\alpha x_\alpha \tilde{C}_{p\alpha}}{R}\right)d\ln\,T \tag{23.5-53}$$

Since x_α and $\tilde{C}_{p\alpha}$ are assumed constant, this equation may be integrated from (p_1, T_1) to (p_2, T_2) to get

$$T_2 = T_1\left(\frac{p_2}{p_1}\right)^{R/\Sigma_\alpha x_\alpha \tilde{C}_{p\alpha}} \tag{23.5-54}$$

We may now combine this expression with Eqs. 23.5-48 and 51 to obtain the desired expression for the gas velocity at plane 2:

$$v_2 = \left\{2T_1\left[1 - \left(\frac{p_2}{p_1}\right)^{R/\Sigma_\alpha x_\alpha \tilde{C}_{p\alpha}}\right]\frac{\Sigma_\alpha x_\alpha \tilde{C}_{p\alpha}}{\Sigma_\alpha x_\alpha M_\alpha}\right\}^{1/2} \tag{23.5-55}$$

By substituting numerical values into Eqs. 23.5-54 and 55, we obtain (see Problem 23A.1) $T_2 = 920K$ and $v_2 = 1726$ ft/s. It may be seen that this treatment is very similar to that presented in Example 15.5-4. It is also subject to the restriction that the throat velocity must be subsonic; that is, the pressure in the nozzle throat cannot fall below that fraction of p_1 required to produce sonic velocity at the throat (see Eq. 15B.6-2). If the ambient pressure falls below this critical value of p_2, the throat pressure will remain at the critical value, and there will be a shock wave beyond the nozzle exit.

Approximation (b): Assumption of very rapid reaction. We may proceed here as in part *(a)*, except that the mole fractions x_α must now be considered functions of the temperature defined by the equilibrium relation

$$\frac{(x_{H_2O})(x_{CO})}{(x_{H_2})(x_{CO_2})} = K_x(T) \tag{23.5-56}$$

and the stoichiometric relations

$$x_{H_2O} = x_{CO} \qquad x_{H_2} = x_{CO_2} \qquad \sum_{\alpha=1}^{4} x_\alpha = 1 \tag{23.5-57, 58, 59}$$

The quantity $K_x(T)$ in Eq. 23.5-56 is the known equilibrium constant for the reaction. It may be considered as a function only of temperature, because of the assumed ideal gas behavior and because the number of moles present is not affected by the chemical reaction. Equations 23.5-57 and 58 follow from the stoichiometry of the reaction and the composition of the gas originally placed in the tank.

The expression for the final temperature is now considerably more complicated. For this reaction, where $\sum_\alpha x_\alpha M_\alpha$ is constant, Eqs. 23.5-49 and 50 may be combined to give

$$R \ln \frac{p_2}{p_1} = \int_{T_1}^{T_2} \left(\frac{d\tilde{H}}{dT}\right) d\ln T \tag{23.5-60}$$

where, with the heat capacities approximated as constants,

$$\frac{d\tilde{H}}{dT} = \sum_{\alpha=1}^{4} [\tilde{H}_\alpha^\circ + \tilde{C}_{p\alpha}(T - T^\circ)]\frac{dx_\alpha}{dT} + \sum_{\alpha=1}^{4} x_\alpha \tilde{C}_{p\alpha} \tag{23.5-61}$$

In general, the integral in Eq. 23.5-60 must be evaluated numerically, since the x_α and the dx_α/dT are all complicated functions of temperature governed by Eqs. 23.5-56 to 59. Once T_2 has been determined from Eq. 23.5-61, however, v_2 may be obtained by use of Eqs. 23.5-48 and 51. By substituting numerical values into these expressions, we obtain (see Problem 23B.2) $T_2 = 937K$ and $v_2 = 1752$ ft/s.

We find, then, that both the exit temperature and the velocity from the nozzle are greater when chemical equilibrium is maintained throughout the expansion. The reason for this is that the equilibrium shifts with decreasing temperature in such a way as to release heat of reaction to the system. Such a release of energy will occur with decreasing temperature in any system at chemical equilibrium, regardless of the reactions involved. This is one consequence of the famous rule of Le Châtelier. In this case, the reaction is endothermic as written and the equilibrium constant decreases with falling temperature. As a result, CO and H_2O are partially reconverted to H_2 and CO_2 on expansion, with a corresponding release of energy.

It is interesting that in rocket engines the exhaust velocity, hence the engine thrust, are also increased if rapid equilibration can be obtained, even though the combustion reactions are strongly exothermic. The reason for this is that the equilibrium constants for these reactions increase with falling temperature so that the heat of reaction is again released on expansion. This principle has been suggested as a method for improving the thrust of rocket engines. The increase in thrust potentially obtainable in this way is quite large.

This example was chosen for its simplicity. Note in particular that if a change in the number of moles accompanies the chemical reaction, then the equilibrium constant, and hence the enthalpy, are functions of the pressure. In this case, which is quite common, the variables p

and T implicit in Eq. 23.5-60 cannot be separated, and a step-by-step integration of this equation is required. Such integrations have been performed, for example, for the prediction of the behavior of supersonic wind tunnels and rocket engines, but the calculations involved are too lengthy for presentation here.

§23.6 USE OF THE MACROSCOPIC BALANCES TO SOLVE UNSTEADY-STATE PROBLEMS

In §23.5 the discussion was restricted to steady state. Here we move on to the transient behavior of multicomponent systems. Such behavior is important in a large number of practical operations, such as leaching and drying of solids, chromatographic separations, and chemical reactor operations. In many of these processes heats of reaction as well as mass transfer must be considered. A complete discussion of these topics is outside the scope of this text, and we restrict ourselves to several simple examples. More extensive discussions may be found elsewhere.[1]

EXAMPLE 23.6-1

Start-Up of a Chemical Reactor

It is desired to produce a substance B from a raw material A in a chemical reactor of volume V equipped with a stirrer that is capable of keeping the entire contents of the reactor fairly homogeneous. The formation of B is reversible, and the forward and reverse reactions may be considered first order, with reaction-rate constants k_{1B}''' and k_{1A}''', respectively. In addition, B undergoes an irreversible first-order decomposition, with a reaction-rate constant k_{1C}''', to a third component C. The chemical reactions of interest may be represented as

$$A \rightleftarrows B \rightarrow C \tag{23.6-1}$$

At zero time, a solution of A at a concentration c_{A0} is introduced to the initially empty reactor at a constant mass flow rate w.

Develop an expression for the amount of B in the reactor, when it is just filled to its capacity V, assuming that there is no B in the feed solution and neglecting changes of fluid properties.

SOLUTION

We begin by writing the unsteady-state macroscopic mass balances for species A and B. In molar units these may be expressed as

$$\frac{dM_{A,\text{tot}}}{dt} = \frac{wc_{A0}}{\rho} - k_{1B}'''M_{A,\text{tot}} + k_{1A}'''M_{B,\text{tot}} \tag{23.6-2}$$

$$\frac{dM_{B,\text{tot}}}{dt} = -(k_{1A}''' + k_{1C}''')M_{B,\text{tot}} + k_{1B}'''M_{A,\text{tot}} \tag{23.6-3}$$

Next we eliminate $M_{A,\text{tot}}$ from Eq. 23.6-3. First we differentiate this equation with respect to t to get

$$\frac{d^2M_{B,\text{tot}}}{dt^2} = -(k_{1A}''' + k_{1C}''')\frac{dM_{B,\text{tot}}}{dt} + k_{1B}'''\frac{dM_{A,\text{tot}}}{dt} \tag{23.6-4}$$

In this equation, we replace $dM_{A,\text{tot}}/dt$ by the right side of Eq. 23.6-2, and then use Eq. 23.6-3 to eliminate $M_{A,\text{tot}}$. In this way we obtain a linear second-order differential equation for $M_{B,\text{tot}}$ as a function of time:

$$\frac{d^2M_{B,\text{tot}}}{dt^2} + (k_{1A}''' + k_{1B}''' + k_{1C}''')\frac{dM_{B,\text{tot}}}{dt} + k_{1B}'''k_{1C}'''M_{B,\text{tot}} = \frac{k_{1B}'''wc_{A0}}{\rho} \tag{23.6-5}$$

---

[1] W. R. Marshall, Jr., and R. L. Pigford, *The Application of Differential Equations to Chemical Engineering Problems,* University of Delaware Press, Newark, Del. (1947); B. A. Ogunnaike and W. H. Ray, *Process Dynamics, Modeling, and Control,* Oxford University Press (1994).

This equation is to be solved with the initial conditions

I.C. 1: at $t = 0$, $\qquad\qquad\qquad M_{B,\text{tot}} = 0$ $\qquad\qquad$ (23.6-6)

I.C. 2: at $t = 0$, $\qquad\qquad\qquad \dfrac{dM_{B,\text{tot}}}{dt} = 0$ $\qquad\qquad$ (23.6-7)

This equation can be integrated to give

$$M_{B,\text{tot}} = \frac{wc_{A0}}{\rho k_{1C}'''}\left(\frac{s_-}{s_+ - s_-}\exp(s_+ t) - \frac{s_+}{s_+ - s_-}\exp(s_- t) + 1\right) \qquad (23.6\text{-}8)$$

where

$$2s_\pm = -(k_{1A}''' + k_{1B}''' + k_{1C}''') \pm \sqrt{(k_{1A}''' + k_{1B}''' + k_{1C}''')^2 - 4\,k_{1B}''' k_{1C}'''} \qquad (23.6\text{-}9)$$

Equations 23.6-8 and 9 give the total mass of B in the reactor as a function of time, up to the time at which the reactor is completely filled. These expressions are very similar to the equations obtained for the damped manometer in Example 7.7-2 and the temperature controller in Example 15.5-2. It can be shown, however, that s_+ and s_- are both real and negative, and therefore $M_{B,\text{tot}}$ cannot oscillate (see Problem 23B.3).

EXAMPLE 23.6-2

Unsteady Operation of a Packed Column

There are many industrially important processes in which mass transfer takes place between a fluid and a granular porous solid: for example, recovery of organic vapors by adsorption on charcoal, extraction of caffeine from coffee beans, and separation of aromatic and aliphatic hydrocarbons by selective adsorption on silica gel. Ordinarily, the solid is held fixed, as indicated in Fig. 23.6-1, and the fluid is allowed to percolate through it. The operation is thus inherently unsteady, and the solid must be periodically replaced or "regenerated," that is, returned to its original condition by heating or other treatment. To illustrate the behavior of such fixed-bed mass transfer operations, we consider as a physically simple case, the removal of a solute from a solution by passage through an adsorbent bed.

In this operation, a solution containing a single solute A at mole fraction x_{A1} in a solvent B is passed at a constant volumetric flow rate w/ρ through a packed tower. The tower packing consists of a granular solid capable of adsorbing A from the solution. At the start of the

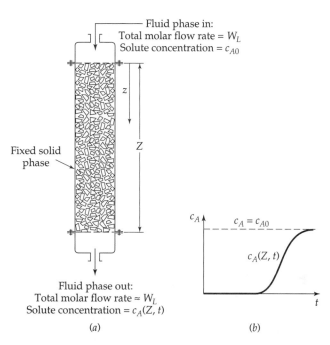

Fluid phase in:
Total molar flow rate = W_L
Solute concentration = c_{A0}

z

Z

Fixed solid phase

c_A

$c_A = c_{A0}$

$c_A(Z, t)$

Fluid phase out:
Total molar flow rate ≈ W_L
Solute concentration = $c_A(Z, t)$

t

(a)

(b)

Fig. 23.6-1. A fixed-bed absorber: (*a*) pictorial representation of equipment; (*b*) a typical effluent curve.

percolation, the interstices of the bed are filled with pure liquid B, and the solid is free of A. The percolating fluid displaces this solvent evenly so that the solution concentration of A is always uniform over any cross section. For simplicity, it is assumed that the equilibrium concentration of A adsorbed on the solid is proportional to the local concentration of A in the solution. It is also assumed that the concentration of A in the percolating solution is always small and that the resistance of the porous solid to intraparticle mass transport is negligible.

Develop an expression for the concentration of A in the column as a function of time and of distance down the column.

SOLUTION

Paralleling the treatment of the gas absorber in Example 23.5-2, we think of the two phases as being continuous and existing side by side as pictured in Fig. 23.6-2. We again define the contact area per unit packed volume of column as a. Now, however, one of the phases is stationary, and unsteady-state conditions prevail. Because of this locally unsteady behavior, the macroscopic mass balances are applied locally over a small column increment of height Δz. We may use Eq. 23.1-3 and the assumption of dilute solutions to state that the molar rate of flow of solvent, W_B, is essentially constant over the length of the column and the time of operation. We now proceed to use Eq. 23.1-3 to write the mass conservation relations for species A in each phase for a column increment of height Δz.

For the *solid phase* in this increment of column we may apply Eq. 22.3-3 locally, keeping in mind that now $M_{A,\text{tot}}$ depends on both z and t:

$$\frac{dM_{A,\text{tot}}}{dt} = W_{A0} \tag{23.6-10}$$

or

$$(1 - \varepsilon)S\Delta z \frac{\partial c_{As}}{\partial t} = (k_x^0 a)(x_A - x_{A0})S\Delta z \tag{23.6-11}$$

Here use has been made of Eq. 22.1-14, and the symbols have the following meaning:

ε = volume fraction of column occupied by the liquid

S = cross-sectional area of (empty) column

c_{AS} = moles of adsorbed A per unit volume of the solid phase

x_A = bulk mole fraction of A in the liquid phase

x_{A0} = interfacial mole fraction of A in the fluid phase, assumed to be in equilibrium with c_{As}

k_x^0 = fluid-phase mass transfer coefficient, defined in Eq. 22.1-14, for small mass-transfer rates

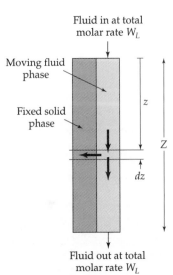

Fluid in at total
molar rate W_L

Moving fluid
phase

Fixed solid
phase

z

Z

dz

Fluid out at total
molar rate W_L

Fig. 23.6-2. Schematic model for a fixed-bed absorber, showing a differential element over which a mass balance is made.

Note that, in writing Eq. 23.6-11, we have neglected convective mass transfer through the solid–fluid interface. This is reasonable if x_{A0} is much smaller than unity. We have also assumed that the particles are small enough so that the concentration of the solution surrounding any given particle is essentially constant over the particle surface.

For the *fluid phase*, in the column increment under consideration, Eq. 23.1-3 becomes

$$\frac{dM_{A,tot}}{dt} = -\Delta W_A + W_{A0} \tag{23.6-12}$$

or

$$\varepsilon c S \Delta z \frac{\partial x_A}{\partial t} = -W_B \Delta z \frac{\partial x_A}{\partial z} - (k_x^0 a)(x_A - x_{A0}) S \Delta z \tag{23.6-13}$$

Here c is the total molar concentration of the liquid. Equation 23.6-13 may be rewritten by the introduction of a modified time variable, defined by

$$t' = t - \left(\frac{\varepsilon c S}{W_B}\right) z \tag{23.6-14}$$

It may be seen that, for any position in the column, t' is the time measured from the instant that the percolating solvent "front" has reached the position in question. By rewriting Eqs. 23.6-13 and 11 in terms of t', we get

$$\left(\frac{\partial x_A}{\partial z}\right)_{t'} = -\frac{(k_x^0 a)S}{W_B}(x_A - x_{A0}) \tag{23.6-15}$$

$$\left(\frac{\partial c_{As}}{\partial t'}\right)_z = \frac{(k_x^0 a)}{(1 - \varepsilon)}(x_A - x_{A0}) \tag{23.6-16}$$

Equations 23.6-15 and 16 combine the equations of conservation of mass for each phase with the assumed mass transfer rate expression. These two equations are to be solved simultaneously along with the interphase equilibrium distribution, $x_{A0} = mc_{As}$, in which m is a constant. The boundary conditions are

B.C. 1:   at $t' = 0$, $c_{As} = 0$ for all $z > 0$ (23.6-17)

B.C. 2:   at $z = 0$, $x_A = x_{A1}$ for all $t' > 0$ (23.6-18)

Before solving these equations, it is convenient to rewrite them in terms of the following dimensionless variables:

$$X(\zeta, \tau) = \frac{x_A}{x_{A1}} \qquad Y(\zeta, \tau) = \frac{mc_{As}}{x_{A1}} \qquad \zeta = \frac{(k_x^0 a)S}{W_B} z \qquad \tau = \frac{(k_x^0 a)m}{(1 - \varepsilon)} t' \quad (23.6\text{-}19, 20, 21, 22)$$

In terms of these variables, the differential equations and boundary conditions take the form

$$\frac{\partial X}{\partial \zeta} = -(X - Y); \qquad \frac{\partial Y}{\partial \tau} = +(X - Y) \tag{23.6-23, 24}$$

with the boundary conditions $Y(\zeta, 0) = 0$ and $X(0, \tau) = 1$.

The solution[2] to Eqs. 23.6-23 and 24 for these boundary conditions is

$$X = 1 - \int_0^\zeta e^{-(\tau + \bar{\zeta})} J_0(i\sqrt{4\tau\bar{\zeta}}) d\bar{\zeta} \tag{23.6-25}$$

Here $J_0(ix)$ is a zero-order Bessel function of the first kind. This solution is presented graphically in several available references.[3]

---

[2] This result was first obtained by A. Anzelius, *Z. angew. Math. u. Mech.*, **6**, 291–294 (1926), for the analogous problem in heat transfer. One method of obtaining this result is outlined in Problem 23D.1. See also H. Bateman, *Partial Differential Equations of Mathematical Physics*, Dover, New York (1944), pp. 123–125.

[3] See, for example, O. A. Hougen and K. M. Watson, *Chemical Process Principles*, Part III, Wiley, New York (1947), p. 1086. Their y/y_0, $b\tau$, and aZ correspond to our X, τ, and ζ.

EXAMPLE 23.6-3

The Utility of Low-Order Moments

For many complex systems, complete descriptions are either infeasible or unnecessary, and it is sufficient to obtain only a few basic characteristics. Specifically, we may ask how one may determine the system volume V and the volume flow rate Q through it from observations of short tracer pulses of mass m introduced at the inlet and then measured at the outlet. Consider for this purpose the macroscopically steady flow through a closed system of arbitrary geometry, but with a single inlet and outlet, such as that suggested in Fig. 7.0-1, except that there are no moving surfaces. The flow and diffusional behavior are arbitrary, except that the tracer distribution must be described by the diffusion equation (Eq. 19.1-7 with Eq. 17.7-3 inserted for the mass flux)

$$\frac{\partial \rho_T}{\partial t} = -(\nabla \cdot \rho_T \mathbf{v}) + (\nabla \cdot \mathcal{D}_{TS} \nabla \rho_T) \tag{23.6-26}$$

in which ρ_T is the local tracer concentration and $\mathcal{D}_{TS}$ is the pseudobinary diffusivity for the tracer moving through the solution that fills the system. Turbulent systems may be included by using time-smoothed quantities and an effective turbulent diffusivity.

In developing the macroscopic balances we shall need to use the condition that there is no flow or diffusion through the walls of the enclosure

$$(\mathbf{n} \cdot \mathbf{v}) = 0 \quad \text{and} \quad (\mathbf{n} \cdot \nabla \rho_T) = 0 \tag{23.6-27, 28}$$

and that the diffusive flux of the tracer is small compared to the convective flux at the inlet and outlet to the system

$$\mathcal{D}_{TS}(\mathbf{n} \cdot \nabla \rho_T) << \rho_T (\mathbf{n} \cdot \mathbf{v}) \tag{23.6-29}$$

Here $\mathbf{n}$ is the outwardly directed unit normal vector. We take the inlet tracer concentration to be zero up to $t = 0$ and also after some finite time $t = t_0$. In practice the concentration pulse duration should be quite short.

SOLUTION

The analysis[4] is based on the moments $I^{(n)}$ of the tracer concentration with respect to time, defined by:

$$I^{(n)}(\mathbf{r}) = \int_{-\infty}^{+\infty} \rho_T(\mathbf{r}, t) t^n dt \tag{23.6-30}$$

We now multiply Eq. 23.6-26 by t^n and integrate with respect to time over the range of nonzero exit tracer concentration

$$\int_0^\infty \frac{\partial \rho_T}{\partial t} t^n dt + \left(\nabla \cdot \mathbf{v} \int_0^\infty \rho_T t^n dt\right) = \left(\nabla \cdot \mathcal{D}_{TS} \nabla \int_0^\infty \rho_T t^n dt\right) \tag{23.6-31}$$

When the first term is integrated by parts and we make use of the notation introduced in Eq. 23.6-30, we get

$$\begin{cases} -nI^{(n-1)} = -(\nabla \cdot \mathbf{v} I^{(n)}) + (\nabla \cdot \mathcal{D}_{TS} \nabla I^{(n)}) & (n \geq 1) \\ 0 = -(\nabla \cdot \mathbf{v} I^{(n)}) + (\nabla \cdot \mathcal{D}_{TS} \nabla I^{(n)}) & (n = 0) \end{cases} \tag{23.6-32, 33}$$

for all systems that give finite moments. We now have a hierarchy of equations for the $I^{(n)}$ in terms of the lower-order moments, and the structure of these equations is very convenient.

In physical terms, it was first noted by Spalding[5] that Eq. 23.6-32 has the same form as the diffusion equation with chemical reaction, Eq. 19.1-16, but with the concentration replaced by $I^{(n)}$ and the reaction term replaced by $nI^{(n-1)}$. Hence we can integrate these equations over the entire volume of the flow system and thereby develop a new set of macroscopic balances.

[4] E. N. Lightfoot, A. M. Lenhoff, and R. I. Rodrigues, *Chem. Eng. Sci.*, **36**, 954–956 (1982).
[5] D. B. Spalding, *Chem. Eng. Sci.*, **9**, 74–77 (1958).

We begin by integrating Eq. 23.6-33, for $n = 0$, over the entire volume of the flow system between planes 1 and 2:

$$0 = -\int_V (\nabla \cdot \mathbf{v}I^{(0)})dV + \int_V (\nabla \cdot \mathscr{D}_{TS}\nabla I^{(0)})dV \tag{23.6-34}$$

The volume integrals may be converted into surface integrals by using the Gauss divergence theorem to get

$$0 = -\int_S (\mathbf{n} \cdot \mathbf{v}I^{(0)})dS + \int_S (\mathbf{n} \cdot \mathscr{D}_{TS}\nabla I^{(0)})dS \tag{23.6-35}$$

where $S = S_f + S_1 + S_2$. The integral over the fixed surface S_f is zero according to Eq. 23.6-27 and 28, and the integrals over the inlet and outlet planes S_1 and S_2 can be simplified, so that we get

$$0 = \int_{S_1} vI^{(0)}dS - \int_{S_2} vI^{(0)}dS \tag{23.6-36}$$

Here we have made use of Eq. 23.6-29 to drop the diffusive terms at planes 1 and 2. If we assume that $I^{(0)}$ is constant over a cross section, we may remove it from the integral, and then we get

$$0 = \langle v_1 \rangle S_1 I_1^{(0)} - \langle v_2 \rangle S_2 I_2^{(0)} \tag{23.6-37}$$

For an incompressible fluid, the volume rate of flow, Q, is constant, so that $\langle v_1 \rangle S_1 = \langle v_2 \rangle S_2$, and

$$I_1^{(0)} = I_2^{(0)} \tag{23.6-38}$$

That is, $I^{(0)}$ evaluated at plane 1 is the same as $I^{(0)}$ at plane 2, and at every point in the system. It is standard notation to abbreviate this quantity as M_0, the zeroth (absolute) moment. Equation 23.6-38 is analogous to Eq. 23.1-2 for a steady-state system with no mass transport across the walls. Next we evaluate $I_1^{(0)}$ for the introduction of a mass m of tracer over a time interval that is very small with respect to the mean tracer residence time $t_{res} = V/Q$:

$$I_1^{(0)} = \int_0^\infty \rho_T(\mathbf{r}, t)|_{S_1}dt = \left(\frac{Q}{V}\int_0^{Q/V} \rho_T(\mathbf{r}, t)|_{S_1}dt\right)\frac{V}{Q} = \left(\frac{m}{V}\right)\frac{V}{Q} = = \frac{m}{Q} \tag{23.6-39}$$

The replacement, in the second step, of the upper limit by Q/V is permitted by the finite duration of the tracer pulse. From the last two equations we then get

$$Q = \frac{m}{I_2^{(0)}} = \frac{m}{M_0} \tag{23.6-40}$$

This provides the possibility of measuring blood flow rate from the mass of an injected tracer and the value of $I_2^{(0)} = M_0$. The latter can be obtained by means of a catheter inserted into the blood vessel or by NMR techniques.

This simple formula[6] was first introduced in 1829 and has been extensively used since 1897 for measurement of blood-flow rates,[7] including cardiac output.[8] It is also widely used for many environmental systems, such as rivers, and also for systems in the process industries.

Next we turn to Eq. 23.6-32, and integrate it over the volume of the flow system, once again making use of the fact that the diffusive term over the inlet and outlet is much smaller than the convective term. This gives

$$-n\int_V I^{(n-1)}(\mathbf{r}, t)dV = -\int_V (\nabla \cdot \mathbf{v}I^{(n)})\, dV \tag{23.6-41}$$

---

[6] E. Hering, *Zeits. f. Physik*, **3**, 85–126 (1829).

[7] G. N. Stewart, *J. Physiol.* (London), **22**, 159–183 (1897).

[8] K. Zierler, *Ann. Biomed. Eng.*, **28**, 836–848 (2000).

or, if $I^{(n)}$ is assumed constant over a cross-section,

$$-n\left(\frac{1}{V}\int_V I^{(n-1)}(\mathbf{r}, t)dV\right) = \frac{1}{V}\left(\langle v_1\rangle S_1 I_1^{(n)} - \langle v_2\rangle S_2 I_2^{(n)}\right) \tag{23.6-42}$$

Then, defining the quantity in parentheses on the left side as the volume average, we get finally[3]

$$-n[I^{(n-1)}]_{\text{vol avg}} = \frac{Q}{V}\left(I_1^{(n)} - I_2^{(n)}\right) \tag{23.6-43}$$

Now, if we set $n = 1$, we get the following:

$$-I^{(0)} = \frac{Q}{V}\left(I_1^{(1)} - I_2^{(1)}\right) = \frac{1}{t_{\text{res}}}\left(I_1^{(1)} - I_2^{(1)}\right) \tag{23.6-44}$$

If the tracer is injected as a *delta-function input*, so that $I_1^{(1)} = 0$, we can use the notation $I_2^{(1)} = M_1$ (the first moment), and the last equation becomes

$$\frac{I_2^{(1)}}{I^{(0)}} = t_{\text{res}} \quad \text{or} \quad \frac{M_1}{M_0} = t_{\text{res}} \tag{23.6-45}$$

This result, used in conjunction with Eq. 23.6-40, has long been applied by cardiologists for determining blood volume. It has since found many other environmental and process-industry calculations.

Higher moments have also proven useful, in particular the central moments

$$\mu_n = \int_{-\infty}^{\infty} \rho_T(t - t_{\text{res}})^n dt \tag{23.6-46}$$

These are commonly applied for the special case of an impulsive tracer input. Then the normalized second central moment, or variance, is

$$\sigma^2 = \frac{\mu_2}{M_0^2} \tag{23.6-47}$$

This is the square of the standard deviation, when the exit tracer profile is a Gaussian distribution. The third central moment is a measure of the asymmetry about t_{res}, and the fourth a measure of the kurtosis. In practice, the fourth moment is nearly impossible to determine accurately from experimental data, and obtaining even the third proves to be quite difficult.

Use of the second moment has found some very important applications in studying tracer dynamics of biological tissue,[9] and again the large literature in the medical field has been extended to many other applications. It is also interesting to note the additivity relationships in serially connected systems. Thus M_0 is invariant to the number of included subsystems, and M_1, μ_2, and μ_3 are additive, but higher-order moments are not.

QUESTIONS FOR DISCUSSION

1. How are the macroscopic balances for multicomponent mixtures derived? How are they related to the equations of change?
2. In Eq. 23.1-1, how are homogeneous and heterogeneous reactions accounted for? What is the physical meaning of $w_{\alpha 0}$?
3. Give a specific example of a system in which the last term in Eq. 23.1-4 is zero.
4. In using Table 23.5-1 one normally specifies the directions of the streams (that is, whether they are input or output streams). How could one proceed if the flow directions change with time?

---

[9] F. Chinard, *Ann. Biomed Eng.*, **28**, 849–859 (2000).

5. Summarize the calculation procedures for the enthalpy per unit mass, $\hat{H} = \hat{U} + p\hat{V}$, in Eq. 23.3-1 and the partial molar enthalpy in Eq. 23.3-1a. What are these quantities for ideal gas mixtures?

6. Review the derivation of the mechanical energy balance in §7.8. What would have to be changed in that derivation, if one wishes to apply it to a nonisothermal, reacting mixture in a flow system with no mass transfer surfaces?

7. To what extent does this chapter provide the background for studying unit operations, such as absorption, extraction, distillation, and crystallization?

8. What changes would have to be made in this chapter to describe processes in a space ship or on the surface of the moon?

PROBLEMS   **23A.1.** **Expansion of a gas mixture: very slow reaction rate.** Estimate the temperature and velocity of the water–gas mixture at the discharge end of the nozzle in Example 23.5-4 if the reaction rate is very slow. Use the following data: $\log_{10} K_x = -0.15$, $\tilde{C}_{p,H_2} = 7.217$, $\tilde{C}_{p,CO_2} = 12.995$, $\tilde{C}_{p,H_2O} = 9.861$, $\tilde{C}_{p,CO} = 7.932$ (all heat capacities are in Btu/lb-mole · F. Is the nozzle exit pressure equal to the ambient pressure?

Answers: 920K, 1726 ft/s; yes, the nozzle flow is subsonic.

23A.2. **Height of a packed-tower absorber.** A packed tower of the type described in Example 23.5-2 is to be used for removing 90% of the cyclohexane from a cyclohexane–air mixture by absorption into a nonvolatile light oil. The gas stream enters the bottom of the tower at a volumetric rate of 363 ft³/min, at 30°C, and at 1.05 atm pressure. It contains 1% cyclohexane by volume. The oil enters the top of the tower at a rate of 20 lb-mol/hr, also at 30°C, and it contains 0.3% cyclohexane on a molar basis. The vapor pressure of cyclohexane at 30°C is 121 mm Hg, and solutions of it in the oil may be considered to follow Raoult's law.

(a) Construct the operating line for the column.

(b) Construct an equilibrium curve for the range of operation encountered here. Assume the operation to be isothermal and isobaric.

(c) Determine the interfacial conditions at each end of the column.

(d) Determine the required tower height using Eq. 23.5-24 if $k_x^0 a = 0.32$ moles/hr · ft³, $k_y^0 a = 14.2$ moles/hr · ft³, and the tower cross section S is 2.00 ft².

(e) Repeat part (d), using Eq. 23.5-25.

Answer: **(d)** ca. 62 ft;   **(e)** 60 ft

23B.1. **Effective average driving forces in a gas absorber.** Consider a packed-tower gas absorber of the type discussed in Example 23.5-2. Assume that the solute concentration is always low and that the equilibrium and operating lines are both very nearly straight. Under these conditions, both $k_y^0 a$ and $k_x^0 a$ may be considered constant over the mass-transfer surface.

(a) Show that $(Y_A - Y_{Ae})$ varies linearly with Y_A. Note that Y_A is the bulk mole ratio of A in the gas phase and Y_{Ae} is the equilibrium gas-phase mole ratio over a liquid of bulk composition X_A (see Fig. 22.4-2).

(b) Repeat part (a) for $(Y_A - Y_{A0})$.

(c) Use the results of parts (a) and (b) to show that

$$W_{Ag,0} = (k_y^0 a)ZS(Y_{A0} - Y_A)_{\ln} \tag{23B.1-1}$$

$$W_{Ag,0} = (K_y^0 a)ZS(Y_{Ae} - Y_A)_{\ln} \tag{23B.1-2}$$

The overall mass transfer coefficient K_y^0 is defined by Eq. 22.4-4. Note that this part of the problem may be solved by analogy with the development in Example 15.4-1.

23B.2. **Expansion of a gas mixture: very fast reaction rate.** Estimate the temperature and velocity of the water–gas mixture at the discharge end of the nozzle in Example 23.5-4 if the reaction rate may be considered infinitely fast. Use the data supplied in Problem 23A.1 as well as the following: at 900K, $\log_{10} K_x = -0.34$; $\tilde{H}_{H_2} = +6340$; $\tilde{H}_{H_2O}(g) = -49,378$; $\tilde{H}_{CO} = -16,636$;

$\tilde{H}_{CO_2} = -83{,}242$ (all enthalpies are given in cal/g-mole). For simplicity, neglect the effect of temperature on heat capacity, and assume that $\log_{10} K_x$ varies linearly with temperature between 900 and 1000K. The following simplified procedure is recommended:

(a) It may be seen in advance that T_2 will be higher than for slow reaction rates, and hence greater than 920K (see Problem 23A.1). Show that, over the temperature range to be encountered, $\tilde{H}$ varies very nearly linearly with the temperature according to the expression $(d\tilde{H}/dT)_{avg} \approx 12.40$ cal/gm-mol $\cdot$ K.

(b) Substitute the result in (a) into Eq. 23.5-41 to show that $T_2 \approx 937$K.

(c) Calculate $\tilde{H}_1$ and $\tilde{H}_2$, and show by use of Eq. 23.5-29 that $v_2 = 1750$ ft/s.

23B.3. Startup of a chemical reactor.

(a) Integrate Eq. 23.6-5 along with the given initial conditions to show that Eq. 23.6-8 correctly describes $M_{B,tot}$ as a function of time.

(b) Show that s_+ and s_- in Eq. 23.6-9 are real and negative. *Hint:* Show that

$$(k_{1A}''' + k_{1B}''' + k_{1C}''')^2 - 4\,k_{1B}'''k_{1C}''' = (k_{1A}''' - k_{1B}''' + k_{1C}''')^2 + 4k_{1A}'''k_{1B}''' \tag{23B.3-1}$$

(c) Obtain expressions for $M_{A,tot}$ and $M_{C,tot}$ as functions of time.

23B.4. Irreversible first-order reaction in a continuous reactor. A well-stirred reactor of volume V is initially completely filled with a solution of solute A in a solvent S at concentration c_{A0}. At time $t = 0$, an identical solution of A in S is introduced at a constant mass flow rate w. A small constant stream of dissolved catalyst is introduced at the same time, causing A to disappear according to an irreversible first-order reaction with rate constant k_1''' sec^{-1}. The rate constant may be assumed independent of composition and time. Show that the concentration of A in the reactor (assumed isothermal) at any time is

$$\frac{c_A}{c_{A0}} = \left(1 - \frac{wt_0}{\rho V}\right)e^{-t/t_0} + \frac{wt_0}{\rho V} \tag{23B.4-1}$$

in which $t_0^{-1} = [(w/\rho V) + k_1''']$.

23B.5. Mass and enthalpy balances in an adiabatic splitter. One hundred pounds of 40% by mass of superheated aqueous ammonia with a specific enthalpy of 420 Btu/lb is to be flashed adiabatically to a pressure of 10 atm. Calculate the compositions and masses of the liquid and vapor produced. For the purposes of this problem you may assume that at thermodynamic equilibrium

$$\log_{10} Y_{NH_3} = 1.4 + 1.53 \log_{10} X_{NH_3} \tag{23B.5-1}$$

where Y_{NH_3} and X_{NH_3} are the *mass* ratios of ammonia to water. The enthalpies of saturated liquid and vapor at 10 atm may be assumed to be

$$\hat{H} = 1210 - 465y_{NH_3} - 115y_{NH_3}^{12} \tag{23B.5-2}$$

Btu/lb of saturated vapor, and

$$\hat{h} = 330 - 950x_{NH_3} + 740x_{NH_3}^2 \tag{23B.5-3}$$

Btu/lb of saturated liquid. Here x_{NH_3} and y_{NH_3} are *mass* fractions of ammonia.
Answer: $P = 36.5$ lbs, $y_P = 0.713$, $\hat{H}_p = 877$ Btu/lb$_m$; $W = 63.6$ lb$_m$, $x_W = 0.22$, $\hat{h}_W = 157$ Btu/lb$_m$

23B.6. Flow distribution in an ideal cascade. Determine the upflowing and downflowing stream flows of individual stages for the ideal cascade described in Example 23.5-3. Express your results as fractions of the feed rate, and start from the bottom of the cascade. Use 12 stages as the closest integer providing the desired separation. It is suggested that you begin by calculating the upflowing and downflowing stream compositions and then use the mass balances

$$D_{n-1} = U_n + W; \qquad x_{n-1}D_{n-1} = y_nU_n + x_WW \tag{23B.6-1}$$

below the feed plate and the corresponding balances above it. Use 10 stages with the bottoms (W) composition equal to a mole fraction of 0.1.

23B.7. Isotope separation and the value function. You wish to compare an existing isotope fractionator that processes 50 moles/hr of a feed containing 1.0 mole% of the desired isotope to a product of 90% purity and a waste of 10% with another that processes 50 moles/hr of 10 mole% material to product and waste of 95% and 2%, respectively. Which fractionation is more effective? Assume the Dirac separative capacity to be an accurate measure of effectiveness.

23C.1. Irreversible second-order reaction in an agitated tank. Consider a system similar to that discussed in Problem 23B.4, except that the solute disappears according to a second-order reaction; that is, $R_{A,\text{tot}} = -k_2'''Vc_A^2$. Develop an expression for c_A as a function of time by the following method:

(a) Use a macroscopic mass balance for the tank to obtain a differential equation describing the evolution of c_A with time.

(b) Rewrite the differential equation and the accompanying initial condition in terms of the variable

$$u = c_A + \frac{w}{2\rho Vk_2'''}\left(1 + \sqrt{1 + \frac{4\rho Vk_2'''c_{A0}}{w}}\right) \tag{23C.1-1}$$

The nonlinear differential equation obtained in this way is a *Bernoulli differential equation.*

(c) Now put $v = 1/u$ and perform the integration. Then rewrite the result in terms of the original variable c_A.

23C.2. Protein purification (Fig. 23C.2). It is desired to purify a binary protein mixture using an ideal cascade of individual ultrafiltration stages of the type shown in the figure. The larger of the two membrane units is the source of separation and each protein flux across the membrane is expressed by

$$N_i = c_i vS_i \tag{23C.2-1}$$

where N_i is the transmembrane protein flux of species i, c_i is its concentration in the upstream solution (assumed to be well mixed), v is the transmembrane superficial velocity, and S_i is a protein-specific *sieving factor.* The smaller membrane unit is used solely to maintain a solvent balance and can be ignored for the purposes of this problem.

(a) Show that the enrichment of protein 1 relative to 2 is given by

$$Y_1 = \alpha_{12}X_1 \tag{23C.2-2}$$

where Y_1 and X_1 are the mole ratios of protein 1 to protein 2 in the product and waste streams, respectively, and $\alpha_{12} = S_1/S_2$.

(b) Determine the number of stages required in an ideal cascade to produce 99% pure protein 1 from a 90% feed in 95% yield as a function of α_{12}. It is suggested that α_{12} be varied from 2 to 200.

(c) Calculate the output concentrations, yield, and stream flow rates for a three-stage cascade, with $\alpha_{12} = 40$, and with a feed of 90% purity to the middle stage.

(d) Compare the Dirac separative capacity of this three-stage cascade with that of a single unit with the same molar ratio of product to feed.

23C.3. Physical significance of the zeroth and first moments. Consider some simple flow systems, such as plug flow and well-stirred vessels, individually and in series or parallel arrangements. Show that flow rates and volumes can be obtained from the moments defined in Example 23.6-3.

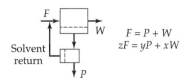

$F = P + W$
$zF = yP + xW$

Fig. 23C.2. A membrane-based binary splitter.

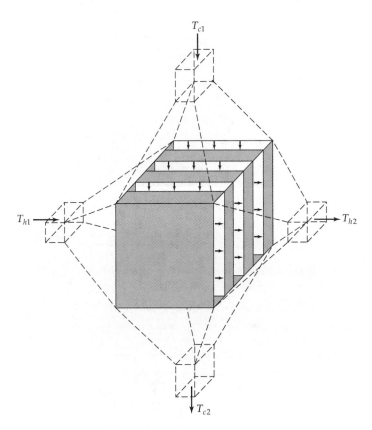

Fig. 23C.4. A schematic representation of a "sandwich-type" cross-flow heat exchanger.

T_{c1}

T_{h1} ⟶   ⟶ T_{h2}

T_{c2}

23C.4. Analogy between the unsteady operation of an adsorption column and a cross-flow heat exchanger[1] (Fig. 23C.4). In the heat exchanger shown in the figure, the two fluid streams flow at right angles to one another, and the heat flux parallel to the wall is neglected. Here exchange of heat is clearly less than for a countercurrent exchanger of the same surface and overall heat transfer coefficient under otherwise identical conditions. The heat flow in these exchangers may be expressed for constant U_{loc} as

$$Q = U_{\text{loc}} A \Delta T_{\ln} Y \tag{23C.4-1}$$

Here Q is the total rate of heat transfer, A is the heat transfer surface area, and $\Delta T_{\ln}$ is the logarithmic mean of $(T_{h1} - T_{c1})$ and $(T_{h1} - T_{c1})$, as defined in the figure. Note that T_{h2} and T_{c2} are the flow-averaged temperatures of the two exit streams. We may then regard Y as the ratio of heat transferred in cross flow to that which would be transferred in counterflow.

Use Eq. 23.6-27 to write an expression for Y as a function of the stream rates, physical properties, heat transfer area, and local overall heat transfer coefficient. Express the result in terms of definite integrals, and assume $\hat{C}_{ph}$, $\hat{C}_{pc}$, and U_{loc} to be constant.

23D.1. Unsteady-state operation of a packed column. Show that Eq. 23.6-25 is a valid solution of Eqs. 23.6-23 and 24. The following approach is recommended:

(a) Take the Laplace transform of Eqs. 23.6-23 and 24 with respect to τ. Eliminate the transform of Y from the resulting expressions. Show that the transform of X may be written for the given boundary conditions as

$$\overline{X} = \frac{1}{s} e^{-[s/(s+1)]\zeta} \tag{23D.1-1}$$

in which $\overline{X}$ is the Laplace transform of X.

---

[1] W. Nusselt, *Tech. Math. Therm.*, **1**, 417 (1930); D. M. Smith, *Engineering*, **138**, 479 (1934).

(b) Rewrite this expression in the form

$$\overline{X} = \frac{1}{s} - \int_0^\zeta e^{-\overline{\zeta}}\left(\frac{1}{s+1}\right)e^{[\overline{\zeta}/(s+1)]}d\overline{\zeta} \qquad (23D.1\text{-}2)$$

Invert this expression to obtain Eq. 23.6-25 by making use of the identity

$$\mathcal{L}\{e^{a\tau}F(\tau)\} = \overline{F}(s-a) \qquad (23D.1\text{-}3)$$

where $\mathcal{L}\{F(\tau)\} = \overline{F}(s)$.

23D.2. **Additivity of the lower moments.** Consider a pair of flow systems meeting the requirements of Example 23.6-3 arranged in series. Show that (i) the zeroth moment is the same at the system inlets and outlets of the first and second systems, and (ii) the first absolute moment and the second and third central moments, but not the fourth central moment, are additive. *Suggestion:* For the second and higher moments it is helpful to recognize that the output from the second unit, following a pulse input to the first may be obtained by the use of the convolution integral

$$c(t) = \int_0^t h_1(t-\tau)h_2(\tau)d\tau \equiv h_1{*}h_2 \qquad (23D.2\text{-}1)$$

where h is a system response to a pulse input. A simple way of proceeding is to recognize that the Laplace transform of $c(t)$ may be written as

$$\mathcal{L}\{c(t)\} \equiv F(s) = h_1(s)h_2(s) \qquad (23D.2\text{-}2)$$

It then follows that

$$F'(s) = h_1'(s)h_2(s) + h_1(s)h_2'(s) \qquad (23D.2\text{-}3)$$

and similarly for the higher derivatives. Now it may also be shown that

$$M_0 = F(0); \qquad M_1 = -\frac{F'(0)}{F(0)} \qquad (23D.2\text{-}4,5)$$

$$\mu_2 = \frac{F''(0)}{F(0)} - \frac{[F'(0)]^2}{[F(0)]^2} \qquad (23D.2\text{-}6)$$

$$\mu_3 = -\frac{F'''(0)}{F(0)} + 3\frac{F'(0)F''(0)}{[F(0)]^2} - 2\frac{[F'(0)]^3}{[F(0)]^3} \qquad (23D.2\text{-}7)$$

$$\mu_4 = \frac{F^{(iv)}(0)}{F(0)} - 4\frac{F'(0)F'''(0)}{[F(0)]^2} + 6\frac{F''(0)[F'(0)]^2}{[F(0)]^3} - 3\frac{[F'(0)]^4}{[F(0)]^4} \qquad (23D.2\text{-}8)$$

23D.3. **Start-up of a chemical reactor.** Rework Example 23.6-1 by use of Laplace transforms of Eqs. 23.6-2 and 3.

23D.4. **Transient behavior of N reactors in series.**[2] There are N identical chemical reactors of volume V connected in series, each equipped with a perfect stirrer. Initially, each tank is filled with pure solvent S. At zero time, a solution of A in S is introduced to the first tank at a constant volumetric flow rate Q and a constant concentration $c_A(0)$. This solution also contains a small amount of a dissolved catalyst, introduced just prior to discharge into the first tank, which causes the following first-order reactions to occur:

$$A \underset{k_{1BA}'''}{\overset{k_{1AB}'''}{\rightleftharpoons}} B \underset{k_{1CB}'''}{\overset{k_{1BC}'''}{\rightleftharpoons}} C \qquad (23D.4\text{-}1)$$

The rate constants in these reactions are assumed constant throughout the system. Let $h = Q/V$, the inverse of the "effective residence time" in each tank. Obtain an expression for $c_\alpha(n)$, the concentration of chemical species α in the nth tank at any time t.

---

[2] A. Acrivos and N. R. Amundson, *Ind. Eng. Chem.*, **47**, 1533–1541 (1955).

Chapter 24

Other Mechanisms
for Mass Transport

§24.1● **The equation of change for entropy**

§24.2● **The flux expressions for heat and mass**

§24.3○ **Concentration diffusion and driving forces**

§24.4○ **Applications of the generalized Maxwell–Stefan equations**

§24.5○ **Mass transfer across selectively permeable membranes**

§24.6○ **Diffusion in porous media**

In Chapter 1 we stated that the molecular transport of momentum is related to the velocity gradient by Newton's law of viscosity. In Chapter 8 we gave Fourier's law, which says that molecular heat transport occurs because of a gradient in temperature. However, when we discussed mixtures in Chapter 19, we pointed out an extra contribution to the molecular heat flux that accounts for the amount of enthalpy transported by the interdiffusion of the various species. In Chapter 17 we gave Fick's (first) law of diffusion, which says that molecular mass transport occurs as the result of a concentration gradient. We indicated there that other driving forces may contribute to the mass flux. The purposes of this chapter are to describe the most important of these additional driving forces and to illustrate some applications.

Important among these forces are the gradients of electrical potential and pressure, which govern the behavior of ionic systems and permselective membranes as well as ultracentrifuges. *Electrokinetic* phenomena in particular are rapidly gaining in importance. Induced dipoles can produce separations, such as *dielectrophoresis* and *magnetophoresis*, which are useful in specialized applications. In addition, we shall find that temperature gradients can cause mass fluxes by a process known as *thermal diffusion*[1] or the *Soret effect*, and that concentration gradients can produce heat transfer by the *diffusion-thermo*,[2] or the *Dufour*, effect. Finally, it is important to realize that in systems containing three or more components, the behavior of any one species is influenced by the concentration gradients of all other species present.

Fortunately the wide range of behavior resulting from these various driving forces can be described compactly via the framework provided by nonequilibrium thermody-

---

[1] The effect was first observed in liquids by C. Ludwig, *Sitzber. Akad. Wiss. Wien* **20**, 539 (1856), but is named after Ch. Soret, *Arch. Sci. Phys. Nat.*, *Genève*, **2**, 48–61 (1879); **4**, 209–213 (1880); *Comptes Rendus Acad. Sci.*, *Paris*, **91**, 289–291 (1880). The first observations in gases were made by S. Chapman and F. W. Dootson, *Phil. Mag.*, **33**, 248–253 (1917).

[2] L. Dufour, *Arch. Sci. Phys. Nat. Genève*, **45**, 9–12 (1872); *Ann. Phys.* (5) **28**, 490–492 (1873).

namics;[3] this topic is summarized in §§24.1 and 2. This discussion concludes with the generalized Maxwell–Stefan equations. In the remaining sections we show how various specializations of these equations can be used to provide convenient descriptions of selected diffusional processes.

Those who do not wish to read the first two sections can go directly to the later sections, where the essential results of nonequilibrium thermodynamics are summarized.

§24.1   THE EQUATION OF CHANGE FOR ENTROPY

Nonequilibrium thermodynamics makes use of four postulates above and beyond those of equilibrium thermodynamics:[1]

1. The equilibrium thermodynamic relations apply to systems that are not in equilibrium, provided that the gradients are not too large (*quasi-equilibrium postulate*).

2. All fluxes in the system may be written as linear relations involving all the forces (*linearity postulate*).

3. No coupling of fluxes and forces occurs if the difference in tensorial order of the flux and force is an odd number (*Curie's postulate*).[2]

4. In the absence of magnetic fields the matrix of the coefficients in the flux–force relations is symmetric (*Onsager's reciprocal relations*).[3]

In this and the following section we will use these postulates, which arose from a need to describe various observed phenomena and also from kinetic theory developments. Note that the nonequilibrium theory we are using excludes consideration of non-Newtonian fluids.[4]

In Problem 11D.1 we saw how to derive Jaumann's entropy balance equation,

$$\rho \frac{D\hat{S}}{Dt} = -(\nabla \cdot \mathbf{s}) + g_S \tag{24.1-1}$$

in which $\hat{S}$ is the entropy per unit mass of a multicomponent fluid, $\mathbf{s}$ is the entropy-flux vector, and g_S is the rate of entropy production per unit volume. At this point we do not know what $\mathbf{s}$ and g_S are, and hence our first task is to find expressions for these quantities

---

[3] The discussion here is for multicomponent systems. A discussion for binary systems can be found in L. Landau and E. M. Lifshitz, *Fluid Mechanics*, 2nd edition, Pergamon Press (1987), Chapter VI. See also R. B. Bird, *Korean J. Chem. Eng.*, **15**, 105–123 (1998).

[1] S. R. de Groot and P. Mazur, *Non-Equilibrium Thermodynamics*, North-Holland, Amsterdam (1962). See also H. B. Callen, *Thermodynamics and an Introduction to Thermostatistics*, Wiley, New York (1985), Chapter 14.

[2] P. Curie, *Oeuvres*, Paris (1903), p. 129.

[3] Nobel laureate **Lars Onsager** (1903–1976) studied chemical engineering at the Technical University of Trondheim; after working with Peter Debye in Zürich for two years, he held teaching positions at several universities before moving on to Yale University. His contributions to the thermodynamics of irreversible processes are to be found in L. Onsager, *Phys. Rev.*, **37**, 405–426 (1931); **38**, 2265–2279 (1931). A summary of experimental verifications of the Onsager reciprocal relations has been given by D. G. Miller, in *Transport Phenomena in Fluids* (H. J. M. Hanley, ed.), Marcel Dekker, New York (1969), Chapter 11.

[4] To describe nonlinear viscoelastic fluids one has to generalize the thermodynamic theory, as described in A. N. Beris and B. J. Edwards, *Thermodynamics of Flowing Systems with Internal Microstructure*, Oxford University Press (1994); M. Grmela and H. C. Öttinger, *Phys. Rev.*, **E56**, 6620–6632 (1997); H. C. Öttinger and M. Grmela, *Phys. Rev.*, **E56**, 6633–6655 (1997); B. J. Edwards, H. C. Öttinger, and R. J. J. Jongschaap, *J. Non-Equilibrium Thermodynamics*, **27**, 356–373 (1997); H. C. Öttinger, *Phys. Rev.*, **E57**, 1416–1420 (1998); H. C. Öttinger, *Applied Rheology*, **9**, 17–26 (1999).

in terms of the fluxes and gradients in the system. To do this we have to use the assumption that the equations of equilibrium thermodynamics are valid locally (the "quasi-equilibrium postulate"), which means that equations such as

$$d\hat{U} = Td\hat{S} - pd\hat{V} + \sum_{\alpha=1}^{N} \frac{\overline{G}_\alpha}{M_\alpha} d\omega_\alpha \tag{24.1-2}$$

can be used in a system that is not too far from equilibrium. In this equation $\overline{G}_\alpha$ is the partial molar Gibbs free energy and M_α the molecular weight of species α. We now apply this relation to a fluid element moving with the mass average velocity $\mathbf{v}$. Then we can replace the differential operators by substantial derivative operators. In that form, Eq. 24.1-2 enables us to express $D\hat{S}/Dt$ in terms of $D\hat{U}/Dt$, $D(1/\rho)/Dt$, and $D\omega_\alpha/Dt$. Then the equation of change for internal energy [Eq. (D) of Table 19.2-4], the overall equation of continuity [Eq. (A) of Table 19.2-3], and the equation of continuity for species α [Eq. (B) of Table 19.2-3] can be used for the three substantial derivatives that have been introduced. Thus, after considerable rearranging, we find

$$\mathbf{s} = \frac{1}{T}\left(\mathbf{q} - \sum_{\alpha=1}^{N} \frac{\overline{G}_\alpha}{M_\alpha} \mathbf{j}_\alpha \right) \tag{24.1-3}$$

$$g_S = -\left(\mathbf{q} \cdot \frac{1}{T^2} \nabla T \right) - \sum_{\alpha=1}^{N} \left(\mathbf{j}_\alpha \cdot \left[\nabla\left(\frac{1}{T}\frac{\overline{G}_\alpha}{M_\alpha} \right) - \frac{1}{T}\mathbf{g}_\alpha \right] \right) - \left(\boldsymbol{\tau}:\frac{1}{T}\nabla\mathbf{v} \right) - \sum_{\alpha=1}^{N} \frac{1}{T}\frac{\overline{G}_\alpha}{M_\alpha} r_\alpha \tag{24.1-4}$$

The entropy production has been written as a sum of products of fluxes and forces. However, there are only $N - 1$ independent mass fluxes $\mathbf{j}_\alpha$, and, because of the Gibbs-Duhem equation, there are also only $N - 1$ independent forces. When we take into account this lack of independence,[5] we may rewrite the entropy flux and the entropy production in the following form:

$$\mathbf{s} = \frac{1}{T}\mathbf{q}^{(h)} + \sum_{\alpha=1}^{N} \frac{\overline{S}_\alpha}{M_\alpha} \mathbf{j}_\alpha \tag{24.1-5}$$

$$Tg_S = -(\mathbf{q}^{(h)} \cdot \nabla \ln T) - \sum_{\alpha=1}^{N} \left(\mathbf{j}_\alpha \cdot \frac{cRT}{\rho_\alpha}\mathbf{d}_\alpha \right) - (\boldsymbol{\tau}:\nabla\mathbf{v}) - \sum_{\alpha=1}^{N} \frac{\overline{G}_\alpha}{M_\alpha} r_\alpha \tag{24.1-6}$$

in which $\mathbf{q}^{(h)}$ is the heat flux with the diffusional enthalpy flux subtracted off,

$$\mathbf{q}^{(h)} = \mathbf{q} - \sum_{\alpha=1}^{N} \frac{\overline{H}_\alpha}{M_\alpha} \mathbf{j}_\alpha \tag{24.1-7}$$

and

$$cRT\mathbf{d}_\alpha = c_\alpha T\nabla\left(\frac{\overline{G}_\alpha}{T} \right) + c_\alpha\overline{H}_\alpha\nabla \ln T - \omega_\alpha\nabla p - \rho_\alpha\mathbf{g}_\alpha + \omega_\alpha \sum_{\beta=1}^{N} \rho_\beta\mathbf{g}_\beta$$

$$= c_\alpha RT\nabla \ln a_\alpha + (\phi_\alpha - \omega_\alpha)\nabla p - \rho_\alpha\mathbf{g}_\alpha + \omega_\alpha \sum_{\beta=1}^{N} \rho_\beta\mathbf{g}_\beta \tag{24.1-8}$$

The second form in Eq. 24.1-8 is obtained[5] by using the relation $d\overline{G}_\alpha = RTd \ln a_\alpha$, where a_α is the activity. In the operation $\nabla \ln a_\alpha$, the derivative is to be taken at constant T and p, and the quantity $\phi_\alpha = c_\alpha\overline{V}_\alpha$ is the volume fraction of species α. The $\mathbf{d}_\alpha$ introduced here are called *diffusional driving forces*, and they account for the *concentration diffusion* (term

---

[5] For the intermediate steps, see C. F. Curtiss and R. B. Bird, *Ind. Eng. Chem. Research*, **38**, 2515–2522 (1999), errata **40**, 1791 (2001).

with $\nabla \ln a_\alpha$), *pressure diffusion* (term with ∇p), and *forced diffusion* (term with $\mathbf{g}_\alpha$). The $\mathbf{d}_\alpha$ are so defined that $\Sigma_\alpha \mathbf{d}_\alpha = 0$.

The entropy production in Eq. 24.1-6, which is a sum of products of fluxes and forces, is the starting point for the nonequilibrium thermodynamics development. According to the "linearity postulate" each of the fluxes in Eq. 24.1-6 ($\mathbf{q}^{(h)}$, $\mathbf{j}_\alpha$, $\boldsymbol{\tau}$, and $\overline{G}_\alpha / M_\alpha$) can be written as a linear function of all the forces (∇T, $\mathbf{d}_\alpha$, $\nabla \mathbf{v}$, and r_α). However, because of "Curie's postulate," each of the $\mathbf{j}_\alpha$ must depend linearly on all of the $\mathbf{d}_\alpha$ as well as on ∇T, and $\mathbf{q}^{(h)}$ must depend linearly on ∇T as well as on all the $\mathbf{d}_\alpha$, but neither $\mathbf{j}_\alpha$ nor $\mathbf{q}^{(h)}$ can depend on $\nabla \mathbf{v}$ or r_α. Similarly the stress tensor $\boldsymbol{\tau}$ will depend on the tensor $\nabla \mathbf{v}$, and also on the scalar driving forces r_α multiplied by the unit tensor. Since the "coupling" between $\boldsymbol{\tau}$ and the chemical reactions has not been studied, we omit any further consideration of it. In the next section we discuss the coupling among all the vector forces and vector fluxes and the consequences of applying the "Onsager reciprocal relations."

§24.2 THE FLUX EXPRESSIONS FOR HEAT AND MASS

We now employ the "linearity postulate" to obtain for the vector fluxes

$$\mathbf{q}^{(h)} = -a_{00} \nabla \ln T - \sum_{\beta=1}^{N} \frac{cRTa_{0\beta}}{\rho_\beta} \mathbf{d}_\beta \tag{24.2-1}$$

$$\mathbf{j}_\alpha = -a_{\alpha 0} \nabla \ln T - \rho_\alpha \sum_{\beta=1}^{N} \frac{cRTa_{\alpha\beta}}{\rho_\alpha \rho_\beta} \mathbf{d}_\beta \qquad \alpha = 1, 2, \ldots, N \tag{24.2-2}$$

In these equations the quantities a_{00}, $a_{0\beta}$, $a_{\alpha 0}$, and $a_{\alpha\beta}$ are the "phenomenological coefficients" (that is, the transport properties). Because the $\mathbf{j}_\alpha$ and $\mathbf{d}_\alpha$ are not all independent, it must be required that $a_{\alpha\beta} + \Sigma_\gamma a_{\alpha\gamma} = 0$, where the sums are over all γ (except $\gamma = \beta$) from 1 to N. Now according to the Onsager reciprocal relations, $a_{\alpha 0} = a_{0\alpha}$ and $a_{\alpha\beta} = a_{\beta\alpha}$ for all values of α and β from 1 to N.

Next we relate the phenomenological coefficients to the transport coefficients. First we relabel $a_{\alpha 0}$ and $a_{0\alpha}$ as D_α^T, the *multicomponent thermal diffusion coefficients*. These have the property that $\Sigma_\alpha D_\alpha^T = 0$. Then we define the *multicomponent Fick diffusivities*,[1] $\mathbb{D}_{\alpha\beta}$, by $\mathbb{D}_{\alpha\beta} = -cRTa_{\alpha\beta}/\rho_\alpha \rho_\beta$. These diffusivities are symmetric ($\mathbb{D}_{\alpha\beta} = \mathbb{D}_{\beta\alpha}$) and obey the relations $\Sigma_\alpha \omega_\alpha \mathbb{D}_{\alpha\beta} = 0$. Then Eq. 24.2-2 becomes

$$\mathbf{j}_\alpha = -D_\alpha^T \nabla \ln T + \rho_\alpha \sum_{\beta=1}^{N} \mathbb{D}_{\alpha\beta} \mathbf{d}_\beta \qquad \alpha = 1, 2, \ldots, N \tag{24.2-3}$$

for the multicomponent mass fluxes. These are the *generalized Fick equations*. When the second form of Eq. 24.1-8 is substituted into Eq. 24.2-3 we see that there are four contributions to the mass-flux vector $\mathbf{j}_\alpha$: the concentration diffusion term (containing the activity gradient), the pressure diffusion term (containing the pressure gradient), the forced diffusion term (containing the external forces), and the thermal diffusion term (proportional to the temperature gradient).

---

[1] C. F. Curtiss, *J. Chem. Phys.*, **49**, 2917–2919 (1968); see also D. W. Condiff, *J. Chem. Phys.*, **51**, 4209–4212 (1969), and C. F. Curtiss and R. B. Bird, *Ind. Eng. Chem. Research*, **39**, 2515–2522 (1999); errata **41**, 1791 (2001). The $\mathbb{D}_{\alpha\beta}$ used here are the negatives of the Curtiss $\tilde{D}_{\alpha\beta}$, which, in turn, are different from the $D_{\alpha\beta}$ used by J. O. Hirschfelder, C. F. Curtiss, and R. B. Bird, *Molecular Theory of Gases and Liquids*, Wiley, New York (1954), second corrected printing (1964), Chapter 11.

Equation 24.2-3 can be turned "wrong-side out"[1,2] and solved for the driving forces $\mathbf{d}_\alpha$:

$$\mathbf{d}_\alpha = -\sum_{\beta \neq \alpha} \frac{x_\alpha x_\beta}{\mathcal{D}_{\alpha\beta}}\left(\frac{D_\alpha^T}{\rho_\alpha} - \frac{D_\beta^T}{\rho_\beta}\right)(\nabla \ln T) - \sum_{\beta \neq \alpha} \frac{x_\alpha x_\beta}{\mathcal{D}_{\alpha\beta}}\left(\frac{\mathbf{j}_\alpha}{\rho_\alpha} - \frac{\mathbf{j}_\beta}{\rho_\beta}\right) \qquad (\alpha = 1, 2, \ldots, N) \qquad (24.2\text{-}4)$$

These are the *generalized Maxwell–Stefan equations*, a special case of which was given in Eq. 17.9-1. The $\mathcal{D}_{\alpha\beta}$ are called the *multicomponent Maxwell-Stefan diffusivities*, and they have been proven to be symmetric;[3] their relation to the $\mathbb{D}_{\alpha\beta}$ will be discussed presently.

When the expression for $\mathbf{d}_\alpha$ in Eq. 24.2-4 is substituted into Eq. 24.2-1, we get

$$\mathbf{q}^{(h)} = -\left[a_{00} + \sum_{\alpha=1}^{N}\sum_{\substack{\beta=1\\\beta\neq\alpha}}^{N} \frac{cRTx_\alpha x_\beta}{\rho_\alpha}\frac{D_\alpha^T}{\mathcal{D}_{\alpha\beta}}\left(\frac{D_\beta^T}{\rho_\beta} - \frac{D_\alpha^T}{\rho_\alpha}\right)\right]\nabla \ln T$$

$$+ \sum_{\alpha=1}^{N}\sum_{\substack{\beta=1\\\beta\neq\alpha}}^{N} \frac{cRTx_\alpha x_\beta}{\rho_\alpha}\frac{D_\alpha^T}{\mathcal{D}_{\alpha\beta}}\left(\frac{\mathbf{j}_\alpha}{\rho_\alpha} - \frac{\mathbf{j}_\beta}{\rho_\beta}\right) \qquad (24.2\text{-}5)$$

The thermal conductivity of a mixture is defined to be the coefficient of proportionality between the heat-flux vector and the temperature gradient when there are no mass fluxes in the system. Thus, the quantity in brackets is, by general agreement, the thermal conductivity k times the absolute temperature T. If we combine this result with the definition in Eq. 24.1-7, w get for the final expression for the heat flux:[3]

$$\mathbf{q} = -k\nabla T + \sum_{\alpha=1}^{N} \frac{\overline{H}_\alpha}{M_\alpha}\mathbf{j}_\alpha + \sum_{\alpha=1}^{N}\sum_{\substack{\beta=1\\\beta\neq\alpha}}^{N} \frac{cRTx_\alpha x_\beta}{\rho_\alpha}\frac{D_\alpha^T}{\mathcal{D}_{\alpha\beta}}\left(\frac{\mathbf{j}_\alpha}{\rho_\alpha} - \frac{\mathbf{j}_\beta}{\rho_\beta}\right) \qquad (24.2\text{-}6)$$

We see that the heat flux vector $\mathbf{q}$ consists of three terms: the heat conduction term (containing the thermal conductivity), the heat diffusion term (containing the partial molar enthalpies and the mass fluxes), and finally the Dufour term (containing the thermal diffusion coefficients and the mass fluxes). The heat diffusion term, already encountered in Eq. 19.3-3, is generally important in diffusing systems. The Dufour term is usually small and can usually be neglected.

Equations 24.2-3, 4, and 6 are the principal results of nonequilibrium thermodynamics. We now have the mass- and heat-flux vectors expressed in terms of the transport properties and the fluxes.

Next we discuss the relation between the matrix of Fick diffusivities $\mathbb{D}_{\alpha\beta}$ and that of the Maxwell–Stefan diffusivities $\mathcal{D}_{\alpha\beta}$. Both matrices are symmetric and of order $N \times N$, and both have $\frac{1}{2}N(N-1)$ independent elements. The $\mathcal{D}_{\alpha\beta}$ are obtained thus:[3]

$$\mathcal{D}_{\alpha\beta} = \frac{x_\alpha x_\beta}{\omega_\alpha \omega_\beta}\frac{\sum_{\gamma\neq\alpha}\mathbb{D}_{\alpha\gamma}(\text{adj } B_\alpha)_{\gamma\beta}}{\sum_{\gamma\neq\alpha}(\text{adj } B_\alpha)_{\gamma\beta}} \qquad \alpha, \beta = 1, 2, \ldots, N \qquad (24.2\text{-}7)$$

in which $(B_\alpha)_{\beta\gamma} = -\mathbb{D}_{\beta\gamma} + \mathbb{D}_{\alpha\gamma}$—that is, the $\beta\gamma$-component of a matrix called B_α, which is of order $(N-1) \times (N-1)$—and adj B_α is the matrix adjoint to B_α. For binary and ternary

---

[2] H. J. Merk, *Appl. Sci. Res.*, **A8**, 73–99 (1959); E. Helfand, *J. Chem. Phys.*, **33**, 319–322 (1960). **Hendrik Jacobus Merk** (1920–1988) performed the inversion of the mass-flux expressions when he was a graduate student in engineering physics at the Technical University of Delft; from 1953 to 1987 he was a professor at the same institution.

[3] C. F. Curtiss and R. B. Bird, *Ind. Eng. Chem. Research*, **38**, 2515–2522 (1999); errata, **40**, 1791 (2001).

Table 24.2-1 Summary[1] of Expressions for the $\mathbb{D}_{\alpha\beta}$ in Terms of the $\mathcal{D}_{\alpha\beta}$. [Note: Additional entries may be generated by cyclic permutation of the indices. Formulas for four-component systems are given in the references.]

Binary:
$$\mathbb{D}_{11} = -\frac{\omega_2^2}{x_1 x_2}\mathcal{D}_{12} \tag{A}$$

$$\mathbb{D}_{22} = -\frac{\omega_1^2}{x_1 x_2}\mathcal{D}_{12} \tag{B}$$

$$\mathbb{D}_{12} = \mathbb{D}_{21} = \frac{\omega_1\omega_2}{x_1 x_2}\mathcal{D}_{12} \tag{C}$$

Ternary:
$$\mathbb{D}_{11} = -\frac{\dfrac{(\omega_2+\omega_3)^2}{x_1\mathcal{D}_{23}} + \dfrac{\omega_2^2}{x_2\mathcal{D}_{13}} + \dfrac{\omega_3^2}{x_3\mathcal{D}_{12}}}{\dfrac{x_1}{\mathcal{D}_{12}\mathcal{D}_{13}} + \dfrac{x_2}{\mathcal{D}_{12}\mathcal{D}_{23}} + \dfrac{x_3}{\mathcal{D}_{13}\mathcal{D}_{23}}} \tag{D}$$

$$\mathbb{D}_{12} = \frac{\dfrac{\omega_1(\omega_2+\omega_3)}{x_1\mathcal{D}_{23}} + \dfrac{\omega_2(\omega_1+\omega_3)}{x_2\mathcal{D}_{13}} - \dfrac{\omega_3^2}{x_3\mathcal{D}_{12}}}{\dfrac{x_1}{\mathcal{D}_{12}\mathcal{D}_{13}} + \dfrac{x_2}{\mathcal{D}_{12}\mathcal{D}_{23}} + \dfrac{x_3}{\mathcal{D}_{13}\mathcal{D}_{23}}} \tag{E}$$

systems, the explicit interrelations are given in Tables 24.2-1 and 2. In Eq. (C) of Table 24.2-1, it can be seen that for a binary mixture the $\mathbb{D}_{\alpha\beta}$ and $\mathcal{D}_{\alpha\beta}$ differ by a factor that is a function of the concentration. However, they do have the same sign, which explains why the plus sign was chosen in Eq. 24.2-3 instead of a minus sign.

We are now in a position to present the three final results of this section that are useful as starting points for solving diffusion problems. For *multicomponent diffusion in gases or liquids*, combining Eqs. 24.1-8 and 24.2-4 gives

$$\sum_{\substack{\beta=1\\\beta\neq\alpha}}^{N} \frac{x_\alpha x_\beta}{\mathcal{D}_{\alpha\beta}}(\mathbf{v}_\alpha - \mathbf{v}_\beta) = -x_\alpha(\nabla \ln a_\alpha)_{T,p} - \frac{1}{cRT}\left[(\phi_\alpha - \omega_\alpha)\nabla p - \rho_\alpha \mathbf{g}_\alpha + \omega_\alpha \sum_{\beta=1}^{N}\rho_\beta \mathbf{g}_\beta\right]$$
$$- \sum_{\substack{\beta=1\\\beta\neq\alpha}}^{N} \frac{x_\alpha x_\beta}{\mathcal{D}_{\alpha\beta}}\left(\frac{D_\alpha^T}{\rho_\alpha} - \frac{D_\beta^T}{\rho_\beta}\right)(\nabla \ln T) \qquad (\alpha = 1, 2, \ldots, N) \tag{24.2-8}$$

This equation has been written in terms of the difference of molecular velocities, $\mathbf{v}_\alpha - \mathbf{v}_\beta$. Equations (D) to (I) of Table 17.8-1 may then be used to write this equation in terms of any desired mass or molar fluxes.

Table 24.2-2 Summary[1] of Expressions for the $\mathcal{D}_{\alpha\beta}$ in Terms of the $\mathbb{D}_{\alpha\beta}$. [Note: Additional entries can be generated by cyclic permutation of the indices. See the original references for four-component systems.]

Binary:
$$\mathcal{D}_{12} = \frac{x_1 x_2}{\omega_1\omega_2}\mathbb{D}_{12} = -\frac{x_1 x_2}{\omega_2^2}\mathbb{D}_{11} = -\frac{x_1 x_2}{\omega_1^2}\mathbb{D}_{22} \tag{A}$$

Ternary:
$$\mathcal{D}_{12} = \frac{x_1 x_2}{\omega_1\omega_2}\frac{\mathbb{D}_{12}\mathbb{D}_{33} - \mathbb{D}_{13}\mathbb{D}_{23}}{\mathbb{D}_{12} + \mathbb{D}_{33} - \mathbb{D}_{13} - \mathbb{D}_{23}} \tag{B}$$

If one wishes to designate one species γ as being special (for example, the solvent), then Eq. 24.2-8 can be rewritten thus (see Problem 24C.1):

$$\sum_{\substack{\beta=1 \\ \text{all } \beta}}^{N} \frac{x_\alpha x_\beta}{D_{\alpha\beta}}(\mathbf{v}_\gamma - \mathbf{v}_\beta) = -x_\alpha(\nabla \ln a_\alpha)_{T,p} - \frac{1}{cRT}\left[(\phi_\alpha - \omega_\alpha)\nabla p - \rho_\alpha \mathbf{g}_\alpha + \omega_\alpha \sum_{\beta=1}^{N}\rho_\beta \mathbf{g}_\beta\right]$$

$$- \sum_{\substack{\beta=1 \\ \text{all } \beta}}^{N} \frac{x_\alpha x_\beta}{\mathcal{D}_{\alpha\beta}}\left(\frac{D_\gamma^T}{\rho_\gamma} - \frac{D_\beta^T}{\rho_\beta}\right)(\nabla \ln T) \qquad (\alpha = 1, 2, \ldots, N) \qquad (24.2-9)$$

Note that in Eq. 24.2-8 there are $N(N - 1)/2$ symmetric diffusivities, $\mathcal{D}_{\alpha\beta}$, and that the $\mathcal{D}_{\alpha\alpha}$ do not appear and are hence not defined. However, in Eq. 24.2-9, there are $N(N + 1)/2$ symmetric diffusivities, but the $\mathcal{D}_{\alpha\alpha}$ (N of them) now appear, and therefore we have to supply an auxiliary relation $\Sigma_\alpha(x_\alpha/\mathcal{D}_{\alpha\beta}) = 0$, in which the sum is over all α. Equation 24.2-9, with the auxiliary relation, is equivalent to Eq. 24.2-8, and both of these generalized Maxwell–Stefan equations are equivalent to the generalized Fick equations of Eq. 24.2-3, together with its auxiliary relation.

For *multicomponent diffusion in gases at low density*, the activity may be replaced by the mole fraction, and furthermore, to a very good approximation, the $\mathcal{D}_{\alpha\beta}$ may be replaced by $\mathscr{D}_{\alpha\beta}$. These are the *binary* diffusivities for all pairs of species in the mixture. Since the $\mathscr{D}_{\alpha\beta}$ vary only slightly with concentration, whereas the $\mathbb{D}_{\alpha\beta}$ are highly concentration-dependent, it is preferable to use the Maxwell–Stefan form (Eq. 24.2-4) rather than the Fick form (Eq. 24.2-3).

For *binary diffusion in gases or liquids*, Eq. (C) of Table 24.2-1 and Eq. 17B.3-1 may be used to simplify Eq. 24.2-8 as follows:

$$\mathbf{J}_A^* = -c\mathcal{D}_{AB}\left[x_A\nabla \ln a_A + \frac{1}{cRT}[(\phi_A - \omega_A)\nabla p - \rho\omega_A\omega_B(\mathbf{g}_A - \mathbf{g}_B)] + k_T\nabla \ln T\right] \qquad (24.2-10)$$

In this equation we have introduced the *thermal diffusion ratio*, defined by $k_T = -D_A^T/\rho\tilde{D}_{AB} = +(D_A^T/\rho\tilde{D}_{AB})(x_A x_B/\omega_A\omega_B)$. Other quantities encountered are the *thermal diffusion factor* α_T and the *Soret coefficient* σ_T, defined by $k_T = \alpha_T x_A x_B = \sigma_T x_A x_B T$. For gases α_T is almost independent of composition, and σ_T is the quantity preferred for liquids. When k_T is positive, species A moves toward the colder region, and when it is negative, species A moves toward the warmer region. Some sample values of k_T for gases and liquids are given in Table 24.2-3.

For binary mixtures of dilute gases, it is found by experiment that the species with the larger molecular weight usually goes to the colder region. If the molecular weights are about equal, then usually the species with the larger diameter moves to the colder region. In some instances there is a change in the sign of the thermal diffusion ratio as the temperature is lowered.[4]

In the remainder of the chapter, we explore some of the consequences of the mass-flux expressions in Eqs. 24.2-8, 9, and 10.

EXAMPLE 24.2-1

Thermal Diffusion and the Clusius–Dickel Column

In this example we discuss the diffusion of species under the influence of a temperature gradient. To illustrate the phenomenon we consider the system shown in Fig. 24.2-1, two bulbs joined together by an insulated tube of small diameter and filled with a mixture of ideal gases A and B. The bulbs are maintained at constant temperatures T_1 and T_2, respectively, and the diameter of the insulated tube is small enough to eliminate convection currents substantially. Ultimately the system arrives at a steady state, with gas A enriched at one end of the tube and depleted at the other. Obtain an expression for $x_{A2} - x_{A1}$, the difference of the mole fractions at the two ends of the tube.

---

[4] S. Chapman and T. G. Cowling, *The Mathematical Theory of Non-Uniform Gases*, 3rd edition, Cambridge University Press (1970), p. 274.

Table 24.2-3 Experimental Thermal Diffusion Ratios
for Liquids and Low-Density Gas Mixtures

Liquids:[a]

Components A–B	T (K)	x_A	k_T
$C_2H_2Cl_4$–n-C_6H_{14}	298	0.5	1.08
$C_2H_4Br_2$–$C_2H_4Cl_2$	298	0.5	0.225
$C_2H_2Cl_4$–CCl_4	298	0.5	0.060
CBr_4–CCl_4	298	0.09	0.129
CCl_4–CH_3OH	313	0.5	1.23
CH_3OH–H_2O	313	0.5	−0.137
cyclo-C_6H_{12}–C_6H_6	313	0.5	0.100

Gases:

Components A–B	T (K)	x_A	k_T
Ne–He[b]	330	0.80	0.0531
		0.40	0.1004
N_2–H_2[c]	264	0.706	0.0548
		0.225	0.0663
D_2–H_2[d]	327	0.90	0.0145
		0.50	0.0432
		0.10	0.0166

[a] R. L. Saxton, E. L. Dougherty, and H. G. Drickamer, *J. Chem. Phys.*, **22**, 1166–1168 (1954); R. L. Saxton and H. G. Drickamer, *J. Chem. Phys.*, **22**, 1287–1288 (1954); L. J. Tichacek, W. S. Kmak, and H. G. Drickamer, *J. Phys. Chem.*, **60**, 660–665 (1956).

[b] B. E. Atkins, R. E. Bastick, and T. L. Ibbs, *Proc. Roy. Soc. (London)*, **A172**, 142–158 (1939).

[c] T. L. Ibbs, K. E. Grew, and A. A. Hirst, *Proc. Roy. Soc. (London)*, **A173**, 543–554 (1939).

[d] H. R. Heath, T. L. Ibbs, and N. E. Wild, *Proc. Roy. Soc. (London)*, **A178**, 380–389 (1941).

SOLUTION

After steady state has been achieved, there is no net motion of either A or B, so that $\mathbf{J}_A^* = 0$. If we take the tube axis to be in the z direction, then from Eq. 24.2-10 we get

$$\frac{dx_A}{dz} + \frac{k_T}{T}\frac{dT}{dz} = 0 \tag{24.2-11}$$

Here the activity a_A has been replaced by the mole fraction x_A, as is appropriate for an ideal gas mixture. Usually the degree of separation in an apparatus of this kind is small. We may therefore ignore the effect of composition on k_T and integrate this equation to get

$$x_{A2} - x_{A1} = -\int_{T_1}^{T_2} \frac{k_T}{T}\,dT \tag{24.2-12}$$

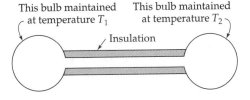

This bulb maintained at temperature T_1 This bulb maintained at temperature T_2

Insulation

Fig. 24.2-1. Steady-state binary thermal diffusion in a two-bulb apparatus. The mixture of gases A and B tends to separate under the influence of the thermal gradient.

Because the dependence of k_T on T is rather complicated, it is customary to assume k_T constant at the value for some mean temperature T_m. Equation 24.2-12 then gives (approximately)

$$x_{A2} - x_{A1} = -k_T(T_m) \ln \frac{T_2}{T_1} \tag{24.2-13}$$

The recommended[5] mean temperature is

$$T_m = \frac{T_1 T_2}{T_2 - T_1} \ln \frac{T_2}{T_1} \tag{24.2-14}$$

Equations 23.2-13 and 14 are useful for estimating the order of magnitude of thermal diffusion effects.

Unless the temperature gradient is very large, the separation will normally be quite small. Therefore it has been advantageous to combine the thermal diffusion effect with free convection between two vertical walls, one heated and the other cooled. The heated stream then ascends, and the cooled one descends. The upward stream will be richer in one of the components—say, A—and the downward stream will be richer in B. This is the principle of the operation of the *Clusius–Dickel column*.[6-8] By coupling many of these columns together in a "cascade" it is possible to perform a separation. During World War II this was one of the methods used for separating the uranium isotopes by using uranium hexafluoride gas. The method has also been used with some success in the separation of organic mixtures, where the components have very nearly the same boiling points, so that distillation is not an option.

The thermal diffusion ratio can also be obtained from the Dufour (diffusion-thermo) effect, but the analysis of the experiment is fraught with problems and experimental errors difficult to avoid.[9]

EXAMPLE 24.2-2

Pressure Diffusion and the Ultra Centrifuge

Next we examine diffusion in the presence of a pressure gradient. If a sufficiently large pressure gradient can be established, then a measurable separation can be effected. One example of this is the ultracentrifuge, which has been used to separate enzymes and proteins. In Fig. 24.2-2 we show a small cylindrical cell in a very high-speed centrifuge. The length of the cell,

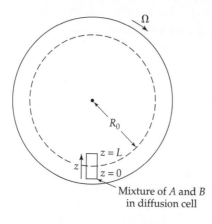

Mixture of A and B in diffusion cell

Fig. 24.2-2. Steady-state pressure diffusion in a centrifuge. The mixture in the diffusion cell tends to separate by virtue of the pressure gradient produced in the centrifuge.

---

[5] H. Brown, *Phys. Rev.*, **58**, 661–662 (1940).

[6] K. Clusius and G. Dickel, Z. *Phys. Chem.*, **B44**, 397–450, 451–473 (1939).

[7] K. E. Grew and T. L. Ibbs, *Thermal Diffusion in Gases*, Cambridge University Press (1952); K. E. Grew, in *Transport Phenomena in Fluids* (H. J. M. Hanley, ed.), Marcel Dekker, New York (1969), Chapter 10.

[8] R. B. Bird, *Advances in Chemical Engineering*, **1**, 155–239 (1956), §4.D.2; *errata*, **2**, 325 (1958).

[9] S. Chapman and T. G. Cowling, *The Mathematical Theory of Nonuniform Gases*, 3rd edition, Cambridge University Press (1970), pp. 268–271.

L, is short with respect to the radius of rotation R_0, and the solution density may be considered a function of composition only. Determine the distribution of the two components at steady state in terms of their partial molar volumes and the pressure gradient. The latter is obtained from the equation of motion as

$$\frac{dp}{dz} = -\rho g_\Omega \approx -\rho \Omega R_0 \tag{24.2-15}$$

For simplicity, we assume that the partial molar volumes and the activity coefficients are constant over the range of conditions existing in the cell.

SOLUTION

At steady state $\mathbf{j}_A = 0$, and the relevant terms in Eq. 24.2-10 give for species A

$$\frac{dx_A}{dz} + \frac{M_A x_A}{RT}\left(\frac{\overline{V}_A}{M_A} - \frac{1}{\rho}\right)\frac{dp}{dz} = 0 \tag{24.2-16}$$

Inserting the appropriate expression for the pressure gradient and then multiplying by $(\overline{V}_B/x_A)dz$, we get for species A

$$\overline{V}_B \frac{dx_A}{x_A} = -\overline{V}_B \frac{g_\Omega}{RT}(\rho \overline{V}_A - M_A)dz \tag{24.2-17}$$

Then we write a similar equation for species B, which is

$$\overline{V}_A \frac{dx_B}{x_B} = -\overline{V}_A \frac{g_\Omega}{RT}(\rho \overline{V}_B - M_B)dz \tag{24.2-18}$$

Subtracting Eq. 24.2-18 from Eq. 24.2-17 we get

$$\overline{V}_B \frac{dx_A}{x_A} - \overline{V}_A \frac{dx_B}{x_B} = \frac{g_\Omega}{RT}(M_B \overline{V}_A - M_A \overline{V}_B)dz \tag{24.2-19}$$

We now integrate this equation from $z = 0$ to some arbitrary value of z, taking account of the fact that the mole fractions of A and B at $z = 0$ are x_{A0} and x_{B0}, respectively. This gives

$$\overline{V}_B \int_{x_{A0}}^{x_A} \frac{dx_A}{x_A} - \overline{V}_A \int_{x_{B0}}^{x_B} \frac{dx_B}{x_B} = \frac{M_B \overline{V}_A - M_A \overline{V}_B}{RT} \int_0^z g_\Omega dz \tag{24.2-20}$$

If g_Ω is treated as constant over the range of integration, then we get

$$\overline{V}_B \ln \frac{x_A}{x_{A0}} - \overline{V}_A \ln \frac{x_B}{x_{B0}} = \frac{M_B \overline{V}_A - M_A \overline{V}_B}{RT} g_\Omega z \tag{24.2-21}$$

Then we take the exponential of both sides to find

$$\left(\frac{x_A}{x_{A0}}\right)^{\overline{V}_B}\left(\frac{x_{B0}}{x_B}\right)^{\overline{V}_A} = \exp\left[(M_B \overline{V}_A - M_A \overline{V}_B)\left(\frac{g_\Omega z}{RT}\right)\right] \tag{24.2-22}$$

This describes the steady-state concentration distribution for a binary system in a constant centrifugal force field. Note that, since this result contains no transport coefficients at all, the same result can be obtained by an equilibrium thermodynamics analysis.[10] However, if one wishes to analyze the time-dependent behavior of a centrifugation, then the diffusivity for the mixing A–B will appear in the result, and the problem cannot be solved by equilibrium thermodynamics.

---

[10] E. A. Guggenheim, *Thermodynamics*, North-Holland, Amsterdam (1950), pp. 356–360.

§24.3   CONCENTRATION DIFFUSION AND DRIVING FORCES

In Chapter 17 we wrote Fick's first law by stating that the mass (or molar) flux is proportional to the gradient of the mass (or mole) fraction, as summarized in Table 17.8-2.

On the other hand, in Eq. 24.2-10 it appears that the thermodynamics of irreversible processes dictates using the activity gradient as the driving force for concentration diffusion. In this section we show that either the activity gradient or the mass (or mole) fraction gradient driving force may be used, but that each choice requires a different diffusivity. These two diffusivities are related, and we illustrate this for a binary mixture.

When we drop the pressure-, thermal-, and forced-diffusion terms from Eq. 24.2-10, we get

$$\mathbf{J}_A^* = -c\mathcal{D}_{AB}x_A\nabla \ln a_A \tag{24.3-1}$$

This may be rewritten by making use of the fact that the activity coefficient is a function of x_A to obtain

$$\mathbf{J}_A^* = -c\mathcal{D}_{AB}\left(\frac{\partial \ln a_A}{\partial \ln x_A}\right)_{T,p}\nabla x_A \tag{24.3-2}$$

The activity may be written as the product of the activity coefficient and the mole fraction ($a_A = \gamma_A x_A$) so that

$$\mathbf{J}_A^* = -c\mathcal{D}_{AB}\left[1 + \left(\frac{\partial \ln \gamma_A}{\partial \ln x_A}\right)_{T,p}\right]\nabla x_A \tag{24.3-3}$$

If the mixture is "ideal," then the activity coefficient is equal to unity, Eq. 24.3-3 becomes the same as Eq. (B) of Table 17.8-2, and $\mathcal{D}_{AB} = \mathscr{D}_{AB}$.

If the mixture is "nonideal," one can express the binary diffusivity $\mathscr{D}_{AB}$ as

$$\mathscr{D}_{AB} = \mathcal{D}_{AB}\left(\frac{\partial \ln a_A}{\partial \ln x_A}\right)_{T,p} = \mathcal{D}_{AB}\left(1 + \left(\frac{\partial \ln \gamma_A}{\partial \ln x_A}\right)_{T,p}\right) \tag{24.3-4}$$

then Eq. 24.3-2 and 3 become

$$\mathbf{J}_A^* = -c\mathscr{D}_{AB}\nabla x_A \tag{24.3-5}$$

which is one of the forms of Fick's law (see Eq. (B) of Table 17.8-2). In order to measure $\mathcal{D}_{AB}$, one has to have measurements of the activity as a function of concentration, and for this reason $\mathcal{D}_{AB}$ has not been popular.

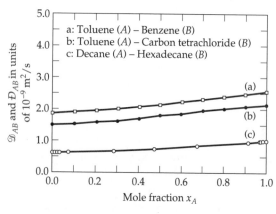

Fig. 24.3-1. Diffusivity in ideal liquid mixtures at 25°C [P. W. M. Rutten, *Diffusion in Liquids*, Delft University Press (1992), p. 31].

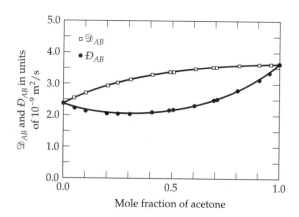

Fig. 24.3-2. Diffusivity in a nonideal liquid mixture (acetone–chloroform (at 25°C) [P. W. M. Rutten, *Diffusion in Liquids*, Delft University Press (1992), p. 32].

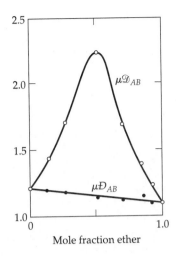

Fig. 24.3-3. Effect of activity on the product of viscosity and diffusivity for liquid mixtures of chloroform and diethyl ether [R. E. Powell, W. E. Roseveare, and H. Eyring, *Ind. Eng. Chem.*, **33**, 430–435 (1941)].

For ideal mixtures $\mathscr{D}_{AB}$ and $Ð_{AB}$ are identical, and are nearly linear functions of the mole fraction as shown in Fig. 24.3-1. For nonideal mixtures $\mathscr{D}_{AB}$ and $Ð_{AB}$ are different nonlinear functions of the mole fraction; an example is shown in Fig. 24.3-2. However, the product $\mu Ð_{AB}$ has been found for *some* nonideal mixtures to be very nearly linear in the mole fraction, whereas $\mu \mathscr{D}_{AB}$ is not (see Fig. 24.3-3). There is no compelling reason to prefer one diffusivity over the other. Most of the diffusivities reported in the literature are $\mathscr{D}_{AB}$ and not $Ð_{AB}$.

§24.4   APPLICATIONS OF THE GENERALIZED MAXWELL–STEFAN EQUATIONS

The generalized Maxwell–Stefan equations were given in Eq. 24.2-4 in terms of the diffusional driving forces $\mathbf{d}_\alpha$, and the expression for $\mathbf{d}_\alpha$ was given in Eq. 24.1-8. When these are combined we get the Maxwell–Stefan equations in terms of the activity gradient, the pressure gradient, and the external forces acting on the various species, given (Eqs. 24.2-8 or 9):

$$-\mathbf{d}_\alpha = \sum_{\substack{\beta=1 \\ \text{all } \beta}}^{N} \frac{x_\alpha x_\beta}{Ð_{\alpha\beta}} (\mathbf{v}_\gamma - \mathbf{v}_\beta) + \text{thermal diffusion terms}$$

$$= -x_\alpha (\nabla \ln a_\alpha)_{T,p} - \frac{1}{cRT}\left[(\phi_\alpha - \omega_\alpha)\nabla p - \rho_\alpha \mathbf{g}_\alpha + \omega_\alpha \sum_{\beta=1}^{N} \rho_\beta \mathbf{g}_\beta \right]$$

$$\alpha = 1, 2, 3, \ldots, N \tag{24.4-1}$$

The thermal diffusion terms have not been displayed here, since they will not be needed in this section. The symbols $\phi_\alpha = c_\alpha \bar{V}_\alpha$ and ω_α designate, respectively, the volume fraction and mass fraction of species α. As explained in §§24.1 and 2, several auxiliary relations have to be kept in mind:

$$\sum_{\alpha=1}^{N} \mathbf{d}_\alpha = 0; \qquad Ð_{\alpha\beta} = Ð_{\beta\alpha}; \qquad \sum_{\alpha=1}^{N} \frac{x_\alpha}{Ð_{\alpha\beta}} = 0 \tag{24.4-2, 3, 4}$$

The first of these relations follows from the definition of the $\mathbf{d}_\alpha$, the second is a consequence of the Onsager reciprocal relations, and the third is needed because of the introduction of an especially designated species γ. The choice as to which species is designated as γ is arbitrary; often setting γ equal to α is convenient. The choice depends on the nature of the system under study, and this point will be illustrated in the examples that follow.

In all previous chapters, the only external force that has been considered has been the gravitational force. In this section we set the external force per unit mass $\mathbf{g}_\alpha$ equal to a sum of forces

$$\mathbf{g}_\alpha = \mathbf{g} - \left(\frac{z_\alpha F}{M_\alpha}\right)\nabla\phi + \delta_{\alpha m}\frac{1}{\rho_m}\nabla p \qquad (24.4\text{-}5)$$

Here $\mathbf{g}$ is the gravitational acceleration, z_α is the elementary charge on species α (for example, -1 for the chloride ion Cl^-), $F = 96485$ abs.-coulombs/g-equivalent is the Faraday constant, ϕ is the electrostatic potential, and the subscript m on the Kronecker delta $\delta_{\alpha m}$ refers to any mechanically restrained matrix, such as a permselective membrane.

In sum, for solving multicomponent diffusion problems in isothermal systems, we now have N mass-flux equations (of which only $N-1$ are independent), the species equations of continuity, and the equation of motion. This set of equations has proven to be useful for solving wide classes of mass transfer problems, and we discuss some of these in the following examples.

Of course, in order to solve multicomponent diffusion problems one needs the Maxwell–Stefan diffusivities $Đ_{\alpha\beta}$ that occur in Eq. 24.4-1. Very few measurements have been made of these quantities, which require the simultaneous measurement of the activity as a function of concentration. Among the few examples of such measurements are those made by Rutten.[1]

EXAMPLE 24.4-1

Centrifugation of Proteins

Protein molecules are large enough that they can be concentrated by centrifugation against the dispersive tendencies of Brownian motion, and this process has proven useful for molecular weight determination as well as for small-scale preparative separations. Show how the behavior of protein molecules in a centrifugal field can be predicted, and the kind of information that can be obtained from their behavior in a centrifuge tube (see Fig. 24.4-1). As we shall see in Example 24.4-3, we may treat the protein and its attendant counter-ions as a single large electrically neutral molecule. Choose the protein as species γ and begin with the mass-flux equation for it. The small ionic species needed for protein stability play no significant role in this development and can be ignored.

SOLUTION

We consider here a pseudobinary system of a single globular protein P in a solvent W, which is primarily water, and initially we restrict the discussion to a dilute solution rotating in a tube perpendicular to the axis of rotation (Fig. 24.,4-1a) at a constant angular velocity Ω. For such a system $x_W \approx 1$ and the solute flow field with respect to stationary axes will be that of rigid-body rotation—namely, $\mathbf{v}_W = \boldsymbol{\delta}_\theta\Omega r$.

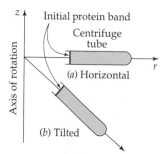

(a) Horizontal

(b) Tilted

Fig. 24.4-1. Ultracentrifugation of proteins, with two possible orientations of the centrifuge tube.

---

[1] Ph. W. M. Rutten, *Diffusion in Liquids*, Delft University Press, Delft, The Netherlands (1992).

Then the radial diffusion of the protein is described by the r-component of the simplified Maxwell–Stefan equation

$$\frac{x_P}{\mathcal{D}_{PW}} v_{Pr} = -\left(1 + \frac{\partial \ln \gamma_P}{\partial \ln x_P}\right)\frac{\partial x_P}{\partial r} - \frac{1}{cRT}(\phi_P - \omega_P)\frac{\partial p}{\partial r} \tag{24.4-6}$$

We see immediately that the protein will move in the positive radial direction if its mass fraction is greater than its volume fraction—that is, if it is denser than the solvent. If Eq. 24.4-6 is multiplied through by $c\mathcal{D}_{PW}$, we get

$$N_P = -c\mathcal{D}_{PW}\left[\left(1 + \frac{\partial \ln \gamma_P}{\partial \ln x_P}\right)\frac{\partial x_P}{\partial r} + \frac{1}{cRT}(\phi_P - \omega_P)\frac{\partial p}{\partial r}\right]$$

$$= -c\mathcal{D}_{PW}\frac{\partial x_P}{\partial r} - \frac{\mathcal{D}_{PW}}{RT}(\phi_P - \omega_P)\frac{\partial p}{\partial r} \tag{24.4-7}$$

in which the usual pseudobinary Fickian diffusivity $\mathcal{D}_{PW}$ is introduced. The diffusivity in Eq. 24.4-7 can be estimated from Eq. 17.4-3 as

$$\mathcal{D}_{PW} = \frac{\kappa T}{6\pi\mu_W R_P f_P} \tag{24.4-8}$$

in which R_P is the radius of a sphere having the volume of the protein molecule, μ_W is the solvent (water) viscosity, and f_P is a hydrodynamic shape factor (that is, a correction factor to account for the nonsphericity of the protein molecule).

From the equation of motion for the solution, we get the pressure gradient in terms of the angular velocity of the ultracentrifuge, thus

$$\frac{\partial p}{\partial r} = \rho \frac{v_\theta^2}{r} = \rho\Omega^2 r \tag{24.4-9}$$

The term $\rho\Omega^2 r$ will not vary significantly over the length of the centrifugation tube, which is small compared with the radius of the ultracentrifuge rotor.

Now we want to get an appreciation of the molecular weight dependence of the pressure-gradient term in Eq. 24.4-7. To do that we introduce the following approximations, valid in the dilute solution limit, common in protein processing:

$$\phi_P = c_P\overline{V}_P = x_P c\overline{V}_P \approx x_P \frac{\overline{V}_P}{\overline{V}_W} = x_P \frac{M_P}{M_W}\frac{\hat{V}_P}{\hat{V}_W} \tag{24.4-10a}$$

$$\omega_P = \frac{\rho_P}{\rho} = \frac{c_P M_P}{cM} = x_P \frac{M_P}{(x_P M_P + x_W M_W)} \approx x_P \frac{M_P}{M_W} \tag{24.4-10b}$$

Here $\hat{V}_P = \overline{V}_P/M_P$ is the partial specific volume of the protein. The partial specific volume of the solvent may be taken as 1 ml/g without significant error, and $\hat{V}_P$ for globular proteins is usually in the neighborhood of 0.75 ml/g. We see then that the decisive factor in permitting effective centrifugation is the ratio of molecular weights rather than the specific volumes, as the latter are not greatly different for the two species.

When Eqs. 24.4-7, 8, 10, and 11 are combined, the protein flux takes the form

$$N_P = -c\mathcal{D}_{PW}\frac{\partial x_P}{\partial r} + x_P(N_P + N_W)$$

$$= -c\mathcal{D}_{PW}\frac{\partial x_P}{\partial r} + c_P v_{\text{migr}} \tag{24.4-11}$$

which somewhat resembles Fick's first law. Here, the "migration velocity" for the protein is

$$v_{\text{migr}} = -\frac{\mathcal{D}_{PW}}{cRT}\left[\frac{M_P}{M_W}\left(\frac{\hat{V}_P}{\hat{V}_W} - 1\right)\right]\rho\Omega^2 r \tag{24.4-12}$$

Note that the radial molar flux of water greatly exceeds that of protein, and that the convective protein flux $c_P v_{\text{migr}}$ is very small.

Next we substitute the molar flux of Eq. 24.4-11 into the species equation of continuity

$$\frac{\partial c_P}{\partial t} = -\frac{\partial N_P}{\partial r} \tag{24.4-13}$$

or, for constant $\mathscr{D}_{PW}$

$$\frac{\partial c_P}{\partial t} = \mathscr{D}_{PW}\frac{\partial^2 c_P}{\partial r^2} - c_P v_{\text{migr}} \tag{24.4-14}$$

which is the equation we want to solve for several specific situations.

(a) Transient Behavior. We first consider migration of a thin protein band under conditions where fractional changes in r are small and no significant amount of protein reaches the far end of the tube. Then we can introduce a new independent variable $u = r - v_{\text{migr}}t$ that enables us to transform $c_P(r, t)$ into $c_P(u, t)$. The diffusion equation becomes

$$\left(\frac{\partial c_P}{\partial t}\right)_u = \mathscr{D}_{PW}\left(\frac{\partial^2 c_P}{\partial u^2}\right)_t \tag{24.4-15}$$

along with the initial condition

$$\text{At } t = 0, \qquad c_P = C\delta(u) \tag{24.4-16}$$

where C is a constant that tells how much protein is contained in the band, and the boundary conditions

$$\text{As } u \to \pm\infty, \qquad\qquad c_P \to 0 \tag{24.4-17}$$

Equation 24.4-17 is a long-tube approximation widely used in this application.

Equations 24.4-15 to 17 describe a Gaussian distribution of the protein about its center of mass, resulting from diffusion and moving with the velocity v_{migr}. The migration velocity can be measured, and this measurement yields a product of protein diffusivity and molecular weight. The broadness of the band in turn provides an independent measure of the diffusivity, and thus, combined with knowledge of the migration velocity and specific volumes, the molecular weight.[2] If the molecular weight is known, for example, from mass spectrometry, the shape factor f_P can be determined. This, in turn, is a useful measure of protein shape.

(b) Steady Polarization. We next consider long-time behavior when the protein has been concentrated at the end of the tube and has attained a steady state. Under these circumstances, there is no radial motion and Eqs. 24.4-6, 9, and 10 give

$$-\left(1 + \frac{\partial \ln \gamma_P}{\partial \ln x_P}\right)\frac{d \ln x_P}{dr} = \frac{1}{cRT}\left(\frac{M_P}{M_W}\right)\left(\frac{\hat{V}_P}{\hat{V}_W} - 1\right)\frac{dp}{dr} \tag{24.4-18}$$

The concentration gradient may be measured, and all other quantities except for M_P may be determined independently of the centrifugation process. Protein activity coefficients may, for example, be obtained from osmotic pressure data. Therefore the molecular weight of the protein may be unambiguously determined. Only mass spectrometry can provide better accuracy, and it is not suitable for all proteins.

(c) Preparative Operation. The speed of centrifugal separation can be greatly increased by tilting the tube as in Fig. 24.4-1b. Here the protein is forced toward the outer boundary of the tube by centrifugal action, and the resulting density gradient causes an axial bulk

---

[2] R. J. Silbey and R. A. Alberty, *Physical Chemistry*, 3rd edition, Wiley, New York (2001), p. 801.

transport by free convection, a process similar to that used for larger particles in disk centrifuges.[3]

EXAMPLE 24.4-2	Show that the results of the last example are equivalent to treating the proteins as small hydrodynamic particles.
Proteins as Hydrodynamic Particles	

SOLUTION

If, in the previous example, we had not used the simplifications in Eqs. 24.4-9 and 10, we would have obtained for the migration velocity in steady-state operation

$$v_{\text{migr}} = -\left(\overline{V}_P - \frac{\omega_P}{c_P}\right)\frac{dp}{dr}\frac{\text{Ð}_{PW}}{RT} \tag{24.4-19}$$

If now we restrict ourselves to dilute solutions so that the activity coefficient is very close to unity, we can set Ð_{PW} to $\mathscr{D}_{PW}$ and use Eq. 24.4-8 for the diffusivity and Eq. 24.4-9 for the pressure gradient. Then the migration velocity becomes

$$v_{\text{migr}} = -\left(\overline{V}_P - \frac{\omega_P}{c_P}\right)(\rho_W\Omega^2 r)\frac{1}{RT}\left(\frac{\kappa T}{6\pi\mu_W R_P f_P}\right) \tag{24.4-20}$$

Next we recognize that $\Omega^2 r = g_{\text{eff}}$ (an effective body force per unit mass resulting from the centrifugal field) and that $\kappa/R = \tilde{N}$ (Avogadro's number), and we get

$$v_{\text{migr}} = -\left(\overline{V}_P\rho_W - \frac{\rho_P}{c_P}\right)\frac{g_{\text{eff}}}{\tilde{N}}\left(\frac{1}{6\pi\mu_W R_P f_P}\right) \tag{24.4-21}$$

where we have used the approximation $\omega_P = \rho_P/(\rho_P + \rho_W) \approx \rho_P/\rho_W$ for a dilute protein solution. Next we set $\overline{V}_P \approx (\frac{4}{3}\pi R_P^3)\tilde{N}$, the volume per mole of protein, and $\rho_P/c_P \approx (\frac{4}{3}\pi R_P^3)(\rho^{(P)})\tilde{N}$, the mass per mole of protein; here $\rho^{(P)}$ is the pure protein density. When these quantities are inserted into Eq. 24.4-21 we get

$$v_{\text{migr}} = \frac{2R_P^2(\rho^{(P)} - \rho_W)g_{\text{eff}}}{9\mu_W f_P} \tag{24.4-22}$$

Comparison with Eq. 2.6-17 shows that the migration velocity for a nonspherical protein in a centrifugal field is the same as the terminal velocity for a sphere in a corresponding gravitational field (divided by the factor f_P to account for deviation from sphericity).

One may also start with an equation of motion for a particle P initially at rest in a suspension sufficiently dilute that particle–particle interactions are negligible. Then the particle velocity relative to a large body of quiescent fluid F is

$$\mathbf{v}_P - \mathbf{v}_F = -\mathscr{D}_{PF}\left(\nabla \ln n_P + \frac{1}{\kappa T}\left[V_P\left(1 - \frac{\rho^{(P)}}{\rho}\right)(\nabla p)_\infty - \mathbf{F}_{em}\right.\right.$$
$$\left.\left. + (\rho^{(P)} + \tfrac{1}{2}\rho)V_P\frac{d\mathbf{v}_P}{dt} + 6\sqrt{\pi\mu\rho}R_P^2\int_0^t \frac{d\mathbf{v}_P}{d\tau}\frac{1}{\sqrt{t-\tau}}d\tau\right]\right) \tag{24.4-23}$$

Here n_P is the number concentration of particles, V_P and R_P are the particle volume and radius, and the subscript ∞ refers to the conditions "far" from the particle (that is, outside the hydrodynamic boundary layer). Equation 24.4-23 is the equation of motion with an added term for Brownian motion, which is important, for example, in aerosol collection.[4] The symbol $\mathbf{F}_{em}$ stands for the electromagnetic force per particle.

---

[3] See, for example, *Perry's Chemical Engineers' Handbook*, McGraw-Hill, New York, 7th edition (1997), p. **18**–113.

[4] See L. D. Landau and E. M. Lifshitz, *Fluid Mechanics*, Pergamon Press, Oxford (1987), pp. 90–91, Problem 7.

The diffusivity $\mathcal{D}_{PF}$ in this example corresponds to $\mathcal{D}_{PW}$ of Example 24.4-1, and it may be seen that there is a very close analogy between the molecular and particulate descriptions. There are, in fact, only three significant differences:

1. The thermodynamic activity coefficient is considered to be unity for the particle.
2. The instantaneous acceleration of the molecule is neglected.
3. The effects of past history (that is, the Basset force given by the integral in Eq. 24.4-23) are neglected for the molecule.

In practice, activity coefficients tend to approach unity in dilute solutions, and the Basset forces tend to be small even for large particles. However, the instantaneous effects of acceleration can be appreciable for particles greater than about one micron in diameter.

EXAMPLE 24.4-3

Diffusion of Salts in an Aqueous Solution

Consider now for simplicity a 1-1 electrolyte M^+X^-, such as sodium chloride, diffusing in a system such as that shown in Fig. 24.4-2. Here well-mixed reservoirs at two different salt concentrations are joined by a constriction in which diffusional transfer between the two reservoirs takes place. The potentiometer shown in the figure measures the potential difference $\Delta\phi$ between the electrodes, without drawing any current from the system. Show how the generalized Maxwell–Stefan equations can be used to describe the diffusional behavior.

SOLUTION

The salt (S) is considered to be fully dissociated, so that the system is being regarded as ternary, with M^+, X^-, and water as the three species. We neglect the pressure diffusion term: the reference pressure cRT in Eq. 24.4-1 is approximately 1350 atmospheres under normal ambient conditions, and the pressure differences occurring in systems such as that pictured are of negligible importance.

The assumption of electroneutrality and no current flow provide the following constraints:

$$x_{M^+} = x_{X^-} = x_S = 1 - x_W \tag{24.4-24}$$

$$N_{M^+} = N_{X^-} = N_S \tag{24.4-25}$$

Here the mole fractions of the cation M^+ and the anion X^- are equal to that of the salt S.

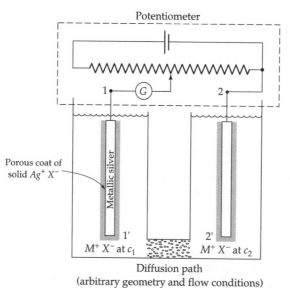

Fig. 24.4-2. Salt diffusion and diffusion potentials. The symbol G denotes a galvonometer.

We may then select species γ in Eq. 24.4-1 to be species α, and use Eqs. 24.4-5 and 24 to obtain for the cation and the anion

$$\frac{1}{c\mathcal{D}_{M^+W}}(x_W N_{M^+} - x_{M^+} N_W) = -x_{M^+}\nabla \ln a_{M^+} + \frac{1}{cRT}\left(\rho_{M^+}\mathbf{g}_{M^+} - \omega_{M^+}\sum_\beta \rho_\beta \mathbf{g}_\beta\right) \tag{24.4-26}$$

$$\frac{1}{c\mathcal{D}_{X^-W}}(x_W N_{X^-} - x_{X^-} N_W) = -x_{X^-}\nabla \ln a_{X^-} + \frac{1}{cRT}\left(\rho_{X^-}\mathbf{g}_{X^-} - \omega_{X^-}\sum_\beta \rho_\beta \mathbf{g}_\beta\right) \tag{24.4-27}$$

Next we use the expression for $\mathbf{g}_\alpha$ in Eq. 24.4-5, as well as Eqs. 24.4-24 and 25, to get:

$$\frac{1}{c D_{M^+W}}(x_W N_S - x_S N_W) = -\left(\frac{\partial \ln a_{M^+}}{\partial \ln x_S}\right)\nabla x_S - \left(\frac{x_S}{RT}\right)F\nabla\phi \tag{24.4-28}$$

$$\frac{1}{c\mathcal{D}_{X^-W}}(x_W N_S - x_S N_W) = -\left(\frac{\partial \ln a_{X^-}}{\partial \ln x_S}\right)\nabla x_S + \left(\frac{x_S}{RT}\right)F\nabla\phi \tag{24.4-29}$$

Note that the ion–ion diffusivity does not appear, because there is no velocity difference between the two ions when there is no current.

The electrostatic potential ϕ may be eliminated between these two equations by adding them together. The resulting flux expression

$$N_S = -\left(\frac{1}{c\mathcal{D}_{M^+W}} + \frac{1}{c\mathcal{D}_{X^-W}}\right)^{-1}\left(\frac{\partial \ln (a_{M^+}a_{X^-})}{\partial \ln x_S}\right)\nabla x_S + x_S(N_S + N_W) \tag{24.4-30}$$

may be put into the form of Fick's law

$$N_S = -c\mathcal{D}_{SW}\nabla x_S + x_S(N_S + N_W) \tag{24.4-31}$$

by introducing the definition of the concentration-based diffusivity

$$\mathcal{D}_{SW} = 2\left(\frac{\mathcal{D}_{M^+W}\mathcal{D}_{X^-W}}{\mathcal{D}_{M^+W} + \mathcal{D}_{X^-W}}\right)\left(1 + \frac{\partial \ln \gamma_S}{\partial \ln x_S}\right) \tag{24.4-32}$$

and, since $a_S = a_{M^+}a_{X^-} = x_S^2\gamma_\pm^2$ and $\gamma_\pm = \sqrt{\gamma_{M^+}\gamma_{X^-}}$,

$$\gamma_S = \gamma_{M^+}\gamma_{X^-} \tag{24.4-33}$$

which is the mean ionic activity coefficient.

The ion-water diffusivities may in turn be estimated from limiting equivalent conductances in the form

$$\lambda_{\alpha\infty} = \lim_{x_\alpha \to 0}\frac{z_\alpha \mathcal{D}_{\alpha W}F^2}{RT} \tag{24.4-34}$$

As a practical matter, diffusivities vary much less with concentration than do conductances, and salt diffusivities can be estimated with fair accuracy up to about 1N concentrations from limiting conductances. A basic reason for this is that ion–ion diffusional interactions, which always occur when a current flows, become appreciable at even modest salt concentrations (see Problem 24C.3).

Eq. 24.4-32 shows that the slower ion tends to dominate in determining the salt diffusivity, and this fact is the justification for treating the protein as a large neutral molecule in Example 24.4-1. Soluble proteins are nearly always charged, but they and their attendant *counter-ions* behave like a neutral salt, and its diffusivity is dominated by the protein moiety, which in turn acts very much like a hydrodynamic particle.

In a concentration gradient, the faster of the two ions tends to get ahead of the slower. However, this results in the formation of a potential gradient tending to speed the slower ion and slow the faster one. It can be shown (see Problem 24B.2) that this so-called *junction potential* is described by

$$\frac{d\phi}{dx_S} = -\frac{RT}{F}\frac{1}{x_S}\left[\frac{(\partial \ln a_{M^+}/\partial \ln x_S)\mathcal{D}_{M^+W} - (\partial \ln a_{X^-}/\partial \ln x_S)\mathcal{D}_{X^-W}}{\mathcal{D}_{M^+W} + \mathcal{D}_{X^-W}}\right] \tag{24.4-35}$$

However, these potentials cannot be measured directly, as the electrodes needed to complete

However, these potentials cannot be measured directly, as the electrodes needed to complete the electric circuit affect the measurement (see Problem 24C.3). One can obtain an approximate value through the use of potassium chloride salt bridges.[5]

This elementary example is only a very bare introduction to a complex and important subject. The interested reader is referred to the large literature on electrochemistry.[6]

EXAMPLE 24.4-4

Departures from Local Electroneutrality: Electro-Osmosis[6]

It is already clear from the preceding discussion of diffusion potential that local departures from electroneutrality do exist in diffusing electrolytes, and they are not always negligible. To examine this situation, consider a long tube of circular cross section containing an electrolyte, at least one component of which is adsorbed on the tube wall. This adsorption results in a fixed surface charge and a region of net charge, the *diffuse double layer*, in the solution adjacent to the tube wall. This net charge will produce an electric field within the tube that varies with radial, but not axial, position. If a potential difference is applied across the ends of the tube, the result will be a fluid flow, known as *electro-osmosis*. Conversely, if a hydrodynamic pressure is used to produce a flow, it will result in a potential difference, known as a *streaming potential*, developing across the ends of the tube. These phenomena are representative of a class known as electrokinetic phenomena. Develop an expression for the electro-osmotic flow developed in the absence of an axial pressure gradient.

SOLUTION

Our first problem is now to develop an expression for the electrostatic potential distribution, after which we can calculate the electro-osmotic flow.

The starting point for the electrostatic potential calculation is the Poisson equation

$$\nabla^2 \phi = -\frac{\rho_e}{\varepsilon} \tag{24.4-36}$$

Here ρ_e is the electrical charge density

$$\rho_e = F \sum_{\alpha=1}^{N} z_\alpha c_\alpha \tag{24.4-37}$$

and ε is the dielectric permittivity of the solution. For the problem at hand, Eq. 24.4-36 reduces to

$$\frac{1}{r} \frac{d}{dr}\left(r \frac{d\phi}{dr} \right) = -\frac{\rho_e}{\varepsilon} \tag{24.4-38}$$

Now, following Newman[6] we assume that the concentration of charge follows a Boltzmann distribution

$$\frac{c_\alpha}{c_{\alpha\infty}} = \exp\left(-\frac{z_\alpha F \phi}{RT} \right) \approx 1 - \frac{z_\alpha F \phi}{RT} \tag{24.4-39}$$

and use a truncated Taylor expansion, known as the Debye–Hückel approximation, so that we can obtain an explicit solution. Here the subscript ∞ can be considered to indicate the centerline of the tube, because, as we shall see, the charge density drops off very rapidly with the distance from the tube wall. For the same reason we may neglect the wall curvature and assume that the net charge at the centerline is zero so that

$$\frac{d^2\phi}{dy^2} = \frac{\phi}{\lambda^2} \qquad \text{where } \lambda = \left(\frac{F^2}{\varepsilon RT} \sum_{\alpha=1}^{N} z_\alpha^2 c_{\alpha\infty} \right)^{-1/2} \tag{24.4-40, 41}$$

[5] R. A. Robinson and R. H. Stokes, *Electrolyte Solutions*, revised edition, Butterworth, London (1965), p. 571. This venerable reference contains a great detail of useful data.

[6] See, for example, J. S. Newman, *Electrochemical Systems*, 2nd edition, Prentice-Hall, Englewood Cliffs, N.J. (1991). Example 24.4-4 is taken from p. 215.

Here $y = R - r$ is the distance measured into the fluid from the wall, and λ is the *Debye length*, which tends to be very small. Thus for a 1-1 electrolyte

$$\lambda \approx \frac{3.0}{\sqrt{c_S}} \tag{24.4-42}$$

where the units of Debye length λ and the salt concentration c_S are Ångströms and molarity, respectively. Thus for a 0.1 N solution, the Debye length is only about 10 Å. As a result, departures from neutrality can usually be neglected in macroscopic systems. Similarly, concentration imbalances are very small for junction potentials, which are typically no more than tens of millivolts (see also Problem 24C.4).

We now need boundary conditions to integrate Eq. 24.4-38, and the first is just the assumption of electroneutrality at large distances from the wall:

B.C. 1: As $\dfrac{y}{\lambda} \to \infty$, $\qquad\qquad\qquad \phi \to 0 \tag{24.4-43}$

The second is obtained from Gauss's law (see Newman[6], p. 75), assuming there is no potential gradient within the solid surface itself,

B.C. 2: At $y = 0$, $\qquad\qquad\qquad \dfrac{d\phi}{dy} = -\dfrac{q_e}{\varepsilon} \tag{24.4-44}$

where q_e is the surface charge per unit area. Integration of Eq. 24.4-40 then gives

$$\phi = -\frac{q_e}{\lambda} e^{-y/\lambda} \tag{24.4-45}$$

Newman[6] gives a more rigorous development that allows for surface curvature, but for any tube of radius greater than tens of nanometers, this is really not necessary.

We are now ready to put these results into the equation of motion, and we shall here assume steady laminar flow, so that

$$0 = \mu \frac{1}{r} \frac{d}{dr}\left(r \frac{dv_z}{dr}\right) - \frac{dp}{dr} + \rho_e E_z \tag{24.4-46}$$

in which the axial electric field strength is

$$E_z = \frac{\partial \phi}{\partial z} \tag{24.4-47}$$

Neglecting the pressure gradient and using Eq. 24.4-36 to eliminate ρ_e, we find

$$0 = \mu \frac{1}{r} \frac{d}{dr}\left(r \frac{dv_z}{dr}\right) + \varepsilon \frac{1}{r} \frac{d}{dr}\left(r \frac{d\phi}{dr}\right) E_z \tag{24.4-48}$$

Now, if curvature is again neglected, this equation may be integrated to give

$$v_z = \left(\frac{\lambda q_e}{\mu}\right) E_z (1 - e^{-y/\lambda}) \tag{24.4-49}$$

The quantity in the first set of parentheses may be considered to be an experimentally determined property of the system, and $\exp(-y/\lambda)$ is negligible over the bulk of the tube cross section for essentially all tubes. Thus the velocity is uniform except very near the wall.

Such electro-osmotic flows are being widely used in microscopic flow reactors and separators—for example, in diagnostic devices—and they offer the advantage of negligible convective dispersion. Note that the velocity is effectively independent of the tube radius. Thus electro-osmosis especially useful in tubes of small radii, where large pressure gradients would otherwise be required to produce the same flow velocity.

EXAMPLE 24.4-5

Additional Mass Transfer Driving Forces

We have now covered all of the mass transfer mechanisms normally considered in a nonequilibrium thermodynamic framework, but there are other possibilities that have proven significant. Here we consider three: the force on a charged particle moving across a magnetic field, and the forces of electrical or magnetic induction. These contain nonlinear terms—that is, products of species velocities and force fields—and therefore they are, strictly speaking, outside the scope of irreversible thermodynamics. However, it has been found permissible to add them to the body forces appearing in Eq. 24.4-1. Develop a specific form for the resulting equation, and show how it can be used to describe mass transfer processes affected by one or more of these additional forces.

SOLUTION

We begin by defining an extended driving force for mass transfer, $\mathbf{d}_{\alpha,ext}$, to include these additional forces:

$$\mathbf{d}_{\alpha,ext} = \mathbf{d}_\alpha + \frac{x_\alpha z_\alpha F}{RT} [\mathbf{v}_\alpha \times \mathbf{B}] + \frac{x_\alpha}{RT} \left\{ \Gamma_\alpha^{el}[\mathbf{E} \cdot \nabla \mathbf{E}] + \Gamma_\alpha^{mag}[\mathbf{B} \cdot \nabla \mathbf{B}] \right\} \tag{24.4-50}$$

Here $\mathbf{B}$ is the magnetic induction, $\mathbf{E} = \nabla \phi$ the electric field, Γ_α^{el} the *electric susceptibility*, and Γ_α^{mag} the *magnetic susceptibility*.

The origin of the terms containing $[\mathbf{v}_\alpha \times \mathbf{B}]$ and $[\mathbf{E} \cdot \nabla \mathbf{E}]$ in Eq. 24.4-50 is in the Lorentz relation

$$\mathbf{F} = q_0(\mathbf{E} + [\mathbf{v} \times \mathbf{B}]) \tag{24.4-51}$$

where q_0 is the electric charge. This is shown explicitly in Eq. 24.4-51 for a charged particle moving through a magnetic field (see Problem 24B.1), but only indirectly for the electric induction $[\mathbf{E} \cdot \nabla \mathbf{E}]$, which is based on the interaction of a *nonuniform field* with an electric dipole.

To show the origin of the $[\mathbf{E} \cdot \nabla \mathbf{E}]$ term in Eq. 24.4-50, consider, for example, the one-dimensional situation pictured in Fig. 24.4-3. An electric field will tend to align dipoles that are normally randomized by Brownian motion, and, if the field is nonuniform, there will be a net force on an aligned dipole of magnitude

$$F_z = q_0 \left[\left(E_0 + \frac{l}{2} \frac{\partial E}{\partial z} \right) - \left(E_0 - \frac{l}{2} \frac{\partial E}{\partial z} \right) \right] = q_0 l \frac{\partial E}{\partial z} \tag{24.4-52}$$

where q_0 is the magnitude of the charge at either end of the dipole and l is the distance between the two centers of charge.

In some cases—for example, the zwitterion form of amino acids—one can determine both q_0 and l from molecular theory. However, for particles and most molecules, one finds only *induced dipoles*: a partial charge separation resulting from the presence of the field. Under the conditions of interest here, only a small fraction of intrinsic dipoles is aligned with the field, and both the fractional alignment of these and the strength of the induced dipoles are

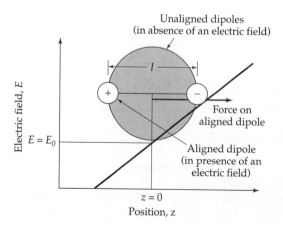

Fig. 24.4-3. Origin of the dielectrophoretic force given in Eq. 24.4-52.

normally assumed to be proportional to the field strength. All these factors are collected into what is usually an experimentally determined quantity, the *electric susceptibility*. The origin of the magnetophoresis term is analogous. We now turn to brief discussions of applications of these new separation mechanisms.

The behavior of ions moving across a magnetic field is the basis of classic mass spectrometry, although time-of-flight mass spectrometers are also in widespread use. Both types of spectrometers are highly developed and find extensive applications for analyzing mixtures from simple inorganic gases to complex nonvolatile biological molecules such as proteins. In fact, where applicable, they provide the most accurate means available for determining protein molecular weight, often within one dalton for a molecular weight typically of the order of tens of thousands.

Both dielectro- and magnetophoresis have long been used on a large process scale for removing small particles suspended in fluids. Nonuniform fields are achieved in the case of dielectrophoresis by using a packing of small dielectric particles, such as glass beads, between electrodes (see, for example, Problem 24B.1). Because particles always move toward the stronger field, one can use alternating current, usually at some tens of kilovolts, and thus avoid electrode reactions. Current flows are extremely small and can normally be neglected. In magnetophoresis, a nonuniform field is achieved by placing ferromagnetic meshes between poles of an electromagnet, which can of course work only with paramagnetic or ferromagnetic materials. A classic example is the removal of color bodies consisting of magnetic iron oxides to whiten clay.

New uses for dielectrophoresis have been developing very rapidly in the fields of biology,[7] advanced materials,[8] including nanotechnology, and environmental monitoring.[9] They include classification, quantitative analysis, and manipulation, including the formation of ordered arrays.

Many of these applications require major extensions of Eq. 24.4-50 to include quadrupolar and even octopolar forces.[10] Moreover, there are strong interactions between electrical forces and hydrodynamics, and both device and particle shape can have profound effects.[11]

§24.5   MASS TRANSFER ACROSS SELECTIVELY PERMEABLE MEMBRANES

Membranes may be viewed physically as thin sheets, usually separating two bulk phases and controlling mass transfer between them. In addition, the membrane is typically kept stationary against external pressure gradients and internal viscous drag by some mechanical constraint, typically a wire mesh or equivalent structure. Membranes consist of

    [7] C. Polk, *IEEE Transactions on Plasma Science*, **28**, 6–14 (2000); J. Suehiro et al., *J. Physics D: Applied Physics*, **32**, 2814–2320 (1999); J. P. H. Bert, R. Pethig, and M. S. Talary, *Trans. Inst. Meas. Control*, **20**, 82–91 (1998); A. P. Brown, W. B. Betts, A. B. Harrison, and J. G. O'Neill, *Biosensors and Bioelectronics*, **14**, 341–351 (1999); O. D. Velev and E. W. Kaler, *Langmuir*, **15**, 3693–3698 (1999); T. Yamamoto, et al., *Conference Record, IAS Annual Meeting (IEEE Industry Applications Society)*, **3**, 1933–1940 (1998); M. S. Talary, et al., *Med. and Bio. Eng. and Computing*, **33**, 235–237 (1995); H. Morgan and N. G. Green, *J. Electrostatics*, **42**, 279–293 (1997).

    [8] L. Cui and H. Morgan, *J. Micromech. Microeng.*, **10**, 72–79 (2000); M. Hase et al., *Proc. Intl. Soc. Optical Eng.*, **3673**, 133–140 (1999); C. A. Randall, *IEEE Intl. Symp. on Applications of Ferroelectrics*, Piscataway, N.J. (1996).

    [9] P. Baron, *ASTM Special Technical Publication*, 147–155 (1999); R. J. Han, O. R. Moss, and B. A. Wong, *Aerosol Sci. Tech.*, 241–258 (1994).

    [10] C. Reichle et al., *J. Phys. D: Appl. Phys.*, **32**, 2128–2135 (1999); A. Ramos et al., *J. Electrostatics*, **47**, 71–81 (1999); M Washizu and T. B. Jones, *J. Electrostatics*, **33**, 187–198 (1994); B. Khusid and A. Acrivos, *Phys. Rev. E*, **54**, 5428–5435 (1996).

    [11] S. Kim and S. J. Karrila, *Microhydrodynamics*, Butterworth-Heinemann, Boston (1991); D. W. Howard, E. N. Lightfoot, and J. O. Hirschfelder, *AIChE Journal*, **22**, 794–798 (1976).

an insoluble, selectively permeable matrix m and one or more mobile permeating species $\alpha, \beta, \ldots$. Mathematically they are defined by three constraints:

1. Negligible curvature

$$\delta \ll R_{\text{curv}} \tag{24.5-1}$$

where δ is the membrane thickness, and R_{curv} is the membrane surface radius of curvature. It follows that mass transport is unidirectional and perpendicular to the membrane surface.

2. Immobility of the matrix

$$v_m = 0 \tag{24.5-2}$$

where v_m is the velocity of the matrix, which serves as the coordinate reference.

3. Pseudosteady behavior

$$\frac{\partial c_\alpha}{\partial t} = 0 \tag{25.4-3}$$

where α is any contained species, including the matrix m. This really means that the diffusional response times within the membrane are short compared to those in the adjacent solutions.

We now wish to show how these constraints can be used to specialize the Maxwell–Stefan equations and to produce compact but reliable descriptions of transport in membranes.

We begin by recognizing that the matrix must be considered to be one of the diffusing species, and we choose to use the Maxwell–Stefan equations only for the mobile species. We may then use Eqs. 24.4-1 and 5 and write for a mixture of N mobile species:

$$\sum_{\substack{\beta=1 \\ \beta \neq \alpha}}^{N} \frac{x_\alpha x_\beta}{Ð_{\alpha\beta}} (x_\alpha - x_\beta) = -x_\alpha \nabla_{T,p} \ln a_\alpha - x_\alpha \left(\frac{\overline{V}_\alpha}{RT} \right) \nabla p - x_\alpha z_\alpha \left(\frac{F}{RT} \right) \nabla \phi \tag{24.5-4}$$

Note that α has been chosen as the reference species in the equation for each α and that the force holding the membrane stationary—that is, the last term in Eq. 24.4-5—has resulted in the elimination of the mass-fraction term in the expression for pressure diffusion.[1]

We next note that, from a thermodynamic point of view, the number of components is the number of independent *mobile* species in the solutions bathing the membrane, because it is the external solution that determines the state of the membrane at equilibrium. We also recognize that, for most situations, the effective molecular weight of the matrix cannot be determined. We thus define the internal system as including only the mobile species and define mole fractions of these species to sum to unity. However, since the interaction of each species with the membrane is quite significant, we also define $Ð'_{\alpha M}$ by

$$\frac{x_M}{Ð_{\alpha M}} = \frac{1}{Ð'_{\alpha M}} \tag{24.5-5}$$

Equation 24.5-5 completes the specialization of the Maxwell–Stefan equations for membrane transport, but we still have to select a generally applicable set of boundary conditions.

These conditions are obtained by requiring the total "potential" of each species to be continuous across the boundary

$$\left(\frac{\overline{V}_{\alpha,\text{avg}}}{RT} \right) (p_m - p_e) + z_\alpha \left(\frac{F}{RT} \right) (\phi_m - \phi_e) = 0 \tag{24.5-6}$$

---

[1] E. M. Scattergood and E. N. Lightfoot, *Trans. Faraday Soc.*, **64**, 1135–1146 (1968).

Here the subscripts m and e refer to conditions within the membrane and in the external solution, respectively. The activity a_α is to be calculated at the composition of the membrane phase but at the pressure of the external solution, and

$$\overline{V}_{\alpha,\text{avg}} = \frac{1}{p_m - p_e} \int_{p_e}^{p_m} \overline{V}_\alpha \, dp \qquad (24.5\text{-}7)$$

In practice, the solutes are normally considered incompressible and $\overline{V}_\alpha$ to be constant across the interface.

Often conditions inside the membrane are very difficult, or even impossible, to determine, and Eq. 24.5-6 is primarily useful under these circumstances to obtain a qualitative understanding of membrane behavior. Partly for this reason complete descriptions of membrane transport are rare (see, however, Scattergood and Lightfoot[1]). Highly simplified, but often directly useful, introductions to membrane transport are available in a variety of sources.[2-4] One venerable approximation, found especially useful by biologists, is that of Kedem and Katchalsky.[5]

However, rapid progress is being made in obtaining fundamental data, and much of this is reported in the *Journal of Membrane Science*. One important area is that of microporous membranes.[6] One can also expect advances in modeling behavior. It has long been known[7] that the generalized Lorentz reciprocal theorem for creeping flows[8] provides a sound basis for extending hydrodynamic diffusion theory of §17.4 to multicomponent diffusion in microporous membranes. Recently developed computational techniques[9] should make the necessary computations tractable enough to provide real predictive power. These techniques can also be used to develop self-assembling structures,[10] which offer new possibilities for highly selective membranes.

This field offers an extremely wide variety of membrane types and of mass transfer processes taking place in them. One can distinguish between biological[11] and synthetic[12] membranes, but there are very wide ranges of composition and behavior within each of these categories. Among the synthetic group there are "homogeneous" membranes, in which the matrix acts as a true solvent for permeating species, and "microporous" membranes, in which the permeating species are confined to matrix-free regions, as well as mixtures of the two types. These factors are important from a materials standpoint, but the formalisms needed to describe their transport behavior are much the same for all.

---

[2] E. L. Cussler, *Diffusion: Mass Transfer in Fluid Systems*, 2nd edition, Cambridge University Press (1997), p. 580.

[3] W. M. Deen, *Analysis of Transport Phenomena*, Oxford University Press (1998), p. 597.

[4] J. D. Seader and E. J. Henley, *Separation Process Principles*, Wiley, New York (1998).

[5] O. Kedem and A. Katchalsky, *Biochem. Biophys. Acta*, **27**, 229 (1958).

[6] K. Kaneko, *J. Membrane Sci.*, **96**, 59–89 (1994); K. Sakai, *J. Membrane Sci.*, **96**, 91–130 (1994); S. Nakao, *J. Membrane Sci.*, **96**, 181–165 (1994).

[7] E. N. Lightfoot, J. B. Bassingthwaighte, and E. F. Grabowski, *Ann. Biomed. Eng.*, **4**, 78–90 (1976).

[8] J. Happel and H. Brenner, *Low Reynolds Number Hydrodynamics*, Prentice-Hall (1965), Martinus Nijhoff (1983), p. 62, p. 85.

[9] S. Kim and S. J. Karrila, *Microhydrodynamics: Principles and Selected Applications*, Butterworth-Heinemann, Boston (1991).

[10] I. Mustakis, S. C. Clear, P. F. Nealey, and S. Kim, *ASME Fluids Engineering Division Summer Meeting*, FEDSM, June 22–26 (1997).

[11] B. Alberts et al., *The Molecular Biology of the Cell*, Garland, New York (1999), Chapters 10 and 11.

[12] W. S. W. Ho and K. K. Sirkar, *Membrane Handbook*, Van Nostrand Reinhold, New York (1992), p. 954; R. D. Noble and S. A. Stern, *Membrane Separations Technology*, Membrane Science and Technology Series, **2**, Elsevier (Amsterdam), p. 718; R. van Reis and A. L. Zydney, "Protein Ultrafiltration" in *Encyclopedia of Bioprocess Technology* (M. C. Flickinger and S. W. Drew, eds.), Wiley, New York (1999), pp. 2197–2214; L. J. Zeman and A. L. Zydney, *Microfiltration and Ultrafiltration*, Marcel Dekker, New York (1996).

There are also a wide variety of process conditions in widespread use. Here we consider only a few examples to illustrate commonly encountered situations.

EXAMPLE 24.5-1

Concentration Diffusion Between Preexisting Bulk Phases

Consider "solute" A diffusing through a membrane placed between binary solutions of solute A in solvent B *under the influence of concentration gradients alone*. This is a commonly encountered situation, including dialysis, blood oxygenation, and many gas-separation systems.[13] There are many variants, including *facilitated diffusion* (see Problem 24C.8). Hemodialysis is a special case, where pressure differences are used to drive water across the membrane, but concentration diffusion of solutes is of primary interest from the present standpoint. Assume for the moment that the flux of solvent, N_B, is already known. Develop an analog to Fick's first law for this system.

SOLUTION

We are primarily concerned here with solute A, and the Maxwell–Stefan equation for it takes the form

$$\frac{x_A v_A}{\mathcal{D}'_{AM}} + \frac{x_A x_B}{\mathcal{D}_{AB}}(v_A - v_B) = -\left[1 + \left(\frac{\partial \ln \gamma_A}{\partial \ln x_A}\right)_{T,p}\right]\frac{dx_A}{dz} \tag{24.5-8}$$

This may be rearranged to give

$$N_A = -c\left(\frac{\mathcal{D}'_{AM}\mathcal{D}_{AB}}{\mathcal{D}'_{AM} + \mathcal{D}_{AB}}\right)\left[1 + \left(\frac{\partial \ln \gamma_A}{\partial \ln x_A}\right)_{T,p}\right]\frac{dx_A}{dz} + x_A(N_A + N_B)\left(\frac{\mathcal{D}'_{AM}}{\mathcal{D}'_{AM} + \mathcal{D}_{AB}}\right) \tag{24.5-9}$$

This is reminiscent of Fick's law, with the first term on the right corresponding to the Fickian diffusive flux and the second to the convective term. However, the effective diffusivity now contains a membrane contribution, and the convective term is now weighted by a ratio of diffusivities. This situation corresponds to the situation pictured in Fig. 24.5-1. The arrow pointing to the right represents the diffusion relative to the solvent, modified by the interaction with the matrix, while the arrow pointing to the left represents the "drag" of the membrane, which tends to reduce the transport relative to the convection occurring in the absence of the membrane matrix—namely, $x_A(N_A + N_B)$. Note that, in general, there are mass transfer boundary layers on both sides of the membrane.

There are several limiting situations of interest. If the membrane interacts only very weakly with the solute, $\mathcal{D}'_{AM} \gg \mathcal{D}_{AB}$. Then

$$N_A = -c\mathcal{D}_{AB}\left[1 + \left(\frac{\partial \ln \gamma_A}{\partial \ln x_A}\right)_{T,p}\right]\frac{dx_A}{dz} + x_A(N_A + N_B) \tag{24.5-10}$$

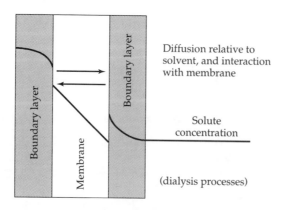

Fig. 24.5-1. Intramembrane mass transport.

[13] W. J. Koros and G. K. Fleming, *J. Membrane Sci.*, **83**, 1–80 (1993).

which is exactly Fick's law. However, it must be remembered that both molar concentration and diffusivity are those in the membrane phase. We next look at a limiting situation, where

$$x_A(N_A + N_B) << N_A \tag{24.5-11}$$

and the distribution between membrane and solution is linear, so that

$$c_{Ae} = K_D c_{Am} \tag{25.5-12}$$

where the subscripts e and m refer to the external solution and membrane phases, respectively, and K_D is the distribution coefficient for the two phases. We may then write

$$N_A = P(c_{Ae0} - c_{Ae\delta}) \tag{24.5-13}$$

where

$$P = K_D D_{A,\text{eff}} \tag{24.5-14}$$

is known as the *membrane permeability*, and

$$D_{A,\text{eff}} = \left(\frac{Ð'_{AM} Ð_{AB}}{Ð'_{AM} + Ð_{AB}} \right) \left[1 + \left(\frac{\partial \ln \gamma_A}{\partial \ln x_A} \right)_{T,p} \right] \tag{24.5-15}$$

The subscripts $_{Ae}0$ and $_{Ae}\delta$ refer to the solute concentrations at the "upstream" and "downstream" sides of the membrane.

EXAMPLE 24.5-2

Ultrafiltration and Reverse Osmosis

Now consider the filtration processes, in which it is desired to remove a solvent selectively relative to a solute by pressure-driven flow across a solute-rejecting membrane. Applications include ultrafiltration and reverse osmosis, the former dealing with macromolecular and the latter with small solutes.[14] Microfiltration and nanofiltration are formally similar, but the particulate nature of the entities being removed presents additional complications we do not wish to consider here.[12] Develop a framework for describing solvent flow rate and filtrate composition as functions of driving pressure.

SOLUTION

Inevitably some solute moves through the membrane along with the solvent, as indicated in Fig. 24.5-1, and it will now be necessary to consider the Maxwell–Stefan equations for both species. However, membrane filtration is a complex process requiring a great deal of information to obtain a complete a priori description, and we therefore begin with an overview of characteristic behavior using Fig. 24.5-2 as a point of departure. Here both the flow through the membrane and the composition of the filtrate are shown schematically as functions of the

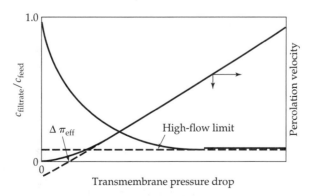

Fig. 24.5-2. Ultrafiltration: flow and solute rejection.

---

[14] R. J. Petersen, *J. Membrane Sci.*, **83**, 81–150 (1993).

transmembrane pressure drop. First we note that the flow increases with pressure drop, slowly at first, but approaching a linear relation asymptotically, and the asymptote crosses the line of zero velocity at a finite pressure drop, $\Delta\pi_{\text{eff}}$. The ratio of the filtrate solute concentration to the feed concentration drops with increasing pressure drop, from unity toward an asymptote,[15] which is normally much lower than unity. Our main concern in this brief introduction is to explain this characteristic behavior in terms of key thermodynamic and transport behavior.

This situation differs fundamentally from that just described, in that pressure diffusion now comes into play, and in that the downstream solution is produced by the transmembrane mass transfer. Hence the downstream ratio of solute to solvent is the same as the ratio of the corresponding mass transfer rates. There is, therefore, no boundary layer on the downstream side of the membrane, and it is an almost universal practice to use a composite structure. Such composite membranes consist of a very thin selective layer on the upstream face, and a comparatively thick, highly porous, nonselective backing that provides mechanical strength. This backing can be ignored in the present example.

We begin by focusing on the intramembrane behavior for which the Maxwell–Stefan equations, modified from Eq. 24.5-4, assume the forms

$$\frac{x_S x_W}{\mathcal{D}_{SW}}(v_S - v_W) + \frac{x_S v_S}{\mathcal{D}'_{SM}} = -\frac{1}{RT}x_S\nabla_{T,p}(RT\ln a_S) - \frac{c_S\overline{V}_S}{cRT}\nabla p \tag{24.5-16}$$

$$\frac{x_S x_W}{\mathcal{D}_{SW}}(v_W - v_S) + \frac{x_W v_W}{\mathcal{D}'_{WM}} = -\frac{1}{RT}x_W\nabla_{T,p}(RT\ln a_W) - \frac{c_W\overline{V}_W}{cRT}\nabla p \tag{24.5-17}$$

Here the subscripts S and W refer to the partially rejected solute and solvent (usually water) respectively. The terms $x_\alpha \overline{V}_\alpha$ have been replaced by $c_\alpha \overline{V}_\alpha / c$, to make the presence of the volume fractions

$$\phi_\alpha = c_\alpha \overline{V}_\alpha \tag{24.5-18}$$

explicit. In addition, the first term on the right has been rewritten as a reminder that the derivative represents the gradient of the partial molar free energy

$$d\overline{G}_\alpha = RTd\ln a_\alpha \tag{24.5-19}$$

with composition, temperature, and pressure being held constant.

We begin by examining flow behavior, and to do this we add Eqs. 24.5-16 and 17 to get a relation between species transport rates and *intramembrane* pressure gradient

$$\left(\frac{N_S}{c\mathcal{D}'_{SM}} + \frac{N_W}{c\mathcal{D}'_{WM}}\right) = -\frac{1}{cRT}\nabla p_m \tag{24.5-20}$$

The subscript m on pressure is a reminder that we have so far calculated only the pressure drop inside the membrane. Here advantage has been taken of the Gibbs–Duhem equation

$$\sum_\alpha \overline{G}_\alpha dx_\alpha = 0 \tag{24.5-21}$$

and the fact that the volume fractions sum to unity.

To obtain the directly measurable difference between upstream and downstream solution pressures, we must go back to Eq. 24.5-6, which takes the form

$$p_e - p_m = \frac{RT}{\overline{V}_S}\ln\frac{a_{Sm}}{a_{Se}} = \pi_e - \pi_m \tag{24.5-22}$$

---

[15] Actually there is a continuing, very slow drop of filtrate solute concentration even at very high pressure drops, presumably resulting from the compression of the membrane. However, we shall not attempt to deal with this small effect here.

Looking at the upstream side of the membrane, for example, Eq. 24.5-22 states that a finite pressure drop across the membrane interface is required to drive the solute against an increase in the thermodynamic activity. Then the measurable transmembrane pressure drop is

$$\Delta p_{ext} = \frac{cRT}{\delta}\left(\frac{N_S}{c\mathcal{D}'_{SM}} + \frac{N_W}{c\mathcal{D}'_{WM}}\right) + \Delta\pi_{eff} \tag{24.5-23}$$

where δ is the membrane thickness, and

$$\Delta\pi_{eff} = (\pi_{e0} - \pi_{e\delta}) - (\pi_{m0} - \pi_{m\delta}) \tag{24.5-24}$$

where the subscripts 0 and δ refer to the upstream and downstream sides of the membrane, respectively. The intramembrane osmotic pressures are seldom known, but they are substantially smaller than the corresponding solution values (see Problems 24C.7 and 8). At the present state of understanding, Eq. 24.5-24 explains why there is a finite intercept to the asymptotic flow behavior, and an elimination of the membrane contributions provides an upper limit to it. It also provides some insight into intramembrane behavior from experimental observations, but not an a priori prediction of the intercept.

Next we eliminate the pressure gradient from Eq. 24.5-16 with the aid of Eq. 24.5-23

$$N_S\left(\frac{x_W}{c\mathcal{D}_{WS}} - \frac{\phi_S}{c\mathcal{D}'_{SM}}\right) - N_W\left(\frac{x_S}{c\mathcal{D}_{SW}} + \frac{\phi_S}{c\mathcal{D}'_{WM}}\right) = -\left(1 + \frac{\partial \ln \gamma_s}{\partial \ln x_s}\right)\frac{dx_s}{dz} \tag{24.5-25}$$

This expression can be integrated to obtain the solute concentration profile (see, for example, Problems 24C.7 and 8). In general the concentration profile shows an increasingly negative slope in the flow direction, and this feature becomes more pronounced as the flow rate through the membrane increases—that is, as the transmembrane pressure drop becomes larger.

At very low flow rates, N_S and N_W are relatively small and diffusion is relatively fast. There is only a small drop in solute concentration across the membrane, and the result is the poor rejection seen for low pressure drops in Fig. 24.5-2. This behavior is suggested by the zero-flow concentration profile in Fig. 24.5-1.

At very high flow rates, on the other hand, concentration gradients are large and diffusion is weak, except very close to the downstream boundary of the membrane, where a very large negative concentration gradient develops. Near the upstream boundary, the two mass-flux terms are large compared to their difference, and one may neglect concentration gradients in calculating the mass-flux ratio:

$$\frac{N_S}{N_W} \approx \frac{x_S}{x_W}\left(\frac{(1/c\mathcal{D}_{SW}) + (\overline{V}_S/\mathcal{D}'_{WM})}{(1/c\mathcal{D}_{SW}) + (\overline{V}_W/\mathcal{D}'_{SM})}\right) \tag{24.5-26}$$

The bases of solute exclusion now become clear:

1. Thermodynamic exclusion, defined by the ratio x_S/x_W.

2. Frictional differentiation, defined by differences in the interaction terms with the membrane $(\overline{V}_S/\overline{V}_W)(\mathcal{D}'_{SM}/\mathcal{D}'_{WM})$.

Both effects are used in practice and are illustrated in the problems.

EXAMPLE 24.5-3	Consider now membranes containing immobilized charges consisting of polyelectrolyte gels. Such gels contain repeated covalently bound ionic groups, as pictured in Fig. 24.5-3. The membrane interior may then be viewed as a solution containing spatially bound *fixed charges*, mobile *counter-ions*, *invading electrolyte*, and water. For simplicity assume that the fixed
Charged Membranes and Donnan Exclusion[16]	

---

[16] H. Strathmann, "Electrodialysis," Section V in *Membrane Handbook* (W. W. S. Ho and K. K. Sirkar, eds.), Van Nostrand Reinhold, New York (1992).

Membrane   Molecular structure

Fig. 24.5-3. A sulfonic acid based ion-exchange membrane.

charges are anions, written as X^-, and the counter-ions are cations, written as M^+. The external solution is aqueous M^+X^-, and it provides the invading electrolyte M^+X^-. Show how the presence of fixed charges produces an exclusion of invading electrolyte.

SOLUTION

This system is dominated by the behavior at the membrane boundary, and we therefore return to Eq. 24.5-6, written for the water and for the salts S or M^+X^-. The expression for water is

$$\ln \frac{a_{We}}{a_{Wm}} = \frac{\overline{V}_W}{RT}(p_m - p_e) = \frac{\overline{V}_W}{RT}\Delta\pi \qquad (24.5\text{-}27)$$

where the subscripts e and m refer to the external solution and the membrane, respectively. Since the intramembrane electrolyte concentration is always higher than the external, thus resulting in a lower internal chemical activity for water, the membrane interior is at a higher pressure than the external solution (see, for example, Problem 24B.4).

The corresponding equation for the salt S yields

$$\frac{a_{Se}}{a_{Sm}} = \frac{x_{M^+e}x_{X^-e}}{x_{M^+m}x_{X^-m}}\frac{\gamma_{Se}}{\gamma_{Sm}} = \exp\left(\frac{\overline{V}_S}{RT}\Delta\pi\right) \qquad (24.5\text{-}28)$$

or

$$\frac{c_{M^+e}c_{X^-e}}{c_{M^+m}c_{X^-m}} = K_D = \frac{\gamma_{Sm}}{\gamma_{Se}}\left(\frac{c_m}{c_e}\right)^2 \exp\left(\frac{\overline{V}_S}{RT}\Delta\pi\right) \qquad (24.5\text{-}29)$$

It follows that

$$c_{Sm}^2 + c_{Sm}c_{X^-m} = \frac{1}{K_D}c_{Se}^2 \qquad (24.5\text{-}30)$$

and therefore that the concentration of salt in the membrane phase is less than that in the solution. This suppression of invading electrolyte by the presence of fixed charges is known as *Donnan exclusion* (see Problem 24B.3).

The preponderance of counter-ions, here M^+, inside the membrane, tends to cause them to diffuse out to the external solution, whereas the co-ions, here X^-, tend to diffuse into the membrane. The result is the development of an electrical potential difference between the membrane and external solution. This is normally estimated, by neglecting osmotic effects and assuming activity coefficients of unity, as

$$\phi_m - \phi_e = \left(-\frac{z_{M^+}F}{RT}\right)\ln\frac{x_{M^+m}}{x_{M^+e}} \qquad (24.5\text{-}31)$$

Equations 24.5-27 to 30 also apply to the relations between solutions on opposite sides of a membrane containing a partially excluded solute on one side, which now corresponds to the membrane phase of the above development. Equation 24.5-31 is often used for lack of knowledge of the neglected effects.

Thus Eq. 24.5-31 is particularly widely used by biologists to explain the origin of the ubiquitous potentials observed across biological membranes.[11] However, the means by which biological membranes can produce and control ion selectivity are extremely sophisticated and are only beginning to be understood.[17]

§24.6   MASS TRANSPORT IN POROUS MEDIA

Porous media are important in many mass transfer applications, some of which, such as catalysis[1] have already been touched on in this text (§18.7), and they exhibit a very wide variety of morphologies.[2,3] Adsorptive processes, such as chromatography, usually take place in granular beds and the absorbent particles themselves are often porous solids. Secondary recovery of crude petroleum typically involves mass transfer in porous rock, and freeze drying, or *lyophilization*, of foods and pharmaceuticals[4] depends on the transport of water vapor through a porous layer of dried solids. Related transport processes occur throughout the large field of particle technology,[5] and, as already indicated in §24.5, some membranes may be considered as microporous structures. Microporous structures abound in living organisms and contribute importantly to both water and solute distribution.[3]

Discussion of porous solids also brings us full circle, back to the discussions of momentum transfer with which this text began. Many of the models used to describe mass transfer in porous media are hydrodynamic in origin, and sometimes the concepts of mass and momentum transfer become blurred.

Predicting the transport of liquids and gases in porous media is a difficult and challenging problem, and no completely satisfactory theory is available. Mass is transported in a porous medium by a variety of mechanisms: (i) by ordinary diffusion, described by the Maxwell–Stefan equations; (ii) by Knudsen diffusion; (iii) by viscous flow according to the Hagen–Poiseuille equation; (iv) by surface diffusion—that is, the creeping of adsorbed molecules along the surfaces of the pores; (v) by thermal transpiration, which is the thermal analog of viscous slip; and (vi) by thermal diffusion. In this discussion, we neglect the last three of these mechanisms.

This problem has been attacked by many investigators,[6] and summarized by others.[7] We give here the principal results of their work. Available models are based either on

---

[17] B. Hill, *Ionic Channels of Excitable Membranes*, Sinauer Associates, Sunderland, Mass. (1992); F. M. Ashcroft, *Ion Channels and Disease: Channelopathies, Academic Press*, New York (1999); D. J. Aidley, *The Physiology of Excitable Cells*, Cambridge University Press (1998).

[1] (a) R. Aris, *The Mathematical Theory of Diffusion and Reaction in Permeable Catalysts*, Vols. 1 and 2 Oxford University Press (1975); (b) O. Levenspiel, *Chemical Reaction Engineering*, 3rd edition, Wiley, New York (1999).

[2] M. Sahimi, *Flow and Transport in Porous Media and Fractured Rock*, Verlagsgesellschaft, Weinheim, Germany (1995); V. Staněk, Fixed Bed Operations, Ellis Horwood, Chichester, England (1994).

[3] F. E. Curry, R. H. Adamson, Bing-Mei Fu, and S. Weinbaum, *Bioengineering Conference* (Sun River, Oregon), ASME, New York (1997).

[4] (a) L. Rey and J. C. May, "Freeze-Drying/Lyophilization of Pharmaceutical and Biological Products" in *Drugs and the Pharmaceutical Sciences* (J. Swarbrick, ed.), Marcel Dekker, New York (1999); (b) P. Sheehan and A. I. Liabis, *Biotech. and Bioeng.*, **60**, 712–728 (1998).

[5] M. Rhodes, *Introduction to Particle Technology*, Wiley, New York (1998).

[6] J. Hoogschagen, *J. Chem. Phys.*, **21**, 2096 (1953), *Ind. Eng. Chem.*, **47**, 906–913 (1955); D. S. Scott and F. A. L. Dullien, *AIChE Journal*, **8**, 113–117 (1962); L. B. Rothfeld, *AIChE Journal*, **9**, 19–24 (1963); P. L. Silveston, *AIChE Journal*, **10**, 132–133 (1964); R. D. Gunn and C. J. King, *AIChE Journal*, **15**, 507–514 (1969); C. Feng and W. E. Stewart, *Ind. Eng. Chem. Fund.*, **12**, 143–147 (1973); C. F. Feng, V. V. Kostrov, and W. E. Stewart, *Ind. Eng. Chem. Fund.*, **13**, 5–9 (1974).

[7] E. A. Mason and R. B. Evans, III, *J. Chem. Ed.*, **46**, 358–364 (1969); R. B. Evans III, L. D. Love, and E. A. Mason, *J. Chem. Ed.*, **46**, 423–427 (1969); R. Jackson, *Transport in Porous Catalysts*, Elsevier, Amsterdam (1977); R. E. Cunningham and R. J. J. Williams, *Diffusion in Gases and Porous Media*, Plenum Press, New York (1980); Chapter 6 of this book gives a summary of the history of the subject of diffusion.

cylindrical channels or aggregates of spheroidal particles, and we shall review a few representative examples here. We shall also restrict the discussion to two limiting situations within the pores of the solid matrix:

(i) *Free-molecule flow of gases*, in which the molecular diameters are short and mean free paths are long relative to the characteristic dimensions of the pores. Under these conditions, there is no significant interaction between the intrapore species.

(ii) *Continuum flow of gases or liquids*, in which both the diameters and the spacing of the intrapore molecules are short compared to the pore dimensions. Here the intrapore fluid can be described by the generalized hydrodynamic theory,[8] and the generalized Maxwell–Stefan equations for multicomponent diffusion can be used.

There are also phenomena for gas transport, known as the *slip-flow* phenomena, in which the mean free paths are comparable to the pore dimensions,[9] but we shall not discuss these here.

Free-Molecule Transport

Transport of rarefied gases is an example of Knudsen flow, already presented in Problem 2B.9. For a long capillary tube of radius a the Knudsen formula takes the form

$$N_A = -\frac{8a}{3}\frac{1}{\sqrt{2\pi M_A RT}}\frac{dp_A}{dz} = -\frac{8a}{3}\sqrt{\frac{RT}{2\pi M_A}}\frac{dc_A}{dz} \tag{24.6-1}$$

Here p_A is the partial pressure of species A in any mixture. Note that Eq. 24.6-1 states that the transport of any individual species under these limiting conditions is unaffected by the presence of others. Thus the total molar flow rate W_A in a tube is proportional to the cube of the tube radius and to the inverse square root of the molecular weight. This dependence on molecular weight is known as *Graham's law*.

Equation 24.6-1 can be rewritten as

$$N_A = -D_{AK}\frac{dc_A}{dz} \tag{24.6-2}$$

which defines the "Knudsen diffusivity" D_{AK}. However, this must be considered as a binary diffusivity for species A relative to the porous medium that is *not* consistent with Fick's law, because the molar flux contains no convective term. As a result D_{AK} is not a state property, containing as it does, the tube radius a. To allow for the tortuous nature of the channels in a porous medium and the limited cross-sectional area available for flow, the flux expression must be further modified by writing

$$\langle N_A \rangle = D_{AK}^{\text{eff}}\frac{dc_A}{dz} \tag{24.6-3}$$

where

$$D_{AK}^{\text{eff}} = (\varepsilon/\tau)D_{AK} \tag{24.6-4}$$

and $\langle N_A \rangle$ is the molar flux based on the total cross section of the porous medium. In this expression ε is the fractional void space in the porous material, and τ is a tortuosity fac-

[8] E. N. Lightfoot, J. B. Bassingthwaighte, and E. F. Grabowski, *Ann. Biomed. Eng.*, **4**, 78–90 (1976).
[9] R. Jackson, *Transport in Porous Catalysts*, Elsevier, Amsterdam (1977); R. E. Cunningham and R. J. J. Williams, *Diffusion in Gases and Porous Media*, Plenum Press, New York (1980).

tor. Although models exist[1a,10] for estimating the magnitude of τ, it must normally be determined experimentally.[11]

As an alternative to Eq. 24.6-4 for the effective Knudsen diffusivity, one may treat the aggregate as a collection of large immobile spheres (or "giant gas molecules"), and use the Chapman–Enskog kinetic theory.[12] Problem 24B.6 shows that this approach yields predictions very similar to those of Eq. 24.6-4. There is remarkable model insensitivity.

EXAMPLE 24.6-1

Knudsen Diffusion

Two large well-stirred reservoirs, each of volume V, are joined by a short duct of cross-sectional area S and length L, filled with a porous solid as indicated in Figure 24.6-1. Initially reservoir 1 is filled with hydrogen at uniform pressure p_0 and reservoir 2 with nitrogen, also at p_0. The entire system is maintained at a constant temperature. At time $t = 0$ a small valve in the duct is opened, and the two reservoirs are allowed to equilibrate with each other. Develop an expression for the total pressure in each reservoir as a function of time, assuming that the flow of each gas through the connecting duct follows Eq. 24.6-1, and that the ideal gas law holds throughout the system.

SOLUTION

We begin by assuming quasi-steady-state behavior in the duct so that, for either gas, the rate of transfer from reservoir 1 to reservoir 2 is given by

$$W_A = \frac{(8/3)(\varepsilon/\tau)aS}{L\sqrt{2\pi M_A RT}}(p_{A1} - p_{A2}) \equiv K_A(p_{A1} - p_{A2}) \tag{24.6-5}$$

where W_A is the molar rate of flow of species A (either nitrogen or hydrogen) and a is the effective radius of the pores in the plug joining the two reservoirs. Now a macroscopic mass balance for reservoir 2 gives

$$V\frac{dc_{A2}}{dt} = \frac{V}{RT}\frac{dp_{A2}}{dt} = K_A(p_{A1} - p_{A2}) \tag{24.6-6}$$

or

$$\frac{dp_{A2}}{dt} = \left(K_A\frac{RT}{V}\right)(p_{A1} - p_{A2}) \tag{24.6-7}$$

Now a mass balance over the whole system yields

$$p_{A1} + p_{A2} = p_0 \tag{24.6-8}$$

The initial conditions are that at time $t = 0$,

$$p_{H1} = p_0 \qquad p_{N1} = 0 \qquad p_{H2} = 0 \qquad p_{N2} = p_0 \tag{24.6-9}$$

These initial conditions complete the specification of the system behavior, and we see that the distributions of the two gases are independent of each other.

For *nitrogen* we can define the dimensionless variables $\psi = p_N/p_0$ and $\tau = (RTK_N/V)t$. Then we may write Eq. 24.6-7 for nitrogen in compartment 2

$$\frac{d\psi_{N2}}{d\tau} = 1 - 2\psi_{N2} \tag{24.6-10}$$

with the initial condition $\psi_{N2}(0) = 1$. The solution to this problem is then

$$\psi_{N2} = \tfrac{1}{2}(1 + e^{-2\tau}) \qquad \psi_{N1} = \tfrac{1}{2}(1 - e^{-2\tau}) \tag{24.6-11}$$

---

[10] W. E. Stewart and M. F. L. Johnson, *J. Catalysis*, **4**, 248–252 (1965).

[11] J. B. Butt, *Reaction Kinetics and Reactor Design*, 2nd edition, Marcel Dekker, New York (1999), p. 500, Table 7.4.

[12] R. B. Evans III, G. M. Watson, and E. A. Mason, *J. Chem. Phys.*, **35**, 2076–2083 (1961).

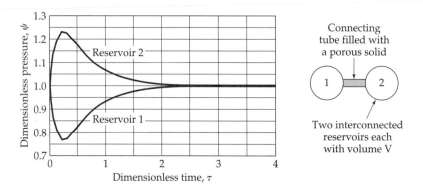

Fig. 24.6-1. Knudsen flow.

For *hydrogen* we note that $K_H = \sqrt{28/2}K_N \approx 3.74K_N$. Hence the differential equation for hydrogen is

$$\frac{d\psi_{H2}}{d\tau} = 3.74(1 - 2\psi_{H2}) \tag{24.6-12}$$

with the initial condition $\psi_{H2}(0) = 0$. The solution to the differential equation is then

$$\psi_{H2} = \tfrac{1}{2}(1 - e^{-7.48\tau}) \qquad \psi_{H1} = \tfrac{1}{2}(1 - e^{-7.48\tau}) \tag{24.6-13}$$

The results are plotted in Fig. 24.6-1.

The ratio $N_A/N_B = -\sqrt{M_B/M_A}$ of molar fluxes obtained here was first observed by Graham[13] in 1833 and rediscovered by Hoogschagen[6] in 1953. Though derived here for Knudsen flow, this relation is valid also for isobaric diffusion well outside the Knudsen region. It has been derived from kinetic theory by several investigators and verified experimentally in tubes and porous media[6,7,13,14] up to very large ratios of passage width to mean free path. Two sets of confirmatory data are shown in Table 24.6-1. In both sets of experiments,[13,14] an apparatus similar to that in Fig. 24.6-1 was used, and various tests gases were used against air. The flux ratios $N_{\text{gas}}/N_{\text{air}}$ were initial values, when each reservoir contained only air or the test gas.

Continuum Transport

To date, fluid mechanical modeling of intrapore transport has been limited to binary solutions in which molecules of the minor constituent (solute) are large compared to those of the solvent. Models for this situation are based on hydrodynamic diffusion theory extended to porous structures.[8] Descriptions are obtained by solving the creeping flow equations of motion for spheres (representing the solute) through a continuum (representing the solvent) in closed channels.[15] Important effects include partial exclusion of solute at the channel entrance and selective interaction with the channel wall. Result to date are limited to single solutes, but the rapid development of computational techniques[16] should permit extension to more complex systems. Hydrodynamic diffusion

---

[13] T. Graham, *Phil. Mag.*, **2**, 175, 269, 351 (1833). **Thomas Graham** (1805–1869), son of a prosperous manufacturer, attended the University of Glasgow from 1819 to 1826; in 1837, he was named professor of chemistry, University College, London, became a Fellow of the Royal Society in 1834, and in the same year was named Master of the Mint.

[14] E. A. Mason and B. Kronstadt, IMP-ARO(D)-12, University of Maryland, Institute for Molecular Physics, March 20, 1967.

[15] Z.-Y Yan, S. Weinbaum, and R. Pfeffer, *J. Fluid Mech.*, **162**, 415–438 (1986).

[16] S. Kim and S. J. Karrila, *Microhydrodynamics: Principles and Selected Applications*, Butterworth-Heinemann, Boston (1991).

Table 24.6-1 Experimental Verification of Graham's Law [T. Graham, *Phil. Mag.*, **2**, 175, 269, 351 (1833); E. A. Mason and B. Kronstadt, IMP-ARO(D)-12, University of Maryland, Institute for Molecular Physics, March 20, 1967].

Gas	N_{gas}/N_{air} Graham[a]	N_{gas}/N_{air} Mason and Kronstadt[b]	$\left(\dfrac{M_{air}}{M_{gas}}\right)^{1/2}$
H_2	3.83		3.791
He		2.66 ± 0.01	2.690
CH_4	1.344	1.33 ± 0.01	1.3437
C_2H_4	1.0191		1.0162
CO	1.0149		1.0169
N_2	1.0143	1.02 ± 0.01	1.0168
O_2	0.9487	0.960 ± 0.005	0.9514
H_2S	0.95		0.9219
Ar		0.855 ± 0.011	0.8516
N_2O	0.82		0.8112
CO_2	0.812		0.8113
SO_2	0.68		0.6724

calculations can be used for microporous membranes, but only if there are no significant intermolecular forces between the solutes and the pore walls.

Modeling viscous flow in these systems has already been discussed in §6.4, and it is common practice to describe such a flow, for the low Reynolds numbers of most interest here, by the Blake–Kozeny expression (Eq. 6.4-9) [see however Rhodes[5] (Chapter 5), Sahimi[2] (Chapter 6), Staněk[2] (Chapter 3)]:

$$v_0 = -\frac{D_p^2}{150\mu} \frac{\varepsilon^3}{(1-\varepsilon)^2} \nabla p \tag{24.6-14}$$

Here v_0 is the superficial mass-average velocity. Note from the discussion of §19.2 that the velocity used here is the mass-average velocity of the fluid through the porous material.

To obtain macroscopic descriptions we may use the generalized Maxwell–Stefan equations (Eq. 24.5-4), and we shall restrict ourselves here to concentration- and pressure-driven flow. Moreover, when the mobile species are small relative to pore dimensions, the boundary conditions simplify to continuity of species concentration and pressure at the interface between the external and "intrapore" fluid.

EXAMPLE 24.6-2

Transport from a Binary External Solution

Simplify the Maxwell–Stefan equations for the diffusion of a binary dilute solution, of a large solute species A in a solvent B, through a macroporous medium M, a matrix with pores large compared to the diameters of both mobile species, but small enough that lateral concentration gradients within each pore may be neglected.

SOLUTION

We begin by determining the pressure-flow relationship and note that we have two ways of doing this: the Blake–Kozeny equation (Eq. 24.6-14), and the diffusion-based result (Eq. 24.5-20) of the previous section.

For high velocities through the porous material and pores large relative to molecular dimensions, it is the mass-average velocity that must be proportional to the pressure gradient, and we may assume that the Blake–Kozeny equation governs the flow. We begin by rewriting Eq. 24.6-14) as

$$\left(\frac{150\mu(1-\varepsilon)^2}{D_p^2\varepsilon^3}\right)(\omega_A v_A + \omega_B v_B) = -\frac{dp}{dz} \tag{24.6-15}$$

and Eq. 24.5-20 as

$$cRT\left(\frac{v_A}{\mathcal{D}'_{AM}} + \frac{v_B}{\mathcal{D}'_{BM}}\right) = -\frac{dp}{dz} \tag{24.6-16}$$

Equating the coefficients of v_A and v_B in these two equations then yields descriptions of $\mathcal{D}'_{AM}$ and $\mathcal{D}'_{BM}$, respectively. If the pores are small relative to the molecular dimensions and the mass-average velocity is not large relative to the diffusional velocities $v_\alpha - v$, one is in a still poorly studied flow region, and one must resort either to experiment or to an appropriate molecular model.[17]

To determine the rate of solute transport, we turn to Eq. 24.5-25 *noting that the diffusivities of that section already include the factor* ε/τ. However, if the pore dimensions are very large with respect to the effective diameters of the solute and solvent molecules, the ratio $\mathcal{D}_{AB}/\mathcal{D}'_{AM}$ will be very small. We can thus obtain

$$N_A x_B - N_B x_A = -c(\varepsilon/\tau)\mathcal{D}_{AB}^{\text{ext}}\nabla x_A \tag{24.6-17}$$

in which

$$\mathcal{D}_{AB}^{\text{ext}} = \mathcal{D}_{AB}^{\text{ext}}\left(1 + \frac{\partial \ln \gamma_A}{\partial \ln x_A}\right) \tag{24.6-18}$$

where the superscript ext refers to conditions in the external solution of the same composition as the pore fluid. Equation 24.6-17 may in turn be rewritten as

$$N_A = -c(\varepsilon/\tau)\mathcal{D}_{AB}^{\text{ext}}\nabla x_A + (N_A + N_B) \tag{24.6-19}$$

which is Fick's first law modified for void fraction and tortuosity. It is widely used.

Exactly as in unconfined fluids, one cannot determine net flow, or pressure drop, from diffusional considerations alone. One needs a flux ratio or equivalent. A specific example is supplied by freeze-drying, where water vapor must diffuse through a porous region of dried solid and where inert gases may be assumed stagnant. This region is also interesting in that conditions can vary from simple continuum diffusion, as here, through the slip-flow region, and on to the Knudsen region.[4b]

It must be remembered that Eq. 24.6-19 and the equations leading up to it represent only the direct effect of molecular diffusion. The convective dispersion resulting from interparticle mixing and local departures from rectilinear flow must be added when using the volume-averaged convective equation (see §20.5, Butt[12] §5.2.5, and Levenspiel[1b] §13.2).

QUESTIONS FOR DISCUSSION

1. How does equilibrium thermodynamics have to be supplemented in order to study non-equilibrium systems, such as those that involve velocity, temperature, and concentration gradients?
2. What new transport coefficients arise in multicomponent mixtures and what do they describe?
3. To what extent does this chapter explain the origin of Eq. 19.3-3? Is that equation completely correct?
4. Is Eq. 24.1-6 really the starting point for the derivation of the complete expressions for the fluxes? Discuss its origin.

---

[17] Z.-Y. Yan, S. Weinbaum, and R. Pfeffer, *J. Fluid Mech.*, **162**, 415–438 (1986).

5. How are the thermal diffusion coefficient, the thermal diffusion ratio, and the Soret coefficient defined? Can the signs of these quantities be predicted a priori?

6. How can one start with Eq. 24.2-8 and obtain Eq. 17.9-1? What restrictions have to be placed on Eq. 17.9-1?

7. What is the proper driving force for diffusion: the gradient of the concentration, the gradient of the activity, or some other quantity?

8. Discuss the Clusius–Dickel column for isotope separation.

9. To describe the steady-state operation of an ultracentrifuge it is not necessary to know any transport properties. Does this seem odd?

10. What various physical phenomena need to be understood in order to describe diffusion in porous media?

PROBLEMS   **24A.1. Thermal diffusion.**

(a) Estimate the steady-state separation of H_2 and D_2 occurring in the simple thermal diffusion apparatus shown in Fig. 24.2-1 under the following conditions: T_1 is 200K, T_2 is 600K, the mole fraction of deuterium is initially 0.10, and the effective average k_T is 0.0166.

(b) At what temperature should this average k_T have been evaluated?

Answers: (a) The mole fraction of H_2 is higher by 0.0183 in the hot bulb

   (b) 330K

24A.2. Ultracentrifugation of proteins. Estimate the steady-state concentration profile when a typical albumin solution is subjected to a centrifugal field 50,000 times the force of gravity under the following conditions:

Cell length $= 1.0$ cm

Molecular weight of albumin $= 45{,}000$

Apparent density of albumin in solution $= M_A/\overline{V}_A = 1.34$ g/cm^3

Mole fraction of albumin (at $z = 0$), $x_{A0} = 5 \times 10^{-6}$

Apparent density of water in the solution $= 1.00$ g/cm^3

Temperature $= 75°$F

Answer: $x_A = 5 \times 10^{-6} \exp(-22.7z)$, with z in cm

24A.3. Ionic diffusivities. The limiting (that is, at zero concentration) equivalent ionic conductances, in dimensions of cm^2/ohm $\cdot$ g-equiv for the following ions at 25°C are:[1] Na$^+$, 50.10; K$^+$, 73.5; Cl$^-$, 76.35. Calculate the corresponding ionic diffusivities from the definion

$$\mathcal{D}_{iw} = \frac{RT}{F^2} \frac{\lambda_{i0}}{|z_i|} \tag{24A.3-1}$$

Note that $F = 96{,}500$ coulombs/g-equiv, $RT/F = 25.692$ mv at 25°C, and 1 coulomb $= 1$ ampere $\cdot$ s.

24B.1. The dimensions of the Lorentz force. Show how the Lorentz force on a charge moving through a magnetic field corresponds to the first term added to the linear $\mathbf{d}_\alpha$ of Eq. 25.4-51 and gives a consistent set of units for this quantity. *Suggestion:* Note that $cRT\mathbf{d}_\alpha$ represents the motive force for diffusive motion of species α per unit volume and that the usual dimensions of the magnetic induction are 1 Weber $= 1$ Newton-second/Coulomb-meter.

24B.2. Junction potentials. Consider two well-mixed reservoirs of aqueous salt at 25°C, as in Fig. 24.4-2, separated by a stagnant region. Salt concentrations are 1.0 N on the left (1) and 0.1 N on the right (2). Estimate junction potentials for NaCl and for KCl using the ion diffusivities

[1] R. A. Robinson and R. H. Stokes, *Electrolyte Solutions*, revised edition, Butterworths, London (1965), Table 6.1.

of Problem 24A.3. Assume constant ion activity coefficients. Which compartment will be the more positive? Why?

24B.3. Donnan exclusion. The sulfonic acid membrane used by Scattergood[2] had the following equilibrium internal composition when immersed in 0.1 N NaCl:

Organic sulfonic acid polymer	c_{X^-}	= 1.03 g-equiv/liter
Water	c_w	= 13.2 g-equiv/liter
Chloride ion	c_{Cl^-}	= 0.001 g-equiv/liter
Sodium ion	c_{Na^+}	= 1.031 g-equiv/liter

Calculate the distribution coefficient of sodium chloride

$$K_D = \frac{(x_{Na^+} x_{Cl^-})_{external}}{(x_{Na^+} x_{Cl^-})_{membrane}} \tag{24B.3-1}$$

Note that the concentration of water in the external solution is about 55.5 g-mol/liter.

Answer: 0.064

24B.4. Osmotic pressure. Typical sea water, containing 3.45% by weight of dissolved salts, has a vapor pressure 1.84% below that of pure water. Estimate the minimum possible transmembrane pressure required to produce pure water, if the membrane is ideally selective.

Answer: about 25 atm

24B.5. Permeability of a perfectly selective filtration membrane. Develop an expression for the hydraulic permeability of the perfectly selective membrane described in Example 22.8-5 in terms of the diffusional parameters introduced in §24.5.

Answer: $K_H = Đ'_{wm}/RT\delta$, where δ is the membrane thickness

24B.6. Model insensitivity. In modeling a porous medium as a parallel network of channels one must allow both for the toruous nature ("tortuosity" τ) of real systems and also the restriction of the transport to the fraction ε of the cross section that is available for flow. Equation 24.6-3 then must be modified to

$$\mathbf{N}_A = -\frac{8}{3} \frac{\varepsilon a}{\tau} \frac{1}{\sqrt{2\pi M_A RT}} \nabla p \tag{24B.6-1}$$

An alternate approach is to consider the transport process to be a diffusion of species A through an immobilized set of giant molecules[3] (these particles comprising the porous medium). This model yields the expression

$$\mathbf{N}_A = -\frac{\pi}{4}\left(\frac{1 + \frac{1}{8}\pi}{1 - \varepsilon}\right) \frac{\varepsilon a}{\tau} \frac{1}{\sqrt{2\pi M_A RT}} \nabla p \tag{24B.6-2}$$

Compare these two equations, noting that the value of ε is often about 0.4.

24C.1. Expressions for the mass flux.

(a) Show how to transform the left side of Eq. 24.2-8 into the left side of Eq. 24.2-9. First rewrite the former as follows:

$$\sum_{\substack{\beta=1 \\ \beta \neq \alpha}}^{N} \frac{x_\alpha x_\beta}{Đ_{\alpha\beta}}(\mathbf{v}_\gamma - \mathbf{v}_\beta) + x_\alpha(\mathbf{v}_\alpha - \mathbf{v}_\gamma) \sum_{\substack{\beta=1 \\ \beta \neq \alpha}}^{N} \frac{x_\beta}{Đ_{\alpha\beta}} \tag{24C.1-1}$$

Rewrite the second term as a sum over *all* β, and then add a term to compensate for the modification of the sum. Note that this change has introduced into the sum a term containing $Đ_{\alpha\alpha}$,

[2] E. M. Scattergood and E. N. Lightfoot, *Trans. Faraday Soc.*, **64**, 1135–1146 (1968).

[3] R. B. Evans, III, G. M. Watson, and E. A. Mason, *J. Chem. Phys.*, **35**, 2076–2083 (1961).

which was not defined because it was not needed. Now, we are at liberty to define $Ð_{\alpha\alpha}$ in any way we choose, and the choice we make is

$$\frac{x_\alpha}{Ð_{\alpha\alpha}} = -\sum_{\substack{\beta=1 \\ \beta\neq\alpha}}^{N} \frac{x_\beta}{Ð_{\alpha\beta}} \quad \text{or} \quad \sum_{\substack{\beta=1 \\ \text{all }\beta}}^{N} \frac{x_\beta}{Ð_{\alpha\beta}} = 0 \qquad (24C.1\text{-}2, 3)$$

This choice enables us to obtain the left side of Eq. 24.2-9, and also the auxiliary relation given after Eq. 24.2-9 is, in fact, just Eq. 24C.1-3 above.

(b) Next repeat the above derivation by replacing $\mathbf{v}_\beta$ by $[\mathbf{v}_\beta + (D_\beta^T/\rho_\beta)\nabla \ln T]$, and verify that both the diffusion terms and the thermal diffusion terms of Eq. 24.2-8 may be transformed into the corresponding terms in Eq. 24.2-9.

24C.2. **Differential centrifugation.** The lysing (bursting) of *E. coli* cells has produced a dilute suspension of *inclusion bodies*, hard insoluble aggregates of a desired protein, unlysed cells, and unwanted dissolved proteins. For purposes of this problem all may be considered as spheres with the properties indicated here.

	Cells	Inclusion bodies	Proteins
Mass or equivalent	1.89×10^{-12} g	2.32×10^{-15} g	50 kilodaltons
Density (g/ml)	1.07	1.3	1.3

Can these materials be effectively separated by centrifugation? Explain.

24C.3. **Transport characteristics of sodium chloride.** In the accompanying table[1] equivalent conductance, diffusivity, and thermodynamic activity coefficients are given for sodium chloride at 25°C. The first two are given as functions of the molarity (M), and the third for molality (m). It may be assumed for the purposes of this problem that $M/m = 1 - 0.019m$. Limiting ionic equivalent conductances (that is, at infinite dilution) are 50.10 and 76.35, respectively. The salt equivalent conductance in turn is defined as

$$\Lambda_S = \lambda_{Na^+} + \lambda_{Cl^-} = K_{sp}/c_S \qquad (24C.3\text{-}1)$$

where the specific conductance $K_{sp} = L/AR$, where R is the resistance of a volume of solution of length L and cross-sectional area A. Use these data to discuss the sensitivity of the solution behavior to the three diffusivities $Ð_{Na^+,W}$, $Ð_{Cl^-,W}$, and $Ð_{Na^+Cl^-}$ needed to describe this response to solution concentration.

Electrochemical characteristics of aqueous NaCl solution at 25°C

Molar concentration	Equivalent conductance (cm²/ohm-equiv)	Diffusivity cm²/s × 10⁵	Molal concentration	Activity coefficient
0	126.45	1.61	0	1
0.00055	124.51			
0.001	123.74	1.585		
0.005	120.64			
0.01	118.53			
0.02	115.76			
0.05	111.06	1.507		
0.1	106.74	1.483	0.1	0.778
0.2	101.71	1.475	0.2	0.735
			0.3	0.710
			0.4	0.693
0.5	93.62	1.474	0.5	0.681

Electrochemical characteristics of aqueous NaCl solution at 25°C *(continued)*

Molar concentration	Equivalent conductance (cm²/ohm-equiv)	Diffusivity cm²/s × 10⁵	Molal concentration	Activity coefficient
			0.6	0.673
			0.7	0.667
			0.8	0.659
			0.9	0.657
1	85.76	1.484	1.0	0.657
			1.2	0.654
			1.4	0.655
1.5	79.86	1.495		
			1.6	0.657
			1.8	0.662
2	74.71	1.516	2.0	0.668
			2.5	0.688
3	65.57	1.563	3.0	0.714
			3.5	0.746
4	57.23		4.0	0.783
			4.5	0.826
5	49.46		5.0	0.874

24C.4. Departures from electroneutrality. Following Newman, estimate the departures from electroneutrality in the stagnant region between the reservoirs of Problem 24B.2 as follows. First calculate the electric field gradient $d^2\phi/dz^2$, where z is the distance measured from reservoir 1 toward reservoir 2, assuming that the salt concentration in g-moles/liter is given by

$$c_s = 1.0 - 0.9\frac{z}{L} \tag{24C.4-1}$$

where L is the length of the stagnant region. Then put the result into Poisson's equation

$$\nabla^2\phi = \frac{F}{\varepsilon}\sum_{i=1}^{N} z_i c_i \tag{24C.4-2}$$

Here ε is the dielectric constant, and F/ε may be taken to be 1.392×10^{16} volt-cm/g-equiv (see Newman[4], pp. 74 and 256), which corresponds to a relative dielectric constant of 78.303. For this problem, the summation reduces to $(c_+ - c_-)$.

24C.5. Dielectrophoretic driving forces. When an electric potential is imposed across an uncharged nonconducting medium, one may write

$$(\nabla \cdot \varepsilon\mathbf{E}) = 0 \tag{24C.5-1}$$

where ε is the dielectric constant.

Show how this equation can be used to calculate the distribution of electric field $\mathbf{E}$ in the region between two coaxial cylindrical metal electrodes of outer and inner radii R_2 and R_1, respectively. You may neglect variations in the dielectric constant. Toward which electrode will particles of positive susceptibility migrate, and how will their migration velocity vary with position?

---

[4] J. S. Newman, *Electrochemical Systems*, 2nd edition, Prentice-Hall, New York (1991), p. 256.

24C.6. **Effects of small inclusions in a dielectric medium.** The production of field nonlinearities by embedded particles can be illustrated by considering the limiting case of a single particle of radius R in an otherwise uniform field. The field distribution in both the external medium and the particle are defined by Laplace's equation, $\nabla^2 \phi = 0$, and by the boundary condition on the sphere surface (here the indices s and c stand for sphere and continuum).

$$\varepsilon_s(\boldsymbol{\delta}_r \cdot \nabla \phi_s) = \varepsilon_c(\boldsymbol{\delta}_r \cdot \nabla \phi_c) \tag{24C.6-1}$$

Develop expressions for ϕ_c and ϕ_s, if $\phi_c \to Ar\cos\theta$ for large r.

24C.7. **Frictionally induced selective filtration.** Describe the glucose rejection behavior of a cellophane[5,6] that shows no thermodynamic rejection. You may assume glucose mole fraction in the feed to the membrane to be 0.01 and the following properties:

$$K_D = 1.0; \qquad \overline{V}_s/\overline{V}_w = 4; \qquad \mathcal{D}'_{wm}/\mathcal{D}'_{gm} = 100; \qquad \mathcal{D}'_{wm}/\mathcal{D}'_{gw} = 25$$

Here the subscripts g, w, and m refer to glucose, water, and the membrane matrix, respectively. *Partial answer:* The high-flow-limiting mole fraction of sugar in the filtrate is 0.00242.

24C.8. **Thermodynamically induced selective filtration.** Describe the behavior of the hypothetical membrane for which $K_D = 1.0$, solute activity coefficients are unity, and $\mathcal{D}'_{sm}/\mathcal{D}'_{wm} = \overline{V}_w/\overline{V}_s$. *Partial answer:* The high-flow limiting product solute concentration is 0.1 times that in the feed.

24C.9. **Facilitated transport.** Consider here the transport of a solute S across a homogeneous membrane from one external solution to another as a complex CS with a carrier C unable to leave the membrane phase. The solute S may be considered to be insoluble in the membrane and convection to be negligible (see Fig. 24C.8). Assume further that:

1. Equilibrium exists at both membrane surfaces according to

$$c_{CS} = K_D c_C c_S \tag{24C.9-1}$$

   where the concentration of S is that in the external solution, and those of C and CS are in the membrane.[7]

2. Both C and CS follow the simple rate expression $N_i = D_{im}\Delta c_i$.

Develop a general expression for the transport rate of S in terms of the total amount of carrier plus carrier complex present in the membrane, the solution concentrations of S, the quantity K_D, and the diffusivities. What is the maximum rate of transport of S (that is, as its concentration at the left of the diagram becomes very high and that at the right is zero)?

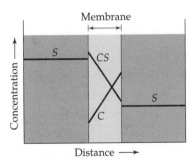

Fig. 24C.8. Elementary facilitated transport. Concentration profiles for the solute (S), the carrier (C), and the complex (CS).

[5] B. Z. Ginzburg and A. Katchalsky, *J. Gen. Physiol.*, **47**, 403–418 (1963).

[6] T. G. Kaufmann and E. F. Leonard, *AIChE Journal*, **14**, 110–117 (1968).

[7] See, however, J. D. Goddard, J. S. Schultz, and R. J. Bassett, *Chem. Eng. Sci.*, **25**, 665–683 (1970), and W. D. Stein, *The Movement of Molecules across Cell Membranes*, Academic Press, New York (1984).

24D.1. Entropy flux and entropy production.

(a) Verify that Eqs. 24.1-3 and 4 follow from Eqs. 24.1-1 and 2.

(b) Show that one can go backward from Eqs. 21.4-5 through 8 to Eqs. 24.1-3 and 4. To do this it may be necessary to use one form of the Gibbs–Duhem equation,

$$\sum_{\alpha=1}^{N} \rho_\alpha \nabla\left(\frac{1}{T}\frac{\overline{G}_\alpha}{M_\alpha}\right) - \frac{1}{T}\nabla p + \sum_{\alpha=1}^{N} \rho_\alpha \frac{\overline{H}_\alpha}{M_\alpha}\frac{1}{T^2}\nabla T = 0 \tag{24D.1-1}$$

Postface

Of all the messages we have tried to convey in this long text, the most important is to recognize the *key role of the equations of change,* developed in Chapters 3, 11, and 19. Written at the microscopic continuum level, they are the key link between the very complex motions of individual molecules and the observable behavior of most systems of engineering interest. They can be used to determine velocity, pressure, temperature, and concentration profile, as well as the fluxes of momentum, energy, and mass, even in complicated time-dependent systems. They are applicable to turbulent systems, and even when complete a priori solutions prove infeasible, simplify the efficient use of data through dimensional analysis. Integrated forms of the equations of change provide the macroscopic balances.

No introductory text can, however, meet the needs of every reader. We have attempted, therefore, to provide *a solid basis in the fundamentals* needed to tackle presently unforeseen applications of transport phenomena in an intelligent way. We have also given extensive references to sources where additional information can be found. Some of these references contain specialized data or introduce powerful problem-solving techniques. Others show how transport analysis can be incorporated into equipment and process design.

We have therefore concentrated on relatively simple examples that illustrate the characteristics of the equations of change and the kinds of questions they are capable of answering. This has required largely neglecting the very powerful numerical techniques available for solving difficult problems. Fortunately, there are now many *monographs on numerical techniques* and packaged programs of greater or lesser generality. Graphics programs are also available, which greatly simplify the presentation of data and simulations.

It should also be recognized that great advances are being made in the molecular theory of transport phenomena, ranging from improved techniques for predicting the transport properties to the development of new materials. *Molecular dynamics and Brownian dynamics simulation techniques* are proving to be very powerful for understanding such varied systems as ultra-low density gases, thin films, small pores, interfaces, colloids, and polymeric liquids.

Simple models of *turbulent transport* have been included, but these are only a modest introduction to a large and important field. Highly sophisticated techniques have been developed for specialized areas, such as predicting the forces and torques on aircraft, the combustion processes in automobiles, and the performance of fluid mixers. It is hoped that the interested reader will not stop with our very limited introductory discussion.

Conversely, we have greatly expanded our coverage of *boundary-layer phenomena,* because its importance and power are now being recognized in many applications. Once primarily the province of aerodynamicists, boundary-layer techniques are now widely used in many fields of heat and mass transfer, as well as in fluid mechanics. Applications abound in such varied fields as catalysis, separation processes, and biology.

Of great and increasing importance is *non-Newtonian behavior,* encountered in the preparation and use of films, lubricants, adhesives, suspensions, and emulsions. Biological examples are exceedingly important, ranging from the operation of the joints to drag-reducing slimes on marine animals, and down to the very basic problem of digesting foodstuffs.

No music and no oral communication would be possible without *compressible flow,* an area we have neglected because of space limitations. Compressible flow is also of critical importance in the design of airplanes, re-entry vehicles in our space program, and in predicting meteorological phenomena. The awesome destructive power of tornadoes is a challenging example of the latter.

Some problems involving transport phenomena in *chemically reacting systems* have been presented. For simplicity, we have taken the chemical kinetics expressions to be of rather idealized forms. For problems in combustion, flame propagation, and explosion phenomena more realistic descriptions of the kinetics will be needed. The same is true in biological systems, and the understanding of the functioning of the human body will require much more detailed descriptions of the interactions among chemical kinetics, catalysis, diffusion, and turbulence.

In basic terms, each of us is internally powered by the close equivalent of fuel cells, with current carried primarily by cations, in particular protons, rather than electrons. There are also *complex electrical transport phenomena* taking place in the now ubiquitous microelectronic devices such as computers and cellular phones. We have provided a very modest introduction to electrotransport, but again the reader is urged to go on to more specialized sources.

No engineering project can be conceived, let alone completed, purely through use of the descriptive disciplines, such as transport phenomena and thermodynamics. Engineering, in the last analysis, depends heavily on heuristics to supplement incomplete knowledge. Transport phenomena can, however, prove immensely helpful by providing useful approximations, starting with order-of-magnitude estimates, and going on to successively more accurate approximations, such as those provided by boundary-layer theory. It is therefore important, perhaps in a second reading of this text, to seek *shape- and model-insensitive descriptions* by examining the numerical behavior of our model systems.

R. B. B.
W. E. S.
E. N. L.

Appendix A

Vector and Tensor Notation[1]

The physical quantities encountered in transport phenomena fall into three categories: *scalars*, such as temperature, pressure, volume, and time; *vectors*, such as velocity, momentum, and force; and (second-order) *tensors*, such as the stress, momentum flux, and velocity gradient tensors. We distinguish among these quantities by the following notation:

$$s = \text{scalar (lightface Italic)}$$
$$\mathbf{v} = \text{vector (boldface Roman)}$$
$$\boldsymbol{\tau} = \text{second-order tensor (boldface Greek)}$$

In addition, boldface Greek symbols with one subscript (such as $\boldsymbol{\delta}_i$) are vectors.

For vectors and tensors, several different kinds of multiplication are possible. Some of these require the use of special multiplication signs to be defined later: the single dot ($\cdot$), the double dot ($:$), and the cross ($\times$). We enclose these special multiplications, or sums thereof, in different kinds of parentheses to indicate the type of result produced:

$$(\ ) = \text{scalar}$$
$$[\ ] = \text{vector}$$
$$\{\ \} = \text{second-order tensor}$$

No special significance is attached to the kind of parentheses if the only operations enclosed are addition and subtraction, or a multiplication in which $\cdot$, $:$, and $\times$ do not appear. Hence $(\mathbf{v} \cdot \mathbf{w})$ and $(\boldsymbol{\tau}:\nabla\mathbf{v})$ are scalars, $[\nabla \times \mathbf{v}]$ and $[\boldsymbol{\tau} \cdot \mathbf{v}]$ are vectors, and $\{\mathbf{v} \cdot \nabla\boldsymbol{\tau}\}$ and

---

[1] This appendix is very similar to Appendix A of R. B. Bird, R. C. Armstrong, and O. Hassager, *Dynamics of Polymeric Liquids, Vol. 1, Fluid Mechanics*, 2nd edition, Wiley-Interscience, New York (1987). There, in §8, a discussion of nonorthogonal coordinates is given. Also in Table A.7-4, there is a summary of the del operations for bipolar coordinates.

$\{\boldsymbol{\sigma} \cdot \boldsymbol{\tau} + \boldsymbol{\tau} \cdot \boldsymbol{\sigma}\}$ are second-order tensors. On the other hand, $\mathbf{v} - \mathbf{w}$ may be written as $(\mathbf{v} - \mathbf{w})$, $[\mathbf{v} - \mathbf{w}]$, or $\{\mathbf{v} - \mathbf{w}\}$, since no dot or cross operations appear. Similarly $\mathbf{vw}$, $(\mathbf{vw})$, $[\mathbf{vw}]$, and $\{\mathbf{vw}\}$ are all equivalent.

Actually, scalars can be regarded as zero-order tensors and vectors as first-order tensors. The multiplication signs may be interpreted thus:

Multiplication Sign	Order of Result
None	Σ
$\times$	$\Sigma - 1$
$\cdot$	$\Sigma - 2$
$:$	$\Sigma - 4$

in which Σ represents the sum of the orders of the quantities being multiplied. For example, $s\boldsymbol{\tau}$ is of the order $0 + 2 = 2$, $\mathbf{vw}$ is of the order $1 + 1 = 2$, $\boldsymbol{\delta}_1 \boldsymbol{\delta}_2$ is of the order $1 + 1 = 2$, $[\mathbf{v} \times \mathbf{w}]$ is of the order $1 + 1 - 1 = 1$, $(\boldsymbol{\sigma}:\boldsymbol{\tau})$ is of the order $2 + 2 - 4 = 0$, and $\{\boldsymbol{\sigma} \cdot \boldsymbol{\tau}\}$ is of the order $2 + 2 - 2 = 2$.

The basic operations that can be performed on scalar quantities need not be elaborated on here. However, the laws for the algebra of scalars may be used to illustrate three terms that arise in the subsequent discussion of vector operations:

a. For the multiplication of two scalars, r and s, the order of multiplication is immaterial so that the *commutative* law is valid: $rs = sr$.

b. For the successive multiplication of three scalars, q, r, and s, the order in which the multiplications are performed is immaterial, so that the *associative* law is valid: $(qr)s = q(rs)$.

c. For the multiplication of a scalar s by the sum of scalars p, q, and r, it is immaterial whether the addition or multiplication is performed first, so that the *distributive* law is valid: $s(p + q + r) = sp + sq + sr$.

These laws are not generally valid for the analogous vector and tensor operations described in the following paragraphs.

§A.1   VECTOR OPERATIONS FROM A GEOMETRICAL VIEWPOINT

In elementary physics courses, one is introduced to vectors from a geometrical standpoint. In this section we extend this approach to include the operations of vector multiplication. In §A.2 we give a parallel analytic treatment.

Definition of a Vector and Its Magnitude

A vector $\mathbf{v}$ is defined as a quantity of a given magnitude and direction. The magnitude of the vector is designated by $|\mathbf{v}|$ or simply by the corresponding lightface symbol v. Two vectors $\mathbf{v}$ and $\mathbf{w}$ are equal when their magnitudes are equal and when they point in the same direction; they do not have to be collinear or have the same point of origin. If $\mathbf{v}$ and $\mathbf{w}$ have the same magnitude but point in opposite directions, then $\mathbf{v} = -\mathbf{w}$.

Addition and Subtraction of Vectors

The addition of two vectors can be accomplished by the familiar parallelogram construction, as indicated by Fig. A.1-1a. Vector addition obeys the following laws:

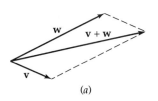
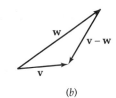

Fig. A.1-1. (*a*) Addition of vectors; (*b*) subtraction of vectors.

Commutative:	$(\mathbf{v} + \mathbf{w}) = (\mathbf{w} + \mathbf{v})$	(A.1-1)
Associative:	$(\mathbf{v} + \mathbf{w}) + \mathbf{u} = \mathbf{v} + (\mathbf{w} + \mathbf{u})$	(A.1-2)

Vector subtraction is performed by reversing the sign of one vector and adding; thus $\mathbf{v} - \mathbf{w} = \mathbf{v} + (-\mathbf{w})$. The geometrical construction for this is shown in Fig. A.1-1b.

Multiplication of a Vector by a Scalar

When a vector is multiplied by a scalar, the magnitude of the vector is altered but its direction is not. The following laws are applicable

Commutative:	$s\mathbf{v} = \mathbf{v}s$	(A.1-3)
Associative:	$r(s\mathbf{v}) = (rs)\mathbf{v}$	(A.1-4)
Distributive	$(q + r + s)\mathbf{v} = q\mathbf{v} + r\mathbf{v} + s\mathbf{v}$	(A.1-5)

Scalar Product (or Dot Product) of Two Vectors

The scalar product of two vectors $\mathbf{v}$ and $\mathbf{w}$ is a scalar quantity defined by

$$(\mathbf{v} \cdot \mathbf{w}) = vw \cos \phi_{vw} \qquad (A.1\text{-}6)$$

in which ϕ_{vw} is the angle between the vectors $\mathbf{v}$ and $\mathbf{w}$. The scalar product is then the magnitude of w multiplied by the projection of $\mathbf{v}$ on $\mathbf{w}$, or vice versa (Fig. A.1-2a). Note that the scalar product of a vector with itself is just the square of the magnitude of the vector

$$(\mathbf{v} \cdot \mathbf{v}) = |\mathbf{v}|^2 = v^2 \qquad (A.1\text{-}7)$$

The rules governing scalar products are as follows:

Commutative:	$(\mathbf{u} \cdot \mathbf{v}) = (\mathbf{v} \cdot \mathbf{u})$	(A.1-8)
Not Associative:	$(\mathbf{u} \cdot \mathbf{v})\mathbf{w} \neq \mathbf{u}(\mathbf{v} \cdot \mathbf{w})$	(A.1-9)
Distributive:	$(\mathbf{u} \cdot \{\mathbf{v} + \mathbf{w}\}) = (\mathbf{u} \cdot \mathbf{v}) + (\mathbf{u} \cdot \mathbf{w})$	(A.1-10)

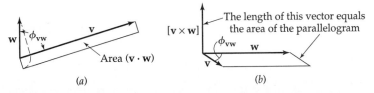

Fig. A.1-2. Products of two vectors: (*a*) the scalar product; (*b*) the vector product.

Vector Product (or Cross Product) of Two Vectors

The vector product of two vectors $\mathbf{v}$ and $\mathbf{w}$ is a vector defined by

$$[\mathbf{v} \times \mathbf{w}] = \{vw \sin \phi_{vw}\}\mathbf{n}_{vw} \qquad (A.1\text{-}11)$$

in which $\mathbf{n}_{vw}$ is a vector of unit length (a "unit vector") perpendicular to both $\mathbf{v}$ and $\mathbf{w}$ and pointing in the direction that a right-handed screw will move if turned from $\mathbf{v}$ toward $\mathbf{w}$ through the angle ϕ_{vw}. The vector product is illustrated in Fig. A.1-2b. The magnitude of the vector product is just the area of the parallelogram defined by the vectors $\mathbf{v}$ and $\mathbf{w}$. It follows from the definition of the vector product that

$$[\mathbf{v} \times \mathbf{v}] = 0 \qquad (A.1\text{-}12)$$

Note the following summary of laws governing the vector product operation:

Not Commutative:	$[\mathbf{v} \times \mathbf{w}] = -[\mathbf{w} \times \mathbf{v}]$	$(A.1\text{-}13)$
Not Associative:	$[\mathbf{u} \times [\mathbf{v} \times \mathbf{w}]] \neq [[\mathbf{u} \times \mathbf{v}] \times \mathbf{w}]$	$(A.1\text{-}14)$
Distributive:	$[\{\mathbf{u} + \mathbf{v}\} \times \mathbf{w}] = [\mathbf{u} \times \mathbf{w}] + [\mathbf{v} \times \mathbf{w}]$	$(A.1\text{-}15)$

Multiple Products of Vectors

Somewhat more complicated are multiple products formed by combinations of the multiplication processes just described:

(a) $rs\mathbf{v}$ (b) $s(\mathbf{v} \cdot \mathbf{w})$ (c) $s[\mathbf{v} \times \mathbf{w}]$
(d) $(\mathbf{u} \cdot [\mathbf{v} \times \mathbf{w}])$ (e) $[\mathbf{u} \times [\mathbf{v} \times \mathbf{w}]]$ (f) $([\mathbf{u} \times \mathbf{v}] \cdot [\mathbf{w} \times \mathbf{z}])$
(g) $[[\mathbf{u} \times \mathbf{v}] \times [\mathbf{w} \times \mathbf{z}]]$

The geometrical interpretations of the first three of these are straightforward. The magnitude of $(\mathbf{u} \cdot [\mathbf{v} \times \mathbf{w}])$ can easily be shown to represent the volume of a parallelepiped with edges defined by the vectors $\mathbf{u}$, $\mathbf{v}$, and $\mathbf{w}$.

EXERCISES

1. What are the "orders" of the following quantities: $(\mathbf{v} \cdot \mathbf{w})$, $(\mathbf{v} - \mathbf{u})\mathbf{w}$, $(\mathbf{ab}:\mathbf{cd})$, $[\mathbf{v} \cdot \rho\mathbf{wu}]$, $[[\mathbf{a} \times \mathbf{f}] \times [\mathbf{b} \times \mathbf{g}]]$?

2. Draw a sketch to illustrate the inequality in Eq. A.1-9. Are there any special cases for which it becomes an equality?

3. A mathematical plane surface of area S has an orientation given by a unit normal vector $\mathbf{n}$, pointing downstream of the surface. A fluid of density ρ flows through this surface with a velocity $\mathbf{v}$. Show that the mass rate of flow through the surface is $w = \rho(\mathbf{n} \cdot \mathbf{v})S$.

4. The angular velocity $\mathbf{W}$ of a rotating solid body is a vector whose magnitude is the rate of angular displacement (radians per second) and whose direction is that in which a right-handed screw would advance if turned in the same direction. The position vector $\mathbf{r}$ of a point is the vector from the origin of coordinates to the point. Show that the velocity of any point in a rotating solid body is $\mathbf{v} = [\mathbf{W} \times \mathbf{r}]$, relative to an origin located on the axis of rotation.

5. A constant force $\mathbf{F}$ acts on a body moving with a velocity $\mathbf{v}$, which is not necessarily collinear with $\mathbf{F}$. Show that the rate at which $\mathbf{F}$ does work on the body is $W = (\mathbf{F} \cdot \mathbf{v})$.

§A.2 VECTOR OPERATIONS IN TERMS OF COMPONENTS

In this section a parallel analytical treatment is given to each of the topics presented geometrically in §A.1. In the discussion here we restrict ourselves to rectangular coordinates and label the axes as 1, 2, 3 corresponding to the usual notation of x, y, z; only right-handed coordinates are used.

Many formulas can be expressed compactly in terms of the *Kronecker delta* δ_{ij} and the *permutation symbol* ε_{ijk}. These quantities are defined thus:

$$\begin{cases} \delta_{ij} = +1, & \text{if } i = j \\ \delta_{ij} = 0, & \text{if } i \neq j \end{cases}$$

(A.2-1)
(A.2-2)

$$\begin{cases} \varepsilon_{ijk} = +1, & \text{if } ijk = 123, 231, \text{ or } 312 \\ \varepsilon_{ijk} = -1, & \text{if } ijk = 321, 132, \text{ or } 213 \\ \varepsilon_{ijk} = 0, & \text{if any two indices are alike} \end{cases}$$

(A.2-3)
(A.2-4)
(A.2-5)

Note also that $\varepsilon_{ijk} = (1/2)(i - j)(j - k)(k - i)$.

Several relations involving these quantities are useful in proving some vector and tensor identities

$$\sum_{j=1}^{3} \sum_{k=1}^{3} \varepsilon_{ijk} \varepsilon_{hjk} = 2\delta_{ih} \tag{A.2-6}$$

$$\sum_{k=1}^{3} \varepsilon_{ijk} \varepsilon_{mnk} = \delta_{im}\delta_{jn} - \delta_{in}\delta_{jm} \tag{A.2-7}$$

Note that a three-by-three determinant may be written in terms of the ε_{ijk}

$$\begin{vmatrix} a_{11} & a_{12} & a_{13} \\ a_{21} & a_{22} & a_{23} \\ a_{31} & a_{32} & a_{33} \end{vmatrix} = \sum_{i=1}^{3} \sum_{j=1}^{3} \sum_{k=1}^{3} \varepsilon_{ijk} a_{1i} a_{2j} a_{3k} \tag{A.2-8}$$

The quantity ε_{ijk} thus selects the necessary terms that appear in the determinant and affixes the proper sign to each term.

The Unit Vectors

Let δ_1, δ_2, δ_3 be the "unit vectors" (i.e., vectors of unit magnitude) in the direction of the 1, 2, 3 axes[1] (Fig. A.2-1). We can use the definitions of the scalar and vector products to tabulate all possible products of each type

$$\begin{cases} (\delta_1 \cdot \delta_1) = (\delta_2 \cdot \delta_2) = (\delta_3 \cdot \delta_3) = 1 \\ (\delta_1 \cdot \delta_2) = (\delta_2 \cdot \delta_3) = (\delta_3 \cdot \delta_1) = 0 \end{cases}$$

(A.2-9)
(A.2-10)

$$\begin{cases} [\delta_1 \times \delta_1] = [\delta_2 \times \delta_2] = [\delta_3 \times \delta_3] = 0 \\ [\delta_1 \times \delta_2] = \delta_3; \qquad [\delta_2 \times \delta_3] = \delta_1; \qquad [\delta_3 \times \delta_1] = \delta_2 \\ [\delta_2 \times \delta_1] = -\delta_3; \qquad [\delta_3 \times \delta_2] = -\delta_1; \qquad [\delta_1 \times \delta_3] = -\delta_2 \end{cases}$$

(A.2-11)
(A.2-12)
(A.2-13)

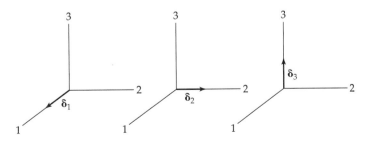

Fig. A.2-1. The unit vectors δ_i; each vector is of unit magnitude and points in the *i*th direction.

---

[1] In most elementary texts the unit vectors are called *i, j, k*. We prefer to use δ_1, δ_2, δ_3 because the components of these vectors are given by the Kronecker delta. That is, the component of δ_1 in the 1-direction is δ_{11} or unity; the component of δ_1 in the 2-direction is δ_{12} or zero.

All of these relations may be summarized by the following two relations:

$$(\boldsymbol{\delta}_i \cdot \boldsymbol{\delta}_j) = \delta_{ij}$$

(A.2-14)

$$[\boldsymbol{\delta}_i \times \boldsymbol{\delta}_j] = \sum_{k=1}^{3} \varepsilon_{ijk} \boldsymbol{\delta}_k$$

(A.2-15)

in which δ_{ij} is the Kronecker delta, and ε_{ijk} is the permutation symbol defined in the introduction to this section. These two relations enable us to develop analytic expressions for all the common dot and cross operations. In the remainder of this section and in the next section, in developing expressions for vector and tensor operations all we do is to break all vectors up into components and then apply Eqs. A.2-14 and 15.

Expansion of a Vector in Terms of its Components

Any vector $\mathbf{v}$ can be completely specified by giving the values of its projections v_1, v_2, v_3, on the coordinate axes 1, 2, 3 (Fig. A.2-2). The vector can be constructed by adding vectorially the components multiplied by their corresponding unit vectors:

$$\mathbf{v} = \boldsymbol{\delta}_1 v_1 + \boldsymbol{\delta}_2 v_2 + \boldsymbol{\delta}_3 v_3 = \sum_{i=1}^{3} \boldsymbol{\delta}_i v_i$$

(A.2-16)

Note that a *vector associates a scalar with each coordinate direction*.[2] The v_i are called the "components of the vector $\mathbf{v}$" and they are scalars, whereas the $\boldsymbol{\delta}_i v_i$ are vectors, which when added together vectorially give $\mathbf{v}$.

The magnitude of a vector is given by

$$|\mathbf{v}| = v = \sqrt{v_1^2 + v_2^2 + v_3^2} = \sqrt{\sum_i v_i^2}$$

(A.2-17)

Two vectors $\mathbf{v}$ and $\mathbf{w}$ are equal if their components are equal: $v_1 = w_1$, $v_2 = w_2$, and $v_3 = w_3$. Also $\mathbf{v} = -\mathbf{w}$, if $v_1 = -w_1$, and so on.

Addition and Subtraction of Vectors

The sum or difference of vectors $\mathbf{v}$ and $\mathbf{w}$ may be written in terms of components as

$$\mathbf{v} \pm \mathbf{w} = \sum_i \boldsymbol{\delta}_i v_i \pm \sum_i \boldsymbol{\delta}_i w_i = \sum_i \boldsymbol{\delta}_i (v_i \pm w_i)$$

(A.2-18)

Geometrically, this corresponds to adding up the projections of $\mathbf{v}$ and $\mathbf{w}$ on each individual axis and then constructing a vector with these new components. Three or more vectors may be added in exactly the same fashion.

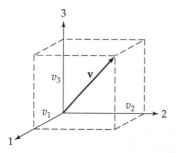

Fig. A.2-2. The components v_i of the vector $\mathbf{v}$ are the projections of the vector on the coordinate axes 1, 2, and 3.

---

[2] For a discussion of the relation of this definition of a vector to the definition in terms of the rules for transformation of coordinates, see W. Prager, *Mechanics of Continua*, Ginn, Boston (1961).

Multiplication of a Vector by a Scalar

Multiplication of a vector by a scalar corresponds to multiplying each component of the vector by the scalar:

$$s\mathbf{v} = s\left\{\sum_i \boldsymbol{\delta}_i v_i\right\} = \sum_i \boldsymbol{\delta}_i [s v_i] \tag{A.2-19}$$

Scalar Product (or Dot Product) of Two Vectors

The scalar product of two vectors $\mathbf{v}$ and $\mathbf{w}$ is obtained by writing each vector in terms of components according to Eq. A.2-16 and then performing the scalar-product operations on the unit vectors, using Eq. A.2-14

$$(\mathbf{v} \cdot \mathbf{w}) = \left(\left\{\sum_i \boldsymbol{\delta}_i v_i\right\} \cdot \left\{\sum_j \boldsymbol{\delta}_j w_j\right\}\right) = \sum_i \sum_j (\boldsymbol{\delta}_i \cdot \boldsymbol{\delta}_j) v_i w_j$$

$$= \sum_i \sum_j \delta_{ij} v_i w_j = \sum_i v_i w_i \tag{A.2-20}$$

Hence the scalar product of two vectors is obtained by summing the products of the corresponding components of the two vectors. Note that $(\mathbf{v} \cdot \mathbf{v})$ (sometimes written as $\mathbf{v}^2$ or as v^2) is a scalar representing the square of the magnitude of $\mathbf{v}$.

Vector Product (or Cross Product) of Two Vectors

The vector product of two vectors $\mathbf{v}$ and $\mathbf{w}$ may be worked out by using Eqs. A.2-16 and 15:

$$[\mathbf{v} \times \mathbf{w}] = \left[\left\{\sum_j \boldsymbol{\delta}_j v_j\right\} \times \left\{\sum_k \boldsymbol{\delta}_k w_k\right\}\right]$$

$$= \sum_j \sum_k [\boldsymbol{\delta}_j \times \boldsymbol{\delta}_k] v_j w_k = \sum_i \sum_j \sum_k \varepsilon_{ijk} \boldsymbol{\delta}_i v_j w_k$$

$$= \begin{vmatrix} \boldsymbol{\delta}_1 & \boldsymbol{\delta}_2 & \boldsymbol{\delta}_3 \\ v_1 & v_2 & v_3 \\ w_1 & w_2 & w_3 \end{vmatrix} \tag{A.2-21}$$

Here we have made use of Eq. A.2-8. Note that the ith-component of $[\mathbf{v} \times \mathbf{w}]$ is given by $\sum_j \sum_k \varepsilon_{ijk} v_j w_k$; this result is often used in proving vector identities.

Multiple Vector Products

Expressions for the multiple products mentioned in §A.1 can be obtained by using the preceding analytical expressions for the scalar and vector products. For example, the product $(\mathbf{u} \cdot [\mathbf{v} \times \mathbf{w}])$ may be written

$$(\mathbf{u} \cdot [\mathbf{v} \times \mathbf{w}]) = \sum_i u_i [\mathbf{v} \times \mathbf{w}]_i = \sum_i \sum_j \sum_k \varepsilon_{ijk} u_i v_j w_k \tag{A.2-22}$$

Then, from Eq. A.2-8, we obtain

$$(\mathbf{u} \cdot [\mathbf{v} \times \mathbf{w}]) = \begin{vmatrix} u_1 & u_2 & u_3 \\ v_1 & v_2 & v_3 \\ w_1 & w_2 & w_3 \end{vmatrix} \tag{A.2-23}$$

The magnitude of $(\mathbf{u} \cdot [\mathbf{v} \times \mathbf{w}])$ is the volume of a parellelepiped defined by the vectors $\mathbf{u}, \mathbf{v}, \mathbf{w}$ drawn from a common origin. Furthermore, the vanishing of the determinant is a necessary and sufficient condition that the vectors $\mathbf{u}, \mathbf{v}$, and $\mathbf{w}$ be coplanar.

The Position Vector

The usual symbol for the position vector—that is, the vector specifying the location of a point in space—is $\mathbf{r}$. The components of $\mathbf{r}$ are then x_1, x_2, and x_3, so that

$$\mathbf{r} = \sum_i \boldsymbol{\delta}_i x_i \tag{A.2-24}$$

This is an irregularity in the notation, since the components have a symbol different from that for the vector. The magnitude of $\mathbf{r}$ is usually called $r = \sqrt{x_1^2 + x_2^2 + x_3^2}$, and this r is the radial coordinate in spherical coordinates (see Fig. A.6-1).

EXAMPLE A.2-1

Proof of a Vector Identity

SOLUTION

The analytical expressions for dot and cross products may be used to prove vector identities; for example, verify the relation

$$[\mathbf{u} \times [\mathbf{v} \times \mathbf{w}]] = \mathbf{v}(\mathbf{u} \cdot \mathbf{w}) - \mathbf{w}(\mathbf{u} \cdot \mathbf{v}) \tag{A.2-25}$$

The i-component of the expression on the left side can be expanded as

$$[\mathbf{u} \times [\mathbf{v} \times \mathbf{w}]]_i = \sum_j \sum_k \varepsilon_{ijk} u_j [\mathbf{v} \times \mathbf{w}]_k = \sum_j \sum_k \varepsilon_{ijk} u_j \left\{ \sum_l \sum_m \varepsilon_{klm} v_l w_m \right\}$$

$$= \sum_j \sum_k \sum_l \sum_m \varepsilon_{ijk} \varepsilon_{klm} u_j v_l w_m = \sum_j \sum_k \sum_l \sum_m \varepsilon_{ijk} \varepsilon_{lmk} u_j v_l w_m \tag{A.2-26}$$

We may now use Eq. A.2-7 to complete the proof

$$[\mathbf{u} \times [\mathbf{v} \times \mathbf{w}]]_i = \sum_j \sum_l \sum_m (\delta_{il}\delta_{jm} - \delta_{im}\delta_{jl}) u_j v_l w_m = v_i \sum_j \sum_m \delta_{jm} u_j w_m - w_i \sum_j \sum_l \delta_{jl} u_j v_l$$

$$= v_i \sum_j u_j w_j - w_i \sum_j u_j v_j = v_i(\mathbf{u} \cdot \mathbf{w}) - w_i(\mathbf{u} \cdot \mathbf{v}) \tag{A.2-27}$$

which is just the i-component of the right side of Eq. A.2-25. In a similar way one may verify such identities as

$$(\mathbf{u} \cdot [\mathbf{v} \times \mathbf{w}]) = (\mathbf{v} \cdot [\mathbf{w} \times \mathbf{u}]) \tag{A.2-28}$$

$$([\mathbf{u} \times \mathbf{v}] \cdot [\mathbf{w} \times \mathbf{z}]) = (\mathbf{u} \cdot \mathbf{w})(\mathbf{v} \cdot \mathbf{z}) - (\mathbf{u} \cdot \mathbf{z})(\mathbf{v} \cdot \mathbf{w}) \tag{A.2-29}$$

$$[[\mathbf{u} \times \mathbf{v}] \times [\mathbf{w} \times \mathbf{z}]] = ([\mathbf{u} \times \mathbf{v}] \cdot \mathbf{z})\mathbf{w} - ([\mathbf{u} \times \mathbf{v}] \cdot \mathbf{w})\mathbf{z} \tag{A.2-30}$$

EXERCISES

1. Write out the following summations:

    (a) $\displaystyle\sum_{k=1}^{3} k^2$ (b) $\displaystyle\sum_{k=1}^{3} a_k^2$ (c) $\displaystyle\sum_{j=1}^{3}\sum_{k=1}^{3} a_{jk} b_{kj}$ (d) $\left(\displaystyle\sum_{j=1}^{3} a_j\right)^2 = \displaystyle\sum_{j=1}^{3}\sum_{k=1}^{3} a_j a_k$

2. A vector $\mathbf{v}$ has components $v_x = 1$, $v_y = 2$, $v_z = -5$. A vector $\mathbf{w}$ has components $w_x = 3$, $w_y = -1$, $w_z = 1$. Evaluate:

    (a) $(\mathbf{v} \cdot \mathbf{w})$

    (b) $[\mathbf{v} \times \mathbf{w}]$

    (c) The length of $\mathbf{v}$

    (d) $(\boldsymbol{\delta}_1 \cdot \mathbf{v})$

    (e) $[\boldsymbol{\delta}_1 \times \mathbf{w}]$

    (f) ϕ_{vw}

    (g) $[\mathbf{r} \times \mathbf{v}]$, where $\mathbf{r}$ is the position vector.

3. Evaluate: (a) $([\boldsymbol{\delta}_1 \times \boldsymbol{\delta}_2] \cdot \boldsymbol{\delta}_3)$ (b) $[[\boldsymbol{\delta}_2 \times \boldsymbol{\delta}_3] \times [\boldsymbol{\delta}_1 \times \boldsymbol{\delta}_3]]$.

4. Show that Eq. A.2-6 is valid for the particular case $i = 1, h = 2$.
   Show that Eq. A.2-7 is valid for the particular case $i = j = m = 1, n = 2$.

5. Verify that $\sum_{j=1}^{3} \sum_{k=1}^{3} \varepsilon_{ijk}\alpha_{jk} = 0$ if $\alpha_{jk} = \alpha_{kj}$.

6. Explain carefully the statement after Eq. A.2-21 that the ith component of $[\mathbf{v} \times \mathbf{w}]$ is
   $\sum_{j} \sum_{k} \varepsilon_{ijk}v_{j}w_{k}$.

7. Verify that $([\mathbf{v} \times \mathbf{w}] \cdot [\mathbf{v} \times \mathbf{w}]) + (\mathbf{v} \cdot \mathbf{w})^2 = v^2w^2$ (the "identity of Lagrange").

§A.3 TENSOR OPERATIONS IN TERMS OF COMPONENTS

In the last section we saw that expressions could be developed for all common dot and cross operations for vectors by knowing how to write a vector $\mathbf{v}$ as a sum $\sum_{i} \boldsymbol{\delta}_{i}v_{i}$, and by knowing how to manipulate the unit vectors $\boldsymbol{\delta}_{i}$. In this section we follow a parallel procedure. We write a tensor $\boldsymbol{\tau}$ as a sum $\sum_{i} \sum_{j} \boldsymbol{\delta}_{i}\boldsymbol{\delta}_{j}\tau_{ij}$, and give formulas for the manipulation of the unit dyads $\boldsymbol{\delta}_{i}\boldsymbol{\delta}_{j}$; in this way, expressions are developed for the commonly occurring dot and cross operations for tensors.

The Unit Dyads

The unit vectors $\boldsymbol{\delta}_{i}$ were defined in the preceding discussion and then the *scalar products* $(\boldsymbol{\delta}_{i} \cdot \boldsymbol{\delta}_{j})$ and *vector products* $[\boldsymbol{\delta}_{i} \times \boldsymbol{\delta}_{j}]$ were given. There is a third kind of product that can be formed with the unit vectors—namely, the *dyadic products* $\boldsymbol{\delta}_{i}\boldsymbol{\delta}_{j}$ (written without multiplication symbols). According to the rules of notation given in the introduction to Appendix A, the products $\boldsymbol{\delta}_{i}\boldsymbol{\delta}_{j}$ are tensors of the second order. Since $\boldsymbol{\delta}_{i}$ and $\boldsymbol{\delta}_{j}$ are of unit magnitude, we will refer to the products $\boldsymbol{\delta}_{i}\boldsymbol{\delta}_{j}$ as *unit dyads*. Whereas each unit vector in Fig. A.2-1 represents a single coordinate direction, the unit dyads in Fig. A.3-1 represent *ordered* pairs of coordinate directions.

(In physical problems we often work with quantities that require the simultaneous specification of two directions. For example, the flux of x-momentum across a unit area of surface perpendicular to the y direction is a quantity of this type. Since this quantity is sometimes not the same as the flux of y-momentum perpendicular to the x direction, it is evident that specifying the two directions is not sufficient; we must also agree on the order in which the directions are given.)

The dot and cross products of unit vectors were introduced by means of the geometrical definitions of these operations. The analogous operations for the unit dyads are introduced formally by relating them to the operations for unit vectors

$$(\boldsymbol{\delta}_{i}\boldsymbol{\delta}_{j}:\boldsymbol{\delta}_{k}\boldsymbol{\delta}_{l}) = (\boldsymbol{\delta}_{j} \cdot \boldsymbol{\delta}_{k})(\boldsymbol{\delta}_{i} \cdot \boldsymbol{\delta}_{l}) = \delta_{jk}\delta_{il} \tag{A.3-1}$$

$$[\boldsymbol{\delta}_{i}\boldsymbol{\delta}_{j} \cdot \boldsymbol{\delta}_{k}] = \boldsymbol{\delta}_{i}(\boldsymbol{\delta}_{j} \cdot \boldsymbol{\delta}_{k}) = \boldsymbol{\delta}_{i}\delta_{jk} \tag{A.3-2}$$

$$[\boldsymbol{\delta}_{i} \cdot \boldsymbol{\delta}_{j}\boldsymbol{\delta}_{k}] = (\boldsymbol{\delta}_{i} \cdot \boldsymbol{\delta}_{j})\boldsymbol{\delta}_{k} = \delta_{ij}\boldsymbol{\delta}_{k} \tag{A.3-3}$$

$$\{\boldsymbol{\delta}_{i}\boldsymbol{\delta}_{j} \cdot \boldsymbol{\delta}_{k}\boldsymbol{\delta}_{l}\} = \boldsymbol{\delta}_{i}(\boldsymbol{\delta}_{j} \cdot \boldsymbol{\delta}_{k})\boldsymbol{\delta}_{l} = \delta_{jk}\boldsymbol{\delta}_{i}\boldsymbol{\delta}_{l} \tag{A.3-4}$$

$$\{\boldsymbol{\delta}_{i}\boldsymbol{\delta}_{j} \times \boldsymbol{\delta}_{k}\} = \boldsymbol{\delta}_{i}[\boldsymbol{\delta}_{j} \times \boldsymbol{\delta}_{k}] = \sum_{l=1}^{3} \varepsilon_{jkl}\boldsymbol{\delta}_{i}\boldsymbol{\delta}_{l} \tag{A.3-5}$$

$$\{\boldsymbol{\delta}_{i} \times \boldsymbol{\delta}_{j}\boldsymbol{\delta}_{k}\} = [\boldsymbol{\delta}_{i} \times \boldsymbol{\delta}_{j}]\boldsymbol{\delta}_{k} = \sum_{l=1}^{3} \varepsilon_{ijl}\boldsymbol{\delta}_{l}\boldsymbol{\delta}_{k} \tag{A.3-6}$$

These results are easy to remember: one simply takes the dot (or cross) product of the nearest unit vectors on either side of the dot (or cross); in Eq. A.3-1 two such operations are performed.

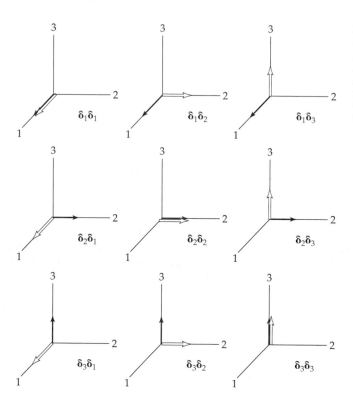

Fig. A.3-1. The unit dyads $\boldsymbol{\delta}_i\boldsymbol{\delta}_j$. The solid arrows represent the first unit vector in the dyadic product, and the hollow vectors the second. Note that $\boldsymbol{\delta}_1\boldsymbol{\delta}_2$ is not the same as $\boldsymbol{\delta}_2\boldsymbol{\delta}_1$.

Expansion of a Tensor in Terms of Its Components

In Eq. A.2-16 we expanded a vector in terms of its components, each component being multiplied by the appropriate unit vector. Here we extend this idea and define[1] a (second-order) tensor as *a quantity that associates a scalar with each ordered pair of coordinate directions* in the following sense:

$$
\begin{aligned}
\boldsymbol{\tau} = {} & \boldsymbol{\delta}_1\boldsymbol{\delta}_1\tau_{11} + \boldsymbol{\delta}_1\boldsymbol{\delta}_2\tau_{12} + \boldsymbol{\delta}_1\boldsymbol{\delta}_3\tau_{13} \\
& + \boldsymbol{\delta}_2\boldsymbol{\delta}_1\tau_{21} + \boldsymbol{\delta}_2\boldsymbol{\delta}_2\tau_{22} + \boldsymbol{\delta}_2\boldsymbol{\delta}_3\tau_{23} \\
& + \boldsymbol{\delta}_3\boldsymbol{\delta}_1\tau_{31} + \boldsymbol{\delta}_3\boldsymbol{\delta}_2\tau_{32} + \boldsymbol{\delta}_3\boldsymbol{\delta}_3\tau_{33} \\
= {} & \sum_{i=1}^{3}\sum_{j=1}^{3}\boldsymbol{\delta}_i\boldsymbol{\delta}_j\tau_{ij}
\end{aligned}
\tag{A.3-7}
$$

The scalars τ_{ij} are referred to as the "components of the tensor $\boldsymbol{\tau}$."

There are several special kinds of second-order tensors worth noting:

1. If $\tau_{ij} = \tau_{ji}$, the tensor is said to be *symmetric*.

2. If $\tau_{ij} = -\tau_{ji}$, the tensor is said to be *antisymmetric*.

3. If the components of a tensor are taken to be the components of $\boldsymbol{\tau}$, but with the indices transposed, the resulting tensor is called the *transpose* of $\boldsymbol{\tau}$ and given the symbol $\boldsymbol{\tau}^\dagger$:

$$
\boldsymbol{\tau}^\dagger = \sum_i\sum_j\boldsymbol{\delta}_i\boldsymbol{\delta}_j\tau_{ji}
\tag{A.3-8}
$$

---

[1] Tensors are often defined in terms of the transformation rules; the connections between such a definition and that given above is discussed by W. Prager, *Mechanics of Continua*, Ginn, Boston (1961).

4. If the components of the tensor are formed by ordered pairs of the components of two vectors $\mathbf{v}$ and $\mathbf{w}$, the resulting tensor is called the *dyadic product of* $\mathbf{v}$ and $\mathbf{w}$ and given the symbol $\mathbf{vw}$:

$$\mathbf{vw} = \sum_i \sum_j \boldsymbol{\delta}_i \boldsymbol{\delta}_j v_i w_j \qquad (A.3\text{-}9)$$

Note that $\mathbf{vw} \neq \mathbf{wv}$, but that $(\mathbf{vw})^\dagger = \mathbf{wv}$.

5. If the components of the tensor are given by the Kronecker delta δ_{ij}, the resulting tensor is called the *unit tensor* and given the symbol $\boldsymbol{\delta}$:

$$\boldsymbol{\delta} = \sum_i \sum_j \boldsymbol{\delta}_i \boldsymbol{\delta}_j \delta_{ij} \qquad (A.3\text{-}10)$$

The magnitude of a tensor is defined by

$$|\boldsymbol{\tau}| = \tau = \sqrt{\tfrac{1}{2}(\boldsymbol{\tau}:\boldsymbol{\tau}^\dagger)}$$

$$= \sqrt{\tfrac{1}{2}\sum_i \sum_j \tau_{ij}^2} \qquad (A.3\text{-}11)$$

Addition of Tensors and Dyadic Products

Two tensors are added thus:

$$\boldsymbol{\sigma} + \boldsymbol{\tau} = \sum_i \sum_j \boldsymbol{\delta}_i \boldsymbol{\delta}_j \sigma_{ij} + \sum_i \sum_j \boldsymbol{\delta}_i \boldsymbol{\delta}_j \tau_{ij} = \sum_i \sum_j \boldsymbol{\delta}_i \boldsymbol{\delta}_j (\sigma_{ij} + \tau_{ij}) \qquad (A.3\text{-}12)$$

That is, the sum of two tensors is that tensor whose components are the sums of the corresponding components of the two tensors. The same is true for dyadic products.

Multiplication of a Tensor by a Scalar

Multiplication of a tensor by a scalar corresponds to multiplying each component of the tensor by the scalar:

$$s\boldsymbol{\tau} = s\left\{\sum_i \sum_j \boldsymbol{\delta}_i \boldsymbol{\delta}_j \tau_{ij}\right\} = \sum_i \sum_j \boldsymbol{\delta}_i \boldsymbol{\delta}_j (s\tau_{ij}) \qquad (A.3\text{-}13)$$

The same is true for dyadic products.

The Scalar Product (or Double Dot Product) of Two Tensors

Two tensors may be multiplied according to the double dot operation

$$(\boldsymbol{\sigma}:\boldsymbol{\tau}) = \left(\left\{\sum_i \sum_j \boldsymbol{\delta}_i \boldsymbol{\delta}_j \sigma_{ij}\right\} : \left\{\sum_k \sum_l \boldsymbol{\delta}_k \boldsymbol{\delta}_l \tau_{kl}\right\}\right) = \sum_i \sum_j \sum_k \sum_l (\boldsymbol{\delta}_i \boldsymbol{\delta}_j : \boldsymbol{\delta}_k \boldsymbol{\delta}_l) \sigma_{ij} \tau_{kl}$$

$$= \sum_i \sum_j \sum_k \sum_l \delta_{il} \delta_{jk} \sigma_{ij} \tau_{kl} = \sum_i \sum_j \sigma_{ij} \tau_{ji} \qquad (A.3\text{-}14)$$

in which Eq. A.3-1 has been used. Similarly, we may show that

$$(\boldsymbol{\tau}:\mathbf{vw}) = \sum_i \sum_j \tau_{ij} v_j w_i \qquad (A.3\text{-}15)$$

$$(\mathbf{uv}:\mathbf{wz}) = \sum_i \sum_j u_i v_j w_j z_i \qquad (A.3\text{-}16)$$

The Tensor Product (the Single Dot Product) of Two Tensors

Two tensors may also be multiplied according to the single dot operation

$$\{\boldsymbol{\sigma} \cdot \boldsymbol{\tau}\} = \left\{\left(\sum_i \sum_j \boldsymbol{\delta}_i \boldsymbol{\delta}_j \sigma_{ij}\right) \cdot \left(\sum_k \sum_l \boldsymbol{\delta}_k \boldsymbol{\delta}_l \tau_{kl}\right)\right\} = \sum_i \sum_j \sum_k \sum_l [\boldsymbol{\delta}_i \boldsymbol{\delta}_j \cdot \boldsymbol{\delta}_k \boldsymbol{\delta}_l] \sigma_{ij} \tau_{kl}$$

$$= \sum_i \sum_j \sum_k \sum_l \delta_{jk} \boldsymbol{\delta}_i \boldsymbol{\delta}_l \sigma_{ij} \tau_{kl} = \sum_i \sum_l \boldsymbol{\delta}_i \boldsymbol{\delta}_l \left(\sum_j \sigma_{ij} \tau_{jl}\right) \tag{A.3-17}$$

That is, the *il*-component of $\{\boldsymbol{\sigma} \cdot \boldsymbol{\tau}\}$ is $\sum_j \sigma_{ij} \tau_{jl}$. Similar operations may be performed with dyadic products. It is common practice to write $\{\boldsymbol{\sigma} \cdot \boldsymbol{\sigma}\}$ as $\boldsymbol{\sigma}^2$, $\{\boldsymbol{\sigma} \cdot \boldsymbol{\sigma}^2\}$ as $\boldsymbol{\sigma}^3$, and so on.

The Vector Product (or Dot Product) of a Tensor with a Vector

When a tensor is dotted into a vector, we get a vector

$$[\boldsymbol{\tau} \cdot \mathbf{v}] = \left[\left\{\sum_i \sum_j \boldsymbol{\delta}_i \boldsymbol{\delta}_j \tau_{ij}\right\} \cdot \left\{\sum_k \boldsymbol{\delta}_k v_k\right\}\right] = \sum_i \sum_j \sum_k [\boldsymbol{\delta}_i \boldsymbol{\delta}_j \cdot \boldsymbol{\delta}_k] \tau_{ij} v_k$$

$$= \sum_i \sum_j \sum_k \boldsymbol{\delta}_i \delta_{jk} \tau_{ij} v_k = \sum_i \boldsymbol{\delta}_i \left\{\sum_j \tau_{ij} v_j\right\} \tag{A.3-18}$$

That is, the *i*th component of $[\boldsymbol{\tau} \cdot \mathbf{v}]$ is $\sum_j \tau_{ij} v_j$. Similarly, the *i*th component of $[\mathbf{v} \cdot \boldsymbol{\tau}]$ is $\sum_j v_j \tau_{ji}$. Clearly, $[\boldsymbol{\tau} \cdot \mathbf{v}] \neq [\mathbf{v} \cdot \boldsymbol{\tau}]$ unless $\boldsymbol{\tau}$ is symmetric.

Recall that when a vector $\mathbf{v}$ is multiplied by a scalar s, the resultant vector $s\mathbf{v}$ points in the same direction as $\mathbf{v}$ but has a different length. However, when $\boldsymbol{\tau}$ is dotted into $\mathbf{v}$, the resultant vector $[\boldsymbol{\tau} \cdot \mathbf{v}]$ differs from $\mathbf{v}$ in *both* length and direction; that is, the tensor $\boldsymbol{\tau}$ "deflects" or "twists" the vector $\mathbf{v}$ to form a new vector pointing in a different direction.

The Tensor Product (or Cross Product) of a Tensor with a Vector

When a tensor is crossed with a vector, we get a tensor:

$$\{\boldsymbol{\tau} \times \mathbf{v}\} = \left\{\left(\sum_i \sum_j \boldsymbol{\delta}_i \boldsymbol{\delta}_j \tau_{ij}\right) \times \left(\sum_k \boldsymbol{\delta}_k v_k\right)\right\} = \sum_i \sum_j \sum_k [\boldsymbol{\delta}_i \boldsymbol{\delta}_j \times \boldsymbol{\delta}_k] \tau_{ij} v_k$$

$$= \sum_i \sum_j \sum_k \sum_l \varepsilon_{jkl} \boldsymbol{\delta}_i \boldsymbol{\delta}_l \tau_{ij} v_k = \sum_i \sum_l \boldsymbol{\delta}_i \boldsymbol{\delta}_l \left\{\sum_j \sum_k \varepsilon_{jkl} \tau_{ij} v_k\right\} \tag{A.3-19}$$

Hence, the *il*-component of $\{\boldsymbol{\tau} \times \mathbf{v}\}$ is $\sum_j \sum_k \varepsilon_{jkl} \tau_{ij} v_k$. Similarly the *lk*-component of $\{\mathbf{v} \times \boldsymbol{\tau}\}$ is $\sum_i \sum_j \varepsilon_{ijl} v_i \tau_{jk}$.

Other Operations

From the preceding results, it is not difficult to prove the following identities:

$$[\boldsymbol{\delta} \cdot \mathbf{v}] = [\mathbf{v} \cdot \boldsymbol{\delta}] = \mathbf{v} \tag{A.3-20}$$

$$[\mathbf{uv} \cdot \mathbf{w}] = \mathbf{u}(\mathbf{v} \cdot \mathbf{w}) \tag{A.3-21}$$

$$[\mathbf{w} \cdot \mathbf{uv}] = (\mathbf{w} \cdot \mathbf{u})\mathbf{v} \tag{A.3-22}$$

$$(\mathbf{uv}:\mathbf{wz}) = (\mathbf{uw}:\mathbf{vz}) = (\mathbf{u} \cdot \mathbf{z})(\mathbf{v} \cdot \mathbf{w}) \tag{A.3-23}$$

$$(\boldsymbol{\tau}:\mathbf{uv}) = ([\boldsymbol{\tau} \cdot \mathbf{u}] \cdot \mathbf{v}) \tag{A.3-24}$$

$$(\mathbf{uv}:\boldsymbol{\tau}) = (\mathbf{u} \cdot [\mathbf{v} \cdot \boldsymbol{\tau}]) \tag{A.3-25}$$

EXERCISES   1.   The components of a symmetric tensor τ are

$$
\begin{array}{lll}
\tau_{xx} = 3 & \tau_{xy} = 2 & \tau_{xz} = -1 \\
\tau_{yx} = 2 & \tau_{yy} = 2 & \tau_{yz} = 1 \\
\tau_{zx} = -1 & \tau_{zy} = 1 & \tau_{zz} = 4
\end{array}
$$

The components of a vector $\mathbf{v}$ are

$$
v_x = 5 \qquad v_y = 3 \qquad v_z = -2
$$

Evaluate

(a) $[\tau \cdot \mathbf{v}]$ (b) $[\mathbf{v} \cdot \tau]$ (c) $(\tau{:}\tau)$

(d) $(\mathbf{v} \cdot [\tau \cdot \mathbf{v}])$ (e) $\mathbf{v}\mathbf{v}$ (f) $[\tau \cdot \delta_1]$

2.   Evaluate

(a) $[[\delta_1\delta_2 \cdot \delta_2] \times \delta_1]$ (c) $(\delta{:}\delta)$

(b) $(\delta{:}\delta_1\delta_2)$ (d) $\{\delta \cdot \delta\}$

3.   If α is symmetrical and β is antisymmetrical, show that $(\alpha{:}\beta) = 0$.

4.   Explain carefully the statement after Eq. A.3-17 that the *il*-component of $\{\sigma \cdot \tau\}$ is $\sum_j \sigma_{ij}\tau_{jl}$.

5.   Consider a rigid structure composed of point particles joined by massless rods. The particles are numbered $1, 2, 3, \ldots, N$, and the particle masses are m_ν ($\nu = 1, 2, \ldots, N$). The locations of the particles with respect to the center of mass are $\mathbf{R}_\nu$. The entire structure rotates on an axis passing through the center of mass with an angular velocity $\mathbf{W}$. Show that the angular momentum with respect to the center of mass is

$$
\mathbf{L} = \sum_\nu m_\nu[\mathbf{R}_\nu \times [\mathbf{W} \times \mathbf{R}_\nu]] \tag{A.3-26}
$$

Then show that the latter expression may be rewritten as

$$
\mathbf{L} = [\mathbf{\Phi} \cdot \mathbf{W}] \tag{A.3-27}
$$

where

$$
\mathbf{\Phi} = \sum_\nu m_\nu\big[(\mathbf{R}_\nu \cdot \mathbf{R}_\nu)\delta - \mathbf{R}_\nu\mathbf{R}_\nu\big] \tag{A.3-28}
$$

is the *moment-of-inertia tensor*.

6.   The kinetic energy of rotation of the rigid structure in Exercise 5 is

$$
K = \sum_\nu \tfrac{1}{2}m_\nu(\dot{\mathbf{R}}_\nu \cdot \dot{\mathbf{R}}_\nu) \tag{A.3-29}
$$

where $\dot{\mathbf{R}}_\nu = [\mathbf{W} \times \mathbf{R}_\nu]$ is the velocity of the νth particle. Show that

$$
K = \tfrac{1}{2}(\mathbf{\Phi}{:}\mathbf{W}\mathbf{W}) \tag{A.3-30}
$$

§A.4   VECTOR AND TENSOR DIFFERENTIAL OPERATIONS

The vector differential operator ∇, known as "nabla" or "del," is defined in rectangular coordinates as

$$
\nabla = \delta_1 \frac{\partial}{\partial x_1} + \delta_2 \frac{\partial}{\partial x_2} + \delta_3 \frac{\partial}{\partial x_3} = \sum_i \delta_i \frac{\partial}{\partial x_i} \tag{A.4-1}
$$

in which the δ_i are the unit vectors and the x_i are the variables associated with the 1, 2, 3 axes (i.e., the x_1, x_2, x_3 are the Cartesian coordinates normally referred to as x, y, z). The symbol ∇ is a vector-operator—it has components like a vector but it cannot stand alone;

it must operate on a scalar, vector, or tensor function. In this section we summarize the various operations of ∇ on scalars, vectors, and tensors. As in §§A.2 and A.3, we decompose vectors and tensors into their components and then use Eqs. A.2-14 and 15, and Eqs. A.3-1 to 6. Keep in mind that in this section equations written out in component form are valid only for rectangular coordinates, for which the unit vectors $\boldsymbol{\delta}_i$ are constants; curvilinear coordinates are discussed in §§A.6 and 7.

The Gradient of a Scalar Field

If s is a scalar function of the variables x_1, x_2, x_3, then the operation of ∇ on s is

$$\nabla s = \boldsymbol{\delta}_1 \frac{\partial s}{\partial x_1} + \boldsymbol{\delta}_2 \frac{\partial s}{\partial x_2} + \boldsymbol{\delta}_3 \frac{\partial s}{\partial x_3} = \sum_i \boldsymbol{\delta}_i \frac{\partial s}{\partial x_i} \tag{A.4-2}$$

The vector thus constructed from the derivatives of s is designated by ∇s (or grad s) and is called the *gradient* of the scalar field s. The following properties of the gradient operation should be noted.

Not Commutative: $\nabla s \neq s\nabla$ (A.4-3)

Not Associative: $(\nabla r)s \neq \nabla(rs)$ (A.4-4)

Distributive: $\nabla(r + s) = \nabla r + \nabla s$ (A.4-5)

The Divergence of a Vector Field

If the vector $\mathbf{v}$ is a function of the space variables x_1, x_2, x_3, then a scalar product may be formed with the operator ∇; in obtaining the final form, we use Eq. A.2-14:

$$(\nabla \cdot \mathbf{v}) = \left(\left\{ \sum_i \boldsymbol{\delta}_i \frac{\partial}{\partial x_i} \right\} \cdot \left\{ \sum_j \boldsymbol{\delta}_j v_j \right\} \right) = \sum_i \sum_j (\boldsymbol{\delta}_i \cdot \boldsymbol{\delta}_j) \frac{\partial}{\partial x_i} v_j$$

$$= \sum_i \sum_j \delta_{ij} \frac{\partial}{\partial x_i} v_j = \sum_i \frac{\partial v_i}{\partial x_i} \tag{A.4-6}$$

This collection of derivatives of the components of the vector $\mathbf{v}$ is called the *divergence* of $\mathbf{v}$ (sometimes abbreviated div $\mathbf{v}$). Some properties of the divergence operator should be noted

Not Commutative: $(\nabla \cdot \mathbf{v}) \neq (\mathbf{v} \cdot \nabla)$ (A.4-7)

Not Associative: $(\nabla \cdot s\mathbf{v}) \neq (\nabla s \cdot \mathbf{v})$ (A.4-8)

Distributive: $(\nabla \cdot \{\mathbf{v} + \mathbf{w}\}) = (\nabla \cdot \mathbf{v}) + (\nabla \cdot \mathbf{w})$ (A.4-9)

The Curl of a Vector Field

A cross product may also be formed between the ∇ operator and the vector $\mathbf{v}$, which is a function of the three space variables. This cross product may be simplified by using Eq. A.2-15 and written in a variety of forms

$$[\nabla \times \mathbf{v}] = \left[\left\{ \sum_j \boldsymbol{\delta}_j \frac{\partial}{\partial x_j} \right\} \times \left\{ \sum_k \boldsymbol{\delta}_k v_k \right\} \right]$$

$$= \sum_j \sum_k [\boldsymbol{\delta}_j \times \boldsymbol{\delta}_k] \frac{\partial}{\partial x_j} v_k = \sum_i \sum_j \sum_k \varepsilon_{ijk} \boldsymbol{\delta}_i \frac{\partial}{\partial x_j} v_k$$

$$= \begin{vmatrix} \boldsymbol{\delta}_1 & \boldsymbol{\delta}_2 & \boldsymbol{\delta}_3 \\ \dfrac{\partial}{\partial x_1} & \dfrac{\partial}{\partial x_2} & \dfrac{\partial}{\partial x_3} \\ v_1 & v_2 & v_3 \end{vmatrix}$$

$$= \boldsymbol{\delta}_1 \left\{ \frac{\partial v_3}{\partial x_2} - \frac{\partial v_2}{\partial x_3} \right\} + \boldsymbol{\delta}_2 \left\{ \frac{\partial v_1}{\partial x_3} - \frac{\partial v_3}{\partial x_1} \right\} + \boldsymbol{\delta}_3 \left\{ \frac{\partial v_2}{\partial x_1} - \frac{\partial v_1}{\partial x_2} \right\} \tag{A.4-10}$$

The vector thus constructed is called the *curl* of **v**. Other notations for $[\nabla \times \mathbf{v}]$ are curl **v** and rot **v**, the latter being common in the German literature. The curl operation, like the divergence, is distributive but not commutative or associative. Note that the ith component of $[\nabla \times \mathbf{v}]$ is $\sum_j \sum_k \varepsilon_{ijk}(\partial/\partial x_j)v_k$.

The Gradient of a Vector Field

In addition to the scalar product $(\nabla \cdot \mathbf{v})$ and the vector product $[\nabla \times \mathbf{v}]$ one may also form the dyadic product $\nabla\mathbf{v}$:

$$\nabla\mathbf{v} = \left\{ \sum_i \boldsymbol{\delta}_i \frac{\partial}{\partial x_i} \right\}\left\{ \sum_j \boldsymbol{\delta}_j v_j \right\} = \sum_i \sum_j \boldsymbol{\delta}_i\boldsymbol{\delta}_j \frac{\partial}{\partial x_i} v_j \tag{A.4-11}$$

This is called the *gradient* of the vector **v** and is sometimes written grad **v**. It is a second-order tensor whose ij-component[1] is $(\partial/\partial x_i)v_j$. Its transpose is

$$(\nabla\mathbf{v})^\dagger = \sum_i \sum_j \boldsymbol{\delta}_i\boldsymbol{\delta}_j \frac{\partial}{\partial x_j} v_i \tag{A.4-12}$$

whose ij-component is $(\partial/\partial x_j)v_i$. Note that $\nabla\mathbf{v} \neq \mathbf{v}\nabla$ and $(\nabla\mathbf{v})^\dagger \neq \mathbf{v}\nabla$.

The Divergence of a Tensor Field

If the tensor $\boldsymbol{\tau}$ is a function of the space variables x_1, x_2, x_3, then a vector product may be formed with operator ∇; in obtaining the final form we use Eq. A.3-3:

$$[\nabla \cdot \boldsymbol{\tau}] = \left[\left\{ \sum_i \boldsymbol{\delta}_i \frac{\partial}{\partial x_i} \right\} \cdot \left\{ \sum_j \sum_k \boldsymbol{\delta}_j\boldsymbol{\delta}_k \tau_{jk} \right\} \right] = \sum_i \sum_j \sum_k [\boldsymbol{\delta}_i \cdot \boldsymbol{\delta}_j\boldsymbol{\delta}_k] \frac{\partial}{\partial x_i} \tau_{jk}$$

$$= \sum_i \sum_j \sum_k \boldsymbol{\delta}_{ij}\boldsymbol{\delta}_k \frac{\partial}{\partial x_i} \tau_{jk} = \sum_k \boldsymbol{\delta}_k \left\{ \sum_i \frac{\partial}{\partial x_i} \tau_{ik} \right\} \tag{A.4-13}$$

This is called the *divergence* of the tensor $\boldsymbol{\tau}$, and is sometimes written div $\boldsymbol{\tau}$. The kth component of $[\nabla \cdot \boldsymbol{\tau}]$ is $\sum_i (\partial/\partial x_i)\tau_{ik}$. If $\boldsymbol{\tau}$ is the product $s\mathbf{vw}$, then

$$[\nabla \cdot s\mathbf{vw}] = \sum_k \boldsymbol{\delta}_k \left\{ \sum_i \frac{\partial}{\partial x_i} (sv_iw_k) \right\} \tag{A.4-14}$$

The Laplacian of a Scalar Field

If we take the divergence of a gradient of the scalar function s, we obtain

$$(\nabla \cdot \nabla s) = \left(\left\{ \sum_i \boldsymbol{\delta}_i \frac{\partial}{\partial x_i} \right\} \cdot \left\{ \sum_j \boldsymbol{\delta}_j \frac{\partial s}{\partial x_j} \right\} \right)$$

$$= \sum_i \sum_j \boldsymbol{\delta}_{ij} \frac{\partial}{\partial x_i} \frac{\partial s}{\partial x_j} = \left\{ \sum_i \frac{\partial^2}{\partial x_i^2} s \right\} \tag{A.4-15}$$

The collection of differential operators operating on s in the last line is given the symbol ∇^2; hence in rectangular coordinates

$$(\nabla \cdot \nabla) = \nabla^2 = \frac{\partial^2}{\partial x_1^2} + \frac{\partial^2}{\partial x_2^2} + \frac{\partial^2}{\partial x_3^2} \tag{A.4-16}$$

This is called the *Laplacian* operator. (Some authors use the symbol Δ for the Laplacian operator, particularly in the older German literature; hence $(\nabla \cdot \nabla s)$, $(\nabla \cdot \nabla)s$, $\nabla^2 s$, and Δs

---

[1] *Caution:* Some authors define the ij-component of $\nabla\mathbf{v}$ to be $(\partial/\partial x_j)v_i$.

are all equivalent quantities.) The Laplacian operator has only the distributive property, as do the gradient, divergence, and curl.

The Laplacian of a Vector Field

If we take the divergence of the gradient of the vector function $\mathbf{v}$, we obtain

$$[\nabla \cdot \nabla \mathbf{v}] = \left[\left\{ \sum_i \boldsymbol{\delta}_i \frac{\partial}{\partial x_i} \right\} \cdot \left\{ \sum_j \sum_k \boldsymbol{\delta}_j \boldsymbol{\delta}_k \frac{\partial}{\partial x_j} v_k \right\} \right]$$

$$= \sum_i \sum_j \sum_k [\boldsymbol{\delta}_i \cdot \boldsymbol{\delta}_j \boldsymbol{\delta}_k] \frac{\partial}{\partial x_i} \frac{\partial}{\partial x_j} v_k$$

$$= \sum_i \sum_j \sum_k \delta_{ij} \boldsymbol{\delta}_k \frac{\partial}{\partial x_i} \frac{\partial}{\partial x_j} v_k = \sum_k \boldsymbol{\delta}_k \left(\sum_i \frac{\partial^2}{\partial x_i^2} v_k \right) \tag{A.4-17}$$

That is, the kth component of $[\nabla \cdot \nabla \mathbf{v}]$ is, in Cartesian coordinates, just $\nabla^2 v_k$. Alternative notations for $[\nabla \cdot \nabla \mathbf{v}]$ are $(\nabla \cdot \nabla)\mathbf{v}$ and $\nabla^2 \mathbf{v}$.

Other Differential Relations

Numerous identities can be proved using the definitions just given:

$$\nabla rs = r\nabla s + s\nabla r \tag{A.4-18}$$

$$(\nabla \cdot s\mathbf{v}) = (\nabla s \cdot \mathbf{v}) + s(\nabla \cdot \mathbf{v}) \tag{A.4-19}$$

$$(\nabla \cdot [\mathbf{v} \times \mathbf{w}]) = (\mathbf{w} \cdot [\nabla \times \mathbf{v}]) - (\mathbf{v} \cdot [\nabla \times \mathbf{w}]) \tag{A.4-20}$$

$$[\nabla \times s\mathbf{v}] = [\nabla s \times \mathbf{v}] + s[\nabla \times \mathbf{v}] \tag{A.4-21}$$

$$[\nabla \cdot \nabla \mathbf{v}] = \nabla(\nabla \cdot \mathbf{v}) - [\nabla \times [\nabla \times \mathbf{v}]] \tag{A.4-22}$$

$$[\mathbf{v} \cdot \nabla \mathbf{v}] = \tfrac{1}{2}\nabla(\mathbf{v} \cdot \mathbf{v}) - [\mathbf{v} \times [\nabla \times \mathbf{v}]] \tag{A.4-23}$$

$$[\nabla \cdot \mathbf{v}\mathbf{w}] = [\mathbf{v} \cdot \nabla \mathbf{w}] + \mathbf{w}(\nabla \cdot \mathbf{v}) \tag{A.4-24}$$

$$(s\boldsymbol{\delta}:\nabla \mathbf{v}) = s(\nabla \cdot \mathbf{v}) \tag{A.4-25}$$

$$[\nabla \cdot s\boldsymbol{\delta}] = \nabla s \tag{A.4-26}$$

$$[\nabla \cdot s\boldsymbol{\tau}] = [\nabla s \cdot \boldsymbol{\tau}] + s[\nabla \cdot \boldsymbol{\tau}] \tag{A.4-27}$$

$$\nabla(\mathbf{v} \cdot \mathbf{w}) = [(\nabla \mathbf{v}) \cdot \mathbf{w}] + [(\nabla \mathbf{w}) \cdot \mathbf{v}] \tag{A.4-28}$$

EXAMPLE A.4-1

Proof of a Tensor Identity

Prove that for *symmetric* $\boldsymbol{\tau}$:

$$(\boldsymbol{\tau}:\nabla \mathbf{v}) = (\nabla \cdot [\boldsymbol{\tau} \cdot \mathbf{v}]) - (\mathbf{v} \cdot [\nabla \cdot \boldsymbol{\tau}]) \tag{A.4-29}$$

SOLUTION

First we write out the right side in terms of components:

$$(\nabla \cdot [\boldsymbol{\tau} \cdot \mathbf{v}]) = \sum_i \frac{\partial}{\partial x_i} [\boldsymbol{\tau} \cdot \mathbf{v}]_i = \sum_i \sum_j \frac{\partial}{\partial x_i} \tau_{ij} v_j \tag{A.4-30}$$

$$(\mathbf{v} \cdot [\nabla \cdot \boldsymbol{\tau}]) = \sum_j v_j [\nabla \cdot \boldsymbol{\tau}]_j = \sum_j \sum_i v_j \frac{\partial}{\partial x_i} \tau_{ij} \tag{A.4-31}$$

The left side may be written as

$$(\boldsymbol{\tau}:\nabla \mathbf{v}) = \sum_i \sum_j \tau_{ji} \frac{\partial}{\partial x_i} v_j = \sum_i \sum_j \tau_{ij} \frac{\partial}{\partial x_i} v_j \tag{A.4-32}$$

the second form resulting from the symmetry of $\boldsymbol{\tau}$. Subtraction of Eq. A.4-31 from Eq. A.4-30 will give Eq. A.4-32.

Now that we have given all the vector and tensor operations, including the various ∇ operations, we want to point out that the dot and double dot operations can be written down at once by using the following simple rule: *a dot implies a summation on adjacent indices*. We illustrate the rule with several examples.

To interpret $(\mathbf{v} \cdot \mathbf{w})$, we note that $\mathbf{v}$ and $\mathbf{w}$ are vectors, whose components have one index. Since both symbols are adjacent to the dot, we make the indices for both of them the same and then sum on them: $(\mathbf{v} \cdot \mathbf{w}) = \Sigma_i v_i w_i$. For double dot operations such as $(\boldsymbol{\tau}:\nabla\mathbf{v})$, we proceed as follows. We note that $\boldsymbol{\tau}$, being a tensor, has two subscripts, whereas ∇ and $\mathbf{v}$ each have one. We therefore set the second subscript of $\boldsymbol{\tau}$ equal to the subscript on ∇ and sum; then we set the first subscript of $\boldsymbol{\tau}$ equal to the subscript on $\mathbf{v}$ and sum. Hence we get $(\boldsymbol{\tau}:\nabla\mathbf{v}) = \Sigma_i\Sigma_j \tau_{ji}(\partial/\partial x_i)v_j$. Similarly, $(\mathbf{v} \cdot [\nabla \cdot \boldsymbol{\tau}])$ can be written down at once as $\Sigma_i\Sigma_j v_j(\partial/\partial x_i)\tau_{ij}$ by performing the operation in the inner enclosure (the brackets) before the outer (the parentheses).

To get the ith component of a vector quantity, we proceed in exactly the same way. To evaluate $[\boldsymbol{\tau} \cdot \mathbf{v}]_i$ we set the second index of the tensor $\boldsymbol{\tau}$ equal to the index on $\mathbf{v}$ and sum to get $\Sigma_j \tau_{ij}v_j$. Similarly, the ith component of $[\nabla \cdot \rho\mathbf{v}\mathbf{v}]$ is obtained as $\Sigma_j(\partial/\partial x_j)(\rho v_j v_i)$. Becoming skilled with this method can save a great deal of time in interpreting the dot and double dot operations in Cartesian coordinates.

EXERCISES

1. Perform all the operations in Eq. A.4-6 by writing out all the summations instead of using the Σ notation.

2. A field $\mathbf{v}(x, y, z)$ is said to be *irrotational* if $[\nabla \times \mathbf{v}] = 0$. Which of the following fields are irrotational?

   (a) $v_x = by$ $v_y = 0$ $v_z = 0$
   (b) $v_x = bx$ $v_y = 0$ $v_z = 0$
   (c) $v_x = by$ $v_y = bx$ $v_z = 0$
   (d) $v_x = -by$ $v_y = bx$ $v_z = 0$

3. Evaluate $(\nabla \cdot \mathbf{v})$, $\nabla\mathbf{v}$, and $[\nabla \cdot \mathbf{v}\mathbf{v}]$ for the four fields in Exercise 2.

4. A vector $\mathbf{v}$ has components

$$v_i = \sum_{j=1}^{3} \alpha_{ij}x_j$$

   with $\alpha_{ij} = \alpha_{ji}$ and $\sum_{i=1}^{3} \alpha_{ii} = 0$; the α_{ij} are constants. Evaluate $(\nabla \cdot \mathbf{v})$, $[\nabla \times \mathbf{v}]$, $\nabla\mathbf{v}$, $(\nabla\mathbf{v})^\dagger$, and $[\nabla \cdot \mathbf{v}\mathbf{v}]$. (*Hint:* In connection with evaluating $[\nabla \times \mathbf{v}]$, see Exercise 5 in §A.2.)

5. Verify that $\nabla^2(\nabla \cdot \mathbf{v}) = (\nabla \cdot (\nabla^2\mathbf{v}))$, and that $[\nabla \cdot (\nabla\mathbf{v})^\dagger] = \nabla(\nabla \cdot \mathbf{v})$.

6. Verify that $(\nabla \cdot [\nabla \times \mathbf{v}]) = 0$ and $[\nabla \times \nabla s] = 0$.

7. If $\mathbf{r}$ is the position vector (with components x_1, x_2, x_3) and $\mathbf{v}$ is any vector, show that
   (a) $(\nabla \cdot \mathbf{r}) = 3$
   (b) $[\nabla \times \mathbf{r}] = 0$
   (c) $[\mathbf{r} \times [\nabla \cdot \mathbf{v}\mathbf{v}]] = [\nabla \cdot \mathbf{v}[\mathbf{r} \times \mathbf{v}]]$ (where $\mathbf{v}$ is a function of position)

8. Develop an alternative expression for $[\nabla \times [\nabla \cdot s\mathbf{v}\mathbf{v}]]$.

9. If $\mathbf{r}$ is the position vector and r is its magnitude, verify that

   (a) $\nabla\dfrac{1}{r} = -\dfrac{\mathbf{r}}{r^3}$ (c) $\nabla(\mathbf{a} \cdot \mathbf{r}) = \mathbf{a}$ if $\mathbf{a}$ is a constant vector

   (b) $\nabla f(r) = \dfrac{1}{r}\dfrac{df}{dr}\mathbf{r}$

10. Write out in full in Cartesian coordinates

(a) $\frac{\partial}{\partial t}\rho\mathbf{v} = -[\nabla \cdot \rho\mathbf{v}\mathbf{v}] - \nabla p - [\nabla \cdot \boldsymbol{\tau}] + \rho g$

(b) $\boldsymbol{\tau} = -\mu\{\nabla\mathbf{v} + (\nabla\mathbf{v})^{\dagger} - \frac{2}{3}(\nabla \cdot \mathbf{v})\boldsymbol{\delta}\}$

§A.5 VECTOR AND TENSOR INTEGRAL THEOREMS

For performing general proofs in continuum physics, several integral theorems are extremely useful.

The Gauss–Ostrogradskii Divergence Theorem

If V is a closed region in space enclosed by a surface S, then

$$\int_V (\nabla \cdot \mathbf{v})dV = \int_S (\mathbf{n} \cdot \mathbf{v})dS \qquad (A.5\text{-}1)$$

in which $\mathbf{n}$ is the outwardly directed unit normal vector. This is known as the *divergence theorem* of Gauss and Ostrogradskii. Two closely allied theorems for scalars and tensors are

$$\int_V \nabla s \, dV = \int_S \mathbf{n}s \, dS \qquad (A.5\text{-}2)$$

$$\int_V [\nabla \cdot \boldsymbol{\tau}] \, dV = \int_S [\mathbf{n} \cdot \boldsymbol{\tau}]dS \qquad (A.5\text{-}3)[1]$$

The last relation is also valid for dyadic products $\mathbf{vw}$. Note that, in all three equations, ∇ in the volume integral is just replaced by $\mathbf{n}$ in the surface integral.

The Stokes Curl Theorem

If S is a surface bounded by the closed curve C, then

$$\int_S (\mathbf{n} \cdot [\nabla \times \mathbf{v}]) \, dS = \oint_C (\mathbf{t} \cdot \mathbf{v})dC \qquad (A.5\text{-}4)$$

in which $\mathbf{t}$ is a unit tangential vector in the direction of integration along C; $\mathbf{n}$ is the unit normal vector to S in the direction that a right-hand screw would move if its head were twisted in the direction of integration along C. There is a similar relation for tensors.[1]

The Leibniz Formula for Differentiating a Volume Integral[2]

Let V be a closed moving region in space enclosed by a surface S; let the velocity of any surface element be $\mathbf{v}_S$. Then, if $s(x, y, z, t)$ is a scalar function of position and time,

$$\frac{d}{dt}\int_V s \, dV = \int_V \frac{\partial s}{\partial t} dV + \int_S s(\mathbf{v}_S \cdot \mathbf{n})dS \qquad (A.5\text{-}5)$$

---

[1] See P. M. Morse and H. Feshbach, *Methods of Theoretical Physics*, McGraw-Hill, New York (1953), p. 66.

[2] M. D. Greenberg, *Foundations of Applied Mathematics*, Prentice-Hall, Englewood Cliffs, N.J. (1978), pp. 163–164.

This is an extension of the *Leibniz formula* for differentiating a single integral (see Eq. C.3-2); keep in mind that $V = V(t)$ and $S = S(t)$. Equation A.5-5 also applies to vectors and tensors.

If the integral is over a volume, the surface of which is moving with the local fluid velocity (so that $\mathbf{v}_S = \mathbf{v}$), then use of the equation of continuity leads to the additional useful result:

$$\frac{d}{dt} \int_V \rho s \, dV = \int_V \rho \frac{Ds}{Dt} \, dV \qquad \text{(A.5-6)}$$

in which ρ is the fluid density. Equation A.5-6 is sometimes called the *Reynolds transport theorem*.

EXERCISES

1. Consider the vector field

$$\mathbf{v} = \boldsymbol{\delta}_1 x_1 + \boldsymbol{\delta}_2 x_3 + \boldsymbol{\delta}_3 x_2$$

   Evaluate both sides of Eq. A.5-1 over the region bounded by the planes $x_1 = 0$, $x_1 = 1$; $x_2 = 0$, $x_2 = 2$; $x_3 = 0$, $x_3 = 4$.

2. Use the same vector field to evaluate both sides of Eq. A.5-4 for the face $x_1 = 1$ in Exercise 1.

3. Consider the time-dependent scalar function:

$$s = x + y + zt$$

   Evaluate both sides of Eq. A.5-5 over the volume bounded by the planes: $x = 0$, $x = t$; $y = 0$, $y = 2t$; $z = 0$, $z = 4t$. The quantities x, y, z, t are dimensionless.

4. Use Eq. A.5-4 (with $\mathbf{v}$ replaced by $\boldsymbol{\tau}$) to show that, when $\tau_{ki} = \sum_j \epsilon_{ijk} x_j$,

$$2 \int_S \mathbf{n} \, dS = \oint_C [\mathbf{r} \times \mathbf{t}] dC$$

   where r is the position vector locating a point on C with respect to the origin.

5. Evaluate both sides of Eq. A.5-2 for the function $s(x, y, z) = x^2 + y^2 + z^2$. The volume V is the triangular prism lying between the two triangles whose vertices are $(2, 0, 0)$, $(2, 1, 0)$, $(2, 0, 3)$, and $(-2, 0, 0)$, $(-2, 1, 0)$, $(-2, 0, 3)$.

§A.6   VECTOR AND TENSOR ALGEBRA IN CURVILINEAR COORDINATES

Thus far we have considered only Cartesian coordinates x, y, and z. Although formal derivations are usually made in Cartesian coordinates, for working problems it is often more natural to use curvilinear coordinates. The two most commonly occurring curvilinear coordinate systems are the *cylindrical* and the *spherical*. In the following we discuss only these two systems, but the method can also be applied to all *orthogonal* coordinate systems—that is, those in which the three families of coordinate surfaces are mutually perpendicular.

We are primarily interested in knowing how to write various differential operations, such as ∇s, $[\nabla \times \mathbf{v}]$, and $(\boldsymbol{\tau} : \nabla \mathbf{v})$ in curvilinear coordinates. It turns out that we can do this in a straightforward way if we know, for the coordinate system being used, two things: (a) the expression for ∇ in curvilinear coordinates; and (b) the spatial derivatives of the unit vectors in curvilinear coordinates. Hence, we want to focus our attention on these two points.

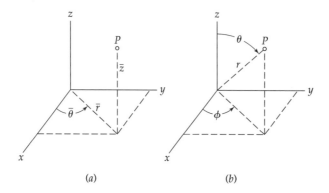

(a)   (b)

Fig. A.6-1. (*a*) Cylindrical coordinates[1] with $0 \leq \bar{r} < \infty, 0 \leq \bar{\theta} < 2\pi, -\infty < \bar{z} < \infty$. (*b*) Spherical coordinates with $0 \leq r < \infty, 0 \leq \theta \leq \pi, 0 \leq \phi < 2\pi$. Note that $\bar{r}$ and $\bar{\theta}$ in cylindrical coordinates are *not* the same as r and θ in spherical coordinates. Note carefully how the position vector **r** and its length r are written in the three coordinate systems:

Rectangular: $\mathbf{r} = \boldsymbol{\delta}_x x + \boldsymbol{\delta}_y y + \boldsymbol{\delta}_z z;$ $r = \sqrt{x^2 + y^2 + z^2}$

Cylindrical: $\mathbf{r} = \boldsymbol{\delta}_r \bar{r} + \boldsymbol{\delta}_z \bar{z};$ $r = \sqrt{\bar{r}^2 + \bar{z}^2}$

Spherical: $\mathbf{r} = \boldsymbol{\delta}_r r;$ $r = r$

Cylindrical Coordinates

In cylindrical coordinates, instead of designating the coordinates of a point by x, y, z, we locate the point by giving the values of r, θ, z. These coordinates[1] are shown in Fig. A.6-1a. They are related to the Cartesian coordinates by

$$\begin{cases} x = r \cos \theta & \text{(A.6-1)} \\ y = r \sin \theta & \text{(A.6-2)} \\ z = z & \text{(A.6-3)} \end{cases} \qquad \begin{array}{ll} r = +\sqrt{x^2 + y^2} & \text{(A.6-4)} \\ \theta = \arctan (y/x) & \text{(A.6-5)} \\ z = z & \text{(A.6-6)} \end{array}$$

To convert derivatives of scalars with respect to x, y, z into derivatives with respect to r, θ, z, the "chain rule" of partial differentiation[2] is used. The derivative operators are readily found to be related thus:

$$\begin{cases} \dfrac{\partial}{\partial x} = (\cos \theta) \dfrac{\partial}{\partial r} + \left(-\dfrac{\sin \theta}{r} \right) \dfrac{\partial}{\partial \theta} + (0) \dfrac{\partial}{\partial z} & \text{(A.6-7)} \\[3mm] \dfrac{\partial}{\partial y} = (\sin \theta) \dfrac{\partial}{\partial r} + \left(\dfrac{\cos \theta}{r} \right) \dfrac{\partial}{\partial \theta} + (0) \dfrac{\partial}{\partial z} & \text{(A.6-8)} \\[3mm] \dfrac{\partial}{\partial z} = (0) \dfrac{\partial}{\partial r} + (0) \dfrac{\partial}{\partial \theta} + (1) \dfrac{\partial}{\partial z} & \text{(A.6-9)} \end{cases}$$

---

[1] *Caution:* We have chosen to use the familiar r, θ, z-notation for cylindrical coordinates rather than to switch to some less familiar symbols, even though there are two situations in which confusion can arise: (a) occasionally one has to use cylindrical and spherical coordinates in the same problem, and the symbols r and θ have different meanings in the two systems; (b) occasionally one deals with the position vector **r** in problems involving cylindrical coordinates, but then the magnitude of **r** is not the same as the coordinate r, but rather $\sqrt{r^2 + z^2}$. In such situations, as in Fig. A.6-1, we can use overbars for the cylindrical coordinates and write $\bar{r}, \bar{\theta}, \bar{z}$. For most discussions bars will not be needed.

[2] For example, for a scalar function $\chi(x, y, z) = \psi(r, \theta, z)$:

$$\left(\frac{\partial \chi}{\partial x} \right)_{y,z} = \left(\frac{\partial r}{\partial x} \right)_{y,z} \left(\frac{\partial \psi}{\partial r} \right)_{\theta,z} + \left(\frac{\partial \theta}{\partial x} \right)_{y,z} \left(\frac{\partial \psi}{\partial \theta} \right)_{r,z} + \left(\frac{\partial z}{\partial x} \right)_{y,z} \left(\frac{\partial \psi}{\partial z} \right)_{r,\theta}$$

Note that we are careful to use different symbols χ and ψ, since χ is a different function of x, y, z than ψ is of $r, \theta,$ and z!

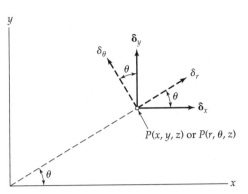

Fig. A.6-2. Unit vectors in rectangular and cylindrical coordinates. The z-axis and the unit vector $\boldsymbol{\delta}_z$ have been omitted for simplicity.

With these relations, derivatives of any scalar functions (including, of course, components of vectors and tensors) with respect to x, y, and z can be expressed in terms of derivatives with respect to r, θ, and z.

Having discussed the interrelationship of the coordinates and derivatives in the two coordinate systems, we now turn to the relation between the unit vectors. We begin by noting that the unit vectors $\boldsymbol{\delta}_x$, $\boldsymbol{\delta}_y$, $\boldsymbol{\delta}_z$ (or $\boldsymbol{\delta}_1$, $\boldsymbol{\delta}_2$, $\boldsymbol{\delta}_3$ as we have been calling them) are independent of position—that is, independent of x, y, z. In cylindrical coordinates the unit vectors $\boldsymbol{\delta}_r$ and $\boldsymbol{\delta}_\theta$ will depend on position, as we can see in Fig. A.6-2. The unit vector $\boldsymbol{\delta}_r$ is a vector of unit length in the direction of increasing r; the unit vector $\boldsymbol{\delta}_\theta$ is a vector of unit length in the direction of increasing θ. Clearly as the point P is moved around on the xy-plane, the directions of $\boldsymbol{\delta}_r$ and $\boldsymbol{\delta}_\theta$ change. Elementary trigonometrical arguments lead to the following relations:

$$\begin{cases} \boldsymbol{\delta}_r = (\cos\ \theta)\boldsymbol{\delta}_x + (\sin\ \theta)\boldsymbol{\delta}_y + (0)\boldsymbol{\delta}_z & \text{(A.6-10)} \\ \boldsymbol{\delta}_\theta = (-\sin\ \theta)\boldsymbol{\delta}_x + (\cos\ \theta)\boldsymbol{\delta}_y + (0)\boldsymbol{\delta}_z & \text{(A.6-11)} \\ \boldsymbol{\delta}_z = (0)\boldsymbol{\delta}_x + (0)\boldsymbol{\delta}_y + (1)\boldsymbol{\delta}_z & \text{(A.6-12)} \end{cases}$$

These may be solved for $\boldsymbol{\delta}_x$, $\boldsymbol{\delta}_y$, and $\boldsymbol{\delta}_z$ to give

$$\begin{cases} \boldsymbol{\delta}_x = (\cos\ \theta)\boldsymbol{\delta}_r + (-\sin\ \theta)\boldsymbol{\delta}_\theta + (0)\boldsymbol{\delta}_z & \text{(A.6-13)} \\ \boldsymbol{\delta}_y = (\sin\ \theta)\boldsymbol{\delta}_r + (\cos\ \theta)\boldsymbol{\delta}_\theta + (0)\boldsymbol{\delta}_z & \text{(A.6-14)} \\ \boldsymbol{\delta}_z = (0)\boldsymbol{\delta}_r + (0)\boldsymbol{\delta}_\theta + (1)\boldsymbol{\delta}_z & \text{(A.6-15)} \end{cases}$$

The utility of these two sets of relations will be made clear in the next section.

Vectors and tensors can be decomposed into components with respect to cylindrical coordinates as was done for Cartesian coordinates in Eqs. A.2-16 and A.3-7 (i.e., $v = \boldsymbol{\delta}_r v_r + \boldsymbol{\delta}_\theta v_\theta + \boldsymbol{\delta}_z v_z$). Also, the multiplication rules for the unit vectors and unit dyads are the same as in Eqs. A.2-14 and 15 and A.3-1 to 6. Consequently the various dot and cross product operations (but *not* the differential operations!) are performed as described in §§A.2 and 3. For example,

$$(\mathbf{v} \cdot \mathbf{w}) = v_r w_r + v_\theta w_\theta + v_z w_z \tag{A.6-16}$$

$$[\mathbf{v} \times \mathbf{w}] = \boldsymbol{\delta}_r(v_\theta w_z - v_z w_\theta) + \boldsymbol{\delta}_\theta(v_z w_r - v_r w_z)$$
$$+ \boldsymbol{\delta}_z(v_r w_\theta - v_\theta w_r) \tag{A.6-17}$$

$$\{\boldsymbol{\sigma} \cdot \boldsymbol{\tau}\} = \boldsymbol{\delta}_r\boldsymbol{\delta}_r(\sigma_{rr}\tau_{rr} + \sigma_{r\theta}\tau_{\theta r} + \sigma_{rz}\tau_{zr})$$
$$+ \boldsymbol{\delta}_r\boldsymbol{\delta}_\theta(\sigma_{rr}\tau_{r\theta} + \sigma_{r\theta}\tau_{\theta\theta} + \sigma_{rz}\tau_{z\theta})$$
$$+ \boldsymbol{\delta}_r\boldsymbol{\delta}_z(\sigma_{rr}\tau_{rz} + \sigma_{r\theta}\tau_{\theta z} + \sigma_{rz}\tau_{zz})$$
$$+ \text{etc.} \tag{A.6-18}$$

Spherical Coordinates

We now tabulate for reference the same kind of information for spherical coordinates r, θ, ϕ. These coordinates are shown in Figure A.6-1b. They are related to the Cartesian coordinates by

$$
\begin{cases}
x = r \sin \theta \cos \phi & \text{(A.6-19)} \\
y = r \sin \theta \sin \phi & \text{(A.6-20)} \\
z = r \cos \theta & \text{(A.6-21)}
\end{cases}
\qquad
\begin{aligned}
r &= +\sqrt{x^2 + y^2 + z^2} & \text{(A.6-22)} \\
\theta &= \arctan(\sqrt{x^2 + y^2}/z) & \text{(A.6-23)} \\
\phi &= \arctan(y/x) & \text{(A.6-24)}
\end{aligned}
$$

For the spherical coordinates we have the following relations for the derivative operators:

$$
\begin{cases}
\dfrac{\partial}{\partial x} = (\sin \theta \cos \phi) \dfrac{\partial}{\partial r} + \left(\dfrac{\cos \theta \cos \phi}{r} \right) \dfrac{\partial}{\partial \theta} + \left(-\dfrac{\sin \phi}{r \sin \theta} \right) \dfrac{\partial}{\partial \phi} & \text{(A.6-25)} \\[2ex]
\dfrac{\partial}{\partial y} = (\sin \theta \sin \phi) \dfrac{\partial}{\partial r} + \left(\dfrac{\cos \theta \sin \phi}{r} \right) \dfrac{\partial}{\partial \theta} + \left(\dfrac{\cos \phi}{r \sin \theta} \right) \dfrac{\partial}{\partial \phi} & \text{(A.6-26)} \\[2ex]
\dfrac{\partial}{\partial z} = (\cos \theta) \dfrac{\partial}{\partial r} + \left(-\dfrac{\sin \theta}{r} \right) \dfrac{\partial}{\partial \theta} + (0) \dfrac{\partial}{\partial \phi} & \text{(A.6-27)}
\end{cases}
$$

The relations between the unit vectors are

$$
\begin{cases}
\boldsymbol{\delta}_r = (\sin \theta \cos \phi)\boldsymbol{\delta}_x + (\sin \theta \sin \phi)\boldsymbol{\delta}_y + (\cos \theta)\boldsymbol{\delta}_z & \text{(A.6-28)} \\
\boldsymbol{\delta}_\theta = (\cos \theta \cos \phi)\boldsymbol{\delta}_x + (\cos \theta \sin \phi)\boldsymbol{\delta}_y + (-\sin \theta)\boldsymbol{\delta}_z & \text{(A.6-29)} \\
\boldsymbol{\delta}_\phi = (-\sin \phi)\boldsymbol{\delta}_x + (\cos \phi)\boldsymbol{\delta}_y + (0)\boldsymbol{\delta}_z & \text{(A.6-30)}
\end{cases}
$$

and

$$
\begin{cases}
\boldsymbol{\delta}_x = (\sin \theta \cos \phi)\boldsymbol{\delta}_r + (\cos \theta \cos \phi)\boldsymbol{\delta}_\theta + (-\sin \phi)\boldsymbol{\delta}_\phi & \text{(A.6-31)} \\
\boldsymbol{\delta}_y = (\sin \theta \sin \phi)\boldsymbol{\delta}_r + (\cos \theta \sin \phi)\boldsymbol{\delta}_\theta + (\cos \phi)\boldsymbol{\delta}_\phi & \text{(A.6-32)} \\
\boldsymbol{\delta}_z = (\cos \theta)\boldsymbol{\delta}_r + (-\sin \theta)\boldsymbol{\delta}_\theta + (0)\boldsymbol{\delta}_\phi & \text{(A.6-33)}
\end{cases}
$$

And, finally, some sample operations in spherical coordinates are

$$
\begin{aligned}
(\boldsymbol{\sigma}:\boldsymbol{\tau}) = {} & \sigma_{rr}\tau_{rr} + \sigma_{r\theta}\tau_{\theta r} + \sigma_{r\phi}\tau_{\phi r} \\
& + \sigma_{\theta r}\tau_{r\theta} + \sigma_{\theta\theta}\tau_{\theta\theta} + \sigma_{\theta\phi}\tau_{\phi\theta} \\
& + \sigma_{\phi r}\tau_{r\phi} + \sigma_{\phi\theta}\tau_{\theta\phi} + \sigma_{\phi\phi}\tau_{\phi\phi}
\end{aligned}
\qquad \text{(A.6-34)}
$$

$$
(\mathbf{u} \cdot [\mathbf{v} \times \mathbf{w}]) =
\begin{vmatrix}
u_r & u_\theta & u_\phi \\
v_r & v_\theta & v_\phi \\
w_r & w_\theta & w_\phi
\end{vmatrix}
\qquad \text{(A.6-35)}
$$

That is, the relations (not involving ∇!) given in §§A.2 and 3 can be written directly in terms of spherical components.

EXERCISES

1. Show that

$$
\int_0^{2\pi} \int_0^\pi \boldsymbol{\delta}_r \sin \theta \, d\theta \, d\phi = 0
$$

$$
\int_0^{2\pi} \int_0^\pi \boldsymbol{\delta}_r \boldsymbol{\delta}_r \sin \theta \, d\theta \, d\phi = \tfrac{4}{3}\pi\boldsymbol{\delta}
$$

where $\boldsymbol{\delta}_r$ is the unit vector in the r direction in spherical coordinates.

2. Verify that in spherical coordinates $\boldsymbol{\delta} = \boldsymbol{\delta}_r\boldsymbol{\delta}_r + \boldsymbol{\delta}_\theta\boldsymbol{\delta}_\theta + \boldsymbol{\delta}_\phi\boldsymbol{\delta}_\phi$.

§A.7 DIFFERENTIAL OPERATIONS
IN CURVILINEAR COORDINATES

We now turn to the use of the ∇-operator in curvilinear coordinates. As in the previous section, we work out in detail the results for cylindrical and spherical coordinates. Then we summarize the procedure for getting the ∇-operations for any orthogonal curvilinear coordinates.

Cylindrical Coordinates

From Eqs. A.6-10, 11, and 12 we can obtain expressions for the spatial derivatives of the unit vectors $\boldsymbol{\delta}_r$, $\boldsymbol{\delta}_\theta$, and $\boldsymbol{\delta}_z$:

$$\frac{\partial}{\partial r}\boldsymbol{\delta}_r = 0 \qquad \frac{\partial}{\partial r}\boldsymbol{\delta}_\theta = 0 \qquad \frac{\partial}{\partial r}\boldsymbol{\delta}_z = 0 \tag{A.7-1}$$

$$\frac{\partial}{\partial \theta}\boldsymbol{\delta}_r = \boldsymbol{\delta}_\theta \qquad \frac{\partial}{\partial \theta}\boldsymbol{\delta}_\theta = -\boldsymbol{\delta}_r \qquad \frac{\partial}{\partial \theta}\boldsymbol{\delta}_z = 0 \tag{A.7-2}$$

$$\frac{\partial}{\partial z}\boldsymbol{\delta}_r = 0 \qquad \frac{\partial}{\partial z}\boldsymbol{\delta}_\theta = 0 \qquad \frac{\partial}{\partial z}\boldsymbol{\delta}_z = 0 \tag{A.7-3}$$

The reader would do well to interpret these derivatives geometrically by considering the way $\boldsymbol{\delta}_r$, $\boldsymbol{\delta}_\theta$, $\boldsymbol{\delta}_z$ change as the location of P is changed in Fig. A.6-2.

We now use the definition of the ∇-operator in Eq. A.4-1, the expressions in Eqs. A.6-13, 14, and 15, and the derivative operators in Eqs. A.6-7, 8, and 9 to obtain the formula for ∇ in cylindrical coordinates

$$\begin{aligned}
\nabla &= \boldsymbol{\delta}_x \frac{\partial}{\partial x} + \boldsymbol{\delta}_y \frac{\partial}{\partial y} + \boldsymbol{\delta}_z \frac{\partial}{\partial z} \\
&= (\boldsymbol{\delta}_r \cos\theta - \boldsymbol{\delta}_\theta \sin\theta)\left(\cos\theta \frac{\partial}{\partial r} - \frac{\sin\theta}{r}\frac{\partial}{\partial \theta}\right) \\
&\quad + (\boldsymbol{\delta}_r \sin\theta + \boldsymbol{\delta}_\theta \cos\theta)\left(\sin\theta \frac{\partial}{\partial r} + \frac{\cos\theta}{r}\frac{\partial}{\partial \theta}\right) + \boldsymbol{\delta}_z \frac{\partial}{\partial z}
\end{aligned} \tag{A.7-4}$$

When this is multiplied out, there is considerable simplification, and we get

$$\nabla = \boldsymbol{\delta}_r \frac{\partial}{\partial r} + \boldsymbol{\delta}_\theta \frac{1}{r}\frac{\partial}{\partial \theta} + \boldsymbol{\delta}_z \frac{\partial}{\partial z} \tag{A.7-5}$$

for *cylindrical* coordinates. This may be used for obtaining all differential operations in cylindrical coordinates, provided that Eqs. A.7-1, 2, and 3 are used to differentiate any unit vectors on which ∇ operates. This point will be made clear in the subsequent illustrative example.

Spherical Coordinates

The spatial derivatives of $\boldsymbol{\delta}_r$, $\boldsymbol{\delta}_\theta$, and $\boldsymbol{\delta}_\phi$ are obtained by differentiating Eqs. A.6-28, 29, and 30:

$$\frac{\partial}{\partial r}\boldsymbol{\delta}_r = 0 \qquad \frac{\partial}{\partial r}\boldsymbol{\delta}_\theta = 0 \qquad \frac{\partial}{\partial r}\boldsymbol{\delta}_\phi = 0 \tag{A.7-6}$$

$$\frac{\partial}{\partial \theta}\boldsymbol{\delta}_r = \boldsymbol{\delta}_\theta \qquad \frac{\partial}{\partial \theta}\boldsymbol{\delta}_\theta = -\boldsymbol{\delta}_r \qquad \frac{\partial}{\partial \theta}\boldsymbol{\delta}_\phi = 0 \tag{A.7-7}$$

$$\frac{\partial}{\partial \phi}\boldsymbol{\delta}_r = \boldsymbol{\delta}_\phi \sin\theta \qquad \frac{\partial}{\partial \phi}\boldsymbol{\delta}_\theta = \boldsymbol{\delta}_\phi \cos\theta \qquad \frac{\partial}{\partial \phi}\boldsymbol{\delta}_\phi = -\boldsymbol{\delta}_r \sin\theta - \boldsymbol{\delta}_\theta \cos\theta \tag{A.7-8}$$

Use of Eqs. A.6-31, 32, and 33 and Eqs. A.6-25, 26, and 27 in Eq. A.4-1 gives the following expression for the ∇-operator:

$$\nabla = \mathbf{\delta}_r \frac{\partial}{\partial r} + \mathbf{\delta}_\theta \frac{1}{r} \frac{\partial}{\partial \theta} + \mathbf{\delta}_\phi \frac{1}{r \sin \theta} \frac{\partial}{\partial \phi} \tag{A.7-9}$$

in *spherical* coordinates. This expression may be used for obtaining differential operations in spherical coordinates, provided that Eqs. A.7-6, 7, and 8 are used for differentiating the unit vectors.

General Orthogonal Coordinates

Thus far we have discussed the two most-used curvilinear coordinate systems. We now present without proof the relations for any orthogonal curvilinear coordinates. Let the relation between Cartesian coordinates x_i and the curvilinear coordinates q_α be given by

$$\begin{cases} x_1 = x_1(q_1, q_2, q_3) \\ x_2 = x_2(q_1, q_2, q_3) \\ x_3 = x_3(q_1, q_2, q_3) \end{cases} \quad \text{or} \quad x_i = x_i(q_\alpha) \tag{A.7-10}$$

These can be solved for the q_α to get the inverse relations $q_\alpha = q_\alpha(x_i)$. Then[1] the unit vectors $\mathbf{\delta}_i$ in rectangular coordinates and the $\mathbf{\delta}_\alpha$ in curvilinear coordinates are related thus:

$$\begin{cases} \mathbf{\delta}_\alpha = \sum_i h_\alpha \left(\frac{\partial q_\alpha}{\partial x_i}\right) \mathbf{\delta}_i = \sum_i \frac{1}{h_\alpha} \left(\frac{\partial x_i}{\partial q_\alpha}\right) \mathbf{\delta}_i \tag{A.7-11} \\[2ex] \mathbf{\delta}_i = \sum_\alpha h_\alpha \left(\frac{\partial q_\alpha}{\partial x_i}\right) \mathbf{\delta}_\alpha = \sum_\alpha \frac{1}{h_\alpha} \left(\frac{\partial x_i}{\partial q_\alpha}\right) \mathbf{\delta}_\alpha \tag{A.7-12} \end{cases}$$

in which the "scale factors" h_α are given by

$$h_\alpha^2 = \sum_i \left(\frac{\partial x_i}{\partial q_\alpha}\right)^2 = \left[\sum_i \left(\frac{\partial q_\alpha}{\partial x_i}\right)^2\right]^{-1} \tag{A.7-13}$$

The spatial derivatives of the unit vectors $\mathbf{\delta}_\alpha$ can then be found to be

$$\frac{\partial \mathbf{\delta}_\alpha}{\partial q_\beta} = \frac{\mathbf{\delta}_\beta}{h_\alpha} \frac{\partial h_\beta}{\partial q_\alpha} - \mathbf{\delta}_{\alpha\beta} \sum_{\gamma=1}^{3} \frac{\mathbf{\delta}_\gamma}{h_\gamma} \frac{\partial h_\alpha}{\partial q_\gamma} \tag{A.7-14}$$

and the ∇-operator is

$$\nabla = \sum_\alpha \frac{\mathbf{\delta}_\alpha}{h_\alpha} \frac{\partial}{\partial q_\alpha} \tag{A.7-15}$$

The reader should verify that Eqs. A.7-14 and 15 can be used to get Eqs. A.7-1 to 3, A.7-5 and A.7-6 to 9.

From Eqs. A.7-15 and 14 we can now get the following expressions for the simplest of the ∇-operations:

$$(\nabla \cdot \mathbf{v}) = \frac{1}{h_1 h_2 h_3} \sum_\alpha \frac{\partial}{\partial q_\alpha} \left(\frac{h_1 h_2 h_3}{h_\alpha} v_\alpha\right) \tag{A.7-16}$$

$$\nabla^2 s = \frac{1}{h_1 h_2 h_3} \sum_\alpha \frac{\partial}{\partial q_\alpha} \left(\frac{h_1 h_2 h_3}{h_\alpha^2} \frac{\partial s}{\partial q_\alpha}\right) \tag{A.7-17}$$

---

[1] P. Morse and H. Feshbach, *Methods of Theoretical Physics*, McGraw-Hill, New York (1953), p. 26 and p. 115.

$$[\nabla \times \mathbf{v}] = \frac{1}{h_1 h_2 h_3} \begin{vmatrix} h_1\boldsymbol{\delta}_1 & h_2\boldsymbol{\delta}_2 & h_3\boldsymbol{\delta}_3 \\ \dfrac{\partial}{\partial q_1} & \dfrac{\partial}{\partial q_2} & \dfrac{\partial}{\partial q_3} \\ h_1 v_1 & h_2 v_2 & h_3 v_3 \end{vmatrix} \qquad (A.7\text{-}18)$$

In the last expression, the unit vectors are those belonging to the curvilinear coordinate system. Additional operations may be found in Morse and Feshbach.[1]

The scale factors introduced above also arise in the expressions for the volume and surface elements $dV = h_1 h_2 h_3 dq_1 dq_2 dq_3$ and $dS_{\alpha\beta} = h_\alpha h_\beta dq_\alpha dq_\beta (\alpha \neq \beta)$; here $dS_{\alpha\beta}$ is a surface element on a surface of constant γ, where $\gamma \neq \alpha$ and $\gamma \neq \beta$. The reader should verify that the volume elements and various surface elements in cylindrical and spherical coordinates can be found in this way.

In Tables A.7-1, 2, and 3 we summarize the differential operations most commonly encountered in Cartesian, cylindrical, and spherical coordinates.[2] The curvilinear coordinate expressions given can be obtained by the method illustrated in the following two examples.

EXAMPLE A.7-1

Differential Operations in Cylindrical Coordinates

Derive expressions for $(\nabla \cdot v)$ and ∇v in cylindrical coordinates.

SOLUTION

(a) We begin by writing ∇ in cylindrical coordinates and decomposing v into its components

$$(\nabla \cdot \mathbf{v}) = \left(\left\{ \boldsymbol{\delta}_r \frac{\partial}{\partial r} + \boldsymbol{\delta}_\theta \frac{1}{r} \frac{\partial}{\partial \theta} + \boldsymbol{\delta}_z \frac{\partial}{\partial z} \right\} \cdot \left\{ \boldsymbol{\delta}_r v_r + \boldsymbol{\delta}_\theta v_\theta + \boldsymbol{\delta}_z v_z \right\} \right) \qquad (A.7\text{-}19)$$

Expanding, we get

$$(\nabla \cdot \mathbf{v}) = \left(\boldsymbol{\delta}_r \cdot \frac{\partial}{\partial r} \boldsymbol{\delta}_r v_r \right) + \left(\boldsymbol{\delta}_r \cdot \frac{\partial}{\partial r} \boldsymbol{\delta}_\theta v_\theta \right) + \left(\boldsymbol{\delta}_r \cdot \frac{\partial}{\partial r} \boldsymbol{\delta}_z v_z \right)$$
$$+ \left(\boldsymbol{\delta}_\theta \cdot \frac{1}{r} \frac{\partial}{\partial \theta} \boldsymbol{\delta}_r v_r \right) + \left(\boldsymbol{\delta}_\theta \cdot \frac{1}{r} \frac{\partial}{\partial \theta} \boldsymbol{\delta}_\theta v_\theta \right) + \left(\boldsymbol{\delta}_\theta \cdot \frac{1}{r} \frac{\partial}{\partial \theta} \boldsymbol{\delta}_z v_z \right)$$
$$+ \left(\boldsymbol{\delta}_z \cdot \frac{\partial}{\partial z} \boldsymbol{\delta}_r v_r \right) + \left(\boldsymbol{\delta}_z \cdot \frac{\partial}{\partial z} \boldsymbol{\delta}_\theta v_\theta \right) + \left(\boldsymbol{\delta}_z \cdot \frac{\partial}{\partial z} \boldsymbol{\delta}_z v_z \right) \qquad (A.7\text{-}20)$$

We now use the relations given in Eqs. A.7-1, 2, and 3 to evaluate the derivatives of the unit vectors. This gives

$$(\nabla \cdot \mathbf{v}) = (\boldsymbol{\delta}_r \cdot \boldsymbol{\delta}_r) \frac{\partial v_r}{\partial r} + (\boldsymbol{\delta}_r \cdot \boldsymbol{\delta}_\theta) \frac{\partial v_\theta}{\partial r} + (\boldsymbol{\delta}_r \cdot \boldsymbol{\delta}_z) \frac{\partial v_z}{\partial r} + (\boldsymbol{\delta}_\theta \cdot \boldsymbol{\delta}_r) \frac{1}{r} \frac{\partial v_r}{\partial \theta}$$
$$+ (\boldsymbol{\delta}_\theta \cdot \boldsymbol{\delta}_\theta) \frac{1}{r} \frac{\partial v_\theta}{\partial \theta} + (\boldsymbol{\delta}_\theta \cdot \boldsymbol{\delta}_z) \frac{1}{r} \frac{\partial v_z}{\partial \theta} + \frac{v_r}{r} (\boldsymbol{\delta}_\theta \cdot \boldsymbol{\delta}_\theta) + \frac{v_\theta}{r} (\boldsymbol{\delta}_\theta \cdot (-\boldsymbol{\delta}_r))$$
$$+ (\boldsymbol{\delta}_z \cdot \boldsymbol{\delta}_r) \frac{\partial v_r}{\partial z} + (\boldsymbol{\delta}_z \cdot \boldsymbol{\delta}_\theta) \frac{\partial v_\theta}{\partial z} + (\boldsymbol{\delta}_z \cdot \boldsymbol{\delta}_z) \frac{\partial v_z}{\partial z} \qquad (A.7\text{-}21)$$

Since $(\boldsymbol{\delta}_r \cdot \boldsymbol{\delta}_r) = 1$, $(\boldsymbol{\delta}_r \cdot \boldsymbol{\delta}_\theta) = 0$, and so on, the latter simplifies to

$$(\nabla \cdot \mathbf{v}) = \frac{\partial v_r}{\partial r} + \frac{1}{r} \frac{\partial v_\theta}{\partial \theta} + \frac{v_r}{r} + \frac{\partial v_z}{\partial z} \qquad (A.7\text{-}22)$$

which is the same as Eq. A of Table A.7-2. The procedure is a bit tedious, but is *is* straightforward.

---

[2] For other coordinate systems see the extensive compilation of P. Moon and D. E. Spencer, *Field Theory Handbook*, Springer, Berlin (1961). In addition, an orthogonal coordinate system is available in which one of the three sets of coordinate surfaces is made up of coaxial cones (but with noncoincident apexes); all of the ∇-operations have been tabulated by the originators of this coordinate system, J. F. Dijksman and E. P. W. Savenije, *Rheol. Acta*, **24**, 105–118 (1985).

Table A.7-1 Summary of Differential Operations Involving the ∇-Operator in Cartesian Coordinates (x, y, z)

$$(\nabla \cdot \mathbf{v}) = \frac{\partial v_x}{\partial x} + \frac{\partial v_y}{\partial y} + \frac{\partial v_z}{\partial z} \tag{A}$$

$$(\nabla^2 s) = \frac{\partial^2 s}{\partial x^2} + \frac{\partial^2 s}{\partial y^2} + \frac{\partial^2 s}{\partial z^2} \tag{B}$$

$$(\boldsymbol{\tau}\!:\!\nabla\mathbf{v}) = \tau_{xx}\left(\frac{\partial v_x}{\partial x}\right) + \tau_{xy}\left(\frac{\partial v_x}{\partial y}\right) + \tau_{xz}\left(\frac{\partial v_x}{\partial z}\right)$$

$$+ \tau_{yx}\left(\frac{\partial v_y}{\partial x}\right) + \tau_{yy}\left(\frac{\partial v_y}{\partial y}\right) + \tau_{yz}\left(\frac{\partial v_y}{\partial z}\right)$$

$$+ \tau_{zx}\left(\frac{\partial v_z}{\partial x}\right) + \tau_{zy}\left(\frac{\partial v_z}{\partial y}\right) + \tau_{zz}\left(\frac{\partial v_z}{\partial z}\right) \tag{C}$$

$$[\nabla s]_x = \frac{\partial s}{\partial x} \tag{D}$$

$$[\nabla s]_y = \frac{\partial s}{\partial y} \tag{E}$$

$$[\nabla s]_z = \frac{\partial s}{\partial z} \tag{F}$$

$$[\nabla \times \mathbf{v}]_x = \frac{\partial v_z}{\partial y} - \frac{\partial v_y}{\partial z} \tag{G}$$

$$[\nabla \times \mathbf{v}]_y = \frac{\partial v_x}{\partial z} - \frac{\partial v_z}{\partial x} \tag{H}$$

$$[\nabla \times \mathbf{v}]_z = \frac{\partial v_y}{\partial x} - \frac{\partial v_x}{\partial y} \tag{I}$$

$$[\nabla \cdot \boldsymbol{\tau}]_x = \frac{\partial \tau_{xx}}{\partial x} + \frac{\partial \tau_{yx}}{\partial y} + \frac{\partial \tau_{zx}}{\partial z} \tag{J}$$

$$[\nabla \cdot \boldsymbol{\tau}]_y = \frac{\partial \tau_{xy}}{\partial x} + \frac{\partial \tau_{yy}}{\partial y} + \frac{\partial \tau_{zy}}{\partial z} \tag{K}$$

$$[\nabla \cdot \boldsymbol{\tau}]_z = \frac{\partial \tau_{xz}}{\partial x} + \frac{\partial \tau_{yz}}{\partial y} + \frac{\partial \tau_{zz}}{\partial z} \tag{L}$$

$$[\nabla^2 \mathbf{v}]_x = \frac{\partial^2 v_x}{\partial x^2} + \frac{\partial^2 v_x}{\partial y^2} + \frac{\partial^2 v_x}{\partial z^2} \tag{M}$$

$$[\nabla^2 \mathbf{v}]_y = \frac{\partial^2 v_y}{\partial x^2} + \frac{\partial^2 v_y}{\partial y^2} + \frac{\partial^2 v_y}{\partial z^2} \tag{N}$$

$$[\nabla^2 \mathbf{v}]_z = \frac{\partial^2 v_z}{\partial x^2} + \frac{\partial^2 v_z}{\partial y^2} + \frac{\partial^2 v_z}{\partial z^2} \tag{O}$$

$$[\mathbf{v} \cdot \nabla\mathbf{w}]_x = v_x\left(\frac{\partial w_x}{\partial x}\right) + v_y\left(\frac{\partial w_x}{\partial y}\right) + v_z\left(\frac{\partial w_x}{\partial z}\right) \tag{P}$$

$$[\mathbf{v} \cdot \nabla\mathbf{w}]_y = v_x\left(\frac{\partial w_y}{\partial x}\right) + v_y\left(\frac{\partial w_y}{\partial y}\right) + v_z\left(\frac{\partial w_y}{\partial z}\right) \tag{Q}$$

$$[\mathbf{v} \cdot \nabla\mathbf{w}]_z = v_x\left(\frac{\partial w_z}{\partial x}\right) + v_y\left(\frac{\partial w_z}{\partial y}\right) + v_z\left(\frac{\partial w_z}{\partial z}\right) \tag{R}$$

$$\{\nabla\mathbf{v}\}_{xx} = \frac{\partial v_x}{\partial x} \tag{S}$$

$$\{\nabla\mathbf{v}\}_{xy} = \frac{\partial v_y}{\partial x} \tag{T}$$

$$\{\nabla\mathbf{v}\}_{xz} = \frac{\partial v_z}{\partial x} \tag{U}$$

$$\{\nabla\mathbf{v}\}_{yx} = \frac{\partial v_x}{\partial y} \tag{V}$$

$$\{\nabla\mathbf{v}\}_{yy} = \frac{\partial v_y}{\partial y} \tag{W}$$

$$\{\nabla\mathbf{v}\}_{yz} = \frac{\partial v_z}{\partial y} \tag{X}$$

$$\{\nabla\mathbf{v}\}_{zx} = \frac{\partial v_x}{\partial z} \tag{Y}$$

$$\{\nabla\mathbf{v}\}_{zy} = \frac{\partial v_y}{\partial z} \tag{Z}$$

$$\{\nabla\mathbf{v}\}_{zz} = \frac{\partial v_z}{\partial z} \tag{AA}$$

$$\{\mathbf{v}\cdot\nabla\boldsymbol{\tau}\}_{xx} = (\mathbf{v}\cdot\nabla)\tau_{xx} \tag{BB}$$
$$\{\mathbf{v}\cdot\nabla\boldsymbol{\tau}\}_{xy} = (\mathbf{v}\cdot\nabla)\tau_{xy} \tag{CC}$$
$$\{\mathbf{v}\cdot\nabla\boldsymbol{\tau}\}_{xz} = (\mathbf{v}\cdot\nabla)\tau_{xz} \tag{DD}$$
$$\{\mathbf{v}\cdot\nabla\boldsymbol{\tau}\}_{yx} = (\mathbf{v}\cdot\nabla)\tau_{yx} \tag{EE}$$
$$\{\mathbf{v}\cdot\nabla\boldsymbol{\tau}\}_{yy} = (\mathbf{v}\cdot\nabla)\tau_{yy} \tag{FF}$$
$$\{\mathbf{v}\cdot\nabla\boldsymbol{\tau}\}_{yz} = (\mathbf{v}\cdot\nabla)\tau_{yz} \tag{GG}$$
$$\{\mathbf{v}\cdot\nabla\boldsymbol{\tau}\}_{zx} = (\mathbf{v}\cdot\nabla)\tau_{zx} \tag{HH}$$
$$\{\mathbf{v}\cdot\nabla\boldsymbol{\tau}\}_{zy} = (\mathbf{v}\cdot\nabla)\tau_{zy} \tag{II}$$
$$\{\mathbf{v}\cdot\nabla\boldsymbol{\tau}\}_{zz} = (\mathbf{v}\cdot\nabla)\tau_{zz} \tag{JJ}$$

where the operator $(\mathbf{v}\cdot\nabla) = v_x\frac{\partial}{\partial x} + v_y\frac{\partial}{\partial y} + v_z\frac{\partial}{\partial z}$

Table A.7-2 Summary of Differential Operations Involving the ∇-Operator in Cylindrical Coordinates (r, θ, z)

$$(\nabla \cdot \mathbf{v}) = \frac{1}{r}\frac{\partial}{\partial r}(rv_r) + \frac{1}{r}\frac{\partial v_\theta}{\partial \theta} + \frac{\partial v_z}{\partial z} \tag{A}$$

$$(\nabla^2 s) = \frac{1}{r}\frac{\partial}{\partial r}\left(r\frac{\partial s}{\partial r}\right) + \frac{1}{r^2}\frac{\partial^2 s}{\partial \theta^2} + \frac{\partial^2 s}{\partial z^2} \tag{B}$$

$$(\boldsymbol{\tau}:\nabla\mathbf{v}) = \tau_{rr}\left(\frac{\partial v_r}{\partial r}\right) + \tau_{r\theta}\left(\frac{1}{r}\frac{\partial v_r}{\partial \theta} - \frac{v_\theta}{r}\right) + \tau_{rz}\left(\frac{\partial v_r}{\partial z}\right)$$
$$+ \tau_{\theta r}\left(\frac{\partial v_\theta}{\partial r}\right) + \tau_{\theta\theta}\left(\frac{1}{r}\frac{\partial v_\theta}{\partial \theta} + \frac{v_r}{r}\right) + \tau_{\theta z}\left(\frac{\partial v_\theta}{\partial z}\right)$$
$$+ \tau_{zr}\left(\frac{\partial v_z}{\partial r}\right) + \tau_{z\theta}\left(\frac{1}{r}\frac{\partial v_z}{\partial \theta}\right) + \tau_{zz}\left(\frac{\partial v_z}{\partial z}\right) \tag{C}$$

$$[\nabla s]_r = \frac{\partial s}{\partial r} \tag{D}$$

$$[\nabla s]_\theta = \frac{1}{r}\frac{\partial s}{\partial \theta} \tag{E}$$

$$[\nabla s]_z = \frac{\partial s}{\partial z} \tag{F}$$

$$[\nabla \times \mathbf{v}]_r = \frac{1}{r}\frac{\partial v_z}{\partial \theta} - \frac{\partial v_\theta}{\partial z} \tag{G}$$

$$[\nabla \times \mathbf{v}]_\theta = \frac{\partial v_r}{\partial z} - \frac{\partial v_z}{\partial r} \tag{H}$$

$$[\nabla \times \mathbf{v}]_z = \frac{1}{r}\frac{\partial}{\partial r}(rv_\theta) - \frac{1}{r}\frac{\partial v_r}{\partial \theta} \tag{I}$$

$$[\nabla \cdot \boldsymbol{\tau}]_r = \frac{1}{r}\frac{\partial}{\partial r}(r\tau_{rr}) + \frac{1}{r}\frac{\partial}{\partial \theta}\tau_{\theta r} + \frac{\partial}{\partial z}\tau_{zr} - \frac{\tau_{\theta\theta}}{r} \tag{J}$$

$$[\nabla \cdot \boldsymbol{\tau}]_\theta = \frac{1}{r^2}\frac{\partial}{\partial r}(r^2\tau_{r\theta}) + \frac{1}{r}\frac{\partial}{\partial \theta}\tau_{\theta\theta} + \frac{\partial}{\partial z}\tau_{z\theta} + \frac{\tau_{\theta r} - \tau_{r\theta}}{r} \tag{K}$$

$$[\nabla \cdot \boldsymbol{\tau}]_z = \frac{1}{r}\frac{\partial}{\partial r}(r\tau_{rz}) + \frac{1}{r}\frac{\partial}{\partial \theta}\tau_{\theta z} + \frac{\partial}{\partial z}\tau_{zz} \tag{L}$$

$$[\nabla^2\mathbf{v}]_r = \frac{\partial}{\partial r}\left(\frac{1}{r}\frac{\partial}{\partial r}(rv_r)\right) + \frac{1}{r^2}\frac{\partial^2 v_r}{\partial \theta^2} + \frac{\partial^2 v_r}{\partial z^2} - \frac{2}{r^2}\frac{\partial v_\theta}{\partial \theta} \tag{M}$$

$$[\nabla^2\mathbf{v}]_\theta = \frac{\partial}{\partial r}\left(\frac{1}{r}\frac{\partial}{\partial r}(rv_\theta)\right) + \frac{1}{r^2}\frac{\partial^2 v_\theta}{\partial \theta^2} + \frac{\partial^2 v_\theta}{\partial z^2} + \frac{2}{r^2}\frac{\partial v_r}{\partial \theta} \tag{N}$$

$$[\nabla^2\mathbf{v}]_z = \frac{1}{r}\frac{\partial}{\partial r}\left(r\frac{\partial v_z}{\partial r}\right) + \frac{1}{r^2}\frac{\partial^2 v_z}{\partial \theta^2} + \frac{\partial^2 v_z}{\partial z^2} \tag{O}$$

$$[\mathbf{v} \cdot \nabla\mathbf{w}]_r = v_r\left(\frac{\partial w_r}{\partial r}\right) + v_\theta\left(\frac{1}{r}\frac{\partial w_r}{\partial \theta} - \frac{w_\theta}{r}\right) + v_z\left(\frac{\partial w_r}{\partial z}\right) \tag{P}$$

$$[\mathbf{v} \cdot \nabla\mathbf{w}]_\theta = v_r\left(\frac{\partial w_\theta}{\partial r}\right) + v_\theta\left(\frac{1}{r}\frac{\partial w_\theta}{\partial \theta} + \frac{w_r}{r}\right) + v_z\left(\frac{\partial w_\theta}{\partial z}\right) \tag{Q}$$

$$[\mathbf{v} \cdot \nabla\mathbf{w}]_z = v_r\left(\frac{\partial w_z}{\partial r}\right) + v_\theta\left(\frac{1}{r}\frac{\partial w_z}{\partial \theta}\right) + v_z\left(\frac{\partial w_z}{\partial z}\right) \tag{R}$$

$$\{\nabla \mathbf{v}\}_{rr} = \frac{\partial v_r}{\partial r} \tag{S}$$

$$\{\nabla \mathbf{v}\}_{r\theta} = \frac{\partial v_\theta}{\partial r} \tag{T}$$

$$\{\nabla \mathbf{v}\}_{rz} = \frac{\partial v_z}{\partial r} \tag{U}$$

$$\{\nabla \mathbf{v}\}_{\theta r} = \frac{1}{r} \frac{\partial v_r}{\partial \theta} - \frac{v_\theta}{r} \tag{V}$$

$$\{\nabla \mathbf{v}\}_{\theta\theta} = \frac{1}{r} \frac{\partial v_\theta}{\partial \theta} + \frac{v_r}{r} \tag{W}$$

$$\{\nabla \mathbf{v}\}_{\theta z} = \frac{1}{r} \frac{\partial v_z}{\partial \theta} \tag{X}$$

$$\{\nabla \mathbf{v}\}_{zr} = \frac{\partial v_r}{\partial z} \tag{Y}$$

$$\{\nabla \mathbf{v}\}_{z\theta} = \frac{\partial v_\theta}{\partial z} \tag{Z}$$

$$\{\nabla \mathbf{v}\}_{zz} = \frac{\partial v_z}{\partial z} \tag{AA}$$

$$\{\mathbf{v} \cdot \nabla \boldsymbol{\tau}\}_{rr} = (\mathbf{v} \cdot \nabla)\tau_{rr} - \frac{v_\theta}{r}(\tau_{r\theta} + \tau_{\theta r}) \tag{BB}$$

$$\{\mathbf{v} \cdot \nabla \boldsymbol{\tau}\}_{r\theta} = (\mathbf{v} \cdot \nabla)\tau_{r\theta} + \frac{v_\theta}{r}(\tau_{rr} - \tau_{\theta\theta}) \tag{CC}$$

$$\{\mathbf{v} \cdot \nabla \boldsymbol{\tau}\}_{rz} = (\mathbf{v} \cdot \nabla)\tau_{rz} - \frac{v_\theta}{r}\tau_{\theta z} \tag{DD}$$

$$\{\mathbf{v} \cdot \nabla \boldsymbol{\tau}\}_{\theta r} = (\mathbf{v} \cdot \nabla)\tau_{\theta r} + \frac{v_\theta}{r}(\tau_{rr} - \tau_{\theta\theta}) \tag{EE}$$

$$\{\mathbf{v} \cdot \nabla \boldsymbol{\tau}\}_{\theta\theta} = (\mathbf{v} \cdot \nabla)\tau_{\theta\theta} + \frac{v_\theta}{r}(\tau_{r\theta} + \tau_{\theta r}) \tag{FF}$$

$$\{\mathbf{v} \cdot \nabla \boldsymbol{\tau}\}_{\theta z} = (\mathbf{v} \cdot \nabla)\tau_{\theta z} + \frac{v_\theta}{r}\tau_{rz} \tag{GG}$$

$$\{\mathbf{v} \cdot \nabla \boldsymbol{\tau}\}_{zr} = (\mathbf{v} \cdot \nabla)\tau_{zr} - \frac{v_\theta}{r}\tau_{z\theta} \tag{HH}$$

$$\{\mathbf{v} \cdot \nabla \boldsymbol{\tau}\}_{z\theta} = (\mathbf{v} \cdot \nabla)\tau_{z\theta} + \frac{v_\theta}{r}\tau_{zr} \tag{II}$$

$$\{\mathbf{v} \cdot \nabla \boldsymbol{\tau}\}_{zz} = (\mathbf{v} \cdot \nabla)\tau_{zz} \tag{JJ}$$

where the operator $(\mathbf{v} \cdot \nabla) = v_r \frac{\partial}{\partial r} + \frac{v_\theta}{r}\frac{\partial}{\partial \theta} + v_z \frac{\partial}{\partial z}$

Table A.7-3 Summary of Differential Operations Involving the ∇-Operator in Spherical Coordinates (r, θ, ϕ)

$$(\nabla \cdot \mathbf{v}) = \frac{1}{r^2}\frac{\partial}{\partial r}(r^2 v_r) + \frac{1}{r \sin \theta}\frac{\partial}{\partial \theta}(v_\theta \sin \theta) + \frac{1}{r \sin \theta}\frac{\partial v_\phi}{\partial \phi} \tag{A}$$

$$(\nabla^2 s) = \frac{1}{r^2}\frac{\partial}{\partial r}\left(r^2 \frac{\partial s}{\partial r}\right) + \frac{1}{r^2 \sin \theta}\frac{\partial}{\partial \theta}\left(\sin \theta \frac{\partial s}{\partial \theta}\right) + \frac{1}{r^2 \sin^2 \theta}\frac{\partial^2 s}{\partial \phi^2} \tag{B}$$

$$(\boldsymbol{\tau}:\nabla\mathbf{v}) = \tau_{rr}\left(\frac{\partial v_r}{\partial r}\right) + \tau_{r\theta}\left(\frac{1}{r}\frac{\partial v_r}{\partial \theta} - \frac{v_\theta}{r}\right) + \tau_{r\phi}\left(\frac{1}{r \sin \theta}\frac{\partial v_r}{\partial \phi} - \frac{v_\phi}{r}\right)$$

$$+ \tau_{\theta r}\left(\frac{\partial v_\theta}{\partial r}\right) + \tau_{\theta\theta}\left(\frac{1}{r}\frac{\partial v_\theta}{\partial \theta} + \frac{v_r}{r}\right) + \tau_{\theta\phi}\left(\frac{1}{r \sin \theta}\frac{\partial v_\theta}{\partial \phi} - \frac{v_\phi}{r}\cot \theta\right)$$

$$+ \tau_{\phi r}\left(\frac{\partial v_\phi}{\partial r}\right) + \tau_{\phi\theta}\left(\frac{1}{r}\frac{\partial v_\phi}{\partial \theta}\right) + \tau_{\phi\phi}\left(\frac{1}{r \sin \theta}\frac{\partial v_\phi}{\partial \phi} + \frac{v_r}{r} + \frac{v_\theta}{r}\cot \theta\right) \tag{C}$$

$$[\nabla s]_r = \frac{\partial s}{\partial r} \tag{D}$$

$$[\nabla s]_\theta = \frac{1}{r}\frac{\partial s}{\partial \theta} \tag{E}$$

$$[\nabla s]_\phi = \frac{1}{r \sin \theta}\frac{\partial s}{\partial \phi} \tag{F}$$

$$[\nabla \times \mathbf{v}]_r = \frac{1}{r \sin \theta}\frac{\partial}{\partial \theta}(v_\phi \sin \theta) - \frac{1}{r \sin \theta}\frac{\partial v_\theta}{\partial \phi} \tag{G}$$

$$[\nabla \times \mathbf{v}]_\theta = \frac{1}{r \sin \theta}\frac{\partial v_r}{\partial \phi} - \frac{1}{r}\frac{\partial}{\partial r}(r v_\phi) \tag{H}$$

$$[\nabla \times \mathbf{v}]_\phi = \frac{1}{r}\frac{\partial}{\partial r}(r v_\theta) - \frac{1}{r}\frac{\partial v_r}{\partial \theta} \tag{I}$$

$$[\nabla \cdot \boldsymbol{\tau}]_r = \frac{1}{r^2}\frac{\partial}{\partial r}(r^2 \tau_{rr}) + \frac{1}{r \sin \theta}\frac{\partial}{\partial \theta}(\tau_{\theta r}\sin \theta) + \frac{1}{r \sin \theta}\frac{\partial}{\partial \phi}\tau_{\phi r} - \frac{\tau_{\theta\theta} + \tau_{\phi\phi}}{r} \tag{J}$$

$$[\nabla \cdot \boldsymbol{\tau}]_\theta = \frac{1}{r^3}\frac{\partial}{\partial r}(r^3 \tau_{r\theta}) + \frac{1}{r \sin \theta}\frac{\partial}{\partial \theta}(\tau_{\theta\theta}\sin \theta) + \frac{1}{r \sin \theta}\frac{\partial}{\partial \phi}\tau_{\phi\theta} + \frac{(\tau_{\theta r} - \tau_{r\theta}) - \tau_{\phi\phi}\cot \theta}{r} \tag{K}$$

$$[\nabla \cdot \boldsymbol{\tau}]_\phi = \frac{1}{r^3}\frac{\partial}{\partial r}(r^3 \tau_{r\phi}) + \frac{1}{r \sin \theta}\frac{\partial}{\partial \theta}(\tau_{\theta\phi}\sin \theta) + \frac{1}{r \sin \theta}\frac{\partial}{\partial \phi}\tau_{\phi\phi} + \frac{(\tau_{\phi r} - \tau_{r\phi}) + \tau_{\phi\theta}\cot \theta}{r} \tag{L}$$

$$[\nabla^2\mathbf{v}]_r = \frac{\partial}{\partial r}\left(\frac{1}{r^2}\frac{\partial}{\partial r}(r^2 v_r)\right) + \frac{1}{r^2 \sin \theta}\frac{\partial}{\partial \theta}\left(\sin \theta \frac{\partial v_r}{\partial \theta}\right) + \frac{1}{r^2 \sin^2 \theta}\frac{\partial^2 v_r}{\partial \phi^2} - \frac{2}{r^2 \sin \theta}\frac{\partial}{\partial \theta}(v_\theta \sin \theta) - \frac{2}{r^2 \sin \theta}\frac{\partial v_\phi}{\partial \phi} \tag{M}$$

$$[\nabla^2\mathbf{v}]_\theta = \frac{1}{r^2}\frac{\partial}{\partial r}\left(r^2 \frac{\partial v_\theta}{\partial r}\right) + \frac{1}{r^2}\frac{\partial}{\partial \theta}\left(\frac{1}{\sin \theta}\frac{\partial}{\partial \theta}(v_\theta \sin \theta)\right) + \frac{1}{r^2 \sin^2 \theta}\frac{\partial^2 v_\theta}{\partial \phi^2} + \frac{2}{r^2}\frac{\partial v_r}{\partial \theta} - \frac{2 \cot \theta}{r^2 \sin \theta}\frac{\partial v_\phi}{\partial \phi} \tag{N}$$

$$[\nabla^2\mathbf{v}]_\phi = \frac{1}{r^2}\frac{\partial}{\partial r}\left(r^2 \frac{\partial v_\phi}{\partial r}\right) + \frac{1}{r^2}\frac{\partial}{\partial \theta}\left(\frac{1}{\sin \theta}\frac{\partial}{\partial \theta}(v_\phi \sin \theta)\right) + \frac{1}{r^2 \sin^2 \theta}\frac{\partial^2 v_\phi}{\partial \phi^2} + \frac{2}{r^2 \sin \theta}\frac{\partial v_r}{\partial \phi} + \frac{2 \cot \theta}{r^2 \sin \theta}\frac{\partial v_\theta}{\partial \phi} \tag{O}$$

$$[\mathbf{v} \cdot \nabla \mathbf{w}]_r = v_r\left(\frac{\partial w_r}{\partial r}\right) + v_\theta\left(\frac{1}{r}\frac{\partial w_r}{\partial \theta} - \frac{w_\theta}{r}\right) + v_\phi\left(\frac{1}{r \sin \theta}\frac{\partial w_r}{\partial \phi} - \frac{w_\phi}{r}\right) \tag{P}$$

$$[\mathbf{v} \cdot \nabla \mathbf{w}]_\theta = v_r\left(\frac{\partial w_\theta}{\partial r}\right) + v_\theta\left(\frac{1}{r}\frac{\partial w_\theta}{\partial \theta} + \frac{w_r}{r}\right) + v_\phi\left(\frac{1}{r \sin \theta}\frac{\partial w_\theta}{\partial \phi} - \frac{w_\phi}{r}\cot \theta\right) \tag{Q}$$

$$[\mathbf{v} \cdot \nabla \mathbf{w}]_\phi = v_r\left(\frac{\partial w_\phi}{\partial r}\right) + v_\theta\left(\frac{1}{r}\frac{\partial w_\phi}{\partial \theta}\right) + v_\phi\left(\frac{1}{r \sin \theta}\frac{\partial w_\phi}{\partial \phi} + \frac{w_r}{r} + \frac{w_\theta}{r}\cot \theta\right) \tag{R}$$

$$\{\nabla \mathbf{v}\}_{rr} = \frac{\partial v_r}{\partial r} \tag{S}$$

$$\{\nabla \mathbf{v}\}_{r\theta} = \frac{\partial v_\theta}{\partial r} \tag{T}$$

$$\{\nabla \mathbf{v}\}_{r\phi} = \frac{\partial v_\phi}{\partial r} \tag{U}$$

$$\{\nabla \mathbf{v}\}_{\theta r} = \frac{1}{r}\frac{\partial v_r}{\partial \theta} - \frac{v_\theta}{r} \tag{V}$$

$$\{\nabla \mathbf{v}\}_{\theta\theta} = \frac{1}{r}\frac{\partial v_\theta}{\partial \theta} + \frac{v_r}{r} \tag{W}$$

$$\{\nabla \mathbf{v}\}_{\theta\phi} = \frac{1}{r}\frac{\partial v_\phi}{\partial \theta} \tag{X}$$

$$\{\nabla \mathbf{v}\}_{\phi r} = \frac{1}{r \sin \theta}\frac{\partial v_r}{\partial \phi} - \frac{v_\phi}{r} \tag{Y}$$

$$\{\nabla \mathbf{v}\}_{\phi\theta} = \frac{1}{r \sin \theta}\frac{\partial v_\theta}{\partial \phi} - \frac{v_\phi}{r} \cot \theta \tag{Z}$$

$$\{\nabla \mathbf{v}\}_{\phi\phi} = \frac{1}{r \sin \theta}\frac{\partial v_\phi}{\partial \phi} + \frac{v_r}{r} + \frac{v_\theta}{r} \cot \theta \tag{AA}$$

$$\{\mathbf{v} \cdot \nabla \boldsymbol{\tau}\}_{rr} = (\mathbf{v} \cdot \nabla)\tau_{rr} - \left(\frac{v_\theta}{r}\right)(\tau_{r\theta} + \tau_{\theta r}) - \left(\frac{v_\phi}{r}\right)(\tau_{r\phi} + \tau_{\phi r}) \tag{BB}$$

$$\{\mathbf{v} \cdot \nabla \boldsymbol{\tau}\}_{r\theta} = (\mathbf{v} \cdot \nabla)\tau_{r\theta} + \left(\frac{v_\theta}{r}\right)(\tau_{rr} - \tau_{\theta\theta}) - \left(\frac{v_\phi}{r}\right)(\tau_{\phi\theta} + \tau_{r\phi} \cot \theta) \tag{CC}$$

$$\{\mathbf{v} \cdot \nabla \boldsymbol{\tau}\}_{r\phi} = (\mathbf{v} \cdot \nabla)\tau_{r\phi} - \left(\frac{v_\theta}{r}\right)\tau_{\theta\phi} + \left(\frac{v_\phi}{r}\right)[(\tau_{rr} - \tau_{\phi\phi}) + \tau_{r\theta} \cot \theta] \tag{DD}$$

$$\{\mathbf{v} \cdot \nabla \boldsymbol{\tau}\}_{\theta r} = (\mathbf{v} \cdot \nabla)\tau_{\theta r} + \left(\frac{v_\theta}{r}\right)(\tau_{rr} - \tau_{\theta\theta}) - \left(\frac{v_\phi}{r}\right)(\tau_{\theta\phi} + \tau_{\phi r} \cot \theta) \tag{EE}$$

$$\{\mathbf{v} \cdot \nabla \boldsymbol{\tau}\}_{\theta\theta} = (\mathbf{v} \cdot \nabla)\tau_{\theta\theta} + \left(\frac{v_\theta}{r}\right)(\tau_{r\theta} + \tau_{\theta r}) - \left(\frac{v_\phi}{r}\right)(\tau_{\theta\phi} + \tau_{\phi\theta}) \cot \theta \tag{FF}$$

$$\{\mathbf{v} \cdot \nabla \boldsymbol{\tau}\}_{\theta\phi} = (\mathbf{v} \cdot \nabla)\tau_{\theta\phi} + \left(\frac{v_\theta}{r}\right)\tau_{r\phi} + \left(\frac{v_\phi}{r}\right)[\tau_{\theta r} + (\tau_{\theta\theta} - \tau_{\phi\phi}) \cot \theta] \tag{GG}$$

$$\{\mathbf{v} \cdot \nabla \boldsymbol{\tau}\}_{\phi r} = (\mathbf{v} \cdot \nabla)\tau_{\phi r} - \left(\frac{v_\theta}{r}\right)\tau_{\phi\theta} + \left(\frac{v_\phi}{r}\right)[(\tau_{rr} - \tau_{\phi\phi}) + \tau_{\theta r} \cot \theta] \tag{HH}$$

$$\{\mathbf{v} \cdot \nabla \boldsymbol{\tau}\}_{\phi\theta} = (\mathbf{v} \cdot \nabla)\tau_{\phi\theta} + \left(\frac{v_\theta}{r}\right)\tau_{\phi r} + \left(\frac{v_\phi}{r}\right)[\tau_{r\theta} + (\tau_{\theta\theta} - \tau_{\phi\phi}) \cot \theta] \tag{II}$$

$$\{\mathbf{v} \cdot \nabla \boldsymbol{\tau}\}_{\phi\phi} = (\mathbf{v} \cdot \nabla)\tau_{\phi\phi} + \left(\frac{v_\phi}{r}\right)[(\tau_{r\phi} + \tau_{\phi r}) + (\tau_{\theta\phi} + \tau_{\phi\theta}) \cot \theta] \tag{JJ}$$

where the operator $(\mathbf{v} \cdot \nabla) = v_r \frac{\partial}{\partial r} + \frac{v_\theta}{r}\frac{\partial}{\partial \theta} + \frac{v_\phi}{r \sin \theta}\frac{\partial}{\partial \phi}$

(b) Next we examine the dyadic product ∇v:

$$\nabla \mathbf{v} = \left\{\boldsymbol{\delta}_r \frac{\partial}{\partial r} + \boldsymbol{\delta}_\theta \frac{1}{r}\frac{\partial}{\partial \theta} + \boldsymbol{\delta}_z \frac{\partial}{\partial z}\right\}\{\boldsymbol{\delta}_r v_r + \boldsymbol{\delta}_\theta v_\theta + \boldsymbol{\delta}_z v_z\}$$

$$= \boldsymbol{\delta}_r\boldsymbol{\delta}_r \frac{\partial v_r}{\partial r} + \boldsymbol{\delta}_r\boldsymbol{\delta}_\theta \frac{\partial v_\theta}{\partial r} + \boldsymbol{\delta}_r\boldsymbol{\delta}_z \frac{\partial v_z}{\partial r} + \boldsymbol{\delta}_\theta\boldsymbol{\delta}_r \frac{1}{r}\frac{\partial v_r}{\partial \theta} + \boldsymbol{\delta}_\theta\boldsymbol{\delta}_\theta \frac{1}{r}\frac{\partial v_\theta}{\partial \theta} + \boldsymbol{\delta}_\theta\boldsymbol{\delta}_z \frac{1}{r}\frac{\partial v_z}{\partial \theta}$$

$$+ \boldsymbol{\delta}_\theta\boldsymbol{\delta}_\theta \frac{v_r}{r} - \boldsymbol{\delta}_\theta\boldsymbol{\delta}_r \frac{v_\theta}{r} + \boldsymbol{\delta}_z\boldsymbol{\delta}_r \frac{\partial v_r}{\partial z} + \boldsymbol{\delta}_z\boldsymbol{\delta}_\theta \frac{\partial v_\theta}{\partial z} + \boldsymbol{\delta}_z\boldsymbol{\delta}_z \frac{\partial v_z}{\partial z}$$

$$= \boldsymbol{\delta}_r\boldsymbol{\delta}_r \frac{\partial v_r}{\partial r} + \boldsymbol{\delta}_r\boldsymbol{\delta}_\theta \frac{\partial v_\theta}{\partial r} + \boldsymbol{\delta}_r\boldsymbol{\delta}_z \frac{\partial v_z}{\partial r} + \boldsymbol{\delta}_\theta\boldsymbol{\delta}_r\left(\frac{1}{r}\frac{\partial v_r}{\partial \theta} - \frac{v_\theta}{r}\right) + \boldsymbol{\delta}_\theta\boldsymbol{\delta}_\theta\left(\frac{1}{r}\frac{\partial v_\theta}{\partial \theta} + \frac{v_r}{r}\right)$$

$$+ \boldsymbol{\delta}_\theta\boldsymbol{\delta}_z \frac{1}{r}\frac{\partial v_z}{\partial \theta} + \boldsymbol{\delta}_z\boldsymbol{\delta}_r \frac{\partial v_r}{\partial z} + \boldsymbol{\delta}_z\boldsymbol{\delta}_\theta \frac{\partial v_\theta}{\partial z} + \boldsymbol{\delta}_z\boldsymbol{\delta}_z \frac{\partial v_z}{\partial z} \tag{A.7-23}$$

Hence, the rr-component is $\partial v_r/\partial r$, the $r\theta$-component is $\partial v_\theta/\partial r$, and so on, as given in Table A.7-2.

| EXAMPLE A.7-2 | Find the r-component of $[\nabla \cdot \boldsymbol{\tau}]$ in spherical coordinates. |

Differential Operations in Spherical Coordinates

SOLUTION

Using Eq. A.7-9 we have

$$[\nabla \cdot \boldsymbol{\tau}]_r = \left[\left\{\boldsymbol{\delta}_r \frac{\partial}{\partial r} + \boldsymbol{\delta}_\theta \frac{1}{r}\frac{\partial}{\partial \theta} + \boldsymbol{\delta}_\phi \frac{1}{r \sin \theta}\frac{\partial}{\partial \phi}\right\} \cdot \{\boldsymbol{\delta}_r\boldsymbol{\delta}_r\tau_{rr} + \boldsymbol{\delta}_r\boldsymbol{\delta}_\theta\tau_{r\theta} + \boldsymbol{\delta}_r\boldsymbol{\delta}_\phi\tau_{r\phi}\right.$$

$$\left.+ \boldsymbol{\delta}_\theta\boldsymbol{\delta}_r\tau_{\theta r} + \boldsymbol{\delta}_\theta\boldsymbol{\delta}_\theta\tau_{\theta\theta} + \boldsymbol{\delta}_\theta\boldsymbol{\delta}_\phi\tau_{\theta\phi} + \boldsymbol{\delta}_\phi\boldsymbol{\delta}_r\tau_{\phi r} + \boldsymbol{\delta}_\phi\boldsymbol{\delta}_\theta\tau_{\phi\theta} + \boldsymbol{\delta}_\phi\boldsymbol{\delta}_\phi\tau_{\phi\phi}\}\right] \tag{A.7-24}$$

We now use Eqs. A.7-6, 7, 8 and Eq. A.3-3. Since we want only the r-component, we select only those terms that contribute to the coefficient of $\boldsymbol{\delta}_r$:

$$\left[\boldsymbol{\delta}_r \frac{\partial}{\partial r} \cdot \boldsymbol{\delta}_r\boldsymbol{\delta}_r\tau_{rr}\right] = [\boldsymbol{\delta}_r \cdot \boldsymbol{\delta}_r\boldsymbol{\delta}_r]\frac{\partial \tau_{rr}}{\partial r} = \boldsymbol{\delta}_r \frac{\partial \tau_{rr}}{\partial r} \tag{A.7-25}$$

$$\left[\boldsymbol{\delta}_\theta \frac{1}{r}\frac{\partial}{\partial \theta} \cdot \boldsymbol{\delta}_\theta\boldsymbol{\delta}_r\tau_{\theta r}\right] = [\boldsymbol{\delta}_\theta \cdot \boldsymbol{\delta}_\theta\boldsymbol{\delta}_r]\frac{1}{r}\frac{\partial}{\partial \theta}\tau_{\theta r} + \text{other term} \tag{A.7-26}$$

$$\left[\boldsymbol{\delta}_\phi \frac{1}{r \sin \theta}\frac{\partial}{\partial \phi} \cdot \boldsymbol{\delta}_\phi\boldsymbol{\delta}_r\tau_{\phi r}\right] = [\boldsymbol{\delta}_\phi \cdot \boldsymbol{\delta}_\phi\boldsymbol{\delta}_r]\frac{1}{r \sin \phi}\frac{\partial}{\partial \phi}\tau_{\phi r} + \text{other term} \tag{A.7-27}$$

$$\left[\boldsymbol{\delta}_\theta \frac{1}{r}\frac{\partial}{\partial \theta} \cdot \boldsymbol{\delta}_r\boldsymbol{\delta}_r\tau_{rr}\right] = \frac{\tau_{rr}}{r}\left[\boldsymbol{\delta}_\theta \cdot \left\{\frac{\partial}{\partial \theta}\boldsymbol{\delta}_r\right\}\boldsymbol{\delta}_r\right] + \frac{\tau_{rr}}{r}\left[\boldsymbol{\delta}_\theta \cdot \boldsymbol{\delta}_r\left\{\frac{\partial}{\partial \theta}\boldsymbol{\delta}_r\right\}\right]$$

$$= \frac{\tau_{rr}}{r}[\boldsymbol{\delta}_\theta \cdot \boldsymbol{\delta}_\theta\boldsymbol{\delta}_r] = \boldsymbol{\delta}_r \frac{\tau_{rr}}{r} \tag{A.7-28}$$

$$\left[\boldsymbol{\delta}_\phi \frac{1}{r \sin \theta}\frac{\partial}{\partial \phi} \cdot \boldsymbol{\delta}_r\boldsymbol{\delta}_r\tau_{rr}\right] = \frac{\tau_{rr}}{r \sin \theta}\left[\boldsymbol{\delta}_\phi \cdot \left\{\frac{\partial}{\partial \phi}\boldsymbol{\delta}_r\right\}\boldsymbol{\delta}_r\right]$$

$$= \frac{\tau_{rr}}{r \sin \theta}[\boldsymbol{\delta}_\phi \cdot \boldsymbol{\delta}_\phi \sin \theta \, \boldsymbol{\delta}_r] = \boldsymbol{\delta}_r \frac{\tau_{rr}}{r} \tag{A.7-29}$$

$$\left[\boldsymbol{\delta}_\theta \frac{1}{r}\frac{\partial}{\partial \theta} \cdot \boldsymbol{\delta}_\theta\boldsymbol{\delta}_\theta\tau_{\theta\theta}\right] = \boldsymbol{\delta}_r\left(-\frac{\tau_{\theta\theta}}{r}\right) + \text{other term} \tag{A.7-30}$$

$$\left[\boldsymbol{\delta}_\phi \frac{1}{r \sin \theta}\frac{\partial}{\partial \phi} \cdot \boldsymbol{\delta}_\theta\boldsymbol{\delta}_r\tau_{\theta r}\right] = \boldsymbol{\delta}_r \frac{\tau_{\theta r} \cos \theta}{r \sin \theta} \tag{A.7-31}$$

$$\left[\boldsymbol{\delta}_\phi \frac{1}{r \sin \theta}\frac{\partial}{\partial \phi} \cdot \boldsymbol{\delta}_\phi\boldsymbol{\delta}_\phi\tau_{\phi\phi}\right] = \boldsymbol{\delta}_r\left(\frac{-\tau_{\phi\phi}}{r}\right) + \text{other terms} \tag{A.7-32}$$

Combining the above results we get

$$[\nabla \cdot \boldsymbol{\tau}]_r = \frac{1}{r^2} \frac{\partial}{\partial r} (r^2 \tau_{rr}) + \frac{\tau_{\theta r}}{r} \cot \theta + \frac{1}{r} \frac{\partial}{\partial \theta} \tau_{\theta r} + \frac{1}{r \sin \theta} \frac{\partial \tau_{\phi r}}{\partial \phi} - \frac{\tau_{\theta\theta} + \tau_{\phi\phi}}{r} \qquad \text{(A.7-33)}$$

Note that this expression is correct whether or not $\boldsymbol{\tau}$ is symmetric.

EXERCISES

1. If $\mathbf{r}$ is the instantaneous position vector for a particle, show that the velocity and acceleration of the particle are given by (use Eq. A.7-2):

$$\mathbf{v} = \frac{d}{dt} \mathbf{r} = \boldsymbol{\delta}_r \dot{r} + \boldsymbol{\delta}_\theta r \dot{\theta} + \boldsymbol{\delta}_z \dot{z} \qquad \text{(A.7-34)}$$

$$\mathbf{a} = \boldsymbol{\delta}_r (\ddot{r} - r\dot{\theta}^2) + \boldsymbol{\delta}_\theta (r\ddot{\theta} + 2\dot{r}\dot{\theta}) + \boldsymbol{\delta}_z \ddot{z} \qquad \text{(A.7-35)}$$

in cylindrical coordinates. The dots indicate time derivatives of the coordinates.

2. Obtain $(\nabla \cdot \mathbf{v})$, $[\nabla \times \mathbf{v}]$, and $\nabla \mathbf{v}$ in spherical coordinates, and $[\nabla \cdot \boldsymbol{\tau}]$ in cylindrical coordinates.

3. Use Table A.7-2 to write down directly the following quantities in cylindrical coordinates:
   (a) $(\nabla \cdot \rho \mathbf{v})$, where ρ is a scalar   (b) $[\nabla \cdot \rho \mathbf{v}\mathbf{v}]_r$, where ρ is a scalar
   (c) $[\nabla \cdot p\boldsymbol{\delta}]_\theta$, where p is a scalar   (d) $(\nabla \cdot [\boldsymbol{\tau} \cdot \mathbf{v}])$
   (e) $[\mathbf{v} \cdot \nabla \mathbf{v}]_\theta$ (f) $\nabla \mathbf{v} + (\nabla \mathbf{v})^\dagger$

4. Verify that the entries for $\nabla^2 v$ in Table A.7-2 can be obtained by any one of the following methods:
   (a) First verify that, in cylindrical coordinates the operator $(\nabla \cdot \nabla)$ is

$$(\nabla \cdot \nabla) = \frac{\partial^2}{\partial r^2} + \frac{1}{r} \frac{\partial}{\partial r} + \frac{1}{r^2} \frac{\partial^2}{\partial \theta^2} + \frac{\partial^2}{\partial z^2} \qquad \text{(A.7-36)}$$

   and then apply the operator to $\mathbf{v}$.
   (b) Use the expression for $[\nabla \cdot \boldsymbol{\tau}]$ in Table A.7-2, but substitute the components for ∇v in place of the components of $\boldsymbol{\tau}$, so as to obtain $[\nabla \cdot \nabla v]$.
   (c) Use Eq. A.4-22:

$$\nabla^2 \mathbf{v} = \nabla(\nabla \cdot \mathbf{v}) - [\nabla \times [\nabla \times \mathbf{v}]] \qquad \text{(A.7-37)}$$

   and use the gradient, divergence, and curl operations in Table A.7-2 to evaluate the operations on the right side.

§A.8   INTEGRAL OPERATIONS IN CURVILINEAR COORDINATES

In performing the integrations of §A.5 in curvilinear coordinates, it is important to understand the construction of the volume elements, as is shown for cylindrical coordinates in Fig. A.8-1 and for spherical coordinates in Fig. A.8-2.

In doing *volume integrals*, the simplest situations are those in which the bounding surfaces are surfaces of the coordinate system. For *cylindrical coordinates*, a typical volume integral of a function $f(r, \theta, z)$ would be of the form

$$\int_{z_1}^{z_2} \int_{\theta_1}^{\theta_2} \int_{r_1}^{r_2} f(r, \theta, z)r \, dr \, d\theta \, dz \qquad \text{(A.8-1)}$$

and for *spherical coordinates* a typical volume integral of a function $g(r, \theta, \phi)$ would be

$$\int_{\phi_1}^{\phi_2} \int_{\theta_1}^{\theta_2} \int_{r_1}^{r_2} g(r, \theta, \phi)r^2 \, dr \, \sin \theta \, d\theta \, d\phi \qquad \text{(A.8-2)}$$

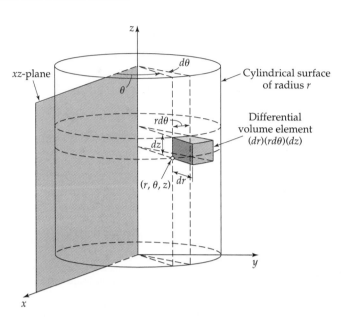

Fig. A.8-1. Differential volume element $r\,dr\,d\theta\,dz$ in cylindrical coordinates, and differential line elements dr, $r\,d\theta$, and dz. The differential surface elements are: $(r\,d\theta)(dz)$ perpendicular to the r direction (intermediate shading); $(dz)(dr)$ perpendicular to the θ direction (darkest shading); and $(dr)(r\,d\theta)$ perpendicular to the z direction (lightest shading).

Since the limits in these integrals ($r_1, r_2, \theta_1, \theta_2$, etc.) are constants, the order of the integration is immaterial.

In doing *surface integrals*, the simplest situations are those in which the integration is performed on one of the surfaces of the coordinate system. For *cylindrical coordinates* there are three possibilities:

On the surface $r = r_0$:
$$\int_{z_1}^{z_2} \int_{\theta_1}^{\theta_2} f(r_0, \theta, z) r_0 \, d\theta \, dz \tag{A.8-3}$$

On the surface $\theta = \theta_0$:
$$\int_{z_1}^{z_2} \int_{r_1}^{r_2} f(r, \theta_0, z) r \, dr \, dz \tag{A.8-4}$$

On the surface $z = z_0$:
$$\int_{\theta_1}^{\theta_2} \int_{r_1}^{r_2} f(r, \theta, z_0) r \, dr \, d\theta \tag{A.8-5}$$

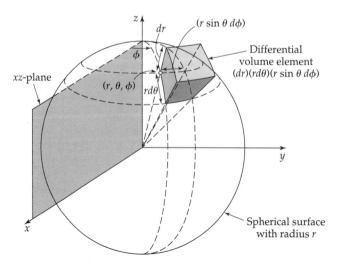

Fig. A.8-2. Differential volume element $r^2 \sin\theta\, dr\, d\theta\, d\phi$ in spherical coordinates, and the differential line elements dr, $r\,d\theta$, and $r\sin\theta\,d\phi$. The differential surface elements are: $(r\,d\theta)(r\sin\theta\,d\phi)$ perpendicular to the r direction (lightest shading); $(r\sin\theta\,d\phi)(dr)$ perpendicular to the θ direction (darkest shading); and $(dr)(r\,d\theta)$ perpendicular to the ϕ direction (intermediate shading).

Similarly, for *spherical coordinates*:

On the surface $r = r_0$:

$$\int_{\phi_1}^{\phi_2} \int_{\theta_1}^{\theta_2} g(r_0, \theta, \phi) r_0^2 \sin \theta \, d\theta \, d\phi \tag{A.8-6}$$

On the surface $\theta = \theta_0$:

$$\int_{\phi}^{\phi_2} \int_{r_1}^{r_2} g(r, \theta_0, \phi) \sin \theta_0 \, r^2 \, dr \, d\phi \tag{A.8-7}$$

On the surface $\phi = \phi_0$:

$$\int_{\theta_1}^{\theta_2} \int_{r_1}^{r_2} g(r, \theta, \phi_0) r^2 \, dr \, \sin \theta \, d\theta \tag{A.8-8}$$

The reader should try making sketches to show exactly what areas are described by each of the above six surface integrals.

If the area of integration in a surface integral is not one of the surfaces of the coordinate system, then a book on differential and integral calculus should be consulted.

§A.9   FURTHER COMMENTS ON VECTOR–TENSOR NOTATION

The boldface notation used in this book is called *Gibbs notation*.[1] Also widely used is another notation referred to as Cartesian tensor notation.[2] As shown in Table A.9-1, a few examples suffice to compare the two systems. The two outer columns are just two different ways of abbreviating the operations described explicitly in the middle column in Cartesian coordinates. The rules for converting from one system to another are as follows.

To convert from expanded notation to Cartesian tensor notation:

1. Omit all summation signs (the "Einstein summation convention")

2. Omit all unit vectors and unit dyads.

3. Replace $\partial / \partial x_i$ by ∂_i.

Table A.9-1

Gibbs notation	Expanded notation in terms of unit vectors and unit dyads	Cartesian tensor notation
$(\mathbf{v} \cdot \mathbf{w})$	$\sum_i v_i w_i$	$v_i w_i$
$[\mathbf{v} \times \mathbf{w}]$	$\sum_i \sum_j \sum_k \varepsilon_{ijk} \boldsymbol{\delta}_i v_j w_k$	$\varepsilon_{ijk} v_j w_k$
$[\nabla \cdot \boldsymbol{\tau}]$	$\sum_i \sum_j \boldsymbol{\delta}_i \dfrac{\partial}{\partial x_j} \tau_{ji}$	$\partial_j \tau_{ji}$
$\nabla^2 s$	$\sum_i \dfrac{\partial^2}{\partial x_i^2} s$	$\partial_i \partial_i s$
$[\nabla \times [\nabla \times \mathbf{v}]]$	$\sum_i \sum_j \sum_k \sum_m \sum_n \boldsymbol{\delta}_i \varepsilon_{ijk} \varepsilon_{kmn} \dfrac{\partial}{\partial x_j} \dfrac{\partial}{\partial x_m} v_n$	$\varepsilon_{ijk} \varepsilon_{kmn} \partial_j \partial_m v_n$
$\{\boldsymbol{\tau} \times \mathbf{v}\}$	$\sum_i \sum_j \sum_k \sum_l \varepsilon_{jkl} \boldsymbol{\delta}_i \boldsymbol{\delta}_l \tau_{ij} v_k$	$\varepsilon_{jkl} \tau_{ij} v_k$

[1] J. W. Gibbs, *Vector Analysis*, Dover Reprint, New York (1960).
[2] W. Prager, *Mechanics of Continua*, Ginn, Boston (1961).

To convert from Cartesian tensor notation to expanded notation:

1. Supply summation signs for all repeated indices.
2. Supply unit vectors and unit dyads for all nonrepeated indices; in each term of a tensor equation the unit vectors must appear in the same order in the unit dyads.
3. Replace ∂_i by $\partial / \partial x_i$.

The Gibbs notation is compact, easy to read, and devoid of any reference to a particular coordinate system; however, one has to know the meaning of the dot and cross operations and the use of boldface symbols. The Cartesian tensor notation indicates the nature of the operations explicitly in Cartesian coordinates, but errors in reading or writing subscripts can be most aggravating. People who know both systems equally well prefer the Gibbs notation for general discussions and for presenting results, but revert to Cartesian tensor notation for doing proofs of identities.

Occasionally *matrix notation* is used to display the components of vectors and tensors with respect to designated coordinate systems. For example, when $v_x = \dot{\gamma}y$, $v_y = 0$, $v_z = 0$, $\nabla \mathbf{v}$ can be written in two ways:

$$\nabla \mathbf{v} = \boldsymbol{\delta}_y \boldsymbol{\delta}_x \dot{\gamma} = \begin{pmatrix} 0 & 0 & 0 \\ \dot{\gamma} & 0 & 0 \\ 0 & 0 & 0 \end{pmatrix} \tag{A.9-1}$$

The second "=" is not really an "equals" sign, but has to be interpreted as "may be displayed as." Note that this notation is somewhat dangerous since one has to infer the unit dyads that are to be multiplied by the matrix elements—in this case, $\boldsymbol{\delta}_x \boldsymbol{\delta}_x$, $\boldsymbol{\delta}_x \boldsymbol{\delta}_y$, and so on. If we had used cylindrical coordinates, $\nabla \mathbf{v}$ would be represented by the matrix

$$\nabla \mathbf{v} = \begin{pmatrix} \dot{\gamma} \sin \theta \, \cos \theta & -\dot{\gamma} \sin^2 \theta & 0 \\ \dot{\gamma} \cos^2 \theta & -\dot{\gamma} \sin \theta \, \cos \theta & 0 \\ 0 & 0 & 0 \end{pmatrix} \tag{A.9-2}$$

where the matrix elements are understood to be multiplied by $\boldsymbol{\delta}_r \boldsymbol{\delta}_r$, $\boldsymbol{\delta}_r \boldsymbol{\delta}_\theta$, and so on, and then added together.

Despite the hazard of misinterpretation and the loose use of "=," the matrix notation enjoys widespread use, the main reason being that the "dot" operations correspond to standard matrix multiplication rules. For example,

$$(\mathbf{v} \cdot \mathbf{w}) = (v_1 \quad v_2 \quad v_3) \begin{pmatrix} w_1 \\ w_2 \\ w_3 \end{pmatrix} = v_1 w_1 + v_2 w_2 + v_3 w_3 \tag{A.9-3}$$

$$[\boldsymbol{\tau} \cdot \mathbf{v}] = \begin{pmatrix} \tau_{11} & \tau_{12} & \tau_{13} \\ \tau_{21} & \tau_{22} & \tau_{23} \\ \tau_{31} & \tau_{32} & \tau_{33} \end{pmatrix} \begin{pmatrix} v_1 \\ v_2 \\ v_3 \end{pmatrix} = \begin{pmatrix} \tau_{11}v_1 + \tau_{12}v_2 + \tau_{13}v_3 \\ \tau_{21}v_1 + \tau_{22}v_2 + \tau_{23}v_3 \\ \tau_{31}v_1 + \tau_{32}v_2 + \tau_{33}v_3 \end{pmatrix} \tag{A.9-4}$$

Of course such matrix multiplications are meaningful only when the components are referred to the same unit vectors.

Appendix **B**

The Fluxes and the Equations of Change

§B.1 NEWTON'S LAW OF VISCOSITY

$$[\tau = -\mu(\nabla \mathbf{v} + (\nabla \mathbf{v})^\dagger) + (\tfrac{2}{3}\mu - \kappa)(\nabla \cdot \mathbf{v})\delta]$$

Cartesian coordinates (x, y, z):

$$\tau_{xx} = -\mu\left[2\,\frac{\partial v_x}{\partial x}\right] + (\tfrac{2}{3}\mu - \kappa)(\nabla \cdot \mathbf{v}) \qquad \text{(B.1-1)}^a$$

$$\tau_{yy} = -\mu\left[2\,\frac{\partial v_y}{\partial y}\right] + (\tfrac{2}{3}\mu - \kappa)(\nabla \cdot \mathbf{v}) \qquad \text{(B.1-2)}^a$$

$$\tau_{zz} = -\mu\left[2\,\frac{\partial v_z}{\partial z}\right] + (\tfrac{2}{3}\mu - \kappa)(\nabla \cdot \mathbf{v}) \qquad \text{(B.1-3)}^a$$

$$\tau_{xy} = \tau_{yx} = -\mu\left[\frac{\partial v_y}{\partial x} + \frac{\partial v_x}{\partial y}\right] \qquad \text{(B.1-4)}$$

$$\tau_{yz} = \tau_{zy} = -\mu\left[\frac{\partial v_z}{\partial y} + \frac{\partial v_y}{\partial z}\right] \qquad \text{(B.1-5)}$$

$$\tau_{zx} = \tau_{xz} = -\mu\left[\frac{\partial v_x}{\partial z} + \frac{\partial v_z}{\partial x}\right] \qquad \text{(B.1-6)}$$

in which

$$(\nabla \cdot \mathbf{v}) = \frac{\partial v_x}{\partial x} + \frac{\partial v_y}{\partial y} + \frac{\partial v_z}{\partial z} \qquad \text{(B.1-7)}$$

---

[a] When the fluid is assumed to have constant density, the term containing $(\nabla \cdot \mathbf{v})$ may be omitted. For monatomic gases at low density, the dilatational viscosity κ is zero.

§B.1   NEWTON'S LAW OF VISCOSITY (continued)

Cylindrical coordinates (r, θ, z):

$$\tau_{rr} = -\mu\left[2\frac{\partial v_r}{\partial r}\right] + (\tfrac{2}{3}\mu - \kappa)(\nabla \cdot \mathbf{v}) \tag{B.1-8^a}$$

$$\tau_{\theta\theta} = -\mu\left[2\left(\frac{1}{r}\frac{\partial v_\theta}{\partial \theta} + \frac{v_r}{r}\right)\right] + (\tfrac{2}{3}\mu - \kappa)(\nabla \cdot \mathbf{v}) \tag{B.1-9^a}$$

$$\tau_{zz} = -\mu\left[2\frac{\partial v_z}{\partial z}\right] + (\tfrac{2}{3}\mu - \kappa)(\nabla \cdot \mathbf{v}) \tag{B.1-10^a}$$

$$\tau_{r\theta} = \tau_{\theta r} = -\mu\left[r\frac{\partial}{\partial r}\left(\frac{v_\theta}{r}\right) + \frac{1}{r}\frac{\partial v_r}{\partial \theta}\right] \tag{B.1-11}$$

$$\tau_{\theta z} = \tau_{z\theta} = -\mu\left[\frac{1}{r}\frac{\partial v_z}{\partial \theta} + \frac{\partial v_\theta}{\partial z}\right] \tag{B.1-12}$$

$$\tau_{zr} = \tau_{rz} = -\mu\left[\frac{\partial v_r}{\partial z} + \frac{\partial v_z}{\partial r}\right] \tag{B.1-13}$$

in which

$$(\nabla \cdot \mathbf{v}) = \frac{1}{r}\frac{\partial}{\partial r}(rv_r) + \frac{1}{r}\frac{\partial v_\theta}{\partial \theta} + \frac{\partial v_z}{\partial z} \tag{B.1-14}$$

[a] When the fluid is assumed to have constant density, the term containing $(\nabla \cdot \mathbf{v})$ may be omitted. For monatomic gases at low density, the dilatational viscosity κ is zero.

Spherical coordinates (r, θ, φ):

$$\tau_{rr} = -\mu\left[2\frac{\partial v_r}{\partial r}\right] + (\tfrac{2}{3}\mu - \kappa)(\nabla \cdot \mathbf{v}) \tag{B.1-15^a}$$

$$\tau_{\theta\theta} = -\mu\left[2\left(\frac{1}{r}\frac{\partial v_\theta}{\partial \theta} + \frac{v_r}{r}\right)\right] + (\tfrac{2}{3}\mu - \kappa)(\nabla \cdot \mathbf{v}) \tag{B.1-16^a}$$

$$\tau_{\phi\phi} = -\mu\left[2\left(\frac{1}{r \sin \theta}\frac{\partial v_\phi}{\partial \phi} + \frac{v_r + v_\theta \cot \theta}{r}\right)\right] + (\tfrac{2}{3}\mu - \kappa)(\nabla \cdot \mathbf{v}) \tag{B.1-17^a}$$

$$\tau_{r\theta} = \tau_{\theta r} = -\mu\left[r\frac{\partial}{\partial r}\left(\frac{v_\theta}{r}\right) + \frac{1}{r}\frac{\partial v_r}{\partial \theta}\right] \tag{B.1-18}$$

$$\tau_{\theta\phi} = \tau_{\phi\theta} = -\mu\left[\frac{\sin \theta}{r}\frac{\partial}{\partial \theta}\left(\frac{v_\phi}{\sin \theta}\right) + \frac{1}{r \sin \theta}\frac{\partial v_\theta}{\partial \phi}\right] \tag{B.1-19}$$

$$\tau_{\phi r} = \tau_{r\phi} = -\mu\left[\frac{1}{r \sin \theta}\frac{\partial v_r}{\partial \phi} + r\frac{\partial}{\partial r}\left(\frac{v_\phi}{r}\right)\right] \tag{B.1-20}$$

in which

$$(\nabla \cdot \mathbf{v}) = \frac{1}{r^2}\frac{\partial}{\partial r}(r^2 v_r) + \frac{1}{r \sin \theta}\frac{\partial}{\partial \theta}(v_\theta \sin \theta) + \frac{1}{r \sin \theta}\frac{\partial v_\phi}{\partial \phi} \tag{B.1-21}$$

[a] When the fluid is assumed to have constant density, the term containing $(\nabla \cdot \mathbf{v})$ may be omitted. For monatomic gases at low density, the dilatational viscosity κ is zero.

§B.2 FOURIER'S LAW OF HEAT CONDUCTION[a]

$$[\mathbf{q} = -k\nabla T]$$

Cartesian coordinates (x, y, z):

$$q_x = -k\frac{\partial T}{\partial x} \tag{B.2-1}$$

$$q_y = -k\frac{\partial T}{\partial y} \tag{B.2-2}$$

$$q_z = -k\frac{\partial T}{\partial z} \tag{B.2-3}$$

Cylindrical coordinates (r, θ, z):

$$q_r = -k\frac{dT}{\partial r} \tag{B.2-4}$$

$$q_\theta = -k\frac{1}{r}\frac{\partial T}{\partial \theta} \tag{B.2-5}$$

$$q_z = -k\frac{\partial T}{\partial z} \tag{B.2-6}$$

Spherical coordinates (r, θ, φ):

$$q_r = -k\frac{\partial T}{\partial r} \tag{B.2-7}$$

$$q_\theta = -k\frac{1}{r}\frac{\partial T}{\partial \theta} \tag{B.2-8}$$

$$q_\phi = -k\frac{1}{r\sin\theta}\frac{\partial T}{\partial \phi} \tag{B.2-9}$$

[a] For mixtures, the term $\Sigma_\alpha(\bar{H}_\alpha/M_\alpha)\mathbf{j}_\alpha$ must be added to $\mathbf{q}$ (see Eq. 19.3-3).

§B.3 FICK'S (FIRST) LAW OF BINARY DIFFUSION[a]

$$[\mathbf{j}_A = -\rho\mathcal{D}_{AB}\nabla\omega_A]$$

Cartesian coordinates (x, y, z):

$$j_{Ax} = -\rho\mathcal{D}_{AB}\frac{\partial\omega_A}{\partial x} \tag{B.3-1}$$

$$j_{Ay} = -\rho\mathcal{D}_{AB}\frac{\partial\omega_A}{\partial y} \tag{B.3-2}$$

$$j_{Az} = -\rho\mathcal{D}_{AB}\frac{\partial\omega_A}{\partial z} \tag{B.3-3}$$

Cylindrical coordinates (r, θ, z):

$$j_{Ar} = -\rho\mathcal{D}_{AB}\frac{\partial\omega_A}{\partial r} \tag{B.3-4}$$

$$j_{A\theta} = -\rho\mathcal{D}_{AB}\frac{1}{r}\frac{\partial\omega_A}{\partial\theta} \tag{B.3-5}$$

$$j_{Az} = -\rho\mathcal{D}_{AB}\frac{\partial\omega_A}{\partial z} \tag{B.3-6}$$

Spherical coordinates (r, θ, ϕ):

$$j_{Ar} = -\rho\mathcal{D}_{AB}\frac{\partial\omega_A}{\partial r} \tag{B.3-7}$$

$$j_{A\theta} = -\rho\mathcal{D}_{AB}\frac{1}{r}\frac{\partial\omega_A}{\partial\theta} \tag{B.3-8}$$

$$j_{A\phi} = -\rho\mathcal{D}_{AB}\frac{1}{r\sin\theta}\frac{\partial\omega_A}{\partial\phi} \tag{B.3-9}$$

[a] To get the molar fluxes with respect to the molar average velocity, replace $\mathbf{j}_A$, ρ, and ω_A by $\mathbf{J}_A^*$, c, and x_A.

§B.4 THE EQUATION OF CONTINUITY[a]

$$[\partial\rho/\partial t + (\nabla\cdot\rho\mathbf{v}) = 0]$$

Cartesian coordinates (x, y, z):

$$\frac{\partial\rho}{\partial t} + \frac{\partial}{\partial x}(\rho v_x) + \frac{\partial}{\partial y}(\rho v_y) + \frac{\partial}{\partial z}(\rho v_z) = 0 \tag{B.4-1}$$

Cylindrical coordinates (r, θ, z):

$$\frac{\partial\rho}{\partial t} + \frac{1}{r}\frac{\partial}{\partial r}(\rho r v_r) + \frac{1}{r}\frac{\partial}{\partial\theta}(\rho v_\theta) + \frac{\partial}{\partial z}(\rho v_z) = 0 \tag{B.4-2}$$

Spherical coordinates (r, θ, ϕ):

$$\frac{\partial\rho}{\partial t} + \frac{1}{r^2}\frac{\partial}{\partial r}(\rho r^2 v_r) + \frac{1}{r\sin\theta}\frac{\partial}{\partial\theta}(\rho v_\theta \sin\theta) + \frac{1}{r\sin\theta}\frac{\partial}{\partial\phi}(\rho v_\phi) = 0 \tag{B.4-3}$$

[a] When the fluid is assumed to have constant mass density ρ, the equation simplifies to $(\nabla\cdot\mathbf{v}) = 0$.

§B.5 THE EQUATION OF MOTION IN TERMS OF τ

$$[\rho D\mathbf{v}/Dt = -\nabla p - [\nabla \cdot \boldsymbol{\tau}] + \rho \mathbf{g}]$$

Cartesian coordinates (x, y, z):[a]

$$\rho\left(\frac{\partial v_x}{\partial t} + v_x \frac{\partial v_x}{\partial x} + v_y \frac{\partial v_x}{\partial y} + v_z \frac{\partial v_x}{\partial z}\right) = -\frac{\partial p}{\partial x} - \left[\frac{\partial}{\partial x}\tau_{xx} + \frac{\partial}{\partial y}\tau_{yx} + \frac{\partial}{\partial z}\tau_{zx}\right] + \rho g_x \quad \text{(B.5-1)}$$

$$\rho\left(\frac{\partial v_y}{\partial t} + v_x \frac{\partial v_y}{\partial x} + v_y \frac{\partial v_y}{\partial y} + v_z \frac{\partial v_y}{\partial z}\right) = -\frac{\partial p}{\partial y} - \left[\frac{\partial}{\partial x}\tau_{xy} + \frac{\partial}{\partial y}\tau_{yy} + \frac{\partial}{\partial z}\tau_{zy}\right] + \rho g_y \quad \text{(B.5-2)}$$

$$\rho\left(\frac{\partial v_z}{\partial t} + v_x \frac{\partial v_z}{\partial x} + v_y \frac{\partial v_z}{\partial y} + v_z \frac{\partial v_z}{\partial z}\right) = -\frac{\partial p}{\partial z} - \left[\frac{\partial}{\partial x}\tau_{xz} + \frac{\partial}{\partial y}\tau_{yz} + \frac{\partial}{\partial z}\tau_{zz}\right] + \rho g_z \quad \text{(B.5-3)}$$

[a] These equations have been written without making the assumption that $\boldsymbol{\tau}$ is symmetric. This means, for example, that when the usual assumption is made that the stress tensor is symmetric, τ_{xy} and τ_{yx} may be interchanged.

Cylindrical coordinates (r, θ, z):[b]

$$\rho\left(\frac{\partial v_r}{\partial t} + v_r \frac{\partial v_r}{\partial r} + \frac{v_\theta}{r}\frac{\partial v_r}{\partial \theta} + v_z \frac{\partial v_r}{\partial z} - \frac{v_\theta^2}{r}\right) = -\frac{\partial p}{\partial r} - \left[\frac{1}{r}\frac{\partial}{\partial r}(r\tau_{rr}) + \frac{1}{r}\frac{\partial}{\partial \theta}\tau_{\theta r} + \frac{\partial}{\partial z}\tau_{zr} - \frac{\tau_{\theta\theta}}{r}\right] + \rho g_r \quad \text{(B.5-4)}$$

$$\rho\left(\frac{\partial v_\theta}{\partial t} + v_r \frac{\partial v_\theta}{\partial r} + \frac{v_\theta}{r}\frac{\partial v_\theta}{\partial \theta} + v_z \frac{\partial v_\theta}{\partial z} + \frac{v_r v_\theta}{r}\right) = -\frac{1}{r}\frac{\partial p}{\partial \theta} - \left[\frac{1}{r^2}\frac{\partial}{\partial r}(r^2\tau_{r\theta}) + \frac{1}{r}\frac{\partial}{\partial \theta}\tau_{\theta\theta} + \frac{\partial}{\partial z}\tau_{z\theta} + \frac{\tau_{\theta r} - \tau_{r\theta}}{r}\right] + \rho g_\theta \quad \text{(B.5-5)}$$

$$\rho\left(\frac{\partial v_z}{\partial t} + v_r \frac{\partial v_z}{\partial r} + \frac{v_\theta}{r}\frac{\partial v_z}{\partial \theta} + v_z \frac{\partial v_z}{\partial z}\right) = -\frac{\partial p}{\partial z} - \left[\frac{1}{r}\frac{\partial}{\partial r}(r\tau_{rz}) + \frac{1}{r}\frac{\partial}{\partial \theta}\tau_{\theta z} + \frac{\partial}{\partial z}\tau_{zz}\right] + \rho g_z \quad \text{(B.5-6)}$$

[b] These equations have been written without making the assumption that $\boldsymbol{\tau}$ is symmetric. This means, for example, that when the usual assumption is made that the stress tensor is symmetric, $\tau_{r\theta} - \tau_{\theta r} = 0$.

Spherical coordinates (r, θ, ϕ):[c]

$$\rho\left(\frac{\partial v_r}{\partial t} + v_r \frac{\partial v_r}{\partial r} + \frac{v_\theta}{r}\frac{\partial v_r}{\partial \theta} + \frac{v_\phi}{r\sin\theta}\frac{\partial v_r}{\partial \phi} - \frac{v_\theta^2 + v_\phi^2}{r}\right) = -\frac{\partial p}{\partial r}$$
$$- \left[\frac{1}{r^2}\frac{\partial}{\partial r}(r^2\tau_{rr}) + \frac{1}{r\sin\theta}\frac{\partial}{\partial \theta}(\tau_{\theta r}\sin\theta) + \frac{1}{r\sin\theta}\frac{\partial}{\partial \phi}\tau_{\phi r} - \frac{\tau_{\theta\theta} + \tau_{\phi\phi}}{r}\right] + \rho g_r \quad \text{(B.5-7)}$$

$$\rho\left(\frac{\partial v_\theta}{\partial t} + v_r \frac{\partial v_\theta}{\partial r} + \frac{v_\theta}{r}\frac{\partial v_\theta}{\partial \theta} + \frac{v_\phi}{r\sin\theta}\frac{\partial v_\theta}{\partial \phi} + \frac{v_r v_\theta - v_\phi^2\cot\theta}{r}\right) = -\frac{1}{r}\frac{\partial p}{\partial \theta}$$
$$- \left[\frac{1}{r^3}\frac{\partial}{\partial r}(r^3\tau_{r\theta}) + \frac{1}{r\sin\theta}\frac{\partial}{\partial \theta}(\tau_{\theta\theta}\sin\theta) + \frac{1}{r\sin\theta}\frac{\partial}{\partial \phi}\tau_{\phi\theta} + \frac{(\tau_{\theta r} - \tau_{r\theta}) - \tau_{\phi\phi}\cot\theta}{r}\right] + \rho g_\theta \quad \text{(B.5-8)}$$

$$\rho\left(\frac{\partial v_\phi}{\partial t} + v_r \frac{\partial v_\phi}{\partial r} + \frac{v_\theta}{r}\frac{\partial v_\phi}{\partial \theta} + \frac{v_\phi}{r\sin\theta}\frac{\partial v_\phi}{\partial \phi} + \frac{v_\phi v_r + v_\theta v_\phi\cot\theta}{r}\right) = -\frac{1}{r\sin\theta}\frac{\partial p}{\partial \phi}$$
$$- \left[\frac{1}{r^3}\frac{\partial}{\partial r}(r^3\tau_{r\phi}) + \frac{1}{r\sin\theta}\frac{\partial}{\partial \theta}(\tau_{\theta\phi}\sin\theta) + \frac{1}{r\sin\theta}\frac{\partial}{\partial \phi}\tau_{\phi\phi} + \frac{(\tau_{\phi r} - \tau_{r\phi}) + \tau_{\phi\theta}\cot\theta}{r}\right] + \rho g_\phi \quad \text{(B.5-9)}$$

[c] These equations have been written without making the assumption that $\boldsymbol{\tau}$ is symmetric. This means, for example, that when the usual assumption is made that the stress tensor is symmetric, $\tau_{r\theta} - \tau_{\theta r} = 0$.

§B.6 EQUATION OF MOTION FOR A NEWTONIAN FLUID WITH CONSTANT ρ AND μ

$$[\rho D\mathbf{v}/Dt = -\nabla p + \mu\nabla^2\mathbf{v} + \rho\mathbf{g}]$$

Cartesian coordinates (x, y, z):

$$\rho\left(\frac{\partial v_x}{\partial t} + v_x\frac{\partial v_x}{\partial x} + v_y\frac{\partial v_x}{\partial y} + v_z\frac{\partial v_x}{\partial z}\right) = -\frac{\partial p}{\partial x} + \mu\left[\frac{\partial^2 v_x}{\partial x^2} + \frac{\partial^2 v_x}{\partial y^2} + \frac{\partial^2 v_x}{\partial z^2}\right] + \rho g_x \quad \text{(B.6-1)}$$

$$\rho\left(\frac{\partial v_y}{\partial t} + v_x\frac{\partial v_y}{\partial x} + v_y\frac{\partial v_y}{\partial y} + v_z\frac{\partial v_y}{\partial z}\right) = -\frac{\partial p}{\partial y} + \mu\left[\frac{\partial^2 v_y}{\partial x^2} + \frac{\partial^2 v_y}{\partial y^2} + \frac{\partial^2 v_y}{\partial z^2}\right] + \rho g_y \quad \text{(B.6-2)}$$

$$\rho\left(\frac{\partial v_z}{\partial t} + v_x\frac{\partial v_z}{\partial x} + v_y\frac{\partial v_z}{\partial y} + v_z\frac{\partial v_z}{\partial z}\right) = -\frac{\partial p}{\partial z} + \mu\left[\frac{\partial^2 v_z}{\partial x^2} + \frac{\partial^2 v_z}{\partial y^2} + \frac{\partial^2 v_z}{\partial z^2}\right] + \rho g_z \quad \text{(B.6-3)}$$

Cylindrical coordinates (r, θ, z):

$$\rho\left(\frac{\partial v_r}{\partial t} + v_r\frac{\partial v_r}{\partial r} + \frac{v_\theta}{r}\frac{\partial v_r}{\partial \theta} + v_z\frac{\partial v_r}{\partial z} - \frac{v_\theta^2}{r}\right) = -\frac{\partial p}{\partial r} + \mu\left[\frac{\partial}{\partial r}\left(\frac{1}{r}\frac{\partial}{\partial r}(rv_r)\right) + \frac{1}{r^2}\frac{\partial^2 v_r}{\partial \theta^2} + \frac{\partial^2 v_r}{\partial z^2} - \frac{2}{r^2}\frac{\partial v_\theta}{\partial \theta}\right] + \rho g_r \quad \text{(B.6-4)}$$

$$\rho\left(\frac{\partial v_\theta}{\partial t} + v_r\frac{\partial v_\theta}{\partial r} + \frac{v_\theta}{r}\frac{\partial v_\theta}{\partial \theta} + v_z\frac{\partial v_\theta}{\partial z} + \frac{v_r v_\theta}{r}\right) = -\frac{1}{r}\frac{\partial p}{\partial \theta} + \mu\left[\frac{\partial}{\partial r}\left(\frac{1}{r}\frac{\partial}{\partial r}(rv_\theta)\right) + \frac{1}{r^2}\frac{\partial^2 v_\theta}{\partial \theta^2} + \frac{\partial^2 v_\theta}{\partial z^2} + \frac{2}{r^2}\frac{\partial v_r}{\partial \theta}\right] + \rho g_\theta \quad \text{(B.6-5)}$$

$$\rho\left(\frac{\partial v_z}{\partial t} + v_r\frac{\partial v_z}{\partial r} + \frac{v_\theta}{r}\frac{\partial v_z}{\partial \theta} + v_z\frac{\partial v_z}{\partial z}\right) = -\frac{\partial p}{\partial z} + \mu\left[\frac{1}{r}\frac{\partial}{\partial r}\left(r\frac{\partial v_z}{\partial r}\right) + \frac{1}{r^2}\frac{\partial^2 v_z}{\partial \theta^2} + \frac{\partial^2 v_z}{\partial z^2}\right] + \rho g_z \quad \text{(B.6-6)}$$

Spherical coordinates (r, θ, ϕ):

$$\rho\left(\frac{\partial v_r}{\partial t} + v_r\frac{\partial v_r}{\partial r} + \frac{v_\theta}{r}\frac{\partial v_r}{\partial \theta} + \frac{v_\phi}{r\sin\theta}\frac{\partial v_r}{\partial \phi} - \frac{v_\theta^2 + v_\phi^2}{r}\right) = -\frac{\partial p}{\partial r}$$
$$+ \mu\left[\frac{1}{r^2}\frac{\partial^2}{\partial r^2}(r^2 v_r) + \frac{1}{r^2\sin\theta}\frac{\partial}{\partial \theta}\left(\sin\theta\frac{\partial v_r}{\partial \theta}\right) + \frac{1}{r^2\sin^2\theta}\frac{\partial^2 v_r}{\partial \phi^2}\right] + \rho g_r \quad \text{(B.6-7)}^a$$

$$\rho\left(\frac{\partial v_\theta}{\partial t} + v_r\frac{\partial v_\theta}{\partial r} + \frac{v_\theta}{r}\frac{\partial v_\theta}{\partial \theta} + \frac{v_\phi}{r\sin\theta}\frac{\partial v_\theta}{\partial \phi} + \frac{v_r v_\theta - v_\phi^2\cot\theta}{r}\right) = -\frac{1}{r}\frac{\partial p}{\partial \theta}$$
$$+ \mu\left[\frac{1}{r^2}\frac{\partial}{\partial r}\left(r^2\frac{\partial v_\theta}{\partial r}\right) + \frac{1}{r^2}\frac{\partial}{\partial \theta}\left(\frac{1}{\sin\theta}\frac{\partial}{\partial \theta}(v_\theta\sin\theta)\right) + \frac{1}{r^2\sin^2\theta}\frac{\partial^2 v_\theta}{\partial \phi^2} + \frac{2}{r^2}\frac{\partial v_r}{\partial \theta} - \frac{2\cot\theta}{r^2\sin\theta}\frac{\partial v_\phi}{\partial \phi}\right] + \rho g_\theta \quad \text{(B.6-8)}$$

$$\rho\left(\frac{\partial v_\phi}{\partial t} + v_r\frac{\partial v_\phi}{\partial r} + \frac{v_\theta}{r}\frac{\partial v_\phi}{\partial \theta} + \frac{v_\phi}{r\sin\theta}\frac{\partial v_\phi}{\partial \phi} + \frac{v_\phi v_r + v_\theta v_\phi\cot\theta}{r}\right) = -\frac{1}{r\sin\theta}\frac{\partial p}{\partial \phi}$$
$$+ \mu\left[\frac{1}{r^2}\frac{\partial}{\partial r}\left(r^2\frac{\partial v_\phi}{\partial r}\right) + \frac{1}{r^2}\frac{\partial}{\partial \theta}\left(\frac{1}{\sin\theta}\frac{\partial}{\partial \theta}(v_\phi\sin\theta)\right) + \frac{1}{r^2\sin^2\theta}\frac{\partial^2 v_\phi}{\partial \phi^2} + \frac{2}{r^2\sin\theta}\frac{\partial v_r}{\partial \phi} + \frac{2\cot\theta}{r^2\sin\theta}\frac{\partial v_\theta}{\partial \phi}\right] + \rho g_\phi \quad \text{(B.6-9)}$$

---

[a] The quantity in the brackets in Eq. B.6-7 is *not* what one would expect from Eq. (M) for $[\nabla \cdot \nabla\mathbf{v}]$ in Table A.7-3, because we have added to Eq. (M) the expression for $(2/r)(\nabla \cdot \mathbf{v})$, which is zero for fluids with constant ρ. This gives a much simpler equation.

§B.7   THE DISSIPATION FUNCTION Φv FOR NEWTONIAN FLUIDS (SEE EQ. 3.3-3)

Cartesian coordinates (x, y, z):

$$\Phi_v = 2\left[\left(\frac{\partial v_x}{\partial x}\right)^2 + \left(\frac{\partial v_y}{\partial y}\right)^2 + \left(\frac{\partial v_z}{\partial z}\right)^2\right] + \left[\frac{\partial v_y}{\partial x} + \frac{\partial v_x}{\partial y}\right]^2 + \left[\frac{\partial v_z}{\partial y} + \frac{\partial v_y}{\partial z}\right]^2 + \left[\frac{\partial v_x}{\partial z} + \frac{\partial v_z}{\partial x}\right]^2 - \frac{2}{3}\left[\frac{\partial v_x}{\partial x} + \frac{\partial v_y}{\partial y} + \frac{\partial v_z}{\partial z}\right]^2 \quad (B.7\text{-}1)$$

Cylindrical coordinates (r, θ, z):

$$\Phi_v = 2\left[\left(\frac{\partial v_r}{\partial r}\right)^2 + \left(\frac{1}{r}\frac{\partial v_\theta}{\partial \theta} + \frac{v_r}{r}\right)^2 + \left(\frac{\partial v_z}{\partial z}\right)^2\right] + \left[r\frac{\partial}{\partial r}\left(\frac{v_\theta}{r}\right) + \frac{1}{r}\frac{\partial v_r}{\partial \theta}\right]^2 + \left[\frac{1}{r}\frac{\partial v_z}{\partial \theta} + \frac{\partial v_\theta}{\partial z}\right]^2 + \left[\frac{\partial v_r}{\partial z} + \frac{\partial v_z}{\partial r}\right]^2$$

$$-\frac{2}{3}\left[\frac{1}{r}\frac{\partial}{\partial r}(rv_r) + \frac{1}{r}\frac{\partial v_\theta}{\partial \theta} + \frac{\partial v_z}{\partial z}\right]^2 \quad (B.7\text{-}2)$$

Spherical coordinates (r, θ, φ):

$$\Phi_v = 2\left[\left(\frac{\partial v_r}{\partial r}\right)^2 + \left(\frac{1}{r}\frac{\partial v_\theta}{\partial \theta} + \frac{v_r}{r}\right)^2 + \left(\frac{1}{r\sin\theta}\frac{\partial v_\phi}{\partial \phi} + \frac{v_r + v_\theta\cot\theta}{r}\right)^2\right]$$

$$+ \left[r\frac{\partial}{\partial r}\left(\frac{v_\theta}{r}\right) + \frac{1}{r}\frac{\partial v_r}{\partial \theta}\right]^2 + \left[\frac{\sin\theta}{r}\frac{\partial}{\partial \theta}\left(\frac{v_\phi}{\sin\theta}\right) + \frac{1}{r\sin\theta}\frac{\partial v_\theta}{\partial \phi}\right]^2 + \left[\frac{1}{r\sin\theta}\frac{\partial v_r}{\partial \phi} + r\frac{\partial}{\partial r}\left(\frac{v_\phi}{r}\right)\right]^2$$

$$-\frac{2}{3}\left[\frac{1}{r^2}\frac{\partial}{\partial r}(r^2 v_r) + \frac{1}{r\sin\theta}\frac{\partial}{\partial \theta}(v_\theta\sin\theta) + \frac{1}{r\sin\theta}\frac{\partial v_\phi}{\partial \phi}\right]^2 \quad (B.7\text{-}3)$$

§B.8   THE EQUATION OF ENERGY IN TERMS OF q

$$[\rho\hat{C}_p DT/Dt = -(\nabla \cdot \mathbf{q}) - (\partial\ln\rho/\partial\ln T)_p Dp/Dt - (\boldsymbol{\tau}:\nabla\mathbf{v})]$$

Cartesian coordinates (x, y, z):

$$\rho\hat{C}_p\left(\frac{\partial T}{\partial t} + v_x\frac{\partial T}{\partial x} + v_y\frac{\partial T}{\partial y} + v_z\frac{\partial T}{\partial z}\right) = -\left[\frac{\partial q_x}{\partial x} + \frac{\partial q_y}{\partial y} + \frac{\partial q_z}{\partial z}\right] - \left(\frac{\partial\ln\rho}{\partial\ln T}\right)_p\frac{Dp}{Dt} - (\boldsymbol{\tau}:\nabla\mathbf{v}) \quad (B.8\text{-}1)^a$$

Cylindrical coordinates (r, θ, z):

$$\rho\hat{C}_p\left(\frac{\partial T}{\partial t} + v_r\frac{\partial T}{\partial r} + \frac{v_\theta}{r}\frac{\partial T}{\partial \theta} + v_z\frac{\partial T}{\partial z}\right) = -\left[\frac{1}{r}\frac{\partial}{\partial r}(rq_r) + \frac{1}{r}\frac{\partial q_\theta}{\partial \theta} + \frac{\partial q_z}{\partial z}\right] - \left(\frac{\partial\ln\rho}{\partial\ln T}\right)_p\frac{Dp}{Dt} - (\boldsymbol{\tau}:\nabla\mathbf{v}) \quad (B.8\text{-}2)^a$$

Spherical coordinates (r, θ, φ):

$$\rho\hat{C}_p\left(\frac{\partial T}{\partial t} + v_r\frac{\partial T}{\partial r} + \frac{v_\theta}{r}\frac{\partial T}{\partial \theta} + \frac{v_\phi}{r\sin\theta}\frac{\partial T}{\partial \phi}\right) = \left[\frac{1}{r^2}\frac{\partial}{\partial r}(r^2 q_r) + \frac{1}{r\sin\theta}\frac{\partial}{\partial \theta}(q_\theta\sin\theta) + \frac{1}{r\sin\theta}\frac{\partial q_\phi}{\partial \phi}\right] - \left(\frac{\partial\ln\rho}{\partial\ln T}\right)_p\frac{Dp}{Dt} - (\boldsymbol{\tau}:\nabla\mathbf{v})$$

$$(B.8\text{-}3)^a$$

[a] The viscous dissipation term, $-(\boldsymbol{\tau}:\nabla\mathbf{v})$, is given in Appendix A, Tables A.7-1, 2, 3. This term may usually be neglected, except for systems with very large velocity gradients. The term containing $(\partial\ln\rho/\partial\ln T)_p$ is zero for fluid with constant ρ.

§B.9   THE EQUATION OF ENERGY FOR PURE NEWTONIAN FLUIDS WITH CONSTANT[a] ρ AND k

$$[\rho \hat{C}_p DT/Dt = k\nabla^2 T + \mu\Phi_v]$$

Cartesian coordinates (x, y, z):

$$\rho \hat{C}_p \left(\frac{\partial T}{\partial t} + v_x \frac{\partial T}{\partial x} + v_y \frac{\partial T}{\partial y} + v_z \frac{\partial T}{\partial z} \right) = k \left[\frac{\partial^2 T}{\partial x^2} + \frac{\partial^2 T}{\partial y^2} + \frac{\partial^2 T}{\partial z^2} \right] + \mu\Phi_v \qquad (B.9\text{-}1)^b$$

Cylindrical coordinates (r, θ, z):

$$\rho \hat{C}_p \left(\frac{\partial T}{\partial t} + v_r \frac{\partial T}{\partial r} + \frac{v_\theta}{r} \frac{\partial T}{\partial \theta} + v_z \frac{\partial T}{\partial z} \right) = k \left[\frac{1}{r} \frac{\partial}{\partial r} \left(r \frac{\partial T}{\partial r} \right) + \frac{1}{r^2} \frac{\partial^2 T}{\partial \theta^2} + \frac{\partial^2 T}{\partial z^2} \right] + \mu\Phi_v \qquad (B.9\text{-}2)^b$$

Spherical coordinates (r, θ, ϕ):

$$\rho \hat{C}_p \left(\frac{\partial T}{\partial t} + v_r \frac{\partial T}{\partial r} + \frac{v_\theta}{r} \frac{\partial T}{\partial \theta} + \frac{v_\phi}{r \sin\theta} \frac{\partial T}{\partial \phi} \right) = k \left[\frac{1}{r^2} \frac{\partial}{\partial r} \left(r^2 \frac{\partial T}{\partial r} \right) + \frac{1}{r^2 \sin\theta} \frac{\partial}{\partial \theta} \left(\sin\theta \frac{\partial T}{\partial \theta} \right) + \frac{1}{r^2 \sin^2\theta} \frac{\partial^2 T}{\partial \phi^2} \right] + \mu\Phi_v \qquad (B.9\text{-}3)^b$$

[a] This form of the energy equation is also valid under the less stringent assumptions k = constant and $(\partial \ln \rho / \partial \ln T)_p Dp/Dt = 0$. The assumption ρ = constant is given in the table heading because it is the assumption more often made.

[b] The function Φ_v is given in §B.7. The term $\mu\Phi_v$ is usually negligible, except in systems with large velocity gradients.

§B.10   THE EQUATION OF CONTINUITY FOR SPECIES α IN TERMS[a] OF j_a

$$[\rho D\omega_\alpha/Dt = -(\nabla \cdot j_\alpha) + r_\alpha]$$

Cartesian coordinates (x, y, z):

$$\rho \left(\frac{\partial \omega_\alpha}{\partial t} + v_x \frac{\partial \omega_\alpha}{\partial x} + v_y \frac{\partial \omega_\alpha}{\partial y} + v_z \frac{\partial \omega_\alpha}{\partial z} \right) = - \left[\frac{\partial j_{\alpha x}}{\partial x} + \frac{\partial j_{\alpha y}}{\partial y} + \frac{\partial j_{\alpha z}}{\partial z} \right] + r_\alpha \qquad (B.10\text{-}1)$$

Cylindrical coordinates (r, θ, z):

$$\rho \left(\frac{\partial \omega_\alpha}{\partial t} + v_r \frac{\partial \omega_\alpha}{\partial r} + \frac{v_\theta}{r} \frac{\partial \omega_\alpha}{\partial \theta} + v_z \frac{\partial \omega_\alpha}{\partial z} \right) = - \left[\frac{1}{r} \frac{\partial}{\partial r} (r j_{\alpha r}) + \frac{1}{r} \frac{\partial j_{\alpha\theta}}{\partial \theta} + \frac{\partial j_{\alpha z}}{\partial z} \right] + r_\alpha \qquad (B.10\text{-}2)$$

Spherical coordinates (r, θ, ϕ):

$$\rho \left(\frac{\partial \omega_\alpha}{\partial t} + v_r \frac{\partial \omega_\alpha}{\partial r} + \frac{v_\theta}{r} \frac{\partial \omega_\alpha}{\partial \theta} + \frac{v_\phi}{r \sin\theta} \frac{\partial \omega_\alpha}{\partial \phi} \right) = \left[\frac{1}{r^2} \frac{\partial}{\partial r} (r^2 j_{\alpha r}) + \frac{1}{r \sin\theta} \frac{\partial}{\partial \theta} (j_{\alpha\theta} \sin\theta) + \frac{1}{r \sin\theta} \frac{\partial j_{\alpha\phi}}{\partial \phi} \right] + r_\alpha \qquad (B.10\text{-}3)$$

[a] To obtain the corresponding equations in terms of J_α^* make the following replacements:

Replace	ρ	ω_α	j_α	v	r_α
by	c	x_α	J_α^*	v^*	$R_\alpha - x_\alpha \sum_{\beta=1}^{N} R_\beta$

§B.11   THE EQUATION OF CONTINUITY FOR SPECIES A IN TERMS OF ω_A FOR CONSTANT^a $\rho\mathscr{D}_{AB}$

$$[\rho D\omega_A/Dt = \rho\mathscr{D}_{AB}\nabla^2\omega_A + r_A]$$

Cartesian coordinates (x, y, z):

$$\rho\left(\frac{\partial\omega_A}{\partial t} + v_x\frac{\partial\omega_A}{\partial x} + v_y\frac{\partial\omega_A}{\partial y} + v_z\frac{\partial\omega_A}{\partial z}\right) = \rho\mathscr{D}_{AB}\left[\frac{\partial^2\omega_A}{\partial x^2} + \frac{\partial^2\omega_A}{\partial y^2} + \frac{\partial^2\omega_A}{\partial z^2}\right] + r_A \tag{B.11-1}$$

Cylindrical coordinates (r, θ, z):

$$\rho\left(\frac{\partial\omega_A}{\partial t} + v_r\frac{\partial\omega_A}{\partial r} + \frac{v_\theta}{r}\frac{\partial\omega_A}{\partial \theta} + v_z\frac{\partial\omega_A}{\partial z}\right) = \rho\mathscr{D}_{AB}\left[\frac{1}{r}\frac{\partial}{\partial r}\left(r\frac{\partial\omega_A}{\partial r}\right) + \frac{1}{r^2}\frac{\partial^2\omega_A}{\partial \theta^2} + \frac{\partial^2\omega_A}{\partial z^2}\right] + r_A \tag{B.11-2}$$

Spherical coordinates (r, θ, ϕ):

$$\rho\left(\frac{\partial\omega_A}{\partial t} + v_r\frac{\partial\omega_A}{\partial r} + \frac{v_\theta}{r}\frac{\partial\omega_A}{\partial \theta} + \frac{v_\phi}{r\sin\theta}\frac{\partial\omega_A}{\partial \phi}\right) = \rho\mathscr{D}_{AB}\left[\frac{1}{r^2}\frac{\partial}{\partial r}\left(r^2\frac{\partial\omega_A}{\partial r}\right) + \frac{1}{r^2\sin\theta}\frac{\partial}{\partial \theta}\left(\sin\theta\frac{\partial\omega_A}{\partial \theta}\right) + \frac{1}{r^2\sin^2\theta}\frac{\partial^2\omega_A}{\partial \phi^2}\right] + r_A \tag{B.11-3}$$

^{a} To obtain the corresponding equations in terms of x_A, make the following replacements:

Replace	ρ	ω_α	$\mathbf{v}$	r_α
by	c	x_α	$\mathbf{v}^*$	$R_\alpha - x_\alpha\sum_{\beta=1}^{N}R_\beta$

Appendix C

Mathematical Topics

In this appendix we summarize information on mathematical topics (other than vectors and tensors) that are useful in the study of transport phenomena.[1]

§C.1 SOME ORDINARY DIFFERENTIAL EQUATIONS AND THEIR SOLUTIONS

We assemble here a short list of differential equations that arise frequently in transport phenomena. The reader is assumed to be familiar with these equations and how to solve them. The quantities a, b, and c are real constants and f and g are functions of x.

Equation	Solution	
$\dfrac{dy}{dx} = \dfrac{f(x)}{g(y)}$	$\int g\,dy = \int f\,dx + C_1$	(C.1-1)
$\dfrac{dy}{dx} + f(x)y = g(x)$	$y = e^{-\int f\,dx}\left(\int e^{\int f\,dx}g\,dx + C_1\right)$	(C.1-2)
$\dfrac{d^2y}{dx^2} + a^2y = 0$	$y = C_1\cos ax + C_2\sin ax$	(C.1-3)
$\dfrac{d^2y}{dx^2} - a^2y = 0$	$y = C_1\cosh ax + C_2\sinh ax$ or	(C.1-4a)
	$y = C_3 e^{+ax} + C_4 e^{-ax}$	(C.1-4b)
$\dfrac{1}{x^2}\dfrac{d}{dx}\left(x^2\dfrac{dy}{dx}\right) + a^2y = 0$	$y = \dfrac{C_1}{x}\cos ax + \dfrac{C_2}{x}\sin ax$	(C.1-5)

---

[1] Some useful reference books on applied mathematics are: M. Abramowitz and I. A. Stegun, *Handbook of Mathematical Functions*, Dover, New York, 9th printing (1973); G. M. Murphy, *Ordinary Differential Equations and Their Solutions*, Van Nostrand, Princeton, N.J. (1960); J. J. Tuma, *Engineering Mathematics Handbook*, 3rd edition, McGraw-Hill, New York (1987).

$$\frac{1}{x^2}\frac{d}{dx}\left(x^2\frac{dy}{dx}\right) - a^2y = 0$$

$$y = \frac{C_1}{x}\cosh ax + \frac{C_2}{x}\sinh ax \text{ or} \tag{C.1-6a}$$

$$y = \frac{C_3}{x}e^{+ax} + \frac{C_4}{x}e^{-ax} \tag{C.1-6b}$$

$$\frac{d^2y}{dx^2} + a\frac{dy}{dx} + by = 0$$

Solve the equation $n^2 + an + b = 0$, and get the roots $n = n_+$ and $n = n_-$. Then (a) if n_+ and n_- are real and unequal,

$$y = C_1\exp(n_+x) + C_2\exp(n_-x) \tag{C.1-7a}$$

(b) if n_+ and n_- are real and equal to n,

$$y = e^{nx}(C_1x + C_2) \tag{C.1-7b}$$

(c) if n_+ and n_- are complex: $n_\pm = p \pm iq$,

$$y = e^{px}(C_1\cos qx + C_2\sin qx) \tag{C.1-7c}$$

$$\frac{d^2y}{dx^2} + 2x\frac{dy}{dx} = 0$$

$$y = C_1\int_0^x \exp(-\bar{x}^2)\,d\bar{x} + C_2 \tag{C.1-8}$$

$$\frac{d^2y}{dx^2} + 3x^2\frac{dy}{dx} = 0$$

$$y = C_1\int_0^x \exp(-\bar{x}^3)\,d\bar{x} + C_2 \tag{C.1-9}$$

$$\frac{d^2y}{dx^2} = f(x)$$

$$y = \int_0^x\int_0^{\bar{x}} f(\bar{\bar{x}})\,d\bar{\bar{x}}\,d\bar{x} + C_1x + C_2 \tag{C.1-10}$$

$$\frac{1}{x}\frac{d}{dx}\left(x\frac{dy}{dx}\right) = f(x)$$

$$y = \int_0^x \frac{1}{\bar{x}}\int_0^{\bar{x}} \bar{\bar{x}}f(\bar{\bar{x}})\,d\bar{\bar{x}}\,d\bar{x} + C_1\ln x + C_2 \tag{C.1-11}$$

$$\frac{1}{x^2}\frac{d}{dx}\left(x^2\frac{dy}{dx}\right) = f(x)$$

$$y = \int_0^x \frac{1}{\bar{x}^2}\int_0^{\bar{x}} \bar{\bar{x}}^2f(\bar{\bar{x}})\,d\bar{\bar{x}}\,d\bar{x} - \frac{C_1}{x} + C_2 \tag{C.1-12}$$

$$\frac{d^2y}{dx^2} = h(y)$$

$$x = \int_0^y \frac{d\bar{y}}{\sqrt{2\int_0^{\bar{y}} h(\bar{\bar{y}})\,d\bar{\bar{y}} + C_1}} + C_2 \tag{C.1-13}$$

$$x^3\frac{d^3y}{dx^3} + ax^2\frac{d^2y}{dx^2} + bx\frac{dy}{dx} + cy = 0 \tag{C.1-14}$$

$y = C_1x^{n_1} + C_2x^{n_2} + C_3x^{n_3}$, where the n_k are the roots of the equation $n(n-1)(n-2) + an(n-1) + bn + c = 0$, provided that all roots are distinct.

Notes:

[a] In Eqs. C.1-4 and C.1-6 the decisions as to whether to use the exponential forms or the trigonometric (or hyperbolic) functions are usually made on the basis of the boundary conditions on the problem or the symmetry properties of the solution.

[b] Equations C.1-5 and C.1-6 are solved by making the substitution $y(x) = u(x)/x$ and then solving the resulting equation for $u(x)$.

[c] In Eqs. C.1-8 to C.1-13, it may be convenient or necessary to change the lower limits of the integrals to some value other than zero.

§C.2   EXPANSIONS OF FUNCTIONS IN TAYLOR SERIES

In physical problems we often need to describe a function $y(x)$ in the neighborhood of some point $x = x_0$. Then we expand the function $y(x)$ in a "Taylor series about the point $x = x_0$":

$$y(x) = y|_{x=x_0} + \frac{1}{1!}\left(\frac{dy}{dx}\bigg|_{x=x_0}\right)(x - x_0) + \frac{1}{2!}\left(\frac{d^2y}{dx^2}\bigg|_{x=x_0}\right)(x - x_0)^2$$

$$+ \frac{1}{3!}\left(\frac{d^3y}{dx^3}\bigg|_{x=x_0}\right)(x - x_0)^3 + \cdots \tag{C.2-1}$$

The first term gives the value of the function at $x = x_0$. The first two terms give a straight-line fit of the curve at $x = x_0$. The first three terms give a parabolic fit of the curve at $x = x_0$, and so on. The Taylor series is often used when only the first several terms are needed to describe the function adequately.

Here are a few Taylor series expansions of standard functions about the point $x = 0$:

$$e^{\pm x} = 1 \pm \frac{x}{1!} + \frac{x^2}{2!} \pm \frac{x^3}{3!} + \cdots \tag{C.2-2}$$

$$\ln(1 + x) = x - \frac{x^2}{2} + \frac{x^3}{3} - \frac{x^4}{4} + \cdots \tag{C.2-3}$$

$$\operatorname{erf} x = \frac{2x}{\sqrt{\pi}}\left(1 - \frac{x^2}{1!3} + \frac{x^4}{2!5} - \frac{x^6}{3!7} + \cdots\right) \tag{C.2-4}$$

$$\sqrt{1 \pm x} = 1 \pm \frac{1}{2}x - \frac{1 \cdot 1}{2 \cdot 4}x^2 \pm \frac{1 \cdot 1 \cdot 3}{2 \cdot 4 \cdot 6}x^3 - \cdots \tag{C.2-5}$$

Further examples may be found in calculus textbooks and handbooks. Taylor series can also be written for functions of two or more variables.

§C.3   DIFFERENTIATION OF INTEGRALS (THE LEIBNIZ FORMULA)

Suppose we have a function $f(x, t)$ that depends on a space variable x and the time t. Then we can form the integral

$$I(t) = \int_{\alpha(t)}^{\beta(t)} f(x, t)\, dx \tag{C.3-1}$$

which is a function of t [see Fig. C.3-1(a)]. If we want to differentiate this function with respect to t without evaluating the integral, we can use the Leibniz formula

$$\frac{d}{dt}\int_{\alpha(t)}^{\beta(t)} f(x, t)\, dx = \int_{\alpha(t)}^{\beta(t)} \frac{\partial}{\partial t} f(x, t)\, dx + \left(f(\beta, t)\frac{d\beta}{dt} - f(\alpha, t)\frac{d\alpha}{dt}\right) \tag{C.3-2}$$

Figure C.3-1(b) shows the meanings of the operations performed here: the first term on the right side gives the change in the integral because the function itself is changing with

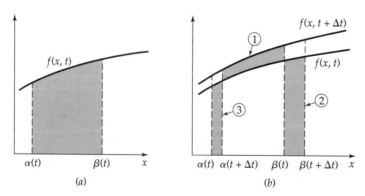

Fig. C.3-1. (a) The shaded area represents $I(t) = \int_{\alpha(t)}^{\beta(t)} f(x, t)dx$ at an instant t (Eq. C.3-1). (b) To get dI/dt, we form the difference $I(t + \Delta t) - I(t)$, divide by Δt, and then let $\Delta t \to 0$. The three shaded areas correspond to the three terms on the right side of Eq. C.3-2.

time; the second term accounts for the gain in area as the upper limit is moved to the right; the second term accounts for the gain in area as the upper limit is moved to the right; and the third term shows the loss in area as the lower limit is moved to the right. This formula finds many uses in science and engineering. The three-dimensional analog is given in Eq. A.5-5.

§C.4 THE GAMMA FUNCTION

The gamma function appears frequently as the result of integrations:

$$\Gamma(n) = \int_0^\infty x^{n-1} e^{-x} dx \tag{C.4-1}$$

$$\Gamma(n) = \int_0^1 \left(\ln \frac{1}{x}\right)^{n-1} dx \tag{C.4-2}$$

$$\Gamma(n + 1) = \int_0^\infty \exp(-x^{1/n}) \, dx \tag{C.4-3}$$

Several formulas for gamma functions are important:

$\Gamma(n + 1) = n\Gamma(n)$ (used to define $\Gamma(n)$ for negative n)   (C.4-4)

$\Gamma(n) = (n - 1)!$ (when n is an integer greater than 0)   (C.4-5)

Some special values of the gamma function are:

$$\Gamma(1) = \Gamma(2) = 1 \tag{C.4-6}$$
$$\Gamma(\tfrac{1}{2}) = \sqrt{\pi} = 1.77245 \ldots \tag{C.4-7}$$
$$\Gamma(\tfrac{3}{2}) = \tfrac{1}{2}\Gamma(\tfrac{1}{2}) = \tfrac{1}{2}\sqrt{\pi} = 0.88622 \ldots \tag{C.4-8}$$
$$\Gamma(\tfrac{1}{3}) = 2.67893 \ldots \tag{C.4-9}$$
$$\Gamma(\tfrac{4}{3}) = \tfrac{1}{3}\Gamma(\tfrac{1}{3}) = 0.89297 \ldots \tag{C.4-10}$$
$$\Gamma(\tfrac{7}{3}) = \tfrac{4}{3}\Gamma(\tfrac{4}{3}) = 1.19063 \ldots \tag{C.4-11}$$

The gamma function is displayed in Fig. C.4-1.

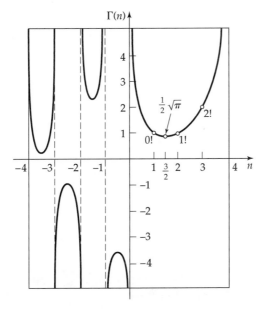

Fig. C.4-1. The gamma function.

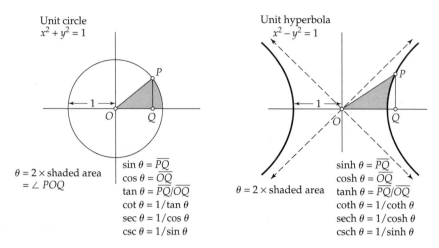

$\theta = 2 \times$ shaded area
$\quad = \angle POQ$

$\sin \theta = \overline{PQ}$
$\cos \theta = \overline{OQ}$
$\tan \theta = \overline{PQ}/\overline{OQ}$
$\cot \theta = 1/\tan \theta$
$\sec \theta = 1/\cos \theta$
$\csc \theta = 1/\sin \theta$

$\theta = 2 \times$ shaded area

$\sinh \theta = \overline{PQ}$
$\cosh \theta = \overline{OQ}$
$\tanh \theta = \overline{PQ}/\overline{OQ}$
$\coth \theta = 1/\coth \theta$
$\text{sech } \theta = 1/\cosh \theta$
$\text{csch } \theta = 1/\sinh \theta$

Fig. C.5-1. Comparison of circular and hyperbolic functions.

§C.5   THE HYPERBOLIC FUNCTIONS

The hyperbolic sine (sinh x), the hyperbolic cosine (cosh x), and the hyperbolic tangent (tanh x) arise frequently in science and engineering problems. They are related to the hyperbola in very much the same way that the circular functions are related to the circle (see Fig. C.5-1). The circular functions (sin x and cos x) are periodic, oscillating functions, whereas their hyperbolic analogs are not (see Fig. C.5-2).

The hyperbolic functions are related to the exponential function as follows:

$$\cosh x = \tfrac{1}{2}(e^x + e^{-x}); \qquad \sinh x = \tfrac{1}{2}(e^x - e^{-x}) \tag{C.5-1, 2}$$

The corresponding relations for the circular functions are:

$$\cos x = \tfrac{1}{2}(e^{ix} + e^{-ix}); \qquad \sin x = \tfrac{1}{2}(e^{ix} - e^{-ix}) \tag{C.5-3, 4}$$

One can derive a variety of standard relations for the hyperbolic functions, such as

$$\cosh^2 x - \sinh^2 x = 1 \tag{C.5-5}$$
$$\cosh(x \pm y) = \cosh x \cosh y \pm \sinh x \sinh y \tag{C.5-6}$$
$$\sinh(x \pm y) = \sinh x \cosh y \pm \cosh x \sinh y \tag{C.5-7}$$

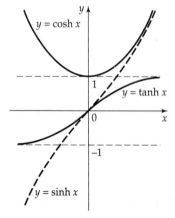

Fig. C.5-2. Comparison of the shapes of the hyperbolic functions.

$$\cosh ix = \cos x; \qquad \sinh ix = i \sin x \qquad (C.5\text{-}8, 9)$$

$$\frac{d \cosh x}{dx} = \sinh x; \qquad \frac{d \sinh x}{dx} = \cosh x \qquad (C.5\text{-}10, 11)$$

$$\int \cosh x \, dx = \sinh x; \qquad \int \sinh x \, dx = \cosh x \qquad (C.5\text{-}12, 13)$$

It should be kept in mind that $\cosh x$ and $\cos x$ are both even functions of x, whereas $\sinh x$ and $\sin x$ are odd functions of x.

§C.6   THE ERROR FUNCTION

The error function is defined as

$$\operatorname{erf} x = \frac{\displaystyle\int_0^x \exp(-\bar{x}^2) \, d\bar{x}}{\displaystyle\int_0^\infty \exp(-\bar{x}^2) \, d\bar{x}} = \frac{2}{\sqrt{\pi}} \int_0^x \exp(-\bar{x}^2) \, d\bar{x} \qquad (C.6\text{-}1)$$

This function, which arises naturally in numerous transport problems, is monotone increasing, going from $\operatorname{erf} 0 = 0$ to $\operatorname{erf} \infty = 1$, and has the value of 0.99 at about $x = 2$. The Taylor series expansion for the error function about $x = 0$ is given in Eq. C.2-4. It is also worth noting that $\operatorname{erf} (-x) = -\operatorname{erf} x$, and that

$$\frac{d}{dx} \operatorname{erf} u = \frac{2}{\sqrt{\pi}} \exp(-u^2) \frac{du}{dx} \qquad (C.6\text{-}2)$$

by applying the Leibniz formula to Eq. C.6-1.

The closely related function $\operatorname{erfc} x = 1 - \operatorname{erf} x$ is called the "complementary error function."

Appendix D

The Kinetic Theory of Gases

In Chapters 1, 9, and 17 we gave a brief account of the use of mean free path arguments to get approximate expressions for the transport properties. Then we gave the rigorous results from the Chapman–Enskog development for dilute monatomic gases. In this appendix we give a brief description of the Chapman–Enskog theory, just enough to show what the theory involves and to show how it gives a sense of unity to the subject of transport phenomena in gases. The reader who wishes to pursue the subject further can consult the standard references.[1]

§D.1   THE BOLTZMANN EQUATION[2]

The starting point for the kinetic theory of low-density, nonreacting mixtures of monatomic gases is the *Boltzmann equation* for the velocity distribution function $f_\alpha(\dot{\mathbf{r}}_\alpha, \mathbf{r}, t)$. The quantity $f_\alpha(\dot{\mathbf{r}}_\alpha, \mathbf{r}, t)\, d\dot{\mathbf{r}}_\alpha d\mathbf{r}$ is the probable number of molecules of species α, which at time t are located in the volume element $d\mathbf{r}$ at position $\mathbf{r}$, and have velocities within the range $d\dot{\mathbf{r}}_\alpha$ about $\dot{\mathbf{r}}_\alpha$. The Boltzmann equation, which describes how f_α evolves with time, is

$$\frac{\partial}{\partial t} f_\alpha = -\left(\frac{\partial}{\partial \mathbf{r}} \cdot \dot{\mathbf{r}}_\alpha f_\alpha\right) - \left(\frac{\partial}{\partial \dot{\mathbf{r}}_\alpha} \cdot \mathbf{g}_\alpha f_\alpha\right) + J_\alpha \tag{D.1-1}$$

---

[1] J. H. Ferziger and H. G. Kaper, *Mathematical Theory of Transport Processes in Gases*, North-Holland, Amsterdam (1972); S. Chapman and T. G. Cowling, *The Mathematical Theory of Non-Uniform Gases*, 3rd edition, Cambridge University Press (1970); J. O. Hirschfelder, C. F. Curtiss, and R. B. Bird, *Molecular Theory of Gases and Liquids*, 2nd corrected printing, Wiley, New York (1964), Chapter 7; E. M. Lifshitz and L. P. Pitaevskii, *Physical Kinetics*, Pergamon, Oxford (1981), Chapter 1.

[2] L. Boltzmann, *Sitzungsberichte Keiserl. Akad. der Wissenschaften*, **66** (2), 275–370 (1872); C. Cercignani, *The Boltzmann Equation and Its Applications*, Springer-Verlag, New York (1988). C. F. Curtiss, *J. Chem. Phys.*, **97**, 1416–1423, 7679–7686 (1992), found it necessary to modify the Boltzmann equation to account for the possibility of orbiting pairs of molecules; the modification, important only at very low temperatures, was found to give much better agreement with the limited low-temperature experimental data.

in which $\partial/\partial\mathbf{r}$ is identical to the ∇-operator, and $\partial/\partial\dot{\mathbf{r}}_\alpha$ is a similar operator involving velocities rather than positions. The quantity $\mathbf{g}_\alpha$ is the external force per unit mass acting on a molecule of species α, and J_α is a very complicated five-fold integral term accounting for the change in f_α because of molecular collisions. The J_α term involves the intermolecular potential energy function (e.g., the Lennard-Jones potential) and the details of the collision trajectories. The Boltzmann equation may be thought of as a continuity equation in the six-dimensional position-velocity space, and J_α serves as a source term. The velocity distribution function is "normalized" to the number density of species α; that is, $\int f_\alpha(\dot{\mathbf{r}}_\alpha, \mathbf{r}, t)d\dot{\mathbf{r}}_\alpha = n_\alpha(\mathbf{r}, t)$.

§D.2   THE EQUATIONS OF CHANGE

When the Boltzmann equation is multiplied by some molecular property $\psi_\alpha(\dot{\mathbf{r}}_\alpha)$ and then integrated over all molecular velocities, the *general equation of change* is obtained:

$$\frac{\partial}{\partial t}\int\psi_\alpha f_\alpha d\dot{\mathbf{r}}_\alpha = -\left(\frac{\partial}{\partial\mathbf{r}}\cdot\int\dot{\mathbf{r}}_\alpha\psi_\alpha f_\alpha d\dot{\mathbf{r}}_\alpha\right) + \int\left(\mathbf{g}_\alpha\cdot\frac{\partial\psi_\alpha}{\partial\dot{\mathbf{r}}_\alpha}\right)f_\alpha d\dot{\mathbf{r}}_\alpha + \int\psi_\alpha J_\alpha d\dot{\mathbf{r}}_\alpha \qquad \text{(D.2-1)}$$

An integration by parts is performed to get this result, and use is made of the fact that f_α is zero at infinite velocities. If ψ_α is a quantity that is conserved during a collision (see §0.3), then the term containing J_α can be shown to be zero.[1]

Now let ψ_α be successively the conserved quantities for monatomic molecules: the mass m_α, the momentum $m_\alpha\dot{\mathbf{r}}_\alpha$, and the energy $\frac{1}{2}m_\alpha(\dot{\mathbf{r}}_\alpha\cdot\dot{\mathbf{r}}_\alpha)$. When these are substituted for ψ_α into Eq. D.2-1, and when a sum over all species α is performed for the second and third of these, we get the *equations of change* for mass of α, momentum, and energy as follows:

$$\frac{\partial}{\partial t}\rho_\alpha = -(\nabla\cdot\rho_\alpha\mathbf{v}) - (\nabla\cdot\mathbf{j}_\alpha) \qquad \text{(D.2-2)}$$

$$\frac{\partial}{\partial t}\rho\mathbf{v} = -[\nabla\cdot\rho\mathbf{v}\mathbf{v}] - [\nabla\cdot\boldsymbol{\pi}] + \sum_\alpha\rho_\alpha\mathbf{g}_\alpha \qquad \text{(D.2-3)}$$

$$\frac{\partial}{\partial t}(\tfrac{1}{2}\rho v^2 + \rho\hat{U}) = -(\nabla\cdot(\tfrac{1}{2}\rho v^2 + \rho\hat{U})\mathbf{v}) - (\nabla\cdot\mathbf{q}) - (\nabla\cdot[\boldsymbol{\pi}\cdot\mathbf{v}]) + \sum_\alpha((\mathbf{j}_\alpha + \rho_\alpha\mathbf{v})\cdot\mathbf{g}_\alpha) \qquad \text{(D.2-4)}$$

In the last of these equations the internal energy per unit volume is defined to be

$$\rho\hat{U} = \tfrac{3}{2}n\kappa T = \int\tfrac{1}{2}m_\alpha(\dot{\mathbf{r}}_\alpha - \mathbf{v})^2 f_\alpha d\dot{\mathbf{r}}_\alpha \qquad \text{(D.2-5)}$$

Thus we see that the equations of continuity, motion, and energy are direct consequences of the conservation laws for mass, momentum, and energy discussed in Chapter 0. Equations D.2-2 to 4 should be compared with Eqs. 19.1-7, 3.2-9, and Eq. (B) and footnote (b) of Table 19.2-4, which were derived by continuum arguments.

§D.3   THE MOLECULAR EXPRESSIONS FOR THE FLUXES

At the same time the equations of change are obtained, the molecular expressions for the fluxes are generated as integrals over the distribution function:

$$\mathbf{j}_\alpha(\mathbf{r}, t) = m_\alpha\int(\dot{\mathbf{r}}_\alpha - \mathbf{v})f_\alpha d\dot{\mathbf{r}}_\alpha \xrightarrow{\text{at equilibrium}} 0 \qquad \text{(D.3-1)}$$

$$\boldsymbol{\pi}(\mathbf{r}, t) = \sum_\alpha m_\alpha\int(\dot{\mathbf{r}}_\alpha - \mathbf{v})(\dot{\mathbf{r}}_\alpha - \mathbf{v})f_\alpha d\dot{\mathbf{r}}_\alpha \xrightarrow{\text{at equilibrium}} p\boldsymbol{\delta} \qquad \text{(D.3-2)}$$

$$\mathbf{q}(\mathbf{r}, t) = \sum_\alpha\tfrac{1}{2}m_\alpha\int(\dot{\mathbf{r}}_\alpha - \mathbf{v})^2(\dot{\mathbf{r}}_\alpha - \mathbf{v})f_\alpha d\dot{\mathbf{r}}_\alpha \xrightarrow{\text{at equilibrium}} 0 \qquad \text{(D.3-3)}$$

In these expressions the fluxes involve integrals over the products of mass, momentum, and energy with the "diffusion velocity" ($\dot{\mathbf{r}}_\alpha - \mathbf{v}$) of species α. Note the similarity between the structure of these *molecular fluxes* (or "diffusive fluxes") and that of the convective fluxes of mass $\rho_\alpha \mathbf{v}$, momentum $\rho \mathbf{v}\mathbf{v}$, and kinetic energy $\frac{1}{2}\rho v^2 \mathbf{v}$ appearing in the equations of change, where $\mathbf{v}$ is the local instantaneous mass average velocity of the gas mixture. Thus the molecular fluxes represent the diffusive movement of mass, momentum, and energy above and beyond that described by the convective fluxes. Note also that the molecular theory automatically generates the molecular work term $-(\nabla \cdot [\boldsymbol{\pi} \cdot \mathbf{v}])$ in the energy equation.

§D.4   THE SOLUTION TO THE BOLTZMANN EQUATION

If the gas mixture were at rest, the velocity distribution function would be given by the Maxwell–Boltzmann distribution function (known from equilibrium statistical mechanics). Then we would find, as shown in §D.3, that $\mathbf{j}_\alpha = 0$, that $\boldsymbol{\pi} = p\boldsymbol{\delta} = n\kappa T\boldsymbol{\delta}$, and that $\mathbf{q} = 0$. The derivation of $p = n\kappa T$ is given in Problem 1C.3.

On the other hand, when there are concentration, velocity, and temperature gradients, the distribution function is given as the Maxwell–Boltzmann distribution multiplied by a "correction factor":

$$f_\alpha(\dot{\mathbf{r}}_\alpha, \mathbf{r}, t) = n_\alpha \left(\frac{m_\alpha}{2\pi\kappa T}\right)^{3/2} \exp[-m_\alpha(\dot{\mathbf{r}} - \mathbf{v})^2/2\kappa T] \, (1 + \phi_\alpha(\dot{\mathbf{r}}_\alpha, \mathbf{r}, t) + \cdots) \quad \text{(D.4-1)}$$

where $\phi_\alpha \ll 1$. In this expression n_α, $\mathbf{v}$, and T are functions of position $\mathbf{r}$ and time t. Since the deviations from equilibrium result from the temperature, velocity, and concentration gradients, $\phi_\alpha(\dot{\mathbf{r}}_\alpha, \mathbf{r}, t)$ can be represented, near equilibrium, as a linear function of the various gradients,

$$\phi_\alpha = -(\mathbf{A}_\alpha \cdot \nabla \ln T) - (\mathbf{B}_\alpha : \nabla \mathbf{v}) + n \sum_\beta (\mathbf{C}_{\alpha\beta} \cdot \mathbf{d}_\beta) \quad \text{(D.4-2)}$$

in which the vector $\mathbf{A}_\alpha$, the tensor $\mathbf{B}_\alpha$, and the vectors $\mathbf{C}_{\alpha\beta}$, all functions of $\dot{\mathbf{r}}_\alpha$, $\mathbf{r}$, and t, are given as the solutions of integrodifferential equations.[1] The quantities $\mathbf{d}_\alpha$ are "generalized diffusional driving forces" that include concentration gradients, the pressure gradient, and external force differences, defined as

$$\mathbf{d}_\alpha = \nabla x_\alpha + (x_\alpha - \omega_\alpha)\nabla \ln p - (\rho/p)\omega_\alpha(\mathbf{g}_\alpha - \Sigma_\beta \omega_\beta \mathbf{g}_\beta)$$

$$= \frac{1}{cRT}\left[(\nabla p_\alpha - \rho_\alpha \mathbf{g}_\alpha) - \omega_\alpha \left(\nabla p - \sum_\beta \rho_\beta \mathbf{g}_\beta\right)\right] \quad \text{(D.4-3)}$$

in which x_α, ω_α, and p_α are the mole fraction, mass fraction, and partial pressure, respectively. Equation D.4-3, valid only for a mixture of monatomic gases at low density, is generalized for other fluids in the discussion of the thermodynamics of irreversible processes in §24.1.

§D.5   THE FLUXES IN TERMS OF THE TRANSPORT PROPERTIES

When Eqs. D.4-1 to 3 are substituted into Eqs. D.3-1 to 3, we get the expressions for the fluxes in terms of $\mathbf{d}_\alpha$, $\nabla\mathbf{v}$, and ∇T:

$$\mathbf{j}_\alpha(\mathbf{r}, t) = +\rho_\alpha \sum_\beta \mathbb{D}_{\alpha\beta}\mathbf{d}_\beta - D_\alpha^T \nabla \ln T \quad \text{(D.5-1)}$$

$$\boldsymbol{\pi}(\mathbf{r}, t) = p\boldsymbol{\delta} - \mu[\nabla \mathbf{v} + (\nabla \mathbf{v})^\dagger - \tfrac{2}{3}(\nabla \cdot \mathbf{v})\boldsymbol{\delta}] \quad \text{(D.5-2)}$$

$$\mathbf{q}(\mathbf{r}, t) = -k\nabla T + \sum_\alpha \frac{\overline{H}_\alpha}{M_\alpha}\mathbf{j}_\alpha + \sum_\alpha \sum_\beta \frac{cRTx_\alpha x_\beta}{\rho_\alpha} \frac{D_\alpha^T}{\mathcal{D}_{\alpha\beta}}\left(\frac{\mathbf{j}_\alpha}{\rho_\alpha} - \frac{\mathbf{j}_\beta}{\rho_\beta}\right) \quad \text{(D.5-3)}$$

In these equations the transport properties appear: the viscosity μ, the thermal conductivity k, the multicomponent thermal diffusion coefficients D_α^T, and the multicomponent Fick diffusivities $\mathbb{D}_{\alpha\beta}$ (the $\text{Ð}_{\alpha\beta}$ are the Maxwell–Stefan diffusivities, closely related to the $\mathbb{D}_{\alpha\beta}$). Thus the kinetic theory predicts the "cross effects": the transport of mass resulting from a temperature gradient (thermal diffusion) and the transport of energy resulting from a concentration gradient (the diffusion-thermo effect).

The pressure term in Eq. D.5-2 comes from the first term in the expansion in Eq. D.4-1 (that is, the Maxwell–Boltzmann distribution), and the viscosity term comes from the second term (that is, the ϕ_α term containing the gradients). The kinetic theory of monatomic gases at low density predicts that the dilatational viscosity will be zero.

§D.6 THE TRANSPORT PROPERTIES IN TERMS OF THE INTERMOLECULAR FORCES

The transport properties in Eqs. D.5-1 to 3 are given by the kinetic theory as complicated multiple integrals involving the intermolecular forces that describe binary collisions in the gas mixture. Once an expression has been chosen for the intermolecular force law (such as the Lennard-Jones (6–12) potential of Eq. 1.4-10), these integrals can be evaluated numerically. For a pure gas, the three transport properties—self-diffusivity, viscosity, and thermal conductivity—are then given by:

$$\mathscr{D} = \frac{3}{8} \frac{\sqrt{\pi m \kappa T}}{\pi \sigma^2 \Omega_\mathscr{D}} \frac{1}{\rho}; \qquad \mu = \frac{5}{16} \frac{\sqrt{\pi m \kappa T}}{\pi \sigma^2 \Omega_\mu}; \qquad k = \frac{25}{32} \frac{\sqrt{\pi m \kappa T}}{\pi \sigma^2 \Omega_k} \hat{C}_V \quad \text{(D.6-1, 2, 3)}$$

The dimensionless "collision integrals" $\Omega_\mu = \Omega_k \approx 1.1\Omega_\mathscr{D}$ contain all the information about the intermolecular forces and the binary collision dynamics. They are given in Table E.2 as functions of $\kappa T/\varepsilon$. If we set the collision integrals equal to unity, we then get the transport properties for a gas composed of rigid spheres.

Thus the transport properties, needed in the equations of change, have been obtained from kinetic theory in terms of the two parameters σ and ε of the intermolecular potential energy function. From these expressions we get $\text{Pr} = \hat{C}_p \mu/k = \frac{2}{5}(\hat{C}_p/\hat{C}_V) = \frac{2}{5}(\frac{5}{3}) = \frac{2}{3}$ and $\text{Sc} = \mu/\rho\mathscr{D} = \frac{5}{6}(\Omega_\mathscr{D}/\Omega_\mu) \approx \frac{3}{4}$, these values being quite good for pure monatomic gases.

§D.7 CONCLUDING COMMENTS

The above discussion emphasizes the close connections among mass, momentum, and energy transport, and it is seen how all three transport phenomena can be explained in terms of a molecular theory for low-density, monatomic gases. It is also important to see that the continuum equations of continuity, motion, and energy can all be derived from one starting point—the Boltzmann equation—and that the molecular expressions for the fluxes and transport properties are generated in the process. In addition, the discussion of the dependence of the fluxes on the driving forces is very closely related to the irreversible thermodynamics approach in Chapter 24.

This appendix has dealt only with low-density, monatomic gases. Similar discus-

---

[3] C. F. Curtiss, *J. Chem. Phys.*, **24**, 225–241 (1956); C. Muckenfuss and C. F. Curtiss, *J. Chem. Phys.*, **29**, 1257–1277 (1958); L. A. Viehland and C. F. Curtiss, *J. Chem. Phys.*, **60**, 492–520 (1974); D. Russell and C. F. Curtiss, *J. Chem. Phys.*, **60**, 514–520 (1974).

[4] J. H. Irving and J. G. Kirkwood, *J. Chem. Phys.*, **18**, 817–829 (1950); R. J. Bearman and J. G. Kirkwood, *J. Chem. Phys.*, **28**, 136–145 (1958).

[5] C. F. Curtiss and R. B. Bird, *Adv. Polymer Sci.*, **125**, 1–101 (1996); *Proc. Nat. Acad. Sci.*, **93**, 7440–7445 (1996); *J. Chem. Phys.*, **106**, 9899–9921 (1997), **107**, 5254–5267 (1997), **111**, 10362–10370 (1999).

sions are available for polyatomic gases,[3] monatomic liquids,[4] and polymeric liquids.[5] In kinetic theories for monatomic liquids, the expressions for the momentum and heat fluxes contain terms similar to those in Eqs. D.3-2 and 3, but also contributions associated with forces between molecules; for polymers, one has the latter contribution, but also additional forces within the polymer chain. In all of these theories one can derive the equations of change from an equation for a distribution function and then get formal expressions for the transport properties.

Appendix E

Tables for Prediction of Transport Properties

Table E.1 Lennard-Jones (6–12) Potential Parameters and Critical Properties

Substance	Molecular Weight M	Lennard-Jones parameters σ (Å)	ε/κ (K)	Ref.	Critical properties[g,h] T_c (K)	p_c (atm)	$\tilde{V}_c$ (cm³/g-mole)	μ_c (g/cm·s × 10⁶)	k_c (cal/cm·s·K × 10⁶)
Light elements:									
H_2	2.016	2.915	38.0	a	33.3	12.80	65.0	34.7	—
He	4.003	2.576	10.2	a	5.26	2.26	57.8	25.4	—
Noble gases:									
Ne	20.180	2.789	35.7	a	44.5	26.9	41.7	156.	79.2
Ar	39.948	3.432	122.4	b	150.7	48.0	75.2	264.	71.0
Kr	83.80	3.675	170.0	b	209.4	54.3	92.2	396.	49.4
Xe	131.29	4.009	234.7	b	289.8	58.0	118.8	490.	40.2
Simple polyatomic gases:									
Air	28.964[i]	3.617	97.0	a	132.4[i]	37.0[i]	86.7[i]	193.	90.8
N_2	28.013	3.667	99.8	b	126.2	33.5	90.1	180.	86.8
O_2	31.999	3.433	113.	a	154.4	49.7	74.4	250.	105.3
CO	28.010	3.590	110.	a	132.9	34.5	93.1	190.	86.5
CO_2	44.010	3.996	190.	a	304.2	72.8	94.1	343.	122.
NO	30.006	3.470	119.	a	180.	64.	57.	258.	118.2
N_2O	44.012	3.879	220.	a	309.7	71.7	96.3	332.	131.
SO_2	64.065	4.026	363.	c	430.7	77.8	122.	411.	98.6
F_2	37.997	3.653	112.	a	—	—	—	—	—
Cl_2	70.905	4.115	357.	a	417.	76.1	124.	420.	97.0
Br_2	159.808	4.268	520.	a	584.	102.	144.	—	—
I_2	253.809	4.982	550.	a	800.	—	—	—	—
Hydrocarbons:									
CH_4	16.04	3.780	154.	b	191.1	45.8	98.7	159.	158.
$CH{\equiv}CH$	26.04	4.114	212.	d	308.7	61.6	112.9	237.	—
$CH_2{=}CH_2$	28.05	4.228	216.	b	282.4	50.0	124.	215.	—
C_2H_6	30.07	4.388	232.	b	305.4	48.2	148.	210.	203.
$CH_3C{\equiv}CH$	40.06	4.742	261.	d	394.8	—	—	—	—
$CH_3CH{=}CH_2$	42.08	4.766	275.	b	365.0	45.5	181.	233.	—
C_3H_8	44.10	4.934	273.	b	369.8	41.9	200.	228.	—
$n{-}C_4H_{10}$	58.12	5.604	304.	b	425.2	37.5	255.	239.	—

i—C_4H_{10}	58.12	5.393	b	408.1	36.0	263.	295.	239.	—
n—C_5H_{12}	72.15	5.850	b	469.5	33.2	311.	326.	238.	—
i—C_5H_{12}	72.15	5.812	b	460.4	33.7	306.	327.	—	—
$C(CH_3)_4$	72.15	5.759	b	433.8	31.6	303.	312.	—	—
n—C_6H_{14}	86.18	6.264	b	507.3	29.7	370.	342.	248.	—
n—C_7H_{16}	100.20	6.663	b	540.1	27.0	432.	352.	254.	—
n—C_8H_{18}	114.23	7.035	b	568.7	24.5	492	361.	259.	—
n—C_9H_{20}	128.26	7.463	b	594.6	22.6	548.	351.	265.	—
Cyclohexane	84.16	6.143	d	553.	40.0	308.	313.	284.	—
Benzene	78.11	5.443	b	562.6	48.6	260.	387.	312.	—
Other organic compounds:									
CH_4	16.04	3.780	b	191.1	45.8	98.7	154.	159.	158.
CH_3Cl	50.49	4.151	c	416.3	65.9	143.	355.	338.	—
CH_2Cl_2	84.93	4.748	c	510.	60.	—	398.	—	—
$CHCl_3$	119.38	5.389	e	536.6	54.	240.	340.	410.	—
CCl_4	153.82	5.947	e	556.4	45.0	276.	323.	413.	—
C_2N_2	52.034	4.361	e	400.	59.	—	349.	—	—
COS	60.076	4.130	e	378.	61.	—	336.	—	—
CS_2	76.143	4.483	e	552.	78.	170.	467.	404.	—
CCl_2F_2	120.91	5.116	b	384.7	39.6	218.	280.	—	—

[a] J. O. Hirschfelder, C. F. Curtiss, and R. B. Bird, *Molecular Theory of Gases and Liquids*, corrected printing with notes added, Wiley, New York (1964).

[b] L. S. Tee, S. Gotoh, and W. E. Stewart, *Ind. Eng. Chem. Fundamentals*, **5**, 356–363 (1966). The values for benzene are from viscosity data on that substance. The values for other substances are computed from Correlation (iii) of the paper.

[c] L. Monchick and E. A. Mason, *J. Chem. Phys.*, **35**, 1676–1697 (1961); parameters obtained from viscosity.

[d] L. W. Flynn and G. Thodos, *AIChE Journal*, **8**, 362–365 (1962); parameters obtained from viscosity.

[e] R. A. Svehla, *NASA Tech. Report R-132* (1962); parameters obtained from viscosity. This report provides extensive tables of Lennard-Jones parameters, heat capacities, and calculated transport properties.

[f] Values of the critical constants for the pure substances are selected from K. A. Kobe and R. E. Lynn, Jr., *Chem. Rev.*, **52**, 117–236 (1962); *Amer. Petroleum Inst. Research Proj.*, **44**, Thermodynamics Research Center, Texas A&M University, College Station, Texas (1966); and *Thermodynamic Functions of Gases*, F. Din (editor), Vols. 1–3, Butterworths, London (1956, 1961, 1962).

[g] Values of the critical viscosity are from O. A. Hougen and K. M. Watson, *Chemical Process Principles*, Vol. 3, Wiley, New York (1947), p. 873.

[h] Values of the critical thermal conductivity are from E. J. Owens and G. Thodos, *AIChE Journal*, **3**, 454–461 (1957).

[i] For air, the molecular weight M and the pseudocritical properties have been computed from the average composition of dry air as given in COESA, U.S. *Standard Atmosphere 1976*, U.S. Government Printing Office, Washington, D.C. (1976).

Table E.2 Collision Integrals for Use with the Lennard-Jones (6–12) Potential for the Prediction of Transport Properties of Gases at Low Densities[a,b,c]

$\kappa T/\varepsilon$ or $\kappa T/\varepsilon_{AB}$	$\Omega_\mu = \Omega_k$ (for viscosity and thermal conductivity)	$\Omega_{\mathcal{D},AB}$ (for diffusivity)	$\kappa T/\varepsilon$ or $\kappa T/\varepsilon_{AB}$	$\Omega_\mu = \Omega_k$ (for viscosity and thermal conductivity)	$\Omega_{\mathcal{D},AB}$ (for diffusivity)
0.30	2.840	2.649	2.7	1.0691	0.9782
0.35	2.676	2.468	2.8	1.0583	0.9682
0.40	2.531	2.314	2.9	1.0482	0.9588
0.45	2.401	2.182	3.0	1.0388	0.9500
0.50	2.284	2.066	3.1	1.0300	0.9418
0.55	2.178	1.965	3.2	1.0217	0.9340
0.60	2.084	1.877	3.3	1.0139	0.9267
0.65	1.999	1.799	3.4	1.0066	0.9197
0.70	1.922	1.729	3.5	0.9996	0.9131
0.75	1.853	1.667	3.6	0.9931	0.9068
0.80	1.790	1.612	3.7	0.9868	0.9008
0.85	1.734	1.562	3.8	0.9809	0.8952
0.90	1.682	1.517	3.9	0.9753	0.8897
0.95	1.636	1.477	4.0	0.9699	0.8845
1.00	1.593	1.440	4.1	0.9647	0.8796
1.05	1.554	1.406	4.2	0.9598	0.8748
1.10	1.518	1.375	4.3	0.9551	0.8703
1.15	1.485	1.347	4.4	0.9506	0.8659
1.20	1.455	1.320	4.5	0.9462	0.8617
1.25	1.427	1.296	4.6	0.9420	0.8576
1.30	1.401	1.274	4.7	0.9380	0.8537
1.35	1.377	1.253	4.8	0.9341	0.8499
1.40	1.355	1.234	4.9	0.9304	0.8463
1.45	1.334	1.216	5.0	0.9268	0.8428
1.50	1.315	1.199	6.0	0.8962	0.8129
1.55	1.297	1.183	7.0	0.8727	0.7898
1.60	1.280	1.168	8.0	0.8538	0.7711
1.65	1.264	1.154	9.0	0.8380	0.7555
1.70	1.249	1.141	10.0	0.8244	0.7422
1.75	1.235	1.128	12.0	0.8018	0.7202
1.80	1.222	1.117	14.0	0.7836	0.7025
1.85	1.209	1.105	16.0	0.7683	0.6878
1.90	1.198	1.095	18.0	0.7552	0.6751
1.95	1.186	1.085	20.0	0.7436	0.6640
2.00	1.176	1.075	25.0	0.7198	0.6414
2.10	1.156	1.058	30.0	0.7010	0.6235
2.20	1.138	1.042	35.0	0.6854	0.6088
2.30	1.122	1.027	40.0	0.6723	0.5964
2.40	1.107	1.013	50.0	0.6510	0.5763
2.50	1.0933	1.0006	75.0	0.6140	0.5415
2.60	1.0807	0.9890	100.0	0.5887	0.5180

[a] The values in this table, applicable for the Lennard-Jones (6–12) potential, are interpolated from the results of L. Monchick and E. A. Mason, *J. Chem. Phys.*, **35**, 1676–1697 (1961). The Monchick–Mason table is believed to be slightly better than the earlier table by J. O. Hirschfelder, R. B. Bird, and E. L. Spotz, *J. Chem. Phys.*, **16**, 968–981 (1948).

[b] This table has been extended to lower temperatures by C. F. Curtiss, *J. Chem. Phys.*, **97**, 7679–7686 (1992). Curtiss showed that at low temperatures, the Boltzmann equation needs to be modified to take into account "orbiting pairs" of molecules. Only by making this modification is it possible to get a smooth transition from quantum to classical behavior. The deviations are appreciable below dimensionless temperatures of 0.30.

[c] The collision integrals have been curve-fitted by P. D. Neufeld, A. R. Jansen, and R. A. Aziz, *J. Chem. Phys.*, **57**, 1100–1102 (1972), as follows:

$$\Omega_\mu = \Omega_k = \frac{1.16145}{T^{*0.14874}} + \frac{0.52487}{\exp(0.77320T^*)} + \frac{2.16178}{\exp(2.43787T^*)} \tag{E.2-1}$$

$$\Omega_{\mathcal{D},AB} = \frac{1.06036}{T^{*0.15610}} + \frac{0.19300}{\exp(0.47635T^*)} + \frac{1.03587}{\exp(1.52996T^*)} + \frac{1.76474}{\exp(3.89411T^*)} \tag{E.2-2}$$

where $T^* = \kappa T/\varepsilon$.

Appendix F

Constants and Conversion Factors

§F.1   Mathematical constants

§F.2   Physical constants

§F.3   Conversion factors

§F.1   MATHEMATICAL CONSTANTS

$$\pi = 3.14159\ldots$$
$$e = 2.71828\ldots$$
$$\ln 10 = 2.30259\ldots$$

§F.2   PHYSICAL CONSTANTS[1]

Gas law constant (R)	8.31451	$J/g\text{-mol} \cdot K$
	8.31451×10^3	$kg \cdot m^2/s^2 \cdot kg\text{-mol} \cdot K$
	8.31451×10^7	$g \cdot cm^2/s^2 \cdot g\text{-mol} \cdot K$
	1.98721	$cal/g\text{-mol} \cdot K$
	82.0578	$cm^3\ atm/g\text{-mol} \cdot K$
	4.9686×10^4	$lb_m\ ft^2/s^2 \cdot lb\text{-mol} \cdot R$
	1.5443×10^3	$ft \cdot lb_f/lb\text{-mol} \cdot R$
Standard acceleration of gravity (g_0)	9.80665	m/s^2
	980.665	cm/s^2
	32.1740	ft/s^2
Joule's constant (J_c) (mechanical equivalent of heat)	4.1840	J/cal
	4.1840×10^7	erg/cal
	778.16	$ft \cdot lb_f/Btu$
Avogadro's number ($\tilde{N}$)	6.02214×10^{23}	$molecules/g\text{-mol}$
Boltzmann's constant ($\kappa = R/\tilde{N}$)	1.38066×10^{-23}	J/K
	1.38066×10^{-16}	erg/K
Faraday's constant (F)	96485.3	$C/g\text{-equivalent}$
Planck's constant (h)	6.62608×10^{-34}	$J \cdot s$
	6.62608×10^{-27}	$erg \cdot s$
Stefan–Boltzmann constant (σ)	5.67051×10^{-8}	$W/m^2 \cdot K^4$
	1.3553×10^{-12}	$cal/s \cdot cm^2 K^4$
	1.7124×10^{-9}	$Btu/hr \cdot ft^2 R^4$
Electron charge (e)	1.60218×10^{-19}	C
Speed of light in a vacuum (c)	2.99792×10^8	m/s

[1] E. R. Cohen and B. N. Taylor, *Physics Today* (August 1996), pp. BG9–BG13; R. A. Nelson, *Physics Today* (August 1996), pp. BG15–BG16.

§F.3   CONVERSION FACTORS

In the tables that follow, to convert any physical quantity from one set of units to another, multiply it by the appropriate table entry. For example, suppose that p is given as $10 \text{ lb}_f/\text{in.}^2$, and we wish to have p in poundals/ft². From Table F.3-2 the result is

$$p = (10)(4.6330 \times 10^3) = 4.6330 \times 10^4 \text{ poundals/ft}^2$$

The entries in the shaded rows and columns are those that are needed for converting from and to SI units.

In addition to the tables, we give a few of the commonly used conversion factors here:

Given a quantity in these units:	Multiply by:	To get quantity in these units:
Pounds	453.59	Grams
Kilograms	2.2046	Pounds
Inches	2.5400	Centimeters
Meters	39.370	Inches
Gallons (U.S.)	3.7853	Liters
Gallons (U.S.)	231.00	Cubic inches
Gallons (U.S.)	0.13368	Cubic feet
Cubic feet	28.316	Liters
Kelvins	1.800000	Degrees Rankine
Degrees Rankine	0.555556	Kelvins

Table F.3-1   Conversion Factors for Quantities Having Dimensions of F or ML/t^2

Given a quantity in these units ↓	Multiply by table value to convert to these units→	$N = kg \cdot m/s^2$ (Newtons)	$g \cdot cm/s^2$ (dynes)	$lb_m \cdot ft/s^2$ (poundals)	lb_f
$N = kg \cdot m/s^2$	(Newtons)	1	10^5	7.2330	2.24881×10^{-1}
$g \cdot cm/s^2$	(dynes)	10^{-5}	1	7.2330×10^{-5}	2.24881×10^{-6}
$lb_m \cdot ft/s^2$	(poundals)	1.3826×10^{-1}	1.3826×10^4	1	3.1081×10^{-2}
lb_f		4.4482	4.4482×10^5	32.1740	1

Table F.3-2 Conversion Factors for Quantities Having Dimensions of F/L^2 or M/Lt^2 (pressure, momentum flux)

Given a quantity in these units ↓ / Multiply by table value to convert to these units →	Pa = N/m² (kg/m·s²)	dyne/cm² = g/cm·s²	poundals/ft² (lb_m/ft·s²)	lb_f/ft²	lb_f/in² (psia)[a]	atm	mm Hg	in. Hg
Pa = N/m² = kg/m·s²	1	10	6.7197×10^{-1}	2.0886×10^{-2}	1.4504×10^{-4}	9.8692×10^{-6}	7.5006×10^{-3}	2.9530×10^{-4}
dyne/cm² = g/cm·s²	10^{-1}	1	6.7197×10^{-2}	2.0886×10^{-3}	1.4504×10^{-5}	9.8692×10^{-7}	7.5006×10^{-4}	2.9530×10^{-5}
poundals/ft² = lb_m/ft·s²	1.4882	1.4882×10^{1}	1	3.1081×10^{-2}	2.1584×10^{-4}	1.4687×10^{-5}	1.1162×10^{-2}	4.3945×10^{-4}
lb_f/ft²	4.7880×10^{1}	4.7880×10^{2}	32.1740	1	6.9444×10^{-3}	4.7254×10^{-4}	3.5913×10^{-1}	1.4139×10^{-2}
lb_f/in.²	6.8947×10^{3}	6.8947×10^{4}	4.6330×10^{3}	144	1	6.8046×10^{-2}	5.1715×10^{1}	2.0360
atm	1.0133×10^{5}	1.0133×10^{6}	6.8087×10^{4}	2.1162×10^{3}	14.696	1	760	29.921
mm Hg	1.3332×10^{2}	1.3332×10^{3}	8.9588×10^{1}	2.7845	1.9337×10^{-2}	1.3158×10^{-3}	1	3.9370×10^{-2}
in. Hg	3.3864×10^{3}	3.3864×10^{4}	2.2756×10^{3}	7.0727×10^{1}	4.9116×10^{-1}	3.3421×10^{-2}	25.400	1

[a] This unit is preferably abbreviated "psia" (pounds per square inch absolute) or "psig" (pounds per square inch gage). Gage pressure is absolute pressure minus the prevailing barometric pressure. Sometimes the pressure is reported in "bars"; to convert from bars to pascals, multiply by 10^5, and to convert from bars to atmospheres, multiply by 0.98692.

Table F.3-3 Conversion Factors for Quantities Having Dimensions of FL or ML^2/t^2 (energy, work, torque)

Given a quantity in these units ↓ / Multiply by table value to convert to these units →	J = kg·m²/s²	ergs = g·cm²/s²	foot poundals lb_m ft²/s²	ft·lb_f	cal	Btu	hp-hr	kw-hr
J = kg·m²/s²	1	10^{7}	2.3730×10^{1}	7.3756×10^{-1}	2.3901×10^{-1}	9.4783×10^{-4}	3.7251×10^{-7}	2.7778×10^{-7}
ergs = g·cm²/s²	10^{-7}	1	2.3730×10^{-6}	7.3756×10^{-8}	2.3901×10^{-8}	9.4783×10^{-11}	3.7251×10^{-14}	2.7778×10^{-14}
foot poundals = lb_m ft²/s²	4.2140×10^{-2}	4.2140×10^{5}	1	3.1081×10^{-2}	1.0072×10^{-2}	3.9942×10^{-5}	1.5698×10^{-8}	1.1706×10^{-8}
ft·lb_f	1.3558	1.3558×10^{7}	32.1740	1	3.2405×10^{-1}	1.2851×10^{-3}	5.0505×10^{-7}	3.7662×10^{-7}
thermochemical calories[a]	4.1840	4.1840×10^{7}	9.9287×10^{1}	3.0860	1	3.9657×10^{-3}	1.5586×10^{-6}	1.1622×10^{-6}
British thermal units	1.0550×10^{3}	1.0550×10^{10}	2.5036×10^{4}	778.16	2.5216×10^{2}	1	3.9301×10^{-4}	2.9307×10^{-4}
Horsepower hours	2.6845×10^{6}	2.6845×10^{13}	6.3705×10^{7}	1.9800×10^{6}	6.4162×10^{5}	2.5445×10^{3}	1	7.4570×10^{-1}
kilowatt hours	3.6000×10^{6}	3.6000×10^{13}	8.5429×10^{7}	2.6552×10^{6}	8.6042×10^{5}	3.4122×10^{3}	1.3410	1

[a] This unit, abbreviated "cal," is used in some chemical thermodynamic tables. To convert quantities expressed in International Steam Table calories (abbreviated "I. T. cal") to this unit, multiply by 1.000654.

Table F.3-4 Conversion Factors for Quantities Having dimensions[a] of M/Lt or Ft/L^2 (viscosity, density times diffusivity)

Given a quantity in these units → Multiply by table value to convert to these units →	Pa·s (kg/m·s)	g/cm·s (poises)	centipoises	$lb_m/ft·s$	$lb_m/ft·hr$	$lb_f·s/ft^2$
Pa·s = kg/m·s	1	10	10^3	6.7197×10^{-1}	2.4191×10^3	2.0886×10^{-2}
g/cm·s = (poises)	10^{-1}	1	10^2	6.7197×10^{-2}	2.4191×10^2	2.0886×10^{-3}
centipoises	10^{-3}	10^{-2}	1	6.7197×10^{-4}	2.4191	2.0886×10^{-5}
$lb_m/ft·s$	1.4882	1.4882×10^1	1.4882×10^3	1	3600	3.1081×10^{-2}
$lb_m/ft·hr$	4.1338×10^{-4}	4.1338×10^{-3}	4.1338×10^{-1}	2.7778×10^{-4}	1	8.6336×10^{-6}
$lb_f·s/ft^2$	4.7880×10^1	4.7880×10^2	4.7880×10^4	32.1740	1.1583×10^5	1

[a] When moles appear in the given and the desired units, the conversion factor is the same as for the corresponding mass units.

Table F.3-5 Conversion Factors for Quantities Having Dimensions of ML/t^3T or F/tT (thermal conductivity)

Given a quantity in these units → Multiply by table value to convert to these units →	W/m·K or kg·m/s³·K	g·cm/s³·K or erg/s·cm·K	$lb_m ft/s^3 F$	$lb_f/s·F$	cal/s·cm·K	Btu/hr·ft·F
W/m·K = kg·m/s³·K	1	10^5	4.0183	1.2489×10^{-1}	2.3901×10^{-3}	5.7780×10^{-1}
g·cm/s³·K	10^{-5}	1	4.0183×10^{-5}	1.2489×10^{-6}	2.3901×10^{-8}	5.7780×10^{-6}
$lb_m ft/s^3 F$	2.4886×10^{-1}	2.4886×10^4	1	3.1081×10^{-2}	5.9479×10^{-4}	1.4379×10^{-1}
$lb_f/s·F$	8.0068	8.0068×10^5	3.2174×10^1	1	1.9137×10^{-2}	4.6263
cal/s·cm·K	4.1840×10^2	4.1840×10^7	1.6813×10^3	5.2256×10^1	1	2.4175×10^2
Btu/hr·ft·F	1.7307	1.7307×10^5	6.9546	2.1616×10^{-1}	4.1365×10^{-3}	1

Table F.3-6 Conversion Factors for Quantities Having Dimensions of L^2/t (momentum diffusivity, thermal diffusivity, molecular diffusivity)

Given a quantity in these units → Multiply by table value to convert to these units →	m²/s	cm²/s	ft²/hr	centistokes
m²/s	1	10^4	3.8750×10^4	10^6
cm²/s	10^{-4}	1	3.8750	10^2
ft²/hr	2.5807×10^{-5}	2.5807×10^{-1}	1	2.5807×10^1
centistokes	10^{-6}	10^{-2}	3.8750×10^{-2}	1

Table F.3-7 Conversion Factors for Quantities Having Dimensions of M/t^3T or F/LtT (heat transfer coefficients)

Given a quantity in these units → Multiply by table value to convert to these units →	W/m²K (J/m²·s·K) kg/s³K	W/cm²K	g/s³K	lb_m/s³F	lb_f/ft·s·F	cal/cm²s·K	Btu/ft²hr·F
W/m²K = kg/s³K	1	10^{-4}	10^3	1.2248	3.8068×10^{-2}	2.3901×10^{-5}	1.7611×10^{-1}
W/cm²K	10^4	1	10^7	1.2248×10^4	3.8068×10^2	2.3901×10^{-1}	1.7611×10^3
g/s³K	10^{-3}	10^{-7}	1	1.2248×10^{-3}	3.8068×10^{-5}	2.3901×10^{-8}	1.7611×10^{-4}
lb_m/s³F	8.1647×10^{-1}	8.1647×10^{-5}	8.1647×10^2	1	3.1081×10^{-2}	1.9514×10^{-5}	1.4379×10^{-1}
lb_f/ft·s·F	2.6269×10^1	2.6269×10^{-3}	2.6269×10^4	32.1740	1	6.2784×10^{-4}	4.6263
cal/cm²s·K	4.1840×10^4	4.1840	4.1840×10^7	5.1245×10^4	1.5928×10^3	1	7.3686×10^3
Btu/ft²hr·F	5.6782	5.6782×10^{-4}	5.6782×10^3	6.9546	2.1616×10^{-1}	1.3571×10^{-4}	1

Table F.3-8 Conversion Factors for Quantities Having Dimensions[a] of M/L^2t or Ft/L^3 (mass transfer coefficients k_x or k_ω)

Given a quantity in these units → Multiply by table value to convert to these units →	kg/m²s	g/cm²s	lb_m/ft²s	lb_m/ft²hr	lb_f s/ft³
kg/m²s	1	10^{-1}	2.0482×10^{-1}	7.3734×10^2	6.3659×10^{-3}
g/cm²s	10^1	1	2.0482	7.3734×10^3	6.3659×10^{-2}
lb_m/ft²s	4.8824	4.8824×10^{-1}	1	3600	3.1081×10^{-2}
lb_m/ft²hr	1.3562×10^{-3}	1.3562×10^{-4}	2.7778×10^{-4}	1	8.6336×10^{-6}
lb_f s/ft³	1.5709×10^2	1.5709×10^1	32.1740	1.1583×10^5	1

[a] When moles appear in the given and the desired units, the conversion factor is the same as for the corresponding mass units.

Notation

Numbers in parentheses refer to equations, sections, or tables in which the symbols are defined or first used. Dimensions are given in terms of mass (M), length (L), time (t), temperature (T), and dimensionless (—). Boldface symbols are vectors or tensors (see Appendix A). Symbols that appear infrequently are not listed.

A = area, L^2

a = absorptivity (16.2-1),—

a = interfacial area per unit volume of packed bed (6.4-4), L^{-1}

a_α = activity of species α (24.1-8),—

C_p = heat capacity at constant pressure (9.1-7), ML^2/t^2T

C_V = heat capacity at constant volume (9.3-6), ML^2/t^2T

c = speed of light (16.1-1), L/t

c = total molar concentration (§17.7), moles/L^3

c_α = molar concentration of species α, (§17.7), moles/L^3

D = diameter of cylinder or sphere, L

D_p = particle diameter in packed bed, (6.4-6), L

$\mathscr{D}_{AB}$ = binary diffusivity for system A–B (17.1-2), L^2/t

$\mathscr{D}_{\alpha\beta}$ = binary diffusivity for the pair α–β in a multicomponent system (17.9-1), L^2/t

$Ð_{\alpha\beta}$ = Maxwell–Stefan multicomponent diffusivity (24.2-4), L^2/t

$\mathbb{D}_{\alpha\beta}$ = Fick multicomponent diffusivity (24.2-3), L^2/t

D_α^T = multicomponent thermal diffusion coefficient (24.2-3), M/Lt

d = molecular diameter (1.4-3), L

$\mathbf{d}_\alpha$ = diffusional driving force for species α (24.1-8), L^{-1}

$E_{tot} = U_{tot} + K_{tot} + \Phi_{tot}$ = total energy in a macroscopic system (15.1-2), ML^2/t^2

E_c = compression term in mechanical energy balance (7.4-3), ML^2/t^3

E_v = viscous dissipation term in mechanical energy balance (7.4-4), ML^2/t^3

e = 2.71828 . . .

e = emissivity (16.2-3),—

$\mathbf{e}$ = combined energy flux vector (9.8-5), M/t^3

$F_{12}, \overline{F}_{12}$ = direct, indirect view factor (16.4-9), (16.5-15),—

$\mathbf{F}_{s\to f}$ = force exerted by the solid on the fluid (7.2-1), ML/t^2

f = friction factor (or drag coefficient) (6.1-1),—

$G = H - TS$ = Gibbs free energy (24.1-2), ML^2/t^2

$G = \langle \rho v \rangle$ = mass velocity (6.4-8), M/L^2t

$\mathbf{g}$ = gravitational acceleration (3.2-8), L/t^2

$\mathbf{g}_\alpha$ = body force per unit mass acting on species α (Table 19.2-1), L/t^2

$H = U + pV$ = enthalpy (9.8-6), ML^2/t^2

h = Planck's constant (14.1-2), ML^2/t

h = elevation (2.3-10), L

$h, h_1, h_{ln}, h_{loc}, h_a, h_m$ = heat transfer coefficients (14.1-1 to 6), M/t^3T

$i = \sqrt{-1}$ (4.1-43),—

$\mathbf{J}_\alpha, \mathbf{J}_\alpha^*$ = molar fluxes (Table 17.8-1), moles/L^2t

$\mathbf{j}_\alpha, \mathbf{j}_\alpha^*$ = mass fluxes (Table 17.8-1), M/L^2t

j_H, j_D = Chilton–Colburn j-factors (14.3-19, Table 22.2-1),—

K = kinetic energy (7.4-1), ML^2/t^2

K_x, K_y = two-phase mass transfer coefficients (22.4-4), moles/tL^2

$\tilde{\kappa}$ = $R/\tilde{N}$ = Boltzmann's constant (1.4-1), ML^2/t^2T

k = thermal conductivity (9.1-1 and 24.2-6), ML/t^3T

k_x = single-phase mass transfer coefficients (22.1-7, 22.3-4, Table 22.2-1), moles/tL^2

k_x^0, k_y^0 = mass transfer coefficient for small mass-transfer rates and small species concentration (22.1-9, 22.4-2), moles/tL^2

$k_x^{\bullet}$ = mass transfer coefficient for high net mass-transfer rates (22.8-2a), moles/tL^2

k_T = thermal diffusion ratio (24.2-10),—

k_e = electrical conductivity (9.5-1), $\text{ohm}^{-1}\,\text{cm}^{-1}$

k_n'' = heterogeneous chemical reaction rate coefficient (18.0-3), $\text{moles}^{1-n}/L^{2-3n}t$

k_n''' = homogeneous chemical reaction rate coefficient (18.0-2), $\text{moles}^{1-n}/L^{3-3n}t$

L = length of film, tube, or slit (2.2-22), L

$\mathbf{L}_{\text{tot}}$ = total angular momentum within a macroscopic system (7.3-1), ML^2/t

l = mixing length (5.4-4), L

l_0 = characteristic length in dimensional analysis (3.7-3), L

M = molar mean molecular weight (Table 17.7-1), M/mole

M_α = molecular weight of species α (Table 17.7-1), M/mole

$M_{\alpha,\text{tot}}$ = total number of moles of species α in macroscopic system (23.1-3), moles

m = mass of a molecule (1.4-1), M

m, n = parameters in power law viscosity model (8.3-3), M/Lt^{2-n},—

$m_{\alpha,\text{tot}}$ = total mass of species α in macroscopic system (23.1-1), M

N = rate of shaft rotation (3.7-28), t^{-1}

N = number of species in a multicomponent mixture (17.7-1),—

$\tilde{N}$ = Avogadro's number, (g-mole)$^{-1}$

$\mathbf{N}_\alpha$ = combined molar flux vector for species α (17.8-2), moles/L^2t

$\mathbf{n}$ = unit normal vector (A.5-1),—

$\mathbf{n}_\alpha$ = combined mass flux vector for species α (17.8-1), M/L^2t

n = molecular concentration or number density (1.4-2), L^{-3}

$\mathbf{P}_{\text{tot}}$ = total momentum in a macroscopic flow system (7.2-1), ML/t

$\mathscr{P}$ = $p + \rho gh$ = modified pressure (for constant ρ and g) (2.3-10), M/Lt^2

$\mathscr{P}_0$ = characteristic pressure used in dimensional analysis (3.7-4), M/Lt^2

p = fluid pressure, M/Lt^2

Q = rate of heat flow across a surface (9.1-1, 15.1-1), ML^2/t^3

$Q_{\overrightarrow{12}}$ = radiant energy flow from surface 1 to surface 2 (16.4-5), ML^2/t^3

Q_{12} = net radiant energy interchange between surface 1 and surface 2 (16.4-8), ML^2/t^3

$\mathbf{q}$ = heat flux vector (9.1-4), M/t^3

q_0 = interfacial heat flux (10.8-14), M/t^3

R = gas constant (in $p\tilde{V} = RT$), ML^2/t^2T mole

R = radius of a cylinder or a sphere, L

R_α = molar rate of production of species α by homogeneous chemical reaction (18.0-2), moles/tL^3

R_h = mean hydraulic radius (6.2-16), L

$\mathfrak{R}$ = real part (of complex quantity) (4.1-43)

$\mathbf{r}$ = position vector (3.4-1), L

$r = \sqrt{x^2 + y^2}$ = radial coordinate in cylindrical coordinates, L

$r = \sqrt{x^2 + y^2 + z^2}$ = radial coordinate in spherical coordinates, L

r_α = mass rate of production of species α by homogeneous chemical reaction (19.1-5), M/tL^3

S_1, S_2 = cross-sectional area at planes 1 and 2 (7.1-1), L^2

S = entropy (11D.1-1, 24.1-1), ML^2/t^2T

T = absolute temperature, T

$\mathbf{T}_{s \to f}$ = torque exerted by a solid boundary on the fluid (7.3-1), ML^2/t^2

$\mathbf{T}_{ext}$ = external torque acting on system (7.3-1), ML^2/t^2

$T_1 - T_0$ = characteristic temperature difference used in dimensional analysis (11.5-5), T

t = time, t

U = internal energy (9.7-1), ML^2/t^2

U = overall heat-transfer coefficient (10.6-15), M/t^3T

$\bar{u}$ = arithmetic mean molecular speed (1.4-1), L/t

$\mathbf{u}$ = unit vector in direction of flow (7.2-1),—

V = volume, L^3

$\mathbf{v}$ = mass average velocity (17.7-1), L/t

$\mathbf{v}^*$ = molar average velocity (17.7-2), L/t

$\mathbf{v}_\alpha$ = velocity of species α (17.1-3, Table 17.7-1), L/t

v_0 = characteristic velocity in dimensional analysis (3.7-4), L/t

v_s = speed of sound (9.4-2, 11C.1-4), L/t

v_* = $\sqrt{\tau_0/\rho}$ = friction velocity (5.3-2), L/t

W = molar rate of flow across a surface, (23.1-4), moles/t

W_α = molar rate of flow of species α across a surface (23.1-3), moles/t

W_m = rate of doing work on a system by the surroundings via moving parts (7.4-1), ML^2/t^3

w = mass rate of flow across a surface (2.2-21), M/t

w_α = mass flow rate of species α across a surface (23.1-1), M/t

x_α = mole fraction of species α (Table 17.7-1),—

x, y, z = Cartesian coordinates

y = distance from wall (in boundary layer theory and turbulence) (§4.4), L

y_α = mole fraction of species α (22.4-2),—

Z = wall collision frequency (1.4-2), $L^{-2}t^{-1}$

z_α = ionic charge, (24.4-5), equiv/mole

alpha $\alpha = k/\rho\hat{C}_p$ = thermal diffusivity (9.1-7), L^2/t

beta β = thermal coefficient of volume expansion (10.9-6), T^{-1}

β = velocity gradient at a surface (12.4-6), s^{-1}

gamma $\gamma = C_p/C_V$ = heat capacity ratio (11.4-56),—

$\dot{\gamma} = \nabla\mathbf{v} + (\nabla\mathbf{v})^\dagger$ = rate-of-deformation tensor (8.3-1), t^{-1}

delta $\Delta X = X_2 - X_1$ = difference between exit and entry values

δ = falling-film thickness (2.2-22), boundary layer thickness (4.4-14), L

$\boldsymbol{\delta}$ = unit tensor (1.2-2, A.3-10),—

$\boldsymbol{\delta}_i$ = unit vector in the i direction (A.2-9),—

δ_{ij} = Kronecker delta (A.2-1),—

epsilon ε = fractional void space (6.4-3),—

$\varepsilon, \varepsilon_{AB}$ = maximum attractive energy between two molecules (1.4-10, 17.3-13), ML^2/t^2

ε_{ijk} = permutation symbol (A.2-3),—

zeta	ζ	= composition coefficient of volume expansion (19.2-2 and Table 22.2-1),—
eta	η	= non-Newtonian viscosity (8.2-1), M/Lt
	η', η''	= components of the complex viscosity (8.2-4), M/Lt
	$\overline{\eta}$	= elongational viscosity (8.2-5), M/Lt
	η_0	= zero shear rate viscosity (8.3-4), M/Lt
theta	θ	= $\arctan(y/x)$ = angle in cylindrical coordinates (A.6-5),—
	θ	= $\arctan(\sqrt{x^2 + y^2}/z)$ = angle in spherical coordinates (A.6-23),—
kappa	κ	= dilatational viscosity (1.2-6), M/Lt
	$\kappa, \kappa_0, \kappa_1, \kappa_2$	= dimensionless constants used in turbulence (5.3-1, 5.4-3, 5.4-5, 5.4-6)
lambda	$\Lambda, \Lambda_v, \Lambda_T, \Lambda_\omega$	= diffusivity ratios (20.2-29),—
	λ	= wavelength of electromagnetic radiation (16.1-1), L
	λ	= mean free path (1.4-3), L
	$\lambda, \lambda_1, \lambda_2, \lambda_k, \lambda_H$	= time constants in rheological models (§8.4 to §8.6), t
mu	μ	= viscosity (1.1-1), M/Lt
nu	ν	= μ/ρ = kinematic viscosity (1.1-3), L^2/t
	ν	= frequency of electromagnetic radiation (16.1-1), t^{-1}
xi	ξ	= composition coefficient of volume expansion (Table 22.2-1),—
pi	$\Pi, \Pi_v, \Pi_T, \Pi_\omega$	= dimensionless profiles (4.4-25, 12.4-21, 20.2-28),—
	π	= 3.14159 . . .
	$\boldsymbol{\pi}$	= $\boldsymbol{\tau} + p\boldsymbol{\delta}$ = molecular momentum flux tensor, molecular stress tensor (1.2-2, 1.7-1), M/Lt^2
rho	ρ	= density, M/L^3
	ρ_α	= mass of species α per unit volume of mixture (Table 17.7-1), M/L^3
sigma	σ	= Stefan–Boltzmann constant, $M/t^3 T^4$
	σ	= surface tension (3.7-12), M/t^2
	σ, σ_{AB}	= collision diameter (1.4-10, 17.3-11), L
tau	$\boldsymbol{\tau}$	= (viscous) momentum flux tensor, (viscous) stress tensor (1.2-2), M/Lt^2
	τ_0	= magnitude of shear stress at fluid–solid interface (5.3-1), M/Lt^2
phi	Φ	= potential energy (3.3-2), ML^2/t^2
	Φ_v	= viscous dissipation function (3.3-3), t^{-2}
	$\boldsymbol{\phi}$	= $\boldsymbol{\pi} + p\mathbf{vv}$ = combined momentum flux tensor (1.7-1), M/Lt^2
	ϕ	= $\arctan y/x$ = angle in spherical coordinates (A.6-24),—
	ϕ	= electrostatic potential (24.4-5), volts
	φ	= intermolecular potential energy (1.4-10), ML^2/t^2
psi	Ψ_1, Ψ_2	= first, second normal stress coefficient (8.2-2, 3), M/L
	Ψ_v	= viscous dissipation function (3.3-3), t^{-2}
	ψ	= stream function (Table 4.2-1), dimensions depend on the coordinate system
omega	$\Omega_\mu, \Omega_k, \Omega_\mathscr{D}$	= collision integrals (1.4-14, 9.3-13, 17.3-11),—
	ω_α	= mass fraction of species α (17.1-2, Table 17.7-1),—
	$\omega_{A1} - \omega_{A0}$	= characteristic mass fraction difference used in dimensional analysis (19.5-7),—

Overlines

$\tilde{X}$	= per mole
$\hat{X}$	= per unit mass
$\overline{X}$	= partial molar (19.3-3, 24.1-2)
$\overline{X}$	= time smoothed (5.1-4)
$\breve{X}$	= dimensionless (3.7-3)

Brackets

$\langle X \rangle$	= average value over a flow cross section
$(X), [X], \{X\}$	= used in vector–tensor operations when the brackets enclose dot or cross operations (Appendix A)

$$\llbracket\ \rrbracket\ =\ \text{dimensionless groupings}$$
$$[=]\ =\ \text{has the dimensions of}$$

Superscripts

$$X^\dagger\ =\ \text{transpose of a tensor}$$
$$X^{(t)}\ =\ \text{turbulent (5.2-8)}$$
$$X^{(v)}\ =\ \text{viscous (5.2-9)}$$
$$X'\ =\ \text{fluctuating quantity (5.2-1)}$$

Subscripts

A, B = species A and B in binary systems

$\alpha, \beta, \ldots$ = species in multicomponent systems

a = arithmetic-mean driving force or associated transfer coefficient (14.1-3)

b = bulk or "cup-mixing" value for an enclosed stream (10.8-33, 14.1-2)

c = evaluated at the critical point (1.3-1)

ln = logarithmic-mean driving force or associated transfer coefficient (14.1-4)

loc = local driving force or associated transfer coefficient (14.1-5)

m = mean transfer coefficient for a submerged object (14.1-6)

r = reduced, relative to critical value (§1.3)

tot = total amount of entity in a macroscopic system

0 = evaluated at a surface

$1, 2$ = evaluated at cross-sections 1 and 2 (7.1-1)

Named dimensionless groups designated with two letters

Br = Brinkman number (10.4-9, Table 11.5-2)

Ec = Eckert number (Table 11.5-2)

Fr = Froude number (3.7-11)

Gr = Grashof number (10.9-18, Table 11.5-2)

$\text{Gr}_\omega, \text{Gr}_x$ = Diffusional Grashof number (19.5-13, Table 22.2-1)

Ha = Hatta number (20.1-41)

Le = Lewis number (17.1-9)

Ma = Mach number (11.4-71)

Nu = Nusselt number (14.3-10 to 15)

Pé = Péclet number (Table 11.5-2)

Pr = Prandtl number (9.1-8, Table 11.5-2)

Ra = Rayleigh number (Table 11.5-2)

Re = Reynolds number (3.7-10)

Sc = Schmidt number (17.1-8)

Sh = Sherwood number (22.1-5)

We = Weber number (3.7-12)

Mathematical operations

D/Dt = substantial derivative (3.5-2), t^{-1}

$\mathscr{D}/\mathscr{D}t$ = corotational derivative (8.5-2), t^{-1}

∇ = del operator (A.4-1), L^{-1}

$\ln x$ = the logarithm of x to the base e

$\log_{10} x$ = the logarithm of x to the base 10

$\exp x = e^x$ = the exponential function of x

$\text{erf } x$ = error function of x (4.1-14, §C.6)

$\Gamma(x)$ = the (complete) gamma function (12.2-24, §C.4)

$\Gamma(x, u)$ = the incomplete gamma function (12.2-24)

$O(\ldots)$ = "of the order of"

Author Index

(Bold-face page numbers indicate that biographical data are given)

Subject Index

•••MOLECULAR FLUX EXPRESSIONS (SEE APPENDIX B.1, B.2, AND B.3)

Momentum (ρ = constant, Newtonian fluid):
$$\boldsymbol{\pi} = p\boldsymbol{\delta} - \mu(\nabla\mathbf{v} + (\nabla\mathbf{v})^\dagger) \qquad \text{or} \qquad \pi_{ij} = p\delta_{ij} - \mu\left(\frac{\partial v_j}{\partial x_i} + \frac{\partial v_i}{\partial x_j}\right)$$

Heat (pure fluid only):
$$\mathbf{q} = -k\nabla T \qquad \text{or} \qquad q_i = -k\frac{\partial T}{\partial x_i}$$

Mass (for a binary mixture of A and B):
$$\mathbf{j}_A = -\rho\mathscr{D}_{AB}\nabla\omega_A \qquad \text{or} \qquad j_{Ai} = -\rho\mathscr{D}_{AB}\frac{\partial\omega_A}{\partial x_i}$$

•••CONVECTED FLUX EXPRESSIONS (SEE §§1.7, 9.7, 17.7)

Momentum:
$$\rho\mathbf{v}\mathbf{v} \qquad \text{or} \qquad \rho v_i v_j$$

Energy:
$$\rho(\hat{U} + \tfrac{1}{2}v^2)\mathbf{v} \qquad \text{or} \qquad \rho(\hat{U} + \tfrac{1}{2}v^2)v_i$$

Mass:
$$\rho\omega_A\mathbf{v} \qquad \text{or} \qquad \rho\omega_A v_i$$

•••COMBINED FLUX EXPRESSIONS

Momentum:
$$\boldsymbol{\phi} = \rho\mathbf{v}\mathbf{v} + \boldsymbol{\pi} = \rho\mathbf{v}\mathbf{v} + p\boldsymbol{\delta} + \boldsymbol{\tau} \qquad\qquad \text{(Eq. 1.7-2)}$$

Energy:
$$\mathbf{e} = \rho(\hat{U} + \tfrac{1}{2}v^2)\mathbf{v} + \mathbf{q} + [\boldsymbol{\pi}\cdot\mathbf{v}] \qquad\qquad \text{(Eq. 9.8-5)}$$
$$= \rho(\hat{H} + \tfrac{1}{2}v^2)\mathbf{v} + \mathbf{q} + [\boldsymbol{\tau}\cdot\mathbf{v}] \qquad\qquad \text{(Eq. 9.8-6)}$$

Mass:
$$\mathbf{n}_A = \rho\omega_A\mathbf{v} + \mathbf{j}_A \qquad\qquad \text{(Eq. 17.8-1)}$$

Note: The quantity $[\boldsymbol{\pi}\cdot\mathbf{v}]$ is the molecular work flux (see §9.8), and $\boldsymbol{\pi} = p\boldsymbol{\delta} + \boldsymbol{\tau}$ (see Table 1.2-21). All fluxes obey the same sign convention: they are positive when the entity being transported is moving from the negative side of a surface to the positive side.